CARBON-NEUTRAL FUELS AND ENERGY CARRIERS

GREEN CHEMISTRY AND CHEMICAL ENGINEERING

Series Editor: Sunggyu Lee
Ohio University, Athens, Ohio, USA

Proton Exchange Membrane Fuel Cells: Contamination and Mitigation Strategies
Hui Li, Shanna Knights, Zheng Shi, John W. Van Zee, and Jiujun Zhang

Proton Exchange Membrane Fuel Cells: Materials Properties and Performance
David P. Wilkinson, Jiujun Zhang, Rob Hui, Jeffrey Fergus, and Xianguo Li

Solid Oxide Fuel Cells: Materials Properties and Performance
Jeffrey Fergus, Rob Hui, Xianguo Li, David P. Wilkinson, and Jiujun Zhang

Efficiency and Sustainability in the Energy and Chemical Industries: Scientific Principles and Case Studies, Second Edition
Krishnan Sankaranarayanan, Jakob de Swaan Arons, and Hedzer van der Kooi

Nuclear Hydrogen Production Handbook
Xing L. Yan and Ryutaro Hino

Magneto Luminous Chemical Vapor Deposition
Hirotsugu Yasuda

Carbon-Neutral Fuels and Energy Carriers
Nazim Z. Muradov and T. Nejat Veziroğlu

CARBON-NEUTRAL FUELS AND ENERGY CARRIERS

Edited by

Nazim Z. Muradov
University of Central Florida
Orlando, FL, USA

T. Nejat Veziroğlu
International Association for Hydrogen Energy
Miami, FL, USA

CRC Press
Taylor & Francis Group
Boca Raton London New York

CRC Press is an imprint of the
Taylor & Francis Group, an **informa** business

CRC Press
Taylor & Francis Group
6000 Broken Sound Parkway NW, Suite 300
Boca Raton, FL 33487-2742

First issued in paperback 2017

© 2012 by Taylor & Francis Group, LLC
CRC Press is an imprint of Taylor & Francis Group, an Informa business

No claim to original U.S. Government works

Version Date: 20110602

ISBN 13: 978-1-4398-1857-2 (hbk)
ISBN 13: 978-1-138-07331-9 (pbk)

Visit the Taylor & Francis Web site at
http://www.taylorandfrancis.com

and the CRC Press Web site at
http://www.crcpress.com

Contents

v

Preface

Carbon is the backbone of life on Earth. Starting from the discovery of fire, our civilization depends vitally on carbon for its energy; carbon made possible the industrial revolution and the rise in the standard of living we currently enjoy. On the other hand, recently, the word "carbon" has acquired a rather negative connotation mostly due to its association with carbon-bearing fossil fuels and, especially, with the product of their combustion—carbon dioxide (CO_2). Fossil fuels are currently being blamed for the ever-increasing levels of CO_2 in the atmosphere, climate change, air pollution, smog, oil spills, and other forms of environmental damage to the biosphere.

Today, more than 29 billion tons of man-made carbon dioxide emissions from a variety of sources (mostly from power generation, transportation and industrial sectors) are annually being pumped into the atmosphere. Some ominous signs of the adverse effects of these emissions on the Earth's ecosystems and climate have already started manifesting themselves in the form of retreating glaciers, hotter summers, stronger storms, increasing floods, etc.

In the face of ever-increasing levels of fossil fuel-derived CO_2, there have been attempts to estimate the tolerable limits of atmospheric CO_2 concentrations in terms of the global mean temperature rise. Analysis of scientific literature on climate impacts indicates that, in the majority of cases, a doubling of preindustrial atmospheric CO_2 concentration and an associated temperature increase of 2°C above the preindustrial level are considered a critical point beyond which a potentially catastrophic impact on the planet's ecosystem might occur. The danger is that beyond this 2°C point in the temperature increase, there exists not only elevated risk of extreme climate impacts and events, but also an increased possibility of strong positive carbon cycle feedbacks that would lead to an even stronger climate impact with the real potential of reaching a "tipping point" (or a runaway situation). These are real concerns that have to be addressed to avert the irreversible changes in the environment and climate.

There is a growing consensus that in order to maintain the present standard of living and establish an environmentally sustainable energy future, new approaches to managing energy resources and fuels have to be developed and implemented. Concerns over an insecure energy supply and the adverse environmental impact of carbonaceous fuels have triggered considerable efforts worldwide to find carbon-free or low-carbon alternatives to conventional fossil fuels. In particular, it is understood that many existing challenges can be solved in conjunction with the development and implementation of carbon-neutral energy systems (i.e., systems that do not increase atmospheric CO_2 concentration).

This book emphasizes the vital role of carbon-neutral energy sources, transportation fuels, and associated technologies in the establishment of a sustainable energy future. From a bird's-eye view, it tries to answer the following question: Will humankind be able to secure enough clean energy and fuels to further prosper and live in harmony with the environment against a backdrop of an ever-growing demand for energy, rapidly depleting resources of affordable fossil fuels, and deteriorating global ecological situation? To answer this question, we solicited help from world-renown experts in a wide range of research areas from photochemistry and fuel-generating enzymes to nuclear and automotive technologies. Without their valuable contributions, a project of this magnitude would not have been possible. We would like to thank all the lead authors and coauthors for their time and efforts, as well as their valuable suggestions and recommendations.

Due to its multidisciplinary nature, this comprehensive sourcebook is organized into 15 chapters, each reflecting a specific area of up-to-date research and development activities. It begins with an introductory chapter where we analyze the Earth's carbon budget with a special emphasis on the effect of human activities on the global carbon cycle, as well as proposed measures to stabilize the atmospheric CO_2 concentration at an acceptable level. The chapter provides a brief overview of a wide range of state-of-the-art technologies for the utilization of carbon-free energy resources and advanced transportation fuels and assesses their technical and commercial potential to meet the target of 10–30 TW of carbon-neutral power by mid-century. Although the technical potential of carbon-free energy resources is by far adequate to meet all humankind energy needs for the foreseeable future, it is realized that switching from fossil fuels currently providing about 80% of primary energy to non-fossil-based energy systems may require too many changes and, consequently, too much time to accomplish that transition. On the other hand, some experts believe that we cannot afford a long transition period because of potentially irreversible changes in global climate caused by the extensive use of fossil fuels.

In Chapter 2, Professor Carl-Jochen Winter (Germany) provides a comprehensive analysis of two main carbon-free energy carriers: electricity and hydrogen. As one of the pioneers of the hydrogen movement, Professor Winter provides the historical and technological background of the hydrogen economy concept. In this wide-ranging chapter, he describes various aspects of future hydrogen economy: hydrogen production, storage and distribution systems, its environmental and climatic relevance, and traditional and new application areas.

Chapter 3 (lead author—Dr. Kenneth Schultz, United States) focuses on the role of nuclear power in the production of carbon-neutral energy and fuels. Advantageously, nuclear energy does not inherently involve any direct use of fossil fuels or generation of CO_2 or other greenhouse gases. Thus, nuclear energy has the potential to make major contributions to the production of carbon-neutral fuels and energy carriers by providing a major carbon-free source of primary energy. In this chapter, a team of authors from a number of leading U.S. research institutions explore how nuclear energy can contribute to the production of carbon-neutral fuels using state-of-the-art nuclear technologies.

Several chapters are dedicated to the efficient utilization of renewable energy resources (solar, wind, geothermal, etc.) and the storage of intermittent renewable energy. In Chapter 4, Professor Gabriele Centi and coauthors (Italy) discuss the role of catalysis and photocatalysis in solving complex scientific problems related to solar-powered conversion of water and CO_2 to carbon-neutral fuels. In Chapter 5, Professor Arif Hepbasli (Turkey) emphasizes the importance of conducting exergetic analysis of renewable energy sources such as solar, wind, and geothermal energy from the viewpoint of designing optimal sustainable energy systems of the future. Electrochemical reduction of CO_2 to fuels is discussed by Professor Serguei Lvov and coauthors (United States) in Chapter 6. The cost-effective around-the-clock utilization of intermittent renewable resources such as solar, wind, wave, and tidal energy would not have been possible without efficient energy storage systems. This aspect is discussed by Professor Bent Sørensen (Denmark) in Chapter 7, which provides an overview of state-of-the-art systems for the storage of intermittent renewable energy resources.

A large section of the book is devoted to different aspects of bioenergy and biofuels production and utilization, and the potential role of bio-inspired energy systems. In Chapter 8, Professor Helmut Tributsch (Germany) discusses the concept of "energy-bionics," inferring that mankind should follow the energy example of nature on a high technological level. This chapter provides specific examples of scientific and technological concepts that aim at reaching this goal. Chapter 9 written by Professor Georg Schaub and Dr. Kyra Pabst (Germany) discusses on thermochemical technologies for the production of synthetic hydrocarbons from lignocellulosic biomass. Fundamental aspects of fermentative bio-hydrogen production are discussed by Professor Debabrata Das and Dr. Chitralekha Nag Dasgupta (India) in Chapter 10. In Chapter 11, a team of U.S. researchers led by Dr. Paul King addresses biomimetic and photobiological aspects of solar fuel production. Dr. Levin and coauthors (Canada) discuss and analyze the prospects of practical applications and outlook for fermentative biofuels such as bioethanol and biobutanol in Chapter 12. The current status and technological aspects of biodiesel production from oily biomass and algae are the focus of Chapter 13 written by Dr. Maximino Manzanera (Spain).

It is universally accepted that the world will continue to rely on fossil fuels to supply the bulk of its primary energy for many decades; thus, finding ways to curb the fossil-derived CO_2 emissions is a major challenge. The concept of fossil fuel decarbonization is considered by many as a feasible and potentially cost-effective near-to-mid-term solution for drastically reducing man-made CO_2 emissions originating from coal, petroleum, and natural gas. In Chapter 14, Dr. Nazim Muradov (United States) assesses the current state of knowledge and technological development with regard to the fossil decarbonization concept and its potential role in the portfolio of carbon mitigation options. The transportation sector consumes a significant fraction of primary energy resources (mostly petroleum) and emits enormous amounts of CO_2 emissions; therefore, it is important to develop and implement effective measures that would help in reducing oil consumption and cut CO_2 emissions in transportation. In Chapter 15, Dr. Sandy Thomas (United States) uses a dynamic computer simulation model to compare various vehicle and fuel options over the entire twenty-first century with regard to greenhouse gas emissions, oil consumption, and urban air pollution.

In summary, this book provides a comprehensive overview of carbon-neutral energy sources and fuels that will play an increasingly important role in the near-to-mid-term future; it highlights the enormous technical, economic, and environmental challenges facing their introduction to the world marketplace. This multidisciplinary reference book is unique in its scope and the diversity of topics covered, and it will be handy to all scientists, engineers, and students working and studying in practically all areas related to alternative energy sources and fuels. It will be a good supplement to textbooks on alternative fuels, renewable energy sources, and hydrogen economy. Nontechnical readership may also find this book useful from the viewpoint of introduction to the field of alternative energy and transportation fuels.

We realize that there is a divergence of opinions and viewpoints especially on such complex issues as energy, environment, and climate, and, therefore, wherever possible, the authors have tried to present a balanced view of the subject. We have made every effort to avoid overlapping in this multiauthored book, but, undoubtedly, some redundancy may still remain from one chapter to the other. For these and any other remaining flaws present in this book, we take full responsibility.

We hope that this book will contribute to an improved understanding and appreciation of the energy and environmental sustainability problems. It is very important to recognize the scope of problems and available options early in order to adequately plan long-term strategies for the transition to sustainable carbon-neutral energy systems, because failure to do that will have an enormous negative impact on our planet and future generations.

NAZIM MURADOV
University of Central Florida

NEJAT VEZIROǦLU
International Association for Hydrogen Energy

Green Chemistry and Chemical Engineering

The subjects and disciplines of chemistry and chemical engineering have encountered a new landmark in the way of thinking about, developing, and designing chemical products and processes. This revolutionary philosophy, termed "green chemistry and chemical engineering," focuses on the designs of products and processes that are conducive to reducing or eliminating the use and/or generation of hazardous substances. In dealing with hazardous or potentially hazardous substances, there may be some overlaps and interrelationships between environmental chemistry and green chemistry. While environmental chemistry is the chemistry of the natural environment and the pollutant chemicals in nature, green chemistry proactively aims to reduce and prevent pollution at its very source. In essence, the philosophies of green chemistry and chemical engineering tend to focus more on industrial application and practice rather than academic principles and phenomenological science. However, as both a chemistry and chemical engineering philosophy, green chemistry and chemical engineering derives from and builds upon organic chemistry, inorganic chemistry, polymer chemistry, fuel chemistry, biochemistry, analytical chemistry, physical chemistry, environmental chemistry, thermodynamics, chemical reaction engineering, transport phenomena, chemical process design, separation technology, automatic process control, and more. In short, green chemistry and chemical engineering is the rigorous use of chemistry and chemical engineering for pollution prevention and environmental protection.

The Pollution Prevention Act of 1990 in the United States established a national policy to prevent or reduce pollution at its source whenever feasible. And adhering to the spirit of this policy, the Environmental Protection Agency launched its Green Chemistry Program in order to promote innovative chemical technologies that reduce or eliminate the use or generation of hazardous substances in the design, manufacture, and use of chemical products. The global efforts in green chemistry and chemical engineering have recently gained a substantial amount of support from the international community of science, engineering, academia, industry, and governments in all phases and aspects.

Some of the successful examples and key technological developments include the use of supercritical carbon dioxide as green solvent in separation technologies; the application of supercritical water oxidation for the destruction of harmful substances; process integration with carbon dioxide sequestration steps; solvent-free synthesis of chemicals and polymeric materials; the exploitation of biologically degradable materials; the use of aqueous hydrogen peroxide for efficient oxidation; the development of hydrogen proton exchange membrane fuel cells for a variety of power generation needs; advanced biofuel productions; the devulcanization of spent tire rubber; the avoidance of the use of chemicals and processes causing the generation of volatile organic compounds; the replacement of traditional petrochemical processes by microorganism-based bioengineering processes; the replacement of chlorofluorocarbons with nonhazardous alternatives; advances in the design of energy-efficient processes; the use of clean, alternative, and renewable energy sources in manufacturing; and much more. This list, even though it is only a partial compilation, is undoubtedly growing exponentially.

This book series by CRC Press/Taylor & Francis Group is designed to meet the new challenges of the twenty-first century in the disciplines of chemistry and chemical engineering by publishing books and monographs based on cutting-edge research and development to the effect of reducing adverse impacts on the environment by chemical enterprise. And in achieving this, the series will detail the development of alternative sustainable technologies that will minimize the hazard and maximize the efficiency of any chemical choice. It aims at providing the readers in academia and industry with an authoritative information source in the field of green chemistry and chemical engineering. The publisher and its series editor are fully aware of the rapidly evolving nature of the subject and its long-lasting impact on the quality of human life in both the present and future. As such, the team is committed to making this series the most comprehensive and accurate literary source in the field of green chemistry and chemical engineering.

Sunggyu Lee

Editors

Nazim Z. Muradov is the principal research scientist at the Florida Solar Energy Center, University of Central Florida, Cocoa, Florida. He holds a DSc in physical chemistry (1990), a PhD in kinetics and catalysis (1975), and an MS in petrochemical engineering (1970). Dr. Muradov's main areas of research include thermocatalytic and photocatalytic hydrogen production systems, solar-powered water-splitting cycles, fossil fuel decarbonization, biomass-derived fuels, reformers for fuel cell applications, radiant detoxification of hazardous wastes, nanostructured carbon materials, and catalytic processing of hydrocarbons.

Dr. Muradov is a member of the board of directors of the International Association for Hydrogen Energy (IAHE) and a member of the board of trustees and the scientific council of the Madrid Institute for Advanced Studies, IMDEA Energia (Spain). He has also been an associate editor of the *International Journal of Hydrogen Energy* since 2006.

Dr. Muradov has authored and coauthored about 200 publications, 36 patents, a book and several book chapters. His paper "From hydrocarbon to hydrogen-carbon to hydrogen economy" (coauthored with T. Nejat Veziroğlu) was identified by Thomson Reuters' *Essential Science Indicators* to be one of the most cited papers in the research area of hydrogen economy, and it was featured as a fast-moving front paper on the *ScienceWatch*® Web site.

Dr. Muradov is a recipient of the University of Central Florida (Institutes and Centers) Distinguished Researcher of the Year Award (1996) and the UCF Research Incentive Award (2003). In 2010, he was granted the honorary title of the IAHE Fellow.

Dr. T. Nejat Veziroğlu, a native of Turkey, graduated from the City and Guilds College, the Imperial College of Science and Technology, University of London, with degrees in mechanical engineering (ACGI, BSc), advanced studies in engineering (DIC.), and heat transfer (PhD).

In 1962, after doing his military service in the Ordnance Section, serving in some Turkish government agencies, and heading a private company, Dr. Veziroğlu joined the Engineering Faculty at the University of Miami. In 1965, he became the director of graduate studies and initiated the first PhD program in the School of Engineering and Architecture. He served as chairman of the Department of Mechanical Engineering from 1971 through 1975, established the Clean Energy Research Institute in 1973, and was the associate dean for research from 1975 through 1979. He took a three-year leave of absence (2004–2007) and founded UNIDO-ICHET (United Nations Industrial Development Organization—International Centre for Hydrogen Energy Technologies) in Istanbul, Turkey. On May 15, 2009, he attained the status of professor emeritus at the University of Miami.

Dr. Veziroğlu organized the first major conference on hydrogen energy: The Hydrogen Economy Miami Energy (THEME) Conference, Miami Beach, March 18–20, 1974. At the opening of this conference, he proposed the hydrogen energy system as a permanent solution for the depletion of fossil fuels and the environmental problems caused by their utilization. Soon after, the International Association for Hydrogen Energy (IAHE) was established, and Dr. Veziroğlu was elected president. As president of IAHE, he initiated the biennial world hydrogen energy conferences in 1976 and the biennial world hydrogen technologies conventions in 2005.

In 1976, Dr. Veziroğlu started the publication of the *International Journal of Hydrogen Energy* (*IJHE*) as its editor-in-chief in order to publish and disseminate hydrogen energy–related research and development results from around the world. IJHE has continuously grown—it now publishes 24 issues a year. Dr. Veziroğlu has also published about 350 papers and scientific reports, edited 160 volumes of books and proceedings, and has coauthored the book *Solar Hydrogen Energy: The Power to Save the Earth*.

Dr. Veziroğlu has memberships in 18 scientific organizations; has been elected to the grade of fellow in the British Institution of Mechanical Engineers, the American Society of Mechanical Engineers, and the American Association for the Advancement of Science; and is the founding president of the International Association for Hydrogen Energy. He has been the recipient of several international awards. He was presented the Turkish Presidential Science Award in 1974, made an honorary professor in Xian Jiaotong University of China in 1981, awarded the I. V. Kurchatov Medal by the Kurchatov Institute of Atomic Energy of the USSR in 1982 and the Energy for Mankind Award by the Global Energy Society in 1986, and elected to the Argentinean Academy of Sciences in 1988. In 2000, he was nominated for the Nobel Prize in Economics for conceiving the hydrogen economy and striving toward its establishment.

Chapter Lead Contributors

Gabriele Centi
Department of Industrial Chemistry and
    Materials Engineering

and

Laboratory of Catalysis for Sustainable
    Production and Energy
Inter-University Consortium for Science
    and Technology of Materials
University of Messina
Messina, Italy

Debabrata Das
Department of Biotechnology
Indian Institute of Technology
Kharagpur, India

Arif Hepbasli
Department of Mechanical Engineering
College of Engineering
King Saud University
Riyadh, Saudi Arabia

Paul W. King
National Renewable Energy Laboratory
U.S. Department of Energy
Golden, Colorado

David B. Levin
Department of Biosystems Engineering
University of Manitoba
Winnipeg, Manitoba, Canada

Serguei N. Lvov
Department of Energy and Mineral
    Engineering
The Pennsylvania State University
University Park, Pennsylvania

Maximino Manzanera
Institute of Water Research
University of Granada
Granada, Spain

Nazim Z. Muradov
Florida Solar Energy Center
University of Central Florida
Cocoa, Florida

Georg Schaub
Engler-Bunde-Institut
Karlsruhe Institute of Technology (KIT)
University of Karlsruhe
Karlsruhe, Germany

Kenneth R. Schultz
General Atomics
San Diego, California

Bent Sørensen
Department of Environmental, Social and
    Spatial Change
Roskilde University
Roskilde, Denmark

C.E. (Sandy) Thomas
Clean Energy Consultant
Alexandria, Virginia

Helmut Tributsch (Retired)
Institute for Physical and Theoretical
    Chemistry
Free University Berlin
Berlin, Germany

T. Nejat Veziroğlu
International Association for Hydrogen
    Energy
Miami, Florida

Carl-Jochen Winter
ENERGON Carl-Jochen Winter e.K.
Überlingen, Germany

1

Energy Options in a Carbon-Constrained World: An Advent of Carbon-Neutral Technologies

Nazim Z. Muradov and T. Nejat Veziroğlu

CONTENTS

1.1 Introduction

Carbon is the basis and the backbone of life on the Earth. It is assumed that life can only exist on planets that have carbon and water (this widely accepted assumption is referred to as *carbon–water chauvinism*). Due to its unique electronic structure carbon atom can easily form strong bonds with other atoms forming vitally important molecules from the simplest molecule methane to such complex molecules as DNA and RNA that control the continuation of life on the Earth (the human body contains 18.5 wt.% of carbon). Through so-called carbon cycle (discussed in the following sections) carbon atoms can be recycled countless times over millions of years.

Our civilization vitally depends on carbon for its energy (starting from the discovery of fire). Carbon is the major constituent of all fossil fuels that made possible the industrial revolution and the rise in the standard of living we currently enjoy. Of particular importance are hydrocarbons, the compounds containing carbon and hydrogen atoms bound together by exceptionally strong C–C and C–H bonds. Liquid hydrocarbon fuels such as gasoline, kerosene, and diesel fuel are ideally suited for surface and air transportation and can be credited to a large extent for the accelerated technological progress during the last 100 years. There is a popular notion that if Mother Nature did not provide us with liquid hydrocarbons humankind would have to spend enormous amounts of wealth and resources to synthesize them. We are also increasingly relying on carbon-based commodities and industrial products. For example, advanced polymers and carbon-reinforced composite materials widely used in building and construction are based on carbon-derived feedstocks; fertilizers are also produced from carbonaceous compounds.

On the other hand, carbon-bearing fuels are currently being blamed for the ever increasing levels of air pollution, smog, oil spills, and other forms of environmental damage to the Earth's ecosystems and for causing health-related problems. Some worrying signs of the adverse effects of fossil-derived greenhouse gas (GHG) emissions on climate have already started manifesting themselves in the form of retreating glaciers, unusually hot summers, stronger hurricanes, increasing floods, etc. These are real concerns that have to be addressed to avert the irreversible changes in our planet's environment.

Although the world's 165 countries are signatories to the United Nations Framework Convention on Climate Change (UNFCC), the treaty that calls for the stabilization of atmospheric GHG, the business-as-usual forecasts project a steady increase in the man-made GHG emissions over the next 100 years. Such projections are based on a growing global economy and abundant fossil fuel resources, which are reflected in the following equation for the amount of GHG emissions from human activities (NETL 2006):

$$GHG\,emissions = economic\,activity(\$) \times \left(\frac{energy\,use}{economic\,activity(\$)} \right) \left(\frac{GHG\,emissions}{energy\,use} \right) \quad (1.1)$$

Note that in Equation 1.1, the first factor (economic activity) is projected to steadily increase (since no country in the world will be willing to constrain its economic growth). Furthermore, it is very unlikely that accessible fossil fuel resources will be left unused, irrespective of the environmental situation evolved (this is especially true for developing countries with considerable fossil fuel resources). This leaves us with the two ratios containing "energy use" member in Equation 1.1 that can be manipulated to decrease GHG emissions. The possible carbon mitigation pathways include a wide range of measures such as the following:

- Energy conservation
- Increase in the efficiency of power plants, industrial processes, automobiles, appliances, etc.
- Expanding nuclear power sector
- Use of renewable energy resources (solar, wind, geothermal, biomass, etc.)
- Use of fossil fuels with CO_2 capture and storage (or sequestration)

Concerns over dwindling fossil energy resources and their adverse environmental impact have triggered considerable worldwide efforts to find carbon-free or low-carbon alternatives to conventional fossil fuels. In this chapter, the authors analyze different energy options in a carbon-constrained world that could become a basis of sustainable energy systems of the future. The basic principles of carbon cycle on the Earth, major carbon reservoirs and carbon fluxes between them, as well as the assessment of anthropogenic carbon emissions, and their effect on biosphere and climate system are discussed. Technical and economic potential of different carbon mitigation options to achieve the atmospheric CO_2 stabilization, while not jeopardizing energy security, sustainable development, and environmental well-being, are elucidated. Finally, the prospects of emergency carbon mitigation options (e.g., geoengineering) for preventing dangerous climatic runaway situations are analyzed in this chapter.

1.2 Earth Carbon Budget

1.2.1 Major Carbon Reservoirs on the Earth

Despite the vital importance of carbon for sustaining life on our planet, its abundance on the Earth is very low: the lithosphere has only 0.032 wt.% of carbon (comparing to, e.g., 5 wt.% for iron) and the atmosphere currently contains 0.039 vol.% (or 392 parts per million [ppm]) of carbon in the form of CO_2. Carbon is stored on the Earth in the following carbon reservoirs:

- CO_2 in the atmosphere
- Dissolved CO_2 and carbonates in the ocean
- Soil organic matter
- Fossil fuel deposits
- Sedimentary rock deposits (e.g., carbonates) in the lithosphere
- Living and dead organisms in the biosphere

Figure 1.1 depicts the relative abundance of the major carbon reservoirs on our planet. It is evident that inorganic deposits of carbon in the lithosphere such as limestone, dolomite, chalk, and other carbonates constitute the largest reservoir of carbon on the Earth. These minerals are the most thermodynamically stable forms of carbon, which explains their abundance on our planet. Organic forms of carbon including soil organic matter (e.g., humus), carbon in the biosphere (e.g., plants, living organisms) represent significantly lesser share of the total carbon inventory. Carbon in the form of fossil fuels (e.g., natural gas (NG), oil, coal, peat, tar) is estimated at about 4000–5000 Gt; however, if we add the potentially recoverable resources of methane hydrates from the ocean floor, this figure would increase by almost 1 order of magnitude (Lackner 2003). To put these numbers in perspective, the values for the "oxygen limits" are also included in the figure (the term "oxygen limits" relates to the amount of fossil carbon that would use up all oxygen available in the Earth's atmosphere during its combustion).

As seen from Figure 1.1, the overall amount of carbon in fossil fuels is relatively low compared, for example, to the lithosphere, but it is precisely fossil fuels that are believed to be linked to the increase in the atmospheric CO_2 levels. From this viewpoint, it is interesting to estimate how high the atmospheric CO_2 concentration would rise if the global resources of fossil fuels are totally burned to CO_2. Figure 1.2 depicts the equivalent amount of CO_2 corresponding to the worldwide resources of major types of fossil fuels: coal, oil,

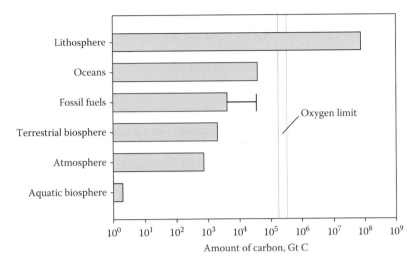

FIGURE 1.1
Major reservoirs of carbon on the Earth. The amount of carbon contained in carbon reservoirs (the error bar includes methane hydrate reserves). (Adapted from U.S. DOE LBL. U.S. DOE Lawrence Berkley Laboratory, http//:esd.lbl.gov/climate/ocean/fertilization.html [accessed January 15, 2010], 2010; Lackner, K., *Science*, 300, 1677, 2003.)

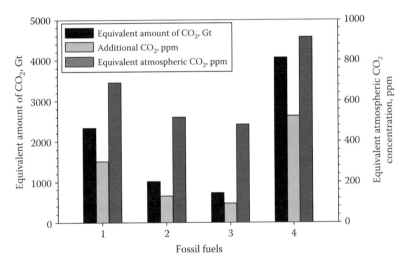

FIGURE 1.2
Worldwide fossil resources and equivalent CO_2 emissions (fossil fuel resources include recoverable reserves, and do not include methane hydrates and nonconventional fuels). 1—coal, 2—oil, 3—natural gas, 4—the sum of all fossil fuels. (Data from NETL, *Carbon Sequestration Technology Roadmap and Program Plan*, U.S. DOE, Office of Fossil Energy, NETL, 2006.)

and NG shown as black bars separately (1–3) and as a sum total (4). Light and dark gray bars represent the corresponding increases in the atmospheric CO_2 concentration over the current level of 392 ppm CO_2 and the total concentration of CO_2 in the atmosphere, respectively, as a result of burning each type of fossil fuel and all of them. Evidently, if all resources of fossil fuels are completely combusted, the atmospheric CO_2 concentration will rise to 917 ppm (this value is a net of any carbon absorption by natural sinks). Although this scenario may never materialize, several burning questions might arise: are these levels of CO_2 in the atmosphere tolerable, and, if not, what can be done about it? Further discussion may provide some answers to these and other related questions, and it is worthwhile to start with the introduction to the global carbon cycle.

1.2.2 Global Carbon Cycle

Over geological time, photosynthetic CO_2 fixation has exceeded respiratory oxidation of organic carbon, which resulted in the accumulation of oxygen and simultaneous reduction of CO_2 in the Earth's atmosphere accompanied by the burial of organic carbon in marine sediments (Falkowski et al. 2000). The carbon cycle has eventually evolved on our planet, which is an extremely complex phenomenon involving the rotation of carbon atoms in nature. Essentially, it is a series of cyclic processes by which carbon is exchanged between four major carbon reservoirs: geosphere, atmosphere, hydrosphere, and biosphere. The amount of carbon in each reservoir and rate at which carbon is exchanged between them could vary seasonally, or annually, or over centuries depending on a variety of factors that are yet to be fully understood. Although most reported schemes dealing with the carbon cycle emphasize the critical role of CO_2 as an exchange "currency" between the reservoirs, the role of methane (CH_4) should not be underestimated.

Figure 1.3 provides a simplified scheme of the natural carbon cycle. It shows unperturbed carbon exchanges between the ocean, the atmosphere, and land in the form of arrows. Carbon fluxes between the reservoirs in Gt C year^{-1} are shown by numbers between the

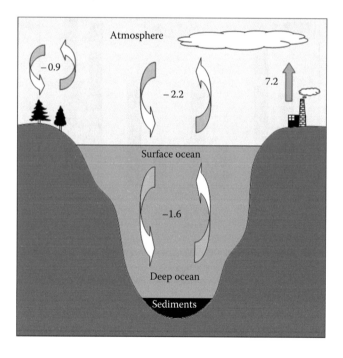

FIGURE 1.3
Simplified schematic representation of global carbon cycle. Positive flux is an input to the atmosphere (fossil emissions + cement), and negative fluxes are losses from the atmosphere (sinks). (Data from The Intergovernmental Panel on Climate Change 4th Assessment Report, Climate Change 2007, *The Physical Science Basis*, Cambridge University Press, Cambridge, U.K., Table 7.1, p. 516, 2007.)

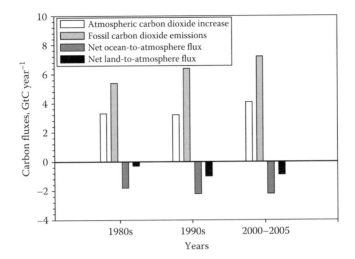

FIGURE 1.4
Changes in carbon fluxes between the atmosphere, land, and ocean during the 1980–2005 period (fossil CO_2 emissions also include emissions from cement-manufacturing plants). Positive fluxes are inputs to the atmosphere, and negative fluxes are losses from the atmosphere (sinks). (Data from The Intergovernmental Panel on Climate Change 4th Assessment Report, Climate Change 2007, *The Physical Science Basis*, Cambridge University Press, Cambridge, U.K., Table 7.1, p. 516, 2007.)

arrows. The positive flux is an input to the atmosphere (fossil-derived emissions plus emissions from cement-manufacturing plants), and negative fluxes are losses from the atmosphere (sinks). Figure 1.4 shows the evolution of net carbon fluxes from 1980s to 2005. The steady increase in the amounts of fossil-generated CO_2 emissions and CO_2 levels in the atmosphere is evident. The flux of anthropogenic CO_2 to the atmosphere to a certain extent is offset by the carbon fluxes from the atmosphere to the ocean and land.

1.2.2.1 Ocean Carbon Cycle

Over millions of years, CO_2 has been continuously exchanging between the atmosphere and the ocean. The estimates indicate that before the industrial revolution the amount of carbon contained in the ocean exceeded that of the atmosphere and land by a factor of 60 and 20, respectively (IPCC PSB 2007). The natural CO_2 exchange between the atmosphere and the ocean is a relatively slow process: the estimated timescale for reaching equilibrium between the atmosphere and the ocean surface is roughly 1 year (it depends on many factors such as wind speed, temperature, precipitation, heat flux, etc.). According to a number of studies (Turley et al. 2006) the rate-limiting step of the overall atmosphere–ocean exchange process is not an air–sea gas exchange, but the mixing of the surface waters with the intermediate and deep ocean, which is a very slow process. Although the ocean can theoretically absorb up to 70%–80% of the projected anthropogenic CO_2 emissions, it would take several centuries to do so due to the slow surface-deep ocean exchange rate (Turley et al. 2006).

Ocean carbon mainly exists in three forms: dissolved inorganic carbon (DIC), dissolved organic carbon (DOC), and particulate organic carbon (POC) (living and dead) in an approximate ratio of DIC:DOC:POC = 2000:38:1 (Falkowski et al. 2000; Sarmiento and Gruber 2006; IPCC PSB 2007). The rate of CO_2 exchange between the atmosphere and the ocean is controlled by three so-called carbon pumps: a solubility pump, an organic carbon

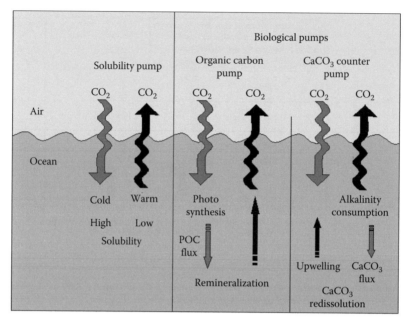

FIGURE 1.5
Simplified representation of three main ocean carbon pumps governing natural CO_2 fluxes in the ocean. (Modified from Heinze, C. et al., *Paleooceanography*, 6, 395, 1991.)

pump, and $CaCO_3$ "counter pump" (the last two pumps are often defined as the forms of a "biological pump") (a simplified scheme of the carbon pumps is shown in Figure 1.5) (Heinze et al. 1991).

CO_2 enters the surface ocean by diffusion and dissolution processes followed by the well-understood series of reactions leading to the formation of bicarbonate (HCO_3^-) and carbonate (CO_3^{2-}) ions (collectively, dissolved CO_2, bicarbonate and carbonate ions are designated as DIC):

$$CO_{2(gas)} + H_2O \leftrightarrow (HCO_3^-)_{aq} + H^+ \leftrightarrow (CO_3^{2-})_{aq} + 2H^+ \tag{1.2}$$

$$CO_{2(gas)} + H_2O + (CO_3^{2-})_{aq} \leftrightarrow (HCO_3^-)_{aq} + H^+ + (CO_3^{2-})_{aq} \leftrightarrow 2(HCO_3^-)_{aq} \tag{1.3}$$

The approximate ratio between the three forms of DIC in the ocean is as follows: $(CO_2)_{aq}$: $(HCO_3^-)_{aq}$: $(CO_3^{2-})_{aq} \approx 1{:}100{:}10$ (IPCC PSB 2007). As seen from the above equations, the addition of CO_2 to seawater results in an increase in the $(HCO_3^-)_{aq}$ concentration and the ocean acidity. The slight alkalinity of the ocean surface (pH varies from 7.9 to 8.25, due to slow dissolution of minerals) greatly facilitates the overall process of CO_2 uptake. The residence time of DIC in the surface ocean, relative to the exchange with the atmosphere and the deeper (intermediate) ocean layers, is less than a decade (IPCC PSB 2007).

Phytoplankton fixes CO_2 through photosynthesis in the surface layers of the ocean, and eventually it sinks as dead organisms, as part of the organic carbon pump action. Due to the intrinsic nature of this photobiological reaction, the efficiency of the organic carbon pump is controlled by the availability of light and "building blocks" such as phosphate, nitrate, silicate, iron-based nutrients, and micronutrients. But before the dead phytoplankton reaches the intermediate layers (about 1000 m deep), most of it is respired or re-mineralized (through the bacterial decay) and recirculated back to the surface as DIC (through the ocean circulation—upwelling) resulting in CO_2 outgassing. A small fraction of the particles reaches the deep ocean sediments where it is buried. The significant amount of organic carbon enters the ocean from rivers and as a result of various marine metabolic processes. The residence time of the dissolved organic matter in the ocean varies from days to thousands of years (the latter being predominant).

The certain fraction of DIC is concurrently fixed with calcium cations (Ca^{2+}) to form calcium carbonate ($CaCO_3$), which is used by shelled organisms (e.g., corals, clams, plankton, some forms of algae) during their growth. After death, these organisms sink to the ocean floor and eventually form carbonate deposits and sedimentary rocks. The decrease in the ocean pH (see above equations) inhibits the bio-calcification processes that are responsible for the formation of corals, phytoplankton, and zooplankton, and leads to the dissolution of $CaCO_3$ and calcerous sediments at the ocean floor. The process of carbonate-based sediment dissolution in response to the atmospheric CO_2 increase occurs on a timescale from one thousand to hundreds of thousands of years, with calcium carbonate dissolution accounting for 60%–70% of the anthropogenic CO_2 emissions uptake (IPCC PSB 2007). Between 22% and 33% of CO_2 absorption can be attributed to the ocean water column, which occurs on a timescale of a hundred to a thousand years (the balance of 7%–8% of CO_2 is due to terrestrial weathering cycles involving silicate carbonates) (Archer and Buffet 2005). Because of very slow rates of carbonate dissolution and silicate weathering, it would take several thousand years for atmospheric CO_2 concentration to reach a new equilibrium value.

Collectively, the solubility and biological pumps control a vertical gradient of carbon (as DIC) between the surface ocean (where its concentration is relatively low) and the deep ocean (with high DIC concentration) layers, and, thus, regulate the exchange of carbon (as CO_2) between the ocean and the atmosphere. It was estimated that intermediate and deep ocean waters mix on a timescale of decades-to-centuries and millennia, respectively (IPCC PSB 2007). A variety of factors can affect the efficiency of the solubility and biological pumps such as surface ocean temperature, salinity, ocean circulation, nutrient supply, to name just a few.

1.2.2.2 Terrestrial Carbon Cycle

Biosphere releases carbon (as CO_2) via plants' and animals' respiration and detritus food chain (i.e., decomposition of organic matter). The carbon flux (mainly, through respiration and the gross primary production [GPP]) between the terrestrial biosphere (which includes vegetation, soil, and detritus) and the atmosphere is estimated at 120 Gt C year^{-1}. Although the above fluxes could vary from year to year, they are assumed to be in balance when averaged over long periods of time. A relatively small amount of carbon (about 1 Gt C year^{-1}) is transported from land to the ocean by rivers in the form of DIC or suspended particles (Richey 2004). Other important, although relatively small, natural carbon fluxes include rock weathering, sediment accumulation, volcanic activity, and conversion of terrestrial organic matter into inert forms of carbon in soils. These carbon fluxes, when averaged over decades, are estimated not to exceed 0.1 Gt C year^{-1} (IPCC PSB 2007).

1.2.3 Interaction between the Earth's Carbon Cycle and Climate System

Interactions between the natural carbon cycle and the climate system are extremely complex, because they involve a large number of physical, chemical, photochemical, biological, and biogeochemical processes that are, in many cases, not very well quantified or even defined. It is universally accepted that the Earth's climate is controlled by the Sun's radiation and a variety of physical phenomena involving the Earth's atmosphere and surface such as absorption, reflection, and emission of radiant energy. The climate system itself is a complex multifaceted system that is governed by myriad interactions between atmosphere, land surface, oceans and seas, snow and glaciers, deserts and terrestrial biosphere. Climate is usually defined in terms of average temperature, precipitation, and wind over a period of time (typically, over period of 30 years) (IPCC PSB 2007). The climate system is constantly changing influenced by its own internal dynamics and a number of external factors such as variations in solar irradiance, volcanic eruptions, and other changes in the atmosphere.

In the absence of natural (e.g., volcanoes) or man-made external changes, the Earth's climate is solely determined by incident solar radiation, which is equal to about 1370 W m^{-2} (i.e., the amount of energy received by 1 m^2 of the top of the Earth's atmosphere facing the Sun in 1 s during daytime; if averaged over the entire planet, the amount of received radiant energy would be 342 W m^{-2}) (IPCC PSB 2007). When averaged over a long period of time, this amount of radiant energy is balanced by a number of energy absorption and emission processes resulting in an equilibrium or zero energy balance. Figure 1.6 depicts

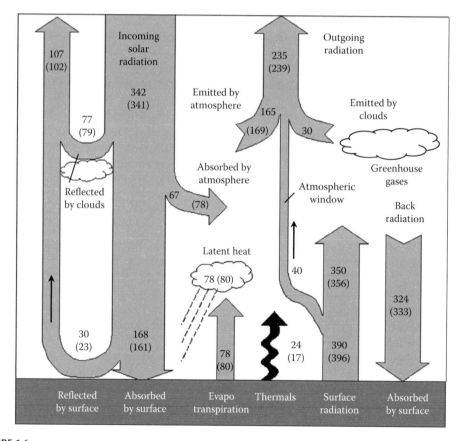

FIGURE 1.6
Earth's global annual mean energy balance (values shown on the diagram are in W m^{-2}). (Data from Kiehl, J. and Trenberth, K., *Bull. Am. Meteorl. Soc.*, 78, 197, 1997; Trenberth, K. et al., *Bull. Am. Meteorol. Soc.*, 90, 311, 2008, doi: 10.1175/2008BAMS2634.1.)

the estimate of the Earth's annual and global mean energy balance detailing the variety of energy absorption, reflection, and emission processes (Kiehl and Trenberth 1997). In 2008, in the light of new observations and analyses, the authors updated the Earth's global annual mean energy budget (the refined values are shown in Figure 1.6 in brackets) (Trenberth et al. 2008). It can be seen that roughly a third of the incoming solar radiation is reflected back to space by the Earth's atmosphere, clouds, aerosols, and land surface (mainly, by snow, ice, and deserts). Although aerosols typically do not contribute much to the overall radiation reflection process (because their abundance is controlled by rains), they become a significant factor during major volcanic eruptions.

The main portion of solar radiant energy ($235–239\,W\,m^{-2}$) is absorbed by the Earth's atmosphere and surface. In order to maintain the energy balance our planet has to release the same amount of energy back to the space, which it does by emitting long-wavelength or infrared (IR) radiation. The Earth's surface absorbs about half (i.e., $161–168\,W\,m^{-2}$) of the incoming solar radiation energy, which it transfers to the atmosphere by means of three processes: (1) warming the air that is in contact with the surface (thermals); (2) evapotranspiration (i.e., energy absorbed during evaporation of water from oceans, seas, and land surface, and then recovered as latent heat during water vapor condensation in clouds); and (3) emission of long-wavelength radiation that is absorbed by clouds and atmospheric gases. Absorption of the long-wavelength (IR) radiation by certain atmospheric gases possessing heat-trapping properties is attributed to a greenhouse effect (thus, these gases are called "greenhouse gases"), which will be discussed in more details in the next section. Trenberth et al. (2008) estimated that an imbalance of $0.9\,W\,m^{-2}$ originates from the enhanced greenhouse effect.

The radiation balance on the Earth's surface could be fundamentally altered by changing (1) the incoming solar radiation, (2) the fraction of reflected solar radiation (i.e., its albedo), and (3) the fraction of long-wavelength radiation from the Earth back to space (IPCC PSB 2007). Evidently, the first option would require actions that are beyond the realm of existing technical capabilities of humans (like, changing the Earth's orbit or altering the Sun radiance). On the other hand, the second option might become a reality in not so distant future, for example, via changes in terrestrial ecosystems (vegetation, etc.) or altering the cloud cover, or adding/controlling atmospheric aerosols (the last two and other so-called geoengineering approaches will be discussed later in the chapter). The third option directly deals with the changes in the GHG concentrations in the atmosphere. The climate system is very sensitive to these changes, and it could react via a variety of direct and indirect feedback mechanisms.

The feedback mechanism could be of positive or negative nature, depending on whether it amplifies or negates, respectively, the effect of the change. The positive feedbacks (or "feedback loops") are of a particular concern, because they could easily lead to "runaway" situations. One important example of the positive feedback mechanism relates to a so-called ice-albedo feedback (IPCC PSB 2007), which functions as follows. The rising levels of GHG in the atmosphere trap increasingly more heat and make the Earth's atmosphere warmer causing snow, glaciers, and polar ice caps to melt. This would result in exposing increasingly broader areas with "dark" land and ocean surfaces that much better absorb solar heat than snow-covered surfaces. The resulting increase in the atmosphere temperature would melt more snow and ice, and so on. The consequence of this positive feedback could be an uncontrolled increase in temperature on the Earth's surface in response to relatively low (in an absolute value) increase in GHG atmospheric concentration.

Another example of the feedback mechanism (both negative and positive) involves clouds, which play an important role in regulating the Earth's climate. Clouds are effective absorbers of long wavelength radiation, and, as such, greatly contribute to the greenhouse effect (i.e., via the positive feedback mechanism since more clouds would result in the more pronounced greenhouse effect and the Earth surface warming). However, clouds also very effectively reflect solar radiation, thus, causing a negative feedback effect, since more clouds result in more reflection and cooling the Earth. Even minor changes in clouds' consistency and location could cause either a positive or negative feedback effect. Because of vital importance of the climate feedback phenomenon and its implications to a very survival of humankind, this issue is at the center of intensive worldwide research efforts.

There are close and multiple interactions between the Earth's carbon cycle and its climate. In particular, some components of the carbon cycle, most importantly, the ocean, biosphere, and human activities can affect the levels of GHG in the atmosphere and, thus, the climate system. Any changes in the land-to-atmosphere or the ocean-to-atmosphere carbon fluxes could potentially affect the CO_2 concentration in the atmosphere and, thus, indirectly influence the climate system. For example, vegetation takes up CO_2 from the atmosphere and sequesters it in the form of carbohydrates, thus, diminishing the greenhouse effect. On the other end of the spectrum, human activities cause CO_2 levels in the atmosphere to increase (via burning fossil fuels and land use changes), thus, amplifying the greenhouse effect and warming the climate. Various feedback mechanisms could potentially determine the extent to which rising CO_2 concentrations would affect the climate. One of the most important feedback effects relates to CO_2–water vapor feedback loop. Water vapor level in the atmosphere may increase in response to rising concentration of atmospheric CO_2 and resulting warming. Consequently, this will cause additional increase in water vapor concentration and the CO_2 flux from the ocean to the atmosphere, further intensifying the greenhouse effect and warming. It was estimated that the effect of the water vapor feedback could double the intensity of the greenhouse effect due to added CO_2 alone (IPCC PSB 2007).

1.2.4 Greenhouse Effect and Carbon Cycle

All objects in nature, including the Sun and the Earth, are emitting electromagnetic radiation. The amount of energy radiated by the Earth's surface is a function of the surface temperature to the fourth power according to the Stefan-Boltzmann law:

$$E = \varepsilon \sigma T^4 \tag{1.4}$$

where
  E is the total energy radiated by a body (e.g., in W m^{-2})
  ε is emissivity coefficient
  σ is Stefan-Boltzmann coefficient
  T is absolute temperature

If we calculate the temperature of a body that emits 235 W m^{-2} of radiant energy based on this equation, we would come up with the temperature of about minus 18°C–19°C (Kreith 2000), which is far below the average temperature on the Earth's surface (14°C). That temperature gap of about 32°C–33°C is attributed to a natural greenhouse effect caused by heat-trapping atmospheric gases, or GHG, predominantly CO_2 and water vapor (note that nitrogen and oxygen—two major constituents of the atmosphere—do not exert such

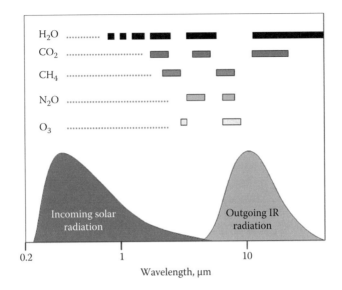

FIGURE 1.7
Simplified representation of a greenhouse effect. (Data from Goody, R., *Atmospheric Radiation: I. Theoretical Basis*, Clarendon, Oxford, U.K., 1964, 436pp; Okabe, H., *Photochemistry of Small Molecules*, John Wiley & Sons, New York, 1978.)

effect). GHG absorb and reflect radiant energy within the atmosphere, which in turn emits most of this long-wavelength radiation energy back to the Earth's surface and a smaller fraction out to space.

Figure 1.7 provides a schematic representation of the greenhouse effect mechanism. The Earth receives solar energy in the form of irradiation with the wavelengths varying in the range of about 0.2–4 μm and greater, which includes a small fraction of ultraviolet (UV) light (0.2–0.4 μm), visible light region (0.4–0.8 μm), and IR (the balance) region. The left curve shows the simplified representation of the spectrum of the incoming solar radiation, which closely follows the spectrum of a black body heated to about 5500 K (the peak of the spectral curve is in the visible area at about 0.5–0.6 μm). As discussed above, the Earth radiation spectrum is mostly a function of temperature, and the associated black body radiation curve extends from wavelengths of 1–3 μm to about 70–80 μm with the peak at about 10 μm (the right curve in the Figure 1.7). Due to structural and electronic properties of H_2O_{gas}, CO_2, CH_4, N_2O, O_3 they are almost transparent to most of the direct sunlight, but very efficiently (up to 80%) absorb outgoing IR radiation directed from the Earth's surface to space. The radiation-absorbing properties of these gases are shown in the Figure 1.7 in the form of horizontal bars the length of which is proportional to the absorption bandwidth of the corresponding molecules in the IR area of the spectrum (the molecules could have several bands, only most important of them are shown).

The common feature of H_2O_{gas}, CO_2, CH_4, N_2O, O_3 gases is that their molecules contain at least three atoms, which allows for a much greater number of fundamental vibrations within their molecules in response to IR photo-excitation, compared to simpler two-atom molecules (which, typically, do not exert the greenhouse effect). The number of fundamental vibrations is determined by the following "rule of thumb" formula:

$$V = 3N - 5 \quad \text{(for linear molecules, e.g., } CO_2\text{)} \tag{1.5}$$

$$V = 3N - 6 \quad \text{(for nonlinear molecules, e.g., } H_2O, CH_4, O_3) \tag{1.6}$$

where
V is the number of possible fundamental vibrations
N is the number of atoms in a molecule

Based on the above formula, H_2O molecule has three vibrations, two stretching and one bending vibration, and CO_2 molecule has four vibrations, two stretching (symmetric and asymmetric) and two bending (in-plane and out-of-plane) vibrations. Figure 1.8 depicts schematically the selected types of vibrations characteristic of CO_2 and H_2O molecules, and the corresponding IR radiation absorption bands. Larger molecules, for example, hydrochlorofluorocarbons, have much larger number of vibrations, and typically they are stronger absorbers of IR radiation. Returning to the Figure 1.7, it can be seen that H_2O_{gas}, CO_2, CH_4, N_2O, O_3 gases efficiently uptake IR radiation in their respective absorption areas and convert it into thermal vibrational energy (note that the radiation absorbing capacity of gases are determined not only by their absorption band, but also by the gas concentration in the atmosphere; consequently, the contribution of H_2O and CO_2 to the heat-trapping effect is much greater than that of other gases due to their relative abundance in the atmosphere). Collectively, these gases cover most of the spectrum emitted by the Earth's surface, leaving a narrow window where CO_2 and H_2O absorption is relatively weak, permitting some of the thermal radiation to escape into the space and, thus, preventing a thermal runaway situation. Carbon cycle via the control mechanism of the amount of CO_2 in the atmosphere potentially has a profound effect on the intensity of the greenhouse effect.

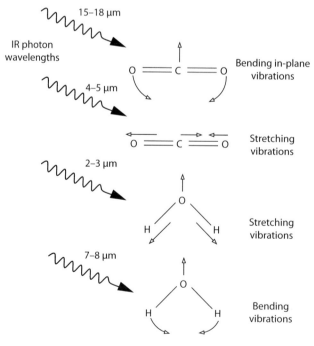

FIGURE 1.8
Characteristic vibrations of CO_2 and H_2O molecules and the corresponding IR radiation absorption bands.

1.3 Effect of Human Activities on Carbon Cycle and Climate System

Human activities contribute to the changes in the carbon cycle and the climate system by causing changes in the carbon flux to the atmosphere and perturbing the fine balance between incoming solar radiation and outgoing IR radiation due to alterations in the amount of GHG, aerosols, and cloudiness in the Earth's atmosphere. The anthropogenic CO_2 emissions comprise two fluxes: (1) CO_2 originated from the use of fossil fuels including cement manufacture (about 80% of total) and (2) CO_2 flux related to the land use changes (e.g., deforestation, agricultural development, etc.) (remaining 20% of the total) (IPCC PSB 2007). In particular, the combined carbon emissions from fossil fuels and cement-manufacturing plants increased from 5.4 ± 0.3 Gt C year^{-1} in the 1980s to 6.5 ± 0.4 Gt C year^{-1} in the 1990s (about 0.7% year^{-1}) to 7.2–7.8 Gt C year^{-1} in 2005 (about 3% year^{-1}) (BP 2006; Marland et al. 2006). Cumulatively, CO_2 emissions due to global fossil fuel burning and cement production have increased by 70% over the last three decades. CO_2 emissions from cement manufacturing are rather small (about 3%) compared to that from fossil fuels combustion. In the land use changes category, CO_2 emissions are estimated at the range of 0.5–2.7 Gt C year^{-1}, contributing from 6% to 39% of the emissions growth rate (Brovkin et al. 2007). This amounts to the contribution of 12–35 ppmv of the total atmospheric CO_2 increase from preindustrial period to the year 2000. Tropical deforestation is a main contributor to the increase in CO_2 flux to the atmosphere due to the land use change.

Although the carbon fluxes that have originated from human activities constitute only a small fraction of the gross natural fluxes within the atmosphere-ocean-land system, they still have caused the appreciable changes in the major carbon reservoirs compared to the preindustrial period. As a result, anthropogenic carbon might induce certain perturbations to the natural carbon cycle, extent of which depends on the specifics of the carbon reservoir. In the ocean reservoir, for example, the biological pump does not directly absorb and store anthropogenic carbon, but rather does it via marine biological cycling of carbon facilitated by high CO_2 concentrations (IPCC PSB 2007). The efficiency and the speed with which the ocean takes up anthropogenic CO_2 are controlled by the rate of the movement of surface waters and their mixing with deeper ocean layers. Much slower (thousands of years) processes of anthropogenic CO_2 absorption by the ocean involve its neutralization by dissolved $CaCO_3$ from sediments. It was estimated that about half of the amount of CO_2 added to the atmosphere will be removed via the carbon cycle within 30 years, and 20% may stay in the atmosphere for thousands of years (Archer 2005).

1.3.1 Concept of Radiative Forcing

The concept of radiative forcing (RF) has been introduced for the quantitative comparison of the impact of different natural and man-made drivers on the climate system; in other words, RF is a measure of how the energy balance in the Earth-atmosphere system is affected when factors that influence climate are altered. RF is defined as the rate of radiative energy change per unit area of the globe (measured at the top of the atmosphere), and it is expressed in W m^{-2} (IPCC PSB 2007). RF can be correlated with the global mean equilibrium temperature change (ΔT_S) at the Earth's surface as follows:

$$\Delta T_S = \lambda RF \tag{1.7}$$

where λ is the climate sensitivity parameter.

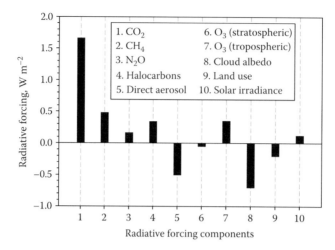

FIGURE 1.9
Changes in the radiative forcing values related to major natural and man-made factors from the start of industrial era (ca. 1750) until 2005. (Data from IPCC PSB, The Intergovernmental Panel on Climate Change 4th Assessment Report, Climate Change 2007, *The Physical Science Basis*, Cambridge University Press, Cambridge, U.K., Table 2.12, p. 204, 2007.)

According to the definition, the RF value is positive when energy of the Earth-atmosphere system increases (i.e., warming effect) in response to a factor(s). Correspondingly, the RF is negative if a factor causes the energy of the Earth-atmosphere system to decrease and, consequently, the system to cool down. The RF values for some major natural and man-made factors are shown in Figure 1.9. Note that the RF values depicted in Figure 1.9 reflect RF changes since the beginning of the industrial era (ca. 1750) until 2005.

Natural RFs mainly result from changes in solar irradiance and major volcanic eruptions. Solar source forcings arise from several direct and indirect factors: (1) a gradual increase in a solar output, (2) the natural cyclic changes in solar irradiance that occur over 11 year periods, and (3) an indirect effect on the forcing via the changes in the abundance of some GHG in the atmosphere (e.g., stratospheric ozone) (IPCC PSB 2007). The resulting net solar forcing value is slightly positive. Major (explosive) volcanic eruptions spewing immense amounts of sulfate-type aerosols into the atmosphere create short-lived (typically, 2–3 years) negative forcing (this forcing is not reflected in Figure 1.9 due to the irregularity and relative rareness of these events).

Human activities greatly contribute to the changes in RF over the last two centuries. It is evident from Figure 1.9 that the RF values for all GHG (including CO_2, N_2O, CH_4, water vapor, and halocarbons) are positive, with the RF for CO_2 having the greatest increase in value. The RF for ozone depends on where the ozone molecules are located: in troposphere (positive RF), or in stratosphere (negative RF); the net balance for ozone RF is positive, that is, it does contribute to warming. The forcing effect of aerosols is rather complex because they could directly or indirectly affect the system. The direct forcing effect involves reflection and absorption of solar and long-wavelength radiation in the atmosphere, whereas indirect effect relates to the changes aerosol particles exert on the optical properties of clouds. The net effect of all aerosol types results in negative RF values. Cumulatively, long-lived GHG have significant RF value of $2.63 \pm 0.26\,W\,m^{-2}$ (RF for tropospheric ozone is $0.35\,W\,m^{-2}$).

Since the beginning of the industrial era, human activities not only altered the abundance of atmospheric GHG, but also changed the land cover over the vast areas on the

Earth's surface, mainly through agriculture and deforestation. These activities directly (e.g., via altering the reflectivity of the land surface) and indirectly (via changing the atmospheric concentrations of CO_2 and CH_4) resulted in appreciable changes in the forcings values. As seen from the Figure 1.9, the changes in the land use result in a negative forcing.

A new type of forcing has emerged over the last half a century, namely due to aircraft operations. During flights, aircrafts leave linear trails of condensation (contrails) that reflect solar radiation and absorb long-wavelength radiation. It was estimated that the contrails noticeably increased the Earth's cloudiness and have caused slight (about 0.01 W m^{-2}) positive increase in the RF value (IPCC PSB 2007) (due to its small value, this forcing is not shown in Figure 1.9). The RF impact of aviation in 2000 (both negative and positive) is summarized as follows (in mW m^{-2}) (Sausen et al. 2005):

Related to CO_2 emissions	+25.3
NO_x-assisted O_3 production	+21.9
NO_x-assisted ambient CH_4 reduction	−10.4
H_2O formation	+2.0
Sulfate particles	−3.5
Soot particles	+2.5
Contrails	+10.0
Overall	+47.8

This estimate does not include the cirrus cloud enhancement caused by aviation (for which no best estimate could be made; however, some sources give the range of 10–80 mW m^{-2}) (IPCC Mitigation 2007). Thus, taking into account the estimate of the total net anthropogenic RF in 2005 of 1.6 W m^{-2} (range of 0.6–2.4 W m^{-2}), the share of aviation would amount to about 3% (range of 2%–8%) of the total (IPCC Mitigation 2007).

Figure 1.10 depicts timescales representing the length of time during which a given RF term would persist in the atmosphere after associated emissions or changes are ceased

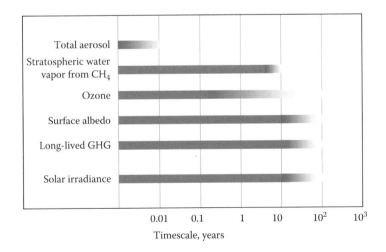

FIGURE 1.10
Timescales for different radiative forcing terms persisting in the atmosphere after associated emissions and changes are ceased. (Data from The Intergovernmental Panel on Climate Change 4th Assessment Report, Climate Change 2007, *The Physical Science Basis*, Cambridge University Press, Cambridge, U.K., Table 2.20, p. 203, 2007.)

(IPCC PSB 2007). The data presented in this figure indicate that the lifetime of various agents in the atmosphere could last from hours to days for aerosols, to tens to hundred years for long-lived GHG and surface albedo (including land use changes). The timescale for CO_2 is not presented in Figure 1.10 because of the high degree of uncertainty associated with the complex and prolonged processes responsible for the removal of CO_2 from the atmosphere.

Summarizing this section, Figure 1.9 shows that the changes in RF resulting from human activities far exceed that from natural sources. The implications of this comparative assessment data are that human activities can potentially impact the climate system more profoundly compared to natural sources.

1.3.2 Global Warming Potential of GHG

Global warming potential (GWP) is a metric tool (or an index) that is widely used to compare the capacity of different greenhouse agents to contribute to global warming. According to an international convention, GWP of CO_2 is a unity (i.e., $GWP_{CO_2} = 1$). GWP of a greenhouse agent is controlled by the following factors:

- Its capacity to absorb IR radiation
- Its spectral properties (i.e., which region of the spectrum relates to its maximum absorbing capacity)
- Its atmospheric lifetime

Thus, greater IR absorption capacity and longer atmospheric lifetime would yield greater values of the GWP index. The calculation of GWP values is based on the time-integrated global mean RF value of the agent (i) 1 kg of which was pulse-emitted to the atmosphere relative to 1 kg of the reference gas CO_2 (IPCC 1990). Thus, GWP of the agent (i) can be defined as follows:

$$GWP_i = \frac{\int_0^{TH} RF_i(t)dt}{\int_0^{TH} RF_{CO_2}(t)dt} = \frac{\int_0^{TH} a_i[C_i(t)]dt}{\int_0^{TH} a_{CO_2}[C_{CO_2}(t)]dt} \tag{1.8}$$

where
  TH is time horizon
  RF_i is the global mean RF of the agent i
  a_i is the RF per unit mass increase in atmospheric abundance of the agent i
  $[C_i(t)]$ is the time-dependent abundance of the agent i
  The denominator includes the corresponding values for the reference gas CO_2

The GWP values of the numerator and denominator are called the absolute GWP (AGWP) of the agent i and CO_2, respectively. The AGWP values for CO_2 for 20, 100, and 500 year time horizon are ($\times 10^{-14}$ W m^{-2} year (kg CO_2)$^{-1}$): 2.47, 8.69, and 28.6, respectively (with the uncertainty of $\pm 15\%$) (IPCC PSB 2007). Figure 1.11 summarizes the GWP values for selected (well-mixed) GHG for the time horizons of 20, 100, and 500 years (uncertainty of the GWP values are $\pm 35\%$) (IPCC/TEAP 2005). It can be seen that the GWP values vary in the range from a few units to tens of thousands. Note that GWP of methane is 25–72 times greater than that of CO_2 over the time horizon of 20–100 years. Indirect GWP

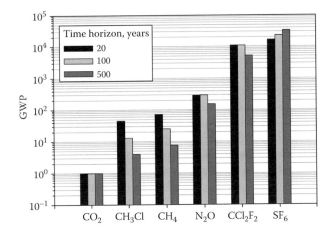

FIGURE 1.11
Global warming potentials for selected greenhouse gases for the time horizon of 20, 100, and 500 years. (Data from The Intergovernmental Panel on Climate Change 4th Assessment Report, Climate Change 2007, *The Physical Science Basis*, Cambridge University Press, Cambridge, U.K., Table 2.14, p. 212, 2007.)

relate to the effects of degradation products or the changes in the concentrations of GHG resulted from the presence of degradation products. For example, the indirect radiative effects of methane relate to changes in tropospheric ozone and stratospheric water vapor levels, and CO_2 concentration.

1.3.3 Radiatively Important Gases

This section discusses radiatively and chemically important gases from the viewpoint of a link between human activities, the carbon cycle, and the climate system.

1.3.3.1 Atmospheric CO_2

The atmospheric CO_2 concentration is one of the major parameters of the global carbon inventory, and it has been measured with great accuracy for many decades. The first continuous measurements of atmospheric CO_2 concentrations were conducted by Keeling at Mauna Loa, Hawaii since 1958 (a high-precision IR gas analyzer was used for this purpose). This was followed by continuous in situ CO_2 measurements at other sites in both hemispheres, for example, Baring Head (New Zealand), Cape Grim (Australia), and the South Pole. The selection of the above locations was dictated by the lack of nearby CO_2 sources or sinks. A totality of direct and indirect atmospheric CO_2 measurements indicate that it has increased from 275 to 285 ppmv during the preindustrial period to 379 ppmv in 2005 (i.e., the growth by 36% over 250 years) (IPCC PSB 2007). In June 2010, Mauna Loa Observatory reported atmospheric CO_2 concentration of 392.04 ppm (i.e., 13 ppm or 3.4% increase in just last 5 years) (CO_2 Now 2010). Most importantly, the measurements confirmed that the rate of CO_2 growth is accelerating, that is, the first and the second 50 ppmv growth were achieved after 200 and 30 years, respectively. The average rate of increase during the 1955–1995 period was 1.4 ppm year^{-1}, during 1995–2005 decade it was 1.9 ppm year^{-1}, and during the last 5 years (2005–2010) it jumped to 2.6 ppm year^{-1}. This is in a correlation with the rate of increase in the amount of carbon in the atmosphere from 3.2 ± 0.1 Gt C year^{-1} in the 1990s to 4.1 ± 0.1 Gt C year^{-1} in the period of 2000–2005 (these values are the net effect of several carbon fluxes between major reservoirs). Based on the observations over the 1995–2005 decade and assuming carbon emissions of 7 Gt C year^{-1} and a remaining airborne fraction of 60%, the global atmospheric CO_2 growth was projected at 1.9 ppmv year^{-1} (IPCC PSB 2007) (apparently, these projections have proved to be rather conservative).

To assess short- and long-term changes in the atmospheric carbon content, the "airborne fraction" parameter (which is the ratio of atmospheric increase in CO_2 concentration to fossil fuel emissions) has been introduced. For the last half a century, the airborne fraction has averaged at 0.55, which indicates that the oceans and terrestrial ecosystems have cumulatively removed about 45% of anthropogenic CO_2 from the atmosphere over this period (IPCC PSB 2007). Another indication of the link between anthropogenic CO_2 emissions and atmospheric CO_2 concentration is the existence of an interhemispheric CO_2 gradient. It was determined that there is a gradient in the atmospheric CO_2 concentrations in the Northern and Southern Hemispheres, which correlates with the CO_2 emissions due to fossil fuel combustion. In particular, the predominance of fossil-derived CO_2 emissions in the Northern Hemisphere (compared to the Southern Hemisphere) causes the occurrence of CO_2 gradient in the amount of about 0.5 ppm per Gt C year^{-1} (IPCC PSB 2007).

RF index for CO_2 in 2005 is estimated at $RF_{CO_2} = 1.66 \pm 0.17$ W m^{-2} (corresponding to the atmospheric CO_2 concentration of 379 ± 0.65 ppm), which is the largest RF among all major forcing factors shown in Figure 1.9. Over the 1995–2005 decade RF_{CO_2} increased by about 0.28 W m^{-2} (or about 20% increase), which represents the largest increase in RF_{CO_2} for any decade since the beginning of the industrial era, and this change is also the largest among all forcing agents. Furthermore, the data show that the changes in the land use greatly contributed to the RF_{CO_2} value in the amount of about 0.4 W m^{-2} (since 1850). It is implied that the remaining three quarters of RF_{CO_2} can be attributed to burning fossil fuels, cement manufacturing, and other industrial emitters of CO_2 (IPCC PSB 2007).

Summarizing, the consistent values of the airborne fraction parameter over decades combined with the existence of a relationship between the interhemispheric CO_2 gradient and the amount of fossil-derived emissions, combined with the recent dramatic changes in RF_{CO_2} values over the last decades unequivocally suggest that *there is a link* between the increases in the atmospheric CO_2 level and man-made CO_2 emissions.

1.3.3.2 Methane

Methane (CH_4) possesses the second largest (after CO_2) RF value of $RF_{CH_4} = 0.48 \pm 0.05$ W m^{-2} among GHG (see Figure 1.9), and it is, next to CO_2, the most important carbon-bearing compound that could potentially affect the global climate system. As pointed out in the previous section, methane is 25–72 times more potent GHG than CO_2 (depending on the length of the time horizon considered: 100 or 20 years). The data indicate that over the last thousands of years atmospheric methane concentration varied in the range of 400–700 ppb (Spahni et al. 2005). The measurements of CH_4 levels in the atmosphere of both Northern and Southern Hemispheres conducted in 2005 gave the value of $1,774.62 \pm 1.68$ ppb (IPCC PSB 2007). More recent (2007) measurements of the atmospheric CH_4 concentrations conducted in the Northern (Mace Head, Ireland) and Southern (Cape Grim, Tasmania) Hemispheres gave the values of 1865 and 1741 ppb, respectively (Blasing 2009). It should be noted that although over the last quarter of a century the methane concentration increased by about 30%, its growth rate substantially slowed down in the late 1990s for reasons yet to be understood.

Atmospheric methane concentration is controlled by the balance between its sources and sinks. Figure 1.12 depicts sources and sinks of methane for the period of 1983–2004.

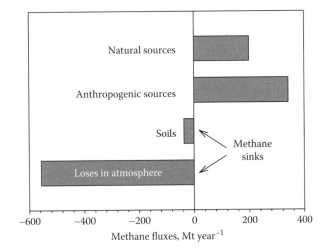

FIGURE 1.12
Sources and sinks of methane for the period of 1983–2004. (Data from The Intergovernmental Panel on Climate Change 4th Assessment Report, Climate Change 2007, *The Physical Science Basis*, Cambridge University Press, Cambridge, U.K., Table 7.6, p. 542, 2007.)

As seen from Figure 1.12, methane is emitted by both natural and man-made sources with the latter being a predominant source. The major natural sources of methane are wetlands, permafrost, vegetation, termites, oceans, methane hydrates, and geothermal sources (mud volcanoes, marine and land seepage, etc.) (emissions from the geological sources are rather significant and estimated at 40–60 Mt CH_4 year^{-1}). The amount of methane (in carbon equivalent) in wetlands and permafrost soils, including subsea permafrost is about twice of the amount of CO_2 in the atmosphere (Heimann 2010). The man-made sources include NG processing facilities, fossil power production, landfills, ruminant animals (cattle, sheep, etc.), rice agriculture, biomass processing, and combustion. It was reported that methane emission from living vegetation accounts for 10%–30% of the total methane emissions (Keppler et al. 2006).

Most biogenic sources of methane involve a complex sequence of primary and secondary fermentation processes with the intermediate formation of carboxylic acids (primarily, acetic acid, CH_3COOH) that are finally converted to methane by methanogenic *Archaea* (IPCC PSB 2007):

$$CH_3COOH \rightarrow CH_4 + CO_2 \qquad (1.9)$$

$$CO_2 + 4H_2 \rightarrow CH_4 + 2H_2O \qquad (1.10)$$

The residence time of methane in the atmosphere is estimated at 8.4 years. Taking into consideration a conversion factor of 2.78 Mt (CH_4) per ppb, and the atmospheric methane concentration of 1774 ppb, the amount of atmospheric methane in 2005 was estimated at 4932 Mt (CH_4). With an annual increase of about 0.6 Mt (CH_4) year^{-1} over the period of 2000–2005, the overall average methane emissions amount to 582 Mt (CH_4) year^{-1} (IPCC PSB 2007).

Methane is primarily removed from the atmosphere via chemical and photochemical chain reactions, which involve extremely reactive hydroxyl radicals (OH•). The primary source of tropospheric OH$^\bullet$-radicals is the UV-light-assisted ozone photodissociation in the presence of water vapor (Hofzumahaus et al. 2004) as follows:

$$O_3 + h\nu \rightarrow O_2 + O(^1D) \quad \text{(radiation wavelengths below 350 nm)} \tag{1.11}$$

where
  O(^{1}D) is an electronically excited state of oxygen atom
  hν is a light photon

$$O\left(^1D\right) + H_2O \rightarrow 2\,OH^\bullet \tag{1.12}$$

$$OH^\bullet + O_3 \rightarrow HO_2^\bullet + O_2 \tag{1.13}$$

$$HO_2^\bullet + O_3 \rightarrow OH^\bullet + 2O_2 \tag{1.14}$$

Additional sources of OH$^\bullet$ are the reactions of UV photolysis of nitrous acid (HONO), nitric acid, and peroxides:

$$HONO + h\nu \rightarrow OH^\bullet + NO \quad \text{(radiation wavelengths of 300–400 nm)} \tag{1.15}$$

$$HNO_3 + h\nu \rightarrow OH^\bullet + NO_2 \quad \text{(radiation wavelengths below 330 nm)} \tag{1.16}$$

$$H_2O_2 + h\nu \rightarrow 2OH^\bullet \quad \text{(radiation wavelengths below 320 nm)} \tag{1.17}$$

OH$^\bullet$-radicals, and to some extent O(^{1}D), attack methane molecules converting them first to oxygenated compounds (e.g., formaldehyde, CHOH) and then to CO_2 and water as follows:

$$OH^\bullet + CH_4 \rightarrow CH_3^\bullet + H_2O \tag{1.18}$$

$$O(^1D) + CH_4 \rightarrow CH_3^\bullet + OH^\bullet \tag{1.19}$$

$$CH_3^\bullet + O_2 + h\nu \rightarrow \cdots \rightarrow CHOH + O_2 + h\nu \rightarrow \cdots \rightarrow CO_2 + H_2O \tag{1.20}$$

Other important methane sinks include biological oxidation in soil and the loss to the stratosphere. The reaction with chlorine radicals was proposed as an additional sink amounting to about 19 Mt (CH$_4$) year^{-1} (Allan et al. 2005). The methane sink strength is estimated in the International Panel on Climate Change (IPCC) Fourth Assessment Report at 581 Mt (CH$_4$) year^{-1} (IPCC PSB 2007). Since most of the methane would eventually be transformed to CO$_2$ via a variety of oxidation routes, methane can be considered part of the global carbon cycle.

A possible effect of methane on the climate system and vice versa is in the focus of intensive studies, because of the high sensitivity of methane biochemistry to temperature and water level changes. Several authors reported a significant increase in CH$_4$ emissions from northern peatlands due to permafrost melting (Wickland et al. 2005). In particular, simulation studies predicted that CH$_4$ emissions from Scotland wetlands would likely increase by 17%, 30%, and 60% if climate warms up by 1.5°C, 2.5°C, and 4.5°C, respectively (Chapman and Thurlow 1996). Modeling studies by Shindel et al. (2004) project that doubling in the atmospheric CO_2 levels would result in 3.4°C warming and, consequently, in 78% increase in CH_4 emissions from wetlands (Shindel et al. 2004).

A recent study published in the *Science* magazine reported that a large amount of methane is seeping from East Siberian Arctic Shelf (ESAS) sediments into the atmosphere (Shakhova et al. 2010). The researchers believe that the sustained release of methane to the atmosphere from thawing Arctic permafrost in all likelihood could be attributed to a positive feedback effect. It was estimated that the annual outgassing from the shallow ESAS areas is about 8 Tg (CH_4) year^{-1}, which is of the same magnitude as the total methane emissions from all oceans (Tg = 10^{12} g). Although this amount of carbon seems to be insignificant when compared to global emissions of about 440 Tg C (as methane) year^{-1} (Heimann 2010), it is very important to determine whether the extensive venting of methane from ESAS is a stable process or it signals the beginning of massive methane release period, which would have unpredictable consequences for climate. The finding also adds a new important variable for climate change models that deals with the potential effects of oceanic warming (Wilson 2010).

Another reason for the concern is that the changes in climate could potentially affect the stability of methane hydrates on the seafloor, where an estimated amount of $\sim 4 \times 10^6$ Tg of CH_4 is stored (Buffet and Archer 2004). Modeling studies suggested that the increase in deep water temperature of 3°C would release about 85% of methane from methane hydrates (Buffet and Archer 2004). The models project that the release of CO_2 as a result of human activities would cause the release of about 2000 Gt of methane from hydrates on a timescale from one thousand to hundred thousands of years (Archer and Buffet 2005). Thus, the above data clearly demonstrate that the release of methane from methane hydrates is a potentially very significant and worrisome positive feedback to the increase in anthropogenic CO_2 emissions.

1.3.3.3 Other Greenhouse Gases and Aerosols

Other major GHG that are of concern with regard to their potential impact on the climate system include (in the order of decreasing RF values): halocarbons, ozone, N_2O, and water vapor. The detailed description of the specifics of their climate changing action is out of the scope of this chapter. In this section, we just briefly describe the agents, their major sources, and recent trends in their abundance in the atmosphere. Because of their tangible effect on the optical properties of the atmosphere, aerosols are also included in this section.

1.3.3.3.1 Halocarbons

The term halocarbons encompasses several types of synthetic molecules typically containing carbon and halogen atoms (sometimes, sulfur and halogens), for example, chlorofluorocarbons (CFC), hydrofluorocarbons (HFC), hydrochlorofluorocarbons (HCFC), perfluorocarbons (PFC), halons, and sulfur hexafluoride (note that very small amounts of some halocarbons, e.g., methyl bromides and chlorides, perfluoromethane occur naturally). The concentrations of these compounds in the atmosphere ranges from few to hundreds ppb, for example, for CFC-12 it equaled to 536–538 ppb (measured in 2006–2007) (Blasing 2009), and their atmospheric lifetime ranges from 45 to 100 years for CFCs and thousands of years for perfluorocarbons. The totality of these compounds have relatively large RF index of RF = 0.337 W m^{-2} (IPCC PSB 2007). The halocarbons were produced by the chemical industry since 1930s for a variety of applications (e.g., refrigeration), and they were leaking since then causing their concentration in the atmosphere to steadily increase up until the 1990s when CFCs were phased out under the Montreal Protocol. Note that the primary objective of this ban was not related to heat-trapping properties of halocarbons, but rather to their ozone-layer-depleting capacity. Since halocarbons are relatively chemically inert and do not react in the troposphere, they diffuse to the stratosphere, where they are exposed to solar UV radiation and photodissociate with the release of chemically reactive chlorine atoms (Cl):

$$CCl_2F_2 + hv \rightarrow CClF_2 + Cl \quad \text{(radiation wavelengths below 230 nm)} \tag{1.21}$$

The produced Cl atoms initiate free radical chain reactions leading to the consumption of ozone:

$$Cl + O_3 \rightarrow ClO + O_2 \tag{1.22}$$

$$ClO + O \rightarrow Cl + O_2 \tag{1.23}$$

Due to the slow rates of diffusion and photodissociation processes, it would take about 10 years from the time of the halocarbons release for the ozone concentration to be affected (Okabe 1978). As a result of the ban on halocarbons, the changes in their RF values are rather insignificant over the period of 1998–2005 (only a slight positive change of +1%, which is mostly due to production of HCFCs, HFCs, and PFCs that will be phased out by 2030) (IPCC PSB 2007).

1.3.3.3.2 Ozone

Ozone (O_3) plays an important role in protecting living things on the Earth's surface from the ruinous effect of solar UV radiation. Due to the presence of ozone in the stratosphere, UV radiation with wavelengths below 300 nm does reach the Earth's surface. Ozone molecules are continuously produced (mostly via photochemical reactions) and decomposed (mostly via free radical chain reactions) in the atmosphere. The equilibrium concentration of ozone in the troposphere is 32 ppb (Blasing 2009). The following reactions control the ozone concentration in the stratosphere (Okabe 1978):

$$O_2 + hv \rightarrow 2O \quad \text{(radiation wavelengths below 240 nm)} \tag{1.24}$$

$$O + O_2 + M \rightarrow O_3 + M \quad \text{(where M is } N_2 \text{ or } O_2) \tag{1.25}$$

$$O + O_3 \rightarrow 2O_2 \tag{1.26}$$

$$O_3 + hv \rightarrow O + O_2 \quad \text{(radiation wavelengths below 900 nm)} \tag{1.27}$$

The maximum equilibrium concentration of ozone is established at an altitude of about 25 km, depending on the intensity of the solar flux and the rates of formation and destruction of ozone (see Figure 1.13).

Man-made emissions of such gases as carbon monoxide, some hydrocarbons, and nitrogen oxides cause the ozone concentration to increase in the troposphere (positive RF) (this explains the fact that the ozone concentration is considerably higher in urban than in rural areas). At the same time, anthropogenic halocarbon emissions reaching the stratosphere have the capacity to destroy the ozone layer (see above), causing negative RF effect. The net RF due to ozone presence in the Earth's atmosphere is positive (i.e., being conducive

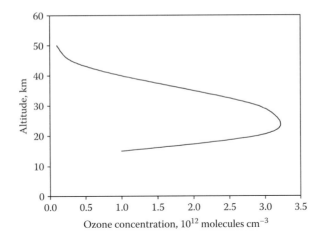

FIGURE 1.13
Ozone concentration in the stratosphere as a function of the altitude. (Data from Okabe, H., *Photochemistry of Small Molecules*, John Wiley & Sons, New York, Table VIII-1B, p. 341, 1978.)

to climate warming). It was estimated that the tropospheric ozone concentration has increased by 20%–50% since the beginning of the industrial era; thus, it could be linked to human activities (IPCC PSB 2007).

1.3.3.3.3 Nitrous Oxide (N$_2$O)

N$_2$O emissions could originate from both natural and man-made sources (about an equal contribution). Its concentration in the atmosphere was determined to be 319 ± 0.12 ppb (2005) and 321–322 ppb (2007), which is an increase of 5–8 ppb since 1998 (IPCC PSB 2007; Blasing 2009). The main natural sources of N$_2$O emissions to the atmosphere include biochemical processes in soils (especially, tropical soils) and the oceans, and photochemical oxidation of biogenic ammonia in the atmosphere. Man-made N$_2$O emissions to the atmosphere mainly originate from biomass burning, agricultural activities (i.e., biochemical transformation of nitrogen-based fertilizers in soils to N$_2$O), cattle, and chemical industry (e.g., nylon manufacture). The N$_2$O lifetime in the atmosphere is roughly 114 years (N$_2$O is removed from the atmosphere mainly through its oxidative transformation to nitrogen oxides and nitric acid). N$_2$O has a relatively small RF index of 0.16 W m^{-2} (2005), and its value has increased by 11% since 1998 (IPCC PSB 2007; Blasing 2009).

1.3.3.3.4 Water Vapor

Water vapor is one of the key climate variables and, at the same time, is the most abundant GHG in the atmosphere. It is estimated that water vapor accounts for the lion's share of IR radiation absorption in the atmosphere and, hence, about 60% of the natural greenhouse effect (for clear skies) (Kiehl and Trenberth 1997). Despite the abundance of water vapor in the atmosphere and its significance as a greenhouse agent, human activities have an insignificant *direct* effect on the atmospheric concentration of water vapor (note that the concentration of water vapor in the lower atmosphere is typically represented as vapor pressure, relative humidity, or a dew point). This is reflected in a relatively low (positive) RF value for stratospheric water vapor. However, human activities could profoundly *indirectly* affect the atmospheric water vapor concentration and, ultimately, the climate system, via a number of mechanisms. For example, the accumulation of other GHG could warm up the atmosphere and increase its water vapor content by facilitating the evaporation process. It was determined that for every 1°C increase in global temperature, the specific humidity rises in average by about 4.9% (over the ocean and land surfaces the increases are 5.7% and 4.3% per each 1°C, respectively) (IPCC PSB 2007). Another indirect mechanism of water vapor formation in the stratosphere involves photochemical oxidation of anthropogenic methane or other hydrogen-containing compounds originating from human activities (e.g., ammonia). Due to the sensitivity of water vapor content to temperature, a variety of models predict that this factor could provide the largest positive feedback with regard to the climate system (Held and Soden 2000).

Recently, an international group of atmospheric scientists from the U.S. National Oceanic and Atmospheric Administration (NOAA) and other institutions published a paper in the *Science* magazine, where they demonstrated a direct link between the changes in the concentration of stratospheric water vapor and mean surface temperature (Solomon et al. 2010). In particular, the authors argued that stratospheric water vapor concentrations probably increased between 1980 and 2000 causing a decade-long (i.e., 1990–2000) surface warming by about 30% (compared with the estimates neglecting this change). Since the year 2000, however, stratospheric water vapor concentration dropped by about 10%, which explains the trend in global surface temperature remaining almost flat since the late 1990s despite continuing increases in the emissions of other GHG, for example, CO_2, CH_4, N_2O, and halocarbons. These observations indicate that stratospheric water vapor is an important driver of global surface temperature change, and the terms related to water vapor impact should be fully represented in climate models.

1.3.3.3.5 Aerosols

Aerosols represent extremely small (about 1–10 µm in size) airborne particles of different chemical nature that enter the atmosphere as a result of both natural processes and human activities. The aerosols could potentially alter the optical properties of the atmosphere, for example, via reflecting, scattering, or absorbing incoming solar radiation and, thus, affecting the energy balance of the system. Naturally occurring aerosols include all kinds of mineral dust that became airborne by wind blowing over the land surface, sea salt aerosols released during stormy weather, sulfate aerosols from volcanic eruptions, micron-sized particles of biological origin, and others. The reported estimates for the dust and sea salt aerosols are 1,000–3,000 and 16,300 Tg year^{-1}, respectively (Textor et al. 2005).

Man-made aerosols mostly relate to fossil fuel and biomass burning, transportation systems (e.g., diesel fuel derived soot), coal and ore surface mining, etc. Three types of aerosols are predominant among a variety of man-made aerosols, namely, sulfate, organic carbon, and black carbon aerosols. Sulfate aerosols are mainly produced by the reactions involving fossil-based SO_2 emissions, and to the lesser extent natural dimethyl sulfide emissions (from marine phytoplankton) and SO_2 of volcanic origin. Globally, the amount of sulfate aerosols is estimated at the range of 91.7–125.5 Tg(S) year^{-1}, with the negative RF value of -0.4 ± 0.02 W m^{-2} (IPCC PSB 2007). Organic and black carbon aerosols mainly originate from combusting fossil fuels, biofuels, and other industrial processes. These two types of carbon-based aerosols have different optical properties: while organic aerosols weakly absorb solar radiation, black carbon is known as a strong absorber of solar radiation. The overall emission of organic and black carbon aerosols due to fossil fuel and biofuel combustion is estimated at 9.7 and 8 Tg C year^{-1}, respectively (Bond et al. 2004; Ito and Penner 2005). The amount of black carbon aerosols increased by a factor of about 3 between 1950 and 2000, and since then has fallen due to emission control measures (Ito and Penner 2005). RF values for organic aerosols are estimated between -0.10 and -0.24 W m^{-2} (depending on the type of fuel combusted: fossil fuel or biofuel), whereas RF value for black carbon from fossil fuels is estimated at +0.2 W m^{-2} (IPCC PSB 2007). Overall, the presence of aerosols in the atmosphere causes negative RF (see Figure 1.9).

1.3.4 Sources of Anthropogenic CO$_2$ Emissions

Figure 1.14 shows the quantities and relative distribution of major GHG (CO$_2$, CH$_4$, N$_2$O, and halocarbons or F-gases) emissions (worldwide data for the year 2005) (WEO 2009). CO$_2$ makes up more than three quarters of the total amount of GHG emissions, with the most of CO$_2$ coming from energy-related sources followed by CO$_2$ from land use change and forestry (LUCF) and, finally, industrial CO$_2$ (which includes nonenergy uses of fossil fuels, gas flaring, industrial emissions). Methane sources (including coal mines, gas leakages, and fugitive emissions) also make a sizable contribution (15.1%) to the overall GHG emissions. N$_2$O (from industry and waste) and F-gases (which mainly include HFC, PFC, and SF$_6$ and other halocarbons) are less important contributors. Since CO$_2$ is a major player among all anthropogenic GHG, the following detailed discussion of different emission sources will be exclusively focused on CO$_2$.

CO$_2$ is emitted by a large number of diverse sources: large power plants and industrial facilities, as well as a myriad of relatively small residential and mobile (transportation) sources. There are many ways to categorize CO$_2$ emission sources, for example, by industrial sector, scale of emissions, CO$_2$ content, and geographical distribution. Some of these CO$_2$ source classifications are presented in this section (only anthropogenic CO$_2$ sources are considered).

1.3.4.1 Classification by CO$_2$ Source Type

There are four major types of CO$_2$ sources; they are presented in Figure 1.15. In the fuel-combustion-type sources, carbon directly reacts with oxygen of air producing CO$_2$, for example, in case of coal-burning power plants (for simplicity, coal is presented as carbon):

$$C + O_2 \rightarrow CO_2 \qquad (1.28)$$

or gasoline-burning transportation:

$$C_nH_m + \left(\frac{n+m}{4}\right)O_2 \rightarrow nCO_2 + \frac{m}{2}H_2O \qquad (1.29)$$

where C$_n$H$_m$ is a generic formula for gasoline.

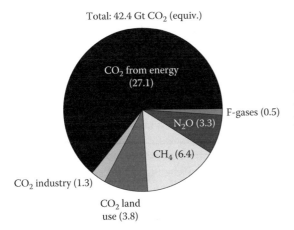

Total: 42.4 Gt CO$_2$ (equiv.)

CO$_2$ from energy (27.1)

F-gases (0.5)

N$_2$O (3.3)

CH$_4$ (6.4)

CO$_2$ industry (1.3)

CO$_2$ land use (3.8)

FIGURE 1.14
World anthropogenic greenhouse gas emissions by source in 2005 (in Gt CO$_2$-equiv.). F-gases: hydrofluorocarbons, perfluorocarbons, and SF$_6$ (from industry). Methane emissions include coal mines, gas leakages, and fugitive emissions. N$_2$O originates from industry and wastes. Land use includes CO$_2$ emissions due to land use change and forestry. (Data from WEO, *World Energy Outlook*, International Energy Agency (IEA), IEA Publications, Paris, France, p. 170, 2009.)

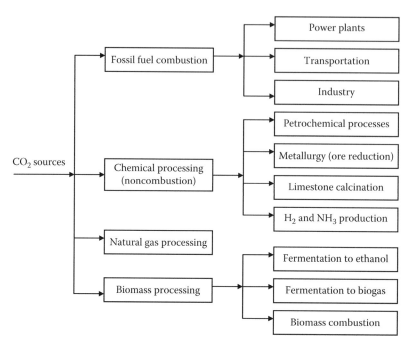

FIGURE 1.15
CO_2 classification by CO_2 source type.

CO₂ could also be emitted from industrial sources that do not involve fuel combustion, for example, as a by-product of a chemical process. Important industrial processes of steam methane reforming (SMR), iron ore reduction by coal, and limestone calcination are examples of such CO_2 sources (overall processes are shown here):

$$CH_4 + 2H_2O \rightarrow CO_2 + 4H_2 \quad \text{(steam methane reforming)} \tag{1.30}$$

$$2Fe_2O_3 + 3C \rightarrow 4Fe + 3CO_2 \quad \text{(iron ore reduction)} \tag{1.31}$$

$$CaCO_3 \rightarrow CaO + CO_2 \quad \text{(limestone calcination)} \tag{1.32}$$

In most of the cases where CO_2 is formed as a by-product, the additional CO_2 is also emitted as a result of fuel combustion (e.g., SMR, limestone calcinations, etc.).

CO_2 is commonly present in NG where its concentrations range from several tenths to several tens volume percent (e.g., CO_2 concentration in NG from Natuna gas field, Indonesia, is 71 vol.%). When CO_2 concentration is above a certain threshold value dictated by economics of NG pipeline transport or specifics of the cryogenic CO_2 liquefaction process, it has to be removed. Biomass, although considered as a carbon-neutral source, is also a significant source of CO_2 emissions via a variety of fermentations, gasification, and combustion processes. Fermentative ethanol plants in the United States and Brazil emit CO_2 emissions in the order of 0.1–0.14 Mt CO_2 year⁻¹ (each) (Kheshgi and Prince 2005). CO_2 emissions from largest (more than 0.1 Mt CO_2 year⁻¹) U.S. stationary sources are presented in Figure 1.16. Power generation and industry are dominant (86%) stationary sources of CO_2 emissions.

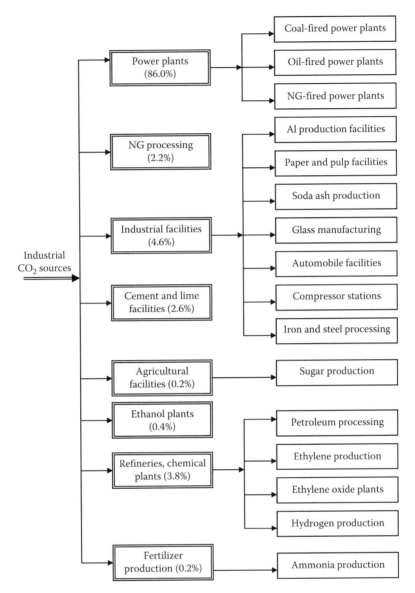

FIGURE 1.16
U.S. stationary CO_2 point sources by the industry type. (Adapted from Jensen, M., CO_2 point source emission estimation methodologies summary, NETL, U.S. DOE Carbon Capture and Sequestration Program, *Regional Carbon Sequestration Partnership's Annual Meeting*, December 12, 2007.)

1.3.4.2 Classification by Industry Sector

Figure 1.17 summarizes the quantity of CO_2 emissions and the relative distribution of CO_2 sources by the industry sector (the data for 1971 and 2007 are presented). Electricity/ heat generation and transport are two major sources of CO_2 emissions: their combined share of CO_2 emissions grew from about half in 1971 to about two-thirds of the global CO_2 emissions in 2007 (IEA Statistics 2009). The emissions from electricity/heat generation increased at faster rates compared to other sectors and their share grew from 27% in 1971 to 41% in 2007. The recent worldwide drastic increase in electricity demand is particularly

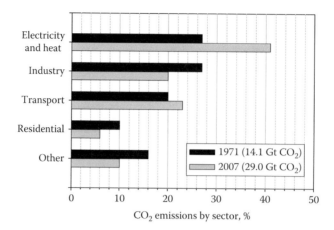

FIGURE 1.17
World CO_2 emissions in 1971 and 2007 organized by sectors. "Other" include agriculture/forestry, commercial and public services, and energy industry (other than electricity and heat). (Data from IEA Statistics, *International Energy Agency Statistics, CO_2 Emissions from Fossil Fuel Combustion*, 2009 edn., IEA, Paris, France, p. 11, 2009.)

troublesome. To put the numbers in a perspective: in order to satisfy the exploding global electricity demand caused by the introduction of flat-screen TVs, game consoles, personal computers, and other electronic gadgets, the equivalent of 560 coal-fired power plants will have to be built over the next two decades (The Week 2009). Over 70% of world electricity and heat generation is provided by fossil fuels, mainly, coal, oil, and gas.

In a contrast to the fuel-flexible electricity/heat generation sector, the transportation sector almost exclusively relies on oil (94% in 2007) (IEA Statistics 2009). CO_2 emission from oil consumption in transport almost doubled since 1971 reaching 60% of the overall oil-related CO_2 emissions in 2007. The increase in demand for transport can be attributed to the relative economic growth (before the economic downturn in 2008), especially, in China and India (in general, car ownership grows with the per capita income). Fuel prices could also be factor determining the choice of vehicle and distance traveled. For example, in the United States, over several decades, relatively low fuel prices encouraged the use of larger cars and longer travel distances (in average, 25,000 km per person per year) (IEA Statistics 2009), whereas in Europe, higher fuel prices contributed to the trend of using smaller cars with higher fuel economy and, thus, lesser CO_2 emissions. IEA projects that the transportation sector will grow by 45% by 2030 (IEA Statistics 2009). In order to reduce CO_2 emissions from this sector, a number of measures need to be implemented such as an improvement in a vehicle fuel economy, a shift from individual to public transportation, and a switch to low-carbon fuels (including, biofuels and hydrogen) and advanced electric, plug-in hybrids and fuel cell (FC) vehicles. These measures will reduce the adverse environmental impact of transport and ease the oil demand.

1.3.4.3 Classification by Type of Fuel

The amount of CO_2 emitted from an individual point source depends on the type of fuel used, which in turn is dictated by the specifics of the industrial sector utilizing this fuel. Figure 1.18 compares the shares of coal, oil, and gas in global CO_2 emissions in 1973 and 2007. Although coal represents only about a quarter of the world primary energy supply in 2007, it accounts for 42% of the global CO_2 emissions due to its high carbon content

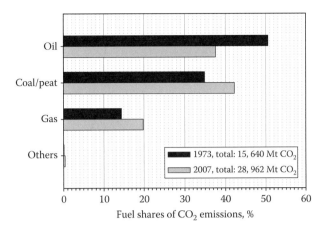

FIGURE 1.18
World CO_2 emissions in 1973 and 2007 organized by fuel. (Adapted from IEA KWES, *International Energy Agency, Key World Energy Statistics*, 2009 edn., IEA, Paris, France, www.iea.org [accessed May 5, 2010], p. 44, 2009.)

relative to its energy content (compared to gas, coal produces about twice as much CO_2 emissions per unit of generated power) (IEA Statistics 2009). Currently, coal power plants are the dominant type of power generation facilities and, will remain as such until at least 2020, and, thus, will maintain their leading role as a major CO_2 emitter in the power sector. Among countries heavily relying on coal for power production are Australia, China, India, Poland, and South Africa that produce between 68% and 95% of their electricity and heat from coal (IEA Statistics 2009). From 1973 to 2007, the share of oil decreased and the share of gas increased (in response to fuel-switching measures). Iron and steel manufacturing sector predominantly utilizes coal and petroleum coke in a blast furnace operation causing significant CO_2 emissions from this industrial sector. Cement industry, which is one of the largest CO_2 emitters, can potentially utilize different types of fuel depending on the local specifics, for example, United States, China, and India are primarily using coal, whereas Mexico is using oil and gas, Europe—a variety of fuels, including nonfossil fuels (IPCC CCS 2005). Oil refining and petrochemical sector is widely using oil and gas as primary fuel (use of petroleum coke as a supplemental fuel is also practiced). Although biomass is not utilized on a large scale worldwide as a fuel source, some countries (Brazil and Scandinavian countries) consider it an important nonfossil fuel. It is projected that by 2030, demand for electricity will almost double compared with 2007 due to the rapid increase in the standard of living in developing countries and further electrification of industrial and residential sectors.

1.3.4.4 Classification by Scale of Emissions

The International Energy Agency (IEA) has developed a database for CO_2 stationary sources that includes power plants, refineries, cement-manufacturing plants, and other major industrial sources of CO_2 emissions globally amounting to about 14,000 sources emitting summarily over $13\,Gt\,CO_2\,year^{-1}$ (IEA GHG 2002). The distribution of CO_2 sources by scale of emissions is represented in Figure 1.19. Of the 14,000 CO_2 emitting sources, about 56% (or about 8000) are stationary sources with the emission level greater than $0.1\,Mt\,CO_2\,year^{-1}$ and the remaining sources (about 6000) are the emitters with the individual capacity of less than $0.1\,Mt\,CO_2\,year^{-1}$; 85% of the overall CO_2 emissions are emitted by large stationary

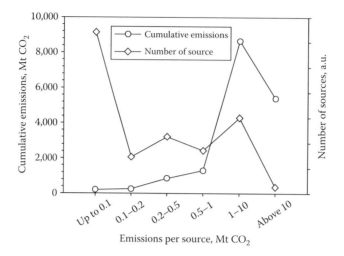

FIGURE 1.19
Sources of CO_2 emissions organized by the scale of emissions (cumulative emissions and number of sources). (Data from IEA, *International Energy Agency, GHG, Building the Cost Curves for CO_2 Storage, Part 1: Sources of CO_2,* PH4/9, 48 pp., July 2002.)

sources emitting more than $1\,Mt\,CO_2\,year^{-1}$. At the lower end, the sources emitting between 0.1 and $0.5\,Mt\,CO_2\,year^{-1}$ account for less than 10% of overall emissions. There are only 330 sources emitting more than $10\,Mt\,CO_2\,year^{-1}$.

1.3.4.5 Classification by CO_2 Content

CO_2 concentration in gas streams of industrial CO_2 emitters varies in a wide range as follows (on a dry basis): from 3 to 4 vol.% in a gas turbine exhaust to about 65 vol.% in the vent gases of NG processing facilities to almost 100 vol.% in the off-gases of fermentation plants. Figure 1.20 summarizes the data on typical CO_2 concentrations in exhaust gases from a

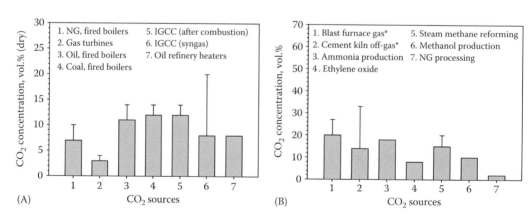

FIGURE 1.20
CO_2 concentration in gaseous streams from industrial sources of (A) fuel combustion type, (B) noncombustion type. * Accounts for CO_2 from fuel combustion. (Data from Metz, B., Davidson, O., de Coninck, H., Loos, M., and Meyer, A. (Eds.), *IPCC Special Report on Carbon Dioxide Capture and Storage,* Prepared by Working Group III of the Intergovernmental Panel on Climate Change, Cambridge University Press, Cambridge, U.K., Tables 2.1 and 2.2, pp. 79–80, 2005.)

variety of industrial sources that include both fuel combustion (A) and noncombustion (B) type CO_2 sources. CO_2 content in flue gases depends on the type of fuel (e.g., coal, oil, NG) and oxidizing gas (e.g., air, O_2-enriched air, oxygen) used. Typically, flue gases from combustion of NG in air have the lowest CO_2 concentrations, and from coal, coke, or residual oil combustion in O_2 or O_2-enriched air have the highest CO_2 concentrations.

In general, off-gases from noncombustion-type industrial CO_2 sources (Figure 1.20B) contain CO_2 in somewhat higher concentrations compared to flue gases from the combustion-type sources. Most importantly, CO_2 partial pressure in these gaseous streams is relatively high (typically, 0.1–0.5 MPa), which is about 1 order of magnitude greater than that in the flue gases of combustion-type sources (typically, 0.005–0.014 MPa), which makes them more suitable for a subsequent recovery of CO_2 (IPCC CCS 2005; IEA GHG 2002). Thus, the majority of CO_2-emitting sources produce gas streams with CO_2 concentration below 15 vol.%, and only a small fraction (less than 2%) generates by-product CO_2 with purity of 95 vol.% and higher (IPCC CCS 2005). The off-gases from sugar-to-ethanol fermentation plants contain almost pure CO_2 (with insignificant amount of organic impurities).

1.3.4.6 *Classification by Geographical Distribution of CO_2 Sources*

The databases on geographical distribution of large stationary CO_2 sources provide useful information with regard to potential CO_2 capture and storage opportunities. Figure 1.21 shows regional distribution of energy-related CO_2 emissions. It can be seen that geographically CO_2 emitting sources are predominantly clustered in four regions: Asia (30%), North America (24%), transitional economies (which include Central and Eastern Europe and former USSR republics) (13%), and OECD (Organization for Economic Cooperation and Development) West (12%). These four regions make up over 90% of the overall CO_2 emissions from power and industrial sectors (other regions account for about 6% of global CO_2 emissions) (IPCC CCS 2005). With regard to the distribution of CO_2 emissions from stationary sources, Asia is the largest emitter (5.6 Gt CO_2 year^{-1}, or 41% of the total), followed by North America (20%) and OECD Europe (13%) (the remaining regions contribute less than 10% of the total CO_2 emissions).

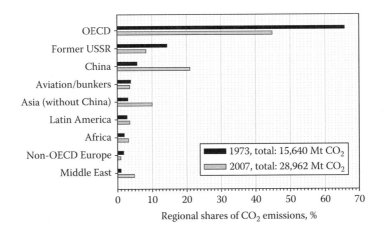

FIGURE 1.21
Geographical distribution of energy-related CO_2 emissions in 1973 and 2007. (Data from IEA KWES, *International Energy Agency, Key World Energy Statistics*, 2009 edn., Paris, France, www.iea.org [accessed May 5, 2010], p. 44, 2009.)

Almost two-thirds (65%) of the world CO_2 emissions in 2007 originated from 10 countries (in a descending order): China, United States, Russian Federation, India, Japan, Germany, Canada, United Kingdom, S. Korea, and Iran (IEA Statistics 2009). Combined, United States and China produce 11.8 Gt CO_2 or 41% of global CO_2 emissions (in 2007, China overtook United States as the world's largest CO_2 emitter). Future scenarios predict that the global CO_2 emissions would increase reaching up to 29.3–44.2 Gt CO_2 in 2020, and 22.5–83.7 Gt CO_2 in 2050 (IPCC CCS 2005). Geographically, there will be a significant redistribution of CO_2 emission sources throughout the world between 2000 and 2030 with China, South Asia, and Latin America getting most of the gain, whereas the share of OECD countries is expected to be reduced. The emission distribution by sector, however, will not dramatically change: power generation over the next 50 years will remain a main CO_2 source closely followed by transportation and industry.

1.3.5 Fossil Fuels as a Main Source of Anthropogenic CO_2 Emissions

Fossil fuels as an important part of the Earth's carbon budget exert a significant impact on the global carbon cycle; therefore, they merit a special consideration. Fossil fuels can be defined as carbon-rich materials that produce thermal energy during exothermic oxidation reaction with oxygen of air, and release CO_2 as a result of this reaction. Main types of fossil fuels are NG, oil, coal, and peat. According to a widely accepted theory of the fossil fuel origin, on a geological timescale of hundreds of million years, the remains of marine organisms, swamp plants, and incompletely decayed plant matter buried under thick layers of rocks and exposed to high temperature and pressure were transformed to coal, oil, or petroleum (a mixture of light and heavy hydrocarbons). Because of a nonuniformity of the "feedstock" and different conditions of the transformation process, no two coals or oils or gases have the same chemical composition.

From historical perspectives, all types of carbonaceous fuels, including biomass and fossil fuels, have originated from solar-powered photosynthesis of biological matter that was converted to fuels over a geological time period. These carbonaceous fuels are differentiated based on the timescale required for their formation: millions to hundreds of million years for coal, oil, and NG, hundreds to thousands of years for peat and months to tens of years for biomass. On the scale of a human life span, coal, oil, and NG are defined as fossil or *nonrenewable* fuels, whereas biomass (and associated biofuels) as *renewable* fuels (IPCC CCS 2005). This classification to some extent is arbitrary because peat is considered a fossil fuel; although, in terms of its formation timescale it is close to biomass (some plants are many hundreds of years old). In this section, fossil fuels as a main source of anthropogenic CO_2 emissions, and the issues related to current and future fossil energy supply are discussed.

1.3.5.1 Natural Gas (Conventional and Unconventional)

NG is the lightest and cleanest (from an environmental viewpoint) form of fossil fuels. In most cases, NG occurs near crude oil reservoirs forming a gas cap between oil and a capping (impervious) rock. Being gas, it can migrate through the porous layers of the Earth's crust and accumulate in locations with favorable temperature and pressure conditions.

The lightest hydrocarbon, methane, is the main component of NG, typically amounting to 70–90 vol.%. Other light hydrocarbons (ethane, propane, butane) are also present in NG along with N_2, He, CO_2, H_2S, and water vapor. Most undesirable components in NG are H_2S (due to its toxicity and chemical aggressiveness) and CO_2 (due to its capacity to lower NG heating value). These components are usually removed from NG before its transportation or liquefaction. Liquefied petroleum gas (mostly, consisting of propane and butane) can be recovered during NG processing.

NG is widely used in a number of important industrial and residential applications, such as power generation, transportation fuel, industrial and residential heating, chemical feedstock for production of fertilizers, rubber, plastics, etc. Over the last half century, world NG production steadily increased. For example, NG production increased from 1.226×10^{12} m^3 (or trillion cubic meter [TCM]) in 1973 to 3.149 TCM in 2008 (IEA KWES 2009). In 2008, the major NG producers were Russia (20.9% of the world total), United States (18.5%), Canada (5.6%), and Iran (3.3%). The major net importers of NG are Japan, United States, Germany, and Italy. Figure 1.22 depicts the shares of gas consumption by different sectors in 1973 and 2007. Note that the share of industry dropped significantly from 54.0% to 35.6%, whereas the shares of transportation and nonenergy use doubled and tripled, respectively, over this period. The demand for NG is projected to continue to grow at about 1.5% year^{-1} (WEO 2009). The biggest increase in production is expected to occur in the Middle East (where the bulk of the gas reserves is located), whereas North America and Eurasia will be the biggest consumers of NG. Gas-fired new power stations using combined-cycle technology are projected to account for the most of the increase in NG demand.

Global proven gas reserves (2008) are estimated at more than 180 TCM, which could meet the gas demand to 2030 and beyond (this estimate includes both conventional and unconventional resources) (WEO 2009). (The term "proven reserves" is defined as the amount of gas that have been discovered and for which there is 90% probability that it can be economically extracted.) Note that the estimates of the proven reserves tend to increase over time due to the evolution of more advanced methods of their assessment, for example, the estimates of proven gas reserves more than doubled since 1980. Three countries, Russia, Iran, and Qatar, possess over half of the proven reserves. The estimate for the ultimately recoverable conventional gas resources is at 471 TCM (this category includes the total volume of gas that is both technically and economically recoverable) (WEO 2009).

Unconventional sources of NG include shale gas, tight sand gas, coal-bed methane, and methane hydrates. A schematic representation of different types of conventional and

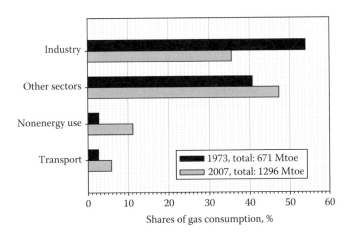

FIGURE 1.22
Shares of world total final natural gas consumption in 1973 and 2007. "Other sectors" include agriculture, residential, commercial and public services, and others. (Data from IEA KWES, *International Energy Agency, Key World Energy Statistics*, 2009 edn., Paris, France, www.iea.org [accessed May 5, 2010], p. 34, 2009.)

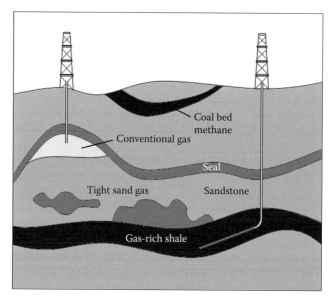

FIGURE 1.23
Geologic nature of major sources of natural gases. (Modified from EIA NG, U.S. DOE Energy Information Agency, http://www.eia.doe.gov/oil_gas/natural_gas/special/ngresources/ngresources.html [accessed March 29, 2010], 2010.)

unconventional NG sources is depicted in Figure 1.23. In contrast to conventional NG, which occurs as a result of migration and trapping of gas by an overlying impermeable formation (seal), shale gas originates from a gas-rich shale rock. Typically, shale gas is found in rocks of Paleozoic and Mesozoic age. Tight sand gas occurs when gas migrates into a sandstone formation and gets trapped there due to inability to further migrate upward. Coal-bed methane is formed during the transformation of organic matter to coal over a geological time span. The production of unconventional gas (shale, sandstone, and coal beds) is rapidly picking up the pace: it rose to more than 50% of U.S. total gas production (Kerr 2010).

In general, shale gas is more costly to produce than conventional NG from wells due to the use of expensive hydraulic fracturing and horizontal drilling equipment; nevertheless, it became an increasingly important source of gaseous fuel in North America and other areas over the past decade, making up nearly 12% of global gas production in 2008 (United States is the largest producer of shale gas with the capacity of 0.3 TCM of gas) (WEO 2009). In 2000, shale gas constituted about 1% of the U.S. gas supply; in 2010 its share increased to 20% (Kerr 2010). During hydraulic fracturing, 12,000–16,000 m^3 of water (plus fine sand and chemicals) are pumped down the well at a pressure of 70 atm, creating cracks that extend for about 1 km from the well. The most economically successful shale gas project is the Barnett Shale project in Texas (its production increased 3000% from 1998 to 2007).

The amount of shale gas reserves is still debated. Although U.S. Energy Information Administration (EIA) is projecting the increase in gas production in the United States by about 20% by 2030, other estimates are much more optimistic, for example, U.S. Geological Survey estimates that the shale gas development would double the U.S. NG reserves (only Barnett Shale contains 1.8×10^9 m^3 of NG) (Hargreaves 2010). IEA estimates of the unconventional gas reserves (without methane hydrates) are 1.15×10^{12} m^3 with half of that amount in shale deposits and another half in tight sandstone and coal beds (Daltorio 2010).

Besides the Barnett shale in north-east Texas, there are other formations that offer a massive potential: the Haynesville shale in Louisiana, the Marcellus Formation in western New York, to name just a few (Tullo 2009). Since 2005, NG prices in the United States are lower relative to crude oil (Tullo 2009). Some overly optimistic analysts predict that shale gas could fuel the United States for the next century, but more cautious estimates indicate that U.S. and Canadian production will likely peak sometime between 2020 and 2040 (Kerr 2010). But, regardless of the estimates, it is clear that the gas reserves are much greater than previously thought, and considering that gas is a much cleaner fuel than coal or oil, the environmental implications of the widespread use of shale gas are difficult to underestimate. But on the other hand, the widespread development of shale gas could cause different types of environmental problems, in particular, the real possibility of groundwater contamination by escaping gas and chemicals, the occurrence of spills of drilling fluids, brines, and chemicals around the wells, etc. Some states in the United States (e.g., New York, Pennsylvania) are moving to strengthen pertinent regulations and tighten licensing requirements for shale gas development (Kerr 2010).

Gas hydrates (or methane hydrates) are another form of the unconventional gas resource, although, much less developed than shale gas or coal bed methane. Vast resources of methane hydrates exist in subsurface sediments in permafrost and in deep oceans; it is believed that upon a successful development, they might become a major source of energy for the foreseeable future. Methane hydrates are formed from water and methane molecules under high-pressure and low-temperature conditions and represent an ice-like solid compound (gas hydrate can be easily transformed to gas if one or both parameters are altered such that hydrate molecules move out of the thermodynamic stability zone). Gas hydrates typically occur in shallow sediments in cold regions (Arctic area), or in deep-water (depths greater than 500–600 m) marine sediments at low temperature and high-pressure conditions.

The resources of methane hydrates vary widely from 1000–5000 TCM (Milkov 2004) to 5,600,000 TCM (Mielke 2000), which is 2–4 orders of magnitude greater than estimated conventional gas resources. National Resources Canada estimates that the worldwide amount of methane in gas hydrates is at least 10^4 Gt of carbon (NR Canada 2006). This is about twice the amount of carbon held in all fossil fuels on the Earth. U.S. Geological Survey estimates that there is about 2.4 TCM of technically recoverable gas hydrates in Northern Alaska alone (USGS 2008). The Mackenzie River delta in Canada contains some of the most concentrated deposits of hydrates in the world. The resources of methane hydrates in Mackenzie Delta and Beaufort Sea regions of Canada are estimated at 8.8–10.2 TCM (Counsel of Canadian Academies 2008). Other countries like United States, Russia, India, Japan, and China also have substantial marine gas hydrate deposits.

However, most of the methane hydrate resources are not commercially recoverable with present-day technologies. United States and Canada lead the worldwide efforts on the commercial development of methane hydrates. Among the most publicized gas hydrate projects is the Mallik Gas Hydrate Research Well Project (also known as The Mallik 2002 Consortium). The primary objectives of the program were to advance fundamental geological, geophysical, and geochemical studies of the Mallik field and to undertake, for the first time, the production testing of the concentrated methane hydrate reservoir (NETL 2010). The Mallik project started in 1998 in an area underlain by over 600 m of permafrost, and was sponsored by the Geological Survey of Canada, JAPEX, The Japanese National Oil Company, USGS, and NETL. The participants successfully drilled three wells to test a major gas hydrate accumulation and conducted the first modern production test of methane hydrates. The initial success of the project led to the formation of another international consortium that started in 2001–2002 and included, besides the original members, participants from Germany, India, and BP-Chevron-Burlington joint venture Group. The research program involved the drilling of a 1200 m deep main production research well and two nearby observation wells. It is worth mentioning another methane hydrate project: Integrated Ocean Drilling Program Expedition 311, which started drilling gas hydrate cores offshore Vancouver Island during the fall of 2005 (NR Canada 2006). Recently, U.S. NIST has developed and reported on-line a database of the properties of gas hydrates, which includes the data from the 2002 Mallik research well in Canada (an international geophysical experiment exploring the feasibility of using them as energy resources) (C&EN October 12, 2009).

1.3.5.2 Oil (Conventional and Unconventional)

Oil is the blood of industry and it is considered by many as the world's most important primary energy source (although its share in the world total final consumption dropped from 48.1% in 1973 to 42.6 in 2007) (IEA KWES 2009). Figure 1.24 compares the shares of oil consumption by a sector in 1973 and 2007. Transportation sector has been and remains the major oil consumer with its share significantly increasing from 45.4% in 1973 to 61.2% in 2007 at the expense of other sectors (except nonenergy use) (IEA KWES 2009). The industrial use of oil dropped from 19.9% to 9.2% over the same period. The first step in the utilization of oil products by consumers is its preprocessing and refining at large refineries, which transform crude oil into a variety of products including motor fuels (gasoline, diesel), aviation fuels (jet fuel, kerosene), heating oil, feedstocks for petrochemical and chemical industries, etc. Figure 1.25 breaks down the world refinery products production. Due to recent technological advances in petrochemical industry, the share of heavy fuel oil has dropped more than in half from 33.8% in 1973 to 15.7% in 2007 (IEA KWES 2009). The share of gasoline did not change much over this period, whereas the share of middle distillates increased appreciably from 26.0% to 33.4% (middle distillates are hydrocarbons produced in the middle range of refinery distillation, which includes kerosene, light and heavy diesel oil, heating oil). The share of aviation fuels increased from 4.2% to 6.3% during the

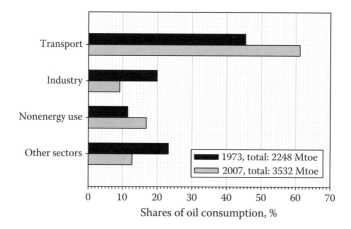

FIGURE 1.24
World oil consumption by sector in 1973 and 2007. "Other sectors" include agriculture, residential, commercial and public services, and others. (Data from IEA KWES, *International Energy Agency, Key World Energy Statistics*, 2009 edn., IEA, Paris, France, www.iea.org [accessed May 5, 2010], p. 33, 2009.)

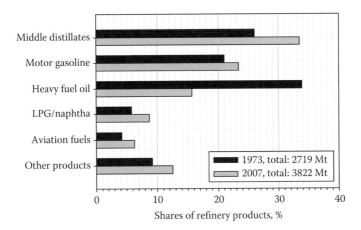

FIGURE 1.25
World refinery products production in 1973 and 2007. (Data from IEA KWES, *International Energy Agency, Key World Energy Statistics*, 2009 edn., IEA, Paris, France, www.iea.org [accessed May 5, 2010], p. 20, 2009.)

same period. In the United States, gasoline is the primary transportation fuel (9 million barrel per day), followed by diesel fuel (2.9 million barrels per day [MBD]), and jet fuel (1.6 MBD) (EIA 2006).

Oil demand is projected to grow by about 1% from 85.2 MBD in 2007 to 88.4 MBD in 2015 and 105.2 MBD in 2030 (WEO 2009) (a barrel is equal to ~159 L). Although there was a drop in demand in 2008–2009 due to the economic downturn, it is projected to progressively recover from 2010 onward. The transport sector will be the main driver of the oil demand increase in all regions between 2007 and 2030 accounting for 97% of the increase in the world oil use (WEO 2009). Most of the projected increase would come from OPEC: their overall output of conventional crude oil, NG liquids, and unconventional oil (predominantly, gas-to-liquids) will rise from 36.3 MBD in 2008 to about 40 MBD in 2015 to 54 MBD in 2030 (WEO 2009). This would bring the OPEC's share of the world oil production from 44% now to 52% in 2030. Non-OPEC conventional oil production has already peaked or is expected to peak between 2010 and 2030. However, the decline in the non-OPEC conventional oil output will be offset by the increase in unconventional oil production.

The report from the HIS Cambridge Energy Research Associates states that oil demand in developed (OECD) countries has probably already peaked and will not exceed the pre-recession (2008) levels, mostly, due to the combination of several factors: (1) demographics and socioeconomics (vehicle ownership rates in developed countries have reached a "saturation" level), (2) more fuel-efficient vehicles (by 2016, mileage of cars and light trucks is projected to increase by 42% and 30%, respectively), and (3) introduction of new more energy-efficient technologies) (Jackson 2009; Schmit 2009). At the same time, the global demand for oil from 2010 to 2020 is projected to increase by almost 14%, mostly due to developing countries, predominantly, China. The report notes that China had 12 million vehicles in 2005, which will grow to 110 million by 2030. Developing countries accounted for about 39% of the global oil demand in 1990; this number will increase to 51% by 2020, whereas the share of developed countries will drop from 61% in 1990 to 49% in 2020.

Unconventional oil resources include oil sands (also known as tar sands and bituminous sands, although, there might be some distinguishing features among them), extra-heavy oil, shale oil, gas-to-liquids, and coal-to-liquids. Canadian (Alberta) oil sands and Venezuela's Orinoco Belt bituminous sands are typical and the most recognized types of unconventional oil resources. Oil sands represent a thick, viscous mixture of heavy organic matter (bitumen), sand, clay, and water. The Canadian oil sands proven reserves are estimated at 178 billion barrels, which puts Canada after only Saudi Arabia in terms of proven oil reserves (WEO 2009). Some estimates of Canadian oil sands reserves go as high as 1 trillion barrels (Koerner 2007). In the United States, tar sand resources are mostly concentrated in the state of Utah (eastern part), and their recoverable reserves are somewhat less than that of Canadian tar sands: 12–20 billion barrels of oil (ANL 2007).

Advantageously, tar sands in Alberta can be recovered by open pit mining techniques, which substantially reduce their cost. The process of oil recovery from tar sands, however, is extremely laborious and energy intensive: 2 ton of the tar sand yields only one barrel of oil, and it requires large amounts of steam and water (2–4.5 volumes of water per one volume of oil, although, most of water is recycled). Roughly, 75% of oil (bitumen) can be recovered from the sand. There are concerns about a potential environmental impact of the oil sands industry because the technology requires the immense amount of water and NG, which results in 20% more CO_2 emissions than conventional oil on a "well-to-wheel" basis. Any future carbon regulations may adversely affect the competitiveness of the process. Canadian oil sands industry was hard-hit by the 2008 recession and relatively low oil prices (since its profitability relies on oil prices around $75–$80 per barrel), and many projects have been suspended and canceled (WEO 2009). If the current challenges could be overcome, oil sands have the potential to contribute to global energy security via diversification of oil supply. Increasingly higher crude oil price will stimulate the increase in the output of Canadian oil sands, and other unconventional oil sources. The global unconventional oil production is projected to increase from 1.8 MBD in 2008 to 7.4 MBD in 2030 (WEO 2009).

The subject of "peak oil" and available oil reserves is one of the most burning and highly debated topics among energy experts, geologists, economists, and politicians alike. In general, it is realized that the issue is not about running out of oil, as much as about the rates: the rates of oil discovery and production vs. the rates of oil consumption and the demand growth and the rate of technological change (Greene 2010). In his recent book, Gorelick argues that despite the finiteness of the global fossil resources, they tend to become less scarce over time, because the declining rates of new oil discovery have been offset by increasing technical capacities to recover a greater fraction of oil from the existing reservoirs (Gorelick 2010).

Nevertheless, the issue of oil reserves prognostication remains a serious problem: indeed, the more accurately we can predict how much of technically recoverable oil still remains underground, the better we can plan and be prepared for inevitability of depletion of this precious resource. But the problem is that if you ask a geologist, an oil company executive, an economist, an environmentalist, or a politician this simple question, how long can we rely on oil to power our economy, you will get completely different estimates from a few years to a hundred years. And the reason for that is that the answer to this question is associated with the analyses of a host of different factors, criteria, and assumptions, some of them of technical and some of political nature. Nevertheless, there are at least three reputable sources of information on the oil and NG reserves: *BP Statistical Review of World Energy, Oil & Gas Journal* (PennWell Corp.), and *World Oil* (Gulf Publishing Company) that are widely referred to in the literature. Figure 1.26 shows most recent estimates of the world proved oil reserves according to these three sources (the estimates include both conventional and unconventional oil resources) (EIA Oil Reserves 2010) (note that "proved

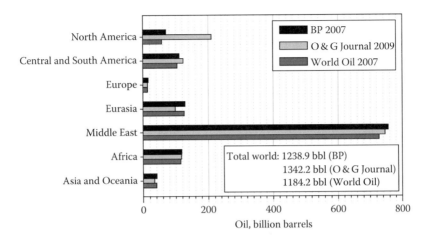

FIGURE 1.26
World proved reserves of oil. (Data from EIA Oil Reserves, U.S. DOE Energy Information Agency, World proved reserves of oil, www.eia.doe.gov/emeu/international/reserves.html [accessed April 2, 2010], 2010.)

reserves" are defined here as quantities of oil that are recoverable under existing economic and operating conditions with a reasonable certainty as indicated by the analysis of geologic and engineering data). As seen from this figure, there is a general agreement between the estimates from these sources (except, for North America). Oil reserves in the Middle East far exceed that of the other parts of the world with the rest of the oil reserves more or less evenly split between the regions, except for Asia/Oceania and Europe that possess much smaller fraction of the estimated oil reserves.

1.3.5.3 Coal and Peat

Coal accounts for 27% of the world primary energy demand. Different types of coal differ by their rank (i.e., its degree of maturity), which is determined by the stage coal reached during so-called coalification process, which is a sequence of the following transformation processes (Carpenter 1988):

$$Peat \rightarrow lignite/brown\,coal \rightarrow sub\text{-}bituminous\,coal \rightarrow bituminous\,coal \rightarrow anthracite$$

This rank-based coal classification system is currently used in North America and some other countries around the world. Coal consumption increased from about 600 Mtoe year^{-1} in 1971 to almost 800 Mtoe year^{-1} in 1988, then dropped to about 500 Mtoe year^{-1} in 2000, and recently started climbing up again reaching 729 Mtoe year^{-1} in 2007 (ton oil equivalent [toe] which equals to 42 GJ; million ton equivalent [Mtoe]). Most of the coal consumption is in power generation and industrial sectors (80%), with a small percentage in transport (0.5%) and nonenergy use (4.4%) sectors (IEA KWES 2009). Most analyses predict that power generation sector will remain for some time the main consumer of coal. Coal demand grows at an average of 1.9% and under the "business as usual" scenario it will increase from 4.549 Gt of coal equivalent in 2007 to 6.98 Gt in 2030 (WEO 2009). China and India will be responsible for about half to two-thirds of coal consumption by 2030.

1.4 Stabilization of Atmospheric CO₂: Prospects and Implications

1.4.1 Scenario of 2°C

In the face of ever-increasing levels of man-made GHG emissions, in particular, fossil-derived CO_2 in the atmosphere, there have been attempts to estimate the tolerable limits of atmospheric CO_2 concentrations in terms of the global mean temperature rise. In practical terms, the question is: what is the threshold level of CO_2 concentration in the atmosphere at which it needs to be stabilized in order to avoid an undesirable increase in the Earth's surface temperature? This turn out to be not an easy task because not only scientific but also some judgmental and even political factors are likely to be involved. The analysis of scientific literature on climate impacts indicates that, in the majority of cases, the temperature increase of 2°C above the preindustrial levels is considered a critical point beyond which a potentially catastrophic impact on the planet's ecosystem might occur (Meinshausen 2006). Note that the target of 2°C was first adopted by European Council in 1996 and later accepted by IPCC and The Group of Twenty (G20) (WEO 2009). The danger is that beyond this 2°C point in the temperature increase there exists not only elevated risk of extreme climate impacts and events, but also an increased possibility of strong positive carbon cycle feedbacks that would lead to even stronger climate impact with the real potential of reaching a "tipping point" or a runaway situation. On top of that, there is still an uncertain, but potentially a devastating impact of a possible methane release from thawing permafrost and/or unstable ocean methane hydrates that could be triggered by warming above certain threshold temperature (Buffet and Archer 2004).

Figure 1.27A shows the range of temperature increases projected at the atmospheric CO_2 stabilization levels between 400 and 650 ppm at the equilibrium (5%–95% range). The increase in the equilibrium atmospheric CO_2 concentration from 400 to 650 ppm results

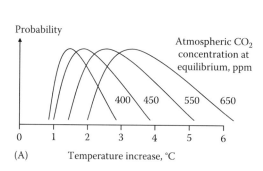

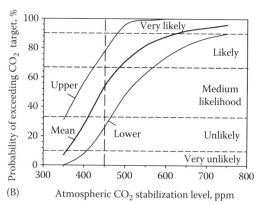

FIGURE 1.27
The probability of temperature increase above preindustrial era in equilibrium for different atmospheric CO_2 equivalence stabilization levels. (A) The range of temperature increases projected at the atmospheric CO_2 stabilization levels between 400 and 650 ppm at the equilibrium (5%–95% range). (B) The probability of exceeding 2°C warming in equilibrium for different atmospheric CO_2 equivalence stabilization levels. The data correspond to the upper-bound, mean, and lower-bound values from 11 sensitivity models. (Data from Wigley, T. and Rapper, S., *Science*, 293, 451, 2001; Meinshausen, M., What does a 2°C target mean for greenhouse gas concentrations? A brief analysis based on multi-gas emission pathways and several climate sensitivity uncertainty estimates, In *Avoiding Dangerous Climate Change*, Eds. Schellnhuber, H., Cramer, W., Nakicenovic, N., Wigley, T., and Yohe, G., Cambridge University Press, Cambridge, U.K., 2006; Murphy. J. et al., *Nature*, 430, 451, 2004.)

in the increase in the probability of the temperature rise from about 1.5°C to about 3.5°C (Wigley and Rapper 2001; Meinshausen 2006). Based on a number of models and climate sensitivity uncertainty estimates, the probability of exceeding 2°C equilibrium warming at different atmospheric CO_2 stabilization levels was determined. Figure 1.27B shows the probability (mean, upper, and lower bounds) of exceeding 2°C warming above the pre-industrial levels in equilibrium for different atmospheric CO_2 equivalence stabilization levels from 350 to 750 ppm (Meinshausen 2006). The models indicated that if the stabilization of atmospheric GHG concentrations would occur at 550 ppm CO_2 (equiv.), the likelihood of exceeding 2°C is very high, ranging between 63% and 99% (mean value—82%) (or "likely" using IPCC terminology). On the other hand, the probability of exceeding 2°C at CO_2 (equiv.) level of 400 ppm or below is estimated at 8%–57% (mean: 28%), which seems to be acceptable (i.e., "unlikely" using IPCC terminology) (for the stabilization level of 350 ppm, the probability would be from 0% to 31%). Note that while some countries support the 550 ppm target, other countries that are particularly vulnerable to an adverse climate impact are advocating much lower target of 350 ppm CO_2 (equiv.) (Hansen et al. 2008). Thus, the 450 ppm CO_2 (equiv.) target seems to be the most reasonable stabilization level, since it relates to 54% (mean) probability of keeping the global temperature increase below 2°C. IPCC stipulates that in order to achieve 450–490 ppm CO_2 (equiv.) target, CO_2 emissions would need to globally drop to 50%–85% below 2000 levels by 2050 (IPCC Mitigation 2007).

In an analytical study (Mignone et al. 2008), the authors examined the implications of delays in carbon mitigation measures using a well-tested model of the ocean carbon cycle. The authors have found, in particular, that when future CO_2 emissions are constrained to decline at the rate of 1% year^{-1}, the peak atmospheric CO_2 concentration (the so-called stabilization frontier) increases at the rate of about 9 ppm year^{-1} (average). The simulation results indicated that the stabilization below a preindustrial doubling (i.e., 550 ppm, which is a commonly cited target in climate policy assessments) would require dedicated mitigation efforts to begin within roughly the next decade (note that the immediate deployment of mitigation measures would place the frontier near 475 ppm). This implies that the delay of more than a decade will not guarantee the stabilization of atmospheric CO_2 concentration below 550 ppm.

In a recent study, Davis et al. (2010) emphasized the importance of overcoming political and technological inertia in curbing man-made CO_2 emissions and slowing climate change. The authors pointed out that the existing energy and transportation infrastructure can be expected to contribute the substantial amount of CO_2 emissions over the next 50 years due to an infrastructural inertia. The commitment to future emissions represented by existing CO_2-emitting devices was estimated by the authors (it was assumed that no additional CO_2 sources would be built, and all the existing CO_2 emitters would be allowed to live out their normal lifetimes). It was estimated that cumulatively 496 Gt (282–701 Gt in lower- and upper-bounding scenarios) of CO_2 from combustion of fossil fuels would be emitted by the existing infrastructure between 2010 and 2060, resulting in CO_2 atmospheric concentration of about 430 ppm, and forcing mean warming of 1.1°C–1.4°C (mean value of 1.3°C) above the preindustrial level. However, there is no doubt, that more CO_2-emitting devices will be built, so this scenario is not realistic, but it offers a means of gauging the threat of climate change from existing devices relative to those that have to be built.

1.4.2 450 Scenario

In this section, the energy and environmental implications of the so-called 450 Scenario (450-S) are analyzed in more details. According to the 450-S, countries take very strong coordinated actions in the energy, industry, and other sectors to limit GHG levels in the atmosphere to 450 ppm CO_2 (equiv.) and, consequently, the global temperature increase to 2°C. Wherever possible, the 450-S will be compared to the Reference Scenario (RS), which corresponds to the "business as usual" approach where no actions are taken to prevent the increase in GHG concentrations and global temperature rise.

1.4.2.1 Implications of 450 Scenario for GHG Emissions

Figure 1.28 shows the projected atmospheric concentrations of CO_2 and all GHG according to the 450- and RSs up to the year 2200 (the ENV-Linkage and MAGICC models were used for the estimates) (WEO 2009). It is evident that the concentration of GHG in the 450-S will be about half of that in the RS, in which the atmospheric concentrations of GHG gases steadily increase during the current century, and reach almost 1000 ppm CO_2 (equiv.) by 2150 (CO_2 reaches 750 ppm). It is noteworthy that in the 450-S, the GHG concentrations peak at 510 ppm CO_2 (equiv.) in 2035, stay flat for about a decade, and then slowly decline to the 450 ppm target level. Figure 1.29 depicts global GHG and energy-related CO_2 emissions trajectories in the 450-S from 1990 to 2050 (WEO 2009). According to the 450-S, the energy-related CO_2 emissions and the cumulative GHG emissions will peak around 2020 at 30.9 and 43.7 Gt CO_2 (equiv.), respectively, followed by a steady decline to 14.5 and 21 Gt CO_2 (equiv.), respectively, by 2050. The rate of the decline is progressively increasing from 1.5% year^{-1} between 2020 and 2030 to about 3% year^{-1} between 2030 and 2050. Compared with the RS, GHG emissions in 450-S would drop by 4 and 14 GT CO_2 (equiv.) in 2020 and 2030, respectively.

Figure 1.30 shows the projected values of global GHG emissions by the type of the GHG agent including both energy and nonenergy-related CO_2, methane, N_2O, and GHG emissions due to the LUCF, but excluding fluorocarbons (F-gases), which have a relatively small share in the overall emissions (WEO 2009). The emissions of CH_4, N_2O, F-gases, and from LUCF stay relatively flat until 2015–2020, and after that they steadily decline and cumulatively amount to 5.1 Gt CO_2 (equiv.) (or about 24% of the total GHG emissions) in 2050. It should be noted that since the cost of the measures to reduce the emissions of the above

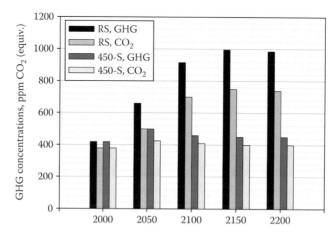

FIGURE 1.28
GHG and CO_2 concentrations according to the Reference (RS) and 450 (R-450) Scenarios for the years 2000–2200. Models used: MAGICC and ENV-Linkages. (Data from *World Energy Outlook*, International Energy Agency (IEA), IEA Publications, Paris, France, pp. 197–199, 2009.)

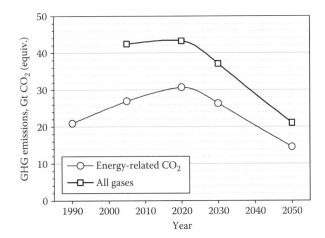

FIGURE 1.29

World GHG and energy-related CO_2 emissions trajectories according to the 450 Scenario for the years 1990–2050. (Data from *World Energy Outlook*, International Energy Agency (IEA), IEA Publications, Paris, France, Table 5.1, p. 200, 2009.)

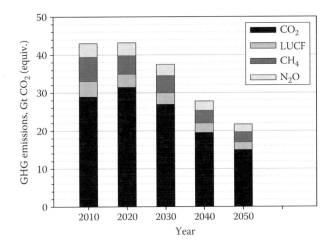

FIGURE 1.30

World current and projected (to 2050) GHG emissions by type in 450 Scenario. CO_2 includes energy and non-energy sources. LUCF-land use change and forestry. F-gases are not shown. (Data from *World Energy Outlook*, International Energy Agency (IEA), IEA Publications, Paris, France, pp. 199–200, 2009.)

gases is relatively small compared to that of energy-related CO_2 emissions, it is projected that the abatement of CH_4, N_2O, LUCF, F-gases (or halocarbons) would account for more than 40% of global GHG abatement by 2020 compared to the RS (by 2050, their abatement potential would be fully utilized) (WEO 2009). Figure 1.31 provides the breakdown of the world energy-related CO_2 emissions by source in 450-S and RS over the period of 1990–2030. As can be seen, power generation sector is the major contributor of CO_2 emissions followed by transport and industry in all considered cases (for simplicity, Figure 1.31 does not show the "other" sources of CO_2 emissions, which in all cases vary within 9%–11% of the total). By 2030, the contributions of power generation and transportation sectors to the overall CO_2 emissions will become rather close. Figure 1.32 compares the trajectories of the

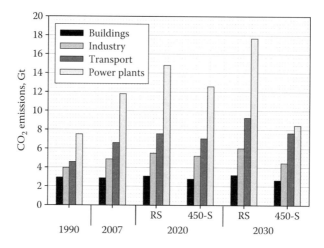

FIGURE 1.31
World energy-related CO_2 emissions by source according to the Reference (RS) and 450 (450-S) Scenarios for the years 1990–2030. (Data from *World Energy Outlook,* International Energy Agency (IEA), IEA Publications, Paris, France, p. 322, 2009.)

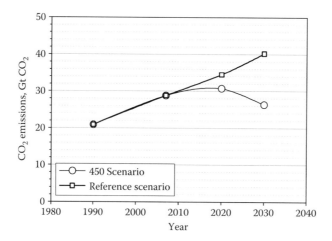

FIGURE 1.32
World energy-related CO_2 emissions trajectories according to the Reference and 450 Scenarios for the years 1990–2030. (Data from *World Energy Outlook,* International Energy Agency (IEA), IEA Publications, Paris, France, p. 322, 2009.)

world energy-related CO_2 emissions over the period of 1990–2030 according to the 450-S and RS. In the 450-S, the peak of global CO_2 emissions will be reached around 2015–2020 (at the value of about 31 Gt CO_2), after which it will gradually decline to 26.4 Gt CO_2 by 2030. The RS will see CO_2 emissions almost linearly increasing over the 1990–2020 period with some appreciable acceleration after 2020.

The above figures imply that in order to stay on the path to the 450 ppm concentration target, the energy-related CO_2 emissions should not exceed 30.7 and 26.6 Gt in 2020 and 2030, respectively (compared to the RS, this would result in the CO_2 savings of 3.85 Gt in 2020). For the practical realization of the 450-S, a set of different abatement measures such an improvement in end-use efficiency, carbon capture and sequestration (CCS), utilization

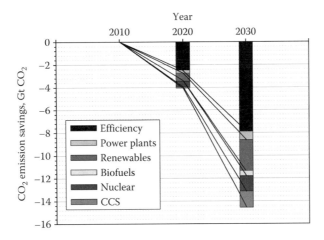

FIGURE 1.33
World energy-related CO_2 emission savings by policy measure in the 450 Scenario. (Data from *World Energy Outlook*, International Energy Agency (IEA), IEA Publications, Paris, France, p. 211, 2009.)

of carbon-free energy sources (e.g., renewables, nuclear), wide use of biofuels, etc., will have to be implemented on a global scale. Figure 1.33 provides the details of the contribution of different abatement measures according to the 450-S in terms of potential CO_2 emission savings over the period of 2010–2030 (compared to the RS). It can be seen that the improvement in the end-use efficiency is the major contributor to energy-related CO_2 emissions abatement measures over the entire period (more than half of total CO_2 savings). This can be attributed to the fact that energy-efficiency improvement measures in buildings and industry typically have short payback periods and negative net abatement costs (WEO 2009). An increased market share of renewables in power and heat generation would result in the total savings of about 20% of CO_2 in 2030. The deployment of additional nuclear-based facilities for power generation and industrial applications would cut CO_2 emissions by about 10% in 2030 compared to the RS. The same percentage of CO_2 savings could be achieved by a widespread deployment of CCS technology. The increased share of biofuels in the transportation sector would result in modest CO_2 savings of 3%, according to the 450-S.

1.4.2.2 Implications of 450 Scenario for Primary Energy Demand and Supply

The 450-S envisages drastic measures and policies aiming at significantly diminishing the primary and final energy demand growth. Figure 1.34 shows the dynamics of the world primary energy demand by fuel and an energy source over the period of 1990–2030 in the 450-S (A), and the difference between the 450-S and the RS in terms of the changes occurring between 2007 and 2030 (B) (WEO 2009). The world primary energy demand reaches nearly 13.6 and 14.4 Gtoe in 2020 and 2030, respectively, which constitutes the reduction of about 6% and 14%, respectively, relative to the RS (WEO 2009). Although the demand in primary energy between 2007 and 2030 will grow by almost 20%, the average annual rate in 450-S will be about half of that in the RS (0.8% and 1.5%, respectively). Fossil fuels will continue to be a major source of energy throughout the projection period; however, their share in the overall energy demand will decline by about 13% by 2030. Coal is the only fuel for which the demand will drop against the current level. Oil will account for the lion's share of the primary energy demand in the 450-S reaching 4.25 Gtoe (or 88.5 MBD)

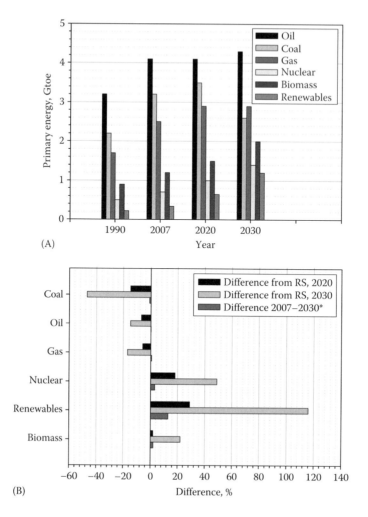

FIGURE 1.34
World primary energy demand by energy source in 1990–2030 time frame in 450 Scenario (A), and difference between 450 and Reference Scenarios (B). Renewables include hydro, solar, wind, geothermal, and others; biomass includes waste. (Data from *World Energy Outlook,* International Energy Agency (IEA), IEA Publications, Paris, France, Table 5.5, p. 212, 2009.)

in 2030 with its share in the total energy demand of 30% (which is about 5% less than in 2007). The projected oil savings will originate from different sources such as the widespread use of electric and plug-in hybrid vehicles (PHV), the introduction to the market of second-generation biofuels, etc. According to the 450-S, road transportation and aviation combined will account for about 70% of oil savings by 2030 with oil share in the transport dropping from 94% in 2007 to 84% in 2030 (WEO 2009).

In response to potential international carbon-restricting policies, energy security, and environmental concerns, the 450-S envisages a significant increase in the share of carbon-free (or zero-carbon) fuels and energy sources in the overall energy demand. In particular, the 2030 demand for nuclear, hydro, non-hydro renewables and biomass/waste will be 49%, 21%, 95%, and 22% higher than in the RS (WEO 2009). The overall share of zero-carbon fuels and energy sources such as nuclear and renewables in the world primary energy demand will increase from 19% in 2007 to 23% in 2020 to 32% in 2030. Biomass consumption

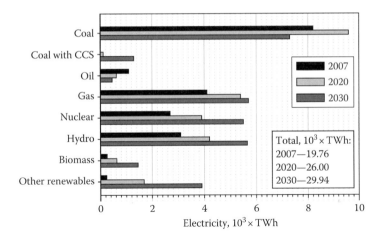

FIGURE 1.35
World electricity generation by the energy source in 450 Scenario in 2007–2030 time period. "Other renewables" include wind offshore and onshore, solar PV, concentrating solar power, geothermal, tide, and wave. (Data from *World Energy Outlook*, International Energy Agency (IEA), IEA Publications, Paris, France, Table 6.1, p. 229, 2009.)

in the 450-S is projected to increase by 350 Mtoe over the RS. Likewise, biofuels production is also projected to significantly increase in the 450-S (in fact, it will more than double relative to the RS). The share of second-generation biofuels will be significantly higher in the 450-S compared to the RS in 2020, and they will become dominant biofuels by 2030.

In the 450-S, global electricity demand will grow from 19,756 TWh in 2007 to almost 30,000 TWh in 2030 (Figure 1.35); however, compared with the RS this will constitute a drop by about 13% (as a result of worldwide energy efficiency increase measures) (WEO 2009). It is projected that carbon-free and low-carbon sources of electricity (e.g., hydro, nuclear, biomass, other renewables) will make up about 60% of the global total by 2030. In contrast, in the RS, only one-third of electricity will be produced from low-carbon sources over the projected period. Production from nuclear plants will meet 18% of the global electricity demand, whereas fossil-based power plants with CCS will be deployed after 2020 with their share being relatively small (5%–6%) by 2030. It can be seen that the growth rate of electricity production from non-hydro renewables (solar, geo, wind, tide, waves) outpaces any other source. The average annual growth rates of wind, solar (PV), solar concentrating power, and geothermal sources are projected at 12.8%, 23.5%, 30%, and 7%, respectively (WEO 2009). Wind (on- and offshore) is projected to produce the lion's share of the renewable electricity, accounting for 26% of all the growth in power generation over the 2007–2030 period. The share of all renewables in electricity generation is projected to increase from 18% in 2007 to 37% in 2030 (the latter includes 19% hydro, 9% wind, 5% biomass, 4% other renewables) (WEO 2009).

Due to lower global primary energy demand and the introduction of the CO_2 emission limits projected in the 450-S, all primary sources, in general, will see a decline in production compared to the RS. According to the 450-S, OPEC oil production will reach 48 MBD in 2030, an increase of 11 MBD against 2008 levels, whereas non-OPEC production will decline from 47 MBD in 2008 to 39 MBD in 2030 (WEO 2009). Likewise, NG production will also slow down by 14%–24% (depending on the region) and see lower prices in the 450-S compared to the RS. Global coal production will follow demand, which will level off around 2020 and then drop to the 2003 level in 2030.

1.4.2.3 Implications of 450 Scenario for Transport

Transportation sector consumes a significant fraction of primary energy resources (about two-thirds of oil), and emits enormous amounts of CO_2 emissions (about a quarter of the total); therefore, it provides an important example of how the 450-S measures would help to reduce the oil consumption and CO_2 emissions. Figure 1.36 depicts the current and projected world transport energy consumption by fuel in the 450-S (A), and the changes relative to the RS over the period of 2007–2030 (B). The total oil consumption will increase from 2161 Mtoe in 2007 to 2306 and 2510 Mtoe in 2020 and 2030, respectively, although it will constitute a drop of 9% and 18%, respectively, compared with the RS. Likewise, NG consumption will drop by 11% and 15% by 2020 and 2030, respectively, compared with the RS. The use of electricity in plug-in hybrid and electric vehicles is projected to increase to about 350 TWh (65 Mtoe) and 835 TWh (122 Mtoe) by 2020 and 2030, respectively, a change of 83% and 165%, respectively, against the RS (WEO 2009). Biofuel consumption will increase from 34 Mtoe in 2007 to 123 and 278 Mtoe by 2020 and 2030, respectively, which are 19% and 109% increase compared with the RS. It is expected that second-generation biofuels produced from ligno-cellulosic biomass and Fischer–Tropsch (FT) liquid fuels would make up most of the projected fuel increase in the road transport and aviation sectors. Biofuels will represent 11%–12% of road transport fuel and 15% of aviation fuel by 2030. According to the 450-S, the 2030 fuel mix will be as follows (%): petroleum fuels—80.2, biofuels—12.0, electricity—6.2, and NG—1.6 (WEO 2009).

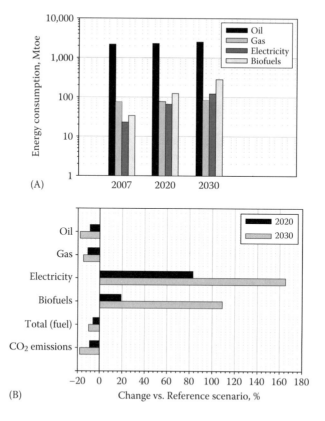

FIGURE 1.36
World transport energy consumption by fuel in 2007, 2020, and 2030 in 450 Scenario (A), and changes in energy consumption by fuel and CO_2 emissions in 450 Scenario vs. Reference Scenario (B). (Data from WEO, *World Energy Outlook,* International Energy Agency (IEA), IEA Publications, Paris, France, Table 6.3, p. 243, 2009.)

The 450-S projects that the implementation of national and international policies and measures would drastically cut CO_2 emissions from transport. In the 450-S CO_2, the emissions in the transportation sector are projected to be reduced by 9% (0.67 Gt) by 2020, and by 18% (1.6 Gt) by 2030, compared to the RS with total emissions reaching 7.1 and 7.7 Gt in 2020 and 2030, respectively (WEO 2009). It can be seen that most of the CO_2 savings (92% and 81% in 2020 and 2030, respectively) would occur in road transport, which could be attributed to a combination of several factors such as the increase in vehicles efficiency, an expanded use of biofuels, and the widespread introduction of plug-in hybrids and electric vehicles. In particular, the efficiency of passenger light duty vehicles is projected to increase by 38% by 2030 (relative to the RS) due to improvements in spark-ignition and diesel engines, non-engine improvements (e.g., tire, lighting), and the market penetration of more efficient engine technologies (e.g., plug-in hybrids and electric vehicles). Accordingly, the 450-S predicts a significant drop in sales of ICE vehicles from 99% (2007) to 52% and 42% in 2020 and 2030, respectively. At the same time, the share of plug-in hybrids and electric vehicles will increase from practically zero in 2007 to 21% and 7%, respectively, in 2030. In contrast, the RS projects only a modest drop (by 6%) in ICE sales by 2030 and an increase in hybrid vehicles sales from 1% in 2007 to 7% in 2030. Note that the RS does not include any plug-in hybrids or electric vehicles sales until 2030. As a result of the shift toward plug-in and electric vehicles, the average on-road CO_2 intensity of new vehicles would drop from 205 gCO_2 km^{-1} (2007) to 125 and 90 gCO_2 km^{-1} in 2020 and 2030, respectively, in the 450-S (WEO 2009).

As could be expected, the increased use of electricity in the transportation sector (especially in light duty vehicles due to the increased share of plug-in hybrids and electric cars) would inevitably result in the additional CO_2 emissions in the power generation sector. According to the 450-S, due an increase in electricity consumption of 880 TWh in transport by 2030 (compared with the RS) the additional CO_2 emissions of 250 Mt are expected to be generated at power plants (WEO 2009). Therefore, it is of utmost importance to implement efficient measures to increase the share of carbon-free energy sources and/or decarbonize fossil fuels used for power generation.

Heavy duty vehicles (e.g., trucks, buses, commercial vehicles) will also see a significant increase in efficiency and reduced CO_2 emissions per kilometer driven according to the 450-S. It is projected that the efficiency of heavy duty vehicles in average will increase by 20% and 34% by 2020 and 2030, respectively, which would result in CO_2 reductions from current (2007) 340 gCO_2 km^{-1} to 270 and 227 gCO_2 km^{-1}, respectively (WEO 2009). As in the case of light duty vehicles, the efficiency improvements would come from both engine and non-engine components, the increased use of biofuels and more efficient logistics. The analysis shows, however, that most of the savings from both light and heavy duty vehicles will in all likelihood be offset by the strong growth in transportation demand, especially, in China and India, which would result in a continued rise in total CO_2 emissions from transport through to 2030 (WEO 2009).

A detailed comparative analysis of various vehicle and fuel options over the entire twenty-first century with regard to GHG emissions, oil consumption, and urban air pollution using dynamic computer simulation model is presented by Dr. Sandy Thomas in Chapter 15.

1.4.2.4 Economics of 450 Scenario

The implementation of the 450-S on a global scale would require a substantial investment in carbon mitigation technologies including an across-the-board enhancement in energy efficiency and the widespread realization of zero- and low-carbon technologies. In this

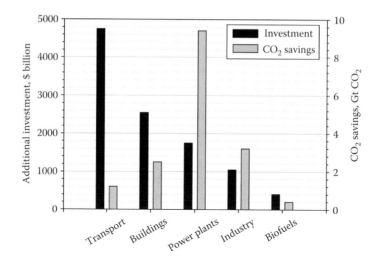

FIGURE 1.37
Cumulative additional investment by sector in the 450 Scenario relative to the Reference Scenario and associated CO_2 savings in the 2010–2030 time period. Investment is expressed in 2008 U.S. dollars. (From WEO, *World Energy Outlook,* International Energy Agency (IEA), IEA Publications, Paris, France, Tables 7.1 and 7.2, p. 259, 2009.)

section, the projections of the additional investment (compared with the RS) that would be required for the stabilization of the atmospheric GHG at the 450 ppm CO_2 (equiv.) are discussed. The five key sectors that accounted for 91% of CO_2 emissions in 2007 and where the most of the additional investment would most likely go are transport, buildings, power generation, industry, and biofuels (WEO 2009). Figure 1.37 depicts cumulative additional investments by a sector in the 450-S relative to the RS and associated CO_2 savings during the 2010–2030 period. It can be seen that the largest increase in the investment ($4750 billion) is in the transportation sector, where money will be spent mainly on purchasing more efficient and less polluting vehicles, for example, hybrids, plug-ins, and electric vehicles. Buildings (residential, commercial, public) is a second largest area of the additional investment amounting to $2550 billion. In this sector, the investment would mainly be directed toward the measures to increase energy efficiency, energy conservation, and the use of renewable resources. The additional investment in power generation sector ($1750 billion) would help to expand the share of renewables (60% of the total investment), nuclear power (16%), and to incorporate CCS technology (7%), especially, in coal-fired power plants. IEA estimates that the total additional investment needed globally for the implementation of the 450-S in the energy sector over the period of 2010–2030 is close to $10,500 billion (WEO 2009). This spending will be directed toward low-carbon energy technologies and energy efficiency efforts to curb the rise of CO_2 levels (C&EN November 16, 2009). The largest CO_2 savings (compared with the RS) are projected in the power generation sector (9.4 Gt CO_2), followed by industry, buildings, transport, and biofuels (overall, about 13 Gt CO_2 savings in the energy sector in 2030). IEA projects that over the 2010–2030 period, the United States and China would have the largest incremental investment in carbon mitigation technologies (about $2050 billion each), followed by EU, India, Japan, and Russia (WEO 2009).

1.4.3 "Stabilization Wedges" Concept

Pacala and Sokolow (2004) have introduced the concept of "Stabilization Wedges" as a useful tool for quantifying the actions that would be required to stabilize atmospheric

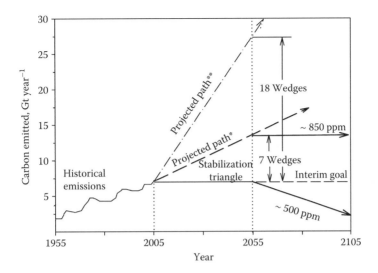

FIGURE 1.38
Schematic representation of the stabilization triangle. Projected path * according to Sokolow (2006). Projected path ** according to Hoffert (2010). (Data from Pacala, S. and Socolow, R., *Science*, 305, 968, 2004; Socolow, R., Stabilization wedges: An elaboration of the concept., In *Avoiding Dangerous Climate Change*, Eds. Schellnhuber, H. et al., Cambridge University Press, Cambridge, U.K., pp. 347–353, 2006; Hoffert, M., *Science*, 329, 1292, 2010.)

GHG concentrations at acceptable levels within 50 years time frame. Figure 1.38 provides the graphical representation of the Stabilization Wedges concept. The authors approximate the baseline (or a business-as-usual) case by a ramp trajectory, which represents a linear growth leading to doubling of global CO_2 (equiv.) emissions by 2055 (i.e., considering only fossil carbon, from about 7 Gt C year^{-1} to about 14 Gt C year^{-1} by mid-century). The flat trajectory on the diagram represents "zero emissions growth," with the target of achieving the removal of 7 Gt C year^{-1} in 2055. The flat and ramp trajectories form the so-called stabilization triangle, which corresponds to the totality of actions required to achieve the stabilization of the atmospheric CO_2 concentration below double the preindustrial levels (500 ppm) over the period of half a century. According to the authors, one "wedge" corresponds to about 1 Gt C year^{-1} of emissions savings in 2055 due to the implementation of a single mitigation strategy (thus, each wedge would reduce CO_2 emissions by 25 Gt C over 50 years). Evidently, in order to achieve the zero emissions growth (or the flat trajectory), there will be need for the introduction of roughly seven wedges. After achieving the interim goal of the zero emissions growth by 2055, a tougher CO_2 target can be implemented that would reduce the global CO_2 (equiv.) emissions rate below the current level. The authors suggested 15 potential wedges or mitigating technologies, some of which already exist on a commercial scale. Although, in most cases, no fundamental breakthroughs are required, these wedges are difficult to implement due to a number of reasons: technical, social, political, and environmental. The wedges can be achieved from implementing energy efficiency increase measures, or from decarbonization of electric power generation and fuel supplies (e.g., by shifting to low-carbon fuels, CCS, use of nuclear or renewable energy resources), or from biological storage in forests and soils (the "Wedges" are described in more details by Professor Winter in Chapter 2).

It is noteworthy that non-carbon energy sources, for example, nuclear or renewables, could produce either a wedge of electricity (by displacing coal electricity), or a wedge of hydrogen fuel (by backing out hydrocarbon fuels in transport). According to the authors of the study, the electricity wedge option would result in much greater (about twice) carbon savings than the hydrogen fuel wedge (Pacala and Socolow 2004). Thus, in most parts of the world, the optimal use of carbon-free energy resources such as nuclear, hydro, solar, wind, etc., will be to provide electricity as a replacement for coal-based electricity (in areas with low share of coal electricity, the fuel-wedge option could be attractive). The down-selection of a particular wedge will depend on many factors and, in most cases, markets would help to determine the "winning" wedges (note that the best wedge in one country or region may not be suitable for another country or region).

However, not all experts entirely agree with the quantitative aspects of the Stabilization Wedges concept. In his recent paper in the *Science* magazine, Hoffert (2010) suggests that the original wedges approach greatly underestimates the needed CO_2 reductions. The author points to the fact that Pacala and Sokolow (2004) built their scenario on a business-as-usual emissions baseline based on the assumption that a shift in the fossil fuels mix will result in the reduction of the carbon-to-energy ratio (i.e., the amount of CO_2 released per unit of energy). The carbon-to-energy ratio did decline during prior shifts from coal to oil to NG. This trend, however, is no longer valid, because as oil and NG production will peak in the near future, coal production will rise, which would effectively increase the carbon-to-energy ratio (e.g., a large number of coal-fired power plants will be built in China, India, and United States). Taking this trend into consideration, Hoffert projects that not 7, but 18 new wedges would be needed to curb carbon emissions by the middle of the century in order to keep future warming below 2°C (see Figure 1.38).

In general, despite some simplifications and scientific, economic, and political uncertainties shrouding certain aspects of the Stabilization Wedges concept, it could be a useful tool providing a guidance as to where to focus the limited resources in order to achieve the stabilization of atmospheric CO_2 concentrations at acceptable levels within the next 50 years, and how to prepare ground for further CO_2 emissions rate reduction in the following half century. In the following sections, some of these stabilization wedges will be considered in more details as part of the carbon mitigation portfolio.

1.5 Carbon-Free and Low-Carbon Alternatives to Fossil Energy

Global primary power consumption today is about 13 TW (TW is equal to 10^{12} W), most of which comes from fossil sources. Several estimates have been reported in the literature with regard to the future power consumption that will satisfy the world's growing demand for energy, support economic growth, and stabilize atmospheric CO_2 levels at an acceptable level. A report by the DOE Basic Energy Sciences workshop estimated that at least 10 TW of carbon-free power has to be generated by the mid-century to avoid an adverse impact of energy sector on the environment and climate (BES 2005). A study by Hoffert et al. (2002), projected that the stabilization of atmospheric CO_2 concentrations at 550, 450, and 350 ppm levels would require generation of 15, 25, and >30 TW, respectively, of carbon-free power by 2050. It is realized that this would be an extremely difficult task requiring the development and practical realization on a global scale of revolutionary technologies in energy production, conversion, storage, and distribution areas.

Three main technological approaches to the introduction of carbon-free energy systems to the global marketplace are most frequently discussed in the literature:

- Decarbonization of fossil energy
- Nuclear energy
- Renewable energy resources

In particular, in his recent paper, Hoffert (2010) estimated that providing by mid-century of 30 TW of carbon-neutral power, perhaps, 10 TW from each of three primary sources: decarbonized or "clean" coal, nuclear, and renewables may be required to sustain economic growth and keep atmospheric CO_2 concentrations below 450 ppm. In the following section, these options are analyzed from a viewpoint of their technological readiness and commercial potential, and whether they could deliver the targeted amount of carbon-free power by mid-century.

1.5.1 Decarbonization of Fossil Energy

The objective of fossil fuel decarbonization is to eliminate or drastically reduce the amount of CO_2 emitted to the atmosphere from the usage of primary fossil resources such as coal, petroleum, and NG. The fossil decarbonization concept offers the advantage of an extension of the fossil fuel era by perhaps another 200 years (purportedly) without an adverse impact of fossil fuels on our planet's ecosystem and climate. It is recognized that as long as fossil fuels are used, fossil decarbonization will be required to decouple economic growth from unacceptable levels of atmospheric emissions of CO_2. Carbon (to be precise, CO_2) capture and sequestration (CCS) is a set of technological measures to prevent anthropogenic CO_2 emissions from reaching the atmosphere by capturing and securely storing CO_2 in different sinks such as geologic formations, the ocean, saline aquifers, terrestrial ecosystems, etc. There is a belief among proponents of the fossil fuel decarbonization option that the concentration of CO_2 in the atmosphere could be stabilized at 450–550 ppmv levels without abandoning the fossil fuel infrastructure. Latest developments in the field of fossil fuel decarbonization, including the state-of-the-art technologies for the large-scale implementation of this technology, as well as environmental, social, and economical aspects of this concept are discussed in details by Muradov in Chapter 14.

1.5.2 Nuclear Option

Nuclear energy is considered an important carbon-free source of energy that could substantially alleviate the potential power shortage problem without disturbing the Earth's fragile carbon balance. In Chapter 3, the team of nuclear energy experts led by Dr. Ken Schultz (General Atomics, San Diego, CA) presents a comprehensive overview of state-of-the-art systems for nuclear-based production of carbon-neutral fuels, in particular, by coupling advanced nuclear reactors with high-temperature electrolysis and thermochemical and hybrid cycles. The objective of this section is to provide a brief introduction to the field with a special emphasis on the carbon mitigation aspects of the nuclear source.

1.5.2.1 Nuclear Energy as a Carbon-Free Source

Nuclear power production is a mature technology: it has been practiced since the middle of the last century, and in some countries it supplies a lion's share of the total electricity demand (e.g., in France, more than 80% of electricity is of a nuclear origin). Nuclear reactors have the highest power density compared with other non-carbon sources of electricity: an average nuclear plant of 1 GW capacity takes up about 6 km^2 of land space (for a comparison, a wind farm of the comparable capacity would occupy an area of 609 km^2) (Percopo 2008). Currently, 440 commercial nuclear power plants are being operated globally (including 104 plants in the United States) generating about 16% of the world's electricity (21% of the U.S. electricity, or about 70% of nonfossil electricity). Most of them are known as Generation II reactors and are based on the 1970s technology and the 1960s materials (Jacoby 2009a).

For almost three decades, no new nuclear plants have been ordered in the United States, and, as a result, the contribution of nuclear power to U.S. electricity supply has been declining as older plants are approaching retirement. The current U.S. administration has announced a rebirth of government support for nuclear power (Johnson 2010). U.S. Nuclear Regulatory Commission (NRC) has recently announced that in the last 2 years, they have received applications to build and operate 28 new nuclear power plants (Jacoby 2009). Nuclear energy R&D spending will be increased by 39%, with special emphasis on the development and implementation of small modular nuclear reactors (about 300 MW$_e$ power output or less, as opposed to a standard 1000 MW$_e$ output typical of modern nuclear plants) with advanced designs. Of particular interest is the "mPower," a light water reactor (Babcock & Wilcox) with the power output of 125 MW$_e$. The reactor is small (about 5 m in diameter and 25 m in height) enough to be manufactured in a factory, shipped to an end-user, and installed underground (the first reactor is scheduled on-line for 2020). The potential end-users include chemical and petrochemical facilities that utilize both electricity and process heat.

There are, however, more cautious assessments of the future role and scope of using nuclear power, especially with regard to available resources of nuclear fuel. In the Report of the Basic Energy Sciences Workshop (organized by the U.S. DOE), it was estimated that producing 10 TW of nuclear power would require construction of a new 1 GW$_{el}$ nuclear plant somewhere in the world every other day for the next 50 years (BES 2005). The proven and recoverable conventional reserves of uranium fuel are assessed at 2.85 and 17.1 million tons, respectively (Price and Blasé 2002). In their recent report, the U.S. National Research Council (NRC) estimated that at the current rate of consumption, uranium reserves might last 50–100 years (NRC 2004); however, if nuclear plants were to provide 10 TW power by the year 2050, uranium resources will last only 30 years (assuming the present-day types of the reactors and ultimately recoverable uranium resources) (Hoffert et al. 2002; Moriarty and Honnery 2007). Jacobson reported that the technical potential of the nuclear energy is between 4.1 and 122 PWh year^{-1} (lower number is for once-through reactors, and high number is for light water and fast-spectrum reactors), which would put the nuclear source for a service in the time frame of 90–300 years (lower and higher numbers are for known and expected reserves, respectively) (Jacobson 2009). The argument put forward by the opponents of nuclear energy boils down to the following: taking into account the nuclear reactor lifetime of 30–40 years, it would be imprudent to base the energy policy on the conventional fission nuclear source without knowing whether there is enough fuel to accomplish the goal. Furthermore, there is some skepticism that such growth in nuclear power might be even feasible because of the severe constraints in cost (currently the price of uranium fuel is at about $4 per barrel of petroleum equivalent, Cowan 2010), nuclear fuel supply, site availability, safety, public acceptance (or opposition), and waste disposal considerations (Moriarty and Honnery 2007).

It was proposed that for the fission nuclear reactors to be a major producer of carbon-neutral energy, a shift to breeder (or "fast") reactors would be necessary, since the breeder reactors could potentially alleviate the uranium resource constraint problem and extend the uranium reserves by a factor of about 30. Advantageously, the breeder reactors can also allow utilizing thorium reserves (estimated at 4.5 million ton) to produce fissile material for use in the nuclear reactors. It was estimated, however, that it may take at least two to three decades before large-scale commercial breeder reactors would contribute to the world's energy needs (Moriarty and Honnery 2007).

Arguments can be found in the literature that there are still ways to significantly expand nuclear fuel resources, for example, via the use of heavy-water reactors that burn unenriched uranium (Cowan 2010) or utilizing practically exhaustible resources of thorium. Thorium was recently "reintroduced" as fuel having a tremendous potential to provide energy without causing massive environmental problems or the potential for further nuclear weapons proliferation (Jacoby 2009; Wetzel 2010). The proponents of the thorium-based technology argue that Th is more abundant, potentially less expensive to process than uranium, and Th-fueled reactors are not conducive to making and collecting materials that can be used to make nuclear bombs. Moreover, radiotoxicity of waste products from the Th usage persists for just tens of years compared with thousands of years typical of the uranium waste. However, some experts are taking a more cautious approach to thorium as a "miracle" nuclear fuel (Swanson 2010). Among the strongest arguments against Th-fuel is that an enormous investment of time and resources would be required before any new type of Th-based nuclear reactor could be licensed for commercial operation. Meanwhile, India is developing its own Th-fueled nuclear industry to take an advantage of the country's large reserves of Th minerals (Jacoby 2009).

Recently, new developments in the nuclear reactor technology showed a potential to significantly expand the role of nuclear energy as a carbon-neutral energy source. In particular, the Generation IV reactors (GEN-IV Initiative) have the goals of improving economics, safety, reliability, and security (including proliferation resistance) of the reactors and the fuel cycle. Most importantly, the Generation IV reactors would significantly expand the range of technological options for producing carbon-neutral fuels such as hydrogen and synthetic fuels and improve the sustainability of the nuclear source to meet the needs of present and future generations. The six leading reactor candidates selected out of several dozens of different reactor concepts are listed in the Table 1.1; they will be given a priority for further development (NRC 2004; Lee et al. 2007). As seen from the Table 1.1, the Generation IV reactors are designed to operate at much higher temperatures than conventional light-water reactors, which would result in a substantial increase in the thermal-to-electrical

TABLE 1.1

Generation IV Nuclear Reactors Suitable for Production of Carbon-Neutral Energy and Fuels

No.	Reactor Type	Temperature, °C	Efficiency, %
1	Very high temperature reactor (VHTR)	1000	50+
2	Supercritical water reactor (SCWR)	400–600	38–45
3	Fast gas-cooled reactor (GFR)	850	48
4	Heavy metal (Pb)-cooled reactor (HMCR)	540–650	Not evaluated
5	Fast sodium-cooled reactor (SCR)	550	Not evaluated
6	Molten salt-cooled reactor (MSR)	700–850	44–50

energy conversion efficiency. U.S. DOE projects that the Generation IV nuclear reactors will be deployed beyond the year 2025 time frame (U.S. DOE GEN IV 2002).

1.5.2.2 Nuclear Fusion Energy

Nuclear fusion energy is considered a long-term option (most experts agree that, in all likelihood, it will contribute to overall energy supply by the end of this century). A multi-national nuclear fusion consortium is building the $12 billion International Thermonuclear Experimental Reactor (ITER) in Cadarache (France) with the objective of demonstrating controlled nuclear fusion by 2030 (the seven countries involved in the project are United States, European Union, Japan, China, Russia, India, and South Korea) (Kim 2010). The project is designed to show that it is possible to harvest energy from magnetically contained plasma heated to extremely high temperatures. A supporting project dubbed "The Broader Approach" funded by Japan and six European nations is located in Rokkasho, Japan (Normile 2010).

In the United States, the $3.5 billion National Ignition Facility (NIF), which was completed in 2009, in the Lawrence Livermore National Laboratory (LLNL) aims at igniting fusion fuel (hydrogen) by firing NIF laser, which produces the world's highest energy pulses (Clery 2010). The laser is designed to produce pulses with the energy of 1.8 MJ in a flash lasting about 20 ns. If the NIF experiments go as planned, the next step will be the development of the Laser Inertial Fusion Engine (LIFE), which is projected to produce up to 500 MW of power. Essentially, it would be similar to the NIF, but it will be working in a continuous mode. Although technically feasible, this approach may be economically challenging due to the high cost of laser diodes.

Osaka University (Japan) researchers also utilize powerful lasers, but attempt a different approach known as fast ignition fusion (Clery 2010). In contrast to the NIF system that uses the same laser to accomplish two necessary functions, namely, the compression and ignition of the fusion fuel, the Japanese researchers use two types of lasers, each optimized for the separate task, which would result in significant economic benefits. In particular, a hydrogen-filled capsule was compressed to the density 600 times exceeding that of a liquid material, and with the second laser the researchers increased temperature to about 10 million degrees Kelvin. Although the researchers were able to detect fusion reactions, ignition itself has not been accomplished. The team now is planning to use more powerful heating laser capable of achieving 60 million degrees Kelvin as part of the project called the Fast Ignition Realization Experiment (FIREX). The FIREX laser will be capable of achieving pulses of power as high as petawatt (10^{15} W) lasting just 10 ps (10^{-11} s), but there are technical challenges ahead, mostly associated with the selection of materials able to withstand such high power impulses.

In Europe, a collaborative nuclear fusion project called the High Power Laser Energy Research (HiPER) is under way (Clery 2010). The objective of the project is to demonstrate power-producing ignition fusion reactor with the plan to start the reactor construction sometime in the mid-decade if NIF experiments are successful. A test-bed fast ignition experiment is currently under construction near Bordeaux, France. There is a possibility that LIFE and HiPER projects will merge in the near future to speed up the development of an ignition fusion power generation unit.

Summarizing the nuclear energy section, it would be interesting to see the real-world competition of nuclear power with other carbon-neutral technologies in the near future. Unlike the 1970s, which saw a real boost in nuclear power in response to the energy crisis, the current global energy field has new dynamic players, for example, solar and wind industries, CCS technologies. These competitors offer the advantage of relatively quick installation, compared with as much as 10 years of construction and $7 billion-plus price tag typical of nuclear power plants installation (Johnson 2010). The outcome of this competition over the next decade will, most likely, define future global energy interplay.

1.5.3 Renewable Energy Sources

Renewable energy sources are defined as resources that are replenished by nature at a sufficiently high rate such that they can be used by humans indefinitely. Renewable energy resources (or for brevity: "renewables") include solar, wind, geothermal, hydropower, ocean thermal, tidal, wave energies, and biomass (including biomass-derived sources such as biogas, landfill gas [LFG], municipal solid waste [MSW], and others). According to the IEA, 3,659,138 GWh of the world's electricity in 2007 was generated by renewable sources (the breakdown is shown in Table 1.2) (IEA 2007). In 2007, renewable energy sources produced about 18% of total global electricity; the share of renewables is projected to increase to 22% (or total of 7640 TWh) in 2030 (WEO 2009). Hydro, wind, and biomass are the largest contributors to the renewable energy mix. Figure 1.39 summarizes data on global power (A) and available energy and technical potential energy (i.e., energy that can feasibly be extracted considering cost and location) (B) for major renewable energy resources (excluding biomass) (BES 2005; Jacobson 2009). As can be seen, in a combination (without biomass), all the renewable sources can provide over 30–40 TW of renewable power, which far exceeds the stated targets for carbon-neutral power by mid-century.

Because of a large diversity of the renewable resources they are typically classified according to their intermittent nature:

- Intermittent sources: solar, wind, tide, wave energy
- Non-intermittent sources: biomass, hydro, geothermal energy, OTEC

TABLE 1.2

The World Gross Electricity Generation by Different Renewable Energy Sources

Source of Renewable Electricity	Electricity Generation, GWh	Percent from Total, %
Solar PV	4,104	0.112
Solar thermal	681	0.019
Wind	173,317	4.737
Geothermal	61,819	1.689
Hydro	3,162,165	86.418
Tide/wave/ocean	550	0.015
Primary solid biomass	158,237	4.324
Liquid biofuels	3,562	0.097
Biogas	26,669	0.729
Municipal waste	56,561	1.546
Industrial waste	11,473	0.314
Total	**3,659,138**	100

Source: From IEA, International Energy Agency, www.iea.org/stats/renewdata.asp?country_code=29 [accessed July 1, 2010], 2007.

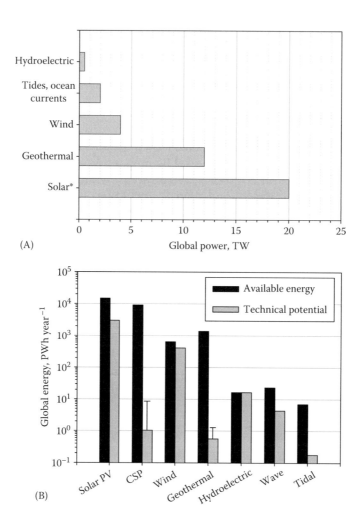

FIGURE 1.39
Worldwide available and technical energy/power potential of renewable energy resources (biomass is not included). *It is assumed that solar conversion systems with 10% efficiency cover 0.16% of the land surface. (A) According to BES (2005) and (B) to Jacobson (2009). (Data from Basic research needs for solar energy utilization, Report of the U.S. DOE Basic Energy Sciences Workshop on Solar Energy Utilization, Publ. Argonne Nat. Lab., 2005; Jacobson, M., *Energy Environ. Sci.*, 2, 148, Table 1, p. 154, 2009.)

Since this chapter puts a special emphasis on the role of carbon in energy systems, we find it useful to also classify the renewable resources according to the presence (or lack) of carbon in the energy source as follows:

- Renewable non-carbogenic energy sources: solar, wind, geothermal, OTEC, etc.
- Renewable carbogenic energy sources: biomass and biomass-derived sources: biogas, LFG, bio-methane, MSW

Figure 1.40 shows the interrelation between two types of the classification of renewable sources. The following is a brief discussion of the renewable non-carbogenic and carbogenic resources.

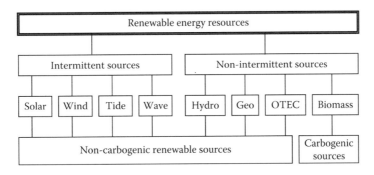

FIGURE 1.40
Classification of renewable energy resources.

1.5.3.1 Solar

Sun is by far the largest of all renewable energy resources. More energy from sun strikes the Earth in 1 h than all the energy consumed by mankind in a year. To produce an electrical equivalent of 10 TW of solar electricity, the required land surface area would be about 220,000 km^2: although very difficult, it is not an insurmountable task (Hoffert et al. 2002). Two major energy technologies taking advantage of this resource are solar photonic (or photovoltaic [PV]) and solar thermal systems.

1.5.3.1.1 Solar PV Systems

Worldwide, about 1700 TW (14,900 PWh year^{-1}) of solar power is theoretically available for solar PV electricity generation (this includes all available land outside of Antarctica), with the estimated technical potential to be below 3000 PWh year^{-1} (Jacobson 2009). The utilization of even 1% of this power would be enough to cover the world's power needs. Total global PV cell production increased from less than 10 MW in 1980 to about 8 GW in 2007 and, in 2008, the total installed capacity of PV systems was about 13 GW (mostly due to dramatic increase in Spain) (BES 2005; WEO 2009). The PV capacities are projected to rise to 200 GW in 2030 (WEO 2009). Due to economic advantages, most PV systems are installed in buildings rather than in central PV power plants (since in the latter case PV competes against grid-electricity prices, which are expected to gradually increase). This trend is likely to remain in future. Commercially available amorphous and polycrystalline silicon PV devices nowadays have efficiencies in the range of 10%–18%. Some top-of-the-line PV cells including single-crystal silicon, GaAs, and CuIn$_{1-x}$Ga$_x$Se$_2$ enjoy solar-to-electric power conversion efficiencies up to 25% under full sunlight, which approaches the theoretical energy conversion limit of 32% for single band-gap devices (the theoretical limit for multi-gap PV cells under full sunlight is about 65%) (Licht 2001; BES 2005).

According to an analytical study by Zweibel (2010), after many years of development, CdTe PV modules have become the lowest-cost producers of solar electricity beating silicon-based PVs (albeit, working at lower efficiency than crystalline Si cells). A key advantage of CdTe systems is that it can be deposited rapidly and over large surfaces (avoiding the need for careful control and adjustment of the vapor-phase deposition stage, which slows the overall process). As a result, CdTe module production costs have dropped substantially from over $2 W^{-1} in 2004 to $0.84 W^{-1} in 2010, which is the lowest among PV systems. These cost reductions bring solar-PV electricity closer to competitiveness with grid electricity. In south-west United States, the cost of solar electricity would be roughly $0.15 kWh^{-1}, and for the rest of the United States—about $0.20 kWh^{-1} (the cost of solar PV electricity is projected to drop to

$0.10 kWh^{-1} by 2020) (Jacobson and Delucchi 2009) (in 2010, the retail price for electricity in the United States was about $0.10 kWh^{-1}). There are concerns, however, that the rather limited Te supply may jeopardize long-term perspectives of the CdTe PV cells' contribution to power generation: if all PV electricity in 2009 (about 7 GW) is to be produced by the CdTe cells, about 640 million tons of Te would have been required, which equals its present annual production worldwide (note that the crustal abundance of Te and Pt is close) (Zweibel 2010). Future developments and technological advancements may alleviate this challenge, for example, by decreasing the film thickness and increasing efficiency (the current efficiency of CdTe modules is 11%; it could be further increased to 15%–16.5% and even more for commercial systems). A study published in *Scientific American* projected that PV systems as well as concentrated solar systems with enough storage to generate electricity around the clock would be able to produce electricity at about 10 cents kWh^{-1} by 2020 (Scientific American 2010).

Despite the 2008–2009 recession, makers of polysilicon and solar cells commissioned new capacities. Hemlock Semiconductor (joint venture with Dow Corning) started up 8500 t year^{-1} polycrystalline silicon facility in Hemlock (MI), and Germany's Wacker Chemie planned a US$1 billion silicon plant in Cleveland (TN) (C&EN December 21, 2009). Recently, DOW Chemical displayed its line of Powerhouse solar shingles (C&EN October 25, 2009). The shingles integrate low-cost Cu-In-Ga-Se (CIGS) PV cells into a roofing material. It is expected that the PV shingles will be commercially available in a matter of 1–2 years.

1.5.3.1.2 Solar Thermal Systems

Concentrated solar power (CSP) systems take advantage of sunlight beams concentrated by different types of mirrors on an absorber of solar radiation which is heated to high temperatures (depending on the concentration ratio). CSP is a proven technology for electricity and heat generation and it has been around for centuries (with regard to heat generation). Available and technical potential energy for CSP systems are estimated at 9,250–11,800 and 1.05–7.8 PWh year^{-1}, respectively (Jacobson 2009). Note that these values are less than that for solar-PV systems, which can be attributed to the fact that the land area required to produce unit electricity is about one-third greater for CSP compared to PV. In the past few years, there has been a surge in the number of projects utilizing CSP technologies. If in 2007, about 0.4–1 TWh of solar electricity was produced using CSP installations, by the year 2030 the CSP capacities are projected to reach up to 124 TWh (WEO 2009). Most of the new installations will be located in sunny areas such as California, Arizona (USA), Southern Europe, Northern Africa, etc., where they can better compete with conventional technologies.

The state-of-the-art solar concentrators can provide solar flux concentrations in the following ranges (Steinfeld and Meier 2004):

Trough (or parabolic) concentrators	30–100 suns
Tower systems	500–5,000 suns
Dish systems	1,000–10,000 suns

Trough-type concentrating systems is the most mature CSP technology. It utilizes parabolic trough solar concentrators that focus solar radiation on a tubular receiver with a heat-transfer fluid (e.g., synthetic oil or molten salt) that could sustain high temperatures (350°C–550°C) to produce steam in a heat exchanger to drive turbines. Nine power plants with the total surface of parabolic trough collectors of 2 km^2 have been built in the Mojave Desert (California) and have been in a full commercial operation for about 20 years without a failure (Collins 2004). These power plants called the Solar Electric Generating Systems (SEGS) generate 354 MW$_e$ of power and provide it to a local utility company (enough to power more than 200,000 households). In 2010, 5 MW CSP plant "Archemede" using molten salts for energy storage was inaugurated near Syracuse (Sicily, Italy); the plant will be integrated into existing combined cycle gas power plant (Guardian 2010). The solar plants have already achieved impressive cost reductions, and the costs are expected to fall to 5–8 cents kWh^{-1} in about a decade or so (and become competitive with fossil and nuclear electricity which is projected to be at 8 cents kWh^{-1} by that time) (Teske 2004; Jacobson and Delucchi 2009). Although grid-connected solar power is still expensive, stand-alone solar technologies are often the least-cost source of power in remote locations distant from the electric grid. Shinnar and Citro (2006) compared the cost of electricity produced by CSP and conventional power plants (e.g., coal, nuclear) and found that even near-term solar thermal electricity would be cheaper than clean coal power plants combined with CCS. The key challenges for solar PV systems and SEGS remain reducing the costs through improvements in the technology and materials, increases in scale, etc. According to some estimates, each square meter of the surface of a solar field is an equivalent of avoiding the annual production of 250–400 kg of CO_2 (Teske 2004).

Solar tower (or central receiver) systems use a circular array of individually tracking mirrors (called heliostats) to concentrate solar radiation onto a central receiver mounted on the top of the tower. As in the case of parabolic concentrators, heat is transferred to power generators through a heat-transfer media or working fluid (which, in most cases, is a molten salt). The 10 MW$_e$ Solar Two solar tower demonstration facility with a molten salt receiver and storage has been built and operated in Barstow, California. Due to higher operational temperatures, the tower-type systems have higher solar energy conversion efficiencies compared with trough concentrators. A recent addition to the family of solar tower systems is the Solar Platform near Seville, Spain, constructed by Solucar: the first (PS10) and second (PS20) solar power plants started operations in 2006 and 2009, respectively. With the overall price tag of €1.2 billion, the plant will be completed by 2013 when it will produce about 300 MW of solar power enough to power the city of Seville (Power Technology 2010). Solar dish systems utilize parabolic dish concentrators that focus sunlight on a thermal receiver. Solar tower and parabolic dish concentrators under direct insolation of 1 kW m^{-2} can produce extremely high flux densities in the range of 1,000–10,000 kW m^{-2} and achieve temperatures up to 1,300°C–1,600°C (Kodama 2003).

The conversion of solar thermal energy to work or Gibbs free energy of the reaction products is limited by the Carnot efficiency and irradiative losses from the receiver, and for an idealized system the overall efficiency can be defined as follows (Kodama 2003):

$$\eta_S = \frac{\alpha_s IC - \varepsilon\sigma T^4}{IC} \cdot \left(\frac{T - T_L}{T}\right) \tag{1.33}$$

where
η_s is an overall idealized system efficiency
I is intensity of solar radiation
T and T_L are the operating temperature of the receiver and the temperature of cold thermal reservoir, respectively
C is the concentration ratio of the solar concentrator
α_s and ε are the absorptance of solar radiation and the emittance of the receiver, respectively
σ is the Stefan–Boltzmann constant

Figure 1.41 depicts the maximum efficiency of an idealized system as a function of the operating temperature for the range of concentration ratios from $C = 500$ to 15,000 assuming

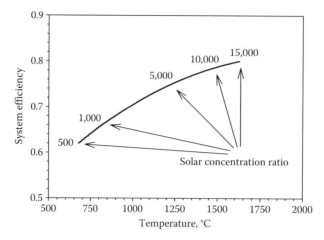

FIGURE 1.41

Solar conversion efficiency and temperature as a function of solar concentrating ratio. (Data from Kodama, T., *Prog. Energy Comb. Sci.*, 29, 567, 2003; Abanades, S. et al., *Energy*, 31, 2805, 2006.)

that receiver is a black body (thus, $\alpha_s = \varepsilon = 1$), there are no conduction or convection losses in the receiver, $T_L = 25°C$ and radiation intensity $I = 1\,kW\,m^{-2}$ (Kodama 2003; Abanades et al. 2006). As seen from Figure 1.41, the system efficiencies of about 80% could theoretically be obtained at the concentration ratio of 10,000–15,000 (at corresponding temperatures of about 1500°C–1700°C). Note that these values are higher than the theoretical energy conversion limit of 32% for single band gap and 65% for multi-band gap PV systems. Thus, potentially, solar thermal systems can provide higher solar-to-work conversion efficiencies than solar photo-conversion systems such as PV units. However, real-world efficiencies of concentrating solar thermal systems are expected to be significantly lower than their optimum values due to a variety of irradiative losses, nonequilibrium conditions, etc. It was determined that the optimal range of operating temperatures for converting concentrated solar radiation into work or fuel using an actual solar receiver would be 527°C–1427°C under solar flux intensities of 1,000–10,000 (Steinfeld and Schubnell 1993).

In the long-term future, space-based solar power systems may play a significant role in the overall energy supply. The solar flux in space is eight times greater than that on the Earth (due to clouds and Earth rotation) (Hoffert et al. 2003). It is proposed that solar electricity will be transmitted to the Earth by microwave energy with the efficiency of 50%–60%. A PV array the size of Manhattan on a geostationary orbit (800 km) would transmit power to $10 \times 13\,km^2$ surface rectenna with $5\,GW_e$ power output (Hoffert et al. 2003).

1.5.3.2 Wind

About 2% of solar energy is converted to wind through atmospheric circulations. The globally available and technically achievable potential of wind energy is estimated at 630–700 and 410 PWh year^{-1}, respectively (the theoretically available wind power is about 72 TW) (Jacobson 2009). In 2007, about 94.1 GW of wind power was installed worldwide, which amounted to just over 1% of the global electricity. The countries with most installed wind capacity (in GW) are Germany (22.2), United States (16.8), and Spain (15.1) (the data for 2008). The share of wind electricity is projected to increase to 4.5% of the global electricity generation by the year 2030 (WEO 2009), which will make wind second to hydropower source of renewable electricity.

Currently, wind is considered the most cost-effective form of renewable energy with an annual growth of 40.5% (2005) (Moriarty and Honnery 2007). The cost of kilowatt-hour of wind-generated energy has dropped from $0.40 kW^{-1} in early 1980s to less than $0.05 kWh^{-1}, and in some locations to $0.03 kWh^{-1} (Rifkin 2002). The European Wind Association forecasts that wind energy will produce 10% of worldwide electricity by 2020 (Evans-Pritchard 2008). In Denmark and some of the northern regions of Germany, wind is now providing 14%–19% of the total electricity output (Jacobson 2009). Spain has installed 15,000 MW of wind power that provides about 28% of its electricity needs (Evans-Pritchard 2008). The world's largest offshore wind farm opened recently off the British coast with 100 wind turbines capable of supplying enough electricity (300 MW$_{el}$) for 200,000 homes a year (Wire Staff, CNN 2010). Energy analysts forecast that the cost of wind energy will continue to drop significantly as the technology further improves. In spite of continued progress in reducing the cost of wind power, the intermittent nature of the wind resource remains a serious drawback. There is a need for an expanded transmission capacity that could deliver wind power from remote rural and offshore sites to major load centers and for improved technologies to store wind energy.

1.5.3.3 Geothermal Energy

The available resources of geothermal energy are immense: they are second to solar energy (1390 PWh year^{-1} [Tester et al. 2005]). However, most of this energy lies deep under the Earth crust, which makes it very difficult and expensive to extract. Thus, the technical potential of geothermal source is estimated at only 0.57–1.21 PWh year^{-1} (Jacobson 2009). Based on the capacity estimates of currently known geothermal basins, the resources of geothermal energy equal roughly 80 times the world's oil resources. The geothermal resources in the United States alone are estimated to exceed 70 million quads (quad is a unit of energy equal to 1.055×10^{18} J; for the reference, one quad is equivalent to the amount of power produced by about 34 nuclear plants during a year) (Rifkin 2002). At present, approximately 2300 MW of geothermal capacity is installed in the United States, most of which is located in California (CA). The world's largest geothermal plant located in The Geysers (CA) has a total installed capacity of 1224 MW$_{el}$ and it generates 2.2% of the state's electricity needs (NCEP 2004). Iceland and Hawaii could potentially build their hydrogen economies based to a large extent on geothermal energy. The outlook for this technology depends on three factors: resource availability, reducing costs, and technology improvement.

The geothermal resource showed a significant cost reduction during the last two decades, approaching that of the wind resource. It was reported that the cost of geothermal energy has come down from 15–16 cents kWh^{-1} in 1985 to 4–6 cents kWh^{-1} today (because of more industry experience, improved drilling technology, and the economy of scale), and it is projected to further drop to 4 cent kWh^{-1} and less by 2020 [Jacobson and Delucchi 2009]. "Hot dry rock" technology can make use of another form of geothermal energy resource, and it is designed to produce base-load electricity at any flat location. It utilizes the increasing temperature of the rock as we move from the surface deep to the center of the Earth (Figure 1.42). At the depth of about 8 km, the rock temperature exceeds that of the boiling point of water. However, no commercial hot rock system is in use anywhere, and the estimated costs of electricity produced by this method are several times higher than that of conventional geothermal systems (Moriarty and Honnery 2007). Exergetic aspects of solar, wind, and geothermal energy systems are presented by Professor Arif Hepbasli in Chapter 5.

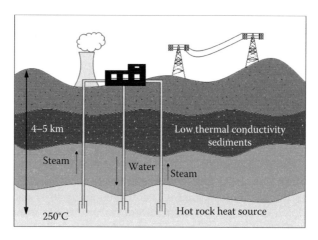

FIGURE 1.42
Schematic representation of a hot rock geothermal source.

1.5.3.4 Hydroelectric, Wave, Tidal, and OTEC

1.5.3.4.1 Hydroelectric

Although both available and technical potential energy for hydroelectric source are rather low (16.5 PWh year^{-1}) compared to solar, wind, and geothermal sources, it is the greatest source of renewable electricity today with installed capacities of 3078 TWh in 2007 (WEO 2009). Hydroelectric power plants today produce a significant fraction (up to 10%) of the electrical power generated in the United States. The total installed hydropower capacity in the United States is approximately 80 GW; much of it is concentrated in the Pacific Northwest. Countries with the largest hydropower capacities are China, Canada, Brazil, United States, Russia, Norway, Venezuela, and Egypt. Although a modest growth in the total output of hydroelectric power is projected (its capacity will rise to 4680 TWh in 2030), its role in carbon-free power production would be rather insignificant.

1.5.3.4.2 Wave and Tidal Energy

Wave and tidal energy sources are examples of underdeveloped renewable resources. Their available potential reaches about 23.6 and 7 PWh year^{-1}, and technical potentials amount to 4.4 and 0.18 PWh year^{-1}, respectively (Jacobson 2009). In 2005, wave- and tidal-based power plants generated 1.4 and 565 GWh of electricity, respectively. The combined capacities of wave and tidal energy are projected to increase to about 13 TWh year^{-1} by 2030 (WEO 2009). For the tidal source to be economical, it has to have the tide height of at least 3 m at a suitable collection area, where turbines can produce electricity from the streams flowing in both directions. Wave energy potential is estimated based on the consideration that 2% of our planet's coastline (800,000 km) have or exceeds wave power density of 30 kW m^{-1}, which translates to about 480 GW (or 4.2 PWh year^{-1}) of technically feasible power output (Jacobson 2009). It is projected that in 2020, wave energy would cost 4 cents kWh^{-1} (Jacobson and Delucchi 2009; Scientific American 2010).

1.5.3.4.3 OTEC

Ocean thermal energy conversion (OTEC) systems utilize the difference between the water temperatures at the surface and depth of warm tropical regions of the ocean (i.e., about 20°C temperature difference between surface and 1000 m depth). Although the overall heat-to-work energy efficiency of the OTEC system is relatively low (about 2%–3%), the immense mass of warm water would potentially allow the OTEC plant to economically produce electricity or hydrogen at suitable locations (O'Bockris and Veziroğlu 2007). Currently there is a renewed interest in OTEC. The techno-economic analysis suggests that the cost of electrical energy produced by an OTEC plant would be between half and one-third that of PV electricity (O'Bockris and Veziroğlu 2007).

1.5.3.5 Biomass and Bioenergy

Biomass is a product of the photosynthesis reaction where solar energy drives the endothermic reaction of water with CO_2 to produce high-energy compounds such as starch. Thus, biomass can be defined as an organic matter available on a renewable (or recurring) basis. As a source of energy, biomass can be produced in nature (e.g., trees, grasses, algae, crops, plants, aquatic plants agricultural products, forestry wastes) or in man-made industrial settings (e.g., sludge, animal wastes and residues, organic waste materials and MSW, production of micro-algae using off-gases from power plants). It should be emphasized that as a CO_2 mitigation technology, biomass growth is limited by a relatively low solar-to-chemical energy conversion efficiency, typically, 0.4%–1% for agricultural biomass, or about $1 W m^{-2}$ (Larson 1993) (solar-to-chemical energy conversion efficiency is defined as a ratio of chemical energy of biomass to energy of incident solar irradiation). Some types of biomass (e.g., specially designed micro-algae and aquatic biomass) can reach somewhat higher energy conversion efficiencies of 2% (the theoretical maximum efficiency is about 11%) (NRC 2004). Due to the low solar energy conversion efficiency, the area covered by biomass for CO_2 conversion applications (based on today's technology) would be beyond practical realization. For example, assuming an average daily intensity of solar irradiation of $200 W m^{-2}$, the area covered by algae absorbing CO_2 from 100 MW coal power plant would be as large as $50 km^2$ (IPCC CCS 2005).

Currently, biomass provides more than 10% of global energy supply (IEA Renewables 2006). Sustainable annual biomass potential for bioenergy production in the United States amounts to about 1.366 billion of dry tons, of which 0.998 billion tons come from agriculture (crop residues, perennial crops, grains-to-biofuels, process residues) and 0.368 billion tons from forest residues (manufacturing residue, logging debris, fuel wood, urban wood waste) (note that except for grains-to-biofuels, almost all of biomass relates to cellulosic biomass) (Forsberg 2009). Global technical potential of biomass as an energy source in 2050 is projected at 33–1135 EJ year^{-1}, depending on land availability and biomass yield (Hoogwijk et al. 2003). However, some experts (e.g., Hoffert et al. 2002) argue that photosynthesis has very low power density (about $0.6 W m^{-2}$) for biomass to significantly contribute as an energy resource. According to a recent study, 10 TW power from biomass would require more than 10% of the Earth's land surface, which is comparable to all human agriculture (Hoffert et al. 2002). Biomass production requires significant amount of water (about 1000–3000 t H_2O per t of biomass) and nutrients (NRC 2004). These factors may potentially result in a significant environmental impact that needs to be carefully considered.

Biomass can be converted to different types of energy. The use of biomass to produce energy (heat, electricity) and fuels (i.e., biofuels such as bioethanol, bio-butanol, biodiesel, "green" gasoline, FT synfuels) has gained a significant interest worldwide recently. Of particular importance are dedicated energy crops, for example, short rotation woody crops, switch grass that are solely grown for production of bioenergy. The surge in biomass-related activities is mainly fueled by three factors, namely, bioenergy is:

- A domestic resource (which would potentially alleviate the dependence on imported oil)

- A carbon-neutral source of energy (although CO_2 is released during the energy use of biomass: directly or in the form of biofuels, the equivalent amount of CO_2 is captured from the atmosphere during its growth, a so-called closed carbon loop)
- Compatible with the fossil fuel infrastructure (e.g., biomass can be coprocessed with coal, and biofuels can be delivered, distributed, and used in existing engines with minimal changes)

Different aspects of bioenergy use and biofuels production are thoroughly described and analyzed in Chapters 8–13 of this book. The objective of this section is to provide some perspectives on the scale of emissions from biomass energy use and its potential impact on CO_2 mitigation strategies.

1.5.3.5.1 *Direct Use of Biomass as a Primary Energy Source*

Direct use of biomass for heat and power production has been practiced in many countries for years. These plants are rather small with the typical capacity of about $30 MW_e$ generating about 0.2–0.3 Mt CO_2 year^{-1} (IPCC CCS 2005). There are about 213 bioenergy plants operating in North America and Brazil, emitting about 73 Mt CO_2 year^{-1} (average emissions per a source: 0.34 Mt CO_2 year^{-1}) (IEA GHG 2002). The CO_2 concentration in the off-gases of the bioenergy plants varies in the range of 3–8 vol.%, which is similar to that from NG-powered power plants. In general, the local availability of biomass feed (mainly, crop and forestry residues) determines the size of the plants. In the places where large and concentrated sources of biomass are available (e.g., near pulp mills, sugar processing plants), large biomass-powered plants could be built. In Sweden, for example, seven combined heat and power plants using pulp mills with the average output of $130 MW_e$ (equiv.) are under operation (IPCC CCS 2005). In addition to biomass-powered plants, there is a great number of power plants worldwide coprocessing biomass and coal.

Future projections point to the increasing role of large bioenergy plants utilizing dedicated (possibly genetically engineered) energy crops. The studies indicated that the plant capacities of several hundreds of MW_e are feasible for bioenergy plants using dedicated energy crops (in many cases, the economy of scale outweighs the additional cost of biomass transportation) (e.g., Dornburg and Faaij 2001). In another analytical study, the authors analyzed various scenarios involving building of large (about $1000 MW_e$) integrated bioenergy plants producing both electricity and synthetic fuels from dedicated energy crops in the United States (Greene 2004). It is very difficult to reliably estimate the scale of future CO_2 emissions from bioenergy plants, which would depend on the extent of the development and utilization of dedicated energy crops and other biomass sources. It is likely that the capacity of bioenergy plants could reach the scales suitable for CO_2 capture and storage, which would enhance the role of biomass as carbon negative energy source.

1.5.3.6 *Storage of Intermittent Renewable Energy*

The major problem with the efficient utilization of solar and wind energy sources relates to their intermittent nature (i.e., daily and seasonal variations) and a nonuniform distribution over the land. A reliable and affordable energy storage system is a prerequisite for the efficient use of these intermittent renewable energy sources and their integration into the nationwide energy system. Thus, the energy storage systems will play an extremely important role and become an integral part of the future renewable-based infrastructure. Available energy storage options have different performance characteristics with regard to their response time, maximum storage capacity, lifetime, etc. In general, intermittent renewable energy can be stored in the form of electrical, chemical (thermochemical), thermal,

TABLE 1.3

Current Intermittent Energy Storage Options

Energy Storage Method	Type of Energy Storage
Batteries	Electrical/electrochemical
Supercapacitors	
Electromagnetic	
Compressed air	Mechanical
Hydrostatic	
Flywheels	
Reversible fuel cell, H_2	Chemical/thermochemical
Reversible chemical reactions	
Thermal (phase change, heat pumps)	Thermal

and mechanical (or kinetic) energy. The current energy storage options are summarized in Table 1.3 (Sørensen 1984; Muradov and Veziroğlu 2008). The conventional electrical energy storage systems such as rechargeable batteries and ultracapacitors are expensive and do not have a sufficient capacity (per unit of volume or weight) to be considered for very large energy systems. From this viewpoint, compressed air, hydrostatic storage, and reversible FC are more promising for the integration into a large renewable energy-based network.

Compressed air (CAES) and hydrostatic energy storage have recently emerged as an efficient and cost-effective means of storing large quantities of electrical energy. CAES systems have successfully been operated in Germany and United States (Alabama) for about a couple of decades. Recent studies by Electric Power Research Institute (EPRI) indicate that the cost of CAES is about half of that of lead-acid batteries and it would add 3–4 cents kWh^{-1} to the cost of PV electricity (Zweibel et al. 2007). Most important task is to find the sites suitable for the pressurized air storage. The EPRI study showed that suitable geologic formations exist in 75% of the United States, often close to metropolitan areas (Zweibel et al. 2007). Zweibel et al. estimate that the storage of 15 billion cubic meters of air, under the pressure of 73 atm, would be required for storing of renewable energy by the year 2050 (i.e., about 1 order of magnitude less than the total storage capacity of NG currently stored in 400 underground reservoirs) (Zweibel et al. 2007). Hydrostatic (or pumped hydro) storage is another well-developed technology for storing large quantities of electrical energy. For example, Denmark uses intermittent power to pump water into Norway's hydropower impoundments, and stored energy of water is later released through turbines (Hoffert 2010). The power output, energy efficiency of the storage system and its cost per kW stored depend on the difference in height between lower and upper levels.

In the *electrolyzer/FC* storage option, water is electrolyzed to hydrogen and oxygen at times of low energy demand using electricity from an intermittent renewable energy source (e.g., solar, wind). Hydrogen is stored in a suitable storage reservoir. At times of a high demand, hydrogen and air electrochemically react in the FC producing electricity. A reversible FC can be used instead of the electrolyzer/FC combination (since the electrochemical reaction is reversible). The storage capacity of this system is determined by the size of the hydrogen storage reservoir. In principle, gaseous hydrogen storage options are similar to those for air (e.g., underground caverns, abandoned mines and aquifers).

Thermal storage systems are designed for storing high-temperature solar heat (e.g., from solar concentrators) and involve several types of storage media, predominantly, thermal oil and molten salts. Thermal oils have a disadvantage of relatively low decomposition temperature (300°C) and inflammability. Molten salts (e.g., $NaNO_2$–$NaNO_3$–KNO_3 mix) have been used as an efficient high-temperature (250°C–600°C) heat-transfer medium since 1930s (e.g., at refineries), but the main problem with that storage medium relates to its high melting point (about 220°C) (which is required to always keep the salt preheated). The first commercial solar thermal plant (50 MW) to incorporate molten salt storage operates in Spain. The feasibility of ionic liquids as a thermal storage medium and heat-transfer fluids in a solar thermal power plant was investigated (Wu et al. 2001).

In *thermochemical storage systems,* thermal energy is stored in chemical bonds by means of reversible thermochemical reactions. Depending on the targeted application, these reversible chemical reactions can be used for long-duration energy storage, or long-distance energy transport (e.g., "chemical heat pump" storage). One example of the thermochemical energy storage is based on a reversible reforming/methanation system (also known as Adam/Eve concept), where the reaction $CH_4 + H_2O \leftrightarrow 3H_2 + CO$ is driven to the right by heating the reactor to 900°C by a solar concentrator.

More detailed information on the state-of-the-art short- and long-term energy storage systems and their role in efficient utilization of intermittent energy is presented by Professor Bent Sørensen in Chapter 7.

1.5.4 Life-Cycle GHG Emissions from Renewable, Nuclear, and Decarbonized Fossil Energy Sources

It would be a mistake to assume that renewable, nuclear and decarbonized (via CCS) fossil sources are completely "clean" and do not produce any GHG emissions. The release of GHG emissions could occur at many stages of the deployment and exploitation of the renewable and nuclear resources, for example, construction, installation, operation and maintenance, accidents, decommissioning. Thus, the life-cycle analysis approach should be considered when evaluating climate-relevant emissions from the renewable and nuclear sources. Figure 1.43 summarizes the data on the life-cycle GHG emissions per kWh of electricity generated from nuclear and a variety of renewable energy sources, excluding biomass (Jacobson 2009).

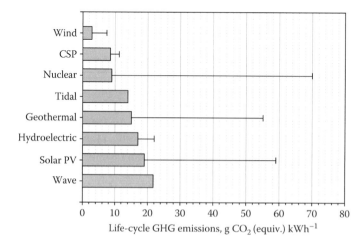

FIGURE 1.43
Life-cycle GHG emissions (in CO_2 equiv.) for nuclear and renewable electricity generation sources. (Data from Jacobson, M., *Energy Environ. Sci.,* 2, 148, Table 3, p. 154, 2009.)

The life-cycle GHG emissions from wind electricity are the lowest among all renewables considered. Assuming 5 MW turbines operating over a lifetime of 30 years, the resulting life-cycle GHG emissions from wind-generated electricity are estimated at 2.8–7.4 g CO_2 (equiv.) kWh^{-1}. Solar CSP generates 8.5–11.3 g CO_2 (equiv.) kWh^{-1} (a CSP plant lifetime is assumed 40 years). The life-cycle GHG emissions from tidal and wave-driven power plants are estimated at 14 and 21.7 g CO_2 (equiv.) kWh^{-1}, respectively (at a lifetime of 15 years). Hydroelectric power plants generate most of GHG emissions during the dam construction phase, but since the plants could last up to 100 years, their life-cycle emissions are relatively low: 17–22 g CO_2(equiv.) kWh^{-1}. Hydroelectric plants located in hot climatic zones could also emit sizable amounts of biogas due to a microbial decay processes. In many cases, the geothermal sources release dissolved or gaseous CO_2 from a well along with hot water. When combining these CO_2 emissions with GHG generated during the construction of geothermal power stations, the life-cycle GHG emissions could reach 15.1–55 g CO_2(equiv.) kWh^{-1}. Solar PV systems have the largest life-cycle GHG emissions. Depending on the payback times and the level of solar insolation, the life-cycle GHG emissions for commercial PV systems varies in the range of 19–59 g CO_2 (equiv.) kWh^{-1} for the insolation of 1300–1700 kWh m^{-2} year^{-1} (Jacobson 2009).

The opponents of expanding nuclear power sector point to an often-overlooked issue related to the amount of life-cycle GHG as a result of using a nuclear source, in particular, during mining of nuclear fuel (uranium), its enrichment, transport, and waste disposal as well as the construction, operation, and decommissioning of the nuclear reactors. Literature sources provide different estimates of life-cycle GHG emissions from nuclear-generated electricity (in g CO_2(equiv.) kWh^{-1}): 9–21 (WNA 2009), 24.2 (Price and Blasé 2002), 16–55 (Fthenakis and Kim 2007), 9–70 (Jacobson 2009). In its report, World Nuclear Association (WNA) estimated that life-cycle CO_2 emissions from nuclear power make up about 1%–2% of that of coal-fired power, and if extremely low-grade ores are envisaged, this figure would rise to about 3% of coal-derived CO_2 emissions (or 6% of gas-derived emissions) (WNA 2009); other authors, however, claim that the amount of CO_2 emissions from nuclear electricity would rise exponentially with the decrease in uranium content of the processed ore (van Lewen and Smith 2005). Interestingly, the values of life-cycle GHG emissions from the nuclear sources are comparable to those from renewables.

Intensive worldwide efforts are underway to eliminate or drastically reduce the amount of CO_2 emitted from fossil fuel (particularly, coal) fired power plants, and one area where many experts are pinning their hopes on is CCS. CCS is a way of preventing anthropogenic CO_2 emissions from reaching the atmosphere by capturing and securely storing CO_2 in such sinks as geologic formations (e.g., deep coal seams, depleted oil fields, and gas reservoirs), the ocean, terrestrial ecosystems, etc. There have been estimates in the literature on the life-cycle GHG emissions from coal power plants coupled with CCS technology (coal-CCS), which include emissions due to the mining and transport of coal, the construction, operation, and decommissioning of coal power plants as well as CO_2 release during its capture and long-term storage (e.g., due to leakage). Overall (world average) life-cycle GHG emissions from coal-fired power plants is a sum of indirect emissions: 176–289 g CO_2 (equiv.) kWh^{-1}, and direct emissions due to coal combustion: 790–1017 g CO_2(equiv.) kWh^{-1} (e.g., for gas-fired plants these numbers are 77–113 and 362–575 g CO_2-equiv kWh^{-1}, respectively) (WNA 2009) (direct CO_2 emissions from fossil-based power plants in the United States are as follows: 951, 894, and 600 g CO_2 kWh^{-1} for coal, petroleum, and NG-powered plants, respectively) (EPA 2000). According to IPCC estimates, the addition of the CCS technology to coal-fired plants could reduce the direct GHG emissions by 85%–90% (IPCC CCS 2005). Thus, the net life-cycle GHG emissions of coal-CCS power plants would amount to about 225–440 g CO_2(equiv.) kWh^{-1}, which is the highest among all electricity-generating options considered. If one adds the "opportunity-cost" emissions due to planning-to-operation delays and possible CO_2 leakage over extended storage time (e.g., 500 years), the value of total GHG emissions from coal-CCS electricity rises to 308–571 g CO_2 (equiv.) kWh^{-1} (Jacobson 2009).

Life-cycle GHG emissions from biofuels, for example, corn and cellulosic ethanol, is a highly debated topic due to the extreme complexity of the system, large number of variables, and differences in the assumptions. Most of the studies accounted for the GHG emissions due to fertilizing, cultivating, watering, harvesting, crops and fuel transporting, etc. and the emissions due to combusting these fuels in vehicles. It was argued that the estimates should also include emissions of soot, NO_x, CO, and accumulation of CO_2 in the atmosphere between biofuel use and regrowth, and emissions due to converting land from one crop to another, etc. (Jacobson 2004). In particular, a study that accounted for land use changes reported that converting gasoline to ethanol (E85) would increase life-cycle GHG emissions by over 90% for corn ethanol and about 50% for cellulosic ethanol (Searchinger et al. 2008). Other authors reported drastically different estimates of life-cycle GHG emissions due to biofuel use; for example, Delucchi (2006) estimated that conversion of gasoline to corn-ethanol E90 fuel could reduce GHG emissions by 2.4%, whereas ethanol produced from switchgrass would reduce U.S. GHG emissions by 52.5%.

Despite the fact that life-cycle GHG emissions from nuclear and renewable energy resources are not zero, in this chapter, we still call them "carbon-free" sources to distinguish them from "low-carbon" sources such as fossil fuels coupled with CCS technology.

1.6 Carbon-Neutral Fuels and Energy Carriers

While curbing CO_2 emissions from power generation plants and large industrial sources is under an intensive technological development, it is realized that CO_2 capture from a myriad of small distributed CO_2 sources (e.g., from transportation and residential sectors) is neither economically nor even technically feasible. Today, transportation consumes a lion's share of the primary energy sources, and it is almost entirely powered by petroleum-based fuels (the share of NG-, LPG-, alcohol-, and all-electric-driven cars is still very small). The transition to carbon-free or low-carbon systems would be particularly challenging in the transportation sector because of its enormous complexity and magnitude (there are close to 1 billion vehicles worldwide, with an annual production rate of about 50 million cars). It is projected that the share of transportation in the U.S. total energy consumption will increase from about 27% in 2000 to about 32% in 2025. If this trend will hold true for the rest of the world, 3–10 TW of carbon-neutral (or low-carbon) transportation fuels will be needed by the mid-century. Several options are technically feasible and currently are under different stages of development, including the use of

- Electricity produced from carbon-free primary sources
- Non-carbon-bearing fuels, such as hydrogen and ammonia
- Fuels that are less carbon intensive than conventional petroleum-based fuels (e.g., synthetic liquid hydrocarbons [SLH], substitute NG, methanol)
- Biofuels (biodiesel, bioethanol, biobutanol, "green" gasoline, etc.)

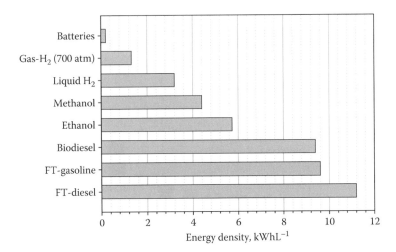

FIGURE 1.44
Comparison of volumetric energy densities of various fuels and power sources.

In the first three cases, nuclear, renewable, or fossil sources coupled with CCS should be used as primary energy sources. Figure 1.44 compares volumetric energy densities of the fuels and energy carriers that are considered as alternatives for petroleum-based fuels in transportation. Biodiesel and SLH have the highest volumetric energy density comparable to that of petroleum-based fuels. The following is a brief discussion of carbon-free fuels and energy carriers, and implications of their use to future energy systems.

1.6.1 Electricity

Electricity is a versatile energy carrier that is increasingly substituting fossil fuels in many areas such as transportation, household appliances, industrial processes (e.g., resistive heating vs. fuel combustion), etc. Most energy projections indicate that this trend will continue in the foreseeable future with a special emphasis on utilizing carbon-free or low-carbon footprint electricity produced from renewable or nuclear energy sources or large centralized fossil-based power plants equipped with CCS. Currently, nuclear and hydro-electric plants are the only major producers of carbon-free electricity (in the United States, their ratio is about 3 to 1, with a relatively small share of other renewable sources). Figure 1.45 summarizes costs of different forms of renewable electricity and compares them to that of fossil and nuclear electricity. In some areas, certain forms of renewable (e.g., wind or geothermal) electricity are cost-competitive with the fossil-derived electricity at today's market prices. The cost of PV, CSP, wave, tidal electricity is still high. The projected cost of wind, wave, geothermal, CSP, and PV electricity in 2020 is (in $ per kWh in 2007 dollars) <0.04, 0.04, 0.04–0.07, 0.8, and 0.10, respectively (Jacobson and Delucchi 2009). The 2009 U.S. NRC report predicts that 10% and 20% of electricity might be generated by renewables by 2020 and 2035, respectively. The report indicates that going beyond 20% would require major scientific advances and a fundamentally different electricity system, mostly, due to the specifics of usage of such intermittent and unpredictable sources as solar and wind energy (Williams 2009).

Electricity is an ideal energy carrier for vehicles due to a number of advantages (Yang 2008):

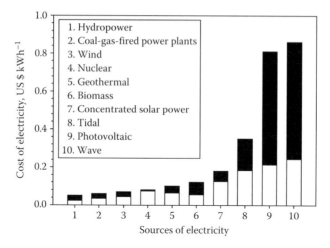

FIGURE 1.45
Cost of renewable electricity relative to the cost of nuclear and fossil electricity. (Adapted from Koonin, S., Environmental stewardship and energy innovation, *Naval Energy Forum*, McLean, VA, October 2009; Jacobson, M. and Delucchi, M., *Sci. Am.* November, 58, 2009.)

- Electricity can be used on board with very high efficiency
- Quiet operation
- Zero point-of-use emissions
- Relatively easy management (e.g., by capture and storage) of GHG emissions (since electricity comes from large centralized power plants)
- It is much less expensive (per km driven) than equivalent amount of gasoline
- It can be produced from a diversity of primary sources and technologies

Electric vehicles (EV) are powered by chemical energy stored in batteries. Battery technologies for the vehicular application have been steadily improving over the last two to three decades as new types of batteries such as Ni-Cd, nickel metal hydride, Li-ion, Li-poly, and zinc-air batteries have been introduced to the market. However, some technical barriers to a broad consumer acceptance of the EV still remain: (1) Relatively low volumetric energy density of batteries (see Figure 1.44) and, hence, limited driving range (about 50–100 km per charge); (2) charging time (hours); (3) lifetime (in general, batteries do not last as long as internal combustion engines [ICE]); and (4) disposal problem. The use of PHV solves the driving range and charging time problems. PHV can be plugged into an electrical grid and drive more than 100 km, and when the battery is depleted the PHV switches to gasoline or diesel engine. Since most cars and light trucks are driven about 50 km day^{-1} or less, the PHV can practically eliminate the gasoline use for a significant fraction of the population. Furthermore, the cost of electricity equivalent to a gallon (about 3.8 L) of gasoline would be less that $1.00 at current electric rates (Duncan 2007).

Although electricity could potentially replace fossil fuels in all energy-related areas, many analytical studies predict that in some application areas (e.g., aviation, heavy trucks) the direct use of liquid hydrocarbon fuels will remain a preferred option. In fact, despite almost century-long worldwide intensive efforts focused on the development of electric cars, they still have only a niche market. On the other hand, the successful market penetration of PHV may boost and expand the role of electricity in transport.

1.6.2 Hydrogen

Hydrogen is considered an ultimate carbon-neutral fuel provided it is produced from water and carbon-free primary energy sources. As a clean fuel and energy carrier, hydrogen is a focus of the comprehensive review by Professor Winter (Chapter 2) and several other chapters dedicated to more specific aspects of hydrogen production and utilization. Therefore, in this section, only a brief discussion of opportunities for an early market penetration of hydrogen will be presented.

One of the most attractive features of hydrogen as fuel is that it could be efficiently converted to electricity and produced from water by means of a reversible electrochemical reaction (thus, avoiding Carnot limitations) involving FC and electrolyzers, respectively:

$$H_2 + \frac{1}{2}O_2 \overset{\text{Fuel cell}}{\underset{\text{Electrolyzer}}{\Longleftrightarrow}} \text{Electricity} + H_2O \qquad (1.34)$$

Thus, early market opportunities for hydrogen may be in transportation (e.g., FC vehicles) and small distributed applications, for example, residential and commercial buildings, where the capture of CO_2 emissions is economically not feasible. This way, replacing hydrocarbon fuels with hydrogen would eliminate CO_2 emissions at the point of energy use. Currently, the main obstacles to the widespread use of hydrogen in vehicles and other areas is the lack of efficient onboard hydrogen storage systems and hydrogen delivery infrastructure, in general, and relatively high cost of FC. These technical challenges are in the focus of numerous national hydrogen R&D programs throughout the world.

It should be noted that in some applications, it may not be necessary to convert hydrogen into electricity as hydrogen can be directly used as fuel, for example, in internal combustion engines (albeit, with lower energy conversion efficiency) and heating appliances. From this viewpoint, the use of suitable H_2–CH_4 mixtures (e.g., Hythane™, where H_2 concentration is 20–30 vol.%) could facilitate early introduction hydrogen fuel to the market. In the area of large-scale power generation, NG-powered turbines can easily be adjusted to operate on hydrogen or hydrogen-methane mixtures (note that although Hythane fuel would produce certain amount of CO_2 during its combustion in ICE or a turbine, the amount of NO_x and other harmful impurities in the emissions would be significantly lower compared to pure NG combustion).

Although hydrogen can be used in ICE, the real benefits of using hydrogen fuel are linked to the use of FC that can deliver higher fuel-to-wheel efficiency. It is recognized that the success of hydrogen-powered transportation will depend on the development and commercialization of efficient onboard hydrogen storage systems and durable and inexpensive FC. Most automakers focus their activities on proton exchange membrane (PEM) FC, which has the advantages of a fast start, a high power density, a convenient temperature range, and other features that satisfy the requirements of light-duty vehicles. The cost of PEM FC, however, is currently prohibitively high (about \$1000 kW^{-1}). There are several prototype vehicles powered by PEM FC from several automakers (e.g., Ford, Toyota, GM and others), but the widespread introduction of FC vehicles into the market would require overcoming a number of significant technical challenges such as improving FC durability, reducing their cost, drastic reduction in Pt-catalyst usage (or finding a replacement for Pt), etc. (the price reduction by a factor of 20–50 would be necessary for PEM FC to compete with ICE engines).

Globally, hydrogen plants consume about 2% of primary energy (Socolow 2006). If CO_2 capture and storage are successfully established in the market as a means of CO_2 mitigation, large centralized NG- or coal-based hydrogen-manufacturing plants may emerge as a source of low-carbon hydrogen. It was estimated that a hydrogen plant with the average capacity of about $1000\,MW_{th}$ (or $720\,t\,day^{-1}\,H_2$) would be adequate to keep the hydrogen and CO_2 capture/storage costs low (this plant could support about 2 million FC vehicles driving $14,000\,km\,year^{-1}$) (IPCC CCS 2005). In this case, CO_2 storage rates for SMR and coal gasification plants would be 1.7 and 3.1 million tons CO_2 per year, respectively (assuming 80% average capacity factor). Thus, fossil-based production of hydrogen in a few very large centralized plants equipped with CCS for the use of hydrogen in a very large number of relatively small mobile (e.g., cars) and distributed end-users could provide a means for significant reductions in life-cycle CO_2 emissions (compared to conventional petroleum-based energy systems). Among the renewable hydrogen production systems, currently, wind energy can generate hydrogen at the lowest price, and this tendency will most likely remain in the future. In a recent study, Moriarty and Honnery (2007) reported a comparative assessment of different renewable sources for hydrogen production; they concluded that wind-powered hydrogen production may become a basis of the future hydrogen economy.

1.6.3 Ammonia

Due to lack of carbon in its molecule, relatively high energy content, and adequate combustion characteristics, ammonia (NH_3) is considered by its proponents a good fuel choice in a carbon-constrained world. Attractive features of ammonia as carbon-free fuel are as follows:

- It yields only water and nitrogen as combustion products when burned in ICE:

$$2NH_3 + \frac{3}{2}O_2 \rightarrow 3H_2O + N_2 \tag{1.35}$$

- It is a high-octane fuel (an octane number of 110).
- Compared to hydrogen, NH_3 lacks the storage, delivery, and other logistic problems (it can be stored as a liquid under moderate pressure at ambient temperatures, similar to liquefied petroleum gas).
- Ammonia is a major chemical commodity, thus, the facilities for its storage, safe handling, transportation, and distribution are available worldwide.

It is noteworthy that the use of ammonia as fuel is not a new idea; in fact, it was used during WWII to power buses in Belgium (due to severe fuel shortages). Ammonia could be used as a fuel directly (in ICE) or via onboard decomposition to hydrogen and nitrogen (in ICE and FC). Recently, the interest in ammonia as fuel was boosted by hopes that it could become a carbon-free alternative to petroleum fuels in transportation. In Canada, Chevrolet Impala was converted to run on ammonia as fuel (Iowa Energy Center 2007). There have been proposals to use an ammonia cracker in combination with an alkaline FC for a vehicular application (Hacker and Kordesh 2003). As of now, however, there is no widespread commercial interest in ammonia as a transportation fuel, mainly, because of its high toxicity.

Ammonia is one of the major products of chemical industry: over 100 million tons of NH_3 per year is produced worldwide. The United States consumes about 20 million tons of ammonia per year with 90% of this amount either applied directly as a fertilizer or converted to solid fertilizers such as ammonium phosphate or urea (Wind 2009). Currently, ammonia is produced by a high-temperature (450°C–600°C), high-pressure (150–180 atm) catalytic synthesis reaction between hydrogen and nitrogen (from atmosphere) using the Haber–Bosch process (Fe- or Ru-based catalysts are used in the process). About 70% of hydrogen used for ammonia synthesis is produced from NG via SMR process, and the rest from coal. Although the current production of ammonia is responsible for only 2% of global CO_2 emissions, fossil-derived ammonia cannot be considered carbon-free fuel (even though it does not release CO_2 during its use as fuel). In future, new less energy-intensive technologies for ammonia manufacturing might be developed, especially, with the use of a carbon-free hydrogen feedstock. The authors of the National Renewable Ammonia Architecture concept project that the first steps in the introduction of ammonia as fuel would occur in farm utility vehicles (Wind 2009). According to the concept, hydrogen used in the ammonia synthesis will be produced by electrolysis using renewable energy resources (solar, wind, etc.).

1.6.4 Synthetic Fuels

Due to ever-increasing anxiety over the depletion of liquid hydrocarbon resources, synthetic liquid fuels are in the focus of intensive R&D and commercial activities worldwide. The main driving forces behind these efforts are twofold: the advantages of using existing petroleum-based fuel infrastructure and car engines with minimal changes, and concerns that other alternative fuels (e.g., hydrogen) may not enter the market in the nearest decade or two. The term *synthetic fuel* typically covers a wide range of liquid, liquefied, and gaseous fuels that could be synthesized from synthesis gas or syngas (H_2–CO mixture): FT hydrocarbons, methanol, synthetic methane (or substitute natural gas [SNG]), dimethyl ether (DME), C_2–C_5 alcohols, Mobil-gasoline, and others.

Currently, considerable activities on production of synthetic fuels from coal are carried out in China. Existing and under construction plants include six 600,000 t year^{-1} methanol-manufacturing plants, two 800,000 t year^{-1} DME plants and two or more FT liquids plants (methanol would be used as fuel and for production of chemicals and DME; the main application of DME is a cooking fuel) (IPCC CCS 2005). A coal gasification plant for the production of synthetic methane is under operation in North Dakota (USA) since 2000. The plant produces about 1.5 million tons per year of CO_2 as a by-product of the gasification process, which is captured and pipelined over 300 km to the Weyburn oil field in Canada, where it is utilized for enhanced oil recovery. In this section, we briefly discuss commercial status of most important synthetic fuels.

Production of SLH (also called FT hydrocarbons) from coal and NG (coal-to-liquid and gas-to-liquid processes, respectively) are well-established technologies. The technology was first introduced and utilized on a large scale by the Germans during WWII fuel shortages. Currently, the most notable technology for producing SLH from coal is the SASOL technology developed and operated (since 1955) in South Africa, where about 20 million tons of coal (in carbon equivalent) is being processed annually producing liquid hydrocarbon fuels and value-added chemicals. In this process, close to 32% of carbon is converted to products, 40% is vented in the form of diluted CO_2 streams, and 28% is released as a concentrated CO_2 stream. Thus, assuming that all CO_2 from the FT plant can be captured and stored, this would result in substantial reductions in CO_2 emissions (by more than half) at the point of the energy use of liquid fuels (compared to petroleum fuels). The commercial manufacture of FT liquid fuels from NG began first in Malaysia in early 1990s with the plant capacity of 12,500 barrels day^{-1}), followed by several plants in Qatar (plants capacities ranging from 30,000 to 140,000 barrels day^{-1}) (IPCC CCS 2005). In contrast to coal-based production of liquid fuels, gas-to-liquid plants produce only diluted CO_2 streams that would be more challenging to process in order to recover CO_2. However, production and use of the coal-derived synfuel generates CO_2 emissions at twice the rate of using gasoline made from crude oil. A coal-fed facility in South Africa is the planet's single biggest point source of CO_2, emitting 20 million tons of CO_2 a year (Kintisch 2008). Thus, the environmental impact of coal-based synfuel production could be disastrous, unless the plants combined with CCS systems are implemented.

In order to be considered carbon-neutral fuels, SLH have to be generated from biomass-based feedstocks or CO_2 from atmosphere. Although, certain aspects of the coal-based technology could be applied to biomass, no widespread large-scale commercial production of SLH from biomass has been implemented yet. Currently, the interest in the biomass-to-liquid (BTL) technology is rapidly increasing. This is due to recent improvements to the technology and the realization that it can provide economical means of converting biomass into liquid fuels with desirable characteristics. SLH enjoy certain advantages over bioethanol and petroleum-based fuels. While during manufacturing of bioethanol and biodiesel only part of the biomass feedstock is utilized (e.g., starch, sugar, oil, cellulose), in the case of SLH production the whole plant could be utilized. Furthermore, in contrast to alcohols that contain significant amount of oxygen (which dramatically reduces energy density of the fuel) SLH contain only carbon and hydrogen; for example, gravimetric energy density of ethanol and synthetic gasoline are 7,102 kcal kg^{-1} and about 11,000 kcal kg^{-1}, respectively (higher the energy density of fuel, the greater the distance that could be traveled by the car on a unit volume of fuel). SLH are most similar to conventional petroleum-based fuels by physical and chemical characteristics; thus, no changes in the fuel infrastructure or car engines will be necessary. On the other hand, compared to petroleum-based fuels, SLH are much cleaner since they contain no sulfurous or other harmful impurities, and, barrel for barrel, they emit less CO_2 and lower levels of NO_x emissions and particulate matter than petroleum fuels (Kintisch 2008). More detailed information on the technological and economical aspects of SLH production from lignocellulosic biomass are presented in Chapter 9 written by Professor Georg Schaub and Dr. Kyra Pabst.

Synthetic NG is produced via a catalytic reaction (called methanation) involving syngas with the H_2/CO molar ratio of 3:1, as follows:

$$3H_2 + CO \rightarrow CH_4 + H_2O \quad (Ni\,catalyst) \quad \Delta H^\circ = -250.6 \,kJ/mol\,(CH_4) \qquad (1.36)$$

This is a strongly exothermic reaction (a reverse reaction to SMR), and the process heat is typically recovered via a variety of process integration schemes. Several commercial coal-to-gas processes currently employ the methanation process: Comflux, HICOM, and others (Lee et al. 2007). Although SNG could be produced via biomass gasification route and potentially used as a transportation fuel (the way compressed NG is currently used), in all likelihood, it will be in disadvantage to SLH due to its relatively low energy storage density.

Methanol is another product that could be produced from biomass-generated syngas, as follows:

$$2H_2 + CO \rightarrow CH_3OH \quad (Cu-Zn\,catalyst) \quad \Delta H^\circ = -92 \,kJ/mol \qquad (1.37)$$

Methanol has certain attractive features as a transportation fuel such as low cost, its storage as a liquid at ambient temperature, good combustion characteristics, etc. Although methanol contains about half the energy of gasoline (per unit volume), it has high octane number of 99 (compensating for the lower energy density). On the downside, methanol is highly toxic, and its widespread use may potentially raise some environmental concerns (e.g., possible groundwater contamination). If necessary, methanol could be converted to more high energy density fuel, gasoline, according to Mobil-process (via dimethyl ether), as follows:

$$2CH_3OH \rightarrow CH_3OCH_3 + H_2O \tag{1.38}$$

$$CH_3OCH_3 \rightarrow \text{Mobil-gasoline} \quad (\text{ZSM-5 catalyst}) \tag{1.39}$$

The process was first commercially practiced in New Zealand and operated until 1996.

1.6.5 Biofuels

The term *biofuel* covers a wide range of gaseous (e.g., biogas, LFG, bio-hydrogen) and liquid (bioethanol, biobutanol, biodiesel, "green" gasoline, etc.) fuels produced from biomass via a variety of fermentative and thermochemical processes. Figure 1.46 depicts main technological (biological and thermochemical) routes to producing carbon-neutral alternative fuels from biomass (including biogas). LFG and biogas are generated by anaerobic digestion or fermentation of any biodegradable organic matter including manure, sewage sludge, MSW, etc. These gases represent very complex gaseous mixtures containing methane and CO_2 as major components along with small amounts of N_2, O_2, H_2S, and a variety of organic and element-organic compounds. If not utilized or treated, LFG and biogas can

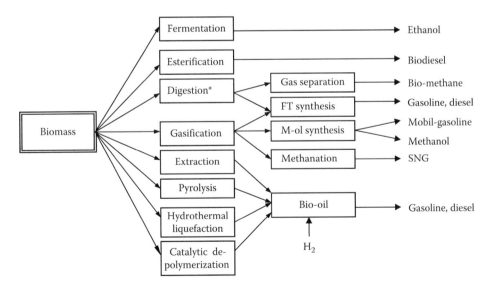

FIGURE 1.46
Main technological routes to production of biofuels from biomass. *Anaerobic digestion of biomass to biogas followed by its reforming to syngas, or recovery of methane. M-ol—methanol, SNG—substitute natural gas. (From Muradov, N. and Veziroğlu, T.N., *Int. J. Hydrogen Energy*, 33, 6804, 2008. With permission.)

potentially create significant environmental damage, because methane is a much more potent GHG than CO_2 (see Section 1.3). Although LFG and biogas are fuels of medium calorific value suitable for electricity generation (via ICE, turbines, or FC), and their resources are vast and widely available throughout the world, they remain mostly unused (though they are becoming increasingly popular in recent years).

Liquid biofuels (e.g., bioethanol, biodiesel) are compatible with the existing fuel infrastructure and vehicle technology and can be conveniently blended with existing transportation fuels and can make a considerable contribution to reducing our dependence on fossil fuels and curbing GHG emissions. It should be emphasized, however, that the economic, environmental, and social benefits of the current generation of biofuels vary significantly, depending on the type of biofuel, feedstock supply, and local economic specifics. Currently, bioethanol and biodiesel are two main commercially produced biofuels. Brazil and United States are the two largest producers of bioethanol in the world, and their production capacities are expected to increase in the foreseeable future.

Bioethanol as transportation fuel is typically used in a mixture with petroleum-based fuels in a wide range of proportions (5%–10% in United States and Europe and up to 85% in Brazil). Recently, U.S. EPA suggested that it would increase the allowable content of ethanol in gasoline to 15% (called E15) (C&EN December 7, 2009). Moving to E15 would add about 7 billion additional gallons of ethanol to the domestic ethanol market. However, one has to take into consideration that the volumetric energy content of ethanol is lower than that of gasoline by a factor of 1.6, which would translate into shorter distance driven on the same tank of fuel and greater CO_2 emission per mile driven compared to gasoline.

Fermentation plants are sources of appreciable quantities of CO_2 emissions: an average plant produces about 0.2 Mt CO_2 year^{-1} (IEA GHG 2002). Advantageously, bioethanol plants produce off-gas with high CO_2 concentration (almost 100 vol.%) that makes it easy to capture and store CO_2. The scale of future global production of bioethanol and related CO_2 emissions will depend on many factors such as improvements in biomass conversion technologies, the competition with other land use (e.g., for food), water availability, the competition with other alternative fuels (IPCC CCS 2005). When ethanol is derived from corn, the well-to-wheel GHG reduction with respect to gasoline typically varies from 10% to 30%, whereas for sugarcane-based ethanol (e.g., from Brazil), the GHG reduction could reach up to 90% (IEA Statistics 2009). The use of oilseed-derived biodiesel results in the well-to-wheel GHG reduction of 40%–60% against conventional petro-diesel.

Currently, relatively high cost of biofuels is a major barrier to their broader introduction to the market. Without government subsidies, only sugarcane-derived ethanol produced in Brazil is competitive with petroleum-based hydrocarbon fuels (of course, the situation may change in the near future since oil prices will continue to go up) (IEA Statistics 2009). Several countries have either mandated or promoted biofuel blending standards in order to diversify fuel supplies for transport. For example, in Brazil, gasoline contains 20%–25% of ethanol, and cars purchased after 2008 can run either on 100% ethanol or ethanol–gasoline blends (IEA Statistics 2009). United States, European Union, Canada, Australia are also mandating the use of biofuels.

In 2007, U.S. administration introduced the Energy Independence and Security Act (EISA), which contains new standards for vehicle fuel economy, as well as the provisions that promote the use of renewable fuels including biofuels (U.S. DOE Biomass 2010). Figure 1.47 depicts renewable fuels production requirements according to the

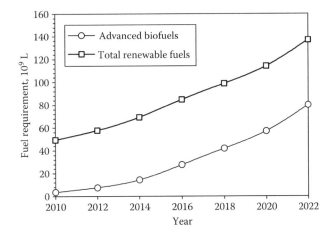

FIGURE 1.47
Production requirements for U.S. domestic alternative fuels under the Renewable Fuels Standard (RFS). Advanced biofuels include cellulosic biofuels and biomass-based diesel. (From U.S. DOE Algal Biofuels, National Algal Biofuels Technology Roadmap, U.S. DOE, Biomass Program, Washington DC, May 2010.)

new Renewable Fuels Standard (RFS). In particular, under RFS, the EISA mandated the production of

- 15 billion gallons of corn starch-based ethanol by 2015
- 500 millions gallons of biodiesel starting in 2009 and peaking at 1 billion gallons by 2012
- 16 billion gallons of cellulosic ethanol by 2022

Biofuels will face significant challenges in meeting their EISA targets. In particular, EISA requires that GHG emissions across biofuels life cycle to be at least 50% less than emissions produced by petroleum-based fuels (gasoline and diesel fuel). Reducing the cost of biofuels is also a significant hurdle to overcome. A study from the U.S. Government Accountability Office (GAO-09-449) points to a number of significant roadblocks to the increased production and use of biofuels (C&EN October 12, 2009). Under the U.S. 2007 renewable fuel law, it is required to increase the amounts of ethanol and other biofuels to 36 billion gallons by 2022.

Many experts, however, have serious questions about the potential impact of these high ethanol production levels to food and livestock prices, as well as water quality. There are also concerns that the United States may face problems with regard to usability of fuel blends containing high content of ethanol in U.S. vehicles. The problem stems from the fact that higher concentration of ethanol in fuel blends could make them too corrosive for most current vehicles and for the fuel distribution system. The GAO report recommends that the life-cycle environmental impacts of the increased production of biofuels need to be carefully assessed and alternative bio-refinery-based fuels be considered. Recent U.S. National Research Council (NRC) report finds that health and non-climate-related damages from corn-grain-based ethanol are similar or slightly worth than those from gasoline because of the energy required to produce and convert corn to fuel (Johnson 2009). The damage cost for ethanol from cellulosic biomass is significantly less compared to corn feedstock.

More detailed information on the technological aspects of fermentative fuels (ethanol, butanol) production and prospects of their practical application is presented in Chapter 12 by Dr. David Levin and co-authors.

Production of biofuels from algae is another area of intensive R&D activities. Many believe that algae farms might provide a salvation from the energy starvation. The year 2009 saw significant boost in algae-related R&D and technology validation projects. Oil and petrochemical companies after years of sitting on the sidelines, entered the alternative fuel game. ExxonMobil have announced that it would invest US$600 million to develop algae-derived biofuels with California-based Synthetic Genomics (C&EN December 21, 2009). Another oil major, Total (France), invested an undisclosed amount in biofuels startup company Gevo, which opened a pilot facility in St. Josef, MO (Reisch 2009). Besides, biofuels, Gevo is also looking into the development of chemical coproduct streams such as isobutanol, polyethylene terephthalate, and polymethyl methacrylate. Universal Oil Products (UOP, USA) has heavily invested in the development of "green" gasoline, diesel, and jet-range hydrocarbon fuels by integrated thermochemical processing of biomass (U.S. DOE EERE 2010). UOP is planning to build a demonstration unit with the capacity of $1\,t\,day^{-1}$ in Hawaii (USA) to convert cellulosic biomass and algae residues to hydrocarbon fuels via integration of fast pyrolysis and hydrodeoxygenation processes. If the project proves to be successful, UOP will expand the operation to a commercial unit with the capacity of 190 million liters of liquid fuels per year. Royal Dutch/Shell significantly expanded its R&D program with Codexis to ferment ethanol from nonfood raw materials. In the heart of the program are Codexis-developed biocatalysts that break down cellulosic biomass, converting them to sugars. The technological aspects of biofuels production from oily biomass, including algae, are outlined in Chapter 13 by Dr. Manzanera.

1.7 Carbon Mitigation Strategies: On the Road to Carbon-Free Energy

1.7.1 Pathways to Carbon-Neutral Energy Systems

Although the prospect of renewable-based economy is seen by many as the best hope of mankind for the sustainable energy future, it is realized that in view of the existing technological and economic hurdles it may take several decades before it could be fully (or to a sufficiently large extent) implemented. Considering that fossil fuels currently provide about 80% of primary energy, switching from fossil to nonfossil-based energy system may require too many changes and, consequently, too much time to accomplish that transition. On the other hand, some experts believe that we cannot afford too long a transition period because of potentially irreversible changes in global climate caused by the extensive use of fossil fuels (Moriarty and Honnery 2007).

The analyses of the three carbon mitigation options—fossil decarbonization and the use of nuclear and renewables resources—indicate that neither of them could provide the required amount of carbon-neutral energy without major changes in the existing infrastructure, technological breakthroughs in many areas, or taking a considerable environmental risk (Muradov and Veziroğlu 2008). In principle, the utilization of large resources of coal, oil shale, and tar sands coupled with CCS could provide all the necessary power for at least a couple of hundred years, but CO_2 sequestration is a new technology without a sufficiently long technical and environmental track record. It is realized that real ecological responses to ocean or geological CO_2 sequestration cannot be adequately examined or predicted, given current levels of understanding, and more research and field demonstrations are needed to confirm practical considerations (e.g., long-term safety, possible leakage, economics, permanence). Professor Weinberg (revered as the father of light-water nuclear reactors) termed CO_2 sequestration a Faustian Bargain (meaning the temptation of a quick technological fix, and a vigilance in assuring that stored CO_2 will not be subsequently released to the biosphere) (Spreng et al. 2007). Nuclear source may provide a significant, but not decisive, share of carbon-free energy, and there is a hope that a widespread implementation of advanced nuclear (e.g., breeder, high-temperature) reactors would provide a means for large-scale production of electricity and carbon-neutral fuels in a reasonable time frame.

The technical potential of all renewable resources combined is well over 30–40 TW of carbon-free power (see Figure 1.39). In the past few years, the cost of PV modules, solar thermal systems, and wind electricity has dropped dramatically and is still continuously dropping, opening the way for their large-scale deployment (wind electricity is already competitive with fossil electricity in some areas). Since solar and wind energy sources are of an intermittent nature, there will be a need for efficient and cost-effective electrical energy storage and transmission systems. Solar electricity generated in the U.S. Southwest or wind electricity generated in the U.S. Northeast would be transmitted over high-voltage transmission lines to a variety of energy storage facilities throughout the country (e.g., compressed air, hydrogen, or hydro-storage, depending on the local availability), where it will be converted to electricity on-demand (Muradov and Veziroğlu 2008). Of the non-intermittent renewable sources, only geothermal resource could provide more than 10 TW of base-load electricity; however, this source is nonuniformly distributed over the land and, therefore, efficient energy storage and transmission systems will also be required to overcome this problem. When fully implemented, this new energy network will come with a big price tag: the additional electrical load would require the expansion of the existing U.S. national grid by about 100% at the cost of $250–$300 billion (EPRI 2004).

Jacobson (Stanford University, U.S.A) analyzed possible solutions to energy security and climate change as well as water supply, air pollution, the land and ocean use, and other issues from the viewpoint of natural energy resources availability and their impact on the environment and civilization in general (e.g., mortality, undernutrition, wildlife, nuclear proliferation) (Jacobson 2009). Nine electric power sources: solar PV, concentrating solar power (CSP), wind, geothermal, hydroelectric, wave, tidal, nuclear and coal with CCS, and two liquid fuels (corn and cellulosic ethanol, both as E85 fuel) options were considered. In order to put electric and liquid fuel sources on an equal footing, 12 combinations of energy sources and new-technology vehicles, for example, battery-electric vehicles (BEV), hydrogen fuel cell vehicles (HFCV), and flex-fuel vehicles run on E85, were evaluated. These combinations were ranked with respect to 11 impact categories such as resource abundance, life-cycle GHG emissions, footprint, mortality, water consumption, effects on wildlife, energy supply disruption, chemical and thermal pollution, and others. Figure 1.48 outlines changes in U.S. GHG emissions (in CO_2-equivalent) upon converting all on-road (both light and heavy duty) vehicles from gasoline to renewable and nuclear electricity-powered BEVs, and bio-ethanol produced from corn and cellulosic biomass (only high estimates are shown for renewable electricity, whereas both low and high estimates are presented for bio-ethanol) (Jacobson 2009). It was determined that running BEVs on wind-generated electricity would provide the greatest reduction in U.S. CO_2 emissions by about 32.5%–32.7%, followed by slightly fewer percent reductions projected for CSP, PV, geothermal tidal/wave, hydro and nuclear sources. The coal-CCS-BEV combination shows the smallest percent reduction in CO_2 emissions (17.7%–26.4%) among the electricity-driven vehicles (mostly, due to high life-cycle emissions and a potential CO_2 leakage from a storage site).

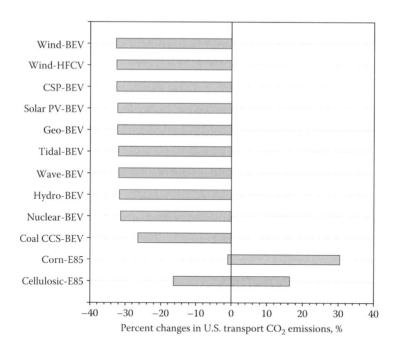

FIGURE 1.48
Percent changes in U.S. GHG emissions (in CO_2 equiv.) upon replacing all onroad (light and heavy duty) vehicles with carbon-neutral technologies. For all cases, except E-85, maximum estimates are presented. For corn and cellulosic-based E85 fuels minimum and maximum estimates are presented and 100% penetration is assumed. (Modified from Jacobson, M., *Energy Environ. Sci.*, 2, 148, 2009.)

Converting vehicles from gasoline to corn-based ethanol fuel either would not provide any benefit with regard to CO_2 reduction, or even would increase the emissions by up to 30.4% (depending on the assumptions). Due to a rather substantial dispersion of the available literature data on the estimates of CO_2 emissions changes using cellulosic ethanol, the figures presented in the diagram range from negative value of −16.4% (i.e., CO_2 reduction) to positive value of 16.4% (i.e., increase in CO_2 emissions).

Four technology tiers clearly emerged from weighing and ranking each combination as follows: Tier 1 (highest-ranked) includes wind-BEV and wind-HFCV combination. This was followed by Tier 2: CSP-BEV, geo-BEV, PV-BEV, tidal-BEV, wave-BEV, and Tier 3: hydroelectric-BEV, nuclear-BEV, and coal-CCS-BEV. The lowest ranked Tier 4 included corn and cellulosic based E85 biofuels, which showed poor ranking with regard to climate impact, air pollution, land use, wildlife damage, and chemical waste categories (it is noteworthy that cellulosic E85 ranked lower than corn-E85, mainly due to its potentially larger land footprint and higher upstream air pollution emissions). The author's recommendations can be summarized as follows: use of wind, CSP, PV, geothermal, tidal, wave, and hydroelectric energy to provide electricity for transportation (in particular, BEVs and HFCVs), residential and industrial sectors will be the most beneficial among all options considered (especially, with regard to energy security, climate impact, and air pollution). Coal-CCS and nuclear options look much less advantageous, and the biofuel options provide no certain benefit and potentially have the greatest negative impacts. To put the numbers in a perspective, the United States could replace all 2007 on-road vehicles with BEVs powered by 73,000–144,000 5 MW wind turbines (which is less than 300,000 airplanes the United States produced during WWII), reducing CO_2 emissions by 32.7% in 2020.

A number of recent studies by Shinnar and Citro (2006), Muradov and Veziroğlu (2005 and 2008), Agarwal et al. (2007) and Forsberg (2009) analyzed different scenarios of the replacement of fossil resources by alternative carbon-free energy sources in different sectors (i.e., industrial, commercial, residential, and transportation). Shinnar and Citro (2006), outlined a scenario of fossil fuels replacement by alternative carbon-free sources and evaluated CO_2 reductions resulting from this action. The authors estimated that 72% of the current use of fossil fuels could be replaced by electricity from alternative carbon-free sources (nuclear, renewables) and 26% by biomass-derived fuels (the remaining 2% cannot be replaced by any sources). The analysis indicated that since 80% of gasoline used for passenger cars and light trucks could be saved by switching to plug-in hybrids, 65% of petroleum used in transport could be replaced by carbon-free electricity (used in BEVs and plug-in hybrids) and the remaining 35% by fuels produced from biomass. The associated reductions in CO_2 emissions from replacing fossil energy by carbon-free electricity and bio-derived fuels are estimated at 76% and 21%, respectively (the maximum percentage of CO_2 emission reduction could reach up to 97%).

According to Shinnar and Citro (2006) analysis, about a quarter of the petroleum fuel usage cannot be replaced by carbon-free electricity in certain areas: those include petrochemicals, aviation, heavy trucks, military, and some industrial applications that require liquid and gaseous hydrocarbon fuels (e.g., diesel, jet-fuel, NG). Of that amount, about half (or about 12% of the total fossil-based fuels used in transportation) is to be in the form of high-energy density liquid fuels (especially for such applications as aviation, heavy trucks, etc.). Among traditional biofuels (e.g., fermentative ethanol and biodiesel), only biodiesel could potentially fit in some of these application areas due its relatively high volumetric energy content (see Figure 1.44). In aviation, the stringent requirements of a high power/weight ratio, an operating temperature range, and fuel specifications leave liquid hydrogen and liquid hydrocarbons (i.e., synthetic jet-fuel) as the only acceptable fuel options (although biodiesel has already been used in aircrafts, it was blended with a conventional jet-fuel; the use of pure biodiesel in jet-engines may require more developmental efforts) (Muradov and Veziroğlu 2008). It is noteworthy that liquid hydrogen occupies four times the volume of jet-fuel, resulting in larger fuel tanks and, consequently, greater air resistance and increased fuel consumption (by about 9%–14%) (Forsberg 2007). Thus, biomass-derived SLH look like the most promising alternatives to petroleum-based fuels in terms of carbon-neutrality and energy density (besides, they are suitable for a great variety of applications including petrochemistry). The question is: do we have enough biomass resources to produce the required amount of SLH? The analyses indicated that the traditional ways of producing biomass-derived fuels (via both fermentative and thermochemical approaches) would not yield the required quantity of transportation fuels without taking a prohibitively large land area and adversely impacting food production.

Recently, a number of researchers reported on the development of alternative approaches to production of carbon-neutral fuels from biomass, which would allow a significant increase in the yield of SLH per unit of biomass feedstock (Bossel et al. 2005; Shinnar and Citro 2006; Agarwal et al. 2007; Muradov and Veziroğlu 2008; Forsberg 2009; Zeman and Keith 2008). These approaches, although differing in technical details, have a common idea of SLH production by using biomass as a source of carbon and water as a source of hydrogen with the energy input from non-carbogenic energy sources (e.g., nuclear, solar, wind) (Figure 1.49). According to the reported estimates, this approach would yield three to four times more synthetic fuels, compared to conventional routes of biofuels production (e.g., via gasification, on the unit of biomass energy basis) (Shinnar and Citro 2006; Agarwal et al. 2007).

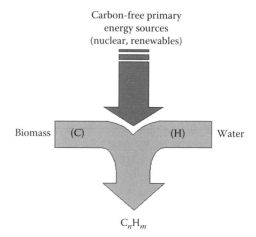

FIGURE 1.49
Production of carbon-neutral hydrocarbon fuels from biomass and water using carbon-free primary energy sources.

In a recent analytical study, Forsberg (2009) outlined a nuclear-hydrogen-biomass concept as an attractive approach to future carbon-neutral energy systems. The author argues that due to high energy intensity of biomass-to-liquid fuel processes, there would be insufficient amount of biomass to meet U.S. liquid transportation fuel demands. However, with the use of nuclear energy to provide heat, electricity, and hydrogen for processing biomass (especially, cellulosic biomass) to fuels, the production of liquid hydrocarbon-based fuels (e.g., diesel fuel) per unit of biomass could be dramatically increased to the extent that U.S. liquid fuel requirements would be met. The proposed nuclear-hydrogen-biomass system is robust and can be implemented with the current or near-term technologies; in particular, nuclear-powered electrolysis and FT technology could provide a basis for the near-term implementation of the biomass-to-liquid fuels process. Assuming that about 0.95 kg of H_2 is required to produce 1 gal of diesel fuel, it would take 92 1 GW nuclear reactors coupled with electrolyzers to provide hydrogen for the production of 1 million barrels of diesel fuel per day (Agarwal et al. 2007; Forsberg 2009). Furthermore, according to Forsberg, the use of electric and plug-in hybrid electric vehicles would enable nuclear energy to meet a major fraction of U.S. transportation needs. For example, a million barrels of oil per day would allow cars and trucks (with an average fuel economy of 20 mi gal^{-1}) drive about 8.2 million miles per day. Twenty-three 1 GW nuclear reactors would produce enough electricity for electric vehicles to drive the same distance (at electric fuel economy of 0.6 kWh mi^{-1}, or 0.038 kWh km^{-1}). At the electricity cost of \$0.10 kWh^{-1}, it would cost \$0.06 for a vehicle to drive 1 mi (or 1.6 km) (the equivalent cost of gasoline would be \$1.21 gal^{-1} or \$0.32 L^{-1}).

Figure 1.50 depicts the schematic diagram of an open-loop carbon-neutral energy system based on the concept of using biomass as a carbon source, water as a hydrogen source, and nuclear or renewables as an energy input source (Muradov and Veziroğlu 2008). Biomass (energy crops, algae, grass, etc.) takes up CO_2 from the atmosphere during its growth, and it is converted to SLH using non-carbogenic renewable hydrogen produced from water (e.g., via PV-electrolysis, CSP-water-splitting cycle, wind electrolysis, nuclear electrolysis, or nuclear-water-splitting cycle). Three possible routes could be suggested to accomplish the biomass-to-SLH conversion:

1. Biomass gasification followed by FT synthesis coupled with a water-splitting process (oxygen is used in the biomass gasification stage, and hydrogen in the FT synthesis stage).

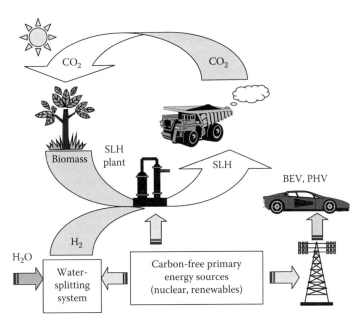

FIGURE 1.50
Schematic of carbon-neutral production of synthetic liquid fuels in an open-loop system. BEV—battery electric vehicle, PHV—plug-in hybrid vehicle, SLH—synthetic liquid hydrocarbon.

2. Hydrothermal biomass liquefaction to bio-oil followed by hydrodeoxygenation of bio-oil to SLH using non-carbogenic hydrogen:

$$CH_nO_m + H_2O \rightarrow C_xH_yO_z + (CH_4, CO, CO_2) \tag{1.40}$$

$$C_xH_yO_z + H_2 \rightarrow C_pH_q + H_2O \tag{1.41}$$

where CH_nO_m, $C_xH_yO_z$, and C_pH_q are generic formulae for biomass, bio-oil, and hydrocarbon, respectively.

This route is particularly attractive for conversion of algal biomass to hydrocarbon fuels because it allows obviating an energy-intensive step of drying an algae feedstock.

3. Biomass pyrolysis to bio-oil followed by hydrodeoxygenation of bio-oil to SLH using non-carbogenic hydrogen (this route would also produce bio-carbon as a by-product of the process):

$$CH_nO_m \rightarrow C_xH_yO_z + (CH_4, CO_x) + C \tag{1.42}$$

$$C_xH_yO_z + H_2 \rightarrow C_pH_q + H_2O \tag{1.41}$$

As of now, neither of these routes has been implemented on a commercial scale (albeit, the first route is very close to commercialization).

Since most cars driven within the cities would preferably be EV or plug-in hydrides operating in an electrical mode, the environmental impact of the SLH-based transportation system would be minimal. At this point, most of the CO_2 emissions would come from the use of SLH by heavy trucks, aviation, industrial, and other sources that would be rather distant from populated areas (thus minimizing the impact). These CO_2 emissions will end up in the atmosphere, from where they will be absorbed by growing biomass, thus closing the loop. This concept would allow using the existing liquid fuel-based infrastructure (while utilizing much cleaner synthetic fuels compared to petroleum-based fuels) without increasing CO_2 concentration in the atmosphere. The key here is to have sufficient amount of carbon-free hydrogen to convert bio-carbon and CO_2 to SLH and maintain the system balanced on carbon.

One should not underestimate the power of a political will and national consensus in a dramatic and rapid move from fossil energy to low-carbon energy resources. France is the most striking example of a country that decisively moved away from oil within about 20–30 years. Before the oil crisis of early 1970s, the entire French economy was based on imported oil. A national consensus was achieved in 1975 that the country had to substantially reduce oil consumption, which led to the creation of three programs: energy conservation, use of nuclear energy, and the deployment of electric-powered high-speed trains (to reduce the use of aircrafts for travels within the range of several hundred km) (Forsberg 2009). The implementation of these programs allowed France to drastically reduce oil consumption within about a quarter of a century. The introduction of similar measures worldwide would drastically ease the heavy reliance on oil, and, thus, prevent potential political and social upheaval and clean up the environment within a reasonable time frame.

1.7.2 Conversion of CO_2 to Fuels

Although the above estimates indicate that the sufficient amount of biomass resources is available for production of carbon-neutral transportation fuels substituting petroleum fuels that cannot be replaced by the use of carbon-free electricity in BEVs and PHVs, other sources of carbon, most importantly, atmospheric CO_2, could also be utilized in the process. If this approach is successful, CO_2 would become an infinite source of carbon for the production of alternative fuels. Although appealing, this approach would face daunting technical challenges with regard to the energy efficient and cost-effective scrubbing of atmospheric CO_2, and its hydrogenation to a desirable range of fuels and products.

Conversion of CO_2 to fuels is an extremely attractive idea that has been actively pursued by researchers worldwide for many decades. Despite numerous technical challenges, certain progress has been achieved in conversion of CO_2 to liquid fuels. Conversion of CO_2 to liquid fuels could be accomplished via at least four routes (Muradov and Veziroğlu 2008):

1. Direct hydrogenation of CO_2 to hydrocarbons

$$2nCO_2 + (m + 4n)H_2 \rightarrow 2C_nH_m + 4nH_2O \qquad (1.43)$$

2. Synthesis of methanol (or other oxygenated fuels), optionally, followed by catalytic (zeolites) conversion of methanol to gasoline (Mobil-process)

$$CO_2 + 3H_2 \xrightarrow[-H_2O]{Catalyst} CH_3OH \qquad (1.44)$$

$$CH_3OH \rightarrow C_nH_m(+CO_x) \tag{1.45}$$

3. Artificial photosynthesis

$$CO_2 + H_2O + solar\ photons \rightarrow 1/n(CH_2O)_n + (O_2) \tag{1.46}$$

$$(CH_2O)_n \rightarrow C_nH_m(+CO_x) \tag{1.47}$$

4. Direct reduction of CO_2 to CO followed by catalytic hydrogenation of CO to fuels

$$CO_2 \rightarrow CO + \frac{1}{2}O_2 \quad \Delta H^\circ = 283\ kJ/mol \tag{1.48}$$

$$CO + 2H_2 \rightarrow CH_3OH\ (or\ C_nH_m) \tag{1.49}$$

The first approach faces technical challenges associated with unfavorable thermodynamics of the reaction and, consequently, low products yields; it would require more fundamental research, especially, in the catalyst development area (Kim et al. 2006). The second option is technologically advanced since it is based on an industrial process for methanol production (32 millions tons of methanol produced worldwide). If necessary, methanol could be further converted to gasoline according to a commercial methanol-to-gasoline Mobile-process. Attempts have been made to mimic Mother Nature and develop "artificial photosynthesis" (bio-inspired) systems, which involve solar-driven catalytic reduction of CO_2 to a variety of products and fuels such as methanol, hydrocarbons, CO, etc. This work is still in an early R&D stage. More detailed discussion of catalytic, photoelectrochemical, electrocatalytic, and bio-mimicry options for CO_2 conversion to different fuels are presented in Chapters 4, 6, 8, 11, and 14.

Direct reduction of CO_2 (reaction 1.48) has been an elusive goal of researchers for many years, since it allows converting CO_2 to two useful products: CO and oxygen. CO could further be converted to hydrogen via water gas shift reaction, or, if necessary, to methanol, or hydrocarbons. The technical feasibility of CO_2 dissociation to CO and O_2 at elevated temperatures (about 2000°C) using a concentrated solar radiation source has been demonstrated recently (Traynor et al. 2002). Although photo-thermal dissociation of CO_2 is technically feasible at the temperature range of the state-of-the-art solar concentrators, 2000°C–2500°C, the close-loop demonstration of the CO_2 cycle with sufficiently high solar energy conversion efficiencies still remains a technical challenge.

Summarizing this section, there is an adequate amount of renewable resources for the replacement of fossil fuels in industrial, commercial, residential, and transportation sectors. This can be done by replacing fossil fuels by electricity from carbon-free sources (e.g., nuclear, renewables) and by advanced biomass-derived liquid hydrocarbon fuels in the areas where electricity cannot be used (e.g., aviation, heavy trucks). The combination of biomass, non-carbogenic hydrogen, and carbon-free primary energy sources could potentially produce the required quantity of carbon-neutral transportation fuels utilizing a reasonable area of the land surface. It was estimated that the cost of replacing 70% of fossil fuels (including coal) is about \$170–\$200 billion per year over the next 30 years; however, a carbon tax of \$45–\$50 t^{-1} of CO_2 would pay for the investment and provide incentives for implementing renewable technologies (Shinnar and Citro 2006).

1.8 Quick Response and Emergency Carbon Mitigation Scenarios

Depending on the ecological situation evolved, there might be a need to not only stabilize but urgently reduce atmospheric CO_2 concentration (e.g., to prevent ecological runaway situations, i.e., an uncontrollable increase in atmospheric CO_2 concentration via positive feedback mechanisms). Traditional approaches like capturing CO_2 from stack gases of power plants or sequestering CO_2 by terrestrial ecosystems may be too slow to make any difference in the rapidly deteriorating situation. For these extreme cases, quick or emergency response measures have been proposed that can supposedly "fix" the problem within a reasonable time frame by reversing the trend and preventing the further slide to the dangerous tipping point. The geoengineering concept is the most highly publicized emergency response option.

1.8.1 Geoengineering

Originally, the term "geoengineering" was proposed by Marchetti (1976) in reference to the problem of CO_2 control in the atmosphere by collecting CO_2 from point sources (e.g., power plants) and injecting it into deep seas taking advantage of natural thermohaline circulations (e.g., the Mediterranean undercurrent entering the Atlantic ocean at Gibraltar) (Marchetti 1976). Currently, the meaning of the term "geoengineering" (sometimes called "climate engineering") is significantly expanded and it is generally defined as a set of measures to deliberately modify or manipulate the Earth's energy balance and climate to counteract the adverse impact of human activities on the global ecosystem. This is a highly controversial concept enjoying support from a number of distinguished scientists, but also facing fierce resistance and criticism from a large group of environmentalists, economists, and scientists having an opposite opinion on the subject. Among most prominent supporters of the geoengineering concept is Nobel Laureate Dr. Paul Crutzen who recently advocated this approach in his article in *Climatic Change* journal (Crutzen 2006). Crutzen and other proponents of the idea argue that the scale and the rate of curbing GHG emissions may not be adequate (due to both technical and political reasons) to avoid the dangerous climate change, so a radical contingency plan in the form of the geoengineering will be needed. The sense of urgency that can be felt in some publications originates from concerns that our planet's climate tipping point may be fast approaching and from the fact that all attempts to control GHG emissions on a global scale so far have been unsuccessful. The opponents of this concept consider the geoengineering a dangerous and untested "quick-fix" approach and are concerned that the excessive reliance on such a radical measure may undercut the resolve to deal with the original cause of the problem, viz., GHG emissions due to human activities (Cicerone 2006). In this section, the pros and cons, and the current status of the geoengineering approach as a mitigation strategy are briefly discussed.

Since the Earth's climate is controlled by two fundamental forces, the amount of solar radiation hitting the Earth and the amount of solar radiation absorbed by the atmosphere, most of the geoengineering proposals generally fall into two main categories:

- Solar radiation management (SRM)
- GHG management

1.8.1.1 Solar Radiation Management

SRM approach aims at reducing the amount of solar radiation striking the Earth surface by deflecting a significant portion of sunlight back to space. It should be noted that this phenomenon is well known and has already been "tested" numerous times (albeit, not as a geoengineering-related activity) since it is a side effect of any major volcano eruption. To some extent, sulfate aerosol particles emitted to the atmosphere during combustion of coal can also backscatter solar radiation, increasing the Earth's albedo. What distinguishes the geoengineering approach from the rather limited effect of these natural and man-made light reflectors is that a special type of aerosol particles with much higher reflectivity can be designed, which are intentionally injected higher into the stratosphere, where they could remain suspended for years. Candidate materials include sulfur-based aerosols, aluminum and alumina particles, and other micro- and nano-sized particles.

It is proposed, for example, that sulfate aerosol precursors such as SO_2, COS, S_2, H_2S are delivered to the stratosphere by balloons or artillery guns, where they are converted to submicron-sized sulfate particles (Crutzen 2006). Once placed in the stratosphere the reflecting particles will remain there for at least 1–2 years (in contrast, in the troposphere, they would last only a week or so). The cooling effect of sulfate aerosols has already been observed during the eruption of Mount Pinatubo volcano in Philippines in 1991 when about 10 million tons of SO_2 was injected into the tropical stratosphere (Bluth et al. 1992). Measurements indicated that due to increased reflection of solar radiation by sulfate particles back to space, the Earth's surface cooled in average about half a degree Celsius in the following year (Lacis and Mishchenko 1995).

It was estimated that to compensate for doubling of CO_2 concentration in the atmosphere, the required stratospheric sulfate loading would be about 5.3 millions tons of a sulfur precursor (Crutzen 2006). According to an analytical study, 1 kg of sulfate particles optimally placed in the stratosphere would negate the warming effect of several hundred tons of CO_2 (Victor et al. 2009). The authors of another aerosol-related proposal estimated that about 10 million tons of dielectric alumina aerosols with the average size of 100 nm would be sufficient to enhance the Earth's albedo by 1% (the use of alumina instead of sulfate aerosols is considered more advantageous from an environmental viewpoint) (IPCC 2001).

The following is the list of alternative geoengineering approaches that are also based on the principle of SRM:

- Brightening clouds by spraying them with sea salt to increase their reflectivity
- Seeding cirrus clouds using airplanes
- Using satellite-guided unmanned "cloud ships" that suck up seawater and spray micro-droplets through giant funnels to form white clouds, which would reflect 1%–2% of the sunlight, thus, potentially canceling out the effect of man-made CO_2 (according to one proposal, 1900 wind-powered ships would travel the seas and create clouds in most suitable locations) (Telegraph, U.K. 2009)
- Putting aluminum mesh structures into the stratosphere
- Releasing a myriad of reflecting micro-balloons into the stratosphere (e.g., it was proposed that micro-balloons with the diameter of 4 mm and thickness of 20 nm are filled with hydrogen and placed at the altitudes of about 25 km where they will have much longer residence times than aerosols; the overall weight of micro-balloons would be about 1 million tons) (IPCC 2001)

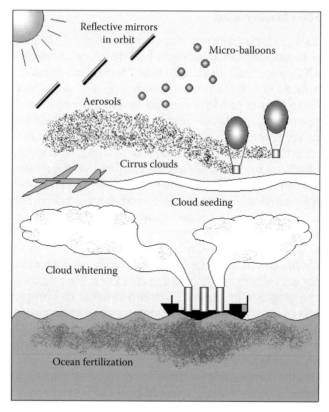

FIGURE 1.51
Schematic representation of selected geoengineering strategies.

- Placing giant mirror arrays into the space orbit
- Whitening roofs, pavements, roads, and other surfaces
- Arctic geoengineering (this approach is aiming at limiting Arctic sea ice loss, which plays an important role in controlling the Earth's albedo)

Figure 1.51 provides a simplified sketch of some of the above-listed geoengineering projects. Currently, no large-scale SRM projects have been undertaken anywhere in the world (except simple local solutions like roof whitening).

1.8.1.2 Greenhouse Gas Remediation

The main objective of proposed GHG remediation-related geoengineering projects is to remove GHG from the atmosphere either directly or indirectly by manipulating natural processes that could facilitate the removal of atmospheric GHGs. Some of these projects are closely associated with CO_2 capture and storage and may not be classified as geoengineering projects per se (among those are the ocean fertilization and other carbon sequestration technologies that are discussed in Chapter 14). A number of proposed GHG remediation geoengineering projects involve the removal of potent GHGs such as methane and CFCs and other halocarbons from the atmosphere. In particular, it was proposed to remove CFCs using powerful lasers that are capable of breaking CFCs into smaller benign molecules (Stix 1993). Removal of methane from the atmosphere may be enhanced by several oxidative processes, for example, via its decomposition by reactive hydroxyl radicals generated by photochemical decomposition of ozone.

1.8.1.3 Economics of Geoengineering

The cost estimates of the proposed geoengineering projects are very sketchy and seldom to come by. Some geoengineering techniques, for example, whitening of roofs, pavements, roads, and some open areas can be accomplished at very little cost (in fact, the projects like roof whitening may even offer some financial payback). Other projects, however, could be extremely costly. The cost estimate for the full-scale implementation of the sulfate-aerosol geoengineering project was reported by Crutzen (2006), based on the information provided by National Academy of Science (NAS 1992). According to this estimate, it would cost annually about $25 billion (in 1992 dollars) to put 1 million tons of sulfates in the stratosphere. Thus, the cost of a continuous deployment of about 2 million tons of sulfate aerosols per year would be about $50 billion or about $50 per world capita, which does not seem to be very high price compared with the cost of other mitigation options (Crutzen 2006).

The Copenhagen Consensus Center and The Royal Society consider the "cloud ship" project to be more cost-effective than other geoengineering projects. According to some recent estimates, the project would cost about $9 billion to test and launch within 25 years, which is much less than the cost of releasing aerosols into the atmosphere ($230 billion) and placing mirrors in space ($395 trillion) (to put these numbers in a perspective, industrialized nations have to commit to spending about $250 billion each year to cut CO_2 emissions) (Telegraph, U.K. 2009). Most experts, however, agree that it is unlikely that cost would play a crucial role in determining whether to deploy geoengineering systems. Most likely, the issues related to risk, politics, and environmental ethics would be more decisive factors in the decision-making process (IPCC 2001).

1.8.1.4 Risk Factors

The opponents of the geoengineering concept have the following ammunition in their arsenal:

- Possible side effects (e.g., aerosols could destroy ozone layer)
- Its effectiveness could be less than expected (the opponents point to the fact that in the ocean fertilization experiments, the estimated amount of CO_2 removed from the atmosphere turned out to be less than expected)
- Low reliability and stability (e.g., due to a possible mechanical failure)
- Difficulties with controlling the geoengineering systems (due to unpredictable impact of certain external events, e.g., solar flares, volcanic eruptions)
- Possibility of weaponization (e.g., the use of cloud machines, creation of droughts)
- Possible adverse impact on terrestrial ecosystems (e.g., plants, wild life)

As of today, there is no a general consensus on the utility of the geoengineering concept: too many technical, environmental, political, and legal issues are involved. Even if a particular technique is developed and successfully tested, its deployment would be extremely challenging due to the global nature of the problem: no one country would be directly responsible for it. Thus, it is likely that for the time being, the geoengineering concept will remain at the stage of research and development of different technological options so they are available when they become an absolute necessity.

1.8.2 Capture of Atmospheric CO_2

Capture of atmospheric CO_2 (CAC) (sometimes also called "air capture") is one of the strategies within the geoengineering category of GHG management, but this topic is so broad and inclusive of so many technologies that it merits a separate consideration. CAC is viewed by many as an ultimate carbon mitigation technology since it can remove CO_2 from any part of the carbon-based economy; thus, the cost of CAC provides an absolute cap on the cost of carbon mitigation (Keith et al. 2006). Secondly, CAC could potentially reduce atmospheric CO_2 concentrations faster than the natural carbon cycle. It is realized that the development and global implementation of efficient CAC systems would alleviate the burden of equipping fossil power plants with expensive CCS systems and remove any restrictions on the amount of CO_2 emitted by transportation and other small sources. Thus, like any other geoengineering approaches, CAC would limit the cost of a worst-case climate scenario and decrease the need for a near-term precautionary abatement (Keith et al. 2006). No wonder that CAC is in the focus of numerous research programs around the world.

There are similarities and differences between CCS and CAC concepts. Similar to CCS, the CAC concept involves the capture of CO_2 and its long-term storage (note that CO_2 storage options in CAC are essentially the same as in CCS). However, CAC has important structural advantages over conventional CCS systems in that CAC removes CO_2 directly from the atmosphere; thus, it manipulates the CO_2 atmospheric concentration rather than stack gases of fossil-based power plants. This allows significantly extending the concept of carbon capture from large centralized industrial CO_2 sources with relatively high CO_2 concentration (5–20 vol.%) to a myriad of small dispersed and mobile CO_2 sources such as cars, airplanes, homes, businesses, etc. (Zeman and Lackner 2004; Keith et al. 2006). Another structural advantage of CAC over CCS is that the CO_2 capture system is decoupled from a CO_2 emission source. This allows fine-tuning the carbon-capturing equipment to a certain fixed CO_2 level such as atmospheric CO_2 concentration (currently, 392 ppmv), and since this value is globally constant, similar CO_2 capture devices could be installed anywhere in the world (Zeman 2007). The last factor is particularly important, because CAC plants can be located close to CO_2 sequestration sites, thus eliminating CO_2 transport cost (with a significant economic gain). This would also allow exploiting the remote CO_2 disposal sites that currently are not considered economically feasible.

The idea of capturing CO_2 from air is not new: some papers dealing with the subject date back to 1940s (the background information on the subject is provided by Spector and Dodge, 1996). Removing CO_2 from air onboard submarines has been practiced for many decades (albeit, at a relatively small scale and not necessarily by cost-effective means). Currently, the interest in this topic is fueled by the concerns over excessive man-made CO_2 emissions and the potential role of CAC as a carbon mitigation strategy. Currently, no large-scale industrial CAC systems exist anywhere in the world; thus, natural photosynthetic systems (plant and algal photosynthesis) are the only practical forms of CO_2 capture from the atmosphere. However, a number of studies suggest that it will be technically and economically feasible to develop practical CAC systems on a timescale relevant to climate policies.

Before we provide an overview of the existing concepts and technological developments in the CAC area, it would be worthwhile to give some theoretical background with regard to physical and economic limits and land constraints on the performance of atmospheric

CO_2 removal systems. Keith et al. (2006) recently assessed the ultimate physical limits on the amount of energy and land required for the implementation of CAC systems. The minimum amount of energy required to extract any gas from a mixture of gases is controlled by thermodynamics of a gas separation process. In particular, the extraction of CO_2 from a gaseous mixture where its partial pressure is p_o and its delivery as a pure CO_2 stream at final pressure of p_f is governed by the enthalpy of mixing (ΔH_{mix}):

$$\Delta H_{mix} = kT \ln\left(\frac{p_f}{p_o}\right) \tag{1.50}$$

where
   k is Boltzmann constant
   T is working temperature
   p_o and p_f are partial pressure of CO_2 in the mixture (i.e., in air) and final pressure of CO_2, respectively

At $p_o = 3.9 \times 10^{-4}$ atm (which corresponds to current atmospheric CO_2 concentration) and $p_f = 1$ atm, the minimum energy needed to capture CO_2 from air would be about $\Delta H_{mix} = 1.6\,GJ\,t^{-1}$ carbon. If we add the energy penalty for compressing CO_2 to 100 atm (required for the geological storage of CO_2) the overall minimum energy requirement for CO_2 capture from air and its sequestration would be about $4\,GJ\,t^{-1}$ carbon (Keith et al. 2006). In practical systems, the energy requirements are expected to be much higher. The energy required for the operation of a CAC device could be provided by fossil, nuclear, or renewable sources. In the former case, the amount of fuel needed to cover energy penalty of the CAC process would constitute about 10%, 8%, and 6% of the carbon-specific energy content of coal, oil, and gas, respectively.

Another important point to consider relates to a relative ease of capturing CO_2 from streams with different CO_2 concentrations: because the enthalpy of mixing is a logarithmic function of partial pressure, the difference in energy penalties for CO_2 capture from air and from much more concentrated sources (e.g., stack gases of power plants) at atmospheric pressure is not as significant as one would think. Indeed, the ratio of the theoretical energy requirements for capturing CO_2 from air ($[CO_2] = 392\,ppmv$) and stack gases containing 1 vol.% of CO_2 at atmospheric pressure is only 1.7, as follows:

$$\frac{\ln(1/3.9 \times 10^{-4})}{\ln(1/10^{-2})} \approx 1.7 \tag{1.51}$$

Thus, in simplistic terms, capturing CO_2 from air would require (at minimum) only roughly—two to three times more energy than capturing it from power plant stack gases (with 90% removal efficiency) (Keith et al. 2006).

Let us now consider minimum land requirements for the practical realization of the CAC systems (generally, the land use is one of the most important constraints for energy technologies). Assuming the use of an efficient CO_2 absorber, the operation of CAC system will be limited by the CO_2 flux from the atmosphere due to the atmospheric motion and turbulent diffusion (Keith et al. 2006). It was estimated that the air capture flux is controlled by CO_2 transport in the atmospheric boundary layer, and it is limited to about $400\,t$ C ha^{-1} year^{-1} (Elliot et al. 2001).

It is interesting to compare CO_2-neutral energy fluxes of the CAC systems associated with fossil-based power plants with that of the carbon-neutral power sources based on

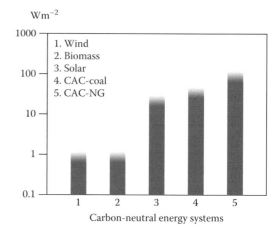

FIGURE 1.52
Carbon-neutral energy fluxes related to fossil- and renewable-based energy systems. (From Keith, D. et al., *Clim. Change*, 74, 17, 2006.)

renewable resources (e.g., wind, solar, biomass). Let us assume that an efficient CAC system is installed to offset CO_2 emissions from coal- or NG-fired power plants. In order to calculate the total power density of the combined zero-net-emission system, we divide the power generated by the fossil-fired plant by the land area required by the CAC plant to capture CO_2 emissions from that plant. The obtained values can be directly compared with the power densities of other carbon-neutral systems. The results are shown in Figure 1.52, which depicts the CO_2-neutral energy fluxes of the CAC systems associated with coal- and NG-fired power plants with that of wind, biomass, and solar-based power generation systems (in W m^{-2} units). It is apparent that the carbon-neutral energy fluxes of the CAC-fossil based systems (which are roughly in 50–100 W m^{-2} range) are greater than that of the alternative systems based on renewable resources (roughly, in 1–20 W m^{-2} range). The authors of the referenced study point to a strong analogy between the land requirements for the CAC and wind power systems (Keith et al. 2006). In particular, both are dependent on the rate of turbulent diffusion in the atmospheric boundary layer, and both must be spaced far enough apart to avoid "overshadowing" and ensure an efficient operation (e.g., in case of wind farms, turbines should be placed 5–10 diameters apart). Thus, in both cases, the footprint of the actual equipment is a small fraction of the required land. The relatively large effective power densities of the CAC-based systems imply that the land use is unlikely to present a constraint on the deployment of the commercial CAC systems.

There are many ideas and proposals currently floating in the literature with regard to technical realization of atmospheric CO_2 capture; all known technological approaches fall into four major categories:

- Alkaline scrubbing solutions, e.g., NaOH, KOH, Ca(OH)$_2$
- Alkali metal carbonate-based scrubbing solutions, e.g., Na$_2$CO$_3$, K$_2$CO$_3$
- Chemical sorbents, e.g., MgO, CaO
- CO_2-selective sorbents, e.g., activated carbons, nano-structured composites, metal-organic frameworks

Although most of the above options have been tested on laboratory scale units, a field demonstration of the technology would be needed to prove their technical feasibility.

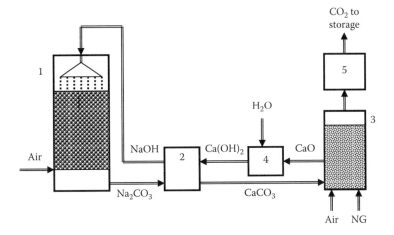

FIGURE 1.53
Simplified process diagram of an atmospheric CO_2 capture system (packed bed tower option). 1—contactor, 2—causticizer, 3—calciner, 4—slaker, 5—CO_2 capture and compression.

The concepts based on CO_2 scrubbing using alkaline solutions are most technologically advanced. One example of such a process is Na–Ca hydroxide-carbonate system reported by Zeman and Lackner (Lackner 2003; Zeman and Lackner 2004). Keith et al. (2006) thoroughly analyzed this system and concluded that the technology could present a scalable near term option for CO_2 removal from air since a CAC plant could be built using already available technological solutions existing in modern paper mills and cement industry. Figure 1.53 provides a simplified block diagram of an air capture system based on the Na–Ca hydroxide-carbonate system. CO_2 is scrubbed from air in a tower-type contactor apparatus (1), where NaOH solution is brought into contact with air at near-ambient conditions (one of the preferable arrangements may include dripping the solution through the tower filled with a packing bed material):

$$CO_2(g) + 2Na^+ + 2OH^- \rightarrow CO_3^{2-} + 2Na^+ + H_2O \quad -110 \text{ kJ/mol-C} \tag{1.52}$$

The resulting sodium carbonate (Na_2CO_3) solution is directed to a causticizer apparatus (2) where it reacts with calcium hydroxide at about 70°C regenerating NaOH and producing solid product $CaCO_3$:

$$CO_3^{2-} + Ca^{2+} \rightarrow CaCO_3(s) \quad 12 \text{ kJ/mol-C} \tag{1.53}$$

This process is quite similar to the kraft recovery process used in the pulp and paper industry. The $CaCO_3$ product is calcined at high temperatures (about 900°C) in a calciner (3) to regenerate CaO and release CO_2:

$$CaCO_3 \rightarrow CaO(s) + CO_2(g) \quad 179 \text{ kJ/mol-C} \tag{1.54}$$

The calcination process has been long practiced at a very large scale for the production of lime and cement and in pulp and paper industry (modern calciners utilize fluidized bed reactors with heat input from NG combustion). It is proposed that the calcining stage will be followed by an amine-based CO_2 capture system to produce a stream of concentrated CO_2. In order to close the loop, CaO reacts with liquid water in a slacker apparatus (4) to produce $Ca(OH)_2$:

$$CaO(s) + H_2O(l) \rightarrow Ca^{2+} + 2OH^- \quad -82 \text{ kJ/mol-C} \tag{1.55}$$

Advantageously, all the chemicals involved in the above system are inexpensive, abundant, and benign materials and all the processes are well understood since they present current industrial practices. The authors of the study (Keith et al. 2006) reported on some physical dimensions and key technological and economic parameters of the spray tower in the air capture system (selected parameters are presented below):

Tower diameter	110 m
Tower height	120 m
CO_2 capture efficiency	50%
NaOH solution flow rate	$1 \text{ m}^3 \text{ s}^{-1}$
CO_2 capture rate	$76,000 \text{ t C year}^{-1}$
Capital cost	$12 million
Operation and maintenance cost	$400,000 year^{-1}

The estimates of the cost components of the entire CAC system are as follows (in $ t^{-1} C):

Spray tower	41
Calcinations	230–373
CO_2 amine capture	49
CO_2 compression	43

Due to high energy intensity of the calcination stage, its cost determines the overall cost of the air capture system, which amounts to about $500 t^{-1} of carbon. Of that amount, about a third relates to capital and maintenance cost, and the remaining two-thirds are the costs of NG and carbon-neutral electricity. The authors argue that the cost could be further brought down via the process optimization and scale-up. The authors of another study dealing with the alkaline scrubber system reported similar dimensions for a full-size commercial device: an open tower about 120 m high and about 100 m in diameter (with the cost of $240 t^{-1} of carbon removed from the atmosphere) (Carnegie Mellon University 2008).

Baciocchi et al. (2006) presented a detailed analysis of the technological aspects of the Na–Ca hydroxide-carbonate air capture system including the process design and material and energy balances of the individual stages of the plant. For their analysis, the authors assumed a CO_2 absorbing column filled with a state-of-the-art Sulzer Mallapak 500Y packing sprayed with 2 M NaOH solution, CO_2 inlet and outlet concentrations of 500 and 250 ppm, respectively. The column performance has been evaluated by numerically solving the material balance of CO_2 in the gas phase:

$$dy = \frac{K_G a_e}{G} P(y - y^*) dz \tag{1.56}$$

where

dy is the mole fraction change taking place in a column section of height dz

K_G is the global mass transfer coefficient

a_e is the packing effective specific area

G is the gas phase molar flux

y is CO_2 molar fraction in the bulk gas phase

y^* is CO_2 molar fraction in an equilibrium with the bulk liquid phase

Based on the calculations, the absorption column for CO_2 capture would have the specific cross section of $18,500 \, m^2 \, t^{-1}$ CO_2 per minute; thus, a unit handling $817,115 \, m^3 \, h^{-1}$ of airflow would have cross section of about $20,000 \, m^2$ (the area equivalent of four soccer fields). The total energy requirements of the process are estimated at $12–17 \, GJ \, t^{-1}$ of CO_2, depending on the process specifics. The fact that these values are comparable with the energy input requirements of the air capture process made the authors of the study question economical and environmental feasibility of the technology. In order to make the process feasible, the authors suggest significantly increasing its energy efficiency and utilizing nonfossil energy resources: nuclear, solar, or other renewables. The excessively high energy cost of this process was also emphasized by other authors (e.g., Herzog 2003).

Zeman (2007) reported the breakdown of energy requirements of an air capture plant based on the same Na–Ca hydroxide-carbonate system as follows (in $kJ \, mol^{-1}$ CO_2):

Air contacting	88
Evaporation	63
Calcinations	256
CO_2 purification/air separation	16
CO_2 compression	19
Total	442

Note that the overall energy requirement estimated by Zeman is somewhat less compared with the estimates provided by Keith et al. ($679 \, kJ \, mol^{-1}$ CO_2) (Keith et al. 2006) and Baciocchi et al. ($516 \, kJ \, mol^{-1}$ CO_2) (Baciocchi et al. 2006). The thermodynamic efficiency of the air capture system was estimated at 6.0%, which is rather low, but would be comparable to other "end of pipe" CO_2 capture technologies (e.g., monoethanolamine scrubbing) if adjusted to comparable CO_2 concentrations. The use of fossil electricity (especially if generated by coal-fired power plants) dramatically reduces the net amount of captured CO_2. A variation of the above system has been reported by Sokolow, who described a process that captures CO_2 in lime water followed by the thermal treatment of the resulting $CaCO_3$ in a calciner producing high pressure CO_2 and regenerated CaO (similar to a cement-manufacturing technology) (Sokolow 2002).

A group of Swiss researchers studied a family of Na- and Ca-based thermochemical cyclic processes for CO_2 capture from air (Nikulshina et al. 2008). For example, in one cycle, air stream containing 500 ppmv of CO_2 was passed over Na_2CO_3 at 50°C to form $NaHCO_3$:

$$Na_2CO_3(s) + CO_2(g) + H_2O(g) \leftrightarrow 2NaHCO_3(s) \quad \Delta H^\circ = -135.5 \, kJ \qquad (1.57)$$

$NaHCO_3$ was thermally decomposed to Na_2CO_3, CO_2, and H_2O at 200°C (the reverse reaction). Na_2CO_3 and H_2O were recycled to the first reactor, and CO_2 was sent to a sequestration site. The total thermal energy requirement for Na-based cycles varied in the range from 390 kJ mol^{-1} CO_2 (for Na_2CO_3–$NaHCO_3$ cycle) to 481 kJ mol^{-1} CO_2 captured (for NaOH–Na_2O–Na_2CO_3 cycle). In $Ca(OH)_2$–$CaCO_3$–CaO thermochemical cycle, the required thermal energy input was much higher: 2485 kJ mol^{-1} CO_2 captured (Nikulshina et al. 2006).

Wind could greatly accelerate CO_2 capture from the atmosphere. Lackner estimated that the size of a CO_2 capture apparatus would be about 1% of the size of a windmill that could displace equal amounts of fossil CO_2 emissions, which implies that the capture system would be cheaper compared with the wind mill system (Lackner 2003). Since the atmosphere mixes rather rapidly, the capture of CO_2 at any given site would be compensated by the release of CO_2 emissions at other even remote sites. One of the proposed technological options is to construct giant filters that would act like flypaper, trapping CO_2 molecules as they drift pass in the wind (Behar 2005). Alkaline solutions would be pumped through porous filters similar to antifreeze flowing through a vehicle's radiator. According to Lackner's estimates, a single commercial size wind scrubber with the height of 70 m and width of 55 m would capture about 90,000 t of CO_2 a year. Some experts, however, are skeptical of the idea; they argue that to capture all anthropogenic CO_2 we would need to cover with the scrubber towers an area at least the size of Arizona (Behar 2005).

1.8.3 Carbon-Negative Systems

The term "carbon-negative system" implies that the amount of atmospheric CO_2 absorbed and sequestered by the system is greater than the CO_2 amount that could ultimately be released back to the atmosphere (e.g., via a leakage, diffusion, degradation, and other processes). At least two options (chemical and biological) for the development of the carbon-negative systems could be envisaged:

1. CAC and its conversion to stable carbonaceous products
2. Pyrolysis of biomass to solid carbonaceous products such as charcoal or bio-carbon

There are growing efforts to develop technologies that utilize CO_2 as a chemical feedstock for the manufacturing of commodity products, polymers, composites, and other materials. This is a challenging task considering the chemical inertness of CO_2 molecule; nevertheless, CO_2 is already being used as a feedstock in the production of a number of important chemicals and materials, such as polycarbonate-based plastics, urea (for fertilizers), salicylic acid (a pharmaceutical ingredient), solvents, etc. However, 115 million tons of CO_2 annually utilized by chemical industry worldwide is not a match for tens of billion tons of atmospheric CO_2 (Ritter 2007). Thus, in order to have a tangible effect on the atmospheric CO_2 reduction, the amount of products produced from CO_2 has to be increased by 2–3 orders of magnitude, which is not realistic due to an unlikely demand. Moreover, not all CO_2-derived products can serve as a long-term sink for CO_2: some of them, for example, urea, pharmaceutical ingredients, will release CO_2 back to the atmosphere via a variety of biochemical and chemical routes in a matter of months or years (Muradov and Veziroğlu 2008).

From this viewpoint, conversion of captured atmospheric CO_2 to solid carbon is more advantageous due to an exceptional chemical inertness of carbon at ambient conditions (coal has been stored underground for millions of years with practically no chemical changes). This approach offers one important advantage over the conventional CCS route in that carbon is permanently locked away off the carbon cycle (it will not "leak" back to the atmosphere as CO_2 stored in geological formations or under the ocean will eventually do). Carbon can be extracted from CO_2 via a variety of chemical reactions involving reducing agents (e.g., hydrogen and some active metals) that are capable of binding oxygen into very stable products (water or oxides).

Although hydrogenation of CO_2 into carbon and water is a well-known reaction (Bosch reaction),

$$2H_2 + CO_2 \rightarrow 2H_2O + C \qquad (1.58)$$

it has attracted a little commercial interest mostly due to a perceived uselessness of the process (why consume expensive hydrogen for converting CO_2 to carbon, if it could be converted to much more valuable products, such as methanol, FT fuels, SNG, etc.). Recently, however, this reaction attracted the attention of researchers as part of a cyclic process aiming at oxygen recovery from CO_2 in a life-support system, particularly, for space exploration applications. The reaction 1.58 is strongly exothermic and is favored by relatively low temperatures and high pressure. At higher temperatures (500°C and above), this reaction is likely to compete with an undesirable reverse water gas shift reaction: $H_2 + CO_2 \rightarrow H_2O + CO$. The practical realization of the reaction (1.58) faces several challenges with regard to unfavorable thermodynamics, relatively slow kinetics (due to the formation of a new phase), and catalyst deactivation problem (due to carbon deposition).

From this viewpoint, an alternative two-step CO_2-to-carbon route looks more promising: first, CO_2 is hydrogenated to CH_4, which is followed by the methane decomposition step (H_2 is recycled), as follows:

$$4H_2 + CO_2 \xrightarrow[-2H_2O]{\text{Catalyst}} CH_4 \longrightarrow 2H_2 + C \qquad (1.59)$$

The first step (the Sabatier reaction) is a well-developed commercial process (production of substitute NG from CO_x); it is typically conducted at 300°C–400°C using supported Ni or Ru catalysts. The second step: methane decomposition to hydrogen and carbon may require more developmental efforts. Although there are two known commercial processes that deal with methane (to be precise, NG) decomposition, namely, the Thermal Black process (currently, not in operation) and Kværner's plasma-driven H&CB process (operated on a limited commercial scale in Norway and Canada), these technologies are very energy intensive and are mainly designed to produce carbon black (with H_2 being a by-product). More information on this topic can be found in Chapter 14.

Figure 1.54 depicts the conceptual scheme of a carbon-negative system based on the first (chemical) approach. Atmospheric CO_2 is captured by a chemical scrubber, and the resulting concentrated CO_2 stream is directed to a CO_2-processing plant. At this plant, CO_2 can be converted to solid carbon (via the reactions 1.57 or 1.58). Carbon produced in the process could be either transformed into structural longlasting materials or sequestered as solid carbon. The stored chemical energy of the carbon product could be recovered sometime in the future (in a less carbon-restrained world), by oxidizing it in a direct carbon fuel cell (DCFC) with the efficiency of 80%–90% (the theoretical efficiency of DCFC is near 100%) (Muradov and Veziroğlu 2005).

According to the second (biological) approach, a certain fraction of carbon can be removed from the natural carbon cycle in the form of bio-carbon (or bio-char, charcoal)

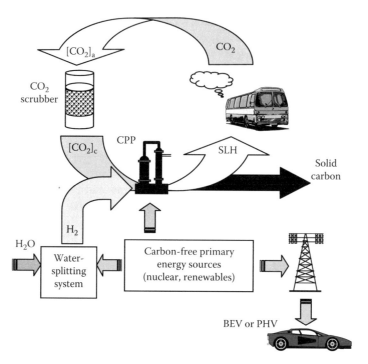

FIGURE 1.54
Schematic representation of a carbon-negative system for production of synthetic fuels and solid carbon. $[CO_2]_a$—atmospheric CO_2, $[CO_2]_c$—concentrated CO_2 stream, BEV—battery electric vehicle, PHV—plug-in hybrid vehicle, CPP—CO_2 processing plant.

via thermal processing (pyrolysis) of biomass. Due to a relatively slow rate of biomass growth, it may take much longer time to remove a given amount of atmospheric CO_2 via the biological option compared with the chemical one. On the other hand, no chemical reagents, huge scrubbers, and associated energy-intensive technological processes would be required if CO_2 is captured by growing biomass: the sun provides all the necessary energy input. In a simplified form, the high-temperature pyrolysis of terrestrial biomass (e.g., plants) process can be presented as follows (Reed 1981):

$$CH_{1.4}O_{0.59} \rightarrow 0.41C + 0.68H_2 + 0.59CO + 0.0005CO_2 + 0.001H_2O + 0.002CH_4 \qquad (1.60)$$

where $CH_{1.4}O_{0.59}$ is a representative formula of terrestrial biomass.

As seen from this equation, about 40% of biomass carbon can be converted to bio-carbon, with the remaining biomass carbon being converted to light gases (CO, CO_2, and CH_4). It is important to mention that some aquatic forms of biomass (e.g., algae, *Lemna minor*, giacint) enjoy much faster growth rates (thus, are able to fastly remove CO_2 from the atmosphere) compared to terrestrial biomass. Recently, Muradov et al. (2010) reported on the pyrolysis of fast-growing biomass—*Lemna minor* (also known as duckweed)—at the temperature range of 400°C–700°C and atmospheric pressure. The following range of products was produced at 400°C (wt.%): bio-char 50.1, pyrolysis gas 12.3, and bio-oil 37.6. Thus, almost half of the dry duckweed's weight ends up in the form of bio-char. Note that pyrolysis gases produced from both terrestrial and aquatic biomass are rich in hydrogen and carbon monoxide. After a preliminary cleanup and conditioning, pyrolysis gas may be suitable for production of carbon-neutral fuels such as hydrogen (via the water gas shift reaction followed by CO_2 removal), or liquid hydrocarbons such as gasoline, diesel (via FT synthesis), or methanol (via methanol synthesis). Alternatively, hydrogen recovered from pyrolysis gas can be used for catalytic hydrodeoxygenation of bio-oil to liquid hydrocarbons. The selection of a particular route for conversion of the biomass pyrolysis products to alternative fuels will be dictated by the specifics of the biomass feedstock and a demand for the particular fuel.

Due to its chemical inertness (at ambient conditions), charcoal can serve as a practically permanent sink for carbon captured from the atmosphere. As an additional benefit, biomass-derived charcoal (bio-char) can be used for the soil amendment applications; thus, bio-char enhances or restores soil fertility while providing means of storing carbon for millennia. In this way, carbon extracted from the captured atmospheric CO_2 helps to remove even more CO_2 from atmosphere via the enhanced growth of biomass. Although the stimulating effect of charcoal on a crop harvest has been known for centuries, recent quantitative studies confirmed that adding charcoal to soil can significantly increase seed germination, plant growth, and crop yields (e.g., crop yields could be increased by up to 200%) (Glaser et al. 2002). It was shown that the application of charcoal to soil increases its nutrient, fertilizer, and water retention capacities, supports microbial communities, and activates the root activity, thus promoting biomass growth (the effect is attributed to higher exchange capacity, surface area, changes in pH, and other factors). Furthermore, the presence of bio-char in soil reduces the emissions of a potent GHG—N_2O to the atmosphere (Lehmann et al. 2006). Thus, the addition of biomass-derived charcoal to soil could potentially have multiple favorable effects on the reduction of atmospheric CO_2, namely, through the enhancement of CO_2 absorption by plants as well as via reduction of CO_2 emissions from soil to the atmosphere. An additional positive effect of adding charcoal to soil is that due to an enhanced nutrient-holding capacity of the carbon-amended soils, the amount of applied fertilizers could be substantially reduced, resulting in an additional economic gain and further reductions in CO_2 emissions into the atmosphere (from fertilizer-manufacturing plants that are major CO_2 producers).

1.9 Conclusions

Our civilization vitally depends on carbon for its energy; carbon made possible the industrial revolution and the rise in the standard of living. On the other hand, fossil fuels are responsible for ever-increasing levels of CO_2 in the atmosphere, air pollution, and other forms of environmental damage that could cause health-related problems. There is a growing consensus among energy industry professionals, policymakers, environmentalists, and general public that in order to maintain the present standard of living and establish an environmentally sustainable energy future, new approaches to managing energy resources and fuels have to be developed and implemented in the near future. In the face of ever-increasing levels of anthropogenic CO_2 emissions, there have been attempts to estimate the tolerable limits of atmospheric CO_2 concentrations in terms of the global mean temperature rise. The analysis of scientific literature on climate impacts indicates that, in the majority of cases, a temperature increase of 2°C above the preindustrial levels is considered a critical point beyond which a potentially catastrophic impact on the planet's ecosystem might occur.

Concerns over dwindling fossil energy resources and their adverse environmental impact have triggered considerable worldwide efforts to find carbon-free or low carbon alternatives to conventional fossil fuels. There is a growing consensus that carbon-neutral energy sources, fuels, and associated technologies (i.e., the technologies that do not increase atmospheric CO_2 concentration) will play the decisive role in the establishment of the sustainable energy future. It was estimated that in order to satisfy the world's growing demand for energy and keep our planet healthy, 10–30 TW of carbon-free power has to be produced by mid-century. Three prominent options discussed in the literature include decarbonization of fossil energy, and the use of nuclear and renewable energy sources.

The analyses of state-of-the-art technologies for utilization of carbon-free energy resources (nuclear, solar, wind, geothermal, and others) and the assessment of their technical potential indicate that these sources can meet the above target by mid-century. However, switching from fossil fuels that currently provide about 80% of primary energy to nonfossil-based energy systems may require too many changes and, consequently, too much time to accomplish that transition. On the other hand, some experts believe that we cannot afford too long a transition period because of potentially irreversible changes in global climate caused by the extensive use of fossil fuels. Recent studies indicated that the combination of two or more carbon-free sources and the use of proper technologies can potentially provide necessary quantities of power and carbon-neutral transportation fuels to make a difference within the reasonable timeframe. Production of synthetic liquid hydrocarbon fuels (e.g., "green" gasoline) by combining nuclear and renewable primary sources, non-carbogenic hydrogen, and biomass (and, eventually, atmospheric CO_2) would allow drastically increasing the yield of bio-derived fuels without competing with food for land and without an adverse environmental impact. Most importantly, these systems can be implemented with current technologies taking advantage of the existing fuel infrastructure; in particular, wind- or nuclear-powered electrolysis combined with FT technology could provide a basis for the near-term implementation of the biomass-to-liquid fuels process, thus, freeing us from a heavy reliance on petroleum.

The removal of CO_2 from atmosphere and its permanent and secure storage, or, alternatively, its economically feasible transformation to stable products (e.g., solid carbon) is an extremely challenging problem. Currently, there are no large-scale atmospheric CO_2 capture units anywhere in the world, but with the intensive R&D efforts and dedicated industrial and governmental support it might be possible to build and validate commercial-size units in the next two to three decades and achieve the necessary cost reduction. Current estimates point to the possibility of capturing CO_2 from air at the cost of $200 t^{-1} C or even less (at this cost, the removal of 50 ppm of CO_2 from the atmosphere would cost close to $20 trillion) (Hansen et al. 2008). The cost seems to be rather high (i.e., much higher than possible future carbon taxes), but the economics of the process might be substantially improved with the development of alternative scrubbing technologies (besides, one has to compare it with other options that will be available 20–50 years from now). IEA estimates that $10.5 trillion in cumulative investments will be needed between now and 2030 to keep atmospheric CO_2 at the 450 ppm level and the global temperature increases below 2°C. The activities in this area will undoubtedly be fueled by a recent announcement by a British billionaire entrepreneur Richard Branson offering a US$25 million prize to anyone who can come up with a technology to remove billions of tons of CO_2 from the Earth's atmosphere.

The transition to carbon-neutral energy systems and transportation fuels is facing an enormous challenge due to an existing infrastructural inertia and a number of factors of political and technological nature. Satisfying growing energy demand without producing immense quantities of carbon emissions and potentially altering climate would require truly extraordinary efforts and government-backed programs with the scale and urgency of the Manhattan (atom bomb) project. One pathway to the sustainable energy future that many experts agree upon involves the worldwide implementation of carbon-neutral technologies based on decarbonized fossil energy, and nuclear and renewable energy sources. It is important to start the transition to carbon–neutral energy systems as soon as possible, while there is still time, and sufficient resources to complete this transition and avoid irreversible changes in our planet's ecosystem.

Acronyms

BEV	battery-electric vehicle
BTL	biomass to liquid
CAC	capture of atmospheric CO_2
CCS	carbon capture and sequestration
CSP	concentrated solar power
DCFC	direct carbon fuel cell
DIC	dissolved inorganic carbon
DOC	dissolved organic carbon
DOE	U.S. Department of Energy
FC	fuel cell
FT	Fischer–Tropsch process
GHG	greenhouse gas
GT	gigaton
GW	gigawatt
GWP	global warming potential
HHV	higher heating value
IEA	International Energy Agency
IPCC	Intergovernmental Panel on Climate Change
LFG	landfill gas
LUCF	land use change and forestry
MBD	million barrels per day
MSW	municipal solid waste
Mtoe	million ton oil equivalent
NG	natural gas
NRC	national research council
NREL	National Renewable Energy Laboratory
OTEC	ocean thermal energy conversion
PEM	proton exchange membrane
PHV	plug-in hybrid vehicle
POC	particulate organic carbon
ppb	part per billion
ppm	parts per million
PV	photovoltaic
RF	radiative forcing

SLH	synthetic liquid hydrocarbons
SMR	steam methane reforming
SNG	synthetic natural gas
SRM	solar radiation management
STH	solar to hydrogen
TCM	trillion cubic meters
toe	ton oil equivalent
TW	terawatt

References

Abanades, S., P. Charvin, G. Flamant, and P. Neveu. 2006. Screening of water-splitting thermochemical cycles potentially attractive for hydrogen production by concentrated solar energy. *Energy* 31: 2805–2822.

Agarwal, R., N. Singh, F. Ribeiro, and N. Delgass. 2007. Sustainable fuel for the transportation sector. *Proc. Natl. Acad. Sci. USA* 104 (12): 4828–4833.

Allan, W. et al. 2005. Interannual variations of ^{13}C in tropospheric methane: Implications for possible atomic chlorine sink in the marine boundary layer. *J. Geophys. Res.* 110: D11306, doi: 10.1029/2004JD005650.

ANL. 2007. About tar sands, http://ostseis.anl.gov/guide/tarsands/index.cfm (accessed September 19, 2007).

Archer, D. 2005. The fate of fossil fuel CO_2 in geologic time. *J. Geophys. Res.* 110 (C9): C09S05, doi: 10.1029/2004JC002625.

Archer, D. and B. Buffet. 2005. Time-dependent response of the global ocean clathrate reservoir to climate and anthropogenic forcing. *Geochem. Geophys. Geosyst.* 6: Q03002, doi: 10.1029/2004GC000854.

Baciocchi, R., G. Storti, and M. Mazzootti. 2006. Process design and energy requirements for the capture of carbon dioxide from air. *Chem. Eng. Process.* 45: 1047–1058.

Behar, M. 2005. Now you CO_2, now you don't. *Popular Sci.* August: 53–58.

BES. 2005. Basic research needs for solar energy utilization, Report of the U.S. DOE Basic Energy Sciences Workshop on Solar Energy Utilization, Publ. Argonne Nat. Lab.

Blasing, T. 2009. Recent greenhouse gas concentrations, Carbon Dioxide Information Analysis Center, Oak Ridge National Laboratory, updated December 2009, http://cdiac.ornl.gov/pns/current_ghg.html (accessed July 9, 2010).

Bluth, G., S. Doiron, C. Schnetzler, A. Krueger, and L. Walter. 1992. Global tracking of SO_2 clouds from the June 1991 Mount Pinatubo eruptions. *Geophys. Res. Lett.* 19: 151–154.

Bockris, J. and N. Veziroğlu 2007. Estimates of the price of hydrogen as a medium for wind and solar sources. *Intern. J. Hydrogen Energy* 32: 1605–1610.

Bond, T. et al. 2004. Technology based global inventory of black and organic carbon emissions from combustion. *J. Geophys. Res.* 109: D14203, doi: 10.1029/2003JD003697.

Bossel, U., B. Eliasson, and G. Taylor. 2005. The future of hydrogen economy: Bright or bleak? www.efcf.com/reports (accessed August 20, 2008).

BP. 2006. British Petroleum, Quantifying energy: BP statistical review of world energy, BP p.l.c. London, U.K., 45pp, http://www.bp.com/productlanding.do?categoryId=6842&contentId=7021390.

Brovkin, V., S. Sitch, W. von Bloh et al. 2004. Role of land cover changes for atmospheric CO_2 increase and climate change during the last 150 years. *Global Change Biol.* 10: 1253–1266.

Buffet, B. and D. Archer. 2004. Global inventory of methane clathrate: Sensitivity to changes in the deep ocean. *Earth Planet. Sci. Lett.* 227: 185–199.

Carnegie Mellon University. 2008. Snatching carbon dioxide from atmosphere, Climate Decision Making Center, Carnegie Mellon University, http://cdmc.epp.cmu.edu (accessed December 12, 2008).

Carpenter, A. 1988. Coal classification. IEA Coal Research Report, IEACR/12, London, U.K., p. 104.

Chapman, S. and M. Thurlow. 1996. The influence of climate on CO_2 and CH_4 emissions from organic soils. *J. Agric. For. Meteorol.* 79: 205–217.

C&EN. October 12, 2009. Biofuels expansion faces challenges. *Chem. Eng. News* 87: 36.

C&EN. October 12, 2009. Database available on gas hydrates. *Chem. Eng. News* 87: 36.

C&EN. October 25, 2009. Dow unveils solar shingle. *Chem. Eng. News* 87: 25.

C&EN. November 16, 2009. Climate change costs top $10 trillion. *Chem. Eng. News* 87: 27.

C&EN. December 7, 2009. Higher ethanol share likely for gasoline. *Chem. Eng. News* 87(49): 32.

C&EN. December 21, 2009. Alternative energy flourishes. *Chem. Eng. News* 87: 28–29.

Cicerone, R. 2006. Geoengineering: Encouraging research and overseeing implementation. *Clim. Change* 77: 221–226.

Clery, D. 2010. Laser fusion energy poised to ignite. *Science* 328: 808–809.

Collins, B. 2004. Utility-scale solar–Today. *Solar Today* July/August: 33–35.

Counsel of Canadian Academies. 2008. Energy from gas hydrates: Assessing the opportunities and challenges for Canada, www.scienceadvice.ca/documents/(2008_07_07)_GH_Report_in_Focus.pdf (accessed September 22, 2008).

CO_2 Now. 2010. http://co2now.org (accessed July 9, 2010).

Cowan, G. March 1, 2010. Letter to editor. *Chem. Eng. News* 7.

Crutzen, P. 2006. Albedo enhancement by stratospheric sulfur injection: A contribution to resolve a policy dilemma? *Clim. Change* 77: 211–219.

Daltorio, T. 2010. Shale natural gas resources: Why Europe doesn't stand much of a chance in this department. Investment U Research, http://www.investmentu.com/IUEL/2010/March/shale-natural-gas-resources.html (accessed March 29, 2010).

Davis, S., K. Caldeira, and H. Damon Matthews. 2010. Future CO_2 emissions and climate change from existing energy infrastructure. *Science* 329: 1330–1333.

Delucchi, M. 2006. Lifecycle analyses of biofuels. Technical Report UCD-ITS-RR-0608, www.its.ucda-vis.edu/publications/2006/UCD-ITS-RR-06-08.pdf (accessed July 8, 2010).

Dornburg, V. and A. Faaij. 2001. Efficiency and economy of wood-fired biomass energy systems in relation to scale regarding heat and power generation using combustion and gasification technologies. *Biomass Bioenergy* 21: 91–108.

Duncan, R. 2007. Plug-in hybrids: Pollution free transport on the horizon. *Solar Today* May/June: 46–48.

EIA NG. 2010. U.S. DOE Energy Information Agency, http://www.eia.doe.gov/oil_gas/natural_gas/special/ngresources/ngresources.html (accessed March 29, 2010).

EIA Oil Reserves. 2010. U.S. DOE Energy Information Agency, World proved reserves of oil, www.eia.doe.gov/emeu/international/reserves.html (accessed April 2, 2010).

Elliot, S., K. Lackner, H. Ziock et al. 2001. Compensation of atmospheric CO_2 buildup through engineered chemical sinkage. *Geophys. Res. Lett.* 28: 1235–1238.

EPA. 2000. Carbon dioxide emissions from generation of electric power in the United States, DOE. EPA Report, Washington, DC.

EPRI. 2004. Power delivery system of the future: A preliminary estimate of cost and benefits, Palo Alto, CA.

Evans-Pritchard, A. 2008. Spain's gain from wind power is plain to see, www.telegraph.co.uk/money/main.jhtml?xml=/money/2008/04/07/ccwind107.xml (accessed November 12, 2008).

Falkowski, P. et al. 2000. The global carbon cycle: A test of our knowledge of Earth as a system. *Science* 290, 291–296.

Forsberg, C. 2007. Future hydrogen markets for large-scale hydrogen production systems. *Int. J. Hydrogen Energy* 32: 431–439.

Forsberg, C. 2009. Meeting US liquid transport fuel needs with a nuclear hydrogen biomass system. *Int. J. Hydrogen Energy* 34: 4227–4236.

Fthenakis, V. and H. Kim. 2007. Greenhouse gas emissions from solar electric and nuclear power: A lifecycle study. *Energy Policy* 35: 2549–2557.

Glaser, B., J. Lehmann, and W. Zech. 2002. Ameliorating physical and chemical properties of highly weathered soils in the tropics with charcoal—A review. *Biol. Fert. Soils* 35: 219.

Goody, R. 1964. *Atmospheric Radiation: I. Theoretical Basis.* Clarendon, Oxford, U.K., 436pp.

Gorelick, S. 2010. *Oil Panic and the Global Crisis. Predictions and Myths.* Wiley-Blackwell, Oxford, U.K.

Greene, N. 2004. Growing energy: How biofuels can help end America's growing oil dependence, NCEP Technical Appendix: Expanding Energy Supply, in the National Commission on Energy Policy, Ending the Energy Stalemate, Washington, DC.

Greene, D. 2010. Oil peak or panic? *Science* 328: 828.

Guardian 2010. Guardian Environmental Week, The world's first molten salt concentrating solar power plant, http://www.guardian.co.uk/environment/2010/jul/22/first-molten-salt-solar-power (accessed January 22, 2011)

Hacker, V. and K. Kordesh K. 2003. Ammonia crackers. In *Handbook of Fuel Cells*, Eds. W. Vielstich et al., J. Wiley & Sons, Chichester, U.K., Vol. 3, pp. 121–127.

Hansen, J., M. Sato, P. Kharecha et al. 2008. Target atmospheric CO_2: Where should humanity aim? *Open Atmos. Sci. J.* 2: 217–231.

Hargreaves, S. 2010. Natural gas: Fuel of then future, http://money.cnn.com/2010/03/29/news/economy/natural_gas/ (accessed March 29, 2010).

Heimann, M. 2010. How stable is the methane cycle? *Science* 327: 1211–1212.

Heinze, C., E. Meier-Reimer, and K. Winn. 1991. Glacial pCO_2 reduction by the world ocean: Experiments with the Hamburg carbon cycle model. *Paleooceanography* 6: 395–430.

Held, I. and B. Soden. 2000. Water vapor feedback and global warming. *Ann. Rev. Energy Environ.* 25: 441–475.

Herzog, H. 2003. Assessing the feasibility of capturing CO_2 from air. Technical Report LFEE 2003-002-WP, Massachusetts Institute of Technology, Cambridge, MA.

Hoffert, M. 2010. Farewell to fossil fuels? *Science* 329: 1292–1294.

Hoffert, M. et al. 2002. Advanced technology paths to global climate stability: Energy for the greenhouse planet. *Science* 298: 981–987.

Hofzumahaus, A. et al. 2004. Photolysis frequency of O_3 to O^1D: Measurements and modeling the IPMMI. *J. Geophys. Res.* 109: D08S90, doi: 1029/2003JD004333.

Hoogwijk, M., A. Faaij, R. van den Broek, G. Bendes, D. Gillen, and W. Turkenburg. 2003. Exploration of the ranges of the global potential of biomass for energy. *Biomass Bioenergy* 25: 119–133.

IEA. 2007. International Energy Agency, www.iea.org/stats/renewdata.asp?country_code=29 (accessed July 1, 2010)

IEA GHG. 2002. *International Energy Agency. GHG, Building the Cost Curves for CO_2 Storage, Part 1: Sources of CO_2,* PH4/9, July 2002, 48pp.

IEA KWES. 2009. *International Energy Agency, Key World Energy Statistics,* 2009 edn., IEA, Paris, France, www.iea.org (accessed May 5, 2010).

IEA Renewables 2006. *International Energy Agency (IEA). Renewables in Global Energy Supply: An IEA Fact Sheet,* IEA/OECD, Paris.

IEA Statistics. 2009. *International Energy Agency Statistics. CO_2 Emissions from Fossil Fuel Combustion,* 2009 edn., IEA, Paris, France.

Iowa Energy Center. 2007. Renewable energy and energy efficiency, education and demonstration, ammonia, http://www.energy.iastate.edu/renewable/ammonia/ammonia/ammonia Mtg07.htm (accessed November, 2008).

IPCC. 1990. The intergovernmental panel on climate change scientific assessment. In *Climate Change,* Eds. J. Houghton et al., Cambridge University Press, Cambridge, U.K., 364pp.

IPCC. 2001. The Intergovernmental Panel on Climate Change. In *Climate Change 2001: IPCC Working Group III: Mitigation.* Online version, http://www.grida.no/publications/other/ipcc_tar/?src=/climate/ipcc_tar/wg3/176.htm (accessed April 14, 2010).

IPCC/TEAP. 2005. *The Intergovernmental Panel on Climate Change Special Report on Safeguarding the Ozone Layer and the Global Climate System: Issues Related to Fluorohydrocarbons and Perfluorocarbons*, Eds. B. Metz et al., Cambridge University Press, Cambridge, NY, 2005, 488pp.

IPCC PSB. 2007. The Intergovernmental Panel on Climate Change 4th Assessment Report, Climate Change 2007. *The Physical Science Basis*, Cambridge University Press, Cambridge, U.K.

IPCC Mitigation. 2007. The intergovernmental panel on climate change. *Climate Change 2007. Mitigation. Contribution of Working Group III to the Fourth Assessment Report of the IPCC*, Eds. P. Bosch, R. Dave, O. Davidson, B. Metz, and L. Meyer, Cambridge University Press, Cambridge, U.K.

Ito, A. and J. Penner. 2005. Historical emissions of carbonaceous aerosols from biomass and fossil fuel burning for the period of 1870–2000. *Global Biochem. Cycles* 19: GB2028, doi: 10.1029/2004GB002374.

Jackson, P. 2009. The future of global oil supply, Understanding the building blocks. Special Report, November 2009, www.cera.com (accessed, May 10, 2010).

Jacobson, M. 2004. The short-term cooling but long term global warming due to biomass burning, *J. Clim.* 17: 2909–2926.

Jacobson, M. 2009. Review of solutions to global warming, air pollution and energy security, *Energy Environ. Sci.* 2: 148–173.

Jacobson, M. and M. Delucchi. 2009. A path to sustainable energy by 2030. *Sci. Am.* November: 58–65.

Jacoby, M. 2009a. Coming back to nuclear energy. *Chem. Eng. News* 87: 14–18.

Jacoby, M. 2009b. Reintroducing thorium. *Chem. Eng. News* 87: 44–46.

Jensen, M. 2007. CO_2 point source emission estimation methodologies summary, NETL, U.S. DOE Carbon Capture and Sequestration Program, *Regional Carbon Sequestration Partnership's Annual Meeting*, Pittsburg, PA, December 12, 2007.

Johnson, J. October 26, 2009. Fossil-fuel costs. *Chem. Eng. News* 87: 6.

Johnson, J. March 8, 2010. New jolt for nuclear power. *Chem. Eng. News* 88: 31–33.

Keith, D., M. Ha-Duong, and J. Stolaroff. 2006. Climate strategy with CO_2 capture from air. *Clim. Change* 74: 17–45.

Keppler, F., J. Hamilton, M. Brass, and T. Rockmann. 2006. Methane emissions from terrestrial plants under aerobic conditions. *Nature* 439: 187–191.

Kerr, R. 2010. Natural gas from shale bursts onto the scene. *Science* 328: 1624–1626.

Kheshgi, H. and R. Prince. 2005. Sequestration of fermentation CO_2 from ethanol production. *Energy* 30: 1865–1871.

Kiehl, J. and K. Trenberth. 1997. Earth's annual global mean energy budget. *Bull. Am. Meteorl. Soc.* 78: 197–208.

Kim, K. 2010. *N. Korea Claims to Achieve Elusive Nuclear Fusion*, Associated Press, http://www.google.com/hostednews/ap/article/ (accessed April 13, 2010).

Kim, J.-S., S. Lee, S. Lee, M. Choi, and K. Lee. 2006. Performance of catalytic reactors for the hydrogenation of CO_2 to hydrocarbons. *Catal. Today* 115: 228–234.

Kintisch, E. 2008. The greening of synfuels. *Science* 320: 306–308.

Kodama, T. 2003. High-temperature solar chemistry for converting solar heat to chemical fuels. *Prog. Energy Combust. Sci.* 29: 567–597.

Koerner, B. 2007. The trillion barrel tar pit. *WIRED Mag.*, hppp://wired.com/wired/archive/12.07/oil.html (accessed September 19, 2007).

Koonin, S. 2009. Environmental stewardship and energy innovation. *Naval Energy Forum*, McLean, VA, October 2009.

Kreith, F. 2000. *The CRC Handbook of Thermal Engineering*, CRC Press, Baca Raton, FL. ISBN 3540663495.

Lacis, A. and M. Mishchenko. 1995. Climate forcing, climate sensitivity and climate response: A radiative modeling perspective on atmospheric aerosols. In *Aerosol Forcing of Climate*, Eds. Charlson, R. and J. Heinztenberg, Wiley, Chichester, U.K., 1995, pp. 11–42.

Lackner, K. 2003. A guide to CO_2 sequestration. *Science* 300: 1677–1678.

Larson, E. 1993. Technology for electricity and fuels from biomass. *Annu. Rev. Energy Environ.* 18: 567–630.

Lee, S. et al. 2007. *Handbook of Alternative Fuel Technologies*, CRC Press, Boca Raton, FL.

Lehmann, J., J. Gaunt, and M. Rondon. 2006. Bio-char sequestration in terrestrial ecosystems—A review. *Mitig. Adapt. Strateg. Global Change* 11, 403–427.

Licht, S. 2001. Multiple bandgap semiconductor/electrolyte solar energy conversion. *J. Phys. Chem. B* 105: 6281–6294.

Marchetti, C. 1976. On geoengineering and CO_2 problem. Report RM-76-17, International Institute for Applied Systems Analysis, Laxenburg, Austria, March 1976.

Marland, G., T. Boden, and R. Andres. 2006. Global, regional and national CO_2 emissions. In *Trends: A Compendium of Data on Global Change*, Carbon dioxide Information Analysis Center, Oak Ridge, TN, http://cdiac.esd.ornl.gov/trends/emis/tre_glob.htm (accessed June 15, 2008).

Meinshausen, M. 2006. What does a 2°C target mean for greenhouse gas concentrations? A brief analysis based on multi-gas emission pathways and several climate sensitivity uncertainty estimates. In: *Avoiding Dangerous Climate Change*, Eds. H. Schellnhuber, W. Cramer, N. Nakicenovic, T. Wigley, and G. Yohe, Cambridge University Press, Cambridge, U.K.

Metz, B., Davidson, O., de Coninck, H., Loos, M., and Meyer, A. (Eds.) 2005. *IPCC Special Report on Carbon Dioxide Capture and Storage*, Prepared by Working Group III of the Intergovernmental Panel on Climate Change, Cambridge University Press, Cambridge, U.K., Tables 2.1 and 2.2, pp. 79–80.

Mielke, J. 2000. Methane hydrates: Energy prospect or natural hazard? CRS Report to Congress: RS20050.

Milkov, A. 2004. Global estimates of hydrate-bound gas in marine sediments: How much is really there. *Earth Sci. Rev.* 66: 183–197.

Mignone, B., R. Socolow, J. Sarmiento, and M. Oppenheimer. 2008. Atmospheric stabilization and the timing of carbon mitigation. *Clim. Change* 88(3–4): 251–265.

Moriarty, P. and D. Honnery. 2007. Intermittent renewable energy: The only future source of hydrogen? *Int. J. Hydrogen Energy* 32: 1616–1624.

Muradov, N., B. Fidalgo, A. Gujar, and A. T-Raissi. 2010. Pyrolysis of fast-growing aquatic biomass *Lemna minor* (duckweed): Characterization of pyrolysis products. *Bioresour. Technol.* 101: 8424–8428.

Muradov N., F. Smith, K. Kallupalayam, and A. T-Raissi. 2008. Production of hydrogen and synthetic hydrocarbon fuels by integrated processing of biomass and biogas. In: *Proceedings XVII World Hydrogen Energy Conference*, Brisbane, Queensland, Australia, 2008.

Muradov, N. and N. Veziroğlu. 2005. From hydrocarbon to hydrogen-carbon to hydrogen economy. *Int. J. Hydrogen Energy* 30: 225–237.

Muradov, N. and T. N. Veziroğlu. 2008. "Green" path from fossil-based to hydrogen economy: An overview of carbon-neutral technologies. *Int. J. Hydrogen Energy* 33: 6804–6839.

Murphy. J., D. Sexton, D. Barnett et al. 2004. Quantification of modeling uncertainties in a large assemble of climate simulations. *Nature* 430: 451–454.

NAS. 1992. National Academy of Science, Policy implications of greenhouse warming: Mitigation, adaptation and the science base, Committee on Science, Engineering and Public Policy, National Academy Press, Washington, DC, 918pp.

NCEP. 2004. Ending the energy stalemate. Report of the National Commission on Energy Policy, Washington, DC.

NETL. 2006. *Carbon Sequestration Technology Roadmap and Program Plan*, U.S. DOE, Office of Fossil Energy, NETL.

NETL. 2010. U.S. DOE National Energy Technology Laboratory. Methane Hydrates, http://www.netl.doe.gov/technologies/oil-gas/future supply/methanehydrates/projects/, last updated February 5, 2010 (accessed June 28, 2010).

Nikulshina, V., N. Ayesa, M. Galvez, and A. Steinfeld. 2008. Feasibility of Na-based thermochemical cycles for the capture of CO_2 from air—Thermodynamic and thermogravimetric analyses. *Chem. Eng. J.* 140: 62–70.

Nikulshina, V., D. Hirsh, M. Mazzotti, and A. Steinfeld. 2006. CO_2 capture from air and co-production of H_2 via the $Ca(OH)_2$-$CaCO_3$ cycle using concentrated solar power-Thermodynamic analysis. *Energy* 31: 1379–1389.

Normile, D. 2010. Fusion "Consolation Prize" gears up for show time. *Science* 328: 1464–1465.

NRC. 2004. *The Hydrogen Economy. Opportunities, Costs, Barriers and R&D Needs*. National Research Council and National Academy of Engineering, National Academies Press, Washington, DC.

NR Canada. 2006. National Resources Canada, Earth Sciences Sector, http://ess.ntcan.gc.ca/2002 2006/ghff/index_e.php (accessed June 28, 2010).

O'Bockris, J. and N. Veziroğlu. 2007. Estimates of the price of hydrogen as a medium for wind and solar resources. *Int. J. Hydrogen Energy* 32: 1605–1610.

Okabe, H. 1978. *Photochemistry of Small Molecules*, John Wiley & Sons, New York.

Pacala, S. and R. Socolow. 2004. Stabilization wedges: Solving the climate problem for the next 50 years with current technologies. *Science* 305: 968–972.

Percopo, B. 2008. U.S. nuclear power's time has come-again. *Power* January: 72.

Power Technology. 2010. Solar Tower, Seville, Spain, http://www.power-technology.com/projects/seville-solar-tower/ (accessed July 23, 2010).

Price, R. and J. Blasé. 2002. Nuclear fuel resources: Enough to last? *NEA News* 20(2): 10–13.

Reed, T. 1981. *Biomass Gasification. Principles and Technology*, Noyes Data Corporation, Park Ridge, NJ.

Reisch, M. October 12, 2009. Chemical from biorefineries. *Chem. Eng. News* 87:28–29.

Richey, J. 2004. Pathways of atmospheric CO_2 through fluvial systems. In: *The Global Carbon Cycle: Integrating Humans, Climate and the Natural World*, Eds. C. Field and M. Raupach, SCOPE 62, Island Press, Washington DC, pp. 329–340.

Rifkin, J. 2002. *The Hydrogen Economy*, J. P. Tarcher/Putnam, New York.

Ritter, S. April 30, 2007. What can we do with CO_2? *Chem. Eng. News* 85:11–17.

Sarmiento, J. and N. Gruber. 2006. *Ocean Biochemical Dynamics*, Princeton University Press, Princeton, NJ.

Sausen, R., I. Isaksen, V. Grewe et al. 2005. Aviation radiative forcing in 2000: An update on IPCC (1999). *Meteorologische Zeitschrift* 114: 555–561.

Schmit, J. October 14, 2009. Fuel-efficient cars help steady oil demand. *USA Today*: 3B.

Scientific American. 2010. Countdown to Copenhagen. *Sci. Am.*, online edition, http://www.flypme-dia.com/issues/sciam/01/ (accessed January 06, 2010).

Searchinger, T., R. Heimlich, R. Houghton et al. 2008. Use of US cropland for biofuels increases greenhouse gases through emissions from land use change. *Science* 319: 1238–1240.

Shakhova, N., I. Semiletov, A. Salyuk, V. Yusupov, D. Kosmach, and O. Gustaffson. 2010. Extensive methane venting to the atmosphere from sediments of the East Siberian Arctic Shelf (ESAS). *Science* 327: 1246–1250.

Shindel, D., B. Walter, and G. Faluvegi. 2004. Impacts of climate change on methane emissions from wetlands. *Geophys. Res. Lett.* 31: L21202, doi: 10.1029/2004GL021009.

Shinnar, R. and F. Citro. 2006. A roadmap to US decarbonization. *Science* 313: 1243–1244.

Socolow, R. 2006. Stabilization wedges: An elaboration of the concept. In *Avoiding Dangerous Climate Change*, Eds. H. Schellnhuber et al., Cambridge University Press, Cambridge, U.K., pp. 347–353.

Sokolow, R. 2002. CO_2 capture: The long term view, *NGO Focus Group Meeting*, Washington, DC.

Solomon, S., K. Rosenlof, R. Portmann et al. 2010. Contributions of stratospheric water vapor to decadal changes in the rate of global warming. *Science* 327: 1219–1223.

Sørensen, B. 1984. Energy storage. *Ann. Rev. Energy* 9: 1–29.

Spahni, R., J. Chapellaz, T. Stocker et al. 2005. Atmospheric methane and nitrous oxide of the late Pleistocene from Antarctic ice cores. *Science* 310: 1317–1321.

Spector, N. and B. Dodge. 1996. Removal of carbon dioxide from atmospheric air. *Trans. Am. Inst. Chem. Eng.* 42: 827–848.

Spreng, D., G. Marland, and A. Weinberg. 2007. CO_2 capture and storage: Another Faustian bargain. *Energy Policy* 35: 850–854.

Steinfeld, A. and A. Meier. 2004. Solar fuels and materials. *Encyclopedia of Energy*, Elsevier, Amsterdam, the Netherlands, Vol. 5, p. 623.

Steinfeld, A. and M. Schubnell. 1993. Optimum aperture size and operating temperature of a solar-cavity-receiver. *Solar Energy* 50: 19–25.

Stix, T. 1993. Removal chlorofluorocarbons from troposphere. *1993 IEEE International Conference on Plasma Science*, IEEE, Vancouver, British Columbia, Canada, p. 135.

Swanson, J. March 1, 2010. Letter to C&EN. *Chem. Eng. News* 88: 6.

Telegraph, U.K. 2009. "Cloud ship" scheme to deflect the Sun's rays is favorite to cut global warming, www.telegraph.co.uk (accessed August 7, 2009).

Teske, S. 2004. Solar thermal power 2020. *Renewable Energy World* 7: 120–123.

Tester, J., E. Drake, M. Driscoll, M. Golay, and W. Peters. 2005. *Sustainable Energy*, MIT Press, Cambridge, MA, 846pp.

Textor, C. et al. 2005. Analysis and quantification of the diversities of aerosol life cycles within AEROCOM. *Atmos. Chem. Phys. Discuss.* 5: 8331–8420.

The Week. 2009, October 2. Noted: 20.

Traynor, A. and R. Jensen. 2002. Direct solar reduction of CO_2 to fuel: First prototype results. *Ind. Eng. Chem. Res.* 41: 1935–1939.

Trenberth, K., J. Fasullo, and J. Kiehl. 2008. Earth's global energy budget. *Bull. Am. Meteorol. Soc.* 90: 311–323, doi: 10.1175/2008BAMS2634.1.

Tullo, A. September 21, 2009. Stepping on the gas. *Chem Eng News*: 2627.

Turley, C., J. Blackford, S. Widdicombe, D. Lowe, P. Nightingale, and A. Rees. 2006. Reviewing the impact of increased atmospheric CO_2 on oceanic pH and the marine ecosystem. In *Avoiding Dangerous Climate Change*, Eds. H. Schellnhuber et al., Cambridge University Press, Cambridge, U.K., pp. 347–353.

U.S. DOE Algal Biofuels. 2010. National Algal Biofuels Technology Roadmap, U.S. DOE, Biomass Program, May 2010, Washington, DC.

U.S. DOE Biomass. 2010. U.S. DOE Biomass program, National Algal Biofuels Technology Roadmap, DOE EERE, Washington, DC.

U.S. DOE EERE. 2010. www1.eere.energy.gov/biomass/pdfs/ibr_arra_uop.pdf (accessed July 2, 2010).

U.S. DOE GEN IV. 2002. U.S. DOE, A technology road map for Generation IV nuclear reactor systems, Washington, DC.

U.S. DOE LBL. 2010. U.S. DOE Lawrence Berkley Laboratory, http//:esd.lbl.gov/climate/ocean/fertilization.html (accessed January 15, 2010).

U.S. EIA. 2006. U.S. DOE Energy Information Agency, Annual Energy Review, Washington, DC.

USGS, 2008. *U.S. Geological Survey, Assessment of Gas Hydrate Resources on the North Slope Alaska, Fact Sheet 2008–3073*, USGS, Washington, DC.

van Lewen, J. and P. Smith. 2005. Nuclear power—The energy balance, www.stotmsmith.nl (accessed December 20, 2005).

Victor, D., M. Morgan, J. Apt, J. Steinbruner, and K. Ricke. 2009. The geoengineering option: A last resort against global warming, *Foreign Affairs*, March/April 2009, http://www.foreignaffairs.com/articles/64829/david-g-victor-m-granger-morgan-jay-apt-john-steinbruner-and-kat/the-geoengineering-option (accessed April 21, 2010).

WEO. 2009. *World Energy Outlook. International Energy Agency (IEA)*, IEA Publications, Paris, France, p. 691.

Wetzel, J. March 1, 2010. Letter to *Chem. Eng. News* 6.

Wickland, K., R. Striegl, J. Neff, and T. Sachs. 2005. Effects of permafrost melting on CO_2 and CH_4 exchange of a poorly drained black spruce lowland. *J. Geophys. Res.* 111: G02011, doi: 10.1029/2005JG000099.

Wigley, T. and S. Rapper. 2001. Interpretation of high projections for global mean warming. *Science* 293: 451–454.

Williams, D. October 12, 2009. Achieving renewable energy. *Chem. Eng. News* 87: 4–6.

Wilson, E. March 8, 2010. Methane from Arctic Ocean. *Chem. Eng. News* 88: 10.

Wind, S. 2009. DK greenroots getting to 350 ppm with renewable ammonia, http://www.dai-lykos.com/story/2009/7/4/750044/-DK-Greenroots-getting-to-350ppm-with-renewable-ammonia/ (accessed January 14, 2010).

Wire Staff, CNN 2010. World's largest offshore wind farm opens in UK. CNN.com, http://www.cnn.com/2010/world/europe/09/23/uk.largest.wind.farm/index.html?hpt=c2 (accessed September 23, 2010).

WNA. 2009. World Nuclear Association (WNA), Comparative carbon dioxide emissions from power generation: Education, www.world-nuclear.org/education/comparativeco2.html (accessed July 08, 2010).

Wu, B., R. Reddy, and R. Rogers. 2001. Novel ionic liquid thermal storage for solar thermal electric power plants. *Proceedings of Solar Forum*, Washington DC.

Yang, C. 2008. Hydrogen and electricity: Parallels, interactions, and convergence. *Int. J. Hydrogen Energy* 33: 1977–1994.

Zeman, F. 2007. Energy and material balance of CO_2 capture from air. *Environ. Sci. Technol.* 41: 7558–7563.

Zeman, F. and K. Lackner. 2004. Capturing carbon dioxide directly from atmosphere. *World Resour. Rev.* 16: 157–172.

Zeman, F., D. Keith. 2008. Carbon neutral hydrocarbons. *Philos. Transact. A Math. Phys. Eng. Sci.* 366: 3901–3918.

Zweibel, K. 2010. The impact of tellurium supply on cadmium telluride photovoltaics. *Science* 328: 699–701.

Zweibel, K., J. Mason, and V. Fthenakis. 2007. A solar grand plan. *Sci. Am.* December, http://www.sciam.com/article.cfm (accessed April 12, 2008).

2

Hydrogen Energy—Abundant, Efficient, Clean: A Debate Over the Energy-System-of-Change*

Carl-Jochen Winter

CONTENTS

---

* *Dedication*: This chapter has two dedications: it is meant to honor both T. Nejat Veziroğlu and the thousands of hydrogen scientists, engineers, entrepreneurs, and policy makers all over the world in whose heads and hands lies the transition to the hydrogen energy economy. It is dedicated to T. Nejat Veziroğlu, who, 36 years ago together with a handful of "hydrogen romantics," started the International Association for Hydrogen Energy (IAHE). This association has so far hosted 17 well-esteemed World Hydrogen Energy Conferences (WHEC) with up to 1000–2000 attendees each, taking place every 2 years on different continent. The 18th WHEC was recently held in 2010 in Essen, Germany. How admirable is Nejat's reputation as a scholar; his sense for continuity, his tireless attitude about never giving up; his inventive skill in establishing the World Hydrogen Energy Conventions scheduled for the odd years in between WHEC; his engagement as editor in chief of Elsevier's distinguished periodical the *International Journal of Hydrogen Energy* (IJHE); and, not least, his recent launch of the T. Nejat Veziroğlu Hydrogen Energy Trust promoting hydrogen energy science and technology and furthering their societal acceptance. This book is also dedicated to the thousands of hydrogen engineers, entrepreneurs, and policy makers around the world who never lose their conviction, their vigor, and their resilience, bringing to market abundant, exergetically efficient, and environmentally and climatically clean hydrogen energy and its technologies. It is up to you, colleagues, to make this happen! What you face is truly a long way to Tipperary. More than one generation will be occupied: Novel energy needs time!

    This work has been published in a special issue of the IJHE (2009). The shortened version of the article is presented here, courtesy of the IJHE.

Both secondary energies, electricity and hydrogen, have much in common: they are technology driven; they are produced from any available primary energy; once produced, they are environmentally and climatically clean over the entire length of their respective conversion chains, from production via storage and transport to dissemination and finally utilization; they are electrochemically interchangeable via electrolyses and fuel cells; they rely on each other, for example, when electrolyzers and liquefiers need electricity or when electricity-providing low-temperature fuel cells need hydrogen as their fuel; in cases of secondary energy transport over longer distances, they compete with each other; in combined fossil fuel cycles, they are produced in parallel exergetically and highly efficiently; hydrogen, in addition to electricity, helps exergizing the energy system and, thus, maximizing the available technical work.

Of course, there are dissimilarities: electricity transports information, hydrogen does not; hydrogen stores and transports energy, electricity does not (in macroeconomic terms). And the most obvious dissimilarity is their market presence, both in capacities and in availability: Electricity is globally ubiquitous (almost), while hydrogen energy is still used in only selected industrial areas and in much smaller capacities.

This chapter consists of 17 sections, 33 figures, 3 tables, and 2 Appendices and describes the up-and-coming hydrogen energy economy, its environmental and climatic relevance, its exergizing influence on the energy system in place, its effect on decarbonizing fossil-fueled power plants, and the introduction of the novel non-heat-engine-related electrochemical energy converter fuel cell in portable electronics, in stationary conversion, and in mobile applications. Hydrogen guarantees environmentally and climatically clean transportation on land, in air and space, and at sea. Hydrogen facilitates the electrification of vehicles with practically no range limits.

2.1 Summary Instead of an Introduction: Inevitably, … It's Hytime!

It takes about 50 years for a new idea to break through and become vogue; no one likes an intruder, particularly when he is upsetting the commonplace.

Hydrogen

- Is the lightest element in the periodic table of elements; its ordinal number is 1.
- Is the most abundant element in the universe.

- Is available on earth only in compounds; freeing hydrogen from these compounds needs energy. That, very briefly, is the root of hydrogen production within the hydrogen energy economy.

- Is considered the "forever fuel," since, like electricity, its secondary energy "running mate," it can be produced from any primary energy fuel such as coal, oil, natural gas, nuclear, all sorts of renewable energies, and from grid electricity. Certainly, because of its environmental and climatic cleanness, "solar hydrogen" (hydrogen made from renewable energies) is the ultima ratio. It is, however, not the precondition for building up the hydrogen energy economy. Hydrogen energy is so far the last missing addition to the continuously further developing energy mix—"until something better comes along" (Scott 2007).

- Is related to electricity, since both compete for the same primary energies; since they may be produced simultaneously with elevated exergetic efficiencies in combined cycles; since they are electrochemically interchangeable via electrolysis and fuel cell; and since hydrogen electrolysis and liquefaction rely on electricity, and electricity depends on hydrogen in low-temperature fuel cells.

- Is environmentally and climatically clean at its point-of-use, and clean over its entire energy conversion chain when produced from renewable electricity or from fossil fuels when carbon capture and storage (CCS) is included. Clean hydrogen energy is compatible with the Kyoto Protocol. Hydrogen energy and its technologies contribute to balancing the environmental and climatic off-balance utilization of the energy system in place.

- Is not oligopolizable, since a "hydrogen energy OPEC" is highly improbable; hydrogen's various primary fuels are obviously much more widely disseminated than OPEC's oil. They are spread over the entire globe, in contrast to the reservoirs of crude oil or natural gas that are concentrated in the "energy strategic ellipse" spreading from the Persian Gulf via Iran, Iraq, and central Asian states to as far as Siberia where the bulk of known fluid fossil fuels are located.

- Is technology driven, because along its complete conversion chain of hydrogen production, storage, transport, dissemination, and utilization, technologies are well understood and marketed or are on the verge of being marketed. Electrolytic production of hydrogen and its production from natural gas, or coal, or heavy fractions of oil are a day-to-day practice. Hydrogen-fueled fuel cells are compact; quiet; clean; and, as chemo-electric converters, and not Carnotian (Sadi Carnot 1796–1832) heat engines, highly efficient. Modularized, their unit capacity ranges from less than watts up to megawatts over more than six orders of magnitude with well-marketable temperature variations; depending on the cell type, the temperatures range from approximately 80°C up to ca. 900°C. They serve as long-life power packages in portable electronics, as stationary combined heat and power (CHP) facilities, as high-temperature topping modules in combined cycles, as prime movers in the electric drive train on board automobiles, and as a replacement for today's miserably inefficient mobile electrical generators. Storage is provided through high-pressure gas tanks, metal hydrides, or dewars containing energy dense cryogenic liquefied hydrogen.

Obviously, hydrogen energy policy is technology politics! With hydrogen technologies, particularly energy import intensive, industrialized nations get a welcome quasi-indigenous energy from the knowledge of their scientists and the skills of their engineers and craftsmen. More energy services from less primary energy is the creed!

- Is inherently securely safe, because long-term and unforeseeable risks are inexistent, since hydrogen energy is without radiotoxicities or radioactivity and its contribution to the anthropogenic greenhouse effect is very small, if any. Since on principle, however, absolutely safe conditions in technical systems are impossible, anywhere and under any condition, hydrogen's specific safety risks need to be thoroughly addressed. Safety-related incidents have been experienced, for instance, in the space launch industry, which is for the time being the only industrial branch that has been using hydrogen energy in large quantities for more than half a century. But it has never had an accident that was *causally* introduced by hydrogen. No less experienced are refineries, and the merchant gas and hydrogen chemistry industries, which are familiar with handling securely and safely very large amounts of hydrogen in their day-to-day practice. There is no arguing with experience and success: addressing hydrogen safety concerns will help hydrogen succeed!

- Is produced worldwide (approximately 50 million tons per annum) and is worth nearly U.S. $280 billion (2006) with an annual addition of approximately 10%, in steam methane reforming (SMR), in partial oxidation of heavy hydrocarbon fractions, and through coal or biomass gasification, or in large electrolyzers. Major hydrogen users are the space flight industry and the electronics industry, glass and food manufacturers, and electrical equipment companies. By far the largest amounts of hydrogen are produced and utilized captively for methanol or ammonia syntheses, and in the refining industry for hydrogen treatment of heavy crude as well as for the production of reformulated gasoline and the de-sulfurization of middle distillate diesel fuel. Worldwide, the amount of captive hydrogen is about seven times that of merchant hydrogen; merchant hydrogen consists of gaseous and liquid hydrogen where the amount of gaseous hydrogen is about six times that of liquid hydrogen.

- Is ubiquitous, since hydrogen energy perpetuates continuously, cleanly, safely, and securely the established world energy trade system; no continent and no nation is excluded as hydrogen producer, hydrogen trader, or hydrogen user. New hydrogen producers and traders will join the club in those areas of the world where so far huge amounts of renewable sources are lying fallow because they cannot be stored or transported. Here hydrogen as a chemical energy carrier makes both possible. Other new hydrogen producers will emerge among the fossil fuel producers of the world when they start decarbonizing and, thus, hydrogenizing their products and start shipping hydrogen instead of fossil fuels. Herewith a switch is predicted from the energy buyer's obligation to clean up the fuels he purchases to the energy seller's cleaning up the fuels prior to selling them. So far, the energy sellers have been freed from any such obligations; traditionally, they simply ship energy raw materials, including their pollutants and potential greenhouse gases.

- Is a clean, secondary energy carrier enabling the "old" energies such as coal, oil, gas, and nuclear fission to join the per se clean renewable energies and electricity, and to continue playing their role in a future environmentally and climatically clean sustainable energy world. Today, electricity and steel keep coal alive; tomorrow hydrogen will help keep clean coal alive. The conversion of coal into hydrogen energy enables coal's return to the transport and household sectors, which it had to leave with the advent of light oil and gas: truly, a renaissance!

- Is sustainable economically, environmentally, and climatically, as well as societally, and from a reversibility standpoint: economically, because of the global ubiquity of the primary renewable fuels from which hydrogen is produced; environmentally and climatically, since electrolytic hydrogen from renewable or nuclear electricity comes from water (from the earth's inventory), and after recombination of hydrogen and oxygen is given back to that inventory; and for hydrogen from fossil fuels, the inclusion of securely safe handling, sequestering, and storing away of carbon is an inevitable necessity. Finally, on principle, energy irreversibilities are not sustainable. Here, striving for much higher exergy efficiencies comes into play: the higher the exergy efficiencies, the lower the conversional irreversibilities are. Turning off the Carnotian heat-engine-related energy system is due; it produces far too much heat of the false temperature at the wrong location where no user asks for it.

- Is on track, because energy sustainability without hydrogen energy is irrational, although there are still many milestones ahead. Energy sustainability is not a momentary value; rather, the road is the destination! Many hydrogen milestones along the road lead to the hydrogen energy economy, and, no doubt, similar to any of the past novel energies, many stumbling blocks will have to be removed. Paving the HYway means more clean energy services from less polluting and less climatically harmful primary fuel! When weighed on the energy sustainability scale, lightweight hydrogen is a heavyweight!

- Is urgent, because we have learned that novel energies need long periods of time—sometimes running into many decades up to half centuries for the irrevocable establishment of a novel energy on the market—and hydrogen energy is no different. Consequently, it's HYtime—it is high time to start the implementation of the hydrogen energy economy and see it through. For the innovation of a novel addition to the global energy system, it so appears that it is almost always too late; humans tend to ignore squeaking wheels, they stay with what they are accustomed to and seldom abandon this attitude without being compelled to do so. Three of these compulsions are clearly foreseeable: (1) the production peaking of the fluid fossil fuels, oil, and gas in the decades to come, and as a consequence, (2) increasing oligopolization of suppliers and skyrocketing prices, and (3) harsh environmental and, in particular climate change challenges.

- Decentralizes the energy system; historically, national energy conversion chains start with energy production at their front end, that is, the conversion of primary energy raw materials into primary and secondary energies, which, after transport, storage, and distribution, are utilized at the back end of the chain. Potentially, through hydrogen energy and fuel cells in stationary or mobile applications, another production link is added to the national chain at its back end. Both ends get the chance to become production ends and to start welcoming the competition. The so far untapped production potential at the back end is immense; in Germany, nearly two-thirds of the national end energy is required for transportation and buildings!

  According to the United Nations' predictions, half of the world's population will be living in urban areas by 2008/2009; by 2050, approximately 70% will be dwelling in megacities that have more (in cases much more) than 10 million inhabitants. As urbanization proceeds, hydrogen and its technologies can help free the agglomerations of fumes, smog, noise, pollutants, and greenhouse gases.

- Exergizes the energy system: whenever energy is converted (produced, handled, stored, transported, disseminated, utilized, etc.) it is split into two parts: energy = exergy + anergy. Exergy is the maximum of the available technical work extracted from energy. By definition, exergy can be converted into any other form of energy, but anergy cannot. The classical Carnotian energy system, followed for over 200 years, produces very high conversional irreversibilities and, thus, too much heat of the false temperature at the wrong place where no potential user asks for it, in power plants, in residential or industrial heating systems, and in automobile engines. On the other hand, hydrogen-supplied fuel cells are exergetically highly efficient, always generating firsthand electricity (=pure exergy), and the exergy content of the remaining heat at the fuel cell's specific temperature, in many cases, exactly meets the requirements of industrial or residential heat demands. It is the irreversibilities of combustion, of heat transfer and energy flow through the installation, which cause exergy destruction and exergy losses. Bringing them down to lower levels in Carnotian systems is welcome, but is becoming increasingly difficult, because only tiny development increments in high-temperature materials, in the physical chemistry of combustion, and in fluid dynamics are slowly asymptotically approaching higher exergy efficiencies; the system immanence tends to its final end. Now, what is urgently needed is a system change to chemo-electric or solar–electric novel systems like hydrogen-supplied fuel cells or solar photovoltaics, which are not Carnotian heat engines. Elevating source-to-sink exergy efficiencies means exergizing energy.

- Is not really anything new: Antoine Lavoisier (1743–1794) and Henry Cavendish (1731–1810) were the first to mention hydrogen in the literature. In 1766, H. Cavendish spoke of "inflammable air" and A. Lavoisier named it "hydrogen" in 1787. Half a century later, around 1839, two friends, the Welshman William Grove (1811–1896) and the German Christian Friedrich Schönbein (1799–1868), published their findings on hydrogen-supplied fuel cells. The fact that between one and two centuries had to pass before hydrogen energy arrived at the verge of taking its inherent place in the world energy market is not a singular phenomenon. Other novel energies and their converters also needed (and will need) time. What the energy world faces is how to construct a bridge from the well-established and economically lucrative *hydrogen economy*, which utilizes hydrogen as a commodity to the *hydrogen energy economy* and its utilization of hydrogen as an energy carrier. In that way, nothing of what has been learned over the centuries-long development of the hydrogen economy is lost; more on the energy aspect is to be added prior to the full establishment of the hydrogen energy economy. The petroleum and natural gas industries, the auto makers, the technical gas industry, and the energy utilities have begun to commit themselves to hydrogen energy; addressing concerns helps both hydrogen and the industry to succeed. Hydrogen energy starts to become a reality!

- Is so far the final addition to energy diversity; it tends to close the anthropogenic energy cycle from the renewable energies of the first solar civilization to coal, to oil, and gas and nuclear fission back to renewable energies, now of the second solar civilization, and finally to hydrogen energy's taking care of renewable energies' storage and transport requirements. Along that line, the anthropogenic energy cycle needed around two centuries and a half. Never in this relatively short period of time did humans use only one form of energy; never did a novel energy fully replace its predecessors and the ever-growing energy demand need them all. After coal, oil, natural gas, and nuclear fission beginning in the eighteenth and continuing into the nineteenth and twentieth centuries, in addition to striving for energy and particularly exergy efficiencies as well as the utilization of renewable energies, the twenty-first century is on the verge of becoming the century of hydrogen energy and its technologies. Hydrogen is the oil and gas of the twentieth and the coal of the nineteenth century. Steadily getting away from oil and natural gas and building up the hydrogen energy economy is the task. We are well on our way and the growing atomic hydrogen/carbon ratio H/C makes it obvious. It tends to a dual triangulation from high carbon to low carbon to no carbon, and, as a consequence, from (almost) no hydrogen to low hydrogen to finally high hydrogen:

Coal : oil : natural gas: hydrogen $= < 1 : 2 : 4 : \infty$

Today, two-thirds of all atoms in fossil fuels burnt are already hydrogen atoms; the trend points to ever-higher hydrogen numbers.

- Dematerializes the energy system; it is the conversion of energy matter that causes environmental and climatic harm and not the conversion of energy; that is why dematerialization of energy is so important. Since the specific atomic masses of hydrogen and carbon are 1 and 12 (g/mol), respectively, the ongoing process of shifting from the carbon-rich/hydrogen-poor hydrocarbon energy economy to the future hydrogen-rich/carbon-poor hydrogen energy economy is accompanied by a continuous dematerialization process. Specifically, energy gets lighter and lighter over time. The "energy era of light" may be the appropriate label for the twenty-first century, because

  a. Efficiencies make more energy services from less weighty primary energy raw materials; renewable energies have no operational primary energy raw materials per se; and hydrogen is the *lightest* element in the periodic table of elements.

  b. Most renewable energies utilize directly or indirectly the *light* of the sun.

  c. Their and hydrogen's utilization *lightens* the burden on the environment and climate.

  d. All of the aforementioned shed *light* on what will become the criterion for the twenty-first century of energy—energy sustainability.

  One thing should not be forgotten, though. In some cases, the flip side of lightness is bulk!

- Helps triggering the next industrial revolution:

  a. The first industrial revolution was triggered by the steam engine.

  b. The second industrial revolution by electrification.

  c. The third industrial revolution is triggered not by only one single technology, but rather a whole range of technologies (see the "Epilogue"), such as energy decentralization, decarbonization, hydrogenation, dematerialization, low weight to weightlessness, biotechnology, information and communication technology, micro-miniaturization, nanotechnology, etc.

- Asks for good global energy governance; it seems that nineteenth and twentieth centuries' thinking and acting in terms of primary energy raw materials is still dominating the world energy scene. How many tons of coal, barrels of oil, cubic meters of gas, or kilograms of uranium are to be traded; at what cost; and with what relevance to the environment and climate change?—these questions still appear in the foreground of argumentations. The twenty-first century, however, sees more and more heavy energy–consuming countries unalterably dependent on fewer and fewer energy suppliers; national energy self-sufficiency is increasingly possible for only a small number of countries. The power of supply oligopolies grows. National energies lose, global energy wins. In this situation, the countries of the world tend to do two things: they revitalize their energy efficiency technologies in order to slow down the need for (imported) primary energy feedstock; and they tend toward good energy governance for the benefit of both energy-rich, though technology-poor, and energy-poor, but technology-rich, nations. Efficient technologies increasingly grow into the role of "national energies," which is a part of the twenty-first century energy thinking and acting. Benjamin Franklin (1706–1790), foreign secretary of the newly independent New England colonies, stated already three centuries ago, "We are bad farmers, because we have too much land." Paraphrased, we get "We are bad energy engineers, because we have too much energy"—too much cheap energy feedstock, stated more precisely, since cheap energy feedstock is the most elusive enemy of energy security!—How prospective Franklin's thinking was!

- Adds value to the energy economy: J. A. Schumpeter's (1883–1950) "Innovations are the driving forces of economic growth" was valid, is valid, and will be valid when hydrogen energy adds value to the energy economy through

    a. Undoubted environmental and climate change benefits

    b. Its avoidance of irreversibilities and, thus, its exergizing ability, providing more technical work from less primary energy

    c. Slow down of physically unavoidable energy value degradation=entropy increase

    d. Its activating "national energy" in the form of energy science and engineering skill, thereby enabling nations to compensate for the imponderabilities of foreign energy markets

    e. The reduction of import dependency and, thus, the avoidance of the price dictates importing countries are suffering under

    f. Stimulating hydrogen technology development and export; "ecological reasoning not only asks for avoidance and renunciation, but also and primarily for unparalleled technology development"!

    g. Helping to decarbonize fossil fuels and, thus, furthering their use until their point of depletion

    h. Switching from the global fossil fuel trade to a global hydrogen trade by decarbonizing fossil fuels and removing pollutants already on the energy seller's side

    i. Decentralizing the national energy scheme, thereby activating so far dormant virtual distributed energy potentials downstream where people live

    j. Making tradable, huge, so far untapped renewable potentials and thereby utilizing the only closed energy material cycle "solar and water from the earth's inventory to hydrogen and water, after hydrogen/oxygen recombination returned to that inventory"; all other cycles are materially open cycles

- k. Not least, professionalization of fallow lying energy potentials at the back end of national conversion chains
- Is not a panacea, but it is considered to be the still lacking cornerstone of anthropogenic sustainable energy building. Economics and ecology are the drivers; hydrogen energy is the enabler. Without hydrogen, expecting energy sustainability is irrational. Hydrogen energy technologies are the choice; hydrogen does not take the energy system to heaven, but it saves it from an environmental and climatic disaster.
- A priori helps to further energy awareness:
  - a. With or without hydrogen, a switch from fossil energies to renewable energies is due; but with hydrogen, huge amounts of renewable sources otherwise lacking storability and transportability can enter the global energy trade
  - b. With or without hydrogen, reduction of irreversibilities in Carnotian energy conversion is due; but with hydrogen, the switch to exergetically efficient combined cycles is facilitated
  - c. With or without hydrogen, decarbonizing fossil fuels is due; but with hydrogen, in combined cycles the exergy efficiency rises significantly
  - d. One item requires hydrogen indispensably; no low-to-medium-temperature fuel cells without hydrogen, either pure hydrogen or hydrogen reformate
- Energy is the forthcoming "terrestrial man to the moon" project which indicates what we are facing; other than the historical man to the moon project, which was the endeavor of one nation, the project of the twenty-first century has to become a worldwide endeavor with no nation excluded, either with its hydrogen production experience, or its hydrogen storage, transport and distribution infrastructure, or its engagement in early adoption of hydrogen energy. A few examples: all nations experienced in the space launch industry are predestined to take over the market sectors of handling, storing, transporting, or combusting large amounts of hydrogen; nations involved in portable electronics will be (and already are) switching to hydrogen-supplied fuel cells to replace low-longevity batteries; or nations with a long history in coal production now face the challenge of hydrogen-supported decarbonization and carbon capture and storage (CCS) for the benefit of the environment and climate; and finally, those nations in the world blessed with a large capacity of renewable energy potentials (hydro, wind, solar, ocean, and others) will sooner or later face the necessity of adding the storable and transportable chemical energy carrier hydrogen in order to enable their renewable sources to contribute to the global energy trade.
- Is the core argument in the Centennial Memorandum of the International Association for Hydrogen Energy (IAHE), submitted to the heads of state of the G8 summits of 2007 in Heiligendamm, Germany, and 2009 in Italy, asking them to give hydrogen energy a top priority in their national and international considerations. The memorandum reads in part: "Hydrogen energy: the abundant clean energy for humankind as a means of mitigating anthropogenic climate change, avoiding environmental challenges, and decelerating the world's ongoing oligopolization of conventional energy raw materials is the permanent solution to the upcoming energy and climate change catastrophe."

2.2 Anthropogenic Climate Change and Hydrogen Energy

It has only been 20 or 30 years since anthropogenic influence on the atmospheric greenhouse effect irretrievably began governing human thinking and acting. Energy, industry, transportation, agriculture, trade, buildings, and, not least, the behavior and attitude of individuals and society and the decisions of policy makers vis-à-vis environment and climate—no area of potential greenhouse interaction is being excluded. From the start of industrialization around 1800, the atmospheric CO_2 concentration has grown from 280 to 380 ppmv (2007) (ppm, parts per million; v, volume). Annually, 36 billion tons of emitted CO_2 (2007) are anthropogenic, of which 29 billion come from fossil fuel combustion and industrial processes; the remaining 7 billion are the consequence of deforestation and agricultural industries. The major emitters are the industrialized countries, with the United States, China, Russia, Japan, India, and Germany at the top, in that order. A rough estimate says that every additionally emitted anthropogenic 30 billion tons of CO_2 raises the atmospheric concentration by ca. 2 ppm. To be added are other greenhouse gases (GHGs) such as CH_4 and N_2O, whose global warming potentials are 21 kg CO_2e/1 kg CH_4 and 310 kg CO_2e/1 kg N_2O, respectively. Other GHGs are diverse fluorine compounds.

The Intergovernmental Panel on Climate Change (IPCC) stated in its Synthesis Report 2007 that "warming of the climate system is unequivocal," and that "eleven of the last twelve years (1995–2006) rank among the twelve warmest years in the instrumental record of global surface temperature (since 1850)" and "rising of the sea level is consistent with warming"; so are the decreases in snow and ice cover, an increase in precipitation in certain areas and a decline in others, the growing intensity of tropical cyclones in the North Atlantic, and increasing ocean acidification due to CO_2 uptake. Figure 2.1 documents an increase from 13.5°C to 14.5°C in global average surface temperature between 1850 and 2000, an increase from −150 to +50 mm in global average sea level between 1870 and 2000, and a decrease from around 37 to 35 million km² in the Northern Hemisphere snow cover between 1920 and 2000.

The IPCC report further states that "global GHG emissions due to human activities have grown since preindustrial times, with an increase of 70% between 1970 and 2004" (Figure 2.2), and "global atmospheric concentrations of CO_2, methane (CH_4) and nitrous oxide (N_2O) have increased markedly as a result of human activities since 1750 and now far exceed preindustrial values determined from ice cores spanning many thousand years." The report concludes that "most of the observed increase in global average temperature since the mid-twentieth century is very likely due to the observed increase in anthropogenic GHG concentrations." The climate process is highly nonlinear.

Figure 2.3 brings best estimates (dots) and likely ranges (bars) of warming assessed for six different scenarios for 2090–2099 relative to 1980–1999. The best estimates indicate temperature rises between 1.8°C and 4°C, with the bars extending over a wide range from around 1°C to 6.4°C. What we learn is that the temperature rise over one century can be dramatic, and that there is still room for reducing the uncertainties of scenarios. Even more dramatic is the message that "anthropogenic warming and sea level rise continue for centuries due to the time scales associated with climate processes and feedbacks, even if GHG concentrations were to be stabilized." What has been concluded over time and again is once again confirmed here. Starting immediately with aggressive mitigation is imperative, although it will not be possible to compensate for the past errors.

Figure 2.4 shows the economic mitigation potentials for seven sectors by 2030 in Gt CO_2e/a over three emission trading certificate values, <20, <50, and <100 U.S. $/t CO_2e.

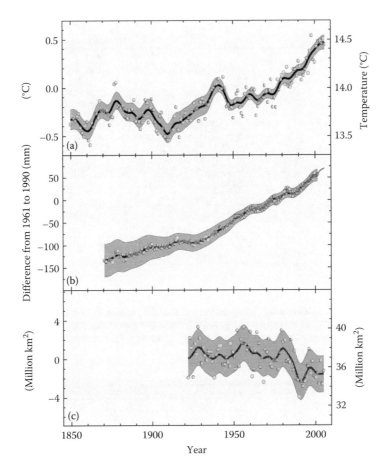

FIGURE 2.1
Changes in (a) global average temperature, (b) global average sea level, and (c) northern hemisphere snow cover. (Adapted from IPCC, *Climate Change 2007*, Metz, B. et al. (Eds.), Intergovernmental Panel on Climate Change, Cambridge University Press, Cambridge, U.K., 2007. With permission.)

The evidence? First, the mitigation potentials of buildings are the highest, followed by agriculture, industry, energy supply, and the other sectors; second, the potentials for non-OECD/EIT countries (EIT economies in transition) are (much) higher than those for OECD countries; and third, it is surprising that the influence of certificate pricing for the sectors transport and waste is almost negligible. If we sum up the <100 U.S. $/t CO_2e potentials of all seven sectors, we end up with some 23 Gt/a CO_2e, and for <20 U.S. $ we are still at 13–14 billion tons. The IPCC comments that "modeling studies show that global carbon price increases from 20 U.S. $ to 80 U.S. $/t CO_2e by 2030 are consistent with stabilization at around 550 ppm CO_2e by 2100." So, the mitigation potential seems real; the time needed, however, is very long. Stabilization cannot be achieved in less than almost one century, and not (only) by deployment of a portfolio of technologies that are either currently available or expected to be commercialized in coming decades. Let us now consider the economic and technology consequences of stabilization in more detail.

Nicholas Stern, former chief economist of the World Bank, published in 2006 his review "The Economics of Climate Change" with these major findings: "Climate change could have very serious impacts on growth and development"; "the costs of stabilizing the climate

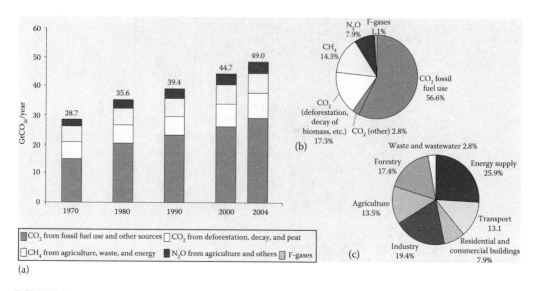

(a)

FIGURE 2.2
Global anthropogenic GHG emissions: (a) global anthropogenic GHG trends, 1970–2004, (b) global anthropogenic GHG emissions in 2004, and (c) GHG emissions by sector in 2004. (Adapted from IPCC, *Climate Change 2007*, Metz, B. et al. (Eds.), Intergovernmental Panel on Climate Change, Cambridge University Press, Cambridge, U.K., 2007. With permission.)

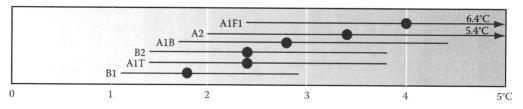

FIGURE 2.3
Warming by 2090–2099 relative to 1980–1999 for non-mitigation scenarios. (Adapted from IPCC, *Climate Change 2007*, Metz, B. et al. (Eds.), Intergovernmental Panel on Climate Change, Cambridge University Press, Cambridge, U.K., 2007. With permission.)

are significant but manageable"; "delay would be dangerous and much more costly"; and "the benefits of strong and early action far outweigh the economic costs of not acting." In detail, the review estimates that if we do not act, the costs and risks of climate change will be equivalent to losing at least 5% (and under certain circumstances 20% or more) of global gross domestic product (GDP) each year, now and forever. On the other side, the cost of measures to reduce emissions to avoid the worst impacts can be limited to one fifth of that amount, that is, to 1% of global GDP each year. Stern claims that the climate change impacts can be substantially reduced if atmospheric GHG concentrations can be stabilized between 450 and 550 ppm CO_2e compared to today's 430 ppm CO_2e with an unaltered yearly rise of 2 ppm (equivalent "e" indicates the climate change effect of the non-CO_2 emissions, CH_4, N_2O, and fluorine compounds compared to CO_2 emissions), of which 380 ppm stem from CO_2 as that GHG with the present comparatively major impact. About 450–550 ppm by 2050, that is, a yearly addition of 2 ppm, is considered "safe" by the political class, the resulting billions for emission mitigation "allowable." Consequently, stabilization in the range of 450–550 ppm by 2050 requires cutting today's global emissions by at least 25%.

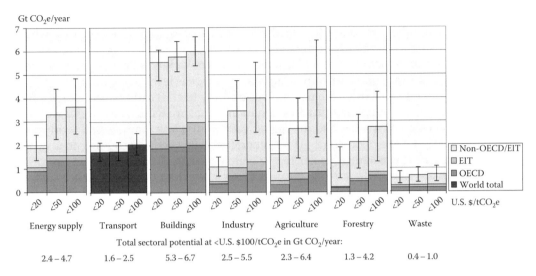

Gt CO$_2$e/year

| Non-OECD/EIT |
| EIT |
| OECD |
| World total |

U.S. \$/tCO$_2$e

Energy supply   Transport   Buildings   Industry   Agriculture   Forestry   Waste

Total sectoral potential at <U.S. \$100/tCO$_2$e in Gt CO$_2$/year:

2.4 – 4.7   1.6 – 2.5   5.3 – 6.7   2.5 – 5.5   2.3 – 6.4   1.3 – 4.2   0.4 – 1.0

FIGURE 2.4
Economic mitigation potentials by sector in 2030 estimated from bottom-up studies. (Adapted from IPCC, *Climate Change 2007*, Metz, B. et al. (Eds.), Intergovernmental Panel on Climate Change, Cambridge University Press, Cambridge, U.K., 2007. With permission.)

This means that the dominant emitters, the industrialized countries, have to reduce their emissions by 60%–80% by 2050, which is identical with the recommendations delivered already in the early 1990s to the German government and parliament, respectively, by the Enquête Commission of the German Bundestag, "Protection of the Earth's Atmosphere"; the recommendations were unanimously agreed upon by both sides of the aisle.

Bluntly, N. Stern speaks of climate change as "the greatest market failure the world has ever seen" interacting with other "market imperfections." He compares the failure with the global challenges of the world wars of the twentieth century and the world recession of the 1920s and 1930s. To fix these imperfections and safeguard them from market failure requires a whole range of countermeasures. So far, however, very little has been and is being done to reduce emissions, although, according to the United Nation's Kyoto Protocol, industrialized countries (still excluding top polluters such as the United States and economies in transition like India and China with their rapid industrialization processes and, thus, their rapid approach to the emission levels of industrialized countries) have agreed upon a reduction of their 1990 emissions by 5.2% in the period between 2008 and 2012. On the contrary, on an average, the world has emitted more GHGs instead of less.

Notwithstanding, let us try to see what reduction possibilities can meet the aforementioned challenges. Seven steps can be envisioned: first, stop deforestation and alter industrial processes in agriculture; second, establish worldwide CO$_2$e emission certificate trading as an economic means of mitigation; third, raise energy efficiencies and in particular exergy efficiencies along all links of energy conversion chains, that is, reduce the irreversibilities in established Carnotian systems and switch to combined cycles with fewer irreversibilities, such as electricity plus heat, electricity plus hydrogen, electricity plus chemical commodities, all of them of maximum technical work (=exergy) extracted from energy; fourth, generate electricity from nonfossil renewable and nuclear sources; fifth, if generated from fossil sources, decarbonize and thus hydrogenize and dematerialize the conversion processes; sixth, introduce hydrogen technologies into the end use sectors where exergetically miserable on-site boilers in buildings and engines in autos need to be replaced by exergetically efficient hydrogen-fueled fuel cells (FCs) or hydrogen-adapted and exergetically optimized ICEs; seventh, cap other emissions like methane, nitric oxides, and fluorine gases, which in part are still small, but rapidly rising CO_2 equivalences; in short, be or at least become conscious of the counteracting ability of innovative technologies and make policy makers and entrepreneurs aware of the unequivocal; energy policy is technology politics!

We have to consider an extraordinary catalogue of low to zero carbon technologies that are afoot, or in labs, or under long-term investigation. Afoot are wind converters approaching unit capacities $\leq 10\,MW$ each and in August 2008 a little less than $100\,GW$ worldwide in total; further afoot are photovoltaic (PV) generators whose efficiencies are on a steady upward trend and approaching, according to cell type, 12%–20%, and in concentrating devices even more than 20%; further, concentrating solar thermal (SOT) power stations with some hundreds of megawatts on line or planned with their appreciable efficiencies, which are much higher than those of even the best PV generators, although they need direct solar light whereas the PV generators work with both direct and diffuse light; and, heat-producing, gasifying, and liquefying biomass facilities, and combined cycles as regular market products in the portfolios of the energy utilities industry.

Technologies in the research and development labs are clean FCs of all sorts for simultaneous production of electricity and heat for portable, stationary, and mobile applications; hydrogen production from high-temperature nuclear heat; hydrogen production from coal, including capture of CO_2 and its secured long-term storage, safely, reliably, and durably; investigations of the geophysical, geochemical, and geobiological behavior of CO_2 underneath the earth's surface or the sea floor.

And finally, long-term investigations deal with metal hydrides of practically applicable energy percentages per weight of some 5–8 wt%; hydrogen in aviation; hydrogen utilization and transport at sea; the combination of liquid hydrogen and high voltage direct current (HVDC) electricity in "supergrids" where hydrogen has two functions, cooling the HVDC line and delivering energy. A thorough collection of hydrogen technologies—today, tomorrow, and later—can be found in Section 2.8.

In a paper by R. H. Sokolow and St. W. Pacala (see References), the overall technological challenge of decarbonization is described. The authors imagine over the next 50 years a triangle between business-as-usual (BAU) emissions [billion tons carbon/year] and stabilization of the 2006 emissions (Figure 2.5), divided into seven "wedges" (1 wedge is equivalent to 1 billion tons of carbon, 7 wedges correspond with today's approximately 7 billion tons of anthropogenic carbon emitted), which have to be reduced over the next 50 years if emissions are to be stabilized.

Figure 2.6 shows 15 wedges (and an open one as an indication of potential future wedges to come) of ways to mitigate emissions, such as efficiency and conservation, power generation, carbon capture and storage (CCS), renewable sources, agriculture, and forestry. The challenges are almost unimaginable! A few examples only: build 1600 new coal-fired power plants of 1000 MW each; increase 40-fold today's wind power potential; operate the 2 billion autos expected by 2050 at 4 L/100 km; add CCS to power plants and generate hydrogen for 1.5 billion cars. Truly, the challenge adopts the character of a marathon, by all means not a sprint! (Ogden 1999).

To conclude this section: Of course, technologies are not the only mitigation means; the behavior of individuals and societies vis-à-vis climate change is also extremely important. Yet, building up the climate consciousness of the public, of industry representatives, and policy makers is a major goal that has not been achieved so far.

The technologies aim at two targets. (1) They raise the energy efficiencies of the established Carnotian conversion system and switch to higher exergy efficient combined cycles; they

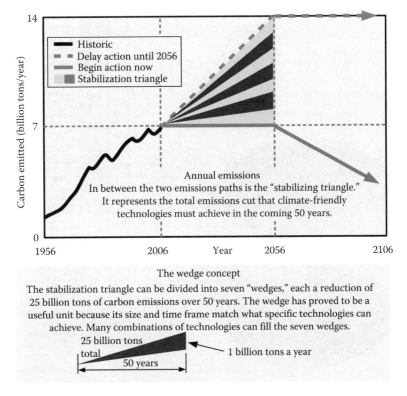

The wedge concept

The stabilization triangle can be divided into seven "wedges," each a reduction of 25 billion tons of carbon emissions over 50 years. The wedge has proved to be a useful unit because its size and time frame match what specific technologies can achieve. Many combinations of technologies can fill the seven wedges.

FIGURE 2.5
Annual carbon emission, 1956–2006 historic, 2006–2106 prospective. (Modified from Sokolow, R.H. and Pacala, St. W., *Sci. Am.*, 9, 28, 2006. With permission.)

produce more energy services from less primary energy raw material; what is not needed to provide more services does not interfere with the greenhouse! And (2), they replace heat engine–related Carnotian conversion technologies with their limited further incremental efficiency potential with hydrogen-supported combined cycles of inherently higher exergy efficiencies. Maximizing an energy system's exergy means aiming at extracting the maximum of technical work from energy and minimizing the interference with the greenhouse.

2.3 Anthropogenic Energy History and Hydrogen Energy

Up until the late eighteenth century, humans exclusively utilized renewable energies of the first solar civilization. Wind blew into the rotors of windmills or the sails of ocean-going ships, solar irradiance helped field crops grow, wood and peat warmed homes, and running water turned waterwheels. Modern energy history covers a rather short period of time, not much more than 200–250 years. It began with the opening of the first industrial coal mine in England in the second half of the eighteenth century: the industrialization of the world began.

Figure 2.7 shows the energy triangle of the primary energy history of the last almost two centuries. All three types of primary energy (raw materials) or feedstock are depicted as isoshares: on the left, coal; on the right, the two fluid hydrocarbons oil and gas; and at

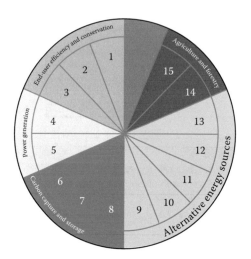

1	End-user efficiency and conservation	Increase fuel economy of two billion cars from 30 to 60 mpg
2		Drive two billion cars not 10,000 but 5000 miles a year (at 30 mpg)
3		Cut electricity use in homes, offices and stores by 25%
4	Power generation	Raise efficiency at 1600 large coal-fired plants from 40% to 60%
5		Replace 1400 large coal-fired plants with gas-fired plants
6	Carbon capture and storage	Install CCS at 800 large coal-fired power plants
7		Install CCS at coal plants that produce *hydrogen* for 1.5 billion vehicles
8		Install CCS at coal-to-syngas plants
9	Alternative energy sources	Add twice todays' nuclear output to displace coal
10		Increase wind power 40-fold to displace coal
11		Increase solar power 700-fold to displace coal
12		Increase wind power 80-fold to make *hydrogen* for cars
13		Drive two billion cars on ethanol, using one sixth of world cropland
14	Agriculture and forestry	Stop all deforestation
15		Expand conservation tillage to 100% of cropland
16		

FIGURE 2.6

Carbon emission saving potentials. (From Sokolow, R.H. and Pacala, St. W., *Sci. Am.*, 9, 28, 2006.). Notes: Wedge (1) Wordfleet: size in 2056 could well be 2 billion cars. Assume they average 10,000 miles/a. Wedge (4) "Large" is 1 GW capacity. Plants run 90% of the time. Wedge (5) Here and below, assume coal plants run at 90% of the time at 50% efficiency. Present coal power output is equivalent to 800 such plants. Wedge (6) Assume 90% of CO_2 is captured. Wedge (7) Assume a car (10,000 miles/year, 60 miles per gal equivalent) requires 170 kg of H_2 a year. Wedge (8) Assume 30 million barrels of synfuels a day, about a third of today's total oil production. Assume half of the carbon originally in the coal is captured. Wedge (10,11) Assume wind and solar produce, on average, 30% of peak power. Thus, replace 2100 GW of 90%-time coal power with 2100 GW(peak) wind or solar plus 1400 GW load-following coal power, for net displacement of 700 GW. Wedge (13) Assume 60 mpg cars. 10,000 miles/year, biomass yield of 15 t a hectare, and negligible fossil fuel inputs. World cropland is 1500 million hectares. Wedge (14) Carbon emissions from deforestation are currently about two billion tons/year. Assume that by 2056 the rate falls by half in the business-as-usual projection and to zero in the flat path. (Modified from Sokolow, R.H. and Pacala, St. W., *Sci. Am.*, 9, 28, 2006. With permission.)

the bottom, the two operationally carbon-free renewable and nuclear energies. The thick dark line is history; it begins outside the right lower corner prior to the turn of the nineteenth century when oil and gas were not yet in use and the renewable energies of the first solar civilization were consecutively replaced by growing amounts of coal. In the later part of the nineteenth and in the twentieth century, oil came into use and helped to further replace the renewable energies, and also began to replace coal. A sharp turn from the nineteenth to the twentieth century in the lower left corner marks this point in history. From then onward through the entire twentieth century, the operationally zero-carbon energies, consisting of a regrowth in the amount of renewable energies, now of the second solar civilization, and, starting in the middle of the century, the appearance of nuclear fission, remained almost constant at ca. 15%. Coal shrank to today's ca. 20%, and oil and gas climbed up to 60%: This is the situation humans were in at the turn of the twenty-first century. Now, where do we go next from here?

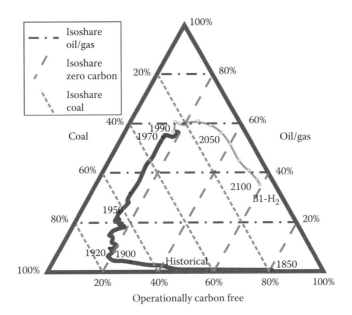

FIGURE 2.7
The energy triangle. (From Barreto, L. et al., *Int. J. Hydrogen Energy*, 28, 267, 2003. With permission.)

One thing seems unavoidable: Experts state that the number of people on earth, some 6.7 billion in 2008, is expected to further grow continuously, though perhaps with a slightly smaller gradient over time, to 9 billion or even more by 2050. For the time being, 80 million, approximately the population of Germany, are added year by year. This growth, combined with the increasing demands of newly industrializing countries, asks for ever more energy! One response out of a whole collection of possible responses (which are not depicted in Figure 2.7) could be what is called "BAU," which means more oil and gas, approximately today's contribution of coal, more nuclear energy, and a little more renewable energy. Since there is not too much room left in the upper corner of Figure 2.7 for a BAU strategy, it seems that this will be of limited (temporal and quantitative) "success." In addition, the extended use of ever more oil and gas, into the 60%–80% range, will of course result in ever stronger (and politically and diplomatically dangerous) dependencies on supply oligopolies. Further, more nuclear power plants, even of the fourth generation with their expected higher safety regime, would require many countries of the world to abandon their hesitant or even negative attitude toward nuclear energy. And hard coal power plants, even those of the admirable exergy efficiencies of 50% of forthcoming modern designs, result in ever more irresponsible GHG emissions, if not privileged with carbon dioxide capture and storage (CCS) equipment; similar effects must be expected with more oil and gas.

Harnessing more renewable sources seems to be the solution now, utilizing the technologies of the second solar civilization like modern wind converters, FCs, solar PV generators, SOT power plants, electrochromic windows, and passive solar buildings. So far, however, because they lack storability and transportability, more or less all renewable energies are utilized only locally, at most regionally; truly large macroeconomic renewable contributions to the world energy trade need the chemical energy carrier hydrogen for storage and transport. One example in Figure 2.7 is the light gray dotted line whose downward gradient could perhaps be steeper, which means less zero-carbon energies over time and more coal, or it could be less steep, requiring more renewables and less coal.

Contributing renewables to the global energy trade is not a question of potential; huge amounts of renewable sources are untapped worldwide. From an energy standpoint, wind in Patagonia blows in vain, solar energy in Australia waits to be harnessed, river giants in Siberia flow "uselessly"—to give only these examples—unless their secondary energies of heat and electricity are converted to hydrogen which can be transported as pressurized GH_2 via continental pipelines or as LH_2 aboard cryogenic tanker ships on transoceanic routes to the major energy-importing countries of the world. Of course, renewably generated electricity can also be sent via overhead lines or perhaps as HVDC to areas with major electricity users, but for distances of 1000 km or more, hydrogen is economically more advantageous, and if at the outlet of the transport system a storable chemical energy carrier is asked for, that is the role for hydrogen, too.

In earlier energy discussions it was heard here and there that the hydrogen energy economy is a question of "if" or "whether" or "possibly." Now, with the knowledge depicted in Figure 2.7 it has become a matter of "how" or "when" or "which way first." In current energy thinking, hydrogen has become indispensable for two major reasons: (1) the earth's huge renewable sources will be tapped only when hydrogen facilitates their contribution to the world energy trade system, and (2) for the decarbonization of fossil fuels, particularly coal, hydrogen is inherent.

At the end of this chapter, we take a look at the end of the twenty-first century: Figure 2.7 depicts the history of human energy in the last 200 years and tries to perpetuate irreversibly the result into the twenty-first century. It seems that a triangle-shaped energy cycle is inscribed in Figure 2.7, beginning with renewable energies of the first solar civilization in the late eighteenth and preceding centuries via coal, oil, gas, nuclear fission, and electricity, and ending after two to three centuries again at renewable energies, now of the second solar civilization, and their energy carrier hydrogen. The path traveled is from no carbon via low and high carbon back to low and no carbon, and consequently from high amounts of renewables to low amounts of renewables and back to high amounts of renewables, and, not least, from no hydrogen use to low hydrogen use to high hydrogen use. In retrospect, the use of fossil fuels proved to be a short interim (compared to human presence on earth), and nuclear an addendum. In foresight, hydrogen energy will complete the "energies-of-light" scheme (see Section 2.4), because carbon with its relative atomic mass of 12 is eliminated, and renewables of no weight and hydrogen with its atomic mass of 1 dominate. The future energy system will become dematerialized and inexhaustible (as long as the sun shines, i.e., for some estimated 4.5 billion years to come).

Figure 2.8 tells us that decarbonization and hydrogenation are nothing really new. The shift from coal via oil and gas to operationally non-carbon renewables, nuclear and hydrogen, is not a jump function, but the final result of a century-long continuous development. In the last 120 years, the carbon tonnage per unit of energy [tons C/kilowatt-years] already decreased by ca. 35%; the dotted line indicates the switchover to less and less carbon and, since the energy demand grows, more and more hydrogen.

Another insight into the continuous shift from solid via liquid to gaseous energy carriers is provided in Figure 2.9. What is seen is the relative percentage of market share over the last century and a half, and a prospective view well into the twenty-first century. The quantity of solids like wood, hay, and coal shrink unremittingly; liquids are about to reach their production maximum after which they will decline, and gases are on a perpetual rise from methane-containing natural gas to prospectively hydrogen. Nothing is said about what the hydrogen is produced from. Of course, in a first phase, the production will remain as it has traditionally been, with hydrogen coming from reformed natural gas, from gasification of coal or partial oxidation of heavy crudes, and from electrolysis

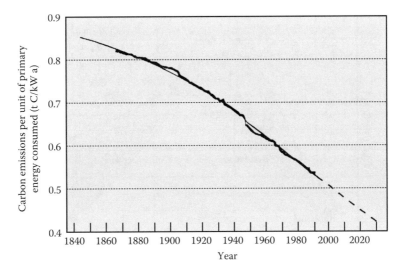

FIGURE 2.8
Energy decarbonization. (Reprinted from *Energy*, 16, CO_2 reduction and removal: Measures for the next century, Nacicenovic, N. and John, A., 1347, Copyright 1991, with permission from Elsevier.)

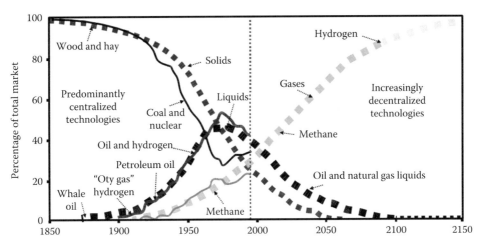

FIGURE 2.9
The global energy systems in transition. (From Hefner, R.A. III., *Int. J. Hydrogen Energy*, 27, 1, 2002. With permission.)

of water (see Section 2.8). Stepwise more and more hydrogen will be introduced from CO_2-sequestered fossil fuels and from renewable and nuclear electricity, and eventually from high-temperature nuclear heat. And more and more decentralized productions from renewable energies will join the game. Whether centralized or decentralized, production will prevail and remains to be seen.

Figure 2.10 brings the value of hydrogen-containing fuels plotted as a function of the relative atomic hydrogen-to-carbon (H/C) ratio, around 1 for solids, around 2–3 for liquids, and around 3 to infinity for gases. Clearly seen are the areas for solids like coal, for liquids like crude oil, the transportation fuels gasoline and diesel and liquefied petroleum gas (LPG), and finally for gases like natural gas containing methane and, not depicted here, hydrogen. What can be learned?

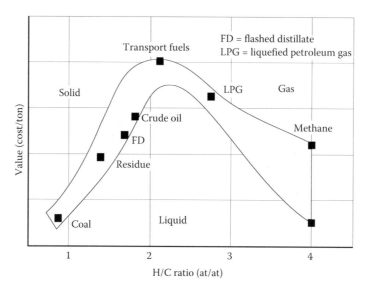

FIGURE 2.10

Value of hydrogen fuels. (From De Jong, K.P. and van Wechem, H.M.H., *Int. J. Hydrogen Energy* 20, 493, 1995. With permission.)

One can read in the literature that energy hydrogenation adds cost. Yes, but this is not generally true. Not surprisingly, coal's cost is the lowest; the switch to liquids is accompanied by a cost increase since the distillation required for transport fuels is particularly cost intensive, but the cost for gases comes down again. The question is, at which cost level will hydrogen enter the picture? If produced from gasified coal it benefits from coal's relatively low cost; the case is similar if it is reformed from natural gas. Electrolytic hydrogen, however, requires inexpensive electricity which, at least in Europe, is hardly imaginable, unless nuclear electrolytic hydrogen or inexpensive hydrogen from nuclear high-temperature heat becomes available and societally accepted. The additional cost for the liquefaction of hydrogen is only justified by applications, which absolutely call for LH_2, such as at filling stations, or on board aircraft, spacecraft, or ocean-going vessels. The liquefaction energy is around 1 kWh per 3 kWh of hydrogen.

2.4 Technological Progress and the World Economy

The waves of technological progress and the waves of the world economy are correlated. Novel technologies, inventions, and systems optimizations are followed by economic prosperity, or, the other way around, technological decline precipitates economic weakness. In some cases, these waves were given names, such as the steam engine age, the electronic era, and the space age. Figure 2.11 gives an indication of exemplary waves experienced during the last 250 years: steam engines, coal, and iron industries in the eighteenth century were succeeded by railway technology and cement industries in the early nineteenth century; the later nineteenth century was extremely inventive and saw electricity, the automobile with its onboard gasoline or diesel engines, and the telephone; plastics, nuclear technologies, aviation and space travel, and electronic technologies were brought to market in the twentieth century.

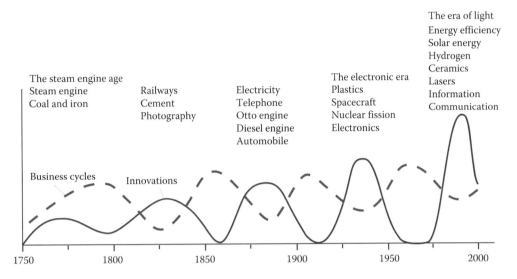

FIGURE 2.11
Waves of technological progress and world business cycles. (Adapted from VDI-Nachrichten, Deutscher Ingenieur-Verlag, Dusseldorf, Germany.)

In the late twentieth and early twenty-first centuries something peculiar happened; more or less, all novel technologies had something in common: they were of low to no weight! Electronic communication technologies replaced weighty letters; miniaturization and micromechanics in electronics dematerialized them; airplanes were produced of low weight carbon fiber plastics instead of weighty metals such as aluminum, titanium, or steel; high-temperature ceramics replaced steel; and in the energy realm, technology-driven efficiencies made more energy services available from less heavy weight energy raw materials; solar primary energies have no weight at all, and hydrogen is the smallest element in the periodic table of elements. Truly, dematerialization is under way.

An "era-of-light" appears to be the appropriate label to characterize the twenty-first century of energy.

Dematerialization of energy prevails. Renewable energy carriers are of light weight; they lack the first link in their energy conversion chain, they have no weighty operational primary energy raw materials per se; energy efficiency reduces the necessary amount of energy raw materials; it contributes to energy services of light weight. Hydrogen has the lowest mass of all elements in the universe; renewable hydrogen energy lightens the burden on environment and climate, and, thus, sheds light on the clean and abundant energy scheme of the twenty-first century. If weighed on the sustainability scale, truly, lightweight hydrogen is a heavyweight.

A general observation corresponds with the era-of-light: In industrialized nations, the percentage of services in their gross national product grows relative to that of production; in some countries it has already reached around 80%, with a tendency to increase still further. Services are of much lower weight than industrial products. Lower weight means easier handling and less transport energy, a step toward sustainability.

A word particularly on energy: In general, energy means energy raw materials and conversion technologies. Two trends can be observed, both pointing to the same direction, technology wins, raw material loses: (1) from coal to oil to natural gas and further to electricity and hydrogen, the weight of the primary energy raw material shrinks; for the two secondary energy carriers, electricity and hydrogen, it is or tends toward zero; and (2) the more efficient the conversion technology, the less weighty material is needed to produce more energy services of low to (almost) no weight. PV generators and solar collectors on roofs or passive solar heating of buildings take an extreme position; they convert weightless solar irradiance directly into electricity or heat without a detour via weighty coal-, oil-, or gas-supplied steam, or gas turbines and electrical generators.

In this context, macroeconomic observations are enlightening. In the early periods of industrialization, nations with indigenous energy raw materials on their own territory were extremely successful; coal and steel were stabile fundaments of their welfare economies. On the other hand, nations without indigenous natural resources did less well; they had to rely on the ingenuity and skill of their scientists, engineers, and craftsmen. Now, the picture has changed radically. Energy raw material poor, but technology-rich nations have developed into the wealthiest nations of the world (such as Switzerland, Japan, and in Germany the southerly federal states of Bavaria and Baden-Wuerttemberg), and the others still struggle to rid themselves of what was once considered beneficial natural wealth: rust belts around the world bear witness! The consequence again and again: "Technologies compete, not fuels!" (Scott 2007).

Accordingly, two recommendations to parliaments and industry, both in Europe, asked for political action: (1) Already in the early 1990s, the German Bundestag's Enquête Commission "Protection of the Earth's Atmosphere" decided unanimously on both sides of the aisle to recommend to the Federal Parliament and Government that the country be run at an energy efficiency of 60% instead of today's 34% (2006). The commission was convinced that its recommendation was not a question of technologies—the technologies were there or would be available—but one of economic viability and political will, and time. And, a few years later, (2) the Federal Institutes of Technology in Zurich and Lausanne, Switzerland, recommended the development of a 2000 watt society. What does that mean? Nothing less than 2000 watt-hours per hour and capita [watt-hours/hour · capita = watt/cap], and nothing more, should be the average energy demand of each inhabitant of Switzerland (including trade and industry). Time, industrial preparedness, and political will are the deciding conditions; technology is and will be available. The situation in other countries is more complicated: the individual energy demand spreads from near zero for the poorest developing countries to 11 kW/cap for North America or particularly energy-rich countries that supply themselves and the world; the world average is around 2 kW/cap. The incentive for much higher energy efficiencies is the more marked the less indigenous energy the country in question can rely on: efficiency technologies are tantamount to energy, they are "energy"! The world average of 2 kW/cap should be approximated from both sides: the developing world coming from the bottom upward, and the industrialized countries from their much higher values downward. Incidentally, efficiency technologies are environmentally and climatically cleaner than the energy they convert. Increasingly stringent environmental and climate change mitigation obligations will force an approach toward the world energy intensity average!

2.5 Hydrogen Energy and Time

Novel energy technologies need time, usually much more than impatience is inclined to tolerate. Many decades, in cases up to half centuries or even centuries are the typical time

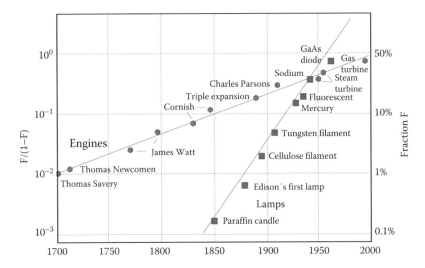

FIGURE 2.12

Efficiency improvement of engines and lamps. (Adapted from Ausubel, J. and Marchetti, C., In *Technological Trajectories and the Human Environment*, J. Ausubel and D. Lanhford (Eds.), National Academy Press, Washington, DC, 1997.)

requirements until their irrevocable market presence. Figure 2.12 brings two examples, heat engines and lamps, and their historical development. What is seen?

Thomas Savery, Thomas Newcomen, and James Watt, all three in Britain, published their steam engine development results in the eighteenth century. Th. Savery's patent on his engine is even dated 1698; he named it "The Miner's Friend" because steam engines in those days helped pump unwanted water out of underground coal mines. The early engines had meager efficiencies; no more than a few percent of the coal's energy content were converted to mechanical energy. Today, around three centuries later, modern steam and gas turbine combined cycles asymptotically approach and even surpass 60% electric (=exergetic) efficiency, and the final figure has not yet been reached. But let us realize that it took three centuries for this development, evolving via many dissimilar designs for steam engines, steam turbines, and later internal combustion engines (ICEs) and gas turbines—three centuries!

The figure also shows the development of lamps. Over one century and a half ago, light came from wax candles with an efficacy of about 0.1%. Edison's first lamp achieved a little less than 1%, and today's gallium arsenide diodes achieve some 50%. But again, 150 years had to pass!

Could it have been accomplished any quicker? Could the development have been accelerated? Most probably not! Figure 2.12 attempts an explanation: for centuries, the development lines for both heat engines and lamps were straight lines in half-logarithmic plotting. It so seems that humans would have had little influence on a potential acceleration. Examples: Otto Hahn split into two the nucleus of the uranium atom in 1938 in Berlin; in the 1940s, Enrico Fermi erected the first nuclear reactor underneath the terraces of a stadium in Chicago; and now, 70 years after the first nuclear reaction a system change to hydrogen fueled up-and-coming chemo-electric FC conversion will join the market, a radically different design than what is plotted in Figure 2.12. The FC is not a Carnotian heat engine; although it simultaneously provides electricity and heat, it is electrically (=exergetically) more efficient than the Carnotian heat engine–generator combination, and it comes in typically very small to small capacities (watts to a few megawatts). Its future development may get a steeper gradient than that of the heat engines, although it has also been known for 170 years already. The aforementioned straight lines for lamps and engines are sigmoidal in character and are thus significant for any other converter under consideration.

Generally, the market approach of a product is regularly depicted by an S-shaped curve. It begins with a slow introduction, turns at a certain point in an "explosion" into market dominance, and then fades out in saturation. Between start and end, many decades may pass by. Let us consider two practical examples: gas turbines and nuclear energy. Gas turbines first: In 1791, a first English patent was granted. Much later, Hans Holzwarth (1877–1953) devoted much of his time to the development of his turbine, with little success. In the 1930s/1940s, Pabst von Ohain's research led to practical gas turbines operated as jet engines aboard the German warplanes Heinkel He 178 and Messerschmitt Me 262. And now, around a century and a half after the mentioned English patent, gas turbines on the ground together with steam turbines in combined cycles generate electricity under commercial conditions with convincing electrical efficiencies around 60%. And now to nuclear: In the 70 years since the first nuclear reaction, 439 reactors have been built worldwide with a capacity of some 393 GW; 32 reactors with 28 GW are being constructed or planned (2007), and the end is not yet visible.

A conjecture may help in understanding why novel energy technologies need so much time. Of course, any novel technology has research and development problems in itself; materials and tools are not at hand; algorithms have to be specifically developed; the origin of failures along the development road must be understood before they can be tackled. Failures and negative research results are positive results too, which is regularly misunderstood by individuals outside the research area. Financial setbacks need to be overcome again and again, whenever other items in the research laboratories are felt to deserve higher priority. Many other parameters have to be accommodated before a research and development result can finally be presented. But, let us make no mistake; in general, up to this point only some 10% of the entire road from initial vision to a stabile market share has been traveled, and it is not realistic to expect much more. What then comes is the market with its own distinct parameters of business activities, marketing, economic viability, micro- and macro-economics, international trade, codes and standards, etc. Consequently, the "classical" novel technology process consumes time, much time!

In the meantime, the "old" technology is not sitting there waiting for its death blow! Its development potential is not zero. The technology is well known; its pros and cons have been experienced, in some cases over very long times; availability, cost, safety, and supply security are givens. Generations of engineers have been educated in the old technologies; they are familiar with their advantages and are conscious of their disadvantages. So, why not develop further what is at hand, instead of taking unknown and insecure routes? Regularly, the novel technology forces the old one toward new frontiers! The competition between the two is open (and in cases fascinating!); sometimes the old technology "wins," sometimes the novel one does. Again, both need rather long periods of time.

Let us give two examples from the transportation sector: Vehicles need to be adapted to the environmental and climatic challenges, operational pollutants have to be reduced, and GHG emissions need to be eliminated. What could be more obvious than switching over from the hydrocarbon fuels now in use to environmentally and climatically clean hydrogen fuel and replacing the more than 100 years old ICE with a modern FC? But, no surprise, the ICE fiercely resists; it defends its survival by completely meeting the strict environmental codes EU1 to 5 (EU6 soon) to the point of measurability limit. It has become more efficient and, as a consequence, it emits fewer GHGs. The EU's call for an average of 120 g of emitted CO_2 per kilometer driven is not out of reach. Or another example: the modern truck (25 t) is already achieving 0.81/100 tkm and 21 g CO_2/tkm. (For comparison, the middle class passenger vehicle with an imposed load of 500 kg is at 171/100 tkm and 403 g CO_2/tkm!) And, what is perhaps the most convincing argument is that the unit price of the marketed ICE is still only some tens of €/kW. The FC is miles away from that price tag, it is working hard to catch up, but, again, it needs time.

As an example for "energy needs time," Figure 2.13 gives a convincing picture of the 200 year long step-by-step development of land transport. It started in the late eighteenth century when carts, horses, and their "fuel" hay or oats were stepwise replaced by steam, and later diesel locomotives and their fuels wood, coal, and later diesel oil. About every 70 years, a novel technology entered the transport market. In the late nineteenth and twentieth century, individual and mass transport of passengers and goods on land and in the air took over most of the market. Again, "technologies compete, not fuels." The arrival of a new transport technology always triggered the market change; the availability of specific fuel followed suit.

In all cases, certain expectations were connected with the novel technology: locomotives allowed for higher speeds and longer distances than horse drawn carts; diesel locomotives were cleaner than coal-fired steam locomotives and offered easier handling of the fuel; air transport made it possible to travel around the globe in reasonable times; and the passenger vehicle stands for individuality. Now, since strict and challenging environmental and climatic cleanliness criteria have to be met, it seems unreasonable to expect the trend to break. Consequently, what follows now? The answer is the FC and a whole range of renewable energy and hydrogen technologies. Supposedly overstated, "no-energy energy supply" is the creed (no-energy means principally no operational primary energy raw

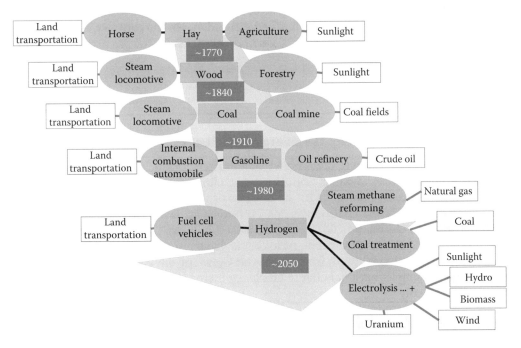

FIGURE 2.13
Land transport. (Adapted from Scott, D.S., *Smelling Land, The Hydrogen Defense against Climate Catastrophe*, Price-Patterson, Canadian Hydrogen Association, Montreal, Canada, 2007.)

materials for solar and solar hydrogen energy technologies). Primary energy raw materials are on the losers' side, and, once again, technologies are on the winners'. Evidently, the beginnings of the up-and-coming technologies were 10–20 years ago, and they will occupy the next 70 years, "until something better comes along" (magnetic levitation? evacuated tubes? submersible ocean travel? … and who knows what all else?).

Eventually, innovations are pushed by "early starters" (regardless of whether they finally become the market winners or not). Do we see circumstances anywhere benefiting early hydrogen technology starters? Yes, we do: there are countries which, within their traditional industrial development scheme, have accumulated certain preferred hydrogen technology skills, manufacturing capabilities, or operation experience. For instance, Japan and the United States: they seem to be well prepared to become (and in some fields already are) active in the large and versatile field of hydrogen-supported portable electronics. Or, space-faring nations around the world have gathered experience for now more than half a century in reforming, liquefying, storing, transporting, handling, and combusting large amounts of hydrogen (and, of course, oxygen), both gaseous and liquefied. And these days, there is no major automobile manufacturer in the world not engaged in the research and development of onboard hydrogen and its utilization in low-temperature mobile FCs or adapted ICEs. Billions are involved, and the times have long passed when cynics quipped that Big Auto's hydrogen and FC development was being financed out of petty cash.

In addition, heavy industry centers of the coal industry, major oil and gas producers, gas traders, and chemical companies have been well aware of hydrogen and its technologies for a very long time. Traditional hydrogen commodity users are considered well prepared to play a role also in the hydrogen energy economy. As a result, there is no need for hydrogen energy to start from scratch; there are lots of active "hydrogen islands": connecting them nationwide is the task! What we face is not too dissimilar from what happened at the turn of the twentieth century when the electricity market began to evolve. Even after a complete century, it has by far not yet reached saturation. On the contrary, for (more than) a century we had one secondary energy carrier—electricity; now we are getting two—electricity and hydrogen.

2.6 Energy Efficiency—No: It's Exergy Efficiency!

The attentive political observer takes note that elevated efficiencies are politically beyond all dispute as measures against anthropogenic climate change. The EU Ministerial Council's formula (2007) reads: "4×20" or 20% less primary energy, a 20% share of renewable energies, and 20% less GHG emissions, each to be achieved by 2020. The G8 summit in Heiligendamm, Germany (2007), even discussed halving GHG emissions by 2050 (a decision confirmed by the G8 summit in Hokkaido, Japan, in 2008), that is, in the extremely short time of only some 40 years from now! And Germany envisaged the ambitious goal (in reach?) of raising its national energy efficiency increase from today's ca. 1% or a little more to 3% annually.

Here is not the place to appraise whether these decisions are technologically and economically justified, whether it is reasonable to assume that they can be realized, and, the most important touchstone of all, whether they will be able to sufficiently restrict the anthropogenic GHG effect in its consequences for humans, their artifacts, fauna, and flora (see Section 2.2).

It is trivial to state that kilowatt-hours not required thanks to more effective conservative energy use as well as elevated efficiencies are environmentally and climatically positively relevant, and that operationally carbon-free renewable and nuclear energies do not contribute to the GHG effect at all.

Not at all trivial, however, is—that is the central point of this section—knowing which efficiency is meant, energy efficiency or exergy efficiency. Exergo-thermodynamics tell us that each energy conversion step along the complete energy conversion chain, link by link from production via storage, transport, dissemination, and finally utilization of energy services, splits up energy into exergy and anergy: energy = exergy + anergy. Obtaining more exergy from energy is the real goal of each energy conversion, because more exergy means more available technical work. This is the ultimate energy challenge and it can be compared with J. W. Gibbs' free energy available to do external work (Josiah Willard Gibbs, 1839–1903). Earlier American literature speaks of "energy availability" meaning available technical work. Exergy can be converted into every other form of energy; anergy cannot. What practical exergy investigators urgently need is to admit that they cannot "unlearn" (uninvent, uninnovate) and to recall the physics of exergo-thermodynamics which was (again) published in the 1960s (e.g., Figures 2.14 and 2.15) and which, blameworthy as it is, have been almost forgotten in the meantime. Since exergy increments within the established prevalent Carnotian system tend asymptotically

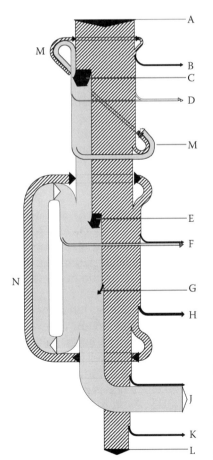

FIGURE 2.14
Exergy/anergy for a steam power plant. A, exergy in (100%); B, unburnt; C, irreversible combustion; D, radiation; E, irreversible heat transfer; F, stack; G, irreversible steam flow; H, mechanical; electrical, magnetical loss; J, condenser; K, in-house use; L, electricity = exergy out (40%); M, air preheat; N, water preheat. (Adapted from Rant, Z., *Brennstoff Wärme Kraft*, 16, 453, 1964.)

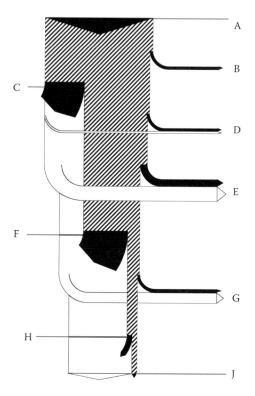

FIGURE 2.15
Exergy/anergy for a coal-fired central heating system. A, exergy: in (100%); B, unburnt; C, irreversible combustion; D, radiation; E, stack; F, irreversible heat transfer; G, distribution; H, irreversible heat transfer II; J, heating energy, exergy: out (63.9%). (Adapted from Rant, Z., *Brennstoff Wärme Kraft*, 16, 453, 1964.)

toward their final maximum, exergy efficiency engineering is increasingly becoming a matter of system change to combined cycles, from electricity only, or heat only, or chemical product only, to simultaneously produced electricity and heat, or electricity and hydrogen, or electricity and chemical commodity, etc. Farsightedness in thermodynamics, or, more precisely, in exergo-thermodynamics is the real goal, not tackling day-to-day efficiency deficits in the applications based on commonly utilized Carnotian energy thermodynamics!

As will be seen in the following, hydrogen energy is an irreplaceable mosaic stone in this picture. Hydrogen helps to bring forward energy conversion systems which avoid conversional irreversibilities and, thus, avoid exergy destruction and exergy losses, or unused exergy and the according build-up of huge amounts of anergy of no anthropogenic usage! Hydrogen helps energy-systems-of-change to approach the maximum of available technical work being extracted from energy; this is the superior parameter that the layout of each energy installation ought to regard—it ought to, but in too many cases it does not, so far.

The situation: After more than 200 years of national energy, the energy efficiency of industrialized Germany, to take that example, is some 34% (2006), and that of the world not much more than 10%. Germany has to introduce 3 kWh of primary energy into the national energy economy in order to provide 1 kWh of energy services after having completed the run through the entire national energy conversion chain, and, bitter to say, for the world the ratio is 10:1! The industrialized country's energy efficiency is really not too impressive at all, but that of the world is devastating, energetically, economically, and ecologically. And this is only one side of the coin. The other side, even more devastating, is that Germany's exergy efficiency is ca. 15%, while that of the world amounts to only a few percent.

In reality, the world's energy system with its deplorably meager exergy efficiency is an anergy system garnished with a small amount of exergy. The system inherently contains large irreversibilities (indicated in Figures 2.14C, F, G and 2.15C, F, H by the large black arrows on the left) and thus produces enormous amounts of heat of no worth to users, conveyed to the ambience and radiated into space. Potential users are elsewhere and the temperature of their heat requirement is a different one. The state-of-the-art energy system is not sustainable for many reasons. Here is one relevant to our exergy argumentation: the energy raw materials in use are finite and their consumption lacks negentropy to compensate for the inherent entropy increase. The case is otherwise with a renewable solar energy system, whose entropy rise is immediately compensated by the negentropy flow from the sun (as long as it shines). Solar hydrogen benefits from these physical facts.

What does all this practically mean? We said that the energy system of the world produces far too much heat of the false temperature at the wrong location (=anergy). However admirable the envisaged electrical efficiencies of modern hard coal stations are, arriving at or even slightly surpassing 50% (=exergy efficiency), thermal power stations still irreversibly provide combustion losses, heat transfer losses, and energy flow losses throughout the system, predominantly because of the installation's exergy destruction (Figure 2.14), and this adds up to huge amounts of anergetic heat of inappropriate temperatures at locations where no user buys them. Or another example: the boiler in a central heating system of a residential or commercial building has a similar problem. It generates flame temperatures of up to 1000°C, although the room radiators require only 60°C–70°C. The energetic efficiency of converting the chemical energy of light oil or natural gas fuel into heat is superb; it reaches almost 100%. The exergetic efficiency, however, of heat delivery of the "right" temperature to the room heating radiators is miserable, only a little more than 10% at most, again because of irreversibilities in combustion, in heat transfer to the heat exchanger, and because of the energy throughput through the entire system (see arrows on the left in Figure 2.15). There is a similar heat problem in the ICE of the automobile: only 20% (30% at the utmost) of the energy content of gasoline or diesel goes into traction (=exergy); the much bigger part is discharged into the environment via irreversible heat exchange in the cooler or irreversible tail pipe exhaust (=anergy). Altogether, the energy system in place is in fact an anergy system that provides quasi as a by-product a little exergy too; it is absolutely sad to have to say that after two centuries and a half of energy engineering!

Certainly, talking is cheap; it is easy for the reviewer to lament the miserable exergetic condition of energy installations in the world without trying to look for ways out of the dilemma. First of all, it is astounding how little the laws of exergo-thermodynamics, known since Gibbs, have so far entered legislative thinking. The laws of parliaments and the laws of nature have grown increasingly divergent, and it cannot be expected that the laws of nature will yield (McKibben 1999). Efficiency increase, that is, more energy services from less primary energy, remains part of the system. However, it is not recognized that exergy efficiency increases have a huge, though dormant, virtual potential which requires a system change to non-heat-engine-related systems! Exergizing the energy system asks for shifting the baseline. "Virtual" means that the potentials are real though hidden, and so far untapped. Through exergy thinking and acting they will be tapped.

Let us try to make this clear by using the three examples mentioned earlier (there are many more in all energy sectors: industry, transport, buildings, trade, etc.). The first example: we spoke about the admirable elevated electrical efficiencies of around 50% of modern coal-fired power stations, verified through asymptotically ever higher temperatures (from a high temperature materials technology point of view more and more difficult to obtain);

these efficiencies remain within the applied system. But with an electrical efficiency of 50% still half of the energy content of the coal is not converted into technical work (=exergy). A system change toward much higher exergy efficiencies, such as the combined production of hydrogen and electricity via air separation, coal gasification, CO_2 and hydrogen separation, and combined cycle power generation, all more or less marketed technologies, is presented in Section 2.9.

The second example: the already mentioned boiler of the central heating system of a building is energetically excellent; almost 100% of the energy content of natural gas or light oil fuel is converted to heat, although to heat of a temperature for which no user exists. Exergetically, however, the boiler is miserable, because it is exergo-thermodynamically simply absurd to generate flame temperatures of up to 1000°C with the objective of supplying room radiator temperatures of some 60°C–70°C. If a hydrogen-fueled low-temperature (<100°C) or middle-temperature (≤200°C) FC is installed in the boiler's stead, it first generates electricity (=pure exergy) from 35% to 40% of the fuel's energy, with the remaining heat still sufficing to warm the building over most of the year. A thought experiment says if the present 15 million boilers in Germany were replaced by FCs of, say, 5 kW(electric) each, an IT-controlled virtual "distributed power station" of 75,000 MW would develop. This comes near the present centrally structured national installations of some 100,000 MW; it approaches an exergetization par excellence of the country's central heating system! (Thought experiments seldom become real, although there is often a certain truth in them. Here, the "truth" says that a distributed competitor of significantly higher exergy efficiencies will have emerged to challenge the central electricity utility system in place: welcome competition which for the time being lies fallow!) All the aforementioned is not inevitable, but it indicates where we ought to be heading.

The third example deals with the ICE in autos. Here, it is not denied that there are still exergo-thermodynamic potentials within the conventional system and that they are being activated stepwise by continuous further development. The Otto and Diesel engines, both inventions of the late nineteenth century and more than 100 years old, are still not fully mature; there is still potential, particularly when the entire auto system is taken into account. However, what is of interest here is the system change toward exergetization: renewable hydrogen or hydrogen from CO_2-sequestered fossil fuels, from nuclear electricity, or, better, nuclear high-temperature heat fed into a hydrogen-optimized ICE or a low-temperature FC. Both are environmentally clean, and without CO_2 emissions along the complete life cycle (well-to-wheel) contributing to the greenhouse effect, they are climatically clean too. For the engineer, the development "race" between the two is highly exciting and the outcome is not decided yet. To make a long story short, despite all the past development aims for steam engines, Stirlings, flywheels, gas turbines, Wankels, and other prime movers aboard automobiles, the FC is the first real alternative in the history of engine technology to truly be taken seriously. Not being a Carnotian heat engine, it is an exergetically highly efficient, clean, quiet, compact, and non-vibrating competitor to the ICE. However, the ICE is not sitting there, waiting to be finished off. Although it is more than 100 years old, it meets the extremely strict legislated codes EU1 to 5 (6) and is thus by definition environmentally clean, and its potential to reduce CO_2 emissions is not zero. Perhaps the most convincing argument vis-à-vis the FC is that the ICE is in around one billion copies in stationary and mobile operations, being reproduced in some 100 million copies per annum, and marketed for some 10 €/kW! From this figure the FC is still miles away, but it is picking up momentum and trying hard to catch up!

Summing up, higher energy efficiencies within the operational system in place are appreciated. The real breakthrough, however, to economic, environmental, and climatic responsibility asks for higher exergy efficiencies and, thus, a system change to combined cycles, which seems expressly suitable for indigenously energy-poor, though technology-rich, industrialized nations. They enjoy an almost inexhaustible and, by the way, renewable "energy" potential in the scientific knowledge of their scientists and the skills of their engineers and craftsmen. Energy technology science and skills are "energy"! Wise energy policy prior to active technology politics provides an entry into the technology-driven hydrogen energy economy and the accordingly necessary system change to the small-to-medium size energy converters FCs of capacities of less than watts to a few megawatts at the back end of national conversion chains. The center of gravity within national chains moves toward their back end. The conventional wisdom of national energy policies is to ensure at economically viable cost the delivery of sufficient amounts of primary energy raw material, which is then converted in the national conversion chain with meager exergy efficiencies stepwise into secondary energies, end energies, and so forth, and finally to energy services. After the system-change to exergetically highly efficient combined cycles, secondary energies become more important than primary energies. Thinking and acting in primary energy raw materials was nineteenth and twentieth century; thinking and acting in exergy-efficient energy conversion technologies is twenty-first century!

An interesting though abstract scheme pointing in the right direction is energy cascading, that is, utilizing heat or cold in steps from higher to lower or from lower to higher temperatures, respectively. Cascading of heat or cold helps to maximize exergy and minimize anergy. The present industrial infrastructure, however, is far from consistent cascading, to put it mildly. Barriers are the geographical dislocation of consecutive users, which does not favor neighborhood industrial structures, dissimilar market requirements, and others. The aforementioned switch to combined cycles producing electricity and hydrogen are cascades of practical application (detailed information on cascading can be found in Section 2.8).

The dispassionate energy economist may now object that such a system change needs decades, if not half centuries to centuries, and trillions worth of investments. Certainly, it is impossible in the twinkling of an eye to systematically convert the energy system in place into something else and pay for it with petty cash. In addition, the longevity of just installed (and still to be installed) investments worth billions (power stations, refineries, pipeline grids, tanker fleets, and the like) are also many decades. It is irrational to expect their dismantling prior to the end of their economic life. But, climate change does not give way to economic considerations. The expectation of being able to reduce the anthropogenic greenhouse effect to a tolerable level simply by continuing developing, perhaps at a slightly accelerated pace, today's energy system is deceptive. The 2°C figure of policy makers as the anthropogenic atmospheric temperature increase considered "allowable" is arbitrary; and even its realization is not at hand. At the latest after future gigantic hurricanes and floods à la "Katrina," at the latest after the melting of land-based Greenland and Antarctic ice and the successive rise in ocean level and flooding of the earth's marshlands where one billion people live, at the latest after arable land which used to feed entire populations has turned into dried-out deserts, all this followed by streams of millions of climate refugees washed ashore where the wealthy "highlanders" live, then at the latest will the call for a system change get louder. To answer the call of being exergo-thermodynamically well-equipped, not more and not less is the objective of this argumentation.

Human imagination is rather finite; its temporal intrusion into the future is only a few years, if that. Regularly, foresightedness is modified by unforeseen surprises, because it is simply an extension of the present. Examples for such surprises are wars, tanker shipwrecks, the intended and fiercely publicly opposed disposal of an oil platform in the North

Sea, diplomatically irritating "playing" with the throughput throttle of an international natural gas pipeline, nuclear reactor accidents, or simply presidential remarks from a major oil-exporting country. The almost immediate consequences are jumps in the price of oil, followed by the other fuels in the global energy trade system, price jumps which hit the nearly unprotected energy buying countries with their extremely high import quota and rather small exergy efficiencies (energy-short Germany's national import quota is 77% (2007); its exergy efficiency a little more than 15%!).

An effective barrier against such surprises is a system change to an exergy-efficient hydrogen energy–supported energy system. One does not have to be a prophet to say that whatever will come, exergy is the insurmountable maximum of technical work, which can be made available from energy! Its limit is set by exergo-thermodynamics. So far, no country of the world has ever touched this limit. The goal is within reach, but it is a long way to Tipperary. Anthropogenic climate change calls for taking this way and seeing it through! Hydrogen energy helps; it crosses the border and enables this system change. Hydrogen exergizes!

2.7 Hydrogen, Electricity: Competitors, Partners?

Electricity and hydrogen have in common that they are secondary energies generated from any primary energy (raw material), none excluded, fossil, nuclear, and renewable. Once generated, they are environmentally and climatically clean along the entire length of their respective energy conversion chains. Both electricity and hydrogen are grid delivered (with minor exceptions); they are interchangeable via electrolysis and FC. Both are operational worldwide, although regionally in absolutely dissimilar capacities.

And their peculiarities? Electricity can store and transport information, hydrogen cannot. Hydrogen stores and transports energy; electricity transports energy but does not store it (in large quantities). For long (i.e., intercontinental) transport routes hydrogen has advantages. The electricity sector is part of the established energy economy. Hydrogen, however, follows two pathways: one where it has been in use materially in the hydrogen economy almost since its discovery in the eighteenth century. To date, it is produced worldwide as a commodity to an amount of some 50 million tons per annum, utilized in methanol or ammonia syntheses, for fat hardening in the food industry, or as a cleansing agent in glass or electronics manufacturing. And, along the other pathway, it serves as an energy carrier in the up-and-coming hydrogen energy economy, which started with the advent of the space launch business after World War II. Essentially, the hydrogen energy economy deals with the introduction of the—after electricity—now second major secondary energy carrier hydrogen, and its conversion technologies. Hydrogen-fueled FCs can replace batteries in portable electronic equipment such as television cameras, laptops, and cellular phones; FCs are being installed in distributed stationary electricity and heat supply systems in capacities of kilowatts to megawatts, and they are operated in transport vehicles on earth, at sea, in the air, and in space. It is never a question of the energy carrier alone, hydrogen or hydrogen reformate. On the contrary, environmentally and climatically clean hydrogen energy technologies along all the links of the energy conversion chain are of overarching importance. Of course, technologies are not energies, but they compare well with "energies." Efficient energy technologies provide more energy services from less primary energy (raw materials). Efficiency gains are "energies"! Especially for energy-poor, but technology-rich, countries, efficiency gains come close to indigenous energy!

A trend is clearly visible: Increasingly, the world is moving from national energies to global energies, and energy technologies serve as their opening valves. CO_2 capture, sequestration, and storage technologies bring hydrogen-producing clean fossil fuels to life, and hydrogen-supported FC technology activates dormant virtual distributed power. Both technologies are key for the hydrogen energy economy which, thus, has the chance of becoming the linchpin of twenty-first century's world energy.

2.7.1 Mechanization, Electrification, and Hydrogenation

The electricity industry began more than 100 years ago with Siemens' electrical generator and Edison's light bulb. Electricity is a success story which is not yet at its end. In industrialized regions, electricity is almost ubiquitous, fitting in locally and temporally, environmentally, and being climatically clean and affordable—more or less.

In the late eighteenth century, James Watt's steam engine initiated the mechanization of industry. A good century later, electrification came into use; it largely replaced mechanization and permeated into almost all energy utilization sectors such as production, households, communication, and railways. Literally and seriously, "electricity is readily available at the socket," really never to be worried about! However, there are weaknesses: blackouts are suffered under, seldom, but once in a while, and many a developing nation's electricity supply with its frequent blackouts is as good as almost inexistent. Further, although battery development is in full swing and progress has been achieved, it is still not easy to operate an automobile over long distances with electricity, much less—if ever—an airplane or a spacecraft.

The question is, can hydrogen be of help wherever it has advantages relative to electricity, wherever electricity is useless because it cannot be stored in large amounts, or wherever electricity and hydrogen together can offer solutions that are inexistent for either one alone? Is it true that after mechanization in the late eighteenth and then in the nineteenth century, after electrification in the late nineteenth and then in the twentieth century, we are now at the start of the twenty-first century on the verge of hydrogenation of the anthropogenic energy system? Answering this question is not too difficult, because we see clear signals. Historically, with the switchover of the anthropogenic energy, centuries from high carbon via low carbon to no carbon, that is, from coal via oil and natural gas to hydrogen, the atomic hydrogen/carbon ratios for coal:oil:natural gas:hydrogen have become $\leq 1:2:4:\infty$. Decarbonization and hydrogenation are continuously increasing, and, since the atomic masses of hydrogen and carbon are 1 and 12, respectively, dematerialization of energy is increasing too. Already today, two-thirds of the fossil fuel atoms burnt are hydrogen atoms; the trend continues.

What is the status of the hydrogen energy economy? There are still only a few industrial sectors where hydrogen serves energetically; all the other areas use it as a commodity. Energetic use includes the space industry, which would even be inexistent without access to the highly energetic recombination of hydrogen and oxygen in the power plants of space launchers; submersibles, where low-temperature high-efficiency hydrogen/oxygen FCs guarantee extended underwater travel and low to zero detectability because the water exhaust is contourless after onboard condensation and possible subsequent utilization as drinking water or for the seamen's sanitary purposes; refineries for the production of reformulated hydrogenized gasoline and desulfurization of diesel; and in cooling systems for large electrical generators.

FCs as replacements for short-life batteries in portable electronics such as laptops, camcorders, and cellular phones, energized with the help of hydrogen or hydrogen-rich methanol cartridges; natural gas or hydrogen-supplied FCs in distributed electricity and heat supply or as replacements for boilers in central heating systems in residential homes and office buildings; FCs as auxiliary power units (APUs) in vehicles or airplanes; hydrogen and ICE s or FCs on board busses or autos; finally, liquefied cryogenic hydrogen instead of kerosene in aviation. All these areas are still in the phase of research and development, or at most in their demonstration phase. The technologically driven hydrogen energy economy—no doubt—is at its beginning and probably has to face decades yet before mass market readiness.

2.7.2 Domains, Partners, and Competitors

Let us now come back to our question posed at the beginning: Electricity, hydrogen—competitors, partners? We distinguish three realms where hydrogen and electricity

a. Have their respective domains

b. Are partners

c. Compete with each other

To (a) belong the aircraft and spacecraft engines; they are/will become undisputedly hydrogen domains, simply because you cannot fly or operate an air- or spacecraft with electricity, the necessary battery sets would be far too heavy and bulky (exceptions: thermionic converters or nuclear reactors for deep space missions). Electricity's domain, on the other hand, lies in the communication sector, in providing light, and in all areas of production electricity is indispensable.

Under (b) we find all the chemo-electric energy converters, the electrolyzers, and the FCs that convert hydrogen efficiently, environmentally, and in a climatically clean manner into heat and electricity in CHP applications in industry, in households, and office buildings; here hydrogen and FCs are an unbeatable combination!

Finally, under (c) we essentially find mobility tasks that can be performed with either electricity or hydrogen: rail transport in Europe is powered by grid delivered electricity; for continental distances, however, as for instance in Canada or in Russia, it may be questionable whether railway electrification through electrolytic hydrogen–powered FCs will not become the economically more viable solution, replacing the traditional overhead electricity contact wire which, for thousands of kilometers, might be the more costly and irksome investment (this idea is D. S. Scott's).

Earlier, the situation in the individual transport realm was not as clear as it is today. As long as there was hope to see on the roads efficient, battery-supported, low-weight, marketable electric vehicles in large numbers, it was not too easy for the hydrogen vehicle to make its point. Now, after many decades of development of long-range auto batteries in the drive train with only rather meager success, the route for hydrogen surface transport in buses, in limousines, and in trucks and lorries is wide open.* However, the decision is still due as to whether there will be an FC or a hydrogen-adapted ICE under the hood, because the "novel" FC has not yet won, and the "old" combustion engine still has potential, consequently, it is not forced to give up. The "race" between the two is highly exciting for

---

* The ongoing electrification of individual city transport of some 50 or 60 km range with the help of newly developed lithium ion batteries is considered not a contradiction. On the contrary, the battery-powered vehicle is an electric vehicle and, thus, in a certain aspect a harbinger of the up-and-coming implementation of the long-range electrical vehicle powered by hydrogen.

the thermodynamicist and the engineer, but it is not decided yet. The FC needs convincing cost, performance, cleanness, and efficiency advantages in order to compete successfully with the more than 100 years of solid experience of the reliable reciprocating piston engine. Cost is the harshest criterion.

One particular partnership development of electricity and hydrogen is worth pointing out: the stationary FC in CHP installations or in the central heating systems of buildings is small and compact with capacities of four orders of magnitude from kilowatts to a few megawatts. As a decentralized energy converter it tri-generates locally and simultaneously electricity and heat and/or cold; consequently, the nation's electrical grid losses are nil (they for the time being sum up to ca. 4% in Germany; in other regions of the world they are sometimes significantly higher!). The distributed FC park with potentially millions of installed FCs compares well with a virtual IT-controlled power station whose capacity easily matches the capacity of the central installations (e.g., for Germany ca. 100,000 MW). Competition between the traditional national energy conversion chain's front-end electricity generation and novel FC–supported back-end generation is foreseen—and welcome. It will be interesting to see which kilowatt-hour of either end of the chain will become the less costly, which the environmentally and climatically cleaner, and which the more reliable!

2.7.3 Exergetization

If the conversion chain's back end of a national energy system becomes a convincing power generator and, thus, a mighty competitor to the established traditional power plant park at the chain's front end, something thermodynamically very important will have occurred: the FCs supplied by hydrogen or hydrogen reformate will have exergized the energy system! Exergizing technology examples were given in Section 2.6 and included the replacement of the home heating boiler with a hydrogen-fueled FC, or the changeover from the not too convincing 20% (at most 30%) exergy efficiency of current autos to FCs or adapted ICEs, both hydrogen fueled and efficient to a non-illusory 50%!

Stationary FCs generate exergetically efficiently and simultaneously electricity and heat, and meet with their FC-specific temperature regime between 80°C and 900°C the exact relative temperature demand of households, industry, and vehicles. Let us never forget that ecological reasoning not only asks for waiving claims and avoiding materialism, but also for unparalleled technology development in order to improve the so far rather poor efficiency of anthropogenic energy use, which was bitterly underestimated for centuries. Hydrogen-supported technology is becoming a harbinger of this development!

Traditionally, electricity is produced at the front end of a national energy conversion chain and used at its back end. There may be a 1000 km between front and back ends. Now, with millions of envisioned FCs' supply at the chain's back end, electricity is also produced there, and that is in the vicinity of electricity users. This is of cardinal importance, because the back end of a national energy conversion chain governs the overall efficiency of a nation, since each kilowatt-hour of energy services not demanded at the chain's end (because of efficiency gains) results in 3 kWh of primary energy (raw materials) not necessary for the nation's economy to be introduced at the front end (Germany's present national energy efficiency ~34%!). In the world, the relation is 1:10 (the present world efficiency ~10%). That is what is meant by the sentence "Hydrogen and FCs exergize the energy system!" They make more electrical and thermal exergy services out of less primary energy. Electricity is pure exergy.

2.7.4 Hydrogen Supply

Wherever energy is discussed, one question is repeatedly asked. Where does the hydrogen come from? There are three answers which would be similarly answered for electricity: from fossil fuels, from renewables, or from nuclear fission. (1) From fossil fuels via reformation or partial oxidation or gasification, preferably from natural gas, like today, or from coal, in future, with capture and sequestration of coproduced carbon dioxide in order to prevent its release into the atmosphere; (2) from renewable electricity via electrolysis, but not before a number of further decades of development and in competition with the direct use of renewable electricity in the power market; or (3) from nuclear fission, if society accepts it.

2.7.5 Thought Experiment

Let us end this discussion with a thought experiment: statistically, Germany's more than 40 million road vehicles are operated only 1 h/day; they are parked for 23 h. Let us imagine that they have FCs under their hoods with a capacity of, say, 50 kW each and are plugged in when parked in the home garages or on employee parking lots. Consequently, only 5% of Germany's car fleet operated at standstill would provide some 100,000 MW, which compares well with the capacity of the central stations on-line today. We said it earlier: thought experiments seldom become real, but contain a certain truth. Here are two truths; the first reads: in the long run, will it really be compatible with the seriously taken energy and transport sustainability so urgently needed to leave useless a whole fleet of "power stations on wheels" (Amory Lovins) with a potential capacity 20 times higher than is in traditional use today, with a price tag as low as some 10 €/kW? (The engineer knows well that a mobile, highly dynamic engine with 10,000 rpm, and a service lifetime of 3,000 h is technically and economically absolutely something different from a stationary power station with 3,000 rpm and a life of 80,000 h prior to the first full maintenance standstill.) And the second truth: Mobile FC vehicles will only be supplied with hydrogen fuel, because any hydrocarbon fuel used instead means the necessity of a 100-million-fold mobile carbon dioxide collecting devices—a technical and economic impossibility!

2.7.6 Secondary Energy Sector Ever More Important

To end this section, none of the aforementioned arguments negates the legitimacy of either electricity or hydrogen; each has its domain; they compete on certain issues, and are partners in some others. Relative to the primary energy (raw materials) sector, the secondary energy sector grows more and more in importance. It begins to dominate the energy scheme of a nation. It will consist in the future of two secondary energy carriers, electricity and hydrogen, developed in tandem!

As mentioned earlier, novel energies need time! It seems almost always too late to start creating consciousness and further awareness. People live and work downstream and ask for reliable, affordable, and clean energy services. Since the hydrogen energy economy moves the center of gravity within the energy conversion chain toward its end, exactly where these people live, professionalization of their supply is needed, not unlike professionalization at the chain's front end where we are accustomed to the professional operation of power plants, refineries, coal mines, and the like. Delay is the foe of success. Consequently, it is HYtime!

2.8 Hydrogen Energy Technologies along Their Entire Conversion Chain

A comparison of materially open-ended and closed energy systems is provided in Figures 2.16 and 2.17.

The traditional system takes something irrevocable from the earth's crust, converts it mechanically, chemically, or nuclearly into something else, and gives it back to the geosphere; often global distances separate the two locations. In the nuclear case, what remains has long, in some cases very long, half-life periods (e.g., plutonium ca. 24,000 years) and is radiotoxic and radioactive. In the case of fossil fuels, the residuals are unavoidably associated with environmental pollutants and the release of GHGs into the atmosphere. Through numerous open ends, the environment and climate are burdened; with the help of additional technologies and systems, optimization engineers try to close up these open ends, sometimes with only poor or no success.

The renewable hydrogen energy system is different: All sorts of technologies convert solar irradiance, wind, hydro (or geothermal heat, tidal or ocean energy, or others) into both the secondary energies heat and/or electricity, which are then used to split de-mineralized water electrolytically (or thermolytically, or otherwise) into hydrogen and oxygen. The oxygen is released to the atmosphere or utilized chemically; the hydrogen in gaseous or liquefied form delivers the energy for the hydrogen energy economy, as a gas in the heat market, re-electrified through FCs or gas turbine/steam turbine combined cycles in the power market, and in all transportation sectors on land, in the air, in space, and at sea. Water is taken from the earth's water inventory; water, after hydrogen usage and recombination with oxygen (from air), is returned to that inventory, physically and chemically unaltered. The locations, though, of water extraction and water return may be global distances apart.

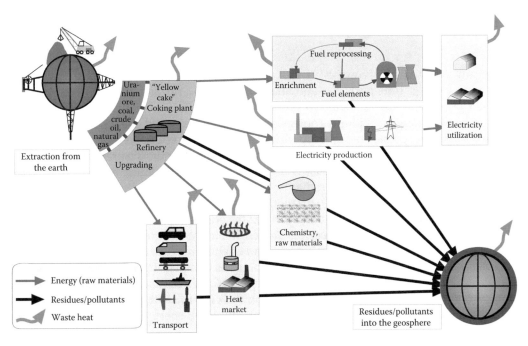

FIGURE 2.16
Materially open-ended energy systems.

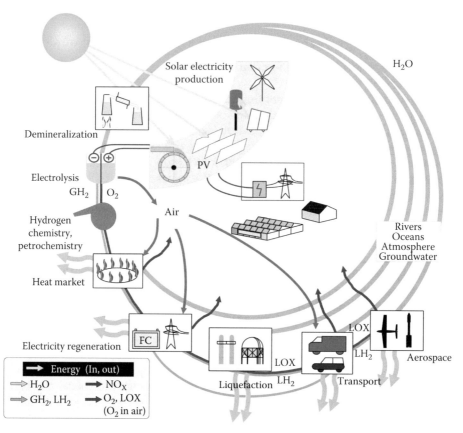

FIGURE 2.17
Materially closed hydrogen energy systems.

Like any other energy conversion chain, the hydrogen chain consists of five links: production, storage, transport, dissemination, and finally utilization. Primary energy raw materials (feedstock) are converted to primary energy, further to secondary energies, end energies, useful energies, with the conversion ending with desired energy services such as warmed or cooled rooms, energy support in transport and production, illuminated living spaces, city streets or workplaces, and all sorts of communication. The renewable energies lack the first chain link (from primary energy raw material to primary energy); they begin with primary energies like solar irradiance, wind energy, upstream hydropower potential, etc. Providing energy services is the sole motivation for the run through of any energy chain; there is no other motivation. The links preceding the energy services have no justification in themselves; they serve to supply services and contribute to meeting the supply conditions in terms of amount and security, cost, safety, and environmental and climatic cleanness.

In countries with high energy imports, sometimes complaints are heard like, "we are energy undersupplied, we have too little energy!" Now, what is really meant is that the amount of energy services for running the country is insufficient, partly because of a lack of primary energy, but mainly because of lamentably small energy and exergy efficiencies. And these are exclusively a matter of technologies.

In Tables 2.1 through 2.3, three technology categories for the hydrogen chain links are listed—state of the art, midterm, and longer term; let us review each of these.

TABLE 2.1

The Hydrogen Energy Technologies: Production

State of the art (with incremental further development)

- Reformation of natural gas
- Gasification of coal
- Partial oxidation of heavy crude oil
- Electrolytic hydrogen from hydropower
- Hydrogen from nuclear electricity

Midterm (ca. 10 years from present)

- Electrolytic renewable hydrogen from wind, PV, solar thermal power, and other renewable sources
- Hydrogen from biomass

Long term (in 20 years or more)

- Hydrogen-supported decarbonization of hydrogen from fossil fuels with CCS
- Hydrogen from coal with the help of high-temperature nuclear heat
- High-temperature electrolysis
- Radiolysis, thermolysis, and photocatalysis of hydrogen

TABLE 2.2

The Hydrogen Energy Technologies: Storage and Transport

State of the art (with incremental further development)

- Hydrogen liquefaction
- Hydrogen cartridges in portable electronics
- Metal hydride containers
- Embrittlement-proof hydrogen pipelines
- Continuous or batchwise GH_2 or LH_2 transport
- Hydrogen in refineries
- Hydrogen in the space business

Midterm (ca. 10 years from present)

- Pick-a-back hydrogen in natural gas pipelines: "NaturalHy"
- 700 bar filament-wound mobile hydrogen tanks
- Vacuum-insulated LH_2 tanks with low boil-off

Long term (in 20 years or more)

- "Supergrid"—an LH_2-cooled superconducting high-capacity cable with simultaneous LH_2 transport
- LH_2 tankship transport
- LH_2 loading and unloading harbor equipment
- Carbon nanostorage

At present, hydrogen is produced predominantly by steam reforming of natural gas, labeled steam methane reforming (SMR), through gasification of coal, and, where cheap electricity is available, through electrolysis of water. Nuclear electricity is used where nuclear operations are societally accepted, for example, in France. With the exception of demonstration projects, in no case where hydrogen is produced from fossil fuels in macroeconomic scales is the coproduced carbon dioxide yet being captured and safely stored away. Renewable hydrogen is nowhere operational in large quantities. A great number of studies and demonstrations are under way, for example, where PV generators or SOT power plants and electrolyzers work together, or where wind electricity must be transported over long distances to the energy user, for example, from off-shore wind parks

TABLE 2.3

The Hydrogen Energy Technologies: Dissemination and Utilization

State of the art (with incremental further development)
- Hydrogen in space transportation
- Spaceborne fuel cells
- Fuel cells in submersibles
- Hydrogen-fueled low-temperature fuel cells in portable, stationary, and mobile applications
- Hydrogen in ICEs and gas turbines
- Hydrogen-fueled mobile APUs

Midterm (ca. 10 years from present)
- Hydrogen/oxygen spinning reserve
- Hydrogen and the high-efficiency ICE
- Hydrogen-filling stations
- Hydrogen in airborne APUs
- Fuel cells replacing airplane ram air turbines
- Geothermal steam temperature rise through mixing with steam from H_2/O_2 recombination

Long term (in 20 years or more)
- Hydrogen jet fuel in air transportation
- Hydrogen as the laminarizing agent in aerodynamics
- Hydrogen and the drive train in sea-going vessels
- Hydrogen propulsion in ICE or fuel cell locomotives

to on-shore users. In all demonstrations, it has been seen that the intermittent renewable energy offer and the electrolyzer's dynamic behavior are fairly closely correlated; the electrolyzer responds rapidly to the varying electricity yields from solar and wind converters.

Clearly, from an environment and climate change standpoint, renewable hydrogen is the ultimate choice, sometimes read as the "primary choice." However, more or less all renewable energy technologies, however admirable their development and market progress over the last years were (and presumably will continue to become), work financially under highly subsidized conditions. Slowly, particularly with galloping oil and gas prices and renewable technologies' further technological development successes, they are approaching market conditions and will get there, realistically, in another few decades.

Consequently, until full market conditions are achieved for renewable technologies, further development toward marketability of hydrogen on the one side and renewable technologies on the other side should be pursued in parallel on a dual carriageway prior to dovetailing their individual results. No reason is seen not to proceed with hydrogen's addition to the energy mix on the marketplace, although renewable technologies are not yet fully market ready. Renewable hydrogen is the ultimate goal, but it is not the precondition for the entry of hydrogen into the market! In the meantime, lots of so far irregularly utilized or even flared hydrogen capacities facilitate their utilization. Figure 2.18 shows the amounts of hydrogen energetically so far not used in Germany, a total of almost 1000 Mm³/a, equivalent to the average consumption of some 7850 FC buses. In addition, hydrogen from some 10,000 sewage plants in the country may be utilized as a transport fuel in the interim.

A strong argument in favor of the utilization of hydrogen energy prior to the maturity of renewable technologies is that "clean coal" via air separation, coal gasification, capture of hydrogen and carbon dioxide, and finally combined cycle electricity generation

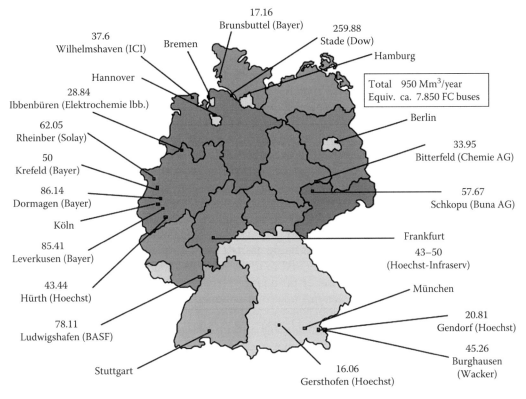

FIGURE 2.18
Hydrogen (Mm³/year) as a by-product of German industry. (Adapted from DWV, DWV–German Hydrogen and Fuel Cell Association 12/2003; Ludwig Boelkow Systemtechnik 1998, Hydrogen as By-Product of German Industry, 1998.)

is inherently connected to hydrogen. Dual benefit is offered by this exergetically highly efficient process: simultaneously cleaning up coal and producing hydrogen energy! The process is not new: it was invented by Friedrich Bergius (1884–1949), who in 1931 received the Nobel Prize for Chemistry for his work on making gasoline from coal. The process is still in industrial use in South Africa (and perhaps elsewhere).

With respect to storage and transport of hydrogen (Table 2.2), a whole collection of technologies are fully operational worldwide for gaseous and liquefied hydrogen (LH₂) or metal hydrides; all that has been learned there over the past century is the welcome preparation for perpetuation into the forthcoming hydrogen energy economy. A special transport method makes headway and deserves particular attention: transport of hydrogen up to a capacity of 10%–15% pick-a-back in operational natural gas pipeline grids without major technical modifications. An ongoing European project labeled "NaturalHY" (www. naturalhy.net) studies the various technology and handling consequences like hydrogen embrittlement of materials, hydrogen loading and off-loading techniques, and change in energy throughput of the natural gas/hydrogen combination. With the addition of hydrogen, the heating value of the mixture decreases, the Wobbe Index, though, remains nearly the same. All in all, it is expected that hydrogen storage and transport in natural gas pipelines will be a welcome inexpensive means of utilizing approved technology also for a novel energy carrier. At least for a first and limited period of time an additional, extremely expensive hydrogen pipeline system investment would not be needed. This applies to centralized hydrogen production or off-ship hydrogen at the front end of the hydrogen conversion chain and to distributed hydrogen utilization at its back end.

Many an argument speaks for centralized vis-à-vis distributed hydrogen production. It is obviously easier to collect and sequester carbon dioxide from a small number of large units than from millions of distributed small ones: here professionals are at work, and the coal, oil, and gas industry and the merchant gas traders well know how to reform natural gas or gasify coal, both in capacities justifying expected low cost. The marine industry and the natural gas traders are well experienced in handling liquefied natural gas (LNG) in tanker ships and harbor equipment, as well as in re-gasifying the cryogenic liquid prior to its introduction into the gas grid. Whatever pros and cons may have emerged, all experience gained is beneficial for the point in time when global hydrogen trade begins. One disadvantage must be faced, though: so far, in all those cases where fossil fuels are involved, almost nowhere is carbon dioxide captured, sequestered, and securely stored away commercially in large amounts in underground storage facilities under an impermeable overhead rock cover. In a few demonstration projects, for example, in the North Sea or in the Gulf of Mexico, practical experience is being gathered.

When renewable hydrogen is generated in distributed installations and utilized on-the-spot in residential energy systems or on board vehicles, no carbon dioxide is involved, and hydrogen pipeline grid costs are nil. As mentioned earlier, time is needed to bring these distributed systems to market. Energy handling at the end of any conversion chain by millions of laypersons is not a sustainable option, but so far a professional regime has nowhere been established similar to what is experienced at the front end of the chain. Professionalization of the chain's end energy services is indispensable. But be that as it may, finally market cost will decide on central or distributed hydrogen production, or both.

A few words on storage: Stationary hydrogen storage is at hand, both for gaseous and liquid hydrogen in high-pressure steel flasks or cryogenic dewars. Large capacity underground storage for gaseous hydrogen (GH_2) in leached salt domes may build on what has been learned from operational underground air or natural gas storage, though special care needs to be taken to prevent leakage of the smallest element of the periodic table of elements: hydrogen! The most challenging venture is the tank on board motor vehicles. For a usual vehicle range of, say, 500 km, the tank for GH_2 requires an inner pressure of 700 bar, which, from a manufacturing and lifelong safety standpoint, is not at all trivial to achieve and maintain. The filling station's pressure will then amount to even ca. 1000 bar, requiring compressor energy. If GH_2 at the filling station is provided by re-gasification of LH_2, then the amount of pressurization energy necessary is smaller. The mobile tank is made of filament-wound carbon fibers with an inner steel or aluminum layer. Because of low cycle fatigue of the tank structure as a consequence of frequent charging and discharging ("breathing" of the tank structure), tanks have to be replaced after certain periods of time, in contrast to the one and only gasoline or diesel tank on duty over a vehicle's entire lifetime. The LH_2 tanks have a double wall structure with an evacuated ring volume and multiple wrinkled aluminum foils to avoid heat transfer from the outside to the inside. That the liquefier requires about 1 kWh per 3 kWh of LH_2 is state of the art. Depending on the tank size, the boil-off rate of modern tank designs is a few % per day or less. The allowable inner pressure of the tank is a few bars, which avoids venting of boil-offs until the pressure allowance is reached. If, however, boil-off occurs, the idea is to avoid venting and rather utilize the boiled off hydrogen in an FC to provide electricity for recharging batteries.

Table 2.3: Local hydrogen dissemination in trucks, trailers, or rail cars with onboard pressurized gaseous or LH_2 is day-to-day practice. Of course, LH_2 is more expensive than pressurized gaseous hydrogen (CGH_2). Eventually, transport and handling, however, of much higher energy density LH$_2$ will offset the higher price. If GH$_2$ is transported pick-a-back in natural gas pipelines, hydrogen separation membranes at the exit points need to be in place.

Three application areas for FCs are visible: (1) in portable electronics fueled with the help of hydrogen or methanol cartridges, (2) in stationary applications, and (3) in the transportation sector in buses, passenger and light duty vehicles and in heavy duty trucks, in aviation, and at sea. It so seems that hydrogenized portable electronics will be the first on the market, because they offer a much longer life than the conventional batteries in use, and, an often heard light side note, because buyers do not care about the cost of the energy involved, which sometimes amounts to some €/kWh! Stationary FCs for industrial use will follow, and finally mobile FCs will be seen on board buses and passenger vehicles.

The most challenging effort in the utilization field is selecting hydrogen/oxygen recombination technology, be it either adapted conventional technology like ICEs or gas turbines, or be it newly developed FCs. Adapted technology has the advantage of familiarity and market success confirmed over many decades, in some cases up to a century; the market price, the behavior under actual long-life conditions, and the operation and maintenance (O&M) requirements are well known; in short, economic viability is a given. The FC (Figure 2.19), on the other hand, is not a Carnotian heat engine, but an exergetically highly efficient chemo-electric converter serving as a prototypical combined cycle in itself, and it is, thus, an example for the aforementioned system change. It generates electricity and heat simultaneously. It promises few irreversibilities (if no reformer is included, the major source of irreversibility); compact design; no moving parts and, thus, no vibrations; and low noise. And, depending on the type of FC, it provides the "right" temperature for heat applications in stationary use: ≤100°C for proton exchange membrane (PEM) FCs, around 200°C for high-temperature PEMs or phosphoric acid FCs, 600°C–650°C for molten carbonate fuel cells (MCFCs), and 700°C–900°C for solid oxide fuel cells (SOFCs). Low- to high-temperature PEMs are exactly what are needed for homes, or buildings, or hospitals, depending under which climate and weather conditions they serve; MCFCs fit the

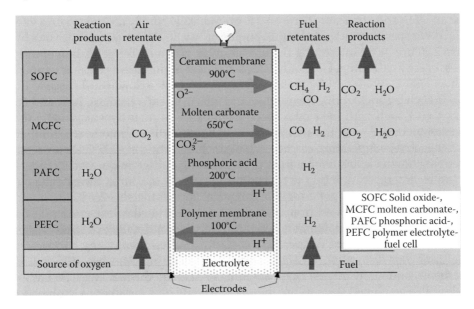

FIGURE 2.19
Various fuel cell systems. SOFC—solid oxide fuel cell, MCFC—molten carbonate fuel cell, PAFC—phosphoric acid fuel cell, and PEFC—polymer electrolyte fuel cell.

exigencies of many small-to-medium size industries, hospitals, and large laboratories, and SOFCs are an excellent topping technology for gas turbine/steam turbine combined cycles.

A thorough exergy analysis of a simulated methane-fueled internally reforming high-temperature SOFC plus bottoming gas turbine in the 100 kW range and heat recovery steam generation is given by A. V. Kotas. The maximum total exergetic efficiency is more than 60%. In case of FCs where the reforming temperatures are insufficient for internal reforming, they require an extra reformer to supply hydrogen reformate from fossil fuels.

An exergy analysis of hydrogen production via steam methane reforming (SMR) is given by A. P. Simpson et al. Approximately 80% of the world's total hydrogen production uses SMR of natural gas. The process consists of the elements natural gas compressor, water pump, reformer, water gas shift reactor, membrane hydrogen separator, air/methane mixer, and a number of heat exchangers; the temperature is 700°C. Most of the exergy destruction occurs in the reformer due to irreversibilities in the fuel/steam mixer and as usual in heat transfer and combustion. The total exergy efficiency and the thermal energy efficiency are ca. 62% and 66%, respectively. Of the 38% of lost exergy, a good 80% is lost within the system; the rest exits with the exhaust stream. Intelligent heat management, varying the steam-to-carbon ratio, and reducing the amount of retentate leaving the membrane separator are means of maximizing the efficiency. A general exergy analysis of energy converters can be found in A. Bejan et al.

The preferred mobile FC is the low-temperature PEM. It fits into the ongoing electrification scheme of the automobile by providing highly efficient electricity for the electric motors in the drive train and for auxiliaries. With only water vapor in the exhaust, PEMs make the vehicle environmentally and climatically clean where it is operated, which is of paramount importance, particularly in the polluted centers of agglomerations. Clean means here locally clean, because whether hydrogen-fueled individual transport is generally clean depends on which primary source hydrogen is produced from, renewable or nuclear electricity, or fossil fuels with or without capture and storage of coproduced CO_2. Replacing the exergetically miserable onboard electric generator with an engine-independent FC may become an interesting first step. Because it is exergetically simply absurd to run a generator of some 5 kW capacity with the help of an engine of, say, 100 kW or (sometimes much) more.

Figure 2.20 brings an interesting though far away vision of exergetization of industrial schemes: energy cascading. Generally, industrial heat requirements start at temperatures as high as 1700°C (or higher) for metalworking and follow a downward cascade through various branches (brickworks, steam power plants, catalytic reactions, heat and cold for buildings, etc.), each with their own specific temperature requirements, finally down to ambient temperature with the remaining heat being radiated into space. Similarly, an upward (negative) temperature cascade is imaginable starting at −155°C for air separation and stepping down via low temperature metal forming, refrigeration, food storage, finally also to ambient temperature (in the context of this article, of course, the cascade ought to start at the liquefaction temperature of hydrogen at approximately −253°C).

The ideal cascade passes heat from one temperature level to the next; exergy destruction is minimized and at ambient temperature the system has optimally extracted exergy from energy! In practice, however, temperature cascading suffers from many barriers, such as geographical user dislocation, dissimilarities of neighboring branches, different market requirements, and many others. Huge amounts of heat are thrown away. So far, exergetically desirable combined systems are in operation only here and there. Combined power and heat cycles in small district heating systems are good examples, and so are inner-company systems, especially in the chemical industry.

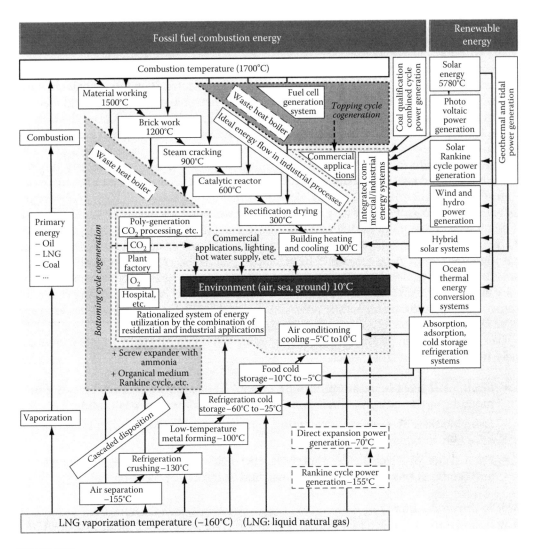

FIGURE 2.20
Energy cascading. (Adapted from Kashiwagi, T., Present status and future prospects of advanced energy technology for solving global environmental problems. *The Global Environmental Technology Seminar JETRO*, Tokyo, Japan, 1990.)

Now, what does this have to do with hydrogen energy? When hydrogen is produced from coal at the start of its conversion chain (see Section 2.9), hydrogen, electricity, and low-temperature heat are being produced simultaneously in an exergy-efficient combined cycle. Similar things can be observed at the back end of hydrogen's conversion chain: exergy-efficient, hydrogen-fueled, low-temperature FCs are in themselves combined cycles simultaneously generating electricity and low-temperature heat. Or a last example, LH$_2$ on board an airplane serves of course as the fuel for the jet engines, but it also cools the outer wing and empennage surfaces of the plane and, thus, laminarizes the air flow, thereby retarding the onset of turbulence flow and, thus, reducing the drag. There are many more examples.

2.9 Hydrogen Production

Hydrogen carries secondary energy. Like electricity, the other secondary energy carrier, it is produced from all thinkable primary energies and electrical energy—coal, oil, natural gas, nuclear electricity, nuclear heat, all sorts of renewable energies, and grid electricity. Hydrogen and electricity are interchangeable via electrolyzer and FC; the electrolyzer makes hydrogen from electricity, the FC makes electricity from hydrogen. Hydrogen from fossil fuels or biomass is a task for chemical process engineers. Typical process technologies are coal gasification, natural gas reforming, or partial oxidation of heavy oil fractions; the required heat is introduced autothermally or allothermally. Obtaining hydrogen from renewable energies is a task for electrochemists. Inexpensive electrolytic hydrogen depends on inexpensive electricity. In order not to weaken hydrogen's inherent character of being environmentally and climatically clean over the entire length of its conversion chain, the chain's first link, the production step from primary energy raw material to primary energy, needs to be clean too. In principle, that is the case for renewable energies and, consequently, also for renewable hydrogen, since they are free of operational primary energy raw materials per se, and it will become the case when hydrogen production from fossil fuels is environmentally nonpolluting, and coproduced CO_2 is sequestered and securely stored away without harming the climate; CCS is inevitable!

The major hydrogen production technologies are those producing hydrogen from fossil fuels, from biomass or from water:

- From fossil fuels by steam reforming of natural gas (SMR), thermal cracking of natural gas, partial oxidation of heavy fractions (POX), or coal gasification
- From biomass by burning, fermenting, pyrolysis, gasification, and follow-on liquefaction, or biological production
- From water by electrolysis, photolysis, thermochemical processes, thermolysis, and combinations of biological, thermal, and electrolytic processes

Prior to the production of electrolytic hydrogen, two key questions must be answered: (1) Why hydrogen, when electricity would do? and (2) Which option: central hydrogen production and dissemination by grid or non-grid transport, or distributed production with low or even zero transport expenses?

To question (1): Is the production of electrolytic hydrogen really necessary, or can the envisaged task be performed, perhaps even better, by electricity itself? The reason for this question is obvious: hydrogen production adds an additional link to the energy conversion chain, and additions add losses and cost and sometimes ecological sequels. Areas where hydrogen is unavoidable are air and space transportation, and surface transportation over up to global distances on land and sea. Possibly, short distance surface transport or transport on long distance rail will be electric (see Section 2.7). In industry, information and communication and all sorts of service businesses as well as mechanical production are the domain of electricity. Hydrogen energy, on the other hand, is key in refineries, in ammonia and methanol syntheses, and for all sorts of hydrogen treatment in industrial chemistry; hydrogen is needed for biomass liquefaction. In buildings, light and electric or electronic appliances are the domain of electricity. Because of its miserable efficiency, Ohmic resistance heating is fading out of use; higher efficiency compressor heat pumps, however, are booming, and low-temperature FC central heating systems depend on hydrogen as the fuel, be it pure hydrogen or reformat.

Now to question (2), central or distributed hydrogen production? The present energy supply system is clearly centralized: electricity or natural gas or light oil are centrally produced in power stations, gas fields, and oil refineries, and the secondary energy carriers electricity, gas, and other energy products are then "diluted" via overhead transmission lines, gas grids, and oil pipelines, or transported on rivers, via road or rail. At the very end of the chain, filling stations supply road vehicles; the local retailer brings fuel oil for buildings' central heating systems or gas for cooking and heating. Now, the supposedly easiest way would perhaps be to mimic the present system when changing over to hydrogen energy: it is well understood, there is a wealth of experience to tap, and those acting are familiar with the technologies. As mentioned earlier, hydrogen could well be transported pick-a-back in the operational natural gas grid, avoiding an extra hydrogen transport system. Further, decarbonizing is much easier when hydrogen from fossil fuels is generated in large systems rather than in distributed installations of much smaller capacities: collecting GHGs from millions of distributed emitters is technically and commercially impossible. On the other hand, indications can no longer be ignored that more and more clean distributed installations such as PVs, wind, and biomass are harnessing local potentials which were previously lying fallow. It might not be illusive to expect that, at least as an addendum to the central system, solar or wind or biomass hydrogen will be produced on the spot of utilization, thus avoiding long and expensive and inefficient transport lines. Time will tell whether vehicles will be filled with electrolytic or reformed hydrogen produced at the forecourt of filling stations; whether the family car will be refilled with electrolytic hydrogen produced with the help of electricity from the roof of their house; or whether district electricity, heat, and hydrogen supply systems will evolve.

Electrolytic production of hydrogen goes back to William Nicholson, who in 1800 reported on the electrolysis of water. Principally, three process versions of electrolytic dissociation of water have been or are being developed:

- Conventional water electrolysis utilizing an alkaline aqueous electrolyte (30 wt.% KOH) with a separation membrane avoiding the remixing of split-up hydrogen and oxygen
- Solid polymer electrolyte (SPE) water electrolysis utilizing a PEM; the technology shows similarities with the polymer electrolyte membrane fuel cell (PEMFC)
- High-temperature steam electrolysis at 700°C–1000°C

The first two technologies are commercially marketed; the third, mainly because of high-temperature material problems, is still far from realization. Figure 2.21 shows for all three the cell voltage over the current density.

For conventional and SPE electrolysis, the respective cell voltages considerably exceed the theoretical decomposition voltage of water electrolysis of 1.23 V at 25°C and 1 bar, which corresponds for the best electrolyzers to an efficiency of 80%–85%. For steam electrolysis, the cell voltage drops significantly. The best present electrolyzers need between 4.3 and 4.9 kWh/Nm³ H_2.

The future of electrolytic hydrogen depends clearly on the price of electricity. So far, major installations with capacities of some 10000 Nm³ H_2/h have only been erected where inexpensive electricity is available, for example, near big hydroelectric dams such as in Aswan, Egypt, or in locations in Norway or Canada, where major amounts of hydrogen are utilized in fertilizer industries.

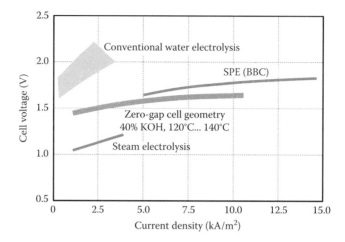

FIGURE 2.21
Electrolysis systems. SPE, solid polymer electrolyte; KOH, potassium hydroxide. (Adapted from Wendt, H., Water splitting methods, In *Hydrogen as an Energy Carrier—Technologies, Systems, Economy,* C.-J. Winter and J. Nitsch (Eds.), Springer Verlag, Berlin, Germany, 1988.)

Since most probably the future average price of electricity will increase rather than decrease (at least in industrialized countries), electrolytic hydrogen may get its market only where the electricity demand is temporarily low, for example, at night, or where base load (nuclear) power station load control ought to be avoided. Another future area where electrolyzers may find their niche is smoothing out the time-dependent solar or wind electricity yield, or in cases where transporting renewable electricity to distant consumers is more expensive than transporting hydrogen. Demonstration units such as the German–Saudi "HYsolar" installation have shown that the electrolyzer is flexible enough to respond to the fluctuating renewable energy yield with respect to both time and capacity (www.hysolar.com). There is hope that solar electricity will soon achieve grid parity with conventional electricity, partly because of technology developments, partly because of rising costs for conventional electricity.

A most important item related to the production of hydrogen in general is its exergizing influence on the cleaning up processes of fossil fuels. Three modern coal plant designs with CCS are under consideration or have begun their demonstration phase. Three typical separations of components from mixtures are key:

- Separation of CO_2 from N_2 in fuel gas decarbonization
- Separation of CO_2 from H_2 in fuel decarbonization
- Separation of O_2 from N_2 in air separation

Systems under consideration or in demonstration or even in routine industrial practice are

1. Integrated gasification combined cycle (IGCC) plants with air separation (in order to get rid of the N_2 ballast and avoid NO_x production), coal gasification, CO shift, desulfurization, CO_2 capture prior to combustion (pre-combustion capture), and utilization of the final syngas $CO + H_2$ in liquefying Fischer–Tropsch synthesis and methanol production, or utilizing the hydrogen for power production in a combined cycle, or for ammonia syntheses, or as feedstock for the hydrogen infrastructure

2. The above plant with air separation and integrated CO_2 capture prior to combustion (labeled "Oxyfuel")

3. A coal plant with CO_2 capture after combustion (post-combustion capture) through amine absorption and thermal steam recovery

All of these designs have advantages and disadvantages; the final "winner" has not yet emerged. IGCC is in operation in a few demonstration plants around the world. Post-combustion has one clear plus, since it seems to be easy to add decarbonization technologies to an existing operable plant, something which is not feasible for the other two designs that have to be built from scratch. All three CCS designs are costly and decrease the overall efficiency of the plant; cost estimates vary between 30 and 50 €/t CO_2 removed, with the major cost item being the in-plant membrane CO_2 capture rather than its subsequent liquefaction, transport, and final storage. The plant efficiency decreases by 8%–12%. Eventually, the cost is hoped to be compensated by reduced CO_2 certificate obligations; even welcome "negative" costs are not illusive when, for enhanced secondary oil or coal-bed methane recovery, CO_2 is injected under pressure into the oil well or the coal bed.

Figure 2.22 brings estimated electricity costs [€-¢/kWh] for dissimilar methods of CO_2 capture in brown coal, hard coal, and natural gas plants for 2020 and 2030 compared to conventional plants without capture; the vertical bars indicate estimation uncertainties. What do we see? Relative to the conventional plant, of course, modern technologies and in particular CO_2 capture raise costs: natural gas combined cycle plants are expected to have the smallest increase, followed by brown coal and hard coal IGCC, in that order. Then comes the monoethanolamine (MEA) options, again with the smallest increase expected for the natural gas plant. The Oxyfuel and Selexol options do not show big cost differences. But all together, the cost rise up to some 5–6 €-¢/kWh is painful. The appraisal is that natural gas is ahead of the others. The question remains as to whether or not the ongoing concentration of natural gas oligopolies will give rise to unacceptable price rises or supply shortages, or both, at least in the long run. In both the other cases, IGCC is leading, followed by not too dissimilar results for MEA, Oxyfuel, and Selexol.

In the table of Figure 2.22, potential CO_2 sinks, both globally and for Germany, are given: depleted oil or gas fields, un-minable coal seams, and saline deep aquifers. Disappointing are the rather limited static ranges of depleted oil and gas fields and coal seams. It appears that the only real long-term potential is offered by deep saline aquifers where salty water absorbs carbon oxide. The large variations in capacity and range, though, point to uncertainties; clarification will only come with practical experience.

CO_2 storage demonstration test sites are in operation in Australia, Europe, Japan, and the United States. In all four locations, CO_2 is stored in deep saline aquifers. In non-power-industry plant operations, the technologies are mature and operational in large scales; emerging technologies promising less cost and higher storage security are under development. The cost is dominated by the specific CO_2 separation technologies (membranes, amines, and others), not by CO_2 compression, liquefaction, and transport. Pipelines or CO_2 vessels similar to those in use for shipping LNG or LPG are day-to-day practice. The distances between major CO_2 producers and potential storage basins around the world are not too large. Nowhere, however, is it known for sure what the biological, chemical, hydraulic, or geological consequences will be over the long term (hundreds of years). It is much too easy to compare CO_2 storage with the millions of years of natural gas, coal, or oil underground storage and expect no cons; time will tell. Other storage possibilities may be un-minable coal seams, depleted natural gas or oil reservoirs, or deep sea "lakes," all of which have their peculiarities. So far too little is known about accompanying degradation of groundwater quality, or about

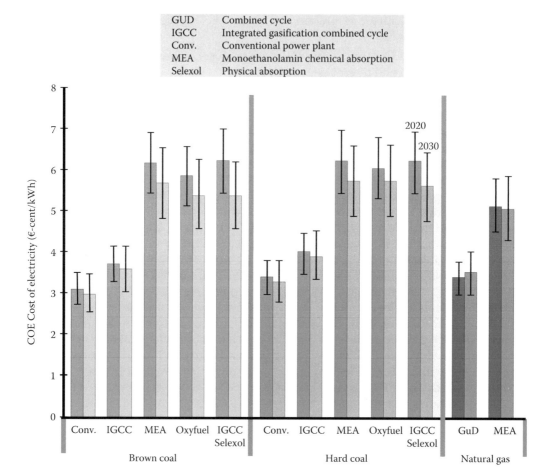

	GUD	Combined cycle
	IGCC	Integrated gasification combined cycle
	Conv.	Conventional power plant
	MEA	Monoethanolamin chemical absorption
	Selexol	Physical absorption

FIGURE 2.22

Cost of electricity for CCS in hard coal, brown coal, and natural gas plants. (Adapted from Linssen, J. et al., Technologien zur Abscheidung von CO_2, BWK Bd. 16, Nr. 7, 2006.)

Storage option	Global		Germany	
	Capacity, Gt CO_2	Static range, years	Capacity, Gt CO_2	Static range, years
Depleted gasfields	690	73	3	8
Depleted oilfields, EOR	120	13	0.1	<1
Deep saline aquifers	400–10000	43–1062	12–28	34–78
Non-minable coal seams, ECBM	40	4	0.4–1.7	<2
EOR-enhanced oil recovery via CO_2 injection		ECBM-enhanced coal bed methane		

potential damage to hydrocarbons or minerals in sedimentary rock, or the acidification influence on deep sea water or subsoil fauna and flora. Carbonization of minerals, on the other hand, is a welcome consequence, as it has been accomplished by nature over millions of years providing stabile and the least mobile carbonates. CO_2 storages will not be absolutely free of leakage; an amount of less than 1% over 100 years is considered "safe."

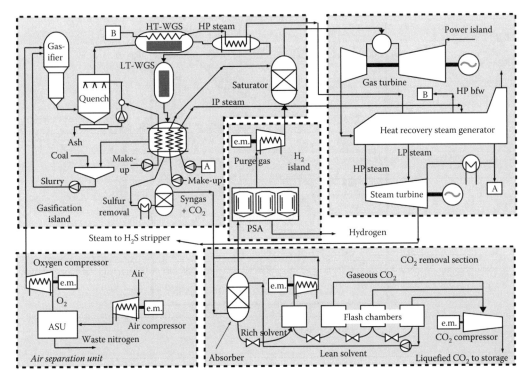

FIGURE 2.23
Combined cycle production of hydrogen, carbon dioxide, electricity, and heat from coal. (From Chiesa, P. et al., *Int. J. Hydrogen Energy*, 30, 747, 2005. With permission.)

Figures 2.23 and 2.24 give the interesting example of exergy-efficient combined production of hydrogen, electricity, and carbon dioxide. What is seen? The conversion of coal into hydrogen occurs in five more or less marketed technology steps (islands): the first step is air separation; the second, oxygen-supported coal gasification; the third and fourth, hydrogen and CO_2 separation; and the fifth and final step adds combined cycle electricity generation. The results are seen in Figure 2.24: with carbon removal of 90%, ready for compression, liquefaction, and transport to the storage site, 58% of the coal's energy content is converted to hydrogen and 4% to electricity—together 62%. If an estimated 10% is reserved for CO_2 capture, transport, and storage, the resulting exergy efficiency is 52%! (For comparison: modern coal-fired power plants are 46% exergy efficient, minus 10% for CCS makes 36%.)

2.10 Hydrogen Handling, Storage, Transport, and Dissemination

Here, thorough practical experience has been accumulated over almost two centuries, starting after the first hydrogen papers of Cavendish and Lavoisier in the late eighteenth century when the balloons were filled with light hydrogen gas. Today, the technical gases industry, hydrogen chemistry, the space launch industry, and refineries are among those most familiar with hydrogen, which they handle as a commodity or as an energy carrier in their day-to-day practice. All that has been learned there can be considered as good preparation for the up-and-coming hydrogen energy economy.

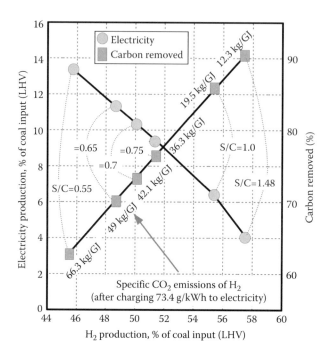

FIGURE 2.24
Exergy-efficient combined hydrogen, electricity, carbon dioxide, and heat production from coal. (From Chiesa, P. et al., *Int. J. Hydrogen Energy*, 30, 747, 2005. With permission.)

We distinguish traded hydrogen and captive hydrogen, with the latter being produced and used in-house in chemical industries or refineries without being traded. Traded hydrogen is mostly in the hands of merchant gas traders who provide hydrogen to glass and electronics manufacturers, to food industries, to electric utilities for cooling big electrical generators, and, not least, to the space industry for onboard FC supply and, in much larger quantities, as the fuel for the LH_2/LOX launcher engines.

Hydrogen is shipped in gaseous or liquefied form using many different means of transportation. Quite a number of CGH pipeline grids are in operation all over the industrialized world. For example, in Germany, a pipeline of 210 km length and ca. 30 bar of inner pressure has been (and still is) in operation since the 1930s; it serves 18 industrial sites. Higher pressure gas comes in steel flasks (200 bar) on road or rail tube trailers, seldom in pressurized pipelines (40 bar) of industrial grids. Liquid hydrogen is transported on the road in lightly pressurized double-walled cryogenic tanks (dewars) at −253°C, even when the customers' demand is not LH_2 but GH_2. The much higher hydrogen transport density of LH_2 justifies liquefaction prior to transport and, consequently, re-gasification after delivery acceptance. Because of the required high purity in the semiconductor business, hydrogen is delivered as LH_2. Economically, one LH_2 tanker truck delivers approximately the same amount of hydrogen as 20 pressurized GH_2 trailers, not only a convincing example of economic viability, but also a welcome emission abatement of diesel truck engine exhausts! An unusual means of transport could have been seen in the early days of the Kennedy Space Center when LH_2 was delivered in vacuum-insulated steel containers on barges from the production sites at the Gulf of Mexico to Cape Canaveral, Florida, using the inner coastal waterways.

Sea-going LH_2 supply from far away places will become necessary when wellhead decarbonization of fossil fuels already at the seller's site is taken seriously, or when sites of immense renewable potential begin contributing to the world's energy trade. Historically, environmentally and climatically cleaning up fossil fuels was (and still is) the obligation of energy buyers. For the time being, the sellers are well off; they simply ship the "dirty" energy carrier. LH_2 tanker transport over ocean distances will be the solution whenever pipeline supply is not feasible. Potential examples are hydrogen from decarbonized coal, say, from Australia or South Africa, or solar hydrogen from Australia or wind-derived hydrogen from Patagonia to Europe, Japan, or North America. All that had been learned from ongoing LNG tanker transport at sea and its cryogenic loading and unloading harbor equipment is a welcome experience for the start of global hydrogen trade.

GH_2 density at ambient temperature and LH_2 density at $-253°C$ are 0.09 and 70.9 kg/m^3, respectively. The volume-related energy density of hydrogen relative to gasoline is approximately 0.3:1, and the weight related density 3:1. In many plants around the industrialized world, liquefaction of hydrogen is more or less a routine practice, in plants of very small capacities up to very big amounts of some 10 t LH_2 per day. Liquefaction is energy intensive: in the classical Claude process around 1 kWh of electricity is needed to liquefy some 3 kWh of hydrogen. Potentially more efficient liquefiers using magneto-caloric magnetization/de-magnetization of rare earth compounds are still deep in their early laboratory phase. Small onboard re-liquefaction installations for boil-offs are (and will be) installed on sea-going LNG tankers (today) and LH_2 tankers (tomorrow); the re-liquefaction temperatures are $-163°C$ and $-253°C$, respectively. Another possibility to avoid boil-off losses is storing the boil-off in metal hydrides underneath the LH_2 spherical balls at the bottom of the ship's hull where the additional weight of the hydride installation simultaneously serves as ballast instead of dead weight water tanks (which, by the way, also transport sometimes very disruptive flora and fauna from the water habitats of one ocean area to those of another).

Hydrogen at ambient temperature consists of 25% ortho- and 75% para-hydrogen, spinning in the same or opposite direction as the nucleus of the hydrogen molecule, respectively. Since at very low temperatures ortho-hydrogen is converted to para-hydrogen accompanied by very high boil-off losses, catalytic conversion of ortho- into para-hydrogen already during the liquefaction process is mandatory.

In practice, stationary storage is realized in on-ground high-pressure containers for GH_2 as well as in vacuum-insulated cryogenic cylinders and balls for LH_2. The ball with the highest LH_2 content so far is located at NASA's Kennedy Space Center at Cape Canaveral, Florida; it serves to fuel the center tank of the space shuttle and contains some 2000 m^3 of LH_2. In the future, GH_2 may well be stored in underground leached salt domes. Practical experience with low to high pressure underground air or NG storage has yielded many lessons learned which may be of help for future underground hydrogen storage. One problem here is how to manage the environmentally critical huge amounts of brine during leaching of the dome. For long distance air, space, and sea transport, only LH_2 meets the requirements.

Near- to long-distance surface transport asks for mobile storage of a different design. Onboard road vehicle storage may well handle hydrogen in all three aggregates, high-pressure gaseous, or LH_2, or metal hydrides. High-pressure GH_2 tanks at 700 bar are filament-wound tanks of cylindrical shape with an inner metal liner; they have a restricted lifetime because of low cycle fatigue damage due to pressure variations when filling and emptying the tank; the tank "breathes." Still, lifetime allowances for the tank are much shorter than the vehicle's life, so more than one tank per vehicle life is needed. Because of their weight and the complicated heat management required when filling or emptying the storage tank, it seems that onboard metal hydride storage has lost the business. The low weight–related energy content of the storage of a few percent, however, is the real reason. Low-pressure and medium temperature metal hydride storage of some 5–8 wt.% H_2 and, thus, an acceptable vehicle range between two fillings would be reasonable. So far, practical laboratory work has not yet succeeded in reaching that goal. Vacuum-insulated double-walled LH_2 containers with inner temperatures of –253°C at a moderate pressure of maximally 3–4 bar, boil-off rates of less than 1% per day, and a content of 7–10 kg LH_2 allowing for a range of 200–300 km is well-developed technology; the technology has been under practical test in some hundreds of demonstration vehicles so far. A leakage-free LH_2-filling receptacle valve is in practical use. One critical point is that if the cylindrical LH_2 container on top of the vehicle's rear axle is bulky, additional length has to be added to the vehicle body. Perhaps a longer-shaped cylindrical tank within the cardan shaft tunnel will be the final solution. Of course, because of minimum boil-off losses, the outer container shape should be as near as possible to the ball shape of minimum specific surface! In an interim phase as long as not too many LH_2 filling stations are in place, the bi-fuel gasoline/hydrogen ICE makes it possible to bridge the gap: whenever a hydrogen-powered vehicle runs out of fuel without a hydrogen station around, one simply switches to gasoline. Temporarily, that is a big advantage of the ICE over the hydrogen-fueled FC car!

To sum up, a wealth of hydrogen storage, transport, and dissemination technology is operational and well understood; safety conditions are known and experienced; and markets are established. Various well-documented stationary storage systems are presented by C. Carpetis (Figure 2.25), and mobile storage criteria—present and expected—are documented by E. Y. Marrero-Alfonso et al. using U.S. Department of Energy and U.S. Department of Defense storage goals. Here specific energy densities by volume and weight are presented for conventional gaseous, liquid, and metal hydride storage systems, but also for future lithium or sodium borates storage media. The respective gravimetric and volumetric data of the latter may increase by a factor of 2–3.

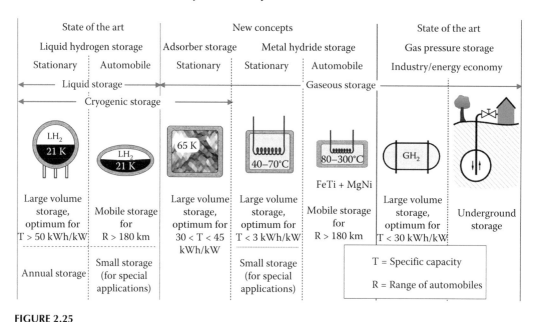

FIGURE 2.25
Hydrogen storage concepts. (Adapted from Carpetis, C., Storage, transport and distribution of hydrogen, In *Hydrogen as an Energy Carrier.* C.-J. Winter and J. Nitsch (Eds.), Springer, Berlin, Germany, 1988.)

Those technologies of the hydrogen energy economy still not yet marketed at large scales are

- Revitalization of hydrogen production from coal, including CO_2 sequestration and storage
- Large-scale electrolysis of high-temperature water vapor
- Solar and wind electrolysis of water developed in tandem with the respective hydrogen technologies
- Simultaneous transport of hydrogen gas in natural gas pipeline grids, including feed-in and phase-out technologies
- LH_2 ocean transport, including cryogenic harbor equipment
- GH_2/LH_2-filling stations for mobile users; LH_2/GH_2 dispensers for stationary users
- Mining of abiogenic hydrogen-rich methane in crystalline gas hydrates from deep sea floors (Clarathene)

2.11 Hydrogen Utilization Technologies

Energy utilization is the final link at the back end of any energy conversion chain where end energy is converted to useful energy and finally to energy utilization services whose efficiencies are decisive for the overall quality of the entire chain. Because each kilowatt-hour of energy services not asked for on the market because of higher efficiencies avoids introducing x ($x > 1$) kilowatt-hours of primary energy raw materials into the national energy economy at the chain's front end. Since the world's energy efficiency is ca. 10% and that of an industrialized country like Germany not much more than 30%, x is 10 for the world and around 3 for industrialized countries.

Usually, four national energy demand areas are distinguished: (1) industry, including energy utilities, (2) transportation, (3) buildings, and (4) small enterprises, trade, and military. A rough estimate for industrialized nations shows that areas (1) through (3) are roughly equal in size, and (4) is small. In Germany, the two areas (2) transportation and (3) buildings sum up to two-thirds of the end energy demand of the nation! Both are located at the end of the nation's energy chain where utilization technologies and their efficiencies are key factors.

Both buildings and transportation demand two forms of energy: one called investive energy for the construction of hardware, its lifelong repair, and dismantling and recycling it at the end of its service life; and the other one called operational energy for the O&M of buildings and transportation vehicles and their infrastructure. The operational energies of buildings, including air conditioning, are provided by various means, such as passive solar, active solar via thermal collectors or PV generators, ambient energy using heat pumps, natural gas or light heating oil for heat supply during wintertime, and hydrogen-supported heat/power blocks for simultaneously supplying heat and electricity. Of course, insulation of the building's envelopes (walls, roof, windows, and cellar ceiling) is of cardinal importance; the better the insulation, the less heat flows from the inside to the outside in winter periods, and from the outside to the inside in summer periods (here, one has to be careful not to interfere with heat gains through passive solar technologies!). "No-energy"

energy supplied buildings ("no" = no commercial energy purchased from the market!), even "neg-energy" buildings that harness more solar energy than is needed to meet their own demand, are not an illusion. Even under the not too favorable weather and climatic conditions of central Europe, residential homes with very little (if not zero) market energy needs have been built and operated over a number of years. The task is not a technological one, it is a question of economic viability! For the time being, no-energy homes are still more expensive than conventional homes or, in other words, the oil or gas price is still too low; but that is only a matter of time.

Transportation comprises infrastructure, vehicles on land, airplanes in the air, and vessels at sea. Land-based infrastructure is provided for (1) individual transport on urban and rural streets, roads, and highways, and (2) mass transport of passengers and goods on rails. Historically, the continental railway infrastructures of the nineteenth century have hardly been further developed, except in Europe and Japan. Here, urbanization and settlement density favor rail systems. New technologies were and are being put on line, making possible high-speed transport at responsible safety levels. Rail electrification via overhead lines is standard. On the basis of the present power mix, no other transport system is environmentally cleaner and emits less GHG per passenger and kilometer traveled.

Inter-infrastructure logistics road/rail/sea, road/rail/air, and road/rail/waterways are effective and time saving. Globalization asks for harbor and airport effectiveness because some 90% of global transport of goods sail on ocean vessels, and more and more air freight crosses the skies. On the other side is the system in North America where individual road transport and mass passenger air transport dominate; only bulk cargo is transported on slow rails. Steam locomotives were largely replaced by diesel locomotives; rail electrification is rare. No other transport system has worse climate consequences than that of North America.

The worldwide system in place depends almost entirely on fossil fuels, or, more precisely, on crude-oil-derived gasoline and diesel; natural gas as fuel or electricity have occupied a small percentage only (a thorough piece on hydrogen and electricity, its parallels, interactions, and convergence is given by Ch. Yang).

In the majority of countries, most of the fuel is imported; heavy users of transport fuel like the United States have in the meantime reached an import quota of more than 50%. The oligopolization of suppliers increases, the number of suppliers shrinks as oil and gas fields are progressively emptied; more of the world's remaining oil and gas resources are in fewer and fewer hands. A geographic "strategic ellipse" has evolved of dominating crude oil and natural gas suppliers to the world where the bulk of resources is located, spreading from the Persian Gulf via Iran, Iraq, and central Asian states to as far as Siberia—a not too comfortable situation for heavily crude-importing countries! To give an example, Germany has to import 77% (2007) of its energy demand, namely 100% of uranium, 60% of hard coal, 84% of natural gas, and almost 100% of oil. Only uncompetitive hard coal, brown coal, very little hydro, and—depending on climate, geography, and topology—renewable energies are available indigenously. Vis-à-vis price dictates the country is almost unprotected. Only one "energy" is securely in its hands, the energy technology knowledge of its scientists and the skill of its engineers and craftsmen. Their task: to harvest more energy services from less (imported) primary energies—the nation's credo: "Technologies compete, not fuels."

Let us now come to the role of hydrogen energy in buildings and transportation. Today, most of the buildings in the Northern Hemisphere are warmed with the help of natural gas or light oil-fueled boilers; they have to provide 60°C or 70°C of room radiator heat. The boiler, particularly a boiler which additionally uses the heat in its condensable exhaust gases, is energetically superb: almost 100% of the fuel's energy content is converted to heat with a flame temperature of around 1000°C. This, however, is exergetically absurd (see Section 2.6), because the irreversibilities in combustion, in heat transfer, the heat exchanger, and the energy flow through the entire system are tremendous and, consequently, the exergy efficiency is very low: exergo-thermodynamically the large temperature difference between flame and radiator temperature is unjustifiable. Instead, let us imagine that the boiler system is replaced by a low to medium temperature FC, which at first is fueled by hydrogen reformate from natural gas, later by pure hydrogen when the hydrogen supply system is operable. The FC follows a combined cycle: firsthand it generates electricity (=pure exergy) with an electrical efficiency of 35%–40%, and the remaining heat has a temperature that compares nicely with the radiator temperature requirement, so it still suffices to warm the house over most of the year; in extreme winter evenings, a small relief boiler is in stand-by mode in order to bridge gaps. The result: the hydrogen/FC system exergizes the energy supply of buildings!

So far only the pros have been discussed, but where are the cons? FCs of that kind are still in their development phase; so far, only around a thousand demonstration units have been constructed and operated. And the FC stack's lifetime is insufficient; it is miles away from the experienced 10–15 years lifetime of conventional boilers. Stack degradation makes its replacement necessary after some thousand hours of operation. Not surprising is that so far the FC's price does not meet market conditions; market-related mass production has not begun yet.

Another situation is seen in industrial FCs with their individual unit capacities of some 100 up to 1000 kW and temperatures around 600°C (MCFC) and 700°C–900°C (SOFC). They are being industrially produced in first lots, although still in small numbers. They serve as tri-generation combined cycles supplying electricity, heat and cold in hospitals, small to medium enterprises, and large building complexes, and a fine specialty, as uninterruptible power suppliers in airports, hospitals, telecommunication control, and computer centers. The German natural gas industry and major electric utility companies have committed to combine their knowledge of FCs of any kind and generate specific operational criteria for their day-to-day business; for details see www.ibz-info.de.

Now to transportation: At sea, the first hydrogen/FC demonstration vessels are being studied and demonstrated, and in aviation studies are under way on FCs replacing today's onboard APUs. Also, studies are pursued on electrification of locomotives using FCs fueled by electrolytic hydrogen; investment costs are expected to be lower than those related to electric overhead lines, particularly for very long distances.

In 2008, approximately 700–800 million passenger vehicles, trucks, and buses operated worldwide in surface transport on streets, roads, and highways, with a reproduction rate of 100 million per year. Gasoline and diesel are the almost exclusively used fuels. The reciprocating piston ICE is the standard prime mover. The onboard electric power system serves auxiliaries, which steadily increase in number and capacity. The present 5 kW for passenger vehicles and 10 kW for trucks and buses do not seem to be the final end of development: vehicle electrification increases. Conventional electrical generators connected to the main engine are an exergetically miserable solution, particularly at vehicle standstill, for example, at a red traffic light when, say, a 100 kW vehicle engine is being run for the benefit of an electrical generator of only 5–10 kW. The same thing happens on an overnight parking lot in hot climates when the air conditioning system of a truck needs power from the main engine with some hundreds of nameplate kilowatts.

The vehicle's electrification upward trend gets support from higher capacity/lower weight nickel metal hydride and lithium-ion battery development with specific energies per volume and weight, respectively, of 180 W-h/L and 80 W-h/kg (nickel metal hydride), and 300 W-h/L and 120–150 W-h/kg (lithium ion), further from plug-in or engine supported hybrids with ICEs and electric motors working in parallel or in series. The motivation is better fuel economy under certain driving conditions (inner city with frequent stop-and-go) and cleaner operations. Plug-in hybrids, however, need electricity from the public grid whose capacity must allow for this additional mandate and whose GHG emissions must be taken into consideration, especially when the bulk of electricity is generated in fossil-fueled power stations—unless affordable renewable electricity is available; the future will tell.

Now to onboard hydrogen energy: We distinguish several discussion lines: (1) Gaseous or LH_2? An attempt to answer this question is found in Section 2.10. (2) Hydrogen ICE or FC as the prime mover? A thorough comparison will be found in Section 2.12. (3) Hydrogen-fueled PEMFC or direct methanol fuel cell? and finally, (4) Which vehicle type first, passenger vehicles, buses, or trucks?

To question (3) about which type of FC will be the final solution: we said earlier that it will be technically (not to speak of economically) impossible to collect and remove CO_2 from hundreds of millions of vehicles disseminated across the world. Consequently, if anthropogenic climate change is truly taken seriously, carbon should not be on board the vehicle, however meager the carbon concentration of the fuel may be—whether natural gas, methanol, or something else (exception: renewable carbon in biofuels). Further, for low-temperature FCs, hydrogen is the fuel of choice; consequently, carbon-containing fuels need reformation to hydrogen. Reformers under the hood bring along additional weight and volume, and, depending on the type of reformed fuel, temperatures of some 700°C–900°C for natural gas or 300°C for methanol. Further, depending on the dynamics of the FC's hydrogen demand, the reformer has to be consecutively accelerated, decelerated, accelerated …, again and again, which is not at all inherently time appropriate for reformer chemistry. Since statistically the FC vehicle is operated only 1 h a day (the average for all vehicles of whichever provenance), the annual asset utilization of the reformer hardly reaches more than 1% per year—an absolutely non-convincing economic solution!

Many a design, experimental, or manufacturing task was (and still is) pursued in auto industry shops and research labs in order to tackle the aforementioned obstacles, because hydrocarbon fuels on board FC vehicles have one convincing advantage: their stationary infrastructures are in operation! Putting all pros and cons together, the final result is clear: the low-temperature FC will be fueled by hydrogen from large-scale central production and/or reformation units outside the vehicle, such as in the forecourts of filling stations; it will not be fueled by onboard reformed hydrogen!

To question (4) about which vehicle type comes first: obviously, in auto industry labs the highest amount of research and development money flows into hydrogen passenger vehicles, either FC or ICE vehicles, buses follow, while trucks are on the waiting list. Hundreds of small-to-medium size passenger demonstration vehicles are on the road; quite a number of buses have been and still are part of the routine services of city transport authorities. CUTE—Clean Urban Transport in Europe, a project of the European Union (www.global-hydrogen-bus-platform.com, www.H2moves.eu), stands for some 30 hydrogen-fueled FC buses operated routinely in European capitals (a few also during the 2008 Olympics in Beijing, China, and in Australia) with the objective to gather technological, economic, and passenger experience. For more details see Section 2.12.

The question remains, how will hydrogen be delivered to the filling station and at what cost? Today, gasoline, diesel, and natural gas are transported to the station via tanker trucks or gas flasks on trailers, seldom via pipelines. A dispenser takes the fuel from the underground storage, a layperson operates the fuel receptacle valve, and the dispensing

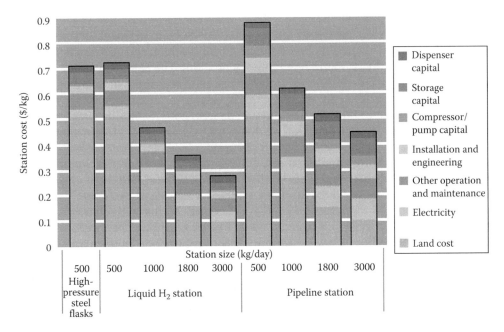

FIGURE 2.26
Hydrogen cost at the filling station. (From Yang, Ch. and Ogden, J., *Int. J. Hydrogen Energy* 32, 268, 2007. With permission.)

time is a few minutes. This is the procedure that future hydrogen-filling stations will have to match.

In Figure 2.26, cost [U.S. $/kg] at a U.S. station is depicted for gaseous and LH_2 against the size of the station [kg H_2/day] (for comparison: 1 kg of hydrogen compares energetically with ca. 1 gal of gasoline = 3.785 L). Three methods of delivery are shown: GH_2 in steel flasks or via pipeline service, and LH_2 in cryogenic tanker trucks. What is seen? Not surprising, smaller stations have the higher cost, because their turnover is limited. But surprisingly, LH_2 costs less than GH_2. Obviously, the higher energy density of LH_2 trucking more than compensates for the liquefaction cost; re-gasification at the station provides GH_2 too. Of the different cost influences, land cost is the highest (which for countries other than the United States might be different). The cost does not include taxes which, of course, differ from country to country. Also excluded is climate change cost, which depends on the fossil or nonfossil primary energies the hydrogen and the necessary electricity are made from, and the diesel fuel the transport trucks are fueled with.

2.12 Hydrogen Energy in Transportation

Historically, transportation on land, in the air, or at sea was (and still is) almost entirely dependent on fossil fuels; only mass railway transport in a few areas of the world (e.g., in Europe or Japan) uses electricity from overhead lines. However, electrification of transportation in general is increasing. For better navigation, more efficient fuel use, and increasing electrical onboard services, sea-going vessels use turbine-generator sets. Aircraft electricity needs have also continuously increased: fly-by-wire, computerized piloting, and general electrical onboard services need ever more electricity which is supplied on the ground by the electrical grid or generated on board by gas turbine–supported APUs; when airborne, electricity is provided by jet engine–mounted generators. The electrical system on trucks and buses has increased to voltages up to 42 V; because of ever more onboard users like air conditioning and operational auxiliaries, capacity increases to more and more kilowatts. The first hybrid cars joined the market a few years ago, with the ability to operate the car in three modes, solely battery supplied, with electricity from an engine-operated generator, or in a combustion engine mode. The motivation for hybrids is better efficiencies under certain (inner city) driving conditions and improved environmental and climatic cleanness. Canadian and Danish engineers studied electrification of long-haul and commuter service train locomotives not depending on overhead electricity lines (Haseli et al., 2008). Their studies compared H_2-ICEs and H_2-PEM FC power trains with conventional diesel engines or coal- or NG-based electric overhead line propulsion. The results are not surprising: CO_2 emissions [kg CO_2/vehicle·km] for conventional propulsion are the highest, those of H_2-PEM FCs are the lowest, with those of H_2-ICEs in between.

Certainly, the electrification trend goes on, also in passenger cars where more and more auxiliaries such as fans and water or fuel pumps are being decoupled from the vehicle's engine and operated electrically at higher efficiencies. Consequently, electricity requirements increase with respect to voltage and capacity. The question, however, is whether future battery development will not only fulfill the above-mentioned requirements, but also those related to the potentially full-electric car! The usual gasoline- or diesel-fueled auto range of some 500 km per tank-filling has by far not yet been reached by the battery-supported electric car. What is still needed is onboard electricity generation parallel to battery supply through engine generation, or through hydrogen-fueled FCs: That is the point!

An exergetically efficient first step is to replace the miserably inefficient engine-operated electrical generator by an engine-independent FC. As long as hydrogen is not taken on board, it may also make sense to install a high-temperature FC and run it on hydrogen reformate from gasoline or diesel. Before the second step is taken, namely the replacement of the ICE by a low-temperature FC and electric drive motors, the decision is due whether the FC is to be fueled with pure hydrogen or with gasoline or diesel reformate. The latter requires a medium-to-high temperature reformer, which is bulky and brings additional weight on board; it has, however, the advantage that the stationary fuel supply system remains unchanged. With respect to climate change, the situation is clear: only hydrogen brings climate change neutrality to transportation, since it will become technologically and economically impossible to collect GHGs from several hundred million (1 billion soon) autos worldwide, with a reproduction rate of around 100 million copies per year. Onboard hydrogen is the choice; there will not be interfuel competition!

A follow-on question arises: How is hydrogen stored on board? Two possibilities have been investigated and developed: high-pressure gaseous hydrogen (CGH_2) tanks and LH_2 containers. (Before metal hydrides of sufficient energy content by weight and volume are market ready, more, perhaps much more, time will be needed.) For today's usual vehicle range of 500 km, the tank's hydrogen pressure in the gaseous case has to be 700 bar, which requires a filling station pressure of around 1000 bar. For the engineer, both numbers are absolutely not trivial (and think of future billions of laypersons who are to handle equipment of such pressures at the stations)! The LH_2 container of modern design and production standard contains some 7–10 kg of LH_2 at an inner pressure of 3–4 bar and a temperature of −253°C; its boil-off rate tends to less than 1% per day. For safety reasons, both gaseous and LH_2-filling hoses and fuelling receptacle valves at the filling station have to be 100%

leakage tight. To date, 56 hydrogen-filling stations are operable in Europe, 26 of them in Germany (for comparison: here some 15,000 gasoline/diesel/NG stations are in place).

All this said, the question has not yet been answered whether the ICE or the FC, both hydrogen supplied, will be the technologically, economically, and ecologically more advantageous prime mover solutions. Let us look at the arguments:

- Although the FC is the much older technology (first publication in 1839), its market presence is still almost nil; until now, it is present in some thousand portable, stationary, and mobile demonstration units worldwide, as well as in space probes and in German submersibles. The case is different for the ICE, which came to market in the late nineteenth century and in the meantime occupies almost the entire mobile and stationary markets; here it is operated in around 1 billion copies with a reproduction rate of ca. 10% p.a.

- Consequently, the cost of the ICE is well known; it is marketed for a few 10 €/kW; worldwide competition is harsh. Engineers and craftsmen in OEM industries and repair shops are technologically well trained; their thinking and acting in favor of the ICE is the consequence of more than 100 years of acquaintance. It is otherwise with the FC: it is at the beginning of its learning curve; historical areas where lessons have been learned are not numerous and limited to certain fields, for example, the space industry, electrochemistry, and, lately, the research and development shops of automobile, electronics, and stationary FC manufacturers. FC market costs are still literally unknown, since so far nowhere have production lots of a size coming near the potential requirements of stationary and mobile markets been practically experienced. And the lifelong operation costs have only been deduced from demonstration units that are rather small in number and only exist in certain application fields.

- Both hydrogen-fueled mobile technologies, the FC and the ICE, are operationally environmentally clean; there is no major difference in their operational behavior vis-à-vis the environment. Both benefit from better vehicle aerodynamics, lower weight construction materials with higher strength and stiffness, and less friction in gears and wheels. The ICE, meeting the European EU5 pollution regulation codes, is literally pollution-free down to the measurability limit, and so is the low-temperature FC (jokers sometimes quip that the gas leaving an ICE exhaust pipe is cleaner than the outside air in certain highly polluted inner city areas of the world, which is, truly, not too far from reality!).

- Climatically, however, the picture is not too clear. On principle, GHG emissions occur during the production process of an energy converter (investive emissions), or they are of operational origin (operational emissions). Both energy converters, the ICE and the FC, may have comparable investive emissions. With respect to the operational emissions, the ICE, because of its lower efficiency, demands more hydrogen fuel per kilometer than the FC, hydrogen fuel which, if produced from fossil fuels emitting non-sequestered CO_2, is more CO_2 intensive than is the smaller amount the FC demands. If produced from renewable energies, the ICE's additional increment of hydrogen fuel demand is irrelevant since the renewable sources hydrogen is produced from are climatically clean. The FC, on the other hand, is not blessed with a stack life as long as the ICE's life, that is, some 3000–4000 h of operation, which is the operational life of the vehicle. Two or three stacks need to be consecutively installed to be commensurate with the life of an ICE.

More stacks per lifetime again bring more investive emissions (not considering cost!). All in all, taking the entire conversion chain into consideration, it is not yet too clear which installation, the hydrogen-fueled ICE or the FC, will be the one which emits less CO_2 lifelong. Further development of both technologies is needed before the matter becomes clear.

- Historically, a great many dissimilar auto engines have been under investigation with the intention to perhaps replace the ICE: the steam engine, the flywheel, the Stirling engine, the Wankel engine, the gas turbine, none has really succeeded, perhaps with the exception of the Wankel engine, which is marketed in small lots in Japan. Now comes the FC, and it seems that for the first time in the history of onboard prime movers it has a real chance to successfully compete with the ICE. Certain FC parameters are encouraging: the FC fits excellently into the ongoing electrification trend; it serves as battery recharging device and as electrical generator in the main drive train; it is clean, efficient, compact, not heavier than the ICE, and fits into a conventional engine compartment without major modifications; it is without moving parts and, thus, vibration-free and noiseless, more or less. The electric motors enjoy the welcome typical characteristics of low price when mass produced, excellent efficiency, acceptable weight and volume, and convincing acceleration; motor/generators bring on board the ability to recuperate brake power. Eventually, the electric motors will be placed in the four wheels of the vehicle (which implies a change in wheel dynamics, though!). The automobile FC industry (Daimler) claims that their 68 kW FC idles within 1 s at 90% capacity, that its hydrogen-to-electric efficiency is 52% (at peak power, the efficiency at usual driving conditions even may reach 58%–59%), and that both its weight and volume are 220 kg or liters, respectively. All in all, the highly exciting "race" between the hydrogen ICE and the FC is not decided yet. Both technologies still have potential for further development; both strive to increase their exergy efficiencies: the ICE by reducing its inherent irreversibilities through utilizing the huge amounts of waste heat in the cooling system and the exhaust, for example, with the help of the Seebeck effect where a voltage is generated when two different materials are hermetically brazed together with heat on one side and cold on the other, or by incorporating an ORC cycle in the high-temperature exhaust stream; and the FC, when using pure hydrogen, by eliminating the reformer with its inherently irreversible (exergy destructing) energy transfer, and by reducing stack degradation in order to stay the entire vehicle lifetime from cradle to grave with one stack—hopefully. "Getting millions of (hydrogen fueled) FCs on the road … will require policy that is as smart as the technology itself" (T. E. Wirth et al.).

- Now, in a thought experiment, let us imagine a future period of time when the majority of vehicles are run by FCs. What does that mean for the industry structure? Today, the automobile engine is in the hands of the mechanical engineer. Casings are foundry products; crankshafts, camshafts, piston rods come from the forging industry; aluminum pistons are manufactured in foundries; the mechanical work is accomplished in the auto industry's shops. All this, more or less, will be subject to change. FCs have no need for crankshafts or engine casings. FCs will also be in the hands of chemical process engineers and electrical engineers. The engine shops of the past will have to close down, while shops for membrane and stack production, for heat exchangers, hydrogen tanks and systems, electrical and electronic equipment will open. An almost complete industrial structure change is foreseen. Of course, this will not be a matter of short notice, a transition phase of many years, presumably decades, is anticipated. Visions seldom become real, but in most cases they indicate the root of the matter. Here an indication is given of a complete revolution of the auto industry's manufacturing infrastructure if the FC is to replace the ICE. Early decisions are due to avoid ruining complete industries: "rust belts" around the world tell an eloquent tale. The tipping point is really two points: each new technology era needs sound technology and market experts; it is up to the latter to investigate in due time the various consequences of introducing the new technology and to act accordingly. It sometimes appears as if exactly here are to be found the reasons for many a failure of past innovations.

In North America, Japan, and Europe, demonstration fleets of hydrogen-fueled FC or ICE buses have been (and still are) on the road operated by municipal transport professionals; they run under conventional conditions with passenger loads and maintain the usual time schedules. One and all, the experiences are positive; no hazard was reported, no major repairs were necessary. The average lay passenger did not even realize that he or she was traveling in a hydrogen-fueled bus, and when informed of the fact, he or she argued in favor of its cleanness and lauded its noiseless operation. And the cons? No surprise, both the cost of the bus and the fuel are high: the one will come down to the levels customary for conventional bus production lots of some hundred thousand copies per year for each brand, and the other must patiently wait for ongoing skyrocketing crude oil prices. In the meantime, gathering demonstration experience continues.

Climate change debates boil hot for the time being in Europe relating to the transportation sector. The EU plans to regulate CO_2e allowables in passenger vehicle transportation to $120\,g\ CO_2e/km$, a limit which will hardly be met by high-capacity limousines and sport utility vehicles. Significantly, the auto industry's voluntary self-commitment to reduce automobile emissions to an average of $140\,g/km$ was clearly missed.

Remedies are offered by a change of fuels to biogenic fuels and hydrogen; in Figure 2.27 automotive well-to-wheel CO_2e emissions [g/kWh] are compared with fuel cost [€/vehicle·km] for a number of biofuels and for gaseous and LH_2, compared to conventional hydrocarbon fuels gasoline and diesel at various crude oil costs per barrel, with or without (German) tax; reference vehicle is a non-hybridized VW Golf. What is seen?

The CO_2e emissions for biofuels, depending on the biomass material they are made from, go down to between 30 and $130\,g/kWh$, and the emissions for hydrogen even to almost zero, both under tax-free conditions at costs that compare rather well with the tax-free costs of gasoline and diesel (especially under exploding barrel prices for crude oil which, as of June 2008, had temporarily reached the historic peak of 139 U.S. $/bbl). Only LH_2 produced with electrical energy from SOT power plants is an outlier; its installed capacity of a few hundred megawatts worldwide might not yet be significant enough to make a reasonable contribution to the huge world automobile fuel market.

All said and considered, we conclude that with the help of biofuels and hydrogen, the GHG emissions of the transportation sector can be reduced to climate change stabilization levels, at costs that do not jeopardize the market. Two remaining questions are: Will the amounts of biomass from which the biofuels are made suffice for at least a good share of the automobile fuel market, without ignoring the life-sustaining priorities of human food and animal feed supply? (see argumentation in Section 2.13). And the second question: In those cases where hydrogen is made from wind, solar, or other electricity-generating renewable sources, is it justified to make hydrogen from electricity when electricity itself is

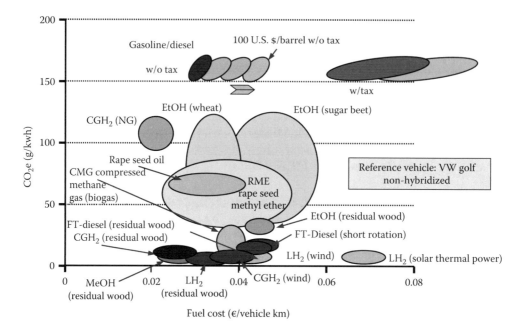

FIGURE 2.27

CO_2(e) Emissions of transportation fuels related to fuel cost. (Adapted from Schindler, J. et al., *Woher kommt die Energie für die Wasserstofferzeugung? Status und Alternativen, Ludwig Bölkow Systemtechnik*. DWV-Deutscher Wasserstoff und Brennstoffzellenverband (Ed.). The German Hydrogen and Fuel Cell Association, 2007.)

asked for on the market? Or in other words, who is going to win the competition between electricity itself and hydrogen made of electricity if the potential gross electricity market size is finite? This question has not yet been satisfactorily answered.

C. E. (Sandy) Thomas presented at the U.S. National Hydrogen Association's annual meeting in 2008 a remarkable paper comparing hydrogen, plug-in-hybrid, and biofuel vehicles; his findings read

- Hydrogen-powered FC vehicles achieve GHG reductions below 1990 levels by 80% or more, hydrogen internal combustion engine hybrid electric vehicles by 60%, and cellulosic (second generation biomass in European terms) plug-in hybrid electrical vehicles 25% at best.
- Urban air pollution would nearly be eliminated with FC vehicles.
- Hydrogen infrastructure cost is not a major issue.
- Hydrogen FC vehicles provide greater cost savings to society than does any other alternative.

It is worthwhile weighing these findings against the aforementioned arguments.

One thing is not questionable: The switch from hydrocarbon fuels to biofuels or hydrogen will not follow a jump function, but rather a continuous process. In an interim period, both types of fuels will share market segments, the novel fuels in slowly increasing, the others in decreasing amounts. At the start, hydrogen will not be produced entirely from renewable sources but from the traditional hydrocarbons with the share of renewables growing in parallel (for details see www.GermanHy.de). An example is given in Figure 2.28, which summarizes the result of a joint (German) project of government, industry and academia, named "Transport energy strategy (VES)." What do we see?

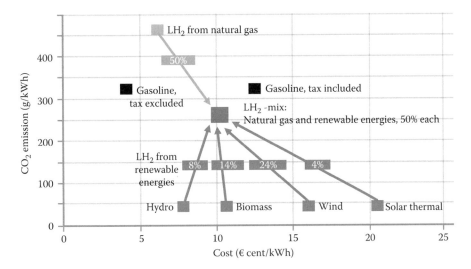

FIGURE 2.28
Costs and CO_2 emissions of LH_2 transportation fuel from natural gas and renewable sources. (Courtesy of VES—Verkehswirtschaftliche Energy Strategie. The German Transport Energy Strategy (VES), Berlin, Germany.)

Depicted are CO_2 emissions in [g/kWh] compared with cost [€-¢/kWh] for an LH_2 automobile, the fuel is natural gas and various renewable sources in a share of 50% each; for comparison the data for gasoline without and with (German) tax are added. As expected, the mix offers a significant reduction in emissions, but also a painful increase in cost: Gasoline without tax is approximately 60% cheaper than the mix (also without tax). That is the cost of GHG mitigation through introduction of renewable sources into automobile fuel! Comforting at best is the reflection that today's gasoline price corresponds with a snapshot of the world crude oil scenario; it seems to be unrealistic to expect that the oil price will ever go sustainably down again: no, rising oil prices benefit renewable hydrogen! (for details see Section 2.11 and Appendix 2.A).

2.13 Hydrogen and Biomass

Biomass is renewable secondary energy. Renewable, because when decaying the carbon it releases to the atmosphere in the form of carbon dioxide taken from the atmosphere during the plants' growth. When energetically used, biomass is biologically or thermochemically treated; it can be burnt to provide heat; it can be gasified or pyrolyzed, fermented, metabolized, or anaerobically digested to low caloric gas which as "bio-natural gas" can be fed into the natural gas grid or combusted in an ICE or FC to deliver electricity and heat in a district heating and electricity supply grid. It can be liquefied, the liquid being added to conventional gasoline or diesel or eventually replacing them. Biomass carries carbon and hydrogen (among other things) and needs additional hydrogen (from wherever) when liquefied. Living biomass has a very low solar-to-biomass efficiency of less than one to a few percent; consequently, biomass needs extraordinarily large surface land areas for the production of a given amount of energy, much larger areas than required by other renewable energy technologies like PV, SOT, or wind energy conversion.

There are many competitive applications for biomass such as food production, pharmaceutical and chemical feedstock or construction material, supply of energy, habitat for a great number of flora and fauna species, fixation of carbon dioxide, storage of water, supply of oxygen, and forest recreation areas for humans. Utilization of biomass depends on a great number of parameters, such as land area, quality of soil, natural or irrigative water supply, insolation, wind for insemination, availability of workforce, energy (e.g., diesel oil for agricultural machinery and transport vehicles, natural gas and electricity for agro-industries), fertilization, pest control, farming skill, and industries producing marketable products. The energy introduced into the different links of the biomass conversion chain prior to its utilization influences the energy-pay-back time, telling us how much operational time is needed in order to regain the amount of energy introduced into the whole procedure; often the pay-back time is so long that it barely justifies the biomass energy harvest. If biomass residues or commercially useless biomass like switchgrass or agricultural and industrial wastes or cellulosic municipal solid or fluid wastes are not converted, but rather "artificial" biomass from short rotation plantations, then the consequences are destruction and often devastation of biotopes that are thousands of years old. Diversity fades away through introduction of monocultures and soil degradation. Eventually, soils are washed away through wind or floods. Noncommercial wood gathering is today's dominating energy supply method in many of the poorest developing countries. It destroys vast land areas around human habitats, and consequently, longer and longer journeys have to be taken to collect it.

Let us take a closer look at gaseous or liquid products made from biomass. We distinguish first and second generations of biomass; let us begin with the first:

- Biogas, after chemical treatment, moisture removal, and pressurization, compares well with natural gas (labeled "bio-natural gas"); the easiest way is feeding it into the existing natural gas grid or, where a grid does not exist, burning it in an ICE or FC to generate electricity, using the heat in an extra district heating system. Biogas has a density of some $80\,kg/m^3$ and a lower heating value (LHV) of $48\,MJ/kg$. With $180\,GJ/hectare$, the yield is reasonable.

- Biodiesel (fatty acid methyl ester) comes from rapeseed or soy oil; its LHV is $37\,MJ/kg$, its density $0.88\,kg/L$.

- Bioethanol from sugarcane (Brazil), corn kernels (United States), or grain ears and sugar beets (Europe) has a density of $0.79\,kg/L$ and an LHV of $27\,MJ/kg$; bioethanol-labeled E85 (upto 85% bioethanol in gasoline) is a common motor fuel in Brazil, the United States, Sweden, and elsewhere; in Germany according to federal law 5% is being added to conventional fuel. A very serious consequence of utilizing bioethanol from corn kernels or grain ears as motor fuel has already been demonstrated in North America and Europe as well as in developing countries where food prices increased significantly after farmers discovered that their income from growing fuel crops was higher than from growing food for humans and fodder for animals. Ethanol from food raw products is not a sustainable solution. Often enough in history when bread prices were raised social uprisings were not far! Bioethanol from cornstalks, straw, wood chips, or wastes and residues may turn out to be sustainable if treated reasonably; this is addressed in the next paragraph which deals with the second generation of biomass. Here, however, productive and long-life enzymes are still not available in industrial scales.

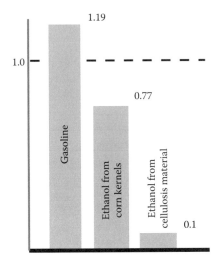

FIGURE 2.29
Amount of fossil energy [MJ] for 1 MJ of fuel. (Adapted from Wald, M.L., *Sci. Am.*, 1, 28, 2007, www.sciam.com.)

- Another critical point is energy-pay-back. Under the Brazilian conditions (weather, climate, soil, labor), the entire sugarcane stalk is processed for glucose and ethanol extraction, and the remaining lignin is converted into process heat in steam plants; the energetic result has a high degree of sustainability (if no extra land had to be provided through deforestation!). The situation is different in the United States and Europe (Figure 2.29) where corn kernels are processed. Here the energy needed for the entire process, mostly natural gas, comes from the market, with the consequence that the energy-pay-back is extremely meager. Cynics speak—not too far from reality—of natural gas plants with a small addition of bioethanol!

Now to the second generation of biomass:

- Biomass-to-liquid (BtL) fuels come from wood chips, bark, straw, stems, stalks, and agricultural, residential, or industrial residues; their energy yield is acceptable; the density is 0.80 kg/L, the LHV ca. 44 MJ/kg. What is still a matter of intensive research and development are the enzymes necessary for breaking up the fibrous material into glucose for further fermentation; a worldwide search for appropriate enzymes is ongoing. If successful, a big leap into the right (sustainable) direction will have been made, because the conflict with food production for humans can be avoided. Simultaneous food and fuel production is the solution when corn kernels and stalks, or grain ears and straw, etc. are harvested and further processed individually.
- Another promising BtL process is flash pyrolysis under oxygen exclusion at a temperature of around 475°C. The product has an LHV half that of light heating oil.

In Figure 2.30, the exergetic efficiency of hydrogen production from biomass is depicted over the biomass' moisture content. From vegetable oil with no moisture via straw, wood, sludge to finally manure, the moisture content increases to 45% and the efficiency shrinks accordingly from 80% to less than 40%; moisture removal is exergy consuming! The largest exergy losses, again because of irreversibilities, occur in the gasifier, followed by losses due to water removal and synthesis gas (syngas) compression.

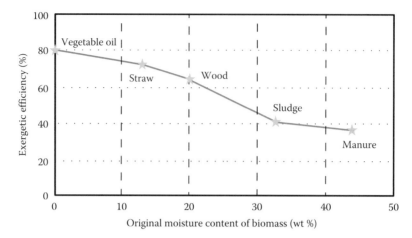

FIGURE 2.30
Hydrogen from biomass. (Adapted from Ptasinski, K.J. et al., Exergy analysis of hydrogen production methods from biomass, *Proceedings of ECOS 2006*, Aghia Pelagia, Crete, Greece, 3, 1601–1618, 2006.)

In conclusion, major biomass capacities and heavy energy users are often spatially dislocated, biomass grows here, and the heavy energy users are there; therefore, diesel oil consuming transportation over sometimes extended distances is mandatory. Two types of biomass are distinguished: on the one side naturally and agriculturally grown biomass, and on the other side wastes, residues, or noncommercial biomass. Land for agricultural or forestry biomass has a great many competing utilizations of which food production is the most important. Energy production from residual or waste biomass and from industrial and residential refuse can fulfill two tasks: being environmentally and climatically responsible, and delivering heat, electricity, or chemical and pharmaceutical commodities. Utilization of biomass leftovers and industrial/residential wastes is not being questioned. It appears, however, that energy utilization of virginally raised biomass is overestimated and may end up in an illusion. In particular, competition between food production and energy usage needs to be avoided; this competition will never be won by energy! Biomass of the first generation is at an impasse. Once again, thinking ahead, trying to anticipate snares prior to taking action is the mandate: energy policies prior to energy politics! Let us avoid barking up the wrong tree, or, even more precise, any tree at all!

In summer 2008, the Commission of the European Union tended to question its earlier regulation of blending gasoline with 10% (first generation) bioethanol by 2020, crediting the use of electric or hydrogen vehicles instead—another step in the right direction of hydrogen in transport!

2.14 Hydrogen Safety

No technology is absolutely safe! What is heard here and there, "this or that is absolutely safe," cannot, in principle, be justified from an engineering standpoint. Each technology is relatively safe; it has its specific safety standard which, of course, applies also to energy technologies and systems; hydrogen energy is not different. In any case, safety is a consequence of the specific science and engineering attributes of the technology in question, and, thus, its risks under operating conditions.

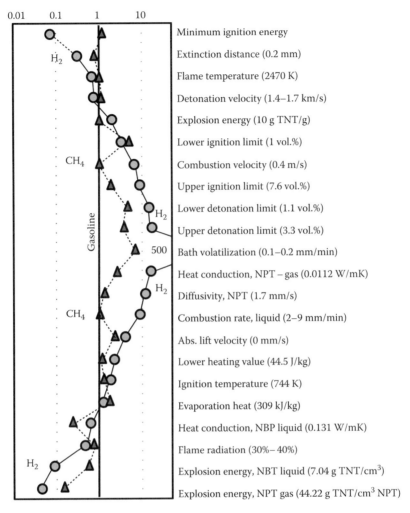

FIGURE 2.31
Safety data of hydrogen and methane compared to gasoline. Data in brackets for gasoline; TNT, tri-nitro-toluene; NPT, normal pressure and temperature (gas); NBT, normal boiling temperature (liquid). (Adapted from Peschka, W. et al., *Liquid Hydrogen: Fuel of the Future*, Springer, Berlin, Germany, 1992.)

In Figure 2.31, selected safety-related data for hydrogen and methane in comparison to gasoline are depicted. Some of the data are not too dissimilar for all three items; others, however, differ significantly. Four categories are particularly interesting for the assessment of hydrogen safety: (1) the diffusivity of hydrogen in air is very high, (2) the ignition energy of an ignitable hydrogen/oxygen mixture is very low, (3) the ignition range is wide, and (4) carbon compounds in hydrogen as well as radioactivity and radiotoxicities are inexistent. Let us discuss.

Like many non-hydrogen gas technologies, hydrogen installations need to be tight in order to prevent leaks or at least keep them as small as ever possible. Since hydrogen is the smallest element in the periodic table of elements and its affinity to oxygen is high, leak tightness is of utmost importance. If, however, leakage occurs or in an accident hydrogen is released to the outside, there is a good chance that no ignitable hydrogen/oxygen mixture is built, or that an ignitable mixture lacks a near-by ignition source, because hydrogen quickly disperses vertically upward into the airy environment; its diffusivity in air is a powerful acceleration source and, thus, a (sort of) safety element. That is the case when hydrogen is handled in open spaces. In closed rooms, precautions need to be taken to avoid hindering the hydrogen from flowing to the outside, such as assuring open outlets in the upper sections of walls, removing barriers of any kind, installing flow accelerators like non-electrically propelled ventilators, and diluting air streams, among other options.

If all precautions against the build-up of an ignitable hydrogen/oxygen mixture have failed, it should be realized that the ignition range for a hydrogen/oxygen compound in comparison to methane or gasoline is much wider, and the ignition energy required for a potential reaction is very small. The frictional electric potential on human skin or a micro-arc from an electric switch may suffice to ignite the mixture. Consequently, in rooms where hydrogen is handled, particular care needs to be taken to avoid even the smallest ignition sources.

Volumes of safety codes and standards have been put together parallel to the decades- or even century-long experience in hydrogen chemistry, in refineries, in the technical gases industry, and in the numerous hydrogen branches where hydrogen is utilized as a commodity. So far, the latest area where safety precautions had to be taken, particularly with respect to LH_2, is the space flight industry where very large amounts of LH_2 (and LOX) are in use as propellants for the jet engines of space launchers. The International Organization for Standardization (ISO) in its Technical Committee ISO TC 197 is establishing the internationally accepted codes and standards for all aspects of the up-and-coming hydrogen energy economy; it is a never-ending effort. A productive source of ongoing European hydrogen safety considerations, theoretical and experimental, is www. Hysafe.org.

Honestly, all that achieved and despite all lessons learned from past safety events— positive and negative—most probably future accidents will occur. One thing, however, gives confidence: none of the accidents in the aerospace business where hydrogen was involved was causally initiated by dysfunctions of the hydrogen system! Two examples: Addison Bain, in his active time head of the hydrogen regime at NASA's Kennedy Space Center at Cape Canaveral, Florida, thoroughly investigated the 1937 Hindenburg zeppelin crash in Lakehurst, New Jersey. The airship was about to land in a thunderstorm atmosphere with high levels of static electrical potential around. Elms fire was observed around the aluminum window frames of the cockpit. At first, it was not the hydrogen inside the ship that caught fire, but rather the zeppelin's hull, which consisted of a weatherized cotton substrate with an aluminized cellulose acetate butyrate dopant—"a cousin to rocket fuel" (Bain). As a consequence, the hydrogen inside was ignited and the airship crashed. And the other example: minutes after lift-off in 1986, the U.S. space shuttle Challenger burnt and burst apart. Again, it was not the hydrogen-filled central tank of the shuttle that caused the accident, but one of the solid fuel boosters mounted outside the hydrogen tank, which, because of a leaking sealing ring, led a hot gas stream onto the insulation material of the hydrogen tank.

Further, there are two positive safety points. Since carbon is not involved in the hydrogen fuel on board space or future land-based or airborne vehicles or vessels at sea, people aboard the vehicles in a potential accident cannot be intoxicated or suffocated. And, since in a future hydrogen energy system radioactivity and radiotoxicities are nonexistent, unforeseen long time (unknown) consequences of potential accidents are impossible.

An example: In 1977, two passenger planes bumped into each other while rolling on the airstrip of the island of Tenerife, Spain; kerosene spills caught fire and burned for some 20 min; fumes, smoke, and toxicants evolved; passengers died from intoxication or suffocation. Could there have been a similar incident with hydrogen fuel? There are significant differences: In hydrogen planes, the LH_2 fuel is compactly stored in double-walled tanks installed above the passenger compartment and surrounded by the plane's fuselage structure; because of limited space availability, no fuel is stored within the wings. If, notwithstanding, LH_2 spills occur, the diffusivity of re-gasified hydrogen in air tends to accelerate the flow rapidly vertically upward. It is not too easy to ignite an LH_2 spillage prior to its gasification. The combustion product is water vapor. Toxicants or suffocating combustion products can only stem from the plane's construction material. As the combustion temperature of the hydrogen/air compound is high, particularly radiated heat injuries can occur, in the worst cases fatal ones. As a consequence of hydrogen's diffusivity, the combustion time is short.

All in all, when fairly, responsibly, and honestly judged, hydrogen incidents cannot be excluded, but due to the specific attributes of hydrogen, their consequences promise fewer fatalities or less severe injuries and material damage. Hydrogen has its specific risks, but its attributes help to alleviate the follow-on effects. No risk can be treated lightly, but engineers who have been working with safety equipment for both hydrocarbons and hydrogen tell us that hydrogen systems, if the safety rules and regulations are strictly adhered to, are safer than the hydrocarbon systems. In the space launching industry, the oxidizing agent is liquefied oxygen LOX which has its own specific safety risks too. Here particularly fat compounds on equipment surfaces need to be removed in order to avoid self-ignition.

At the end of this section, a few general thoughts: the energy systems we are accustomed to are in the hands of professionals and laypersons. The former run coal mines, oil and gas fields, power stations, refineries, pipelines and electricity grids, liquefaction plants, and tanker ships, and they operate buses, trucks, locomotives, airplanes, and spacecraft. In the hands of laypeople are residential energy systems, autos, and electrical and electronic equipment. As an inherent consequence of hydrogen energy, in particular renewable hydrogen energy and its technologies, decentralization of energy increases, such as solar PV generators and thermal collectors on roofs, hydrogen-fueled FCs in the cellars of buildings, and hydrogen from the dispenser at the filling stations, and this equipment ought not remain in the hands of laypersons. Not only safety considerations but also effective and efficient energy utilization ask for indispensable professionalization also at the back end of the energy conversion chain where energy decentralization will be taking place. As a convincing example just one professionalized technology is mentioned, robotized fueling of hydrogen vehicles: The (lay) driver, entering the station, stops at a red light, remains seated in his car and identifies himself and the type of his vehicle by inserting his plastic card. A robot opens the tank lid, inserts the fuel receptacle valve, confirms absolute leak tightness, fills the tank, and finally closes the lid again. The light switches to green. Not even one hydrogen drop was lost. Filling time is similar to what we are accustomed to today, a few minutes. After a while, the driver will be notified that the amount of his purchase has been deducted from his account.

A final thought indirectly related to safety aspects: we already mentioned that an increase in efficiency is urgent so that less primary energy produces more energy services. Less primary energy corresponds to fewer safety risks; what is not utilized is of no safety relevance. Finally, since decarbonization replaces more and more carbon with hydrogen, carbon-related risks tend to vanish.

2.15 Hydrogen Energy: Costs and CO_2 Emissions

Like any other energy, hydrogen energy has to meet a range of criteria before successfully entering the market. The two major, perhaps dominating, criteria are costs and CO_2 emissions. Of course, costs are key for the entry of any energy into a competitive large-scale market, and hydrogen energy is no different. Carbon dioxide is the predominant GHG with the maximum influence on anthropogenic climate change, followed by methane CH_4, nitrous oxide N_2O, and fluorine gases. Correctly, all emissions have to be taken into account along the complete energy conversion chain from the very beginning of primary energy conversion to finally energy services utilization.

In Figure 2.32, costs [€-¢/kWh H_2] and CO_2 emissions [g/kWh H_2] of gaseous and LH_2 energy production are shown. Three primary energies have been taken into account: natural gas in the upper part of the figure, coal in the middle, and renewable and nuclear energies at the bottom. No link of the energy conversion chain was forgotten, from production via storage, transport, and electricity generation in the case of electrolysis. What is seen?

Hydrogen production from natural gas shows, not surprisingly, the lowest cost, although moderate amounts of CO_2 emissions, if not captured and sequestered. Hydrogen from coal has moderately higher costs, but its emissions without sequester are prohibitively high; CO_2 capture and sequester bring them down to acceptable levels. For hydrogen from renewable or nuclear energies the picture changes: now, because renewable energies are not yet fully developed to unsubsidized market levels, costs are prohibitive, and emissions

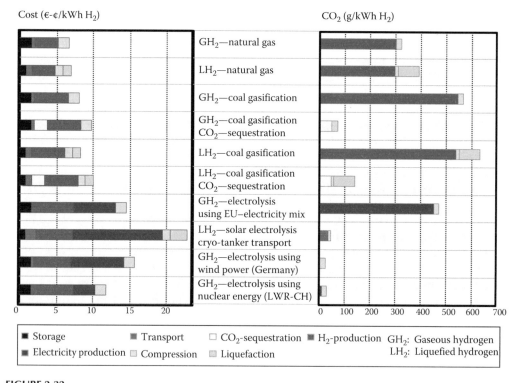

FIGURE 2.32
Costs and CO_2 emissions of hydrogen production. (Adapted from Wietschel, M. and Hasenauer, U., *Energiespiegel*, November 12, 2004.)

tend to zero. Clearly seen are the cost dominance of electricity production with renewable technologies and the unacceptable CO_2 emissions of fossil fuels without capture, sequester, and final storage; in both cases, further technology development is imperative.

A comparison of sequestered emissions in the coal-hydrogen cases with those of renewable hydrogen gives a clear indication of the importance of capture and sequester. Environmentally and climatically no big difference is seen between hydrogen from sequestered fossil fuels and hydrogen from renewable sources; costs, however, differ significantly. The cardinal question remains: if coal-hydrogen including carbon capture, sequester, and storage (CCS) is the climatically clean solution, at least in the interim until unsubsidized renewable hydrogen is market ready, will then the whole CO_2 complex of capture, liquefaction, dehydration, transport, storage, and deposition be economically viable and geo-scientifically responsible in the long term?

Clearly, the retrofitted plant share of an operational power plant, a nuclear station, and a modern, highly efficient fossil plant deliver the lowest mitigation costs. There are a number of CCS methods which, so far, have not yet revealed a priority technology, which is the reason for the wide cost range for CO_2-free (better CO_2-restricted) fossil plants of 30–50 €/t CO_2. Renewable plants are still far from unsubsidized market conditions; further development which lowers their cost is urgent. Hydrogen production within a cap and trade price bandwidth of 5–30 €/t CO_2 seems marketable; of course, use of nuclear plants requires societal acceptance.

2.16 Energy 2050 at a Glance: Concluding Remarks

This manuscript was written around the beginning of 2009; it was written particularly as an accompanying framework text of the forthcoming 18th World Hydrogen Energy Conference, which was held from May 16 to 20, 2010, in Essen, Germany.

Until the mid-twenty-first century, we have before us some 40 years, a very short time in energy economic and technology categories. For illustration, let us recall that the first nuclear reaction was experienced by Otto Hahn in Berlin in 1938; now, after 70 years, nuclear power stands for (only) some 7%–8% of primary energy equivalent worldwide, and a reactor's operational life is 40–60 years. Coal mines need 20–30 years before the first loaded tipper truck arrives at the mine mouth, and their life may exceed 100 years. Electric utility plants (nuclear or fossil) are seldom decommissioned prior to occasional technology re-powering; their total lifetime approaches 50 or more years. Hydropower plants are even operated for around 100 years. In the energy utilization realm, buildings are hardly replaced by new constructions only because of a potential improvement of their energy situation. European cities statistically replace their buildings not earlier than after some 70 years, or after much longer time periods, if at all; exemptions are artifacts of cultural heritage such as cathedrals, cloisters, or castles. Individual mass road transport and commercial aviation massively started only after World War II; in the meantime they have been around for 60–70 years. And finally, novel energy utilization technologies regularly needed also long development time periods: a first gas turbine was patented 200 years ago, but began its triumphal utility not earlier than a few decades ago. The FC was mentioned in literature for the first time already in 1839, but has still not achieved irrefutable mass market success; energy and energy technologies need time; many decades up to half to full centuries are typical!

What applies to energy technologies is even more applicable to primary energy feed-stock or primary energies. The first solar civilization began with humans' advent on earth and is now reduced to merely noncommercial irksome wood or dung collecting in the world's poorest developing countries. Coal's success worldwide started with the opening of the first commercial coal mine in England in the second half of the eighteenth century and is today with approximately 20% of the world's supply still in full swing. Oil and gas, although explored already in the second half of the nineteenth century, really began their worldwide mass market success not much earlier than after World War II and stand now at approximately 60% of the world's total energy demand. It appears that the next two additions to the mix will be energy and exergy efficiency gains, and renewable energies, these now of the second solar civilization. What is clearly seen is the continuous shrinking of national energies and the rise of the now prevailing international energy trade system; only a few nations in the world are 100% energy self-sufficient, though remaining depen-dent on energy technologies from other countries. What is further seen is that energy is not static; ongoing development to energy heterogeneity is the rule. Altogether, energy is asso-ciated with very long lead times, sometimes even centuries: up until late in the eighteenth century, renewable energies of the first solar civilization were exclusively utilized—wood, running water, and wind; the nineteenth century was the century of coal; the twentieth century was supplemented by oil, natural gas, and fissionable uranium. Now at the turn of a new century, there are three more additions: energy and exergy efficiency gains, renew-able energies of the second solar civilization, and the secondary energy carrier hydrogen. Perhaps by the end of the twenty-first century, nuclear fusion would also have arrived.

In Figure 2.33, energy and the well-being of people on earth are correlated. Nearly 80 million more humans live on earth each year (approximately the population of Germany). By 2050, the earth's population is expected to rise to approximately 9 billion. More people means more energy. Industrialization of those regions where more people live means even

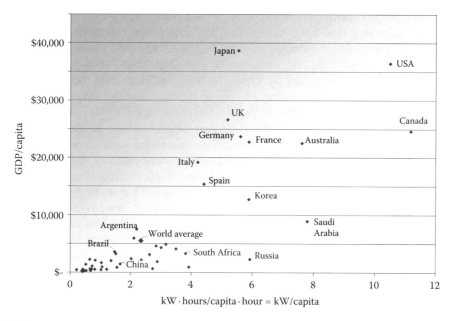

FIGURE 2.33
Energy demand vs. the GDP of nations (2005). (From IEA, *Key World Energy Statistics*, International Energy Agency, Paris, France, 2006.)

more energy. The world's average demand is 2 kWh per capita and hour = 2 kW/capita. The majority of people live below that average; some even have no access to commercial energy at all. All in all, it seems not too farfetched to suppose that the world's energy demand will rise further, only mitigated by rising efficiencies and, thus, relatively shrinking primary energy demand in the industrialized world.

There is also a close correlation between climate change and energy: the production, handling, and utilization of energy has a major influence on climate change (for details see Section 2.2). Three energies whose applications do not emit any of the GHGs influencing the earth's climate may cut the Gordian knot of mitigation: energy and exergy efficiency gains, nuclear fission, and all sorts of renewable energies. The potentials of all three are immense: regularly, modern industrialized nations reduce their annual energy demand by around 1%, Germany even plans aiming at 3% (achievable?), which is not a matter of available technologies, but of economic viability and political will! The forthcoming switch to exergy-efficient combined cycles adds another incentive to demand reduction. By 2050, nuclear fission will still not suffer under uranium supply shortages, although fast breeder reactors may not be operational by then. Both nuclear fission and renewable energies need hydrogen, the former in high-temperature reactors predominately as exergy-efficient, high-temperature heat source for allothermal hydrogen–supported coal decarbonization, the latter in order to facilitate their contribution to the world energy trade, since so far all utilization of nonstorable renewables is restricted to local, at most regional, applications. For a period of up to a half or full century, nuclear fusion will most probably still remain a scientific and engineering research and development venture, an absolutely fascinating and challenging one though.

The three climate neutral energies mentioned relate dissimilarly to energy price levels. Aggressive striving for increased energy and exergy efficiencies is a direct consequence of the exploding price jumps of conventional energies. Cheap energy is the most elusive enemy of energy security, as well of higher efficiency novel energy technologies. At this stage, freeing renewable energies from their remarkably high subsidies will, besides further technological development, be achieved by elevating energy price levels, the matter will quasi resolve itself. And the third climate neutral energy, nuclear fission, plays a price-smoothing role, since its overall cost calculation is predominately technology dependent; only a small cost share stems from the fuel, and in addition, its contribution to climate change is nil.

What does all this mean keeping in mind our projections for 2050? Let us try to avoid the mistake often made when looking into the future, namely, simply extrapolating the present situation. First of all, humans are not too well prepared for what the energy world will encounter down the road. Environmentally and climatically clean energy by 2050 is still far from certain. What ought to be in operation 40 years from now must already be in the pipeline today, or it will not be! What will not be in commercial operation at all is easy to see: nuclear fusion. Coal and nuclear fission will continue to provide energy to the world, not only until 2050 but also well beyond that date, coal perhaps in slowly shrinking relative amounts, fission with smaller growth rates, new coal undisputedly with CCS, and nuclear with fourth generation reactors and their expected higher safety regimes or, even more farsighted, with high-temperature reactors run exergetically efficiently in combined cycle mode, serving not only the electricity, but also the high-temperature heat market. Oil may retain its supply contribution or it will go down; here two tendencies work in opposite directions: on one side, the galloping price trend builds up market barriers that favor competing energy alternatives such as the global hydrogen trade, and on the other side, so far economically nonviable and extremely climatically harmful "crudes" like tar sands or oil shales or untouched sources on or below deep sea floors or under ice cover approach commercial viability. Natural gas is the present champion, and it appears that it will remain so for at least another 40 year, which, diplomatically and commercially, will not go down too well with nations that are heavily dependent on gas imports, to put it mildly. Diversification of suppliers and supplying modes including LH_2 tanker transport is mandatory!

What truly will be novel in the energy arena are the three newcomers: (1) energy and exergy efficiency gains, (2) renewable energies of the second solar civilization, and (3) hydrogen energy. All three are characterized exclusively by technological knowledge and engineering skill, not by the operational feedstock. Renewable energies lack operational energy feedstock, in principle, and for energy and exergy efficiency gains technologies are the key to providing more services from less primary energy raw materials. The energy market dominance switches from today's energy raw material providers to those knowledgeable about energy technology science and engineering. Energy technologies become more important than energy raw materials. The center of gravity within the world's energy conversion chain moves toward the chain's back end: environmentally and climatically clean, efficient secondary energies, end energies, useful energies, and finally energy services will characterize human energy supply systems by 2050. Energy raw materials were prevalent in the nineteenth and twentieth centuries, whereas exergetically efficient conversion technologies of low irreversibilities and, thus, low exergy destruction are the technologies of the twenty-first century!

Hydrogen not only adds to electricity another environmentally and climatically clean secondary energy carrier, it is also indispensable for combined cycle coal utilization; it is essential for tapping exergy-efficient distributed stationary FC power and heat at the back end of national energy conversion chains, and finally, it makes transportation-related climate change neutral.

"Who looks ahead, is the master of the day." Energy is not a matter of tackling day-to-day inconveniencies—it is about foresight, about thinking and acting in long waves, not in jump functions; energy is not for the impatient—a decade is nothing for energy! Securing the energy future is much harder and much more time (and money) consuming than managing its present; for the time being, energy lives on borrowed time. For visionaries, it might be disappointing not to see more of the novel additions to the mix in 40 years' time, and for climatologists the energy approach to climatic cleanness might be unsatisfactorily slow. Three potential influences are—almost—outside the powers of public intervention: oil price jumps as a consequence of the growing oligopolization of suppliers; climate catastrophes; and, perhaps the most serious influence, lacking awareness of the clear indications of forthcoming developments: "We thirst for knowledge, but we are drowning in a sea of information" (N. Postman).

If the next 40 years are taken seriously, humans must accept that energy development is something that is not completed in a jiffy, but requires positiveness, patience, pertinacity, and resilience. Humans must be fully aware of exergo-thermodynamics, aggressively promote renewable energies, and add the so far last lacking leg of the energy triangle, comprised of (1) hydrogen-supported cleaned up fossil fuels, (2) operationally carbon-free renewable energies and responsible nuclear fission, and (3) the two secondary energy carriers, electricity and hydrogen.

What hydrogen energy needs is vigor, not fickleness; major capital, not small change; continuity, not ups and downs; and, most importantly, conviction, not ambivalence! What we face is nothing less than an energy-system-change, comparable to the step into the electricity age that started more than a century ago and has by no means yet come to an end, or to the step into modern transportation, also 100 years old, although billions of people have still never sat in an automobile or booked a flight ticket. Hydrogen energy offers an innovation push; it is a powerful major job engine.

"Visions are more important than knowledge, since knowledge is finite" (Albert Einstein), and Ernst Bloch adds, "Visions need timetables." Here, a timetable has been drafted consisting of up-and-coming novel energy technologies in the energy-system-change ahead, and their respective time frames.

We close these concluding remarks with a retrospection on findings already published in the early 1990s: "Assuming that by the middle of the next century [*note: the twenty-first century*] it will be necessary to reduce CO_2 emissions more than 60%, the development of a hydrogen [*energy*] economy is not only consistent with the call for an energy supply which is as economical as possible. Such CO_2 reduction goals even mandate the utilization of these technologies" (Fraunhofer ISI—Institut für Systemtechnik und Innovationsforschung and PROGNOS).

2.17 Epilogue

> The energy system compares nicely with
> a bike, if not pushed forward, it tumbles!

Niles Eldredge and the late Harvard palaeontologist, Stephen Jay Gould, presented at the Annual Meeting of the Geological Society of America in 1971 a landmark paper introducing into the evolution theory the term "punctuated equilibrium," meaning that in an extremely limited short period of time species rapidly grow into a higher state of their intellectual being, or novel species enter the scene. After the end of that time period, the evolution falls back into its usual almost glacially slow Darwinian pace, until the next punctuated equilibrium comes along.

Technology too is subject to such happenings. Take the short period of a few decades at the turn of the nineteenth and twentieth centuries: Almost all of a sudden the automobile arrived on the road with its propelling reciprocating piston engine; the electrical generator provided electricity for the manufacturing industry and city street lighting; the telephone made communication easy, from the 1920s onward with the help of long distance cables across the Atlantic; and oil and gas fueled the booming industrialization of Europe and the new world.

But after that rather short period of time, "nothing more" happened. Of course, the auto was further developed, its speed increased, fuel consumption went down, the two-seater developed into a four-, even multi-seater, reliability was improved and safety requirements were met; but, in principal, the original configuration only changed minimally: the vehicle still has four wheels, an Otto or Diesel engine still powers it, the vehicle is still made of steel, and mineral oil derivatives still serve as the fuel. Only recently (compared to the more than 100 years of automobile history) have gradual changes turned into principal changes: for the first time in the history of the automobile, hydrogen and the FC ("the power station on wheels") as well as the vehicle's electronification ("the computer on wheels") offer the chance of a next punctuated equilibrium.

The good old "steam telephone" arrived at its life's end: It was replaced by wireless mobile telecommunication via satellites; automatic information exchange between personal computers took over; television is available and provides information 24 h a day; the World Wide Web delivers any information any time at any place, and drastic cost reductions up to 99% make communication and information a mass availability phenomenon: the semiconductor and the transistor were—and still are—the key technologies!

Sadly, firms in established industries usually innovate hesitantly, and mostly only in response to new technologies coming over the horizon that threaten their survival. Correspondingly, Max Planck observed, "The usual way a new scientific truth becomes generally accepted is not that its opponents are persuaded and stand corrected but that its opponents gradually die out and the next generation grows up with that truth from the start"—a bitter recognition? A realistic one!

When will the next energy punctuated equilibrium occur? We do not really know. But what we already know are the names of what will become part of it. They read: de-carbonization of fossil fuels via CCS and, thus, their hydrogenation and dematerialization; storage and transport of large-scale renewable energies via the chemical energy carrier hydrogen, which enables them to take part in the global energy trade system; and exergetization of the energy system and, thus, making use of the maximum extractable technical work from energy. In short, an energy-system-of-change is due with combined cycles and energy converters with low irreversibilities and, thus, minimum exergy destruction and exergy losses! Further names for the next energy punctuated equilibria are decentralization of energy and professionalization of the back end of national energy conversion chains, which, at this stage, is in the hands of the lay population; energy is much too precious to leave it there!

Literally, more or less all expected energy punctuations have to do with hydrogen energy. All together, they are epitomized in the up-and-coming hydrogen energy economy: it is HYtime!

Acknowledgments

Those with whom I have had the pleasure and satisfaction to collaborate over some three to four decades deserve my particular gratitude: Susan Giegerich and the group of hundreds of colleagues in industry, in the German Aerospace Center (DLR), and The Solar and Hydrogen Energy Research Center (ZSW) in Stuttgart and Ulm—both without hesitation willing to be of help whenever and whereever required, both contributing excellence where I failed, and giving hints where my thoughts seemed weak or sometimes even false. Susan, an American citizen, polished my "German English" and corrected didactics. The mostly younger colleagues of the group, energy and aerospace engineers who were experts at personal computing and at computational graphics took almost pastoral care when my electronics did not do what I wanted them to do: Thank you both, and many commonalities to come!

Tax money for R&D in almost all fields of hydrogen energy and its technologies still plays an indispensable role. The funds allocated in industrialized countries around the world have appreciably increased in the recent past. But what is even more important is the identification of those who dedicate the money to the different R&D fields that deal with the issue.

Here, the State Government of North Rhine-Westphalia, Germany, and especially its administration for Economic Affairs and Energy, deserves tribute and recognition: Germany's energy state in an early period of time recognized that the 200 years of the Ruhr area of indigenous coal tend to an end and are to cross the roads from electricity and steel which kept coal alive to electricity, hydrogen energy and stainless steel which will be keeping clean coal alive! It is considered only consequent that the state government appreciably supported the operations of the 18th World Hydrogen Energy Conference 2010 in

Essen, Germany, – in Essen, is the capital of the historic Ruhr area that was awarded the title of The European Capital of Culture for 2010. Hydrogen energy and its technologies are part of that culture: they are clean, abundant, efficient, and accompanied by an energy-system-change to non-heat-engine related conversions. The vigor of hydrogen energy avoids the "rust belts" under which so many areas of the world suffer.

Special thanks go to the editor, Elsevier, who agreed with the publication in its distinguished and well-esteemed periodical series, the *International Journal of Hydrogen Energy* (IJHE), which is also available online at www.sciencedirect.com. As I had already published in the past, I was familiar with the professional approach of the publisher; so, the risk was low, if any: thank you, Elsevier!

Abbreviations, Acronyms, and Glossary

BAU	business as usual
bbl	barrels (of crude oil) 1 barrel rc. 159 L
Bergius, Friedrich	1884–1949, winner of the 1931 Chemistry Nobel Prize
Bio-natural gas	biogas meeting natural gas specifications
BtL	biomass-to-liquid
BZ	Brennstoff-Zelle (fuel cell)
C	carbon
CCS	carbon capture and storage (or carbon capture and sequestration)
CGH_2	compressed gaseous hydrogen
CH_4	methane
CHP	combined heat and power system
CO_2	carbon dioxide
$CO_2(e)$	carbon dioxide equivalent of non-–carbon dioxide compounds
COE	cost of electricity
Dewar	vacuum-insulated container, after James Dewar (1842–1923)
Diesel fuel	after Rudolf Diesel (1858–1913), the inventor of the Diesel engine
EIT	economies in transition
Elms fire	St. Elmo's fire (St. Elmo, also known as Erasmus of Formiae, the patron saint of sailors), electrostatic discharge in an atmospheric electrical field
Entropy	energy devaluation, in closed systems ever increasing, never decreasing
EOR	enhanced oil recovery
ER	equivalence ratio of fuel and oxidizer
FC	fuel cell
Fermi	Enrico Fermi (1901–1954), nuclear scientist
G8	the group of eight highly industrialized nations
GDP	gross domestic product
GH_2	gaseous hydrogen
GHG	greenhouse gas
GT	gas turbine
H_2	hydrogen
H_2O	water

HHV	higher heating value
HT	high temperature
HVDC	high-voltage direct current
IAHE	International Association for Hydrogen Energy
ICE	internal combustion engine
IGCC	integrated gasification coal combustion
IPCC	Intergovernmental Panel on Climate Change
ISAM	integrated starter–alternator motor
ISO TC	International Organization for Standardization, Technical Committee
It's Hytime	it's hydrogen time (www.itshytime.de)
KOH	potassium hydroxide
kW	kilowatt
LH_2	liquefied hydrogen
LHV	lower heating value
LNG	liquefied natural gas
LOX	liquefied oxygen
LPG	liquefied petroleum gas
MCFC	molten carbonate fuel cell
MEA	monoethanolamine
MHR	(high temperature) modular helium reactor
mpgge	miles per gallon gasoline equivalent
MW	megawatt
N_2	nitrogen
N_2O	nitric oxide
$NaBH_4$	sodium borohydride
Negentropy	compensation for entropy increase (e.g., through solar irradiance)
NG	natural gas
NOx	nitrogen oxide
O&M	operation and maintenance
OEM	original equipment manufacturer
Ohmic	resistance heating after Georg Simon Ohm (1789–1854)
ORC	organic Rankine (steam) cycle, after William J. M. Rankine (1820–1872)
Otto engine	after its inventor Nikolaus August Otto (1832–1891)
Pabst von Ohein	Hans Joachim Pabst von Ohein (1911–1998), renowned jet engine engineer
PEMFC	polymer electrolyte membrane fuel cell: hydrogen purity 99.99% (vol.), "4×9"
POX	partial oxidation of hydrocarbons
ppb	parts per billion
ppm	parts per million
ppmv	parts per million, by volume
PV	photovoltaic
Pyrolysis	gasification under oxygen exclusion
Retentate	non-reacted leftover process flow (e.g., in an NG reformer)
rpm	revolutions per minute
S	sulfur
Selexol	physical solvent (of carbon dioxide)

Slipstream hydrogen	inexpensive hydrogen at the filling station from a nearby IGCC plant
SMR	steam methane reforming
SOFC	solid oxide fuel cell
SOT	solar thermal (power plants)
SPE	solid polymer electrolyte
Syngas	synthetic gas consisting of CO, H_2, and (CO_2)
tkm	ton-kilometer
VES	(Verkehrswirtschaftliche Energiestrategie) A Transport Energy Strategy (in German at http://www.bmvbs.de/-,1423.2458/Verkehrswirtschaftliche-Energi.htm)
Wobbe Index	interchangeability of fuel gases
wt %	percent by weight

Appendix 2.A: The German Hydrogen-Autobahn Ring—A Nationwide Project

Time has come to identify the industrialized world a novel addition to the energy mix and its technological capabilities and, thus, demonstrate hydrogen's maturity and economic viability to the public, to trade and industry, and not least to administrators and politicians.

With a peak price of U.S. $ 4/gal* of gasoline at the U.S. filling station in June 2008, energetically equivalent to U.S. $ 4/kg of hydrogen, the commercial viability of hydrogen energy is near, if not already achieved; even more so when at the same time EURO 1.5/L* of gasoline at the dispenser in Germany is taken into account, energetically equivalent to a fantastic (for U.S. citizens, even for Europeans), though real in day-to-day practice, U.S. $8.52/kg of hydrogen!

The hydrogen autobahn ring from Berlin via Hanover, Düsseldorf, Stuttgart, and Munich back to Berlin consists of some 10–15 hydrogen filling stations (one every 200–300 km) designed and constructed by the technical gases industry. The stations are supplied with LH_2 from the two national liquefaction plants located in Ingolstadt and Leuna right alongside the ring, or with GH_2 from all places where today hydrogen is being flared, or from the national hydrogen pipeline running pretty much parallel to the ring from the Ruhr area to Cologne over approximately 250 km.

The first vehicles to be fueled with hydrogen are city buses, light duty vans of small-to-medium size industries or trade companies, and numerous short- to long-range passenger vehicles provided by automakers in Munich, Ingolstadt, Stuttgart, Rüsselsheim, Cologne, and Wolfsburg, all of these locations touched by the envisaged ring. Besides these OEM industries, various hydrogen industries are invited to offer their products to this first of its kind central European hydrogen showcase, thus alerting other markets to join.

---

* In the meantime, the fuel prices went significantly down again; this, however, does not really change the message: increasing supply shortages and growing suppliers' oligopolization tend to enforce a mainstream upward general price trend! Statistically, the average gasoline price at the dispenser in Germany grew from 1950 to 2008 by c. 2 €-¢ per annum.

Appendix 2.B: A Hydrogen Energy Tycoon?

Do we see any Friedrich Krupps, Henry Fords, Werner von Siemens, Cornelius Vanderbilts, and Bill Gates in hydrogen and hydrogen technologies? Do we already see entrepreneurial matadors somewhere in the world who are devoting their thinking and acting, their skills, their financial capital, and their organizational talent to evolving hydrogen markets? To clean hydrogen production, to its different types of storage, to hydrogen transport and trade, to hydrogen utilization technologies? Are we expecting well-known companies to start or be on the verge of starting to become matadors in hydrogen energy businesses?

We do, and we also do not (yet); both answers are true. Of course, there are the space rocket launching companies which would not even exist without hydrogen, in this case liquefied, stored, transported, and combusted hydrogen; there are the industrial chemistry companies utilizing hydrogen as a commodity; there are, of course, the Seven Sisters running their refineries; and there are the methanol or ammonia manufacturers producing their needed hydrogen captively. Sections 2.9, 2.10, and 2.11 covered all of this.

All aforementioned hydrogen businesses have something in common: they belong to "old" well-established markets: hydrogen as space-launching propellant began more than half a century ago, and hydrogen chemistry and trade in technical gases are much older still. What was meant with our question about the hydrogen matadors refers to those who take care of the novel markets-to-come of the forthcoming hydrogen energy economy and here the answer is rather modest!

Section 2.8 introduced three table summaries of hydrogen energy technologies already marketed in small quantities, or in a waiting position, or still in R&D labs and development shops. But is there a matador visible? One whose key technology is the basis for the up-and-coming economically viable hydrogen energy market? Someone like Henry Ford, who started the mass production of reasonably priced autos (the legendary "Tin-Lizzy") and gained a world industrial empire; as did Werner von Siemens, whose electrical generator provided the core solution of generating power at one place and using it somewhere else, the still valid solution of geographically disconnected energy production and utilization.

In our times, we had Geoffrey Ballard who, with a number of colleagues, founded Ballard Power Systems in Burnaby, British Columbia, Canada; and we have almost all big world auto makers who are developing FC vehicles—a little hesitantly, though, since they are in parallel developing other electric vehicles that get their electricity not on board but from outside, like the plug-ins, the hybrids, the pure electric battery vehicles, and combinations thereof. For the industry's policy makers, market developments are still not too clearly foreseeable; perhaps here we get a feeling of the frequent change of FC vehicle market entrance dates which automobile companies used to announce.

For stationary or portable FCs, a wealth of small to very small companies have developed worldwide that are still in their research, development, and demonstration phases delivering small lots of products to a limited number of clients. Normally these companies' financial situation is modest, to say the least, if not risky, since the live of risk capital with interest rates of 30% or even higher. Similar things are true for mobile storage developers. An exception to this general observation are perhaps the big players in electronic devices, who have clearly devoted themselves to portable micro-to-mini FCs for all sorts of portables like cellular phones, camcorders, television cameras, and the like.

How about the major electricity utilities and the coal industry and their inclination to build efficient combined cycle power plants delivering simultaneously both electricity and hydrogen? No, they are still on their usual pathway constructing exergetically excellently efficient coal-fired electricity plants with nearly 50% efficiency or even a little higher. The engineer and the energy economist admire that, no doubt, but let us be realistic; the remaining 50% of the coal's energy content is still being converted to high-temperature exhaust heat with no industrial user around; only in the very rare situations when, say, a cement factory or a steel mill is located in the vicinity does the high temperature exhaust heat perhaps find a market.

Electrolytic wind-hydrogen or solar-hydrogen is even farther away from the market. Still, wind energy converters and solar generators "only" deliver electricity, and when, say, an off-shore wind park needs efficient and reliable electricity transport in order to be connected to its far away on-shore users, HVDC solutions enjoy priority (if the distance and nasty sea floor conditions allow for). The situation changes when very large amounts of wind or solar electricity are planned to contribute to the world energy scene, for example, wind from Patagonia in the far away South of Argentina, or solar energy from Australia, both commissioned to supply Europe or Japan or the United States. In such cases, hydrogen as the transportation means is unavoidable. But, far and wide, no major energy company in the world is following that idea as yet, not to speak of a matador.

The technical gases industry is well prepared to play an important role in the hydrogen energy field. The major companies—Linde, Air Products, Air Liquide, Praxair, and perhaps a few others—are experts in electrolyzers, steam methane reformers (SMRs), liquefiers, hydrogen dispensers, and filling stations. None of them, however, has developed into a champion's role, leaving all the others behind, so far.

Similarly, "Big Oil" is absolutely knowledgeable and experienced in hydrogen and its technologies. Large amounts of captive hydrogen are in use in crude oil refining, for the production of reformulated gasoline or de-sulfurized diesel. But again, no champion has evolved yet.

Having said all this, can a hydrogen energy tycoon realistically be expected? Most probably not. Let us see: Most of the hydrogen energy technologies along their complete conversion chain from production of hydrogen via storage and transport to dissemination and finally utilization go back to inventors who have lived and researched over the past two centuries and a half starting in the later eighteenth century. Mostly as late as in the second half of the twentieth century, their inventions were taken over by developers in national labs or universities, and their results are now under the control of the appropriate industries who simply buy what has left the labs, approaches market readiness, and promises profitable return (see Tables 2.1 through 2.3). The coal, oil, and gas industries are familiar with all aspects of hydrogen production in gasifiers, reformers, partial oxidizers, and other approaches. The electrochemical industry builds and operates electrolyzers. Pipelines hundreds of kilometers long for GH_2 and LH_2 (much shorter) are day-to-day practice. Storage on the ground and underneath are fully operable, taken care by the technical gases or industrial chemistry industries, or by space-launching companies.

In the final link of the hydrogen conversion chain, the utilization link, we see a different picture: The hydrogen-fueled portable mini-to-micro FCs are clearly in the domain of the electronics industries. Small-to-medium size companies have specialized on portable FCs in the kilowatt range for military applications or leisure activities. Deliverers of central heating systems for residential homes or office buildings are active in low-to-medium-temperature FC replacements of the traditional boiler/burner combinations. Here a challenging controversy is to be expected between central heating system companies and electricity utilities. Because with their FCs, the system companies no longer deliver only heat devices, but devices that simultaneously generate heat and electricity. In a country like Germany, for example, with some 15 million boilers/burners replaced by FCs (5–10 kW electric), the distributed power easily sums up to today's full electric power on line! Since this newly evolving competition in the electric power market competes with the traditional power business of the electricity utility companies on line, the matter will become rather touchy! An exciting development is foreseen and as its result one or two matadors may evolve.

The auto manufacturers deserve special attention: It may be that the present major challenges—cost reduction, fuel consumption reduction, change of fuel to carbon poor/hydrogen richer compounds—will be mastered by further-developed ICE vehicles, natural gas or biofuels, and hybridized electric vehicles of various designs. In the long run when the traditional fossil fuels get scarcer (and ever more expensive), the ICE's development potential approaches its limit, and the land surface area dedicated to the production of biofuels is completely exploited, then hydrogen energy, in particular renewable hydrogen, gets to its tipping point.

Let us return to our question: Will we see "A hydrogen energy tycoon?" It seems not too realistic to expect one, at least not in an early period of time. The energy-related industry branches appear to be well prepared to add to their portfolio hydrogen energy and all sorts of hydrogen technologies, as soon as indications of forthcoming profitability favor investments. One thing, however, should not be forgotten: energy is a highly political matter, and so will be hydrogen energy! As mentioned earlier: "The laws of parliaments and the laws of nature have developed increasingly divergent, and it is unreasonable to expect that the laws of nature will yield!"

References

The following are literature sources not expressly cited in the text, though strongly related to the subject presented:

Aceves, S.M. et al. 2006. Vehicular storage of hydrogen in insulated pressure vessels. *Int. J. Hydrogen Energy* 31: 2274–2283.

Ausubel, J. and C. Marchetti. 1997. Elektron: Electrical systems in retrospect and prospect. In *Technological Trajectories and the Human Environment*. J. Ausubel and D. Lanhford (Eds.). Washington, DC: National Academy Press, pp. 115–140.

Ball, M. et al. 2007. Integration of a hydrogen economy into the German energy system: An optimising modelling approach. *Int. J. Hydrogen Energy* 32: 1355–1368.

Barreto, L. et al. 2003. The hydrogen economy in the 21st century: A sustainable development scenario. *Int. J. Hydrogen Energy* 28: 267–284.

Bejan, A. et al. 1996. *Thermal Design and Optimization*. New York: Wiley.

Bose, T. and P. Malbrunot. *Hydrogen, Facing the Energy Challenges of the 21st Century*. John Libbey Eurotext, www.jle.com (accessed December 2008).

Carpetis, C. 1988. Storage, transport and distribution, Chap. 9. In *Hydrogen as an Energy Carrier*. C.-J. Winter and J. Nitsch (Eds.). Berlin, Germany: Springer, pp. 249–289.

Cherry, R.S. 2004. A hydrogen utopia? *Int. J. Hydrogen Energy* 29: 125–129.

Chiesa, P. et al. 2005. Co-production of hydrogen, electricity and CO_2 from coal with commercially ready technology. Part A: Performance and emissions. *Int. J. Hydrogen Energy* 30: 747–767.

Consonni, St. and F. Vigano. 2005. Decarbonized hydrogen and electricity from natural gas. *Int. J. Hydrogen Energy* 30: 701–718.

De Jong, K.P. and H.M.H. van Wechem. 1995. Carbon: Hydrogen carrier or disappearing skeleton? *Int. J. Hydrogen Energy* 20: 493–499.

Dincer, I. 2002. Technical, environmental and exergetic aspects of hydrogen energy systems. *Int. J. Hydrogen Energy* 27: 265–285.

Dunker, R. 2005. Wege zum emissionsfreien Kohlekraftwerk. BWK Bd. 57 (2005) Nr. 11, personal communication with Vattenfall.

Dunn, S. 2002. Hydrogen futures: Toward a sustainable energy system. *Int. J. Hydrogen Energy* 27: 235–264.

DWV. 1998. DWV–German Hydrogen and Fuel Cell Association 12/2003; Ludwig Boelkow Systemtechnik 1998, Hydrogen as By-Product of German Industry.

Edgerton, D. 2006. *The Shock of the Old.* London, U.K.: Profile Books.

Ewan, B.C.R. and R. Allen. 2005. A figure of merit assessment of the routes to hydrogen. *Int. J. Hydrogen Energy* 30: 809–819.

Gould, S.J. and N. Eldredge. 1993. Punctuated equilibrium comes of age. *Nature* 366: 223–227.

Haseli, Y. et al. 2008. Comparative assessment of greenhouse gas mitigation of hydrogen passenger trains. *Int. J. Hydrogen Energy* 33: 1788–1796.

Hefner, R.A. III. 2002. The age of energy gases. *Int. J. Hydrogen Energy* 27: 1–9.

Hijikata, T. 2002. Research and development of international clean energy network using hydrogen energy (WE-NET). *Int. J. Hydrogen Energy* 27: 115–129.

IEA. 2006. *Key World Energy Statistics.* International Energy Agency, Paris, France.

IPCC. 2005. *Carbon Dioxide Capture and Storage.* Metz, B. et al. (Eds.), Intergovernmental Panel on Climate Change, Cambridge, U.K.: Cambridge University Press, pp. 428.

IPCC. 2007. Intergovernmental Panel on Climate Change. *Climate Change 2007.* Metz, B. et al. (Eds.), Cambridge, U.K.: Cambridge University Press.

Iwasaki, W. 2003. Magnetic refrigeration technology for an international clean energy network using hydrogen energy. *Int. J. Hydrogen Energy* 28: 559–567.

Karim, G.A. 2003. Hydrogen as a spark ignition engine fuel. *Int. J. Hydrogen Energy* 28: 569–577.

Kashiwagi, T. 1990. Present status and future prospects of advanced energy technology for solving global environmental problems. *The Global Environmental Technology Seminar JETRO,* Tokyo, Japan, 1990.

Kotas, A.V. 2008. An analysis of SOFC/GT CHP system based on exergetic performance criteria. *Intern. J. Hydrogen Energy* 33: 2566–2577.

Kreutz, Th., R. Williams, St. Consonni, and P. Chiesa. 2005. Co-production of hydrogen and CO_2 from coal with commercially ready technology. Part B: Economic analysis. *Int. J. Hydrogen Energy* 30: 769–784.

Linssen, J. and M. Wallbeck. 2001. Forschungszentrum Jüelich. Ferienkurs Energieforschung, Bwk Bd. 18, Nr. 7 pp. 102–126, ISSN: 1433-5522, ISBN 3-89336-291-6.

Lovins, A.B. 2003. *Twenty Hydrogen Myths,* www.rmi.org/sitepages/art7516.php (accessed November 2007).

Lovins, A.B. and B.D. Williams. 1999. A strategy for the hydrogen transition, paper given at the *10th Annual Meeting.* National Hydrogen Association, Vienna, VA, April 7–9, 1999.

Lutz, A.E., R. Bradshaw, J. Keller, and D. Witmer. 2003. Thermodynamic analysis of hydrogen production by steam reforming. *Int. J. Hydrogen Energy* 28: 159–167.

Lutz, A.E., R. Larson, and J. Keller. 2002. Thermodynamic comparison of fuel cells to the Carnot cycle. *Int. J. Hydrogen Energy* 27: 1103–1111.

Marrero-Alfonso, E.Y. et al. 2007. Minimizing water utilization in hydrolysis of sodium borohydride: The role of sodium metaborate hydrates. *Int. J. Hydrogen Energy* 32: 4723–4730.

Martinez-Frias, J., A. Pham, and S. Aceves. 2003. A natural gas-assisted steam electrolyzer for high-efficiency production of hydrogen. *Int. J. Hydrogen Energy* 28: 483–490.

McDowall, W. and M. Eams. 2007. Towards a sustainable hydrogen economy: A multi-criteria sustainability appraisal of competing hydrogen futures. *Int. J. Hydrogen Energy* 32: 4611–4626.

McKibben, B. 1999. In *Natural Capitalism,* Boston, New York, London.

Nacicenovic, N. and A. John, A. 1991. CO_2 reduction and removal: Measures for the next century. *Energy* 16: 1347–1377.

Nakicenovic, N. 1996. Freeing energy from carbon. *Daedalus* 125: 95–112.

Neelis, M.L. et al. 2004. Exergetic life cycle analysis of hydrogen production and storage systems for automotive applications. *Int. J. Hydrogen Energy* 29: 537–545.

Ogden, J.M. 1999. Developing an infrastructure for hydrogen vehicles: A Southern California case study. *Int. J. Hydrogen Energy* 24: 709–730.

Peschka, W., E. Wilhem, and U. Wilhelm. 1992. *Liquid Hydrogen: Fuel of the Future*. Berlin, Germany: Springer.

Prince-Richard, S., M. Whale, and N. Djilali. 2005. A techno-economic analysis of decentralized electrolytic hydrogen production for fuel cell vehicles. *Int. J. Hydrogen Energy* 30: 1159–1179.

Ptasinski, K.J. et al. 2006. Exergy analysis of hydrogen production methods from biomass. *Proceedings of ECOS 2006*, Aghia Pelagia, Crete, Greece, Vol. 3, pp. 1601–1618.

Rant, Z. 1964. Thermodynamische Bewertung der Verluste bei technischen Energieumwandlungen. *Brennstoff Wärme Kraft* 16: 453.

Romm, J.J. 2004. *The Hydrogen HYPE*. Island Press, New York.

Rosen, M.A. and D.S. Scott, 2003a. Entropy production and exergy destruction: Part I, hierarchy of earth's major constituencies. *Int. J. Hydrogen Energy* 28: 1307–1313.

Rosen, M.A. and D.S. Scott. 2003b. Entropy production and exergy destruction: Part II, illustrative technologies. *Int. J. Hydrogen Energy* 28: 1315–1323.

Ruijven, B. van, D. van Ruuven, and B. de Vries. 2007. The potential role of hydrogen in energy systems with and without climate policy. *Int. J. Hydrogen Energy* 32: 1655–1672.

Schindler, J. et al. 2007. *Woher kommt die Energie für die Wasserstofferzeugung? Status und Alternativen*, *Ludwig Bölkow Systemtechnik*. DWV-Deutscher Wasserstoff und Brennstoffzellenverband (Ed.). The German Hydrogen and Fuel Cell Association, Berlin, Germany.

Scientific American. 2006. Energy's future beyond carbon. *Scientific American*, 295(3): 50–102.

Scott, D.S. 2007. *Smelling Land, the Hydrogen Defense against Climate Catastrophe*, Montreal, Canada: Canadian Hydrogen Association p. 485.

Seymour, E.H., F. Borges, and R. Fernandes. 2007. Indicators of European public research in hydrogen and fuel cells—An input-output analysis. *Int. J. Hydrogen Energy* 32: 3212–3222.

Sheffield, J.W. and C. Sheffield (Eds.). 2007. *Assessment of Hydrogen Energy for Sustainable Development*. Berlin, Germany: Springer. www.springer.com, ISBN 978-1-4020-6442-5 (e-book).

Sigfusson, T.I. 2008. *Planet Hydrogen. The Taming of the Proton*. Oxfordshire, U.K.: Coxmoor Publishing, 13 Number ISBN: 078-1-901892-7, 10 Number: ISBN 1 901892 27 1.

Simpson, A.P. et al. 2007. Exergy analysis of hydrogen production via steam methane reforming. *Int. J. Hydrogen Energy* 32: 4811–4820.

Sokolow, R.H. and St. W. Pacala. 2006. A plan to keep carbon in check. *Scientific American* 9: 28–35.

Sørensen, B., 2007. *Hydrogen & Fuel Cells*. Boston, MA: Elsevier, ISBN: 0-12-655281-9.

Steinberg, M. 1999. Fossil fuel decarbonization technology for mitigating global warming. *Int. J. Hydrogen Energy* 24: 771–777.

Thomas, C.E. (Sandy). 2008. Comparison of transportation options in a carbon-constrained world: Hydrogen, plug-in hybrids, biofuels. Paper given at the *National Hydrogen Association Annual Meeting*, Sacramento, CA, March 31, 2008.

VDI-Nachrichten. 1998. Dusseldorf: Deutscher Ingenieur-Verlag, ISSN: 0042-1758.

Verfondern, K. (Ed.). 2007. *Nuclear Energy for Hydrogen Production*. Energy Technology, Vol. 58. Jülich, Germany: Forschungszentrum Jülich, ISBN 978-3-89336-468-8.

Verne, J. 1874. *Mysterious Island*, Paris.

Verkehrswirtschaftliche Energiestrategie, an initiative of German auto makers in collaboration with the Federal Ministry of Transport, Berlin http://www.germanhy.de/page/index.phy?9654 (accessed August 2008).

Vielstich, W., A. Lamm, and H.A. Gasteiger (Eds.). 2003. *Handbook of Fuel Cells, Fundamentals, Technology, Applications*, 4 Vols. www.wiley.com/hfc

Wald, M.L. 2007. Is ethanol for the long haul? *Scientific American* 1: 28–35. www.sciam.com

Wendt, H. 1988. Water splitting methods, chap. 7. In *Hydrogen as an Energy Carrier*. C.-J. Winter and J. Nitsch (Eds.). Springer, pp. 166–208.

Wietschel, M. and U. Hasenauer. 2004, Facts für die energiopelitik von morgen, Hoffnungsträger Wasserstoff. *Energiespiegel*, November 12, 2004. National swiss paul scherrer Institut, http:// gabe.web.psi.ch/energie-spiegel

Winter, C.-J. 2000. Energy sustainability—The road is the destination. Invited paper given at the *Energy and Sustainability Forum of the Federal Institute of Technology*, Lausanne, Switzerland, March 28, 2000.

Winter, C.-J. 2002a. Wasserstoff aus Biomasse—Status quo und Perspektiven, *Vortrag gehalten auf dem Fachkongress "Kraftstoffe der Zukunft" der Bundesinitiative Bioenergie BBE, 04.-05*, Berlin, Germany, December 2002.

Winter, C.-J. 2002b. Wasserstoff und Kohle—Castor und Pollux, *Vortrag gehalten auf dem Deutschen Wasserstoff Energietag 2002*, Essen, Germany, November 12–14, 2002.

Winter, C.-J. 2003a. Forum hydrogenium. A dispute on energy. Paper given at *HYFORUM 2003, The International Hydrogen Energy Forum 2003*, Beijing, China, October 21–24, 2003.

Winter, C.-J. 2003b. On the HYway—Sustainable assets in Germany's energy state's portfolio. *Int. J. Hydrogen Energy* 28: 477–481.

Winter, C.-J. 2004. The hydrogen energy economy: An address to the World Economic Forum 2004. *Int. J. Hydrogen Energy* 29: 1095–1097.

Winter, C.-J. 2005. Electricity, hydrogen—Competitors, partners? *Int. J. Hydrogen Energy* 30: 1371–1374.

Winter, C.-J. 2007. Energy efficiency, no: It's exergy efficiency! *Int. J. Hydrogen Energy* 32: 4109–4111.

Winter, C.-J. and J. Nitsch. 1988a. *Wasserstoff als Energieträger*. Berlin, Germany: Springer Verlag.

Winter, C.-J. and J. Nitsch. (Eds.). 1988b. *Hydrogen as an Energy Carrier—Technologies, Systems, Economy*. Berlin, Germany: Springer Verlag, ISBN 3-540-18896-7, ISBN 0-387-18896-7.

Winter, C.-J., R.L. Sizmann, and L.L. Vant-Hull (Eds.). 1991. *Solar Power Plants, Fundamentals, Technology, Systems, Economics*. Berlin, Germany: Springer-Verlag, ISBN 3-540-18897-5, ISBN 0-387-18897-5.

Wirth, T.E. et al. 2003. The future of energy policy. *Foreign Affairs* 82: 132–135.

Wit, M.P. de and A. Faaij. 2007. Impact of hydrogen onboard storage technologies in the performance of hydrogen fuelled vehicles. A techno-economic well-to-wheel assessment. *Int. J. Hydrogen Energy* 32: 4859–4870.

Yang, Ch. 2008. Hydrogen and electricity: Parallels, interactions, and convergence. *Int. J. Hydrogen Energy* 33: 1877–1994.

Yang, Ch. and J. Ogden. 2007. Determining the lowest-cost hydrogen delivery mode. *Int. J. Hydrogen Energy* 32: 268–286.

3

![line]

Nuclear Power for the Production of Carbon-Free Energy and Fuels

Kenneth R. Schultz, Maximilian B. Gorensek, J. Stephen
Herring, Michele A. Lewis, Robert Moore, James E. O'Brien,
Paul S. Pickard, Benjamin E. Russ, and William A. Summers

CONTENTS

Nuclear energy does not inherently involve any use of fossil fuels or the production of carbon dioxide or other greenhouse gases (GHGs). Nuclear energy thus has the potential to make major contributions to the production of carbon-neutral fuels and energy carriers by providing a major source of the primary energy for these fuels. Forty percent of the U.S. primary energy consumption goes for the production of electricity, and that electricity provides 19% of the end-use energy consumption in the United States (LLNL 2008). The production of electricity accounts for 40% of the energy-related CO_2 production in the United States, primarily from burning coal and natural gas. Nuclear power provides 21% of the U.S. electricity and virtually no CO_2. Increased use of nuclear energy to produce electricity, displacing coal and natural gas would reduce CO_2 production. However, reducing the other 60% of the U.S. energy-related CO_2 emissions requires addressing the nonelectric energy applications. The purpose of this chapter is to explore how nuclear energy can contribute to the production of carbon-neutral and carbon-free fuels and energy carriers.

3.1 Introduction

Nuclear power does not produce CO_2 and thus offers the potential for a carbon-free source of primary energy. However, the construction of the nuclear power plant and the provision of nuclear fuel for that power plant can involve use of products that required fossil energy to produce. Examples are the use of coal and natural gas to produce steel and concrete to build the plant, and the consumption of electricity produced from fossil fuels to enrich the uranium used in the plant. This issue has been studied in depth for the production of electricity from nuclear energy in comparison with other sources of electricity (Meier 2002) with the conclusion that the life-cycle production of CO_2 from nuclear power is among the lowest, as shown in Figure 3.1.

Thus, nuclear power currently provides a virtually CO_2-free way to produce 20% of the electricity used in the United States. As use of nuclear power is expanded and fossil power generation is reduced, the amount of CO_2 produced as a by-product of electricity generation will decline. If all electricity were produced by renewable and nuclear energy, energy-related CO_2 production could be reduced up to 40%. To address the other 60% requires that we seek to apply nuclear energy to nonelectric applications.

3.1.1 Hydrogen

Combustion of fossil fuels provides 86% of the world's energy (DOE/EIA 2000). Drawbacks to fossil fuel utilization include limited supply, pollution, and carbon dioxide emissions, thought to be responsible for global warming (DOE/EIA 2003). Hydrogen can replace

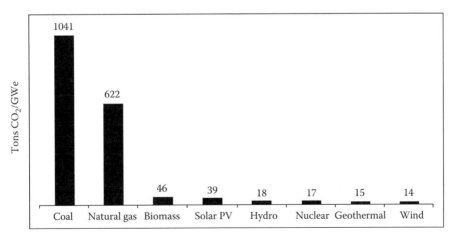

FIGURE 3.1

Comparison of life-cycle CO_2 emissions—tons of carbon dioxide equivalent per GWh. (Adapted from Meier, P.J., Life-cycle assessment of electricity generation systems and applications for climate change policy analysis, PhD thesis, University of Wisconsin, Madison, WI, UWFDM-1181, August 2002.)

fossil fuels in transportation, reducing vehicle emissions of CO_2, NO_X, and SO_X and making possible fuel cell vehicles with more than double the mileage of conventional engines. Hydrogen is an environmentally attractive end-use fuel that has the potential to displace fossil fuels. However, contemporary hydrogen production is primarily based on fossil fuels, which negates many of hydrogen's environmental advantages. This industry produces hydrogen for use in the production of fertilizers and in oil refineries to lighten heavy crude oils and produce cleaner burning fuels, and for other industrial uses, primarily by steam reformation of methane. The fastest growing of these uses is for oil refining, shown in Figure 3.2 (Forsberg and Peddicord 2001). In the United States, this hydrogen industry produces 11 million tons of hydrogen a year with a thermal energy equivalent of 48 GW. In doing so, it consumes 5% of the U.S. natural gas usage and releases 74 million tons of CO_2 per year.

There is currently a great deal of interest in the transition from our present petroleum-based transportation system to the one based on hydrogen. A significant "hydrogen economy" is predicted that will reduce our dependence on petroleum imports and reduce pollution and GHG emissions (DOE/Hydrogen 2002). Hydrogen is an environmentally attractive fuel that has the potential to displace fossil fuels, but hydrogen is an energy *carrier*, not an energy *source*. While hydrogen is the most plentiful element in the Universe, on Earth it occurs naturally chemically bound in water, hydrocarbons, carbohydrates, or other compounds. Energy must be invested to separate the hydrogen.

The transition to a hydrogen economy will require significant expansion in the production of hydrogen—factor of 18 more for transportation energy alone. Clearly, new sources of hydrogen will be needed. Hydrogen produced from water using nuclear energy can be one of the sources and would avoid both use of fossil fuels and GHG emissions.

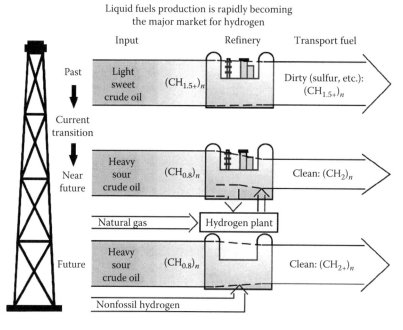

FIGURE 3.2
Use of hydrogen to lighten heavy crude oils is growing rapidly. (From Forsberg, C.W. and K.L. Peddicord, *Nucl. News*, 41, 2001.)

Hydrogen could be produced from nuclear energy by several means. Electricity from nuclear power can separate water into hydrogen and oxygen by electrolysis. Low-temperature electrolysis (LTE) is a proven, commercial technology. The net efficiency is the product of the efficiency of the reactor in producing electricity, times the efficiency of the electrolysis cell, which, at the high pressure needed for distribution and utilization, is about 75%. Our current nuclear power plants, mostly light water reactors (LWRs), have 32% electrical efficiency, so the net efficiency of hydrogen production is about 24%. If the electrolysis were done using steam at a high temperature, nuclear-produced heat could be substituted for some of the electricity and the net heat-to-hydrogen efficiency could be raised to ~50%. Thermochemical water-splitting cycles could get all of their input energy from nuclear-produced heat, using coupled, thermally driven chemical reactions to split water into oxygen and hydrogen, and offer the promise of heat-to-hydrogen efficiencies of ~50%. These processes are discussed in this chapter.

3.1.2 Biofuels

Biofuels, transportation fuels made from various plant products, have the potential to be carbon neutral. The CO_2 that is released into the atmosphere when the biofuel is used originally came from the atmosphere and was captured by the plants that provided the biomass for the biofuel production process. If the energy used to gather, process, and convert the biomass into transport fuel is provided by a carbon-free or carbon-neutral source, the biofuel will be carbon neutral. Forsberg has proposed that the energy to produce biofuels be provided by nuclear power (Forsberg 2009). He notes that the annual sustainable biomass production in the United States is about 1.3 billion dry tons and that it is projected that this biomass could produce about 30% of our transportation fuel needs (Perlack et al. 2005, Koonin 2006, Somerville 2006). Fuel production from biomass is energy intensive and less than half the energy value of the dry biomass ends up in the biofuel, as shown in Figure 3.3.

If the energy to process the biomass into biofuel is provided by nuclear power, the energy content of the resulting fuel is increased by almost a factor of 3, as also shown in Figure 3.3. Forsberg points out that the processes for the conversion of biomass to synthetic gasoline or diesel fuel, which avoid the CO_2 production of the fermentation process used to produce ethanol, require large amounts of hydrogen. Thus, the use of nuclear power to provide heat, electricity, and hydrogen can enable significant improvement in the biomass to biofuel process and can greatly expand the amount of biofuel that could be sustainably produced.

3.1.3 Synthetic Hydrocarbon Fuels: "Synfuel"

Synthetic liquid hydrocarbons have been synthesized for more than three-quarters of a century from nonliquid feedstocks—principally coal but also natural gas in recent years. The leading process is the Fischer–Tropsch (F-T) process, which uses synthesis gas—hydrogen and carbon monoxide—as its feed and produces a synthetic "crude" that undergoes further processing to a range of commercially finished products (Syntroleum 2006). The synthesis gas is produced by coal gasification and by reforming of natural gas. Both of these feed preparation processes (in addition to the cost of the feed) can represent a significant cost component of the entire synthesis process. For coal, synthesis gas is produced according to the gasification reaction (3.1). The water–gas shift reaction can be used to produce additional H_2 as shown in (3.2). The reaction for producing F-T products from

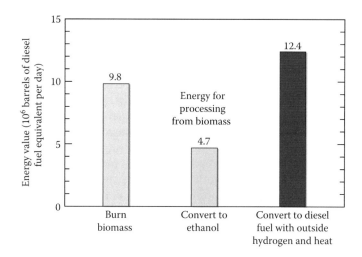

FIGURE 3.3
Energy value of 1.3 billion tons of biomass. (Figure courtesy of C.W. Forsberg.)

synthesis gas (CO and H_2) is (3.3). Thus, the simultaneous F-T and water–gas shift reactions in the reactor lead directly to the complete reaction (3.4).

Synfuel by coal gasification:

$$\text{Gasification} \quad 2C + \frac{1}{2}O_2 + H_2O \rightarrow 2CO + H_2 \tag{3.1}$$

$$\text{Water–gas shift} \quad CO + H_2O \rightarrow H_2 + CO_2 \tag{3.2}$$

$$\text{F-T reaction} \quad CO + 2H_2 \rightarrow -CH_2- + H_2O \tag{3.3}$$

$$\text{Net reaction} \quad 2C + H_2O + \frac{1}{2}O_2 \rightarrow -CH_2- + CO_2 \tag{3.4}$$

Note that two carbons are required to produce one F-T CH_2 product, the other carbon being emitted as carbon dioxide.

The extra hydrogen that is provided by the water–gas shift can be provided by water splitting with energy supplied from a nuclear reactor or other non-CO_2-emitting source. In this case, synthesis gas is still provided by coal gasification as above (3.5) and the extra hydrogen is provided by water splitting (3.6), for the net reaction shown in (3.7).

Synfuel by coal gasification + hydrogen from water splitting:

$$\text{Gasification} \quad C + \frac{1}{4}O_2 + \frac{1}{2}H_2O \rightarrow CO + \frac{1}{2}H_2 \tag{3.5}$$

$$\text{Water splitting} \quad \frac{3}{2}H_2O + \text{Energy} \rightarrow \frac{3}{2}H_2 + \frac{3}{4}O_2 \tag{3.6}$$

$$\text{F-T reaction} \quad CO + 2H_2 \rightarrow -CH_2- + H_2O \tag{3.3}$$

$$\text{Net reaction} \quad C + H_2O + \text{Energy} \rightarrow -CH_2- + \frac{1}{2}O_2 \tag{3.7}$$

In comparison with the conventional gasification and F-T sequence, only half the carbon is required, and there is *no* CO_2 produced in the conversion process. Further, oxygen is provided by the water splitting, which avoids the need for an external oxygen supply to the gasification process and even has some excess oxygen for potential sale. Figure 3.4 illustrates the complete block diagram for the idealized process.

Fossil-fired power plants produce CO_2, which could be captured and converted to CO for the production of synthetic fuels. CO_2 can be converted to CO by the reverse water–gas shift reaction (3.8). CO could then be used in the F-T reaction with additional hydrogen from water splitting to produce synfuel (Schultz and Bogart 2006).

Synfuel by CO_2 capture + H_2 from water splitting:

$$\text{Reverse water–gas shift} \quad CO_2 + H_2 \rightarrow CO + H_2O \tag{3.8}$$

$$\text{F-T reaction} \quad CO + 2H_2 \rightarrow -CH_2- + H_2O \tag{3.3}$$

$$\text{Water-splitting} \quad 3H_2O + \text{Energy} \rightarrow 3H_2 + \frac{3}{2}O_2 \tag{3.9}$$

$$\text{Net reaction} \quad CO_2 + H_2O + \text{Energy} \rightarrow -CH_2- + \frac{3}{2}O_2 \tag{3.10}$$

In this ideal case, no coal is needed at all, and CO_2 is consumed rather than produced. The excess O_2 would be used in the fossil power plant that provides the CO_2, simplifying CO_2 capture. There is currently considerable effort underway on developing CO_2 capture systems for new and extant power plants. The increasing concern with global climate change suggests that there is a reasonable likelihood of such plants operating in the timeframe

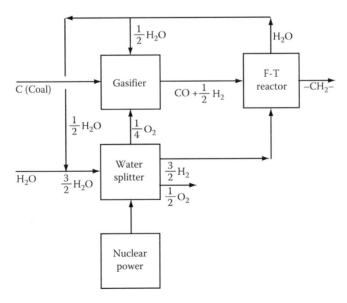

FIGURE 3.4
Coal gasification and F-T processes augmented with externally provided hydrogen produced by water splitting.

associated with synthetic fuel from carbon dioxide. Such a synergistic system, dubbed "twice burned coal" or "recycled coal," has the potential to significantly reduce the current emissions of CO_2 since the carbon in the coal is used once for power production and then again for liquid hydrocarbon fuel synthesis.

Several different techniques have been proposed for capture of the CO_2 needed for this synthesis process from the flue gas of fossil-fired power plants. Absorption of CO_2 by an aqueous solution of potassium carbonate followed by thermal solvent regeneration is a commercially available process. Membrane gas absorption in amine solvents flowing in permeable membranes with thermal recovery has attractive features. For coupling with nuclear production of hydrogen and synthetic fuel production, capture of the CO_2 by oxy-firing the coal plant appears the best choice (Schultz et al. 2009). The oxygen needed to oxy-fire the coal plant is provided by the water-splitting process that produces the hydrogen. The complete system is shown in Figure 3.5.

The overall impact of alternative sources of transportation fuels can be seen by examining the CO_2 flows resulting from energy consumption (LLNL 2002). The 2002 U.S. petroleum-based transportation economy released 1811 million metric tons (MMt) of CO_2 per year. Approximately another 100 MMt of CO_2 were released in the production and processing of that petroleum. If that fuel were produced from coal by gasification and F-T processing, 1113 MMt of carbon from coal would be needed and 2046 MMt of CO_2 would be produced. Added to the 1811 MMt that is released when the fuel is burned, the net CO_2 release would be 3857 MMt/year. Comparing these numbers to the current use of coal for electricity production of 565 MMt/year of carbon as coal and release of 2070 MMt/year of CO_2, it is apparent that reliance on coal-based synfuels for transportation fuel needs

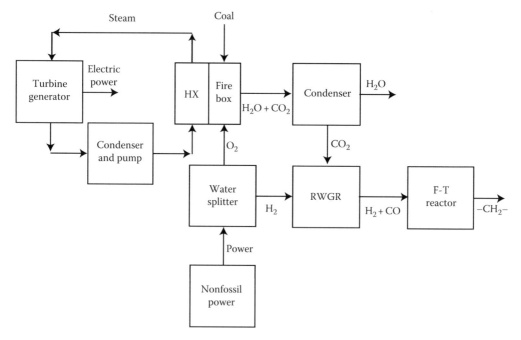

FIGURE 3.5
Oxyfuel coal plant and F-T processes augmented with externally provided oxygen and hydrogen produced by water splitting. RWGR is a reverse water gas shift reactor. HX is a heat exchanger. *Note*: For simplification, the reaction products are not balanced in this figure.

TABLE 3.1

Fuel Needed and CO_2 Released for Alternate Transportation Fuel Sources

Units: MMt/Year	Transportation Fuel from			
	Oil	Coal	Coal + H_2 from Water	CO_2 + H_2 from Water
Oil needed	612	—	—	—
Coal needed	—	1113	556	0
H_2 needed	—	—	130	260
CO_2 produced	~100	2046	104	−1811
CO_2 released on use	1811	1811	1811	1811
Net CO_2 released	1911	3857	1905	0

Current carbon use as coal/CO_2 produced: 565/2070 MMt/year; H_2 use: 11 MMt/year.

would require tripling the amount of coal used and increasing by half the total amount of CO_2 produced.

If a CO_2-free source of hydrogen, such as nuclear or solar energy, is provided, the production of synfuels could be done using 556 MMt/year of carbon as coal and producing 1905 MMt/year of CO_2. This would mean doubling the current consumption of coal but with no increase in the current production of CO_2. If the carbon needed for synfuel were provided from CO_2 captured from flue gas of current coal-fired power plants, the mass flows match well. About 565 MMt/year of carbon is used and released in the form of CO_2, and about 565 MMt/year is needed for synfuel production. All the transportation fuel currently used could be provided using CO_2 captured from current coal-fired power plants. This would require no additional coal use and would actually cut current release of CO_2 in half. Both these scenarios would require a significant increase in the amount of hydrogen that would have to be produced, and would require development of non-CO_2 emitting techniques, such as water splitting, for its production. These alternate scenarios are summarized in Table 3.1. The dramatic difference in CO_2 production of the scenarios is illustrated in Figure 3.6.

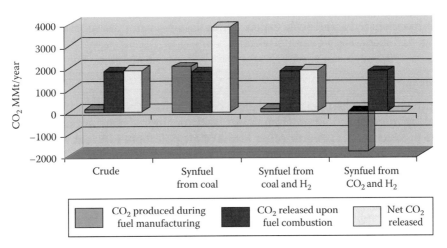

FIGURE 3.6
CO_2 released by alternative sources of transportation fuels, MMt/year.

This hydrogen synfuel concept would allow us to significantly reduce our use of petroleum and cut our CO_2 emissions in half, while still using our existing hydrocarbon-based transportation infrastructure. The hydrogen-production infrastructure needed for synfuel production could be used to produce hydrogen for direct application via fuel cells in the future. A hydrogen–synfuel economy would provide a bridge to a pure hydrogen economy.

3.1.4 Nuclear Production of Hydrogen

A common thread for all the processes for the production of reduced-carbon and carbon-neutral fuels is the need for a large-scale source of hydrogen at reasonable cost. Hydrogen is used to upgrade heavy crude petroleum and get more fuel per unit carbon released. Hydrogen can make significant improvements in the efficiency of producing biofuels from biomass. Hydrogen can be used to make synthetic hydrocarbon fuels, eventually "recycling" the CO_2 from fossil power plants to make transportation fuels. And hydrogen by itself alone is the ultimate carbon-free transportation fuel.

Nuclear power can provide the large-scale source of hydrogen at reasonable cost, which is needed to make these visions into a reality. The rest of this chapter describes the status of R&D on the nuclear production of hydrogen. The following sections describe high-temperature electrolysis (HTE), thermochemical cycles in general, focusing then on the high-temperature sulfur–iodine (S-I) and hybrid sulfur (HyS) cycles, and the H_2SO_4 decomposition process that is common to both. The chapter closes with a look at advanced thermochemical cycles that could allow the utilization of nuclear power at lower temperatures for an efficient production of hydrogen.

3.2 High-Temperature Electrolysis

The Idaho National Laboratory (INL) has been the lead lab for HTE research and development, initially under the Department of Energy (DOE) Nuclear Hydrogen Initiative and recently as part of the DOE Next-Generation Nuclear Plant (NGNP) project. The INL HTE program has included small-scale experiments, detailed computational modeling, system modeling, and technology demonstration.

3.2.1 Background

High-temperature nuclear reactors have the potential for substantially increasing the efficiency of hydrogen production from water, with negligible consumption of fossil fuels and little production of GHGs or other forms of air pollution. Efficient water splitting for hydrogen production can be accomplished via HTE or thermochemical processes, using high-temperature nuclear process heat. In order to achieve high efficiencies, both processes require high-temperature operation. Thus, these hydrogen-production technologies are tied to the development of advanced high-temperature nuclear reactors. High-temperature electrolytic water splitting supported by nuclear process heat and electricity has the potential to produce hydrogen with overall thermal-to-hydrogen efficiencies of 50% or higher, based on high heating value. This efficiency is near that of the thermochemical processes (Yildiz and Kazimi 2006, O'Brien 2008, O'Brien et al. 2008).

A research program is under way at the INL to simultaneously address the technical and scale-up issues associated with the implementation of solid-oxide electrolysis cell technology for efficient hydrogen production from steam. A progression of electrolysis cell and stack testing activities, at increasing scales, along with a continuation of supporting research activities in the areas of materials development, single-cell testing, detailed computational fluid dynamics analysis, and system modeling, is underway. The INL successfully completed a 1080 h test of the HTE integrated laboratory scale (ILS) technology demonstration experiment during the fall of 2008. The HTE ILS achieved a maximum hydrogen-production rate in excess of 5.7 N m³/h, with a power consumption of 18 kW. During May through September of 2009, the HTE program operated a 10-cell stack for 2580 h continuously, with an average cell degradation rate of 8.2% per 1000 h.

The HTE program also includes an investigation of the feasibility of producing syngas by simultaneous electrolytic reduction of steam and carbon dioxide (co-electrolysis) at high temperatures using solid-oxide cells. Syngas, a mixture of hydrogen and carbon monoxide, can be used for the production of synthetic liquid fuels via F-T processes. This concept, coupled with nuclear energy, provides a possible path to reduce GHG emissions and increase energy independence, without a major infrastructure shift that would be required for a purely hydrogen-based transportation systems (Mogensen et al. 2008, O'Brien et al. 2009a, Stoots and O'Brien 2009). Furthermore, if the carbon dioxide feedstock is obtained from biomass, the entire concept would be carbon neutral.

3.2.1.1 Thermodynamics of High-Temperature Electrolysis

Consider a control volume surrounding an isothermal electrolysis process. In this case, both heat and work interactions cross the control volume boundary. The first law for this process is given by

$$Q - W = \Delta H_R \tag{3.11}$$

For reversible operation,

$$Q_{rev} = T\Delta S_R \tag{3.12}$$

such that

$$W_{rev} = \Delta H_R - T\Delta S_R = \Delta G_R \tag{3.13}$$

The thermodynamic properties appearing in Equation 3.13 are plotted in Figure 3.7 as a function of temperature for the H_2-H_2O system from 0°C to 1000°C at standard pressure. This figure is often cited as a motivation for HTE versus LTE. It shows that the Gibbs free energy change, ΔG_R, for the reacting system decreases with increasing temperature, while the product of temperature and the entropy change, $T\Delta S_R$, increases. Therefore, for reversible operation, the electrical work requirement decreases with temperature, and a larger fraction of the total energy required for electrolysis, ΔH_R, can be supplied in the form of heat, represented by $T\Delta S_R$. Since heat-engine-based electrical work is limited to a production thermal efficiency of 50% or less, decreasing the work requirement results in higher overall thermal-to-hydrogen production efficiencies. Note that the total energy requirement, ΔH_R, increases only slightly with temperature and is very close in magnitude to the

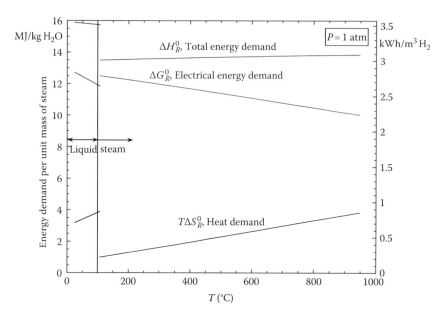

FIGURE 3.7
Standard-state ideal energy requirements for electrolysis as a function of temperature.

lower heating value of hydrogen. The ratio of ΔG_R to ΔH_R is about 93% at 100°C, decreasing to only about 70% at 1000°C. Operation of the electrolyzer at high temperatures is also desirable from the standpoint of reaction kinetics and electrolyte conductivity, both of which improve dramatically at higher operating temperatures. Potential disadvantages of high-temperature operation include the limited availability of very high-temperature process heat and materials issues such as corrosion and degradation.

The solid-oxide electrolysis cell is a solid-state electrochemical device consisting of an oxygen-ion-conducting electrolyte (e.g., yttria-stabilized zirconia [YSZ] or scandia-stabilized zirconia [ScSZ]) with porous electrically conducting electrodes deposited on either side of the electrolyte. A cross section of a planar design is shown in Figure 3.8. The design depicted in the figure shows an electrolyte-supported cell with a nickel cermet cathode and a perovskite anode such as strontium-doped lanthanum manganite (LSM). In an electrolyte-supported cell, the electrolyte layer is thicker than either of the electrodes. The flow fields conduct electrical current through the stack and provide flow passages for the process gas streams. The separator plate or bipolar plate separates the process gas streams. It must also be electrically conducting and is usually metallic, such as a ferritic stainless steel.

As shown in the figure, a mixture of steam and hydrogen at 750°C–950°C is supplied to the cathode side of the electrolyte (note that cathode and anode sides are opposite to their fuel-cell-mode roles). The half-cell electrochemical reactions occur at the triple-phase boundary near the electrode–electrolyte interface, as shown in the figure. Oxygen ions are drawn through the electrolyte by an applied electrochemical potential. The ions liberate their electrons and recombine to form molecular O_2 on the anode side. The inlet steam–hydrogen mixture composition may be as much as 90% steam, with the remainder hydrogen. Hydrogen is included in the inlet stream in order to maintain reducing conditions at the cathode. The exiting mixture may be as much as 90% H_2. It is not desirable to attempt to operate the stack with a higher outlet composition of hydrogen because

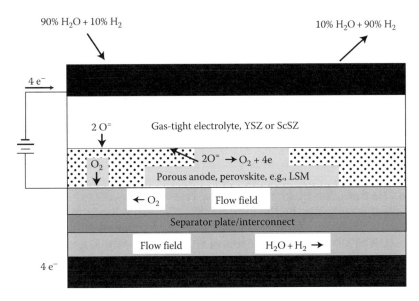

FIGURE 3.8
Cross section of a planar HTE stack.

there is a risk of local steam starvation, which can lead to significantly reduced cell performance. Furthermore, based on detailed analyses of large-scale HTE systems, the overall hydrogen-production efficiency is almost flat above about 50% steam utilization (O'Brien et al. 2009b). Product hydrogen and residual steam is passed through a condenser or membrane separator to purify the hydrogen.

3.2.1.2 Cell Materials

The solid-oxide electrolysis cell is a solid-state electrochemical device consisting of an oxygen-ion-conducting electrolyte (e.g., YSZ or ScSZ) with porous electrically conducting electrodes deposited on either side of the electrolyte. The standard electrolyte material is formed by doping zirconia (ZrO_2) with, for example, 8 mol% of yttria (yttrium oxide, Y_2O_3). A dopant composition of 8% or higher yields a "fully stabilized" electrolyte. The dopant serves two purposes. It "stabilizes" the cubic (or fluorite) crystal structure over a wide temperature range. Undoped zirconia exhibits a monoclinic crystal structure at room temperature and a tetragonal phase above 1170°C. Zirconia doped with yttria is called YSZ. In addition to stabilizing the crystal structure, when trivalent Y is substituted for tetravalent Zr, holes (unfilled positions) in the oxygen sub-lattice are introduced at the same time. This makes it possible for oxygen ions to move through the solid by hopping from hole to hole in the lattice. YSZ is, therefore, a good oxygen-ion conductor. Other compounds such as scandia (Sc_2O_3) can also be used as the dopant. ScSZ has a significantly higher ionic conductivity than YSZ but is more expensive. Other potential electrolyte materials include samaria-doped ceria and calcia-doped lanthanum gallate.

The most common steam–hydrogen electrode material is porous nickel–zirconia cermet, but a nickel–ceria cermet can also be used. In the electrolysis mode, this electrode serves as the cathode. Because the cathode contains nickel metal, reducing conditions must be maintained on this electrode during cell operation. This is typically accomplished by including ~10% mole fraction hydrogen in the inlet flow. The nickel in the cathode acts as a catalyst for steam reduction. The zirconia in the cermet provides ionic conductivity. Porosity allows the steam to migrate to the active electrochemical reaction site and hydrogen to migrate away from the sites. The active reaction sites correspond to what is typically termed the "triple-phase boundary" where the electronic, ionic, and gas phases coexist. These sites occur at locations where a pore structure intersects with nickel and zirconia particles.

Several materials have been studied for the air–oxygen electrode. This electrode must operate in a highly oxidizing environment. The most common material used is strontium-doped lanthanum manganite $La_{0.8}Sr_{0.2}MnO_3$, or LSM. This material provides good electronic conductivity and good catalytic activity and tolerance to the oxidizing environment. LSM is an example of a class of materials called perovskites, which have the general chemical formula ABO_3 where "A" and "B" are two cations of very different sizes (A much larger than B), and O is the anion that bonds to both. These perovskites exhibit p-type electrical conductivity that is enhanced by the introduction of lower-valence dopant cations such as Sr^{2+} to replace La^{3+} cations. Strontium-doped (LSF) and strontium + cobalt-doped lanthanum ferrites (LSCF) have also received lots of attention recently. These materials are more catalytically active than LSM and therefore yield generally better performance, especially at temperatures below 800°C.

3.2.1.3 Electrolysis Cell Designs and Stack Configurations

Several basic cell designs have been developed for solid-oxide fuel cell (SOFC) applications including electrolyte-supported, electrode-supported, and porous ceramic or metal substrate-supported cells. A full discussion of these various cell designs and the various fabrication techniques is beyond the scope of this chapter. Common cell characteristics include a dense gas-tight electrolyte layer, with porous electrodes on either side. In an electrolyte-supported cell, the electrolyte layer is thicker than either of the electrodes and must have sufficient mechanical strength to withstand any stresses. However, as a result of the relatively thick electrolyte, ionic resistance across the electrolyte is large for this design. The best performing SOFC cells of recent design are the anode-supported cells in which the mechanical strength is provided by a thick (~1.5 mm) layer of anode (usually nickel-YSZ cermet) material (Williams et al. 2006). Thin electrolyte and cathode layers are deposited on the anode material by screen printing or other techniques. This design has exhibited very high performance in SOFC tests. Some researchers have suggested that the best performance for the electrolysis mode of operation could be obtained using air-side (e.g., LSM) electrode-supported cells (Ni et al. 2006). A wealth of additional information on materials, configurations, and designs of solid-oxide electrochemical systems is available in reference (Singhal and Kendall 2003).

The majority of the electrolysis stack testing that has been performed at INL to date has been with planar stacks fabricated by Ceramatec, Inc. of Salt Lake City, UT. An exploded view of the internal components of one of these stacks is shown in Figure 3.9. The cells have an active area of 64 cm². The stacks are designed to operate in cross flow, with the steam–hydrogen gas mixture flowing from front to back in the figure and air flowing from right to left. Airflow enters at the rear though an air inlet manifold and exits at the front directly into the furnace. The steam/hydrogen inlet and outlet manifolds are visible in the photograph. The power lead attachment tabs, integral with the upper and lower interconnect plates are also visible in Figure 3.9.

Stack operating voltages were measured using wires that were directly spot-welded onto these tabs. The interconnect plates are fabricated from ferritic stainless steel. Each interconnect includes an impermeable separator plate (~0.46 mm thick) with edge rails and

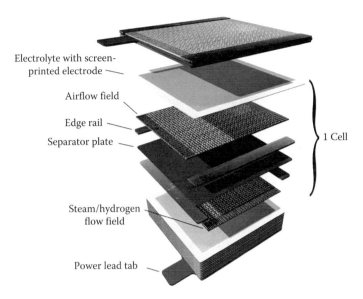

Electrolyte with screen-printed electrode

Airflow field

Edge rail

Separator plate

1 Cell

Steam/hydrogen flow field

Power lead tab

FIGURE 3.9
Exploded view of electrolysis stack components.

two corrugated "flow fields," one on the air side and one on the steam/hydrogen side. The height of the flow channel formed by the edge rails and flow fields is 1.0 mm. Each flow field includes 32 perforated flow channels across its width to provide uniform gas-flow distribution. The steam/hydrogen flow fields are fabricated from nickel foil. The air-side flow fields are ferritic stainless steel. The interconnect plates and flow fields also serve as electrical conductors and current distributors. To improve performance, the air-side separator plates and flow fields are pre-surface-treated to form a rare-earth stable conductive oxide scale. A perovskite rare-earth coating is also applied as a bond layer to the separator-plate oxide scale by either screen printing or plasma spraying. On the steam/hydrogen side of the separator plate, a thin (~10 µm) nickel metal coating is applied as a bond layer.

The stack electrolytes are ScSZ, about 140 µm thick. The air-side electrodes (anode in the electrolysis mode), are a strontium-doped manganite. The electrodes are graded, with an inner layer of manganite/zirconia (~13 µm) immediately adjacent to the electrolyte, a middle layer of pure manganite (~18 µm), and an outer bond layer of cobaltite. The steam/hydrogen electrodes (cathode in the electrolysis mode) are also graded, with a nickel–zirconia cermet layer (~13 µm) immediately adjacent to the electrolyte and a pure nickel outer layer (~10 µm).

Planar stacks can also be assembled using electrode-supported cells. Advanced technology SOFC stacks based on anode-supported cell technology have been developed by several manufacturers under the Solid-State Energy Conversion Alliance (SECA) (Williams et al. 2006) program.

3.2.2 Experimental Program

3.2.2.1 Small-Scale Tests

The experimental program at INL includes a range of test activities designed to characterize the performance of solid-oxide cells operating in the electrolysis mode. Small-scale activities are intended to examine candidate electrolyte, electrode, and interconnect materials with single cells and small stacks. Initial cell and stack performance and long-term degradation characteristics have been examined. Larger scale experiments are designed to demonstrate the technology and to address system-level issues such as hydrogen recycle and heat recuperation.

A schematic of the experimental apparatus used for single-cell testing is presented in Figure 3.10. The schematic for stack testing is similar. Primary components include gas supply cylinders, mass-flow controllers (MFCs), a humidifier, online dew point and CO_2 measurement stations, temperature and pressure measurement devices, a high-temperature furnace, a solid-oxide electrolysis cell (SOEC), and a gas chromatograph (GC). Nitrogen is used as an inert carrier gas. Carbon dioxide and related instrumentation is included for co-electrolysis experiments. Inlet flow rates of nitrogen, hydrogen, carbon dioxide, and air are established by means of precision MFCs. Hydrogen is included in the inlet flow as a reducing gas in order to prevent oxidation of the Nickel cermet electrode material. Airflow to the stack is supplied by the shop air system, after passing through a two-stage extractor/ dryer unit. The cathode-side inlet gas mixture, consisting of hydrogen, nitrogen, and possibly carbon dioxide (for co-electrolysis tests), is mixed with steam by means of a heated humidifier. The dew point temperature of the nitrogen/hydrogen/CO_2/steam gas mixture exiting the humidifier is monitored continuously using a precision dew point sensor. All gas lines located downstream of the humidifier are heat-traced in order to prevent steam condensation. Inlet and outlet CO_2 concentrations are also monitored continuously using online infrared CO_2 sensors, when applicable.

For single-button-cell testing, an electrolysis cell is bonded to the bottom of a zirconia tube, using a glass seal. During testing, the tube is suspended in the furnace. The cells are electrolyte-supported with a ScSZ electrolyte, about 150 μm thick. The outside electrode, which is exposed to air, acts as the cathode in fuel cell mode and the anode in electrolysis mode. This electrode is a doped manganite. The inside steam–hydrogen electrode (electrolysis cathode) material is a nickel cermet. Both button-cell electrodes incorporate a

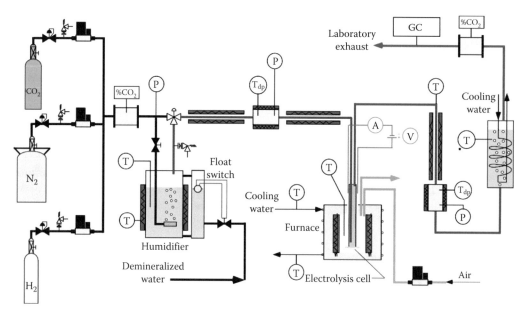

FIGURE 3.10
Schematic of single-cell co-electrolysis test apparatus.

platinum wire mesh for current distribution and collection. The button cells include both an active cell area (2.5 cm^2 for the cell shown) and a reference cell area. The active cell area is wired with both power lead wires and voltage taps. The reference cell area is wired only with voltage taps, allowing for continuous monitoring of open-cell potential. The power lead and voltage wires are routed to the far end of the zirconia tube via several small-diameter alumina tubes fixed to the outside of the zirconia manifold tube. A type-K stainless-steel sheathed thermocouple is mounted on the manifold tube and bent around in front of the button cell in order to allow for continuous monitoring of the button-cell temperature. The inlet gas mixture enters this tube, directing the gas to the steam/hydrogen/CO_2 side (inside) of the cell. The cell is maintained at an appropriate operating temperature (800°C–850°C) via computer-based feedback control. The furnace also preheats the inlet gas mixture and the air sweep gas. Oxygen produced by electrolysis is captured by the sweep gas stream and expelled into the laboratory. The product stream exits the zirconia tube and is directed toward the downstream dew point and CO_2 sensors and then to a condenser through a heat-traced line. The condenser removes most of the residual steam from the exhaust. The final exhaust stream is vented outside the laboratory through the roof. Rates of steam and CO_2 electrolysis are monitored by the measured change in inlet and outlet steam and CO_2 concentration as measured by the online sensors. In addition, a GC has been incorporated into the facility downstream of the condenser to precisely quantify the composition of the dry constituents in the electrolysis product stream (including any CH_4 that may be produced).

   Results of initial single (button)-cell HTE tests completed at the INL were documented in detail in reference (O'Brien et al. 2005a). Button cell tests are useful for basic performance characterization of electrode and electrolyte materials and of different cell designs (e.g., electrode-supported, integrated planar, and tubular). Results of initial short-stack HTE tests performed at INL are provided in O'Brien et al. (2006, 2007).

3.2.2.2 Co-Electrolysis of Carbon Dioxide and Steam

Representative co-electrolysis results are presented in Figure 3.11, from Stoots et al. (2009b). This figure shows the outlet gas composition (dry basis) from a 10-cell electrolysis stack as a function of stack current. The solid data symbols represent measurements obtained from GC. The lines represent predictions based on a chemical equilibrium co-electrolysis model (Stoots et al. 2009b). The open data symbols show the cold inlet mole fractions of CO_2, H_2, and CO (zero). Note that these values are different than the zero-current outlet compositions shown in the figure. Even without any electrolysis, the reverse-shift reaction occurs in the stack at 800°C, resulting in the production of some CO and consumption of CO_2 and H_2. During co-electrolysis, the mole fractions of CO_2 and steam (not shown in Figure 3.11) decrease with current, while the mole fractions of H_2 and CO increase. For the conditions chosen for these tests, the ratio of H_2 to CO is close to the desired 2-to-1 value for syngas production.

3.2.2.3 Large-Scale Demonstration: The INL Integrated Laboratory-Scale Facility

One of the objectives of the HTE program is technology scale-up and demonstration. To this end, the HTE group developed a 15 kW HTE test facility, termed the ILS HTE test facility. Details of the design and initial operation of this facility are documented in Housley et al. (2007), Stoots and O'Brien (2008), and Stoots et al. (2009a). A condensed description of the facility will be provided here. The ILS includes three electrolysis modules, each

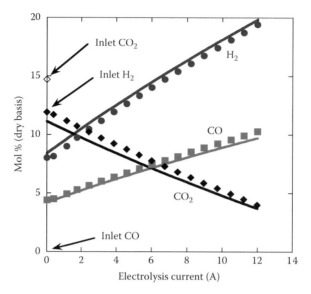

FIGURE 3.11
Outlet gas composition as a function of current density for co-electrolysis experiments, 10-cell stack.

consisting of four stacks of 60 cells, yielding 240 cells per module and 720 cells total. The cells are similar to those discussed earlier. Each electrolysis module utilizes an independent support system supplying electrical power for electrolysis, a feedstock gas mixture of hydrogen and steam (and sometimes nitrogen), a sweep gas, and appropriate exhaust handling. Each module includes a controlled inlet flow of deionized water, a steam generator, a controlled inlet flow of hydrogen, a superheater, inlet and outlet dew point measurement stations, a condenser for residual steam, and a hydrogen vent. All three modules were located within a single hot zone. Heat recuperation and hydrogen product recycle were also incorporated into the facility.

An exploded view of one of the ILS module assemblies including the recuperative heat exchanger, base manifold unit, and four-stack electrolysis unit is presented in Figure 3.12. For each four-stack electrolysis module, there were two heat exchangers and one base manifold unit. Each base manifold unit has nine flow tubes entering or exiting at its top and only four flow tubes entering or exiting at the bottom of the unit and at the bottom of the heat exchangers, thereby reducing the number of tube penetrations passing through the hot zone base plate from nine to just four. This feature also reduces the thermal load on the hot zone base plate. An internally manifolded plate-fin design concept was selected for this heat recuperator application. This design provides excellent configuration flexibility in terms of selecting the number of flow elements per pass and the total number of passes in order to satisfy the heat transfer and pressure drop requirements. Theoretical counterflow heat exchanger performance can be approached with this design. This design can also accommodate multiple fluids in a single unit. More details of the design of the recuperative heat exchangers are provided in Housley et al. (2008).

Figure 3.12 shows a cut-away design rendering of the three ILS electrolysis modules with their base manifolds and heat exchangers beneath. This illustration also shows the instrumentation wires for intermediate voltage and temperature readings. Each module is instrumented with twelve 1/16″ sheathed thermocouples for monitoring gas temperatures in the electrolysis module manifolds and in the base manifold. These thermocouples are

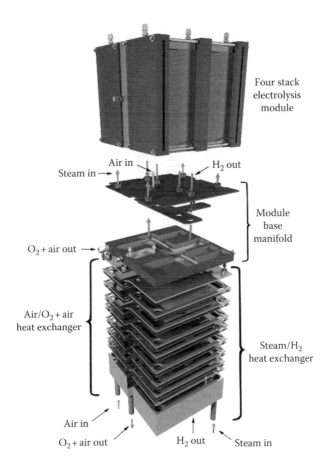

Air in
Steam in →
H_2 out

Four stack
electrolysis
module

Module
base
manifold

O_2 + air out →

Air/O_2 + air
heat exchanger

Steam/H_2
heat exchanger

Air in
O_2 + air out
H_2 out
Steam in

FIGURE 3.12
Exploded view of heat exchanger, base manifold unit, and four-stack electrolysis unit.

attached to the manifolds using compression fittings. There are also 12 miniature 0.020″ diameter inconel-sheathed type-K thermocouples per module, which are used for monitoring internal stack temperatures. Access to the internal region of the stacks is provided via the air outlet face. The internal thermocouples are inserted into the small exit airflow channels. Similarly, seven intermediate voltage tap wires per module are inserted into the airflow channels of the four stacks.

Two compression bars are shown across the top of each module in Figure 3.13. These bars are used to maintain compression on all of the stacks during operation in order to minimize electrical contact resistance between the cells, flow fields, and interconnects. The bars are held in compression via spring-loaded tie-downs located outside of the hot zone under the base plate.

Note that the heat exchangers are partially imbedded in the insulation thickness. The top portion of each heat exchanger is exposed to the hot zone radiant environment, which helps to insure that the inlet gas streams achieve the desired electrolyzer operating temperature prior to entering the stacks. The temperature at the bottom of each heat exchanger will be close to the inlet stream temperature, minimizing the thermal load on the hot zone base plate in the vicinity of the tubing penetrations. A photograph of the three ILS electrolysis modules installed in the hot zone is shown in Figure 3.13.

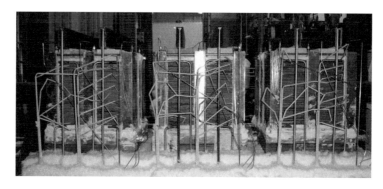

FIGURE 3.13
HTE ILS hot zone with three modules installed.

Over a period of 700 h of test time, module-average ASR values increased by about a factor of 5, from an initial value near 1.5 Ωcm^2. Some of the observed degradation was related to balance-of-plant issues. For example, prior to about 480 h of operation, unanticipated condensation occurred in the hydrogen recycle system, which led to erratic control of the hydrogen recycle flow rate due to the intermittent presence of liquid water in the mass flow controllers. This problem resulted in time periods during which there may have been no hydrogen flow to the ILS stacks, leading to accelerated performance degradation associated with oxidation of the nickel cermet electrodes. Despite the problems with the ILS, we were able to successfully demonstrate large-scale hydrogen production with heat recuperation and hydrogen recycle, as would be required in a large-scale plant. Peak electrolysis power consumption and hydrogen-production rate were 18 kW and 5.7 N m^3/h, respectively, achieved at about 17 h of elapsed test time.

3.2.2.4 Short-Stack Long-Duration Test Results

Long-duration steady-state operations with a 10-cell stack containing an improved materials set began on May 29, 2009 at the thermal neutral voltage of 12.9 V. The stack completed 2500 h at constant input on September 10 at 20:40. A summary graph of the results of the test is shown in Figure 3.14. This graph shows the operating voltage, stack current, and ASR with data recorded every 5 min. The stack operating conditions were kept constant at the same values as those for the polarization curve test discussed above.

Several events disrupted the steady-state testing. It is not clear whether these events affected the long-term performance of the stack. Most of these disruptions involved interruptions of inlet gas flows (N_2 and H_2), caused by problems with the gas mass flow controller electronics. Later in testing (beyond 1200 h), the gas line exiting the humidifier would become occasionally clogged, also interrupting gas flows to the stack. When blockage of gas flow occurred, the stack would not receive sufficient steam and electrical current would drop drastically due to insufficient oxygen ions being available to carry the current through the electrolyte. Around 2100 h elapsed test time, the problem gas line was replaced.

During the first 60 h, the nickel oxide in the hydrogen electrode is being reduced and all of the components of the cell are sintering. The ASR of the stack reached 2.3 Ωcm^2 at 60 h elapsed test time, at which point it stabilized. Beyond 60 h elapsed test time, the average rate of degradation was 8.15% per 1000 h.

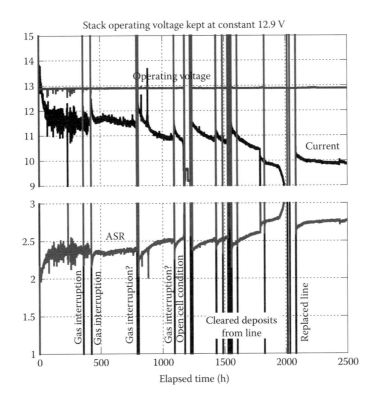

FIGURE 3.14
ASR, current, and voltage for the 2500 h 10-cell stack test.

At the conclusion of 2500 h, the stack had a per-cell ASR of 2.75 Ω cm^2. The hydrogen-production rate as measured by the stack current (Faraday's Law) was 690 sccm (41.4 N L/h, 3.7 g/h). The stack electrical current was 9.89 A.

Although the final determination of the cause of this improved performance must await a detailed examination of the cells after the stacked is disassembled, it is thought that the spinel coating had the greatest effect in reducing the degradation rate. This statement is based upon this test and parametric studies conducted at Ceramatec. The revised formulation of the oxygen electrode had a more minor effect.

Also of interest is that the hydrogen product from the last 700 h of testing was intermittently compressed, stored, and subsequently used to produce methane. A methanation experiment was constructed under the INL Hybrid Energy Systems program. This apparatus consists of a ventilated enclosure, hydrogen compressor, hydrogen storage tank, heated methanation reactor, nickel catalyst, micro gas chromatograph, and various other instrumentation. Operating in a semi-continuous fashion, this experiment has successfully demonstrated in-line methanation of electrolysis products and has provided reaction kinetics information to systems modelers.

The effect of reactor outlet temperature has also been considered. Figure 3.15 shows overall hydrogen-production efficiencies, based on high heating value in this case, plotted as a function of reactor outlet temperature. This figure was also presented in the Fundamentals section of this report. The figure includes a curve that represents 65% of the thermodynamic maximum possible efficiency (i.e., 65% exergetic efficiency) for any thermal water splitting process, assuming heat addition occurs at the reactor outlet

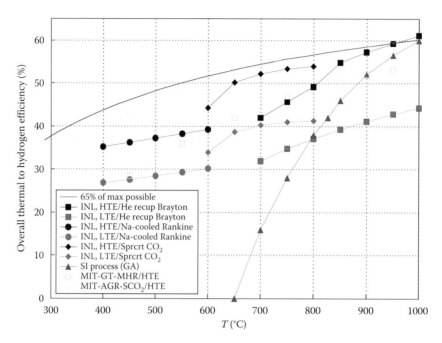

FIGURE 3.15
Overall thermal-to-hydrogen efficiencies (based on HHV) for HTE coupled to three different reactor types, as a function of reactor outlet temperature.

temperature and heat rejection occurs at $T_L = 20°C$ (O'Brien 2008). In order to cover a broad range of possible reactor outlet temperatures, three different advanced-reactor/power-conversion combinations were considered: a helium-cooled reactor coupled to a direct recuperative Brayton cycle, a supercritical CO_2-cooled reactor coupled to a direct recompression cycle, and a sodium-cooled fast reactor coupled to a Rankine cycle. Each reactor/power-conversion combination was analyzed over an appropriate reactor outlet temperature range.

The figure shows results for both HTE and LTE. Results of system analyses performed at MIT (Yildiz and Kazimi 2006) are also shown. The lower MIT curve, labeled MIT-GT-MHR/HTE, represents overall efficiency predictions for a helium-cooled reactor with a direct Brayton cycle power conversion unit. The upper MIT curve, labeled MIT-AGR-SCO$_2$/HTE, represents overall efficiency predictions for a CO_2-cooled advanced gas reactor with a supercritical CO_2 power conversion unit. The results presented in Figure 3.15 indicate that, even when detailed process models are considered, with realistic component efficiencies, heat exchanger performance, and operating conditions, overall hydrogen-production efficiencies in excess of 50% (HHV) can be achieved for HTE with reactor outlet temperatures above 850°C. For reactor outlet temperatures in the range of 600°C–800°C, the supercritical CO_2/recompression power cycle is superior to the He-cooled/Brayton cycle concept. This conclusion is consistent with results presented in Yildiz and Kazimi (2006). The efficiency curve for the S-I process also includes values above 50% for reactor outlet temperatures above 900°C, but it drops off quickly with decreasing temperature, and falls below values for LTE coupled to high-temperature reactors for outlet temperatures below 800°C. Note that even LTE benefits from higher reactor outlet temperatures because of the improved power conversion thermal efficiencies.

3.2.3 Degradation in Solid-Oxide Electrolysis Cells

At present, a complete understanding and reasonable agreement on the causes of degradation and electrochemical mechanisms behind them does not exist. However, continuing research and the consensus of experts in the field (Sohal 2009) indicate a number of probable causes for the long-term degradation that has been seen.

Experimental data on degradation can be classified into three main categories: (1) baseline progressive constant-rate degradation, (2) degradation corresponding to transients caused by thermal or redox (**red**uction and **ox**idation) cycling phenomena occurring in a cell, and (3) degradation resulting from a sudden incident or a failure/malfunction of a component or a control in a stack system. However, there is no clear evidence if different events lead to similar or drastically different electrochemical degradation mechanisms within a cell.

3.2.3.1 SOEC versus SOFC Stacks

Degradation data have been obtained both in single cells as well as in stacks. Degradation mechanisms in a stack are not identical to those in a single cell (Virkar 2007). Also, degradation in a SOEC is not identical to that in a SOFC. Long-term single-cell tests show that SOEC operation exhibits greater degradation rates than SOFC operation. Therefore, SOFC degradation can be used for background information and guidance. But for specific SOEC stack development, all studies have to be done on SOEC stacks. Some researchers observed that higher operating temperatures increase degradation in SOEC, but higher current density does not increase degradation. However, Argonne National Laboratory (ANL) observed higher degradation in higher current flow regions of O_2-electrodes. Such unconfirmed and conflicting opinions need to be resolved during future research.

3.2.3.2 Air/Oxygen Electrode

It is understood that degradation of the O_2-electrode is more severe than that of the H_2-electrode. Therefore, it was proposed to focus initially on the degradation of the O_2-electrodes in a stack. ANL examination of an SOEC stack operated by INL for ~1500 h showed that the O_2-electrode delaminated from the bond layer/electrolyte. However, the causes of the delamination can be termed as speculative because confirmative tests proving the fundamental cause(s) have not been performed. It is thought that high oxygen evolution in over-sintered region can build up high pressure at those locations. In SOEC mode, O_2 has to be pushed out, hence chances of delamination increase. Therefore, the high porosity of O_2-electrode is very important. This opinion was expressed by many participants and hence deserves further examination. Per ANL observations, the delamination occurs in cell areas with high current flows. It was also suggested that chromium poisoning originating from the interconnects or from balance-of-plant piping may get located at the interface or triple phase boundary (TPB). This can result in separation of the bond layer from the O_2-electrode. Deposition of impurities at the TPB and delamination can adversely impact the electrochemical reactions and ionic conductivity in the cell. It was also mentioned that at the electrode–electrolyte interface, forced electron transfer can also form defects. No detailed discussion on the specifics of the phenomenon took place.

3.2.3.3 Air/O$_2$-Electrode Side Bond Layer

An O$_2$-electrode side protective bond layer is often used. Because it is next to the O$_2$-electrode, it encounters similar electrochemical phenomena that lead to cell degradation. GE's work also makes reference to bond layer issues. However, besides ANL's observations, at the workshop, no other studies or data were presented that can demonstrate the bond layer's significance relative to the O$_2$-electrode in terms of overall cell degradation. ANL found an average of 1%–8% (~30% maximum) Cr-contamination in the bond layer, probably originating from interconnects. Cr contaminants were found in association with lanthanum strontium chromite (LSC). In the O$_2$ bond layer, a secondary phase may form. However, there are conflicting opinions about the severity of Cr contamination. ANL observed delamination and weak interface between the O$_2$-electrode and LSC bond layer, which can prevent solid-state Cr from diffusing into the O$_2$-electrode. For this reason, the O$_2$-electrode can remain stable. However, a weak interface is not desirable from electrical conductivity point of view.

3.2.3.4 Electrolyte

In electrolytes, the main cause of degradation is loss of ionic conductivity. Müller et al. (2003) showed that during first 1000 h of testing, yttria- and scandia-doped zirconia (8 mol% Y$_2$O$_3$ Sc-ZrO$_2$/8YSZ) electrolytes showed ~23% performance degradation. For the next 1700 h of testing, the decrease in conductivity was as high as 38%. An increase in tetragonal phase during annealing at the expense of cubic and monoclinic phases was detected for the 3YSZ samples. However, 3YSZ and 4YSZ samples showed much smaller decreases in conductivity after 2000 h of test. Both Steinberger-Wilckens (2008) and Hauch (2007) reported formation of impurities at the TPBs. A substantial amount of SiO$_2$ was detected at the Ni/YSZ H$_2$-electrode–electrolyte interface during electrolysis, while no Si was detected in other reference cells. This Si-containing impurities were probably from albite glass sealing. ANL observed that cubic, tetragonal, and monoclinic phases of ZrO$_2$ remained stable at the present scandia-doping level.

3.2.3.5 Steam/H$_2$-Electrode

Overall, many researchers reported that the contribution of the steam/H$_2$-electrode to SOEC degradation is much less than that attributed to other cell components. ANL also observed Si as a capping layer on steam/H$_2$-electrode. It probably was carried by steam from the seals, which contain Si. SiO$_x$ also emanates from interconnect plates. Mn also diffuses from interconnects, but the significance of Mn diffusion is not known. Hauch (2007) observed contaminants containing Si to segregate to the innermost few microns of the H$_2$-electrode near the electrolyte. The impurities that diffused to and accumulated at the TPBs of the H$_2$-electrode are believed to be the main cause of performance degradation in SOECs (Hauch 2007). In the literature, it has been noted that steam content greater than 30% shows conductivity loss. Therefore, an optimum ratio of steam-H$_2$ mixture and steam utilization percentage needs to be determined.

3.2.3.6 Interconnects

Interconnects can be a source of serious degradation. Sr, Ti, and Si segregate and build up at interfaces. Sr segregates to the interconnect–bond layer interface. Mn segregates to the interconnect surface. Si and Ti segregate to the interconnect–passivation layer interface. Cr contamination can originate from interconnects and it can interact with O_2-electrode surface or even diffuse into the O_2-electrode. Chromium reduction (Cr^{6+} to Cr^{3+}) can take place at electrode–electrolyte interface (Singh et al. 2008). Under the sponsorship of U.S. DOE SECA program, coatings for the interconnects are being developed. Coated stainless steel interconnects have shown reduced degradation rates. GE observed higher degradation with stainless steel current collectors than with Au current collectors (Guan et al. 2006).

3.2.3.7 Contaminants and Impurities

A hydrogen electrolysis plant or a laboratory-scale experiment is always connected to piping, gas storage tanks/cylinders, or other such equipments. These components can be a source of undesirable particles/chemicals, which can get deposited at different locations in a SOECs. It has been shown in previous sections that any foreign particles depositing at the TPB can lead to degradation in the cell performance. The reactant gases can also have some undesirable impurities. It is understood that the balance of plant and gases are merely sources of impurities. The phenomenological causes of the degradation depend on other electrochemical reasons.

3.2.4 Concluding Remarks

High-temperature nuclear reactors have the potential for substantially increasing the efficiency of hydrogen production from water, with no consumption of fossil fuels, no production of GHGs, and no other forms of air pollution. Water splitting for hydrogen production can be accomplished via HTE or thermochemical processes, using high-temperature nuclear process heat. In order to achieve high efficiencies, both processes require high-temperature operation. Thus, these hydrogen-production technologies are tied to the development of advanced high-temperature nuclear reactors. High-temperature electrolytic water splitting supported by nuclear process heat and electricity has the potential to produce hydrogen with overall thermal-to-hydrogen efficiencies of 50% or higher, based on high heating value. This efficiency is near that of the thermochemical processes, but without the severe corrosive conditions of the thermochemical processes and without the fossil fuel consumption and GHG emissions associated with hydrocarbon processes.

An overview of the HTE technology status and the HTE research and development program at the INL has been presented. Large-scale system analyses performed at the INL and elsewhere indicate very promising potential for HTE as a large-scale hydrogen-production technology. The INL HTE experimental program has demonstrated hydrogen production at a variety of scales from single-button cells (~1 W) and short stacks (~500 W) to the successful operation of the 15 kW ILS facility. These experiments also served to demonstrate the straightforward scalability of the HTE technology. The ILS demonstrated a peak hydrogen-production rate in excess of 5.6 m^3/h and operated for over 1000 h the fall of 2008. Although some issues require further research and development, including cell and stack long-term performance degradation, HTE is the only advanced hydrogen-production technology that has successfully demonstrated hydrogen-production rates greater than ~100 L/h.

Industry is currently preparing to begin the mass production of solid-oxide fuel cells, primarily for stationary power application. This commitment indicates that the issues of fuel cell manufacturing cost and long-term performance have reached a level of maturity that industry is comfortable in moving on to commercial deployment. It is likely that once the fundamental mechanisms of high-temperature electrolyzer cell degradation have been identified, long-term performance of solid-oxide electrolyzers will also improve drastically.

3.3 Introduction to Thermochemical Cycles for Nuclear Production of Hydrogen

3.3.1 Thermochemical Cycles

Thermochemical water splitting is the conversion of water into hydrogen and oxygen by a series of thermally driven chemical reactions. The direct thermolysis of water requires temperatures in excess of 2500°C for significant hydrogen generation:

$$H_2O \rightarrow H_2 + \frac{1}{2}O_2 \quad (2500°C\,min) \tag{3.14}$$

At this temperature, only about 10% of the water is decomposed. In addition, a means of preventing the hydrogen and oxygen from recombining upon cooling must be provided or no net production would result. A thermochemical water-splitting cycle accomplishes the same overall result using much lower temperatures. The S-I cycle is a prime example of a thermochemical cycle. It consists of three chemical reactions, which sum to the dissociation of water:

$$I_2 + SO_2 + 2H_2O \rightarrow 2HI + H_2SO_4 \quad (120°C) \quad (Exothermic) \tag{3.15}$$

$$H_2SO_4 \rightarrow SO_2 + H_2O + \frac{1}{2}O_2 \quad (\sim850°C) \quad (Endothermic) \tag{3.16}$$

$$2HI \rightarrow I_2 + H_2 \quad (\sim400°C) \quad (Endothermic) \tag{3.17}$$

$$H_2O \rightarrow H_2 + \frac{1}{2}O_2 \tag{3.14}$$

With a suitable catalyst, the high-temperature reaction (3.16) reaches 10% conversion at only 510°C and 83% conversion at 850°C. Moreover, there is no need to perform a high-temperature separation as the reaction ceases when the stream leaves the catalyst.

Energy, as heat, is input to a thermochemical cycle via one or more endothermic high-temperature chemical reactions. Heat is rejected via one or more exothermic low-temperature reactions. Other thermally neutral chemical reactions may be required to complete the cycle so that all the reactants, other than water, are regenerated. In the S-I cycle, most of the input heat goes into the dissociation of sulfuric acid. Sulfuric acid and hydrogen iodide are formed in the endothermic reaction of H_2O, SO_2, and I_2, and the hydrogen is generated in the mildly endothermic decomposition of hydrogen iodide. The combination of high-temperature endothermic reactions, low-temperature exothermic reactions, and energy neutral closing reactions is not sufficient for a cycle to be thermodynamically realizable. Each reaction must also have favorable ΔG_R (Gibbs free energy of reaction). A reaction is favorable if ΔG is negative, or at least not too positive. For example, each of the four chemical reactions of the UT-3 Cycle, shown in the following equations, has a slightly positive ΔG. The flow of gaseous reactant through the bed of solid reactants sweeps the gaseous products away, resulting in total conversion of the solid reactants to solid products:

$$Br_2(g) + CaO(s) \rightarrow CaBr_2(s) + \frac{1}{2}O_2(g) \quad (672°C) \tag{3.18}$$

$$3FeBr_2(s) + 4H_2O(g) \rightarrow Fe_3O_4(s) + 6HBr(g) + H_2(g) \quad (560°C) \tag{3.19}$$

$$CaBr_2(s) + H_2O(g) \rightarrow CaO(s) + 2HBr(g) \quad (760°C) \tag{3.20}$$

$$Fe_3O_4(s) + 8HBr(g) \rightarrow Br_2(g) + 3FeBr_2(s) + 4H_2O(g) \quad (210°C) \tag{3.21}$$

$$H_2O(g) \rightarrow H_2(g) + \frac{1}{2}O_2(g) \tag{3.14}$$

Sometimes, it is possible to electrochemically force a non-spontaneous reaction; such a process is termed a hybrid thermochemical cycle. The HyS cycle, also known as the Westinghouse Sulfur cycle or as the Ispra Mark 11 cycle has the same high-temperature endothermic reaction as the S-I cycle. The HyS cycle is closed by the electrochemical oxidation of sulfur dioxide to sulfuric acid:

$$H_2SO_4 \rightarrow SO_2 + H_2O + \frac{1}{2}O_2 \quad (850°C) \tag{3.16}$$

$$SO_2 + 2H_2O \rightarrow H_2SO_4 + H_2 \quad (80°C \text{ electrolysis}) \tag{3.22}$$

$$H_2O \rightarrow H_2 + \frac{1}{2}O_2 \tag{3.14}$$

In phase one of the DOE-supported study described in (Brown et al. 2002), General Atomics, Sandia National Laboratories (SNL) and the University of Kentucky carried out a search of the world literature on thermochemical water-splitting cycles for matching to a high-temperature nuclear reactor heat source. They located and catalogued 822 references and identified 115 separate thermochemical water-splitting cycles. They evaluated these against the quantifiable screening criteria shown in Table 3.2 and selected the 25 most promising cycles for detailed technical evaluation. The top 10 cycles from this screening process are shown in Table 3.3. A similar study for matching thermochemical cycles to a solar heat source (McQuillan et al. 2005) evaluated 182 cycles, selected 67 as promising, and chose 14 for detailed analysis.

From Table 3.3, it can be seen that all of the top 10 thermochemical cycles require an upper temperature of 750°C–1000°C. This implies the need for a high-temperature reactor with an outlet temperature much greater than the ~300°C of today's water-cooled reactors. Table 3.3 also shows that three of the top four thermochemical cycles are sulfur-based cycles that employ thermal decomposition of sulfuric acid as the high-temperature step in the cycle. Two of the top four cycles are "hybrid" cycles that include both thermally driven reactions and electrolytic reactions. The HyS hybrid thermochemical cycle and the S-I pure thermochemical cycle have emerged as the leading candidate thermochemical cycles for R&D for application to nuclear energy around the world with recent programs in the United States, Japan, Korea, France, China, and India. The high-temperature sulfuric acid thermal decomposition reaction, the S-I cycle, and the HyS cycle are each discussed in more detail in later sections of this chapter.

TABLE 3.2

Thermochemical Screening Criteria

Desirable Characteristic	Rational	Metric
1. Minimum number of chemical reactions steps	A smaller number indicates a simpler process and lower costs	Number of chemical reactions
2. Minimum number of separation steps	A smaller number indicates a simpler process and lower costs	Number of chemical separations, excluding simple phase separation
3. Minimum number of elements	A smaller number indicates a simpler process and lower costs	Number of elements, excluding oxygen and hydrogen
4. Employ elements that are abundant	Use of abundant elements will lower the cost and permit implementation on a large scale	Score is based on least abundant element in cycle
5. Minimize use of expensive materials by avoiding corrosive chemicals	The effect of materials cost on hydrogen-production efficiency and cost	Score is based on the relative corrosiveness of the process solutions
6. Minimize the flow of solids	Chemical plant costs are considerably higher for solids processing plants	Score is based on minimization of solid flow problems
7. Heat input temperature compatible with materials	Limit on temperature will be material separating the reactor coolant from the process stream	Score is based on the high-temperature heat input being close to that delivered by an advanced nuclear reactor
8. Many papers from many authors and institutions	Cycles that have been thoroughly studied have a lower probability of undiagnosed flaws	Score is based on the number of papers published dealing with the cycle
9. Tested at a moderate or large scale	Processes for which the basic chemistry has not been verified are suspect	Score is based on the degree to which the chemistry has been actually demonstrated
10. Good efficiency and cost data available	A significant amount of engineering design work is necessary to estimate process efficiencies and production costs	Score is based on the degree to which efficiencies and cost have been estimated

Source: From Brown, L.C. et al., High efficiency generation of hydrogen fuels using thermochemical cycles and nuclear power, *AIChE 2002 Spring National Meeting*, New Orleans, LA, March 11–15, 2002.

3.3.2 Nuclear Reactors for Hydrogen Production

SNL evaluated various nuclear reactors for their ability to provide the high-temperature heat needed by high-temperature thermochemical cycles and to be interfaced safely and economically to hydrogen-production processes (Marshall 2002). The recommended reactor technology should require minimal technology development to meet the high-temperature requirement and should not present any significant design, safety, operational, or economic issues.

Furthermore, it was recommended that an intermediate helium loop should be used between the reactor coolant loop and the hydrogen-production system. This will assure that any leakage or permeation (e.g., tritium) from the reactor primary coolant loop will not contaminate the hydrogen-production system or expose hydrogen plant personnel to radiation from the primary loop coolant. It also assures that corrosive process chemicals cannot enter the core of the nuclear reactor. The heat exchanger interface sets the boundary conditions for selection of the reactor system. The principal requirement is the temperature of the highest temperature reaction of the thermochemical cycle, which must account for the temperature drop between the core outlet and the point of application

TABLE 3.3

Top 10 Thermochemical Cycles from Screening Process

Cycle	Name	T/E*	Temperature (°C)	Reaction	Total Score
1	Hybrid Sulfur[a]	T	850	$2H_2SO_4(g) \rightarrow 2SO_2(g) + 2H_2O(g) + O_2(g)$	85
		E	77	$SO_2(g) + 2H_2O(a) \rightarrow H_2SO_4(a) + H_2(g)$	
2	Ispra Mark 13[b]	T	850	$2H_2SO_4(g) \rightarrow 2SO_2(g) + 2H_2O(g) + O_2(g)$	80
		E	77	$2HBr(a) \rightarrow Br_2(a) + H_2(g)$	
		T	77	$Br_2(l) + SO_2(g) + 2H_2O(l) \rightarrow 2HBr(g) + H_2SO_4(a)$	
3	UT-3 Univ. of Tokyo[c]	T	600	$2Br_2(g) + 2CaO \rightarrow 2CaBr_2 + O_2(g)$	79
		T	600	$3FeBr_2 + 4H_2O \rightarrow Fe_3O_4 + 6HBr + H_2(g)$	
		T	750	$CaBr_2 + H_2O \rightarrow CaO + 2HBr$	
		T	300	$Fe_3O_4 + 8HBr \rightarrow Br_2 + 3FeBr_2 + 4H_2O$	
4	Sulfur–Iodine[d]	T	850	$2H_2SO_4(g) \rightarrow 2SO_2(g) + 2H_2O(g) + O_2(g)$	78
		T	450	$2HI \rightarrow I_2(g) + H_2(g)$	
		T	120	$I_2 + SO_2(a) + 2H_2O \rightarrow 2HI(a) + H_2SO_4(a)$	
5	Julich Center EOS[e]	T	800	$2Fe_3O_4 + 6FeSO_4 \rightarrow 6Fe_2O_3 + 6SO_2 + O_2(g)$	68
		T	700	$3FeO + H_2O \rightarrow Fe_3O_4 + H_2(g)$	
		T	200	$Fe_2O_3 + SO_2 \rightarrow FeO + FeSO_4$	
6	Tokyo Inst. Tech. Ferrite[f]	T	1000	$2MnFe_2O_4 + 3Na_2CO_3 + H_2O \rightarrow$ $2Na_3MnFe_2O_6 + 3CO_2(g) + H_2(g)$	64
		T	600	$4Na_3MnFe_2O_6 + 6CO_2(g) \rightarrow$ $4MnFe_2O_4 + 6Na_2CO_3 + O_2(g)$	
7	Hallett Air Products 1965[e]	T	800	$2Cl_2(g) + 2H_2O(g) \rightarrow 4HCl(g) + O_2(g)$	62
		E	25	$2HCl \rightarrow Cl_2(g) + H_2(g)$	
8	Gaz de France[e]	T	725	$2K + 2KOH \rightarrow 2K_2O + H_2(g)$	62
		T	825	$2K_2O \rightarrow 2K + K_2O_2$	
		T	125	$2K_2O_2 + 2H_2O \rightarrow 4KOH + O_2(g)$	
9	Nickel Ferrite[g]	T	800	$NiMnFe_4O_6 + 2H_2O \rightarrow NiMnFe_4O_8 + 2H_2(g)$	60
		T	800	$NiMnFe_4O_8 \rightarrow NiMnFe_4O_6 + O_2(g)$	
10	Aachen Univ Julich 1972[e]	T	850	$2Cl_2(g) + 2H_2O(g) \rightarrow 4HCl(g) + O_2(g)$	59
		T	170	$2CrCl_2 + 2HCl \rightarrow 2CrCl_3 + H_2(g)$	
		T	800	$2CrCl_3 \rightarrow 2CrCl_2 + Cl_2(g)$	

[a] Brecher et al. (1977).
[b] Beghi (1986).
[c] Yoshida et al. (1990).
[d] Besenbruch (1982).
[e] Williams (1980).
[f] Ueda et al. (1974).
[g] Tamaura et al. (1985).

in the hydrogen-production system. A required reactor outlet temperature of 950°C was assumed for the study. This should permit a peak process temperature of 850°C, adequate for all but one of the top 10 cycles.

The reactor coolant becomes a primary consideration for determining which concepts are most appropriate. The reactor/coolant types considered include pressurized water-cooled reactors, boiling water-cooled reactors, alkali liquid metal-cooled reactors, heavy liquid metal-cooled reactors, gas-cooled reactors, organic-cooled reactors, molten

TABLE 3.4

Reactor Selection Requirements and Criteria

Basic requirements

1. Chemical compatibility of coolant with primary loop materials and fuel
2. Coolant molecular stability at operating temperatures in a radiation environment
3. Pressure requirements for primary loop
4. Nuclear requirements: parasitic neutron capture, neutron activation, fission product effects, gas buildup, etc.
5. Basic feasibility, general development requirements, and development risk

Important criteria

1. Safety
2. Operational issues
3. Capital costs
4. Intermediate loop compatibility
5. Other merits and issues

salt-cooled reactors, liquid-core reactors, and gas-core reactors. The reactor types were assessed against the five requirements and five important criteria given in Table 3.4.

Based on this assessment, and upon evaluation of the relative development requirements for candidate reactors, the following conclusions and recommendations were made:

- PWR, BWR, and organic-cooled reactors—not recommended: cannot achieve the high temperatures needed
- Liquid-core and alkali-metal-cooled reactors—significant development risk due to materials concerns at the high temperatures needed
- Heavy-metal and molten salt-cooled reactors—promising, but significant development needed
- Gas-cooled reactors—baseline choice, only modest development needed for helium gas-cooled reactor
- Gas-core reactors—not recommended, too speculative

Helium gas-cooled reactors were recommended as the baseline choice for a reactor heat source for the S-I thermochemical cycle and the HyS cycle for hydrogen production.

An example of a recent design for the helium gas-cooled reactor is General Atomics' Gas Turbine-Modular Helium Reactor (LaBar 2002). This reactor consists of 600 MWt modules that are located in underground soils. The direct-cycle gas turbine power conversion system is located in an adjacent silo, as shown in Figure 3.16. Other designs for high-temperature gas-cooled nuclear reactors include the Pebble Bed Modular Reactor, which is being developed in South Africa by Westinghouse and others (Nichols 2001).

This new generation of reactor has the potential to provide significant advantages compared to the earlier generation reactors that are now providing nuclear power in the United States. Known as Generation IV reactors, they promise to be more efficient, more economical, and to provide additional safety features. The helium-cooled reactor utilizes high-temperature ceramic fuel and a core design, which together provide passive safety. A catastrophic accident is not possible. Under all conceivable accident conditions, the reactor

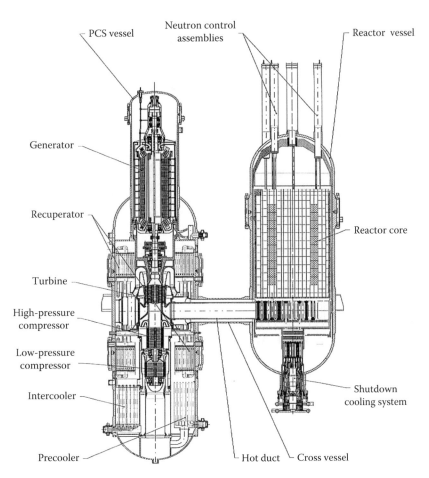

PCS vessel

Neutron control assemblies

Reactor vessel

Generator

Recuperator

Reactor core

Turbine

High-pressure compressor

Low-pressure compressor

Intercooler

Shutdown cooling system

Precooler

Hot duct

Cross vessel

FIGURE 3.16
The GT-MHR is an example of HTGR.

fuel stays well below failure conditions with no actions required by the plant operators or equipment. By avoiding the need for massive active safety back-up systems, the capital cost of the reactor is reduced. For electric power production, the high-temperature fuel also allows high-efficiency power conversion. The gas turbine cycle is projected to achieve 48% thermal efficiency.

The high helium outlet temperature also makes possible the use of the MHR for the production of hydrogen using the S-I and HyS cycles. By replacing the gas turbine system with a primary helium circulator, an intermediate heat exchanger, an intermediate helium loop circulator, an the intermediate loop piping to connect to the hydrogen-production plant, the GT-MHR could be changed into an "H_2-MHR," as shown in Figure 3.17. The MHR is also well suited to provide the electricity and high-temperature steam for HTE as described in the previous section.

3.3.3 Concluding Remarks

The DOE-supported study of thermochemical production of hydrogen by nuclear energy described above identified the S-I thermochemical water-splitting cycle and the HyS

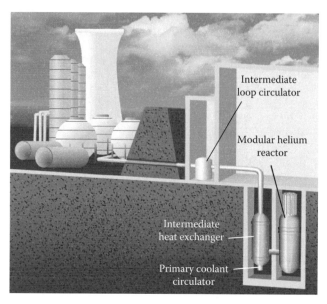

FIGURE 3.17
The H_2-MHR is an HTGR adapted to hydrogen production.

cycle coupled to the modular helium reactor (the H_2-MHR) as attractive candidate systems for hydrogen production. These processes and the high-temperature decomposition of sulfuric acid that is common to both are described in the following sections of this chapter.

The high temperatures needed by the thermochemical cycles listed in Table 3.3 restricted the search for reactors described in Marshall (2002), Brown et al. (2003) to those with very high-temperature capability. If thermochemical cycles could be found that exhibited favorable technical characteristics and reasonable efficiency at lower temperatures, the spectrum of reactors that could provide the heat would be expanded. This could make possible application of reactors, such as the sodium-cooled reactor with outlet temperature around 500°C, for which a significant experience base exists. This question is addressed and lower temperature thermochemical cycles are discussed in the last section of this chapter.

Nuclear production of hydrogen would allow large-scale production of hydrogen while avoiding the release of CO_2. Nuclear production of hydrogen could thus become the enabling technology for the hydrogen economy.

3.4 Sulfuric Acid Decomposition for the Sulfur-Based Thermochemical Cycles

The sulfur-based thermochemical cycles are being considered as promising processes for the large-scale production of H_2 using heat from advanced reactors. The S-I, HyS, and sulfur–bromine cycles all require the decomposition of H_2SO_4 to produce SO_2 as the high-temperature reaction step. In each of the three cycles, the SO_2 is then oxidized in the proceeding reaction step, regenerating H_2SO_4 that is recycled back to the acid decomposition step. The net reaction for each of the thermochemical cycles is the production of H_2 and O_2 with all other chemicals used in the process being recycled.

Sulfur-iodine

$$(1) \quad H_2SO_4 \rightarrow H_2O + SO_2 + \frac{1}{2}O_2 \tag{3.16}$$

$$(2) \quad 2HI \rightarrow I_2 + H_2 \tag{3.17}$$

$$(3) \quad 2H_2O + SO_2 + I_2 \rightarrow H_2SO_4 + 2HI \tag{3.15}$$

Hybrid sulfur

$$(1) \quad H_2SO_4 \rightarrow H_2O + SO_2 + \frac{1}{2}O_2 \tag{3.16}$$

$$(2) \quad 2H_2O + SO_2 \rightarrow H_2SO_4 + H_2 \tag{3.22}$$

Sulfur-bromine

$$(1) \quad H_2SO_4 \rightarrow H_2O + SO_2 + \frac{1}{2}O_2 \tag{3.16}$$

$$(2) \quad SO_2 + Br_2 + 2H_2O \rightarrow 2HBr + H_2SO_4 \tag{3.23}$$

$$(3) \quad 2HBr \rightarrow H_2 + Br_2 \tag{3.24}$$

3.4.1 Sulfuric Acid Decomposition Description

The basic functions required in the sulfuric acid decomposition process are concentration of dilute H_2SO_4, produced in the prior thermochemical cycle reaction step, followed by acid boiling/vaporization, superheating, and decomposition in the presence of a catalyst. For the process to be economical, it is also essential to recuperate as much of the sensible heat in the high-temperature product stream as possible. The products of acid decomposition are separated for subsequent reaction steps. This process is illustrated in Figure 3.18. For most sulfur-based cycles, dilute acid, nominally 20 mol%, is concentrated to approximately 40 mol% to reduce the energy required in the subsequent boiling and superheating stages. For 40 mol% acid, boiling occurs at approximately 400°C. In the superheating stage, H_2SO_4 decomposes to form SO_3 and H_2O vapor at a temperature of approximately 600°C. The vapors are heated further in the catalytic decomposition stage to temperatures of 800°C–900°C to achieve conversion of $SO_{3(g)}$ to $SO_{2(g)}$, with the associated $O_{2(g)}$ and $H_2O_{(g)}$. The catalyst used in the tests conducted in the DOE nuclear hydrogen program was platinum with a TiO_2 ceramic substrate. The decomposition product stream is then passed through a recuperator to recapture the sensible heat. After condensing the unreacted acid and water, the gaseous products are transferred to the next reaction section in the cycle for further processing. Depending on the concentration of the unreacted acid, it is either mixed with fresh feed acid for the acid decomposer or recycled through the acid concentrator.

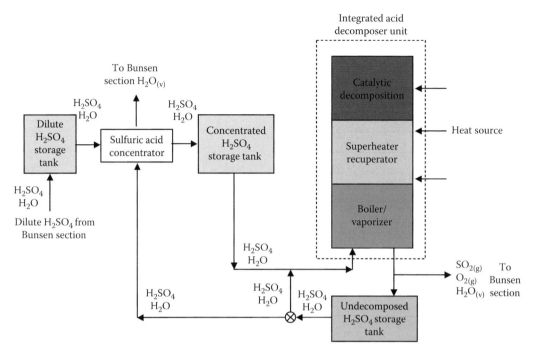

FIGURE 3.18
Schematic diagram of SNL acid decomposition process.

3.4.2 Key Issues

The key technical challenges in developing a viable process for sulfuric acid decomposition are associated with the severe thermal and chemical environment that the system must operate in. The challenges include the following:

- High temperatures up to 900°C
- Highly corrosive (oxidizing) environment
- Operation at pressure
- Sealing technologies for connections between high-temperature components
- Thermal cycling (shock)
- Effectiveness of heat recuperation
- Durability of materials of construction
- Economics of materials of construction

This combination of issues has generally been considered as a key feasibility barrier for the sulfur cycles and the most identifiable technical challenge.

3.4.3 Early Acid Decomposition Studies under the NHI Program

Early experiments were conducted at General Atomics as part of the S-I process development (Norman et al. 1981, Norman et al. 1982) and at Westinghouse as part of the HyS process development (Brecher et al. 1977). These experiments were generally short-duration experiments performed in glass hardware at a laboratory scale to investigate the chemistry and thermodynamics of the decomposition process.

The DOE NE Nuclear Hydrogen Initiative (NHI) was initiated in 2003 to investigate advanced hydrogen-production methods for transportation fuels. NHI thermochemical cycle research focused on the S-I and HyS cycles, and work was initiated at SNL in 2003 to develop options for the high-temperature acid decomposition step in these cycles. Other work has also been conducted related to these cycles. Among the previous efforts are thermodynamic analysis and modeling (Ozturk et al. 1995, Ponyavin et al. 2006), catalyst efficiency (Yannopoulos and Pierre 1984, Ginosar 2005), construction materials (Moore et al. 2008), and decomposer design and prototype construction (Subramanian et al. 2005, Moore et al. 2006, Wilson et al. 2006, 2007, Minatsuki et al. 2007, Connolly et al. 2009).

The corrosive environment presents a range of materials challenges for the different temperature and acid composition regimes, particularly in the acid boiling and condensation regimes. In addition, assuring the integrity of the array of connections and seals for the different components, which were constructed with different materials, added to the challenge. Initial efforts at SNL focused on the use of high-temperature corrosion-resistant alloys (Incoloy, Hastelloy, and Saramet) (Gelbard et al. 2006). These materials were selected based on literature corrosion data for the relevant range of operating conditions. The acid boiler, superheater, and decomposer were separate units constructed of different metals to minimize corrosion in the specific temperature and concentration range for each process. These individual components then required connections and seals to form the complete unit.

Initial tests at SNL using metallic components were successful in producing SO_2 for limited periods; however, significant corrosion was observed for all of the metal construction materials. The most severe corrosion was observed in the process section where condensation of H_2SO_4 acid occurred. Although several materials were evaluated, no long-term solution was identified for metallic components over the range of operating conditions required in the full system. Additionally, problems were encountered with seals and connections. Temperature cycling for the flanges exposed to temperatures of 400°C and above resulted in warping of the flanges and subsequent system leaks.

Alternative corrosion-resistant materials for construction were needed to establish the viability of long-term process operation. Ceramics, glasses, and thermoplastic construction materials satisfied corrosion requirements but in general are more difficult to fabricate and are less robust to thermal cycling and generally involve more difficult assembly. Evaluation of several design options to address the fabrication and interconnect issues led to a new design approach for a more robust acid decomposer based on corrosion-resistant ceramic materials and commercially available ceramic components. This evaluation process led to the adoption of the silicon carbide bayonet acid heat exchanger concept described in the next section.

3.4.4 SNL SiC Bayonet Heat Exchanger

As shown in Figure 3.19, the bayonet decomposer assembly consists of two heat exchanger tubes; a small diameter tube, open at both ends, is placed inside a larger bayonet tube with a closed top. The two tubes are connected at the bottom using a Teflon manifold with Kalrez O-ring seals. The feed acid enters the annular space between the inner heat exchanger tube and outer heat exchanger tube. Heat is supplied to the surface of the outer tube as in a conventional tube and shell heat exchanger. As the acid travels up the annular space, it is heated to boiling in the lower part of the apparatus and the resulting vapors are subsequently superheated to approximately 850°C in the upper section of the apparatus.

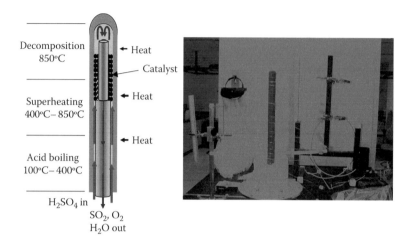

FIGURE 3.19
Schematic of bayonet acid decomposer and 27″ decomposer with ceramic clam shell heaters used in laboratory tests.

A catalyst is placed in the annular space in the top section of the apparatus. As the super-heated acid passes through the catalyst bed, it is decomposed to produce $SO_{2(g)}$, $O_{2(g)}$, and $H_2O_{(g)}$. The decomposition products along with any undecomposed acid enter the inner heat exchanger tube at the top, travel down the inner tube, and exit at the bottom. As the gases travel down the inner tube, they exchange heat with the incoming acid in the outer annular space resulting in effective recuperation.

In the new decomposer design, all high-temperature components (>200°C) are constructed using silicon carbide or other ceramic materials. The bayonet heat exchanger design integrated the boiling, superheat, decomposition, and recuperation stages, which essentially eliminates all high-temperature connections and seals. In addition, most silicon carbide components were commercially available. For the small-scale test of the decomposer, quartz was used for the inner tube based on availability; however, silicon carbide would be the likely choice in larger scale systems (Moore et al. 2007a).

The bayonet design has several distinctive advantages over previous designs:

- Boiler/superheater/catalytic decomposition functions are integrated into a single-unit operation.
- Connections between the boiling, superheating, and decomposition section of the process are eliminated.
- Corrosion is minimized by the use of ceramics at temperatures above 200°C.
- Heat recuperation is integral to the bayonet design.
- Connections to the unit, inlet and outlet, are at the relatively cool end (~150°C) of the unit allowing for the use of routine materials for sealing.

3.4.5 Bayonet Experimental Results

The temperature profile of the silicon carbide acid decomposer as a function of position is given in Figure 3.20. Heat from the ceramic clamshell heaters to the outside of the acid

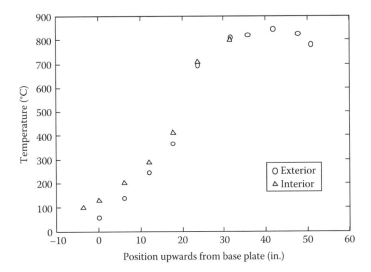

FIGURE 3.20
Temperature profile of the bayonet acid decomposer giving the interior and exterior temperature as a function of position.

decomposer is only supplied to the upper two-thirds of the unit. For the bottom one-thirds section of the acid decomposer, heat to the incoming feed acid is provided primarily by heat recuperated from the gaseous product stream.

The bayonet SiC acid decomposition system was constructed and tested at the ~300 L/h scale as part of the United States French S-I integrated laboratory-scale experiment. The demonstration unit used a 54 in. SiC outer tube, which was electrically heated and thermally insulated. Figure 3.21 shows pictures of the front and back sides of the acid decomposition apparatus as installed in the S-I integrated laboratory-scale experiment at General Atomics.

The bayonet acid decomposer was operated and tested with varying acid feed concentration, feed flowrates, and decomposition pressures to establish the operational parameters for the design. Figure 3.22 is a graph of SO_2 production in L/min at 1, 2, and 3 bar operating pressure with a constant acid feed of 7.1 mol/h. As pressure is increased, the expected decrease in SO_2 production is observed as the higher operating pressures shift the equilibrium from right to left in the equation for SO_2 production, given by

$$SO_3 \rightarrow SO_2 + \frac{1}{2}O_2 \tag{3.25}$$

The effect of varying the acid flow rate to the decomposer on SO_2 production is illustrated in Figure 3.23. The acid feed concentration was 40 mol% and the tests were performed at ambient pressure. Acid feed concentrations varied from 5, 11, and 14 mol/h. The data show the expected increase in SO_2 production as acid feed is increased. At the highest flow rates, heat transfer to the catalyst region can become a limiting factor in the rate of SO_2 production.

At a given temperature and pressure, the theoretical maximum SO_2 production rate for H_2SO_4 decomposition can be calculated using equilibrium data for H_2SO_4 decomposition and an equation of state (Gelbard et al. 2005). Table 3.4 lists the results from experimental decomposition tests at 850°C with varying decomposition pressure along with the theoretical maximum SO_2 production rates. The experimental values are very close to the

(a)   (b)

FIGURE 3.21
SNL acid decomposition section. (a) Liquid acid processing section and (b) acid decomposition and gas processing section.

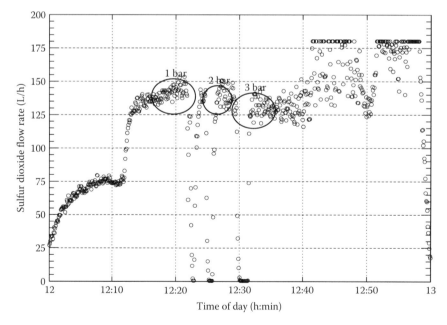

FIGURE 3.22
Production of sulfur dioxide as a function of decomposition pressure at 1, 2, and 3 bar.

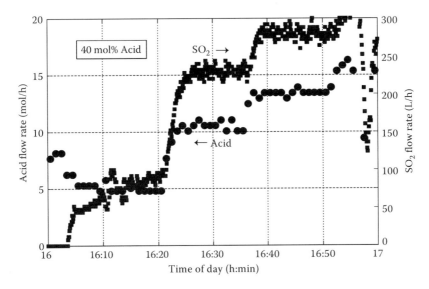

FIGURE 3.23
Production of sulfur dioxide as a function of acid feed rate.

TABLE 3.5

Comparison of Experimental SO_2 Production Rates and Calculated Theoretical Maximum Values

Pressure (bar)	Acid Flow Rate (mol/h)	Theoretical Maximum Conversion Fraction at 850°C	Theoretical Maximum SO_2 Production Rate (L/h)	Experimental SO_2 Production Rate (L/h)
1	3.6	0.86	77	75
1	7.1	0.86	152	145
2	7.1	0.82	145	140
3	7.1	0.80	141	135

theoretical maximum values indicating that near theoretical rates of SO_2 production are achieved in the bayonet decomposer, and the product stream exiting the bayonet decomposer is at or very near to equilibrium composition (Table 3.5).

3.4.6 Bayonet Decomposer Improvements

The bayonet decomposer tested in the ILS experiment was an experimental unit constructed mainly with commercially available components. No effort was made to optimize the design in the ILS work, but several options have been considered to improve efficiency.

The SO_2 production rate in the bayonet is limited by heat transfer from the external source to the acid stream in the catalyst region where significant heat is required to support decomposition. Increasing heat transfer in this region allows a higher production rate from the bayonet heat exchanger. Although numerous methods of improving heat transfer can be imagined (fins on surface, surface roughness, etc.), two of the easiest improvements are made by decreasing the hydraulic channel sizes and/or by increasing mass flow rates sufficiently to enter turbulent flow. Both techniques involve engineering tradeoffs and result in higher flow velocities in the annulus. Increasing the length of the bayonet to provide increased heat transfer areas allows higher flow rates to be used. The relatively short lengths used in the ILS decomposer were sufficient for small-scale experiments but high-capacity units would use longer bayonet tube lengths.

The longer tube lengths that would be used in a larger scale system would permit higher flow rates resulting in a transition from laminar to turbulent flow, which results in significantly improved heat transfer. However, turbulent flow must be achieved at a flow rate that is still consistent with the rate of temperature increase that allows the fluid in the catalyst region to achieve the desired decomposition temperatures. Initial analysis of the possible increase in SO_2 production from a commercial length tube (~4.3 m) due to turbulent heat transfer indicates that about a factor of 10 increase in mass flow rate (SO_2 production) can be achieved.

3.4.7 Process Scale-Up

The proposed scale-up approach for the bayonet decomposer is based on a multi-bayonet version of the ILS design. Multiple bayonet tube assemblies would be manifold to provide the desired capacity in a single-ended tube and shell-type configuration as illustrated in Figure 3.24. The heat source would be helium flow from a high-temperature, gas-cooled reactor. The resulting component would be similar to current commercial bayonet heat exchanger designs. Demonstration of manifold materials and sealing technologies would be required. The multi-tube manifold could be fabricated as a Ta-coated/lined steel chamber on the higher temperature SO_2 side with the possibility of using a Teflon-lined chamber on the lower temperature side. Materials for the higher temperature seals (Helium side seals) for the acid decomposer–Helium interface could be constructed of Grafoil (graphite sheet), or Kalrez (fluoropolymer). Validation of these seals with the selected seal material and design needs to be demonstrated at the intermediate scale (~10 tube assemblies).

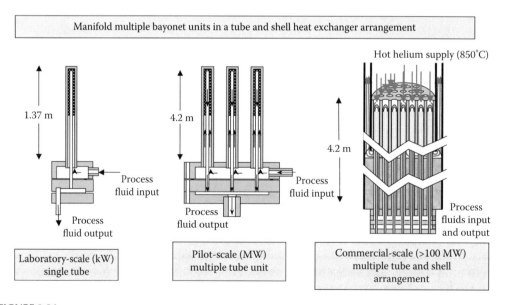

FIGURE 3.24
Scale-up of the bayonet tube acid decomposer. Multiple decomposer tubes are connected through a common manifold. The design is very similar to multi-tube bayonet heat exchanger.

3.4.8 Materials Options for a Scaled-Up Process

For the materials used in the ILS process, no corrosion issues were identified over the time frames of the ILS project. Silicon carbide, glass, alumina, and Teflon operating under the conditions of the tests exhibited no observable corrosion. The types of materials used in the laboratory-scale system are generally available for construction of a large-scale process. Table 3.6 lists material options for a larger scale process. Several potentially applicable materials were identified in the scoping studies but were not used in the ILS. Some of these alternative materials are potential candidates at larger scales, including

- Tantalum and Tantalum alloy-lined/coated steel
- Glass ceramics
- High-temperature glass (up to 300°C)-lined components

Tantalum and tantalum alloys are observed to be very corrosion resistant to H_2SO_4 at high temperatures. However, tantalum is susceptible to corrosion by SO_2 and can suffer from hydrogen embrittlement under certain conditions. Glass ceramics share many properties similar to glass and quartz but can be used to much higher temperature (1000°C), are far less brittle, and cost less than silicon carbide. A relatively new product from the Pfaudler Corporation is a glass lining with a maximum temperature rating of 300°C. The glass-lined pipe used in the ILS is limited to 230°C. Unfortunately, the higher temperature material is currently only available for special chemical reactor linings and is not available for glass-lined pipe. However, it may be more widely available in the future. Each of these materials would require testing before use in the acid decomposition process.

3.4.9 Concluding Remarks

The new SiC bayonet acid decomposer design eliminates many of the previous problems with corrosion and high-temperature seals and provides excellent heat recuperation. The decomposer has been tested at decomposition pressures from 1 to 3 bar and at a decomposition temperatures up to 850°C. The efficiency of H_2SO_4 decomposition to form SO_2 was

TABLE 3.6

Materials Options for the Major Components of a Commercial Scale Acid Decomposer

Unit Operation	Material Candidates	Material Status
Acid concentration	Lined steel (glass, Teflon), ceramics, graphite, and thermoplastics sealing material	Commercially available, demonstrated at large scale. Large-scale acid concentrators are commercially available
Acid decomposer (vaporizer, decomposer, recuperator)	SiC, SiO_2, graphite sealing material	SiC bayonet demonstrated at small scale, but large-scale application uses multiple tubes
SiC bayonet multiple tube manifold	Ta-coated/lined stainless steel Possible ceramic glass- or Teflon-lined steel	Ta-coated/lined components rely on commercial technology but require demonstration
Bayonet tube to He seals	Grafoil, Graphite O rings Teflon, Kalrez	Seals are available in this temperature range but require demonstration
H_2SO_4 piping, tanks	Glass- or Teflon-lined steels Ta-coated steels, ceramics	Glass- and Teflon-lined steels are commercially available and demonstrated in the laboratory-scale process

determined to be 74%–86% conversion, very near the theoretical maximum value based on chemical equilibrium data. The outlet conditions can be modified as needed to interface with either the S-I or HyS cycles. The basic design of the small-scale unit developed and tested in the ILS experiments can be extended to larger scales suitable for commercial applications by utilizing multiple bayonet units in a single-ended tube and shell heat exchanger configuration.

3.5 Sulphur-Iodine Cycle

As noted in Section 3.3, the S-I cycle (or I-S cycle as it is known in Japan) scored very well in the thermochemical cycle screening process. It has emerged as a leading candidate for application of nuclear energy to hydrogen production in programs around the world, including the United States, Japan, Korea, and France. This section gives the status of the S-I cycle development program in the United States (Figure 3.25).

3.5.1 S-I Baseline Process Summary

Water thermally dissociates at significant rates into hydrogen and oxygen at temperatures approaching 4000°C. As indicated in Figure 3.26, the S-I process consists of three primary chemical reactions that accomplish this same result at much lower temperatures. The process involves decomposition of sulfuric acid and hydrogen iodide, and regeneration of these reagents using the Bunsen reaction. Process heat is supplied at temperatures equal to or greater than 700°C to concentrate and decompose sulfuric acid. The exothermic Bunsen

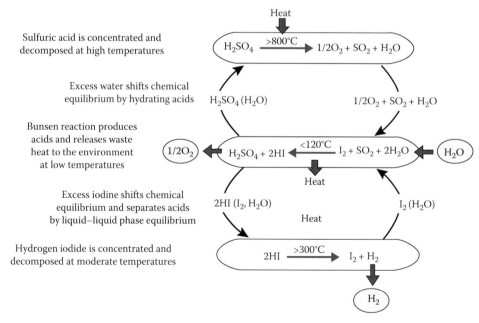

FIGURE 3.25
The S-I thermochemical water-splitting process.

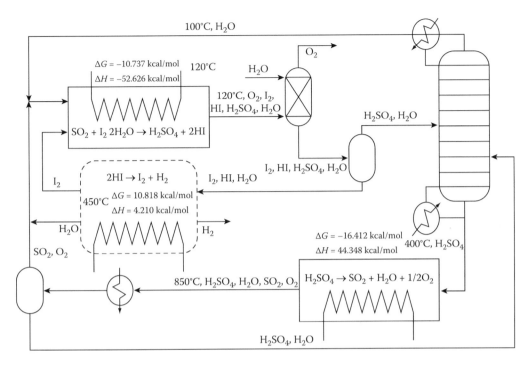

FIGURE 3.26
Simplified S-I process flow schematic.

reaction is performed at temperatures near 120°C and releases waste heat to the environment. Hydrogen is generated during the decomposition of hydrogen iodide, using process heat at temperatures greater than 300°C. Figure 3.27 shows a simplified process flow diagram of the S-I cycle. The product hydrogen gas is produced at a pressure of 4.0 MPa.

3.5.1.1 S-I Process Description

The S-I thermochemical cycle consists of three chemical reactions that result in dissociation of water into hydrogen at oxygen:

$$\text{Bunsen reaction:} \quad I_2 + SO_2 + 2H_2O \rightarrow 2HI + H_2SO_4 \quad (T \sim 120°C) \tag{3.15}$$

$$H_2SO_4 \text{ decomposition:} \quad H_2SO_4 \rightarrow SO_2 + H_2O + \frac{1}{2}O_2 \quad (T > 800°C) \tag{3.16}$$

$$\text{HI decomposition:} \quad 2HI \rightarrow I_2 + H_2 \quad (T > 350°C) \tag{3.17}$$

$$\text{Net} \quad H_2O \rightarrow H_2 + \frac{1}{2}O_2 \tag{3.14}$$

All three reactions are operated under conditions of chemical equilibrium. Energy inputs to the process are heat to the endothermic H_2SO_4 and hydrogen iodide (HI) decomposition reactions and electrical energy required for pumping process fluids and heat pumps. Heat at about 120°C is rejected from the exothermic Bunsen reaction. With the exception of water, all reactants are regenerated and recycled.

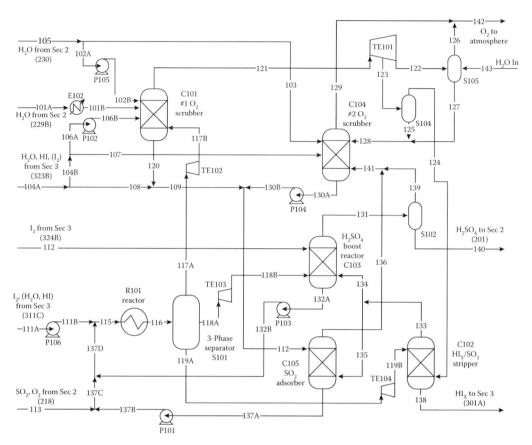

FIGURE 3.27
S-I process Section 1 flow stream description.

The hydrogen-production system design is organized into subsystems according to these three chemical reactions. These subsystems are referred to as Sections 1 through 3. Section 1 includes all the process equipment associated with production of the aqueous sulfuric acid phase and the hydrogen iodide ($HI/I_2/H_2O$) phase. Section 1 also includes equipment to purify the oxygen before release. Section 2 concentrates the aqueous sulfuric acid phase and then decomposes the concentrated acid. The decomposition products and the water removed from concentrating the acid are returned to Section 1. Section 3 concentrates and decomposes HI. Section 3 also includes equipment to purify the product hydrogen gas.

3.5.1.1.1 *Section 1 (Bunsen Reaction)*

The flow sheet for Section 1 is shown in Figure 3.27 and the stream compositions are shown in Table 3.7. This flow sheet is essentially the same as that described previously (Brown et al. 2003) and is based on experimental data obtained in the 1980s and later validated during the ILS experiment during 2006–2009. A key component in Section 1 is the flow reactor R101, which functions as a heat exchanger. Because the kinetics of the Bunsen reaction are very fast, the rate of heat transfer controls the reaction rate. The mixture exiting R101 consists of three immiscible phases, of which two are liquid and one is gas. These phases are separated in component S101. The lower phase, which consists of $HI/I_2/H_2O$ in the approximate molar ratios 2/8/10, is stripped of dissolved SO_2 and H_2SO_4 in the

TABLE 3.7

Section 1 Stream Components

Stream ID	Species Present						Phase	Pressure (bar)	Temperature (°C)
	H_2SO_4	HI	I_2	H_2O	SO_2	O_2			
101A	X			X	X		L	12.00	40.6
101B	X			X	X		L	12.00	86.5
102A	X			X	X	X	L	1.01	38.0
102B	X			X	X	X	L	4.40	38.0
103	X			X			L	1.01	38.0
104A		X	X	X			L	7.00	120.4
104B		X	X	X			L	7.00	120.4
105	X			X	X	X	L	1.01	38.0
106A		X	X	X			L	7.00	120.4
106B		X	X	X			L	12.00	120.4
107		X	X	X			L	1.01	95.4
108		X	X				L	4.20	95.4
109	X	X	X	X	X	X	L	1.85	119.9
111A		X	X	X			L	7.00	118.8
111B		X	X	X			L	12.00	119.9
112		X	X	X			L	7.00	115.6
113				X	X	X	V	7.00	40.1
115	X	X	X	X	X	X	V+L	7.00	115.6
116	X	X	X	X	X	X	V+L	7.00	119.9
117A		X	X	X	X	X	V	7.00	119.9
117B		X	X	X	X	X	V	4.20	81.1
118A	X			X	X		L	7.00	119.9
118B	X			X	X		L	1.85	119.9
119A		X	X	X	X		L	7.00	119.9
119B		X	X	X	X		L	1.85	119.9
120	X	X	X	X	X	X	L	4.20	111.3
121				X		X	V	4.20	111.3
122				X		X	L+V	1.01	111.3
123				X		X	L+V	1.85	15.9
124				X		X	V	1.85	15.9
125				X			L	1.85	15.9
126				X		X	V	1.01	15.9
127				X			L	1.01	15.9
128				X			L	1.01	15.9
129				X		X	V	1.01	39.9
130A	X	X	X	X			L	1.01	119.9
130B	X	X	X	X			L	1.85	119.9
131	X		X	X	X	X	L	1.85	111.4
132A		X	X	X	X		L	1.85	111.4
132B		X	X	X	X		L	7.00	111.4
133		X	X	X	X	X	V	1.85	119.9
134		X	X	X	X	X	V	1.85	119.9
135		X	X	X	X	X	V	1.85	119.9
136		X	X	X	X	X	V	1.85	96.5

(continued)

TABLE 3.7 (continued)

Section 1 Stream Components

Stream ID	Species Present						Phase	Pressure (bar)	Temperature (°C)
	H_2SO_4	HI	I_2	H_2O	SO_2	O_2			
137A	X	X	X	X	X		L	1.85	96.5
137B	X	X	X	X	X		L	7.00	96.5
137C		X	X	X			L	1.85	119.9
137D		X	X	X	X	X	V	1.85	111.4
140	X			X	X		L	1.85	111.4
141		X	X	X	X	X	V	1.85	102.0
142				X		X	V	1.01	24.9
143				X			L	1.01	24.9

packed column C102. Prior to stripping, the pressure is lowered in order to use a recycle O_2 stream as the stripping agent. The SO_2 is directly stripped by the O_2 stream. As the SO_2 is depleted, the reaction equilibrium shifts to produce more SO_2 from H_2SO_4 reacting with HI, which results in simultaneous removal of SO_2 and H_2SO_4. The stripped lower phase exits C102 and is transferred to Section 3, where it is processed to produce the hydrogen product gas. The upper phase exiting S101 (aqueous H_2SO_4) is transferred to boost reactor C103, which is a packed column. A portion of the upper phase exiting C102 (O_2 and SO_2) is also transferred to C103. In this column, the sulfuric acid is concentrated by contacting it with I_2 and SO_2. The molar ratio of H_2SO_4 to H_2O entering C103 is about 1:6 and is increased to about 1:4 at the exit of C103.

The remaining equipment in Section 1 is associated with processing the O_2 stream. The SO_2 is scrubbed from the O_2 stream in packed column C101. A small amount of iodine is added to C101 in order to minimize the amount of water required to remove SO_2 via the Bunsen reaction. The water streams to C101 are supplied from Section 2 and are cooled prior to entering C101 in order to remove the heat of reaction and minimize the amount of heat that is removed by R103. The SO_2 remaining in the upper phase that exits boost reactor C103 is scrubbed from the oxygen in packed column C104. Because all of the SO_2 exiting C102 cannot be used in boost reactor C103, a portion of this stream is transferred to packed column C105 to scrub the SO_2 from the oxygen. This SO_2 is then transferred to R101. Energy recovery turbines are used to recover the work available from the compressed gasses and to provide the O_2 stripping gas at an appropriate pressure.

Because the design point at the outlet of component R101 is essentially fixed in terms of temperature, compositions of both liquid phases, and the partial pressure of SO_2, the only free variable is the system pressure. The system pressure must be below the operating pressure of Section 2 so that the gases can flow from Section 2 without compression. The pressure must also be sufficiently high for efficient operation of the SO_2 stripping column. Operating R101 at higher pressures does provide the benefit of a lower SO_2 to O_2 ratio at the exit of this reactor, which leaves less SO_2 to be removed in the first oxygen scrubber (C101). The operating pressure for R101 has been set at 0.7 MPa.

3.5.1.1.2 Section 2 (Sulfuric Acid Decomposition)

The flow sheet for Section 2 is shown in Figure 3.28 and the stream compositions are shown in Table 3.8. Section 2 concentrates and decomposes the aqueous sulfuric acid phase that is transferred from Section 1. All of the heat from the secondary helium loop system (HLS)

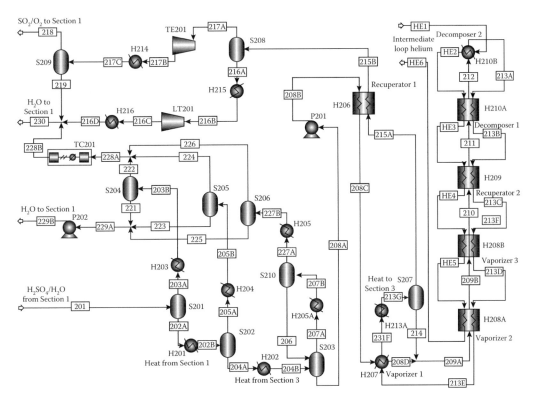

FIGURE 3.28
S-I process Section 2 flow stream description.

from the reactor is transferred to Section 2. As previously discussed, the sulfuric acid vaporizers and decomposers are operated at a pressure slightly lower than that of the secondary helium coolant in order to prevent chemical contamination of the secondary HLS while minimizing pressure differentials across these components.

The H_2SO_4/H_2O stream from Section 1 (201) is concentrated in three successive vacuum flashes. The first flash (S201) is adiabatic but heat is added in the other two flashes. The first flash cools the stream significantly below the temperatures of Section 1 so that heat released in R103 can be used to evaporate water in the second flash (S202) of Section 2. The third vacuum flash (S203) uses heat recovered from Section 3. The vapors from the three flashes are cooled and the condensate returned to Section 1 (229B), as are the condensed vapors from the vacuum pump (228B). The concentrated acid is pumped from vacuum conditions to the pressure of the decomposition subsection using P201 and passed to the countercurrent vapor–liquid recuperator (H206).

The hot liquid from the recuperator is heated and vaporized in three stages. The first two evaporators (H207 and H208A) are flow-through evaporators, with the stream progressively vaporized as it passes through the countercurrent heat exchangers. The condensate from the decomposer product is added between the first and second vaporizers. Any minerals that enter with the deionized water feed to the process and any mobile corrosion products from the entire system will eventually be deposited in the sulfuric acid decomposition process. If the final evaporation occurred while the stream was flowing through tubes, the minerals and corrosion products could deposit on the heat transfer surfaces and

TABLE 3.8

Section 2 Stream Components

Stream ID	H₂SO₄	HI	I₂	H₂O	SO₂	O₂	Phase	Pressure (bar)	Temperature (°C)
	H_2SO_4	HI	I_2	H_2O	SO_2	O_2			
201	X	X		X		X	L	1.85	111.8
202A	X	X		X		X	L	0.1	66.4
202B	X	X		X		X	V+L	0.1	110
203A	X			X		X	V	0.1	66.4
203B	X			X		X	V	0.1	40
204A	X	X				X	L	0.1	110
204B	X	X				X	V+L	0.1	194.2
205A	X			X		X	V	0.1	110
205B	X			X		X	L	0.1	40
206	X	X				X	L	0.1	160
207A	X	X				X	V	0.1	189.9
207B	X	X				X	V+L	0.1	160
208A	X	X				X	L	0.1	189.9
208C	X	X				X	L	70.5	376.9
208D	X	X				X	L	70.5	401.9
209A	X	X				X	L	70.5	403
209B	X	X	X			X	V+L	70.5	563.9
210	X	X	X			X	V	70.5	626.9
213A	X	X	X	X	X	X	V	70.5	900
213G	X	X	X	X	X	X	V+L	70.5	401.9
214	X	X				X	L	70.5	401.9
215A	X	X	X	X	X	X	V	70.5	401.9
215B	X	X		X	X	X	V+L	70.5	250.9
216A	X	X		X	X	X	L	70.5	252.6
216B	X	X		X	X	X	L	70.5	120
216C	X	X		X	X	X	L	2	118.9
216D	X	X		X	X	X	V+L	2	40
217A	X			X	X	X	V	70.5	252.6
217B	X			X	X	X	V+L	7	128.3
217C	X			X	X	X	V+L	7	40
218	X			X	X	X	V	7	40
219	X			X		X	L	7	40
221							L	0.1	40
222	X			X		X	V	0.1	40
223	X			X		X	L	0.1	39.6
224	X					X	V	0.1	39.6
225	X	X				X	L	0.1	40
226							V+L	0.1	40
227A	X	X				X	V	0.1	160
227B	X	X				X	L	0.1	40
228A	X			X		X	V	0.1	40
228B	X			X		X	V+L	7	120
229	X	X		X		X	L	12	40.5
HE1							V	70.6	924
HE6							V	69.6	557.5

contribute to fouling or even plugging. For this reason, the third vaporizer (H208B) is a pool-type unit.

The sulfuric acid vapors are decomposed in two steps:

$$H_2SO_4 \leftrightarrow SO_3 + H_2O \quad (T > 450°C) \tag{3.26}$$

$$SO_3 \rightarrow SO_2 + \frac{1}{2}O_2 \quad (T > 800°C) \tag{3.25}$$

The first reaction is very fast and equilibrium is maintained, shifting to more complete decomposition as the temperature is raised or as SO_3 is removed by the second reaction. The first reaction begins to occur in the final vaporizer (H208B), but most of the vaporization occurs primarily in H209, which is a standard countercurrent heat exchanger.

The second reaction requires a catalyst. The catalytic reaction occurs in two decomposers (H210A and H210B). Conceptually, the decomposers are countercurrent heat exchangers with catalyst on the heat transfer surfaces. The differences between the two decomposers are in how they are heated. H210B is heated with helium from the secondary HTS (HE1). Decomposer H210A, which operates at slightly lower temperatures, is heated with both secondary helium (HE2) and the decomposer product (213A). In addition, both helium and decomposer product are used in parallel to heat the gas–gas recuperator (H209) and vaporizers 2 and 3 (H208A and H208B). Only the decomposer product is used to heat the first vaporizer (H207). Finally, the decomposer is used to provide the heat requirements for Section 3 (H213A). Some condensation occurs in this final heat exchanger. The product gases (215A) are separated (S207) from the condensate and pass to the gas–liquid recuperator (H206). The condensate (214) is recycled to the second vaporizer (H208A). The condenser (H213A) and separator (S207) are physically located above the vaporizers (H208A and H208B) such that the gravitational head exceeds the pressure drop through the decomposition system, which eliminates the need for a pump capable of operating at these extreme conditions.

The two-phase product from the decomposition system (215B) is separated and power is recovered separately from the two phases. Each phase is cooled to as low a temperature as practical before they are transferred to Section 1. Condensate from the cooling process is transferred separately to Section 1.

3.5.1.1.3 Section 3 (Hydrogen Iodide Decomposition)

Two different processes have been investigated for HI decomposition. One process, referred to as extractive distillation, uses phosphoric acid to strip HI from the HI/H_2O/ I_2 (HIx) mixture and to break the HI/H_2O azeotrope. The other process is referred to as reactive distillation and involves reacting the HI/H_2O/I_2 mixture in a reactive bed to affect the separation process and produce hydrogen. Extractive distillation is a proven process but requires significant amounts of energy and many components to perform the extraction, distillation, concentration, reaction, and separation steps of the process (see Figure 3.29). The kinetics for reactive distillation are still relatively unknown, but the process can be performed in a single component without requiring concentration of the acid (see Figure 3.30). For the *n*th-of-a-kind S-I-Based H_2 plant design, the HI decomposition flow sheet is based on the reactive distillation process. One disadvantage of reactive distillation is that there is significant recycle of HI back to Section 1, which increases equipment sizes for Section 1.

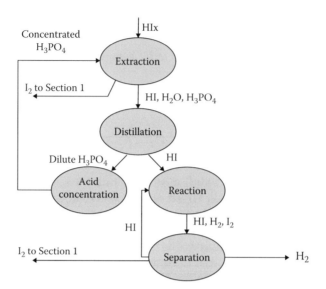

FIGURE 3.29
Extractive distillation process diagram.

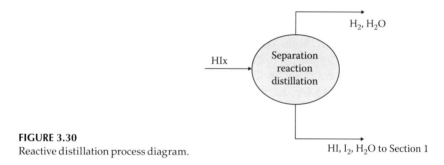

FIGURE 3.30
Reactive distillation process diagram.

The flow sheet for Section 3 is shown in Figure 3.31 and the stream compositions are shown in Table 3.9. The reactive distillation process is performed within a packed-bed distillation column using activated carbon as a catalyst. Significant quantities of heat are required for Section 3, but there are also significant quantities of heat available for recovery within Section 3. Because much of the heat available for recovery is at a temperature lower than required for process operations, heat pumps (using water as the working fluid) are used to transfer the heat from the lower temperature source to the higher temperature sink. The distillation column has a high heat duty in the reboiler and has heat available from the condenser. The first heat pump transfers heat from the condenser to the reboiler at the expense of externally supplied shaft work. Much of the heat is used to break the binding energy of the $HI/I_2/H_2O$ complex. For recycled HI, the heat of mixing is recovered and a second heat pump is used to raise the temperature of this stream to that required for heating the distillation column feed. The remaining heat duty of Section 3 is recovered from the decomposer products of Section 2.

The heat pumps consist of a multi-staged steam compressor, a hot heat exchanger complex that transfers heat to the process and condenses steam, an expansion valve in

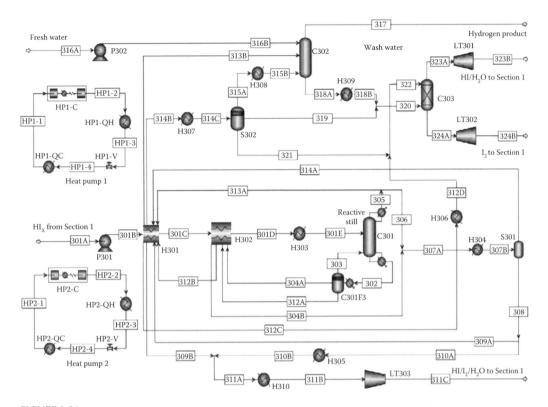

FIGURE 3.31
S-I process Section 3 flow stream description.

which the condensate is partially flashed and the pressure is lowered, and a cold heat exchanger in which the condensate is evaporated. Mechanical energy drives the compressor and much of this mechanical energy adds internal energy to the steam. The hot heat exchanger network includes the interstage coolers of the compressor as well as the subsequent condenser.

Recovery of the heat of mixing of the $HI/I_2/H_2O$ is necessary for process efficiency but does require additional process steps (described below) to produce a stream of pure I_2 that is required for operation of the H_2SO_4 boost reactor in Section 1. The HIx from Section 1 (301A) is pumped up to the operating pressure of the reactive distillation column (C301) and heated to the required feed temperature in a series of heat exchangers (H301, H302, and H303). Most of the required heat is recovered from cooling hot streams, but a portion of the heat is obtained from Heat Pump 2. The distillation column operates with a vapor product (305) consisting of $HI/H_2/H_2O$. The cold side reflux condenser on C301 is HP1-QH, which is the heat input to Heat Pump 1. The hot side of the reboiler on C301 is HP2-QC. The heavier liquid phase is nearly pure iodine. This phase is cooled, washed with water in C303, and returned to Section 1 for use in the H_2SO_4 boost reactor. The lighter liquid phase is cooled to the same temperature as the column overhead and then combined with the column overhead (306) to recover the heat of solution from the undecomposed HI. This heat is removed in H304, which is the hot side of HP2-QC and the heat input to Heat Pump 2. HP2-QH, the heat output of Heat Pump 2, is the hot side of H303 and the final preheat of the still feed. The two-phase stream leaving H304 is separated in S301. The vapor phase contains the hydrogen product gas. This stream is cooled in

TABLE 3.9

Section 3 Stream Components

Stream ID	Species Present				Phase	Pressure (bar)	Temperature (°C)
	H_2O	I_2	HI	H_2			
301A	X	X	X		L	1.85	119.85
301B	X	X	X		L	40.00	123.07
301C	X	X	X		L	40.00	247.17
301D	X	X	X		L	40.00	269.87
301E	X	X	X		V + L	40.00	282.46
302	X	X	X	X	L	40.00	289.69
303	X	X	X	X	V	40.00	289.06
304A	X	X	X		L	40.00	289.06
304B	X	X	X		L	40.00	257.17
305	X	X	X	X	V	40.00	262.17
306	X	X	X	X	V	40.00	262.17
307A	X	X	X	X	V + L	40.00	271.06
307B	X	X	X	X	V + L	40.00	257.17
308	X	X	X	X	L	40.00	257.17
309A	X	X	X	X	L	40.00	257.17
309B	X	X	X	X	L	40.00	139.01
310A	X	X	X	X	L	40.00	257.17
310B	X	X	X	X	L	40.00	139.01
311A	X	X	X	X	L	40.00	139.01
311B	X	X	X	X	L	40.00	120
311C	X	X	X	X	L	7.00	118.78
312A	X	X	X		L	40.00	289.06
312B	X	X	X		L	40.00	257.17
312C	X	X	X		L	40.00	139.01
313A	X	X	X	X	V	40.00	262.17
313B	X	X	X	X	V + L	40.00	139.01
314A	X	X	X	X	V	40.00	257.17
314B	X	X	X	X	V + L	40.00	139.01
314C	X	X	X	X	V + L	40.00	120
315A	X	X		X	V	40.00	120
315B	X	X		X	V + L	40.00	40
316A	X				L	1.01	25
316B	X				L	40.00	27.26
317	X			X	V	40.00	28.82
318A	X	X	X	X	L	40.00	42.97
318B	X	X	X	X	V + L	40.00	110
319	X	X	X	X	L	40.00	120
320	X	X	X	X	V + L	40.00	117.69
321	X	X	X	X	L	40.00	120
322	X	X	X	X	L	40.00	118.27
322D	X	X	X		L	40.00	120
323A	X	X	X	X	L	40.00	120.55
323B	X	X	X	X	L	7.00	120.39

TABLE 3.9 (continued)

Section 3 Stream Components

Stream ID	H₂O	I₂	HI	H₂	Phase	Pressure (bar)	Temperature (°C)
	H_2O	I_2	HI	H_2			
324A	X	X	X	X	L	40.00	118.04
324B	X	X	X	X	L	7.00	115.57
HP1-1	X				V	44.77	257.12
HP1-2	X				V	71.60	288.46
HP1-3	X				L	71.60	287.4
HP1-4	X				V+L	44.77	257.12
HP2-1	X				V	41.21	252.13
HP2-2	X				V	79.50	295.69
HP2-3	X				L	79.50	294.6
HP2-4	X				V+L	41.21	252.13

H308 and the condensate is removed in S302. The hydrogen is then washed with water in C302 to yield the final hydrogen product. The condensate from S302 consists of two liquid phases. The heavier phase is primarily iodine and is washed in C303 for use in the H_2SO_4 boost reactor of Section 1. The lighter phase is mostly water and is combined with the lower phase from C302, which is also mostly water. These two streams function as the wash for C303.

Pure water is not very effective at washing iodine out of hydrogen, but iodine is very soluble in HI. For this reason, a small amount of the distillation overhead (313A) is split from the main flow (305), cooled, and then added back to the wash column (C301) at an intermediate stage.

The liquid phase from S301 is split for heat recovery. Part of the heat is recovered into the column feed and the remainder is cooled in H305, which is the hot side of H202 in Section 2.

3.5.1.2 Plant Performance and Efficiency

The flow sheets for Sections 2 and 3 have been analyzed and optimized using AspenPlus® process simulation software. Because there are still very limited thermophysical property data for the $HI/I_2/H_2O$ vapor equilibrium, the equilibrium conditions for Section 1 are based on previous calculations (Norman et al. 1981). Approximately two-thirds of the fresh water required for hydrogen production is supplied to Section 3 and the remainder is supplied to Section 1. The product hydrogen gas is produced at a pressure of 4.0 MPa.

The hydrogen-production rate for a commercial four-module plant (2400 MWt) is 6436 mol/s = 12.97 kg/s. The corresponding heat rate is 1840.44 MWt, using the higher heating value of hydrogen (141.9 MJ/kg). Shaft work associated with the primary and secondary coolant circulators produces 96 MWt. Shaft work associated with the hydrogen-production system produces 684 MWt. The hydrogen plant is assumed to produce 15 MWe through energy-recovery turbines and house loads are assumed to be 6 MWe. Assuming pump motor efficiencies of 95% and electricity produced at 48% thermal efficiency using

GT-MHRs, the net electricity required by the plant is 812 MWe. The overall plant efficiency is then calculated to be

$$\eta = \frac{\Delta H^0_{H_2O}(T = 25°C)}{Q + W/\eta_{el}}$$

$$\eta = 100 \times \frac{HHV\,H_2}{Q + W/0.4} = \frac{1841.15}{2400 + 812/0.48} = 45.0\%$$

3.5.2 Integrated Loop Experiments

An international collaborative effort between SNL, General Atomics Corporation (GA), and the French Commissariat à l'Energie Atomique (CEA) was undertaken to investigate S-I thermochemical cycles for the production of hydrogen from water. This project was conducted as an International Nuclear Energy Research Initiative (INERI) project supported by the CEA and the U.S. DOE NHI. The integrated experiment was performed at the General Atomics facility in San Diego, CA. The objective of the program was to demonstrate thermochemical hydrogen production at a rate of 100–200 L/h in a process utilizing scalable technology, which enables the design of a larger pilot-scale process in the future. The key issues addressed in this project included selection of construction materials, characterization of individual chemical process operations, and process control and technology scalability. Each participant developed one of the three major process sections of the S-I process.

The three sections of the S-I process were constructed on skids with Lexan enclosures. The process sections were integrated through a chemical storage or buffer skid. With the exception of SO_2, all chemicals used in the S-I process are transferred to and from the chemical storage skid. For safety reasons, SO_2 gas was transferred directly from the acid decomposition skid to the HI-production skid. The chemical storage skid allowed for each section of the S-I process to operate in either the stand-alone or integrated mode.

3.5.2.1 Process Description

3.5.2.1.1 Hydrogen–Iodide Production (Bunsen Section)

The HI production section known as the Bunsen section (Figure 3.32) performs two primary process functions: Separation of O_2 from the $SO_2/O_2/H_2O$ stream received from the acid decomposition section and the reaction of SO_2, H_2O, and I_2 to form HI and H_2SO_4. The $SO_2/O_2/H_2O$ mixture is collected in a buffer tank where the H_2O is cryogenically removed. The remaining SO_2–O_2 mixture is compressed to liquefy the SO_2. The O_2 remains gaseous and is first scrubbed with H_2O to remove any residual SO_2 and then vented. The liquid SO_2 is transferred to the chemical reactor (Bunsen reactor) where it is combined with I_2 and H_2O to produce HI and H_2SO_4. The Bunsen reaction is exothermic and is performed in a countercurrent three-phase reactor at ~120 °C and 6 bar. The heavy phase from the Bunsen reaction exiting the bottom of the reactor contains mainly HI whereas the light phase exiting the top of the reactor contains mainly H_2O and H_2SO_4. The materials of construction used in the process are stainless steel for the SO_2–O_2 separation line and PTFE and tantalum tubing. The Bunsen reactor itself is made from glass-lined, double-walled carbon steel tubing.

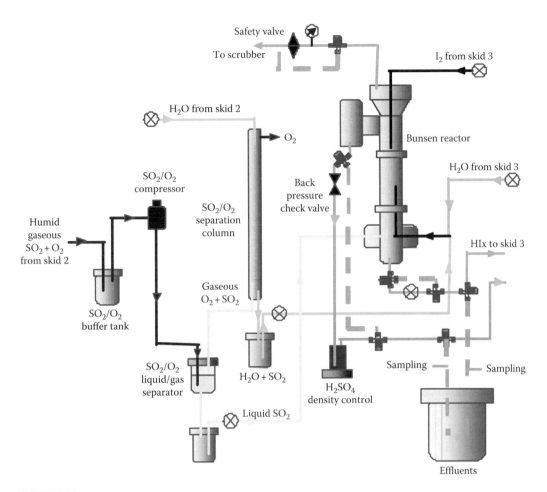

FIGURE 3.32
Schematic diagram of the HI production or Bunsen section of the S-I process.

3.5.2.1.2 Sulfuric Acid Decomposition

A schematic of the H_2SO_4 decomposition process is shown in Figure 3.33. Two main chemical processes in this section are the concentration of dilute H_2SO_4 received from the Bunsen process and the thermal-catalytic decomposition of H_2SO_4 to produce SO_2, H_2O, and O_2. Incoming acid from the Bunsen process is ~20 mol% and is concentrated to 40 mol% in a thin film-type evaporator operated under vacuum. The evaporator is constructed of silicon carbide and Teflon. Preconcentration of the acid before decomposition is not a requirement but allows the acid decomposer to operate at higher capacity. The concentrated acid is fed to the acid decomposer where it is decomposed at 850°C in the presence of a platinum catalyst. The acid decomposer is an all ceramic, bayonet type heat exchanger constructed of concentric silicon carbide tubes with a catalyst placed in the top section of the apparatus. Detailed descriptions of the acid concentrator and decomposer are given elsewhere (Helie et al. 2007, Moore et al. 2007b). Any undecomposed acid is collected at the exit of the acid decomposer for recycle. The product stream, SO_2, H_2O, and O_2, is passed through a water-chilled condenser at 5°C to remove most of the H_2O to avoid corrosion in the current

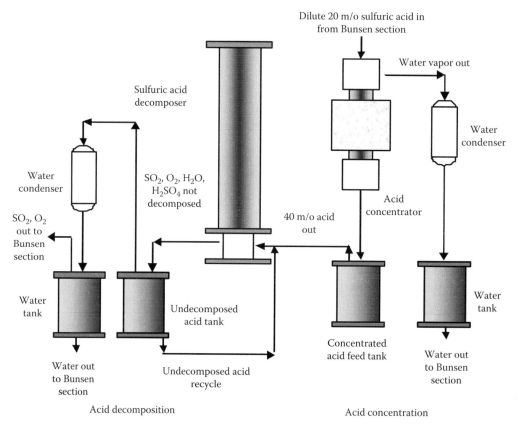

FIGURE 3.33
Schematic diagram of the sulfuric acid decomposition section for the S-I process.

piping. The H_2O is transferred to the chemical storage skid and the SO_2 and O_2 mixture is transferred directly to the Bunsen section.

The acid decomposer is operated at a pressure of 2 bar or greater. Pressure is maintained with a Teflon backpressure regulator placed at the product exit of the process. The backpressure regulator effectively isolates the acid decomposer from any pressure fluctuations created by cycling of the gas compressor in the Bunsen section. PTFE/glass rotameters are used to monitor the flowrate of acid feed and recycle acid to the acid decomposer. The composition of the product gas stream is monitored with an O_2 analyzer and the flowrate is determined with a flowmeter constructed of Ryton.

3.5.2.1.3 HI Decomposition Section

The HI decomposition process system (Figure 3.34) consists of six major chemical processes: I_2 extraction, HI distillation, HI decomposition, iodine recycle, HI recycle, and phosphoric acid concentration. The HI decomposition skid receives HI feed, 2:8:10 molar ratios of $HI:I_2:H_2O$, from the HI production process and first removes the I_2 in a liquid–liquid extraction process utilizing H_3PO_4 acid to break the HI, H_2O azeotrope. The HI is distilled from the remaining H_2O, H_3PO_4 solution and pumped to the reactor for decomposition to I_2 and H_2. As the reaction is thermodynamically limited to 20% conversion, the unreacted HI must be separated and recycled in order to reach the desired H_2 production

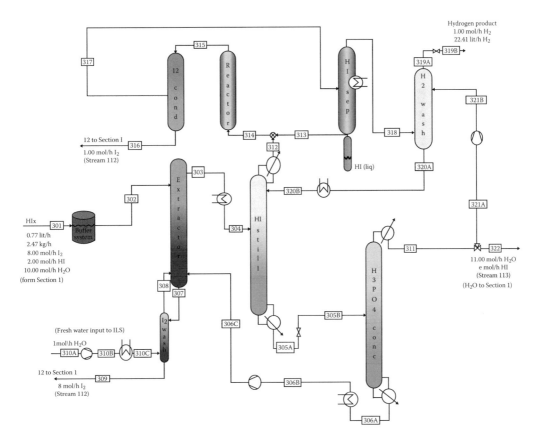

FIGURE 3.34
HI decomposition process for the S-I process.

rates. Any undecomposed HI in the product stream is cryogenically separated from I_2 for recycle of the HI back to the reactor. The cryogenic method can recover up to 90% of the unreacted HI. The material of construction for the plumbing and process vessels is mainly a tantalum—10% tungsten alloy. This material was selected based on its corrosion resistance as well as its flexibility for plumbing. The main process vessels are 2″ diameter tube with the process lines 0.5″ tubing. Tantalum-coated Swagelok fittings were used for process line connections.

3.5.2.1.4 Process Integration

As previously stated, integration of the three sections of the S-I process was through a chemical storage or buffer skid. A picture of the S-I process with the three process sections and the chemical storage skid is given in Figure 3.35. The chemical storage skid, only partially visible in the back, houses Teflon-lined stainless steel tanks for H_2SO_4 and H_2O, and heated glass-lined steel tanks for HI and I_2. A chemical scrubber was also located on this skid for disposal of any unwanted chemicals produced in stand-alone or integrated testing.

3.5.2.2 Concluding Remarks

The objective of testing of the S-I laboratory-scale experiment was to establish the capability for integrated operations and demonstrate hydrogen production from the S-I cycle.

FIGURE 3.35
Integrated laboratory-scale experiment for the S-I process. From the left, the acid decomposition skid, the Bunsen process skid, the chemical storage skid (only partially visible in the back), and the hydrogen iodine decomposition skid.

This objective was achieved with the successful integrated operation of the SNL acid decomposition and CEA Bunsen reactor sections to produce the required heavy and light acid phase product streams, and the subsequent generation of hydrogen from that material in the HI decomposition section. This is the first time the S-I cycle has been constructed using engineering materials and operated at prototypic temperature and pressure to produce hydrogen, demonstrating its viability for nuclear hydrogen production.

3.6 Hybrid Sulfur Cycle

Thermochemical water-splitting cycles can be solely powered by thermal energy (i.e., "pure" thermochemical cycles) or by a combination of both thermal and electrical energy (i.e., "hybrid" thermochemical cycles). A significant advantage of hybrid cycles is the ability to use electrochemical reactors to conduct one or more otherwise non-spontaneous reactions. This can greatly simplify the process chemistry and reduce the number of reaction steps necessary to close the thermochemical cycle. The hybrid sulfur (HyS) cycle is an example of a hybrid thermochemical cycle. It is one of the simplest thermochemical water-splitting processes, comprising only two reactions with all reactants and products in the fluid state. It entails two coupled reactions that cycle sulfur between its +4 (sulfite) and +6 (sulfate) oxidation states. One reaction step is the thermochemical decomposition of sulfuric acid, which is common to all thermochemical processes based on the sulfur cycle chemistry, including the S-I cycle. The second step is the electrochemical oxidation of sulfur dioxide to sulfuric acid with the simultaneous generation of hydrogen.

The HyS cycle was first proposed by Brecher and Wu (1975) at Westinghouse Electric Corp. Known at that time as the "Westinghouse Sulfur Cycle," considerable effort was conducted by Westinghouse in developing the cycle from its inception in the early 1970s until eventual project termination in the mid-1980s. Simultaneous development also took place in Europe at the Joint Research Center in Ispra, Italy, where it was known as the

Mark 11 cycle (Beghi 1986), and at the Nuclear Research Center (KFA) in Jülich, Germany, as well as elsewhere. As a result of decreased worldwide research in advanced nuclear technologies beginning in the early 1980s, development activities essentially ceased for nearly 20 years. Beginning in 2003, the Savannah River National Laboratory (SRNL) initiated renewed development on the HyS cycle under the U.S. DOE, Office of Nuclear Energy's (DOE-NE's) NHI (Sink 2006). Since then, researchers in a number of countries, including the European Union, France, the Republic of Korea, Japan, and South Africa have also initiated research programs on the HyS Process. U.S. work on HyS has remained centered at SRNL (Summers and Gorensek 2006, Summers 2008, 2009), while related work on the electrochemical step has been conducted at the University of South Carolina (USC) (Sivasubramanian et al. 2007, Staser and Weidner 2009).

3.6.1 Process Chemistry

The HyS cycle is comprised of two coupled reaction steps. The thermochemical step, which is common to all sulfur cycles, is the high-temperature decomposition of sulfuric acid into water, sulfur dioxide, and oxygen:

$$H_2SO_4(aq) \xrightarrow{\text{Heat, } T>800°C} H_2O(g) + SO_2(g) + \frac{1}{2}O_2(g) \tag{3.16}$$

This is an equilibrium-limited reaction that requires a catalyst, as well as heat input at relatively high temperatures, such as those that an advanced, high-temperature gas-cooled reactor (HTGR) heat source could supply. The electrochemical step is the SO_2-depolarized electrolysis of water:

$$SO_2(aq) + 2H_2O(l) \xrightarrow{\text{Power, } T \approx 80°C-120°C} H_2SO_4(aq) + H_2(g) \tag{3.22}$$

which takes place at lower temperatures and requires the input of electric power. The combined net effect of the two reactions is the splitting of one mole of water into one mole of hydrogen and one-half mole of oxygen.

The advantage of the SO_2-depolarized electrolysis approach is a substantial reduction in the electric power requirement per unit of hydrogen produced compared to conventional direct water electrolysis:

$$H_2O(l) \xrightarrow{\text{Power}} H_2(g) + \frac{1}{2}O_2(g) \tag{3.14}$$

which is a single-step, water-splitting approach based on the input of electric power only. As is well known, the standard potential at 25°C for water electrolysis is −1.229 V. The standard potential at 25°C for SO_2-depolarized electrolysis, on the other hand, is only −0.158 V (Gorensek et al. 2009a), a reduction of 87%. In both cases, practical cell operating voltages are higher due to overpotentials resulting from irreversibilities caused by kinetic and mass transfer limitations, as well as Ohmic losses. These overpotentials increase as the current flow through a given size cell is increased. Economic considerations lead to tradeoffs between low cell voltages (high efficiency) and high current densities (low capital cost). Alkaline water electrolyzers are typically operated at −1.8 to −2.0 V, while SO_2-depolarized electrolyzers (SDEs) are expected to be capable of achieving −0.6 V at practical current densities (500 mA/cm²). Under these conditions, the HyS cycle requires approximately one-third as much electricity for the electrochemical step as conventional water electrolysis. As long as the energy required to close the cycle by decomposing the sulfuric acid leaving the cell to regenerate the SO_2 reactant is less than the energy needed for the additional electricity used by conventional water electrolysis, the HyS cycle will enjoy a net efficiency advantage. As detailed further in the discussions in the following sections, both theoretical and experimental results indicate a substantial efficiency advantage for the HyS cycle.

3.6.2 SO_2-Depolarized Electrolyzer

The SDE step distinguishes HyS from the other sulfur-based thermochemical cycles. The electrochemical reactor employed to conduct this step is termed the SDE. Simply stated, it consists of a water electrolyzer that utilizes sulfur dioxide at the anode to produce sulfuric acid rather than oxygen, thereby significantly reducing the overall cell potential. The half-cell reaction at the anode,

$$SO_2(aq) + 2H_2O(l) \rightarrow H_2SO_4(aq) + 2H^+ + 2e^- \qquad (3.27)$$

splits water to produce sulfuric acid, protons, and electrons. The hydrogen ions (protons) pass through the electrolyte to the cathode while the electrons travel through the external electric circuit. The result is hydrogen gas being evolved at the cathode via the following half-cell reaction:

$$2H^+ + 2e^- \rightarrow H_2(g) \qquad (3.28)$$

Early work at Westinghouse was based on a parallel plate electrolyzer concept that used a microporous diaphragm separator. The electrolyzer was built up from parallel banks of individual cells connected in series and contained in a pressurized container. Both the catholyte and anolyte contained sulfuric acid at similar concentrations (up to 75 wt% H_2SO_4) and were separately recirculated. The anolyte feed was saturated with dissolved SO_2, while the catholyte (operated at higher pressure) was kept SO_2 free. The pressure differential resulted in a net flow of sulfuric acid through the separator from the cathode to the anode.

When research on the HyS cycle was restarted in 2003 by SRNL, a much different design approach was taken for the SDE reactor. Proton exchange membranes (PEMs), such as Nafion®, have undergone considerable development over the past two decades since the cessation of the Westinghouse work. PEMs, which are also used extensively in the chlor-alkali industry, form the basis for almost all automotive fuel cell development, and they are also emerging for water electrolysis applications. As a result, most current SDE work has been based on PEM technology, allowing for the leveraging of the large investments, technological advancements, and cost reductions made in this technology by the fuel cell developers, automotive companies, and others.

Adopting the fuel cell design approach, SDEs have been constructed using compact membrane–electrode assemblies (MEAs) to allow close contact between the polymer electrolyte and the electrodes. MEAs incorporating PEM electrolyte separators provide for the shortest possible current path between anode and cathode, minimizing IR losses, and increasing cell efficiency. As a result, the PEM-based SDE offers the promise of higher efficiency, smaller footprint, and lower capital costs than the two-compartment cells considered earlier. This helps to overcome one of the major challenges faced by HyS developers in the past, that of creating a cost-effective, scalable electrochemical reactor.

Figure 3.36 illustrates the SDE configuration used by SRNL. The MEA (anode catalyst layer, PEM, and cathode catalyst layer) is at the "heart" of the SDE. A carbon cloth or paper diffusion layer is pressed against each side by a graphite flow field, which is designed to force fluid flow through the diffusion layer. The anolyte is sulfuric acid saturated with dissolved SO_2. Unlike the Westinghouse design, the catholyte is water. This helps the membrane remain hydrated and also serves to sweep out any elemental sulfur that may form at the cathode due to SO_2 crossover. Essentially no undissociated H_2SO_4 remains in solution at typical anolyte concentrations, 30%–75 wt% H_2SO_4. Consequently, both the applied potential and the anionic character of the membrane provide an effective barrier to sulfate and bisulfate ion diffusion, keeping sulfuric acid out of the catholyte. However, neutral species, such as SO_2, may diffuse through the water phase of the membrane and cause undesired side reactions at the cathode, such as the formation of elemental sulfur. Prevention of this phenomenon is one of the major challenges of SDE design.

The use of a PEM electrolyte imposes some limitations on the SDE. The conductivity of the most common PEM material, Nafion, depends on the water content of the membrane; water content, in turn, depends on the activity of water at the membrane surfaces. When in contact with sulfuric acid, it has been shown that the resistivity of Nafion increases with the H_2SO_4 content of the acid. The practical upper limit for the acid concentration in a Nafion-equipped SDE is about 50 wt% H_2SO_4 (Gorensek and Summers 2009). Furthermore, Nafion is limited to an operating temperature less than 100°C. High acid product concentrations are preferred in order to minimize energy requirements in the acid decomposition portion of the HyS cycle, and high temperatures are preferred in order to improve the reaction kinetics and increase the electrolyzer efficiency. Therefore, a PEM other than Nafion, preferably one that has good conductivity and low SO_2 permeability at high acid concentration and is capable of high-temperature operation, is needed to further improve the technology.

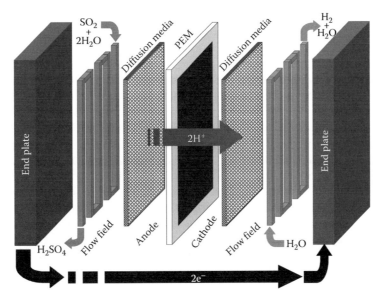

FIGURE 3.36
PEM-type design for SO_2-depolarized electrolyzer.

The SDE design developed by SRNL utilizes a liquid anolyte consisting of sulfuric acid saturated with SO_2. An alternative SDE design has been developed and tested at the USC (Sivasubramanian et al. 2007). In this approach, gaseous SO_2 is fed to the anode while liquid water is fed to the cathode. A large pressure differential from cathode to anode is maintained in order to help create a net water flux across the PEM. Water transported across the membrane reacts with gaseous SO_2 at the anode to form sulfuric acid, which accumulates and is removed as the anode product. As with the liquid-fed design, hydrogen gas is evolved at the cathode and leaves along with excess water. When compared to the all-liquid SDE design, a significantly higher catholyte (water) recirculation rate is needed in the gaseous anolyte approach in order to control the SDE temperature. More importantly, a closed HyS flow sheet incorporating this gaseous anode design will differ in the way the products of the high-temperature sulfuric acid decomposition step are handled compared to a liquid-fed SDE. Since SO_2 and O_2 are quite easily separated based on their solubility difference, a liquid-fed SDE flow sheet can utilize straightforward preferential dissolution to do this separation and recycle the SO_2 to the electrolyzer. The gaseous SO_2 anode flow sheet will be somewhat more complex and require more challenging equipment design. Otherwise, initial flow sheet studies have shown the performance of the two designs to be comparable if the same SDE performance is achieved for each.

3.6.2.1 SDE Component Development

The SDE is composed of several major components, each of which requires careful selection, analysis, characterization, and experimental study. The use of a PEM as the cell electrolyte dictates the general geometry and basic design approach for the SDE. Component-level research has primarily focused on the electrocatalyst for both the anode and cathode and the PEM electrolyte.

3.6.2.1.1 Electrocatalyst Development

Selection of a suitable electrocatalyst is important in order to (1) increase the electrical efficiency by minimizing kinetic overpotential losses and (2) maintain long-term stable cell performance. Prior work by Westinghouse Electric in the 1980s investigated the use of precious metal blacks (Pd and Pt) as electrocatalysts for the SDE reactions (Lu et al. 1980). The results indicated that palladium was a more effective electrocatalyst than platinum for SDE. More recently, SRNL conducted experimental work to compare high surface area catalysts using palladium on carbon (Pd/C) and platinum on carbon (Pt/C) (Colón-Mercado and Hobbs 2007).

Electrochemical characterization of catalysts consisted of cyclic voltammograms (CVs) in the solution purged with nitrogen and linear sweep voltammograms (LSVs) in SO_2-saturated sulfuric acid solutions. Pd/C showed somewhat higher initial activity than Pt/C but was much less stable. After a few cycles, the activity of the Pd/C catalyst decreased significantly, whereas the Pt/C exhibited very good stability and activity for the oxidation of SO_2. Tafel plots showed lower potentials (ca. 100 mV) and much higher exchange currents (ca. 1000 times greater) for the oxidation of SO_2 on Pt/C compared to Pd/C. Furthermore, the activation energy for the oxidation of SO_2 on Pt/C is at least half of that on a Pd/C surface. As a result, SRNL has focused its SDE work on the use of Pt electrocatalysts. Recently, SRNL has conducted experiments to measure the performance of Pt-alloy catalysts created by alloying Pt with various transition metals (Colón-Mercado et al. 2008). The best performance was observed when Pt was alloyed with non-noble metals such as cobalt and chromium. There remains a need to further increase the anode catalyst performance for the oxidation of sulfur dioxide in order to meet the commercial performance goals. This can be accomplished through the development of more active catalysts combined with increased operating temperature, which improves the overall reaction kinetics.

3.6.2.1.2 SDE Electrolyte Development

In addition to the electrocatalyst, the other key component of an SDE is the PEM electrolyte. There are many requirements of a PEM for the successful functioning of the electrolyzer. The PEM must be stable in highly corrosive solution (>30 wt% H_2SO_4 saturated with SO_2) and at high operating temperature (80°C–140°C), allow minimal transport of SO_2, and must maintain high ionic conductivity under these conditions. These requirements allow the electrolyzer to perform at high current density and low cell potential, thus maximizing the energy efficiency for hydrogen production. Lastly, the PEM serves to separate the anolyte from the hydrogen output in order to prevent the production of undesired sulfur-based reaction products and poisoning of the cathode catalyst.

The SRNL has investigated numerous candidates for use as membranes in the SDE (Hobbs 2009). Both commercially available and experimental membranes were selected and evaluated for chemical stability, sulfur dioxide transport, and ionic conductivity characteristics. An array of thicknesses, equivalent weights, chemistry, and reinforcements were considered. Many of the membranes considered are also being utilized or are under development for use in PEM fuel cells, PEM electrolyzers, and/or direct methanol fuel cells (DMFCs). This permits the developers of an SDE to leverage much of the extensive work being done by others for these related PEM applications. Membranes being developed for DMFC applications are particularly relevant since a major challenge for this technology is the reduction of methanol crossover. Methanol, like sulfur dioxide, is a neutral species that is soluble in water. The methanol crossover issue is similar to the SDE problem with the crossover of dissolved SO_2.

A partial list of the membranes selected and characterized by SRNL is given in Table 3.10. All of these membranes exhibited excellent short-term chemical stability in concentrated sulfuric acid solutions at temperatures up to 80°C.

The conductivity and SO_2 transport for each membrane were evaluated using a custom-made permeation cell. Each membrane was also tested for through-plane ionic conductivity using electrochemical impedance spectroscopy. Representative results are shown in Figure 3.37 (Hobbs 2009).

TABLE 3.10

Candidate PEM Materials for the SDE Application

Membrane Description	Source
Perfluorinated sulfonic acid (PFSA)	DuPont
Polybenzimidizole (PBI)	BASF (Germany)
Sulfonated Diels-Alder polyphenylene (SDAPP)	Sandia National Laboratories
Stretched recast PFSA	Case Western Reserve University
Nafion/fluorinated ethylene propylene (FEP) blends	Case Western Reserve University
Treated PFSA	Giner Electrochemical Systems, LLC
Perfluorocyclobutane-biphenyl vinyl ether (BPVE)	Clemson University
Perfluorocyclobutane-biphenyl vinyl ether-hexafluoroisopropylidene (BPVE-6F)	Clemson University

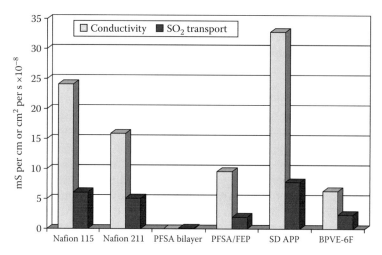

FIGURE 3.37
Proton exchange membrane results showing ionic conductivity and SO$_2$ diffusion. (From Hobbs, D., Challenges and progress in the development of a sulfur dioxide-depolarized electrolyzer for efficient hydrogen production, *Pittcon Conference & Expo 2009*, Chicago, IL, 2009.)

The lowest SO$_2$ transport coefficient was measured for membranes of the PFSA bilayer and high equivalent weight PFSA type. However, these membranes also showed unacceptably low ionic conductivity, negating their use in the SDE. Several of the advanced membranes had promising combined performance (low SO$_2$ transport and high conductivity). Both the PFSA/FEP blend and the BPVE-6F membranes exhibited lower SO$_2$ transport than Nafion, but with somewhat lower ionic conductivity. The SDAPP membrane had not only high conductivity but also a somewhat higher SO$_2$ transport coefficient. Results for the PBI membrane (not shown in Figure 3.37) were also encouraging, with low SO$_2$ transport and reasonable conductivity.

The test results shown in Figure 3.37 were determined at 67°C and atmospheric pressure. However, several of the more promising membranes, such as the PBI, BPVE, and SDAPP, are expected to perform significantly better at higher temperature (120°C–140°C), which is not possible with a PFSA membrane due to its lower temperature limit. The higher temperature operation is also expected to improve reaction kinetics and lower the cell potential. Preliminary testing of these membranes at higher temperature using gaseous SO$_2$ as the anolyte has shown promising results (Staser et al. 2009). Operation with liquid anolyte (SO$_2$ dissolved in sulfuric acid) at the higher temperatures requires a high-pressure test facility in order to maintain adequate SO$_2$ concentrations in the feed. No results for higher temperature/higher pressure testing with a liquid-fed SDE have been reported yet, although this is an obvious future research direction.

3.6.2.2 SDE Testing

Extensive SDE testing has been reported by SRNL using both a button-cell electrolyzer and a larger single-cell electrolyzer (Steimke and Steeper 2005, 2006, Hobbs et al. 2009). Photographs of the single-cell electrolyzer are shown in Figure 3.38.

It is constructed of steel support plates, graphite plates, graphite flow fields, graphite diffusions layers, PVDF gaskets, and PFA-lined steel tubing. Copper current collectors

FIGURE 3.38
The SRNL 60 cm² single-cell SO₂-depolarized electrolyzer.

are sealed from contacting the process fluids. The cell is designed to permit easy replacement of the MEA, consisting of the PEM bonded between platinized carbon anode and platinized carbon cathode layers. The cell has a nominal active cell area of 60 cm². A test facility capable of automated unattended testing of SDE units at pressures up to 600 kPa and temperatures up to 80°C was designed and constructed. Most of the single-cell tests were operated to generate approximately 10–20 L/h of hydrogen. The SDE was operated in the liquid-fed mode, with the anolyte consisting of sulfuric acid saturated with dissolved sulfur dioxide. Water was also added on the cathode side of the cell in order to maintain membrane hydration and to remove any sulfur species that could result from SO_2 crossover. Representative polarization curve results are shown in Figure 3.39 (Steimke et al. 2009a).

The test results show a cell potential of approximately 760 mV at a current density of 500 mA/cm² during operation at 80°C with 30 wt% sulfuric acid. This compares to the SRNL commercial design goal of 600 mV at 500 mA/cm². Operation at higher temperature (120°C–140°C) and use of advanced membranes are expected to lower the cell potential in order to achieve the commercial goal. The advanced membranes will also be required to maintain high conductivity in the presence of high sulfuric acid concentrations.

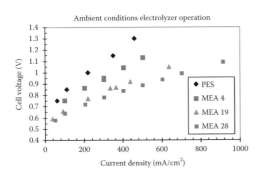
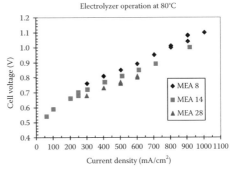

FIGURE 3.39
Representative polarization curve for the SRNL single-cell electrolyzer. (From Steimke, J.L. et al., Results from testing hybrid sulfur single-cell and three-cell stack, *Hybrid Sulfur Electrolyzer Workshop and Information Exchange*, Aiken, SC, April 20–21, 2009, http://srnl.doe.gov/hse_workshop/Steimke%20Electrolyzer%20Performance.pdf [accessed October 21, 2009].)

A major concern of SDE operation is the diffusion of sulfur dioxide through the PEM to the cathode where it can be reduced to hydrogen sulfide and eventually elemental sulfur. Earlier tests performed by SRNL revealed the buildup of a sulfur layer inside of the cell's MEA (Summers 2009). More recent testing indicated that such a sulfur buildup could be avoided through the proper selection of membranes and the choice of operating conditions. In this case, the SO_2 crossover may be reduced to the point where a manageable amount of H_2S (part-per-million level) is generated and leaves the cell with the hydrogen gas, but no sulfur buildup occurs inside the SDE. SRNL conducted two tests of 212 h each using a Nafion membrane verifying this mode of operation (Steimke et al. 2009b). Longer term testing is planned in the future, including operation at higher temperature and pressure using advanced membranes.

3.6.2.3 Larger-Scale SDE Development

The largest reported SDE device tested to date is a multi-cell stack rated at 100 L/h of hydrogen production demonstrated by SRNL in 2008 (Summers 2009). A photograph of the three-cell SDE installed in the SRNL test facility is shown in Figure 3.40. This unit utilized round plates with 160 cm² active area per cell stacked in a bipolar arrangement. It was successfully operated at 80°C and 700 kPa for 72 h with a constant hydrogen output of 86 L/h. Full stack capacity of 100 L/h could not be achieved due to test facility limitations. Development of higher capacity SDE units, consisting of larger area plates in multi-cell stacks, will be necessary in order to construct commercial-scale electrolyzers.

3.6.3 Sulfuric Acid Decomposition

The decomposition of sulfuric acid to release oxygen and regenerate sulfur dioxide is the thermochemical step in the HyS cycle. It is also the high-temperature step of the S-I cycle and all other "sulfur cycle" thermochemical processes. Much of the research and testing for this step has been performed as part of the work on the S-I cycle, and most of the work can be directly applied to HyS process. A detailed discussion of sulfuric acid decomposition for the sulfur-based thermochemical cycles is given in Section 3.4.

FIGURE 3.40
The SRNL three-cell SO_2-depolarized electrolyzer. (From Summers, W.A., Hybrid sulfur thermochemical cycle, *2009 DOE Hydrogen Program and Vehicle Technologies Program Annual Merit Review and Peer Evaluation Meeting*, Arlington, VA, May 18–22, 2009, http://www.hydrogen.energy.gov/pdfs/review09/pd_13_summers.pdf [accessed October 21, 2009].)

3.6.4 HyS Process Flow Sheet

A recent HyS flow sheet has been presented by SRNL and others (Gorensek et al. 2009b). It is based on the use of a PEM-type SDE and a bayonet decomposition reactor. Many earlier efforts to prepare flow sheets for HyS and other thermochemical processes used somewhat simplified and idealized designs. The SRNL-led team, which included experienced industrial partners, attempted to develop a flow sheet reflecting normal industrial practices, including realistic operating conditions, such as heat exchanger approach temperatures, conservative ambient design values, and the need for blow-down and make-up streams. An Aspen Plus flow sheet was prepared as shown in Figure 3.41. A summary of this flow sheet is presented in the following.

The anolyte exiting the SDE was assumed to contain 50 wt% H_2SO_4 at 100°C and 20 bar. Future SDE designs using advanced membranes may allow higher acid concentrations and higher temperature operation. The pressure was selected to provide product hydrogen under pressure, which minimizes downstream compression requirements. The operating pressure was limited, however, by phase equilibrium considerations. Higher pressures favor SO_2 solubility but are limited by SO_2 vapor pressure—two liquid phases could otherwise form in the SDE in some circumstances.

The simplest way to remove unreacted SO_2 and trace O_2 from the SDE product is by dropping the pressure. This requires recompressing the predominantly SO_2 and trace O_2 vapor stream that is out-gassed so that it can be recycled to the SDE feed system. The shaft work required is not excessive and the separation can be made without any additional heat input.

The degassed SDE product needs to be concentrated before introducing it into the bayonet decomposition reactor in order to avoid the large thermal penalty associated with heating excess water in this reactor. Several acid concentration methods are available. All consume energy, some more than others. A recuperative method is clearly needed in order to minimize the energy requirements for the acid concentration portion of the flow sheet. SRNL has chosen vacuum distillation with recuperative preheating/partial vaporization of the feed streams to concentrate the SDE product. Vacuum is maintained using a two-stage steam ejector, keeping temperatures high enough to allow use of cooling water in the condenser, yet low enough to permit metallic materials of construction. Water removed in the concentration process is condensed and recycled to the SDE feed system. The vacuum column bottoms, containing concentrated sulfuric acid at 75 wt% H_2SO_4, can be pumped to the necessary pressure and fed directly to the bayonet reactor.

The bayonet reactor performance is based on a peak process temperature of 875°C and an operating pressure of 8.6 MPa. This assumes that the high-temperature heat is provided by hot helium at 900°C, representing a high-temperature gas reactor outlet temperature of 950°C. At these conditions, fractional conversion of H_2SO_4 in the decomposition reactor is 48.1%, and the effluent temperature at the bayonet exit is 254.7°C with a vapor fraction of 20.4% (assuming 10°C minimum temperature difference for recuperation).

Effluent from the bayonet reactor is readily separated into unreacted H_2SO_4 feed and SO_2/O_2 product by means of a vapor–liquid split. The acid is simply recycled to the vacuum still. The vapor is cooled further and let down to the pressure of the SDE feed system, resulting in a three-phase mixture: wet liquid SO_2, a saturated solution of SO_2 in H_2O, and wet O_2 gas contaminated with SO_2. The O_2 gas can be scrubbed with the water collected in the concentration process to remove most of the SO_2. The two liquid phases can then be combined with that water and with recycled, spent anolyte to form fresh anolyte feed.

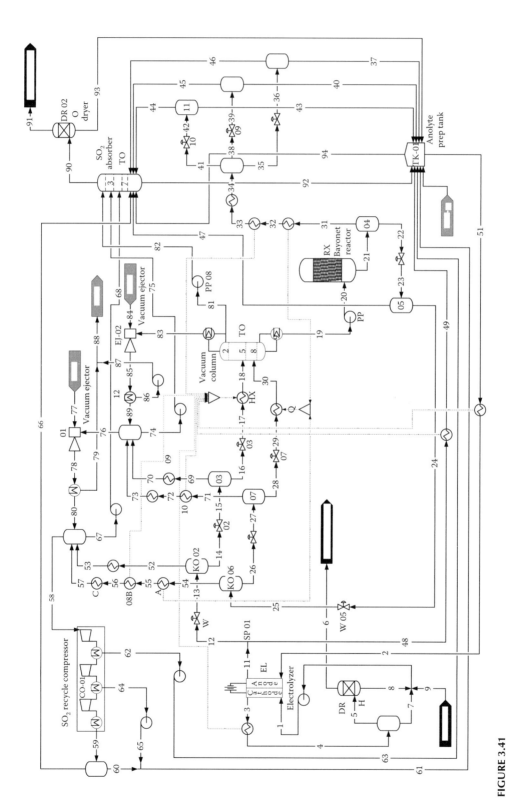

FIGURE 3.41

HyS flow sheet using a PEM SDE and a bayonet decomposition reactor. (From Gorensek, M.B. et al., Hybrid sulfur process reference design and cost analysis, Savannah River National Laboratory, Report No. SRNL-L1200-2008-00002, Aiken, SC (doi: 10.2172/956960), June 12, 2009.)

Integration of the PEM SDE and the bayonet reactor as described above results in a flow sheet that uses only proven technology, with the sole exception of the SDE and the decomposition reactor. Thus, the flow sheet itself does not introduce any additional technical hurdles. This should help give the performance projections greater credibility.

3.6.5 Energy Requirements and System Efficiency

The basis for the stream data associated with the flow sheet was 1.0 kmol/s hydrogen-production rate. Heat exchanger specifications are presented in Table 3.11. External heat input is required in only two places: the bayonet reactor (RX-01) and the Vacuum Column (TO-01) reboiler. The bayonet reactor receives 340.3 kJ/mol H_2 high-temperature external heat, assumed to be supplied from hot helium heated in a high-temperature nuclear reactor. The reboiler, however, does not need such high temperatures since it operates in the 100°C–125°C range. The amount required, only 75.5 kJ/mol H_2, can be supplied with low-pressure steam. Heat rejection to cooling water adds up to 252.9 kJ/mol H_2, with over 80% attributable to the vacuum column condenser. Recuperation takes place

TABLE 3.11

Performance Summary for HyS Flow Sheet Heat Exchangers

Block ID	Duty (MW$_{th}$)	Temperature, °C (K)		Heat Exchanged with
		Inlet	Outlet	
CO-01/Stage 1 Cooler	−2.290	138.02 (411.2)	40.0 (313)	Cooling water
CO-01/Stage 2 Cooler	−9.109	137.79 (410.9)	40.0 (313)	Cooling water
CO-01/Stage 3 Cooler	−0.132	143.76 (416.9)	40.0 (313)	Cooling water
DR-01	−2.045	76.96 (350.1)	40.0 (313)	Cooling water
DR-02	−0.197	41.12 (314.3)	40.0 (313)	Cooling water
HX-01	−17.688	100.00 (373.0)	76.96 (350.1)	HX-02
HX-02	130.486	66.96 (340.1)	94.41 (367.6)	HX-01, HX-04B, HX-05, HX-06, HX08B, HX-10A
HX-03	31.777	100.50 (373.7)	114.62 (387.8)	HX-04A, HX-08A
HX-04A	−30.543	254.70 (527.9)	110.50 (383.7)	HX-03
HX-04B	−8.400	110.50 (383.7)	76.96 (350.1)	HX-02
HX-04C	−6.243	76.96 (350.1)	40.0 (313)	Cooling water
HX-05	−47.903	100.00 (373.0)	80.00 (353.0)	HX-02
HX-06	−18.686	84.71 (357.9)	78.77 (351.9)	HX-02
HX-07	−3.702	84.08 (357.2)	40.0 (313)	Cooling water
HX-08A	−1.234	147.33 (420.5)	110.50 (383.7)	HX-03
HX-08B	−37.440	110.50 (383.7)	76.96 (350.1)	HX-02
HX-08C	−4.021	76.96 (350.1)	40.0 (313)	Cooling water
HX-09	−1.476	80.38 (353.5)	40.0 (313)	Cooling water
HX-10A	−0.368	118.57 (391.7)	76.96 (350.1)	HX-02
HX-10B	−11.503	76.96 (350.1)	40.0 (313)	Cooling water
HX-11	−1.310	93.77 (366.9)	40.0 (313)	Cooling water
HX-12	−0.354	95.05 (368.2)	40.0 (313)	Cooling water
RX-01	340.280	123.63 (396.8)	254.70 (527.9)	High-temperature source
TO-01 Reboiler	75.482	102.74 (375.9)	122.87 (396.0)	Low-temperature source
TO-01 Condenser	−210.542	44.80 (318.0)	40.0 (313)	Cooling water

TABLE 3.12

Power Requirements for HyS
Electrolyzers, Pumps, and
Compressors

Block ID	Work (MW$_e$)
EL-01	115.782
CO-01/Stage 1	1.464
CO-01/Stage 2	1.357
CO-01/Stage 3	0.025
PP-01	0.022
PP-02	1.836
PP-03	0.071
PP-04	0.002
PP-05	0.028
PP-06	0.055
PP-07	0.021
PP-08	0.212
PP-09	0.00003

in exchangers HX-02 and HX-03, accounting for 130.5 and 31.8 kJ/mol H$_2$ heat exchange, respectively. The vacuum ejectors require a small quantity of steam (1.31 kJ/mol H$_2$) for their operation.

Electric power requirements are detailed in Table 3.12. The SDE accounts for the majority (115.8 kJ/mol H$_2$) of the 120.9 kJ/mol H$_2$ electric energy consumed by the process. Most of the remainder, about 2.8 kJ/mol H$_2$, is attributable to SO$_2$ recycle compressor. The actual pumping requirement will be somewhat higher because frictional losses due to flow-through equipment have been ignored. Assuming a thermal-to-electric conversion efficiency of 33% (representing conventional reactor power conversion efficiency), the total electric power requirement is equivalent to a heat input of 366.4 kJ/mol H$_2$. If advanced high-temperature gas reactors were used to provide the electric power (45% conversion efficiency), the equivalent heat input would be 268.7 kJ/mol H$_2$.

The total heat consumed by the process is, therefore, 340.3 + 75.5 + 1.3 = 417.1 kJ/mol H$_2$. Adding the thermal equivalent of the power consumption (assuming a conventional power source), this flow sheet requires a total of 783.5 kJ/mol H$_2$ of primary energy. This compares favorably with a conventional reactor powering direct water electrolysis at 1080 kJ/mol H$_2$ (Gorensek et al. 2009b). Using an HTGR power source for electricity generation as well as thermal energy would reduce the primary energy requirement from 783.5 to 685.8 kJ/mol H$_2$.

The HyS flow sheet presented above combines a PEM SDE with a bayonet decomposition reactor and otherwise uses only proven chemical process technology. If the SDE and high-temperature decomposition reactor perform as projected (600 mV cell potential, 40% conversion, 50 wt% H$_2$SO$_4$ product; 950°C HTGR operating temperature, 50°C drop between the primary and secondary helium coolants, 25°C minimum temperature difference between helium coolant and process stream, 10°C minimum temperature difference for recuperation, adequate heat transfer characteristics), the flow sheet process will split water into H$_2$ and O$_2$ while consuming 340.3 kJ/mol H$_2$ high-temperature heat, 76.8 kJ/mol H$_2$ low-temperature heat, and 120.9 kJ/mol H$_2$ electric power. Should the ultimate source of the electric power be a conventional nuclear reactor (electricity from heat at 33% conversion efficiency), the net thermal efficiency would be 30.9%, LHV basis, excluding the power needed for helium circulators. This is significantly more efficient than alkaline electrolysis coupled with conventional reactor power (22.4%, LHV basis). A higher HyS process net thermal efficiency, 35.3%, LHV basis would be attainable with electric power provided by an HTGR power. This also exceeds HTGR-powered direct water electrolysis (30.6%, LHV basis). On an HHV basis, the HTGR powered HyS process efficiency is 41.7%.

Further process efficiency improvements are possible based on the development of suitable SDE membranes that permit operation at higher temperature and higher acid concentrations.

Preliminary experiments with PBI membranes at USC show that significantly higher acid concentrations, in excess of 65% H_2SO_4 by weight, may be feasible (Weidner 2009). In this case, external energy needed for acid concentration could be reduced or eliminated. For example, the 75.5 kJ/mol H_2 heat duty for the vacuum column could be eliminated (recuperative heating would suffice) and the bayonet heat requirement could possibly be lowered to 321 kJ/mol H_2 (80 wt% acid feed). Using HTGR power with an assumed thermal-to-electric conversion efficiency of 45%, the primary energy requirement would then be reduced to 591 kJ/mol H_2. This would give a net thermal efficiency of 40.9%, LHV basis, or 48.3%, HHV basis. Some reactor system developers have projected a future HTGR power conversion efficiency of 52% (Richards et al. 2006), which would result in a HyS process with a net thermal efficiency of 43.6%, LHV basis (51.6%, HHV basis).

3.6.6 Summary and Technical Challenges

The HyS process has compelling advantages as a thermochemical water-splitting cycle. It has a simple two-step chemistry requiring only hydrogen-, oxygen-, and sulfur-containing species. Plant design studies have shown that it has the potential for high efficiency, and early economic analysis shows it to be cost-competitive when powered by either advanced nuclear reactors or solar central receivers. The two key process components, the SDE and the sulfuric acid decomposition reactor, have both been successfully demonstrated at laboratory-scale of approximately 100 L/h hydrogen output. The next anticipated development step is the construction and operation of an integrated laboratory-scale experiment to demonstrate closed-loop operation of an entire HyS cycle. This would likely be followed by a MW-scale pilot plant prior to scale-up to full commercial plant capacities.

Some of the specific technical challenges remaining to be addressed include the following:

- SDE development
  1. Reduction in cell potential through the use of higher temperature operation and improved electrocatalyst
  2. High-temperature membranes with low SO_2 crossover
  3. Operation with higher H_2SO_4 product
  4. Demonstration of long-time stable operation with minimal operating potential degradation
  5. Scale-up to larger cell sizes and multi-cell modules
- Acid decomposer
  1. Demonstration of a long-life catalyst
  2. Operation with convective-heating using a helium heat source

3. Decomposer design and material selections
4. SiC joining
5. Catalyst replacements methods
- HyS process and system development
1. Integrated system demonstrations at increasingly larger scale
2. Materials of construction validation
3. Capital cost reduction and evaluations

The majority of the work presented here was the result of activities by the SRNL and its partners, with the primary source of funding support provided by the U.S. DOE's NHI program. Increasingly, other entities have begun to perform research on the HyS cycle, and international collaboration has increased. A HyS work package has been added to the research subjects under the framework of very-high temperature reactor (VHTR) system in the Generation IV International Forum. These activities are expected to address the remaining technical challenges for the HyS cycle and to lead to pilot-plant facilities and demonstration of a commercial-scale plant.

3.7 Low-Temperature Thermochemical Cycles: The Cu-Cl Cycle

The U.S. DOE recognized that the higher temperatures required for most thermochemical cycles placed a heavy burden on currently available materials of construction. DOE subsequently supported a program that identified and evaluated thermochemical cycles that could operate with lower temperature heat that could be obtained from new heat sources such as the sodium-cooled and the supercritical water-cooled nuclear reactors. These produce process heat near 550°C. Several potentially high-efficiency cycles were identified. Based on various analyses, the copper–chlorine (Cu-Cl) was selected as one of the most promising. ANL led the effort to investigate the Cu-Cl cycle in 2004 with a focus on completing proof-of-concept for the major reactions in the cycle as well as determining estimates of the process's efficiency and its cost of hydrogen production.

DOE provided a framework for international and national collaborative research. An INERI was established with Atomic Energy of Canada Limited (AECL) and with Commissariat à l'Energie Atomique (CEA) in France. AECL provided seed money to the University of Ontario Institute of Technology (UOIT), which was used to obtain funding from the Ontario Research Foundation. This grant funded research on various aspects of the development effort at several Canadian universities. The Nuclear Energy Research Initiative-Consortium (NERI-C) program funded Pennsylvania State University (lead), the University of South Carolina, and Tulane University to study advanced electrochemical technologies.

3.7.1 Process Chemistry

Two versions of the Cu-Cl cycle were reported in the literature (Dokiya and Kotera 1976, Carty et al. 1981). ANL initially studied both of these versions but determined that combining features of both led to a more efficient cycle with a potentially lower cost for H_2

TABLE 3.13

Reactions in the Cu-Cl Cycle

	Reaction	Temperature (°C)	
Hydrolysis	$2CuCl_2 + H_2O \rightarrow Cu_2OCl_2 + 2HCl$	340–400	(3.29)
Decomposition	$Cu_2OCl_2 \rightarrow \frac{1}{2}O_2 + 2CuCl$	450–530	(3.30)
Electrolysis	$2CuCl + 2HCl \rightarrow 2CuCl_2 + H_2$	100	(3.31)
Net	$H_2O \rightarrow H_2 + \frac{1}{2}O_2$		(3.14)

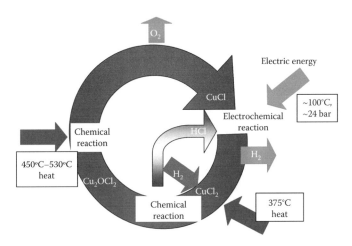

FIGURE 3.42
Schematic of the Cu-Cl cycle.

production and a lower maximum process temperature (Lewis and Masin 2009, Lewis et al. 2009a,b). The major reactions are given in Table 3.13 and a schematic of the process is given in Figure 3.42.

The hydrolysis reaction involves contacting $CuCl_2$ with steam, that is, this is a gas–solid reaction for which mass and heat transfer is critically important. The products of the hydrolysis reaction are copper oxychloride, Cu_2OCl_2, which is found in nature as the mineral melanothallite, and HCl as well as unreacted steam. The HCl and steam mixture is subsequently condensed and transferred to the electrolyzer feed tanks. The Cu_2OCl_2 solid product is transferred to a higher temperature reactor where it decomposes to oxygen and molten CuCl salt. The CuCl salt is cooled and its enthalpy is recovered for use in the process elsewhere. The CuCl is then dissolved in HCl solution prior to transfer to the electrolyzer. By conducting the hydrolysis and decomposition reactions separately, the HCl and O_2 gases are generated in different reactors eliminating any mixing of these two gases. In the electrolysis reaction, Cu(I) is oxidized to Cu(II) at the anode while the hydrogen ion is reduced at the cathode.

3.7.1.1 Hydrolysis Reaction

The two major challenges associated with the hydrolysis reaction, $2CuCl_2(s) + H_2O(g) \rightarrow Cu_2OCl_2(s) + 2HCl(g)$, are (1) identification of a reactor design that provides the needed mass and heat transfer and (2) the need for excess steam to drive the reaction to the right. Both challenges must be met to obtain essentially 100% yields of the desired products, a must for a cyclic process. In addition, to have an efficient and a cost-effective process, the steam to $CuCl_2$ molar ratio must be as low as possible since heating and vaporizing massive amounts of water is energy and capital intensive.

Several types of reactor designs were tested at ANL. The earliest fixed-bed reactors did not provide sufficient mass and heat transfer to obtain high yields of the desired products. When very high flow rates of the carrier gas (as measured by the gas hourly space velocity [GHSV]) were used, yields of the desired products were higher. For example, in the experiments with a GHSV of 43,000/h, Cu_2OCl_2 represented 87–89 wt% of the product while in experiments with GHSVs of 8,900 and 26,000/h Cu_2OCl_2 represented 48 and 66 wt%, respectively of the product. These results confirmed the need for a reactor design that would provide extremely good mass and heat transfer.

The design that appears the most promising is a spray reactor (Ferrandon et al. 2010b). Such a reactor provides high mass and heat transfer and is suitable for injecting liquids into a hot environment. Spray reactors are used commercially in the recovery of HCl from spent pickling solutions. Several different versions of a laboratory-scale spray reactor have now been tested. The ultrasonic nozzle was easier to use and gave a finer spray than the "pneumatic" nebulizer tested (Ferrandon et al. 2010b). Yields were high. The amount of Cu_2OCl_2 was measured as 95–97 wt% with a small amount 3–5 wt% CuCl. The x-ray diffraction (XRD) pattern is shown in Figure 3.43. Note the one-to-one correspondence of the peaks in the laboratory-produced Cu_2OCl_2 and those of melanothallite, the mineral form of Cu_2OCl_2. Note also the peak for CuCl, which is observed in all of our XRD patterns.

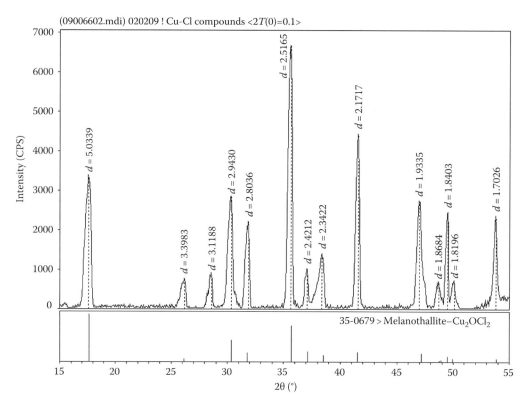

FIGURE 3.43
X-ray diffraction pattern for the product of the hydrolysis reaction using an ultrasonic nozzle, 370°C and 1 bar.

However, the steam-to-$CuCl_2$ ratio required was 17–23, which is too high to obtain a reasonable efficiency for the overall process.

Le Chatelier's Principle suggests that the amount of water required for complete reaction could be reduced by removing the gaseous products and/or by reducing the operating pressure to drive the reaction to the right. The results of an AspenPlus sensitivity study in Figure 3.44 show that reducing the pressure from 1 to 0.5 bar reduces the steam to $CuCl_2$ molar ratio from about 17 to 12 while maintaining the yield of Cu_2OCl_2 near 100%. Experiments to validate this prediction were conducted in the spray reactor/ultrasonic nozzle apparatus to which an aspirator pump was added at the exit of the bubbler. A schematic is shown in Figure 3.45, along with pictures of the ultrasonic nozzle, the three-zone furnace, which houses the reactor, the product Cu_2OCl_2 collected in the bottom of the reactor. Preliminary experiments at subatmospheric pressures indicate that it is possible to reduce the steam-to-$CuCl_2$ ratio to 11–15. Further details of the apparatus and the experimental method will be published (Ferrandon et al. 2010b).

All of the XRD patterns examined thus far contained peaks assigned to CuCl. This peak can be seen in Figure 3.43 at a d-spacing of 3.11 Å. There are two modes for CuCl formation. One is the thermal decomposition of $CuCl_2$, $2CuCl_2 \Leftrightarrow 2CuCl + Cl_2$ and the second is the thermal decomposition of Cu_2OCl_2, $Cu_2OCl_2 \Leftrightarrow 2CuCl + \frac{1}{2}O_2$. Operating conditions must be chosen to prevent the first reaction where chlorine is formed.

Several types of experiments have been completed to determine which of the reactions for CuCl formation is the more important. Conductivity, pH, and UV–VIS spectroscopy were used at the Commissariat à l'Energie Atomique (CEA) to study the hydrolysis reaction. Their experiments are described in detail elsewhere (Ferrandon et al. 2010b). In their experiments, $CuCl_2$ particles were contacted with steam from 190°C to 540°C, with a hold period of 30 min at 390°C. The integrated conductivity of the condensed effluent increased with time corresponding to the sorption of the generated HCl. Doizi noted that the concentration of chloride ion was calculated from the conductivity measurements as 0.0115 mol/L while the pH measurement indicated a hydronium ion concentration of 0.01116 mol/L. The gaseous effluent was analyzed in real time by UV–VIS spectrometry. Chlorine has an absorption band between 300 and 400 nm. No chlorine was observed spectroscopically until the temperature reached 400°C and above. These data suggest that the thermal decomposition of $CuCl_2$ to CuCl and Cl_2 does not occur until 400°C.

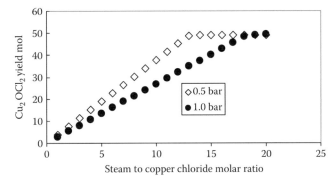

FIGURE 3.44
Results of a sensitivity study showing the dependence of Cu_2OCl_2 yield as a function of two pressures.

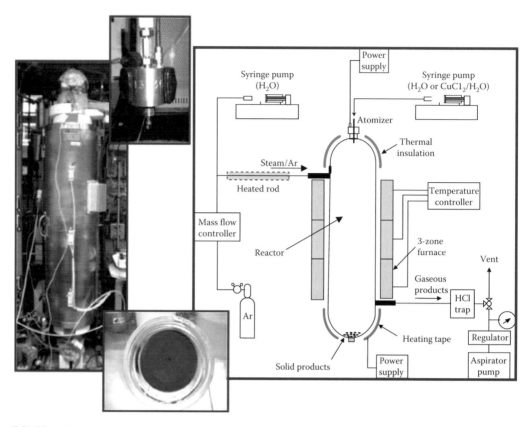

FIGURE 3.45
Components of the hydrolysis reactor: reactor furnace (left most); ultrasonic nozzle (upper center); Cu_2OCl_2 powders collected in the bottom of the reactor (bottom center); schematic of the hydrolysis reactor with the aspirator (right).

3.7.1.2 Decomposition of Cu_2OCl_2 Reaction

The decomposition of Cu_2OCl_2: $Cu_2OCl_2(s) \rightarrow \frac{1}{2}O_2(g) + 2CuCl(s)$ is a relatively simple thermal decomposition reaction. The result of the initial study at Gas Research Institute was that Cu_2OCl_2 decomposed to produce molten CuCl; O_2 production was not confirmed (Carty et al. 1981). Several more recent experiments have now been run to verify that O_2 is the only gaseous product and that the amount of oxygen corresponds to 100% of the theoretical amount (Serban et al. 2004, Ferrandon et al. 2010b). The most recent work involved heating a product from the hydrolysis reactor from room temperature to 700°C and following the evolution of the gaseous products with a mass spectrometry. The mass spectrum for O_2 is shown in Figure 3.46. Using a calibration curve and the integrated area of the peak, it was determined that the area corresponded to 100% of the theoretical amount if the sample contained 95% Cu_2OCl_2, which is consistent with the presence of some CuCl. Note that evolution of O_2 started at about 400°C and was essentially complete around 525°C.

3.7.1.3 Electrolysis Reaction

The objective of the development effort for the electrolysis reaction was to design an electrolyzer that can meet target values for the cell potential (0.6–0.7 V) and the current density

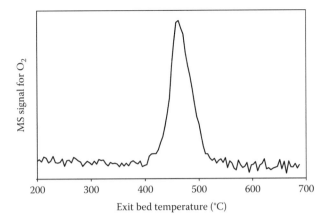

FIGURE 3.46
Mass spectrum of oxygen released during heating of Cu_2OCl_2 produced in-house.

(500 mA/cm²). These were required for a reasonable overall efficiency for the cycle. The electrolysis reaction has been studied at the Atomic Energy of Canada Limited (AECL) as part of the INERI program. Most of the work at AECL is proprietary. The NERI-C program focused on advanced electrochemical techniques and was concerned with development of a thermodynamic model for determining speciation as a function of Eh and pH, anion exchange membrane development, and membrane testing.

Initial experiments at AECL showed that the anode reaction was mass transfer limited (Stolberg et al. 2009). Half-cell tests showed that stirring and/or operating at higher temperatures increased the current density for a given cell voltage. These results were subsequently confirmed by work at Pennsylvania State University (PSU) and USC. The major challenge observed in single-cell tests was that copper crossed over from the anode to the cathode. This crossover was manifested by a large increase in cell voltage. SEM micrographs of used MEAs at PSU showed that copper dendrites had formed in the membrane and/or on the electrode surface and that the dendrites in some cases tore holes in the membrane itself. In other cases, copper deposited on the cathode. AECL recently reported that their latest cell design provided 2–3 days of operation with a stable voltage of 0.6–0.7 V in a flow-through cell at room temperature using a cation exchange membrane (Stolberg 2009). The current density was lower than the target value of 500 mA/cm². However, it is expected that the current density will increase as the operating temperature is raised and further improvements are made.

PSU in the NERI-C program completed a thermodynamic model for determining the speciation of the copper ions in solution. Details of the results of this work can be found elsewhere (Gong et al. 2009) and Chapter 6 of this book. Briefly, it was determined that much of the copper in solution exists as chloro-copper complexes. The concentration of the various species depends on Eh and pH.

3.7.2 Modeling Activities

AspenPlus was used to simulate the process design. The initial work was done at the University of Chicago-Illinois (UIC) (Nankani et al. 2007). A simplified version of the Aspen flow sheet is given in Figure 3.47 and is described in more detail elsewhere (Ferrandon et al. 2008). This is the flow sheet for which the current values for efficiency are based. This flow sheet is currently undergoing revision and further optimization.

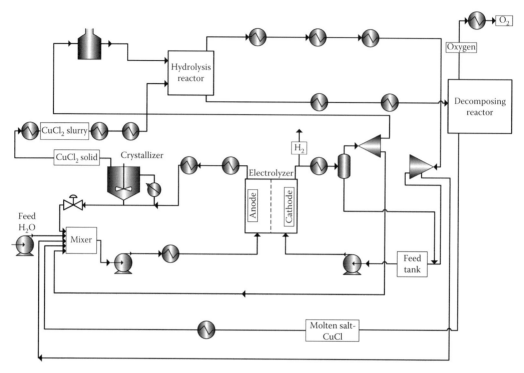

FIGURE 3.47
Flow sheet of Cu-Cl cycle process.

The conceptual design corresponding to the flow sheet includes four unit operations:

1. The hydrolysis/oxychloride decomposition reactors
2. Direct heat exchanger
3. The electrolyzer
4. The crystallizer

A brief description of these operations is provided here. Wherever possible, commercially practiced operations are used.

3.7.2.1 Hydrolysis/Oxychloride Decomposition Reactors

In the hydrolysis reactor, the hot, pressurized (24 bar) $CuCl_2$ slurry is sprayed into a superheated (400°C) steam environment at 0.25 bar where it forms a free jet. As the jet expands, it aspirates the superheated steam into the jet resulting in high mass and heat transfer between the $CuCl_2$ in the jet and the steam. The $CuCl_2$ is converted to Cu_2OCl_2 and HCl. The HCl and unreacted steam exit the hydrolysis reactor, are cooled in feed preheaters, and then fed to the cathode of the electrolyzer via the cathode feed tank. A steam ejector and the volume contraction of the hydrolysis stream, as it condenses, pulls the vacuum on the hydrolysis reactor. Dry, free-flowing solid Cu_2OCl_2 accumulates at the bottom of the hydrolysis reactor. The solid copper oxychloride flows by gravity through an L valve to the oxychloride decomposition reactor. The hydrolysis reactor is very similar to a spray roaster used in the steel industry to recover HCl from $FeCl_2$ generated by the steel pickling process. The flow of solid Cu_2OCl_2 from the hydrolysis reactor to the oxychloride decomposition reactor is similar to the flow of cracking catalyst in a fluid bed catalytic reactor in a refinery.

In the oxychloride decomposition reactor, the Cu_2OCl_2 is heated up to 550°C. Between 450°C and 550°C, the Cu_2OCl_2 decomposes to oxygen and molten CuCl. The oxygen leaves the oxychloride decomposition reactor as a gas and the molten CuCl spills over the weir. The heat in the molten CuCl stream is recovered in a heat exchanger. The molten CuCl enters the direct heat exchanger and is atomized by a spinning disc. As the molten CuCl cools, it heats the (acid-poor) vapor stream from the cathode flash. The vapor stream is then fed to the hydrolysis reactor. During the cooling, the CuCl is granulated. The granulated CuCl is fed back to the anode feed tank via a screw feeder or via gravity flow.

3.7.2.2 Direct Heat Exchanger

The direct heat exchanger has not been demonstration but is similar in concept to the Bateman Granulation system, which is used to make granulated steel or slag.

3.7.2.3 Electrolyzer

We have assumed that the electrolyzer will operate near 100°C at 24 bar in order to produce hydrogen at 300 psi and that the cell potential will be 0.6–0.7 V with a current density of $500 \, mA/cm^2$. The commercial electrolyzer consists of individual electrolyzer cells consisting of a membrane and anodic/cathodic compartments using a modular design of stacks and modules to attain the necessary hydrogen-production throughput. The ion-exchange membrane is located in the middle of the cell and is sandwiched between the two electrodes, the anode and the cathode. Both the anode and cathode are porous, carbon/carbon-felt electrodes typically used in PEM fuel cell design. In the CuCl electrolyzer, the cathode has an electrocatalyst layer to promote the hydrogen generation reaction whereas the anode does not need this layer. Moving outward from the electrode is the bipolar plate, also called a flowplate, which acts as a channel for gas and electrolyte flow into and out of the anodic and cathodic compartments.

3.7.2.4 Crystallizer

The crystallizer is based on a model (Mathias 2007) derived from solubility data in the literature (Novikov et al. 1979). This unit operation has not been demonstrated.

3.7.2.5 Alternate Conceptual Design

An alternate version of the Cu-Cl cycle is under development by six Canadian universities funded by an Ontario Research Foundation (ORF) grant. The lead university is the University of Ontario Institute of Technology. Their work is focused on the development of large laboratory-scale equipment, modeling, materials of construction, exergy analysis, control system design, and reliability assessment (Naterer et al. 2009, Orhan et al. 2010, Rosen et al. 2010).

3.7.2.6 Efficiency Calculations

AspenPlus was used to develop mass and energy balances for the process described above. The simulation provided stream flows and properties as well as heat exchanger duties and work requirements for pumps. Heat usage was optimized with pinch analysis and a heat exchanger network. For a CuCl cycle plant producing 125 MT H_2/day, 191 MW of thermal energy and 100.5 MW of electrical energy are required (Kolb 2008). This calculation includes heat and energy demands for the chemical plant, the power plant, and the intermediate heat exchanger. The energy efficiency of the process is defined as energy out divided by energy in. Using a 40% factor for converting heat to electricity and the low heating value for hydrogen, the efficiency of this process was estimated as

$$\text{Efficiency} = \frac{\text{Mol. of } H_2 \text{ produced} \times \text{LHV}}{\left(\text{Shaft work} + \text{Electrochemical work} + \text{Pinch heat}\right)} \tag{3.32}$$

$$\text{Efficiency} = \frac{125,000 \times 33.3}{\left(24\left(3000/0.4\right) + 97,500/0.4 + 191,000\right)} = 39\% \left(\text{LHV}\right)$$

Using the higher heating value of hydrogen (39.4 kWh/kg), the efficiency is 46%.

This efficiency value is subject to change as more is learned about the chemistry within the cycle. As mentioned above, the operability of the crystallizer has not been demonstrated and target values for the electrolyzer's cell potential and current density are assumed.

3.8 Conclusions

Nuclear energy does not inherently involve any use of fossil fuels or the production of carbon dioxide or other GHGs. Nuclear energy thus has the potential to make major contributions to the production of carbon-free and carbon-neutral fuels and energy carriers by providing a major source of the primary energy for these fuels. The construction of the nuclear power plant and the provision of nuclear fuel for that power plant can involve use of coal and natural gas to produce steel and concrete to build the plant, and electricity used to enrich the uranium used in the plant. The life-cycle production of CO_2 from nuclear power, however, is among the lowest of any source of energy and is less than 2% that of coal. Forty percent of the U.S. primary energy consumption goes for the production of electricity and accounts for 40% of the energy-related CO_2 production in the United States, primarily from burning coal and natural gas. Nuclear power provides 21% of the U.S. electricity and virtually no CO_2. Increased use of nuclear energy to produce electricity, displacing coal and natural gas, would reduce CO_2 production. However, reducing the other 60% of the U.S. energy-related CO_2 emissions requires addressing the nonelectric energy applications.

Nuclear energy can provide the process heat that is used for a variety of industrial processes, including the production of fuels from biomass. Nuclear energy is well suited to provide a large-scale source of hydrogen at reasonable cost. Hydrogen is used to upgrade heavy crude petroleum and get more fuel per unit carbon released. Hydrogen can make significant improvements in the efficiency of producing biofuels from biomass. Hydrogen can be used to make synthetic hydrocarbon fuels, eventually "recycling" the CO_2 from fossil power plants to make transportation fuels. And hydrogen by itself alone is the ultimate carbon-free transportation fuel.

Nuclear energy can produce hydrogen by LTE, by high-temperature steam electrolysis (HTSE), and by a number of thermochemical water-splitting cycles. LTE is a fully commercial technology as is the production of electricity by nuclear power. The net efficiency (higher heating value basis) of producing hydrogen by this means is ~25% for current LWRs, to about 36% for advanced high-temperature reactors. Coupled with HTSE or thermochemical water splitting, these advanced reactors could reach efficiencies above 50%. HTSE is at an advanced development stage and could be applied to a demonstration hydrogen-production plant in the near future. A number of different thermochemical water-splitting cycles are in various stages of research and development. The S-I cycle and the HyS cycle are approaching readiness for a similar demonstration. A number of lower temperature thermochemical cycles are being investigated, with the Cu-Cl cycle being most advanced and looking very promising. This would allow efficient hydrogen production from lower temperature nuclear reactors.

Nuclear energy can play a major role in the production of carbon-free and carbon-neutral fuels for a sustainable energy future. The technologies described in this chapter can help make this carbon-free, sustainable energy future a reality.

Nomenclature

ASR	area-specific resistance, $\Omega \, cm^2$
F	Faraday number, $96487 \, C/mol$
ΔG_R	Gibbs energy of reaction, J/mol
ΔH_f^o	enthalpy of formation, J/mol
H_i	component sensible enthalpy, J/mol
ΔH_R	enthalpy of reaction, J/mol
HHV	higher heating value
i	current density, A/cm^2
I	current, A
j	number of electrons transferred per molecule of hydrogen produced
LHV	lower heating value
$\dot{N}_{H_2}$	molar hydrogen flow rate, mol/s
$\Delta \dot{N}_{H_2}$	molar hydrogen-production rate, mol/s
P	pressure, kPa
q''	heat flux, W/cm^2
Q_H	high-temperature heat addition, J/mol
Q_L	low-temperature heat rejection, J/mol
$\dot{Q}_T$	isothermal heat transfer rate, W
$\dot{Q}$	heat transfer rate, W
R_u	universal gas constant, J/mol K
ΔS_R	entropy of reaction, J/mol K
T	temperature, K
T_o	standard temperature, K
T_L	temperature of heat rejection, K
T_H	temperature of heat addition, K
T_R	reactant temperature, K
T_P	product temperature, K

V	voltage, V
V^o	standard-state open-cell potential, V
V_N	Nernst potential, V
V_{op}	operating voltage, V
V_{tn}	thermal neutral voltage, V
$\dot{W}$	work, rate basis, W
y	mole fraction
η_H	overall thermal-to-hydrogen efficiency
η_{th}	power cycle thermal efficiency
η_e	electrolysis efficiency

Acknowledgments

The authors of this chapter acknowledge the support of their institutions for the preparation of the materials presented here: Argonne National Laboratory, General Atomics, Idaho National Laboratory, Sandia National Laboratories, and Savannah River National Laboratory.

Argonne National Laboratory is a U.S. Department of Energy laboratory managed by UChicago Argonne, LLC under contract number DE-AC02-06CH11357.

This work was supported in part by the U.S. Department of Energy, Office of Nuclear Energy, Next Generation Nuclear Plant program as well as the Idaho National Laboratory, Laboratory Directed Research and Development Program. The Idaho National Laboratory is operated for the U.S. Department of Energy's Office of Nuclear Energy by the Battelle Energy Alliance under contract number DE-AC07-05ID14517.

Sandia National Laboratories is a multi-program laboratory operated by Sandia Corporation, a wholly owned subsidiary of Lockheed Martin company, for the U.S. Department of Energy's National Nuclear Security Administration under contract number DE-AC04-94AL85000.

Savannah River National Laboratory is managed and operated for the U.S. Department of Energy by Savannah River Nuclear Solutions, LLC under contract number DE-AC09-08SR22470.

References

Beghi, G. E. 1986. A decade of research on thermochemical hydrogen at the joint research center, Ispra. *Int. J. Hydrogen Energy* **11**, 761–771.

Besenbruch, G. E. 1982. General atomic sulfur-iodine thermochemical water-splitting process. *Am. Chem. Soc., Div. Pet. Chem., Prepr.* **271**, 48–53.

Brecher, L. E., S. Spewock, and C. J. Warde. 1977. The Westinghouse sulfur cycle for the thermochemical decomposition of water. *Int. J. Hydrogen Energy* **2**, 7–15.

Brecher, L. E. and C. K. Wu. 1975. Electrolytic decomposition of water, U.S. Patent 3,888,750, Westinghouse Electric Corporation, June 10.

Brown, L. C., G. E. Besenbruch, R. D. Lentsch, K. R. Schultz, J. F. Funk, P. S. Pickard, A. C. Marshall, and S. K. Showalter. 2003. *High Efficiency Generation of Hydrogen Fuels Using Nuclear Power.* GA-A24285, Rev. 1, General Atomics, San Diego, CA.

Brown, L. C., G. E. Besenbruch, K. R. Schultz, S. K. Showalter, A. C. Marshall, P. S. Pickard, and J. F. Funk. 2002. High efficiency generation of hydrogen fuels using thermochemical cycles and nuclear power. *AIChE 2002 Spring National Meeting*, New Orleans, LA, March 11–15.

Carty, R. H., M. Mazumder, J. D. Schreider, and J. B. Panborn. 1981. *Thermochemical Hydrogen Production*, Vols. 1–4. Gas Research Institute for the Institute of Gas Technology, GRI-80/0023, Chicago, IL.

Colón-Mercado, H. R., M. C. Elvington, and D. T. Hobbs. May 2008. Component development needs for the hybrid sulfur electrolyzer. Savannah River National Laboratory, Report No. WSRC-STI-2008–00291, Aiken, SC (doi: 10.2172/935436).

Colón-Mercado, H. R. and D. T. Hobbs. 2007. Catalyst evaluation for a sulfur dioxide-depolarized electrolyzer. *Electrochem. Commun.* **9**(11), 2649–2653.

Connolly, S. M., E. Zabolotny, D. F. McLaughlin, and E. J. Lahoda. 2009. Design of a composite sulfuric acid decomposition reactor, concentrator, and preheater for hydrogen generation processes. *Int. J. Hydrogen Energy* **34**(9), 4074–4080.

DOE/EIA. 2000. International Energy Outlook: DOE/EIA-0484 (2000), The Energy Information Administration of the Department of Energy, www.eia.doe.gov

DOE/EIA. 2003. Impacts of the Kyoto protocol on U.S. energy markets and economic activity: SR/OIAF/98-03, The Energy Information Administration (EIA) of the Department of Energy.

DOE/Hydrogen. February 2002. A national vision of America's transition to a hydrogen economy—To 2030 and beyond, National Hydrogen Vision Meeting Document, U.S. Department of Energy.

Dokiya, D. and Y. Kotera. 1976. Hybrid cycle with electrolysis using a Cu-Cl system. *Int. J. Hydrogen Energy* **1**, 117–121.

Ferrandon, M., M. Lewis, F. Alvarez, and E. Shafirovich. 2010a. Hydrolysis of $CuCl_2$ in the Cu-Cl thermochemical cycle for hydrogen production: Experimental studies using a spray reactor with an ultrasonic atomizer. *Int. J. Hydrogen Energy* **35**(5), 1895–1904.

Ferrandon, M., M. Lewis, D. Tatterson, A. Gross, D. Doizi, L. Croizé, V. Dauvois, J. L. Roujou, Y. Zanella, and P. Carles. 2010b. Hydrogen production by the Cu-Cl thermochemical cycle: Investigation of the key step of hydrolyzing $CuCl_2$ to Cu_2OCl_2 and HCl using a spray reactor, *Int. J. Hydrogen Energy* **35**(3), 992–1000.

Ferrandon, M. S., M. A. Lewis, D. F. Tatterson, and A. Zdunek. November 2008. Overview of the development and status of the thermochemical Cu-Cl cycle. *2008 AIChE Annual Meeting*, Philadelphia, PA.

Forsberg, C. W. 2009. Meeting U.S. liquid transport fuel needs with a nuclear hydrogen biomass system. *Int. J. Hydrogen Energy* **34**, 4227–4236.

Forsberg, C. W. and K. L. Peddicord. 2001. Hydrogen production as a major nuclear energy application. *Nucl. News*, 41–45.

Gelbard, F., J. C. Andazola, G. E. Naranjo, C. E. Velasquez, and A. R. Reay. 2005. High pressure H_2SO_4 decomposition experiments for the sulfur-iodine thermochemical cycle, Sandia National Laboratories, SAND2005-5598, Albuquerque, NM.

Gelbard, F., R. C. Moore, M. E. Vernon, E. J. Parma, D. A. Rivera, J. C. Andazola, G. E. Naranjo, C. E. Velasquez, and A. R. Reay. 2006. Pressurized sulfuric acid decomposition experiments for the sulfur-iodine thermochemical cycle, *World Hydrogen Energy Conference*, Lyon, France, June 13–16, 2006.

Ginosar, D. M. 2005. Activity and stability of catalysts for the high temperature decomposition of sulfuric acid. Technical Report, Idaho National Laboratory, Idaho Falls, ID.

Gong, Y., E. Chalkova, N. Akinfiev, V. Balashov, M. Fedkin, and S. N. Lvov. 2009. CuCl-HCl electrolyzer for hydrogen production via Cu-Cl thermochemical cycle. *ECS Trans.* **19**(10), 21–32.

Gorensek, M. B., J. A. Staser, T. G. Stanford, and J. W. Weidner. 2009a. A thermodynamic analysis of the SO_2/H_2SO_4 system in SO_2-depolarized electrolysis. *Int. J. Hydrogen Energy* **34**(15), 6089–6095.

Gorensek, M. B. and W. A. Summers. 2009. Hybrid sulfur flowsheets using PEM electrolysis and a bayonet decomposition reactor. *Int. J. Hydrogen Energy* **34**(9), 4097–4114.

Gorensek, M. B., W. A. Summers, C. O. Bolthrunis, E. J. Lahoda, D. T. Allen, and R. Greyvenstein. 2009b. Hybrid sulfur process reference design and cost analysis. Savannah River National Laboratory, Report No. SRNL-L1200-2008-00002, June 12, 2009, Aiken, SC (doi: 10.2172/956960).

Guan, J., B. Ramamurthi, J. Ruud, J. Hong, P. Riley, and N. Minh. 2006. High performance flexible reversible solid oxide fuel cell. GE Global Research Center Final Report for DOE Cooperative Agreement DE-FC36-04GO-14351.

Hauch, A. 2007. Solid oxide electrolysis cells—Performance and durability. PhD thesis, Technical University of Denmark, Risø National Laboratory, Roskilde, Denmark.

Helie, M., P. Carles, J. Duhamet, D. Ode, J. Leybros, N. Pons et al. 2007. Sulfur-iodine integrated laboratory-scale experiment. I-NERI Project—2006-001-F Annual Report (2007).

Hobbs, D. 2009. Challenges and progress in the development of a sulfur dioxide-depolarized electro-lyzer for efficient hydrogen production, *Pittcon Conference & Expo 2009*, Chicago, IL.

Hobbs, D. T., W. A. Summers, H. R. Colón-Mercado, M. C. Elvington, J. L. Steimke, T. J. Steeper, D. T. Herman, and M. B. Gorensek. 2009. Recent advances in the development of the hybrid sulfur process for hydrogen production, Abstracts of Papers, *238th ACS National Meeting*, Washington, DC, August 16–20, 2009, p. NUCL-160.

Housley, G., K. Condie, J. E. O'Brien, and C. M. Stoots. 2007. Design of an integrated laboratory scale experiment for hydrogen production via high temperature electrolysis. Paper No. 172431, *ANS Embedded Topical: International Topical Meeting on the Safety and Technology of Nuclear Hydrogen Production, Control, and Management*, June 24–28, 2007, Boston, MA.

Housley, G. K., J. E. O'Brien, and G. L. Hawkes. 2008. Design of a compact heat exchanger for heat recuperation from a high temperature electrolysis system. *2008 ASME International Congress and Exposition*, Paper# IMECE2008-68917, Boston, MA, November 2008.

Kolb, G. 2008. Personal communication, Argonne National Laboratory.

Koonin, S. E. 2006. Getting serious about biofuels. *Science* **311**, 435.

LaBar, M. P. June 2002. The gas-turbine-modular helium reactor: A promising option for near-term deployment, *International Congress on Advanced Nuclear Power Plants*, Hollywood, FL.

Lewis, M. A. and J. G. Masin. 2009. The evaluation of alternative thermochemical cycles—Part II: The down-selection process. *Int. J. Hydrogen Energy* **34**, 4125–4135.

Lewis, M. A., M. S. Ferrandon, D. F. Tatterson, and P. Mathias, 2009a. Evaluation of alternative ther-mochemical cycles—Part III further development of the Cu-Cl cycle. *Int. J. Hydrogen Energy* **34**, 4136–4145.

Lewis, M. A., J. G. Masin, and P. A. O'Hare. 2009b. Evaluation of alternative thermochemical cycles. Part I: The methodology. *Int. J. Hydrogen Energy* **34**, 4115–4124.

LLNL. 2002. Lawrence Livermore National Laboratory, U.S. 2002 Carbon Dioxide Emissions from Energy Consumption, http://eed.llnl.gov/flow/(accessed January 24, 2005).

LLNL. 2008. Lawrence Livermore National Laboratory (LLNL), Energy Flow Charts 2008, https://publicaffairs.llnl.gov/news/energy/energy.html (accessed December 6, 2009).

Lu, W. T. P., R. L. Ammon, and G. H. Parker. July 1980. A study on the electrolysis of sulfur dioxide and water for the sulfur cycle hydrogen production process. Westinghouse Electric Corporation, Advanced Energy Systems Division, NASA Contractor Report, NASA-CR-163517.

Marshall, A. C. August 2002. An assessment of reactor types for thermochemical hydrogen produc-tion. Sandia National Laboratories Report.

Mathias, P. M. May 2007. Aspen plus model for solubilities in the $CuCl_2$-CuCl-HCl-H_2O system. Report to ANL.

McQuillan B. W., L. C. Brown, G. E. Besenbruch, R. Tolman, T. Cramer, B. E. Russ, B. A. Vermillion et al. May 2005. High efficiency generation of hydrogen fuels using solar thermo-chemical splitting of water (solar thermo-chemical splitting for H_2) annual report. General Atomics Report GA-A24972.

Meier, P. J. August 2002. Life-cycle assessment of electricity generation systems and applica-tions for climate change policy analysis. PhD thesis, University of Wisconsin, Madison, WI, UWFDM-1181.

Minatsuki, I., H. Fukui, and K. Ishino, 2007. A development of ceramics cylinder type sulfuric acid decomposer for thermo-chemical iodine-sulfur process pilot plant. *J. Power Energy Syst.* **1**(1), 36–48.

Mogensen, M., S. H. Jensen, A. Hauch, L. B. Chorkendorff, and T. Jacobsen. 2008. Reversible solid oxide cells. *Ceramic Engineering and Science Proceedings*, Vol. 28, no. 4, Advances in Solid Oxide Fuel Cells III—A Collection of Papers Presented at the 31st International Conference on Advanced Ceramics and Composites, 2008, pp. 91–101.

Moore, R. C., F. Gelbard, E. Parma, M. Vernon, R. Lenard, and P. Pickard. 2007a. A laboratory-scale H_2SO_4 decomposer apparatus for use in hydrogen production cycles, *American Nuclear Society Annual Meeting*, Boston, MA, June 2007.

Moore, R. C., F. Gelbard, M. Vernon, E. Parma, and P. Pickard. 2007b. H_2SO_4 Section Performance Assessment and the Next Generation Design, September 15, 2007.

Moore, R. C., B. Russ, W. Sweet, H. Helie, and N. Pons, 2008. Materials and operational issues experienced during testing of the sulfuric acid and hydroiodic acid decomposer sections of the sulfur iodine integrated laboratory-scale experiment. Sandia National Laboratories, December 2008.

Moore, R. C., M. E. Vernon, F. Gelbard, E. Parma, and H. Stone. 2006. Preliminary design of the sulfuric acid section of the integrated lab scale experiment. Nuclear Hydrogen Initiative Milestone Report, Sandia National Laboratories, March 15, 2006.

Müller, A. C., A. Weber, D. Herbstritt, and E. Ivers-Tiffée. 2003. Long term stability of yttria and scandia doped zirconia electrolytes. *Proceedings 8th International Symposium on SOFC*, Singhal, S. C. and Dokiya, M., eds. PV 2003–2007, The Electrochemical Society, Pennington, NJ, pp. 196–199.

Ni, M., M. K. H. Leung, and D. Y. C. Leung. 2006. A modeling study on concentration overpotentials of a reversible solid oxide fuel cell. *J. Power Sources* **163**, 460–466.

Nankani, R. V., M. Kumar, L. E. Wedgewood, and L. C. Nitsche. 2007. Personal communication, Argonne National Laboratory, 2007 and 2008.

Naterer, G. F., S. Suppiah, M. Lewis, K. Gabriel, I. Dincer, M. A. Rosen, M. Fowler et al. 2009. Recent Canadian advances in nuclear-based hydrogen production and the thermochemical Cu-Cl cycle. *Int. J. Hydrogen Energy* **34**, 2901–2917.

Nichols, D. 2001. The pebble bed modular reactor. *Nucl. News*, 35–40.

Norman, J. H., G. E. Besenbruch, L. C Brown, D. R. O'Keefe, and C. L. Allen, 1982. Thermochemical water-splitting cycle, bench-scale investigations and process engineering. Final Report for the period February 1977 through December 31, 1981. GA-A16713, General Atomics, San Diego, CA, May 1982.

Norman, J. H., G. E. Besenbruch, and D. R. O'Keefe, 1981. Thermochemical water-splitting cycle for hydrogen production. Gas Research Institute Report GRI-A 16713.

Novikov, G. I., L. E. Voropaev, P. K. Rud'ko, and I. M. Zharskii, 1979. *Zhurnal Neorganicheskoi Khimii* (in Russian), **24**, 811.

O'Brien, J. E. 2008. Thermodynamic considerations for thermal water splitting processes and high-temperature electrolysis. *2008 ASME International Congress and Exposition*, Paper# IMECE2008-68880, Boston, MA, November 2008.

O'Brien, J. E., M. G. McKellar, and J. S. Herring. 2008. Performance predictions for commercial-scale high-temperature electrolysis plants coupled to three advanced reactor types. *2008 International Congress on Advances in Nuclear Power Plants*, Anaheim, CA, June 8–12, 2008.

O'Brien, J. E., M. G. McKellar, C. M. Stoots, J. S. Herring, and G. L. Hawkes. 2009a. Parametric study of large-scale production of syngas via high temperature electrolysis. *Int. J. Hydrogen Energy* **34**, 4216–4226.

O'Brien, J. E., C. M. Stoots, and G. L. Hawkes. 2005a. Comparison of a one-dimensional model of a high-temperature solid-oxide electrolysis stack with CFD and experimental results. *Proceedings, 2005 ASME International Mechanical Engineering Congress and Exposition*, Orlando, FL, November 5–11.

O'Brien, J. E., C. M. Stoots, J. S. Herring, and J. J. Hartvigsen. 2006. Hydrogen production performance of a 10-cell planar solid-oxide electrolysis stack. *J. Fuel Cell Sci. Technol.* **3**, 213–219.

O'Brien, J. E., C. M. Stoots, J. S. Herring, and J. J. Hartvigsen. 2007. Performance of planar high-temperature electrolysis stacks for hydrogen production from nuclear energy. *Nucl. Technol.* **158**, 118–131.

O'Brien, J. E., C. M. Stoots, J. S. Herring, P. A. Lessing, J. J. Hartvigsen, and S. Elangovan. 2005b. Performance measurements of solid-oxide electrolysis cells for hydrogen production. *J. Fuel Cell Sci. Technol.* **2**, 156–163.

O'Brien, J. E., C. M. Stoots, M. G. McKellar, E. A. Harvego, K. G. Condie, G. K. Housley, J. S. Herring, and J. J. Hartvigsen. 2009b. Status of the INL high temperature electrolysis research program—Experimental and modeling. *Fourth Information Exchange Meeting on Nuclear Hydrogen*, NEA, Oakbrook, IL, April 14–16, 2009.

Orhan, M., I. Dincer, G. F. Naterer, and M. A. Rosen. 2010. Coupling of copper–chlorine hybrid thermochemical water splitting cycle with a desalination plant for hydrogen production from nuclear energy. *Int. J. Hydrogen Energy* **35**:1560–1574.

Ozturk, I. T., A. Hammache, and E. Bilgen, 1995. An improved process for H_2SO_4 decomposition step of the sulfur-iodine cycle. *Energ. Convers. Manage.* **36**(1), 11–21.

Perlack, R. D., L. L. Wright, A. F. Turhollow, R. L. Grahan, B. J. Stocks, and D. C. Erbach. 2005. Biomass as feedstock for a bioenergy and bioproducts industry. DOE/GO-102995-2135, ORNL=TM-2005/66, April 2005.

Ponyavin, V., Y. Chen, T. Mohamed, M. Trabia, and M. Wilson. 2006. Modeling and parametric study of a ceramic high temperature heat exchanger and chemical decomposer. *Proceedings of IMECE2006*, Chicago, IL, November 5–10, 2006.

Richards, M., A. Shenoy, K. Schultz, L. Brown, E. Harvego, M. McKellar, J. P. Coupey, S. M. M. Reza, F. Okamoto, and N. Handa. 2006. H2-MHR conceptual designs based on the sulphur-iodine process and high-temperature electrolysis. *Int. J. Nucl. Hydrogen Prod. Appl.* **1**(1), 36–50.

Rosen, M. A., G. F. Naterer, C. C. Chukwu, R. Sadhankar, and S. Suppiah. 2010. Nuclear-based hydrogen production with a thermochemical copper-chlorine cycle and supercritical water reactor: Equipment scale-up and process simulation. *Int. J. Energy Res.*, (doi: 10.1002/er.1702) April 20, 2010 (in press).

Schultz, K. R. and S. L. Bogart. 2006. Hydrogen and synthetic hydrocarbon fuels—A natural synergy. *National Hydrogen Association Annual Meeting*, Long Beach, CA, March 13–16, 2006.

Schultz, K. R., S. L. Bogart, R. P, Noceti, and A. V. Cugini. April 2009. Synthesis of hydrocarbon fuels using renewable and nuclear energy. *Nucl. Technol.* **166**, 56–63.

Serban M., M. A. Lewis, and J. K. Basco. 2004. Kinetic study for the hydrogen and oxygen production reactions in the copper-chlorine thermochemical cycle. *Conference Proceedings, 2004 AIChE Spring National Meeting: 2004*, pp. 2690–2698.

Singh, P., L. R. Pederson, J. W. Stevenson, D. L. King and G. L. McVay. 2008. Understanding degradation processes in solid oxide fuel cell systems. *Presented at the Workshop on Degradation in Solid Oxide Electrolysis Cells and Strategies for its Mitigation*, October 27, 2008, Fuel Cell Seminar & Exposition, Phoenix, AZ.

Singhal, S. C. and K. Kendall. 2003. *Solid Oxide Fuel Cells*, Elsevier Advanced Technology, Oxford, U.K..

Sink, C. J. 2006. An overview of the U.S. Department of Energy's research and development program on hydrogen production using nuclear energy, *Presentation, AIChE Spring National Meeting*, Orlando, FL, April 23–27, 2006. http://www.aiche-ned.org/conferences/aiche2006spring/session_51/AICHE2006spring-51b-Sink.pdf (accessed October 14, 2009).

Sivasubramanian, P., R. P. Ramasamy, F. J. Freire, C. E. Holland, and J. W. Weidner. 2007. Electrochemical hydrogen production from thermochemical cycles using a proton exchange membrane electrolyzer. *Int. J. Hydrogen Energy* **32**(4), 463–468.

Sohal, M. S. 2009. Degradation in solid oxide cells during high temperature electrolysis. INL External Report, INL/EXT-09-15617, June 2009.

Somerville, C. 2006. The billion tonne biofuels vision. *Science* **311**, 435.

Staser, J. A., K. Norman, C. H. Fujimoto, Hickner, M. A., and J. W. Weidner. 2009. Transport properties and performance of polymer electrolyte membranes for the hybrid sulfur electrolyzer. *J. Electrochem. Soc.* **156**(7), B842–B847.

Staser, J. A. and J. W. Weidner. 2009. Effect of water transport on the production of hydrogen and sulfuric acid in a PEM electrolyzer. *J. Electrochem. Soc.* **156**(1), B16–B21.

Steimke, J. and T. Steeper. 2005. Generation of hydrogen using electrolyzers with sulfur dioxide depolarized anodes. *Conference Proceedings, AIChE Annual Meeting*, Cincinnati, OH, October 30–November 4, 2005, pp. 581d/1–581d/19.

Steimke, J. L. and T. J. Steeper. 2006. Generation of hydrogen using electrolyzer with sulfur dioxide depolarized anode. *Conference Proceedings, AIChE Annual Meeting*, San Francisco, CA, November 12–17, 2006, pp. 126c/1–126c/10.

Steimke, J. L., T. Steeper, and D. Herman. 2009a. Results from testing hybrid sulfur single-cell and three-cell stack. *Hybrid Sulfur Electrolyzer Workshop and Information Exchange*, Aiken, SC, April 20–21, 2009. http://srnl.doe.gov/hse_workshop/Steimke%20Electrolyzer%20Performance pdf (accessed October 21, 2009).

Steimke, J. L., T. J. Steeper, and D. T. Herman. 2009b. Prevention of sulfur formation in hybrid sulfur electrolyzer. Paper presented at the *2009 AIChE Annual Meeting*, Nashville, TN, November 8–13, 2009.

Steinberger-Wilkens, R. 2008. Degradation Issues in SOFCs. Presented at the *Workshop on Degradation in Solid Oxide Electrolysis Cells and Strategies for its Mitigation*, October 27, 2008, Fuel Cell Seminar & Exposition, Phoenix, AZ.

Stolberg, L. May 2009. Personal communication.

Stolberg, L, H. Boniface, S. McMahon, S. Suppiah, and S. York. 2009. Development of the electrolysis reactions involved in the Cu-Cl thermochemical cycles. *Proceedings of the International Conference on Hydrogen Production*, Oshawa, CA.

Stoots, C. M. and J. E. O'Brien. 2008. Initial operation of the high-temperature electrolysis integrated laboratory scale experiment at INL. *International Congress on Advances in Nuclear Power Plants*, Anaheim, CA, June 8–12, 2008.

Stoots, C. M. and J. E. O'Brien. 2009. Results of recent high-temperature co-electrolysis studies at the Idaho National Laboratory. *Int. J. Hydrogen Energy* **34**, 4208–4215.

Stoots, C. M., J. E. O'Brien, K. Condie, L. Moore-McAteer, G. K. Housley, J. J. Hartvigsen, and J. S. Herring. April 2009a. The high-temperature electrolysis integrated laboratory experiment. *Nucl. Technol* **116**(1), 32–42.

Stoots, C. M., J. E. O'Brien, J. S. Herring, and J. J. Hartvigsen. 2009b. Syngas production via high-temperature coelectrolysis of steam and carbon dioxide. *J. Fuel Cell Sci. Technol.* **6**, part 1, paper no. 011014, 1–12.

Subramanian, S., V. Ponyavin, C. R. DeLosier, Y. Chen, and A. E. Hechanova. 2005. Design considerations for a compact ceramic offset strip-fin high-temperature heat exchanger. *Proceedings of the ASME Turbo Expo*, Reno, NV, GT2005-68745, pp. 989–996.

Summers, W. A. 2008. Hybrid sulfur thermochemical process development. DOE 2008 Hydrogen Program Annual Merit Review, Arlington, VA, June 9–13, 2008. http://www.hydrogen.energy. gov/pdfs/review08/pd_26_summers.pdf (accessed October 21, 2009).

Summers, W. A. 2009. Hybrid sulfur thermochemical cycle. *2009 DOE Hydrogen Program and Vehicle Technologies Program Annual Merit Review and Peer Evaluation Meeting*, Arlington, VA, May 18–22, 2009. http://www.hydrogen.energy.gov/pdfs/review09/pd_13_summers.pdf (accessed October 21, 2009).

Summers, W. A. and M. B. Gorensek. 2006. Nuclear hydrogen production based on the hybrid sulfur thermochemical process. *Proceedings of the 2006 International Congress on Advances in Nuclear Power Plants, Embedded Topical Meeting*, Reno, NV, June 4–8, 2006, pp. 2254–2256.

Syntroleum. 2006. Syntroleum, Inc. For an exhaustive review of the Fischer–Tropsch process, Syntroleum, Inc. sponsors a web site http://www.fischer-tropsch.org/ (accessed April 2006).

Tamaura, Y., A. Steinfeld, P. Kuhn, and K. Ehrensberger. 1995. Production of solar hydrogen by a novel, 2-step, water-splitting thermochemical cycle. *Energy* **20**, 325–330.

Ueda, R., H. Tagawa, S. Sato, T. Yasuno, S. Ohno, and M. Maeda. 1974. *Production of Hydrogen from Water Using Nuclear Energy, A Review*. Japan Atomic Energy Research Institute, Tokyo, Japan, p. 69.

Virkar, A. V. 2007. A model for solid oxide fuel cell (SOFC) stack degradation. *J. Power Sources* **172**, 713–724.

Weidner, J. W. 2009. Personal communication, E-mail dated June 17, 2009.

Williams, L. O. 1980. *Hydrogen Power*. Pergamon Press, Amsterdam, the Netherlands.

Williams, M. C., J. P. Strakey, W. A. Surdoval, and L. C. Wilson. 2006. Solid oxide fuel cell technology development in the US. *Solid State Ionics* **177**(19–25), 2039–2044.

Wilson, M. A., C. Lewwinsohn, J. Cutts, Y. Chen, and V. Ponyavin. 2007. Design of a ceramic heat exchanger for sulfuric acid decomposition. *Proceedings of the ICONE-15. The 15th International Conference on Nuclear Engineering*, Nagoya, Japan, April 22–26, 2007.

Wilson, M. A., C. Lewinsohn, J. Cutts, E. N. Wright, and V. Ponyavin. 2006. Optimization of microchannel features in a ceramic heat exchanger for sulfuric acid decomposition. *2006 AIChE Annual Meeting*, San Francisco, CA, 2006.

Yannopoulos, L. N and J. F Pierre, 1984. Hydrogen production process: High-temperature scalable catalysts for the conversion of SO_3 to SO_2, *Int. J. Hydrogen Energy* **9**(5), 383–390.

Yildiz, B. and M. S. Kazimi. 2006. Efficiency of hydrogen production systems using alternative nuclear energy technologies. *Int. J. Hydrogen Energy* **31**, 77–92.

Yoshida, K., H. Kameyama, T. Aochi, M. Nobue, M. Aihara, R. Amir, H. Kondo et al. 1990. A simulation study of the UT-3 thermochemical hydrogen production process. *Int. J. Hydrogen Energy* **15**, 171–178.

4

Solar Production of Fuels from Water and CO₂

Gabriele Centi, Siglinda Perathoner, Rosalba Passalacqua, and Claudio Ampelli

CONTENTS

4.1 Introduction

The need to accelerate the introduction of renewables in the energy pool has been fostered by the recent oil price crisis that created intense social pressure to accelerate the transition from fossil to renewable fuels. At the same time, concerns regarding greenhouse gases have been progressively turned from possibility to reality, and the fraction of world population accessing a massive use of energy is exponentially increasing. Therefore, it is no more an eco-political issue, but an increasing social demand for the need of some fundamental changes in the present energy supply system (Balzani et al., 2008; Dovì et al., 2009).

The actual fraction of renewable energy sources is still often limited to non-ecoefficient options (such as wood combustion) or to routes that further potential increase is close to limit (hydroelectricity). On the other hand, the need to focus on solar energy derives from a simple consideration. The actual average global energy consumption is about 16 TW and it is estimated to increase to about 25 TW by 2050 (IEA, 2009). The amount of sunlight striking the Earth's atmosphere continuously is 1.75×10^5 TW. Considering a 60% transmittance through the atmospheric cloud cover, 1.05×10^5 TW reaches the Earth's surface continuously. If the irradiance on only 1% of the Earth's surface could be converted into electric

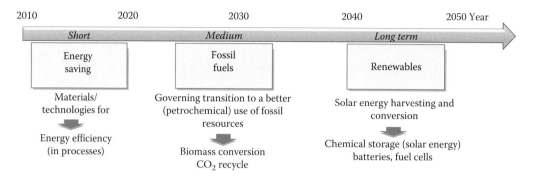

FIGURE 4.1
Simplified scenario for sustainable energy in relation to use of renewable and solar energy. (Adapted from Centi, G. and van Santen, R.A: *Catalysis for Renewables*. 2007. Copyright Wiley-VCH Verlag GmbH & Co. KGaA. Reproduced with permission.)

energy with 10% efficiency, it would provide a resource base of 105 TW, for example, 10 times the estimated world energy increase by 2050. By contrast, the estimated amount of energy extractable by wind is about 2–4 TW, by tides about 2–3 TW, by biomass 5–7 TW, and by geothermal energy 3–6 TW (Lewis et al., 2005). The increase in the use of solar irradiation is thus an unquestionable necessity.

4.1.1 Solar Energy and Sustainable Energy Scenarios

The complexity of the energy problem requires a deeper analysis. The design and development of future fuels requires defining the sustainable energy scenarios and the effective scientific strategies necessary to address the interconnected energy and environment issues (Nakćenović, 2000). A detailed discussion of these aspects is not within the scope of this chapter, but we may briefly recall the need on a short-term perspective of new technologies for energy saving and efficiency, and for use of biomass (Figure 4.1). On a medium term, a better use of renewable resources is needed, including solving the issue of energy storage and transport, and find a sustainable solution to CO_2 emissions, because the full transition to nonfossil fuels will require a longer time. Finally, in the long-term, the renewable energy scenario, based in particular on solar energy, will become predominant (Centi and van Santen, 2007).

Therefore, alternative feedstocks for synthetic fuels, from stranded and unconventional fossil fuel reserves to green biomass uses represent the short-to-medium term solution, which integrates with energy saving and an increasing use of renewable energy (wind, geothermal, and solar energy, particularly using photovoltaic [PV] cells). However, in a medium-to-long term perspective, it is necessary to enhance the role of solar energy in producing fuels directly and the recycling of CO_2.

4.2 Solar Fuels

We define here as solar fuels the direct production of H_2 from water, and the conversion of CO_2 to fuels using solar energy. It should be mentioned that the production of solar fuels has been a long-term dream from the beginning of photochemistry studies.

G. Ciamician, a pioneer of modern photochemistry, in a famous address presented before the VIII International Congress of Applied Chemistry held in New York in 1912, indicated the need to replace "fossil solar energy" (i.e., coal) with the energy that the Earth receives from the Sun every day (Venturi et al., 2005).

After a century, solar fuels are still a grand challenge (Lewis et al., 2005), but significant progress has been made recently and will be soon further made, due to the large recent R&D investments in this field. For example, ARPA-E (U.S. Advanced Research Projects Agency—Energy, which aims to provide access to the funding needed to bring the next generation of energy technologies to fruition) gave out $23.7 million in grants in October 2009 to start-ups and universities experimenting in the relatively new field of direct solar fuels. Similarly, a significant push on R&D activities in this field is present worldwide.

Solar clean fuels and energy carriers of the future often consider only H_2 and electrical energy, although the latter may be not considered a proper fuel. In both cases, storage is difficult and costly. The energy density of batteries is still too low for a cost-efficient use out of the grid, and H_2, being a gas even at very low temperatures and H_2-storage materials being still far from meeting the performances required for application (see also later discussion regarding Figure 4.3), also exhibits considerable limitations in a wider use as an energy vector. In addition, a distribution system exists already for electrical energy (even if energy loss—around 30%—is still a main issue, and in energy-intensive mobile applications, such as cars, the storage is the limiting factor) (Chen et al., 2009). A new energy infrastructure requires to be created for H_2 for use as fuel. Integration into the existing energy infrastructure is a critical aspect for economics and thus to accelerate the shifts to renewable energy. We therefore consider solar H_2 as an intermediate chemical to produce other more suitable energy vectors.

The need of solar-to-chemical conversion (solar fuels) for integration (not as an alternative) to solar-to-electrical energy (in PV cells) needs to be further clarified. There is a peak in energy consumption during the day, when PV cells have the maximum power efficiency. PV cells could thus efficiently integrate the extra energy required during these peaks in consumption, when also the cost in production of energy is higher. Nevertheless, a closer inspection evidences a mismatch between power consumption and solar irradiation during the year and in terms of geographical distribution. The current methods to store electrical energy are not very efficient, and in general a still critical issue is the storage of electrical energy, notwithstanding the R&D progresses in nanomaterials for these applications (Centi and Perathoner, 2009a). Therefore, the conversion of solar-to-chemical energy is a necessary integration to the current solar options for storing and transporting solar energy.

PV cells are close to grid parity (Bhandari and Stadler, 2009; Curtright et al., 2008), while economic parity for solar fuels is still far to be reached. PV cells may be a first step to produce solar fuels, for example, by water electrolysis. Tributsch (2008) recently reviewed the outlook and economics of PV hydrogen generation. In the past, this renewable means of hydrogen production has suffered from low efficiency (2%–6%), which increased the area of the PV array required, and, therefore, the cost of generating hydrogen. Today, the efficiency of the PV-electrolysis system has been optimized by matching the voltage and maximum power output of the photovoltaics to the operating voltage of proton exchange membrane (PEM) electrolyzers (Gibson and Kelly, 2008). The optimization process increased the hydrogen generation efficiency to 12% for a solar-powered PV-PEM electrolyzer that could supply enough hydrogen to operate a fuel cell vehicle.

Tributsch (2008) indicates that PV hydrogen generation today is about 10 times costlier with respect to PV electricity, although analyzing the development of progress in this field he concluded that PV hydrogen generation is expected to have a bright future. When

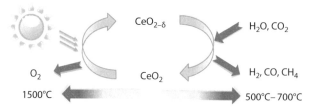

FIGURE 4.2
Scheme of the production of solar fuels by thermochemical reaction using doped cerium-oxide. (Adapted from Chueh, W.C. and Haile, S.M., *ChemSusChem*, 2, 735, 2009.)

more effective integrated solar H_2 cells are developed, for example, based on direct photo-electrolysis (often indicated as photocatalytic water splitting) and not based on a two-step approach, the costs will be further reduced making solar H_2 economically competitive, particularly for decentralized production (Currao, 2007; Nowotny and Sheppard, 2007).

Thermochemical production of H_2 using solar energy and suitable chemical cycles to lower the temperature with respect to pure thermal water decomposition is also an alternative to low-temperature photoelectrolysis using semiconductor materials. Abanades and Flamant (2006) reported, for example, a new thermochemical cycle for H_2 production based on CeO_2/Ce_2O_3 oxides. The thermal reduction of Ce^{IV} to Ce^{III} (endothermic step) is performed in a solar reactor (inert atmosphere) at about 2000°C and a pressure of 100–200 mbar. The hydrogen generation step (water splitting with Ce^{III} oxide) was instead made in the temperature range of 400°C–600°C. These results evidence some of the limits in solar thermal H_2 production: (i) need of very high temperatures and (ii) quite different temperature regions for the two redox steps. This reflects in high costs, low productivities, and materials stability problems. Chueh and Haile (2009) also reported CeO_2 doped with 15% samarium as an efficient system to produce syngas (from water and CO_2) in the absence of a metal catalyst, and methane in the presence of Ni as the catalyst (Figure 4.2). The oxide is pretreated at 1500°C (24 h) in an oxygen-free atmosphere and then interacted with a flow of water and carbon dioxide as a carrier gas to generate hydrogen, methane, and CO.

More complex solutions are also possible. Jin et al. (2009) recently proposed a new approach based on medium-temperature solar thermal energy (150°C–300°C) that can be converted into a high-grade solar fuel by integrating this technique with the endothermic reaction of hydrocarbons. In particular, they proposed a solar/methanol fuel hybrid thermal power plant and a solar-hybrid combined cycle with inherent CO_2 separation using chemical-looping combustion, to generate electricity or hydrogen. They tested the concept in a 5 kW solar receiver/reactor prototype. H_2 was produced through methanol steam reforming. The authors indicated this as a new direction to an efficient utilization of low-grade solar thermal energy, and for the production of cost-effective solar fuels. Even if further data are necessary to understand the viability of the approach, particularly in terms of energy balance and sustainability, a general comment is that the economics in solar fuels is a critical issue that requires having simple and robust solutions without too many and complex steps.

4.2.1 Energy Vectors

There is a considerable interest in H_2 as a future energy vector. Its renewable production using solar energy is thus a target in sustainable energy scenarios and renewable H_2 production would reduce the use of fossil fuels actually necessary for its production.

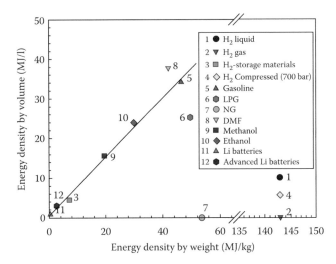

FIGURE 4.3
Energy density per weight versus per volume in a series of liquid and gaseous fuels (from fossil sources, or renewables such as ethanol and DMF), H$_2$ (liquid, gas, compressed at 700 bar, and stored in advanced nanomaterials), and electrical energy (Li batteries, conventional and advanced). NG—natural gas; DMF—dimethylfuran; LPG—liquefied petroleum gas.

On the other hand, the energy density (per unit volume or weight) of gasoline, for example, is by far larger than that possible for H$_2$ and for electrical energy, even considering future possible developments in storage materials. Figure 4.3 reports the energy density per weight or volume in a series of liquid fuels (from fossil or renewable sources), H$_2$ (gas, liquid, compressed or in storage materials), and electrical energy (in conventional or new generation Li batteries). Both high energy densities per weight and per volume are required for practical applications. This graph clearly evidences that electrical energy transport out of the grid or electrical energy storage will remain a main issue in the future as well.

The main drawback of using H$_2$ as a clean energy vector is in transport/storage when compared with liquid fuels. In addition, realistic energy scenarios should consider, besides sustainability, the continuity in usage of the existing technologies (and infrastructure) wherever possible. However, a renewable source of H$_2$ could be used to convert CO$_2$ back to liquid fuels, which can be then easily stored/transported and will integrate into the existing energy infrastructure (Centi and Perathoner, 2004, 2009b). The liquid fuels that can be obtained from CO$_2$ and water solar conversion are thus a valuable carbon-neutral energy vector, because CO$_2$ is recycled. As discussed more in detail later, the issue is that some of the products of CO$_2$ photochemical reduction, such as methane, alcohols, or acids (methanol, formic acid, and oxalic acid) when present in low concentration in solution do not have the characteristics discussed above to be suitable energy vectors. Methane is a gas and has most of the same problems discussed for hydrogen. Methanol or formic acid could be valuable, but their recovery from diluted aqueous solutions may be energetically costly.

Various recent reviews have discussed the role of H$_2$ for a sustainable energy future, but much less have considered alternative energy vectors. Muradov and Veziroğlu (2008) discussed a scenario for the transition from current fossil-based to hydrogen economy that includes two key elements: (i) Changing the fossil decarbonization strategy from one based on CO_2 sequestration to one that involves sequestration and/or utilization of solid carbon and (ii) producing carbon-neutral synthetic fuels from bio-carbon and hydrogen generated from water using carbon-free sources (nuclear, solar, wind, and geothermal). They suggested that this strategy would allow taking advantage of the existing fuel infrastructure without an adverse environmental impact, and it would secure a smooth carbon-neutral transition from fossil-based to future hydrogen economy.

Bockris (2008) also analyzed different alternatives, and evidenced the interest in converting CO_2 with low-cost (renewable) hydrogen to produce methanol, as an energy vector that avoids a radical change of the distribution method. Olah et al. (2006), while analyzing the interrelationship between fuels and energy in their famous book have also evidenced the continuing need for hydrocarbons. Hydrogen economy presents significant shortcomings and thus they emphasized the need of shifting toward a methanol economy. Solar hydrogen (Nowotny and Sheppard, 2007) may be a better and more sustainable alternative when combined with the possibility of forming liquid fuels easily transportable and with high energy densities. In addition, hydrogen is unsafe particularly in transporting energy over long distances. The transformation of hydrogen into a safe transportable chemical is thus desirable for bulk energy transport.

There are some possible alternatives. Ammonia is a transportable form of chemically-bound hydrogen that does not contribute to the carbon footprint of energy supply (Christensen et al., 2005). It can be manufactured with well-established processes (although still not via solar energy routes) and although a transport infrastructure exists to supply the fertilizer demands in the world (ammonia is easily liquefied under moderate pressure), it is not comparable in terms of the dimensions for use as an energy vector. However, there are clear concerns regarding toxicity and the smell, as well as the potent greenhouse effect of ammonia causing severe constraints for leakage. In addition, risks related to boiling liquid expanding vapor explosion (BLEVE) and unconfined vapor cloud explosion (UVCE) are significant. Even if there is a considerable push in promoting ammonia as and energy vector, the drawbacks are relevant, although in part common to other energy systems also. In addition, ammonia as an energy vector would imply changing the current energy infrastructure with large investments.

CO_2 conversion to liquid fuels (methanol or other liquid chemicals), either indirectly using solar H_2, or directly by photocatalytic CO_2 reduction and simultaneous H_2O oxidation using solar energy, appears thus a preferable and more sustainable route. It forms safer chemicals with high energy densities, showing lower risks in storage, and which may be well integrated with the existing energy infrastructure with minimal investments. As discussed later, it will also allow delocalized energy production, with the reduction of some of the problems associated with centralized energy production and the risks associated with energy transport.

Producing solar fuels via recycling CO_2 is thus a carbon-neutral approach to store and transport solar energy, which can be well integrated in the current energy infrastructure.

4.3 Bio-Routes to Solar Fuels

In a broad sense, the area of biofuels may be also considered a bio-route to solar fuels. Solar energy and CO_2 are captured by plants or algae and converted to complex molecules (cellulose, hemicellulose, lignine, starch, lipides, oils, etc.), which are then converted by different possible routes to finally produce biofuels (Centi and van Santen, 2007). However, in a more strict and correct sense, bio-solar fuels refer to only those that can be obtained directly through photosynthesis processes using bio-organisms. Plants and some bacteria convert incident solar radiation into stored chemical energy. Two protein assemblies participate in the process. Photosystem II uses light to split water into oxygen, protons, and electrons (Lubitz et al., 2008). The latter two are used by photosystem I to convert CO$_2$ to carbohydrates. The efficiency is low, typically lower than 1%. Some algae are more efficient in the process, up to about 10%, and from this derives the interest in the production of biofuels from (micro)algae. However, due to the complexity of the biomass, the issue and cost (included in terms of energy) is in the harvesting of the biomass and its transformation to biofuel. This is the main issue in all the actual effort for a bio-based economy.

In order to make feasible bio-routes to solar energy, it is thus necessary to simplify this multistep process, enabling the possibility of direct synthesis of fuels. One approach is to design organisms that can produce fuels like H$_2$, ethanol, or methane directly using sunlight. The light-energy conversion in engineered microorganisms was reviewed recently by Johnson and Schmidt-Dannert (2008). Advances in molecular biology and metabolic engineering have significantly increased the possibilities in this field recently, particularly with the introduction of light-driven proton pumping or anoxygenic photosynthesis into *Escherichia coli* to increase the efficiency of metabolically engineered biosynthetic pathways. Donohue and Cogdell (2006) discussed also how microbiology can contribute toward the provision of clean solutions to the world's energy needs.

In general, in oxygenic photosynthetic microorganisms such as algae and cyanobacteria, water is split into electrons and protons, which during the primary photosynthetic process are redirected by photosynthetic electron transport chain, and ferredoxin, to the hydrogen-producing enzymes hydrogenase or nitrogenase (Allakhverdiev et al., 2009). By these enzymes, e$^-$ and H$^+$ recombine and form gaseous hydrogen. The biohydrogen activity of hydrogenase can be very high but it is extremely sensitive to photosynthetic O$_2$. In contrast, nitrogenase is insensitive to O$_2$, but has lower activity. At present, the efficiency of biohydrogen production is low. However, theoretical expectations suggest that the rates of photon conversion efficiency for H$_2$ bioproduction can be high enough (>10%) (Allakhverdiev et al., 2009). Cyanobacteria (formerly known as blue-green algae) appear rather interesting solar energy converters of hydrogen (H$_2$) (Bothe et al., 2008; Sakurai and Masukawa, 2007) and increasing R&D attention focuses on this topic. There are, however, still many challenges to solve before the use of photobiological H$_2$ production: oxygen sensitivity of the enzyme, preferential energy consumption by other metabolic pathways, and heat tolerance.

The mechanism of action is quite complex and is schematically summarized in Figure 4.4 (Wall et al., 2008). In the chloroplastic membranes of higher plants, the complex Ndh, an analogue of mitochondrial respiratory complex I, catalyzes the entry of electrons in a cyclic transfer pathway around photosystem I. This pathway has proved to be most important in conditions of environmental stress. In cyanobacteria, the ancestors of the photosynthesis organelles (chloroplasts), the operation of this enzyme complex is closely correlated with hydrogen bioproduction capacity. In fact, photosynthesis consists of two processes: light energy conversion to biochemical energy by a photochemical reaction, and CO$_2$ reduction to organic compounds such as sugar phosphates, through the use of this biochemical energy by Calvin-cycle enzymes. Under certain conditions, however, a few groups of microalgae and cyanobacteria consume biochemical energy to produce molecular hydrogen instead of reducing CO$_2$ (Figure 4.4).

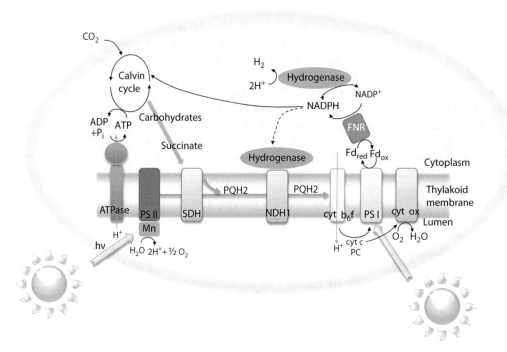

FIGURE 4.4
Scheme of the mechanism of H_2 production in cyanobacteria.

It may also be possible to use solar light to increase the efficiency in electrical energy production in microbial fuel cells (MFCs). Cho et al. (2008) showed the impact of solar-powered electricity generation by *Rhodobacter sphaeroides* in a single-chamber MFCs (Figure 4.5). With light, the maximum power density was $790\,\text{mW}\,\text{m}^{-2}$ when compared to very low values in dark (< than $0.5\,\text{mW}\,\text{m}^{-2}$). MFCs are devices that exploit microbial catabolic activities

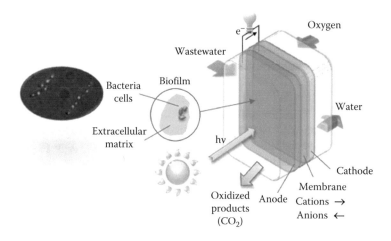

FIGURE 4.5
Scheme of solar-driven microbial (*Rhodobacter sphaeroides*) fuel cells for enhanced electrical energy generation using wastewater from agrofood or agroenergy production.

to generate electricity from a variety of materials, including complex organic waste and renewable biomass (Watanabe, 2008). MFCs with respect to chemical fuel cells may utilize wastewater or diluted solutions coming from biomass processing (agrofood or agroenergy), although sensibility to poisoning and energy density output are a significant limitation. Nevertheless, MFCs will probably find use in the future in the production of sustainable bioenergy by organic waste treatment coupled with electricity generation.

The other possible approach is that of artificial leaves (Birch, 2009). In these leaves, light is used to drive the splitting of water into its component parts to make a fuel. Artificial water-splitting devices, based, for example, on the combination of PV and photoelectro-chemical technology are known from more than a decade to be more efficient in producing hydrogen than photosynthesis. Khaselev and Turner (1998) reported an efficiency of about four times higher using gallium arsenide and gallium indium phosphide as semiconduct-ing materials. H_2 forms on the gallium indium phosphide, while a platinum electrode is used as second counter electrode. The problem was that those semiconductors degraded far too quickly in solution (less than one day). In addition, these materials are costly and risky to manage. Progresses have been made recently in this PEC approach (Currao, 2007), in the oxygen-evolving catalysts (Kanan and Nocera, 2008) as well as in the use of cheaper and more robust semiconductors such as titania (Nowotny, 2008), but still the problem of stability and cost-effectiveness remains.

There is thus an increasing interest in exploring the alternative bio-based approach for producing solar fuel, even taking into account the limits in scaling-up, complexity, cost, and recovery of the fuels. There are two possible approaches: (i) entrapping component of plants in an inorganic matrix to intensify the process and make the system more robust or (ii) fully biomimetic or bio-inspired systems.

To intensify the photosynthesis process, it is possible to immobilize photosynthetically active matter (pigments, proteins, thylakoids, chloroplasts, and whole plant cells) within biologically inert, but biocompatible, matrices and use these hybrid materials to convert solar energy to chemical energy (Rooke et al., 2008).

Su and coworkers (Meunier et al., 2009a,b) reported the performances of thylakoids (the photosynthetic key part of plant cells) encapsulated into a three-dimensional silica net-work in the photocatalytic splitting of H_2O. Compared to the free thylakoid suspension, the bioactivity of entrapped thylakoids can be significantly extended. Since these photo-synthetic structures are very sensitive to ionic strength and osmotic pressure as well as to traces of alcohols, a biocompatible synthesis pathway has been designed to allow the formation of a robust hybrid silica gel without releasing any by-product during its con-struction. In these conditions, the photochemical activity of thylakoids can be preserved for more than 40 days. This method has also been applied to encapsulate more voluminous organelles such as chloroplasts and thus in principle this "artificial leaf" approach could be extended to the immobilization of whole cells (vegetable cells, cyanobacteria, or algae). However, the extension is not so obvious, due to the high sensitivity of these "living" systems, the need to provide nutrients, and to recover the products. In fact, thylakoids pro-duce oxygen from water, while for use in solar fuel production it is necessary to produce H_2 (for example, using cyanobacteria). Carbohydrates are produced from water and CO_2, but their recovery would destroy the system. The results are thus an interesting advance, but to produce solar fuels still a large effort is necessary and the real applicability is still questionable in terms of potential cost-effectiveness.

In the bio-mimicking or -inspired approach, a first issue regards a better understanding of natural photosynthesis. Despite advances in characterizing the active site and protein ligand structure of the enzymes involved in the photosystem, a true understanding of the overall structure functionalities is still missing. This reflects in limitations in the possible design of new systems that work on the same principles as in nature.

A possible approach for biomimetic structures is based on dendrimers, which are branching polymers. Aida and coworkers (Choi et al., 2003; Li and Aida, 2006) proposed the construction of artificial photosynthetic systems based on multiporphyrin dendrimers. Inspired by the wheel-like structure of LH1 and LH2 in photosynthetic purple bacteria, they prepared a series of multiporphyrin dendrimers with a structural feature where all the zinc porphyrin units were placed in the same layer of the dendritic architecture. Upon complexation of fullerene-appended dipyridine ligands, a layer of fullerene was engineered surrounding the zinc porphyrin spherical layer. Photoinduced electron transfer occurred upon excitation of zinc porphyrin moieties and resulted in charge separation among zinc porphyrins and fullerenes. The number of zinc porphyrins and of fullerenes has a great impact on the photoinduced electron transfer properties. The multiporphyrin array is an artificial light-harvesting antenna that mimics the wheel-like antenna complexes of photosynthetic purple bacteria.

Dendrimers based on Ru(II) and Os(II) polypyridine complexes as building blocks and 2,3-bis(2′-pyridyl)pyrazine as bridging ligands were also developed by Serroni and coworkers (2003) as light-harvesting antenna materials. The dendrimers exhibit a huge absorption in the visible region and energy migration patterns whose direction and efficiency depend on the synthetically determined topography of the systems.

The design of these dendrimer-based artificial photosystems harvest the light at the end of the branches and transfer the energy to a central processor for splitting water or carbon dioxide. There are two different classes of dendrimers: compact dendrimers with constant distance between neighboring branching points throughout the macromolecule and extended dendrimers where this distance increases from the system periphery to the center. Supritz et al. (2006) discussed whether energy transport (via Frenkel excitons) occurs in a coherent or incoherent manner. A main existing problem is how efficient is the transfer of the energy in these fast photochemical transformations. The energy transfer within Zn(II)—porphyrin dendrimers—and how it depends on the dendrimer structure was discussed by Larsen et al. (2007).

Several advances have been made on the understanding of dendrimers-based materials as efficient artificial light harvesting systems. However, the main issue is the complexity, the cost of these systems, and weak robustness (lack of long-term photostability, although similar to many other polymers). The possible use of these artificial photosynthesis systems is thus still a distant target. Many other nanostructure systems are possible, based on the large variety of hybrid inorganic-organic systems developed in recent years (Cong and Yu, 2009; Kamat, 2007; Löbmann, 2007). However, their suitability to develop cost-effective artificial photosystems for solar fuel production has to be explored.

There are thus interesting achievements in this field, but in general, the practical applicability of these systems in larger-scale application for secure and sustainable solar-to-fuel technology requires a further substantial improvement of productivity and economics, and robustness of the operations.

4.3.1 Concentrating Solar Biomass Gasification

More than 2.5 billion years of photosynthesis have been required to accumulate the fossil energy stocks that we are using now and there are increasing concerns pointing out that the net productivity of world's vegetation is not high enough to substitute the fossil energy stocks and therefore more efficient innovations to harness solar energy by photosynthetic mechanisms are required (Tikkanen et al., 2009). As discussed in Section 4.1.1, biofuels must be considered more as a transition solution in integration to fossil fuels more than as the solution for renewable energy. Nevertheless, it is clear from life-cycle assessments (Gallezot, 2008; Udo de Haes and Heijungs, 2007) that growing, harvesting, and transport of biomass have a large impact on carbon neutrality of biofuels.

Also, the process of biomass transformation requires significant energy, particularly in the thermochemical conversion of biomass to syngas (gasification). The idea of Hertwich and Zhang (2009) was thus to produce synfuel from biomass using concentrated solar energy as its main energy source. They also used H_2 from water electrolysis with solar power for reverse water gas shift to avoid producing CO_2 during the process. In a process simulation, they compared the solar biofuel concept with two other advanced synfuel concepts: second generation biofuel and coal-to-liquid, both using gasification technologies and capture and storage of CO_2 generated in the fuel production. The solar-driven biofuel requires only 33% of the biomass input and 38% of total land as the second generation biofuel, while still exhibiting a CO_2-neutral fuel cycle. There is a trade-off between reduced biomass feed costs and the increased capital requirements for the solar-driven process; it is attractive at intermediate biomass and CO_2 prices.

4.4 Low- versus High-Temperature Approach in Producing Solar Fuels

There are two main options for using solar energy to produce fuels, the high- or low-temperature approaches. The thermochemical approach uses the high temperatures reached in solar furnaces to produce H_2 (and O_2) from water or CO (and O_2) from CO_2 (Kodama, 2003; Kodama and Gokon, 2007; Licht, 2005; N'Tsoukpoe et al., 2009). Thermochemical cycles are necessary to lower the temperatures necessary. An example is the use of a metal oxide that spontaneously reduces at high temperatures and is then reoxidized by interaction with H_2O to form H_2 or alternatively with CO_2 to produce CO. Nevertheless, temperatures above 1200°C–1400°C are necessary. This creates a number of issues in terms of materials and stability, cost-effectiveness, and productivity. The syngas (CO/H_2) should be then catalytically upgraded to fuels (methanol, FT hydrocarbons). The approach is essentially suitable for solar plants, while may be difficult to adapt in a delocalized production of solar fuels. In addition, scale-up problems are relevant.

4.4.1 High-Temperature Thermochemical Approaches

There are solar peculiarities in comparison to conventional thermochemical processes: high thermal flux density and frequent thermal transitions due to the fluctuating insolation. Therefore, conventional industrial thermochemical processes are generally not suitable for solar-driven processes. Thus, adaptation of the process to solar high-temperature heat is required. The solar high-temperature heat supply changes during the day and is zero during the night. It is possible to store solar high-temperature heat in a thermal storage system and use the stored thermal energy continuously for the process. However, thermal storage at high temperatures is very difficult to achieve on an industrial scale.

The thermal inertia problems and start-up difficulties in an intermittently operated system will be more severe if the chemical plant is larger. It is also difficult for a complicated process with a number of steps (e.g., reaction, separation, and concentration) to be effectively operated with quick response to an intermittent heat supply from concentrated solar radiation.

This is a challenge to solar reactor (heliostats) concepts. Direct irradiation of the redox working material (metal oxides) enables the reactor to respond quickly to the intermittent heat supplied by concentrated solar radiation. Several types of windowed solar reactors are currently being developed—monolithic, foam, rotary-type, CR5 (Counter Rotating Ring Receiver Reactor Recuperator, developed at U.S. Sandia National Laboratories), and internally circulating fluidized bed reactors. In the first two reactors, both the thermal reduction and reoxidation by water steps are carried out in a single reactor, although in a cyclic fashion and at different temperatures, while the redox material is moved in different zones (at different temperatures) in the last two reactors.

Fe_3O_4/FeO and ZnO/Zn are two of the most recognized thermochemical water-splitting cycles:

$$Fe_3O_4 \rightarrow 3FeO + \frac{1}{2}O_2 \tag{4.1a}$$

$$H_2O + 3FeO \rightarrow Fe_3O_4 + H_2 \tag{4.1b}$$

$$ZnO \rightarrow Zn(g) + \frac{1}{2}O_2 \tag{4.2a}$$

$$Zn + H_2O \rightarrow ZnO + H_2 \tag{4.2b}$$

The first solar high-temperature step of the thermal reduction of magnetite (Fe_3O_4) to wustite (FeO) proceeds at temperatures above 1370°C under 1 bar. The second step of reoxidation by water proceeds at temperatures below 540°C. The thermal decomposition of ZnO proceeds endothermically and the temperature for which the $\Delta G°$ equals zero is 1220°C. The reoxidation by water of the metallic zinc thermodynamically proceeds at temperatures below 760°C under 1 bar (Kodama and Gokon, 2007). Therefore, the material has to be continuously cycled between these temperatures, with consequent large stresses and easy sintering.

Heat losses are increased at higher temperatures, and heating the steam feed to a higher temperature requires a higher-temperature heat supply. This results in a low degree of utilization of the heliostat, which can represent over 50% of the capital cost in a solar thermal hydrogen plant. In addition, higher temperatures require higher concentration of the solar flux, with thus larger heliostats and costs. Therefore, lower temperature operation is an important factor.

Scaling up the system creates further engineering difficulties, for example, in making a cluster of the reactors at the top of a solar tower. The scale of one reactor must be limited by the size limitations imposed by its quartz window and the need to have uniformly heated reaction zones. Therefore, a cluster of the reactors will be required for scaled applications. This indicated that the usual scale of economy in processes is not effective in this case, because for larger productions many parallel solar reactors are needed. In addition, an aspect often underestimated is the sophisticated solar tracking necessary during the day to maintain uniform temperature in the solar reactor.

Multistep water-splitting cycles capable of working below about 650°C are possible: thermochemical cycles with more than three steps and thermochemical–electrochemical hybrid cycles with more than two steps (Kodama and Gokon, 2007). However, the promising cycles all suffer from the use of corrosive reactants and difficulties in the separation steps.

Various large projects have explored up to pilot plant experimentation the use of thermo-chemical high-temperature approach to produce solar fuels. The EU project HYDROSOL-II (SES6-CT-2005-020030) had the objective of establishing the basis for mass production of solar hydrogen (Agrafiotis et al., 2005; Roeb et al., 2009). A 100 kW (thermal) pilot plant for solar hydrogen production via a two-step thermochemical water-splitting process was built. The solar thermal reactor uses ceramic monoliths coated with Mn-ferrite as the redox component. The process works by cycling stationary ferrite-coated monoliths through sequential oxidation and reduction steps at 800°C and 1200°C, respectively.

Figure 4.6 illustrates the concept (Roeb et al., 2009). By using two such monoliths in parallel, it is possible to quasi-continuously produce hydrogen. A prototype solar reactor and peripherals suitable to proof the process concept were developed and have been successfully tested in DLR's solar furnace in Cologne (Germany). Repeated cycling operations were possible, but in most cases a decay of the amount of the evolved hydrogen was observed from cycle to cycle, due to inhomogeneous heating of the monolithic absorber in the very first cycles and due to disappearance of the porosity and the associated loss of surface area in later cycles (Roeb et al., 2009).

A variation of the concept is to use these solar concentrators and reactors to drive the CO$_2$ reforming with methane (solar dry reforming of methane) to produce syngas (mixtures of CO/H$_2$), which can be then converted to methanol and Fischer–Tropsch products.

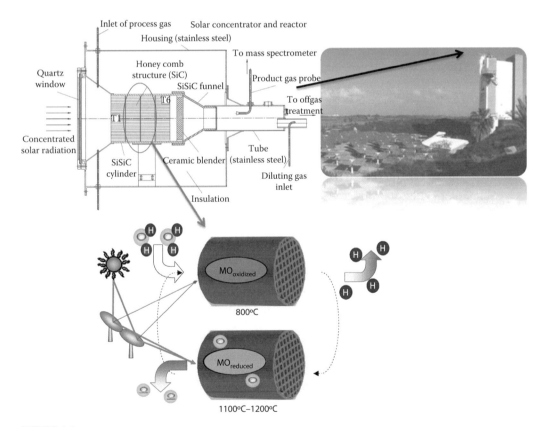

FIGURE 4.6
Scheme of the solar concentrator and reactor, and of the monolithic ceramic honeycombs coated with active redox pair materials used in the HYDROSOL-II project. (Adapted from Roeb, M. et al., *Int. J. Hydrogen Energy*, 34, 4537, 2009.)

The advantage is the possibility of continuous operations, instead of cyclic, while the disadvantage is the need of a methane feed. The solar illumination in this case provides the heat necessary for the endothermic dry reforming of methane with CO_2. Also, in this case, the main issues relate to the difficulty of having a homogeneous temperature in the monolith and problems of materials stability and of carbon formation. Scale-up of the solar system to larger productions of solar fuels remains an issue.

Possible alternatives are also the solar steam reforming of methane, for example, feeding water instead of CO_2, or the solar decomposition of methane to carbon (rejected) and hydrogen. All these reactions are endothermic and the Sun provides the heat for reaction. However, in the steam reforming of methane the water-to-methane ratio should be greater than about three, in order to avoid excessive buildup of carbon over the catalyst. In the conventional catalytic process, with external heat supply, this dilution is not critical, but in solar reformers, where the uniformly heated zone is limited (and thus the possible spatial velocities), the productivity and thus cost-effectiveness are significantly affected. The reaction is controlled by the rate of heat transfer, and thus the parameters to improve the reaction rate are very limited. In solar decomposition of methane, the production of a solid (carbon) is a clear technical issue, particularly for larger-scale applications. Solar dry reforming of methane with CO_2 appears the preferable choice between these three options, although in all cases the problem of catalyst stability (due to inhomogeneities in the heating) is a main issue.

Therefore, although the potential gain in energy efficiency of solar reformers over the conventional catalytic process to generate syngas or H_2 is attractive, the low productivity and absence of scale economy are the main critical issues, together with the problem of materials. The feasibility of concentrated solar power (CSP) for producing solar fuels was proven, but not stability of operations or its economics. Recent economic studies suggest a main role of CSP in electrical energy generation, particularly to integrate the peak demand (Geyer, 2008; Sarangi et al., 2009). The expansion of the market for CSP will probably be an incentive in future for their use in producing solar fuels.

4.4.2 Low-Temperature Approaches

Delocalization of energy production is another important aspect to consider, in particular for a better integration with territory, reduction of the eco-impact, and decrease of fuel transport costs and risks. Low-temperature approaches in producing solar fuels present evident advantages in this respect.

In the low-temperature approach, solar energy is used by a suitable semiconductor to generate by charge separation the electrons and holes, which further react with water and CO_2. The reduction of the latter can be a two-step approach, for example, generation of electricity in PV cells or by wind, and then use of the electrons to reduce carbon dioxide electrochemically/catalytically in a physically separate cell (Tributsch, 2008). Alternatively, a one-step approach is possible by coupling the two processes in a single unit, for example, a photoelectrochemical/catalytic approach (Barton et al., 2008; Centi et al., 2003; Currao, 2007; Kaneko et al., 2006). The physical separation of the two reactions of water oxidation and CO_2 reduction, in a photoanode and electrocathode, respectively, is necessary to increase the efficiency of the two reactions and limit charge recombination.

The same device can also be used for the production of physically separated flows of H_2 and O_2 during water photoelectrolysis. This device could be also used to produce renewable H_2 by photocatalytic reforming of chemicals present in waste streams from agrofood or agrochemical production, such as diluted streams of ethanol, glycerol, etc. There are many diluted waste streams containing ethanol and other organics, which are too diluted to be used as feed to make (i) H$_2$ by catalytic routes, (ii) methane by anaerobic digestion or (iii) in new generation fuel cells (Centi and van Santen, 2007). The photoreforming of these waste streams to produce H$_2$ is thus attractive.

Many studies have been dedicated recently to the water splitting on semiconductor catalysts under solar irradiation. Recent developments have been reviewed in detail by Fierro and coworkers (Navarro Yerga et al., 2009), Kudo and Miseki (2009), and Minero and Maurino (2007). Remarkable progress has been made since the pioneering work by Fujishima and Honda in 1972, but the development of photocatalysts with improved efficiencies for hydrogen production from water using solar energy still faces major challenges. Most of the recent efforts focus on the search for active and efficient photocatalysts, for example, through new materials and synthesis methods. While good quantum efficiencies (>50%) have been obtained with ultraviolet light, the use of visible light still poses major challenges. Often results still report H$_2$ or O$_2$ evolution from an aqueous solution containing a sacrificial reagent, but this could be acceptable only when waste water is used.

Many oxides consisting of metal cations with d^0 and d^{10} configurations, metal (oxy)sulfide and metal (oxy)nitride photocatalysts have been reported, especially during the latest decade (Kudo and Miseki, 2009). These semiconductors have a band gap in the visible region, but stability is still critical. Some of the best results have been reported by Domen (2009). He reported that some typical elements containing oxynitride photocatalysts such as (Ga$_{1-x}$Zn$_x$) (N$_{1-x}$O$_x$) actually work under visible light irradiation to realize the overall water splitting (Hirai et al., 2007; Maeda et al., 2006a). This was the first example of the overall water-splitting reaction under visible light irradiation on a photocatalyst with a band gap less than 3 eV. (Ga$_{1-x}$Zn$_x$) (N$_{1-x}$O$_x$) is a solid solution of GaN and ZnO and has been proved to be relatively stable materials during the overall water-splitting reaction with proper modification. To obtain effective photocatalytic activity of the overall water-splitting reaction, however, it is indispensable to construct hydrogen evolution sites on the particles of (Ga$_{1-x}$Zn$_x$) (N$_{1-x}$O$_x$). Several materials have been known as hydrogen evolution sites such as NiO, RuO$_2$, and Rh-Cr mixed oxide. Among them, a Rh/Cr$_2$O$_3$ core/shell nanostructure has been proven to efficiently reduce protons into H$_2$ molecules and simultaneously prevent the reverse reaction to form H$_2$O from H$_2$ and O$_2$. It is also possible to replace Rh metal with other noble metals such as Pt (Maeda et al., 2006b). The proposed reaction mechanism of H$_2$ evolution on these core/shell photocatalysts is shown in Figure 4.7 (Domen, 2009). The Cr$_2$O$_3$-covered Rh electrode selectively reduces H$^+$ but not oxygen molecules. These results evidence the role of a proper design in nanomaterials for improving the efficiency in solar fuel production.

Many studies recently focused on the investigation of semiconductor nanoparticles having the band gap in the visible region. CdS quantum dots are often used. Ryu et al. (2007), for example, studied the photocatalytic production of H$_2$ in water with visible light using nanocomposite catalysts, which include quantum-sized (Q-sized) CdS, CdS nanoparticles embedded in zeolite cavities (CdS/zeolite), and CdS quantum dots (Q-CdS) deposited on KNbO$_3$ (CdS/KNbO$_3$ and Ni/NiO/KNbO$_3$/CdS). The rate of H$_2$ production in alcohol/water mixtures and other electron donors at $\lambda \geq 400$ nm is the highest with the hybrid catalyst, Ni/NiO/KNbO$_3$/CdS with a measured quantum yield, φ, of 8.8%. Good quantum yields are thus obtained, but with sacrificial agents, and it is known that CdS nanoparticles are not stable in these conditions.

An improvement of these concepts has been reported recently by Zheng et al. (2009). An enhanced visible-light-driven photocatalyst, CuInS$_2$, was prepared by a template solvothermal route. The products show complex hierarchical architectures assembled from

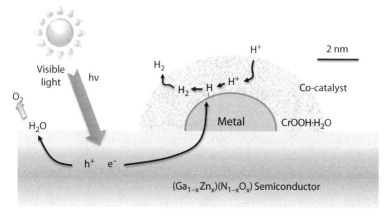

FIGURE 4.7
Schematic model of H_2 evolution at core/shell metal/Cr_2O_3 co-catalyst on $(Ga_{1-x} Zn_x)(N_{1-x}O_x)$. (Adapted from Domen, K., Efficient hydrogen evolution sites of photocatalysts for water splitting, *Presented at the 21st North American Meeting (NAM)*, San Francisco, CA, Paper 2896, 2009.)

interleaving two-dimensional microcrystals and near monodispersity. The involved CuS hierarchitectures form in situ and then act as the self-sacrificed templates, resulting in the obtained $CuInS_2$ inheriting the hierarchical architectures and monodispersity. A hydrogen yield of 59.4 µmol/h·g under visible light irradiation (significantly higher than previous reports) has been attained over the $CuInS_2$ doped with Pt. It is thus possible to increase the photocatalyst performances, but the issue on lack of stability was not solved. It is thus imperative in this field to develop materials that are stable and do not decompose with time.

While in theory similar semiconductors materials could be used also for the photocatalytic conversion of CO_2 to fuels, in practice the problem is more complex and thus the semiconductors active in water splitting (photoelectrolysis) are not effective in CO_2 conversion. This reaction will be discussed more in detail in the following section.

The photoelectrochemical path to water splitting involves instead the separation of the oxidation and reduction processes into half-cell reactions (Currao, 2007). Three are the possible approaches. Arrangements using either PV cells (PV approach) or semiconductor-liquid junctions (SCLJ approach), or a combination of the two (PV/SCLJ approach) can be realized.

The approach based on solid-state photovoltaics is to couple a PV system and an electrolyzer into a single system. Semiconductor layers are connected in series, one behind the other, in a single monolithic device capable of generating the potential needed to split water. These so-called tandem cells or multi-junction cells are modified with, or connected to, H_2-and O_2-producing electrodes, like Pt and RuO_2-modified Pt acting as cathode and anode, respectively. For example, n-p $GaInP_2$/GaAs, n-p $Al_xGa_{1-x}As$/Si, and multiple junction p-i-n amorphous Si cells were used for the photoelectrolysis of water (Currao, 2007).

PV approach was also used in solar-driven water electrolysis at elevated temperatures (Licht, 2003). The basic principle is the decrease of the electrochemical water-splitting potential with increasing temperature. Solar radiation is used for generating the necessary potential by illuminating PV cells as well as for the heat source to facilitate water electrolysis. This permits smaller band gap solar cells to drive the water cleavage at sufficiently low temperatures (500°C) from molten NaOH.

In the SCLJ approach, the water-splitting potential is generated directly at the semiconductor–liquid interface. Fujishima and Honda reported for the first time in 1972 sunlight-assisted electrolysis of water using crystalline TiO_2 photoelectrodes (Fujishima and

Honda, 1972). The photoelectrochemical cell consisted of TiO$_2$ (rutile) as a photoanode and platinum as a cathode. Illumination of the TiO$_2$ electrode led to O$_2$ evolution on the photoanode and H$_2$ evolution on the cathode. The quantum efficiency increased with an increase in alkalinity in the TiO$_2$ photoanode compartment and in acidity in the Pt cathode compartment. This means that the thermodynamic potential of 1.23 V required for water splitting was substantially decreased due to the presence of a large pH gradient between the compartments (ΔpH ~ 13, chemical bias ~ 0.77 V).

Many semiconductor materials have been used to drive the water oxidation and the water reduction at the same time, for example, Ni-doped In-TaO$_4$, InNbO$_4$, Ln$_2$Ti$_2$O$_7$ (Ln = La, Pr, Nd), Y$_2$Ti$_2$O$_7$ and Gd$_2$Ti$_2$O$_7$, and H$_2$La$_{2/3}$Ta$_2$O$_7$. For all these materials, the photocatalytic activity increases significantly when loaded with a reducing and/or oxidizing co-catalyst, such as Pt, RuO$_2$, or NiO. These materials may be considered, on a nanoscale, as small photoelectrochemical cells, but the absence of physical separation between the two stages of O$_2$ and H$_2$ evolution determines (i) an easy recombination, and (ii) the need of downstream separation. In the photoelectrochemical approach, the two stages are instead physically separated.

A photoelectrolysis cell based on two illuminated SCLJ may be also realized. An n-type semiconductor is used for the evolution of O$_2$ and a p-type semiconductor for the evolution of H$_2$. By separating the oxidation and reduction processes into half-cell reactions, a better optimization of each reaction and reducing recombination is possible. Besides, two semiconductors with smaller band gaps can be utilized since each needs only to provide part of the water-splitting potential. The smaller band gap means more absorption in the visible region of the solar spectrum where the Sun has a greater photon flux. As a result, the maximum theoretical efficiency is considerably higher.

The electrons formed in the n-type semiconductor (photoanode) recombine with holes formed in the p-type semiconductor (photocathode) via back contact connections in both materials. This is theoretically possible only if the valence band of the photocathode lies positive (higher electrochemical potential) with respect to the conduction band of the photoanode. This means, that proper selection of both semiconductor electrode characteristics ensures that the energy necessary for water photoelectrolysis is gathered entirely from the illumination, eliminating the necessity of applying energy from an external source.

In a PV/SCLJ approach for overall water splitting, a PV cell is used together with a semiconductor that is in direct electrolyte contact. The PV cell can be combined either with a reduction (photocathode) or with an oxidation (photoanode) photocatalyst.

Using these approaches, good efficiencies in separate H$_2$ and O$_2$ production are possible even with visible light. A triple junction amorphous Si solar cell coated with indium-tin-oxide (ITO) and connected to a Pt cathode showed a good solar to hydrogen efficiency (5%–6%) (Kelly and Gibson, 2006). However, stability is not over one month, mainly due to corrosion of the ITO and semiconductor layers by the electrolyte, even if the photoanode was protected from corrosion by a fluorinated tin oxide (SnO$_2$:F) layer.

This result evidences that two of the major issues in these photoelectrochemical approaches were the complexity and cost of the multistack electrodes and their weak stability during operations. These systems may be adapted with difficulty to the solar conversion of CO$_2$ to fuels. In general, the need is to develop robust and cost-effective systems that may be scaled-up and are suitable for small-to-medium-scale installations.

4.4.3 Centralized versus Distributed Approaches for Solar Fuels

The cost of investment is a critical issue and to accelerate the introduction of solar fuels there is the need of technologies that may be implemented effectively on a small scale also, in order to reduce capital cost. This is the common issue to accelerate the introduction of sustainable and based-on-renewable chemical and energy processes (Centi et al., 2009). In a difficult-to-predict economic scenario, similar to that for renewable energy, technologies that are effective only on a large scale and requiring large investments, will have a lower probability to be implemented and will require longer times for decision/realization.

On the other hand, it is clear that small-to-medium-scale technologies have to be based on a different R&D vision, centered on the use of process intensification, alternative energy technologies, high-performance materials, etc. In fact, it is necessary to break down the scale economy, and thus small-to-medium-scale processes should be based on highly efficient small modular units, with the increase in productivity realized by increasing their number, and not their size (parallel modular approach). This introduces advantages in terms of flexibility, eco-impact, safety, and a faster rate to introduce novel processes. It is also at the heart of the possibility of decentralized (distributed) production, based on the philosophy of producing where is needed (thus avoiding long-distance transport) and with production sites eco-compatible with the environment (e.g., where the local impact is below the capacity of self-depuration of the environment, and thus there are no irreversible decreases in biodiversity and eco-quality indicators), differently from the actual refinery and petrochemical large sites based on scale-economy.

The actual philosophy in several energy companies is that new technologies need a high-cost entry ticket (e.g., high costs of investment) to have suitable revenue on the investment, because this will limit the number of companies that can enter the market and thus limit the possible competition. This is not the correct vision of the problem, not only in terms of sustainability, because the entrance gap has to be in technology, and not in economics. The limit in this vision can be seen from the actual drastic changes in the market due to the aggressive economics of some emerging countries. It is thus knowledge-economy, not money-economy that should drive sustainability.

This reflects also in terms of the suitable technologies for solar fuels. It should be remarked that the energy problem is so large scale and complex that it can be never correct to consider that only one or few technologies have to be developed/applied. There is the need and space for all technologies, even if with different degrees of application. Nevertheless, it may be correct to discuss in terms of the share of R&D investment and the probability of application from a sustainable vision. From this point of view and limiting the discussion to solar fuels, it is clear that centralized technologies have received more attention with respect to distributed applications. On the other hand, it is also true that several of the technologies having potential to be effective on the small-scale such as those discussed in Section 4.2, have often taken into limited account the constrains for applicability.

There is thus the need for reconsidering in a more comprehensive vision the question of solar fuels and their perspectives, also in terms of centralized versus delocalized approaches and related technological constrains.

4.5 Solar Conversion of CO_2 to Liquid Fuels

In the Section 4.4.2, it was evidenced how the conversion of CO_2 to liquid fuels is a key part of the vision for future carbon-neutral fossil fuels. In addition, it was shown how solar H_2 may be considered a part of this vision, which is itself a suitable energy vector. Already various aspects have been addressed on this question of converting carbon dioxide, but it is useful to dedicate additional and more specific comments, because this part has been discussed much less in literature with respect to solar H$_2$ or electrical energy. Some earlier aspects on CO$_2$ conversion to fuels have been reviewed by Centi and Perathoner (2004) discussing the status and perspectives on the heterogeneous catalytic reactions with CO$_2$. A recent review (Centi and Perathoner, 2009c) and an advanced report (Centi et al., 2008a) discuss more closely the opportunities and prospects in the chemical recycling of carbon dioxide to fuels.

The vision of carbon dioxide as the hydrogen-storage material of the future was discussed in an essay by Enthaler (2008), where the cycle between CO$_2$ and formic acid was identified as the preferable option for renewable energy systems, notwithstanding the problems in terms of toxicity and instability associated with formic acid. Various aspects of the chemical fixation of CO$_2$ in constructing a future low-carbon global economy with reference to energy sources, thermodynamic considerations, net carbon emissions, and availability of reagents, were discussed by Tsang and coworkers (Kerry Yu et al., 2008), while recent advances in CO$_2$ were captured by Jones and coworkers (Choi et al., 2009).

4.5.1 Electrochemical/Catalytic Conversion of CO$_2$

Electrochemical utilization of CO$_2$ has been studied for many years. Recent reviews in this field are by DuBois (2006) and Gattrell et al. (2006). There are two main approaches depending whether the conversion of CO$_2$ is studied in aqueous or nonaqueous solutions. Formic acid is the main reaction product in electrolysis of aqueous solutions of CO$_2$. The formation of formic acid is known for over a century, because it competes with hydrogen evolution in the reduction of CO$_2$. A problem in the utilization of CO$_2$ in aqueous solution derives from its low solubility in water at standard temperature and pressure. At the surface of the electrode, very small amounts of CO$_2$ are available for the reaction to proceed. For aqueous solutions, in order to speed the reaction rate, the pressure must be increased.

Numerous studies have been made on the electrochemical reduction of CO$_2$ under high pressure on various electrodes in an aqueous electrolyte. Table 4.1 shows the Faradaic efficiencies of CO$_2$ reduction on Group 8–10 metal electrodes, which have low overpotentials

TABLE 4.1

The Electrochemical Reduction of CO$_2$ under 30 atm Pressure at 163 mA cm^{-2} (25°C, 0.1 mol dm^{-3} KHCO$_3$, and Charge Passed 300°C)

Electrode	E°/V	\multicolumn{7}{c}{Faradaic Efficiency, %}							PCD (CO$_2$ Red)/mA cm^{-2}
		CH$_4$	C$_2$H$_6$	C$_2$H$_4$	CO	HCOOH	H$_2$	CO$_2$ Red	Total	
Fe	−1.63	2.03	0.4	0.16	4.2	28.6	51.6	35.4	87.8	57.7
Cu	−1.54	3.09	0.17	0.38	15.6	21.9	40.9	41.5	88.4	7.6
Rh	−1.41	0.26	0.03	0.01	51.0	19.5	13.1	80.8	93.9	131.7
Ir	−1.55	0.62	0.05	0.05	22.3	22.3	48.3	40.4	88.8	66.0
Ni	−1.59	0.72	0.08	0.11	33.5	31.3	36.0	65.7	91.7	107.1
Pd	−1.56	0.13	0.01	Trace	46.1	35.6	12.8	81.8	94.6	133.3
Pd*	−1.76	0.21	0.01	0.02	35.1	44.0	33.8	79.4	93.2	397.0
Pt	−1.48	0.22	0.02	Trace	6.2	50.4	33.6	56.7	90.3	92.4

Source: Adapted from Hara, K. et al., *J. Electroanal. Chem.*, 391, 141, 1995.

* Current density, 500 mA cm^{-2}.

for hydrogen formation (Hara et al., 1995). Even if this paper is relatively old, no remarkable advances have been made later. Productivity increases substantially at 30 bar of CO_2 with respect to that at 1 bar of CO_2. However, the total cathodic current barely increases with increasing CO_2 pressure. In terms of products, CO, H_2, and formic acid are mainly observed. Traces of methane and C_2 hydrocarbons are observed but in very low amounts (below few percent). In addition, the observed efficiencies were recorded for only few minutes, due to fast deactivation.

Solvents with a high solubility for CO_2 are used in the nonaqueous electrochemical reduction of CO_2. Carbon dioxide concentration in dimethylformamide is about 20 times higher than that in aqueous solutions, and in propylene carbonate and methanol, the CO_2 solubility is about eight and five times higher, respectively. However, a high CO_2 solubility requires large current densities, whereas low electrolytic conductivities lead to high ohmic losses. For this reason, methanol is often used to balance these two aspects. Cu-based foils give the best results as electrodes. Another problem is that very high current densities at the copper cathode are necessary to maximize the formation of hydrocarbons, but a fast deactivation is present in these conditions (Hori et al., 2005). The formation of products from the potential and partial current densities is shown in Figure 4.8.

Hori et al. (2005) using immobilized CuCl on a Cu-mesh electrode reported a Faradaic efficiency of about 70% to C_2H_4, although the electrode deactivates rapidly. In addition, corrosive media (high pressure, metal halides) are used because metal halides are necessary to promote the surface concentration of CO_2 at the electrode.

Note also that Cu is the only metal that gives appreciable amounts of C2 hydrocarbons (Table 4.1). Several critical aspects determine the performances, such as (i) the gas evolution in the electrochemical cells, which reduce electrolyte conductivity and increase ohmic resistance, (ii) the pH and reaction temperature, which has opposite influence on the solubility of CO_2 and selectivity towards C2 products, and (iii) the porosity of electrodes. Using TiO_2 nanotube composite electrodes for the electrochemical reduction of CO_2 to methanol, current efficiencies over 60% were reported (Qu et al., 2005).

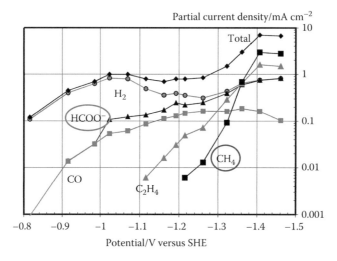

FIGURE 4.8
Dependence of the formation of products from the potential and partial current densities in the CO_2 reduction in methanol under pressure and using Cu-foil electrode. (Adapted from Gattrell, M. et al., *J. Electroanal. Chem.*, 594, 1, 2006.)

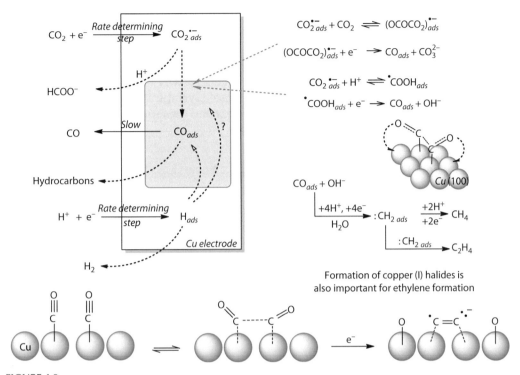

FIGURE 4.9
Overview of the key aspects in the reaction mechanism of CO$_2$ reduction over Cu electrodes. (Adapted from Gattrell, M. et al., *J. Electroanal. Chem.*, 594, 1, 2006.)

The key aspects in the mechanism on Cu electrodes are summarized in Figure 4.9. The initial stage is the formation of a carbon dioxide anion radical $^{\bullet}CO_2^-$, which explains why metal halides are necessary in stabilizing this intermediate and avoid the formation of formate ions, which desorb from the surface. The reduction of this intermediate, or alternatively of that formed by the reaction of CO$_2$ with the carbon dioxide anion radical gives, adsorbed carbon monoxide (CO) as the key intermediate. Chemisorbed carbon monoxide can react with protons and electrons (in the presence hydroxide anions) to give water and chemisorbed methylene (:CH$_2$), which may be either further hydrogenated to CH$_4$ or can react with another methylene intermediate following a Fischer–Tropsch-like chain growth mechanism. Surface science studies on Cu (100) crystals suggest the possible alternative that two vicinal chemisorbed CO molecules react together with simultaneous breaking of respective C–O bonds to form a chemisorbed radical-anion C2 species by one electron transfer (Figure 4.9, bottom reaction mechanism). This species is then the precursor for C2 hydrocarbons. It may be observed that in all cases the proposed mechanisms are rather speculative, because no direct in situ evidences have been provided. Two recent reviews (Gattrell et al., 2006; Jitaru, 2007) discussed in detail the reaction mechanism, as well as the performances of the different electrocatalysts and the dependence of the behavior from the reaction conditions.

Shibata et al. (2008) showed that alkanes and alkenes up to C6 can be obtained by CO$_2$ electroreduction at room temperature and atmospheric pressure by the application of a commercially available Cu-electrode, provided the pretreatment by electropolishing is avoided. The product distribution follows the Schultz–Flory distribution, and, depending on the applied potential, the chain growth probability (α) ranges from 0.23 to 0.31, values lower than those obtained in Fischer–Tropsch synthesis over heterogeneous Co- or Fe-based catalysts.

When the same electrode material was pretreated by electropolishing, it behaved like a pure Cu electrode, giving mainly methane and ethene. It was suggested that the oxygen coverage of the electrodes is a function of the surface crystallinity, but the mechanism of this marked effect on the surface reactivity is unclear and the reproducibility of the results is problematic.

4.5.2 From Liquid to Gas Phase Electrocatalytic Conversion of CO_2

The feasibility of CO_2 conversion to fuels depends on the possibility to form liquid fuels under solventless conditions such as the production of long-chain hydrocarbons and/or alcohols, which can be easily collected without the need to be distilled from liquid solutions (a quite energy-intensive process).

There are very few studies based on this approach, but it was demonstrated that using nanostructured carbon-based electrodes it is possible to reduce CO_2 electrocatalytically in the gas phase using the protons flowing through a membrane. Long-chain hydrocarbons and alcohols up to C9–C10 are formed, with the preferential formation of isopropanol using carbon-nanotube-based electrodes (Centi et al., 2003, 2006, 2007a, 2009). Productivities are still limited, but these results prove the concept of a new approach to recycle CO_2 back to fuels.

The features of the electrode used in this gas-phase electrocatalytic reduction of CO_2 are close to those used in PEM fuel cells (Centi and Perathoner, 2009b), for example, a carbon cloth/Pt on carbon black/Nafion®-assembled electrode (gas diffusion electrode [GDE]). The electrocatalyst is Pt supported on carbon black, which is then deposited on a conductive carbon cloth to allow both electrical contact and the diffusion of gas-phase CO_2 into the electrocatalyst. The Pt particles are at the interphase in contact with Nafion through which protons diffuse. On the Pt nanoparticles, the gas-phase CO_2 reacts with the electrons and protons to produce long-chain hydrocarbons and alcohols, the relative distributions of which depends on the reaction temperature. However, using this electrode assembly, acetone is the major product at 60°C, while at room temperature the productivities are lower, but longer-chain hydrocarbons can be formed (Centi et al., 2007a).

Using a similar GDE configuration, but with carbon nanotubes as the substrate (instead of carbon black) for the electrocatalyst nanoparticles, it is possible to obtain isopropanol as the main product of reaction (Centi and Perathoner, 2009b; Gangeri et al., 2009). Figure 4.10 also evidences that using carbon nanotubes it is possible to also use iron nanoparticles instead of a noble metal, although the latter shows better stability. In addition, it is shown that the use of N-doped carbon nanotubes (N/CNT) allows a further improvement in the productivity to isopropanol.

Not only the nature of carbon is relevant, but also the presence of nanocavities, which could favor the consecutive conversion of intermediates with the formation of C–C bonds. Figure 4.11 reports the comparison of the behavior at 25°C in the electrocatalytic reduction of CO_2 of two 10wt% Fe/CNT samples. In the first catalyst, the iron particles are located on both the inner and the outer surface of the CNT (the usual situation), as shown from transmission electron microscopy data (Centi and Perathoner, 2009d). In the second sample, Fe is localized only inside the CNT. The localization inside leads to an enhanced productivity and influences the product distribution. Larger iron particles are present inside the CNT.

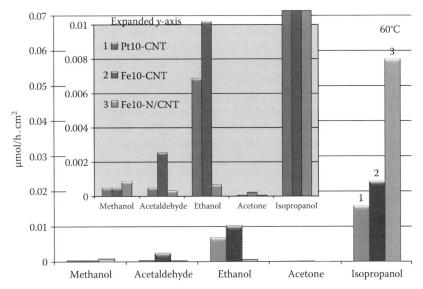

FIGURE 4.10

Products distribution at 60°C in the electrocatalytic reduction of carbon dioxide in gas phase over Nafion 117/ (Pt or Fe(10%)/CNT)20%/carbon cloth gas diffusion membrane electrode. Tests in a semi-batch cell, using a 0.5 M KCO_3 electrolyte on the anode side and operating the cathode in the gas phase with a continuous feed of 50% CO_2 in humidified nitrogen. (Adapted from Centi, G. and Perathoner, S., *Top. Catal.*, 52, 948, 2009; Gangeri, M. et al., *Catal. Today*, 143, 57, 2009.)

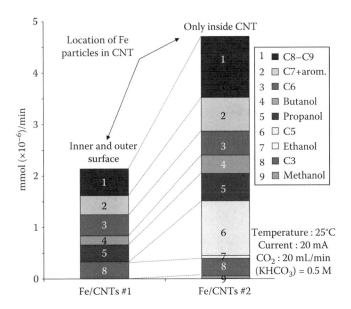

FIGURE 4.11

Comparison of the product distribution in CO_2 electrocatalytic reduction at 25°C on Fe@CNT (10 wt%. Fe): the iron particles are located both on the inner and outer surface of CNT (#1) or only on the inner surface (#2). (Adapted from Centi, G. and Perathoner, S., *Catal. Today*, 150, 151, 2009, doi:10.1016/j.cattod.2009.09.009.)

The reaction mechanism of the formation of isopropanol as the main reaction intermediate is still unclear. The product distribution is quite different from Schultz–Flory distribution, as in the results reported by Shibata et al. (2008) using non-electropolished Cu electrodes. This suggests that the mechanism could be different from that present in Fisher–Tropsch (FT) synthesis. Note that isopropanol and not n-propanol is formed, and that no methane was formed, as expected in the FT mechanism. It could also be observed that comparing the results of Fe10-CNT with those of Fe10-N/CNT (e.g., a similar electrocatalyst, but using N/CNT as the substrate) (Figure 4.10), the formation of ethanol decreases and that isopropanol increases. This indication is not consistent with an FT-like mechanism.

4.5.3 Coupling Photo- and Electrocatalysis

This approach for the electrocatalytic reduction of CO_2 to fuels is sustainable only when the electrons and protons necessary for the reaction are supplied using renewable resources, for example, using solar energy by integrating a photoanode cell on which the photodissociation of water produces the electrons and protons necessary for the electrocatalytic reduction of CO_2. An effective device should thus integrate the photoanode (for water dissociation) to the electrocatalyst (for CO_2 reduction) in a photoelectrocatalytic (PEC) reactor.

It should be evidenced that in the perspective of a practical use of PEC solar cells, the design should be quite different from that used typically in literature. The anode and cathode in the PEC device should be in the form of a thin film separated from a proton-conducting membrane (Nafion, for example) and deposited over a porous conductive substrate that allows the efficient collection/transport of the electrons over the entire film, as well the diffusion of protons to/from the membrane. It is also necessary to allow an efficient evolution of oxygen on the anode side, an efficient diffusion of CO_2, and the evolution of reaction products on the cathode side.

The conceptual scheme of a PEC reactor to convert CO_2 back to fuels at near room temperature and pressure using solar light and water is reported in Figure 4.12 (Centi et al., 2004, 2007a). Using this device, it will be possible in the future to develop "artificial trees" able to capture the CO_2 and convert it to liquid fuels (hydrocarbons, alcohols) (Centi and

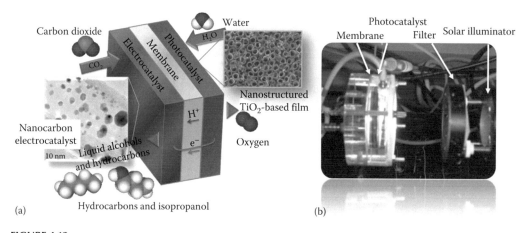

FIGURE 4.12
(a) Simplified scheme of a PEC device for the CO_2 reduction to fuels using solar energy. (b) View of lab-scale PEC device. (Adapted from Centi, G. and Perathoner, S., *Catal. Today*, 150, 151, 2009, doi:10.1016/j.cattod.2009.09.009.)

Perathoner, 2009b). The PEC reactor shows analogies with PEM fuel cells, for example, the R&D of the latter can be used to reduce costs and improve scale-up. One side of the cell is composed of a nanostructured TiO$_2$-based thin film where photoelectrolysis of water using solar light occurs. Protons diffuse through a membrane on the other side of the PEC device and react with CO$_2$ in the presence of electrons (generated in water photodissociation; anode and cathode sides are connected by a wire) and a special electrocatalyst, based on the concept of nanoconfinement, as discussed before. The PEC concept was originally proposed by "Hitachi Green Center" researchers (Ichikawa, 1995; Ichikawa and Doi, 1996).

A specific problem in PEC solar cells is the need to have a specific nanostructure in the photoanode. The use of an array of one-dimensional aligned nanostructures (nanorods, nanotubes, etc.) improves light harvesting and limits charge recombination at the grain boundaries with respect to an assembly of nanoparticles, while maintaining a high geometrical surface area necessary to improve the photoresponse (Ampelli et al., 2008; Centi and Perathoner, 2008b). An optimal contact/interface with the H$^+$-membrane is necessary.

A further general issue is regarding the need to use preparation methods to produce the photoanode that (i) can be cost-effective, (ii) can allow a good control of the nanostructure in terms of one-dimensional array characteristics, vertical alignment, density, and thickness, and (iii) can be easily scalable to larger scale (at least 10×10 cm). The choice of suitable preparation methods having all these characteristics is quite restricted (Centi and Perathoner, 2008b). A good method is the anodic oxidation of titanium thin foils to form ordered arrays of vertically aligned titania nanotubes (Centi et al., 2007b).

An alternative approach was proposed in the U.S. DoE report "Catalysis for energy" (Bell et al., 2007), where a conceptual model of a semiconducting membrane was presented. The channels permit proton communication between cells containing half-reaction catalysts, but do not permit passing of reactants or products through. Molecular catalysts for water oxidation and carbon dioxide reduction are anchored at opposite sides of the electrode. The electrode is designed to absorb solar radiation and facilitate diffusion of the resulting point charges into the respective molecular catalysts responsible for the carbon dioxide reduction and water oxidation chemistry.

An attempt to go in this direction was reported recently by Grimes and coworkers (Varghese et al., 2009) who studied the use of N-doped TiO$_2$ nanotube arrays for the solar conversion of carbon dioxide and water vapor to methane and other hydrocarbons. However, they tested the performances of these titania nanotubes (doped with Pt or Cu) in a batch-type photoreactor. Therefore, a flow-through photocatalytic membrane design was not used as suggested in the cited DoE report. The best results were obtained for an N-doped titania nanotube array annealed at 600°C using Cu as co-catalysts (about half of the surface area covered by copper nanoparticles). Using a 400 nm wavelength high-pass filter, the productivities were low, while illuminating with solar radiation (which has also UV components), H$_2$ and methane, and minor amounts of CO, and other alkanes and alkenes were observed. No alcohols were observed differently from the results in Figures 4.10 and 4.11. Other authors investigating the CO$_2$ photoreduction on TiO$_2$-based catalysts instead observed mainly the formation of oxygenated products (Centi and Perathoner, 2009d). Two recent reviews discussed the development in this area (Kitano et al., 2007; Usubharatana et al., 2006).

The reaction mechanism in the photoreduction of CO$_2$ involves two important species, H$^\bullet$ (hydrogen atom) and $^\bullet$CO$_2^-$ (carbon dioxide anion radical), produced by electron transfer from the conduction band of the excited semiconductor. The solubility of CO$_2$ in water is low, as noted before, and the CO_2 photoreduction process competes with H_2 and H_2O_2 formation, which consumes H^+ and e^-. There are thus intrinsic limitations to improve the productivity to fuels in CO_2 photocatalytic reduction following the conventional approach. It is to be noted that methanol and formaldehyde are the easier products of CO_2 reduction in water solution, but the formed solutions are highly diluted. The recovery of these chemicals is thus expensive.

It may be concluded that the direct CO_2 reduction of semiconductor-based materials is conceptually attractive, but limited in terms of perspectives of applications. The physical separation of the water photodissociation from the CO_2 reduction stages, such as in the PEC approach, coupled to solventless conditions is the requirement for application.

4.5.4 Thermochemical Conversion of CO_2

In thermochemical conversion of CO_2, solar irradiation is not used to generate the reactive species that convert carbon dioxide as in the case of semiconductors (low-temperature approach), but simply to provide the heat of reaction. Dissociation of the C-O bond has a zero ΔG at 3350 K, and thus the approach is suitable only in combination with a redox cycle, based on metal oxides as discussed in the Section 4.5.3, or in more complex thermochemical pathways, which allow reducing the necessary temperature. When this reaction of CO_2 dissociation is coupled with the thermochemical water splitting to produce H_2 and O_2, the CO/H_2 mixture obtained may be then converted by established routes to methanol, FT products, etc.

A critical issue is the fast back reaction both in CO_2 dissociation (to CO and mono-oxygen) and water splitting. Therefore, a very fast and efficient quenching is necessary. On the other hand, the zone of irradiation is limited, particularly in terms of length of the redox material bed, in order to have uniform heating. The conversion per pass is limited, but the rapid quenching downstream does not make the recycle very efficient. There are thus intrinsic large difficulties in having good productivities with this approach, as well as in the scale-up of the system.

Traynor and Jensen (2002) of Los Alamos Renewable Energy (LARE) Corporation were among the first to report prototypes results on the direct solar reduction of CO_2 to CO and oxygen. In the first prototype system, a solar focusing mirror and a secondary concentrator were used to provide high solar intensity around a ceramic rod. This high-temperature, high-solar irradiance environment provided strong heating of CO_2 with the resultant dissociation to CO and oxygen. Quenching of the back reaction was provided by the geometry and gas dynamics of the system and by cool gas quencher jets just downstream. The best-measured net conversion of CO_2 to CO was near 6%, which is compared to a plant design target of 12%. The observed peak conversion of solar energy to chemical energy was 5%. Later, Price et al. (2004) also made a computer modeling to determine the flow, temperature, and reactions occurring in the prototype device. The solar flux heated a zirconia rod at the throat of the device to 2625 K, which in turn heated the surrounding gas by convection. All CO formation reactions occurred in the boundary layer of the zirconia rod and just beyond the region of high solar flux.

LARE announced in 2005 the process called SOLAREC™ (Solar Reduction of Carbon). In the LARE reactor, CO_2 is fed into a reaction chamber sealed at one end by a quartz window 8 cm in diameter. The chamber is fixed at the focal point of a mirrored dish that concentrates sunlight through the chamber's window onto a ceramic rod set inside the chamber. These rods reach a temperature of around 2400°C enough to cause dissociation of carbon dioxide to CO and oxygen, although conversion is low.

One of the drawbacks of this approach is the high operating temperature. High temperatures lead to heavy thermal losses, which in turn can reduce efficiency. Though the Sun's energy is free, an equipment to generate and withstand these temperatures is expensive to build, making efficient operation vital if the process is to be cost-effective.

Sandia National Laboratory (Albuquerque, United States) has developed an alternative system indicated as CR5 (short for counter-rotating ring receiver reactor recuperator), which operates at lower temperatures, although still above 1000°C. Like the LARE reactor, it has a concentrator dish that focuses the Sun's rays. In this case, the high temperatures are generated on one side of a stack of 14 rings made of a cobalt ferrite ceramic, a material that when heated releases oxygen from its molecular lattice without destroying the lattice's integrity. The rings, which are about 30 cm in diameter, rotate at around one revolution per minute inside a sealed double chamber.

Sunlight focused through a window in the hot side of the chamber heats the rings to 1500°C, causing the ceramic lattice to liberate oxygen atoms. As the rings rotate, the heated sections pass to the rear of the chamber, where they cool to around 1100°C as they is bathed in CO_2. At this temperature, the deoxygenated ceramic reacts with the CO_2 molecules to be reoxidized producing CO. As the ring continues to rotate, the reoxygenated section passes back into the hot side of the chamber and the cycle begins again. The material is thus continuously cycled between the two temperatures, with thus very large problems of material stress (fast sintering). In addition, conversion is limited and the control of back reactions difficult. The CR5 was originally developed to produce hydrogen using steam in the cool chamber, but was then converted to split CO_2, because it could offer a more efficient way of capturing solar energy. Burning the CO formed in the solar reactor should deliver 10% of the energy that was required to produce it. They calculate that the prototype will be able to produce about 100 L of CO per hour.

The CR5 Sandia project is called "Sunlight to Petrol" or S2P. The prototype contains 14 cobalt ferrite rings, each about one foot in diameter and turning at one revolution per minute. An 88 m^2 solar furnace will blast sunlight into the unit.

4.6 Conclusions

Producing solar fuels is a hot topic with widespread scientific and industrial interest, as well as great attention from the public (Gray, 2009). Many blogs and Internet forums discuss this topic, as well as various articles in numerous widely circulated newspapers. One of the initiatives of the current Swedish Presidency of the European Union (November 2009) is dedicated to "molecular science for solar fuel," as part of the EU vision for future energy systems to transform the solar energy into a fuel that can be stored and used when needed. At "California Institute of Technology"—Caltech (CA)—a Solar Fuel Center for Chemical Innovation (CCISolar) (www.ccisolar.caltech.edu) was recently created. These are some of the many initiatives that demonstrate the active interest in this field.

Many companies start initiatives in this area. At Massachusetts Institute of Technology, the MIT-spin out "Sun Catalytix" (www.suncatalytix.com) explores two approaches to split water to produce hydrogen. About half of the companies use a biological catalyst and the other half employ chemical catalysts. They commercialize a new, active, versatile, and affordable catalyst that splits water into oxygen and hydrogen fuel, mimicking photosynthesis. However, the real applicability of the system is unclear. "Carbon Sciences,

Inc." (www.carbonsciences.com) claims to have developed a unique, energy-efficient, and highly scalable biocatalytic process that apparently is able to produce directly usable fuels from CO_2 deriving from large emitters, such as a power plant. However, no clear data are provided to understand the real applicability and performance of the system. These two examples are indicative of the large attention on this topic, and the commercial possibilities as well, but also point out the absence often of more complete and reliable data that can provide a good techno-economic assessment of the proposed technologies and solutions.

In this review, various aspects of this topic have been discussed, evidencing not only the possibilities but also the limits of the alternatives. This is a scientific area in very dynamic evolution. It is thus natural that many different possible routes are explored and that it is still unclear what is preferable, where the technological limits are, and which one shows better perspectives. The discussion has evidenced some of these aspects, although in a fast growing field like that of solar fuels it is always difficult to have conclusive indications.

Some of the aspects we remarked are the need to produce easily transportable and stored fuels, which can be integrated into the existing energy infrastructure. In this sense, we strongly suggest that liquid fuels produced from carbon dioxide and water using solar energy is the preferable route, notwithstanding the difficulties in the reaction.

There are many options in this direction, but often experimentation does not consider how to transfer the proposed solution into an application, the problems and costs in the recovery of the fuels, and the problems related to scale-up. Even if solar fuels should be still considered a medium-to-long term solution, it is necessary to consider the aspects indicated above for a correct evaluation of the topics on which to focus the investigations.

Solar fuels are still a grand challenge, but using the words of Gray (2009) "this is the century when we will start paying back with the capital generated through fundamental research in chemistry."

References

Abanades S. and Flamant G. 2006. Thermochemical hydrogen production from a two-step solar-driven water-splitting cycle based on cerium oxides. *Solar Energy*, 80(12): 1611–1623.

Agrafiotis C., Roeb M., Konstandopoulos A.G., Nalbandian L., Zaspalis V.T., Sattler C., Stobbe P., and Steeled A.M. 2005. Solar water splitting for hydrogen production with monolithic reactors. *Solar Energy*, 79(4): 409–421.

Allakhverdiev S.I., Kreslavski V.D., Thavasi V., Zharmukhamedov S.K., Klimov V.V., Nagata T., Nishihara H., and Ramakrishna S. 2009. Hydrogen photoproduction by use of photosynthetic organisms and biomimetic systems. *Photochemical and Photobiological Sciences*, 8(2): 148–156.

Ampelli C., Passalacqua R., Perathoner S., Centi G., Su D.S., Weinberg G. 2008, Synthesis of TiO_2 thin films: Relationship between preparation conditions and nanostructure. *Topics in Catalysis*, 50: 133–144.

Balzani V., Credi A., and Venturi M. 2008. Photochemical conversion of solar energy. *ChemSusChem*, 1(1–2): 26–58.

Barton E.E., Rampulla D.M., and Bocarsly A.B. 2008. Selective solar-driven reduction of CO_2 to methanol using a catalyzed p-GaP based photoelectrochemical cell. *Journal of the American Chemical Society*, 130(20): 6342–6344.

Bell A.T., Gates B.C., and Ray D. 2007. *Basic Research Needs: Catalysis for Energy* (PNNL-17214), Washington, DC: U.S. Department of Energy, Report from a Workshop held in August 6–8, 2007, Bethesda, MD (http://www.sc.doe.gov/bes/reports/list.html) (accessed on February 25, 2011).

Bhandari R. and Stadler I. 2009. Grid parity analysis of solar photovoltaic systems in Germany using experience curves. *Solar Energy*, 83(99): 1634–1644.

Birch H. 2009. The artificial leaf. *Chemistry World*, 6(5): 42–45.

Bockris J.O'M. 2008. Hydrogen no longer a high cost solution to global warming: New ideas. *International Journal of Hydrogen Energy*, 33(9): 2129–2131.

Bothe H., Winkelmann S., and Boison G. 2008. Maximizing hydrogen production by cyanobacteria. *Zeitschrift fuer Naturforschung—Section C Journal of Biosciences*, 63(3–4): 226–232.

Centi G., Cum G., Fierro J.L.G., López Nieto J.M. 2008a. Direct Conversion of Methane, Ethane and Carbon Dioxide to Fuels and Chemicals, *CAP Report*, Spring House, PA: The Catalyst Group Resources (www.catalystgrp.com) (accessed on February 25, 2011).

Centi G. and Perathoner S. 2004. Heterogeneous catalytic reactions with CO_2: Status and perspectives. *Studies in Surface Science and Catalysis*, 153(Carbon Dioxide Utilization for Global Sustainability): 1–8.

Centi G. and Perathoner S. 2008b. Nano-architecture and reactivity of titania catalytic materials. Part 2. Bidimensional nanostructured films. *Catalysis*, 20: 1–39.

Centi G. and Perathoner S. 2009a. The role of nanostructure in improving the performance of electrodes for energy storage and conversion. *European Journal of Inorganic Chemistry*, 26: 3851–3878.

Centi G. and Perathoner S. 2009b. Catalysis: Role and challenges for a sustainable energy. *Topics in Catalysis*, 52(8): 948–961.

Centi G. and Perathoner S. 2009c. Opportunities and prospects in the chemical recycling of carbon dioxide to fuels. *Catalysis Today*, 148(3–4): 191–205, doi:10.1016/j.cattod.2009.07.075.

Centi G. and Perathoner S. 2009d. Problems and perspectives in nanostructured carbon-based electrodes for clean and sustainable energy. *Catalysis Today*, 150(1–2): 151–162, doi:10.1016/j.cattod.2009.09.009.

Centi G., Perathoner S., and Rak Z.S. 2003. Gas-phase electrocatalytic conversion of CO_2 to fuels over gas diffusion membranes containing Pt or Pd nanoclusters. *Studies in Surface Science and Catalysis*, 145(Science and Technology in Catalysis 2002): 283–286.

Centi G., Perathoner S., Wine G., and Gangeri M. 2006. Converting CO_2 to fuel: A dream or a challenge? *Preprints of Symposia—American Chemical Society Division of Fuel Chemistry*. 51: 745–746.

Centi G., Perathoner S., Wine G., and Gangeri M. 2007a. Electrocatalytic conversion of CO_2 to long carbon-chain hydrocarbons. *Green Chemistry*, 9(6): 671–678.

Centi G., Passalacqua R., Perathoner S., Su D.S., Weinberg G., and Schlögl R. 2007b. Oxide thin films based on ordered arrays of 1D nanostructure. A possible approach toward bridging material gap in catalysis. *Physical Chemistry Chemical Physics*, 9: 4930–4938.

Centi G., Trifiró F., Perathoner S., and Cavani F. 2009. *Sustainable Industrial Chemistry*, Weinheim, Germany: Wiley-VCH. (ISBN-10: 3-527-31552-7).

Centi G. and van Santen R.A. 2007. *Catalysis for Renewables*, Weinheim, Germany: Wiley-VCH (ISBN: 978-3-527-31788-2).

Chen H., Cong T.N., Yang W., Tan C., Li Y., and Ding Y. 2009. Progress in electrical energy storage system: A critical review. *Progress in Natural Science*, 19(3): 291–312.

Cho Y.K., Donohue T.J., Tejedor I., Anderson M.A., McMahon K.D., and Noguera D.R. 2008. Development of a solar-powered microbial fuel cell. *Journal of Applied Microbiology*, 104(3): 640–650.

Choi M.-S., Yamazaki T., Yamazaki I., and Aida T. 2003. Bioinspired molecular design of light-harvesting multiporphyrin arrays. *Angewandte Chemie—International Edition*, 43(2): 150–158.

Cong H.-P. and Yu S.-H. 2009. Self-assembly of functionalized inorganic-organic hybrids. *Current Opinion in Colloid and Interface Science*, 14(2): 71–80.

Christensen C.H., Johannessen T., Sørensen R.Z., andNørskov J.K. 2005. Towards an ammonia-mediated hydrogen economy? *Catalysis Today*, 111(1–2): 140–144.

Choi S., Drese J.H., and Jones C.W. 2009. Adsorbent materials for carbon dioxide capture from large anthropogenic point sources. *ChemSusChem*, 2(9): 796–854.

Chueh W.C. and Haile S.M. 2009. Ceria as a thermochemical reaction medium for selectively generating syngas or methane from H_2O and CO_2. *ChemSusChem*, 2(8): 735–739.

Curtright A.E., Morgan M.G., and Keith D.W. 2008. Expert assessments of future photovoltaic technologies. *Environmental Science & Technology*, 42 (24): 9031–9038.

Currao A. 2007. Photoelectrochemical water splitting. *Chimia*, 61(12): 815–819.

Domen K. 2009. Efficient hydrogen evolution sites of photocatalysts for water splitting. *Presented at the 21st North American Meeting* (NAM), San Francisco, CA, Paper 2896.

Donohue T.J. and Cogdell R.J. 2006. Microorganisms and clean energy. *Nature Reviews Microbiology*, 4(11): 800.

Dovì V.G., Friedler F., Huisingh D., and Klemeš J.J. 2009. Cleaner energy for sustainable future. *Journal of Cleaner Production*, 17(10): 889–895.

DuBois D.L., 2006. Electrochemical reactions of carbon dioxide, In *Encyclopedia of Electrochemistry*, Weinheim, Germany: Wiley-VCH (ISBN: 978–3–527–30399–1), Vol. 7A, Ch. 6.2, pp. 202–225.

Enthaler S. 2008. Carbon dioxide—The hydrogen-storage material of the future? *ChemSusChem*, 1(10): 801–804.

Fujishima A. and Honda K. 1972. Electrochemical photolysis of water at a semiconductor electrode, *Nature*, 238: 37–38.

Gallezot P. 2008. Catalytic conversion of biomass: Challenges and issues. *ChemSusChem*, 1(8–9): 734–737.

Gangeri M., Perathoner S., Caudo S., Centi G., Amadou J., Begin D., Pham-Huu C. et al. 2009. Fe and Pt carbon nanotubes for the electrocatalytic conversion of carbon dioxide to oxygenates. *Catalysis Today*, 143(1–2): 57–63.

Gattrell M., Gupta N., and Co A. 2006. A review of the aqueous electrochemical reduction of CO_2 to hydrocarbons at copper. *Journal of Electroanalytical Chemistry*, 594(1): 1–19.

Geyer M. 2008. International market introduction of concentrated solar power-policies and benefits. In *Proceedings of ISES World Congress 2007—Solar Energy and Human Settlement*, Heidelberg, Germany: Springer-Verlag (ISBN 978-3-540-75996-6), pp. 75–82.

Gibson T.L. and Kelly N.A. 2008. Optimization of solar powered hydrogen production using photovoltaic electrolysis devices. *International Journal of Hydrogen Energy*, 33(21): 5931–5940.

Gray H.B. 2009. Powering the planet with solar fuel. *Nature Chemistry*, 1(4): 7.

Hara K., Kudo A., and Sakata T. 1995. Electrochemical reduction of carbon dioxide under high pressure on various electrodes in an aqueous electrolyte. *Journal of Electroanalytical Chemistry*, 391(1–2): 141–147.

Hertwich E.G. and Zhang X. 2009. Concentrating-solar biomass gasification process for a 3rd generation biofuel. *Environmental Science & Technology*, 43(11): 4207–4212.

Hirai T., Maeda K., Yoshida M., Kubota J., Ikeda S., Matsumura M., and Domen K. 2007. Origin of visible light absorption in GaN-rich $(Ga_{1-x}Zn_x)(N_{1-x}O_x)$ photocatalysts. *Journal of Physical Chemistry C*, 111(51):18853–18855.

Hori Y., Konishi H., Futamura T., Murata A., Koga O., Sakurai H., and Oguma K. 2005. "Deactivation of copper electrode" in electrochemical reduction of CO_2. *Electrochimica Acta*, 50(27): 5354–5369.

IEA (International Energy Agency). 2009. *Key World Energy Statistics 2009*. http://www.iea.org/textbase/nppdf/free/2009/key_stats_2009.pdf (accessed on February 25, 2011).

Ichikawa S. 1995. Chemical conversion of carbon dioxide by catalytic hydrogenation and room temperature photoelectrocatalysis. *Energy Conversion and Management*, 36(6–9): 613–616.

Ichikawa S. and Doi R. 1996. Hydrogen production from water and conversion of carbon dioxide to useful chemicals by room temperature photoelectrocatalysis. Catalysis Today, 27: 271–277.

Jin H.G., Hong H., Sui J., and Liu Q.B. 2009. Fundamental study of novel mid- and low-temperature solar thermochemical energy conversion. *Science in China, Series E: Technological Sciences*, 52(5): 1135–1152.

Jitaru M. 2007. Electrochemical carbon dioxide reduction—Fundamental and applied topics (Review), *Journal of University of Chemical Technology and Metallurgy*, 42(4): 333–344.

Johnson E.T., and Schmidt-Dannert C. 2008. Light-energy conversion in engineered microorganisms. *Trends in Biotechnology*, 26(12): 682–689.

Kamat P.V. 2007. Meeting the clean energy demand: Nanostructure architectures for solar energy conversion. *Journal of Physical Chemistry C*. 111(7): 2834–2860.

Kanan M.W. and Nocera D.G. 2008. In situ formation of an oxygen-evolving catalyst in neutral water containing phosphate and Co^{2+}. *Science*, 321: 1072–1075.

Kaneko M., Nemoto J., Ueno H., Gokan N., Ohnuki K., Horikawa M., Saito R. and Shibata T. 2006. Photoelectrochemical reaction of biomass and bio-related compounds with nanoporous TiO_2 film photoanode and O_2-reducing cathode. *Electrochemistry Communication*, 8(2): 336–340.

Kelly N.A. and Gibson T.L. 2006. Design and characterization of a robust photoelectrochemical device to generate hydrogen using solar water splitting. *International Journal of Hydrogen Energy*, 31(12): 1658–1673.

Kerry Yu K.M., Curcic I., Gabriel J., Tsang S.C.E. 2008. Recent advances in CO_2 capture and utilization. *ChemSusChem*, 1(11): 893–899.

Khaselev O. and Turner J.A. 1998. A monolithic photovoltaic-photoelectrochemical device for hydrogen production via water splitting. *Science*, 280: 425–427.

Kitano M., Matsuoka M., Ueshima M. and Anpo M. 2007. Recent developments in titanium oxide-based photocatalysts. *Applied Catalysis A: General*, 325: 1–14.

Kodama T. 2003. High-temperature solar chemistry for converting solar heat to chemical fuels. *Progress in Energy and Combustion Science*, 29(6): 567–597.

Kodama T. and Gokon N. 2007. Thermochemical cycles for high-temperature solar hydrogen production. *Chemical Reviews*. 107(10): 4048–4077.

Kudo A. and Miseki Y. 2009. Heterogeneous photocatalyst materials for water splitting. *Chemical Society Reviews*, 38(1): 253–278.

Larsen J., Brüggemann B., Khoury T., Sly J., Crossley M.J., Sundström V., and Åkesson E. 2007. Structural induced control of energy transfer within Zn(II)—Porphyrin dendrimers. *Journal of Physical Chemistry A*, 111(42): 10589–10597.

Lewis N.S., Crabtree G., Nozik A., Wasielewski M., and Alivisatos P. 2005. *Basic Research Needs for Solar Energy Utilization*. Washington, DC: U.S. Department of Energy. http://authors.library.caltech.edu/8599/1/SEU_rpt05.pdf (accessed on February 25, 2011).

Li O.-S. and Aida T. 2006. Construction of artificial photosynthetic systems based on multiporphyrin dendrimers. *Polymer Preprints, Japan*, 55(1): 405.

Licht S. 2003. Solar water splitting to generate hydrogen fuel: Photothermal electrochemical analysis. *Journal of Physical Chemistry B*, 107(18): 4253–4260.

Licht S. 2005. Thermochemical solar hydrogen generation. *Chemical Communications*, 37(7): 4635–4646.

Löbmann P. 2007. From sol-gel processing to bio-inspired materials synthesis. *Current Nanoscience*, 3(4): 306–328.

Lubitz W., Reijerse E.J., and Messinger J. 2008. Solar water-splitting into H_2 and O_2: Design principles of photosystem II and hydrogenases. *Energy & Environmental Science*, 1(1): 15–31.

Maeda K., Teramura K., Lu D., Takata T., Saito N., Inoue Y., and Domen K. 2006a. Photocatalyst releasing hydrogen from water. *Nature*, 440: 295.

Maeda K., Teramura K., Lu D., Saito N., Inoue Y., and Domen K. 2006b. Noble-metal/Cr_2O_3 core/shell nanoparticles as a cocatalyst for photocatalytic overall water splitting. *Angewandte Chemie International Edition*, 45(46): 7806–7809.

Meunier C.F., Cutsem P.V., Kwon Y.-U., and Su B.-L. 2009a. Investigation of different silica precursors: Design of biocompatible silica gels with long term bio-activity of entrapped thylakoids toward artificial leaf. *Journal of Materials. Chemistry*, 19(24): 4131–4137.

Meunier C.F., Cutsem P.V., Kwon Y.-U., and Su B.-L. 2009b. Thylakoids entrapped within porous silica gel: Towards living matter able to convert energy. *Journal of Materials Chemistry*, 19(11): 1535–1542.

Minero C. and Maurino V. 2007. Solar photocatalysis for hydrogen production and CO_2 conversion. In *Catalysis for Renewables* (G. Centi and R. van Santen, Eds.), Weinheim, Germany: Wiley-VCH, Ch. 16, pp. 351–385.

Muradov N.Z. and Veziroğlu T.N. 2008. "Green" path from fossil-based to hydrogen economy: An overview of carbon-neutral technologies. *International Journal of Hydrogen Energy*, 33(23): 6804–6839.

Nakćenović N. 2000. Energy scenarios (Chapter 9). In *World Energy Assessment: Energy and the Challenge of Sustainability*. UNDP/UN-DESA/World Energy Council. http://stone.undp.org/undpweb/seed/wea/pdfs/chapter9.pdf (accessed on February 25, 2011).

Navarro Yerga R.M., Alvarez Galván M.C., del Valle F., Villoria de la Mano J.A., and Fierro J.L. 2009. Water splitting on semiconductor catalysts under visible-light irradiation. *ChemSusChem*, 2(6): 471–485.

Nowotny J. and Sheppard L.R. 2007. Solar-hydrogen. *International Journal Of Hydrogen Energy*, 32(14): 2607–2608.

Nowotny J. 2008. Titanium dioxide-based semiconductors for solar-driven environmentally friendly applications: Impact of point defects on performance. *Energy & Environmental Science*, 1(5): 565–572.

N'Tsoukpoe K.E., Liu H., Le Pierrès N., and Luo L. 2009. A review on long-term sorption solar energy storage. *Renewable and Sustainable Energy Reviews*, 13(9): 2385–2396.

Olah G.A., Goeppert A., and Surya Prakash G.K. 2006. *Beyond Oil and Gas: The Methanol Economy*, Weinheim, Germany: Wiley-VCH (ISBN: 978-3-527-32422-4).

Price R.J., Morse D.A., Hardy S.L., Fletcher T.H., and Jensen R.J. 2004. Modeling the direct solar conversion of CO_2 to CO and O_2. *Industrial & Engineering Chemistry Research*, 43(10): 2446–2453.

Qu J., Zhang X., Wang Y., and Xie C. 2005. Electrochemical reduction of CO_2 on RuO_2/TiO_2 nanotubes composite modified Pt electrode. *Electrochimica Acta*, 50(16–17): 3576–3580.

Ryu S.Y., Choi J., Balcerski W., Lee T.K., and Hoffmann M.R. 2007. Photocatalytic production of H_2 on nanocomposite catalysts. *Industrial & Engineering Chemistry Research*, 46(23): 7476–7488.

Roeb M., Neises M., Säck J.-P., Rietbrock P., Monnerie N., Dersch J., Schmitz M., and Sattler C. 2009. Operational strategy of a two-step thermochemical process for solar hydrogen production. *International Journal of Hydrogen Energy*, 34(10): 4537–4545.

Rooke J.C., Meunier C., Léonard A., and Su B.-L. 2008. Energy from photobioreactors: Bioencapsulation of photosynthetically active molecules, organelles, and whole cells within biologically inert matrices. *Pure and Applied Chemistry*, 80(11):2345–2376.

Sakurai H. and Masukawa H. 2007. Promoting R & D in photobiological hydrogen production utilizing mariculture-raised cyanobacteria. *Marine Biotechnology*, 9(2): 128–145.

Sarangi S.K., Barpujari B., and Dawar R. 2009. Emerging technology options for clean power generation—Concentrated solar power (CSP). *Presented at Petrotech 2009* (New Delhi, India, January 2009), Paper P09-841. http://www.solarthermalworld.org/files/csp%20india.pdf?download (accessed on February 25, 2011).

Serroni S., Campagna S., Puntoriero F., Loiseau F., Ricevuto V., Passalacqua R., and Galletta M. 2003. Dendrimers made of Ru(II) and Os(II) polypyridine subunits as artificial light-harvesting antennae. *Comptes Rendus Chimie*, 6(8–10): 883–893.

Shibata H., Moulijn J.A., and Mul G. 2008. Enabling electrocatalytic Fischer–Tropsch synthesis from carbon dioxide over copper-based electrodes. *Catalysis Letters*, 123(3–4): 186–192.

Supritz C., Engelmann A., and Reineker P. 2006. Energy transport in dendrimers. *Journal of Luminescence*, 119–120 (Issue SPEC. ISS.): 337–340.

Tikkanen M., Suorsa M., and Aro E.-M. 2009. The flow of solar energy to biofuel feedstock via photosynthesis. *International Sugar Journal*, 111(1323): 156–163.

Traynor A.J. and Jensen R.J. 2002. Direct solar reduction of CO_2 to fuel: First prototype results. *Industrial & Engineering Chemistry Research*, 41: 1935–1939.

Tributsch, H. 2008. Photovoltaic hydrogen generation. *International Journal of Hydrogen Energy*, 33(21): 5911–5930.

Udo de Haes H.A. and Heijungs R. 2007. Life-cycle assessment for energy analysis and management. *Applied Energy*, 84(7–8): 817–827.

Usubharatana P., McMartin D., Veawab A., and Tontiwachwuthikul P. 2006. Photocatalytic process for CO_2 emission reduction from industrial flue gas streams. *Industrial & Engineering Chemistry Research*, 45(8): 2558–2568.

Varghese O.K., Paulose M., LaTempa T.J., and Grimes C.A. 2009. High-rate solar photocatalytic conversion of CO_2 and water vapor to hydrocarbon fuels. *Nano Letters*, 9(2): 731–737.

Venturi M., Balzani V., and Gandolfi M.T. 2005. Fuels from solar energy. A dream of giacomo ciami-
cian, the father of photochemistry. *Proceedings of the 2005 Solar World Congress* (Orlando, FL,
August 2005), American Solar Energy Society. Boulder, Colorado. http://www.gses.it/pub/
Ciamician.pdf

Wall J., Harwood C.S., and Demain A.L. 2008. *Bioenergy*. Washington, DC: ASM Press (ISBN
978-1-55581-478-6).

Watanabe K. 2008. Recent developments in microbial fuel cell technologies for sustainable bioenergy.
Journal of Bioscience and Bioengineering, 106(6): 528–536.

Zheng L., Xu Y., Song Y., Wu C., Zhang M., and Xie Y. 2009. Nearly monodisperse CuInS$_2$ hierarchical microarchitectures for photocatalytic H$_2$ evolution under visible light. *Inorganic Chemistry*,
48(9): 4003–4009.

5

Efficient Utilization of Solar, Wind, and Geothermal Energy Sources through Exergy Analysis

Arif Hepbasli

CONTENTS

5.1 Introduction

Since the beginning of 2007, the European Union has focused its emphasis on developing a new policy that puts energy back at the heart of this union action (Doukas et al., 2009). European energy strategy and policy are strongly driven by the twin objectives of sustainability (including environmental aspects) and security of supply. Implementation of new energy technologies are of big importance for satisfying these objectives, such as renewable energy systems at the supply side and improved energy end-use efficiency at the demand side (EC, 2006; Doukas et al., 2009).

Renewable energy can satisfy at least two orders of magnitude more than the world energy demand without negative environmental impacts. Wind and solar photovoltaics (PVs) are experiencing an exponential growth as their costs decrease. Interest is renewed in solar-thermal power, while geothermal energy deserves more attention (Lior, 2010).

The implementation of renewable energy strategies typically involve three major technological changes, namely, (i) energy savings on the demand side, (ii) efficiency improvements in the energy production, and (iii) replacement of fossil fuels by various sources of renewable energy. Consequently, large-scale renewable energy implementation plans must include strategies for integrating renewable sources into coherent energy systems with associated energy savings and increases in efficiency. First, the major challenge is to expand the amount of renewable energy in the supply system. Renewable energy is considered an important resource in many countries around the world, but on a global scale less than 15% of primary energy supply is renewable energy, and the major part is hydro power and wood fuels in developing countries. Renewable sources, such as wind and solar, only constitute a very small share of the total supply; however, their potential is substantial. In some regions and countries, the share of renewable energy has grown substantially during the last couple of decades. Two major challenges of renewable energy strategies for sustainable development can be identified. One challenge is to integrate a high share of intermittent resources into the energy system, especially the electricity supply, while the other is to include the transportation sector in the strategies (Lund, 2010).

Sustainable development does not make the world "ready" for the future generations, but it establishes a basis on which the future world can be built. A sustainable energy system may be regarded as a cost-efficient, reliable, and environmentally friendly energy system that effectively utilizes local resources and networks. It is not "slow and inert" like a conventional energy system, but it is flexible in terms of new techno-economic and political solutions. The introduction of new solutions is also actively promoted (Alqnne and Saari, 2006). Dincer (2002) reported the linkages between energy and exergy, exergy and the environment, energy and sustainable development, and energy policy making and exergy in detail. The author provided the following key points to highlight the importance of the exergy and its essential utilization in numerous ways:

1. It is a primary tool that best addresses the impact of energy resource utilization on the environment.

2. It is an effective method using the conservation of mass and conservation of energy principles together with the second law of thermodynamics for the design and analysis of energy systems.

3. It is a suitable technique for furthering the goal of more efficient energy–resource use, for it enables the locations, types, and true magnitudes of wastes and losses to be determined.

4. It is an efficient technique revealing whether or not and by how much it is possible to design more efficient energy systems by reducing the inefficiencies in existing systems.

5. It is a key component in achieving a sustainable development.

An exergy analysis (or second law analysis) has proven to be a powerful tool in the simulation thermodynamic analyses of energy systems. In other words, it has been widely used in the design, simulation, and performance evaluation of energy systems. The exergy analysis method is employed to detect and to evaluate quantitatively the causes of the thermodynamic imperfection of the process under consideration. It can, therefore, indicate the possibilities of thermodynamic improvement of the process under consideration, but only an economic analysis can decide the expediency of a possible improvement (Szargut et al., 1998; Rosen and Dincer, 2003a).

The concepts of exergy, available energy, and availability are essentially similar. The concepts of exergy destruction, exergy consumption, irreversibility, and lost work are also essentially similar. Exergy is also a measure of the maximum useful work that can be done by a system interacting with an environment that is at constant pressure P_0 and temperature T_0. The simplest case to consider is that of a reservoir with heat source of infinite capacity and invariable temperature T_0. It has been considered that maximum efficiency of heat withdrawal from a reservoir that can be converted into work is the Carnot efficiency (Rosen and Dincer, 2003b; Rosen et al., 2005).

Dincer et al. (2004) highlighted that, to provide an efficient and effective use of fuels, it is essential to consider the quality and quantity of the energy used to achieve a given objective. In this regard, the first law of thermodynamics deals with the quantity of energy and asserts that energy cannot be created or destroyed, whereas the second law of thermodynamics deals with the quality of energy, that is, it is concerned with the quality of energy to cause change, degradation of energy during a process, entropy generation, and the lost opportunities to do work. More specifically, the first law of thermodynamics is concerned only with the magnitude of energy with no regard to its quality; on the other hand, the second law of thermodynamics asserts that energy has quality as well as quantity. By quality, it means the ability or work potential of a certain energy source having certain amount of energy to cause change, that is, the amount of energy that can be extracted as useful work, which is termed as exergy. First and second law efficiencies are often called energy and exergy efficiencies, respectively. It is expected that exergy efficiencies are usually lower than the energy efficiencies, because the irreversibilities of the process destroy some of the input exergy.

This chapter is organized as follows. It starts with the introductory information on renewable energy resources (RERs) and the exergy concept, followed by energetic and exergetic relations used in the analyses of the RERs. Some important applications of RER systems along with their performance evaluation aspects and indicators are discussed and explained. Finally, some concluding remarks are presented.

5.2 General Relations

5.2.1 "Dead (Reference)" State

Exergy analysis provides a method to evaluate the maximum work extractable from a substance relative to a reference state (i.e., "dead" state). This reference state is arbitrary, but for terrestrial energy conversion the concept of exergy is most effective if it is chosen to reflect the environment on the surface of the Earth. The various forms of exergy are due to random thermal motion, kinetic energy, potential energy associated with a restoring force, or the concentration of species relative to a reference state. In order to establish how much work potential is contained in a given resource, it is necessary to compare it against a state defined to have zero work potential. An equilibrium environment that cannot undergo an energy conversion process to produce work is the technically correct candidate for a reference state (Hermann, 2006).

It should be noticed that exergy is always evaluated with respect to a reference environment (i.e., dead state). When a system is in equilibrium with the environment, the state of the system is called the dead state due to the fact that the exergy is zero. At the dead state, the conditions of mechanical, thermal, and chemical equilibrium between the system and the environment are satisfied: the pressure, temperature, and chemical potentials of the system equal those of the environment, respectively. In addition, the system has no motion or elevation relative to coordinates in the environment. Under these conditions, there is neither possibility of a spontaneous change within the system or the environment nor an interaction between them. The value of exergy is zero. Another type of equilibrium between the system and environment relates to a restricted form of equilibrium, where only the conditions of mechanical and thermal equilibrium (thermo-mechanical equilibrium) must be satisfied. Such state is called the restricted dead state. At the restricted dead state, the fixed quantity of matter under consideration is imagined to be sealed in an envelope impervious to mass flow, at zero velocity and elevation relative to coordinates in the environment, and at the temperature T_0 and pressure P_0 taken often as 25°C and 1 atm (Moran, 1982).

5.2.2 General Energetic and Exergetic Relations

A general expression for the conservation of mass of a control volume is given as (Bejan et al., 1996)

$$\frac{dm_{cv}}{dt} = \sum_{in} \dot{m}_{in} - \sum_{out} \dot{m}_{out} \tag{5.1a}$$

where the left side represents the time rate of change of mass contained within the control volume, and the subscripts *in* and *out* stand for inlet and outlet states, respectively.

In steady state, $dm_{CV}/dt = 0$. Therefore, Equation 5.1a is reduced to

$$\sum_{in} \dot{m}_{in} = \sum_{out} \dot{m}_{out} \tag{5.1b}$$

The energy balance on a rate basis for a control volume with multiple inlets and outlets is written as (Bejan et al., 1996)

$$\frac{dE_{CV}}{dt} = \dot{Q}_{CV} - \dot{W}_{CV} + \sum_{in} \dot{m}_{in}\left(h + \frac{V^2}{2} + gz\right)_{in} - \sum_{out} \dot{m}_{out}\left(h + \frac{V^2}{2} + gz\right)_{out} \tag{5.2a}$$

where

dE_{CV}/dt is the time rate of change of energy

$\dot{Q}_{CV}$ is the heat transfer rate over the boundary of the control volume

$\dot{W}_{CV}$ includes work effects, such as those associated with rotating shafts, displacement of the boundary, and electrical effects

h is the specific enthalpy

V is the velocity

z is the elevation

In a steady state, $dE_{CV}/dt = 0$. Therefore, Equation 5.2a is reduced to

$$\dot{Q}_{CV} - \dot{W}_{CV} = \sum_{out} \dot{m}_{out}\left(h + \frac{V^2}{2} + gz\right)_{out} - \sum_{in} \dot{m}_{in}\left(h + \frac{V^2}{2} + gz\right)_{in} \tag{5.2b}$$

The exergy balance can be expressed in various forms that may be more appropriate for particular applications. A convenient form of the exergy balance for closed systems is given as (Bejan et al., 1996)

$$\frac{dEx}{dt} = \sum_j \left(1 - \frac{T_0}{T_j}\right)\dot{Q}_j - \left(\dot{W} - P_0 \frac{dV}{dt}\right) - \dot{Ex}_{dest} \tag{5.3a}$$

where

dEx/dt is the time rate of change of exergy

$[1 - (T_0/T_j)]\,\dot{Q}_j$ represents the exergy transfer rate associated with heat transfer at the rate $\dot{Q}_j$ occurring at the location on the boundary where the instantaneous temperature is T_j

$\dot{W}$ indicates energy transfer rate by work, and the associated exergy transfer is given by $\dot{W} - P_0\,(dV/dt)$ where dV/dt is the time rate of change of system volume

$\dot{Ex}_{dest}$ accounts for exergy destruction rate due to irreversibilities within the system and is related to entropy generation rate by $\dot{Ex}_{dest} = T_0\,\dot{S}_{gen}$

The closed system exergy balance is used as a basis for extending the exergy balance concept to control volumes, which is the case of greater practical utility. Like mass, energy, and entropy, exergy can be transferred into or out of a control volume where streams of matter enter and exit. Accordingly, the counterpart of Equation 5.3a applicable to control volumes requires the addition of terms accounting for each exergy transfers (Bejan et al., 1996):

$$\frac{dEx_{CV}}{dt} = \sum_j \left(1 - \frac{T_0}{T_j}\right)\dot{Q}_j - \left(\dot{W}_{CV} - P_0 \frac{dV_{CV}}{dt}\right) + \sum \dot{m}_{in}\,ex_{in} - \sum \dot{m}_{out}\,ex_{out} - \dot{Ex}_{dest} \tag{5.3b}$$

where dEx_{CV}/dt is the time rate of change in the exergy of the control volume.

The term $\dot{Q}_j$ represents the heat transfer rate at the location on the boundary of the control volume where the instantaneous temperature is T_j and the associated exergy transfer rate is given by

$$\dot{Ex}_{q,j} = \left(1 - \frac{T_0}{T_j}\right)\dot{Q}_j \tag{5.3c}$$

As the control volume energy rate balance, $\dot{W}_{CV}$ indicates the energy transfer rate by work other than flow work. The associated exergy transfer rate is given by

$$\dot{Ex}_w = \dot{W}_{CV} - P_0 \left(\frac{dV_{CV}}{dt} \right) \tag{5.3d}$$

The term $\dot{W}$ indicates energy transfer rate by work, and the associated exergy transfer is given by $\dot{W} - P_0(dV/dt)$ where dV/dt is the time rate of change of system volume. $\dot{Ex}_{dest}$ accounts for exergy destruction rate due to irreversibilities within the system and is related to entropy generation rate by $\dot{Ex}_{dest} = T_0 \dot{S}_{gen}$.

In steady state, $dEx_{CV}/dt = 0$ and $dV_{CV}/dt = 0$, so Equation 5.3b is reduced to

$$0 = \sum_j \left(1 - \frac{T_0}{T_j} \right) \dot{Q}_j - \dot{W}_{CV} + \sum \dot{m}_{in} \, ex_{in} - \sum \dot{m}_{out} \, ex_{out} - \dot{Ex}_{dest} \tag{5.3e}$$

Expressed in terms of the time rates of exergy transfer and destruction, Equation 5.3e takes the following form (Bejan et al., 1996):

$$0 = \sum_j \dot{Ex}_{q,j} - \dot{W}_{CV} + \sum \dot{Ex}_{in} - \sum \dot{Ex}_{out} - \dot{Ex}_{dest} \tag{5.3f}$$

where
$\dot{Ex}_{in}$ and $\dot{Ex}_{out}$ are exergy transfer rates at inlets and outlets, respectively
$\dot{Ex}_{q,j}$ is given by Equation 5.3c

5.2.3 Exergy Balance Equations

The general exergy balance can be written as follows (Hepbasli, 2008):

$$\sum \dot{Ex}_{in} - \sum \dot{Ex}_{out} = \sum \dot{Ex}_{dest} = \dot{I} \tag{5.4a}$$

$$\dot{Ex}_{heat} - \dot{Ex}_{work} + \dot{Ex}_{mass,in} - \dot{Ex}_{mass,out} = \dot{Ex}_{dest} \tag{5.4b}$$

with

$$\dot{Ex}_{heat} = \sum \left(1 - \frac{T_0}{T_k} \right) \dot{Q}_k \tag{5.4c}$$

$$\dot{Ex}_{work} = \dot{W} \tag{5.4d}$$

$$\dot{Ex}_{mass,in} = \sum \dot{m}_{out} \psi_{out} \tag{5.4e}$$

$$\dot{Ex}_{mass,out} = \sum \dot{m}_{out} \psi_{out} \tag{5.4f}$$

where
$\dot{Q}_k$ is the heat transfer rate through the boundary at temperature T_k at location k
$\dot{W}$ is the work rate

The geothermal brine exergy inputs from the production field of two geothermal district heating systems (GDHS) investigated are calculated from the following equations:

$$\dot{Ex}_{brine} = \dot{m}_w \left[(h_{brine} - h_0) - T_0 (s_{brine} - s_0) \right] \tag{5.5}$$

The exergy destructions in the heat exchanger, pump, and the system itself are calculated from

$$\dot{Ex}_{dest,\, HE} = \dot{Ex}_{in} - \dot{Ex}_{out} = \dot{Ex}_{dest} \tag{5.6}$$

$$\dot{Ex}_{dest,pump} = \dot{W}_{pump} - (\dot{Ex}_{out} - \dot{Ex}_{in}) \tag{5.7}$$

and

$$\dot{Ex}_{dest,system} = \sum \dot{Ex}_{dest,HE} + \sum \dot{Ex}_{dest,pump} \tag{5.8}$$

The exergy efficiency of a heat exchanger is determined by the increase in the exergy of the cold stream divided by the decrease in the exergy of the hot stream on a rate basis as follows:

$$\varepsilon_{HE,1} = \frac{\dot{m}_{cold} (\psi_{cold,out} - \psi_{cold,in})}{\dot{m}_{hot} (\psi_{hot,in} - \psi_{hot,out})} \tag{5.9}$$

Kotas (1995) also derived the following relation to study the effect of operating temperature range on the heat exchanger performance where $T_{1,m}$ and $T_{2,m}$ are the mean temperatures of the streams and it is assumed that $\Delta T < T_{1,m}$ where $\Delta T = T_{1,m} - T_{2,m}$

$$\varepsilon_{HE,2} = 1 - \left(\frac{2 T_0 \Delta T / T_{1,m}^2}{\left((T_{1,m} - T_0) / T_{1,m} \right) + \left(T_0 \Delta T / T_{1,m}^2 \right)} \right) \tag{5.10}$$

5.2.4 Energy and Exergy Efficiency Relations

Basically, the energy efficiency of the system can be defined as the ratio of total energy output to total energy input

$$\eta = \left(\frac{\dot{E}_{output}}{\dot{E}_{input}} \right) \tag{5.11}$$

where in most cases *"output"* refers to *"useful"* one.

Numerous ways of formulating exergetic (or exergy or second-law) efficiency (effectiveness, or rational efficiency) for various energy systems are given in detail elsewhere (Kotas, 1995). In a similar way, exergy efficiency may be defined as the ratio of total exergy output to total exergy input:

$$\varepsilon = \left(\frac{\dot{Ex}_{output}}{\dot{Ex}_{input}} \right) \tag{5.12}$$

where "output" refers to "net output" or "product" or "desired value," and "input" refers to "given" or "used."

5.2.5 Some Exergetic Parameters

Thermodynamics analysis of renewable energy systems may also be performed using the following parameters (Xiang et al., 2004):
  Fuel depletion ratio

$$\delta_i = \left(\frac{\dot{Ex}_{dest,i}}{\dot{F}_T} \right) \tag{5.13}$$

  Relative irreversibility (exergy destruction)

$$\chi_i = \left(\frac{\dot{Ex}_{dest,i}}{\dot{Ex}_{dest,T}} \right) \tag{5.14}$$

  Productivity lack

$$\xi_i = \left(\frac{\dot{Ex}_{dest,i}}{\dot{P}_T} \right) \tag{5.15}$$

  Exergetic factor

$$f_i = \left(\frac{\dot{F}_i}{\dot{F}_T} \right) \tag{5.16}$$

5.2.6 Exergetic Improvement Potential Rate

Van Gool (1997) has also proposed that maximum improvement in the exergy efficiency for a process or system is obviously achieved when the exergy loss or irreversibility $(\dot{Ex}_{in} - \dot{Ex}_{out})$ is minimized. Consequently, he suggested that it is useful to employ the concept of an exergetic *improvement potential* when analyzing different processes or sectors of the economy. This improvement potential in the rate form, denoted $\dot{IP}$, is given by

$$\dot{IP} = (1 - \varepsilon)(\dot{Ex}_{in} - \dot{Ex}_{out}) \tag{5.17}$$

5.2.7 Sustainability Index

A sustainable development requires that the resources should be used efficiently. Exergy analysis method is a very useful tool for maximizing the benefits and using the resources very efficiently. In this analysis, the Sustainability Index (*SI*) method directly related to exergy efficiency (ε) is used to improve and contribute to the sustainable development as follows (Rosen et al., 2008):

$$SI = \left(\frac{1}{1-\varepsilon}\right) \tag{5.18}$$

5.2.8 Advanced Exergy Analysis

Splitting the exergy destruction into endogenous/exogenous and unavoidable/avoidable parts represents a new direction in exergy analysis, which can be called advanced exergy analysis. These splittings improve the accuracy of exergy analysis and our understanding of the thermodynamic inefficiencies, and facilitate an exergoeconomic optimization (Morosuk and Tsatsaronis, 2008).

Only a part of the exergy destruction rate within a component can be avoided. The exergy destruction rate that cannot be reduced due to technological limitations such as availability and cost of materials and manufacturing methods is the unavoidable ($\dot{Ex}_{dest,k}^{UN}$) part of the exergy destruction. The remaining part represents the avoidable ($\dot{Ex}_{dest,k}^{AV}$) part of the exergy destruction. Thus, splitting the exergy destruction into unavoidable and avoidable parts in the *k*th component provides a realistic measure of the potential for improving the thermodynamic efficiency of a component, as given below (Morosuk and Tsatsaronis, 2008):

$$(\dot{Ex}_{dest,k}) = (\dot{Ex}_{dest,k}^{UN}) + (\dot{Ex}_{dest,k}^{AV}) \tag{5.19}$$

Using an appropriate method (Morosuk and Tsatsaronis, 2006), the unavoidable endogenous exergy destruction within the *k*th component may be calculated when all the remaining components operate without irreversibilities. This calculation allows us then to obtain the unavoidable endogenous exergy destruction and subsequently the avoidable endogenous, the unavoidable exogenous, and the avoidable exogenous parts of exergy destruction within the *k*th component. The endogenous avoidable part of the exergy destruction can be reduced through improving the efficiency of the *k*th component. The exogenous avoidable part of the exergy destruction can be reduced by a structural improvement of the overall system or by improving the efficiency of the remaining components and of course by improving the efficiency in the *k*th component. The endogenous unavoidable part of the exergy destruction cannot be reduced because of technical limitations for the *k*th component. The exogenous unavoidable part of the exergy destruction cannot be reduced because of technical limitations in the other components of the overall system for the given structure (Morosuk and Tsatsaronis, 2008).

The advanced exergetic evaluation can be extended to consider components in pairs or larger groups in an effort to completely understand the thermodynamic interactions among the system components. However, all these considerations become more powerful when they are used for the exergoeconomic evaluation of the same system. In such a calculation, not only the exergy destruction, but also the investment cost for each system components is split into avoidable/unavoidable and endogenous/exogenous parts (Morosuk and Tsatsaronis, 2007). An advanced exergoeconomic evaluation should be conducted only when the investment costs for the different cases can be assessed with acceptable accuracy, whereas an advanced exergetic evaluation, as reported here, is always helpful.

5.3 Exergy Analysis of Renewable Energy Resources

5.3.1 Solar Energy Systems

In evaluating the performance of solar energy systems using exergy analysis method, calculation of the exergy of radiation is very crucial. However, its calculation is a problem of unquestionable interest, since exergy represents the maximum quantity of work that can be produced in some given environment (usually the terrestrial environment, considered as an infinite heat source or sink) (Bejan, 1988). Over a period of more than 20 years, many papers including various approaches to the calculation of radiation exergy have been published, as reported by Petela (2003) and Candau (2003). The first relation was proposed by Petela (1964), who gave the maximum efficiency ratio (or exergy-to-energy ratio for radiation) for determining an exergy of thermal emission at temperature T as follows (Petela, 2003, 2005):

$$\psi_{srad,max} = 1 + \frac{1}{3}\left(\frac{T_0}{T}\right)^4 - \frac{4}{3}\left(\frac{T_0}{T}\right) \tag{5.20}$$

where $\psi_{srad,max}$ is exergy of thermal emission at temperature T. T was taken to equal the solar radiation temperature (T_s) with 6000 K in exergetic evaluation of a solar cylindrical–parabolic cooker and a solar parabolic-cooker by Petela (2005) and Ozturk (2004), respectively. Equation 5.20 was also derived by Szargut (2003), who presented a simple scheme of a reversible prime mover (different from that of Petela) transforming the energy of radiation into mechanical or electrical work and took into account that the solar radiation had a composition similar to that of a black body.

The history of Equation 5.20 is as follows (Millan et al., 1997): It was first derived by Petela (1964) and independently by Landsberg and Mallinson (1976) and Press (1976), while Spanner (1964) proposed Petela's equation for the direct solar radiation exergy (2003). As a consequence, it is sometimes called the Landsberg efficiency (Szargut, 2003), the Petela–Landsberg upper bound (Badescu, 1991), or, more appropriately, the Petela–Press–Landsberg and Mallinson formula (Bejan, 1988).

Petela (2003) reported that in the existing literature in the above field, the most discussed formulae for heat radiation exergy were those derived by him (Petela, 1964; Spanner, 1964; Jeter, 1981) given in Equations 5.20 through 5.22 in Table 5.1, respectively. As for the application of these equations to various thermal systems by some investigators, Chaturvedi et al. (1998) used Equation 5.20 in the performance assessment of a variable capacity direct expansion solar-assisted heat pump system for domestic hot water application. Eskin (2000) considered Equation 5.21 in evaluating the performance of a solar process heating system, while Singh et al. (2000) and Ucar and Inalli (2006) considered Equation 5.22 in the exergetic analysis of a solar thermal power plant and the exergoeconomic analysis of a solar-assisted heating system, respectively.

TABLE 5.1

Numerators (Output) and Denominators (Input) of the Limiting Energy Efficiency of Radiation Utilization along with Exergy-to-Energy Ratios for Solar Radiation Reported by Various Researchers

	Researchers			
Items	**Petela (1964)**	**Spanner (1964)**	**Jeter (1981)**	**Nobusawa (1980)**
Input	Radiation energy	Radiation energy	Heat	—
Output	Useful work = radiation exergy	Absolute work	Net work of a heat engine	—
Expression proposed (exergy-to-energy ratio for solar radiation)	$1+\dfrac{1}{3}\left(\dfrac{T_0}{T}\right)^4-\dfrac{4}{3}\left(\dfrac{T_0}{T}\right)$ (5.20a)	$1-\dfrac{4}{3}\left(\dfrac{T_0}{T}\right)$ (5.21)	$1-\left(\dfrac{T_0}{T}\right)$ (5.22)	0.95

Source: Petela, R., *J. Heat Transfer*, 86, 187, 1964; Spanner, D.C., *Introduction To Thermodynamics*, Academic Press, London, U.K., 1964; Svirezhev, Y.M. et al., *Ecol. Model.*, 169, 339, 2003; Jeter, S.M., *Sol. Energy*, 26, 231, 1981; Nobusawa, T., *Introductory Text of Exergy*, Tokyo, Ohmsha (in Japanese), 1980.

In Table 5.1, various limiting energy efficiencies of utilization of the radiation matter are listed according to (Petela, 1964; Spanner, 1964; Jeter, 1981), while Nobusawa (1980) used the most simplified approach (by taking exergy-to-energy ratio of solar radiation to be 0.95) for evaluating the exergy of solar radiation. Petela (2003) explained this as follows: "Bejan's conclusion was that all theories (Spanner's, Jeter's, and Petela's) concerning the ideal conversion of thermal radiation into work, although obtaining different results, were correct. Although really correct, they had incomparable significance for the true estimation of the radiation exergy value. All of the efficiencies assumed work as an output. However, Petela's work was equal to the radiation exergy, Jeter's work was the heat engine cycle work and Spanner's work was an absolute work. As an input, Petela and Spanner assumed the radiation energy, whereas Jeter assumed heat."

Svirezhev et al. (2003) applied a new concept regarding the exergy of solar radiation to the analysis of satellite data describing the seasonal (monthly) dynamics of four components of the global radiation balance: incoming and outgoing short- and long-wave solar radiation. They constructed maps of annual mean exergy and some derivative values using real satellite data. They concluded that exergy could be a good indicator for crucial regions of the oceans, which were indicated where its values were maximal.

5.3.1.1 Solar Collector Applications

The instantaneous exergy efficiency of the solar collector can be defined as the ratio of the increased water exergy to the exergy of the solar radiation (Ozturk, 2004). In other words, it is a ratio of the useful exergy delivered to the exergy absorbed by the solar collector (Singh et al., 2000).

$$\varepsilon_{scol}=\left(\frac{\dot{E}x_u}{\dot{E}x_{scol}}\right) \tag{5.23}$$

with

$$\dot{Ex}_u = \dot{m}_w\left[(h_{w,out}-h_{w,in})-T_0(s_{w,out}-s_{w,in})\right] \tag{5.24a}$$

or

$$Ex_u = \dot{m}_w C_w \left[(T_{w,out} - T_{w,in}) - T_0 \left(Ln \frac{T_{w,out}}{T_{w,in}} \right) \right] \tag{5.24b}$$

$$Ex_u = \dot{Q}_u \left[1 - \left(\frac{T}{T_{w,out} - T_{w,in}} \right) \left(Ln \frac{T_{w,out}}{T_{w,in}} \right) \right] \tag{5.24c}$$

and

$$Ex_{scol} = A I_T \psi_{srad,max} \tag{5.25}$$

where

the subscripts "in" and "out" denote the inlet and outlet of the solar collector, respectively

$\psi_{srad,max}$ may also be calculated using Equations 5.20 through 5.22 given in Table 5.1

The exergy of solar radiation with beam (I_{be}) and diffuse (I_d) components for parabolic collectors is also given by the following equation (Onyegegbu and Morhenne, 1993), which was used by Eskin (1999) in the performance evaluation of a cylindrical parabolic concentrating collector:

$$\psi_{srad,be,d} = I_{be} \left(1 - \frac{4T_0}{3T_s} \right) + I_d \left(1 - \frac{4T_0}{3T_s^*} \right) \tag{5.26}$$

with

$$\frac{T_s}{T_s^*} = 0.9562 + 0.2777 \ln \left(\frac{1}{\kappa} \right) + 0.0511\kappa \tag{5.27}$$

where κ is the dilution factor of diffuse radiation (Landsberg and Tonge, 1979).

Solar water heater is the most popular means of solar energy utilization because of technological feasibility and economic attractiveness compared with other solar energy utilization options (Xiaowua and Bena, 2005). Table 5.2 lists exergy efficiency values for various types of solar collectors, while one application among these will be explained below.

Figure 5.1 illustrates a schematic diagram of the experimental solar water heating (SWH) system tested in Solar Energy Institute of Ege University, Izmir, Turkey (Gunerhan and Hepbasli, 2007). This SWH system consists mainly of three parts, namely, (1) the flat plate solar collector (2 m² aperture area), (2) the heat exchanger with water storage tank, and (3) the circulating pump. Water is circulated through the closed collector loop to a heat exchanger, where its heat is transferred to the potable water. The collector is oriented toward south, inclined at an angle equal to 45° at Bornova in Izmir, Turkey (latitude 38°28′N: longitude 27°15′E).

Figure 5.2 illustrates the exergy efficiency curve for the solar collector investigated to observe the trend of the exergy efficiencies of the solar collector. This curve was proposed by the authors using the similar methodology applied to the thermal efficiency curve for

TABLE 5.2

Exergy Efficiency Values for Various Types of Solar Collectors

Investigator	Type of Solar Heater/Data Used	Location (City/Country)	Technical Specification	Exergy Efficiency
Xiaowuan and Bena (2005)	Domestic-scale/experimental	Yun Nan/China	Top and bottom hot water temperatures: 50°C and 38°C in the storage barrel, respectively; 2.5 m² of collector area; an average ambient temperature of 25°C; an average ambient water temperature of 24°C; 466 W m⁻² of whole-day average solar radiation and 196.4 kg of water storing log capacity for the storage barrel	0.77% for solar water heater
Gunerhan and Hepbasli (2007)	Solar water heating system/experimental	Bornova in Izmir/Turkey (latitude 38°28′N: longitude 27°15′E)	A flat plate solar collector with an aperture area of 2 m²; solar radiation values varying between 662.22 and 851.61 W m⁻²	Varied between 2.02% and 3.37% for the solar collector, 10% and 16.67% for the circulating pump an 16% and 51.72% for the heat exchanger at a dead (reference) state temperature of 32.77°C, while those were found to range from 3.27% to 4.39% for the overall system
Baghernejad and Yaghoubi (2010)	Integrated solar combined cycle/design plant data	Yazd/Iran	Yazd thermo-solar power plant with an estimated full capacity of 467 MW; a solar radiation intensity of about 800 W m⁻²; a solar field consisting of 42 loops (with each loop of 6 collectors from type of from type of LS-3) Assuming a collector fluid entering temperature of 298°C and a mass flow rate of fluid in the collectors with 218 kg/s and an outlet collector fluid temperature of 393°C	27% for solar collector
Akpınar and Koçyiğit (2010)	Flat-plate solar air heater having different obstacles on absorber plates/experimental	Elazig/Turkey (latitude 38.41°N, longitude 39.14°E, altitude 1067 m above sea level)	Four types of absorber plates are used; dimensions of 1.2 × 0.7 × 0.12 m	Varied between 8.32% and 44% for four types of solar air heaters
Alta et al. (2010a)	Flat-plate solar air heater with/without fins/experimental	Antalya/Turkey	Three different types of designed flat-plate solar air heaters, two having fins) and the other without fins; external dimensions of 0.63 × 0.315 × 0.15 m	Varied between 0.20% and 0.83% for three types of solar air heaters
Gupta and Kaushik (2010)	A conceptual direct steam generation solar–thermal power plant/proposed conceptual design	Longitude 5°58′W, latitude 37°24′N	A parabolic trough solar collector field consisting of 7 parallel rows of ET-100 collectors and each row composed of 10 collectors; an aperture width of 5.76 m; a single collector with overall length of 98.5 m; a direct solar irradiance of 875 W m⁻²	44.91% for collector, 69.05% for receiver and 31.02% for collector field

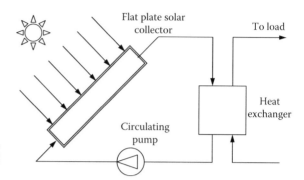

FIGURE 5.1
Schematic of the solar water heating system investigated. (Modified from Gunerhan, H. and Hepbasli, A., *Energ. Buildings*, 39, 509, 2007.)

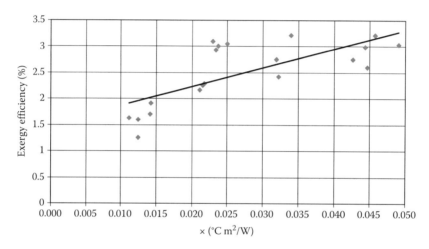

FIGURE 5.2
Exergy efficiency curve for the solar collector investigated. (Adapted from Gunerhan, H. and Hepbasli, A., *Energ. Buildings*, 39, 509, 2007.)

solar collectors. In this regard, the exergy efficiency was correlated with a coefficient of determination (R^2) of 0.54 as follows:

$$\varepsilon = 35.987\, x + 1.5087 \tag{5.28}$$

which is capable of predicting the values of exergy efficiencies of the solar collector under various operation conditions.

Gunerhan and Hepbasli (2007) have also undertaken a parametric study to investigate how varying water inlet temperature to the collector will affect the exergy efficiencies of the SWH system components. It is clear from Figure 5.3 that the exergetic efficiency of the overall system is almost stable with the values of under 4%.

5.3.1.2 Solar Thermal Power Plants

The solar thermal power system, in general, can be considered as consisting of two subsystems, namely, a collector–receiver subsystem and a heat engine subsystem as illustrated in Figure 5.4 (Singh et al., 2000).

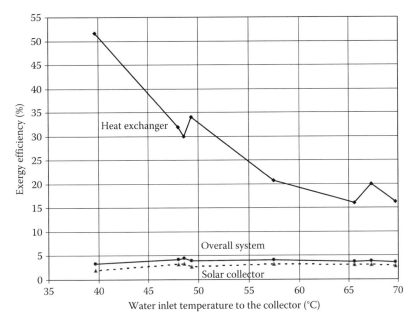

FIGURE 5.3
Variation of exergy efficiencies versus water inlet temperatures to the solar collector. (Adapted from Gunerhan, H. and Hepbasli, A., *Energ. Buildings*, 39, 509, 2007.)

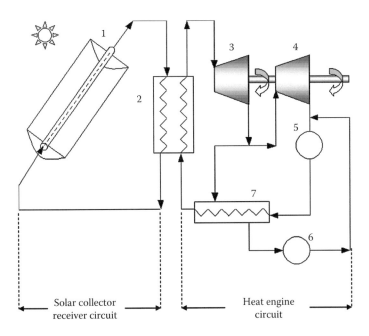

FIGURE 5.4
Schematic of a solar power thermal plant. 1, parabolic through collector; 2, boiler heat exchanger; 3, high pressure turbine; 4, low pressure turbine; 5, condenser; 6, valve; 7, regenerator. (Modified from Singh, N. et al., *Renew. Energ.*, 19, 135, 2000.)

The collector–receiver circuit consists of a number of parabolic trough collectors arranged in modules operating in tracking mode. The main components of the heat engine circuit are a heater (here the boiler heat exchanger), a turbine having two stages, a condenser, a pump, and a regenerator. Exergy analysis is conducted in two main subsystems along with their components, as given by Singh et al. (2000).

The overall exergy efficiency of the collector–receiver circuit is calculated in a similar manner given by Equation 5.23 as follows (Singh et al., 2000):

$$\varepsilon_{o,col\text{-}rec} = \left(\frac{\dot{Ex}_{u,col\text{-}rec}}{\dot{Ex}_{par,s}} \right) \tag{5.29}$$

with the useful exergy of the whole parabolic collector

$$\dot{Ex}_{u,col\text{-}rec} = n_{par,col}\dot{Ex}_u \tag{5.30}$$

and the exergy received by the parabolic collector given as

$$\dot{Ex}_{par,s} = \dot{Q}_s\left(1 - \frac{T_0}{T_s}\right) \tag{5.31}$$

where
$n_{par,col}$ is the number of the parabolic collectors
$\dot{Q}_s$ is the transferred solar heat
T_s is the solar temperature

The exergy received by the parabolic collector may also be calculated using Equations 5.26 through 5.27, and the relations given in Table 5.1. The useful exergy of one parabolic collector ($\dot{Ex}_u$) may be found using Equation 5.24b or c.

The total efficiency of the heat engine subsystem consisting of the boiler heat exchanger and the heat engine cycle is calculated as follows:

$$\varepsilon_{h,eng} = \left(\frac{\dot{W}_{net}}{\dot{Ex}_{h,eng}} \right) \tag{5.32}$$

with the available exergy of the working fluid of the heat engine cycle given as

$$\dot{Ex}_{h,eng} = \dot{Q}_u\left(1 - \frac{T_0}{T_{Ran,0}}\right) \tag{5.33}$$

where
$\dot{Q}_u$ is the useful transferred heat
$\dot{W}_{net}$ is the net work done by the heat engine cycle
$T_{Ran,0}$ is the ambient temperature of the Rankine cycle

The overall exergy efficiency of the solar thermal power system is the product of the two separate efficiencies of the subsystems (Singh et al. 2000):

$$\varepsilon_{h,eng} = \left(\frac{\dot{W}_{net}}{\dot{Ex}_{par,s}} \right)$$ (5.34)

Exergetic studies of two solar thermal power plants are briefly explained below. In this regard, Baghernejad and Yaghoubi (2010) has proposed an Integrated Solar Combined Cycle System as a means of integrating a parabolic trough solar thermal plant with modern combined cycle power plants. They analyzed the Integrated Solar Combined Cycle in Yazd, Iran using design plant data, while energy and exergy analyses for the solar field and combined cycle were carried out to assess the plant performance and to pinpoint sites of primary exergy destruction. The values of energy and exergy efficiencies for the system were found to be 46.17% and 45.6%, respectively. These efficiencies were higher than simple combined cycle power plant without solar contribution and Rankine cycle power plants with parabolic trough technology. Gupta and Kaushik (2010) performed energy and exergy analyses of the different components of a proposed conceptual direct steam generation solar-thermal power plant. They found that the maximum energy loss was in the condenser, followed by solar collector field. The maximum exergy loss was in the solar collector field while in other plant components it was small. The possibilities to further improve the plant efficiency were identified and exploited. For minimum exergy loss in receiver, the inlet temperature of water to the receiver, which was governed by the number of feed water heaters, bleed pressure, and mass fraction of bleed steam, must be optimum. It was reported that the exergy loss in the collector could be reduced by increasing the concentration ratio of the collector, which was limited due to material and design considerations, and the exergy loss in the receiver could be reduced by keeping optimum mean temperature of the receiver or heat energy collection. For the maximum exergetic efficiency of a parabolic-trough collector row, the inlet temperature of collector fluid (water) circulating through the receiver should be optimum for the same exit temperature of collector fluid.

5.3.1.3 Photovoltaics and Hybrid Solar Collectors

Exergetic evaluation of PVs has been performed by some investigators (Fujisawa and Tani, 1997; Saitoh et al., 2003; Sahin et al., 2007; Joshi et al., 2009a; Sarhaddi et al., 2010) in hybrid (PV/thermal) systems as a part of the system.

Electrical energy is not affected by ambient conditions and therefore is equivalent to work. If global irradiance is I, energetic efficiency of the solar cell is η_{scell}, the instantaneous electrical exergy is then as follows (Fujisawa and Tani, 1997):

$$\dot{Ex}_e = \eta_{scell}I = \varepsilon_{scell}I$$ (5.35)

where ε_{scell} is the exergetic efficiency of the solar cell.

The exergy efficiency of a hybrid solar collector that generates both electric power and heat, may be calculated as follows (Saitoh et al., 2003):

$$\varepsilon_{PV/Thermal} = \left(\frac{\eta_{conver}I + \dot{Ex}_{solar}}{\dot{Ex}_I} \right)$$ (5.36)

with

$$Ex\dot{}_{solar} = \left(\frac{T_{fluid} - T_0}{T_{fluid}} \right) \dot{Q}_{solar} \tag{5.37}$$

and

$$Ex\dot{}_I = 0.95I \tag{5.38}$$

where
η_{conver} is the conversion efficiency
I is the global irradiance (W m^{-2})
$Ex\dot{}_{solar}$ is the exergy of heat (W m^{-2})
$Ex\dot{}_I$ is the exergy of global irradiance (W m^{-2})
$\dot{Q}_{solar}$ is the collected solar heat amount per unit time per panel area (W m^{-2})
T_{fluid} is the supply temperature of collector fluid (K)

Fujisawa and Tani (1997) defined the *synthetic exergy of the PV/T collector* as the total value of the electrical and thermal exergies as follows:

$$Ex\dot{}_{PV/T} = Ex\dot{}_e + Ex\dot{}_{thermal} = (\varepsilon_e + \varepsilon_{thermal}) = \varepsilon_{PV/T}I \tag{5.39}$$

with

$$Ex\dot{}_e = \varepsilon_e I \tag{5.40}$$

and

$$Ex\dot{}_{thermal} = \dot{Q}\eta_{Carnot} = \dot{Q}\left(\frac{T_f - T_0}{T_f} \right) = \varepsilon_{thermal}I \tag{5.41}$$

Fujisawa and Tani (1997) designed and constructed a PV/thermal hybrid collector on their university campus. The collector consisted of a liquid heating flat-plate solar collector with mono-Si *PV* cells on substrate of nonselective aluminum absorber plate; the collector area was $1.3 \times 0.5\,m^2$. From the annual experimental evaluation based on exergy, Fujisawa and Tani (1997) concluded that the *PV/T* collector could produce higher output density than a unit *PV* module or liquid heating flat-plate solar collector. Using exergetic evaluation, the best performance of available energy was found to be that of the coverless *PV/T* collector at 80.8 kWh/year, the second to be the *PV* module at 72.6 kWh/year, the third to be the single-covered *PV/T* collector at 71.5 kWh/year, and the worst to be the flat-plate collector at 6.0 kW h/year.

Saitoh et al. (2003) performed field measurements from November 1998 to October 1999 at a low energy house at Hokkaido University. A system diagram of the experimental equipment is illustrated in Figure 5.5. The power generated was measured by giving the variable load for the maximum power point tracking. Two hybrid solar panels were connected in parallel, while the brine was supplied at constant temperature by a fluid supply system with a circulating pump. The volumetric flow rate per panel was fixed at 1 l/min.

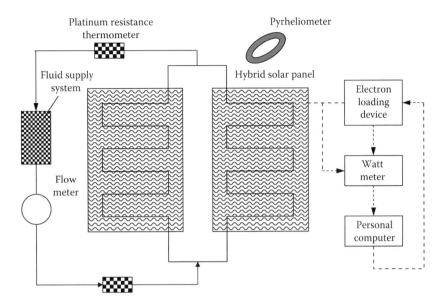

FIGURE 5.5
System diagram of experimental equipment. (Modified from Saitoh, H. et al., *Appl. Therm. Eng.*, 23, 2089, 2003.)

The tilt angle of the panel was 30°, which gave the maximum annual global irradiance. Except for winter, the mean conversion efficiency of array and the collector efficiency were stable at 8%–9% and 25%–28%, respectively. The dependency on solar energy was 46.3%. Energy and exergy efficiency values of single-junction crystalline silicon *PV* system were discussed by Saitoh et al. (2003). The exergy efficiencies of a solar collector, a *PV* array and a hybrid solar collector were found to be 4.4%, 11.2%, and 13.3%, respectively, whereas, energy efficiencies of the same solar-powered devices were 46.2%, 10.7%, and 42.6%, respectively. The solar collector had the lowest value of exergy efficiency, but in terms of energy efficiency the solar collector and hybrid solar collector were relatively more effective compared to the *PV* array.

Sahin et al. (2007) have applied exergy analysis to a *PV* system. They also obtained exergy components and *PV* array exergy efficiency, while they compared energy, electrical, exergy efficiencies under given climatic and operating conditions. Energy efficiency was seen to vary between 7% and 12% during the day. In contrast, exergy efficiencies, which incorporate the second law of thermodynamics and account for solar irradiation exergy values, were lower for electricity generation using the considered *PV* system and varied between 2% and 8%.

Joshi et al. (2009a) have carried out the performance analysis of both *PV* and *PV/T* system in terms of exergy efficiency and reported that the thermal energy due to solar radiation was actually a heat loss to the *PV* system whereas it was a useful heat for a *PV/T* system. They have also shown that the electrical (exergy) efficiency of a *PV* system can be improved if the heat can be removed from the *PV* surface.

Sarhaddi et al. (2010) evaluated the exergetic performance of a solar *PV/T* air collector, while they performed a detailed energy and exergy analysis to calculate the thermal and electrical parameters, exergy components, and exergy efficiency of a typical *PV/T* air collector. They also used an improved electrical model to estimate the electrical parameters of a *PV/T* air collector. Further, a modified equation for the exergy efficiency of a *PV/T* air collector was derived in terms of design and climatic parameters. It was reported that the exergy efficiency used in the previous literature had some deficiencies such as it did not include the exergy of pressure terms and it had a significant error at low solar radiation intensity.

5.3.1.4 Solar Exergy Maps

Joshi et al. (2009b) developed for the first time a solar exergy map concept and conducted a comprehensive case study to show how it is utilized and how it is significant for practical solar applications. Based on the exergy content of the solar radiation, they evaluated the performance of a *PV/T* system for different Indian cities, namely, Bangalore (latitude 12°58′N, longitude 77°38′E), Jodhpur (latitude 26°18′N, longitude 73°04′E), Mumbai (latitude 18°55′N, longitude 72°54′E), New Delhi (latitude 28°35′N, longitude 77°12′E), and Srinagar (latitude 34°08′N, longitude 74°51′E) for a year, and for different cities in the United States, namely, Chicago (latitude 41°50′N, longitude 87°37′W), Las Vegas (latitude 36°10′N, longitude 115°12′W), Miami (latitude 25°46′N, longitude 80°12′W), New York (latitude 40°47′N, longitude 73°58′E), Portland (latitude 43°40′N, longitude 70°15′W), San Antonio (latitude 29°23′N, longitude 98°33′W), San Francisco (latitude 37°47′N, longitude 122°26′W), Tucson (latitude 32°7′N, longitude 110°56′W), and Tulsa (latitude 36°09′N, longitude 95°59′W) for different months of January, April, June, and October. Exergy maps for the exergy of solar radiation were developed, while the exergy efficiency of *PV/T* system was determined for the above-mentioned Indian and American climatic conditions. It was found that the predicted exergy efficiency was in a good agreement with the experimental results for the climatic conditions of New Delhi, India. It was observed that the average exergy efficiency was highest in Bangalore from January (28%) to April (32.6%) and from September (32.5%) to December (32.4%), and it was highest in Srinagar from May (29.5%) to August (26.8%) for Indian climatic conditions; for American climatic conditions, the *PV/T* system gave the best performance in terms of exergy efficiency in Las Vegas (32%) and Tucson (32.5%–31.5%) in April and June.

   Alta et al. (2010b) mapped the national spatial distribution of mean monthly exergy values of solar radiation over Turkey at a resolution of 500 m using universal kriging based on solar radiation data from 152 geo-referenced locations. They also estimated mean exergy value of solar radiation in Turkey at $13.5 \pm 1.74 \, \text{MJ/m}^2\text{day}$, with a mean annual exergy-to-energy ratio of 0.93.

5.3.2 Wind Energy Systems

The exergy of fluid currents is entirely due to the kinetic energy of the fluid. The maximum work is extracted from a moving fluid when the velocity is brought to zero relative to the reference state. The kinetic energy of the fluid and its exergy have the same numerical value (Weston, 2006):

$$\psi_f = \frac{1}{2}\rho V_0^3 \tag{5.42}$$

where
  ρ is the density of the wind
  V_0 is the wind speed

5.3.2.1 Wind Energy Applications

Koroneos et al. (2003) presented two diagrams where utilization of the wind's potential in relation to the wind speed and exergy losses in the different components of a wind turbine (i.e., rotor, gearbox, and generator) were included. The authors concluded that, "According to Betz's law, wind turbine can take advantage of up to 60% of the power of the wind. Nevertheless, in practice, their efficiency is about 40% for quite high wind speeds. The rest of the energy density of the wind not obtainable is exergy loss. This exergy loss appears mainly as heat. It is attributed to the friction between the rotor shaft and the bearings, the heat that the cooling fluid abducts from the gearbox, the heat that the cooling fluid of the generator abducts from it and the thyristors, which assist in smooth starting of the turbine and which lose 1%–2% of the energy that passes through them."

As for the studies conducted on exergetic assessment of wind energy systems, Sahin et al. (2006a) developed a new exergy formulation for wind energy, which was more realistic and also accounted for the thermodynamic quantities enthalpy and entropy. The relation developed for the total exergy for wind energy is as follows, while the input and output variables for the system considered are illustrated in Figure 5.6 (Sahin et al. 2006a).

$$Ex = \dot{E}_{gen} + \dot{m}C_p(T_2 - T_1) + \dot{m}T_{at}\left(C_p \ln\frac{T_2}{T_1} - R\ln\frac{P_2}{P_1} - \frac{\dot{Q}_{loss}}{T_{at}}\right) \tag{5.43}$$

with

$$\dot{Q}_{loss} = \dot{m}C_p(T_{at} - T_{ave}) \tag{5.44}$$

where
$\dot{E}_{gen}$ is the generated electricity (i.e., the total kinetic energy difference)
T_{at} is the atmospheric temperature
P_2 is the pressure at the exit of the wind turbine for a wind speed V_2
P_1 is the pressure at the inlet of the wind turbine for a wind speed V_1
$\dot{Q}_{loss}$ represents heat losses of wind turbine
T_{ave} is the mean value of input and output wind chill temperatures

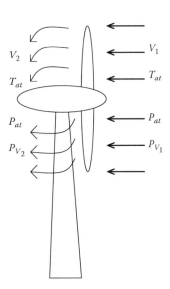

FIGURE 5.6
Wind turbine and representative wind energy input and output variables.
(Modified from Sahin, A.D. et al., *Int. J. Energ. Res.*, 30, 553, 2006.)

Sahin et al. (2006) reported as follows: "The exergy analysis of wind energy shows that there are significant differences between energy and exergy analysis results. According to one classical wind energy efficiency analysis technique, which examines capacity factor, the resultant wind energy efficiency is overestimated relative to what it really is. The capacity factor normally refers to the percentage of nominal power that the wind turbine generates. Average differences between energy and exergy efficiencies are approximately 40% at low wind speeds and up to approximately 55% at high wind speeds."

Ozgener and Ozgener (2007) performed exergy analysis of a wind turbine system (1.5 kW) located in Solar Energy Institute of Ege University (latitude 38.24 N, longitude 27.50 E), Izmir, Turkey. They reported that exergy efficiency changed between 0% and 48.7% at different wind speeds based on a dead state temperature of 25°C and atmospheric pressure of 101.325 kPa, considering pressure differences between state points. Depending on temperature differences between state points, exergy efficiencies were found to be 0%–89%.

The exergetic efficiency of a turbine is defined as a measure of how well the stream exergy of the fluid is converted into useful turbine work output or inverter work output. By using inverter work output as useful work and applying this to wind turbine, the following relations are obtained (Ozgener and Ozgener, 2007).

$$\varepsilon = \left(\frac{\dot{W}_{e,io}}{\dot{Ex}_1 - \dot{Ex}_2} \right) \tag{5.45}$$

with

$$\dot{Ex}_{dest} = (\dot{Ex}_1 - \dot{Ex}_2) - \dot{W}_e \tag{5.46}$$

where
$\dot{W}_{e,io}$ is the power at inverter output
$\dot{Ex}_{dest}$ is the exergy destruction
$\dot{Ex}_1$ and $\dot{Ex}_2$ are exergy rates calculated using Equations 5.47a and 5.47b or Equation 5.48

The total flow exergy of air is calculated from (Wepfer et al. 1979)

$$\psi_{a,t} = (C_{p,a} + \omega C_{p,v})T_0\left[\left(\frac{T}{T_0}\right) - 1 - \ln\left(\frac{T}{T_0}\right)\right] + (1 + 1.6078\omega)R_aT_0 \ln\left(\frac{P}{P_0}\right)$$

$$+ R_aT_0\left\{(1 + 1.6078\omega)\ln\left(\frac{1 + 1.6078\omega_0}{1 + 1.6078\omega}\right) + 1.6078\omega\ln\left(\frac{\omega}{\omega_0}\right)\right\} \tag{5.47a}$$

where the specific humidity ratio is

$$\omega = \frac{\dot{m}_v}{\dot{m}_a} \tag{5.47b}$$

Assuming air to be a perfect gas, the specific physical exergy of air is calculated by the following relation (Kotas, 1995):

$$\psi_{a,per} = C_{p,a}\left(T - T_0 - T_0 \ln\frac{T}{T_0}\right) + R_a T_0 \ln\left(\frac{P}{P_0}\right) \tag{5.48}$$

Pope et al. (2010) performed an energy and exergy analysis of four different wind power systems, including both horizontal and vertical axis wind turbines (VAWTs). They encompassed significant variability in turbine designs and operating parameters through the selection of systems. In particular, two airfoils (NACA 63(2)-215 and FX 63–137) commonly used in horizontal axis wind turbines were compared with two VAWTs. A Savonius design and Zephyr VAWT benefit from operational attributes in wind conditions that are unsuitable for airfoil type designs. Each system was also analyzed with respect to both the first and second laws of thermodynamics. The aerodynamic performance of each system was numerically analyzed by computational fluid dynamics software, FLUENT. A difference in energy and exergy efficiencies of between 50% and 53% was predicted for the airfoil systems, whereas 44%–55% differences are predicted for the VAWT systems.

Baskut et al. (2010) reported exergy efficiency results of the wind turbine power plants. They investigated effects of meteorological variables, such as air density, pressure difference between state points, humidity, and ambient temperature, on exergy efficiency. Some key parameters were also given monthly for the three turbines. Exergy efficiency values varied between 23% and 27%, while temperatures changed from 268.15 to 308.15 K with air densities of 1.368–1.146 kg/m³.

5.3.2.2 Wind Exergy Maps

Sahin et al. (2006b) introduced a new concept in the area of wind energy through new spatiotemporal exergy maps and studied both energetic and exergetic aspects of wind energy in more detail for the first time. They presented energy and exergy efficiencies in the form of geostatistical maps for 21 climatic stations in the province of Ontario, Canada, over a period of 4 months. Exergy maps provided a new and original approach as well as more meaningful and useful information than energy analysis regarding the efficiency, losses, and performance for wind turbines. Energy and exergy efficiencies were estimated using measured generated power data. Output electricity power data for a 100 kW wind turbine, which was more convenient for rural area renewable energy application, with a rotor diameter at 18 m and hub height 30 m were given. Geostatistical spatiotemporal maps for January, April, July, and October were developed and discussed. In particular, the authors determined that exergy efficiencies were lower than energy efficiencies during the month of October, and one of the highest energy efficiency areas, which was observed in western Ontario, was less significant based on exergy efficiency. Without summer topographical heating, the relative differences between these efficiencies were low during October in most parts of Ontario. But wind chill became more appreciable during this month. The results indicated that aerial differences between energy and exergy efficiencies were equal to approximately 20%–24% at low wind speeds and to approximately 10%–15% at high wind speeds.

5.3.3 Geothermal Energy Systems

Geothermal energy is a reliable and promising renewable energy resource. In 1892, the first GDHS began operations in Boise, Idaho, the United States (Mary et al., 2009). Since then, a number of GDHSs installations have been made worldwide. Various investigations on the efficient utilization of geothermal energy resources have also been conducted to attain sustainable development (Hepbasli, 2010).

5.3.3.1 Geothermal District Heating Systems

Hepbasli (2010) comprehensively reviewed GDHSs from three angles: energetic, exergetic, and exergoeconomic analyses and assessments (to the best of our knowledge, this was the first review of this kind in the field). The author gave a historical background on the previously conducted GDHS studies with a special emphasis on the above three aspects, followed by the review and classification of the GDHSs. The following is a description of GDHSs including some important exergetic relations.

Figure 5.7 illustrates a schematic diagram of the simplified GDHS where user system is heated by geothermal energy (Ozgener et al., 2007a,b; Hepbasli, 2010). A GDHS consists mainly of three circuits, namely, (a) energy production circuit (EPC) (geothermal well loop), (b) energy distribution circuit (EDC) (district heating distribution network, and (c) energy consumption circuit (ECC). The actual operational values for a typical GDHS, namely, for the Salihli GDHS, in Manisa, Turkey, given as an illustrative example, are given below (Ozgener et al., 2005a), while the latest values for that are given elsewhere (Ozgener and Ozgener, 2008a). The wellhead temperatures of the production wells vary from 56°C to 115°C, while the volumetric flow rates of the wells range from 0.002 to 0.02 m³/s. Geothermal fluid is sent to the primary plate-type heat exchanger (between the geothermal fluid and the district heating water) and is cooled to about 44°C, as its heat is transferred to the district heating water. The temperatures obtained during the operation of the Salihli GDHS are, on average, 96°C/44°C for the district heating distribution network and 62°C/43°C for the building circuit. By using the control valves for flow rate and temperature at the building main station, the needed amount of water is sent to each housing unit and the heat balance of the system is

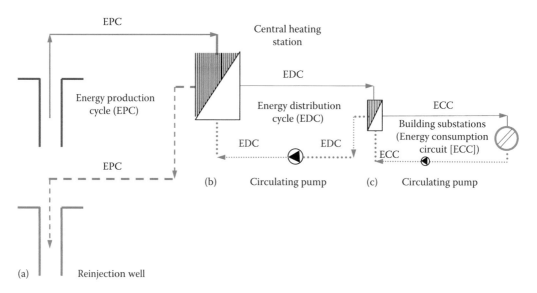

FIGURE 5.7
Basic simplified geothermal district heating energy system schema. (Adapted from Hepbasli, A., *Energ. Convers. Manage.*, 51, 2041, 2010.)

achieved. The geothermal fluid, collected from the four production wells at an average well heat temperature of about 96°C, is pumped to the inlet of the heat exchanger mixing tank and a main collector (from four production wells) with a total mass flow rate of about 47.62 kg/s.

The geothermal brine exergy inputs from the production field of the two GDHSs investigated may be calculated from the following equations (Ozgener, 2005; Hepbasli, 2010):

$$\dot{Ex}_{brine} = \dot{m}_w \left[(h_{brine} - h_0) - T_0(s_{brine} - s_0) \right] \tag{5.49}$$

The exergy destructions in the heat exchanger, pump, and the system itself are calculated from

$$\dot{Ex}_{dest,HE} = \dot{Ex}_{in} - \dot{Ex}_{out} = \dot{Ex}_{dest} \tag{5.50}$$

$$\dot{Ex}_{dest,pump} = \dot{W}_{pump} - (\dot{Ex}_{out} - \dot{Ex}_{in}) \tag{5.51}$$

and

$$\dot{Ex}_{dest,system} = \sum \dot{Ex}_{dest,HE} + \sum \dot{Ex}_{dest,pump} \tag{5.52}$$

The exergy efficiencies of GDHSs with and without reinjected thermal water are calculated as follows, respectively (Ozgener et al., 2004; Ozgener and Ozgener, 2008a):

$$\varepsilon_{system} = \left(\frac{\dot{Ex}_{useful,HE}}{\dot{Ex}_{brine}} \right) = 1 - \left(\frac{\dot{Ex}_{dest,system} + \dot{Ex}_{reinjected} + \dot{Ex}_{natural\,discharged}}{\dot{Ex}_{brine}} \right) \tag{5.53}$$

$$\varepsilon_{system} = \left(\frac{\dot{Ex}_{useful,HE}}{\dot{Ex}_{brine}} \right) = 1 - \left(\frac{\dot{Ex}_{dest,system} + \dot{Ex}_{natural\,discharged}}{\dot{Ex}_{brine}} \right) \tag{5.54}$$

where
$\dot{Ex}_{dest,system}$ is the exergy destructions in the system taking place as the exergy of the fluid lost in the heat exchanger and the pumps
$\dot{Ex}_{reinjected}$ is the exergy loss due to the reinjected water
$\dot{Ex}_{natural\,discharged}$ is the exergy destruction due to the natural direct discharge of the system (due to a significant amount of water leaks; pipeline losses)

Table 5.3 indicates GDHSs in Turkey, which have been energetically and exergetically evaluated. As can be seen from the values given in this table along with the footnote, the exergy efficiencies of the GDHSs ranged from 42.94% to 80.0% at the varying dead state temperatures of 0°C–25°C.

5.3.3.2 Classification of Geothermal Resources Using Specific Exergy Index

Geothermal resources have been classified as low, intermediate or high enthalpy resources according to their reservoir temperatures (Dickson and Fanelli, 2009). The temperature ranges used for these classifications are arbitrary and they are not generally agreed upon. Temperature is used as the classification parameter because it is the most convenient to

TABLE 5.3

Comparison of Various Geothermal District Heating Systems in Terms of Exergy Destructions and Exergy Efficiencies

Name of GDHS	Location/Country	Dead State Temperature (°C)	Date of Data Used	Total Exergy Input (kW)	Exergy Destructions/Losses (% of the Total (Brine) Exergy Input)				Exergy Efficiency (%)	Reference
					Pumps	Heat Exchangers	Thermal Reinjection	Natural Direct Discharge (Pipe Line)		
Balcova GDHS	Izmir/Turkey	13.1	January 1, 2003	9,164.29	3.22	7.24	22.66	24.1	42.94	Ozgener et al. (2004)
Gonen GDHS	Balikesir/Turkey	6	February 1, 2004	2,657.5	14.81	7.11	12.96	1.06	64.06	Oktay and Aslan (2007)
Salihli GDHS	Manisa/Turkey	2.9	February 1, 2004	2,564	2.22	17.90	—	20.44	59.44	Ozgener et al. (2005a)
Balcova GDHS	Izmir/Turkey	13.1[a]	January 2, 2004	14,808.15	1.64	8.57	14.84	28.96	46.00	Ozgener et al. (2005b)
Gonen GDHS	Balikesir/Turkey	10[b]	February 1, 2004	2,333.33	17.45	8.19	7.37	0.86	66.13	Ozgener et al. (2005c)
Salihli GDHS	Manisa/Turkey	25[c]	February 1, 2004	1,524	4.27	31.17	—	8.98	55.58	Ozgener et al. (2007)
Balcova GDHS	Izmir/Turkey	11.4[d]	January 2, 2004	14,390	1.74	8.83	14.18	28.7	46.55	Ozgener et al. (2006)
Bigadic GDHS	Balikesir/Turkey	11[e]	December 2006	2,889	1.7	26.3	—	34	49.00	Oktay and Dincer (2007)
Salihli GDHS	Manisa/Turkey	3.1[f]	2006/2007 heating season	2,600–5,403	1.05–2.19	14.82–17.65	18.14–20.06	1.70–2.23	60.3–62.9	Ozgener and Ozgener (2008a)
Afyon GDHS	Afyon/Turkey	0–20[g]	2006/2007 heating season	2,391.51	23.45	76.55	Not available data	Not available data	53.14–55.28	Ozgener and Ozgener (2008b)
Salihli GDHS	Manisa/Turkey	0–20[h]	2007/2008 heating season	1,025.97	7.90	43.31	Not available data	Not available data	61.1	Ozgener and Ozgener (2008b)

Gonen GDHS	Balikesir/Turkey	4[i]	January 13, 2007	3,203	0.9	8.24	32.49	1.13	50.0	Ozgener and Ozgener (2009)
Edremit GDHS	Balikesir/Turkey	13.4	January 20, 2007	1,927	1.66	6.07	29.94	8.04	54.26	Oktay and Dincer (2008)
Dikili GDHS	Izmir/Turkey	1.9[j]	Project data	10,332.82	0.76	10.34	38.77	—	50.12	Kalinci et al. (2008)

Sources: Ozgener, L. et al., *ASME, J. Energ. Resour. Technol.*, 126, 293, 2004; Ozgener, L. et al., *Int. J. Energ. Res.*, 29, 398, 2005a; Ozgener, L. et al., *Build. Environ.*, 40, 1309, 2005b; Ozgener, L. et al., *Geothermics*, 34, 632, 2005c; Ozgener, L. et al., *Build. Environ.*, 41, 699, 2006; Ozgener, L. et al., *Heat Transfer Eng.*, 28, 357, 2007a; Ozgener, L. et al., *Renew. Sust. Energ. Rev.*, 11, 1675, 2007b; Oktay, Z. and Aslan, A., *Geothermics*, 36, 167, 2007; Oktay, Z. and Dincer, I., *Int. J. Green Energy*, 44, 549, 2007, http://www.informaworld.com/smpp/title~content=t713597260~db=all~tab=issueslist~branches=4; Oktay, Z. and Dincer, I., *ASHRAE Trans.*, 116, 2008; Ozgener, L. and Ozgener, O., *J. Energ. Resour—ASME*, 130, 022302-1-6, 2008a; Ozgener, L. and Ozgener, O., *P. I. Mech. Eng. A. J. Pow.*, 222, 167, 2008b; Ozgener, L. and Ozgener, O., *Appl. Energ.*, 86, 1704, 2009; Kalinci, Y. et al., *Energ. Buildings*, 40, 742, 2008.

[a] The study was conducted for four various dates, namely, January 1, 2003, December 2, 2003, January 2, 2004, and February 2, 2004 at dead state temperatures of 13.1°C, 12°C, 10.4°C, and 6.5°C, respectively.

[b] A parametric study was performed to investigate how varying reference temperature from 0°C to 20°C would affect the exergy efficiencies.

[c] The study was conducted for December 1, 2003, January 1, 2004, and February 1, 2004, at dead state temperatures of 10.8, 9.7, and 2.9, respectively. In addition, two parametric expressions of exergy efficiencies were developed as a function of the reference temperature, by varying the dead state temperature from 0°C to 25°C.

[d] The reference-state temperature value of 11.4°C was obtained from the average ambient temperature data measured between 2000 and 2004 for 2 January in Izmir. The dead state temperature varying from 0°C to 25°C was considered to conduct a parametric study.

[e] A parametric study was performed to investigate how varying reference temperature from 2°C to 24°C would affect the exergy efficiencies. The highest exergy loss (accounting for 37%) occurred from the pipes of the system. This GDHS does not have a reinjection section yet.

[f] This study has been conducted based on a wide range of the monitored data in the heating season for the first time.

[g] The effect of varying dead state temperatures on the exergy efficiencies was investigated, while yearly average data provided for one representative unit were included here. The values for heat exchanger and pumps have been reported to be equal to those for the overall system. Exergy efficiency values have been reported to range from 57.73% to 72.08% at the dead state temperatures of 0°C–20°C (Ozgener and Ozgener, 2008b).

[h] The effect of varying dead state temperatures on the exergy efficiencies was investigated, while yearly average data provided for one representative unit were included here. Exergy efficiency values have been reported to range from 58.25% to 71.60% and from 57.23% to 80.00% at the dead state temperatures of 0°C to 20°C before and after reinjection studies, respectively (Ozgener and Ozgener, 2008b).

[i] The study was conducted based on an exergoeconomic assessment.

[j] A parametric study was performed to investigate how varying reference temperature from 0°C to 20°C would affect the exergy efficiencies. Project data taken from the Dikili municipality were utilized.

measure. In addition, temperature or enthalpy alone can be ambiguous when defining a geothermal resource because two independent thermodynamic properties are required to define the thermodynamic state of fluid. Geothermal energy is already in the form of heat, and from the thermodynamic point of view, work is more useful than heat because not all the heat can be converted to work. Therefore, geothermal resources should be classified by their exergy, a measure of their ability to do work (Lee, 1996, 2001).

Lee (2001) proposed a new parameter, the so-called specific exergy index (*SExI*) for better classification and evaluation as follows:

$$SExI = \left(\frac{h_{brine} - 273.16\, s_{brine}}{1192} \right) \tag{5.55}$$

which is a straight line on an *h-s* plot of the Mollier diagram. Straight lines of *SExI* = 0.5 and *SExI* = 0.05 can therefore be drawn in this diagram and used as a map for classifying geothermal resources by taking into account the following criteria:

- *SExI* < 0.05 for low-quality geothermal resources
- 0.05 ≤ *SExI* < 0.5 for medium-quality geothermal resources
- *SExI* ≥ 0.5 for high-quality geothermal resources

Here, the demarcation limits for these indices are exergies of saturated water and dry saturated steam at 1 bar absolute.

In order to map any geothermal field on the Mollier diagram as well as to determine the energy and exergy values of the geothermal brine, the average values for the enthalpy and entropy are calculated from the following equations (Quijano, 2000):

$$h_{brine} = \left(\frac{\sum_{i}^{n} \dot{m}_{w,i} h_{w,i}}{\sum_{i}^{n} \dot{m}_{w,i}} \right) \tag{5.56}$$

$$s_{brine} = \left(\frac{\sum_{i}^{n} \dot{m}_{w,i} s_{w,i}}{\sum_{i}^{n} \dot{m}_{w,i}} \right) \tag{5.57}$$

5.3.3.3 Some Energetic and Exergetic Indicators

The energetic renewability ratio, exergetic renewability ratio, energetic reinjection ratio, and exergetic reinjection ratio for geothermal systems have been developed by Coskun et al. (2009) as follows:

5.3.3.3.1 Energetic Renewability Ratio

The energetic renewability ratio is defined as the ratio of useful renewable energy obtained from the system to the total energy input (all renewable and nonrenewable together) to the system and given by

$$R_{Ren,E} = \left(\frac{\dot{E}_{useful}}{\dot{E}_{T}} \right) \tag{5.58}$$

5.3.3.3.2 Exergetic Renewability Ratio

The exergetic renewability ratio is defined as the ratio of useful renewable exergy obtained from the system to the total exergy input (all renewable and nonrenewable together) to the system and calculated using

$$R_{Ren,Ex} = \left(\frac{\dot{E}_{useful}}{\dot{Ex}_T} \right) \tag{5.59}$$

Here, total energy and exergy input terms include the wellhead geothermal water energy only in energy and exergy efficiencies. But in the above equations for energetic and exergetic renewability ratios, all renewable and nonrenewable energy and exergy inputs are considered for input.

5.3.3.3.3 Energetic Reinjection Ratio

The energetic reinjection ratio is defined as the ratio of renewable energy discharged to the environment or reinjected to the well from the system to the total geothermal energy supplied to the system and is obtained from

$$R_{Rein,E} = \left(\frac{\dot{E}_{nd}}{\dot{E}_{gw}} \right) \tag{5.60}$$

5.3.3.3.4 Exergetic Reinjection Ratio

The exergetic reinjection ratio is defined as the ratio of renewable exergy discharged to environment or reinjected to the well from the system to the total geothermal exergy supplied to the system and found from

$$R_{Rein,Ex} = \left(\frac{\dot{Ex}_{nd}}{\dot{Ex}_{gw}} \right) \tag{5.61}$$

Hepbasli (2010) has recently proposed the following indices for GDHSs.

5.3.3.3.5 Distribution Cycle Exergetic Ratio

The distribution cycle exergetic ratio (R_{DCEx}) is defined as the ratio of the exergetic capacity of the district heating water circulating through the district heating distribution network (MW) to the exergetic capacity of the geothermal reservoir in the field considered (MW) asfollows:

$$R_{DCEx} = \left(\frac{\dot{Ex}_{dhw}}{\dot{Ex}_{gr}} \right) \tag{5.62}$$

5.3.3.3.6 Energy Consumptions Circuit Exergetic Ratio

The energy consumptions circuit exergetic ratio (R_{ECCEx}) is defined as the ratio of the exergetic capacity of the fluid circulating through the energy consumptions circuit (building substations) district heating distribution network (MW) to the exergetic capacity of the geothermal reservoir in the field considered (MW) as follows:

$$R_{ECCEx} = \left(\frac{\dot{Ex}_{dhw}}{\dot{Ex}_{gr}} \right) \tag{5.63}$$

The average values for the enthalpy and entropy to be used in the calculations may be calculated from Equations 5.14 and 5.15.

5.3.3.3.7 Reservoir-Specific Exergy Utilization Indice

The reservoir-specific exergy utilization indice (*RSExUI*) is defined as the ratio of the exergetic capacity of the geothermal reservoir in the field considered (MJ) to the total heating surface area of all the buildings (m²) and is given by

$$RSExUI \ (MJ/m^2) = \left(\frac{\dot{Ex}_{gr} \ (MJ)}{A_{building} \ (m^2)} \right) \tag{5.64}$$

5.3.3.3.8 Geothermal Brine–Specific Exergy Utilization Indice

The geothermal brine–specific exergy utilization indice (*GBSExUI*) is defined as the ratio of the geothermal brine exergy input value (MJ) to the total heating surface area of all the buildings (m²) and is given by

$$GBSExUI \ (MJ/m^2) = \left(\frac{\dot{Ex}_{gb} \ (MJ)}{A_{building} \ (m^2)} \right) \tag{5.65}$$

5.4 Concluding Remarks

Exergetic aspects of solar, wind, and geothermal energy systems can be summarized as follows:

1. Exergy is a way to a sustainable development. In this regard, exergy analysis is a very useful tool, which can be successfully used in the performance evaluation of RERs as well as all energy-related systems.
2. Exergy efficiency of various types of solar collectors range between 0.77% and 44.91%.
3. Based on an exergetic evaluation of wind energy systems, average differences between energy and exergy efficiencies were obtained to be approximately 40% at low wind speeds and up to approximately 55% at high wind speeds (Sahin et al. 2006a).
4. Wind and solar exergy maps were developed for the first time by Sahin et al. (2006b) and Joshi et al. (2009b), respectively.

5. Although various studies have been undertaken to perform energy and exergy analyses of geothermal systems, the exergy analysis of GDHSs based on the actual field data was performed by Ozgener et al. (2004, 2005a,b,c) for the first time.

6. Geothermal resources should be classified by their exergy, a measure of their ability to do work. The classification with reference to SExI was more meaningful as there was no general agreement on the arbitrary temperature ranges used in the classification of geothermal resources by temperature (Lee 1996, 2001).

7. Based on a comparison of exergy efficiency values for GDHSs installed in Turkey, the exergy efficiencies of the GDHSs ranged from 42.94% to 80.0% at the varying "dead" state temperatures of 0°C to 25°C. As expected, the lower the reference state temperature, the significantly larger the exergy losses in the system pipeline and heat exchangers (Hepbasli, 2010).

8. All the GDHSs energetically and exergetically analyzed and evaluated to date have been installed in Turkey. These analysis methods should be applied to other GDHSs worldwide.

9. Until now, seven GDHSs have been analyzed using energy and exergy analysis methods (Hepbasli, 2010).

10. In conclusion, the author hopes that the analyses and assessments data reported here will provide the investigators with necessary knowledge on how to increase the efficiency of solar, wind, and geothermal energy systems.

Acknowledgments

The author would like to express his appreciation to his wife Fevziye Hepbasli and his daughter Nesrin Hepbasli for their continued patience, understanding, and full support throughout the preparation of this chapter as well as all the other ones.

Nomenclature

A	area (m^2)
C	specific heat (kJ/kgK)
E	energy (kJ)
$\dot{E}$	energy rate (kW)
ex	specific exergy (kJ/kg)
$\dot{Ex}$	exergy rate (kW)
$\dot{F}$	exergy rate of fuel (kW)
g	gravitational constant (m/s^2)
$GBSExUI$	geothermal brine specific utilization indices (MJ/m^2)
h	specific enthalpy (kJ/kg)
I	global irradiance (W/m^2)
$\dot{I}$	rate of irreversibility, rate of exergy consumption (kW)
$\dot{IP}$	rate of improvement potential (kW)
$\dot{m}$	mass flow rate (kg/s)

n	number, index number (dimensionless)
P	pressure (kPa)
$\dot{P}$	exergy rate of the product (kW)
$\dot{Q}$	heat transfer rate (kW)
R	ideal gas constant (kJ/kgK), ratio (dimensionless)
R^2	determination of coefficient
$RSExUI$	reservoir specific exergetic utilization indices (MJ/m^2)
s	specific entropy (kJ/kgK)
$\dot{S}$	entropy rate (kW/K)
$SExI$	specific exergy index (dimensionless)
SI	sustainability index (dimensionless)
t	period between local maxima and minima of the tidal record, time (s)
T	temperature (°C or K)
V	speed, velocity (m/s)
W	work (kJ)
$\dot{W}$	rate of work (or power) (kW)
z	elevation (m)

Greek Letters

κ	dilution factor (dimensionless)
η	energy (first law) efficiency (dimensionless)
ψ	flow (specific) exergy (kJ/kg), maximum efficiency ratio (or exergy-to-energy ratio for radiation (dimensionless)
Δ	interval
ρ	density (kg/m^3)
δ	fuel depletion rate (dimensionless)
ε	exergy (second law) efficiency (dimensionless)
ξ	productivity lack (dimensionless)
χ	relative irreversibility (dimensionless)
ω	specific humidity ratio (kg$_{water}$/kg$_{air}$)

Indices

a	air
at	atmospheric
AV	avoidable
ave	average
be	beam
col	collector
$conver$	conversion
cv	control volume
d	diffuse
$DCEx$	distribution cycle exergetic
$dest$	destroyed, destruction
dhw	district heating water
e	electrical
E	energetic
$ECCEx$	energy consumptions circuit exergetic
eng	engine
Ex	exergetic

f	fluid
gb	geothermal brine
gen	generation, generated
gr	geothermal reservoir
gw	geothermal water
h	heat
HE	heat exchanger
i	successive number of elements
in	input
io	inverter output
k	location
max	maximum
nd	natural direct discharge
out	output
p	constant pressure
par	parabolic
per	perfect
PV	photovoltaic
q	heat
r	reinjected thermal water
Ran	Rankine
Ren	renewability
rec	receiver
Rein	reinjection
s	solar
scell	solar cell
scol	solar collector
t	total
u	useful
UN	unavoidable
v	water vapor
w	water
0	dead (reference) state
1	initial state
2	final state

Abbreviations

GDHS	geothermal district heating system
PV	photovoltaic
RER	renewable energy resource

References

Akpinar, E. K. and Koçyiğit, F. 2010. Energy and exergy analysis of a new flat-plate solar air heater having different obstacles on absorber plates. *Applied Energy* 87:3438–3450.

Alqnne, K. and Saari, A. 2006. Distributed energy generation and sustainable development. *Renewable and Sustainable Energy Reviews* 10(6):539–558.

Alta, D., Bilgili, E., Ertekin, C., and Yaldiz, O. 2010a. Experimental investigation of three different solar air heaters: Energy and exergy analyses. *Applied Energy* 87:2953–2973.

Alta, D., Ertekin, C., and Evrendilek, F. 2010b. Evrendilek. Quantifying spatio-temporal dynamics of solar radiation exergy over Turkey. *Renewable Energy* 35:2821–2828.

Badescu, V. 1991. Maximum conversion efficiency for the utilization of multiply scattered solar radiation. *Journal of Physics D* 24:1882–1885.

Baghernejad, A. and Yaghoubi, M. 2010. Exergy analysis of an integrated solar combined cycle system. *Renewable Energy* 35:2157–2164.

Baskut, O., Ozgener, O., and Ozgener, L. 2010. Effects of meteorological variables on exergetic efficiency of wind turbine power plants. *Renewable and Sustainable Energy Reviews* 14(9):3237–3241, doi:10.1016/j.rser.2010.06.002.

Bejan, A. 1988. *Advanced Engineering Thermodynamics*. Wiley, New York.

Bejan, A., Tsatsaronis, G., and Moran, M. 1996. *Thermal Design & Optimization*. John Wiley & Sons, Inc., New York.

Candau, Y. 2003. On the exergy of radiation. *Solar Energy* 75:241–247.

Chaturvedi, S. K., Chen, I. D. T., and Kheireddine, A. 1998. Thermal performance of a variable capacity direct expansion solar-assisted heat pump. *Energy Conversion and Management* 39(3/4):181–191.

Coskun, C., Oktay, Z., and Dincer I. 2009. New energy and exergy parameters for geothermal district heating systems. *Applied Thermal Engineering* 29(11–12):2235–2242.

Dickson, M. H. and Fanelli, M. 2009. Istituto di Geoscienze e Georisorse, CNR, Pisa, Italy, Prepared on February 2004. (2 March). http://wwwsoc.nii.ac.jp/grsj/iga/bukai-files/What_is_Geothermal_org.pdf

Dincer, I. 2002. The role of exergy in energy policy making. *Energy Policy* 30:137–149.

Dincer, I., Hussain, M. M., and Al-Zaharnah I. 2004. Energy and exergy use in public and private sector of Saudi Arabia. *Energy Policy* 32(141):1615–1624.

Doukas, H., Papadopoulou, A. G., Nychtis, C., Psarras, J., and van Beeck, N. 2009. Energy research and technology development data collection strategies: The case of Greece. *Renewable and Sustainable Energy Reviews* 13:682–688.

EC, European Commission. 2006. *Action Plan for Energy Efficiency: Realizing the Potential*. COM(2006)545 final; October 19.

Eskin, N. 1999. Transient performance analysis of cylindrical parabolic concentrating collectors and comparison with experimental results. *Energy Conversion and Management* 40:175–191.

Eskin, N. 2000. Performance analysis of a solar process heat system. *Energy Conversion and Management* 41:1141–1154.

Fujisawa, T. and Tani, T. 1997. Annual exergy evaluation on photovoltaic-thermal hybrid collector. *Solar Energy Materials and Solar Cells* 47:135–148.

Gunerhan, H. and Hepbasli, A. 2007. Exergetic modeling and performance evaluation of solar water heating systems for building applications. *Energy and Buildings* 39(5):509–516.

Gupta, M. K. and Kaushik, S. C. 2010. Exergy analysis and investigation for various feed water heaters of direct steam generation solar–thermal power plant. *Renewable Energy* 35:1228–1235.

Hepbasli, A. 2008. A key review on exergetic analysis and assessment of renewable energy resources for a sustainable future. *Renewable and Sustainable Energy Reviews* 12(3):593–661.

Hepbasli, A. 2010. A review on energetic, exergetic and exergoeconomic aspects of geothermal district heating systems (GDHSs). *Energy Conversion and Management* 51(10):2041–2061.

Hermann, W. A. 2006. Quantifying global exergy resources. *Energy* 31(12):1685–1702.

Joshi, A. S., Dincer, I., and Reddy, B. V. 2009b. Development of new solar exergy maps. *International Journal of Energy Research* 33(8):709–718.

Joshi, A. S., Tiwari, A., Tiwari, G. N., Dincer, I., and Reddy, B. V. 2009a. Performance evaluation of a hybrid photovoltaic thermal (PV/T) (glass-to-glass) system. *International Journal of Thermal Science* 48:15–64.

Jeter, S. M. 1981. Maximum conversion efficiency for the utilization of direct solar radiation. *Solar Energy* 26(3):231–236.

Kalinci, Y., Hepbasli, A., and Tavman, I. 2008. Determination of optimum pipe diameter along with energetic and exergetic evaluation of geothermal district heating systems: Modeling and application. *Energy and Buildings* 40(5):742–755.

Kotas, T. J. 1995. *The Exergy Method of Thermal Power Plants.* Krieger Publishing Company, Malabar, FL.

Koroneos, C., Spachos, T., and Moussiopoulos, N. 2003. Exergy analysis of renewable energy sources. *Renewable Energy* 28:295–310.

Landsberg, P. T. and Mallinson, J. R. 1976. Thermodynamic constraints. Effective temperatures and solar cells. *CNES*, Toulouse, 27–46.

Landsberg, P. T. and Tonge, G. 1979. Thermodynamics of the conversion of diluted radiation. *Journal of Physics A: Mathematical and General.* 12:551.

Lee, K. C. 1996. Classification of geothermal resources an engineering approach. In *Proceedings of the Twenty-First Workshop on Geothermal Reservoir Engineering*, Stanford University, Stanford, CA. 5 pp.

Lee, K. C. 2001. Classification of geothermal resources by exergy. *Geothermics* 30:431–442.

Lior, N. 2010. Sustainable energy development: The present (2009) situation and possible paths to the future. *Energy* 35:3976–3994.

Lund, H. 2010. The implementation of renewable energy systems. Lessons learned from the Danish case. *Energy* 35:4003–4009.

Millan, M. I., Hernandez, F., and Martin, E. 1997. Letter to the Editor. Comments on "Available solar energy in an absorption cooling process." *Solar Energy* 61(1):61–64.

Moran, M. J. 1982. *Availability Analysis: A Guide to Efficiency Energy Use.* Englewood Cliffs, NJ: Prentice-Hall.

Morosuk, T. and Tsatsaronis, G. 2006. The "Cycle Method" used in the exergy analysis of refrigeration machines: From education to research. In C. Frangopoulos, C. Rakopoulos, G. Tsatsaronis, eds. *Proceedings of the 19th İnternational Conference on Efficiency, Cost, Optimization, Simulation and Environmental İmpact of Energy Systems*, vol. 1. Aghia Pelagia: Crete, Greece; July 12–14, pp. 157–163.

Morosuk, T. and Tsatsaronis, G. 2007. Exergoeconomic evaluation of refrigeration machines based on avoidable endogenous and exogenous costs. In A. Mirandola, O. Arnas, A. Lazzaretto, eds. *Proceedings of the 20th International Conference on Efficiency, Cost, Optimization, Simulation and Environmental Impact of Energy Systems*, vol. 3. Padova: Italy; June 25–28, pp. 1459–1467.

Morosuk, T. and Tsatsaronis, G. 2008. A new approach to the exergy analysis of absorption refrigeration machines. *Energy* 33:890–907.

Nobusawa, T. 1980. *Introductory Text of Exergy*, Tokyo: Ohmsha (in Japanese).

Oktay, Z. and Aslan, A. 2007. Geothermal district heating in Turkey, the Gonen case study. *Geothermics* 36:167–182.

Oktay, Z. and Dincer, I. 2007. Energetic, exergetic, economic and environmental assessments of the Bigadic geothermal district heating system as a potential green solution. *International Journal of Green Energy*, 44(5):549–69, http://www.informaworld.com/smpp/title~content=t713597260~db=all~tab=issueslist~branches=4

Oktay, Z., and Dincer, I. 2008. Energetic, exergetic and environmental assessments of the Edremit geothermal district heating system. *ASHRAE Transaction*, 114:116–127.

Onyegegbu, S. O. and Morhenne, J. 1993. Transient multidimensional second law analysis of solar collectors subjected to time-varying insolation with diffuse components. *Solar Energy* 50(1):85.

Ozgener, L. 2005. Exergoeconomic analysis of geothermal district heating systems (in Turkish). PhD thesis, Natural and Applied Sciences, Mechanical Engineering Science Branch, Ege University, p.102 (Co-supervisors: A. Hepbasli and I. Dincer).

Ozgener, L., Hepbasli, A., and Dincer, I. 2004. Thermo-mechanical exergy analysis of Balcova geothermal district heating system in Izmir. Turkey. *ASME, Journal of Energy Resources Technology* 126:293–301.

Ozgener, L., Hepbasli, A., and Dincer, I. 2005a. Exergy analysis of Salihli geothermal district heating system in Manisa, Turkey. *International Journal of Energy Research* 29(5):398–408.

Ozgener, L., Hepbasli, A., and Dincer, I. 2005b. Energy and exergy analysis of geothermal district heating systems: An application. *Building and Environment* 40(10):1309–1322.

Ozgener, L., Hepbasli, A., and Dincer, I. 2005c. Energy and exergy analysis of the Gonen geothermal district heating system in Turkey. *Geothermics* 34(5):632–645.

Ozgener, L., Hepbasli, A., and Dincer, I. 2006. Effect of reference state on the performance of energy and exergy evaluation of geothermal district heating systems: Balcova example. *Building and Environment* 41(6):699–709.

Ozgener, L., Hepbasli, A., and Dincer, I. 2007a. Parametric study of the effect of reference state on energy and exergy efficiencies of geothermal district heating systems (GDHSs): An application of the Salihli GDHS in Turkey. *Heat Transfer Engineering* 28(4):357–364.

Ozgener, L., Hepbasli, A., and Dincer, I. 2007b. A key review on performance improvement aspects of geothermal district heating systems and applications. *Renewable and Sustainable Energy Reviews* 11(8):1675–1697.

Ozgener, O. and Ozgener, L. 2007. Exergy and reliability analysis of wind turbine systems: A case study. *Renewable and Sustainable Energy Reviews* 11:1811–1826.

Ozgener, L. and Ozgener, O. 2008a. Monitoring of energetic and exergetic performance analysis of Salihli Geothermal District Heating System. *Journal of Energy Resources Technology—Transactions of the ASME* 130(2):022302-1-6.

Ozgener, L. and Ozgener, O. 2008b. Thermomechanical exergy and thermoeconomic analysis of geothermal district heating systems. *Proceedings of the Institution of Mechanical Engineers, Part A—Journal of Power and Energy* 222(A2):167–177.

Ozgener, L. and Ozgener, O. 2009. Monitoring of energy exergy efficiencies and exergoeconomic parameters of geothermal district heating systems (GDHSs). *Applied Energy* 86(9):1704–1711.

Ozturk, H. H. 2004. Experimental determination of energy and exergy efficiency of the solar parabolic-cooker. *Solar Energy* 77:67–71.

Petela, R. 1964. Energy of heat radiation. *Journal of Heat Transfer* 86:187–192.

Petela, R. 2003. Exergy of undiluted thermal radiation. *Solar Energy* 74:469–488.

Petela, R. 2005. Exergy analysis of the solar cylindrical-parabolic cooker. *Solar Energy* 79:221–233.

Pope, K., Dincer, I., and Naterer, G. F. 2010. Energy and exergy efficiency comparison of horizontal and vertical axis wind turbines. *Renewable Energy* 35:2102–2113.

Press, W. H. 1976. Theoretical maximum for energy from direct and diffuse sunlight. *Nature* 264:734–735.

Quijano, J. 2000. Exergy analysis for the Ahuachapan and Berlin Geothermal fields, El Salvador. *Proceedings World Geothermal Congress*, May 28–June 10, Kyushu-Tohoku, Japan.

Rosen, M. A. and Dincer, I. 2003a. Exergy methods for assessing and comparing thermal storage systems. *International Journal of Energy Research* 27(4):415–430.

Rosen, M. A. and Dincer, I. 2003b. Exergy-cost-energy-mass analysis of thermal systems and processes. *Energy Conversion and Management* 4(10):1633–1651.

Rosen, M. A., Dincer, I., and Kanoglu, M. 2008. Role of exergy in increasing efficiency and sustainability and reducing environmental impact. *Energy Policy* 36: 128–137.

Rosen, M. A., Le, M. N., and Dincer, I. 2005. Efficiency analysis of a cogeneration and district energy system. *Applied Thermal Engineering* 25(1):147–159.

Sahin, A. D., Dincer, I., and Rosen, M. A. 2006a. Thermodynamic analysis of wind energy. *International Journal of Energy Research* 30(8):553–566.

Sahin, A. D., Dincer, I., and Rosen, M. A. 2006b. New spatio-temporal wind exergy maps. *Transactions of the ASME, Journal of Energy Resources Technology* 128 (3):194–202.

Sahin, A. D., Dincer, I., and Rosen, M. A. 2007. Thermodynamic analysis of solar photovoltaic cell systems. *Solar Energy Materials and Solar Cells* 91:153–159.

Saitoh, H., Hamada, Y., Kubota, H., Nakamura, M., Ochifuji, K., Yokoyama, S., and Nagano, K. 2003. Field experiments and analyses on a hybrid solar collector. *Applied Thermal Engineering* 23:2089–2105.

Sarhaddi, F., Farahat, S., Ajam, H., and Behzadmehr, A. 2010. Exergetic performance assessment of a solar photovoltaic thermal (PV/T) air collector. *Energy and Buildings* 42:2184–2199.

Singh, N., Kayshik, S. C., and Misra, R. D. 2000. Exergetic analysis of a solar thermal power system. *Renewable Energy* 19(1–2):135–143.

Spanner, D. C. 1964. *Introduction to Thermodynamics*. Academic Press, London, U.K.

Svirezhev, Y. M., Steinborn, W. H., and Pomaz, V. L. 2003. Exergy of solar radiation: Global scale. *Ecological Modeling* 169:339–346.

Szargut, J., Morris, D. R., and Stewart, F. R. 1998. *Exergy Analysis of Thermal, Chemical, and Metallurgical Processes*. Edwards Brothers Inc., Ann Arber, MI.

Szargut, J. T. 2003. Anthropogenic and natural exergy losses (exergy balance of the Earth's surface and atmosphere). *Energy* 28:1047–1054.

Ucar, A. and Inalli, M. 2006. Exergoeconomic analysis and optimization of a solar-assisted heating system for residential buildings. *Building and Environment* 41:1551–1556.

Van Gool, W. 1997. Energy policy: Fairly tales and factualities. In *Innovation and Technology—Strategies and Policies*, O.D.D. Soares, A. Martins da Cruz, G. Costa Pereira, I.M.R.T. Soares, and A.J.P.S. Reis, eds. Kluwer, Dordrecht, the Netherlands, pp. 93–105.

Wepfer, W. J., Gaggioli, R. A., and Obert, E. F. 1979. Proper evaluation of available energy for HVAC. *ASHRAE Transactions* 85(1):214–230.

Weston, A. 2006. Hermann. Quantifying global exergy resources. *Energy* 31(12): 1685–1702.

Xiang, J. Y., Cali, M., and Santarelli, M. 2004. Calculation for physical and chemical exergy of flows in systems elaborating mixed-phase flows and a case study in an IRSOFC plant. *International Journal of Energy Research* 28:101–115.

Xiaowua, W. and Bena, H. 2005. Exergy analysis of domestic-scale solar water heaters. *Renewable and Sustainable Energy Reviews* 9:638–645.

6

Electrochemical Reduction of CO_2 to Fuels

Serguei N. Lvov, Justin R. Beck, and Mark S. LaBarbera

CONTENTS

6.1 Introduction

The electrochemical reduction of carbon dioxide (CO_2) has been studied extensively in a wide variety of conditions and analyzed with numerous techniques. It has been viewed as a potential component of a renewable energy cycle. CO_2 reduction has been examined as a means for reducing climate change due to greenhouse gasses, as well as for the production of liquid fuels from solar, wind, and nuclear energy. The primary focus of current research in CO_2 electroreduction is maximizing the current efficiency, or Faradaic efficiency, toward the production of specific high-value products. To overcome this challenge, new selective electrocatalysts for an efficient CO_2 reduction should be found. This chapter summarizes the advances in materials and techniques reported on the electrochemical reduction of CO_2 from the published papers available to the authors covering the years from 1963 to 2010.

The majority of research on the electrochemical CO_2 reduction has been performed on a laboratory scale in order to investigate and maximize product yields. Most of these studies have been performed with water electrolysis occurring at the anode as the source of hydrogen, although in some tests pure hydrogen was fed. Extensive work has been carried out, and is summarized here, on the effects of catalyst composition, reaction conditions, ionic membrane, analytical techniques, and kinetics and reaction mechanism. This chapter is designed to serve as a summary and reference tool for the reader to gather important information on a variety of details in electrochemical CO_2 reduction, as well as forward them to the relevant publications. The focus of this chapter is the use of metals, metal complexes, polymers, and inorganic heterogeneous catalysts for the electrochemical reduction of CO_2. However, homogenous catalysts and photochemical/photoelectrochemical reduction of CO_2 are not covered in this chapter.

6.2 Fundamentals of Water and CO_2 Electrolysis

The generation of hydrogen from water electrolysis is a well-researched topic and is widely considered an integral part of carbon-neutral economy. This section gives a brief description of the electrolysis of water and CO_2.

The thermodynamics of water electrolysis determine the necessary voltage required to generate $H_2(g)$ and $O_2(g)$ from water electrochemically. The Gibbs energy of reaction, also known as the free energy, is directly proportional to the standard potential of the electrochemical reaction:

$$\Delta_r G = -zFE_{cell,eq} \tag{6.1}$$

where

$\Delta_r G$ and $E_{cell,eq}$ are the Gibbs energy and electrode potential at equilibrium of the electrochemical reaction, respectively

z is the charge number of the half reactions

In order for an electrochemical cell to operate electrolytically, a potential difference must be applied to the cell that is greater in magnitude than $E_{cell,eq}$ (Cohen et al., 2007; Grimes et al., 2008).

For the case of electrolytic water splitting, the following anodic and catholic half reactions dominate:

$$H_2O(l) \rightarrow \frac{1}{2}O_2(g) + 2H^+ + 2e^- \text{ (anodic half reaction occurring at positive electrode)} \tag{6.2}$$

$$2H^+ + 2e^- \rightarrow H_2(g) \text{ (cathodic half reaction occurring at negative electrode)} \tag{6.3}$$

resulting in the total electrolysis reaction:

$$H_2O(l) \rightarrow H_2(g) + \frac{1}{2}O_2(g) \text{ (overall reaction)} \tag{6.4}$$

At 25°C and 1 bar, the $\Delta_r G$ for the water-splitting reaction (6.4) is +237.1 kJ mol⁻¹ (Lide, 2008), giving a decomposition potential (DP) of –1.229 V. This means that a potential more negative than –1.229 V must be applied in an electrolysis cell to generate $H_2(g)$ and $O_2(g)$.

6.2.1 Proton Exchange Membrane Electrolysis

Proton exchange membrane (PEM) electrolyzers, also referred to as polymer electrolyte electrolyzers or solid polymer electrolyzers, are used to generate ultrahigh-purity hydrogen

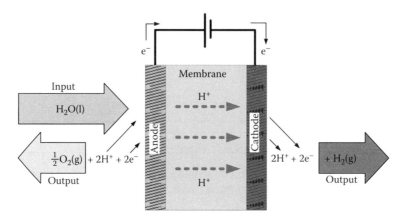

FIGURE 6.1
Electrolysis of water in proton exchange membrane electrolysis cell.

(<99.999%). The PEM water electrolyzer stack generally operates at 60%–70% efficiency, based on the electrical input and the higher heating value of the produced hydrogen. The most common membrane material is Nafion®, a porous sulfonated tetrafluroethylene polymer, though some other polymers have been developed to be used in PEM electrolyzers.

In PEM electrolysis, systems operate by conducting protons between the electrodes through the membrane and recombine them to form gaseous H_2 at the cathode, which is the negatively charged electrode. A schematic of a PEM electrolysis cell is provided in Figure 6.1. Water is fed to the anode, where it is adsorbed onto the catalyst freeing two protons into the membrane, two electrons to a power supply, and producing ½ O_2(g) at the catalyst sites. At the cathode, the protons from the membrane recombine on a catalyst site with the electrons from the power supply forming H_2(g). The anodic and cathodic half reactions are presented by reactions (6.2) and (6.3).

Membrane-based water electrolyzers are the current technology of preference for the production of high-purity hydrogen. Industrial PEM water electrolyzer systems are commonly operated up to 200 kW. The durability of these systems has developed to allow for thousand of hours of continual operation with failures in less than 1% of cells in the stack. The membrane electrolyzer is a compelling technology for further development and widespread deployment allowing a small footprint and distributed applications of varying scale. The membrane-based water electrolyzers are now cost-competitive with the natural gas reforming process, for high purity hydrogen production. Still, there is a need for improving the membrane electrolyzer performance by increasing the activity of the anode catalyst and decreasing the ionic resistance of the membrane. The technology still has a number of difficulties such as effective electrocatalysts for oxygen evolution, robust and high-conducting proton-conductive membranes, and the electrochemical stability of bipolar plates.

Water electrolysis should be considered as a predecessor of an electrolyzer where CO_2 is converted to a hydrocarbon fuel.

6.2.2 Water and CO_2 Electrolysis

A water–CO_2 electrolysis cell shares many of the same features as a water electrolysis cell. An example of such a cell producing synthesis gas (syngas), a mixture of CO(g) + H_2(g), is shown in Figure 6.2.

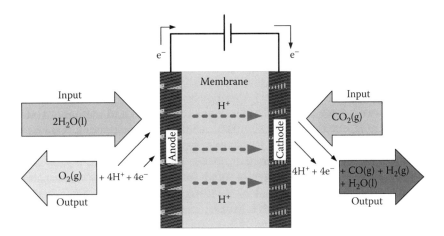

FIGURE 6.2
Electrolysis of CO$_2$ and water to produce syngas.

The overall reaction is given by

$$H_2O(l) + CO_2(g) \rightarrow CO(g) + H_2(g) + O_2(g) \text{ (overall reaction)} \tag{6.5}$$

with the half reactions as

$$2H_2O(l) \rightarrow O_2(g) + 4H^+ + 4e^- \text{ (anodic half reaction)} \tag{6.6}$$

$$4H^+ + 4e^- + CO_2(g) \rightarrow CO(g) + H_2(g) + H_2O(l) \text{ (cathodic half reaction)} \tag{6.7}$$

The Gibbs energy for reduction (6.5) is +494.3 kJ mol^{-1} at 25°C and 1 bar, giving a DP of −1.281 V for H$_2$O(l)/CO$_2$(g) electrolysis to produce syngas and oxygen.

Both water and water–CO$_2$ electrolyzers can use identical processes at the anode, as the formation of H$^+$ ions is the key here. The H$^+$ is transported though the electrolyte, most often by means of some type of proton-conductive membrane. It is at the cathode that the two processes diverge. Further description of the mechanism analysis can be found later in this chapter, but the general agreement has been that adsorbed CO$_2$ reacts with adsorbed H to either form an HCO intermediate, in the case of formate-related products, or water and adsorbed CO in the case of most other hydrocarbons. The adsorbed intermediates can further react with adsorbed H to produce higher hydrocarbons. The adsorption of CO$_2$ and its intermediates strongly affects the resulting product distribution, and for this reason the choice of catalyst material has been found to have a major impact on the resulting products. In the case of CO$_2$ electroreduction, any current that does not contribute to CO$_2$ reduction is considered a waste. Thus, formation of H$_2$ as a product is generally considered an inefficiency, except in the cases where syngas is the desired product.

While direct CO$_2$ electroreduction has been demonstrated in membrane-based cells at current densities up to 0.1 A cm^{-2}, the efficiency of CO$_2$ conversion to fuels should be significantly improved. This can be achieved by developing new catalysts for the electrolyzer cathode with higher selectivity with respect to the electrochemical reduction product and synthesizing novel ion conductive membranes that are stable and highly conductive in both acidic and basic environments.

6.3 Catalysts in Electrochemical CO_2 Reduction

A considerable amount of research has been performed in an attempt to classify a wide variety of catalyst materials under different conditions. Characterization studies have been performed in both aqueous (Hori et al., 1985; Vassiliev et al., 1985c; Azuma et al., 1989, 1990; Noda et al., 1990; Bandi et al., 1993; Hori et al., 1994; Hara et al., 1997) and methanol systems (Ortiz et al., 1995; Saeki et al., 1996; Ohta et al., 1998). Additional nonaqueous systems have been examined as well (Ikeda et al., 1987). Pressure effects have been investigated in a number of studies (Nakagawa et al., 1991; Hara et al., 1995a; Hara and Sakata, 1997b). Gas diffusion electrodes (GDEs) have been a geometry of much interest, with some studies attempting to classify materials by performance (Cook et al., 1990a; Schwartz et al., 1994). In the following section, much of the prevailing research is broken down by the choice of the cathodic catalyst.

6.3.1 Aluminum

The aluminum alloy Cu:Zn:Al, studied in an aqueous solution of Na_2CO_3, $Cu(NO_3)_2$, and $Al(NO_3)_2$, has been shown to electrochemically reduce CO_2 to CH_3OH with high Faradaic efficiency (Kobayashi and Takahashi, 2004).

6.3.2 Boron

B-doped diamond electrodes have been used to electrochemically reduce CO_2 to peroxycarbonate at 70% Faradaic efficiency (Saha et al., 2004).

6.3.3 Carbon

C electrodes have been used in a wide variety of aqueous solutions. Products have included CO, HCOOH, and H_2, while CH_4, C_2H_6, C_2H_4, C_3H_8, C_4H_{10}, and CH_3OH have been observed (Eggins et al., 1988; Hara et al., 1997; Hernandez et al., 1999). A kinetics and mechanism study has also been performed (Eggins et al., 1996).

6.3.4 Cobalt

Cobalt catalyst preferably reduce water to $H_2(g)$ in the electrolysis of high pressure aqueous CO_2 solutions (Hara and Sakata, 1997b).

6.3.5 Copper

Extensive research has been carried out on the use of Cu catalysts of various types using aqueous, methanol, and other nonaqueous electrolytes to study the electrochemical reduction of CO_2. Cu has shown a promising behavior in the production of a variety of hydrocarbons. A summary of the current state of electrochemical CO_2 reduction on Cu electrodes in various electrolyte solutions is provided in the following sections.

6.3.5.1 Aqueous Electrolyte Solutions

The electrochemical reduction of CO_2 with a Cu electrode in aqueous electrolyte solution has been tested with numerous supporting electrolytes as well as in a variety of experimental conditions, including pH, temperature, pressure, and applied potential. Cu electrodes in aqueous environments have shown a strong selectivity in reducing CO_2 to CH_3OH and CH_4 depending on the electrolyte and reaction conditions, with primary side products being $HCOOH$, C_2H_4, and H_2; in all cases, the selectivity toward H_2 evolution was reduced with decreasing temperature and increasing pressure (Hori et al., 1986, 1989a,b, 2002; Cook et al., 1987, 1988b; Kim et al., 1988; Noda et al., 1989; Frese, 1991; Hara et al., 1994; Ohkawa et al., 1994; De Jesús-Cardona et al., 2001; Kaneco et al., 2002a, 2002c, 2003; Ogura et al., 2003; Pettinicchi et al., 2003; Dube and Brisard, 2005; Shibata et al., 2008; Chang et al., 2009). Electrolysis of CO_2 on Cu GDEs had similar product distribution to dissolved CO_2, the primary products being CH_4 and C_2H_4 (Dewulf and Bard, 1988; Cook et al., 1990b; Komatsu et al., 1995; Yano et al., 2002a; Ogura, 2003).

Studies to determine the effect of supporting electrolytes on the products distribution in aqueous solutions have been carried out. Inorganic salts were tested as electrolytes in the reduction of CO_2 on Cu, yielding CO, CH_4, C_2H_4, C_2H_5OH, and C_3H_7OH (Hori et al., 1989a,b).

Supercritical CO_2 in aqueous electrolyte solution has been electrochemically reduced on Cu electrodes. The primary products were CO and H_2 (Zhao et al., 2004).

Kinetic studies have been carried out to determine limiting reaction rates, intermediates, and adsorbed species on a Cu electrode. Kinetic studies of the relationship between surface charge and product selectivity have been carried out (Noda et al., 1989; Momose et al., 2002; Tsubone et al., 2005). The effect of catalyst geometry on reaction rates and product selectivity has been studied using Cu electrodes in aqueous electrolyte solutions (Ohta et al., 1995). Studies have been carried out on adsorbed species and their role in reaction mechanism and rate (DeWulf et al., 1989; Hori et al., 1991; Smith et al., 1997). Cu crystal structure has been shown to affect the selectivity of CO_2 reduction products in aqueous electrolyte solutions (Hori et al., 2002; Takahashi et al., 2002).

Cu electrodes in the presence of Cu halides, such as CuCl and CuBr, in GDE electrodes have shown up to 80% Faradaic efficiency toward C_2H_4 production (Ogura et al., 2003, 2004; Yano et al., 2004). Alloys of Cu:Ag and Ni:Cu were used in the presence of Cu halide electrolytes to reduce CO to C_2H_4 (Ogura et al., 2005).

Cu alloys have shown a particularly high selectivity toward higher hydrocarbons than other electrode materials. Cu:Zn:Al has been shown to yield 97% Faradaic efficiency for CH_3OH, and Cu:Sn electrodes showed very high selectivity to $HCOOH$, while Cu:Zn has shown mixed selectivity to CH_3OH, C_2H_5OH, CH_3COCH_3, and light hydrocarbons (Katoh et al., 1994; Pettinicchi et al., 2003; Kobayashi and Takahashi, 2004; Li and Oloman, 2005, 2006). Cu alloys containing Cu:Ni, Cu:Sn, Cu:Pb, Cu:Zn, Cu:Cd, Cu:Ru, and Cu:Ag have been used to electrolyze CO_2 in aqueous solutions to generate CH_3OH, CO, and $HCOOH$ (Watanabe et al., 1991a,b; Popić et al., 1997). Cu:Ag alloys were tested to determine the optimal ratio of Cu to Ag for C_2 length products such as C_2H_5OH and C_2H_4 (Ishimaru et al., 2000; Yano et al., 2002b). Studies have been carried out to optimize the overall CO_2 conversion by using Fe:Ni:Pd:Cu alloys, yielding primarily CO, $HCOOH$, and CH_4 (Yamamoto et al., 2000; Tryk et al., 2001).

Deactivation of Cu electrodes and Cu alloy electrodes in aqueous electrolyte solutions has been demonstrated during the electroreduction of CO_2. Significant research has been carried out in characterizing and mitigating this poisoning, specifically by implementation of pulse electrolysis (DeWulf et al., 1989; Jermann and Augustynski, 1994; Augustynski et al., 1996; Friebe et al., 1997; Smith et al., 1997; Momose et al., 2002; Hori et al., 2005; Tsubone et al., 2005; Yano et al., 2007; Yano and Yamasaki, 2008).

6.3.5.2 Methanol and Organic Electrolyte Solutions

Methanol-based electrolyte solutions have been extensively studied in the electrochemical reduction of CO_2 on Cu electrodes. The most commonly utilized supporting electrolytes were NaOH, LiOH, and KOH; using all of these electrolytes, similar CO_2 reduction product selectivity has been demonstrated. CH_4 and C_2H_4 have been shown to be the primary products in CO_2 electrolysis in methanol solutions; however, CO, H_2, and HCOOH were also common electroreduction products (Murata and Hori, 1991; Naitoh et al., 1993; Saeki et al., 1994; Mizuno et al., 1995a,b, 1997b; Kaneco et al., 1999, 1999a,b,c, 2000, 2002, 2006, 2006a, 2007a,b; Shibata et al., 2008).

Temperature and pressure dependence of electrochemical CO_2 reduction on Cu electrodes in methanol electrolyte solutions has been studied and has produced well-described results. Decreased temperature and increased pressure both increased the concentration of dissolved CO_2 and increased the Faradaic efficiency toward CH_4 and C_2H_4 production (Naitoh et al., 1993; Saeki et al., 1994; Mizuno et al., 1995a,b, 1997b; Saeki et al., 1995; Kaneco et al., 1999a,b, 2002b, 2006, 2007b).

Alloys consisting of Cu:Re have been utilized as a catalyst in the electrochemical reduction of CO_2 in methanol electrolyte solutions. Reduction products of this alloy catalyst system were primarily H_2, with CO and CH_4 being lesser products (Schrebler et al., 2002). Cu:Ni alloys have been shown to reduce CO_2 in methanol and water solutions, producing mainly C_2H_4, CH_4, CO, and H_2 (Kaneco et al., 2007c).

As with aqueous systems, catalyst deactivation has been reported in Cu electrodes reducing CO_2 in methanol electrolyte systems. The nature of this performance degradation has been characterized, and mitigation of the performance degradation was achieved via pulse electrolysis (Nogami et al., 1994).

A comparison of the effects of cation and anion exchange membranes on the reduction products from CO_2 electroreduction with Cu catalysts in methanol electrolyte solutions was performed. The primary cation exchange membrane reduction product was C_2H_4, while the main anion exchange membrane products were HCOOH and CO (Komatsu et al., 1995).

Some studies have been carried out using organic solvent electrolyte solutions other than methanol. Supercritical CO_2 and water in 1-*n*-butyl-3-methylimidazolium hexafluorophosphate solution (a room temperature ionic liquid) were successfully electrolyzed into CO and H_2 with trace amounts of HCOOH (Zhao et al., 2004). Ethanol solvent was used with LiCl under high pressure to electrolyze CO_2 on Cu catalysts for CH_3OH production (Li and Prentice, 1997).

6.3.6 Gallium

GaAs has been used in the electrochemical reduction of CO_2 to CH_3OH with high Faradaic efficiencies at low current densities in aqueous electrolytes (Canfield and Frese, 1983; Frese and Canfield, 1984).

6.3.7 Gold

Various Au catalysts have been studied in the reduction of CO_2 in both aqueous and methanol solvents. Under acidic conditions, pH 5.2–6.8 with aqueous phosphate buffer solution, CO was the primary product along with small amounts of CH_4, C_2H_4, and C_2H_6 (Hori et al., 1987; Maeda et al., 1987; Fujihira and Noguchi, 1992; Noda et al., 1995). Polycrystalline Au electrodes and Au:Ag:Cu alloys were studied in aqueous electrolyte solutions and were shown to progressively deactivate as electrolysis proceeded. This deactivation was shown to be mitigated by the pre-electrolysis of electrolyte solution (Kedzierzawski and Augustynski, 1994; Augustynski et al., 1996). Porous Au film electrodes were studied in the reduction of CO_2 to CO in aqueous $KHCO_3$ at ambient conditions and $-1.2\,V$ versus saturated calomel electrode (SCE), where Faradaic efficiencies of 75% were obtained (Stevens et al., 2002). The effects of temperature on product selectivity using an Au electrode in a KOH–methanol electrolyte showed the main products to be CO, H_2, and HCOOH, with the selectivity toward CO production greatly improving as temperature dropped from 288 to 249 K (Kaneco et al., 1998b).

Au alloy electrodes have been studied as catalysts in the electrochemical reduction of CO_2 in various electrolyte solutions. Cu:Au alloys were shown to generate light hydrocarbons during the electrochemical reduction of CO_2 in aqueous solutions (Tsubone et al., 2005). Both Au:Pt:C electrodes in acetonitrille solutions and Au:Hg electrodes in dimethyl sulfoxide yielded CO as the primary product with HCOOH formed as a side product (Haynes and Sawyer, 1967; Christensen et al., 1990). Au:Pt alloy in various aqueous electrolyte solutions was shown to reduce CO_2 to CH_3OH, H_2, and HCOOH (Brisard et al., 2001).

Au catalysts have been studied to determine kinetic behavior during the electrochemical reduction of CO_2. Under ambient conditions using an electrolyte solution of tetraethylammonium perchlorate in dimethyl sulfate, CO_2 reduction kinetics have been studied at Au electrodes to find that mass transfer limitations were dominant (Haynes and Sawyer, 1967). Fourier transform infrared (FTIR) analysis of Au electrodes during electrochemical CO_2 reduction in methanol electrolyte solutions has provided a detailed analysis of reaction intermediates (Ortiz et al., 1995).

6.3.8 Indium

In catalysts have been studied in a variety of environments, including both aqueous and methanol solutions, to test the effects of supporting electrolyte, pH, and temperature on selectivity and kinetics. In aqueous solutions, ambient conditions favored CO evolution, while high yields of HCOOH were achieved either with elevated pressure or low pH even at high current densities (Kapusta and Hackerman, 1983; Ito et al., 1985; Ikeda et al., 1987; Mahmood et al., 1987a; Todoroki et al., 1995). In methanol solutions, a wider variety of products have been produced. Under ambient conditions, HCOOH and CO were the primary products, while elevated pressure yielded CH_2O, HCOOH, and CH_3COOH, and by altering the supporting electrolytes at low temperatures, CH_4, C_2H_4, C_2H_6, CO, and HCOOH have been produced (Kaneco et al., 1999b; Aydin and Köleli, 2004). Studies have also been carried out on the effects of morphology of In–TiO_2 catalysts on selectivity (Cueto et al., 2006). FTIR analysis of In electrodes during electrochemical CO_2 reduction in methanol solutions has provided a detailed analysis of reaction intermediates (Ortiz et al., 1995).

6.3.9 Iridium

Single crystal Ir electrodes have been studied in the electrochemical reduction of CO_2 in aqueous electrolyte solutions with a focus on kinetics (Hoshi et al., 1995b).

6.3.10 Iron

Fe electrodes have been used to study the electrochemical reduction of CO_2 in multiple electrolyte solutions. Under high pressure in aqueous electrolytes, Fe electrodes have been shown to reduce CO_2 to CH_4, C_2H_6, C_2H_4, C_3H_8, C_3H_6, C_4H_{10}, CO, HCOOH, and H_2 (Hara et al., 1995b). Fe electrodes treated with Na_2S have been tested in aqueous electrolytes, reducing CO_2 to H_2, HCOOH, and CO (Hara et al., 1997). Stainless steel electrodes have been studied in a tetrabutylammonium perchlorate in acetonitrile solution to produce $(COOH)_2$ (Fischer et al., 1981).

Kinetic studies have been carried out on Fe electrodes during the electrochemical reduction of CO_2. FTIR analysis of CO_2 reduction in aqueous electrolyte solutions on Fe electrodes has been studied to determine the adsorbed species on Fe and reaction mechanism (Koga et al., 1998). Carbon fiber GDEs impregnated with Fe catalyst have been studied, and carbon fibers activated prior to Fe impregnation have shown improved reactivity for CO_2 electroreduction (Yamamoto et al., 2000; Tryk et al., 2001).

6.3.11 Lead

The electrochemical reduction of CO_2 on Pb electrodes has been well reported for aqueous electrolytes. At ambient conditions and low current densities, product yields are almost exclusively HCOOH (Ikeda et al., 1987; Mahmood et al., 1987a; Todoroki et al., 1995; Köleli et al., 2003a; Köleli and Balun, 2004; Subramanian et al., 2007; Innocent et al., 2009). Due to mass transport limitations, carrying out electrolysis at higher current densities yielded significant amounts of H_2 evolution from water electrolysis (Todoroki et al., 1995). Lower temperature and increased pressure reduced mass transport limitations and increased HCOOH yield (Innocent et al., 2009). Methanol solutions produced similar results to aqueous in that HCOOH was the primary product; however, methanol electrolyte solutions also produced some amounts of CO, CH_4, and H_2 (Kaneco et al., 1998c). Supercritical CO_2 solutions have been reduced on Pb electrodes yielding CO and $(COOH)_2$ as the primary reduction products (Abbott and Eardley, 2000). Pb electrodes in the presence of hydroxyl amine and tetramethylammonium chloride in methanol have been shown to generate glyoxalate, glycolate, and oxalate ions (Bewick and Greener, 1970; Eggins et al., 1997).

Pb alloy catalysts have been studied in the electrochemical reduction of CO_2 in aqueous electrolyte solutions. Pb:Sn and Pb:Sn:In electrodes have shown strong selectivity toward producing HCOOH, with selectivity increasing with increasing pressure and acidity and decreasing temperature (Mahmood et al., 1987a; Watanabe et al., 1991a; Mizuno et al., 1995b; Köleli et al., 2002, 2003a). A Pb:In:Sn:Zn electrode in propylene carbonate electrolyte solution has been shown to produce mainly $C_2H_2O_4$, CO, and $C_2H_2O_3$ (Bewick and Greener, 1970; Ito et al., 1985). Pb-rich Cu alloys have been shown to electrochemically reduce CO_2 in aqueous electrolytes to C_2H_3OH, CH_3OH, and C_3H_7OH (Schwartz et al., 1994).

6.3.12 Mercury

Hg pool electrodes have been tested in a variety of conditions to study product selectivity and kinetic behavior. A kinetic study of Hg electrodes was carried out focusing on the effects of pH on electron transfer (Ryu et al., 1972). Testing on the effects of high pressure on Hg pool electrodes has shown 100% Faradaic efficiency toward HCOOH production at pressures above 10 bar (Todoroki et al., 1995). Kinetic studies of Hg electrodes in aqueous solutions have been carried out to describe rate-limiting steps, which were typically found to be mass transfer controlled under standard conditions (Haynes and Sawyer, 1967; Ryu et al., 1972).

The electrochemical reduction of CO_2 to HCOOH in aqueous electrolytes has shown high Faradaic efficiency with Hg electrodes (Russell et al., 1977).

6.3.13 Nickel

The electrochemical kinetics has been studied on Ni catalysts during the reduction of CO_2. The primary reduction products from CO_2 electrochemical reduction on activated Ni GDEs are CO and H_2 at an approximately 1:1 ratio; the mechanism of this reaction has been studied and reported (Williams et al., 1978; Kudo et al., 1993; Hara and Sakata, 1997b; Yamamoto et al., 2002). FTIR analysis has been used to study the adsorbed species and reaction mechanism during electrochemical CO_2 reduction on Ni catalysts (Koga et al., 1998). Ni-impregnated GDE performance has been improved by activating the carbon fiber gas diffusion substrate prior to Ni impregnation (Yamamoto et al., 2000, 2002; Tryk et al., 2001). Cu:Ni catalysts have been shown to generate CH_3OH and HCOOH at low overpotentials during the electrochemical reduction of CO_2 in aqueous solutions (Watanabe et al., 1991a). FTIR analysis of Ni electrodes during electrochemical CO_2 reduction in methanol solutions has provided a detailed analysis of reaction intermediates (Ortiz et al., 1995).

6.3.14 Palladium

Pd has been studied both as an electrochemical catalyst and as a proton-conductive membrane in the electrochemical reduction of CO_2. Pd cathodes in aqueous solutions used in the electrochemical reduction of CO_2 yielded primarily CO and H_2 (Shiratsuchi et al., 1992; Kolbe and Vielstich, 1996; Saeki et al., 1996; Hara and Sakata, 1997b). Pd-impregnated GDE performance has been improved by activating the carbon fiber gas diffusion substrate prior to Pd impregnation (Yamamoto et al., 2000; Tryk et al., 2001).

Pd and Pd:Ru alloys have been shown to reduce CO_2 to HCOOH at high Faradaic efficiencies (Yoshitake et al., 1995a; Furuya et al., 1997; Iwakura et al., 1998). Pd has been studied as a proton conductor as well as an electrocatalyst in the formation of HCOOH (Yoshitake et al., 1995; Kolbe and Vielstich, 1996). The morphology of Pd catalysts has been shown to significantly affect the selectivity of the reduction reaction (Hoshi et al., 1997a). Pd catalyst poisoning from supporting electrolyte cation and mitigation of catalyst poisoning has also been studied (Yoshitake et al., 1995b).

6.3.15 Platinum

Pt and Pt alloy catalysts have been studied in a variety of conditions for use in the electrochemical reduction of CO_2. The primary CO_2 reduction product in aqueous electrolyte media are H_2, CO, HCOOH, CH_3OH, and CH_2O (Huang et al., 1991; Hara and Sakata, 1997a; Hoshi et al., 1999; Brisard et al., 2001). Increased pressure has been shown to alter the product yield, producing mainly CH_4 with C_2H_3OH, C_2H_4, C_2H_6, and HCOOH as side products (Hara et al., 1995; Hara and Sakata, 1997a; Centi et al., 2007). The effect of using acetonitrile instead of water as the electrolyte solvent has been shown to shift product yield using pure Pt and Pt:Au:C electrodes. As the acetonitrile to water ratio increased, the product yield for pure Pt electrodes shifted from HCOOH to $(COOH)_2$ (Tomita et al., 2000). When using Pt:Au:C electrodes in acetonitrile electrolyte solutions, the product yields were CO, H_2, and CO_2 (Christensen et al., 1990).

The effects of Pt and Pt alloy structure on the kinetics of CO_2 reduction have been studied. Detailed studies on the adsorbed species, specifically adsorbed CO, and reaction rates on Pt electrodes reducing CO_2 have been carried out (Huang et al., 1991; Arevalo et al., 1994; Taguchi and Aramata, 1994; Hoshi et al., 1995, 1996a,b, 1997a,b, 1999, 2003;

Marcos et al., 1995, 1997; Mendez et al., 1999; Hoshi and Hori, 2000). Polycrystalline Pt structure kinetics has been studied in comparison to single crystal kinetics during the electrochemical reduction of CO_2 (Huang et al., 1991). Pt:Pd and Pt:Rh catalysts have been studied with respect to reaction kinetics and adsorbed species have been studied using FTIR and stripping voltammetry (Taguchi et al., 1994a; Marcos et al., 1995). Differential electrochemical mass spectrometry (DEMS) has shown that a Pt catalyst in aqueous electrolyte media has adsorbed COOH and CO in bridge and linear configurations (Tomita et al., 2000; Brisard et al., 2001).

A supercritical liquid mixture of CO_2 and 1,1,1,2-tetrafluoroethane was studied in an electrochemical reduction of CO_2 on Pt electrode surfaces. The supercritical phase had the effect of decreasing reduction potential while shifting product yield toward $(COOH)_2$ (Abbott and Eardley, 2000). A study was carried out using Pt electrodes to catalyze the electrochemical reduction of CO_2 using cyclic voltammetry (CV) to analyze kinetic behavior of the system (Abbott and Harper, 1996).

6.3.16 Rhenium

Re, Re:Cu, and Re:Au catalysts were studied in the electroreduction of CO_2 in methanol electrolytes. CO was the main reduction product in all cases, followed by CH_4 and H_2. CH_4 production showed slightly higher Faradaic efficiency with pure Re compared to Re:Cu alloy (Schrebler et al., 2002). Au:Re alloy showed significant mass transport limitations, only yielding CH_4 in well-stirred environments (Schrebler et al., 2001).

6.3.17 Rhodium

Kinetic studies have been carried out on Rh catalysts in the electrochemical reduction of CO_2 in aqueous solutions. DEMS studies were utilized to determine adsorbed species and rate-determining steps of CO_2 electrochemical reduction and the H_2 evolution reaction (Marcos et al., 1995). Single crystal and polycrystalline Rh electrodes were used to electrochemically reduce CO_2 in aqueous solutions with a focus on the kinetics of adsorbed species (Hoshi et al., 1995a; Arevalo et al., 1998).

6.3.18 Ruthenium

Ru catalysts have been studied in aqueous electrolyte solutions for the electrochemical reduction of CO_2. Pure Ru catalysts in aqueous electrolyte solutions showed primarily CO_2 reduction products of CH_4 with some CH_3OH and CO production (Frese and Leach, 1985; Summers and Frese, 1988). Ru deposited on diamond electrodes showed a slightly altered product distribution with the main products being H_2, CO, CH_4, CH_3OH, and HCOOH (Spataru et al., 2003).

Ru alloys have been studied as catalysts in the electrochemical reduction of CO_2 in aqueous electrolyte solutions. Ru-oxide electrodes modified by Cu, Ag, and Cd have been shown to reduce CO_2 electrochemically to CH_3OH and CH_3COOH in aqueous solutions (Bandi, 1990; Bandi and Kühne, 1992; Popić et al., 1997). Ru:Ti alloy using nanotube and nanoparticle geometry was able to produce CH_3OH at more than 60% Faradaic efficiency (Qu et al., 2005). In Ru:Pd alloys, more than 90% Faradaic efficiency toward HCOOH has been reported (Furuya et al., 1997).

6.3.19 Silver

Ag catalysts have been studied for the reduction of CO$_2$ in both aqueous and methanol solvents. An Ag GDE and Nafion membrane were shown to reduce CO$_2$ to CO and H$_2$ in 1:2 CO to H$_2$ ratio, appropriate for methanol synthesis, from –1.7 to –1.75 V versus SCE (Saeki et al., 1996; Hara and Sakata, 1997b; Yano et al., 2002b; Hori et al., 2003; Delacourt et al., 2008). In 0.1 M KOH-methanol solution, dissolved CO$_2$ was reduced at an Ag electrode; results showed the reduction of temperature from 273 to 248 K greatly increased the selectivity of CO formation while reducing H$_2$ formation rates (Kaneco et al., 1998a). Ag:Cu alloy catalysts have been shown to improve Faradaic efficiency compared to pure Ag electrodes in reducing CO$_2$ to C$_2$H$_4$, CO, CH$_3$OH, C$_2$H$_5$OH, and HCOOH in aqueous electrolyte solutions (Watanabe et al., 1991a; Schwartz et al., 1994; Ishimaru et al., 2000; Ogura et al., 2005).

Catalyst deactivation in aqueous environments during electrochemical CO$_2$ reduction has been studied and reported. The mitigation of catalyst poisoning has been demonstrated with electrolyte pre-electrolysis, as well as by performing electroreduction via pulse electrolysis with pure Ag, as well as Ag alloys with Ag and Cu (Augustynski et al., 1996; Shiratsuchi and Nogami, 1996; Ishimaru et al., 2000).

6.3.20 Tin

Sn catalysts have been studied in aqueous electrolyte solutions for the electrochemical reduction of CO$_2$. Pure Sn electrodes have been shown to primarily produce HCOOH and CO while producing significant amounts of H$_2$ under acidic and neutral conditions (Ito et al., 1985; Ikeda et al., 1987; Mahmood et al., 1987a; Köleli et al., 2002, 2003a,c; Li and Oloman, 2007). FTIR analysis of Sn electrodes during electrochemical CO$_2$ reduction in methanol solutions has provided a detailed analysis of reaction intermediates (Ortiz et al., 1995).

Sn and Cu mesh electrodes have been shown to produce HCOOH as the primary product, while selectivity was shown to depend on temperature and pressure (Watanabe et al., 1991a; Katoh et al., 1994; Köleli et al., 2003a,c; Li and Oloman, 2005, 2006). Kinetic studies have been carried out to analyze the mechanism at various pH, pressures, and electrode preparation methods (Kapusta and Hackerman, 1983; Nogami et al., 1993).

6.3.21 Titanium

Ti catalysts have been studied in methanol electrolyte to electrochemically reduce CO$_2$. Ti has been shown to reduce CO$_2$ to HCOOH and CO at ambient conditions, yielding some H$_2$ only at reduced temperature (Mizuno et al., 1997a). Ti and TiO$_2$ morphology has been studied with respect to the selectivity of the products (Cueto et al., 2006; Chu et al., 2008). A comparative study of adsorbed species on TiO$_2$ and Pt-doped TiO$_2$ electrodes in aqueous electrolyte solutions showed significant differences in kinetic behavior between the two electrodes (Koudelka et al., 1984).

6.3.22 Tungsten

An extended electrolysis study of WO$_3$:polyaniline (PAn):polyvinylsulfate catalyst showed strong initial selectivity toward HCOOH. This selectivity remained high but decreased with time (Endo et al., 1997).

6.3.23 Zinc

Zn has been employed as a catalyst in both pure and alloy forms, as well as in aqueous and methanol electrolyte solutions when electrochemically reducing CO_2. Pure Zn catalysts in methanol yielded primarily HCOOH and CO; however, Zn:Cu alloys in methanol electrolytes have been shown to produce some amounts of CH_4 and C_2H_4 (Ikeda et al., 1987; Saeki et al., 1996; Ohya et al., 2009). Zn alloys, Zn:Cu and Zn:Cu:Al, in aqueous electrolytes have shown very promising results in the formation of liquid hydrocarbons, specifically CH_3OH, CH_3CH_2OH, and CH_3COCH_3 (Watanabe et al., 1991a; Pettinicchi et al., 2003; Kobayashi and Takahashi, 2004).

6.3.24 Metal Complex, Polymer, and Inorganic Catalysts

6.3.24.1 Metal Complexes

A variety of Co-containing complexes have been studied as electrochemical catalysts for the reduction of CO_2. A carbon electrode coated with octa-butoxy-phthalocyanine was used in acidic aqueous electrolyte solutions to generate CO and H_2 in a 4:1 ratio (Abe et al., 1996a,b, 1997). Various macrocyclic Co complexes have been studied in the reduction of CO_2 to CO while mitigating H_2 evolution (Kapusta and Hackerman, 1984; Lieber and Lewis, 1984; Mahmood et al., 1987b; Atoguchi et al., 1991a,b; Savinova et al., 1992; Ramos Sende et al., 1995; Aga et al., 1996, 1997; Vasudevan et al., 1996; Zhang et al., 1996; Sonoyama et al., 1999; Shibata and Furuya, 2003). Novel electrode preparation techniques have been studied and reported in developing various Co complex electrodes for electrochemical CO_2 reduction (Tanaka and Aramata, 1997). Co complexes in Prussian blue-polyaniline films on a Pt cathode have been shown to reduce CO_2 into $C_3H_6O_3$, CH_3OH, C_2H_5OH, and $OC(CH_3)_2$ (Ogura et al., 1993).

Numerous Ru complexes have been studied for the electrochemical reduction of CO_2 in aqueous electrolytes. The reaction mechanism of CO_2 electrochemical reduction to CO on various Ru complexes including mono(bipyridine)carbonylruthenium complexes have been reported as yielding high Faradaic efficiency for CO (Collomb-Dunand-Sauthier et al., 1994b; Shibata and Furuya, 2003). Carbon electrodes modified by {Ru^0(2,2'-bipyridine) $(CO)_2\}_n$ have been shown to reduce CO_2 to CO with high Faradaic efficiency and improved electrode stability (Collomb-Dunand-Sauthier et al., 1994a; Ramos Sende et al., 1995).

Fe complexes have been studied in the electrochemical reduction of CO_2 in aqueous electrolytes. Fe-(4,5-dihydroxybenzene-1,3-disulfonate) complex immobilized in Prussian blue films on Pt electrodes have been shown to electrochemically reduce CO_2 with high catalytic activity to the formation of $C_3H_6O_3$ and other organic compounds (Ogura et al., 1994b; Ramos Sende et al., 1995; Vasudevan et al., 1996; Nakayama et al., 1997; Sonoyama et al., 1999; Shibata and Furuya, 2003). In situ FTIR studies have been carried out on bis(1,8-dihydroxynaphthalene-3,6-disulfonato) iron(II) complex fixed in a Prussian blue catalyst film deposited on a Pt electrode, with the primary reduction products as C_1 to C_3 carbon length hydrocarbons (Ogura et al., 1996).

6.3.24.2 Polymer and Inorganic Catalysts

Among conductive polymers, there have been two main materials of focus: polypyrrole and polyaniline. Polypyrrole has been tested on Pt, Au, and Sn-oxide substrates in methanol with HCHO, HCOOH, and CH_3COOH as the main products (Aydin and Köleli, 2004). Polyaniline has also been tested in methanol on a Pt substrate with HCHO, HCOOH, and

CH$_3$COOH as the products (Aydin and Köleli, 2002; Köleli et al., 2003b, 2004). Similarly, there have been two main inorganic catalysts tested: Everitt's salt and Prussian blue. Everitt's salt has predominantly been used in conjunction with homogenous catalysts, and thus is not covered in this chapter. Prussian blue has usually been used as part of a poly-aniline electrode. Some testing has been performed on a Pt GDE, with the main products as CH$_3$OH, CH$_3$CH$_2$OH, HCHO, HCOOH, CH$_3$COOH, and C$_3$H$_6$O$_3$ (Ogura and Yoshida, 1986). More tests with both Prussian blue and polyaniline can be found in the next section.

6.3.24.3 Composites

It has been a common occurrence to see complex, polymer, and inorganic catalysts tested as part of a single electrode. Fe complexes have often been used on Prussian blue–polyaniline electrodes in aqueous KCl with common products found to be CH$_3$OH, HCOOH, CH$_3$COOH, and C$_3$H$_6$O$_3$ (Ogura et al., 1994a, 1995, 1998). Spectroscopy has been used to investigate the electrode properties and reaction mechanism (Ogura et al., 1996; Nakayama et al., 1997). Other complexes have also been tested on Prussian blue and polymer electrodes as well. Co complexes on Pt have yielded C$_3$H$_6$O$_3$, CH$_3$OH, CH$_3$CH$_2$OH, and HCHO (Ogura et al., 1993). Another Co complex was tested on polypyrrole in an acetonitrile–water solution (Zhang et al., 2009). A Ni complex with polypyrrole on both Pt and C in various solutions yielded CO as a primary product (Zhalko-Titarenko et al., 1990). A study was performed with both a polypyrrole–molybdenum blue and Fe complex–Prussian blue electrode tested with and without Fe and Co complexes, where CH$_3$OH, CH$_3$CH$_2$OH, HCHO, and C$_3$H$_6$O$_3$ were found to be the main products (Ogura et al., 1994b).

6.4 Membranes

A significant number of experiments have been performed in standard undivided electro-chemical cells. It should be noted, however, that many experiments were performed using a variety of ionic and nonionic membrane materials.

6.4.1 Nonionic Membranes

A variety of inert materials have been used to divide the cathode and anode cavities, though many of them found use in the earlier years of testing. Some of the more standard materials included glass (Ryu et al., 1972; Russell et al., 1977; Ogura and Fujita, 1987; Ogura and Endo, 1999; De Jesús-Cardona et al., 2001; Köleli et al., 2002; Stevens et al., 2002; Yano et al., 2002a; Ogura et al., 2004), plastic (Mahmood et al., 1987a), and paper (Yamamoto et al., 2002). Due to its permeability to hydrogen, Pd has also seen use as a membrane in addition to its use as a catalyst material (Yoshitake et al., 1995b; Iwakura et al., 1998).

6.4.2 Cation Exchange Membranes

Overall, cation exchange membranes have been the predominant material used in much of the divided cell tests. Many tests simply listed the material as a cationic conductor (Williams et al., 1978; Hori and Suzuki, 1982a; Aurian-Blajeni et al., 1983; Hori et al., 1985, 1989, 1995b; Ogura and Yoshida, 1988; Yoshitake et al., 1995b; Eggins et al., 1997; Ohta et al., 1998; Yano et al., 2002b; Delacourt et al., 2008). Some specific brands have been listed, including Selemion™ (Hori et al., 1987; Yano et al., 2004) and the largest contributor Nafion (Yoshida et al., 1993; Kedzierzawski and Augustynski, 1994; Nagao et al., 1994; Komatsu et al., 1995; Augustynski et al., 1996; Hernandez et al., 1999; Jarzębińska et al., 1999; Magdesieva et al., 2002; Kobayashi and Takahashi, 2004; Centi et al., 2007). A variety of Nafion types have been used, including Nafion 115 (Dewulf and Bard, 1988), Nafion 315 (Ito et al., 1985; Ikeda et al., 1987; Noda et al., 1990), Nafion 423 (Innocent et al., 2009), Nafion 450 (Li and Oloman, 2005, 2006), Nafion 430 (Subramanian et al., 2007), and Nafion 961 (Subramanian et al., 2007). The two most common types used were Nafion 417 (Cook et al., 1987, 1988b, 1989; Azuma et al., 1990; Shiratsuchi et al., 1992; Kyriacou and Anagnostopoulos, 1993; Shiratsuchi et al., 1993; Nogami et al., 1994; Hara et al., 1995, 1995b, 1997a,b; Todoroki et al., 1995; Shiratsuchi and Nogami, 1996; Ishimaru et al., 2000; Köleli and Balun, 2004; Köleli et al., 2004) and Nafion 117 (Cook et al., 1988a, 1990a; Watanabe et al., 1991a; Katoh et al., 1994; Hoshi et al., 1995, 2005; Mizuno et al., 1995, 1997a; Terunuma et al., 1997; Kaneco et al., 1998a,b,c, 1999a,b,c, 2000, 2002a,b, 2003, 2006, 2006a,b, 2007a,b,c; Qu et al., 2005; Li and Oloman, 2007; Yano et al., 2007; Yano and Yamasaki, 2008; Ohya et al., 2009).

6.4.3 Anion Exchange Membranes

Far fewer papers have focused on the use of anion exchange membranes for CO_2 electroreduction (Lee and Tak, 2001). One paper performed an analysis and comparison between cation exchange and anion exchange membranes onto which metals had been deposited for the reduction of CO_2 to CO (Hori et al., 2003).

6.5 Test Conditions

Throughout the various tests performed, numerous reaction conditions have been examined. The following papers are those that focused specifically on the effects of temperature, pressure, and pH on the reaction rates and product distributions.

6.5.1 Temperature

Some papers have examined the effects of various temperatures on CO_2 reduction. The effect has been examined on Cu in aqueous H_2CO_3 solution for CH_4 and C_2H_4 production (Hori et al., 1986) as well as on Ti in KOH in methanol for the production of CO and HCOOH (Mizuno et al., 1997a). However, the majority of the studies on the effect of temperature has been focused on using low-temperature (below ambient) solutions. The general reason for this has been the increased solubility of CO_2 in the electrolyte, which benefits mass transfer. Testing has been performed to compare the effects on various metals (Azuma et al., 1989, 1990). A considerable amount of efforts has been put into examining the effect on Cu electrodes, especially by Kaneco et al. (2002a, 2003). Aqueous bicarbonate solutions were examined, as well as various electrolytes in methanol (Naitoh et al., 1993; Kaneco et al., 1999a,b, 2002). Other tested systems include Ag and Au in aqueous KOH (Kaneco et al., 1998a,b) as well as Pt and Rh in methanol (Mizuno et al., 1995).

6.5.2 Pressure

As with temperature, the effect of pressure on CO_2 reduction has also been widely studied. The general result was similar to the temperature experiments, where high pressure increased CO_2 concentrations and benefited mass transfer. Some researchers investigated the effect of varying pressure on a variety of metals (Hara et al., 1995a), with additional focus put on Group VIII metals (Nakagawa et al., 1991). Other papers focused on specific materials such as polypyrrole (Aydin and Köleli, 2002, 2004). Testing with metals included Cu (Kyriacou and Anagnostopoulos, 1993; Hara et al., 1994; Saeki et al., 1995) and Hg (Paik et al., 1969). Varying pressure was also used to characterize Pt and metal-porphyrin GDE performance (Hara et al., 1995; Sonoyama et al., 1999). High pressure was a specific condition used in numerous tests. Some of the tested electrode materials include C (Hara et al., 1997), Fe (Hara et al., 1995b), Sn (Köleli et al., 2003a,c), Cu (Li and Prentice, 1997), and Pb, Hg, and In (Todoroki et al., 1995). High-pressure GDE tests have also been performed with a variety of metals (Hara and Sakata, 1997a,b).

6.5.3 pH

Electrolysis has been performed in acidic, neutral, and basic conditions, often depending on the desired electrolyte. Some research has focused specifically on the effect of pH on reduction with various metals (Vassiliev et al., 1985c) such as Au (Noda et al., 1995). The pH effect on complexes based on Ru (Chardon-Noblat et al., 1998) and WO_3 (Endo et al., 1997) has been examined. More exotic materials have been examined, such as RuO_2 on B-doped diamond (Spataru et al., 2003). Acidic conditions have been tested on Pb, In, and Sn GDEs for HCOOH production (Mahmood et al., 1987a).

6.5.4 Combination of Operational Conditions

While the above papers have mainly focused on one condition, some researchers have examined the effect of multiple conditions in parallel. Köleli et al. examined high-pressure, high-temperature conditions for Sn and Pb granules in a fixed-bed reactor (Köleli et al., 2003a,c; Köleli and Balun, 2004). Kaneco et al. tested Cu in methanol under low-temperature, high-pressure conditions (Kaneco et al., 2006, 2007b). Some tests have examined the effects of varying both temperature and pressure for Ni (Kudo et al., 1993) and Sn–Cu (Li and Oloman, 2006). The effect of varying temperature on high-pressure reduction on In, Sn, and Pb was examined for HCOOH production (Mizuno et al., 1995b). One test investigated the effect of varying temperature, pressure, and pH for Cu in aqueous $KHCO_3$ (De Jesús-Cardona et al., 2001).

6.6 Test Methods

Throughout the majority of the tests, the two predominant test methods used were electrolysis and CV. Electrolysis was performed either potentiostatically or galvanostatically in order to analyze the product distributions and current densities. CV was often performed to determine multiple system characteristics. It was used to probe reaction mechanisms and kinetics, as well as to determine the potentials for initial electrolysis. In addition to these two methods, a variety of other techniques have also been used.

6.6.1 Electrochemical Impedance Spectroscopy

Electrochemical impedance spectroscopy has generally been used to study kinetics, mechanism, and electrode surface during electroreduction (Hackerman, 1983; Bandi et al., 1993; Kapusta and Köleli et al., 2003b; Isaacs et al., 2005).

6.6.2 Anode Stripping Voltammetry

The test setup and analysis for anodic stripping voltammetry was similar to CV tests, but the purpose was to analyze adsorbed intermediates and reduction products by oxidizing them. One study focused on the effects of Pt crystal structure on CO_2 reduction (Hoshi et al., 2003). Taguchi et al. focused on using the technique to investigate the reduction of CO_2 to adsorbed CO (Taguchi and Aramata, 1994; Taguchi et al., 1994a,b).

6.6.3 Rotating Disk and Rotating Ring-Disk Electrodes

Like anodic stripping, the rotating disk electrode and rotating ring-desk electrode were generally used to analyze intermediates and products through their ensuing oxidation (Zhang et al., 1996). One investigation looked at the production of methanol with respect to crystal structure (Frese and Canfield, 1984). Another study combined the use of rotating electrodes with mass spectrometry for the analysis of CH_4 and C_2H_4 production (Wasmus et al., 1990).

6.6.4 Spectroscopy and FTIR

After electrolysis and CV, spectroscopy appeared to be one of the most common research tools utilized. It was usually performed in combination with other techniques. FTIR spectroscopy was the most common method of spectroscopy used. One of the main advantages of FTIR analysis was that it could often be performed in situ and in combination with other methods (Christensen et al., 1988, 1990; Nikolic et al., 1990; Ogura et al., 1996; Marcos et al., 1997; Nakayama et al., 1997; Arevalo et al., 1998; Isaacs et al., 2005). One of the most important uses of FTIR was for determining intermediates and reaction pathways (Vassiliev et al., 1985a; Arevalo et al., 1994; Christensen et al., 1995; Koga et al., 1998; Ogura et al., 2005). One study examined the relation of products and intermediates toward electrode poisoning (Smith et al., 1997). Particular focus has been put on the analysis of CO formation and adsorption (Huang et al., 1991; Taguchi et al., 1994a,b; Hori et al., 1995a).

6.6.5 Differential Electrochemical Mass Spectrometry

DEMS has been touted as a sensitive analytical tool for studying electrochemical processes (Fujihira and Noguchi, 1992). The primary uses have been the analysis of adsorbates (Kolbe and Vielstich, 1996), intermediates (Brisard et al., 2001), and overall reaction mechanisms (Dube and Brisard, 2005) of CO_2 reduction. It has also been used for examining electrode poisoning by use in parallel with pulse electrolysis (Friebe et al., 1997).

6.6.6 Chronoamperospectroscopy, Chronopotentiometry, and Coulometry

A number of the more classical electrochemical techniques have also been incorporated into CO_2 electroreduction research, though most of their use took place prior to the mid-1990s. These techniques included coulometric analysis (Lieber and Lewis, 1984; Hori et al., 1988), chronopotentiometry (Haynes and Sawyer, 1967; Hori et al., 1989b, 1991), and chronoamperospectroscopy (Abe et al., 1996a,b).

6.6.7 Electrode Reactivation and Pulse Electrolysis

One major issue with CO_2 electroreduction has been electrode deactivation or poisoning over time. For this reason, techniques have been developed to extend catalyst life. This led to the investigation of electrode reactivation through anodic polarization (Kedzierzawski and Augustynski, 1994; Hori et al., 2005). This has often been implemented in the form of pulse electrolysis, where the potential is alternated regularly between cathodic and anodic polarization. This has been performed on Ag (Shiratsuchi and Nogami, 1996), though much of the focus has been on Cu (Shiratsuchi et al., 1993; Nogami et al., 1993, 1994; Ishimaru et al., 2000; Lee and Tak, 2001; Tsubone et al., 2005) and Cu-oxide (Yano et al., 2007; Yano and Yamasaki, 2008).

6.7 Mechanism and Kinetics

The reaction mechanisms and kinetic relations of CO_2 have been an area of great interest in research. Reviews have been conducted on general reaction pathways and those for certain metals (Darensbourg and Kudaroski, 1984; Chaplin and Wragg, 2003). One proposed reaction pathway can be seen in Figure 6.3 (Kaneco et al., 2007c). This mechanism suggests two distinct pathways for reduction. Formation of adsorbed CO_2 was considered a common intermediate for most hydrocarbons, with other compounds resulting from the formation of adsorbed formate. One focus of the research has been on analyzing CO as an initial intermediate in reduction (Hori and Murata, 1990; Taguchi et al., 1994b). The role of hydrogen in the reaction has been investigated through the use of deuterium as a marker (Sobkowski and Czerwinski, 1975; Yoshitake et al., 1995a,b). The effects of crystal structure on reduction have been studied with particular focus on CO and H_2 adsorption. Single crystal Pt was by far the most studied material (Rodes et al., 1994a,b,c; Hoshi et al., 1995, 1996a,b, 1997b, 1999, 2003; Hoshi and Hori, 2000), though testing has also been performed on single crystal Ag (Hoshi et al., 1997c), Pd (Hoshi et al., 1997a), and Ir (Hoshi et al., 1995b). The availability of protons in solution can also affect reduction. This has been investigated

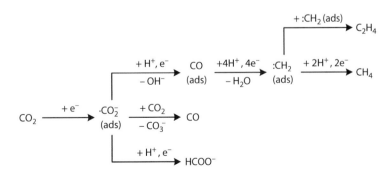

FIGURE 6.3
Proposed mechanism for CO_2 reduction. (Adapted from Kaneco, S. et al., *Bull. Catal. Soc. India*, 6, 74, 2007c.)

in solvents of varying proton availability (Amatore and Saveant, 1981; Eggins and McNeill, 1983; Vassiliev et al., 1985a).

Specific materials have been tested as well. Pt has been a metal of strong interest (Giner, 1963; Sobkowski and Czerwinski, 1974; Vassiliev et al., 1985b; Arevalo et al., 1994; Mendez et al., 1999; Brisard et al., 2001), as has been Cu (Cook et al., 1989; Hori et al., 1991; Saeki et al., 1995; Smith et al., 1997; Dube and Brisard, 2005). Palladium (Yoshitake et al., 1995b; Kolbe and Vielstich, 1996) and metal complexes (Furuya and Koide, 1991; Christensen et al., 1995) have also been examined. Additional materials include Au (Noda et al., 1995), Ni and Fe (Koga et al., 1998), Sn and In (Kapusta and Hackerman, 1983), C (Eggins et al., 1996), Hg (Ryu et al., 1972), Re on Au (Schrebler et al., 2001), and WO_3 (Endo et al., 1997).

6.8 Full Electrochemical Cells

Up to this point, the majority of research has been performed on a laboratory scale, often operating under batch reactor conditions. There have been a number of papers investigating more realistic and large-scale reactor designs, or full electrochemical cells. The most common style has been the filter-press reactor, the style used by most PEM fuel cells and electrolyzers as described earlier. Generally, an ion-conducting membrane, usually proton conducting, separated the anode and cathode, though some designs used a nonconductive membrane. An example of this design can be seen in Yamamoto et al. (2002). In this study, a filter-press configuration was used, though not quite a zero-gap spacing between the electrodes and membrane was demonstrated. A GDE was used in this work. The membrane was a paper filter through which a $0.5\,M$ $KHCO_3$ aqueous electrolyte solution flowed. In this case, the goal was to produce a CO and H_2 syngas product. A similar approach was also performed using a Nafion membrane for syngas production (Delacourt et al., 2008). It was found that inserting a thin layer of pH buffer solution between the membrane and the cathode improved performance.

For most cases, the desired product has been formic acid or formate production. In these reactors, an aqueous electrolyte was used either at both the anode and cathode, which were separated with some form of Nafion membrane, or through the membrane itself. Catholytes have included potassium phosphate buffer (Subramanian et al., 2007) and NaOH (Innocent et al., 2009), while $KHCO_3$ has been used in a PTFE filter membrane (Whipple et al., 2010). The reactor scale has varied from microreactors (Whipple et al., 2010) to continuous flow reactors (Subramanian et al., 2007). A concerted effort has been made by Li and Oloman on the scale-up of formic acid and formate production reactors. Initial work was performed with filter-press reactors using Nafion membranes to convert from batch to continuous reactor design (Li and Oloman, 2005, 2006). Later works focused on an industrial scale reactor using a trickle-bed design (Li and Oloman, 2006, 2007). In both cases, a variety of aqueous catholytes were tested, as well as multiple Sn and Sn–Cu electrodes.

6.9 Reviews

Over the decades of research activities in the CO_2 electroreduction area, a number of reviews have been published. A handful of the them have focused on covering the general aspects of CO_2 reduction (Ayers, 1994; Scibioh and Vijayaraghavan, 1998; Scibioh and Viswanathan, 2004). A scheme has been proposed for classifying electrochemical systems by the electrodes and electrolytes (Jitaru et al., 1997). Models for conditions at the electrode surface have also been proposed (Gupta et al., 2006). Systems with specific materials have been reviewed, including Cu (Gattrell et al., 2006), phthalocyanines (Vasudevan et al., 1996), and metals deposited on Nafion membranes (Enea et al., 1995). Additional focus has been put on hydrocarbon production (Sánchez-Sánchez et al., 2001; Gattrell et al., 2007; Olah et al., 2009). Analyses have been performed on using CO_2 reduction for energy storage (Hori and Suzuki, 1982b) and on the efficiency of producing methanol from both biomass and atmospheric CO_2 (Weimer et al., 1996).

6.10 CO₂ Electroreduction with CuCl Electrolysis

6.10.1 Hydrogen Production

Thermochemical cycles are considered promising technological solutions for high-efficiency production of hydrogen and, in the context of renewable energy concept, can become a crucial part of industrial infrastructure in the sustainable economy. In a typical thermochemical water-splitting cycle, water is involved in a series of high- and low-temperature chemical reactions, and the only output products are hydrogen and oxygen. All other chemical components are recycled. Some of the cycles include an electrolytic step, and in this case, the cycle should be termed a hybrid electrothermochemical cycle. Still, the specific physicochemical conditions will allow performing the electrolysis at a much lower DP and, therefore, be more efficient compared with water electrolysis.

The benchmark method for producing hydrogen from water as the only input has been water electrolysis, which is costly and provides a relatively low efficiency of 18%–24% for the overall process (Lewis et al., 2009). A variety of thermochemical cycles have been proposed since the early 1970s (Dang and Steinberg, 1980; Carty et al., 1981; McQuillan et al., 2005; Petri et al., 2006; Elder and Allen, 2009) as alternatives to water electrolysis. The particular advantage of the thermochemical cycles compared to water electrolysis is the more favorable thermodynamics of the process at elevated temperatures, which allows for less electric energy use and, therefore, lower cost of hydrogen when the thermal energy is relatively inexpensive.

While the operating temperature range for different thermochemical cycles varies up to about 2000°C, the moderate temperature thermochemical cycles, which can be performed at 400°C–600°C, are especially attractive because such conditions create more opportunities for combining the cycles with a number of available heat sources such as solar concentrators and Generation IV nuclear reactors, which have been recently under an intense development. Advantageously, conducting the thermochemical cycles at moderate temperatures (sometimes called alternative thermochemical cycles) allows for the use of a wider range of construction materials. This is a critical issue as material selection becomes crucial as operating temperatures are increased.

The CuCl thermochemical cycle is among the most attractive alternative thermochemical cycles proposed for hydrogen production due to its moderate temperature requirements and high efficiency. This cycle is regarded as one of the most promising approaches due to its moderate temperature, around 550°C, reduced complexity, inexpensive chemicals, and potentially high efficiency. The optimized CuCl thermochemical cycle can be considered as a system of three main interrelated reactions (Dokya and Kotera, 1976; Carty et al., 1981; Lewis et al., 2009): a low-temperature hydrogen-producing electrolysis reaction (6.8) and two high-temperature reactions (6.9) and (6.10):

$$2CuCl \bullet nH_2O(aq) + 2HCl \bullet mH_2O(aq) \rightarrow 2CuCl_2 \bullet (n+m)H_2O(aq) + H_2(g),$$

$$\text{electrolysis } (25°C - 80°C) \tag{6.8}$$

$$2CuCl_2(s) + H_2O(g) \rightarrow CuCl_2 \bullet CuO(s) + 2HCl(g), \text{ hydrolysis } (310°C - 375°C) \tag{6.9}$$

$$CuCl_2 \bullet CuO(s) \rightarrow 2CuCl(s) + (½)O_2(g), \text{ decomposition } (450°C - 530°C) \tag{6.10}$$

In addition, four common processes

$$H_2O(l) \rightarrow H_2O(g), \text{ evaporation} \tag{6.11}$$

$$2CuCl(s) + nH_2O(l) \rightarrow 2CuCl \bullet nH_2O(aq), \text{ dissolution} \tag{6.12}$$

$$2CuCl_2 \bullet (n+m)H_2O(aq) \rightarrow 2CuCl_2(s) + (n+m)H_2O(l), \text{ crystallization} \tag{6.13}$$

$$2HCl(g) + mH_2O(l) \rightarrow 2HCl \bullet mH_2O(aq), \text{ dissolution} \tag{6.14}$$

are needed to complete the cycle so that the only overall chemical input is liquid water and the outputs are H_2 and O_2 gas, while all other chemicals are recycled.

In this technology, the key step is H_2 gas evolution via oxidation of CuCl dissolved in highly concentrated aqueous HCl in an electrolyzer. The general concept of CuCl electrolysis, reaction (6.8), using a proton-conductive membrane is illustrated in Figure 6.4.

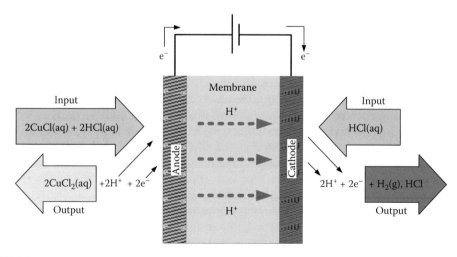

FIGURE 6.4
Conceptual scheme of CuCl electrolysis with a proton-conductive membrane.

The input reagents on the anode side include CuCl, which is produced in the decomposition reaction (6.10) and dissolved in aqueous HCl solution, and the cathodic input usually contains aqueous HCl solution. However, the catholyte could be pure water in a liquid or vapor phase. Upon applying an electric potential difference between the electrodes, the anodic reaction proceeds with oxidation of Cu(I) to Cu(II), which are present as chloride complexes, while the protons are transported through the membrane from the positively charged anode to the negatively charged cathode, where they are electrochemically reduced to H_2 gas, as shown in Figure 6.4.

It should be noted that the anodic reaction, as it is presented in Figure 6.4, only represents one possible mechanism and does not comprehensively reflect the actual anolyte chemistry. As a matter of fact, the real anolyte speciation can be far more complex, and the thermodynamic analysis of the $CuCl$-$CuCl_2$-HCl-H_2O system is required to properly understand the speciation issue.

A CuCl electrolyzer recently developed at Pennsylvania State University (United States) (Balashov et al., 2010) was based on $5\,cm^2$ PEM fuel cell hardware (Electrochem, Inc.) and was using Nafion-based HYDRion MEAs from Ion Power, Inc. designed for water electrolysis. The electrode working area was $5\,cm^2$. The thicknesses of the Nafion membranes in different MEA samples were between 127 and $178\,\mu m$. Two carbon paper diffusion layers (AA 10, Ion Power Inc.) were applied to each side of the membrane without hot pressing. The double diffusion layers on each side of the MEAs were used for a good electrical contact with the graphite bipolar plates, which had flow-through serpentine channels for solution supply and removal of electrolysis products. The compositions of the anolyte solution varied in range with 0.1–$0.3\,mol\,kg^{-1}$ CuCl(aq) dissolved in 1.5–$4\,mol\,kg^{-1}$ HCl(aq). Deionized water was used as the catholyte in this study as the first step in studying the electrolysis process. The anolyte solution was always kept under argon. The hydrogen production in the electrolytic cell was driven by an external applied voltage in the range of 0.3–$1.0\,V$ using a DC power supply. The experimental system allowed the determination of the hydrogen production rate and the total mass of produced gas. The experimental temperature for different runs was in the range from 22°C to 30°C, and electrolysis was studied at two different flow rates of anolyte at 30 and $68\,mL\cdot min^{-1}$, with the catholyte flow rate at $28.5\,mL\cdot min^{-1}$. Linear sweep voltammetry was the main method used in this study to obtain the experimental DP and to observe the current response to the applied voltage during electrolysis. Repeated linear sweep voltammetry measurements were performed with different anolyte concentrations and at different flow rates. The obtained hydrogen production rates correspond to Faraday's law within the experimental errors. The electrolysis was performed for 2 h and, according to the observed correspondence to Faraday's law, the cathodic reaction was represented by the reduction of protons to molecular H_2 at the electrolyzer cathode.

The DPs for several different conditions of CuCl–HCl electrolysis were theoretically calculated at 25°C using a specially developed speciation model for the $CuCl$-$CuCl_2$-HCl-H_2O system. It was found that the calculated DP value for the $0.26\,mol\,kg^{-1}$ CuCl in $2.84\,mol\,kg^{-1}$ HCl anolyte was in good agreement with the experimental DP values for the CuCl electrolysis and was always around $0.4\,V$.

The current efficiency, η_c, of the CuCl electrolyzer was estimated by the comparison between the experimentally produced amount of hydrogen per unit time, and the theoretical rate of hydrogen gas generation at a particular current density was calculated from Faraday's law. The average current efficiency calculated for all experimental points in the CuCl electrolyzer was around 98%. This essentially meant that the hydrogen production reaction $2H^+(aq) + 2e^- \rightarrow H_2(gas)$ was the main process consuming applied current, and no side reactions were apparent during the time of the experiment.

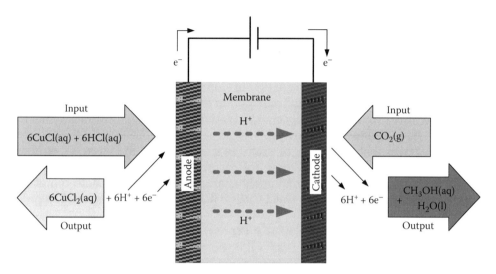

FIGURE 6.5
Conceptual scheme of the CuCl/CO$_2$ electrolyzer.

The voltage efficiency, η_v, was estimated in this study as the percent ratio of the experimentally obtained DP to the applied voltage. Based on the experimental data obtained in this study, it was concluded that the voltage efficiency of the CuCl electrolyzer when pure water was used as catholyte could be 80% at 0.5 V and 0.1 A cm^{-2}, 50% at 0.8 V and 0.35 A cm^{-2}, and 40% at 1.0 V and 0.55 A cm^{-2}.

6.10.2 CO$_2$ Reduction

It is believed that the CuCl thermochemical cycle could also be used to efficiently reduce CO$_2$ to a fuel. A schematic of the possible CuCl(aq)/CO$_2$(g) electrolyzer is presented in Figure 6.5. If such an electrolyzer were developed, it could be run at a much lower applied potential compared to performing electrolysis using only water as the anolyte. The experimental work on this concept is in progress.

6.11 Summary

As seen in this chapter, there has been considerable work done with regard to converting CO$_2$ into valuable chemical compounds and fuels using electrochemistry. Every aspect of the problem from product distributions to catalysis to reaction kinetics and mechanisms has been explored in one way or another. However, one underlying trend is that there has not been much work done on the process to scale up to larger practical-size units. A considerable part of this has been related to the implementation of an appropriate catalyst and ion exchange membrane. Many of the most promising catalysts were often noncommon materials that may not be readily available and thus require custom synthesis. At this point, two main restrictions to the process scale-up appear to be products distribution, the process efficiency, and the required potential or current. The majority of the produced hydrocarbon and oxygenated fuels have been C$_1$ and C$_2$ compounds, with the more advanced products often being formed at much lower concentrations. The high potentials required and low corresponding current densities will need to be improved if the process is to be considered on a scale larger than the laboratory bench.

There is still considerable work to be explored in this field, and the most critical aspect relates to catalyst development. Like fuel cell technology, the choice of catalyst can make all of the difference in a process' feasibility. It can be seen through reviewing the current collection of research work that metal alloys can be customized to provide the desired adsorption properties required for certain products. The catalyst geometry has also been found to impact not only the product distribution but also the cell potential and current density. For this reason, it would be very useful to be able to reproduce many of the studied systems in more efficient cell designs, such as the filter-press style used by the more advanced fuel cells and electrolyzers. This would not only provide a more accurate analysis of the process but would also remove some potential issues such as the solution resistance often found in methanol-based cells. Advanced cell designs would also allow more room to experiment with a variety of catalyst strictures and allow researchers to more accurately observe and analyze the process of CO_2 electroreduction.

References

Abbott A.P. and Eardley C.A. 2000. Electrochemical reduction of CO_2 in a mixed supercritical fluid. *J. Phys. Chem. B* 104(4):775–779.

Abbott A.P. and Harper J.C. 1996. Electrochemical investigations in supercritical carbon dioxide. *J. Chem. Soc., Faraday Trans.* 92(20):3895–3898.

Abe T., Imaya H., Yoshida T., Tokita S., Schlettwein D., Wöhrle D., and Kaneko M. 1997. Electrochemical CO_2 reduction catalysed by cobalt octacyanophthalocyanine and its mechanism. *J. Porphyr. Phthalocyanines* 1(4):315–321.

Abe T., Taguchi F., Yoshida T., Tokita S., Schnurpfeil G., Wöhrle D., and Kaneko M. 1996a. Electrocatalytic CO_2 reduction by cobalt octabutoxyphthalocyanine coated on graphite electrode. *J. Mol. Catal. A Chem.* 112(1):55–61.

Abe T., Yoshida T., Tokita S., Taguchi F., Imaya H., and Kaneko M. 1996b. Factors affecting selective electrocatalytic CO_2 reduction with cobalt phthalocyanine incorporated in a polyvinylpyridine membrane coated on a graphite electrode. *J. Electroanal. Chem.* 412(1–2):125–132.

Aga H., Aramata A., and Hisaeda Y. 1997. The electroreduction of carbon dioxide by macrocyclic cobalt complexes chemically modified on a glassy carbon electrode. *J. Electroanal. Chem.* 437(1–2):111–118.

Amatore C. and Saveant J.M. 1981. Mechanism and kinetic characteristics of the electrochemical reduction of carbon dioxide in media of low proton availability. *J. Am. Chem. Soc.* 103(17):5021–5023.

Arevalo M.C., Gomis-Bas C., and Hahn F. 1998. Reduced CO_2 on a polycrystalline Rh electrode in acid solution: Electrochemical and in situ IR reflectance spectroscopic studies. *Electrochim. Acta* 44(8–9):1369–1378.

Arevalo M.C., Gomis-Bas C., Hahn F., Beden B., Arevalo A., and Arvia A.J. 1994. A contribution to the mechanism of reduced CO_2 adsorbates electro-oxidation from combined spectroelectrochemical and voltammetric data. *Electrochim. Acta* 39(6):793–799.

Atoguchi T., Aramata A., Kazusaka A., and Enyo M. 1991a. Electrocatalytic activity of CO(II) TPP-pyridine complex modified carbon electrode for CO_2 reduction. *J. Electroanal. Chem.* 318(1–2):309–320.

Atoguchi T., Aramata A., Kazusaka A., and Enyo M. 1991b. Cobalt (II)–tetraphenylporphyrin–pyridine complex fixed on a glassy carbon electrode and its prominent catalytic activity for reduction of carbon dioxide. *J. Chem. Soc., Chem. Commun.* 1991(3):156–157.

Augustynski J., Jermann B., and Kedzierzawski P. 1996. Electroreduction of carbon dioxide in aqueous solutions at metal electrodes. *Am. Chem. Soc. Div. Fuel Chem.* 41:1420–1424.

Aurian-Blajeni B., Halmann M., and Manassen J. 1983. Electrochemical measurement on the photoelectrochemical reduction of aqueous carbon dioxide on p-gallium phosphide and p-gallium arsenide semiconductor electrodes. *Solar Energy Mater.* 8(4):425–440.

Aydin R. and Köleli F. 2002. Electrochemical reduction of CO_2 on a polyaniline electrode under ambient conditions and at high pressure in methanol. *J. Electroanal. Chem.* 535(1–2):107–112.

Aydin R. and Köleli F. 2004. Electrocatalytic conversion of CO_2 on a polypyrrole electrode under high pressure in methanol. *Synth. Met.* 144(1):75–80.

Ayers W.M. 1994. An overview of electrochemical carbon dioxide reduction. *Spec. Publ. Royal Soc. Chem.* 153:365–374.

Azuma M., Hashimoto K., Hiramoto M., Watanabe M., and Sakata T. 1989. Carbon dioxide reduction at low temperature on various metal electrodes. *J. Electroanal. Chem.* 260(2):441–445.

Azuma M., Hashimoto K., Hiramoto M., Watanabe M., and Sakata T. 1990. Electrochemical reduction of carbon dioxide on various metal electrodes in low-temperature aqueous KHCO media. *J. Electrochem. Soc.* 137:1772.

Balashov V.N., Schatz R.S., Chalkova E., Akinfiev N.N., Fedkin M.V., and Lvov S.N. 2010. CuCl electrolysis for hydrogen production in the Cu-Cl thermochemical cycle. *J. Electrochem. Soc.* 158:B266–B275.

Bandi A. 1990. Electrochemical reduction of carbon dioxide on conductive metallic oxides. *J. Electrochem. Soc.* 137:2157.

Bandi A. and Kühne H.M. 1992. Electrochemical reduction of carbon dioxide in water: Analysis of reaction mechanism on ruthenium-titanium-oxide. *J. Electrochem. Soc.* 139:1605.

Bandi A., Schwarz J., and Maier C.U. 1993. Adsorption of CO_2 on transition metals and metal oxides. *J. Electrochem. Soc.* 140:1006.

Bewick A and Greener G.P. 1970. The electroreduction of CO_2 to glycollate on a lead cathode. *Tetrahedron Lett.* 11(5):391–394.

Brisard G.M., Camargo A.P.M., Nart F.C., and Iwasita T. 2001. On-line mass spectrometry investigation of the reduction of carbon dioxide in acidic media on polycrystalline Pt. *Electrochem. Commun.* 3(11):603–607.

Canfield D. and Frese Jr. K.W. 1983. Reduction of carbon dioxide to methanol on n- and p-GaAs and p-InP: Effect of crystal face, electrolyte and current density. *J. Electrochem. Soc.* 130(8):1772–1773.

Carty R.H., Mazumder M.M., Schreider J.D., and Pangborn J.B. 1981. Thermochemical hydrogen production, Vols. 1–4. Gas Research Institute for the Institute of Gas Technology, Chicago, IL, GRI-80/0023.

Centi G., Perathoner S., Winè G., and Gangeri M. 2007. Electrocatalytic conversion of CO_2 to long carbon-chain hydrocarbons. *Green Chem.* 9(6):671–678.

Chang T.Y., Liang R.M., Wu P.W., Chen J.Y., and Hsieh Y.C. 2009. Electrochemical reduction of CO_2 by Cu_2O-catalyzed carbon clothes. *Mater. Lett.* 63(12):1001–1003.

Chaplin R.P.S. and Wragg A.A. 2003. Effects of process conditions and electrode material on reaction pathways for carbon dioxide electroreduction with particular reference to formate formation. *J. Appl. Electrochem.* 33(12):1107–1123.

Chardon-Noblat S., Deronzier A., Ziessel R., and Zsoldos D. 1998. Electroreduction of CO_2 catalyzed by polymeric [Ru (bpy)(CO)$_2$] n films in aqueous media: Parameters influencing the reaction selectivity. *J. Electroanal. Chem.* 444(2):253–260.

Christensen P.A., Hamnett A., Higgins S.J., and Timney J.A. 1995. An in-situ Fourier Transform Infrared study of CO_2 electroreduction catalysed by Ni (0)—4,4′-dimethyl-2,2′-bipyridine and Ni (0)—1,10-phenanthroline complexes. *J. Electroanal. Chem.* 395(1–2):195–209.

Christensen P.A., Hamnett A., and Muir A.V.G. 1988. An in-situ FTIR study of the electroreduction of CO_2 by copc-coated edge graphite electrodes. *J. Electroanal. Chem.* 241(1–2):361–371.

Christensen P.A., Hamnett A., and Muir N.A. 1990. CO_2 reduction at platinum, gold and glassy carbon electrodes in acetonitrile: An in-situ FTIR study. *J. Electroanal. Chem.* 288(1–2):197–215.

Chu D., Qin G., Yuan X., Xu M., Zheng P., and Lu J. 2008. Fixation of CO₂ by electrocatalytic reduction and electropolymerization in ionic liquid-H₂O solution. *ChemSusChem.* 1(3):205–209.

Cohen E.R., Cvitaš T., Frey J.G., Holmstrom B., Kuchitsu K., Marquardt R., Mills, I. et al. 2007. *Quantities, Units and Symbols in Physical Chemistry.* 3rd edn. IUPAC. RSC Publishing, Cambridge, U.K.

Collomb-Dunand-Sauthier M.N., Deronzier A., and Ziessel R. 1994a. Electrocatalytic reduction of CO₂ in water on a polymeric [{Ru⁰(bpy)(CO)₂}ₙ] (bpy = 2,2'-bipyridine) complex immobilized on carbon electrodes. *J. Chem. Soc., Chem. Commun.* (2):189–191.

Collomb-Dunand-Sauthier M.N., Deronzier A., and Ziessel R. 1994b. Electrocatalytic reduction of carbon dioxide with mono (bipyridine) carbonylruthenium complexes in solution or as polymeric thin films. *Inorg. Chem.* 33(13):2961–2967.

Cook R.L., MacDuff R.C., and Sammells A.F. 1987. Efficient high rate carbon dioxide reduction to methane and ethylene at in situ electrodeposited copper electrode. *J. Electrochem. Soc.* 134(9):2375–2376.

Cook R.L., MacDuff R.C., and Sammells A.F. 1988a. Ambient temperature gas phase CO₂ reduction to hydrocarbons at solid polymer electrolyte cells. *J. Electrochem. Soc.* 135(6):1470–1471.

Cook R.L., MacDuff R.C., and Sammells A.F. 1988b. On the electrochemical reduction of carbon dioxide at in situ electrodeposited copper. *J. Electrochem. Soc.* 135:1320.

Cook R.L., MacDuff R.C., and Sammells A.F. 1989. Evidence for formaldehyde, formic acid, and acetaldehyde as possible intermediates during electrochemical carbon dioxide reduction at copper. *J. Electrochem. Soc.* 136(7):1982–1984.

Cook R.L., MacDuff R.C., Sammells A.F. 1990a. Gas-phase CO₂ reduction to hydrocarbons at metal/solid polymer electrolyte interface. *J. Electrochem. Soc.* 137(1):187–189.

Cook R.L., MacDuff R.C., and Sammells A.F. 1990b. High rate gas phase CO₂ reduction to ethylene and methane using gas diffusion electrodes. *J. Electrochem. Soc.* 137(2):607–608.

Cueto L., Hirata G., and Sánchez E. 2006. Thin-film TiO₂ electrode surface characterization upon CO₂ reduction processes. *J. Sol. Gel. Sci. Technol.* 37(2):105–109.

Dang V.D. and Steinberg M. 1980. Application of the fusion reactor to thermochemical-electrochemical hybrid cycles and electrolysis for hydrogen production from water. *Energy Convers. Manag.* 20:85.

Darensbourg D.J. and Kudaroski R. 1984. Metal-induced transformations of carbon dioxide. Carbon-carbon bond-forming processes involving anionic group VIB metal derivatives, and the x-ray structure of [PNP][cis-MeW (CO) 4PMe3]. *J. Am. Chem. Soc.* 106(12):3672–3673.

De Jesús-Cardona H., del Moral C., and Cabrera C.R. 2001. Voltammetric study of CO₂ reduction at Cu electrodes under different KHCO₃ concentrations, temperatures and CO₂ pressures. *J. Electroanal. Chem.* 513(1):45–51.

Delacourt C., Ridgway P.L., Kerr J.B., and Newman J. 2008. Design of an electrochemical cell making syngas (CO-H₂) from CO₂ and H₂O reduction at room temperature. *J. Electrochem. Soc.* 155:B42.

Dewulf D.W. and Bard A.J. 1988. The electrochemical reduction of CO₂ to CH₄ and C₂H₄ at Cu/Nafion electrodes (solid polymer electrolyte structures). *Catal. Lett.* 1(1):73–79.

DeWulf D.W., Jin T., and Bard A.J. 1989. Electrochemical and surface studies of carbon dioxide reduction to methane and ethylene at copper electrodes in aqueous solutions. *J. Electrochem. Soc.* 136:1686.

Dokya M. and Kotera Y. 1976. Hybrid cycle with electrolysis using Cu–Cl system. *Int. J. Hydrogen Energy* 1:117.

Dube P. and Brisard G.M. 2005. Influence of adsorption processes on the CO₂ electroreduction: An electrochemical mass spectrometry study. *J. Electroanal. Chem.* 582(1–2):230–240.

Eggins B.R., Bennett E.M., and McMullan E.A. 1996. Voltammetry of carbon dioxide. Part 2. Voltammetry in aqueous solutions on glassy carbon. *J. Electroanal. Chem.* 408(1–2):165–171.

Eggins B.R., Brown E.M., McNeill E.A., and Grinshaw J. 1988. Carbon dioxide fixation by electrochemical reduction in water to oxalate and glyoxylate. *Tetrahedron Lett.* 29(8):945–948.

Eggins B.R., Ennis C., McConnell R., and Spence M. 1997. Improved yields of oxalate, glyoxylate and glycolate from the electrochemical reduction of carbon dioxide in methanol. *J. Appl. Electrochem.* 27(6):706–712.

Eggins B.R. and McNeill J. 1983. Voltammetry of carbon dioxide: Part I. A general survey of voltammetry at different electrode materials in different solvents. *J. Electroanal. Chem.* 148(1):17–24.

Elder R. and Allen R. 2009. Nuclear heat for hydrogen production: Coupling a very high/high temperature reactor to a hydrogen production plant. *Prog. Nucl. Energy* 51:500.

Endo N., Miho Y., and Ogura K. 1997. Hydrogenation of CO_2 on the cathodized tungsten trioxide/polyaniline/polyvinylsulfate-modified electrode in aqueous solution. *J. Mol. Catal. A Chem.* 127(1–3):49–56.

Enea O., Duprez D., and Amadelli R. 1995. Gas phase electrocatalysis on metal/Nafion membranes. *Catal. Today* 25(3–4):271–276.

Fischer J., Lehmann T., and Heitz E. 1981. The production of oxalic acid from carbon dioxide and water. *J. Appl. Electrochem.* 11(6):743–750.

Frese Jr. K.W. 1991. Electrochemical reduction of CO at intentionally oxidized copper electrodes. *J. Electrochem. Soc.* 138:3338.

Frese Jr. K.W. and Canfield D. 1984. Reduction of CO_2 on n-GaAs electrodes and selective methanol synthesis. *J. Electrochem. Soc.* 131:2518.

Frese Jr. K.W. and Leach S. 1985. Electrochemical reduction of carbon dioxide to methane, methanol, and CO on Ru electrodes. *J. Electrochem. Soc.* 132:259.

Friebe P., Bogdanoff P., Alonso-Vante N., and Tributsch H. 1997. A real-time mass spectroscopy study of the (electro) chemical factors affecting CO_2 reduction at copper. *J. Catal.* 168(2):374–385.

Fujihira M. and Noguchi T. 1992. A highly sensitive analysis of electrochemical reduction products of CO_2 on gold by new differential electrochemical mass spectroscopy (DEMS). *Chem. Lett.* 21(10):2043–2046.

Furuya N. and Koide S. 1991. Electroreduction of carbon dioxide by metal phthalocyanines. *Electrochim. Acta* 36(8):1309–1313.

Furuya N., Yamazaki T., and Shibata M. 1997. High performance Ru–Pd catalysts for CO_2 reduction at gas-diffusion electrodes. *J. Electroanal. Chem.* 431(1):39–41.

Gattrell M., Gupta N., and Co A. 2006. A review of the aqueous electrochemical reduction of CO_2 to hydrocarbons at copper. *J. Electroanal. Chem.* 594(1):1–19.

Gattrell M., Gupta N., and Co A. 2007. Electrochemical reduction of CO_2 to hydrocarbons to store renewable electrical energy and upgrade biogas. *Energy Convers. Manag.* 48(4):1255–1265.

Giner J. 1963. Electrochemical reduction of CO_2 on platinum electrodes in acid solutions. *Electrochim. Acta* 8(11):857–865.

Grimes C.A., Oomman K., Varghese K., and Ranjan S. 2008. *Light, Water, Hydrogen.* Springer, New York.

Gupta N., Gattrell M., and MacDougall B. 2006. Calculation for the cathode surface concentrations in the electrochemical reduction of CO_2 in $KHCO_3$ solutions. *J. Appl. Electrochem.* 36(2):161–172.

Hara K., Kudo A., and Sakata T. 1995a. Electrochemical reduction of carbon dioxide under high pressure on various electrodes in an aqueous electrolyte. *J. Electroanal. Chem.* 391(1–2):141–147.

Hara K., Kudo A., and Sakata T. 1995b. Electrochemical reduction of high pressure carbon dioxide on Fe electrodes at large current density. *J. Electroanal. Chem.* 386(1–2):257–260.

Hara K., Kudo A., and Sakata T. 1997a. Electrochemical CO_2 reduction on a glassy carbon electrode under high pressure. *J. Electroanal. Chem.* 421(1–2):1–4.

Hara K., Kudo A., Sakata T., and Watanabe M. 1995c. High efficiency electrochemical reduction of carbon dioxide under high pressure on a gas diffusion electrode containing Pt catalysts. *J. Electrochem. Soc.* 142:L57.

Hara K. and Sakata T. 1997a. Electrocatalytic formation of CH_4 from CO_2 on a Pt gas diffusion electrode. *J. Electrochem. Soc.* 144(2):539–545.

Hara K. and Sakata T. 1997b. Large current density CO_2 reduction under high pressure using gas diffusion electrodes. *Bull. Chem. Soc. Jpn.* 70(3):571–576.

Hara K., Tsuneto A., Kudo A., and Sakata T. 1994. Electrochemical reduction of CO_2 on a Cu electrode under high pressure. *J. Electrochem. Soc.* 141(8):2097–2103.

Hara K., Tsuneto A., Kudo A., and Sakata T. 1997b. Change in the product selectivity for the electrochemical CO_2 reduction by adsorption of sulfide ion on metal electrodes. *J. Electroanal. Chem.* 434(1–2):239–243.

Haynes L.V. and Sawyer D.T. 1967. Electrochemistry of carbon dioxide in dimethyl sulfoxide at gold and mercury electrodes. *Anal. Chem.* 39(3):332–338.

Hernandez R.M., Marquez J., Marquez O.P., Choy M., Ovalles C., Garcia J.J, and Scharifker B. 1999. Reduction of carbon dioxide on modified glassy carbon electrodes. *J. Electrochem. Soc.* 146:4131.

Hori Y., Ito H., Okano K., Nagasu K., and Sato S. 2003. Silver-coated ion exchange membrane electrode applied to electrochemical reduction of carbon dioxide. *Electrochim. Acta* 48(18):2651–2657.

Hori Y., Kikuchi K., Murata A., and Suzuki S. 1986. Production of methane and ethylene in electrochemical reduction of carbon dioxide at copper electrode in aqueous hydrogencarbonate solution. *Chem. Lett.* 15(6):897–898.

Hori Y., Kikuchi K., and Suzuki S. 1985. Production of CO and CH_4 in electrochemical reduction of CO_2 at metal electrodes in aqueous hydrogencarbonate solution. *Chem. Lett.* 14(11):1695–1698.

Hori Y., Koga O., Yamazaki H., and Matsuo T. 1995a. Infrared spectroscopy of adsorbed CO and intermediate species in electrochemical reduction of CO_2 to hydrocarbons on a Cu electrode. *Electrochim. Acta* 40(16):2617–2622.

Hori Y., Konishi H., Futamura T., Murata A., Koga O., Sakurai H., and Oguma K. 2005. "Deactivation of copper electrode" in electrochemical reduction of CO_2. *Electrochim. Acta* 50(27):5354–5369.

Hori Y. and Murata A. 1990. Electrochemical evidence of intermediate formation of adsorbed CO in cathodic reduction of CO_2 at a nickel electrode. *Electrochim. Acta* 35(11–12):1777–1780.

Hori Y., Murata A., Ito S., Yoshinami Y., and Koga O. 1989a. Nickel and iron modified copper electrode for electroreduction of CO_2 by in-situ electrodeposition. *Chem. Lett.* 18(9):1567–1570.

Hori Y., Murata A., Kikuchi K., and Suzuki S. 1987. Electrochemical reduction of carbon dioxide to carbon monoxide at a gold electrode in aqueous potassium hydrogen carbonate. *J. Chem. Soc., Chem. Commun.* (10):728–729.

Hori Y., Murata A., and Takahashi R. 1989b. Formation of hydrocarbons in the electrochemical reduction of carbon dioxide at a copper electrode in aqueous solution. *J. Chem. Soc., Faraday Trans. 1* 85(8):2309–2326.

Hori Y., Murata A., Takahashi R., and Suzuki S. 1988. Enhanced formation of ethylene and alcohols at ambient temperature and pressure in electrochemical reduction of carbon dioxide at a copper electrode. *J. Chem. Soc., Chem. Commun.* (1):17–19.

Hori Y., Murata A., and Yoshinami Y. 1991. Adsorption of CO, intermediately formed in electrochemical reduction of CO_2, at a copper electrode. *J. Chem. Soc., Faraday Trans.* 87(1):125–128.

Hori Y. and Suzuki S. 1982a. Electrolytic reduction of carbon dioxide at mercury electrode in aqueous solution. *Bull. Chem. Soc. Jpn.* 55(3):660–665.

Hori Y. and Suzuki S. 1982b. Cathodic reduction of carbon dioxide for energy storage. *J. Res. Inst. Catal. Hokkaido Univ.* 30(2):81.

Hori Y., Takahashi I., Koga O., and Hoshi N. 2002. Selective formation of C2 compounds from electrochemical reduction of CO_2 at a series of copper single crystal electrodes. *J. Phys. Chem. B* 106(1):15–17.

Hori Y., Wakebe H., Tsukamoto T., and Koga O. 1994. Electrocatalytic process of CO selectivity in electrochemical reduction of CO_2 at metal electrodes in aqueous media. *Electrochim. Acta* 39(11–12):1833–1839.

Hori Y., Wakebe H., Tsukamoto T., and Koga O. 1995b. Adsorption of CO accompanied with simultaneous charge transfer on copper single crystal electrodes related with electrochemical reduction of CO_2 to hydrocarbons. *Surf. Sci.* 335(1–3):258–263.

Hoshi N. and Hori Y. 2000. Electrochemical reduction of carbon dioxide at a series of platinum single crystal electrodes. *Electrochim. Acta* 45(25–26):4263–4270.

Hoshi N., Ito H., Suzuki T., and Hori Y. 1995a. CO_2 reduction on Rh single crystal electrodes and the structural effect. *J. Electroanal. Chem.* 395(1–2):309–312.

Hoshi N., Kato M., and Hori Y. 1997c. Electrochemical reduction of CO_2 on single crystal electrodes of silver Ag (111), Ag (100) and Ag (110). *J. Electroanal. Chem.* 440(1–2):283–286.

Hoshi N., Kawatani S., Kudo M., and Hori Y. 1999. Significant enhancement of the electrochemical reduction of CO_2 at the kink sites on Pt (S)-[n (110) × (100)] and Pt (S)-[n (100) × (110)]. *J. Electroanal. Chem.* 467(1–2):67–73.

Hoshi N., Mizumura T., and Hori Y. 1995. Significant difference of the reduction rates of carbon dioxide between Pt (111) and Pt (110) single crystal electrodes. *Electrochim. Acta* 40(7):883–887.

Hoshi N., Noma M., Suzuki T., and Hori Y. 1997a. Structural effect on the rate of CO_2 reduction on single crystal electrodes of palladium. *J. Electroanal. Chem.* 421(1–2):15–18.

Hoshi N., Sato E., and Hori Y. 2003. Electrochemical reduction of carbon dioxide on kinked stepped surfaces of platinum inside the stereographic triangle. *J. Electroanal. Chem.* 540:105–110.

Hoshi N., Suzuki T., and Hori Y. 1997b. Catalytic activity of CO_2 reduction on Pt single-crystal electrodes: Pt (S)-[n (111) × (111)], Pt (S)-[n (111) × (100)], and Pt (S)-[n (100) × (111)]. *J. Phys. Chem. B* 101(42):8520–8524.

Hoshi N., Suzuki T., and Hori Y. 1996a. CO_2 reduction on Pt (S)-(n (111) × (111) I single crystal electrodes affected by the adsorption of sulfuric acid anion. *J. Electroanal. Chem.* 416(1):61–66.

Hoshi N., Suzuki T., and Hori Y. 1996b. Step density dependence of CO_2 reduction rate on Pt (S)-[n (111) × (111)] single crystal electrodes. *Electrochim. Acta* 41(10):1647–1654.

Hoshi N., Uchida T., Mizumura T., and Hori Y. 1995b. Atomic arrangement dependence of reduction rates of carbon dioxide on iridium single crystal electrodes. *J. Electroanal. Chem.* 381(1–2):261–264.

Huang H., Fierro C., Scherson D., and Yeager E.B. 1991. In situ Fourier transform infrared spectroscopic study of carbon dioxide reduction on polycrystalline platinum in acid solutions. *Langmuir* 7(6):1154–1157.

Ikeda S., Takagi T., and Ito K. 1987. Selective formation of formic acid, oxalic acid, and carbon monoxide by electrochemical reduction of carbon dioxide. *Bull. Chem. Soc. Jpn.* 60(7):2517–2522.

Innocent B., Liaigre D., Pasquier D., Ropital F., Léger J.M., and Kokoh K.B. 2009. Electro-reduction of carbon dioxide to formate on lead electrode in aqueous medium. *J. Appl. Electrochem.* 39(2):227–232.

Isaacs M., Armijo F., Ramírez G., Trollund E., Biaggio S.R., Costamagna J., and Aguirre M.J. 2005. Electrochemical reduction of CO_2 mediated by poly-M-aminophthalocyanines (M = Co, Ni, Fe): Poly-co-tetraaminophthalocyanine, a selective catalyst. *J. Mol. Catal. A, Chem.* 229(1–2):249–257.

Ishimaru S., Shiratsuchi R., and Nogami G. 2000. Pulsed electroreduction of CO_2 on Cu-Ag alloy electrodes. *J. Electrochem. Soc.* 147:1864.

Ito K., Ikeda S., Yamauchi N., Iida T., and Takagi T. 1985. Electrochemical reduction products of carbon dioxide at some metallic electrodes in nonaqueous electrolytes. *Bull. Chem. Soc. Jpn.* 58(10):3027–3028.

Iwakura C., Takezawa S., and Inoue H. 1998. Catalytic reduction of carbon dioxide with atomic hydrogen permeating through palladized Pd sheet electrodes. *J. Electroanal. Chem.* 459(1):167–169.

Jarzębińska A., Rowiński P., Zawisza I., Bilewicz R., Siegfried L., and Kaden T. 1999. Modified electrode surfaces for catalytic reduction of carbon dioxide. *Anal. Chim. Acta* 396(1):1–12.

Jermann B. and Augustynski J. 1994. Long-term activation of the copper cathode in the course of CO_2 reduction. *Electrochim. Acta* 39(11–12):1891–1896.

Jitaru M., Lowy D., Toma M., Toma B., and Oniciu L. 1997. Electrochemical reduction of carbon dioxide on flat metallic cathodes. *J. Appl. Electrochem.* 27(8):875–889.

Kaneco S., Hiei N., Xing Y., Katsumata H., Ohnishi H., Suzuki T., and Ohta K. 2002a. Electrochemical conversion of carbon dioxide to methane in aqueous $NaHCO_3$ solution at less than 273 K. *Electrochim. Acta* 48(1):51–55.

Kaneco S., Hiei N., Xing Y., Katsumata H., Ohnishi H., Suzuki T., and Ohta K. 2003. High-efficiency electrochemical CO_2-to-methane reduction method using aqueous $KHCO_3$ media at less than 273 K. *J. Solid State Electrochem.* 7(3):152–156.

Kaneco S., Iiba K., Yabuuchi M., Nishio N., Ohnishi H., Katsumata H., Suzuki T., and Ohta K. 2002b. High efficiency electrochemical CO_2-to-methane conversion method using methanol with lithium supporting electrolytes. *Ind. Eng. Chem. Res.* 41:21.

Kaneco S., Iiba K., Hiei N., Ohta K., Mizuno T., and Suzuki T. 1999a. Electrochemical reduction of carbon dioxide to ethylene with high faradaic efficiency at a Cu electrode in CsOH/methanol. *Electrochim. Acta* 44(26):4701–4706.

Kaneco S., Iiba K., Katsumata H., Suzuki T., and Ohta K. 2006. Electrochemical reduction of high pressure CO_2 at a cu electrode in cold methanol. *Electrochim. Acta* 51(23):4880–4885.

Kaneco S., Iiba K., Katsumata H., Suzuki T., and Ohta K. 2007a. Effect of sodium cation on the electrochemical reduction of CO$_2$ at a copper electrode in methanol. *J. Solid State Electrochem.* 11(4):490–495.

Kaneco S., Iiba K., Katsumata H., Suzuki T., and Ohta K. 2007b. Electrochemical reduction of high pressure carbon dioxide at a cu electrode in cold methanol with CsOH supporting salt. *Chem. Eng. J.* 128(1):47–50.

Kaneco S., Iiba K., Ohta K., and Mizuno T. 1999a. Electrochemical reduction of carbon dioxide on copper in methanol with various potassium supporting electrolytes at low temperature. *J. Solid State Electrochem.* 3(7):424–428.

Kaneco S., Iiba K., Ohta K., and Mizuno T. 2000. Reduction of carbon dioxide to petrochemical intermediates. *Energy Sources* 22(2):127.

Kaneco S., Iiba K., Ohta K., Mizuno T., and Saji A. 1998a. Electrochemical reduction of CO$_2$ at an Ag electrode in KOH-methanol at low temperature. *Electrochim. Acta* 44(4):573–578.

Kaneco S., Iiba K., Ohta K., Mizuno T., and Saji A. 1998b. Electrochemical reduction of CO$_2$ on Au in KOH methanol at low temperature. *J. Electroanal. Chem.* 441(1–2):215–220.

Kaneco S., Iiba K., Suzuki S., Ohta K., and Mizuno T. 1999c. Electrochemical reduction of carbon dioxide to hydrocarbons with high faradaic efficiency in LiOH/methanol. *J. Phys. Chem. B* 103(35):7456–7460.

Kaneco S., Iwao R., Iiba K., Ohta K., and Mizuno T. 1998c. Electrochemical conversion of carbon dioxide to formic acid on Pb in KOH/methanol electrolyte at ambient temperature and pressure. *Energy* 23(12):1107–1112.

Kaneco S., Iwao R., Iiba K., Itoh S.I., Ohta K., and Mizuno T. 1999b. Electrochemical reduction of carbon dioxide on an indium wire in a KOH/methanol-based electrolyte at ambient temperature and pressure. *Environ. Eng. Sci.* 16(2):131–137.

Kaneco S., Katsumata H., Suzuki T., and Ohta K. 2006a. Electrochemical reduction of carbon dioxide to ethylene at a copper electrode in methanol using potassium hydroxide and rubidium hydroxide supporting electrolytes. *Electrochim. Acta* 51(16):3316–3321.

Kaneco S., Katsumata H., Suzuki T., and Ohta K. 2006b. Electrochemical reduction of CO$_2$ to methane at the Cu electrode in methanol with sodium supporting salts and its comparison with other alkaline salts. *Energy Fuels* 20(1):409–414.

Kaneco S., Sakaguchi Y., Katsumata H., Suzuki T., and Ohta K. 2007c. Cu-deposited nickel electrode for the electrochemical conversion of CO$_2$ in water/methanol mixture media. *Bull. Catal. Soc. India* 6:74.

Kaneco S., Yabuuchi M., Katsumata H., Suzuki T., and Ohta K. 2002c. Electrochemical reduction of CO$_2$ to methane in methanol at low temperature. *Fuel Chem. Div. Preprints* 47(1):71.

Kapusta S. and Hackerman N. 1983. The electroreduction of carbon dioxide and formic acid on tin and indium electrodes. *J. Electrochem. Soc.* 130:607.

Kapusta S. and Hackerman N. 1984. Carbon dioxide reduction at a metal phthalocyanine catalyzed carbon electrode. *J. Electrochem. Soc.* 131:1511.

Katoh A., Uchida H., Shibata M., and Watanabe M. 1994. Design of electrocatalyst for CO$_2$ reduction. *J. Electrochem. Soc.* 141(8):2054–2058.

Kedzierzawski P. and Augustynski J. 1994. Poisoning and activation of the gold cathode during electroreduction of CO. *J. Electrochem. Soc.* 141:L58.

Kim J.J., Summers D.P., and Frese Jr. K.W. 1988. Reduction of CO$_2$ and CO to methane on Cu foil electrodes. *J. Electroanal. Chem.* 245(1–2):223–244.

Kobayashi T. and Takahashi H. 2004. Novel CO$_2$ electrochemical reduction to methanol for H$_2$ storage. *Energy Fuels* 18(1):285–286.

Koga O., Matsuo T., Yamazaki H., and Hori Y. 1998. Infrared spectroscopic observation of intermediate species on Ni and Fe electrodes in the electrochemical reduction of CO$_2$ and CO to hydrocarbons. *Bull. Chem. Soc. Jpn.* 71(2):315–320.

Kolbe D. and Vielstich W. 1996. Adsorbate formation during the electrochemical reduction of carbon dioxide at palladium—A DEMS study. *Electrochim. Acta* 41(15):2457–2460.

Köleli F., Atilan T., and Palamut N. 2002. Electrochemical reduction of CO_2 on granule electrodes in a fixed-bed reactor in aqueous medium. *Fresenius Environ. Bull.* 11(6):278–283.

Köleli F., Atilan T., Palamut N., Gizir A.M., Aydin R., and Hamann C.H. 2003a. Electrochemical reduction of CO_2 at Pb-and Sn-electrodes in a fixed-bed reactor in aqueous K_2CO_3 and $KHCO_3$ media. *J. Appl. Electrochem.* 33(5):447–450.

Köleli F., Röpke T., and Hamann C.H. 2003b. Electrochemical impedance spectroscopic investigation of CO_2 reduction on polyaniline in methanol. *Electrochim. Acta* 48(11):1595–1601.

Köleli F., Yesilkaynak T., and Balun D. 2003c. High pressure-high temperature CO_2 electro-reduction on Sn granules in a fixed-bed reactor. *Fresenius Environ. Bull.* 12(10):1202–1206.

Köleli F. and Balun D. 2004. Reduction of CO_2 under high pressure and high temperature on Pb-granule electrodes in a fixed-bed reactor in aqueous medium. *Appl. Catal. A, Gen.* 274(1–2):237–242.

Köleli F., Röpke T., and Hamann C.H. 2004. The reduction of CO_2 on polyaniline electrode in a membrane cell. *Synth. Met.* 140(1):65–68.

Komatsu S., Tanaka M., Okumura A., and Kungi A. 1995. Preparation of Cu-solid polymer electrolyte composite electrodes and application to gas-phase electrochemical reduction of CO_2. *Electrochim. Acta* 40(6):745–753.

Koudelka M., Monnier A., and Augustynski J. 1984. Electrocatalysis of the cathodic reduction of carbon dioxide on platinized titanium dioxide film electrodes. *J. Electrochem. Soc.* 131:745.

Kudo A., Nakagawa S., Tsuneto A., and Sakata T. 1993. Electrochemical reduction of high pressure CO_2 on Ni electrodes. *J. Electrochem. Soc.* 140:1541.

Kyriacou G.Z. and Anagnostopoulos A.K. 1993. Influence CO_2 partial pressure and the supporting electrolyte cation on the product distribution in CO_2 electroreduction. *J. Appl. Electrochem.* 23(5):483–486.

Lee J. and Tak Y. 2001. Electrocatalytic activity of Cu electrode in electroreduction of CO_2. *Electrochim. Acta* 46(19):3015–3022.

Lewis M.A., Masin J.G., and O'Hare P.A. 2009 Evaluation of alternative thermochemical cycles, Part I: The methodology. *Int. J. Hydrogen Energy* 34:4115.

Li H. and Oloman C. 2005. The electro-reduction of carbon dioxide in a continuous reactor. *J. Appl. Electrochem.* 35(10):955–965.

Li H. and Oloman C. 2006. Development of a continuous reactor for the electro-reduction of carbon dioxide to formate–Part 1: Process variables. *J. Appl. Electrochem.* 36(10):1105–1115.

Li H. and Oloman C. 2007. Development of a continuous reactor for the electro-reduction of carbon dioxide to formate–Part 2: Scale-up. *J. Appl. Electrochem.* 37(10):1107–1117.

Li J. and Prentice G. 1997. Electrochemical synthesis of methanol from CO_2 in high-pressure electrolyte. *J. Electrochem. Soc.* 144:4284.

Lide D.R. 2008. *Handbook of Chemistry and Physics*, 89th edn. CRC Press, Boca Raton, FL.

Lieber C.M. and Lewis N.S. 1984. Catalytic reduction of carbon dioxide at carbon electrodes modified with cobalt phthalocyanine. *J. Am. Chem. Soc.* 106(17):5033–5034.

Maeda M., Kitaguchi Y., Ikeda S., and Ito K. 1987. Reduction of carbon dioxide on partially-immersed au plate electrode and Au-SPE electrode. *J. Electroanal. Chem.* 238(1–2):247–258.

Magdesieva T.V., Yamamoto T., Tryk D.A., and Fujishima A. 2002. Electrochemical reduction of CO_2 with transition metal phthalocyanine and porphyrin complexes supported on activated carbon fibers. *J. Electrochem. Soc.* 149:D89.

Mahmood M.N., Masheder D., and Harty C.J. 1987a. Use of gas-diffusion electrodes for high-rate electrochemical reduction of carbon dioxide. I. Reduction at lead, indium-and tin-impregnated electrodes. *J. Appl. Electrochem.* 17(6):1159–1170.

Mahmood M.N., Masheder D., and Harty C.J. 1987b. Use of gas-diffusion electrodes for high-rate electrochemical reduction of carbon dioxide. II. Reduction at metal phthalocyanine-impregnated electrodes. *J. Appl. Electrochem.* 17(6):1223–1227.

Marcos M.L., Gonzalez-Velasco J., Bolzan A.E., and Arvia A.J. 1995. Comparative electrochemical behaviour of CO_2 on Pt and Rh electrodes in acid solution. *J. Electroanal. Chem.* 395(1–2):91–98.

Marcos M.L., Velasco J.G., Hahn F., Beden B., Lamy C., and Arvia A.J. 1997. In situ FTIRS study of 'reduced' CO_2 on columnar-structured platinum electrodes in different acid media. *J. Electroanal. Chem.* 436(1–2):161–172.

McQuillan B.W., Brown L.C., Besenbruch G.E., Tolman R., Cramer T., and Russ B.E. 2005. High efficiency generation of hydrogen fuels using solar thermo-chemical splitting of water. Annual Report GA-A24972, San Diego, CA.

Mendez E., Martins M.E., and Zinola C.F. 1999. New effects in the electrochemistry of carbon dioxide on platinum by the application of potential perturbations. *J. Electroanal. Chem.* 477(1):41–51.

Mizuno T., Kawamoto M., Kaneco S., and Ohta K. 1997a. Electrochemical reduction of carbon dioxide at Ti and hydrogen-storing Ti electrodes in KOH–methanol. *Electrochim. Acta* 43(8):899–907.

Mizuno T., Ohta K., Kawamoto M., and Saji A. 1997b. Electrochemical reduction of CO_2 on Cu in 0.1 M KOH-methanol. *Energy Sources* 19:249–258.

Mizuno T., Naitoh A., and Ohta K. 1995a. Electrochemical reduction of CO_2 in methanol at −30°C. *J. Electroanal. Chem.* 391(1–2):199–201.

Mizuno T., Ohta K., Sasaki A., Akai T., Hirano M., and Kawabe A. 1995b. Effect of temperature on electrochemical reduction of high-pressure CO_2 with In, Sn, and Pb electrodes. *Energy Sources, Part A: Recovery, Util. Environ. Effects* 17(5):503–508.

Momose Y., Sato K., and Ohno O. 2002. Electrochemical reduction of CO_2 at copper electrodes and its relationship to the metal surface characteristics. *Surf. Interface Anal.* 34(1):615–618.

Murata A. and Hori Y. 1991. Product selectivity affected by cationic species in electrochemical reduction of CO_2 and CO at a Cu electrode. *Bull. Chem. Soc. Jpn.* 64(1):123–127.

Nagao H., Mizukawa T., and Tanaka K. 1994. Carbon-carbon bond formation in the electrochemical reduction of carbon dioxide catalyzed by a ruthenium complex. *Inorg. Chem.* 33(15):3415–3420.

Naitoh A., Ohta K., Mizuno T., Yoshida H., Sakai M., and Noda H. 1993. Electrochemical reduction of carbon dioxide in methanol at low temperature. *Electrochim. Acta* 38(15):2177–2179.

Nakagawa S., Kudo A., Azuma M., and Sakata T. 1991. Effect of pressure on the electrochemical reduction of CO_2 on group VIII metal electrodes. *J. Electroanal. Chem.* 308(1–2):339–343.

Nakayama M., Iino M., and Ogura K. 1997. In situ infrared spectroscopic investigations on the electrochemical properties of Prussian blue—Polyaniline-modified electrodes with various anionic Fe (II) complexes working as a mediator for the electroreduction of CO_2. *J. Electroanal. Chem.* 440(1–2):251–257.

Nikolic B.Z., Huang H., Gervasio D., Lin A., Fierro C., Adzic R.R., and Yeager E.B. 1990. Electroreduction of carbon dioxide on platinum single crystal electrodes: Electrochemical and in situ FTIR studies. *J. Electroanal. Chem.* 295(1–2):415–423.

Noda H., Ikeda S., Oda Y., Imai K., Maeda M., and Ito K. 1990. Electrochemical reduction of carbon dioxide at various metal electrodes in aqueous potassium hydrogen carbonate solution. *Bull. Chem. Soc. Jpn.* 63(9):2459–2462.

Noda H., Ikeda S., Oda Y., and Ito K. 1989. Potential dependencies of the products on electrochemical reduction of carbon dioxide at a copper electrode. *Chem. Lett.* (2):289–292.

Noda H., Ikeda S., Yamamoto A., Einaga H., and Ito K. 1995. Kinetics of electrochemical reduction of carbon dioxide on a gold electrode in phosphate buffer solutions. *Bull. Chem. Soc. Jpn.* 68(7):1889–1895.

Nogami G., Aikoh Y., and Shiratsuchi R. 1993. Investigation of fixation mechanism of carbon dioxide on oxide semiconductors by current transients. *J. Electrochem. Soc.* 140:1037.

Nogami G., Itagaki H., and Shiratsuchi R. 1994. Pulsed electroreduction of CO_2 on copper electrodes-II. *J. Electrochem. Soc.* 141(5):1138.

Ogura K. 2003. Electrochemical and selective conversion of CO_2 to ethylene. *Denki Kagaku Oyobi Kogyo Butsuri Kagaku* 71(8):676–680.

Ogura K. and Endo N. 1999. Electrochemical reduction of CO_2 with a functional gas-diffusion electrode in aqueous solutions with and without propylene carbonate. *J. Electrochem. Soc.* 146(10):3736.

Ogura K., Endo N., and Nakayama M. 1998. Mechanistic studies of CO_2 reduction on a mediated electrode with conducting polymer and inorganic conductor films. *J. Electrochem. Soc.* 145(11):3801–3809.

Ogura K., Endo N., Nakayama M., and Ootsuka H. 1995. Mediated activation and electroreduction of CO_2 on modified electrodes with conducting polymer and inorganic conductor films. *J. Electrochem. Soc.* 142:4026.

Ogura K. and Fujita M. 1987. Electrocatalytic reduction of carbon dioxide to methanol. VII: With quinone derivatives immobilized on platinum and stainless steel. *J. Mol. Catal.* 41(3):303–311.

Ogura K., Higasa M., Yano J., and Endo N. 1994a. Electroreduction of CO_2 to C2 and C3 compounds on bis (4,5-dihydroxybenzene-1,3-disulphonato) ferrate (II)-fixed polyaniline/Prussian blue-modified electrode in aqueous solutions. *J. Electroanal. Chem.* 379(1–2):373–377.

Ogura K., Mine K., Yano J., and Sugihara H. 1993. Electrocatalytic generation of C2 and C3 compounds from carbon dioxide on a cobalt complex-immobilized dual-film electrode. *J. Chem. Soc., Chem. Commun.* (1):20–1.

Ogura K., Nakayama M., and Kusumoto C. 1996. In situ Fourier transform infrared spectroscopic studies on a metal complex-immobilized polyaniline/Prussian blue-modified electrode and the application to the electroreduction of CO. *J. Electrochem. Soc.* 143:3606.

Ogura K., Oohara R., and Kudo Y. 2005. Reduction of CO_2 to ethylene at three-phase interface effects of electrode substrate and catalytic coating. *J. Electrochem. Soc.* 152(12):213.

Ogura K., Sugihara H., Yano J., and Higasa M. 1994b. Electrochemical reduction of carbon dioxide on dual-film electrodes modified with and without cobalt (II) and iron (II) complexes. *J. Electrochem. Soc.* 141:419.

Ogura K., Yano H., and Shirai F. 2003. Catalytic reduction of CO_2 to ethylene by electrolysis at a three-phase interface. *J. Electrochem. Soc.* 150:D163.

Ogura K., Yano H., and Tanaka T. 2004. Selective formation of ethylene from CO_2 by catalytic electrolysis at a three-phase interface. *Catal. Today* 98(4):515–521.

Ogura K. and Yoshida I. 1986. Electrocatalytic reduction of carbon dioxide to methanol in the presence of 1,2-dihydroxybenzene-3,5-disulphonatoferrate(III) and ethanol. *J. Mol. Catal.* 34(1):67–72.

Ogura K. and Yoshida I. 1988. Electrocatalytic reduction of CO_2 to methanol: Part 9: Mediation with metal porphyrins. *J. Mol. Catal.* 47(1):51–57.

Ohkawa K., Noguchi Y., Nakayama S., Hashimoto K., and Fujishima A. 1994. Electrochemical reduction of carbon-dioxide on hydrogen-storing materials. 3. The effect of the absorption of hydrogen on the palladium electrodes modified with copper. *J. Electroanal. Chem.* 367(1–2):165–173.

Ohta K., Hasimoto A., and Mizuno T. 1995. Electrochemical reduction of carbon dioxide by the use of copper tube electrode. *Energy Convers. Manag.* 36(6–9):625–628.

Ohta K., Kawamoto M., Mizuno T., and Lowy D.A. 1998. Electrochemical reduction of carbon dioxide in methanol at ambient temperature and pressure. *J. Appl. Electrochem.* 28(7):717–724.

Ohya S., Kaneco S., Katsumata H., Suzuki T., and Ohta K. 2009. Electrochemical reduction of CO_2 in methanol with aid of CuO and Cu_2O. *Catal. Today* 148(3–4):329.

Olah G.A., Goeppert A., and Prakash G.K.S. 2009. Chemical recycling of carbon dioxide to methanol and dimethyl ether: From greenhouse gas to renewable, environmentally carbon neutral fuels and synthetic hydrocarbons. *J. Org. Chem.* 74(2):487–498.

Ortiz R., Marquez O.P., Marquez J., and Gutierrez C. 1995. FTIR spectroscopy study of the electrochemical reduction of CO_2 on various metal electrodes in methanol. *J. Electroanal. Chem.* 390(1–2):99–107.

Paik W., Andersen T.N., and Eyring H. 1969. Kinetic studies of the electrolytic reduction of carbon dioxide on the mercury electrode. *Electrochim. Acta* 14(12):1217–1232.

Petri M.C., Yildiz B., and Klickman A.E. 2006. US work on technical and economic aspects of electrolytic, thermochemical, and hybrid processes for hydrogen production at temperatures below 550°C. *Int. J. Nucl. Hydrogen Prod. Appl.* 1(1):79.

Pettinicchi S., Boggetti H., Lopez de Mishima B.A., Mishima H.T., Rodriguez J., and Pastor E. 2003. Electrochemical reduction of carbon dioxide on copper alloy. *J. Argent. Chem. Soc.* 91(1/3):107–118.

Popić J.P., Avramov-Ivić M.L., and Vuković N.B. 1997. Reduction of carbon dioxide on ruthenium oxide and modified ruthenium oxide electrodes in 0.5 M NaHCO$_3$. *J. Electroanal. Chem.* 421(1–2):105–110.

Qu J., Zhang X., Wang Y., and Xie C. 2005. Electrochemical reduction of CO$_2$ on RuO$_2$/TiO$_2$ nanotubes composite modified Pt electrode. *Electrochim. Acta* 50(16–17):3576–3580.

Ramos Sende J.A., Arana C.R., Hernandez L., Potts K.T., Keshevarz K.M., and Abruna H.D. 1995. Electrocatalysis of CO$_2$ reduction in aqueous media at electrodes modified with electropolymerized films of vinylterpyridine complexes of transition metals. *Inorg. Chem.* 34(12):3339–3348.

Rodes A., Pastor E., and Iwasita T. 1994a. Structural effects on CO$_2$ reduction at Pt single-crystal electrodes. I: The Pt (110) surface. *J. Electroanal. Chem.* 369(1–2):183–191.

Rodes A., Pastor E., and Iwasita T. 1994b. Structural effects on CO$_2$ reduction at Pt single-crystal electrodes. II: Pt(111) and vicinal surfaces in the [011] zone. *J. Electroanal. Chem.* 373(1–2):167–175.

Rodes A., Pastor E., and Iwasita T. 1994c. Structural effects on CO$_2$ reduction at Pt single-crystal electrodes. 3. Pt (100) and related surfaces. *J. Electroanal. Chem.* 377(1–2):215–225.

Russell P.G., Kovac N., Srinivasan S., and Steinberg M. 1977. The electrochemical reduction of carbon dioxide, formic acid, and formaldehyde. *J. Electrochem. Soc.* 124:1329.

Ryu J., Andersen T.N., and Eyring H. 1972. Electrode reduction kinetics of carbon dioxide in aqueous solution. *J. Phys. Chem.* 76(22):3278–3286.

Saeki T., Hashimoto K., Fujishima A., Kimura N., and Omata K. 1995. Electrochemical reduction of CO$_2$ with high current density in a CO$_2$-methanol medium. *J. Phys. Chem.* 99(20):8440–8446.

Saeki T., Hashimoto K., Kimura N., Omata K., and Fujishima A. 1996. Electrochemical reduction of CO$_2$ with high current density in a CO$_2$ methanol medium at various metal electrodes. *J. Electroanal. Chem.* 404(2):299–302.

Saeki T., Hashimoto K., Noguchi Y., Omata K., and Fujishima A. 1994. Electrochemical reduction of liquid CO$_2$. *J. Electrochem. Soc.* 141:L130.

Saha M.S., Furuta T., and Nishiki Y. 2004. Conversion of carbon dioxide to peroxycarbonate at boron-doped diamond electrode. *Electrochem. Commun.* 6(2):201–204.

Sánchez-Sánchez C.M., Montiel V., Tryk D.A., Aldaz A., and Fujishima A. 2001. Electrochemical approaches to alleviation of the problem of carbon dioxide accumulation. *Pure Appl. Chem.* 73(12):1917–1928.

Savinova E.R., Yashnik S.A., Savinov E.N., and Parmon V.N. 1992. Gas-phase electrocatalytic reduction of CO$_2$ to CO on carbon gas-diffusion electrode promoted by cobalt phthalocyanine. *React. Kinet. Catal. Lett.* 46(2):249–254.

Schrebler R., Cury P., Herrera F., Gomez H., and Cordova R. 2001. Study of the electrochemical reduction of CO$_2$ on electrodeposited rhenium electrodes in methanol media. *J. Electroanal. Chem.* 516(1–2):23–30.

Schrebler R., Cury P., Suarez C., Munoz E., Gomez H., and Cordova R. 2002. Study of the electrochemical reduction of CO$_2$ on a polypyrrole electrode modified by rhenium and copper–rhenium microalloy in methanol media. *J. Electroanal. Chem.* 533(1–2):167–175.

Schwartz M., Vercauteren M.E., and Sammells A.F. 1994. Fischer–Tropsch electrochemical CO reduction to fuels and chemicals. *J. Electrochem. Soc.* 141:3119.

Scibioh M.A. and Vijayaraghavan V.R. 1998. Electrocatalytic reduction of carbon dioxide: Its relevance and importance. *J. Sci. Ind. Res.* 57(3):111–123.

Scibioh M.A. and Viswanathan B. 2004. Electrochemical reduction of carbon dioxide: A status report. *Proc. Indian Natl. Sci. Acad.* 70:407.

Shibata M. and Furuya N. 2003. Simultaneous reduction of carbon dioxide and nitrate ions at gas-diffusion electrodes with various metallophthalocyanine catalysts. *Electrochim. Acta* 48(25–26):3953–3958.

Shibata H., Moulijn J.A., and Mul G. 2008. Enabling electrocatalytic Fischer–Tropsch synthesis from carbon dioxide over copper-based electrodes. *Catal. Lett.* 123(3):186–192.

Shiratsuchi R., Aikoh Y., and Nogami G. 1993. Pulsed electroreduction of CO$_2$ on copper electrodes. *J. Electrochem. Soc.* 140:3479.

Shiratsuchi R., Hongo K., Nogami G., and Ishimaru S. 1992. Reduction of CO_2 on Fluorine-doped SnO_2 thin-film electrodes. *J. Electrochem. Soc.* 139:2544.

Shiratsuchi R. and Nogami G. 1996. Pulsed electroreduction of CO_2 on silver electrodes. *J. Electrochem. Soc.* 143:582.

Smith B.D., Irish D.E., Kedzierzawski P., and Augustynski J. 1997. A surface enhanced Roman scattering study of the intermediate and poisoning species formed during the electrochemical reduction of CO on copper. *J. Electrochem. Soc.* 144:4288.

Sobkowski J. and Czerwinski A. 1974. Kinetics of carbon dioxide adsorption on a platinum electrode. *J. Electroanal. Chem.* 55(3):391–397.

Sobkowski J. and Czerwinski A. 1975. The comparative study of $CO_2 + H_{ads}$ reaction on platinum electrode in H_2O and D_2O. *J. Electroanal. Chem.* 65(1):327–333.

Sonoyama N., Kirii M., and Sakata T. 1999. Electrochemical reduction of CO_2 at metal-porphyrin supported gas diffusion electrodes under high pressure CO_2. *Electrochem. Commun.* 1(6):213–216.

Spataru N., Tokuhiro K., Terashima C., Rao T.N., and Fujishima A. 2003. Electrochemical reduction of carbon dioxide at ruthenium dioxide deposited on boron-doped diamond. *J. Appl. Electrochem.* 33(12):1205–1210.

Stevens G.B., Reda T., and Raguse B. 2002. Energy storage by the electrochemical reduction of CO_2 to CO at a porous Au film. *J. Electroanal. Chem.* 526(1–2):125–133.

Subramanian K., Asokan K., Jeevarathinam D., and Chandrasekaran M. 2007. Electrochemical membrane reactor for the reduction of carbon dioxide to formate. *J. Appl. Electrochem.* 37(2):255–260.

Summers D.P. and Frese Jr. K.W. 1988. Electrochemical reduction of carbon dioxide. Characterization of the formation of methane at ruthenium electrodes in carbon dioxide saturated aqueous solution. *Langmuir* 4(1):51–57.

Taguchi S. and Aramata A. 1994. Surface-structure sensitive reduced CO_2 formation on Pt single crystal electrodes in sulfuric acid solution. *Electrochim. Acta* 39(17):2533–2538.

Taguchi S., Aramata A., and Enyo M. 1994a. Reduced CO_2 on polycrystalline Pd and Pt electrodes in neutral solution: Electrochemical and in situ Fourier transform IR studies. *J. Electroanal. Chem.* 372(1):161–170.

Taguchi S., Ohmori T., Aramata A., and Enyo M. 1994b. Adsorption of CO and CO_2 on polycrystalline Pt at a potential in the hydrogen adsorption region in phosphate buffer solution: An in situ Fourier transform IR study. *J. Electroanal. Chem.* 369(1–2):199–205.

Takahashi I., Koga O., Hoshi N., and Hori Y. 2002. Electrochemical reduction of CO_2 at copper single crystal Cu (S)-[n (111) × (111)] and Cu (S)-[n (110) × (100)] electrodes. *J. Electroanal. Chem.* 533(1–2):135–143.

Tanaka H. and Aramata A. 1997. Aminopyridyl cation radical method for bridging between metal complex and glassy carbon: Cobalt (II) tetraphenylporphyrin bonded on glassy carbon for enhancement of CO_2 electroreduction. *J. Electroanal. Chem.* 437(1–2):29–35.

Terunuma Y., Saitoh A., and Momose Y. 1997. Relationship between hydrocarbon production in the electrochemical reduction of CO_2 and the characteristics of the Cu electrode. *J. Electroanal. Chem.* 434(1–2):69–75.

Todoroki M., Hara K., Kudo A., and Sakata T. 1995. Electrochemical reduction of high pressure CO_2 at Pb, Hg and in electrodes in an aqueous $KHCO_3$ solution. *J. Electroanal. Chem.* 394(1–2):199–203.

Tomita Y., Teruya S., Koga O., and Hori Y. 2000. Electrochemical synthesis and engineering-electrochemical reduction of carbon dioxide at a platinum electrode in acetonitrile-water mixtures. *J. Electrochem. Soc.* 147(11):4164–4167.

Tryk D.A., Yamamoto T., Kokubun M., Hirota K., Hashimoto K., Okawa M., and Fujishima A. 2001. Recent developments in electrochemical and photoelectrochemical CO_2 reduction: Involvement of the $(CO)_2$-dimer radical anion. *Appl. Organomet. Chem.* 15(2):113.

Tsubone K., Tanaka F., Komatsu M., Adachi Y., and Aihara M. 2005. Carbon dioxide recycling by pulsed electroreduction-influence of the surface state of Cu electrode to Faradaic efficiency. *Bull. Fac. Human Environ. Sci.* 36:13–21.

Vassiliev Y.B., Bagotzky V.S., Khazova O.A., and Mayorova N.A. 1985a. Electroreduction of carbon dioxide: Part II. The mechanism of reduction in aprotic solvents. *J. Electroanal. Chem.* 189(2):295–309.

Vassiliev Y.B., Bagotzky V.S., Osetrova N.V., Khazova O.A., and Mayorova N.A. 1985c. Electroreduction of carbon dioxide. I: The mechanism and kinetics of electroreduction of CO_2 in aqueous solutions on metals with high and moderate hydrogen overvoltages. *J. Electroanal. Chem.* 189(2):271–294.

Vassiliev Y.B., Bagotzky V.S., Osetrova N.V., and Mikhailova A.A. 1985b. Electroreduction of carbon dioxide: Part III. Adsorption and reduction of CO_2 on platinum. *J. Electroanal. Chem.* 189(2):311–324.

Vasudevan P., Phougat N., and Shukla A.K. 1996. Metal phthalocyanines as electrocatalysts for redox reactions. *Appl. Organomet. Chem.* 10(8):591–604.

Wasmus S., Cattaneo E., and Vielstich W. 1990. Reduction of carbon dioxide to methane and ethylene—An on-line MS study with rotating electrodes. *Electrochim. Acta* 35(4):771–775.

Watanabe M., Shibata M., Kato A., Azuma M., and Sakata T. 1991a. Design of alloy electrocatalysts for CO_2 reduction. *J. Electrochem. Soc.* 138:3382.

Watanabe M., Shibata M., Katoh A., Sakata T., and Azuma M. 1991b. Design of alloy electrocatalysts for CO_2 reduction: Improved energy efficiency, selectivity, and reaction rate for the CO_2 electroreduction on Cu alloyelectrodes. *J. Electroanal. Chem.* 305(2):319–328.

Weimer T., Schaber K., Specht M., and Bandi A. 1996. Methanol from atmospheric carbon dioxide: A liquid zero emission fuel for the future. *Energy Convers. Manag.* 37(6–8):1351–1356.

Whipple D.T., Finke E.C., and Kenis P.J.A. 2010. Microfluidic reactor for the electrochemical reduction of carbon dioxide: The effect of pH. *Electrochem. Solid State Lett.* 13(9):109–111.

Williams R., Crandall R.S., and Bloom A. 1978. Use of carbon dioxide in energy storage. *Appl. Phys. Lett.* 33:381.

Yamamoto T., Tryk D.A., Fujishima A., and Ohata H. 2002. Production of syngas plus oxygen from CO_2 in a gas-diffusion electrode-based electrolytic cell. *Electrochim. Acta* 47(20):3327–3334.

Yamamoto T., Tryk D.A., Hashimoto K., Fujishima A., and Okawa M. 2000. Electrochemical reduction of CO_2 in the micropores of activated carbon fibers. *J. Electrochem. Soc.* 147:3393.

Yano J., Morita T., Shimano K., Nagami Y., and Yamasaki S. 2007. Selective ethylene formation by pulse-mode electrochemical reduction of carbon dioxide using copper and copper-oxide electrodes. *J. Solid State Electrochem.* 11(4):554–557.

Yano H., Shirai F., Nakayama M., and Ogura K. 2002a. Efficient electrochemical conversion of CO_2 to CO, C_2H_4 and CH_4 at a three-phase interface on a cu net electrode in acidic solution. *J. Electroanal. Chem.* 519(1–2):93–100.

Yano H., Shirai F., Nakayama M., and Ogura K. 2002b. Electrochemical reduction of CO_2 at three-phase (gas | liquid | solid) and two-phase (liquid | solid) interfaces on Ag electrodes. *J. Electroanal. Chem.* 533(1–2):113–118.

Yano H., Tanaka T., Nakayama M., and Ogura K. 2004. Selective electrochemical reduction of CO_2 to ethylene at a three-phase interface on copper (I) halide-confined Cu-mesh electrodes in acidic solutions of potassium halides. *J. Electroanal. Chem.* 565(2):287–293.

Yano J. and Yamasaki S. 2008. Pulse-mode electrochemical reduction of carbon dioxide using copper and copper oxide electrodes for selective ethylene formation. *J. Appl. Electrochem.* 38(12):1721–1726.

Yoshitake H., Kikkawa T., Muto G., and Ota K. 1995a. Poisoning of surface hydrogen processes on a Pd electrode during electrochemical reduction of carbon dioxide. *J. Electroanal. Chem.* 396(1–2):491–498.

Yoshitake H., Kikkawa T., and Ota K. 1995b. Isotopic product distributions of CO_2 electrochemical reduction on a D flowing-out Pd surface in protonic solution and reactivities of "subsurface" hydrogen. *J. Electroanal. Chem.* 390(1–2):91–97.

Yoshida T., Tsutsumida K., Teratani S., Yasufuku K., and Kaneko M. 1993. Electrocatalytic reduction of CO_2 in water by [Re (bpy)(CO)₃Br] and [Re (terpy)(CO)₃Br] complexes incorporated into coated Nafion membrane(bpy = 2, 2′-bipyridine; terpy = 2, 2′: 6′, 2′-terpyridine). *J. Chem. Soc., Chem. Commun.* (7):631–633.

Zhalko-Titarenko O.V., Lazurskii O.A., and Pokhodenko V.D. 1990. Electrocatalytic reduction of carbon dioxide on electrodes modified by polypyrrole with an immobilized complex of nickel with 1, 4, 8, 11-tetraazacyclotetradecane. *Theor. Exp. Chem.* 26(1):40–44.

Zhang J., Pietro W.J., and Lever A.B.P. 1996. Rotating ring-disk electrode analysis of CO_2 reduction electrocatalyzed by a cobalt tetramethylpyridoporphyrazine on the disk and detected as CO on a platinum ring. *J. Electroanal. Chem.* 403(1–2):93–100.

Zhang A., Zhang W., Lu J., Wallace G.G., and Chen J. 2009. Electrocatalytic reduction of carbon dioxide by cobalt-phthalocyanine-incorporated polypyrrole. *Electrochem. Solid State Lett.* 12:E17.

Zhao G., Jiang T., Han B., Li Z., Zhang J., Liu Z., He J., and Wu W. 2004. Electrochemical reduction of supercritical carbon dioxide in ionic liquid 1-n-butyl-3-methylimidazolium hexafluorophosphate. *J. Supercritical Fluids* 32(1–3):287–291.

7

Energy Storage and Other Ways of Handling Intermittent Energy Production from Renewable Sources

Bent Sørensen

CONTENTS

7.1 Introduction

A convenient reference case for discussing energy storage options is oil, a substance formed by millions of years of geological processing of biomass (Sørensen, 1984). Currently, world economy is based on the transient usage of this unique and finite resource, available for a century or two at nearly zero extraction cost and offering very high energy storage density and convenient conversions to and from the store that may be a simple metal or composite container. An example of filling this type of store is given in Figure 7.1. The crude oil has been modestly refined to diesel oil, capable of furnishing a fuel-to-wheel efficiency of 27% when used in a common-rail internal combustion engine based on the diesel cycle to accomplish the chemical-to-mechanical work conversion step when extracting fuel from the store (Sørensen, 2005). Few alternative energy storage systems are able to compete with the filled oil tank in terms of performance. Below I give an overview of the current status of different storage devices followed by a more detailed discussion of some key options, including not only those involving storage facilities but also other ways of handling intermittent energy production, for example, by making use of energy exchange and trade arrangements.

7.2 Overview of Energy Storage Options

An energy storage system may consist of all or some of the following components (see Figure 7.2): a device for accepting input energy and converting it to the form required by

FIGURE 7.1
Reference storage system: The filled oil container: the rate of energy transfer at a filling station is as high as 30–50 MW. (Photo courtesy of B. Sørensen.)

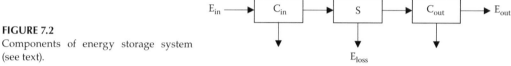

FIGURE 7.2
Components of energy storage system (see text).

the store, the store itself, possibly associated with a loss of energy with time, and finally an output device, converting the stored energy to the form required for subsequent use (cf. Jensen and Sørensen, 1984). All steps in this chain may have less than 100% efficiency, and a goal for constructing energy storage devices is to keep the losses to a minimum or to avoid them, by combining components, by keeping the energy form unchanged, and by storing the energy using a time-independent (loss-free) facility.

Table 7.1 shows some of the storage methods resembling the storage in the automobile fuel tank of Figure 7.1. The high density of stored energy in solid or liquid fuels is exhibited. For gaseous fuels such as hydrogen the density is high only by mass, not by volume, which points to one of the difficulties in using hydrogen as a storage option, particularly in vehicles. The energy storage devices listed in Table 7.2 are particularly suited for storage of heat at various temperatures. For the higher temperature levels, conversion into mechanical or electric energy may be feasible, but for the lower levels, the second law of thermodynamics makes the losses in such attempted applications too large for acceptance. Cycle efficiencies are not given, because the losses from the stores are heat losses that depend on insulation materials used, on shape of the store, and on ambient temperatures as function of time, factors that have to be considered for each proposed store by taking into account the peculiar data pertaining to that store at its anticipated location. The heat stores listed in Table 7.2 may be interesting for solar thermal systems such as building space heating or hot-water supply systems, where the need for seasonal storage in most non-equatorial climates poses a difficult problem.

In Table 7.3, the storage of energy originally in a high-quality form, with the aim to recover it in the same or another high-quality form, is considered. An additional option for high-quality energy storage is the storage in the form of hydrogen mentioned in Table 7.1. All these options, possibly except that of elevated water, are currently considered too expensive for large-scale energy storage, but as will be mentioned below, some of them are approaching viability for medium size storage, for example, in road

TABLE 7.1

Fuel-Type Energy Storage

Storage Form	Energy Density		Cycle Efficiency
	kJ kg⁻¹	MJ m⁻³	

Storage Form	kJ kg⁻¹	MJ m⁻³	Cycle Efficiency
Conventional fuels			
Crude oil	42,000	37,000	
Coal	32,000	42,000	
Dry wood	12,500[a]	10,000	
Synthetic fuels			
Hydrogen, gas	120,000	10	0.4–0.6
Hydrogen, liquid	120,000	8,700	
Hydrogen, metal hydride	2,000–9,000	5,000–15,000	
Methanol	21,000	17,000	
Ethanol	28,000	22,000	

Sources: Adapted from Jensen, J., *Energy Storage*, Butterworth Publications, London, U.K., 1980; Jensen, J. and Sørensen, B.: *Fundamentals of Energy Storage.* 1984. Copyright Wiley-VCH Verlag GmbH & Co. KGaA; *Renewable Energy*, 4th edn. Sørensen, B., Copyright (2010) from Elsevier.

For hydrogen gas, the density is quoted at ambient pressure and temperature.

[a] Oven-dry wood may reach values up to 20,000 kJ kg⁻¹.

TABLE 7.2

Thermal Energy Storage

Storage Form	Energy Density	
	kJ kg⁻¹	MJ m⁻³

Storage Form	kJ kg⁻¹	MJ m⁻³
Thermal—low quality		
Water, 100°C → 40°C	250	250
Rocks, 100°C → 40°C	40–50	100–140
Iron, 100°C → 40°C	~30	~230
Thermal—high quality		
Rocks, e.g., 400°C → 200°C	~160	~430
Iron, e.g., 400°C → 200°C	~100	~800
Inorganic salts, heat of fusion, over 300°C	>300	>300

Sources: Adapted from Jensen, J., *Energy Storage*, Butterworth Publications, London, U.K., 1980; Jensen, J. and Sørensen, B.: *Fundamentals of Energy Storage.* 1984. Copyright Wiley-VCH Verlag GmbH & Co. KGaA; *Renewable Energy*, 4th edn. Sørensen, B., Copyright (2010) from Elsevier.

The temperature range in which the store is designed to work is indicated.

vehicles, and may be heading for viability for large power plant usage. The pumped hydro option is already viable in a number of places, but requires appropriate types of geological formations not only suited for holding water, but also environmentally acceptable for such use. The same is true for natural elevated water reservoirs, currently used in hydropower schemes. If environmentally acceptable, these can provide the equivalent

TABLE 7.3

High-Quality Energy Storage, Divided into Methods Based on Mechanical and Electrochemical Storage Processes

Storage Form	Energy Density		Cycle Efficiency
	kJ kg^{-1}	MJ m^{-3}	
Mechanical			
Pumped hydro, 100 m head	1	1	0.65–0.8
Compressed air		~15	0.4–0.5
Flywheels, steel	30–120	240–950	
Flywheels, advanced	>200	>100	~0.95
Electrochemical			
Lead–acid	40–140	100–900	0.7–0.8
Nickel–cadmium	~350	~350	<0.7
Lithium ion and other advanced batteries	400–700	300–1400	0.7–0.8
Superconducting		~100	~0.85

Sources: Adapted from Jensen, J., *Energy Storage*, Butterworth Publications, London, U.K., 1980; Jensen, J. and Sørensen, B.: *Fundamentals of Energy Storage*. 1984. Copyright Wiley-VCH Verlag GmbH & Co. KGaA; *Renewable Energy*, 4th edn. Sørensen, B., Copyright (2010) from Elsevier.

The cycle efficiencies given are typical of currently available systems. For compressed air energy storage, both electricity and heat inputs are included on equal terms in estimating the cycle efficiency.

of energy storage by suitable management of the water flows (holding back flow when intermittent power production is large and accelerating it when there is insufficient power production).

7.3 Dealing with Energy Intermittency

Several renewable energy sources flow intermittently or at varying rates. Solar radiation is zero at night and during day it depends on incident angle, cloud cover, and air turbidity. Furthermore, its seasonal distribution varies just the wrong way, if compared to space heating demand. Wind energy varies according to weather systems passing over a given location, with recurrence times typically between 1 and 3 weeks. On the other hand, its seasonal variations are fairly modest and generally follow those of power demand. Hydro is generally stable, but varies seasonally as dictated by the components of water balance. For run-of-the-river hydro, the variations may change over a couple of weeks, while for reservoir-based hydro, there are no short-term problems, but only the general question of having sufficient precipitation (and snow-melt where relevant) coming into the reservoirs, normally relative to some average-year situation for which the system is designed. Wave energy, being essentially accumulated wind energy, shows even larger deviations from its average level. Only biofuels have behavior like fossil fuels that can be stored in containers and used when called for.

Biomass and hydro storage are already examples of energy stores directly associated with renewable energy. One may add secondary devices, such as those listed in Tables 7.1 through 7.3. The cost of using each of these storage forms vary widely, and several of the options are only feasible in particular consumer segments (e.g., the lithium-ion batteries having first been used only for consumer products with a very small energy demand, such as watches, but are currently climbing up the energy ladder to products like video cameras, laptops, and lawn mowers), aiming at the automobile market.

The disadvantage of energy stores is that they often have to store energy in a form different from both the initial one and from the energy required by the final user (cf. Figure 7.2). For example, storing electricity in underground hydrogen caverns requires three stages, each involving equipment (at a cost) and each having an efficiency less than 100%: first, electricity is turned into hydrogen by alkaline electrolysis or some other reverse fuel cell principle, then it is put into the cavern (excavated at a cost) using pumps and perhaps cooling devices, and, finally, electricity is regenerated from hydrogen using either some Carnot engine (combustion followed by driving say a steam turbine and generator) or a fuel cell. Losses through such chains of conversion are the reason that efforts are put into the alternative possibility of altering the way energy is used, for example, using the hydrogen stored on the basis of excess wind power directly for vehicles in the transportation sector, thus avoiding having to go back to electricity and in this way avoiding a loss that today is 50% or more, no matter how it is done (advanced steam engines or fuel cells).

Several of the technologies mentioned above are presently in a stage of early development and far from ready to enter the market, neither by the currently used costing methods nor by an idealized life-cycle costing where indirect costs and costs of impacts elsewhere in our society such as environmental or climate costs are all included (usually involving some highly uncertain numerical value). For example, fuel cells promise higher efficiency and the ability to operate on fuels created from renewable sources without pollution. Demonstration fleets of vehicles have been equipped with current state-of-the-art fuel cells, but they cost more than all the rest of the vehicle. Evidently, the costs must go down when it comes to mass production. The efficiency will eventually grow from the current one not much above 30% to somewhere closer to the theoretical maximum of about 65%. But what about operational life? Current cells have significant degradation after 1–2 years and the "research goal" of the auto manufacturers (and that was stated before the current financial problems) is only 5 years. Consumer cars today have an average lifetime exceeding 15 years, so two or three changes of the fuel cell unit in the car is required, this being for the key component for which currently the cost of the preinstalled fuel cell is already prohibitively high. To the subsequent fuel cell replacements should be added the cost of installing the replacements. Some 10 years ago, there were automobile manufacturers who told us that there would be a major penetration of fuel cell vehicles by 2004. That did not happen, and today you hear instead that electric vehicles probably are better alternatives or hybrids if you need a car with mileage range of more than 100–150 km.

The cause for such new belief in battery-operated vehicles is that advanced lithium-technology batteries can now finally be built at sizes relevant for consumer vehicles. Like fuel cells, the cost of such batteries is much higher than that of the rest of the car, but with a potential for coming down. I would comment on this ongoing war between fuel cell and battery vehicles as follows: Batteries and fuel cells are two very similar technologies, which differ only in that the chemicals capable of producing energy when allowed to interact are stored within the cell in a battery, but outside it in a fuel cell. In both cases, modern devices have a polymer membrane or electrolyte (with a finite lifetime) and two electrodes prone to degradation and shrouding by debris deposited on their surfaces. During the 1970s and 1980s, advanced batteries were the center of attention, but it was never possible to reach the around 5 year lifetime of the old lead-acid technology. Therefore, batteries were abandoned for automotive applications during the 1990s, and research instead centered on the fuel cells and notably the low-temperature proton-exchange membrane cells. Fuel cell research, at first enthusiastic, slowed down during the first decade of the third millennium, for exactly the same reasons as those halting the battery development 25 years earlier. However, in the meantime, small batteries have improved their lifetime to about the 5 year goal and are beginning to expand in the direction of a larger capacity.

The interesting thing about the two technologies is that they have some complementary properties that make hybrid technologies very promising. Even with the low weight of lithium-ion batteries relative to metal-hydride and lead-acid batteries, the increased energy demand for the additional mass of the batteries begins to be a limiting factor as ranges above 100 km are contemplated. For fuel cells, the cost increases with nameplate rating, and to match a fossil car in accelerating uphill, even a small car will need some 85 kW of fuel cells, although the average power level during all driving is only 5 kW. This points to the battery-fuel cell hybrid as the ideal solution: The fuel cell could be rated slightly above the average power level and the electric engine could operate on batteries with an electric motor to provide the maximum power output when required. A further advantage is that this hybrid car would be a natural continuation of the diesel-battery hybrids likely to gain acceptance in the near-future oil crunch era (diesel, not gasoline, because of the recent breakthrough in diesel engines making them substantially more efficient than any gasoline-driven Otto engine).

Energy storage solutions vary widely in range of capacity and in the amount of energy that can be stored. The characteristic storage requirements of different renewable energy systems, used to cover some specific types of load, will often point to the type of storage that might be suitable. The vehicle example used above is looking to batteries and compressed hydrogen containers as a replacement for present-day fossil-fuel containers. Methanol as a means of getting closer to the current technology was tried in the 1990s, but onboard reformation of methanol to hydrogen turned out more complex than anticipated. If the fuel cell development continues to slow down, the advanced biofuels (so-called second-generation technologies, based only on biomass residues, with no use of biomass parts useful for food production) may well take over.

Competition to secondary energy storage may come not only from storable biofuels but also from new ways of arranging the entire energy system. Obviously, if demand management could change demand profiles to approach the variations in renewable energy availability, it would solve some of the problems: You put your clothes into the washing machine but it only starts when receiving a signal through the power line that there is available excess of wind power. This is fine if the lull is not extending for 2 weeks. Another target would be the many rechargeable batteries used in portable equipments. With more standardization, one could recharge enough batteries during high-wind periods to last for a 2 week long lull. However, with the present distribution of total demand, the load management is unlikely to contribute more than about 10% to the solution of the intermittency problem that would arise if renewable energy were to take a dominant share in energy supply. Also biofuels would be covering less than 100% of the energy needs in the transportation sector, for resource availability reasons, even with a substantial efficiency improvement. So what are the other avenues available?

The variations in solar and wind energy are not the same at different geographic locations. Solar energy varies with longitude and some leveling may be achieved with transmission between time zones. For wind energy, dispersed location within a region takes care of short-term variations in the production of individual turbines, but long-distance transmission can take care of at least part of the main wind production variations, as sites separated by more than 500 km usually experience different weather systems. Interregional power transmission is already in place in many countries, but they will probably have to be reinforced in capacity (at a known cost) in order to serve as a mitigating option for renewable energy intermittency, in addition to the present uses of the lines (providing peak and emergency power). Any remaining mismatch between supply and demand must be covered by active energy storage (Sørensen, 2009).

7.4 Short-Term Energy Storage

Energy storage options may be divided in terms of the length of storage that they offer. Long-term stores such as hydro reservoirs may store energy for many months and have an important impact on the overall energy budget, while small-time storage, for example, capacitors, can help avoiding flickering of the light when power-network companies switch between different suppliers of electricity. Of course, there is a smooth transition between short- and long-term storage devices. For instance, batteries in lawn mowers or electric vehicles have to be recharged every day, while batteries using the same technology in watches and calculators can last for years before having to be recharged or replaced.

The potter's wheel (Sørensen, 1984) is an example of a classical application of short-term storage of the kinetic energy of a rotating plate. It stabilizes the rotational speed and thus helps perfect the roundness of the pots produced. Similar uses of inertia are made in flywheels attached to irregularly rotating shafts, such as those driven by explosion-type engines, with the purpose of smoothing the shaft-power output from such devices. The storage time associated with such use of flywheels is a fraction of a second. Use of flywheels has also been suggested for more substantial energy storage, for example, by placing large flywheels horizontally underground (to reduce accident risks), for instance to smooth out variations in the renewable energy input from power sources such as wind.

What flywheels may do for mechanical energy can be done for electric energy by capacitors, and probably at a lower cost. The electric current is used to build up assemblies of charge, which can either be used to create delays in a circuit, or by switching the current on and off to actually store energy that can later be regained directly in the form of electric energy.

The advantage of use of superconducting materials is that Ohmic losses could be avoided in storing electromagnetic energy (Table 7.3). For example, a magnet could be charged by a superconducting coil with near-zero heat losses. Because known superconductors are superconducting only at very low temperatures, the system has to be cooled, and likely at a substantial energy cost. Demonstration facilities already built, for example, by the U.S. Department of Defense and by the European Organization for Nuclear Research (CERN) scientific center in Switzerland, allow slow charging but extremely fast release of the energy stored, which would be essential in some applications. The installations are costly and there is a substantial economy of scale, which makes superconducting storage rather a large- and medium-term option. This option is envisaged as possibly playing a role in connection with intercontinental power exchange through superconducting cables, reasoning that if the cryogenic technologies are already used in transmission, one could add benefits by also using cryogenics for superconducting energy storage aimed at load management or handling of intermittent renewable resources such as solar electricity from desert locations (e.g., those of North Africa, as seen from a European perspective, see Sørensen, 2010).

For batteries, the intermediate storage form is chemical, not electrical, which causes both losses and problems of electrode and electrolyte degradation. There are two broad classes of batteries used or considered: the conventional batteries with the chemicals stored inside the battery, and the flow batteries, also called reversible fuel cells, where the chemical substances are stored outside the unit. At present, fuel cells of the alkaline type is dominating the electrolysis market, while the reverse process, hydrogen to electricity, is only used in vehicles demonstrating a possible future use of another type of fuel cells (polymer electrolyte membrane [PEM]) for road vehicles. Any kind of fuel cell or battery is characterized by the choice of electrode material and the choice of electrolyte. The development has been in the direction away from liquid and toward solid electrolytes, and although the lifetime problems are similar for battery and fuel cell technologies, it seems that batteries are doing better than fuel cells at the moment: While neither PEM nor high-temperature fuel cells have yet reached the promised efficiency and durability, the advanced battery types, such as lithium-ion batteries, first out-competed lead-acid and nickel-metal technologies in the market for small consumer appliances (watches, mobile phones, etc.), then moved a bit up to laptops, then a bit more up to lawn mowers and hedge saws, and finally seems to begin to reach the price and performance level required for hybrid vehicles. This means that electrochemical storage technologies, which used to be considered small scale and short term (such as bicycle lights or flashlights), except possibly automobile starter batteries, which could reach a useful life of some 5 years, are now readying themselves for breakthrough as a solution to storage requirements on nearly any scale, including power utility applications, say in case of intermittent wind inputs to the utility system. Any problems? Well, the price is not quite there yet, and after all, the round-trip efficiency still lurks around 75%. The difficulty is that losses are determined by electrochemical potential jumps across the storage device that cannot always be optimized in both directions, for storing and for retrieving power.

The examples so far have been for storing and retrieving high-quality energy, and it should perhaps be mentioned in passing, that for nonrenewable energy sources such as fuels, those in solid or fluid form can be stored directly, whereas the gaseous ones (natural gas and in the future perhaps hydrogen produced from renewable sources) need to be compressed, forcing acceptance of the associated energy expenditure and the subsequent possibility of losses by leaking. The systems requirements make this more interesting for large-scale stores.

This brief survey will now give some examples of heat storage. Low-quality storage, particularly if we are talking about low-temperature heat used in households and commerce, is typically under 100°C and sometimes substantially below this level. A heat store could be a hot-water tank of a few cubic meters, serving a household for a winter day or 2–3 summer days before needing a reheat, or it could be a communal pebble bed store assisting a building-integrated solar heating system. Moving from heat capacity stores to latent heat stores, an advanced system might use the energy absorption and release associated with phase change in a suitable material such as Glauber's salt (Na_2SO_4 with 10 crystal water molecules per salt molecule). Finally, one can move from physical change to chemical reactions, selecting some that involve large amounts of energy added or released, depending on the direction of the reaction. A number of such systems have been tested in various configurations (from integration into the heating system of a detached residence to that of an apartment building or cluster of houses). Because of the cost of solar collectors, the present pricing in areas with a space heating demand favors rather small systems aimed only at supplementary energy during a (possibly extended) summer period. This case, as well as that of pure hot-water supply in areas nearer to the equator, has made storage in water tanks the preferred technology. The problem with solar heating is that solar radiation has a seasonal variation opposite to that of space-heating demand, and, furthermore, the efficiency of solar thermal collectors depends on the inlet temperature of the water circuit through the collector. This makes it difficult to select an optimal size for the heat store, because a small system has low efficiency (high inlet temperature) during summer where the solar resource is largest, and a large store has too low an outlet temperature during the winter to be able to provide useful heat in winter. One might add a heat pump to remedy this, but that is another capital-intensive component in a system already quite costly.

7.5 Long-Term Energy Storage

During a transition period, the simplest way to deal with intermittent energy sources such as wind and solar radiation, when they reach a level of penetration where they sometimes produce more energy than the total demanded, and at other times less than required, may be to use the existing fossil fuel system for as long as it can be operated. This means keeping the old fossil-fired power stations on standby for use in case of insufficient renewable energy production, and to accept that surpluses are sometimes lost, if they cannot be exported and used by neighboring regions. This procedure is not acceptable in the long run because of its negative effect on the cost of renewable energy, and because the deficit periods can be quite extended (e.g., for solar energy in winter), as a result, the use of fossil energy may remain too high for meeting greenhouse gas reduction targets.

One solution is to replace the backup use of fossil fuels by renewable energy fuels that can be stored, such as biomass, synthetic hydrocarbon fuels, or hydrogen. While use of hydrogen in fuel cells or for conventional combustion does not emit carbon dioxide, it is not the case for biomass combustion or use of synthetic hydrocarbons, unless carbon-capturing devices are part of the system. It is conventional to consider biomass as carbon neutral, but this is an assertion that has to be considered in more detail: A growing plant assimilates CO_2 from the atmosphere and stores the carbon in its fibers. When the leaves of the plant wither or the plant dies, decomposers transform the residues in a way that releases the CO_2 back into the atmosphere. If instead, humans take the residues (straw, wood, etc.) and burn them, the CO_2 is also released, so there is no difference. The reason for calling the cycle "carbon neutral" is that the net release of CO_2 is zero, but the impacts on climate further depends on the delay between assimilation and reemission. Plants that grew millions of year ago to subsequently become transformed into fossil fuels cannot be considered carbon neutral, because the carbon assimilated millions of years ago is immaterial for current climate problems, in contrast to the CO_2 from burning the fuels. The same is true for cutting down and burning a forest that had its initial growth period 50–200 years ago. The surplus of CO_2 assimilation stops when the forest has come into a state of equilibrium, where addition of biomass equals losses by whatever mechanism is making the leaves or detached parts decay. This assimilation surplus period is 50–100 years depending on the type of forest, and, after that, burning the standing crop wood (as distinct from the residues) is definitely climate negative. On the other hand, if the wood is used for furniture or as building material with a lifetime of 50–100 years, then there is no immediate change in carbon accounting when going from a forest in equilibrium to some nonenergy use of the wood.

For short-rotation crops, the time duration between carbon-assimilating growth and carbon-releasing decomposition or combustion is so short that such crops may be considered carbon neutral. However, these crops are often food crops, so again one should use only the residues for energy extraction (combustion associated with air pollution or second-generation biofuels, with different and possibly less pollutant emissions). Converting biomass into biofuels has the additional advantage that most nutrients are retained in the "residues from residues," that is, the char from gasification or the slurry from biogas production, and thus may be returned to the fields where they can replace chemical fertilizers.

Is a 50–100 year delay of CO_2 emissions sufficient? Will climate-changing human activities really have stopped within the next 40–50 years? Climate-negotiating politicians try to avoid touching the fossil fuel industry here and now, and public acceptance that we have to do something has a tendency to diminish, when citizens realize that it does not work without lifestyle changes away from ever-increasing amounts of material goods. Renewable energy sources do not allow infinite material growth, which is the paradigm behind the economic organization of most current societies. Nonmaterial growth is fine, more art and culture, more bicycle trips in the national parks, more social engagement, and the like, but if this is not to be a job threat, it requires solidarity in distribution policy, a feature not contained in the prevailing eighteenth-century liberalistic policy preferences. Changing the economic paradigm is a tall order. It is certainly easier to accept the climate changes and let those who have the money move away from floods and desertification fires and tropical diseases. It is just a pity that there are no viable alternatives to renewable energy in the long run (in other places, I have explained why I do not believe nuclear energy is an alternative, see e.g., Sørensen, 2005).

More intricate systems can be used to convert excess solar or wind energy into fuels. Hydrogen (produced by electrolysis) is an obvious choice, if there is a demand for hydrogen, say in the transportation sector, but also if electricity has to be regenerated, hydrogen stored in underground reservoirs may be a viable option. Alternative fuels (or rather fuels closer to those we have become accustomed to during the last 100 years) containing carbon can only be produced from solar or wind if there is a source of carbon. This could be CO_2 capture from current fossil fuel operations, or direct solar thermal or electrochemical conversion of atmospheric CO_2.

A lot of technical development is required for these solutions to become viable. In the medium term, there may be more modest energy storage challenges that can be solved by already existing technology. One example of such a storage technology is compressed air storage. While compression in the case of hydrogen is just for convenience, as the storage is of chemical energy and hydrogen has a very low density making storage volumes too big without compression, then for air, the energy stored is that of compression. Excess wind or solar power in the form of electricity is used to compress the air. This is usually done in several stages in order to minimize the associated heating (as the heat energy would get lost after prolonged storage). Power is regained when needed by passing the compressed air through a turbine. The expansion is associated with a strong cooling, and heat has to be added in order to release the air at near-ambient conditions. The first such installation is rated at 290 MW and has been operated since 1981 at Huntorf in North Germany, with a 0.3 million cubic meter underground cavern. A subsequent 110 MW plant has been operated by Alabama Electric Corporation in the United States since 1991. It recuperates heat from the compression process and stores the heat for the turbine stage of new power production.

Storage cycle efficiencies are typically 40%–50%, for the original use as peak-load plants (charging the store at night, discharging during peak load hours, cf. Sørensen, 2010). The longer storage times needed for renewable energy operation will lower the round-trip efficiency, due to larger heat losses.

For elevated water stores, there may either be a pumping upward and regeneration of power by letting the water down to the lower level through an ordinary hydro turbine, or no upward pumping but a coordinated operation of the reservoir-based hydro system and the wind or solar electricity system requiring storage, such that hydro covers all the demands from customers within the two areas combined, when there is no wind or solar production, and hydro production is reduced when wind/solar production exceeds demand. Obviously, this type of operation requires strong transmission links between the two subsystems. Only few locations in the world are blessed with hydro installations with suitable large (seasonal or annual) reservoirs, and they were typically built long ago, when issues such as flooding heritage-level mountain landscapes were not raising opposition as they are today. It is thus necessary to look for other long-term storage options.

Batteries were discussed above and underground hydrogen storage will be covered in the next section. Further ideas would comprise reversible chemical reactions with high levels of energy input and output, but none of these ideas are close to commercial use. The scaling up of advanced battery technologies from the smaller-scale applications that are already in place is quite a new development and will need further price reductions to become really interesting. Hydrogen storage is in principle also technologically ready, but is also on the expensive side and awaiting the possible success in the development of fuel cells, for two reasons. One is that the round-trip efficiency using either reversible high-temperature fuel cells both to convert electricity to hydrogen and hydrogen to electricity, or the use of conventional alkaline fuel cells (an established technology called "electrolysers") to produce hydrogen but advanced fuel cells to regenerate electricity, is higher than using conventional combustion-based turbines to regenerate electric power. The other is the hope that in the future, hydrogen will gain a larger place in our energy systems as energy carrier, for example, for use in vehicles, so that at least some of the energy stored after conversion to hydrogen need not go through another loss-prone conversion back to electricity, but can be used directly in vehicles or stationary fuel cells.

Stationary use of fuel cells has the advantage that the heat generated in all the conversion processes can be used to satisfy heat demands. This could be used in district heating systems (for central stores) or in individual buildings, in case the fuel cell technology becomes decentralized (in which case low-temperature fuel cells such as PEM cells will be used). Waste heat generated at all conversion steps makes it possible to cover low-temperature heat needs without dedicated heat plants. For vehicles, the most likely technology in an energy system with hydrogen as an everywhere available carrier would be battery/fuel cell hybrids as a natural replacement for the battery/diesel hybrids expected to dominate the near future.

7.6 Special Opportunities

The global potential for offshore wind parks is very large (Sørensen, 2008a). However, large surges of surplus power are likely to appear, spread over the year with intervals of several weeks, and unfortunately the same is true for lulls. Wind follows the weather patterns of front systems passing over the region in question. A particular study for Denmark and surrounding countries shows that having wind turbines dispersed over the country, occupying offshore sites in the North Sea, the Baltic Sea, and the Kattegat Sea (bordered by Sweden, Norway, and Denmark), provides a smoothing of wind power production making the power availability curve change from having above-zero power 70% of the time (for one wind turbine) to having it 100% of the time, for the system as a whole. The power duration curve is very different from the nearly flat one characterizing the current coal-fired plants. It drops more or less linearly from twice the average to zero (Sørensen, 2010). Some smoothing can be achieved by trade with neighboring countries, using the organization Nordpool's power auction facility, but prices often vary unfavorably for renewable systems, as the power exchange rules have been set on the basis of conventional power production systems (Meibom et al. 1999).

Denmark has natural gas storage systems inaugurated during the 1980s, when the first gas pipeline from the North Sea to mainland Denmark was constructed. At that time, fear was expressed that the line could become severed by Russian submarines or even trawlers. Should this happen, repair might, in the worst case, take 6 months, according to the estimates made at that time. Therefore, the Danish government decided that an underground store should be built that could hold 6 months of natural gas use. This was accomplished by two separate underground facilities. One was based on a salt intrusion, that is, a dome of salt reaching nearly to the surface. Compared to other types of stores, a salt dome store is cheap to construct: You drill a vertical hole and flush water through it at low pressure (to minimize energy costs) for say 2 years, thereby creating an underground cavern. When completed, it may be sealed with a tightening gel, or steel canisters may be lowered into the cavity, now allowing gas storage at pressures of around 23 MPa (measured at extraction). The Danish store at Lille Thorup holds 4.2×10^8 m^3 of natural gas. The second underground store built was an aquifer store, taking advantage of an upward bending of an underground aquifer to store gas at the top, with the water in the aquifer preventing the gas from flowing away. This facility, at Stenlille, holds 3.5×10^8 m^3 of natural gas at an exit pressure of 17 MPa. The glaciation-deposit geology of Denmark with 2–4 km down to the bedrock allows these two low-cost stores to be built. In regions with solid rock near the surface, more costly drilling is required for making the cavity, but sometimes this cost can be avoided, for example, if an abandoned mine can be used for gas storage.

The natural gas stores can equally be used for hydrogen, with somewhat lower pressures due to the higher permeability of hydrogen (Sørensen et al. 2001). Electricity may be regenerated at times of insufficient wind, at round-trip efficiencies above 50%, especially if the price of solid oxide fuel cell (SOFC) generators on the market would reduce, but probably also with the new generation of hydrogen-optimized gas turbines. Surplus wind power would be converted to hydrogen by a reverse-operation fuel cell, either the conventional alkaline type or one of the new concepts. SOFCs would receive competition from the decentralized PEM fuel cells that according to one vision (Figure 7.3) might be installed in homes and buildings as replacement for earlier gas boilers, under the motto "be your own power producer." As often as possible, the hydrogen would be used directly, thereby avoiding the storage cycle losses. This would be possible, if hydrogen hybrid vehicles (hydrogen fuel cell plus batteries, as replacement for the earlier common-rail diesel engine plus batteries) would reach a sizeable proportion in the transportation sector.

Perhaps an even more exiting possibility would be to consider what would happen if Denmark produced more wind energy than it could use domestically on average. There is plenty of wind to take from: According to the Sørensen (2008a) study, more than twice the total Danish energy consumption (not just the electricity consumption) could be covered

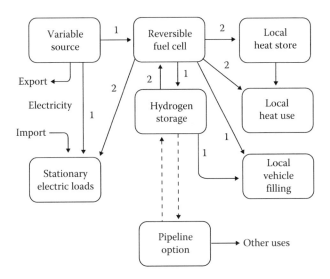

FIGURE 7.3
Decentralized energy system with local hydrogen stores in buildings and vehicles. Arrows "1" denote flow direction when there is a power surplus, arrows "2" when there is a power deficit. (From Sørensen, B. et al., Scenarier for samlet udnyttelse af brint som energibærer i Danmarks fremtidige energisystem. Final public domain report for project 1763/99-0001 under the Danish Energy Agency Hydrogen Programme. Roskilde University IMFUFA Texts No. 390, 2001.)

by wind turbines sweeping an area equal to only 0.02% of the Danish land area. There would be a surplus to export, most likely to Germany, who themselves have much less wind potential than Denmark and where a Denmark-Germany transmission connectivity is already established and ready for reinforcement. Later, the transmission network could be expanded to accept the substantial wind potentials in Sweden, Norway, and Finland, which like in Denmark by far exceed the total domestic energy needs of each country. The Nordic countries could be a major supplier of energy to Germany, in the form of electricity, piped hydrogen, or both (Sørensen, 2008b). Generally, as far as wind is concerned, many countries near coasts could supply surpluses to inner continental countries (Sørensen, 2010).

References

Jensen, J., 1980. *Energy Storage.* Butterworth Publications, London, U.K.

Jensen, J., Sørensen, B., 1984. *Fundamentals of Energy Storage.* Wiley, New York.

Meibom, P., Svendsen, T., Sørensen, B., 1999. Trading wind in a hydro-dominated power pool system. *Int. J. Sustain. Dev.* 2, 458–483.

Sørensen, B., 1984. Energy storage. *Ann. Rev. Energy* 9, 1–29.

Sørensen, B., 2005. *Hydrogen and Fuel Cells.* Elsevier Academic Press, Burlington, VT (2nd edn. expected by December 2011).

Sørensen, B., 2008a. A new method for estimating off-shore wind potentials. *Int. J. Green Energy* 5, 139–147.

Sørensen, B., 2008b. A renewable energy and hydrogen scenario for northern Europe. *Int. J. Energy Res.* 32, 471–500.

Sørensen, B., 2009. Energy storage column, 4 essays: living with intermittency; 2024: an energy storage tale; short-term energy storage: adaptable flexibility; storing energy: a look to the future. In: *Renewable Energy Focus* 10(2), 30–32; 10(3), 26–27; 10(5), 32–34; 11(1,2010), 56–57.

Sørensen, B., 2010. *Renewable Energy*. 4th edn. Elsevier Academic Press, Burlington, VT.

Sørensen, B., Petersen, A., Juhl, C., Pedersen, T., Ravn, H., Søndergren, C., Simonsen, P., et al., 2001. Scenarier for samlet udnyttelse af brint som energibærer i Danmarks fremtidige energisystem. Final public domain report for project 1763/99–0001 under the Danish Energy Agency Hydrogen Programme. Roskilde University IMFUFA Texts No. 390.

8

Energy Bionics: The Bio-Analogue Strategy for a Sustainable Energy Future*

Helmut Tributsch

CONTENTS

The remarkable energy strategy of nature has not only provided a sustainable basis for survival of living beings in nearly all climatic zones of our earth, it has even optimized our ecosystem out of an entirely hostile primitive environment. The discipline of energy bionics tries to understand the principles involved and to adapt them for the benefit of mankind using technically practical materials and processes. Can the future energy strategy of an industrialized world function as a high-technological continuation of natural evolution of biological energy technology? It turns out that the natural strategy of splitting

---

* The idea was previously discussed in the book: H. Tributsch; Erde, wohin gehst du? Solare Bionik-Strategie: Energie-Zukunft nach dem Vorbild der Natur, Shaker–Media, Aachen, 2008.

water with solar light for hydrogen, which is then attached to carbon-carrying molecules is a demanding technology but highly practical. It provides the basis for a sustainable fuel and material technology that mankind could gradually approach using wave-, wind-, and photovoltaic energy for hydrogen production and gasified biomass as carbon carriers. Additional scientific challenges in relation to energy, which man has to tackle, include photocatalytic water splitting, solar cells of nanomaterials based on kinetic charge separation, solar-powered water technology based on water cohesion, catalysis of multielectron transfer for fuel generation, as well as numerous mechanisms for energy storage, confinement, and conservation. Advanced nanomaterial technology can provide many advantages for energy applications as well. On the long term, there is no alternative to artificial carbon dioxide fixation. It is suggested to take advantage of sulfide-oxidizing bacteria, supplied by a solar generated sulfide energy source, for an artificial biomass industry similar to deep-sea ecosystems. Chemical industry must also learn to combine carbon dioxide with hydrogen. A significant advantage of the projected bionic energy strategy is that our fuel-producing and fuel-supplying industry can essentially remain the same while the fuel becomes sustainable.

It is suggested to adopt the energy strategy of nature as a long-term goal, to improvise temporarily but to take the right long-term decisions, and to stepwise learn the missing biological energy technologies by aimed research in the course of the current century.

8.1 Implementing Energy Strategies: What We Know

While the long-term need for sustainable fuel technology is incontestable, practical steps toward that aim are modest and characterized by conflicting opinions on the feasibility and economical reality of preferred options. A full-scale hydrogen economy is faced with the high costs for hydrogen production from photovoltaic electricity and for the infrastructure, especially for transport, if hydrogen is selected as the preferred fuel. Vegetable oils from agricultural crops will have a future but will compete with agriculture for food production in a world with increasing population and living standards. Large-scale efforts to recapture carbon dioxide for recycling into underground deposits suffer from the problem that they do not lead to a sustainable situation and that the long-term stability and safety of such technology will require careful studies and much practical experience. While it is generally accepted that regenerative energy sources could ultimately provide energy for all mankind, their development is presently slower compared with the growth of world consumption. Worldwide energy consumption will be more than double during the next 50 years, which is especially due to the economical growth in presently less industrialized nations. Renewable energy sources like photovoltaic energy show high growth rates, of the order of 30%; however, they start from a low market volume and still are very dependent on massive public subsidies.

When discussing the possible development of renewable energy technologies, one has to consider some typical patterns, which have been observed during the introduction of other energy technologies. Generally, it is seen that shifting consumption patterns typically requires an extensive and time-consuming infrastructure development. It has also been observed that the faster a new technology grows, the more it pushes up the costs of raw materials or other inputs. This is, for example, presently observed in the photovoltaic silicon industry, where up to now cheap waste silicon from electronic industry has been used for the fabrication of photovoltaic cells. Now there is a shortage of this silicon, and the price of the raw material goes up. The second example is the price of indium, which is needed for the fabrication of CIS ($CuInS_2/Se_2$) solar cells. Especially since indium is also increasingly used in modern flat-panel electronic monitors, the price has recently increased more than fivefold. Also, intelligent innovative energy concepts do not automatically succeed. We have recently witnessed a dynamic development and much optimism with fuel cell technology. In this context, it is interesting to note that already more than 100 years ago (in 1897), Nobel laureate Wilhelm Ostwald recommended fuel cell research as an important strategy toward the protection of earth resources. Even though he was scientifically correct, it was apparently more practical (lucrative?) to develop the polluting, noisy, and less-efficient combustion engines. On the other hand, sometimes energy technologies develop even though they start from a higher price level. Oil, for example, was twice as costly as coal when the advantage of liquid fuel started to replace the solid fuel. New fuels or energy technologies apparently succeed when they provide new and superior features. Government efforts to develop a modern energy technology do, on the other hand, not necessarily succeed, as, for example, seen with nuclear energy technology, which was pushed in many countries but was not widely accepted by the population. An important issue for the success of a new technology is of course also timing. Many additional factors have to fit in support of a favorable development. Today, increasing emphasis exists on the cleanliness of energy systems for health and environmental reasons. Additionally, increased flexibility, availability, and safety are important issues. When planning the introduction of a new energy technology, it has also to be considered that it typically takes two to three decades before a primary energy form reaches a 1% share of the global market. A similar time period is also needed to develop a new energy technology. One well-known example is the silicon photovoltaic technology. Within 50 years, the thermodynamic limits of efficiency for energy conversion have gradually been approached by laboratory cells, and industrial cells follow with one-third smaller efficiencies. A similar trend has been known from the development of the steam engine and of wind turbines. When discussing the possible development and future of renewable energy technologies, it is definitely necessary to consider such experience, which, at the first glimpse, makes it very difficult to seriously predict possible developments in the energy field. This uncertainty is reflected in available studies of the possible energy future. While energy options are quite well discussed [1], there is no agreement about the potential of renewable energies. There are serious studies that attribute a 30%–50% or even smaller share of renewable energy technologies to the world energy consumption by the year 2050 [2–6]. There are other studies that project a fully renewable industrial society [7–9], however, on the basis of substantial energy saving. Other studies support the renewable energy strategy, while pinpointing realistic challenges [10–12]. It is remarkable that a previously optimistic, well-known environmentalist feels that nuclear energy cannot be avoided in a future energy scenario [13]. The big problem is that if the effect of fossil energy systems on the climate development is supported by further evidence [14], presently discussed energy strategies may not be sufficient for safeguarding our ecological environment.

Considering the conflicting experiences with the development of energy systems during the last two centuries, it appears to be difficult to adopt a straightforward rational strategy toward an optimal solution and to generate sufficient confidence and credibility to stimulate the required investment along the suggested direction.

However, this contribution will attempt to develop the concept for an energy strategy based on the 100 million years long experience of evolution in biological systems. Nature has not only succeeded in evolving regenerative energy technologies for sustaining life but has also gradually improved the living conditions and the atmosphere on earth by liberating oxygen from water and generating moisture in the environment. Our suggestion and hypothesis is that, if we try to understand nature and follow with our technological efforts along the same strategic pathway of natural evolution, we will suffer from less mistakes and progress faster toward a sustainable future of industrial society. In other words, if we agree on a reasonable sustainable strategy for the future, modeled according to nature, we could focus our scientific and technological efforts and make much more pointed progress. It should be analyzed here to what extent our proposal could become a realistic alternative.

8.2 Our Energy Reality

With respect to the energy future, mankind is faced with a complex reality: The fossil energy economy relies on global energy reserves, which may at the long term entirely unbalance the world climate by liberating excessive carbon dioxide. When oil and natural gas should run short, oil shale and hydrated methane may fill the gap. The supporters of nuclear energy refer to a, in their opinion, well-tested technology, which operates approximately 450 reactors worldwide. They are confronted with a lot of resistance from the public and depend on governments for insurance against accidents and reprocessing of nuclear waste. There is also a security issue. Proponents of large-scale nuclear energy are faced with the difficulty to convince the world, that more than 15,000 reactors and a functioning reprocessing and disposal infrastructure for nuclear waste must be built. This number is based on the assumption that 45,000 GW (GW is gigawatt) of energy will be needed by the end of the century, and at least one-third will have to be supplied by nuclear reactors, which will supply not only electricity but also heat and hydrogen. The estimated exploitable uranium reserves, now for 47 years, would then reduce to 1.5 years.

Also, the dream of fusion technology is faced with unanswered questions. For 50 years, fusion research has aimed at igniting and sustaining small suns on earth, and it may take another 50 years before this may work out. Because of the high neutron fluxes involved, and the presence of a wide spectrum of materials involved, fusion reactors are expected to produce a similar amount of radioactive waste as conventional nuclear reactors. Nobody knows, by the way, how a confining reactor wall will look like that should resist an intensive neutron bombardment at temperatures around 1000°C. Since no heat exchange rods can be used like in conventional nuclear reactors, fusion reactors must be very large and will be expensive. Where should the enormous waste heat go when large numbers of fusion reactors work?

The only sustainable energy option, solar energy in a broader sense including secondary sources such as wind energy, biomass, and ocean thermal energy, has recently seen a lot of progress. Supported by significant subsidies, photovoltaics cells experiences an annual growth of 30%–40%, and has reached a yearly world production of nearly 2 GW, measured at maximum solar intensity. Impressive is the contribution of the much cheaper wind power. Worldwide 160 GW of wind power has already been installed. Nevertheless, in spite of a lot of optimism, the reality is that solar energy technology is growing much slower than energy consumption worldwide. While today the worldwide energy consumption is approximately 15,000 GW, it is expected to increase to 45,000 GW by the end of this century. Solar energy is growing too slow from a too low level, which explains why few decision makers consider it a reliable option for the nearer future. These considerations show that there is presently no realistic alternative against climate change [14] and that there is an urgent need for action.

Why has solar energy, which supplies 7000 times more energy than mankind presently needs, not developed to a more realistic alternative? Our interpretation is that our present approach to solar energy is only partial and has been too narrow and too much based on man's short-term technical experience. Silicon solar cells, wind machines, and solar heat collectors, which are the key technologies for technical solar energy harvesting, all have been invented by man during his comparatively short industrial endeavor. These inventions, though technologically impressive, may open a too narrow and too specialized area of applications for satisfying our energy needs in a sustainable world.

On the other hand, living nature has, for hundreds of million years, evolved a wide spectrum of energy technologies, which are highly interconnected and have successfully sustained life on earth and positively shaped our fragile ecosystem. It may be that we need the long-term "experience" of nature for shaping the energy future of our industrialized world. To what extent do we need to adopt essential features of its energy strategy to get our industrial world on a reliable path toward sustainability?

8.3 Bionics Strategy and What Technology Has Already Learned from Nature

Bionics is a research direction that aims at transferring technology from living nature into modern industry [15–18]. In contrast to biomimetic research, it does not insist on copying largely unstable molecules or materials to understand and demonstrate the function of biological mechanisms. Bionics recognizes that it cannot copy biological technology since self-organized biological matter cannot yet be reproduced and handled by man. However, it is interested in identifying the fundamental working principles of biological–technical systems and is aiming at getting equivalent industrial models working with technical materials. The aim of bionics is the technical function of systems similar to those in biology. They can be even more efficient because they can be constructed for a more specialized function. In this respect, bionics has to take account of significant differences between biological systems and technical ones, as Figure 8.1 explains for the special case of bionics, related to the field of energy technology, which is the subject of this study. It is a quite complex and interdisciplinary area of technology dealing with the interconversion and storage of energy, with materials used in energy systems, and with strategies aimed at conserving and handling energy. In order to distinguish this area of bionics from other areas such as, for example, those dealing with structural or mechanical engineering, flight technology, interface technology, or signal processing, the author suggests to name it "energy bionics," it is bionics dealing with energy matters.

Biological systems are typically complex. Several different principles are sometimes applied and have been optimized by the same biological system. The interesting biological energy principle has first to be identified and understood. Then a technical concept is developed. For this aim, some applied research is typically needed. As to be shown later, there are cases where the biological principle can only be understood after systematic

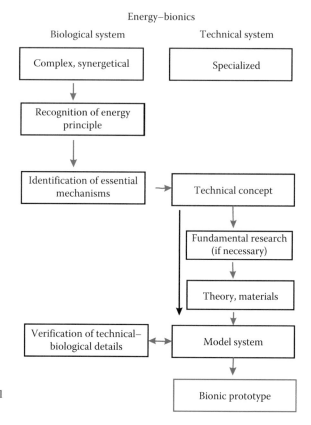

FIGURE 8.1
Flow sheet comparing biological and technical
systems and explaining bionic approaches.

fundamental research. Nature has applied energy principles that man has not yet explored
or even recognized. On the basis of new insights, a bionic model of the identified energy
system is designed. To verify whether it is really relevant, additional studies of the
corresponding biological system may be necessary. On the basis of new results, a bionic
prototype of the energy system is then conceived and tested. A bionic prototype may be
built up of modern technical materials, but should work on the basis of the physical chemical principles applied in the biological system.

When we intend to consider and develop energy technologies according to bionic points
of view, and when we intend to follow the experience of nature in such a highly transdis-
ciplinary area as sustainable energy utilization, it is important to know that bionics is not
a first-time adventure of mankind. After awaking from his animal past and diversifying
through his upright movement, man has probably always again solved technical problems
through observation of nature. He obviously has learned to know and control fire as a
natural phenomenon. Maybe he has recognized the advantage of rolling movements by
observing dung beetles during their pill moving activity. Man may have learned sewing
and weaving from weaver birds, and may have recognized the use of textiles when seeing similar structures as natural protections around the stems of palm trees. He may have
acquired the widely distributed adobe building technique with clay, mixed with straw,
from the potter birds or from certain insects, which use the same technique. The functions
of nearly all of our mechanical tools are reflected in the biting tools, bills, and claws of animals. From native people, we know that they identified new medicines by observing what
sick animals prefer to consume.

Modern industry has needed some time to rediscover nature as a model for techno-logical innovation [19]. Today, ships get specialized hull surfaces tailored according to the principles applied to the shark skin. Or they get a pear-like front structure copied from dolphin noses. On turbulence-prone areas of the Airbus 340, foils are applied, which mimic the grooves in the skin of a shark. Ultralight shot-proof protections take advantage from extraordinarily tension-resistant polymer fibers, adapted from spiderwebs. Self-cleaning surfaces copy the strategy of the lotus leaf surface [20], and highly effective glues take advantage from a gluing technique that certain sea snails apply when sticking to rocks. Modern insecticides implement the successful chemical defense strategy of certain plants. The burdock has shown us how to fasten objects in an elegant and simple way. Tree frogs have learned to avoid gliding off from slippery plant surfaces. Their honeycomb-like lam-ina on the sole of their feet already decorate commercial winter tires for cars. Innumerable architectonic and structural patterns of our modern world have an origin as discoveries of bionics. Especially, ultralight constructions in nature have significantly stimulated modern architectural and structural designs. Sandwich- and honeycomb structures, like hollow structures or mechanically highly optimized branching structures, or foam-like materials are self-evident details in innumerable, often very modest green plants. If bionic discov-eries would not be left to coincidences and to the curiosity of individuals, but would be systematically explored according to scientific principles and sorted according to require-ment, the advantage of bionic knowledge would really become evident. Modern industry is slowly recognizing this opportunity. The Mercedes Benz Bionic Car has, for example, been developed in analogy to a tropical fish, and has turned out to be one of the most efficient compact cars in the world with a gasoline consumption of 4.3 L per 100 km. The principle of a minimum material use for highest possible strength and efficiency, which nature has fully considered, is today already applied in many technical constructions. Building parts are calculated with a computer until, within an optimization strategy mod-eled according to evolutionary conditions, they are highly functional and material saving.

Our civilization is faced with the challenge of developing, within half a century, an affordable, sustainable, and highly diversified and complex energy economy. Nature has, in the course of evolution, gained incredible experience because the energy front has always been decisive for survival. It has adjusted to ice ages as well as to desert conditions and strongly varying environmental parameters and has even managed to steadily bal-ance and optimize the surrounding world.

What did nature accomplish in terms of energy technology, which includes efficient energy production as well as efficient energy conversion and storage? Let us imagine a primitive man with all the abilities that we still admire during Olympic games. He requires on the average 100 W power, or 2000 kcal over 24 h. A crocodile, a powerful reptile with a long evolutionary history, needs only one-tenth of this power to perform. Six million years ago, a giant bird, Argentavis, lived where Argentina is situated today. With its wings with air-filled bones and 1.5 m long feathers, it spanned 7 m and had the dimension of a sports aircraft, but had a weight of only 70 kg. Scientists estimated that he needed only 170 W for his predominantly sailing flight activities. This is remarkable since an ordinary car is con-structed for a power of 50–100 kW. And we have to consider that the present-day industrial man is consuming 60–120 times more energy than a primitive man in order to satisfy his modern living requirements.

Nature has, of course, in order to fill the niches of energy distribution structures, evolved living beings with all kinds of energy needs. There are the koalas or the sloth bears that are turning over extremely little energy, or the beasts of prey and the hummingbirds, which are dependent on a very high energy turnover. Nature also knows large-scale technology such as photosynthesis of plants, which provides fuels and materials on a gigantic scale. And there are energy technologies in nature, which mankind has not yet implemented, because it does not yet understand the mechanisms involved.

8.4 What Is the Energy Strategy of Nature?

Nature uses solar energy to split water. This is a quite energy-intensive process during which an energy-poor product, oxygen, is split off from water in such a way that an energy-rich product, hydrogen, becomes available. In nature, hydrogen, mobilized via this process, is not utilized as a gas, but directly attached to carbon compounds. In this way, the energy becomes available in the form of chemical products that allows the synthesis of all kinds of energy carriers and chemicals under moderate environmental conditions. Nature simultaneously uses the energy cycle in order to activate and turn over all necessary materials and chemicals. This is, as we will see, a very useful strategy, which made life, as we know it, on the whole only possible. While plants use solar energy to drive photosynthesis via carbon dioxide fixation for the generation of carbohydrates, animals use these as energy carriers while liberating again carbon dioxide via their own intricate energy cycle.

Figure 8.2 shows these energy-converting processes in a scheme. Light captured by chlorophyll in the leaf structures can separate electrons from positive charges, which induce the oxidation of water. In two subsequent light-induced excitation steps, the electrons are supplied with additional energy and would then energetically be able to reduce protons to hydrogen. Nature did not follow the path of a pure hydrogen technology, but attaches, as already mentioned, the hydrogen to an organic carbon compound. This compound (NADPH) and a second one (adenosine triphosphate [ATP]), which is produced parallel to the photosynthetic membrane processes via an equally light-driven proton current, serve the plant to fix carbon dioxide from air via a complicated process. This is accomplished by a complex sequence of chemical steps, the so-called Calvin cycle. It is shown in Figure 8.2 that green leaves consume water and liberate oxygen. The processes in the animal world function exactly complementary to the plant world. This shows the scheme on the right side of Figure 8.2. Via the fairly complicated Krebs cycle, the energy-rich carbohydrates are converted in such a way into energy carriers and materials, so that animals can satisfy their complex needs. Essential is that energy-rich electron-transferring molecules as well as the typical biological fuel ATP are generated. These two elements are sufficient to power a variety of energetic machines. They can generate mechanical work, synthesize chemical products, and induce biological light or electricity. In this way, carbon dioxide and water are generated by the animal kingdom, while carbohydrates and oxygen are consumed. Altogether, the biological energy system in Figure 8.2 shows that it is sustainable in relation to the use of both energy and chemical species. It is, however, most important to realize that, in order to operate this successful energy cycle, nature is applying a series of technologies, which our present civilization is either not yet properly handling or entirely ignoring. Seven of these are indicated in Figure 8.2. They comprise photocatalytic water splitting, a special, kinetic strategy for generating electricity from light (nanophotovoltaics), techniques to fix carbon dioxide, and a highly developed catalyst technology for energy conversion. Nature has also evolved a very peculiar energy conversion technology, based on protonic currents and gradients, handles efficient fuel cells that work at ambient temperature and operates a solar-powered water technology that uses water in the tensile

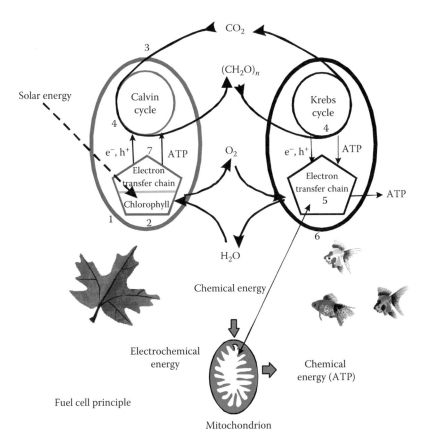

FIGURE 8.2
Scheme explaining solar-powered energy and material flows in nature with the complementary cycles of plant life (left) and animal kingdom (right). The inserted numbers list major supporting energy technologies: 1—water splitting with solar light, 2—nanophotovoltaics, 3—carbon dioxide fixation, 4—energy conversion catalysis, 5—proton energy cycles, 6—fuel cell principle, and 7—tensile water technology.

state, where water can be pulled. Mankind will have to learn a lot in order to understand and adopt these neglected or overlooked energy technologies. Their discussion and analysis will later be an essential part of our assessment of a bionic energy strategy.

Figure 8.2 appears to be an abstract scheme, but it is reflected in a very practical reality. Let us take as an example a hummingbird, sucking nectar from a flowering bush. The green leaves of the bush are generating the fuel from sunlight. It is upgraded and distributed. The hummingbird takes it up from flowers, converts it, and stores it. It uses it for its living activities and for its sophisticated and energy-intensive flight mechanism and as an energy reserve for the yearly migration over the Gulf of Mexico to South America. Biological energy conversion systems have a high standard that the technical energy systems also require.

8.5 Let Us Continue Evolution on a High-Technological Level

Man does presently not handle the light-induced catalytic splitting of water, like it occurs during the primary processes of photosynthesis. He cannot handle photosynthetic carbon dioxide reduction either. In addition, his knowledge on catalytic processes is too modest to be able to operate similar complex processes of energy conversion as nature did evolve during evolution. But humanity could for the beginning try to follow up and imitate the principal energy ideas with less-advanced technologies. Man could, on the basis of presently available sustainable technologies, at least try to start on the right track already now. And he could then, step-by-step, and with improving scientific understanding and technological know-how, gradually approach the biological energy strategy over the coming decades. This would have the advantage that little time would be lost and man could already now efficiently invest into a sustainable biology-inspired energy future. In the area of energy technology, mankind would simply decide to integrate energy developments into the million years old energy strategy of nature by concentrating scientific and technological efforts on closing the gaps and developing this unique strategy further. The sensitive equilibrium of our environment is simply too precious to try a drastically different, man-invented energy path and there is an additional opportunity to be considered: By imitating nature according to bionic principles, man could exploit the limits of natural laws even better than nature. Compared to complex biological systems, he could design his technological developments much simpler and in a much more specialized way. He can, of course, also use materials and mechanisms, which, for some reasons, nature did not develop and apply. Here, of course, materials, prepared at high temperatures, which nature did not exploit, provide remarkable opportunities.

The energy cycles, in analogy to the biological ones of Figure 8.2, which could be approached by man already in the nearer future are shown in Figure 8.3. In this context, it has to be pointed out that it may take a century to really reach a standard comparable with the natural example because of the variety of biological technologies we still have to learn (some of these are listed as numbers in Figure 8.2). One of these technologies is the important process of light-induced catalytic water splitting. Why we do not yet master this technology is discussed later. Because we presently cannot make technical use of it, we have to liberate hydrogen from water via sustainably generated electricity. It is well known that water decomposes into hydrogen and oxygen, when electric current is passed through it. Our industrial technology today knows several techniques to produce electricity in a sustainable way. Besides the elegant photovoltaic technology, in which light is converted directly into electricity without moving mechanical parts, we know solar thermal generation of electricity, or electricity generation by wind or waves. Electricity could, of course, also be generated by classical hydroelectric installations, ocean currents, ocean thermal gradients, or chimney effects in upwind power stations. Also, geothermal electricity could be used or hydrogen could directly be produced in high-temperature solar installations.

In Figure 8.3, it is shown that the commercially most economic energy for water electrolysis, which we still will identify, is applied for hydrogen generation. Like in photosynthesis, this hydrogen should now be attached to a carbon-carrying compound, so that we obtain a valuable chemical energy carrier. The carbon carrier has, of course, to be obtained in a sustainable way. Since our industrial technology does not yet handle carbon dioxide fixation from the air, this carbon carrier must be derived from agricultural industry or forestry. It must be natural biomass. This biomass has then to be upgraded with hydrogen to obtain significantly more energy-rich compounds.

For the herein proposed coupling of hydrogen to biomass, the Fischer–Tropsch process provides the right catalytic technology. Via this process, it would be possible to synthesize practically any carbon-containing product desirable, as we can do now starting from fossil fuels.

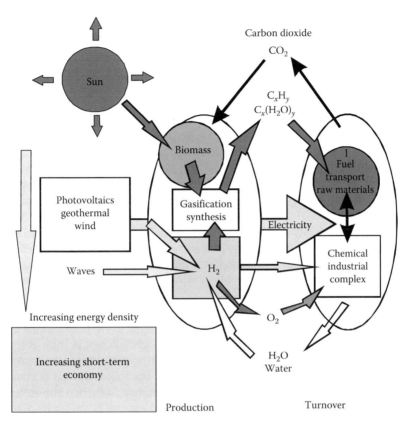

FIGURE 8.3
Possible technical energy circuits as bio-analogue strategy (compare Figure 8.2) for a sustainable energy economy.

As well known, the Fischer–Tropsch process has already been used in Germany during World War II at a large scale in order to synthesize gasoline from coal and water. This occurs at high temperatures during which water, the hydrogen donor, is decomposed and hydrogen is allowed to react with carbon monoxide. In the here-proposed process of Figure 8.3, the carbon compound will not be fossil carbon, but biomass, which already contains a certain concentration of hydrogen. It will be significantly upgraded by the attachment of additional hydrogen. It can be seen that the left ellipsoid in Figure 8.3, which describes bionic energy generation according to the biological model (Figure 8.2), in this case indeed reflects a sustainable energy technology, which uptakes carbon dioxide and water, and releases chemical energy carriers and oxygen.

The produced fuels and chemicals should now be used for the production-, transport-, and living activities of our industrialized society, which are symbolized by the right ellipse in Figure 8.3. One realizes that here organic fuels and oxygen are consumed and water together with carbon dioxide is released into the environment. Of course, there will be smaller details, which would be different in the technical energy diagram (Figure 8.3), compared to the biological one (Figure 8.2). Electricity, for example, and hydrogen will directly flow from the energy-generating (left) ellipse to the energy-consuming one (right ellipse), because they are technically useful there and we handle the corresponding technologies. Altogether, the proposed simple bionic energy diagram will be as sustainable with respect to energy and materials as the biological one. However, the energy processes involved are still intricate and before all have to be made economical.

On the first sight, it is immediately realized what is needed most urgent—a most economical source for sustainable hydrogen and also for a carbon carrier. As a sustainable carbon source today, only biomass from agriculture and forestry is available in reasonable quantities. A maximum amount of carbon compounds can be obtained from gasification of biomass at high temperatures. This is today a largely developed technology that is being improved in the direction of different specialized applications. Via skilled processing, the biomass is heated in such a way that mostly hydrogen and carbon monoxide are generated as synthetic gases. In this gas mixture, all materials of biomass are present and can react with additional hydrogen. Via the catalytic technology of the Fischer–Tropsch process, practically any energy-rich product, gasoline as well as diesel fuel, can be synthesized now.

It is presently economically not feasible to generate hydrogen with photovoltaic current on a large scale. Photovoltaic electricity is presently still 10 times more expensive than electricity from fossil fuels. Only public subsidizes and special conditions such as a larger distance from the public grid make photovoltaic energy competitive. Wind energy is clearly cheaper but not widely available.

Incident solar energy drives thermal processes and intensive water evaporation, which produces the large variety of known weather phenomena. They may induce a significant concentration of solar energy. Incident diluted solar light energy is thus eventually concentrated in the form of mechanical wind energy. Smaller-sized technical equipments are required for the harvest of more concentrated energy. For this reason, it is cheaper to harvest sustainable solar energy in the form of wind energy and the economic advantage further increases with increasing wind velocity. Wind energy is still subsidized in many countries but it is nearly competitive with fossil energy.

Water has a higher density than air. Therefore, classical hydroelectricity and electricity from wave energy are economically still more attractive. This is well known for hydroelectric energy, which is widely applied. But the availability of suitable sites for hydroelectric plants is limited. It is for this reason that our attention should be focused on wave energy for hydrogen production. In the open ocean, the height of waves is correlated with the speed of wind.

8.6 Wave Energy as a Start into Economic Hydrogen Production

It can be seen from Figure 8.4, in which the approximate price for electricity in Euro cents is plotted for different regenerative energy systems, that electricity from wind energy is significantly more cost efficient than that from the present photovoltaic systems. In inhabited areas, it is relatively easy to be introduced into the public grid. However, problems with land and seashore use arise due to conflicts with public acceptance, as well as with irregular wind patterns. Our aim here is to discuss whether there is still a more efficient and economically promising sustainable energy source for hydrogen.

Simple physical considerations indicate that harvesting energy from a diluted medium such as air needs larger installations and is thus more costly than harvesting the same amount of energy from a medium, which has a higher density such as water. This conclusion is supported by the fact that water power installations are economical today. But

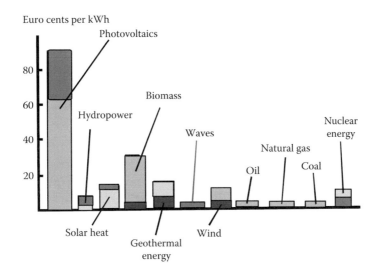

FIGURE 8.4

Energy costs in Euro cents per kWh for different energy technologies. Different shades distinguish between costs under different conditions.

in many countries, water power has already been exploited to a large extent. We have, therefore, to look at the ocean where wind power is generating waves. The height of these waves is correlated with the wind power. The exploitation of wave power has a long history (e.g., [21–24]). There are several construction concepts that are being investigated and applied [22,23]. A typical strategy is to build wave power installations near the coast so that the power can be transported on land via electric cables. There are several economical estimates that place the cost of wave power, which is typically estimated for waves 2 m in height, significantly lower than that for wind energy. This is, as already mentioned, reasonable because water has a much higher density and the necessary technical installations are proportionally smaller. As Figure 8.4 suggests, wave energy systems will, after their technology has matured, consequently become the most economic sustainable form of energy. Up to now, around 41 wave power projects have become known. At present, technical solutions, usually the mechanical energy of waves, are transferred via a hydraulic system to an electricity generator for electric power generation. The wave energy systems are anchored near or at the coast.

If wave energy should serve for massive hydrogen production, the wave energy installations cannot be limited to the coast region near inhabited land where wave patterns are often irregular and the sea has to be sufficiently shallow for anchoring. A wave power technology for hydrogen production has to be developed for areas of the sea where wave patterns are regular, present all over the year, and wave height reaches a maximum, since the energy contained in waves increases with the square of wave height. According to the TOPEX-Poseidon significant wave height maps of the world oceans [25], there are large areas where waves reach heights between 4 and 6 m practically throughout the year. Those areas are mostly confined to the region south of the 40th degree latitude (roaring forties) or, to a smaller extent, to the extreme north toward arctic latitudes. Especially in the southern latitude, the waves are present all over the year. The ocean in these high-wave regions is typically deep and the wave installations should, therefore, function as huge buoys, which are harvesting mechanical energy from a swimming body, which is periodically displaced by the wave height from the stagnant rest of the construction under water.

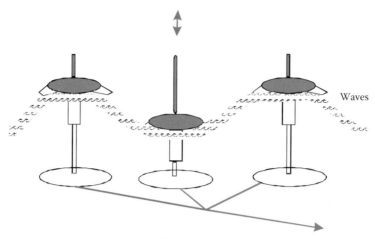

Mechanical ---> Electrical energy --> Hydrogen

FIGURE 8.5
Scheme of a hydrogen-producing wave buoy that is constructed in such a way that the swimmer is periodically moving with respect to a submerged inertial countermass that also stays in position due to its large water resistance.

Such structures have been proposed for electricity production via hydraulic systems and electromotors, and they could be redesigned for hydrogen generation. In order to adapt such installations for massive hydrogen generation for offshore production at a low cost, they have to be as simple and as durable as possible. Our suggestion is, as shown in Figure 8.5, to omit the hydraulic system to transform the vertical movement directly into electricity via a linear electricity generator in which a magnet passes through a coil thus generating direct current electricity with different polarities for the up- and downward movements [24]. A simple mechanical switch could invert one phase to produce direct electricity for hydrogen evolution. In order to avoid that the underwater counterweight for the periodical movement is displaced by wave power, it has to be either sufficiently heavy and inert or would have to spread out like an umbrella to generate a high resistance against the upward force in the sea water. This latter concept would consume less material.

The direct electrolysis of sea water into hydrogen and oxygen has already been studied by Bockris and others in greater detail quite some time ago [1]. The problem of anodic chlorine evolution parallel to oxygen evolution can be essentially suppressed by choosing appropriate electrodes and adequate limiting current densities. In recent years, some significant progress has been made in spite of little research on this subject. It was found that manganese-containing electrodes could release oxygen from sea water at high current densities with 99.6% efficiency [26]. Research in this direction should be intensified. It should not be overlooked that the photosynthetic system had to solve the same problem with its manganese-containing oxygen evolution center.

Stepwise, hydrogen energy buoys could be linked together to larger buoy fields via horizontal structural elements or cables under water, along which hydrogen could be collected and transferred to storage ships from where hydrogen could be carried on by ships or via undersea pipelines. One hydrogen-producing element (1 buoy) would have a diameter of approximately 20 m and would follow a vertical periodical movement of 4–5 m. The estimated power would be 1 MW. If one imagines how large ships are lifted by high waves, one can imagine the forces involved in large wave power. One thousand buoy elements

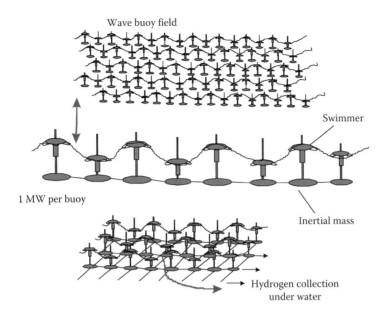

Wave buoy field

Swimmer

1 MW per buoy

Inertial mass

Hydrogen collection
under water

FIGURE 8.6
Arrangement of wave buoys in buoy fields, which produce hydrogen via mechanical movements and water electrolysis. Hydrogen is collected in the deeper, quieter water regions.

clustered together would operate as floating 1 GW hydrogen energy fields (Figure 8.6). They could have their own propagation and positioning system.

The idea to install wave energy buoys in violent seas may meet psychological barriers. Heavy storms may be expected to threaten the installations and to make logistics very difficult. Therefore, it should be emphasized that the buoys would barely look out of the water so that storms would not have a significant impact on them. The updrifting, vertically mobile floating bodies could be streamlined to minimize horizontal wind pressure. Altogether, one would essentially be dealing with underwater power stations, which take advantage of circular water movements within the upper 10–20 m of the ocean. Hydrogen would be collected under water where the ocean is calm. If hydrogen pipelines are not installed for medium- or long-range transport, liquid hydrogen-carrying transport submarines could be considered. A simplified view of such a wave power field including auxiliary installations is shown in Figure 8.6.

The technology needed for such floating hydrogen installations is basically available but would have to be optimized by aimed research and development. By keeping the technology as simple as possible, the impact of the harsh environment on corrosion and lifetime could be kept on an acceptable level. The buoy elements could be constructed in mass production in installations of the shipping industry and then transported to the desired location for transfer into the sea where they are interlinked by hydrogen-transporting underwater tubes. Figure 8.7 schematically shows how such floating hydrogen platforms could be integrated into the worldwide flow of chemical fuels. From the floating, hydrogen-producing platforms in the southern and northern seas, the hydrogen would first be transported (by pipeline or by ship) to areas with abundant biomass or biomass plantations. This could be Brazil and the Congo in the southern Atlantic and Norway or Canada in the northern Atlantic, southern Chile in the southern Pacific and Siberia, Canada or the northwestern United States in the northern Pacific. Here the gasified biomass would

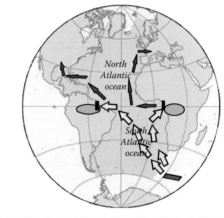

FIGURE 8.7
World map explaining an example of logistics for hydrogen generation from the sea. The use of hydrogen from a high-wave ocean region for refinement of gasified biomass and its introduction into the existing infrastructure for fuel processing and distribution are indicated.

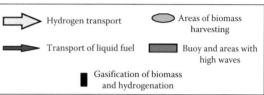

be combined with hydrogen for a transportable liquid fuel, which then could be shipped to industrial areas for further refining to gasoline, diesel, or chemicals as desired. The entire expensive infrastructure for the present distribution of fuels and chemicals could be maintained but the new fuel would be sustainable. This is a surprising and interesting result of our analysis. To be able to essentially maintain our complex and expensive infrastructure for fuel processing and distribution would be a significant advantage for an energy strategy.

8.7 Economy of Hydrogen from High-Wave Ocean Regions

Quite generally, it can be stated that, owing to the higher density of water as compared to air of similar speed, wave energy contains two to three orders of magnitude more kinetic energy compared to wind. This allows much smaller devices to produce a similar amount of power in a fraction of the space. Wind energy varies with a cube of airspeed but wave energy with the square of wave height. Therefore, much higher average power can be produced from high waves per unit of time. An important additional economical advantage is that wave energy can be exploited by modular systems over extended ocean areas. The costs of electric energy from experimental wave energy installations have dropped from 30 cents per kWh (around 1980) to less than 5 cents per kWh (in 2005). The various parameters influencing these costs (location, type of installation, dependence of wave presence on seasons, and length of electric cables needed) have amply been considered in the literature. Since wave power installations have typically been constructed or are being projected near populated areas such as Europe, Central America, or Hawaii, the energy output from the waves is a limiting factor. In these areas, and according to wave height data [25], wave heights above 2 m are untypical. When transferring, as we suggest, wave power installations into high-wave regions such as south of the 40th degree of latitude, the wave height will double and the energy contained will be fourfold. Since, according to the TOPEX-Poseidon significant wave height maps [25], such waves will be essentially present all around the year, another factor of 2 will have to be considered. This means that eight times more primary energy will be available. In the ocean areas concerned, the wave frequency will be lower, approximately 10 s compared to 5 s in calmer oceans. We consequently still have an energetic advantage amounting to a factor of 4. If we allow another factor of 2 for additional costs (hydrogen collection, infrastructure, rough sea-engineering technology), hydrogen from high-wave oceans could still become economical if sufficient time (—one to two decades) is left for research and technological development. In the long term, when technology is optimized, costs are expected to drop further.

Now let us make a purely theoretical calculation and consider that the world electricity consumption is approximately 2000 GW. The additional need for fuel energy may be several times higher (world energy consumption is around 15,000 GW). In this case, the world would need approximately 10,000 wave power fields of 1 GW size with a surface area of 10 km^2 each. The total area of high-wave ocean to be dotted with such installations would be of the order of 100.000 km^2. This would be a rectangle of 100×1000 km. When looking at the world map, we can immediately recognize that such an area is easy to accommodate in the oceans south of Africa alone where anyway, not far away, the ships from the gulf area are presently passing by with their loads of oil. In a sustainable energy economy, mankind will, of course, never need to occupy such large ocean areas for energy production because sustainable solar energy can be harvested all over the world with many different kinds of technologies. The challenge is, however, to make these technologies economically competitive. Wind energy, and, in the long term, also the worldwide most available photovoltaic electricity are expected to become gradually economical for massive hydrogen production.

8.8 Role of Biomass as Energy Carrier

Within an energy strategy modeled according to living nature (Figure 8.3), only biomass from natural photosynthesis is presently available as a carbon-containing sustainable energy carrier. Today, it globally contributes 18% to the total energy consumption. Most of this comes from developing countries. Since biomass is simultaneously relevant for food production, and, since, with the ongoing population growth increasingly fertile land is lost to urbanization and erosion, natural biomass is a valuable good. Worldwide energy consumption has grown to such an extent that photosynthetic biomass production can, in a sustainable energy economy, only supply a limited fraction of energy. A fundamental problem of natural biomass is that its production is not very efficient. It is true that fast-growing plants like corn or sugarcane convert, during their most dynamic growth period, 3%–4% of the incident solar energy into biomass. A similar amount they consume for their own living activities. But, in the yearly average, with three harvests per year, the overall efficiency is only 0.5%. Forests even grow with only 0.1% or smaller efficiency. Nature did not evolve plants for high energy conversion efficiency only, but for other purposes too. Europe is a suitable continent for considerations on biomass availability. In contrast to wide areas of Africa, Asia, and Australia, it is supplied with fertile land and sufficient rain. Forests cover a significant fraction of the land, in overpopulated Germany, for example, one-third of the surface area. Nevertheless, the European Community (EC) was, in 2003, only able to supply 69 Mtoe in the form of biomass for energy purposes (1 Mtoe = energy equivalent to 1 million tons of oil). This corresponds to 4% of the total energy consumption. (In the United States at the same time, biomass contributed 3.2% to the total energy consumption.) An EC study projected that this contribution may be increased until 2010, with great efforts, to 180 Mtoe, and by 2030 to 210–250 Mtoe. The quantity of 300 Mtoe would be the maximum contribution the EC could provide, which would be limited by serious impact on the environment. This means that the EC may not be able to supply more than 16% of its present energy consumption in the form of biomass. This is by far not sufficient to replace fossil fuels and there is consequently not sufficient biomass to act as carbon source within the bionic fuel strategy. In addition, biomass will also have to substitute fossil fuels as the raw material for chemical industry, which consumes roughly one-third of fossil energy. An additional factor is food production, which will have to increase to feed the growing world population. It will compete with biomass production for energy and fertile land. This short analysis shows that mankind will have to explore and implement artificial techniques for fixation of carbon dioxide from the air.

8.9 Need for Artificial Carbon Dioxide Fixation

There have been significant scientific efforts toward carbon dioxide fixation from the air. The main idea followed was to do this electrochemically, which means by transferring stepwise electrons to carbon dioxide from an electrode. As a consequence, protons are added to the carbon compound yielding alcohol and finally methane. Many metals and organic compounds were tested as catalysts, but the results were not encouraging. Yields were low and catalysts subject to deterioration. The main problem turned out to be the high activation energy for the transfer of the first electron to carbon dioxide. It generates a significant energy loss.

Can we learn from nature how to fix carbon dioxide? Definitively, nature in photosynthesis does not fix carbon dioxide via electron transfer. It rather exposes carbon dioxide to a hydrogen donor and catalyzes the attachment of hydrogen. There are also bacteria that do not depend on solar energy but use hydrogen as energy source. They succeed in attaching hydrogen to carbon dioxide. When bacteria have evolved chemical strategies for attaching hydrogen to carbon dioxide, scientists may learn and imitate them with advanced chemical technology. This could, for example, become an important long-term development challenge for our chemical industry.

8.10 Archaic Bacteria: The Search for the Seeds
for a Futuristic Agriculture

There is still another interesting possibility for an efficient technologically feasible carbon dioxide fixation from the air. This idea, which is also based on bionic considerations, could lead to a futuristic agriculture, which could be operated on infertile soils and on the ocean. It fulfils the important condition that it may fix carbon dioxide at least one order of magnitude more efficiently than natural photosynthesis. The author has published the concept for this strategy, which exploits the carbon dioxide fixation ability of specialized, archaic bacteria, on the basis of first experiments a long time ago [27] and has since added a variety of supporting experimental information [28,29].

They show that a technology based on such a principle would be technically feasible but its technological application would, like similar innovations for large-scale applications, need substantial scientific and technological efforts over several decades. But the efficient artificial fixation of carbon dioxide from the air is such an important enterprise that there is no alternative to such efforts, if a large-scale bio-analogue fuel production should be achieved.

Now let us consider the background of such a bionic strategy. Besides natural photosynthesis on the earth surface, which depends on sunlight, there is a second biomass system that exists in the dark of the deep sea around sources of chemical energy. The latter is liberated by geothermal and volcanic activity along the fracture lines between continental masses (see Figure 8.8).

Beginning with the early seventies of the last century, deep diving exploration vessels like the diving boat Alvin have documented unique living forms, ecological communities, and biomass systems. Typical are the 1–2 m long reddish worms, sticking to the ground and slowly moving in the ocean currents, mussels, crabs, and eel-like creatures. Altogether, already more than 500 new species have been detected in these ecosystems. We are dealing with biomass systems that are associated with the so-called black smokers, sporadically developing and vanishing hot springs that contain large amounts of sulfide compounds. The water under high pressures can be several hundred degrees hot and mixes with the few degrees Celsius cool ocean water. The sulfur compounds that appear as black clouds are energy rich and consist of dissolved and solid sulfur species. They can be used as an energy source by primitive microorganisms like the archaea, which are the

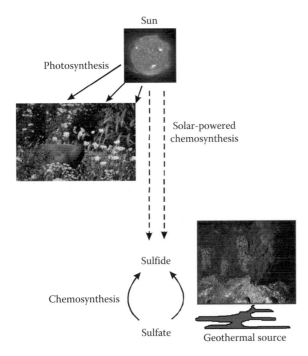

FIGURE 8.8
Scheme comparing photosynthesis (left) with chemosynthetic biomass generation (right) in deep sea. The proposed solar chemosynthetic process (center) for bacterial biomass generation via the sulfate–sulfide cycle combines the two technologies.

forerunners of our present bacteria. The energy conversion occurs around the black smokers at temperatures up to 100°C. These microorganisms are able to fix carbon dioxide and develop their living activities using the sulfide energy by oxidizing the sulfide to sulfate. We speak of chemosynthesis in contrast to photosynthesis, which is powered by solar energy. But there is an important difference. Chemical energy can be used more easily by these microorganisms than by light-collecting plants or algae for carbon dioxide fixation. This is obviously related to the difficulty in extracting energy from light-driven processes. Here, species are involved that only have a short lifetime of, maybe, 10^{-7} s. During such a short period, excitation energy has to be passed on to generate stable energy-storing compounds. This is a complicated and not especially an effective process. Energy-rich stable molecules can, on the other hand, provide energy for an unlimited period, until the carbon dioxide-fixing reaction is initiated and completed. Energy conversion processes in the deep sea, which are based on chemosynthesis, are therefore very efficient. The bacteria that grow on the chemical energy, which pours forth from the ocean underground, can in turn be consumed by many exotic species of this astonishing ecosystem. Ultimately, geothermal energy from the underground, heat, and pressure, is converting energy-poor sulfur compounds, sulfates, into energy-rich sulfur compounds, sulfides. These pour out from the ocean floor and power the deep-sea ecosystem like sunlight on the earth surface lets plants grow.

Now the essential ideas in relation with the proposed bionic solar technology for carbon dioxide fixation should be analyzed. Since the energy source for the bacteria is simple sulfides like iron sulfide, pyrite, it is imaginable to use solar energy for the conversion of sulfates into sulfides. This way the chemosynthetic process of carbon dioxide fixation by bacteria is coupled to the solar technical generation of iron sulfide (Figure 8.8). The sulfide of iron is preferred here because iron is nontoxic and entirely compatible with life processes, in contrast to most other metals. By utilizing sulfide produced from sulfate via solar energy, the bacteria energetically ultimately grow on solar energy, fixing carbon dioxide like algae in sunlight:

$$\text{solar energy} + 2\text{FeSO}_4 \rightarrow \text{FeS}_2 + \text{Fe} + 4\text{O}_2 \xrightarrow{\text{bacteria} + \text{CO}_2 + \text{H}_2\text{O}} 2\text{FeSO}_4$$

$$+ \text{bacterial biomass (C, H, O)} \tag{8.1}$$

It has been shown that this process can be at least 10 times more efficient for biomass generation than photosynthesis [27,29]. But many details on the mechanism of bacterial interaction with sulfide particles for energy turnover remain to be investigated [30,31], and bacteria with the fastest turnover of energy have to be identified. For this purpose, the mechanisms of sulfide oxidation for energy extraction have to be explored and optimized.

How such a solar-powered biomass technology, modeled after the deep-sea biomass system, would look like is shown in Figure 8.9. Also a picture of iron sulfide-oxidizing bacteria, *Acidothiobacillus caldus* and *Leptospirillum ferrooxidans*, is shown. Solar energy in a desert is converted into electricity and heat, which is used to generate iron sulfide from iron sulfate. A possible process has been explored [28]. It applies a thermo-electrochemical process proceeding with carbon electrodes. The generated sulfide energy is supplied to bacteria, which, in appropriate tanks, grow to biomass by oxidizing it to sulfate. The biomass formed is then pumped away for fuel generation via a pipeline. This biomass from bacteria could be gasified like biomass from photosynthesis, and, by adding sustainable hydrogen, converted into high-quality liquid fuels. Such a biomass technology would not consume fertile land, and would use little water, which could even be largely recycled.

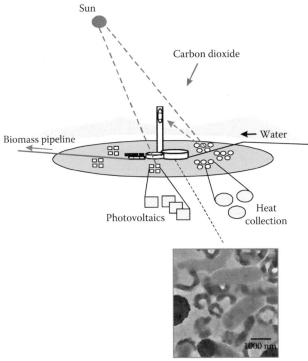

FIGURE 8.9
Production scheme for solar-powered bacterial biomass. Solar electricity and solar heat are used to generate sulfide from sulfate for bacteria cultivation and carbon dioxide reduction in large tanks.

Of course, it would also be possible to operate such a technology on platforms on the sea using wind and ocean power as sustainable energy.

Quite generally seen, the elaborate fuel-generating and -converting molecular mechanisms of numerous archaic bacteria provide remarkable additional opportunities for bionic energy research. There are, for example, various species of methane-producing archaea, ancient bacteria, which attach hydrogen to carbon dioxide and release methane as a by-product of their metabolism. If hydrogen would be obtained from solar energy utilizing bacteria, such as purple bacteria (*Rhodobacter capsulatus*) and transferred to methane liberating bacteria this would yield an interesting sustainable fuel system. Other bacterial energy cycles could also be taken into consideration.

If we summarize our considerations on artificial biomass generation (Figures 8.8 and 8.9), we arrive at the conclusion that it will be possible on the long term to develop biomass technologies, which provide significantly higher production efficiencies compared with photosynthetic biomass buildup, and they will not compete with agriculture for food production. The key reason is the expected much higher industrial efficiency of solar thermal–electric processes compared to photosynthetic processes.

8.11 Elements of a Bio-Analogue Fuel Economy

Since in the future we will be able to count on both our present biomass and a futuristic new biomass technology for carbon-carrying molecules, we can imagine a bio-analogue,

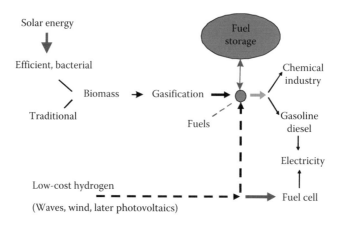

FIGURE 8.10
Simplified scheme for the illustration of a bio-analogue technical fuel supply for our world. Sustainable hydrogen and natural together with artificial biomass are the bases, and large-dimensioned fuel storage will finally allow humanity a certain control of the world climate.

bionic fuel cycle as depicted in Figure 8.10. Biomass from natural and artificial sources will be gasified, and, in this way, yield an optimal amount of carbon compounds. These will be upgraded by the attachment of sustainable hydrogen. This way, all energy carriers and also the large variety of chemicals for the chemical industry can be synthesized. Gasoline and diesel will be the main products. Besides these, hydrogen will also be used directly for electricity generation in fuel cells, steel production, and a variety of chemical processes. In addition, as already indicated in Figure 8.3, the direct generation of electricity with photovoltaic cells, wind machines, and via other technologies will also play a relevant role. Important for the proposed energy cycle of Figure 8.10 is an energy sink, a storage system for the fuel that has been generated in a sustainable way. This means that surplus fuel is temporarily to be taken out of the cycle in large quantities in order to be used again when needed later. Besides, the use of sustainable hydrogen in relation to carbon-containing fuel could be increased or decreased, as needed. An essential advantage of this energy strategy would then be the possibility to control the carbon dioxide concentration in the atmosphere.

Sustainable fuel will be deposited underground in large quantities, when carbon dioxide is to be extracted from the atmosphere. This would be necessary to gradually neutralize an advanced greenhouse effect, or to compensate a possible large volcanic eruption. In the case of an approaching ice age, on the other hand, carbon dioxide would be accumulated in the atmosphere to support solar heating.

The sustainable fuel cycle of Figure 8.10 is a first bionic prototype of the biological energy cycle (Figure 8.2), and was derived on the basis of the analysis of a more accurate model (Figure 8.3). The latter showed that we are far from understanding and handling biological energy conversion processes, with relevant, required energy technologies indicated in the form of numbers within Figure 8.2. Since we are not yet able to split water with light and since we are not yet able to fix carbon dioxide in a sustainable way, we have to aim at hydrogen from ocean waves. We also have to rely on gasification of photosynthetic biomass to obtain carbon-containing energy carriers before we may acquire the knowledge to fix carbon efficiently by solar generation of the energy source of sulfide-oxidizing bacteria. The advantage of this bionic approach would be that we start with a blueprint of an energy strategy that allows us to essentially maintain our complex fuel infrastructure. Only the present fossil fuel will be replaced with sustainable fuel.

In order to make sustainable fuel cheap and widely available on the long term, we have however to learn the identified energy technologies from nature and integrate them into the bionic energy strategy. Only by adapting these technologies we will be able to approach nature's energy standard. In the following sections, we have reviewed these missing technologies and pinpointed the scientific and technological challenges we have to face.

8.12 Nature as an Example for Advanced Energy Catalysis

For our agriculture, ammonia is needed as a starting material for fertilizer production. The combination of hydrogen and nitrogen according to the Haber-Bosch process requires high temperatures and pressures. Nature handles the process of nitrogen fixation at ambient conditions. Various green plants, such as beans or alders, keep nitrogen-fixing bacteria in their roots to do the job.

When we breathe oxygen, we allow the energy-rich products, supplied via our food, use this sink for electrons for the conversion of chemical energy into electrochemical one. In order to facilitate this process, nature uses a catalyst composed of an iron atom and a copper atom in an appropriate organic environment. In a technical fuel cell, the process is similar. Electrons from a fuel like hydrogen are transferred via the electrodes and the electric circuit to oxygen for reduction to water. The required catalyst is platinum particles and higher temperatures are needed to make the process efficient. Nature is able to perform all the essential energy conversion processes at ambient temperature. The catalysts applied involve iron (proton reduction), copper (oxygen reduction), manganese (oxygen evolution), or molybdenum (nitrogen fixation), typically clustered with other elements. In contrast, our technology uses noble metals, platinum, palladium, rhodium, or ruthenium typically at quite high temperatures.

What nature has accomplished with its catalyst strategy is the ability to interconvert all required energy forms, to generate an entire family of energy carriers, and to do this at ambient temperatures. It has evolved catalysts with a high degree of incorporated information, which guides the complicated reaction processes. This was the precondition for making living beings so incredibly dynamic with respect to their energy turnover.

What is the key trick for replacing a noble element catalyst for one made of a common transition metal? The problem with common transition metals like iron, copper, or cobalt is that, when exchanging electrons with oxygen, they tend to form oxides that drastically change their properties. Nature takes these elements and embeds them into an organic environment, such as a porphyrin or heme ring. This leaves the electron transfer properties largely unchanged, but avoids the irreversible reaction with molecular oxygen.

One can make useful technical catalysts, starting from organic-biological catalyst blueprints. When a biological catalyst center like a cobalt-porphyrin is heated with additional metals and chemicals, the carbon environment carbonizes to graphite, which is chemically very stable (Figure 8.11). The Co-catalyst center, however, essentially survives. Additional centers are established between the graphite layers and cooperate with the Co-center. Reasonable oxygen reduction catalysts for fuel cells result [32,33,34]. The catalyst centers are of molecular nature, similar to their biological example, and in contrast to our established technical oxygen reduction catalysts, which function as nanoparticles. By simulating biological energy conversion catalysts and building up theoretical understanding

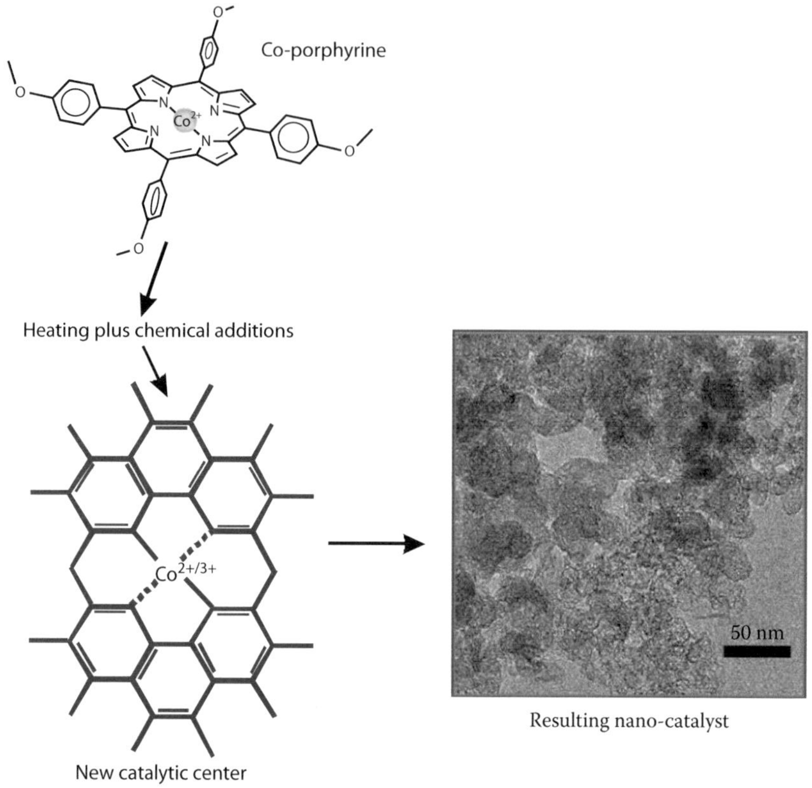

FIGURE 8.11
Bio-analogue catalysts for energy systems made by modifying biological catalytic centers, by heating and use of chemicals, into similar ones within a graphitic environment.

of the mechanisms involved, science may gradually learn this art of guiding essential electron transfer processes along least energy-consuming pathways.

Multielectron transfer catalysis is not the only important energy achievement of biological systems. They have also optimized energy storage and evolved a wide variety of well-catalyzed processes that take care of an efficient interconversion of chemical energy. Understanding the way, for example, how chemical energy is stored and mobilized, at ambient temperature, in the form of brown fat, which has a similar energy content as gasoline, is a big challenge for bionics. Also the widely used biological fuel ATP should attract special attention as it works entirely different from our technical fuels. When these fuels are turned over, quite strong chemical bonds are broken and again formed. With ATP, no relevant strong bond is broken, but the entire chemical structure of the molecule and its aqueous environment is involved when energy is turned over and a phosphate group is released. In principle, during energy release, a significant amount of previously nonavailable energy, entropic energy, becomes available for the reaction. Why did nature choose such a strange mechanism for the most important and mostly turned over biological fuel? Our interpretation is that the aim is to make the energy conversion well functioning at ambient temperature. Derivable from the principle of least action, irreversible thermodynamics (linear range) requires from a chemical reaction to follow the law of minimum entropy production to proceed efficiently. It should not produce much disorder in the environment. As a consequence, a minimum of entropic energy ($T\Delta S$) may globally be generated, when the turnover in entropy can internally be supplied from entropic energy already present in and released from the reacting chemicals. In other words, if previously nonavailable, entropic, energy is made available, the corresponding decrease in entropy will reduce the total entropy turnover of the reaction. This possibility is apparently considered in the energy-releasing ATP hydrolysis reaction, which guarantees efficient energy conversion at ambient and body temperature. Energy bionics recognizes in biological chemical energy technology a very relevant area for research.

8.13 Photovoltaic Nanocells of Plants: The Advantage of Kinetic Charge Separation

A solar cell is, in principle, a simple membrane that is only transparent for photo-generated charge carriers in one direction (Figure 8.12a). Our present silicon solar cells have evolved to quite sophisticated electronic devices. They generate a thermodynamic gradient in the form of an imprinted electric field, generated via the p-, n-doping transition (Figure 8.12b). This electric field has to be sustained for the lifetime of the solar cell. This is typically more than two decades, during which the gradient should not destroy itself by field-driven diffusion processes. It is therefore not surprising that only exceptionally stable materials like silicon, or gallium arsenide, or unusually tolerant materials like the self-repairing copper-indium-selenide became practically successful as solar cell materials. The present technical strategy of photovoltaics is working well, but after half a century of development it still needs substantial subsidies toward commercialization.

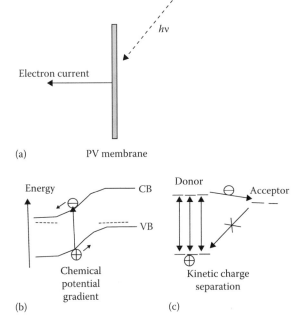

FIGURE 8.12
(a) A solar cell is just a membrane that allows electrons only to pass in one direction. The mechanism for rectification can be based on an inbuilt thermodynamic potential gradient (as in classical solar cells via p-n doping) (b) or it can be based on kinetic rectification (as in the photosynthetic membrane and in dye or polymer solar cells) (c).

Nature also converts light into electricity. It happens in the photosynthetic membrane. But nature is not applying a permanently imprinted thermodynamic gradient for charge separation. It rather organizes and tailors macromolecules in such a way that electrons can only be transferred into one direction (Figure 8.12c). The advantages are that kinetically determined solar cells can relatively be easily assembled at low temperatures from abundant particulate and molecular raw materials. This principle functions in the so-called nanosolar cells, which do not allow the buildup of thermodynamic gradients. Such nanosolar cells are, for example, the purely empirically established polymer [35] and dye solar cells [36,37]. The molecular nanosolar cell has, as a bionic technology, a significant potential for development because it promises simpler and cheaper solar cells and opens a largely unexploited research field with respect to the principles involved.

It is interesting to note that the established technical principle of photovoltaics has long been considered to be the only existing one. This belief was supported by the derivation of the diode equation from the Planck radiation formula. It was concluded that the phenomenon is controlled by reversible thermodynamics [38]. However, a photovoltaic cell is a system that not only converts energy but also generates entropy in the form of heat and can better be described by irreversible thermodynamics. In fact, it could be shown that photovoltaic principles can be derived from the fundamental principle of least action [39], which leads to the irreversible thermodynamic principle of minimum entropy production. Interestingly, both photovoltaic mechanisms, the classical technical one applied in silicon cells and the nanosolar cell principle applied in the photosynthetic membrane, can be derived and identified. While the useful electric power W_c in a classical solar cell is proportional to the product of thermodynamic current flux j_i and thermodynamic potential gradient, $\nabla \mu_i$,

$$W_c \sim j_i \nabla \mu_i \tag{8.2}$$

the useful work from a kinetic solar cell W_k depends on the product of chemical reaction rate w_i and chemical affinity A_i:

$$W_k \sim w_i A_i \tag{8.3}$$

The chemical affinity A_i is equivalent to the deviation of a chemical system from equilibrium and proportional to the Gibbs free energy turnover. The chemical affinity is approaching zero when a system reaches equilibrium that is when illumination stops. w_i describes the reaction rate of the process involved. By providing an electron transfer mechanism in which k_1 of the forward photoreaction is much larger than k_2 of the reverse reaction (such as is the case with the fullerene in polymer solar cells), one allows both w_i and A_i to become large.

The nanosolar cell principle in the photosynthetic membrane evolved 2–3 billion years ago and conserved the same principle of charge transfer rectification in all later biological applications among more advanced photosynthetic species. The reason is apparently the high efficiency of this process. Excitation energy is activated within a radical pair for transfer of an electron. A nuclear spin polarization occurs and under spin conservation, the electron spin is inverted simultaneously (Figure 8.13). While this occurs some energy is lost so that the entire process is a combined quantum-classical one. The consequence is that, if the electron should react back into the ground state, it would have to overcome a 0.2 eV high activation barrier while simultaneously having to invert its spin. Such a process has, of course, a very low probability. The reverse reaction rate constant that enters

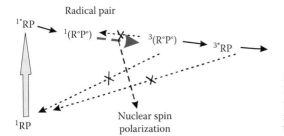

FIGURE 8.13
During the primary photosynthetic process, electron transfer is rectified via a photodynamically induced nuclear spin polarization. The electron cannot react back.

into chemical affinity A and into the reaction rate w is correspondingly low, leading to a correspondingly high output in photovoltaic energy W.

Today, understanding of electron transfer is essentially based on Marcus theory [40], which describes it as a reversible tunneling process subject to weak interaction. This essentially excludes unidirectional electron transfer, as needed for kinetic solar cells. Stronger interactions requiring molecular feedback processes must be involved for obtaining electronic rectification. Electron transfer processes involving irreversible mechanisms and feedback have been calculated [41,42]. The results show that for directional electron transfer, the electron, while being transferred, has to exert an influence on the molecular environment, which improves the rate of electron transfer. A feedback reaction of this type is a process well defined in time. There is a time before the feedback and a time after. Quantum physics does not accept a time concept, and time-dependent quantum processes only arise from interactions with classical perturbation. If such a mechanism of feedback interaction is calculated via the electron density of an individual electron cloud as a continuous model considering infinitesimal feedback steps, the self-organized behavior of an individual electron, while interacting with the molecular environment, can be calculated. It turns out that molecular feedback for electron transfer may not only make the process directional. It can also become significantly faster. Such an electron transfer mechanism, which was named "stimulated electron transfer," supports kinetic electronic charge separation because it is directional and efficient, leading to a large w in relation (8.3). The challenge is that it has to be learned how appropriate feedback mechanisms during electron transfer can be stimulated and implemented in the form of suitable electron transfer bridges. First inspirations may be obtained from molecular biology where numerous elements of electron transfer chains dynamically react with respect to electron transfer. The amino acid cysteine and similar thiols may be examples. L-Cysteine or thioglycolic acid adsorbed to a semiconductor particle via the thiol sulfur supports anodic, but not cathodic, electron exchange. An electron extracted from the thiol sulfur leads to a reorganization of electron density over the remaining molecule, which supports regeneration of the electron. But the opposite reaction pathway for an electron is blocked. This peculiar nonsymmetrical electron transfer behavior of cysteine has been continuously used in biological systems for electron transfer bridges in ferredoxins since early evolution. We need to handle directional electron transfer in order to build kinetically determined solar cells in a creative way. In present empirically developed nanosolar cells, in polymer and dye solar cells, the peculiar properties of fullerene molecules and the nonsymmetrical properties of the I^-/I_3^- redox couple are essentially responsible for the kinetic rectification of electron transfer. The same species are, however, also responsible for some disadvantages such as chemical instability.

The dye solar cell (Figure 8.14) has started as a bionic prototype for chlorophyll-sensitized electricity-generating photosynthetic membranes [36] more than three and a half decades

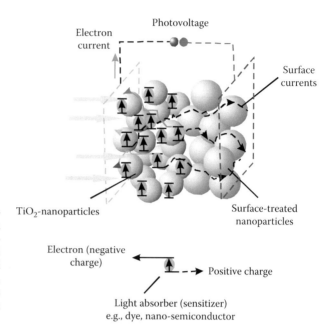

FIGURE 8.14
A dye or nanosolar cell operates with light-absorbing dots (dyes or nanoparticles) on interfaces of nanoparticles. Charge separation is not determined by an imprinted thermodynamic gradient (as in classical Si-solar cells), but by a kinetic mechanism. For chemical–structural reasons, electrons have a preference to move in one direction only.

ago, but in spite of the significant progress made [37], it could not yet be commercialized due to instability problems. It is interesting to note that the species, which are involved in kinetic charge separation, the I^-/I_3^- redox system, is also essentially involved in instability problems. In polymer solar cells, the same is true for the fullerene molecule. It is suggested that the scientific basis of kinetic charge separation is essentially lacking, which shows the challenge and need of bionic energy research.

8.14 Photocatalytic Water Splitting

Fuel production via light-generated electricity and water splitting is a fundamental quantum solar energy conversion process, which is applied in nature, and, thus, via fossil fuels and biomass, responsible for most of our present energy supply. Significant advantages of this technology are elegant energy conversion mechanisms, applicability at environmental temperatures, and potentially high-energy conversion efficiencies. The key problem with oxygen liberation from water is that the most efficient thermodynamic potential at $E_0 = 1.23$ V is described as a four-electron extraction process. This means that the transfer of all four electrons is subject to a similar low energy turnover. If the extraction of one of these electrons from water requires significantly more energy, this will make the entire process correspondingly less favorable. Extraction of four electrons with a comparably low energy turnover is, however, not accomplished in practice unless a highly sophisticated catalyst is handling the electron transfer process. If the conditions only allow extraction of one electron from water, an oxidation potential of $E_0 = 2.8$ V will, for example, apply. This is approximately observed at the titanium dioxide electrode when a photo-generated hole is reacting via the energetically low valence band of this photocatalytic material. The energy conversion pathway in this case is consequently very unfavorable leading to radical intermediates. A multielectron transfer process, which is energetically needed, has to be catalyzed by a molecular environment, which safeguards the transfer of four electrons with approximately an equivalent energy turnover. This is a significant challenge that nature in the photosynthetic membrane has solved via a manganese complex. Two pairs of manganese ion centers in an oxide-calcium environment absorb four positive charges, which, in a highly catalytic process, are then neutralized by the extraction of four electrons from water. During the last three decades, all imaginable organic manganese cluster compounds have been synthesized and tested for water splitting. None of these has shown a behavior, which is nearly comparable with the properties of the biological manganese complex. A puzzling property of this complex is, by the way, that it is molecularly quite weakly bonded and that the manganese centers can easily be extracted and replaced. Because in photosynthesis water oxidation is occurring at a much lower overpotential than comparable processes in electrochemistry, we have suggested that the manganese compound extracts electrons via a kinetically self-organized mechanism. The first electron transferred induces a change, which favors the second, which in turn favors the third, and the fourth. If the set of equations is solved and the feedback action is varied and optimized, it is found that the kinetic formalism can be reduced to one equation describing how the first electron is "slaving" the others to enable a self-organized mechanism of multielectron transfer [41,42]. In such a process, of course, dynamic molecular changes are expected, since the manganese atoms thereby change their oxidation states and their coordination. Such necessary molecular rearrangements can, in principle, easily be handled via manganese complexes since manganese can readily change its coordination structure. If this concept proves to be correct, water-splitting catalysts will have to be identified in the form of three-dimensional, disordered, and loosely arranged structures that are able to change their morphology and can easily be penetrated by protons. And a very characteristic property of such a catalyst will be that it is able to handle electronic charges in a highly nonlinear way. Understanding and controlling such cooperative electron transfer processes is the main challenge toward photocatalytic water splitting within the bionic energy approach.

In photosynthesis, the electrons, taken from water, are energetically raised in two steps via photoinduced processes (Figure 8.15). Nature has selected two steps to make better use of the available solar energy. The electrons are raised until the electrons could, in principle, reduce protons to hydrogen. But nature prefers to attach hydrogen to a carbon compound, which it uses as an energy carrier.

Few illuminated materials have, by the way, the properties to react with water. Nature handles all major electrocatalytic processes of energy conversion via transition metal centers involving manganese, molybdenum, iron, cobalt, or copper. This, of course, suggests that coordination chemical mechanisms are involved. Indeed, semiconducting materials that supply electrons or holes via energy bands derived from transition metals show very specific interfacial properties by allowing a simultaneous exchange of coordinated ligands. If the semiconductor ruthenium disulfide is illuminated and polarized as an anode, oxygen evolution occurs with a quantum efficiency exceeding 60% [43]. The energy gap of this material is too small to allow water splitting without applied electric supporting potential. However, this material shows that when electrons are turned over in such a way that water species adsorb to the interface forming hydroxide–oxide complexes, a stepwise oxidation of water may occur. In this case, the dynamically changing ruthenium-hydroxide-oxide-peroxide complex provides the catalytic environment for reasonably favorable multielectron transfer processes. This is apparently also the reason why ruthenium oxide as a metal-based catalyst shows very favorable properties for water oxidation too. Photoinduced coordination chemistry is, therefore, a key to water oxidation. However, if abundant transition

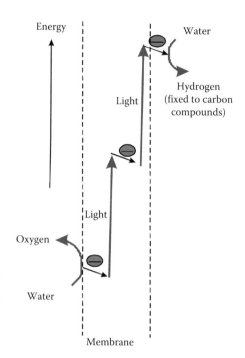

FIGURE 8.15
Photocatalytic water splitting is a four-electron transfer process (using eight photons for maximum solar energy capture) from water to protons leading to a hydrogenated carbon compound (in biology) or hydrogen (in technology). The main challenge is avoidance of energy-wasting radical chemical intermediates.

metals should be put to work, which cannot reach very high oxidation states, transition metal clusters will be required to accommodate four positive charges for water oxidation. This is obviously the way nature proceeded by evolving the manganese complex, which contains four manganese atoms.

A key precondition for oxygen evolution is, as we have seen, electron extraction via transition metal d-states, which are chemically able to react with water in a specialized catalytic environment. Similar properties are expected from calcium-containing manganate perovskites, which are presently under investigation. They provide holes on d-states and simultaneously are also characterized by a manganese-oxide-calcium molecular environment. Such a model system may help us to learn more about parameters, which nature has optimized toward the highly efficient manganese complex. It is expected that the ability of calcium manganates to get involved in cooperative charge carrier dynamics [44–46] will turn out to be crucial for water oxidation [47].

The efficiency of hydrogen generation via photovoltaic water electrolysis is [48]

$$\eta_{\text{phel}} = \frac{\Delta H_{\text{H}_2\text{O}}}{\Delta H_{\text{ph}} \left(1 + \dfrac{\text{losses}}{\Delta G_{\text{H}_2\text{O}}}\right)} \tag{8.4}$$

where
   $\Delta H_{\text{H}_2\text{O}}$ and ΔH_{ph} are the enthalpy changes due to water electrolysis and the photoreaction, respectively
   $\Delta G_{\text{H}_2\text{O}}$ is the free energy turnover during water electrolysis

From Equation 8.4, it can be deduced that the efficiency for hydrogen generation via photoelectrolysis can be maximized by making the technical–electrochemical losses small as compared to the Gibbs free energy change for water electrolysis ΔG_{H_2O}. If this can be done, the efficiency is simply the ratio of the enthalpy for water electrolysis and the enthalpy for the photo-generated charge carriers, which is determining the efficiency η_{phel} for hydrogen generation. It is recognized that by well matching the corresponding enthalpies and reducing trivial technical–electrochemical losses, the efficiencies for the photovoltaic hydrogen generation can approach 100%. In other words, nearly all photovoltaic energy generated by the solar cell can be converted into hydrogen energy, provided the catalytic conditions for water splitting are so favorable that the optimization needed can indeed be approached.

Equation 8.4 derived for photovoltaic hydrogen generation is significantly more optimistic than previous efficiency estimations for photovoltaic water splitting, which consider different cumulative secondary losses [49]. The relevance of relation (8.4) has been tested experimentally by using a combined silicon aluminum-gallium-arsenide tandem solar cell, which generated photovoltaic energy with 20% efficiency. It was directly fitted with catalytic metallic contacts consisting of platinum black and ruthenium dioxide for hydrogen evolution and oxygen evolution, respectively. Placed in contact with water, hydrogen and oxygen evolved resulting in an effective efficiency for hydrogen generation of 18.3%. This means that more than 90% of photovoltaic energy could indeed be converted into chemical energy of hydrogen [48]. For the development of practical, economic catalysts for water splitting, however, multielectron transfer has to be thoroughly studied [50]. This shows that the prospects of photovoltaic hydrogen are quite promising, but the scientific challenges for developing a commercially functioning technology are quite significant and complex, as a review of this subject shows [51].

8.15 Handling Water within a Bionic Energy Economy

Water is an integral element of the biological energy cycle (Figure 8.2) since it reacts to participate in energy conversion. Altogether, nature handles water in a sustainable way since it is taken from the environment and also given back to it. Nature has also learned to take advantage of the availability of water from its energy system. In deserts with extreme water shortage, there are numerous small animals for which it is sufficient to draw water from metabolism, that is, from their energy system. An example is the gerbil that never has to drink in spite of a dry grain diet. It obtains its water from splitting fat and from oxidizing hydrogen, which is attached to organic molecules. With this strategy, the gerbil could conquer vast desert regions, where other mammals were not able to survive.

Such a water recovery from the fuel cycle is evidently also possible within the bionic energy cycle (Figure 8.3). Especially when hydrogen could be generated economically on a large scale from wave or wind energy, there is an opportunity to recover water in a straightforward way for additional use. Would an American cover his 12 kW energy consumption from solar hydrogen, which turns over with oxygen to water, he would get 66 L of water daily. A European with approximately half the energy consumption would still liberate 33 L of water daily. By recovering water in the liquid form, they would, within this strategy, simultaneously reduce their energy consumption by 14%. This comes from the fact that less energy is needed when water is not transformed into vapor. Such advantage is already applied in the so-called condensing boiler.

The amount of 33–66 L of water daily from energy consumption is sufficient to cover the water needs of a man, considering that energy consumption will further increase. The

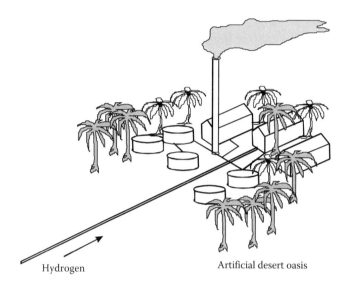

FIGURE 8.16
Water should, as a product of hydrogen energy use, be recycled for domestic and agricultural use around desert-based industrial centers.

average per capita water consumption today in Africa is 47 L and in Asia 85 L per day. It is, however, significantly larger in Germany (129 L) and the United States (295 L).

Man will continue to be engaged in energy-intensive activities, and may even intensify energy consumption. Where he brings sustainable energy he will also have water. This may help him to live in deserts where it would also be useful to transfer high-energy activities.

Hydrogen, which is generated in stormy oceans or later by wind energy in remote areas, will be pumped via pipelines into desert regions like the Sahara, the Arabian desert, or into the interior of Australia, where industrial colonies would be established. If metallurgical industries, like steel, titanium, or aluminum industries, or other high-energy technologies would be operated, people would simultaneously enjoy a reasonable degree of water supply from industrial and their own energy consumption (Figure 8.16). Starting from such water oases, it may be even possible to recover some desert land by planting and irrigating resistant plants.

8.16 Solar Energy and Tensile Water

More and more frequently, it is claimed that water will play a strategic role in our world. Already now there are countries that are competing and quarreling for water. This is not surprising. Not only people but also their cattle need water. For the production of 1 kg of wheat, 1000 L of water is needed. And wheat is not an especially water-consuming plant.

It appears that we are quite helpless in facing the water shortage, or we accept high-energy consumption for desalination of sea water. And doing so, we are faced with a fascinating natural water technology, which works at ambient technology and is practically unknown to our industrial society. High-growing trees like sequoia or eucalyptus can pump water more than 100 m high via solar evaporation. And well-adapted trees like mangroves can desalinate seawater or extract water from extremely dry or salty soil. Barely 5 m high eucalyptus trees in western Australia have been found to extract water with a solar-powered pulling force of 40–50 bar from the dry underground. In order to understand what this technology means for our ecosystem on earth, we should try to imagine how our world would look like without this astonishing technology. Trees would be only a few meters high. On dry and salty lands of our continents, there would be nearly no vegetation. It is, for example, known that in the Amazon rain forest, more than 50% of humidity is due to evaporation from trees.

The cohesion–tension mechanisms for the ascent of sap in plants, introduced between 1894 and 1895 [52,53], has been discussed and explored for more than one century. It is presently very actively supported by many plant physiologists [54–60] and has found access into practically all textbooks on plant physiology. Highlights for this theory have been measurements of sap ascent in boiled or poisoned shoots, the measurements of significant negative pressures in tall trees [61,62], and determinations of tree trunk circumferences that vary with evaporation activity. These findings support a mechanism by which water is pulled through the xylem conduits of trees by tension gradients produced by water loss through evaporation. Water could simply be seen as following a water potential gradient from more positive potentials in the underground environment of the roots to very negative potentials in the atmosphere around the treetops. Reversible thermodynamic mechanisms, based on thermodynamic quantities such as the water potential, are somehow activated and driven by water loss through evaporation. The tensile state of water that develops is essentially assumed to be a metastable side phenomenon of evaporative water loss from the tree capillary structure. This mechanism can, in principle, be simulated with a simple model, in which an evaporation body of ceramics is pulling a mercury column, in order to reduce the dimension of the experimental setup (Figure 8.17). The effective experimental height reached is, with less than 25 m, however, significantly lower than that reached by trees.

The ability of water to sustain tension can be understood by considering the properties of ice as compared with water. Ice is less dense than water and thus floating on it. Nevertheless, it is much stronger and can sustain tension. When water is diluted during evaporation from tree tissues, it similarly builds up higher strength because more hydrogen bonds are formed.

A critical classical question was how nature could have placed such an important technology on a phenomenon, which, within the frame of reversible thermodynamics, is scientifically considered to be highly unstable and sensitive against the rupture of water columns.

This view was confronted more recently with the results of a molecular kinetic model of tensile water behavior in trees by Tributsch et al. [63]. According to this hypothesis, tensile water is not simply a highly metastable state, on which plants would have to rely on, but a state of dynamic self-organization, driven by solar energy. When two tensile water molecules V' are interacting to generate an additional one from ordinary water A (feedback-induced activation of additional hydrogen bond)

$$A + 2V' \xrightleftharpoons[k_1,k_2]{} 3V' \tag{8.5}$$

and when it is additionally assumed that tensile water can be evaporated to vapor molecules B and react back through condensation,

$$V' \xrightleftharpoons[k_3,k_4]{} B \tag{8.5a}$$

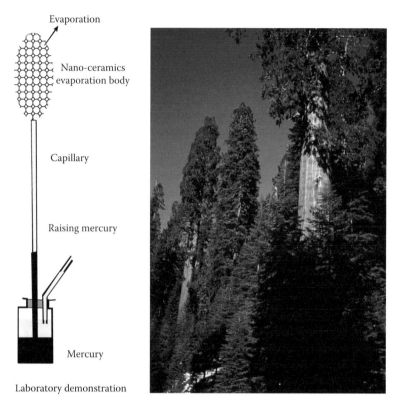

FIGURE 8.17
Under the expansion of water in micro- and nanostructured environment, due to solar evaporation, additional hydrogen bonds can be activated to provide tensile strength. It can be used to pull water to the tops of sequoia trees (figure to the right).

a stationary equation of the following form is reached:

$$V'^3 - V'^2 \frac{k_1 A}{k_2} + \frac{k_3}{k_2} V' - \frac{k_4 B}{k_2} = 0 \qquad (8.6)$$

It has exactly the same form of the empirical van der Waals equation for real gases including water, which also describes the buildup of negative pressures. However, the interpretation is different. Equations 8.5, 8.5a, and 8.6 describe a process of dynamic self-organization of water subject to a through-flow of energy (from solar evaporation). The autocatalytic mechanism involved has its origin in the dynamics of hydrogen bonding, activated by stretching water (Equation 8.5), and can explain the buildup of the tensile state, cavitation, oscillations, and a bistable state of evaporation, actually observed in trees [63]. One is dealing with a vapor machine subject to irreversible thermodynamics. The buildup of the tensile state and the transport of water in a tree can be compared with the (much more complicated) self-organization of a living organism. This explains also the relative stability of the system, as long as solar energy is supplying it with energy.

The tensile water technology is an energy technology, which is an integral part of nature's well-balanced energy strategy. Our world would look entirely different without this energy technology and would be much less favorable for sustaining life. This shows for one special energy technology that our present scientific knowledge is far from being sufficient for safeguarding our fragile environment into a sustainable future.

8.17 Household and Industrial Waste: With Hydrogen Back to Raw Materials

Within the biological energy-material cycle, most waste products are recycled. We say most because also nature knows final deposits for residues. Let us think of the mussel deposits in sediments that now form huge mountains, or the deposits of carbon and mineral oil. In our modern industrial society, attempts are made to recycle waste materials, but the largest portion of the waste is simply burned. Not only heat but also a huge amount of oxygen compounds are produced. This way one multiplies the waste and makes it even less useful and poorer in energy. Plastic burns to carbon dioxide, nitrogen compounds yield nitrogen oxides, iron yields rust, and silicon yields quartz sand.

In nature, waste is again upgraded. When termites, for example, digest decaying wood, and convert it with complex chemistry, they grow on it and can themselves later be eaten by other animals. In fact, remaining energy is extracted and used to upgrade chemical compounds. The carbon dioxide formed is, of course, equally recycled by the photosynthetic process.

Ultimately, one would have to add energy to waste products to make them useful again. This would succeed with civilization waste in an optimal way if waste would be treated with hydrogen at elevated temperatures in order to form hydrogen-containing products [64]. The hydrogen would transform the useless, energy-poor waste products into interesting raw materials. From plastic materials, methane or oil-like products would result, the nitrogen contained would yield ammonia, from rust pure iron would be formed, and from silicon the technically useful gas silane would be produced. Of course, technologies for separation and refining would also be needed. But nearly all products could be reused. The remaining, undesired products could be further transformed using concentrated solar radiation [64]. Solar energy could also provide the heat needed for hydrogenation processes.

Upgrading urban and chemical waste with hydrogen, heat and radiation from solar energy could become a technology inseparable from a sustainable energy future.

8.18 Energy Materials and Sustainability

Our industrial society is suffering from the constantly changing building materials. Plastic in numerous variations, asbestos, foam materials, and rock wool testify for experiments that are not yet completed. Much later we found out that materials pose problems with health, the environment, or degradation. Compared with nature, our experience is not sufficient because we do not take the time to wait until we understand the materials. It is therefore useful to look at nature's material strategy. Nature also made many inventions of new materials, but it had the opportunity to test them over and over. Only

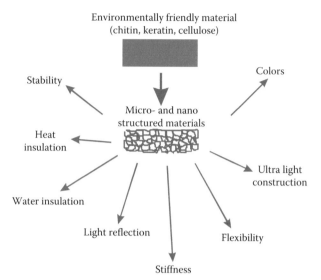

FIGURE 8.18
Nature uses few environment-friendly materials such as chitin, keratin, and cellulose that are tailored by micro- and nanostructuring toward functional perfection.

very useful inventions survived. Nature essentially limits its energy materials to a few environment-friendly, recyclable materials like chitin, keratin, and cellulose. Starting from these, nature forms, through nano- and microstructuring, an entire palette of functionally highly developed layers and structures (Figure 8.18).

These are, for example, the elaborate furs and feather clothes, which protect animals. These are also the elastic wings of birds, which take advantage of the power of the wind. Or they form the water-conducting structures of trees, which extract humidity from the ground with the help of solar energy. They also form the iridescent surfaces of beetles or butterfly wings, which show their colors because of regular microscopic patterns, which generate an interference of light. They are built of the same material as the fine feathers of the snowy owl, which protect this bird against the arctic cold, and simultaneously suppress flight noise. They make up the dense fur of the sea otter, which does not allow the water to permeate toward the animals skin.

Nano- and microstructuring of selected, useful materials for most diverse energy applications is a challenge, which also our civilization will have to face.

8.19 Nature's Ionic Electricity-Generation Technology

When looking back at the development of electricity-generation industry, with its impact on signal transmission, illumination, machines, and information technology including computers, we remain impressed by its enormous potential. Energy transmission and conversion here are based on driving electronic currents through electronic materials (Figure 8.19, top). These electronic materials give access to a diversity of applications and are still in the focus of dynamic scientific development. Remarkably, nature is basing essentially the equivalent technology on ionic currents and ionic materials (Figure 8.19, bottom). Ionic processes are mediating signal transmission and brain function. Proton currents are propelling rotating motors such as the flagellar propagation mechanism of bacteria. Electric energy is stored in proton gradients and proton currents are involved in

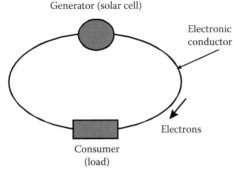

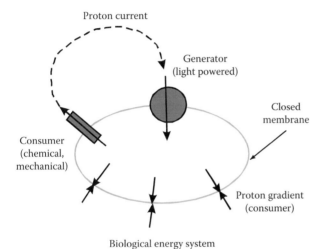

FIGURE 8.19

Technical power systems convert energy via electronic currents and biological systems via proton currents. Both, in terms of energy-converting systems and energy-storing materials, proton-conducting systems provide ample new opportunities.

the chemical fuel generating and converting infrastructure of metabolic circuits. Living nature has demonstrated the use of ionic energy technology within the largely aqueous environment of biological systems. The potential of ionic technologies of energy conversion is significant and we are just beginning to think about it, for example, on how to construct light-driven ion pumps [65]. Proton current technology offers new opportunities in many areas of energy applications. At the long term, industrial society should turn attention to it.

8.20 Energy Bionics: A Never-Ending Lesson

In all fields of energy technology, the wealth of bionic energy examples is really large and may occupy scientists during many decades to come. But energy bionics is only a very young research discipline and still has to find its path. Its philosophy is that by looking closer at nature we are learning faster than by merely relying on independent basic and applied research (which is also needed). Nature has developed an incredible technical fantasy to obtain energy from the sun and the environment, or to protect itself from surplus

FIGURE 8.20
In nearly every field of energy technology, nature sets standards: (a) solar architecture (termites, India), (b) thermal insulation (snowy owl), (c) daylight systems (window plants, Namibia), and (d) cooling systems (seals on skeleton coast of Namibia).

heat. In Figure 8.20a, few examples are shown just to indicate some areas of interest. Figure 8.20a shows a termite building from southern India. Insect populations have evolved sophisticated examples of air conditioning using ambient energy. The highly specialized feather coat of the snowy owl (Figure 8.20b) does not only provide a high standard of insulation against heat loss but also avoids any noise during flight and provides the white color as camouflage for the arctic environment. Figure 8.20c shows the window plant that relies on nonimaging optics, an only recently developed area of optics, for collecting, filtering, and distributing daylight. The seals from the Namibian coast (Figure 8.20d) usually avoid heat loss from their large fins by cooling the blood flow there via heat exchange systems. But on a hot desert beach, they apparently invert the heat flow and use their fins for cooling.

In addition, transparent insulation, an insulation that passes light but retains the heat by cutting off heat flow, is an ancient invention of nature. It is applied in the pelt of the polar bear as well as in hairy alpine and arctic plants that allow light scattering toward their surface, but avoid heat loss by providing many air pockets for insulation. One example is the catkin of the willow, which for this reason is flowering so early in the year. The further north the willow is growing, the smaller it becomes, and on the islands of Spitsbergen, it is just creeping over the ground to seek protection from wind and cold. Here the willow has invented an additional strategy against the cold. With little material, the down-like fibres which this plant sets free, it forms a kind of transparent blanket as protection. It allows the light to pass, but holds the air relatively still, so that heat has difficulty in escaping from the plant. Such ultralight transparent heat-insulating blankets have not yet found application in our building technology.

Recently, it has been studied why tree barks are cool even in the hottest summer [66]. They have to stay cool because the water under tension that is transported in the tree trunk

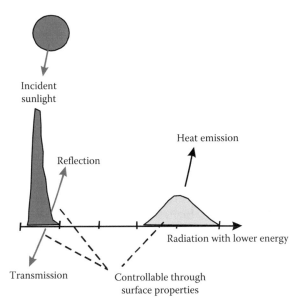

FIGURE 8.21
By balancing sunlight absorption and heat emission by selecting materials with appropriate optical properties, nature can efficiently heat or cool.

below the bark does not tolerate excessive heat. The barks date back 300 million years of evolution and are obviously optimized with respect to heat exchange. Figure 8.21 visualizes incident higher visible and infrared energy radiation and the emitted low-energy heat in an energy diagram and explains how this is done.

In order to keep the surface cool, the incident solar energy is maximally reflected and the generated heat is maximally emitted in the infrared region at the temperature the object reaches. In order to maximally emit heat energy in the infrared, a surface must, for thermodynamic reasons, also optimally absorb radiation energy there. This means that the barks must show strong absorption in the infrared (wave length 5–10 μm) where heat is emitted according to the Planck's law. Barks have indeed adapted to keep cool. In contrast, one can imagine that in order to heat up, the situation should be opposite. Incoming radiation should be maximally absorbed and infrared emission of heat should be inhibited. This can be done by selecting surface materials that cannot absorb radiation in the respective infrared region. Nature is apparently considering such conditions to keep living beings warm. But who is controlling the infrared properties of our buildings and technological structures in order to heat or cool them?

At the long-term, mankind will handle a large variety of sustainable energy technologies and energy tricks, which can complement each other. Only this synergetic cooperation of various technologies will ultimately lead bio-analogue strategies to full performance. We want to explain how this works with the example of the winter survival strategy of the beaver in northern forests. He survives as warm-blooded animal in his "wooden hut" in the middle of a dammed-up lake at winter temperatures of –50°C. How is he doing it? He is smearing his building from the inside with wet clay, which is freezing. This way, a real ice igloo is created. It is known that via its round form and the high degree of insulation provided by ice, an igloo provides an efficient protection against the cold. Then the beaver takes advantage of his excellent fur and amplifies its function by crowding tightly with the entire family. The damming of the lake around the beaver hut has in addition the advantage that water serves as heat storage and also liberates latent heat during the freezing process. When it gets very cold and when the fat reserve becomes smaller, the beaver leaves his hut under water in order to eat plant food, stored in the neighborhood. All technologies together make up the high-energy standard of his survival strategy.

Our industrial society also needs to learn a variety of energy technologies for dealing with the energy challenge.

8.21 Bionic Energy Strategy Compared with Other Visions

The concept that mankind should, at the long term, follow nature's example in dealing with a sustainable technology for fuels and chemicals has guided us to a straightforward model for handling the energy problem of the world. Sustainable carbon compounds from biomass gasification should be upgraded by hydrogen from a sustainable source to yield the desired fuels and chemicals. The search for high-density secondary solar energy pointed attention to wave power in violent oceans. The thus generated hydrogen is then to be attached to a carbon carrier derived from gasified biomass for fuel production. The presently available expensive infrastructure of the fossil fuel industry for the refining and distribution of oil products can thus essentially be maintained. When dealing with the challenge toward a sustainable energy for our world, we have to aim at technological innovation at the large scale. There is no easy way toward a sustainable energy economy. And, other energy options for the future pay a similar and even larger price. As a reference, fusion research should be mentioned. It has already lasted for 50 years and may need another 50 years before the first prototypes of power stations can be constructed. They are expected to be large and expensive and will be faced with problems like tritium toxicity, heat pollution, radioactive waste from neutron activation, and the problem of nuclear proliferation (plutonium breading in high neutron fluxes). Since many materials are involved and high neutron flux densities are generated, fusion reactors are expected to involve similar amounts of radioactive waste as conventional fission reactors. Also, the heat-exchanging confinement of the fusion plasma is an unsolved problem. Fusion reactors will also just produce heat, and the technologies leading to fuel production are costly and not very elegant. Compared to this long-term initiative, the here-suggested strategy for sustainable fuel production is, in comparison, straightforward, entirely compatible with environmental considerations, and well fitting into our present understanding of industrial chemical complexes. The other large-scale competing energy technology is conventional nuclear fission with improved reactors and maybe from nuclear breeders that will extend the use of uranium. This available nuclear technology is presently considered to be the most likely energy option besides solar technology by energy strategy planners. It also involves far-reaching consequences for the world and for mankind. More than 15,000 fission or breeder reactors will be needed, which will have to be replaced every 50 years. A worldwide, expensive, and highly sensitive infrastructure will be required for radioactive waste reprocessing, plutonium handling, and final waste depositing. There will be risks for nuclear technology from natural disasters like large earthquakes, terrorist attacks, and political turmoil. The last half century of nuclear technology with only 500 reactors and military interest in nuclear materials of just a few countries have already caused the contamination of 2% of the world's land surface for long periods. What will be the situation in 1000 years after 200,000 fission reactors may have been built and operated?

The prospect of affordable hydrogen from high-wave oceans would justify a 10- to 20-year effort to build up and optimize prototypes of hydrogen buoys (compare Figure 8.22).

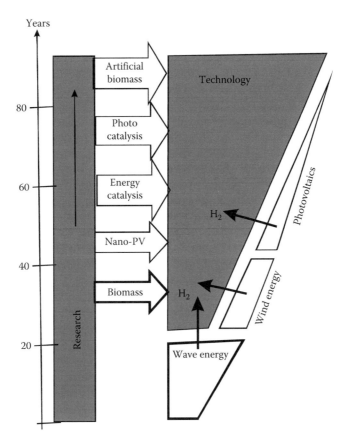

FIGURE 8.22
Timetable for the implementation of bionic solar energy technology by starting with wave-generated hydrogen and gasified biomass, and systematically learning required complementary technologies.

If the expected economic feasibility can be confirmed and investment promises to pay off, the future of this proposed fuel-generating energy technology could be straightforward. Mass production of energy buoys in the shipbuilding industry could be initiated and a lot of work could thereby be created worldwide. There would be additional favorable side effects. The growing need of biomass for fuel production with hydrogen will increase the presently too low price for wood and energy plant crops. This could stimulate the cultivation of energy crops worldwide and farmers could adjust to a long-term sustainable fuel boom. It would become economically more attractive to cultivate land for long-term sustainable use. The infrastructure for fuel distribution would essentially remain the same and industry presently active in the fossil fuel market could easily shift to sustainable fuels with increasing costs for oil and natural gas explorations. These companies could gradually shift such efforts in the direction of ocean-based platforms for hydrogen production.

There are sufficient ocean areas available toward the Antarctic and Arctic regions, where large, 4–6 m waves are present practically all the year round. They do not limit the supply of cheap hydrogen for sustainable fuel. However, it may be expected that, at the longer term, biomass will become too scarce. Then artificial or hybrid technologies for carbon dioxide fixation, which guarantee higher solar energy conversion efficiencies than traditional biomass should be exploited. One example, which already has been shown to work, consists in supplying solar energy synthesized simple energy carriers (e.g., FeS_x) to sulfide-oxidizing bacteria such as they exist in the deep sea around sulfide-emitting thermal vents (black smokers). Biomass efficiencies exceeding the efficiency of natural photosynthesis by one order of magnitude are feasible. Such artificial solar-powered chemosynthetic processes for producing bacterial biomass could be operated not only in desert regions with little water but also on sea platforms [27–29].

The floating hydrogen-generating buoy fields represent a technology that is inherently not polluting the sea. In contrast, such a technological infrastructure could even support ecosystems. The underwater connecting grid between the buoys could serve as a huge platform for fish farming, which should be productive in cold seas. A 1 GW hydrogen-generating platform could be a rectangle with edges 3–5 km in length. With many such platforms installed, there would be a large area available for cultivating fish and seafood, which anyhow will be needed for the still-growing world population.

The concept for sustainable energy supply proposed here is quite different from the concepts proposed after the first energy crisis 30 years ago. At that time, the vision was that photovoltaic panels in desert regions would supply photovoltaic electricity for water electrolysis and a hydrogen economy based on pure hydrogen. Today, we know that photovoltaic cells are still much too expensive for large-scale installations and that changing the infrastructure for fuel distribution will cost a very high price. In addition, it is also known that sandstorms pose problems and that a silicon solar cell heating from 10°C in the morning to 70°C hours later will lose one-third of their efficiency. The here-proposed strategy concentrating on high-density secondary solar energy offers a more cost-efficient and realistic approach, offering a smoother transition to a larger scale and economically sustainable energy future. Later, with more research into bio-analogue kinetically determined solar energy conversion, significantly cheaper photovoltaic cells and other solar energy-converting systems will add to a much larger diversity of energy-converting and hydrogen-producing technologies (Figure 8.22).

The strategy proposed here for wave-generated hydrogen energy involves technologies that can be developed in a reasonably straightforward way from established technological know-how. Still necessary research and development tasks in fields like construction engineering, biocorrosion resistance, seawater electrolysis, biomass gasification, and others appear to be realistic enterprises. A significant worldwide research initiative is simultaneously to be installed in order to investigate and implement all the additional technologies, which we eventually need in order to approach nature's elegant energy strategy (Figure 8.22). It may take from one half to one century to reach a reasonable standard and market penetration with nanophotovoltaics, catalysis of energy conversion, photocatalysis, and artificial biomass. However, the author sees no excuse for postponing the development of a bio-analogue sustainable energy technology into the more distant future while going on stressing the environment with fossil fuel technology and risking increasing political confrontations that face diminishing oil and gas reserves.

Altogether, the scientific–technological challenge may be explained via the scheme shown in Figure 8.23. The presently functioning, relatively modest research and development effort in solar energy technology is expressed by the upper section. The additional three sections have to be added as part of the bio-analogue, bionic approach. The left segment describes the efforts to get cheap hydrogen from high ocean waves and remote wind resources. The right segment describes the efforts needed for biomass upgrading and development of artificial biomass technology. The bottom segment summarizes efforts needed to investigate and implement bio-analogue energy technologies like energy

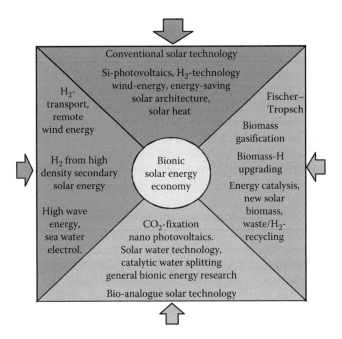

FIGURE 8.23
Scheme visualizing research needs toward a bionic solar energy economy. While present solar energy technologies (upper segment) should be further developed, cheap hydrogen from wave and wind energy (left segment) will enable a dynamic fuel industry based on upgrading conventional and artificial biomass (right segment). A variety of additional bionic technologies, ranging from catalytic water splitting to nanophotovoltaics and tensile water technology will gradually improve the standard and economics.

conversion catalysis, solar catalytic water splitting, nanophotovoltaics, carbon dioxide fixation, and a larger variety of biological energy experiences. All these segments together only will, at the long term (probably one century), lead to a functioning bionic energy economy. Figure 8.23 also shows why the present solar energy efforts are not sufficiently convincing for decision makers. The upper segment alone would not have helped nature to implement its well-functioning energy strategy. It needed much more technologies, and, most important, the experience to integrate them into a working, sustainable flow system for energy and materials.

8.22 Summary and Discussion

On the basis of reasonable arguments, it would be highly desirable if industrial civilization would decide to ultimately continue nature's sustainable energy strategy on a highly technological level. It consists in splitting water and attaching the hydrogen to carbon compounds for the generation of a wide spectrum of fuels and chemicals. In a bio-analogue technological strategy, the carbon source will be photosynthetically derived from biomass, and, at the long term, necessarily also from artificial photosynthetic sources. The sustainable generation of cheap hydrogen will be the biggest near-term challenge. The different possibilities are discussed and the conclusion is reached that hydrogen from stormy oceans, generated by buoys harvesting the energy of big waves, could become economical within a comparatively short period of two to three decades. The necessary technology is basically available or could be elaborated with adequate investments. The products of synthesis of carbon compounds from biomass with hydrogen can take advantage of the presently functioning infrastructure for fossil fuel distribution and would perfectly fit into the competence of petrochemical industry. This is a significant strategic advantage since its huge and elaborate infrastructure would be available and could continue to function. A comparatively smooth transition to a sustainable fuel economy is expected. Several new technologies will have to be learned from nature and adapted to modern industrial standards. A strategic plan of necessary steps toward such a goal is given, which may be reached within one century (Figure 8.22). Detailed research and technological challenges have been listed in Ref. [67].

The political difficulties, encountered with the development of largely untested, mostly nuclear, new energy technologies, discussed at the beginning, may be largely avoided if the need to follow the experience of evolution could be convincingly communicated. The recently founded International Renewable Energy Agency (IRENA) could support scientific and industrial activities in individual countries and coordinate sustainable energy efforts internationally. But a high standard of research, with experienced specialized laboratories performing standard control measurements, will be necessary to avoid setbacks in credibility and funding after incorrect statements of progress in renowned journals, as it repeatedly happened in difficult, and therefore little attractive research fields like photocatalytic water splitting [68,69]. The prospects of a bionic energy future appear to be very promising, even though mankind would be faced with significant scientific challenges. People would understand the need to develop these efforts, considering that presently most research is not aimed at securing the survival but is serving purely commercial or military aims. The fact that the processing and distribution infrastructure for fuels will essentially remain the same, while the fuels will change from fossil to sustainable, offers straightforward opportunities for financing the industrial development of the bionic energy technology via carbon credits.

Carbon dioxide-emitting industries will pay for credits as long as they are nominally consuming nonsustainable fuel. With accumulating payments, they will become nominally and increasingly sustainable using the same chemical fuel. Simultaneously, the fuel-producing industry will use the money for increasing the fraction of sustainable fuel.

The actual global carbon cycle in our world ecosystem in gigatons per year is shown in Figure 8.24a. It is seen how fossil energy consumption and deforestation is presently contributing to the increase of carbon dioxide in the atmosphere.

Considering that energy consumption during the present century will increase by a factor of 3, and considering a full-grown bionic energy economy by the end of the century, we are expecting a carbon cycle as depicted in Figure 8.24b. More carbon will be cycling in the atmosphere and its carbon dioxide content will be slightly higher, but a sustainable situation would have been reached. It will not be very different from a situation where much more forests would be present on the earth because energy industry is modeled according to plant life.

The bionic energy system will, as already indicated, in the long term provide the means to control the carbon content in the atmosphere and thus the climate. This is done via a fuel sink, a huge underground storage system, as indicated in Figure 8.10, which can be filled or emptied, depending on the climate requirements. When less carbon is desired in the atmosphere, more carbon-free hydrogen will be used for energy purposes.

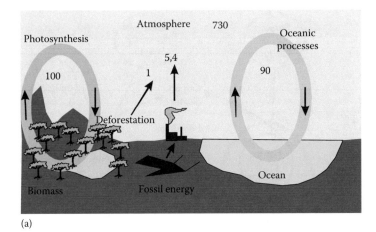

(a)

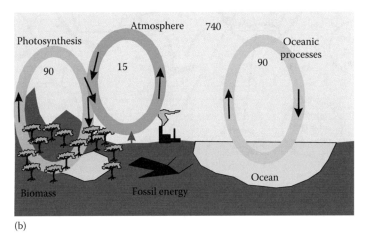

(b)

FIGURE 8.24

The carbon cycle (in gigatons per year) for our present world (a) and after implementation of bionic energy economy at the end of the present century with three times more energy consumption (b).

Altogether, mankind should, via the bionic energy strategy, ultimately approach the sustainable biological energy system as shown in Figure 8.25. The core functions, as depicted by the two ellipses that represent energy production and consumption, can superficially be imitated by the bionic energy scheme of Figure 8.10, which will be the first target.

But nature is much more sophisticated with diverse additional technologies (spheres in Figure 8.25) that are implemented to support and amplify the core activities. As our analysis has shown, in certain areas entirely new research disciplines will have to be initiated in order to develop similar skills.

Mankind should study and apply these technologies, and add others from human creativity, which fit into the energy strategy of biological evolution. The bionic strategy is in scientific and technological respect highly challenging, but, considering the potential of industrial civilization, it is a realistic goal. It is perhaps the only energy strategy where we can be sure not to be faced with fundamental mistakes and setbacks, because nature has tested it for so many million years.

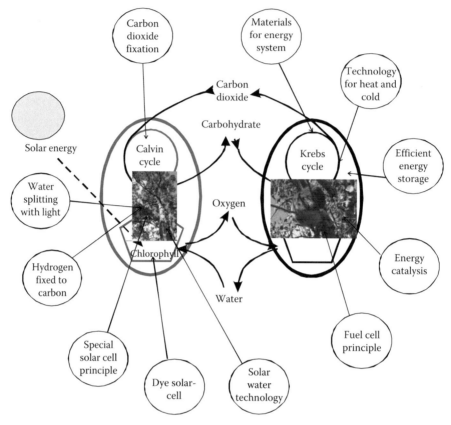

FIGURE 8.25

The complex energy and materials flow system of living nature (Figure 8.2) is recalled indicating the most important supporting energy technologies. In principle, man should associate his energy-generating systems to plant life and his energy-consuming urban and industrial activities to animal life without interfering with the global ecological balance.

References

1. Bockris, J.O.M. 1980. *Energy Options*, Australian & New Zealand Book Company, Sydney, New South Wales, Australia.
2. Shell International Limited. 2001. *Global Business Environment (PXG), Energy Needs, Choices and Possibilities, Scenarios to 2050*, London 2001.
3. IEA. 2004. International Energy Agency, *World Energy Outlook* 2004, http://www.iea.org/textbase/nppdf/free/weo2004.pdf
4. Deutsche Physikalische Gesellschaft. 2005. *Klimaschutz und Energieversorgung in Deutschland* 1990–2020, Bad Honnef, Germany, September 2005.
5. Heinloth, K. 1997. *Die Energiefrage*, Vieweg Verlag, Wiesbaden, Germany.
6. Kleinknecht, K. 2007. *Wer im Treibhaus sitzt-Wie wir der Klima-und Energiefalle entkommen*, Piper, München, Germany.
7. Scheer, H. 1999. *Solare Weltwirtschaft, Strategie für die Ökologische Moderne*, Verlag Antje Kunstmann, München, Germany.

8. Scheer, H. 2005. *Energieautonomie. Eine neue Politik für erneuerbare Energien*, Verlag Antje Kunstmann, München, Germany.

9. Lehmann, H., Reetz, T. 1995. *Zukunftsenergien. Strategien einer neuen Energiepolitik*. Berlin, Germany.

10. Nitsch, J., Leitstudie. 2007. Ausbaustrategie Erneuerbare Energie. Erstellt für das Bundesministerium für Umwelt (Germany), Naturschutz und Reaktorsicherheit.

11. Welt im Wandel. 2003. Energiewende zur Nachhaltigkeit, Wissenschaftlicher Beirat der Bundesregierung, Globale Umweltveränderungen, Springer Verlag, Berlin, Germany, ISBN: 3-540-40160-1.

12. von Weizsäcker, E.U., Lovins, A.B., Lovins, L.H. 1995. Factor Four: Doubling Wealth, Halving Resource Use, Earthscan Publications Ltd. 1999.

13. Lovelock, J. 2006. *The Revenge of Gaia. Why the Earth Is Fighting Back*. Penguin Books, London, U.K.

14. IPCC. 2007. *Intergovernmental Panel on Climate Change, Special Report*, Carbon dioxide capture and storage. Metz, B., Davidson, O., deConinck, H., Loos, M., Meyer, L. (Eds.), Cambridge University Press, Cambridge, U.K., ISBN: 92-9169-119-4.

15. Nachtigall, W. 1974. *Phantasie der Schöpfung. Faszinierende Entdeckungen der Biologie und Biotechnik*. Hoffmann und Campe, Hamburg, Germany.

16. Tributsch, H. 1976. Wie das Leben leben lernte. *Physikalische Technik in der Natur*, DVA, Stuttgart, Germany.

17. Nachtigall, W., Blüchel, G. 2001. *Das grosse Buch der Bionik. Neue Technologien nach dem Vorbild der Natur*, DVA, Stuttgart, Germany.

18. Blüchel, G, Malik, F. (Herausgeber). 2006. Faszination Bionik, Die Intelligenz der Schöpfung, Malik Management Zentrum, St. Gallen, Switzerland, www.bionik.de

19. Wienecke-Janz, D. (Ed.), 2008. *Spektrum: Bionik, Vorbild Natur in Leben und Technik*, Wissen Media Verlag, Gütersloh, Germany, 2008.

20. Barthlott, W., Neinhuis, C. 1997. Purity of the sacred lotus, or escape from contamination in biological surfaces. *Planta* 202: 1–8.

21. Swissnuclear Dokument. 2007. http:www.swissnuclear.ch/documents/Erneuerbare%20 Energien%20und%20neue%20Nuklearanlagen,%20Teil%206.pdf (accessed October 2007).

22. State of Hawaii, DBEDT, Strategic Industries Division. 2002. *Feasibility of Developing Wave Power as a Renewable Energy Resource of Hawaii*, January 2002, http://www.state.hi.us/dbedt/ert/ wavereport/wavereport.html

23. Hagermann G. 1992. Wave energy resource and economic assessment for the State of Hawaii prepared by SEASUN Power Systems for the Department of Business, Economic Development, and Tourism, Final Report, June 1992.

24. Thorpe, T.W. 1992. A review of wave energy Volume 1 and 2, ETSU report R 72, December 1992; Economic Analysis of Wave Energy Devices, *Third European Conference*, Patras, Greece, 1998.

25. NASA, Jet Propulsion Laboratory, http:/topex-www.jpl.nasa.gov/gallery/science.html (retrieved 2007).

26. Izumiya, K., Akiyama, E, Habazaki, H., Kumagai, N., Kawashima, A., Hashimoto, K. 1998. Anodically deposited manganese oxide and manganese-tungsten oxide electrodes for oxygen evolution from seawater. *Electrochim. Acta* 43(21): 3303–3312.

27. Tributsch, H. 1979. Solar bacterial biomass bypasses efficiency limits of photosynthesis. *Nature* 281: 555–556.

28. Bilal, B.R., Tributsch, H. 1998. Thermo-electrochemical reduction of sulfate to sulfide using a graphite cathode. *J. Appl. Electrochem.* 28: 1073–1081.

29. Tributsch, H. 2003. Coupling bio-geochemical processes to regenerative energy for an industrial carbon cycle. *Hydrometallurgy* 71(1–2): 293–300.

30. Tributsch, H., Rojas-Chapana, J. 2006. Bacterial strategies for obtaining chemical energy by degrading sulfide minerals. In *The Practice, Theory and Microbiology of Biomining*, Rawlings, D.E., Barrie Johnson, D. (Eds.), Springer Verlag, Germany, Chapter 13, pp. 263–280.

31. Rojas-Chapana, J., Tributsch, H. 2004. Interfacial activity and leaching patterns of *Leptospirillum ferrooxidans* on pyrite. *FEMS Microbiol. Ecol.* 47: 19–29.
32. Bogdanoff, P., Herrmann, I., Hilgendorff, M., Dorbandt, I., Fiechter, S., Tributsch, H. 2004. Probing structural effects of pyrolyzed CoTMPP-based electrocatalysts for oxygen reduction via new preparation strategies. *J. New Mater. Electrochem. Syst. (JNMES) Canada* 7: 85–92.
33. Tributsch, H. 2007. Multi-electron transfer catalysis for energy conversion based on abundant transition metals. *Electrochim. Acta* 52(6): 2302–2316.
34. Tributsch, H., Koslowski, U., Dorbandt, I. 2008. Experimental and theoretical modelling of Fe-, Co-, Cu-, Mn-based electrocatalysts for oxygen reduction. *Electrochim. Acta* 53(5): 2198–2209.
35. Sariciftci, N.S., Heeger, A.J. 1997. In *Handbook of Organic Conductive Molecules and Polymers*, Nalwa, H.S. (Ed.), Vol. 1. Wiley, New York.
36. Tributsch, H. 1972. Reaction of excited chlorophyll molecules at electrodes and in photosynthesis. *Photochem. Photobiol.* 16: 261; Tributsch, H. 2004. Dye sensitization solar cells: A critical assessment of the learning curve. *Coord. Chem. Rev.* 248: 1511–1530.
37. O'Regan, B., Grätzel, M. 1991. A low-cost, high-efficiency solar cell based on dye-sensitized colloidal TiO_2 films. *Nature* 353: 373.
38. Würfel, P. 2005. *Physics of Solar Cells. From Principles to New Concepts.* Wiley, Weinheim, Germany.
39. Tributsch, H. 2009. Nano-composite solar cells: The requirement and challenge of kinetic charge separation. *Solid State Electrochem.* 13: 1127–1140; doi: 10.1007/s10008-008-0668-2.
40. Marcus, R.A. 1956. On the theory of oxidation-reduction reactions involving electron transfer. *J. Chem. Phys.* 24: 966; 1991. Theory of electron-transfer rates across liquid-liquid interfaces. 2. Relationships and application. *J. Phys. Chem.* 95: 1050.
41. Tributsch, H., Pohlmann, L. 1992. Far from equilibrium cooperative electron transfer: The energetic advantage. *Chem. Phys. Lett.* 188: 338; 1995. Synergetic molecular approaches towards artificial and photosynthetic water photoelectrolysis. *J. Electroanal. Chem.* 396: 53; 1997. Synergetic electron transfer in molecular electronic and photosynthetic mechanisms. *J. Electroanal. Chem.* 438: 37; 1998. Electron transfer: Classical approaches and new frontiers. *Science* 279: 1891.
42. Pohlmann, L., Tributsch, H. 1992. Stimulated and cooperative electron transfer in energy conversion and catalysis. *J. Theor. Biol.* 155: 443; 1992. A model for cooperative electron transfer. *J. Theor. Biol.* 156: 63; 1994. A model for stimulated and cooperative electron transfer. In *On Self-Organization*, Springer Series in Synergetics, Mishra, R.K. Maaß, D., Zwierlein, E. (Eds.), Vol. 61, Springer Verlag, Berlin, Germany, p. 133; 1997. Self-organized electron transfer. *Electrochim. Acta* 42(18): 2737.
43. Kühne, H.M., Tributsch, H. 1983. Oxygen evolution from water mediated by infrared light on iron doped RuS_2 electrodes. *J. Electrochem. Soc.* 130: 448.
44. Dagotta, E. 2003. *Phase Separation and Colossal Magneto-Resistance.* Springer Series in Solid State Sciences, Springer, Berlin, Germany.
45. Fiebig, M., Miyano, M.K., Tomyoka, Y., Tokura, Y. 1998. *Science* 280: 1925–1928.
46. Jooss, Ch., Wu, L., Beetz, T., Klie, R.F., Beleggia, M., Schofield, M.A., Schramm, S., Hoffmann, J., Zhu, Y., 2007. Polaron melting and ordering as key mechanisms for colossal resistance effects in manganites. *Proc. Natl. Acad. Sci. USA* 104(34): 13597–13602.
47. Leidel, N., Ikeda, S., Jooss, Ch., Fiechter, S., Tributsch, H. 2010. $CaMnO_3$ as model system for the $CaMn_4O_x$ photosynthetic oxygen evolution center. Submitted to *J. Phys. Chem.*
48. Licht, S., Wang, B., Mukerji, S., Soga, T., Umeno, M.,Tributsch, H. 2000. Efficient solar water splitting, exemplified by RuO_2-catalyzed AlGaAs/Si photoelectrolysis. *J. Phys. Chem. B.* 104: 8920.
49. Bolton, J.R., Strickler, S.J., Conolly, J.S., 1985. Limiting and realizable efficiencies of solar photolysis of water. *Nature* 316: 459.
50. Tributsch, H. 2007. Multi-electron transfer catalysis for energy conversion based on abundant transition metals. *Electrochim. Acta* 52: 2302–2316.
51. Tributsch, H. 2008. Photovoltaic hydrogen generation. *Int. J. Hydrogen Energy* 33: 5911–5930; doi: 10.1016/j.ijhydene.2008.08.017.

52. Dixon, H., Joly, J. 1894. On the ascent of sap. *Ann. Bot.* 8: 468–470; On the ascent of sap. *Phil. Trans. Royal Soc. London B* 186: 563–576.

53. Askenasy, E. 1895. *Über das Saftsteigen*, Verhandlungen des Naturhistorischen Medizinischen Vereins, Heidelberg, NF 5, Germany, pp. 325–345.

54. Sperry, J.S., Saliendra, N.Z., Pockman, W.T., Cochard, H.E., Cruziat, P, Davis, S.D., Ewers, F.W., Tyree, M.T. 1996. New evidence for large negative xylem pressures and their measurement by the pressure chamber method. *Plant Cell Environ.* 19: 427–436.

55. Tyree, M.T. 1997. The cohesion-tension theory of sap ascent: Current controversies. *J Exp. Bot.* 48: 1753–1765.

56. Tyree, M.T. 2003. The ascent of water. *Nature* 423: 923.

57. Steudle, E. 1995. Trees under tension. *Nature* 378: 663–664.

58. Steudle, E. 2001. The cohesion-tension mechanism and the acquisition of water by plant roots. *Annu. Rev. Plant Physiol. Mol. Biol.* 52: 847–875.

59. Milburn, J.A. 1996. Sap ascent in vascular plants: Challengers to the cohesion theory ignore the significance of immature xylem and the recycling of Münch water. *Ann. Bot.* 78: 399–407.

60. Cochard, H., Ameglio, T., Cruiziat, P. 2001. The cohesion theory debate continues. *Trends Plant Sci.* 6: 456.

61. Scholander, P.F., Hammel, H.T., Bradstreet, E.D., Hemmingsen, E.A. 1965. Sap pressure in vascular plants. *Science* 148: 339–346.

62. Kappen, L., Lange, O.L., Schulze, E.-D., Eventari, M., Buschbom, U. 1972. Extreme water stress and photosynthetic activity of the desert plant. *Artemsia herba-alba* Asso. *Oecologia* (*Berlin*) 10: 177–182.

63. Tributsch, H., Cermak, J., Nadezhdina, N. 2005. Kinetic studies on the tensile state of water in trees. *J. Phys. Chem. B* 109: 17692–17707.

64. Tributsch, H. 1989. Feasibility of toxic chemical waste processing in large scale solar installations. *Solar Energy* 43(3): 139–143.

65. Tributsch, H. 2000. Light driven proton pumps. *Int. J. Ionics* 3&4: 161–171.

66. Henrion, W., Tributsch, H. 2009. Optical solar energy adaptations and radiative temperature control of green leaves and tree barks. *Solar Energy Mat. Sol. Cells,* 93: 98–107.

67. Tributsch, H. 2008. *Erde, wohin gehst du? Solare Bionik-Strategie: Energie-Zukunft nach dem Vorbild der Natur,* Shaker Media, Aachen, Germany.

68. Duonghong, D., Borgarello, E., Grätzel, M.J. 1981. Dynamics of light-induced water cleavage in colloidal systems. *J. Am. Chem. Soc.* 103: 4685.

69. Khan, S.U.M., Al-Sgahry, M., Ingler, W.B. 2002. Efficient photochemical water splitting by a chemically modified n-TiO$_2$. *Science* 297: 2243.

9

Synthetic Hydrocarbon Fuels from Lignocellulosic Biomass

Georg Schaub and Kyra Pabst

CONTENTS

9.1 Introduction: Background and Driving Forces

In an era of increasing oil prices and climate concerns, biofuels have gained attention as nonfossil secondary energy. Governments have become interested in securing the supply of raw materials and limiting climate change, many innovative proposals have been made and development work has started for innovative candidate fuels.

Chemical upgrading of biomass to high-value fuels such as liquid or gaseous hydrocarbon mixtures appears as an attractive route. These fuels allow the use of present infrastructures such as gas stations for liquid fuels or pipeline systems for natural gas. Lignocellulosic biomass as feedstock helps to avoid competition with food production if residues (or waste) from food-producing agriculture systems are used. In addition, wood, forestry residues, and energy crops can also serve as lignocellulosic feedstocks.

Today's supply of liquid fuels for automobiles worldwide depends almost completely on petroleum (more than 95%, total flow around 2×10^9 t/year). In the near future, the demand for fuels will increase, as the number of automobiles increases with the present growth of economies and prosperity. There will be pressure on the supply of fuels for two reasons.

Petroleum has become more expensive and will increasingly cause political and military conflicts. Its production will pass a maximum for conventional qualities in the foreseeable future, while at the same time CO_2 concentrations in the atmosphere will increase causing climate change effects. Therefore, significant changes regarding the raw material basis and processing technologies are necessary, and new fuels produced preferably from renewable raw materials will be needed to complement or replace the present petroleum-based fuels. Even if—as a mid- or long-term solution—electromobility will appear as a new alternative, liquid fuels for internal combustion systems will remain dominant for mid- to long-distance car applications for a longer period of time.

As for gaseous hydrocarbon mixtures (natural gas), there are extended gas grids with pipeline systems distributing large quantities to households and industries. This infrastructure can be used for distributing bioenergy in the form of substitute natural gas (SNG), either as a heating fuel or as a transport fuel for automobiles.

Given the situation as described, there is presently a debate about which fuels from lignocellulosic biomass with their yield potentials and characteristic advantages appear most attractive. The present contribution discusses various biofuels produced via thermochemical conversion of biomass raw materials, with special emphasis on synthetic hydrocarbon mixtures, either liquid or gaseous. Since the required amounts are very high, both quality and yield aspects are significant.

9.2 Value of Fuels and Lignocellulose as Raw Material

Chemical fuels can be assessed according to various criteria: characteristic properties with respect to combustion/utilization, storage and handling, raw material availability, and production cost (Table 9.1). Energy density, heating value, and specific CO_2 emission values are directly linked to the elemental analysis and contents of the main constituents carbon, hydrogen, and oxygen. Figure 9.1 gives an overview of the hydrogen/carbon and oxygen/carbon ratios of various fuels. Included are both fossil and nonfossil fuels or raw materials and lines for constant higher heating values (HHV; Figure 9.1a). The HHV, according to the reaction enthalpy released during combustion (Equation 9.1), can be approximated on the basis of the elemental composition carbon/hydrogen/oxygen, according to Equation 9.2 (Boie 1953, Meunier 1962). This approximation can also be used to estimate CO_2 emissions per thermal energy (in kg/GJ) released when burning a fuel (Figure 9.1b and Table 9.1).

$$CH_xO_y + \left(1 + \frac{x}{4} - \frac{y}{2}\right)O_2 \rightarrow CO_2 + \frac{x}{2}H_2O \quad \Delta_RH < 0 \tag{9.1}$$

TABLE 9.1

Significant Criteria for the Assessment of Chemical Fuels

Energy density (per mass or volume), heating value
C-content (specific CO_2 emission), minor constituents (process complexity for cleaning)
Application potential, flexibility, handling, combustion characteristics
Transportation characteristics, storage
Efficiency in production/upgrading process
Availability of raw material/resources
Production cost, price, market potential

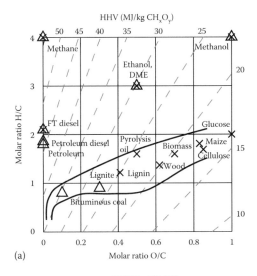

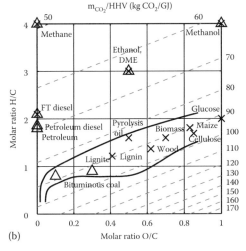

FIGURE 9.1
Molar ratio H/C and O/C of fossil and renewable fuels. (a) With correlation of HHV. (Based on Boie, W., *Energietechnik*, 3, 309, 1953.) (b) With correlation of specific CO_2 emission (based on stoichiometry). Triangle, fossil; cross, renewable; full lines, range of coalification.

$$\Delta_R H^0_{298} = -422.5 - 117.2x + 177.5y \quad (kJ/mol) \tag{9.2}$$

$$HHV = -\frac{\Delta_R H^0_{298}}{M_{CH_xO_y}} \quad (kJ/kg) \tag{9.3}$$

$$\frac{m_{CO_2}}{HHV} = \frac{44,000}{(422.5 + 117.2x - 177.5y)} \quad (kg/GJ) \tag{9.4}$$

Among the biofuels listed in Table 9.2, only ethanol and biodiesel (fatty acid methyl ester, FAME), known as "1st Generation" biofuels, are presently applied in significant quantities. They are most commonly added to petroleum fuels in their pure form or as ether (ethanol as ethyl-tertiary-butyl-ether, ETBE), and fit with actual engine technologies. As they are presently produced from sugar and starch or natural fatty acid esters, respectively, the raw materials compete with other applications (such as food or feed).

TABLE 9.2

Overview of Liquid and Gaseous Fuels Generated from Lignocellulose and Their Properties

	Units	Pyrolysis Oil	FTS Diesel	SNG	Hydrogen	Petro Diesel (Reference)
Chemical compounds		Aromatic, phenolic, alkyl, carboxyl CH_xO_y	Hydrocarbon (alkane) C_nH_{2n+2} $n = 10–20$	Hydrocarbon CH_4	Molecular H_2	Hydrocarbon (alkane, aromatic) C_nH_{xn} $n = 10–20+$ $x = 0.5–2$
Composition	CH_xO_y	$CH_{1.6}O_{0.5}$	$CH_{2.1}$	CH_4	H_2	$CH_{1.86}$
Boiling temperature[a]	°C	>150	160–380 (var)	−161.5	−252.8	180–360 (var)
Mass density[b]	kg/L	1.0	0.77	0.135[c]	0.017[c]	0.84
Energy density[b]	MJ/L	10.0	34.3	6.4[b]	1.9[c]	35.0
Application		FU	ICE	ICE/FC	ICE/FC	ICE
Other significant properties: vaporization, ignition, combustion, chemical stability						

[a] At 1 bar.
[b] At 15°C.
[c] At 200 bar.

Biofuels produced from lignocellulose via either hydrolysis/fermentation or thermo-chemical conversion are far more flexible with respect to the raw materials used and are less critical with respect to competition with food or feed applications. They are called "2nd Generation" biofuels, with some representatives indicated in Table 9.2. Listed are pyrolysis oil and Fischer–Tropsch (FT) diesel fuel as liquid representatives, and SNG and hydrogen as gaseous fuels (with petroleum-derived diesel fuel as reference). Of particular significance may be synthetic liquid hydrocarbon fuels (e.g., FT diesel), which due to their high energy density and handling characteristics are favored in mobile applications (cars, airplanes, and ships). Experience is available from large-scale production, handling, and utilization of FT fuels produced from coal and natural gas feedstocks (Schaub 2006).

Lignocellulose is the most abundant kind of biomass on earth (annual growth rate ca. 150 Gt/year, terrestrial inventory ca. 4000 Gt). It presently serves as raw material source for the pulp and paper industry for cellulose (global cellulose production 150 Mt/year), for natural fibers or for the building sector. As it cannot serve as food for humans or feed for animals, it represents a potential fuel source, either in its raw form (for combustion) or after chemical and biotechnological upgrading. The constituents of lignocellulose are cellulose, hemicellulose, and lignin, with cellulose being predominant for wood, grass, and straw (Table 9.3). Its heating value correlates with the elemental analysis CH_xO_y (Table 9.4). The structural elements determining chemical or biological reactivity include saturated ring systems, typical of cellulose and hemicellulose polysaccharide species, as well as lignin with aromatic rings connected via alkyl groups (Figure 9.2).

Cellulose is a high molecular weight polymer composed of units of D-glucose ($C_6H_{10}O_5$) bound together in a linear structure by ether-type linkages and hydrogen bonds. Hemicellulose is a more complex heterogeneous polymer exhibiting a branched molecular structure based on two-to-four (and occasionally five-to-six) different sugar units such as D-xylose or D-glucose. The degree of polymerization is generally from 50 to 200 as compared to 7000 to 10,000 of native cellulose. Xylan is the predominant hemicellulose of all

TABLE 9.3

Raw Material Lignocellulose and Typical Composition

	Units	Softwood	Hardwood	Straw (Barley, Wheat)	Maize (Gavott/Doge)
Content	wt.%				
Cellulose[a]		47 ± 2	45 ± 2	42–47	34–40
Hemicellulose[a]		27 ± 2	30 ± 5	35–39	45–53
Lignin[a]		28 ± 3	20 ± 4	15–16	11–12
H_2O-soluble		3 ± 2	5 ± 3	7–12	16–18
Molar ratio	—	$CH_{1.37}O_{0.62}$			$CH_{1.81}O_{0.83}$
Heating value HHV[b]	kJ/kg	20.3			18.0

Source: From Howard, J.B. et al., Pyrolysis of biomass, *Proceedings, Joint Meeting of Chemical Engineering, Chemical Industry and Engineering Society of China*, American Institute of Chemical Engineers, Beijing, China, September 19–22, Vol. II, pp. 638–652, 1982.

Cellulose: $CH_{1.69}O_{0.85}$ (HHV = 17.2 MJ/kg); lignin: $CH_{1.22}O_{0.41}$ (HHV = 24.9 MJ/kg).

[a] Percent of extracted biomass (H_2O).

[b] Per mass dry organic fraction CH_xO_y (Boie 1953).

TABLE 9.4

Fossil and Renewable Compounds in Organic Raw Materials and Fuels (Calculated with Equations 9.2 through 9.4)

CH_xO_y	Name	Molar Mass (g/mol)	$\Delta_R H$ (kJ/mol)	HHV (MJ/kg)	LHV (MJ/kg)	$m_{CO_2}/\Delta_R H$ (kg/GJ)
CH_4	Methane	16.0	891.3	55.7	51.2	49.4
CH_2	n-Alkane	14.0	656.9	46.9	43.5	67.0
$CH_{1.8}$	Petroleum	13.8	633.5	45.9	42.6	69.5
$CH_{1.4}$	Bitumen	13.4	586.6	43.8	40.7	75.0
$CH_{0.8}O_{0.1}$	Bituminous coal	14.4	498.5	34.6	32.0	88.3
$CH_{1.6}O_{0.7}$	Biomass	24.8	485.8	19.6	17.2	90.6
$CH_{1.37}O_{0.62}$	Wood	23.3	473.0	20.3	18.0	93.0
$CH_{1.69}O_{0.82}$	Cellulose	26.8	475.0	17.7	15.4	92.6
$CH_{1.22}O_{0.41}$	Lignin	19.8	492.7	24.9	22.4	89.3
$CH_{1.67}O_{0.83}$	Carbohydrate	27.0	470.9	17.5	15.2	93.4
CH_2O	Glucose	30.0	479.4	16.0	13.7	91.8
$CH_{1.81}O_{0.83}$	Maize	27.1	487.3	18.0	15.6	90.3
$CH_{1.85}O_{0.125}$	Lipid	15.9	617.1	38.9	35.8	71.3
$CH_{1.58}O_{0.32}$	Protein	18.7	550.9	29.5	26.7	79.9

$\Delta_R H$, heat of combustion reaction; HHV (LHV), higher (lower) heating value.

hardwoods, accounting for 25–35 wt.%. Xylan is a polymer of the pentose sugar, D-xylose, and can contain some carboxylic acid and methyl-ether groups.

Lignin is an aromatic three-dimensional polymer based primarily on phenylpropane units. It acts as a cementing agent for the cellulose and hemicellulose fibers in wood. The complex chemical structure of lignin has been the subject of many investigations (Freudenberg and Neish 1968, after Howard et al. 1982). There is presently a significant interest in producing base chemicals from lignin, a renewable raw material available in large quantities.

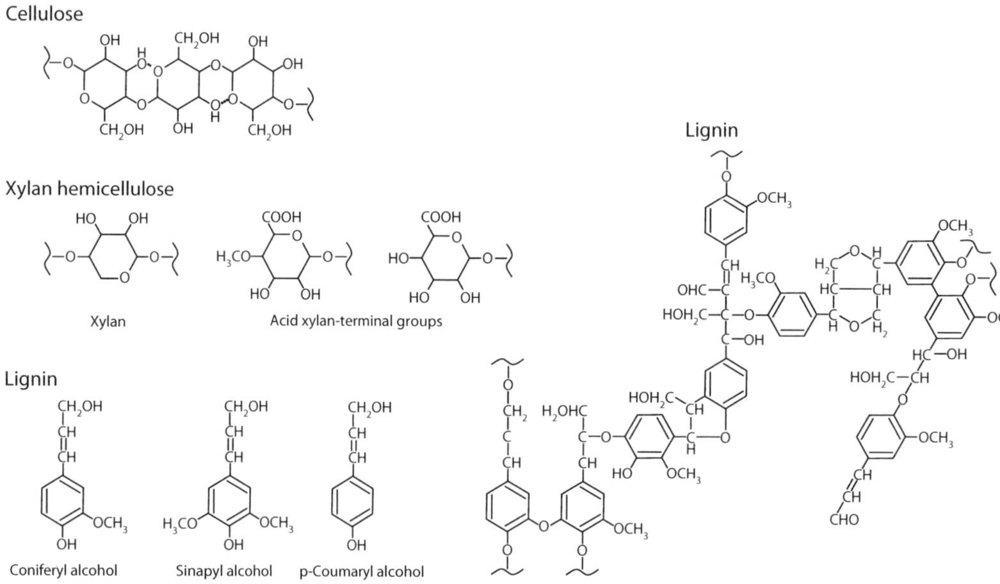

FIGURE 9.2
Structural elements in lignocellulose molecules.

The availability of lignocellulose as raw material is limited by the net productivity of photosynthesis (NPP), which depends on a combination of geographical factors and the kind of vegetation. In particular, the most significant limiting factors for NPP are water availability, intensity of incoming solar radiation, soil quality, type of plant (C_3 or C_4), and temperature (Larcher 2003). Plant growth is a highly unsteady process and follows the cyclic changes of day/night and the seasons. The energy efficiency of photosynthesis (as the ratio of heating value of biomass versus incoming solar radiation) is limited by the albedo of the leaves, quantum yields, and the photochemical efficiency. The values achieved in natural plants are low, that is, with a theoretical maximum of ca. 6%, since other criteria besides energy efficiency dominate in biological systems, for example, repair mechanisms of the cell. The energy efficiencies of photosynthesis achieved in temperate climates are generally below 1%. A look at the global distribution of vegetation indicates that water availability is a significant limiting factor.

The actual crop yield and the resulting heating value data for Germany (as an example of temperate climate zone) are listed in Table 9.5. Higher yield values can be achieved in warmer climates and also for aquatic plants (like water hyacinth or seaweed).

The energy density achievable in agriculture and forestry (Table 9.5) must be seen in comparison with the average intensity of the incoming solar radiation (which has an average of 168 W/m² or 53 GJ/ha × year over time and surface area of the globe, Kiehl and Trenberth, 1997, after IPCC 2007). The carbon flows involved in photosynthesis represent a significant contribution to the global carbon cycle, as well as the oxygen flows generated contribute to sustaining an environment on earth that is beneficial to life. The examples shown in Table 9.5 include purposely grown energy crops, for example, miscanthus and maize. Combinations with multiseasonal wood plants in so-called agroforestry systems may be interesting with respect to biodiversity criteria. Since thermochemical conversion accepts varying biomass qualities and materials as feedstock, agriculture and forestry can operate with significant flexibility.

TABLE 9.5

Annual Dry Matter Production in Temperate Climate Agriculture and Forestry with Common Agriculture Practice, and Chemical Energy Yields (with Heating Value HHV = 16–20 MJ/kg Dry Matter)

	Yield[a] (t/ha × Year)	Energy Yield (GJ/ha × Year)
Wheat (whole plant)	10.5–17.5	168–350
Oilseed rape (whole plant)	8.5–12	136–240
Maize (whole plant)	11–19	176–380
Miscanthus	10–30	160–600
Wood	4–18	64–360

Source: After FNR, Bioenergy—Plants, raw materials, products, Report Fachagentur Nachwachsende Rohstoffe, 2nd edn., www.fnr.de, 2009.

[a] Dry organic matter CH_xO_y.

Achieving a high yield potential by good agricultural practice (up to about 30 t dry matter per hectare) will only be possible in areas with high water availability and high-quality soils. Further addition of water, for example, via irrigation, can be performed only to a limited level, since the supply of drinking water for the world population, the supply of process water for industry, and maintenance of minimum water levels in rivers are all competing with one another.

9.3 Thermochemical Routes for Biomass Conversion to Fuels

9.3.1 Overview

Thermochemical conversion requires thermal or catalytic activation of biomass molecules at elevated temperatures and reaction without or with gaseous reactants (O_2, H_2O, CO_2, H_2). Usually, the subsequent upgrading or conversion of primary products is also carried out at elevated temperatures, often in the presence of a solid catalyst. All these reactions are summarized as thermochemical conversion. Wide experience with fossil organic raw materials (hard coal, lignite, and peat) has grown since the beginning of the industrial time period. This experience with fossil solid feedstocks can be transferred to biomass feedstocks because there are important analogies although ultimate analysis and detailed characteristic properties are different. Chemical pathways, product selectivities, and feedstock reactivity can be understood in terms of chemical composition (Table 9.3, Figure 9.2).

An overview of the most important thermochemical conversion routes for lignocellulosic biomass is given in Figure 9.3. Products are either individual compounds (e.g., methanol, methane) or gaseous and/or liquid mixtures (e.g., synthesis gas and gaseous/liquid hydrocarbons). Idealized formulations of the chemical reactions involved, together with values for heat of reaction per mole of carbon are listed in Table 9.6. Heat effects are quantified on the basis of an empirical correlation for HHV to be calculated with molar ratios H/C and O/C of any organic material (see the caption of Table 9.6, Boie 1953, after Meunier 1962). Also included is the formation reaction of biomass (photosynthesis). For the quantitative treatment, lignocellulosic biomass is assumed to be represented by the stoichiometric formula $CH_{1.6}O_{0.7}$.

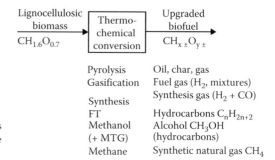

FIGURE 9.3

Overview of thermochemical conversion routes for lignocellulosic biomass feedstocks (see Abbreviations).

TABLE 9.6

Idealized Thermochemical Reactions for the Production of Liquid and Gaseous Fuels from Lignocellulose and Related Enthalpy Changes (Examples), and Photosynthesis/Combustion as Reference

				$\Delta_R H^0_{298}$ per Mole of C[a] (kJ/mol)
Photosynthesis	$CO_2 + H_2O$	$\xrightarrow{h\nu}$	$CH_2O + O_2$	+479.4
	$CO_2 + 0.8\,H_2O$	$\xrightarrow{h\nu}$	$CH_{1.6}O_{0.7} + 1.05\,O_2$	+485.8
Pyrolysis[b]	$CH_{1.6}O_{0.7}$	$\longrightarrow$	$0.3\,C + 0.55\,CH_{1.6}O_{0.455} +$ $0.15\,CO + 0.3\,H_2O$	−17.0
Gasification	$CH_{1.6}O_{0.7} + 0.3\,H_2O$	$\longrightarrow$	$CO + 0.95\,H_2$	+68.7
	$CH_{1.6}O_{0.7} + 0.15\,O_2$	$\longrightarrow$	$CO + 0.8\,H_2$	+25.8
FT synthesis	$CO + 2\,H_2$	$\longrightarrow$	$-(CH_2)- + H_2O$	−185.7
CH_4 synthesis	$CO + 3\,H_2$	$\longrightarrow$	$CH_4 + H_2O$	−250.1
Combustion	$CH_{1.6}O_{0.7} + 1.05\,O_2$	$\longrightarrow$	$CO_2 + 0.8\,H_2O$	−485.8

[a] Heat of reaction based on empirical correlation for HHV of CH_xO_y (Equation 9.2) for defined molecular compounds values from Probstein and Hicks (1990).

[b] Stoichiometry: own estimate based on reported product distribution (Henrich et al. 2009).

9.3.2 Pyrolysis

Thermal decomposition (or pyrolysis) of an organic material occurs if the material is exposed to a temperature high enough for thermal activation and breaking of chemical bonds. It normally means thermal decomposition in the absence of an external reactive gas and a catalyst. The extent of conversion and product distribution are affected by reaction time, temperature–time history, pressure, and solid particle size. There is a strong analogy between pyrolysis of coal particles (Howard 1981) and pyrolysis of lignocellulosic biomass (Howard et al. 1982, Nunn et al. 1985, Bridgewater 2003). Primary products of pyrolysis are very complex mixtures, with aromatic and aliphatic structural elements, wide varieties of molecular size, gases, liquids, and solids.

Of particular interest for the production of liquid fuel products are pyrolysis processes carried out at high heating rates followed by rapid quenching of the products ("fast pyrolysis"), where secondary reactions of medium-size molecules (liquids) are suppressed. An example process uses a solid heat carrier that during mixing heats up the biomass feedstock rapidly to 700°C, with rapid vapor–solid separation and quenching (Henrich and Dinjus 2003). The reaction stoichiometry in Table 9.5 is based on the reported product distributions (Henrich et al. 2009). The liquids generated should contain about 80% of the chemical heat content of the lignocellulosic raw material. Together with the solid char, they are envisaged to serve as a slurry feedstock for subsequent gasification to synthesis gas (to be further processed to synthesis products). If the primary liquid is supposed to be used as a liquid fuel, it has to be (hydro)processed using sophisticated and costly upgrading processes, if today's quality requirements for engine fuels have to be met (Bridgewater 2007, Elliott 2007).

In case a solid fuel for co-combustion in coal burners or entrained flow gasifiers is the preferred product, a variation of the pyrolysis process (called torrefaction) is envisaged. Mild (i.e., low-temperature) pyrolysis should lead to a large fraction of solid char product, easy to grind and to pulverize (Kiel et al. 2009).

9.3.3 Gasification

Gasification aims at the production of gases from solid carbonaceous material such as wood and solid residues from agriculture or forestry. This process has been applied for a long time with coal and peat of various kinds as feedstock (Hebden and Stroud 1981, Higman and van der Burgt 2003). The gasification process occurs via pyrolysis reactions and subsequent heterogeneous (gas–solid) reactions of the solid residue from pyrolysis with O_2, H_2O, CO_2, and H_2 as reactive gases (Johnson 1981, Schaub and Reimert 2003). The separate reactions for the reactants most commonly used (O_2, partial oxidation; H_2O, steam gasification) are represented as endothermic reactions in Table 9.6. The products of the overall reaction are CO/H_2 mixtures containing additional gases:

$$CH_{1.6}O_{0.7} + a\,O_2 + b\,H_2O \rightarrow c\,CO + d\,H_2 + e\,CO_2 + f\,CH_4 \tag{9.5}$$

Feedstock conversion, as well as yield and composition of the product are determined by the combined effects of feedstock properties and reaction conditions, according to chemical thermodynamics, kinetics of individual reactions, and stoichiometry. For example, as in pyrolysis, the yields of liquid byproducts (e.g., tar, phenols) depend on the reaction conditions that prevail during pyrolysis and secondary reactions of the volatile products (temperature, residence time, temperature–time history, etc.).

The most important degree of freedom in the design of gasification reactors is the gasification temperature with oxygen input as the key independent variable in autothermal gasification (Figure 9.4). Higher reactor temperatures result from higher feed rates of oxygen because exothermic combustion reactions become more important. As a result of increased gasification rates, residence time of the feedstock particles may become shorter, thus allowing entrained flow reactor systems with a characteristic residence time of only seconds (as compared to minutes in the case of moving bed reactors and fluidized bed reactors in between). Since temperature strongly affects the secondary reactions of primary pyrolysis products, tar compounds resulting from pyrolysis reactions can be completely converted into gaseous products at a high-enough temperature. Thus, generation of a tar-free gas is possible at the expense of synthesis gas yield, as shown in Figure 9.5. The loss in synthesis gas yield corresponds to increased yields of oxidation products CO_2 and H_2O. Other effects on synthesis gas composition can be attributed to the CO shift reaction (water gas shift) and CH_4 reforming reaction, both being affected by temperature and pressure. Table 9.7 lists the characteristic compositions of synthesis gases from biomass for different gasification reactor systems. An overview of the recent progress in biomass gasification process development was presented by Knoef (2005).

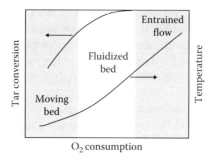

FIGURE 9.4

Reactor principles of different gasification processes and correlation between oxygen input, tar conversion, and gasification temperature in autothermal operation (schematic), analogous to coal gasification data. (Adapted from Schaub, G. and Reimert, R., Gas production from coal, wood and other solid feedstocks, In *Ullmann's Encyclopedia of Industrial Chemistry*, pp. 357–380, 6th edn., Vol. 15, Wiley VCH, Weinheim, Germany, 2003; Unruh, D. et al., *Energy Fuels*, 24, 2634, 2010.)

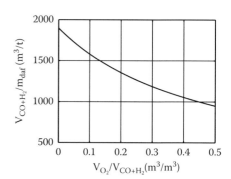

FIGURE 9.5

Effect of oxygen consumption on synthesis gas yield in autothermal gasification of biomass (estimate based on stoichiometry and combustion reactions in Table 9.6).

TABLE 9.7

Characteristic Compositions of Synthesis Gas (Dry) from Biomass in Entrained Flow and Fluidized Bed Gasification Processes

| Gasification Process | Pressure | Medium | $x_{i,dry}$ (vol %) | | | | x_{H_2}/x_{CO} |
			H_2	CO	CO_2	CH_4	
Entrained flow[a]	Elevated	O_2	30	50	17	0.1	0.60
Entrained flow[b]	Atm/elev.	O_2	35.2	41.1	22.0	0.1	0.86
Fluidized bed[c]	Atm	H_2O	35–45	22–25	20–25	5–12	1.5–2
Fluidized bed[d]	Atm	H_2O	16–27	24–43	10–25	9–16	0.6–1.2

Atm/elev., near ambient/elevated pressure.
[a] Henrich and Dinjus (2003).
[b] Althapp (2003).
[c] Hofbauer et al. (2001) and Potetz (2008).
[d] Van der Meijden et al. (2007, 2008).

9.3.4 Synthesis Gas Conversion

Synthesis gas, produced today in large quantities from natural gas, petroleum fractions, and coal, is considered to be one of the important base chemicals. It is either converted to organic chemicals like methanol and further products of C1 chemistry, to waxes, hydrogen, or to liquid or gaseous fuels (hydrocarbons). Today's synthesis gas chemistry

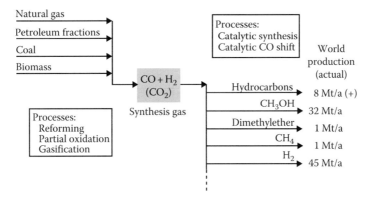

FIGURE 9.6
Present principal production and conversion routes for synthesis gas: Overview. (Based on Arpe, H.J., *Industrielle Organische Chemie*, Wiley VCH, Weinheim, Germany, 2007.)

is represented schematically in Figure 9.6, including raw materials and synthesis gas production processes, and catalytic synthesis processes, with catalytic CO shift conversion in cases where H_2/CO ratios have to be increased. Production routes for hydrocarbons (except for CH_4), either to alkene molecules or to liquid hydrocarbon fuels (diesel fuel, gasoline, kerosene) occurs via FT synthesis reactions (Schulz 1999). This will be discussed in more detail in Section 9.4 and methane synthesis (where no carbon–carbon bonds have to be formed) will be discussed in Section 9.5.

9.3.5 Analogy Biomass and Fossil Fuels Upgrading

In all thermochemical routes for biomass conversion to fuels, the analogy with fossil feedstocks can be seen. Process principles and underlying chemistry, as well as practical research, development, and operational experience gained in fossil fuel upgrading should be extensively used. Thus, biomass conversion can strongly take advantage of the existing technical knowledge from fossil fuel technologies (Elliott 1981, Schmalfeld 2008).

9.4 Liquid Hydrocarbons via Fischer–Tropsch (FT) Synthesis

FT synthesis has gained considerable interest in the past decades for the conversion of natural gas and coal into high-value liquid hydrocarbon fuels (Schulz 1999, 2003, Van der Laan and Beenackers 1999, Bartholomew and Farrauto 2006). The availability of cheap natural gas, for example, as associate gas in oil production, has given new momentum to synthesis technologies first developed in the early- and mid-twentieth century. Biomass as feedstock has been considered as potential feedstock since political aims were defined in the European Union regarding the introduction of biofuels in the transportation sector (Tijmensen et al. 2002, EC Directive 2003, Schaub et al. 2004, Schaub 2006). The particular attraction to biomass for energy use is that synthetic hydrocarbon fuels can be produced from any kind of biomass (energy crops, residues, waste) and the product is compatible with the present liquid fuels supply and distribution infrastructure.

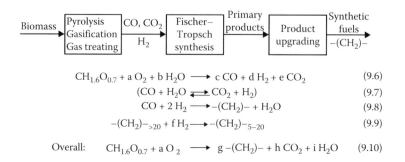

$$CH_{1.6}O_{0.7} + a\,O_2 + b\,H_2O \longrightarrow c\,CO + d\,H_2 + e\,CO_2 \qquad (9.6)$$

$$(CO + H_2O \rightleftharpoons CO_2 + H_2) \qquad (9.7)$$

$$CO + 2\,H_2 \longrightarrow -(CH_2)- + H_2O \qquad (9.8)$$

$$-(CH_2)-_{>20} + f\,H_2 \longrightarrow -(CH_2)-_{5-20} \qquad (9.9)$$

Overall: $$CH_{1.6}O_{0.7} + a\,O_2 \longrightarrow g\,-(CH_2)- + h\,CO_2 + i\,H_2O \qquad (9.10)$$

FIGURE 9.7

Flow scheme for the conversion of lignocellulosic biomass feedstocks to liquid hydrocarbon fuels via FT synthesis.

In the overall process (Figure 9.7), synthesis gas produced from biomass raw materials has to be cleaned from impurities and adjusted to the needs of the subsequent synthesis step. Here, the H_2/CO ratio has to be increased by means of the water-gas-shift reaction. Primary products from FT synthesis commonly are upgraded by a hydrogen treatment (hydrocracking, isomerization) to produce a high-quality product suitable as fuel for modern internal combustion engines. If diesel fuel with molecular carbon numbers in the range of 10 to 20 is the desired product, present strategies achieve maximum yields by a two-step approach. During synthesis, long-chain waxy hydrocarbon molecules (C_{21+}) are produced as intermediates that, during product upgrading, are converted by catalytic hydrocracking to high-quality diesel fuel. The simplified overall reaction (Equation 9.10 in Figure 9.7) indicates the significance of O_2 consumption during gasification: the more O_2 is consumed, the more CO_2 is formed, resulting in lower overall yields of hydrocarbon products.

FT chemistry includes polymerization reactions with $-CH_2-$ as the monomer, produced on a solid catalyst surface. They mainly lead to linear alkane and alkene hydrocarbons, ranging from methane to high molecular weight waxes. The characteristics of these reactions that occur on the surface of the solid catalyst (Equation 9.8 in Figure 9.7) are summarized in Table 9.8.

The reaction mechanism is dominated by a competition between hydrocarbon chain growth and chain termination (i.e., desorption from the catalyst surface, Figure 9.8). FT catalysts are able to favor hydrocarbon chain growth and slow down desorption reactions, with detailed reaction mechanisms depending on the particular catalyst and on reaction conditions (temperature, syngas composition, and pressure) (Schulz 2005). Carbon number distribution of FT products approximately follows a simple statistical model with a chain growth probability parameter α, analogous to a representation of ideal polymerization (Figure 9.9). The higher the α value, the higher is the probability of growth of an alkyl

TABLE 9.8

Characteristics of FT Synthesis Reaction

Polymerization on solid catalyst surface, monomer: $-CH_2-$ ("nontrivial")
Catalyst: Fe, Co (specific interaction with C, surface activation required)
Product mixture: n-alkanes, 1-alkenes (minor: i-alkanes, O-compounds)
T: 180°C–250°C (low-T, Co mainly), 300°C–360°C (high-T, Fe), P: 10–60 bar

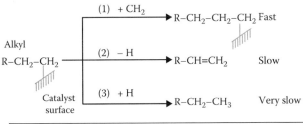

(1) Chain growth, (2) Desorption as alkene, (3) Desorption as alkane

FIGURE 9.8
Reaction pathways for alkyl species on catalyst surface and relative rates at low-temperature conditions. ("Fischer-Tropsch-regime". With kind permission from Springer Science+Business Media: *Topics Catal.*, Major and minor reactions in Fischer-Tropsch synthesis on cobalt catalysts, 26, 2003, 73, Schulz, H.)

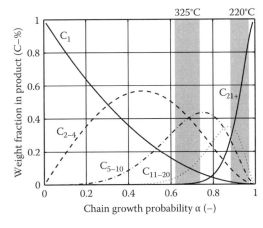

FIGURE 9.9
Idealized composition of primary FT product according to Anderson–Schulz–Flory distribution. (Adapted from Dry, M., *Catal. Today*, 6, 183, 1990; Sie, S.T., *Rev. Chem. Eng.*, 14, 109, 1998.)

species adsorbed on the catalyst surface, as compared to desorption into the gas phase. Low-temperature synthesis (180°C–250°C), which is mainly considered for liquid fuel production, exhibits α values in the range of 0.85–0.96, whereas in high-temperature synthesis (300°C–360°C) α values are in the range of 0.6–0.75.

As catalysts for FT synthesis, Co and Fe are considered to be the most attractive. Both are able to catalyze hydrocarbon chain growth and slow down desorption reactions, although the detailed reaction mechanisms are different (Schulz 2003). The most significant features of both catalysts are summarized in Table 9.9. Considerable effort has been made to improve catalyst activity and selectivity, especially in the case of Co, given the high material cost. One important strategy for catalyst improvement strategies focuses on identifying the optimum cobalt crystallite size/dispersion on particle or monolithic supports.

Reactor design in FT synthesis is dominated by the need to remove the exothermic heat of FT reactions (see Table 9.6). For low-temperature conditions, multi-tubular fixed bed reactors and three-phase slurry reactors are used today in commercial-scale plants, and fluidized bed reactors (circulating or bubbling) for high-temperature conditions, as indicated in Figure 9.10. A systematic approach to reactor design and scale-up must include, besides syngas injection and dispersion strategies, coolant selection and heat transfer management for temperature control, as well as a proper catalyst design (Krishna and Sie 1994). Large reactor size (up to 20,000 barrels per day or 0.5 million tons per year

TABLE 9.9

Significant Features of Cobalt and Iron Catalysts in FT Synthesis (Simplified and General Trends Only)

	Cobalt	Iron
Common FT principle		Chain growth favored compared with (slow) desorption reactions
Evolution of FT regime	Reconstruction of metal surface	Carbide formation
FT reaction rates	Higher	Lower
Min synthesis T	Lower	Higher
Chain growth	Higher	Lower
Increasing T	α decreases, CH_4 becomes dominant product FT regime released	α decreases, no undue CH_4 formation
Increasing p_{H_2}/p_{CO}	Selectivity changes to CH_4 mainly	No undue CH_4 formation
Increasing p_{H_2O}	Less CH_4, longer chains	Reaction rate lower
Alkali promotion	"No" effect	Essential for good activity, stability, selectivity
CO/CO_2 shift activity	Not existing (almost)	Fast reaction
Alkene secondary reactions	Significant	Marginal
Catalyst stability	Higher	Lower

Source: Based on Schaub, G. et al., *Erdöl, Erdgas, Kohle* 120, 327, 2004. With permission.

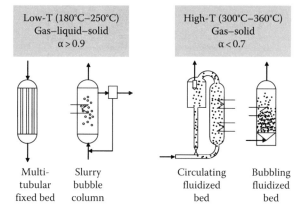

FIGURE 9.10
Multiphase reactors used in industrial FT processes. (Adapted from Sie, S.T. and Krishna, R., *Appl. Catal. A Gen.*, 186, 55, 1999.)

Low-T (180°C–250°C) Gas–liquid–solid $\alpha > 0.9$ — Multi-tubular fixed bed / Slurry bubble column

High-T (300°C–360°C) Gas–solid $\alpha < 0.7$ — Circulating fluidized bed / Bubbling fluidized bed

and possibly higher) together with a complex cooling equipment require significant knowledge in the design, manufacture, and operation of synthesis reactors. In addition, significant advances have been made in the field of reaction engineering science. As an innovative approach, monolithic catalyst configurations have been proposed (Güttel et al. 2008) in order to achieve better mass and heat transfer conditions. Considering the overall process, however, the FT synthesis step is less significant for capital investment than synthesis gas production (see Section 9.6).

The final hydrocarbon fuel products exhibit favorable properties for low-pollutant-forming combustion in modern car engines. FT diesel fuel has practically no aromatics and no sulfur content, ensuring low emissions of NO_x and soot. Table 9.10 summarizes the most important characteristic property ranges for FT and petroleum-derived diesel fuels.

TABLE 9.10

Characteristic Property Ranges for Conventional Petroleum Diesel Fuel (European Standard) and FT Product

	Units	Petroleum Diesel (Reference)		FT Diesel	
		Range[a]	Example[b]	Range[c]	Example[b]
Content					
Total aromatics	wt.%	n.a.	28	<1	0.14
Polycyclic aromatics	wt.%	<11	n.a.	n.a.	n.a.
Sulfur	ppm wt	<10	8	<5	<1
H/C molar ratio	—	n.a.	1.8	>2	2.1
Cetane number	—	>51	53	>60	75
Lower heating value	MJ/kg	n.a.	43.1	>43	43.8
Density (20°C)	kg/L	0.82–0.86[d]	0.83	<0.8	0.77
T 10% volatile[e]	°C	[f]	221	[g]	187
T 15% volatile	°C		354		321

Note: n.l., no limit; n.a., data not available.

[a] DIN EN 590 (2004).
[b] Herrmann et al. (2006).
[c] ReNew (2008) and Sie et al. (1991).
[d] 15°C.
[e] ASTM D86 distillation.
[f] <65% volatile at 250°C, >85% volatile at 350°C.
[g] Variable.

Besides emission reduction, there is potential for fuel consumption reduction, if internal combustion processes are adjusted to optimum-designed ("taylor-made") FT fuels, for example, with respect to rate of vaporization and rate of flame propagation. Achieving final product properties according to desired specifications (distillation curve, cold flow properties, cetane index, etc.) is possible due to today's highly developed petroleum refining processes for hydrocarbon upgrading (hydrocracking, isomerization, etc.). The flexibility of these processes for designing product fuel composition and properties explains the interest of car and engine manufacturers. Different designs of synthesis product upgrading schemes are needed if different combinations of fuel yield/fuel specifications are desired (De Klerk 2009, Leckel 2009).

In addition to FT synthesis, a chemical route to liquid hydrocarbon fuels via methanol exists, which is known as methanol-to-gasoline (MTG, Figure 9.11, Maxwell and Stork 2001). Here, methanol is an intermediate, produced in a high-selectivity synthesis process. The advantageous aspects of this route include (a) methanol as versatile and storable intermediate, to be used also in chemical industry, (b) potential flexibility of the final MTG process to produce attractive petrochemicals (ethene, propene), and (c) process complexity appearing less significant than in FT synthesis.

Synthesis gas treatment and subsequent synthesis are similar for various solid feedstocks and for gasification (biomass, coal). In the case of coal as feedstock, practiced since the 1930s, minor constituents of the product gas that may damage FT catalysts are removed in high-efficiency scrubbing processes. In the case of natural gas as feedstock, no synthesis gas cleaning is required, and no adjustment of H_2/CO ratios is necessary either, because the feedstock contains sufficient hydrogen.

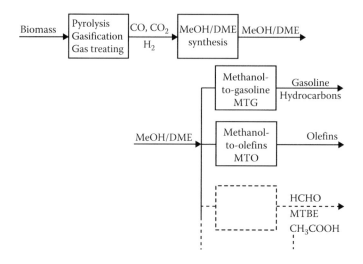

FIGURE 9.11
Flow scheme for the conversion of lignocellulosic biomass to methanol/dimethylether and flexibility to produce hydrocarbon fuels or petrochemicals. (Adapted from Arpe, H.J., *Industrielle Organische Chemie*, Wiley VCH, Weinheim, Germany, 2007.)

9.5 Gaseous Hydrocarbons via Methane Synthesis

Besides liquid hydrocarbon fuels, methane can be produced from synthesis gas, to be used as SNG replacing natural gas. The overall process biomass-to-SNG is represented in Figure 9.12. In contrast to biomass-to-liquid (BtL; Figure 9.7), methane is produced in a one-step synthesis process, typically integrated with CO shift in order to provide enough H_2 for synthesis according to Equation 9.13 in Figure 9.12. Methane synthesis from gases produced by coal gasification was investigated intensively in the 1980s, when SNG based on gasification of coal was considered as a potential long-term replacement for natural gas (Probstein and Hicks 1990). A large-scale plant was built at that time in North Dakota, generating an equivalent of 2 GW (HHV in SNG, Dakotagas 2009). Additionally, methanation is commonly employed as a gas purification step in hydrogen or ammonia production to convert low levels of CO, in order to avoid poisoning effects by CO in subsequent catalytic

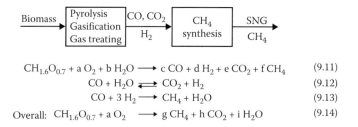

$$CH_{1.6}O_{0.7} + a\,O_2 + b\,H_2O \longrightarrow c\,CO + d\,H_2 + e\,CO_2 + f\,CH_4 \qquad (9.11)$$
$$CO + H_2O \rightleftarrows CO_2 + H_2 \qquad (9.12)$$
$$CO + 3\,H_2 \longrightarrow CH_4 + H_2O \qquad (9.13)$$
$$\text{Overall: } CH_{1.6}O_{0.7} + a\,O_2 \longrightarrow g\,CH_4 + h\,CO_2 + i\,H_2O \qquad (9.14)$$

FIGURE 9.12
Flow scheme for the conversion of lignocellulosic biomass to methane (SNG).

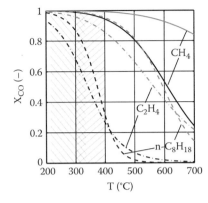

FIGURE 9.13

Equilibrium conversion of CO in synthesis reaction to different products. Full lines, CH_4 (Equation 9.13 in Figure 9.12); dotted lines, C_2H_4; broken lines, $n-C_8H_{18}$, feed ratio $H_2/CO = 3$; gray, 50 bar; black, 1 bar; shaded areas, temperature range FTS 210°C–330°C, SNG 300°C–400°C.

processes. Biomass conversion to SNG has been proposed by various research groups and companies in the recent past (Wokaun 2009).

The methanation reaction can be characterized by (a) the high exothermic heat release, (b) the large quantities of synthesis gas that must be handled (four to five volume units of dry syngas generate one unit of methane), and (c) the large amount of steam formed during methane synthesis. CO conversion is limited by chemical equilibrium and can be carried out at temperatures up to about 525°C. Low temperature and high pressure favor CO conversion (see Figure 9.13). Among the hydrocarbons shown, methane is favored compared to ethene and n-octane. At low enough temperatures (and/or pressures high enough), however, small amounts of longer-chained hydrocarbons may be formed accordingly that could increase the heating value of the product gas (per volume). The reaction must be promoted by a catalyst, most common as active component is nickel supported on an alumina or other oxide support. Typical reaction temperatures are 300°C–400°C and the reaction pressure depends on the pressure applied in gasification on the one hand and pressure requirements of the product gas SNG on the other.

In contrast to the FT reaction (Equation 9.8 in Figure 9.7), the desired product methane exhibits a selectivity of nearly 100% over a wide range of temperatures. Therefore, control of heat removal and thus reaction temperature is not as critical as that in FT synthesis. The most common strategies applied for temperature control include (a) application of cooled recycle gas in fixed bed reactors, (b) application of a multistage concept with intermediate quenching with steam, and (c) application of fluidized bed reactors with internal heat transfer equipment (Müller 2003). The gas recycle alternative appears to be a robust solution. In addition, application of cooled fluidized bed reactors was successfully demonstrated in the 1980s.

In the case of biomass as feedstock, the potential of low gasification temperature (due to relatively high gasification reactivities of biomass) allows for high CH_4 production in the gasifier (see Table 9.7), with oxygen consumption lowered by the heat-contributing exothermic methane synthesis reaction. This leads to higher overall energy efficiencies as compared to FT synthesis (see Section 9.6). Thus, process-wise, there are two advantages of methane synthesis as compared to FT synthesis: (a) higher overall energy conversion efficiency due to primary methane formation and (b) no complex product upgrading required as the desired product methane is formed directly and with high selectivity during synthesis.

9.6 Conversion Efficiencies and Economics

For the production of synthetic hydrocarbon fuels from biomass, there is currently no practical experience with large-scale plants so far. Therefore, overall mass balance and achievable product yields have been estimated on the basis of lab-scale experiments or extrapolations from coal or natural gas as feedstocks. Various studies have recently been published with estimated yield and efficiency values based on flow sheet simulations. These studies indicate that efficiencies in terms of chemical energy and in terms of carbon recovered in the hydrocarbon product may be expected in the range of 30%–50% and 25%–45%, respectively (Table 9.11). This can be interpreted in terms of the molar stoichiometry of the overall FT biofuel process for expected heating value efficiencies (Equations 9.15 and 9.16):

$$100CH_{1.6}O_{0.7} + 45O_2 \xrightarrow{\eta_{HHV}\ 50\%} 40\ -(CH_2)- +60CO_2 + 40H_2O \tag{9.15}$$

$$100CH_{1.6}O_{0.7} + 69O_2 \xrightarrow{\eta_{HHV}\ 30\%} 24\ -(CH_2)- +76CO_2 + 56H_2O \tag{9.16}$$

The real case, as expected according to Table 9.11, will most likely be between these two cases that correspond to synfuel yields of 0.23 and 0.14 t/t, respectively, as liquid hydrocarbon product C_{5-20} per dry and ash-free biomass. A potential improvement for carbon utilization efficiency can be seen for the addition of external hydrogen (in the form of CH_4 or H_2). If hydrocarbon yields are increased in such a way that all carbon present in the feedstock biomass would be converted to hydrocarbons, this would mean a maximum use of biomass carbon for the production of liquid hydrocarbon fuels. As discussed in Section 9.4, the limited carbon or chemical energy efficiency of the overall process can be attributed to the following factors: (a) Overall stoichiometry: how much chemical energy can be

TABLE 9.11

Yield and Efficiency Values (Carbon or Heating Value Ratio of Liquid Hydrocarbon Products and Biomass Feedstock) from BtL Process Studies

	Units	Tijmensen et al. (2002)	Dimmig and Olschar (2003)	Dena (2006)	Leible et al. (2007)
Feed flow dry	t/h	80[a]	100[a]	100[a]	18/900[b]
	GW	0.37[a]	0.5[a]	0.5[a]	0.1/5[a]
Product flow					
$-(CH_2)-$[c]	t/h	10–15	18	15–17[d]	119–139[d]
Gasoline	t/h			6–7	
Diesel	t/h			9–10	
η Carbon[e]	%	26–41	36	34	23–27
η HHV[f]	%	32–51	45	42	29–34

[a] Basic definition.
[b] Calculated with 18 (or 20) MJ/kg HHV.
[c] Fuels C_{5-20}.
[d] Calculated with 44 MJ/kg HHV.
[e] m_C in C_{5-20}/m_C in biomass feed.
[f] HHV of C_{5-20}/HHV of biomass feed.

stored according to the overall stoichiometry in the desired fuel product hydrocarbon, that is, what is the heat of the overall reaction (where exothermicity means a loss in chemical energy)? (b) Different temperature levels: how much heat is required for the endothermic gasification step, where the exothermic synthesis reaction, due to its low temperature level, cannot contribute significantly? (c) Energy requirements for continuous operation: which are the internal energy requirements of the process (e.g., high-pressure application, transport of solids and fluids, size reduction of solids)? (d) Product selectivity: which fraction of the reactants is converted to the desired products (e.g., to C_{5-20} hydrocarbons during FT synthesis and hydrocracking)?

Regarding the energy loss during the sequence of conversion processes in terms of chemical energy, efficiency can be compared for biomass and different fossil feedstocks (Table 9.12). When starting with a solid feedstock (coal as fossil, biomass as renewable feedstock), hydrocarbon synthesis leads to overall efficiencies of about 30%–50%, with a gaseous feedstock (natural gas) of 55%–63%. This has to be compared to gasoline or diesel from petroleum refining as a reference case with efficiencies of about 90%–94%. As a consequence, if biomass is converted to synthetic fuels, the saving effect for fossil CO_2 is reduced by the relatively low conversion efficiency. Replacing petroleum products with relatively high refining efficiencies therefore may be questioned from a viewpoint of maximum CO_2 saving scenarios. This would change drastically if automotive fuels were produced from more difficult solid feedstocks (like coal, tar sands). Well-to-wheel analysis of FT diesel production routes from various feedstocks reflect the complex effects of feedstock characteristics on CO_2 emissions, conversion efficiencies, and production costs (van Vliet et al. 2009). Table 9.12 also shows that bio-SNG production has higher efficiencies than BtL conversion due to low-temperature gasification and better internal heat management (see Section 9.5).

Fuel conversion processes (like synthetic fuel production) can be assessed using energy efficiency. These efficiencies are summarized in Figure 9.14 for synthetic hydrocarbon fuels (liquid and gaseous) and for biomass and coal as feedstocks. For comparison, efficiencies

TABLE 9.12

Overall Molar Yields of Synthetic Hydrocarbon Products from Various Raw Materials as Compared to Natural Hydrocarbons from Petroleum

					η_{HHV} (%)
Natural gas[a]	100 CH_4	$\longrightarrow$	68–80	$-(CH_2)-$	53–63
Biomass[b]	100 $CH_{1.6}O_{0.7}$	$\longrightarrow$	24–40	$-(CH_2)-$	30–50
Coal[c]	100 $CH_{0.8}O_{0.1}$	$\longrightarrow$	28–40	$-(CH_2)-$	35–50
Petroleum[d]	100 $CH_{1.8}$	$\longrightarrow$	94	$-(CH_2)-$	94
Biomass[e]	100 $CH_{1.6}O_{0.7}$	$\longrightarrow$	35–50	CH_4	50–70
Coal[f]	100 $CH_{0.8}O_{0.1}$	$\longrightarrow$	28–35	CH_4	50–60

Note: $-(CH_2)-$, liquid hydrocarbons; CH_4, SNG; η_{HHV}, conversion efficiency based on higher heating value.

[a] Audus et al. (2001) and Sie et al. (1991).
[b] Tijmensen et al. (2002).
[c] Hoogendorn (1976), from Jüntgen et al. (1981).
[d] Own estimate.
[e] Own estimate, based on van der Drift et al. (2005) and Duret et al. (2005).
[f] Own estimate, based on Probstein and Hicks (1990).

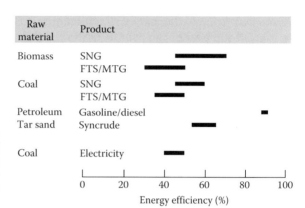

FIGURE 9.14
Overview of typical energy efficiency values of fuel upgrading processes (as ratio of HHVs of product and raw material), including biomass-derived liquid and gaseous hydrocarbons (FTS/MTG and SNG). (Table 9.12 or own estimate.)

for processing petroleum crude, coal, and tar sand, as well as production of electricity by combustion have been included. In this time of climate change concerns, these efficiency values mean a significant assessment criterion.

If biomass crop yields per hectare of land are combined with process efficiencies, the overall biofuel yields per hectare of land can be estimated. In the example of German climatic conditions discussed in Section 9.2 (Table 9.5), synfuel potentials of biomass from biofuel yields per agricultural area are about one-fourth of the present German consumption of automotive fuels (based on utilization of 20% agricultural land for energy crops and of about 5% if 25% of total straw yields are used [Schaub and Vetter 2008]). Comparison of overall fuel yields per area of land indicates that synthetic biofuels achieve higher yields per cultivated land than 1st Generation biofuels (FAMEs or ethanol from sugar or starch).

The future of synthetic biofuels will be strongly affected by economic and political factors. Oil price increases and global conflicts favor the use of biomass raw materials, although their price is not independent of petroleum. Figure 9.15 shows cost estimates for the production of upgraded fuels from biomass, with varying specific capital investment (leading to varying intercept values), and varying energy efficiency (leading to varying slopes). Definitions and assumptions made are summarized in Table 9.13. Parameters of the straight lines are (a) specific investment in Euro per GW as HHV in products and (b) energy efficiency as ratio HHV of product and feedstock, with its inverse value as slope. Characteristic combinations of intercept and slope values include 0/1 for equity line, 2.5/1.1 for petroleum refining as reference, with only limited changes in the structure of feedstock molecules (case I), 2–4/1.15–1.25 for biomass pyrolysis to solid or liquid products (case II),

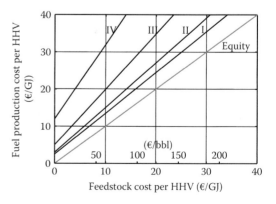

FIGURE 9.15
Production cost of upgraded fuels from biomass (cases II–IV) and from petroleum (case I). Characteristic parameters: Operating cost, capital-related (in €/GJ)/energy efficiency (in % HHV): I (reference): 2.5/91, II: 3/83, III: 5/67, IV: 12/50, definitions and assumptions: Table 9.13, examples for I–IV: see text and Table 9.14.

TABLE 9.13

Definitions and Assumptions Used in Production Cost Estimates of
Synfuels from Biomass (Figure 9.15)

Intercept (€/GJ)	$\dfrac{\text{Total annual cost (w/o feedstock)}}{\text{Annual HHV in product}}$
Total annual cost (w/o feedstock) including capital cost, maintenance, labor, utilities	20% of capital investment/year
Slope	1/energy efficiency
Energy efficiency	HHV in product/HHV in feedstock
Data source (investment, efficiency)	Published data or own estimate

TABLE 9.14

Characteristic Data of Biomass Upgrading Processes Used in Production Cost
Estimates[a]

	Feedstock Capacity		Investment (10⁶ €)	Conversion Efficiency (%)	Product
	(kt/year)	(GW)[b]			
Torrefaction/pyrolysis[c]	80	0.05	15	85	Solid/liquid
Gasification[d]	80	0.05	15	75–80	Fuel gas (H₂)
Bio-SNG[e]	80	0.05	20	60–65	CH₄
BtL/FTS[f]	1,000	0.65	500–1,000	35–50	Liquid synfuels
Oil refining[g]	15,000	21.8	7,000–12,000	91–94	Liquid fuels

[a] Own estimates based on analogies or references.
[b] As HHV in feedstock.
[c] Kiel et al. (2009).
[d] Schulze and Gaensslen (1984).
[e] Zwart and Mäkinen (2009).
[f] Dena (2006).
[g] Own estimate, European standard.

2.5–4/1.2–1.3 for biomass gasification to fuel gas or H_2 (case II), 3–6/1.5 for biomass-to-SNG (case III), and 12–16/2 for BtL hydrocarbon products (case IV) (see Table 9.14).

The highest degree of upgrading (namely to liquid hydrocarbons) can only be viable economically if petroleum feedstock cost becomes very high. For example, biomass feedstock cost of 5 €/GJ correspond to crude oil feedstock cost of about 18 €/GJ, the latter corresponding to a crude oil price of about 120 €/bl. In contrast, pyrolysis or gasification of biomass are less expensive but lead to lower-value pyrolysis oil products or less versatile fuel gas (or hydrogen) products. BtL, due to the lowest energy efficiency as discussed in Section 9.4, has the steepest slope in Figure 9.15. Breakdown of the capital costs for a BtL plant shows shares of 50%/20%/30% for feedstock pretreatment, gasifier, and oxygen plant/gas treatment/synthesis and product upgrading, respectively (Tijmensen et al. 2002).

9.7 Conclusion and Perspectives

The following conclusions can be drawn for the integration of biomass in the energy supply system in the form of synthetic hydrocarbons. Technologies are available for synthetic liquid and gaseous fuel production, both in pilot/demonstration stage with similarities to coal as feedstock. In gasification as the first step in this conversion route, nonspecific, heterogeneous lignocellulosic feedstocks can be used as raw materials, either as energy crops or as residues from other kinds of biomass utilization (e.g., for food or feed applications). It offers a broad flexibility, as any carbon-containing biomass can be used.

9.7.1 Liquid Hydrocarbons (Bio-Synfuel BTL)

- Controlled chemical structure of hydrocarbon compounds in product mixtures allows for improved internal combustion engines with reduced pollutant emission.
- Supply security for liquid fuels can be increased (due to regional supply of raw biomass material).
- Fossil CO_2 emissions can be avoided (if petroleum fuels are replaced).
- Process complexity is high (due to multistep flow scheme, complex H_2 management, high pressure applied, solids handling).
- High production cost due to high plant investment (due to process complexity).
- Energy conversion efficiencies are moderate (30%–50%), due to combination of endothermic (high-temperature) gasification and exothermic (low-temperature) synthesis.
- For market implementation, a demonstration phase is envisaged until 2014 (ReNew 2008). The most favorable areas in Europe are Central France, East Germany, and Poland (due to high land and biomass availability). The first demonstration plant is presently in start up phase.
- Integration of biomass in coal-based synfuel industry is penalized by high additional fossil CO_2 emissions in coal-to-synfuel conversion.

9.7.2 Substitute Natural Gas (Bio-SNG)

- Product SNG is easy to integrate into gas grids for natural gas.
- Energy security can be improved (by replacing natural gas).
- Fossil CO_2 emissions can be avoided (by replacing natural gas).
- Process complexity is lower than with liquid hydrocarbon synthesis (lower number of process steps, no H_2 management required, etc.).
- Specific production cost (per energy unit) is lower than that with liquid synthetic hydrocarbons, decentralized production may therefore be more easily economically viable.
- Energy efficiency is moderate (but higher than that with BtL), due to potential CH_4 formation in (low-temperature) gasification, that is, exothermic heat utilization for biomass gasification.
- Demonstration plants are operating (Güssing) or are in design or construction phase.

Biofuels can provide only a limited part to the solution regarding a secure energy supply, besides other renewable energy resources. Therefore, as efficiency increases, energy saving measures must be taken, for example, in the area of mobility, adjustment of car manufacturing strategies and consumer behavior, and maximizing fuel economy of cars in order to minimize or avoid utilization of fossil fuels.

In the broader context of biomass utilization for energy applications, fossil CO_2 mitigation effects are most significant if biomass replaces coal or tar sand (either in direct combustion or avoiding their CO_2-producing chemical upgrading). Other alternatives (to liquid bio-synfuels) include H_2 production from biomass via gasification for integration in petroleum refining, thus increasing transport fuel yields. For example, today's light heating oil could thus be converted into clean diesel fuel. However, biomass conversion to liquid transport fuels (that can replace petroleum fuels) offers the possibility to increase supply security for liquid fuels and sustain regional agriculture and forestry.

Abbreviations

a,b,c, etc.	stoichiometric factors
A	area
α	chain growth parameter in FT reaction
BtL	biomass-to-liquid
Daf	dry and ash free
Dm	dry matter
DME	dimethylether
EU	European Union
FAME	fatty acid methyl ether
FC	fuel cell
FT	Fischer–Tropsch
FTS	Fischer–Tropsch synfuels
FU	fuel
H	energy efficiency
Ha	hectare (=10,000 m^2)
HHV	higher heating value (product H_2O:liquid)
$\Delta_R H^o_{298}$	(molar) heat of combustion reaction at 298 K and $p_i = 1.013 \times 10^5$ Pa
ICE	internal combustion engine
LHV	lower heating value (product H_2O:gaseous)
M	mass
M	molar mass/weight
MeOH	methanol
MTG	methanol-to-gasoline
MTO	methanol-to-olefins
NPP	net productivity of photosynthesis
SNG	substitute natural gas
V	volume
x, y	molar ratios hydrogen/carbon and oxygen/carbon
x_i	volume fraction of component i in gas mixture

References

Althapp, A. 2003. Synthetic transportation fuels from biomass via Fischer–Tropsch synthesis–Principles and perspectives. Paper presented at the conference *Regenerative Kraftstoffe*, November 13–14, Stuttgart, Germany.

Arpe, H.J. 2007. *Industrielle Organische Chemie*. Weinheim, Germany: Wiley VCH.

Audus, H., G. Choi, A. Heath, and S. Tam. 2001. *Stud. Surf. Sci. Catal.* 136: 519–524.

Bartholomew, C.H. and R.J. Farrauto. 2006. *Fundamentals of Industrial Catalytic Processes*. Chapter 6.5 Fischer-Tropsch Synthesis, pp. 398–484.

Boie, W. 1953. Beiträge zum feuerungstechnischen Rechnen—Teil 1: Ableitung neuer Heizwertformeln mit umfassendem Gültigkeitsbereich. *Energietechnik* 3: 309–316.

Bridgewater, A.V., ed. 2003. *Pyrolysis and Gasification of Biomass and Waste*. Newbury, U.K.: CPL Scientific Press.

Bridgewater, A.V. 2007. Applications for utilisation of liquids produced by fast pyrolysis of biomass. *Biomass Bioenergy* 31(8): 1–7.

Dakotagas. 2009. http://www.dakotagas.com/Miscellaneous/pdf/Mediakit/DGCMediaKit_0509.pdf (accessed June 23, 2009).

De Klerk, A. 2009. Can Fischer–Tropsch syncrude be refined to on-specification diesel fuel? *Energy Fuels* 23: 4593–4604.

Dena (Deutsche Energie-Agentur). 2006. Biomass-to-liquid BtL. Realisierungsstudie (Summary). Final Report.

Dimmig, T. and M. Olschar. 2003. Sekundärenergieträger aus Biomasse—Eine analyse. Final Report, Research Project BMVEL.

DIN EN 590. 2004. European Diesel Fuel Specification.

Dry, M. 1990. The Fischer–Tropsch process—Commercial aspects. *Catal. Today* 6: 183–206.

Duret, A., C. Friedli, and F. Marechal. 2005. Process design of synthetic natural gas (SNG) production using wood gasification. *J. Cleaner Prod.* 13: 1434–1446.

EC Directive 2003/30, followed by Directive 2009/28/EC of the European Parliament on the promotion of the use of energy from renewable sources. *Off. J. Eur. Union* 2009, L 140/16.

Elliott, M.A., ed. 1981. *Chemistry of Coal Utilization*. 2nd Suppl. Vol. New York: John Wiley.

Elliott, D.C. 2007. Historical developments in hydroprocessing bio-oils. *Energy Fuels* 21: 1792–1815.

FNR. 2009. *Bioenergy—Plants, Raw Materials, Products*. Report Fachagentur Nachwachsende Rohstoffe, 2nd edn., http://www.fnr-server.de/ftp/pdf/literatur/pdf_330-bioenergy_2009.pdf (accessed February 23, 2011).

Freudenberg, K. and A.C. Neish. 1968. *Constitution and Biosynthesis of Lignin*. New York: Springer Verlag.

Güttel, R., U. Kunz, and T. Turek. 2008. Reactors for Fischer–Tropsch synthesis. *Chem. Eng. Technol.* 31: 746–754.

Hebden, D. and H.J.F. Stroud. 1981. Coal gasification processes, in *Chemistry of Coal Utilization*, M.A. Elliott, ed., pp. 1599–1752, 2nd Suppl. Vol. New York: John Wiley.

Henrich, E., N. Dahmen, and E. Dinjus. 2009. Cost estimate for biosynfuel production via biosyncrude gasification. *Biofuels, Bioprod. Bioref.* 3: 28–41.

Henrich, E. and E. Dinjus. 2003. Tar-free, high pressure synthesis gas from biomass. *Pyrolysis and gasification of biomass and waste. Proceedings of the Expert Meeting*, Strasbourg, September 30–October 1, 2002. Newbury, U.K.: CPL Press, pp. 511–526.

Herrmann, H.D., N. Pelz, R.R. Maly, J.J. Botha, P.W. Schaberg, and M. Schnell. 2006. Effect of GTL diesel fuels on emissions and engine performance. Paper presented at the *World Gas Conference*, June 5–9, Amsterdam, the Netherlands.

Higman, C. and M. van der Burgt. 2003. *Gasification*. Burlington, MA: Elsevier Science.

Hofbauer, H., R. Rauch, P. Foscolo, and D. Matera. 2001. Hydrogen-rich gas from biomass steam gasification. Paper presented at *1st World Conference on Biomass for Energy and Industry, Proceedings*, Sevilla, Spain, June 5–9, 2000.

Hoogendorn, J.C. 1976. Gas from coal for synthesis of hydrocarbons. Paper presented at American Institute of Mining Engineers, *23rd Annual Meeting*, 1974. Cited by Jüntgen, H., Klein, J., Knoblauch, K., Schröter, H.-J., and Schulze, J. 1981. In *Chemistry of Coal Utilization*, M.A. Elliott, ed., pp. 2071–2158, 2nd Suppl. Vol. New York: Wiley & Sons.

Howard, J.B. 1981. Fundamentals of coal pyrolysis and hydropyrolysis. In *Chemistry of Coal Utilization*, M.A. Elliott, ed., pp. 665–784, 2nd Suppl. Vol. New York: John Wiley.

Howard, J.B., M.R. Hajaligol, J.P. Longwell, T.R. Nunn, and W.A. Peters. 1982. Pyrolysis of biomass, *Proceedings, Joint Meeting of Chemical Engineering, Chemical Industry and Engineering Society of China*, American Institute of Chemical Engineers, Beijing, China, September 19–22, Vol. II, pp. 638–652.

IPCC. 2007. Intergovernmental panel on climate change. *Climate Change 2007, The Physical Science Base*. Cambridge, U.K.: Cambridge University Press.

Johnson, J.L. 1981. Fundamentals of coal gasification. In *Chemistry of Coal Utilization*, M.A. Elliott, ed., pp. 1491–1598, 2nd Suppl. Vol. New York: John Wiley.

Jüntgen, H., J. Klein, K. Knoblauch, H.J. Schröter, and J. Schulze. 1981. Conversion of coal and gases produced from coal into fuels, chemicals, and other products. In *Chemistry of Coal Utilization*, M.A. Elliott, ed., pp. 2071–2158, 2nd Suppl. Vol. New York: John Wiley.

Kiehl, J. and K. Trenberth. 1997. Earth's annual global mean energy budget. *Bull. Am. Meteorol. Soc.* 78: 197–206.

Kiel, J.H.A., F. Verhoeff, H. Gerhauser, W. van Daalen, and B. Meuleman. 2009. BO2-technology for biomass upgrading into solid fuel—An enabling technology for IGCC and gasification-based BtL. Paper presented at the *3rd International Freiberg Conference on IGCC & XtL Technologies*, May 18–21, Dresden, Germany.

Knoef, H.A.M., ed. 2005. *Handbook of Biomass Gasification*. Enschede, the Netherlands: Biomass Technology Group.

Krishna, R. and S.T. Sie. 1994. Strategies for multiphase reactor selection. *Chem. Eng. Sci.* 49: 4029–4055.

Larcher, W. 2003. *Physiological Plant Ecology*. Berlin, Germany: Springer Verlag.

Leckel, D. 2009. Diesel production from Fischer–Tropsch: The past, the present, and new concepts. *Energy Fuels* 23: 2342–2358.

Leible, L., S. Kälber, G. Kappler et al. 2007. Kraftstoff, Strom und Wärme aus Stroh und Waldrestholz— Eine Systemanalytische Untersuchung, Wissenschaftliche Berichte des Forschungszentrums Karlsruhe GmbH, FZKA 7170.

Maxwell, I.E. and W.H.J. Stork. 2001. Hydrocarbon processing with zeolites. In *Studies in Surface Science and Catalysis*, H. van Bekkum, E.M. Flanigen, P.A. Jacobs, and J.C. Jansen, eds., pp. 802–805. Amsterdam, the Netherlands: Elsevier.

Meunier, J. 1962. *Vergasung fester Brennstoffe und oxidative Umwandlung von Kohlenwasserstoffen*. Weinheim, Germany: Verlag Chemie.

Müller, W.D. 2003. Methanation and methane synthesis. In *Ullmann's Encyclopedia of Industrial Chemistry*, pp. 388–395, 6th edn., Vol. 15. Weinheim, Germany: Wiley VCH.

Nunn, T.R., J.B. Howard, J.P. Longwell, and W.A. Peters. 1985. Product composition and kinetics in the rapid pyrolysis of milled wood lignin. *Ind. Eng. Chem. Process Des. Dev.* 24: 844–852.

Potetz, A. 2008. Synthetic biofuels made in Güssing. Paper presented at the *3rd BtL Congress*, October 15–16, Berlin, Germany.

Probstein, R.F. and R.E. Hicks. 1990. *Synthetic Fuels*, Cambridge, MA: pH Press.

ReNew. 2008. Renewable fuels for advanced powertrains. Executive summary. www.RENEW-fuel.com (accessed February 23, 2011).

Schaub, G. 2006. Synthetic fuels and biofuels for the transportation sector—Principles and perspectives. *Oil Gas Eur. Mag.* 1/2006 In *Erdöl, Erdgas, Kohle* 122: OG 34–38.

Schaub, G. and R. Reimert. 2003. Gas production from coal, wood and other solid feedstocks. In *Ullmann's Encyclopedia of Industrial Chemistry*, Barbara Elvers, ed., pp. 357–380, 6th edn., Vol. 15. Weinheim, Germany: Wiley VCH.

Schaub, G., D. Unruh, and M. Rohde. 2004. Synthetische Kraftstoffe aus Biomasse über die Fischer-Tropsch Synthese—Grundlagen und Perspektiven. *Erdöl, Erdgas, Kohle* 120: 327–331.

Schaub, G. and A. Vetter. 2008. Biofuels for automobiles—An overview. *Chem. Eng. Technol.* 31: 721–729.

Schmalfeld, J., ed. 2008. *Die Veredelung und Umwandlung von Kohle—Technologien und Projekte 1979 bis 2000 in* Hamburg, Germany. DGMK, ISBN 978-3-936418-88-0.

Schulz, H. 1999. Short history and present trends of Fischer–Tropsch synthesis. *Appl. Catal. A: Gen.* 186: 3–12.

Schulz, H. 2003. Major and minor reactions in Fischer–Tropsch synthesis on cobalt catalysts. *Topics Catal.* 26: 73–85.

Schulz, H. 2005. Comparing Fischer-Tropsch synthesis on iron and cobalt catalysts. *Prep. Pap.-Am. Chem. Soc., Div. Pet. Chem.* 50: 155–157.

Schulze, J. and H. Gaensslen. 1984. Kosten der Wasserstofferzeugung—neue Vorkalkulationsmethoden und Ergebnisse. *Chem. Ind.* 36: 135–140.

Sie, S.T. 1998. Process development and scale-up: IV case history of the development of a Fischer-Tropsch synthesis process. *Rev. Chem. Eng.* 14: 109–157.

Sie, S.T. and R. Krishna. 1999. Fundamentals and selection of advanced Fischer–Tropsch reactors. *Appl. Catal. A: Gen.* 186: 55–70.

Sie, S.T., M.M.G. Senden, and H.M.H. van Wechem. 1991. Conversion of natural gas to transportation fuels via the shell middle destillate (SMDS). *Catal. Today* 8: 371–394.

Tijmensen, M.J.A., A.P.C. Faaij, C.N. Hamelinck, and M.R.M. van Hardeveld. 2002. *Biomass Bioenergy* 23: 129–152.

Unruh, D., K. Pabst, and G. Schaub. 2010. Fischer–Tropsch synfuels from biomass: Maximizing carbon efficiency and hydrocarbon yield. *Energy Fuels* 24: 2634–2641.

Van der Drift, A., L.P.L.M. Rabon, and H. Boerrigter. 2005. Heat from biomass via synthetic natural gas. Paper presented at the *14th European Biomass Conference & Exhibition*, October 17–21, Paris, France.

Van der Laan, G.P. and A.A.C.M. Beenackers. 1999. Kinetics and selectivity of the Fischer–Tropsch synthesis: A literature review. *Catal. Rev. Sci. Eng.* 413: 255–318.

Van der Meijden, C.M., A. van der Drift, and B.J. Vreugdenhil. 2007. Experimental results from the allothermal biomass gasifier Milena. Paper presented at the *15th European Biomass Conference*, May 7–11, Berlin, Germany.

Van der Meijden, C.M., H.J. Veringa, A. van der Drift, and B.J. Vreugdenhil. 2008. The 800 KWTH allothermal biomass gasifier Milena. Paper presented at the *16th European Biomass Conference*, June 2–6, Valencia, Spain.

Van Vliet, O.P.R., A.P.C. Faaij, and W.C. Turkenburg. 2009. Fischer–Tropsch diesel production in a well-to-wheel perspective: A carbon, energy flow and cost analysis. *Energy Convers. Manag.* 50: 855–876.

Wokaun, A. 2009. Bio-SNG and its role in the future energy system. Paper presented at the *International Conference Bio-SNG '09—Synthetic Natural Gas from Biogas*, May 26–27, Zurich, Switzerland. www.bio-sng.web.psi.ch (accessed September 25, 2009).

Zwart, R. and T. Mäkinen. 2009. R&D needs and recommendations for the commercialisation of the production and use of renewable substitute natural gas from biomass. Report ECN-BKM-2009–341. www.ecn.nl, www.biosng.com (accessed September 25, 2009).

10

Fundamentals of Biohydrogen Production Processes

Chitralekha Nag Dasgupta and Debabrata Das

CONTENTS

10.1 Introduction

Energy has played a very important role in our life since the beginning of human history. Mankind's fuels have continually evolved as better, more efficient, safer, and cleaner fuels. From wood, to animal fat, to coal, to petroleum, to natural gas, to hydrogen, a clear trend to lighter and cleaner fuels is apparent. Fossil fuels have unanimously ruled the world for more than one century. The indiscriminate use of fossil fuels has gone to such an extent that it has not only polluted the environment but also exhausted the limited fuel reserves, necessitating the quest for an alternative energy. Hydrogen (H_2) is most promising in the succession of fuel evolution, with several technical, socioeconomic, and environmental benefits to its credit. It has the highest energy content per unit weight of any known fuel ($142\,kJ\ g^{-1}$) and can be transported for domestic/industrial consumption through conventional means. H_2 gas is safer to handle than domestic natural gas. H_2 is now universally accepted as an environmentally safe, renewable energy resource and an ideal alternative to fossil fuels that does not contribute to the greenhouse effect. Presently, 40% H_2 is produced from natural gas, 30% from heavy oils and naphtha, 18% from coal, 4% from electrolysis, and about 1% from biomass. However, today, biological H_2 production processes are becoming important mainly due to two reasons: utilization of renewable energy resources and usually operation at ambient temperature and atmospheric pressure. The microbial production of hydrogen by fermentation can be broadly classified into two main categories—one is light independent and the other is light dependent. The light-independent fermentation processes, commonly known as dark fermentation, employ both obligate and facultative anaerobic bacteria for the production of H_2 from a variety of potentially utilizable substrates, including refuse and waste products. It generally gives a high rate of H_2 evolution and does not rely on the availability of light sources. In contrast, in photofermentation, small-chain organic acids are used by photosynthetic bacteria as electron donors for the production of H_2 at the expense of light energy. Both green algae and blue-green algae (cyanobacteria) can convert water to hydrogen. This chapter deals with the fundamental of these biohydrogen production processes.

10.2 Microbiology of Hydrogen Production

In nature, some microorganisms use hydrogen as their energy source and some are hydrogen generators. They have their own hydrogen fuel cycle. Microorganisms produce hydrogen mostly in response to anaerobiosis. Hydrogen production can be observed in a wide variety of microorganisms classified in Figure 10.1. Versatility in their habitats and nutritional types help them to withstand extreme conditions like complete darkness, high temperature, and anaerobic condition. Hydrogen production from different microorganisms is depicted in Table 10.1.

10.2.1 Dark Fermentative Bacteria

Dark-adapted bacteria are heterotrophs and produce hydrogen in anoxia. As they are adapted in dark, most suddenly have been found under the soil layers, water, sewage, inside the vegetables, etc. and can withstand oxygen deficiency. They grow on organic substrates (heterotrophic growth). These substrates are degraded by oxidation to provide the building blocks and metabolic energy for growth. Dark fermentative hydrogen production consumes less energy in a reactor setup as these organisms do not require light energy.

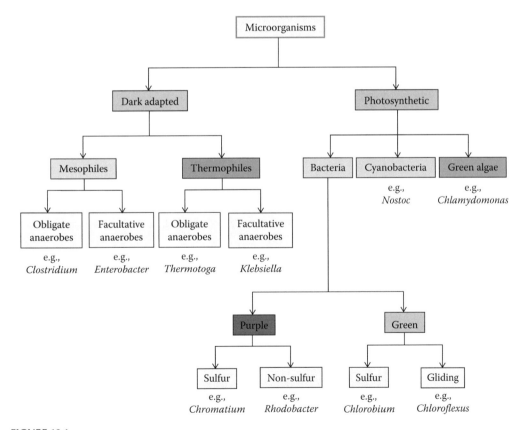

FIGURE 10.1
Schematic representation of different hydrogen-producing microorganisms.

TABLE 10.1

Hydrogen Production of Different Microorganisms

Type of Organism	Name of the Microorganism	Enzyme	Rate of Hydrogen Production	Yield	Conditions	Reference
Dark fermentative bacteria						
Obligate anaerobes	C. pasteurianum	Hydrogenase	$9\,mmol\,L^{-1}\,h^{-1}$	$1.5\,mol\,H_2\,mol^{-1}$ glucose	Glucose	Brosseau and Zajic (1982)
	Clostridium butyricum CGS5	Hydrogenase	$9.3\,mmol\,L^{-1}\,h^{-1}$	$2.78\,mol\,H_2\,mol^{-1}$ sucrose	Sucrose, pH 5.5	Chen et al. (2005)
Facultative anaerobes	E. coli	Formate hydrogenase lyase	$31\,mL\,h^{-1}\,OD_{unit}^{-1}\,L^{-1}$	$1.67\,mol\,H_2\,mol^{-1}$ glucose	Nougat wastewater	Penfold et al. (2003)
	E. cloacae	Hydrogenase	$19.95\,mmol\,L^{-1}\,h^{-1}$	$2.12\,mol\,H_2\,mol^{-1}$ glucose	Glucose, pH 6.5, 37°C	Kumar et al. (2001)
	Citrobacter intermedin	Hydrogenase/ Formate hydrogenase lyase	$3.7\,mmol\,L^{-1}\,h^{-1}$	$1\,mol\,H_2\,mol^{-1}$ glucose	Glucose	Brosseau and Zajic (1982)
Thermophiles obligate anaerobes	T. maritima DSM3109	Hydrogenase	$15.2\ 12\,mmol\,L^{-1}\,h^{-1}$	$1.67\,mol\,H_2\,mol^{-1}$ glucose	pH 6.5–7, 75°C–80°C	Nguyen et al. (2008)
	T. thermosaccharo lyt-icum PSU-2	—	$12.12\,mmol\,L^{-1}\,h^{-1}$	$2.53\,mol\,H_2\,mol^{-1}$ sucrose	60°C, pH 6.25	O-Thong et al. (2008)
Facultative anaerobes	Klebsiella oxytoca HP1	—	—	$3.6\,mol\,H_2\,mol^{-1}$ sucrose	65°C, pH 7.1	Minnan et al. (2005)
Photosynthetic bacteria						
Purple sulfur	T. roseopersicina	Hydrogenase	$1.8\,mmol\,L^{-1}\,h^{-1}$	—	Thiosulfate sulfide	Horwith et al. (2004)
	Chromatium sp. O.U. 1010	—	$11\,\mu L\,mg^{-1}$ dry wt·h^{-1}	—	Paper mill effluent	Rao and Mahmood (1997)

Purple non sulfur	*Rhodobacter spaeroides RV*	Nitrogenase	2.9 mmol L^{-1} h^{-1}	4 mol H_2 mol^{-1} lactic acid	Lactic acid, ammonia	Fascetti and Todini (1995)
Green sulfur	*Chlorobium limicola*	—	—	—	—	—
Green gliding	*C. aurantiacus*	—	—	—	—	—
Cyanobacteria Heterocystous	*Anabaena cylindrical* B 629	Hydrogenase/ nitrogenase	0.103 μmol mg^{-1} dry wt · h^{-1}	—	7000 lux, 5% CO_2	Lambert and Smith (1977)
Nonheterocystous	*Anaebaena variabilis* ATCC 29413	Hydrogenase/ nitrogenase	0.25 μmol mg^{-1} dry wt · h^{-1}	—	150 μE m^{-2} s^{-1}, 1% CO_2	Berberoglu et al. (2008)
	Gleobacter PCC 7421	Hydrogenase/ nitrogenase	1.38 μmol mg^{-1} chl a (protein) h^{-1}	—	20 μE m^{-2} s^{-1}, CO	Moezelaar et al. (1996)
	Oscillotoria limosa	Hydrogenase/ nitrogenase	0.83 μmol mg^{-1} chl a (protein)/h	—	1200 lux, CO_2	Heyer et al. (1989)
	Synechocystis sp PCC 6803	Hydrogenase/ nitrogenase	0.81 μmol mg^{-1} chl a (protein) h^{-1}	—	50 μE m^{-2} s^{-1}, NaHCO$_3$	Burrows et al. (2008)
	Synechococcus PCC 602	Hydrogenase/ nitrogenase	0.66 μmol mg^{-1} dry wt · h^{-1}	—	20 μE m^{-2} s^{-1}, CO	Howarth and Codd (1985)
Green algae	*C. reinhardtii*	Hydrogenase	0.093 mmol L^{-1} h^{-1}	—	100 μE m^{-2} s^{-1}, acetate, S-deprived	Laurinavichene et al. (2006)
	Chlorella sorokiniana Ce	Hydrogenase	0.06 mmol L^{-1} h^{-1}	—	120 μE m^{-2} s^{-1}, acetate, S-deprived	Chader et al. (2009)

It has been reported that *Clostridia* could utilize a large number of carbohydrates such as arabinose, cellobiose, fructose, galactose, lactose, sucrose, xylose, etc. (Taguchi et al. 1993). It suggests that it can a be very effective organism for hydrogen production from industrial waste water (Suzuki and Karube 1981). Some of these can grow in ambient temperature (mesophiles), while some are adapted to high temperatures (thermophiles). Thermophiles are considered to be more promising candidates for hydrogen production than mesophiles as they can

- Produce hydrogen near stoichiometry, that is, 4 mol of hydrogen per mole of glucose
- Avoid contamination of hydrogen-consuming microorganisms as well as other microorganisms
- Use hot industrial effluent directly as substrate

However, for thermophiles, extra energy is needed to maintain the high temperature in the reaction broth. They derive metabolic energy from oxidation of external substrates. Thermophiles are mostly obligate anaerobes found in various geothermally heated regions of the earth such as hot springs, deep-sea hydrothermal vents, as well as decaying plant matter. They can also use a broad range of substrates like cellulose, hemicellulose, and pectin-containing biomass (Van de Werken et al. 2008). By using sweet sorghum juice as carbon and energy substrate, 58% of the theoretical maximum at a maximal production rate of $21 \, \text{mmol} \, L^{-1} \, h^{-1}$ has been achieved by using *Caldicellulosiruptor saccharolyticus* (Claassen et al. 2004).

Microaerophilic or facultative bacteria are always advantageous for experimental work than obligate anaerobes as they are easier to handle. Obligate anaerobes are very sensitive to oxygen and often do not survive in the presence of low oxygen concentrations. Facultative anaerobes are capable of generating ATP by aerobic respiration; however, they are also capable of switching to fermentation in absence of oxygen (anaerobic respiration is also possible). Factors influencing fermentative metabolism are the concentration of oxygen and fermentable material in the environment. Bacteria of family Enterobacteriaceae, for example, *Escherichia coli*, *Enterobacter*, and *Citrobacter*, produce hydrogen from complex carbohydrates in absence of oxygen. In *E. coli*, hydrogen production is mediated by the enzyme formate hydrogen lyase (FHL) using sugar solution as substrate (Penfold et al. 2003). However, in *Enterobacter*, hydrogen production is catalyzed by the enzyme hydrogenase (Mishra et al. 2002). Presence of both FHL activity and other hydrogenases activity has been observed in *Citrobacter* (Kim et al. 2008b). Seol et al. (2008) compared hydrogen production in different strains of Enterobacteriaceae and found that final hydrogen yield ($1.7–1.8 \, \text{mol} \, H_2 \, \text{mol}^{-1}$ glucose) is very similar in the four strains, *C. amalonaticus* Y19, *E. coli* K-12 MG1655, *E. coli* DJT135, and *E. aerogenes*.

10.2.2 Photosynthetic Bacteria

In contrast to dark-adapted bacteria, photosynthetic bacteria require light energy for hydrogen evolution. They are very efficient candidates for hydrogen production.

- The main advantage of using these organisms for hydrogen production is that there is no oxygen evolution during photosynthesis so the anaerobic condition is automatically maintained in the reactor.

Photosynthetic bacteria are of two types—purple and green. Phylogenetic studies have shown that purple and green bacteria are far distantly related (Gibson et al. 1979, Fox et al. 1980). They are probably metabolically most versatile organisms on earth. They can adapt to any probable situation that might arise. In anaerobic conditions, they can live photoheterotrophically by using different organic acids as substrate. This process is known as photofermentation.

- Another advantage of photofermentative microbes is that they can utilize organic acids produced by dark fermentative microbes as the substrate for photofermentation.

Light energy is used for the generation of ATP via cyclic phosphorylation. Most of the photosynthetic bacteria are active diazotrophs (exception green gliding) that fix N_2 by nitrogenase under anoxic nitrogen-limiting conditions (Madigan et al. 1984, Kimble and Madigan 1992, Warthmann et al. 1992), although some of them may contain hydrogenases (Wu and Mandrand 1993). The hydrogen evolution is observed normally by nitrogenase when ATP and electrons are available (Sasikala et al. 1993). Some of them may also possess uptake hydrogenase that reoxidize the H_2 to protons.

10.2.3 Cyanobacteria

Other interesting prokaryotes, cyanobacteria, evolve oxygen during photosynthesis similar to eukaryotes. However, oxygen evolution is the key problem for hydrogen production as the enzymes nitrogenase and hydrogenase are very sensitive to oxygen. Many of them have a specialized cell known as a heterocyst for maintaining the anaerobic condition, for example, *Nostoc*, and *Anabaena*. Reserve carbon is transported inside the heterocyst, oxidized to release electrons that reduce protons inside the heterocyst. Nitrogenase is the key enzyme responsible for hydrogen generation within the heterocyst. However, the non-N_2-fixing cyanobacteria mostly metabolize hydrogen by hydrogenase. Existence of [NiFe]-hydrogenase has been characterized in unicellular non-N_2-fixing *Synchocystis* (Gutekunst et al. 2005). There is another type of hydrogenase, the uptake hydrogenase present in cyanobacteria, which acts in reverse direction and oxidizes the molecular hydrogen. Advantages of using cyanobacteria for hydrogen production are as follows:

- Necessity of very simple reaction broth.
- Vigorous growth.
- Lesser chances of contamination.
- Hydrogen evolution is separated from oxygen evolution either by compartmentalization (heterocysts) or by temporal separation.

10.2.4 Green Algae

Eukaryotes do not have much versatility in their metabolic pathways. They are mostly aerobes, which have evolved to respire with oxygen. Hardly in their life cycle do they exhibit fermentative metabolism. The only potential eukaryotes that can produce hydrogen in anoxia are green algae. They carry out oxygenic photosynthesis like higher plants in the presence of oxygen. However, they can also survive in anoxia in their natural habitats. Hydrogen evolution in green algae is identified as a consequence of anaerobiosis (Happe and Naber 1993) or nutrient deprivation (Melis et al. 2000). Higher plants are rarely exposed to the anoxic conditions. However, most of the algal populations are aquatic and sometimes suffer from anoxia in stagnant water or during algal blooms. Gaffron and his coresearchers first observed that under aerobic condition unicellular green algae, *Scenedesmus obliquus*, is able to generate hydrogen in the presence of light (Gaffron and Rubin 1942). After this discovery, an extensive research on hydrogen metabolism in photosynthetic algae has been done, and it has been found that the photosynthetically generated electrons in absence of oxygen can combine with protons to evolve harmless hydrogen instead of producing toxic fermented products like alcohols and acids. Thus, hydrogen metabolism not only indicates the evolutionary linkage with prokaryotes but also is very crucial and essential for cell survival. It has been found that enzyme hydrogenase is involved in algae for metabolizing hydrogen.

10.3 Biochemistry of Hydrogen Production

Microorganisms are structurally very simple but the great diversity has been observed in their physiology and metabolism. They can readily switch over from one metabolic pathway to another according to the different environmental conditions. This feature allows microorganisms to flourish in all habitats suitable for life on earth. Energy is the basic need to maintain a life, and microorganisms mostly have two different ways to derive energy for them: one is from sunlight and another is from chemical oxidation of substrates. Oxygen (aerobic respiration), some other terminal electron acceptor like nitrate (anaerobic respiration), and partially oxidized intermediates or protons (fermentation, hydrogen production by biophotolysis) could accept the reducing equivalents generated in entire metabolic pathways. How the cell rid off excess electrons depends on the habitat and physiology of particular group of microorganisms. Hydrogen metabolism is very common in bacteria. They generated hydrogen as a by-product of reductive metabolism. Some organism like methanogenic bacteria can consume hydrogen by oxidative metabolism to obtained energy from H–H bond. They are involved in recycling the hydrogen in nature (Zeikus 1977). The chemical equation for methanogenesis can be summarized as follows:

$$4H_2 + CO_2 \rightarrow CH_4 + 2H_2O \tag{10.1}$$

For every methane that is generated, one ATP is also generated (Conrad 1999). They oxidize H_2 to a proton (H^+) and a hydride ion (H^-) via a hydrogenase enzyme. Some hydrogen-producing microorganisms also possess bidirectional hydrogenase and uptake hydrogenase by which they can reconsume hydrogen to maintain the redox potential of the cell. To increase the yield of hydrogen, consumption of hydrogen is not desirable during hydrogen generation. Therefore, the reaction broth in a reactor should devoid of methanogens and uptake hydrogenase activity of the hydrogen-producing organisms to increase the yield of hydrogen.

During heterotrophic metabolism, microorganisms are not capable to fix CO_2 and they use external organic substrates as their carbon source. Mainly two metabolic processes drive heterotrophic metabolism: fermentation and respiration. Under anoxic condition, microorganisms mostly follow the fermentative metabolism to derive energy. In dark condition, energy is derived from partial oxidation of organic substrates and ATP has been generated by substrate-level phosphorylation (dark fermentation). During oxidation of substrate, ferredoxin (Fd), an electron carrier is reduced, which is reoxidized by reducing protons via enzyme hydrogenase or nitrogenase to generate hydrogen.

Phototrophic organisms can generate ATP by using light energy. Electron transport system, membrane-bound ATPase and electron acceptors are involved for this type of ATP generation (photophosphorylation). Photofermentative bacteria can reduce organic acids to mostly produce CO_2 and hydrogen in anoxic environment (photofermentation). Some other photosynthetic bacteria can use electrons from inorganic or organic substrates carried through the photosystem. They are mostly diazotrophs, and hydrogen production is catalyzed by nitrogenase. Microorganisms mostly go through hydrogen metabolism to cope up with anaerobiosis. However, it is not always true that anaerobiosis can induce fermentative hydrogen metabolism. Even within single prokaryote a great versatility in metabolism has been observed. Besides having the normal fermentation and respiration, they can respire in absence of oxygen using other terminal electron acceptor like NO_3 or fumarate (anaerobic respiration), for example, *E. coli* (Stewart 1988). In the presence of oxygen, the facultative microorganisms can switch over in oxidative respiration instead of hydrogen metabolism to generate energy. However, the obligate anaerobes cannot survive in the presence of oxygen. They are only adapted to fermentative metabolism.

The atmosphere of the early Earth was anoxic. Hydrogen production is the consequence of anoxia. Afterward, eukaryotic life developed and atmosphere became aerobic with sufficient level of oxygen. Divergence of metabolism started toward oxygenic photosynthesis and aerobic respiration to generate energy. Cyanobacteria are prokaryotes but have great similarity with eukaryotic world. They do oxygenic photosynthesis with two photosystems. However, they have capability to generate hydrogen in anaerobic condition. In eukaryotic world, green algae like prokaryotes can potentially produce hydrogen to deal with oxygen deficiency. Both cyanobacteria and green algae are mostly having photoautotrophic metabolism. They use water as their electron source and CO_2 as their carbon source. They contain chlorophylls and other pigments to capture light energy. The energy has been used for splitting water (biophotolysis) into protons (H^+), electrons (e^-), and O_2. In normal photosynthesis, those electrons are channelized for reduction of CO_2. But, in anoxia, electrons are mostly used for the reduction of protons to produce molecular hydrogen (direct biophotolysis). The reducing equivalents [NAD(P)H] produced during oxidation of reserve carbon can also serve electrons for the generation of hydrogen (indirect biophotolysis) when the terminal electron acceptor oxygen is absent. Therefore, it is suggested that anaerobiosis plays the key role to divert biochemical pathways toward the hydrogen metabolism. Different biohydrogen production processes are illustrated in Figure 10.2.

10.3.1 Hydrogen Production via Dark-Fermentation Process

Fermentation is a metabolic process aiming at the regeneration of ATP, the metabolic energy currency, and at supplying metabolites for biosynthetic processes without involvement of oxygen. A group of microorganism can produce hydrogen via fermentative conversion of organic substrate or storage products without involvement of light energy. During heterotrophic growth, complex organic polymers are hydrolyzed by hydrolytic enzymes to monomers and are further oxidized to volatile fatty acids, alcohols, CO_2, and H_2. The overall process is represented in Figure 10.3.

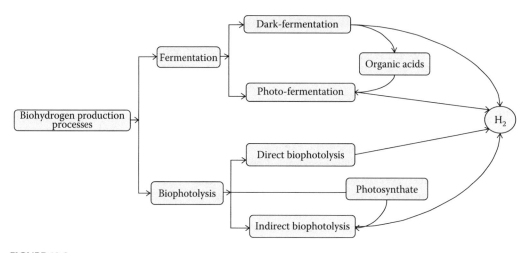

FIGURE 10.2
Different hydrogen production processes.

Generally, the organic substrates are cleaved to produce building blocks of the cell (biosynthesis of carbohydrate, protein, lipids, etc.), and part of the cleavage products is used for energy generation. During crisis, oxidation of endogenous storage products has also been found in microorganism. The oxidative reactions result in ATP regeneration by substrate-level phosphorylation. The oxidized carbon is usually released in the form of carbon dioxide.

The electrons released in these oxidation reactions are transferred to carriers such as nicotinamide adenine dinucleotides (NAD) or Fd. The reduced carriers ($NADH_2$ and FdH_2) are reoxidized by transferring the electrons to protons or other cleavage products or their derivatives, which are thereby reduced and then excreted, from the cells. The fermentation of carbohydrates and various others substances leads to the formation of the following products: ethanol, 2-propanol, 2–3 butanediol, *n*-butanol, formate, acetate, lactate, propanoate, butyrate, capronoate, acetone, carbon dioxide, and hydrogen either as sole products or as a mixture.

Complex organic polymers generate glucose by hydrolysis. Glucose produces pyruvate via glycolytic pathway and generates ATP. There are two types of biochemical reactions leading to the conversion of pyruvate to hydrogen.

One is typical for the obligate anaerobe clostridia (McCord et al. 1971) or facultative and thermophilic bacteria (Zeikus et al. 1979). They are forming products such as acetate or butyrate along with hydrogen. It is observed that during anaerobic fermentation, pyruvate, the product of glucose catabolism, is oxidized to acetyl coenzyme A (acetyl-CoA) employs by pyruvate-ferredoxin oxidoreductase (PFOR) (Uyeda and Rabinowitz 1971) and farther conversion to acetyl phosphate with concomitant generation of ATP and acetate. Oxidation of pyruvate to acetyl-CoA requires reduction of Fd. Reduced Fd is oxidized by Fe-hydrogenase and catalyzes the formation of H_2. The overall reaction of the processes can be described as follows:

$$\text{Pyruvate} + \text{CoA} + 2\,\text{Fd (ox)} \rightarrow \text{Acetyl-CoA} + 2\,\text{Fd (red)} + CO_2 \tag{10.2}$$

Hydrogen is formed when Fd is oxidized by Fd-H_2 oxidoreductase according to the following equation:

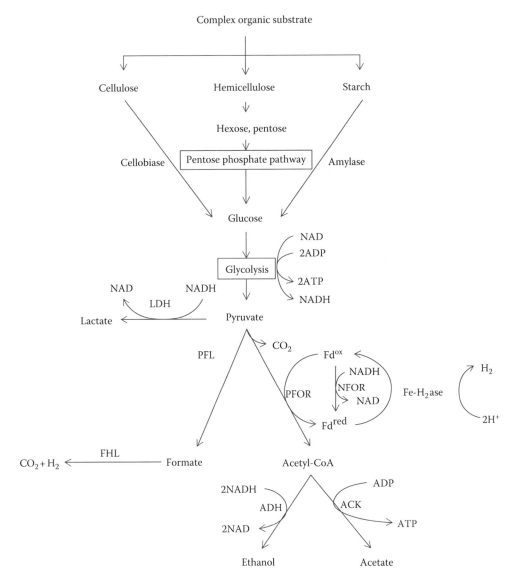

FIGURE 10.3
Overall biochemical pathways for dark fermentative metabolism (LDH—lactate dehydrogenase, PFOR—pyruvate ferredoxin oxidoreductase, PFL—pyruvate formate lyase, FHL—formate hydrogen lyase, ADH—alcohol dehydrogenase, ACK—acetyl CoA kinase, NFOR—NADH: ferredoxin oxidoreductase).

$$2H^+ + Fd\ (red) \rightarrow H_2 + Fd\ (ox) \tag{10.3}$$

The second type of H_2-evolving reaction is typical for some facultative anaerobic bacteria. For example *E. coli* and *Citrobacter* belong to family Enterobacteriaceae. This fermentation is known as mixed-acid fermentation, and formate is the characteristic product. The enzyme pyruvate formate lyase catalyzes the reaction (Knappe and Sawers 1990):

$$Pyruvate + CoA \rightarrow Acetyl\text{-}CoA + formate \tag{10.4}$$

Hydrogen is formed when formate is cleaved by an FHL enzyme system to give CO_2 and H_2. Stephenson and Stickland (1932) first described this pathway in the 1930s:

$$Formate \rightarrow CO_2 + H_2 \qquad (10.5)$$

In the lactic acid fermentation, pyruvate is directly reduced to lactate. No hydrogen is produced when ethanol, lactate, or propionate is the sole fermentation product.

In several types of fermentation, acetyl-CoA is converted to acetoacetyl-CoA. This intermediate serves as an electron acceptor for $NADH_2$ and is converted to yield butyrate, *n*-butanol, acetone, 2-propanol, or capronoate. The synthesis of acetoacetyl-CoA prevents the bacterium from regenerating ATP from acetyl-CoA and therefore results in a low cell yield. Some fermentative bacteria contain an enzyme, $NADH_2$-Fd oxidoreductase which catalyzes the reaction as follows:

$$NADH_2 + 2\ Fd\ (ox) \rightarrow NAD + 2\ Fd\ (red) \qquad (10.6)$$

Hydrogen can be evolved by using reduced Fd via mediation of a hydrogenase. The reduction of Fd by $NADH_2$ enables bacteria to produce hydrogen even from reducing power, which is primarily provided on the level of NAD. However, the reduction of Fd by $NADH_2$ is thermodynamically unfavorable reaction. This means that the reaction proceeds only if the reduced Fd is reoxidized by hydrogenase, and hydrogen is evolved. Hydrogen produced in a reaction broth has inhibitory effect on hydrogen production. Hydrogen yield can be increased if hydrogen is collected from a bioreactor in low partial pressure (Nath and Das 2004).

Carbohydrates, mainly glucose, are the preferred carbon sources for fermentation process, which predominantly give rise to acetic and butyric acids together with hydrogen gas. The overall reaction of hydrogen production via dark fermentation as follows:

$$C_6H_{12}O_6 + 2H_2O \rightarrow 2CH_3COOH + 2CO_2 + 4H_2 \qquad (10.7)$$

$$C_6H_{12}O_6 \rightarrow CH_3CH_2CH_2COOH + 2CO_2 + 2H_2 \qquad (10.8)$$

The end products of glucose fermentation by obligate anaerobic and facultative anaerobic chemoheterotrophs, like *Clostridia*, and Enteric bacteria, are produced through pyruvate. Facultative anaerobic bacteria give 4 mol of hydrogen per mol of glucose, whereas strict anaerobic bacteria give two (Nath and Das 2004). Facultative anaerobes are less sensitive to oxygen. In the presence of oxygen, they can switch over to aerobic respiration. Aerobic respiration rapidly depletes the oxygen in reaction broth and recovers the fermentative hydrogen production activity. As a consequence, a facultative anaerobe is considered a better microorganism than a strict anaerobe in fermentative hydrogen production processes (Oh et al. 2002).

Some facultative anaerobic Enteric bacteria can undergo anaerobic respiration instead of fermentation by using nitrate, fumarate, etc. as terminal electron acceptor. This might hamper the hydrogen production. Reaction broth in a reactor should be devoid of those electron acceptors.

Photofermentative bacteria use organic acids as a substrate and are oxidized completely to produce CO_2 and H_2. Organic acids produced during dark-fermentative metabolism could be used as a substrate of photofermentation. The two-stage process, dark fermentation followed by photofermentation using the spent media of dark fermentation, has been developed for complete oxidation of organic substrate and simultaneous H_2 production (Das 2009).

Anaerobic fermentation enables mass production of hydrogen via relatively simple processes from a wide spectrum of potentially utilizable substrates including refuse and waste products. Moreover, fermentative hydrogen production generally gives higher rate and does not rely on the availability of light sources. Thus, in a reactor, no light source is required. For the dark fermentative thermophiles, heat energy is required to maintain the high temperature of the reaction broth.

10.3.2 Hydrogen Production by Photosynthetic Reaction

Metabolism of dark-adapted microorganisms depends on the availability of substrate. Any time they can starve with depletion of substrate. In substrate crisis, energy can be derived only by limited endogenous catabolism and energy mostly dissipates in a non-reusable form. To sustain life and fulfill all metabolic demands, life evolved to utilize light energy from the sun. The sun is the unlimited source of energy. Microorganism has developed different light-harvesting complexes (LHCs) for capturing photons. Light energy is converted to chemical energy via photophosphorylation. Two principal classes of photosynthetic bacteria, the purple bacteria and the green bacteria carry out photosynthesis with single photosystem (Blankenship 1992).

Purple bacteria contain a photosystem II (PSII)-like reaction center (RC) and thus are incapable of reducing Fd but can generate ATP via cyclic electron flow. The electrons desired for nitrogenase-mediated hydrogen evolution are derived from inorganic/organic substrates. Bacteriochlorophyll (P_{870}) in RC is excited and releases electrons that reduce bacteriopheophytin in the RC. Once reduced, the bacteriopheophytin reduces several intermediate quinone molecules to, finally, a quinone in "quinone pool." Electrons are now transported from the quinone through a series of iron–sulfur proteins (FeS) and cytochromes (Cyts) back to the RC (P_{870}). Cyt bC_1 complex interacts with the quinone pool during photosynthetic electron flow as a proton motive force to derive ATP synthesis (Michel and Deisenhofer 1988) (Figure 10.4).

Purple sulfur bacteria are mostly photoautotrophs and obligate anaerobes. However, in dark, some species are capable of chemolithoautotrophic or chemoorganoheterotrophic growth in the presence of oxygen. Members of Chromatiaceae use sulfide or organic substrate as electron donors (Kampf and Pfennig 1980).

In these bacteria, H_2 evolution is coupled with inorganic (sulfur-containing compounds) or organic substrate-driven reserve electron flow. A reversed electron flow operates in purple bacteria to reduce NAD^+ to NADH (McEwan 1994). The reduced inorganic or organic substrates are oxidized by Cyts and electrons from them eventually end up in quinone pool. However, the energy potential of quinone is insufficiently negative to reduce NAD^+ directly. Therefore, the electrons from the quinone pool are forced backward to reduce NAD^+ to NADH. This energy requiring process is called reversed electron flow (McEwan 1994) (Figure 10.4). Electrons and ATP are used for CO_2 fixation as well as N_2 fixation. They contain Ribulose bisphosphate carboxylage (Rubisco) and phosphoribulokinase (Tabita 1999) and fix CO_2 via Calvin cycle. Most of them do not have complete TCA cycle as they lack oxoglutarate dehydrogenase (Kondratieva 1979). There is partial oxidation of substrate required for biosynthesis of some valuable products.

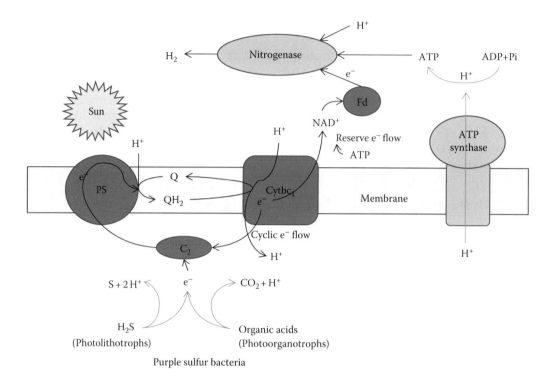

FIGURE 10.4
Cyclic electron flow, hydrogen evolution, and ATP synthesis in purple bacteria.

Hydrogen is evolved when electrons are transferred toward nitrogenase via Fd by expense of ATP. The majority of purple sulfur bacteria are active diazotrophs. They are able to fix N_2 to ammonia (Madigan 2004). Nitrogen fixation is linked with hydrogen production. Under nitrogen starvation most of the purple sulfur bacteria are able to produce molecular hydrogen (Mitsui 1975). No oxygen is evolved in this process. Hydrogen production through nitrogenase is a very expensive process and requires 4 ATP for 1 mol hydrogen generation (Equation 10.9).

$$Fd\,(red) + 4ATP + 2H^+ \rightarrow Fd\,(ox) + 4ADP + 4Pi + H_2 \tag{10.9}$$

where Pi is inorganic orthophosphate.

Some species contain membrane-bound hydrogenase that has hydrogen uptake capability instead of production of hydrogen. Hydrogen serves as photosynthetic electron donor for them. This capability was first detected in *Allochromatium vinosum* strain D (Gaffron 1935). Uptake hydrogenase resists excess loss of protons from the cell and maintains the redox potential of the cell (Equation 10.10).

$$H_2 \rightarrow 2H^+ + 2e^- \tag{10.10}$$

Hydrogen production is not well explored in purple sulfur bacteria. Most of the purple nonsulfur bacteria are well studied in this regards (Kumazawa and Mitsui 1982). Purple nonsulfur bacteria exhibit aerobic respiration in photoautotrophic growth. Hydrogen or organic substrates are the electron donors for photosynthesis. Electrons and ATP (generated by photophosphorylation) are used for fixing CO_2 via Calvin cycle. TCA cycle and electron transport chain operate completely and oxygen is the terminal electron acceptor (Table 10.2). However, in the presence of oxygen, no evolution of hydrogen occurs as oxygen is the acceptor of electron instead of protons. In addition, the activity of enzyme nitrogenase is suppressed in the presence of oxygen.

Hydrogen evolution in purple nonsulfur bacteria has been observed in the absence of oxygen. They depend on external organic substrate for carbon and electron source. In photoheterotrophic growth, they use light energy only for ATP generation via cyclic electron flow (Table 10.2). Reducing equivalents generated during oxidation of substrate reduce Fd followed by nitrogenase to fix nitrogen. Protons are mainly used for the formation of ammonia:

$$N_2 + 6H^+ + 6e^- \rightarrow 2NH_3 \tag{10.11}$$

In nitrogen starvation, nitrogenase catalyzes the formation of molecular hydrogen from protons. The final amount of hydrogen produced is also affected by the activity of uptake hydrogenase (Koku et al. 2002), which should be as low as possible. Highest yield of hydrogen was obtained at 30°C–37°C and with a high ratio of carbon to nitrogen in the medium (Eroglu et al. 1999). Photosynthetic nonsulfur purple bacteria are considered the best means of photobiological hydrogen production (Das and Veziroğlu 2001). The advantage of this process relies on application of organic wastes as the source of organic carbon (photofermentation). This adds extra advantage in economy of hydrogen production. Some success has been found using industrial wastewater as substrate (Yetis et al. 2000). However, some extra afford has to be given for pretreatment of the industrial effluent before using within the reactor as it may be toxic. It could be opaque or having some color that can reduce light diffusion in the reactor. The overall reaction of the photofermentation is given in Table 10.3 and Equation 10.12:

$$CH_3COOH + 2H_2O \rightarrow 4H_2 + 2CO_2 \tag{10.12}$$

Green sulfur bacteria are obligate anaerobic photoautotrophic bacteria. In contrast to purple bacteria, they have the PSI type RC reduced directly Fd. Electron flow around the RC is insufficient for the direct reduction of Fd; it can only generate ATP. Some sulfur compounds donate electrons to RC for the reduction of Fd. Bacteriochlorophyll (P_{840}) absorbs light near 840 nm and resides at a significantly more negative reduction potential in comparison to purple bacteria. Unlike purple bacteria in which the first stable electron acceptor molecule resides at about 0.0 reduction potential, the electron acceptors of green bacteria (FeS proteins) reside at about −0.6 reduction potential and have a much more electronegative reduction potential than NADH (Blankenship 1985). In green bacteria, Fd is reduced by FeS protein and serves directly as electron donor for the H_2 production (Figure 10.5).

Most of them can fix nitrogen by the enzyme nitrogenase, and in N_2-limited condition nitrogenase catalyzes the reduction of protons to hydrogen similar to purple bacteria. When H_2S donates electrons to green bacteria, sulfur globules remain outside the cell. This is unlike purple bacteria where the globules of sulfur remain inside of the bacterial cell (Imhoff 2004, Friedrich et al. 2005). RC donates electrons only to the nitrogenase for hydrogen production coupled with nitrogen fixation but not for CO_2 fixation. They lack Rubisco the key enzyme of Calvin cycle. Green sulfur bacteria fix CO_2 by reductive TCA cycle (Holo and Sirevåg 1986, Hugler et al. 2005) (Table 10.2). Therefore, there is no competitive inhibition of hydrogen production by CO_2 fixation. CO_2-free environment within

TABLE 10.2

Mode of Metabolism of Different Types of Photosynthetic Bacteria during Hydrogen Production

Photosynthetic Bacteria	Sensitivity to O_2	Growth Condition	Carbon Metabolism		During H_2 Production		Electron Donor
			Anabolism	Catabolism	Requirement of Light Energy	Mode of ATP Formation	
Purple sulfur	Obligate anaerobes	Photoautotrophic	Calvin cycle	Partial TCA cycle	To transfer electrons to Fd and ATP formation	Photophosphorylation	Mostly sulfur compounds
Purple nonsulfur	Facultative anaerobes	Photoheterotrophic	(External substrate is used)	Photofermentation	ATP formation	Substrate-level phosphorylation	Organic substrate
Green sulfur	Obligate anaerobes	Photoautotrophic	Reductive TCA cycle	—	ATP formation	Photophosphorylation	Mostly sulfur compounds

TABLE 10.3

Various Biological Hydrogen Production Processes with General Overall Reactions Involved Therein, Broad Classification of Microorganisms Used, and Their Relative Advantages

Process	General Reactions and Broad Classification of Microorganisms Used	Advantages
1. Dark fermentation	$C_6H_{12}O_6 + 2H_2O \rightarrow 2\ CH_3COOH + 2CO_2 + 4H_2$ Fermentative bacteria	(i) It can produce H_2 all day long without light (ii) A variety of carbon sources can be used as substrates (iii) It produces valuable metabolites such as butyric, lactic and acetic acids as by products (iv) It is anaerobic process, so there is no O_2 limitation problem
2. Photofermentation	$CH_3COOH + 2H_2O + light = 4H_2 + 2CO_2$ Purple bacteria, microalgae	(i) A wide spectral light energy can be used by these bacteria (ii) Can use different waste materials
3. Direct biophotolysis	$2H_2O + light = 2H_2 + O_2$ Microalgae	(i) Can produce H_2 directly from water and sunlight (ii) Solar conversion energy increased by 10 folds as compared to trees, crops
4. Indirect biophotolysis	$6H_2O + 6CO_2 + light = C_6H_{12}O_6 + 6O_2$ $C_6H_{12}O_6 + 2H_2O = 4H_2 + 2CH_3COOH + 2CO_2$ $2CH_3COOH + 4H_2O + light = 8H_2 + 4CO_2$ Overall reaction: $12H_2O + light = 12H_2 + 6O_2$ Microalgae, cyanobacteria	(i) Can produce H_2 from water (ii) Has the ability to fix N_2 from atmosphere

a reactor is not required but N_2-free environment is required to inhibit NH_3 formation. Storage product in green sulfur bacteria is mainly glycogen. In dark, they can catabolize storage product to produce different organic acids similar to dark fermentation. It has been suggested that storage product serve as energy source in dark and reducing power in light (Sirevåg and Ormerod 1977).

Very few studies have been done for the hydrogen production from green gliding bacteria. They have PSII type of RC like purple bacteria but having light-harvesting structure chlorosome like green sulfur bacteria. Most of them probably do not have nitrogenase (Heda and Madigan 1986). Hydrogenase was isolated from green gliding bacteria *Chloroflexus aurantiacus* and used as efficient biocatalyst for hydrogen production (Gogotov et al. 1991). This organism can survive in dark only in the presence of oxygen. This organism survives in dark having orange color but in the presence of light it becomes green. *Chloroflexus* uses reduced sulfur compounds such as hydrogen sulfide, thiosulfate or elemental sulfur, and also hydrogen. *C. aurantiacus* is thought to grow photoheterotrophically in nature, but it has the capability of fixing inorganic carbon through photoautotrophic growth. *C. aurantiacus* has been demonstrated to use a novel autotrophic pathway known as the 3-hydroxypropionate pathway (Hügler et al. 2003). The complete electron transport chain for *Chloroflexus* spp. is not yet known.

10.3.3 Hydrogen Production through Biophotolysis

Photosynthetic bacteria exhibit photosynthesis without evolution of oxygen. It is advantageous as the enzymes for hydrogen production are sensitive to oxygen. Cyanobacteria

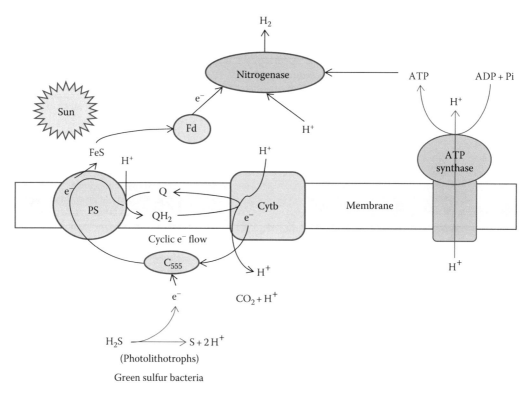

FIGURE 10.5
Cyclic electron flow, hydrogen evolution, and ATP formation in green bacteria.

evolve oxygen during photosynthesis, though these are prokaryotes. Some of them developed separate structure heterocyst to protect nitrogenase from oxygen. Some of them use hydrogenase for hydrogen production in dark. In dark, hydrogenase is reduced by the oxidation of storage product. Only difference to the heterotrophic growth is that they are not using external substrate. All the electrons used for the reduction of Fd are actually coming from water oxidation. Thus, this type of hydrogen production is known as indirect biopholysis instead of dark fermentation. It is also possible to produce hydrogen through direct biophotolysis by sparging the reaction broth continuously with some inert gas. In green algae, hydrogen production through combination of direct and indirect biophotolysis has been achieved by suppressing the oxygen-evolving complex by using sulfur-deprived reaction broth (Melis 2002).

10.3.3.1 Direct Biophotolysis of Water

Hydrogen production through direct biophotolysis is a biological process to split water to produce hydrogen and oxygen by utilizing solar energy. Hydrogen production by direct biophotolysis is mostly observed in green algae and cyanobacteria. Both of them contain PSI and PSII for capturing light energy and exhibit oxygenic photosynthesis like higher plants. In PSII, solar energy is captured by the photosynthetic pigments present in LHC and transferred to the RC of PSII. The LHC is composed of hundreds of molecules of chlorophylls and accessory pigments. Most of the chlorophyll a in a cell is actually involved in light harvesting. These are referred to as antenna chlorophylls. The light energy transferred from antenna chlorophylls excites a special pair of chlorophyll a present in the RC. The magnesium ion (Mg^{2+}) present in the ground state chlorophyll is having no unpaired electron (singlet state). But after getting the solar energy, one electron from 2P orbital transferred to the 3S orbital leading to the triplet state with two unpaired electrons. The excited state of chlorophyll is very unstable. The chlorophyll-a molecule gets rid of this energy by ejecting an electron and generating a strong oxidant (photooxidation) PS680* (E = 0.82 V) capable of splitting water into protons (H^+), electrons (e^-) and O_2. Electron from the water is taken by the oxidized chlorophyll to return into normal state (Equation 10.13, Figures 10.6 and 10.7).

$$2H_2O \rightarrow 4H^+ + 4e^- + O_2 \tag{10.13}$$

Two photons are required for releasing one electron form magnesium. However, as soon as the excitation energy exceeds the capacity of the RC different dangerous by products, that is, triplet state of magnesium, singlet oxygen and superoxide can be formed and lead to severe photodamages. One of the leading photodamage has been observed is oxidation of pigments lead to photobleaching and might be death of cells. However, in cyanobacteria, repigmentation after exposure of low irradiation has been observed (Nultsch and Agel 1986). Thus, photobleaching is not exactly a photodamage phenomenon, rather it is a light adaptation process. There are some protective pathways to escape from generation of harmful intermediates one of them is direct quenching of chlorophyll, that is, transfer of excess energy to the adjacent carotenoid molecules. Carotenoids liberate excess energy by fluorescence and heat. Algal cells directly exposed to light energy for a longer time can overcome photodamages by another phenomenon called photoinhibition, that is, light-dependent inhibition of PSII (Aro et al. 1993). Algal cells that are exposed directly in light

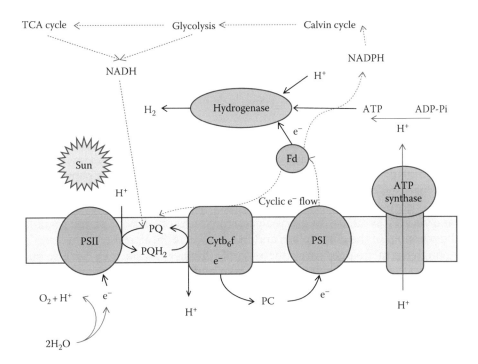

FIGURE 10.6
Direct and indirect biophotolysis of green algae for H_2 production.

FIGURE 10.7
Difference in electron pairing in Mg^{2+} of chlorophyll in ground state (singlet) and in excited state (triplet).

struggling with these problems but the cells floating under the surface cells might not be able to get enough solar energy. Overall, photosynthetic efficiency becomes lower as the surface cells are not able to manifest excess energy and deeper cells are not getting enough solar energy. Hydrogen production through biophotolysis in algae solely depends on light conversion efficiency of PSII, so photobleaching and photoinhibition might hamper and even stop hydrogen production. Hydrogen production in a photobioreactor can be enhanced by increasing overall photosynthetic efficiency that depends on proper light regime, transparency of the reactor, good agitation in the reactor, and proper architect of the reactor. Proper design of a photobioreactor might help to avoid shading effect, photo-damages, and loss of energy by fluorescence and heat.

Electrons are liberated out from the PSII and transferred to PSI through a series of electron carriers, Pheo, Plastoquinone, Cyt b_6f complex, plastocyanine, which is in the increasing order of standard redox potential and hence electron passes spontaneously from negative redox potential to positive redox potential. Electrons are transferred through these electron carriers and Cyt complex to PSI (P700), which further reduces the oxidized Fd (Equation 10.14). Protons from the stroma are drawn into the thylakoid lumen through PQ center along with electron flow-generating proton gradient across the thylakoid membrane (which separates lumen and stroma). Light energy is converted into chemical energy by generating ATP when the protons are released back to the stroma through ATPase. If the cells suffer from shortage of energy, electrons flow back to PQ center from Fd. This cyclic flow of electron generates only ATP (Munekage et al. 2004).

$$2Fd\,(red) + 2H^+ + NADP^+ \rightarrow 2Fd\,(ox) + NADPH + H^+ \tag{10.14}$$

Fd is located in stroma where its reduced form is specific to hydrogenase as well as Fd-NADP$^+$ reductase and shows its bifunctional ability. In aerobic condition, biomass is produced by fixing carbon using reducing power ($NADPH_2$) via Calvin's cycle or pentose phosphate pathway (Horecker 2002) and produce carbohydrates (CH_2O) and/or lipids (Hatch and Slack 1970). Energy is driven from ATP produced by proton gradient across the thylakoid membrane via ATP synthase. In green algae, stored carbohydrate is starch granules present in the pyrenoid of the chloroplast, and in cyanobacteria, stored carbohydrate is glycogen. Although the microalgae carry out oxygenic photosynthesis, they can also encounter anaerobiosis in their natural habitats: Tarns, lake, and sea sediments become anoxic owing to insufficient water stirring or to O_2 deprivation during algal blooms. Photosynthetic organisms survive under anaerobic conditions by switching their metabolism to fermentation (Gaffron and Rubin 1942). A problem regarding the fermentative metabolism is the harmful end products. High concentrations of ethanol or formate are toxic, and organic acids (e.g., lactic oracetic acid) acidify the cell (Kennedy et al. 1992). Through dark fermentation of reserve carbon, O_2 and H_2 production is temporally (light/dark) separated (Miura et al. 1997). This is advantageous for the oxygen-sensitive hydrogenase. In anaerobic dark conditions, PFOR is responsible for decarboxylation (CO_2 evolution) of pyruvate to acetyl-CoA linked with H_2 production via reduction of Fd (Equation 10.15).

$$\text{Pyruvate} \rightarrow \text{Acetyl-CoA} + CO_2 + H_2 \qquad (10.15)$$

However, in the presence of light, the photosynthetically produced O_2 permits oxidative pyruvate degradation and respiratory ATP generation. In anaerobic environment, photosynthetically generated O_2 is not sufficient to activate respiratory chain fully. However, the partially active respiratory chain presumably does not suffice as an electron sink. Furthermore, under this physiological anaerobiosis, the electron-consuming Calvin cycle also fails to function as an electron sink. The hydrogenase plays an important role in physiological anaerobiosis (Happe et al. 2002). Photosynthetically generated electrons reduce protons and evolve molecular hydrogen. This is known as hydrogen production through direct biophotolysis as electrons are coming directly from water. Electrons (e^-) from reduced Fd are used by the [FeFe]-hydrogenase to reduce protons (H^+) and evolve hydrogen (H_2) (Schnackenberg et al. 1996).

Green algae are the only known potential eukaryotes with both oxygenic photosynthesis and a hydrogen metabolism. N_2-fixing cyanobacteria produce hydrogen mainly by nitrogenase, instead of bidirectional hydrogenase. However, in several non-N_2-fixing cyanobacteria, H_2 evolution is also observed through bidirectional hydrogenase (Howarth and Codd 1985, Serebryakova et al. 1998) (Equation 10.16). It has been observed that the unicellular non-N_2-fixing cyanobacteria exhibit relatively low H_2 production rate compared to N_2-fixing strain:

$$2\text{Fd (red)} + 2H^+ \rightarrow H_2 + 2\text{Fd (ox)} \qquad (10.16)$$

Efficiency of hydrogen production by direct biophotolysis is very low as the enzyme hydrogenase is highly sensitive to even moderately low concentration of oxygen. Simultaneous production of oxygen from splitting of water during hydrogen production is the major problem inside a photobioreactor. Another problem may arise to have an explosion within the reactor due to mixing of these gases. Researchers are seeking of a good way out from this problem using different bioengineering approaches. To maintain continuous anoxia, sparging with inert gases might be the one solution. In anoxia or in physiological anaerobiosis, reducing equivalents from fermentation are reoxidized by the reduction of protons.

Melis and coworkers developed sulfur-deprived method for green algae to achieve anaerobic condition within a photobioreactor (Melis 2002). This approach is to generate biomass through photosynthesis then inactivate water splitting activity to generate anoxia. Hydrogen evolves by deriving electrons from oxidation of reserve carbohydrate. After production of biomass if the green algae are shifted to the media deprived of sulfur, PSII gets partially inactivated. Therefore, oxygen evolution reduced to generate anaerobic condition in the reactor. Anaerobiosis activates hydrogenase, and electrons coming partly from water splitting and partly from catabolism of pyruvate are reduced to evolve hydrogen. Thus, hydrogen production by sulfur-deprived method is a combination of direct and indirect biophotolysis. Sulfur is very crucial for biosynthesis of proteins present in PSII. If the algal cells are exposed to continuous strong light that inhibits PSII almost completely, but has little affect on PSI. This phenomenon is known as photoinhibition. Research was stimulated by a paper by Kyle et al. in 1984, showing that photoinhibition is accompanied by selective loss of a 32 kDa protein (later identified as the PSII RC protein D1) followed by activation of the RC through rapid inbuilt repair mechanism. Only when the repair cycle is unable to match the rate of damage to PSII, photoinhibition is observed (Kyle et al. 1984). In absence of sulfur, rebiosynthesis of the D1 protein after loss is inhibited, as the production of sulfur-containing amino acids cysteine or methionine becomes impossible. Thus in absence of sulfur, activity of PSII is hampered but has very little effect on PSI. Endogenous sulfur has been contributed to maintain the physiology of the cell. Partial inhibition of PSII can generate anaerobic condition for the cell within a photobioreactor, as there is less water-oxidation activity to evolve O_2 and the residual O_2 is used by respiration (Wykoff et al. 1998). Anaerobiosis induces the expression of [FeFe]-hydrogenase in algal cells (Happe and Kaminski 2002, Forestier et al. 2003), and sustained hydrogen production can be achieved (Ghirardi et al. 2000, Melis et al. 2000). In this condition, electrons are derived less from the water mostly from the oxidation of reserve carbohydrates. Electrons are derived from the reserve carbon source is reduced PQ pool. The two enzyme complexes, NAD(P) dehydrogenase and ferredoxin-NADP-reductase are involved in electron transfer from NAD(P)H and electron flow around PSI (Nixon 2000). In this situation, as the terminal electron acceptor (O_2) is absent, the excess electrons are deposited through another way, that is, the Fd-hydrogenase pathway (Mus et al. 2005), to produce H_2. Moreover, the presence and absence of O_2 acts as a switch for the hydrogen production. The existence of this anaerobic pathway in aerobic phototrophs probably focused the evolutionary linkage, and this has been exploited by researchers for the production of hydrogen. Significant research has been done to improve H_2 production targeting this NAD(P)H plastoquinone oxido-reductase pathway. A mutant alga that overaccumulate starch significantly increases the rate and duration of H_2 production (Kruse et al. 2005a). Chen et al. (2003) reported another type of mutant with inefficient ability to transport sulfate into the chloroplast. This was obtained by effecting sulfate permease activity and this mutant can evolve H_2 without depleting sulfate in the culture media. Some PSII inhibitors have been also used to inhibit water oxidation activity. After a sufficient dark anaerobic incubation (for induction of hydrogenase) by adding PSII inhibitor 3-(3,4-dichlorophenyl)-1,1-dimethylurea that binds at the Q_B site of PSII and blocks transport of electrons, high rate of H_2 production can be detected (Happe and Naber 1993). Electron flow around PSI can be sustained and enhanced by adding external electron donor like NAD(P)H into the cell culture media and to maintain the cell growth (Mus et al. 2005). Instead of acetate, which is normally used in the media as a carbon source, CO_2 emitted in industries can be coupled to the system as it will be less expensive in large-scale H_2 production as well as helpful for waste removal (Skjånes et al. 2007).

Achieving proper anaerobic conditions within the cell is the main key for H_2 production. Still there are many other physiological barriers. Lee and Greenbaum (2003) achieved enhanced production of H_2 by eliminating the competitive inhibition of photosynthetic H_2 production by CO_2 fixation. They inserted a proton-conductive channel in the thylakoid membrane to transport the protons across the membrane and ceased the photophosphorylation (ATP formation) that is required for CO_2 fixation. Since this designer alga no longer requires a CO_2 (HCO_3^-)-free environment for hydrogen production, this can contribute to increase the photosynthetic efficiency as PSII requires bicarbonate (HCO_3^-) for better activity (Lee and Greenbaum 2003).

Though there are many bottlenecks to achieve theoretical value of hydrogen production by biophotolysis, it might be the best potential method to satisfy the future energy demand because the reducing power and energy are derived from virtually limitless resources, that is, water and sunlight respectively. Also the theoretical energy efficiency is much higher for hydrogen production from biophotolysis (40%) (Prince and Kheshgi

2005) compared to that from biomass. The overall reaction in biophotolysis is as given in Equation 10.17. Advantages of hydrogen production through direct biophotolysis are summarized in Table 10.3.

$$2H_2O + Fd\ (ox) + h\nu \rightarrow Fd\ (red)(4e^-) + 4H^+ \rightarrow Fd\ (ox) + 2H_2 \tag{10.17}$$

10.3.3.2 Indirect Biophotolysis

In direct biophotolysis, electrons are derived directly from water, but in indirect biophotolysis, electrons are donated by the reserve carbon source that has been produced by fixing CO_2 via the Calvin cycle during photosynthesis. Thus, the electrons are not directly coming from water as in case of direct biophotolysis. The stored carbohydrate is oxidized to produce H_2. The general reaction is as follows (Table 10.3):

$$12H_2O + 6CO_2 \rightarrow C_6H_{12}O_6 + 6O_2 \tag{10.18}$$

$$C_6H_{12}O_6 + 12H_2O \rightarrow 12H_2 + 6CO_2 \tag{10.19}$$

Indirect biophotolysis is very common in cyanobacteria. This is a very efficient process to separate O_2 and H_2 evolution phases (Hallenbeck and Benemann 2002) (Equations 10.18 and 10.19). Nonheterocystous cyanobacteria can separate O_2 and H_2 production in time (temporal separation), in a process where H_2 is produced from bidirectional hydrogenase-utilizing stored carbohydrates as electron donor (Schmitz et al. 1995, 2002, Tamagnini et al. 2002).

Many species of heterocystous cyanobacteria contain another enzyme called nitrogenase, which fix nitrogen (N_2) to ammonia (NH_3) and produce H_2 in the process. Nitrogenase can also reduce protons (H^+) by deriving electrons (e^-) from a reserve carbon source produced during photosynthesis (Equations 10.20 and 10.21). In the heterocycts, only PSI is present. The electrons are donated to PSI in the heterocyst coming from reserve carbon transported from the neighbor vegetative cell. These cells are able to maintain an anaerobic atmosphere through compartmentalization (special separation), and thereby have the advantage to avoid the enzyme inactivation by oxygen (Figure 10.8). However, this hydrogen production is energetically burdensome due to biosynthesis and maintenance of heterocysts and the significant ATP requirement of nitrogenase. Another problem is that the presence of uptake hydrogenase in the heterocyst makes the yield of hydrogen lower than its theoretical maximum value as it oxidizes the molecular hydrogen:

$$N_2 + 8H^+ + Fd\ (red)(8e^-) + 16\ ATP \rightarrow 2\ NH_3 + H_2 + Fd\ (ox) + 16ADP + Pi \tag{10.20}$$

$$8H^+ + 8e^- + 16ATP \leftrightarrow 4H_2 + 16ADP + 16Pi \tag{10.21}$$

Recently, the presence of uptake hydrogenase antigens has been reported in both the vegetative cells and heterocysts of *Nostoc punctiforme*, corresponding most probably to an inactive and an active form of the enzyme (Seabra et al. 2009). It was known that nonheterocystous N_2-fixing cynobacteria lack O_2-dependent uptake hydrogenase activity and can evolve more H_2 than heterocystous cyanobacteria (Kashyap et al. 1996); however, according to the recent reports, the uptake hydrogenase has been found in almost all

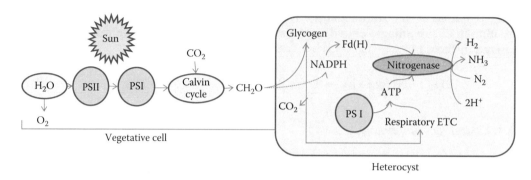

FIGURE 10.8
Indirect biophotolysis in heterocystous cyanobacteria for H_2 production.

N_2-fixing strains examined so far (Tamagnini et al. 2002, Schütz et al. 2004). Heterocyst provides spatial separation of O_2 evolution and H_2 evolution. For the nonheterocystous cyanobacteria in addition to temporal separation (light and dark), high rate of synthesis of nitrogenase to maintain a certain amount of active enzyme under aerobic conditions is observed. Second, it has been seen that, nitrogenase can be converted from active to non-active form after a sudden and short-term exposure to high oxygen concentrations (Stal and Krumbein 1985, 1987).

10.4 Genetic and Metabolic Engineering Approaches for the Improvement of Hydrogen Production

A different group of microorganisms mostly have different molecular mechanisms for the generation of hydrogen. Many proteins are involved for hydrogen generation. Some of them are well characterized in protein as well as gene level. Researchers are targeting the genes involved directly or indirectly in hydrogen generation for maximization of hydrogen production. Many genetic and metabolic engineering works have been done in different hydrogen-producing organisms to enhance the yield. Characterization has been done for the enzymes involved for catalyzing the production of hydrogen including their types, maturation systems, different domains, and their functions. Three different enzyme hydrogenase, nitrogenase, and FHL can catalyze the reduction of protons to hydrogen. The ultimate goal of biohydrogen research is to produce as much hydrogen as possible. Genetic and metabolic modifications can be a very effective and promising method to optimize and redirect the flow of reducing equivalents (electrons) to the enzyme and also the accumulation of protons (H^+), to achieve a satisfactory yield of hydrogen (H_2) in large scale. The various genetic and metabolic engineering strategies for overcoming the bottlenecks of hydrogen production are summarized in Table 10.4.

10.4.1 Enhancing the Uptake of External Substrate for Fermentative Hydrogen Production

In fermentative hydrogen production, external substrates are the sole source of electrons. Genetic modification can be done for better uptake of external substrate.

TABLE 10.4

Different Molecular Approaches for Improvement of Hydrogen Production

Strategies	Advantages	Microorganisms	References
Substrate uptake	Increasing the reducing power	*C. reinhardtii*	Doebbe et al. (2007)
Metabolic redirection by inhibiting or overexpressing the crucial metabolic enzymes	Redirecting the e⁻ flux toward hydrogen-producing enzyme	*E. cloacae* IIT-BT08	Kumar et al. (2001)
		Synechocystis (mutant M55)	Cournac et al. (2004)
		R. capsulatus (IR4)	Vignais et al. (2006)
Enhancing the capacity of deriving e⁻ from carbohydrates	Contributing in hydrogen production in anoxic dark condition	*C. reinhardtii* (Stm6 strain)	Kruse et al. (2005a)
Pigment reduction	Increasing the photoconversion efficiency	*R. sphaeroides* (mutant MTP4 & P3)	Vasilyeva et al. (1999) Kondo et al. (2002)
		Green algae (*C. reinhardtii, Dunaliella salina*)	Polle et al. (2003), Mussgnug et al. (2007)
		Cyanobacteria (*Synechocystis* PCC 6803)	Bernat et al. (2009)
Generating anaerobic environment	Activating hydrogen-producing enzyme	*C. reinhardtii* (*arp* mutant) *C. reinhardtii* (D1 protein mutant)	Melis (2007) Ruhle et al. (2008) Torzillo et al. (2009)
Oxygen-tolerant enzyme	Producing hydrogen in the presence of oxygen	*A. aeolicus* *Chlamydomonas sp.*	Guiral et al. (2006) Ghirardi et al. (2000)
Introducing foreign efficient hydrogen-producing enzyme	Enhancing hydrogen production in that particular microorganism that may be efficient in other criteria	*S. elongates* *R. rubrum*	Miyake and Asada (1997), Asada et al. (2000), King et al. (2006) Kim et al. (2008a)
Eleminating uptake hydrogenase activity	Inhibiting reoxidation of produced hydrogen	*T. roseopersicina* (*hypF*-deficient mutant)	Fodor et al. (2001)
		R. rubrum (*hupL*-deficient mutant)	Ruiyan et al. (2006) Gokhan et al. (2008)
		R. sphaeroides (*hupSL*-deficient mutant)	
Eliminating competitive inhibition by other e⁻ acceptor	Redirecting the e⁻ flux toward the hydrogen-producing enzyme	Green algae (*C. reinhardtii* 137c)	Lee and Greenbaum (2003)

By overexpressing the external substrate transporter proteins, it might be possible to increase the hydrogen production. There are some reports about modification in transporter proteins. Species of *Chlorella* has a hexose uptake protein that is involved in transferring external carbohydrate to the cell. Recently, the HUP1 (hexose uptake protein) from *Chlorella kessleri* was introduced into *Chlamydomonas reinhardtii* to increase the supply of external glucose into the cell. Hydrogen production capacity was increased about 150% by supplying 1 mM glucose to a strain of Stm6 where HUP1 has been inserted (Doebbe et al. 2007).

10.4.2 Redirecting the Electron Pull toward Hydrogen Production

Mostly the microbial communities produce hydrogen by fermentative metabolism of organic substrates. Fermentative hydrogen production depends on the substrate conversion efficiency. All the metabolic engineering has been done for better conversion of substrate to hydrogen. Theoretically, 1 mol glucose can produces 12 mol of hydrogen in absence of oxygen. But this reaction thermodynamics is less favorable ($\Delta G = -9.5$ kcal mol^{-1} glucose under ambient condition) and no ATP has been formed. Most of the electron flask has been redirected toward formation of acetate, yielding -51.6 kcal mol^{-1} (Thauer et al. 1977), theoretically sufficient for the production of several ATPs. Strategy has been taken by the researchers to redirect all electron pull toward proton reduction. Reducing equivalents (NADH) generated during fermentative oxidation are utilized for some other acids and alcohol formation. Tanisho et al. (1989) reported that these excess reducing equivalents could be disposed off via proton reduction mediated by electrons carrier and hydrogenase leading to the formation of H$_2$ in *E. aerogenes*. Redirection of metabolic pathway toward hydrogen generation has been achieved in *Enterobacter cloacae* IIT-BT08 by blocking alcohol and some acid formations. Double mutants, with defects in both alcohol and organic acid formation pathways, had higher H$_2$ yields (3.4 mol mol^{-1} glucose) than the wild-type strain (2.1 mol mol^{-1} glucose) (Kumar et al. 2001, Kumar and Das 2000). Metabolic pathway redirection has done in *E. coli* also by upregulating FHL complex and disrupting the pathways competing for pyruvate (Yoshida et al. 2005). Many different genes are involved in fermentative hydrogen metabolism including the key proteins and accessory proteins. Modification in accessory genes can also help to increase hydrogen production (Fan et al. 2009). Some modifications have also been made in the metabolic pathways to redirect the electron flux toward the hydrogen-producing enzyme. Thus, NAD(P)H-dehydrogenase-deficient mutant M55 of *Synechocystis*, a non-N$_2$-fixing cyanobacteria, accumulate NADPH and evolve H$_2$ only by bidirectional NAD(P)-dependent hydrogenase (Cournac et al. 2004). By improving the synthesis of D-malic enzyme and using D-malate as the sole carbon source in strain IR4 of *R. capsulatus*, a 50% higher hydrogen production has been demonstrated (Vignais et al. 2006).

Along with external electron donors like H$_2$O for cyanobacteria and green algae, and sulfur-containing inorganic compounds or organic acids for photosynthetic bacteria, reducing equivalents can also be derived either from reserve carbon source produced during biomass formation or from an external organic source. In cyanobacteria and green algae, this alternative metabolic process can produce hydrogen via fermentative reactions and/or reducing PQ by NAD(P) dehydrogenase to maintain the NAD/NADH balance and ATP supply (Gfeller and Gibbs 1984, Hemschemeier and Happe, 2005). After random gene insertion in *C. reinhardtii*, a strain named Stm6 with modified respiratory metabolism was isolated. This strain is able to accumulate large amount of starch in the cells, and has low dissolved oxygen concentration. It can produce 5–13 times more hydrogen than the wild type (Kruse et al. 2005b). Supplying a carbon source like acetate or glucose externally, hydrogen production can be increased more than when using inorganic media. Metabolic engineering by genetic modification could be a stable solution for efficient hydrogen production. This approach could rationalize different other problems and parameters for producing hydrogen in a large scale with a pilot reactor.

10.4.3 Enhancement of Photoconversion Efficiency in Photosynthetic Microorganisms

Photosynthetic bacteria are different from dark fermentative bacteria in their source of energy. They convert light energy to chemical energy. Thus, the photosynthetic efficiency is important for the supply of energy for photoheterotrophic and photoautotrophic bacteria. Hydrogen production by photosynthetic bacteria is catalyzed mostly by nitrogenase with the expense of 4 mol of ATP for generation of 1 mol hydrogen. Thus, hydrogen production by nitrogenase is an energy-intensive process and depends much on conversion of light energy to formation of ATP. Photosynthetic efficiency can be increased by modifying the light-harvesting antenna complexes responsible for capturing the solar energy. Reduction of pigments can reduce shelf shading and loss of energy by fluorescence and heat. Reduction of pigments can give additional advantage to the design of a photobiorector for enhancing the photoconversion efficiency. Photofermentative bacteria *Rhodobacter* is well explored for hydrogen production. It has two antenna complexes LH1 and LH2. Mutant *Rhodobacter* (P3 mutant) with 2.7-fold decreases in core antennal (LH1) content and 1.6-fold increases in peripheral antennal (LH2) content has given accelerated H_2 production compared to wild type (Vasilyeva et al. 1999). Using a *Rhodobacter sphaeroides* mutant MTP4 within the plate-type reactor, 50% more hydrogen is produced than its wild-type counterpart *R. sphaeroides* RV (Kondo et al. 2002).

Large antenna complexes of cyanobacteria and green algae help them to grow in low-light condition, but this might not be very useful in normal light conditions. High amounts of pigments in the cell produce self-shading and lower the photosynthetic efficiency in large photobioreactors. Greenbaum has reported maximum light conversion efficiency of 80% for green algae (Greenbaum 1988). The recent research has mostly focused on improving the light absorption capacity. Reduction in pigment content can lead to better penetration of light inside the reactor and reduce the wastage of light energy by fluorescence and heat. It has been observed that hydrogen production could be increased by a mutant algal strain having reduced chlorophyll antenna (Polle et al. 2003). It was also expected that the cyanobacteria deficient in phycobilisomes, which implies uncoupling of electron transport and preferable excitation of PSI by illumination, can enhance the rate of hydrogen production (Bernát et al. 2009). Another promising technology, the RNA interference or RNAi technology has also been used to downregulate the entire family of LHCs in *C. reinhardtii* (Mussgnug et al. 2007).

10.4.4 Improvement of Hydrogen-Producing Enzymes

Hydrogen production is catalyzed mostly by hydrogenase. But in diazotrophs, hydrogen production is mostly coupled with nitrogen fixation and is catalyzed by the enzyme nitrogenase. Both the enzymes are characterized and studied well. Phylogenetically different enzymes, their corresponding genes, maturation systems, catalytic domains, and mechanism of actions are important to better understand the enzyme characteristics and their limitations. These information help researchers to improve hydrogen production by doing modification in molecular level. A brief description about phylogenetically different enzymes and their catalytic domains are discussed in the following.

10.4.4.1 Hydrogenase

Phylogenetic analyses, based on sequence alignments of catalytic subunits of hydrogenases, have led to the identification of three phylogenetically distinct classes of enzymes, [NiFe]-hydrogenases, [FeFe]-hydrogenases, and Fe-S free hydrogenases, initially called metal free and now renamed [Fe]-hydrogenases, each characterized by a distinctive functional core that is conserved within each class (Vignais and Billoud 2007). Although turnover for hydrogen production is much higher for [FeFe]-hydrogenases, [NiFe]-hydrogenases are more oxygen tolerant. [Fe]-hydrogenase is, however, not involved in hydrogen production. [Fe]-hydrogenases catalyze the reduction of CO_2 by H_2 to methane, and the bacteria that rely on such enzymes are mostly methanogens.

10.4.4.1.1 Types of Hydrogenase Involved in Hydrogen Production

[NiFe]-hydrogenases. They are widespread among prokaryotes and found to catalyze both H_2 evolution and uptake, and with low-potential multiheme Cyts such as Cyt c_3 they act as either electron donors or electron acceptors, depending on their oxidation state (Cammack et al. 1994). The revised system of Vignais et al. (2001) subdivides [NiFe]-hydrogenases into four groups depending on their functional role within the organism.

Membrane-bound uptake hydrogenase. They (encoded by *hupSL*) link the oxidation of H_2 to the reduction of anaerobic electron acceptors, such as NO_3^-, SO_4^{2-}, fumarate, or CO_2 (anaerobic respiration), or to O_2 (aerobic respiration), with recovery of energy in the form of a proton motive force. The enzyme has two subunits. Larger one is *hupL*-coded protein, responsible for uptaking hydrogen, and the smaller subunit, coded by *hupS*, is involved after the reduction (Vignais and Colbeau 2004). In the presence of uptake hydrogenase, there is no net H_2 production as the hydrogen formed is usually reoxidized via a Knallgas reaction, under ambient conditions.

Cytoplasmic H_2 sensors. They are predominantly cytoplasmic proteins (lack signal peptide sequence at the N-terminal end) and are not directly involved in energy transducing reactions. Their role is to detect the presence of H_2 in the environment and to trigger a cascade of cellular reactions controlling the synthesis of respiratory [NiFe]-hydrogenases (Vignais and Colbeau 2004).

Bidirectional hydrogenase. They are termed as bidirectional hydrogenases since these enzymes are able to bind to soluble factors like NAD or NADP and cause them to reoxidize anaerobically using protons released from water as electron acceptors. Cyanobacterial [NiFe]-bidirectional hydrogenases have been purified as pentameric enzyme, and the corresponding structural genes, *hoxEFUYH*, are clustered together with three open reading frames of unknown function(s) (Long et al. 2007). Typically, they are composed of two moieties, *hoxY* and *hoxH* genes and the diaphorase moiety, encoded by the *hoxU*, *hoxE*, and *hoxF* genes homologous to some subunits of Complex 1 of mitochondrial and bacterial respiratory chains that contain NAD(P), FMN, and Fe-S-binding sites.

H_2 evolving energy-conserving hydrogenase. The multimeric enzymes (six subunits or more) of this group reduce protons from water to dispose of excess reducing equivalents produced by the anaerobic oxidation of C1 organic compounds of low potential, such as carbon monoxide or formate (Vignais and Colbeau 2004).

10.4.4.1.2 [NiFeSe]-Hydrogenase

Recently, it was found that the enzyme obtained from *Desulfomicrobium baculatum* comprised of one of the terminal cysteine (Cys) ligands to the nickel (Ni) replaced by selenocysteine (Garcin et al. 1999).

10.4.4.1.3 [FeFe]-Hydrogenases

These hydrogenases are found mostly in some facultative bacteria and some strict anaerobes as well as in green algae. They are soluble, monomeric enzymes that can in some cases be cytoplasmic, as is the case in *Clostridium pasteurianum* hydrogenase CpI. [FeFe]-hydrogenases are extremely sensitive to inactivation by O_2 and catalyze both H_2 evolution and uptake. *Desulfovibrio* spp. has periplasmic, heterodimeric [FeFe]-hydrogenase, which can be purified aerobically and catalyzes mainly H_2 oxidation (Nicolet et al. 1999). Green algal hydrogenases are located in the chloroplast and are bidirectional (capable of catalyzing hydrogen oxidation or proton reduction to produce H_2). It is reported that, in algae, anaerobically inducible *hydA* genes encode a special type of highly active [FeFe]-hydrogenase. Gene sequences that encode for polypeptides with the entire essential attributes of an [FeFe]-hydrogenase protein are termed hydB but not in green algae (Vignais et al. 2001).

10.4.4.2 Catalytic Site of Hydrogenase

10.4.4.2.1 [FeFe]-Hydrogenase

Many [FeFe]-hydrogenases are monomeric and consist of catalytic subunit only. However, dimeric, trimeric, and tetrameric are also known (Vignais et al. 2001). The catalytic subunits of FeH_2ase in comparison to [NiFe]-hydrogenases exhibit considerable variations in sizes. Indeed, in addition to containing the conserved domain H cluster of approximately 350-amino acid residues, they often consist of additional domains that accommodate the FeS clusters (responsible for electron transfer between the active sites). The 3D structures have confirmed that the core of the [FeFe]-hydrogenase catalytic site consists of two Fe atoms: Termed Fe1 and Fe2 (Nicolet et al. 2000). These are bridged by CO and a small molecule initially modeled in *Desulfovibrio desulfuricans* as 1,3 propanedithiol on the basis of electron density alone. Each iron atom is also liganded to two diatomic molecules, assigned to one CO and one CN^-. This combination of S, –CN, and –CO ligands tends to make the metal site "softer" so that it behaves more like the zero valent metal (Cammack 2001). The catalytic site is bound to the protein through only one cysteic thiolate ligand to Fe1 and hydrogen bonds involving each of the cyanide ligands. Fe1 has six ligands in a distorted octahederal conformation, whereas Fe2 has five ligands with an empty site in *D. desulfuricans* and an additional water ligand in *C. pasteuranium*.

10.4.4.2.2 [NiFe]-Hydrogenase

The x-ray crystallographic analysis shows that the catalytic site of the protein consists of a binuclear center with one Ni and one Fe atom linked to the protein by four cysteic thiolates. In the oxidized form of the *Desulfovibrio gigas* [NiFe]-hydrogenase, the Ni atom has three close and two distant ligands in a highly distorted square pyramidal conformation, with a vacant axial sixth ligand site, whereas the Fe atom has six ligands in a distorted octahederal conformation. Three of the Fe ligands are diatomic molecules modeled as two cyanides and one CO in *D. gigas* and *Desulfovibrio fructosovoran*; however, in *Desulfovibrio vulgaris*, the iron ligands were modeled as two CO and one SO (Higuchi et al. 1997). The existence of such a binuclear center provides an alternative to an Ni ion to be both catalytic and extensively redox active site.

10.4.4.3 Hydrogen Channels

Since the active site is deeply buried in the protein, the reactants, H_2 and hydrons have to diffuse to the catalytic center. Earlier experiments suggested that these small hydrophobic molecules diffusing inside hydrogenase take advantage of the well-defined preexisting packing defects which arise spontaneously in proteins. The movement of the gas molecules occurs from cavity to cavity as the cavities fluctuate into existence inside the protein indicating that no permanent "channels" as such exist inside the protein. The first x-ray crystallography of NiFe hydrogenase showed that the active site is buried within the large subunit at approximately 3 nm from the surface indicating the possibility of specific channels for H_2 diffusion (Carlos et al. 2001). It is known that Xe binds to the hydrophobic part of the protein and thus has been used as a probe to analyze the gas access channel. The crystal structure of the Xe-bound NiFe hydrogenase from *D. fructosovarans* solved at 6 Å resolution and a molecular dynamics calculation has shown a small cavity that could reach the active site via several channels (Montet et al. 1997). This study shows that there are several channel entrances near the surface of the enzyme, which combine to form one small channel leading to the active site.

10.4.5 Nitrogenase

The bacterial enzyme nitrogenase has agronomic importance as it is responsible for the natural fertilization of soil by fixing N_2 into NH_3. Though the nitrogenase is mainly involved in N_2-fixation, it is also capable of simultaneous H_2 production. Researchers are exploiting this capability of nitrogenase to produce higher amount H_2 in N_2-limited condition.

10.4.5.1 Catalytic Site of Nitrogenase

The nitrogenase enzyme relies on three metal–sulfur complexes to carry out the catalytic function: An Fe_4S_4 cubane unit in the dinitrogenase reductase protein (Fe protein), a P-cluster [Fe_8S_7], and iron–molybdenum/vanadium/or only iron cofactor in the dinitrogenase protein.

10.4.5.1.1 Dinitrogenase

There are three types of dinitrogenase found among the nitrogenases, which vary depending on the metal content. Type one contains molybdenum (Mo) (Thiel 1993), type two contains vanadium (V) instead of Mo (Kentemich et al. 1988, Thiel 1993), and type three has neither Mo nor V but it contains iron (Fe) (Kentemich et al. 1991, Bishop and Premakumar 1992). Nitrogenase with an MoFe cofactor at its active site, which consists of $MoFe_7S_9$ and the organic acid homocitrate, is widely spread and best studied by researchers. It is an $\alpha_2\beta_2$ heterotetramer, encoded by the genes nifD and nifK, having molecular weight of about 220–240 kDa, respectively. It breaks apart the atoms of nitrogen.

10.4.5.1.2 Dinitrogenase Reductase

It is a homodimer of about 60–70 kDa, contains Fe protein, encoded by *nifH*, and plays the specific role of mediating the transfer of electrons from the external electron donor (an Fd or a flavodoxin) to the dinitrogenase, the active site (Orme-Johnson 1992, Flores and Herrero 1995, Masepohl et al. 1997).

10.4.6 Limitations of Enzymes and Molecular Approaches to Overcome

If harnessed properly, hydrogenase-containing microorganisms could be used to supply the next generation economic fuel. However, the major challenges that researchers face in this transit is to overcome the sensitivity of hydrogenase to oxygen. Continuous production of H_2 is difficult in the light since oxygen is one of the obligatory products of water oxidation. Isolation of hydrogenase from cells and its subsequent analysis is also very difficult. Induction of artificial anaerobic conditions adds significant cost to hydrogen production in large scale. The photosynthetic bacteria are anoxygenic during hydrogen production.

Cyanobacteria and green algae, however, produce oxygen due to their PSII activity, which inhibits the hydrogenase responsible for proton reduction, and, as a result, the organisms stop metabolizing hydrogen. Thus, a prolonged enhanced H_2 production can be obtained only by generating anaerobic conditions where hydrogenase can be induced. By changing some physicochemical parameters, for example, by sparging with inert gas inside the reactor or depleting the media of sulfur that stalls the PSII activity, anaerobic environment can be achieved in the presence of light. Research is currently being performed within genetic and metabolic engineering for the generation of anaerobic conditions in photosynthetic organisms. Under normal growth condition, photosynthetic rate is about four- to seven-fold higher than the respiration rate. By using attenuated photosynthesis/respiration ratio (P/R ratio) mutants (*apr* mutants) of *C. reinhardtii*, the P/R ratio drops below one, thereby establishing anaerobic conditions that mimic the physiological status of sulfur-deprived cells (Melis 2007). So by using the *apr* mutant, continuous hydrogen production can be obtained. The sulfur-deprivation method that may cause cell growth inhibition and death can be avoided (Ruhle et al. 2008), and also the method of continuous argon sparging that dilutes the produced hydrogen and is expensive to use. It has also been reported that a sulfur-deprived *C. reinhardtii* D1 mutant that carried a double amino acid substitution is superior to the wild type for hydrogen production. The leucine residue L159 in the D1 protein was replaced by isoleucine, and the asparagine N230 was replaced by tyrosine (L159I-N230Y). This strain is very efficient for prolonged H_2 production and also having lower chlorophyll content and higher respiration rate that contributes in net yield (Torzillo et al. 2009).

To date, only a few hydrogenases are known to display tolerance to oxygen and various strategies have been adopted to generate oxygen-tolerant mutants. Hydrogenase from extremophiles that are naturally tolerant to oxygen may provide further inspiration for the development of such technologies. *Aquifex aeolicus* hydrogenase is a good candidate for biotechnological applications due to their high resistance to aerobic and thermal inactivation (Guiral et al. 2006). Early biochemical studies on oxygen inactivation of [FeFe]-hydrogenase suggests that the level of sensitivity of different enzymes to oxygen varied. It was observed that algal hydrogenase, which is the smallest and the simplest of hydrogenases, encoding only the H-cluster where the active domain is located, is much more sensitive to oxygen inactivation than, for example, CpI hydrogenase. Although this hydrogenase is similar, it harbors additional F-cluster domains (homologous to Fe-S clusters of Fd). It was believed that this domain might provide additional protection to the active site from oxygen inactivation. On the other hand, the conserved amino acid sequence of the H-cluster or the active site suggests that the amino acid composition of this domain is critical to the protection of the active site from oxygen. It is also observed that [FeFe]-hydrogenase are more sensitive to oxygen than [NiFe]-hydrogenases. Efficiency of the enzymes is another important factor affecting the rate of hydrogen photoproduction. In photosynthetic bacteria, net hydrogen production can be increased by improving the efficiency of nitrogenase. Nitrogenase is also sensitive to oxygen, but less than hydrogenase. Nitrogenase is mostly present in anaerobic prokaryotes or involved in anaerobic hydrogen production in cyanobacteria separated from oxygen-producing phase through special (heterocyst) or temporal (light/dark) separation. This reaction happens under nitrogen-limiting conditions and is a highly energy-consuming process. By knocking out *glnB* and *glnK*, the problem of repression of nitrogenase by ammonium ions has been overcome in *R. sphaeroides* (Kim et al. 2008a).

Another possibility for the researchers is through heterologus overexpression of efficient enzymes into the cell. Cyanobacteria have bidirectional [NiFe]-hydrogenase that is inefficient for hydrogen production, and hydrogen is mostly produced by nitrogenase. An efficient [FeFe]-hydrogenase (*hydA*) from *C. pasteurianum* was introduced and expressed in the cyanobacteria *Synechococcus elongatus* and succeeded to achieve enhanced hydrogen production without coexpression of maturation proteins (Miyake and Asada 1997, Asada et al. 2000, King et al. 2006). Clostridial *hydA* has also been cloned into *Rhodospirillum rubrum*, and the native hydrogenase of *R. rubrum hydC* has been overexpressed. In both cases, it was observed that pyruvate was the electron donor for hydrogen production (Kim et al. 2008a). Significant research has been performed to increase the oxygen tolerance of hydrogen-producing enzymes, hydrogenases in particular (Xu et al. 2005). By introducing random and site-directed mutagenesis in *Chlamydomonas*, strains with 10-fold more oxygen tolerance have been obtained (Ghirardi et al. 2000).

However, to achieve the theoretical maximum value for hydrogen production, there is another obstacle in cyanobacteria and photosynthetic bacteria. The produced hydrogen could be reoxidized again by another enzyme known as uptake hydrogenase (NiFe-hydrogenase). Generating mutants deficient of uptake hydrogenase activity dramatically increases the net production of hydrogen. Inactivation of hydrogen uptake activity in *Thiocapsa roseopersicina* (*hypF*-deficient mutant) under nitrogen-fixing condition caused a significant increase in the hydrogen evolution capacity. This mutant is therefore a promising candidate for use in practical biohydrogen-producing systems (Fodor et al. 2001). Ruiyan et al. (2006) also reported that a *R. rubrum* mutant deleted of *hupL* gene encoding the large subunit of uptake hydrogenase produced increased amount of H_2. To date, significant research has been performed to inhibit the uptake hydrogenase activity using different approaches. Gokhan et al. (2008) used a suicide vector for site-directed mutagenesis of uptake hydrogenase (*hupSL*) in *R. sphaeroides*. They obtained 20% more H_2 production than wild type.

An anaerobic environment may not always be sufficient for enzyme activation, as there are other competitors present for capturing electrons. Lee and Greenbaum (2003) have described that there is a competitive inhibition of hydrogen production by CO_2 fixation. CO_2 fixation can be inhibited if ATP requirements are not fulfilled. Genetic insertion of a polypeptide protein channel with a hydrogenase promoter can reduce the proton gradient across the thylakoid membrane by avoiding the ATPase channel where ATP is produced during proton transfer in algal cells.

10.5 Mathematical Modeling of Biological Hydrogen Production Processes

The primary objective of a kinetic model developed for fermentation is the prediction of kinetic behavior of hydrogen fermentation performance. An appropriate model of fermentation, with the technical, economic, and physiological implications would be a powerful instrument to predict and control problem fermentations, and be helpful to understand the fermentation process (Marin 1999). Among the numerous models developed, the majority of the models are biochemically knowledge-based models (same), which consist of a set of mathematical equations describing the phenomena occurring during fermentation. The main advantage of this type of model is that they account for biological phenomena. The model parameters with some biological significance can be obtained, but their structures may be strongly nonlinear, complex, and difficult to verify and validate, and can pose problems in terms of parameters identification. In general, the fermentation kinetic model can be subdivided into a growth model, a substrate model, and a product model.

10.5.1 Determination of Cell Growth Kinetic Parameters

The rate of biomass production and substrate degradation can be determined by the differential technique as follows:

$$\frac{dx}{dt} = \frac{X_{n+1} - X_{n-1}}{t_{n+1} - t_{n-1}} \quad \frac{ds}{dt} = \frac{S_{n+1} - S_{n-1}}{t_{n+1} - t_{n-1}} \tag{10.22}$$

where
  X is the biomass concentration (g L^{-1})
  S is the substrate concentration (g L^{-1})
  n is the sampling number

The specific growth rate (μ) of biomass can be written as

$$\mu = \frac{1}{x} \frac{dx}{dt} \tag{10.23}$$

where dx/dt are calculated for different experimental data points.

Assuming that cell growth follows Monod equation (Bailey and Ollis 1986), specific cell growth rate can be approximated as

$$\mu = \frac{\mu_{max} S}{K_S + S} \tag{10.24}$$

Equation 10.24 may be linearized to find out the kinetic parameters with the help of Lineweaver–Burk plot:

$$\frac{1}{\mu} = \frac{K_S}{\mu_{max} S} + \frac{1}{\mu_{max}} \tag{10.25}$$

Regression analysis was used to find the best fit for a straight line on a plot $1/\mu$ versus $1/S$ to determine the values of maximum specific growth rate (μ_{max}) and saturation constant (K_S). The values of μ_{max} and K_S were $0.4\,h^{-1}$ and $5.5\,g\,L^{-1}$, respectively, using glucose as substrate using *E. cloacae* IIT-BT (Nath et al. 2008). Both the values of μ_{max} and K_S were found lower than those reported ($1.12\,h^{-1}$ and $8.9\,g\,L^{-1}$, respectively).

In case of chemostat (CSTR), under steady-state condition and sterile feed, $D = \mu$. Then Equation 10.25 may be written as

$$\frac{1}{D} = \frac{K_S}{\mu_{max} S} + \frac{1}{\mu_{max}} \tag{10.26}$$

Kinetic models based on material balances for continuous hydrogen production process were constructed to describe the steady-state behavior of the substrate sucrose, biomass, and products (H_2 and VFA) in the anaerobic culture adapted to the enhanced production of hydrogen (Chen et al. 2001). The proposed model was able to interpret the trends of the experimental data. The maximum specific growth rate (μ_{max}), Monod constant (K_S), and yield coefficient for cell growth ($Y_{X/S}$) were estimated as $0.17\,h^{-1}$, $68\,mg\ COD\ L^{-1}$, and $0.1\,g\ g^{-1}$, respectively.

10.5.2 Development of Mathematical Models to Correlate Substrate and Biomass Concentration with Time

Now,

$$\frac{dx}{dt} = \mu x = \frac{\mu_{max}Sx}{K_S + S} \tag{10.27}$$

$Y_{X/S}$ is the cell mass growth/mass of substrate consumed which is given as

$$Y_{X/S} = \frac{dx}{dS} \tag{10.28}$$

Assuming $Y_{X/S}$ constant throughout the fermentation, the rate of glucose degradation can be presented as

$$\frac{dS}{dt} = \frac{\mu_{max}Sx}{Y_{x/s}(K_S + S)} \tag{10.29}$$

$Y_{X/S}$ is calculated from the experimental data by averaging the values of $Y_{x/S}$ obtained at different data points.

$$Y_{x/S} = \frac{(x - x_0)}{(S_0 - S)} \tag{10.30}$$

x in Equation 10.29 is replaced by an expression involving S_o, x_o, and $Y_{x/S}$. It can then be integrated to give an expression for simulated values of S as a function of t:

$$\mu_{max}(x_0 + Y_{X/S}S_0)t = [x_0 + Y_{X/S}(S_0 + K_S)]\ln\left[\frac{x_0 + Y_{X/S}(S_0 - S)}{x_0}\right] - K_SY_{x/S}\ln\frac{S}{S_0} \tag{10.31}$$

From the above expression, simulated substrate profile with t was determined by using Wegstein convergence method of successive substitutions on each iteration (Himmelblau 1996). Also simulated values of cell mass concentration, x were calculated by using the following relation:

$$x = x_0 + Y_{x/S}(S_0 - S) \tag{10.32}$$

10.5.3 Development of Substrate Inhibition Model

Medium constituents such as substrate or product may inhibit the specific growth rate. Since in the present system under discussion, product is gaseous H_2, which escapes and is collected in a gas collector, product inhibition can be neglected. Therefore, substrate inhibition is taken into consideration. Andrews model for substrate inhibition is given by

$$\mu = \frac{\mu_{max}s}{(K_i + s + s^2)/K_p} \tag{10.33}$$

For the present H_2-generating system, a substrate inhibition model of the form

$$\mu = \frac{\mu_{max}S}{K_S + S - K_i S^2} \tag{10.34}$$

is proposed.

10.5.3.1 Calculation of Kinetic Parameters K_S, μ_{max}, and K_i

Equation 10.34 suggests a nonlinear relationship between the specific growth rate (μ) and substrate concentration (S).

For calculating the kinetic parameter values from Equation 10.34, a numerical methods approach, viz., *Method of least squares,* is made use of:

$$S \equiv p \tag{10.35}$$

$$\frac{1}{\mu} \equiv q \tag{10.36}$$

$$\frac{-K_i}{\mu_{max}} \equiv a \tag{10.37}$$

$$\frac{1}{\mu_{max}} \equiv b \tag{10.38}$$

$$\frac{K_S}{\mu_{max}} \equiv c \tag{10.39}$$

Using the above relations, Equation 10.34 can be rewritten in the form

$$q = ap + b + \frac{c}{p} \tag{10.40}$$

Now, when a data set of p and q is available, the coefficients a, b, and c can be calculated by solving the following three equations simultaneously:

$$an + b\sum \frac{1}{P_i} + c\sum \frac{1}{p_i^2} = \sum \frac{q_i}{p_i} \tag{10.41}$$

$$a \sum p_i + bn + c \sum \frac{1}{p_i} = \sum q_i \qquad (10.42)$$

$$a \sum p_i^2 + b \sum p_i + cn = \sum q_i p_i \qquad (10.43)$$

Here, n = number of data of p and q, and i refers to the ith data of p or q.

Once values for a, b, and c are obtained, values of kinetic parameters K_S, μ_{max}, and K_i can be calculated by using Equations 10.37 through 10.39.

10.5.3.2 Simulation of Biomass Concentration and Substrate Concentration Profiles as Functions of Time

Now,

$$\frac{dx}{dT} = \mu x = \frac{\mu_{max} S}{K_S + S - K_i S^2} x \qquad (10.44)$$

Thus,

$$\frac{dS}{dt} = \frac{-1}{Y_{x/S}} \frac{\mu_{max} S}{K_S + S - K_i S^2} x \qquad (10.45)$$

x in Equation 10.45 is replaced by an expression involving S, S_0, x_0, and $Y_{x/S}$ (from Equation 10.30). It can then be integrated to give an expression for simulated values of S as a function of t:

$$\mu_{max}(x_0 + Y_{x/S} S_0)t = [x_o + Y_{x/S}(S_0 + K_S)] \ln\left(\frac{x_0 + Y_{x/S}(S_0 - S)}{x_0}\right) - K_i(x_0 + Y_{x/S} S_0)$$

$$\times \left[S - S_0 + \frac{(x_0 + Y_{x/S} S_0)}{Y_{x/S}} \ln\left(\frac{x_0 + Y_{x/S}(S_0 - S)}{x_0}\right) \right] - K_S Y_{x/S} \ln \frac{S}{S_0} \qquad (10.46)$$

Hence, from the above expression, simulated values for substrate concentration at different times are calculated.

Simulated values for biomass concentration (x) are calculated by substituting the values of the substrate concentration in Equation 10.32. Substrate and biomass concentration profiles have good resemblance as compared to the experimental results (Kumar et al. 2000). The little deviation between the experimental and theoretical curves might be due to the assumption that $Y_{x/S}$ remained constant over the system irrespective of time of fermentation. This was assumed by ignoring the effect of maintenance energy on cell growth and metabolism.

10.5.4 Development of Mathematical Models for Cell Growth Kinetics in Photofermentation Process

To study the growth kinetics of *R. sphaeroides* O.U 001 in the second stage of the two-stage fermentation, logistic equation was used. Logistic equations are a set of equations that characterize growth in terms of carrying capacity. Recently, the logistic model, as a sigmoidal shaped model, has been the most popular one due to its "goodness of fit" and has been widely used in describing the growth of microorganism. The use of logistic model has a benefit of representing the entire growth curve including the lag phase (if present), the exponential growth, and the stationery phases. Usually, the logistic model was used to show the self-regression made by the increase of cell concentration common in batch fermentation. An equation based on the logistic model with growth-associated production of hydrogen in photofermentation has been developed.

The specific growth rate for the logistic model can be represented as

$$\mu = k_c \left[1 - \frac{x}{x_{max}} \right] \tag{10.47}$$

where
μ is the specific growth rate (h^{-1})
k_c is the apparent specific growth rate (h^{-1})
x is the dry cell mass concentration ($g\ L^{-1}$)
x_{max} is a model parameter that is called maximum dry cell mass concentration ($g\ L^{-1}$)

Recalling that the cell growth rate can be expressed as

$$\frac{dx}{dt} = \mu x \tag{10.48}$$

and inserting Equations 10.48 into 10.47 we get:

$$\frac{dx}{dt} = k_c x \left[1 - \frac{x}{x_{max}} \right] \tag{10.49}$$

The integration of Equation 10.49 with the boundary condition $x(0) = x_0$ yields the logistic equation for cell concentration for a batch reactor represented as

$$x = \frac{x_0 e^{k_c t}}{1 - x_0/x_{max}(1 - e^{k_c t})} \tag{10.50}$$

Although Equation 10.50 is implicit in its dependence on substrate concentration, for the batch hydrogen production experiments of this study where the initial substrate concentration and volume of inoculum are kept constant, the logistic model is a fair approximation of the growth kinetics. Equation 10.50 can be rearranged to the following form:

$$x = \frac{x_{max}}{1 + e^{-k_c t}[(x_{max}/x_0) - 1]} \tag{10.51}$$

or

$$x = \frac{a}{1 + be^{-ct}}$$

where

a = x_{max}
c = k_c
b = [(x_{max}/x_0) - 1]

Logistics constants can be determined by using a software program, Curve expert 1.3, for fitting curves.

10.5.5 Cumulative Hydrogen Production by Modified Gompertz Equation

The cumulative hydrogen production in the anaerobic reactor can be modeled by modified Gompertz equation (Lay et al. 1997), which can be written as

$$H(t) = P \exp\left\{-\exp\left[\frac{R_m e}{P}(\lambda - t) + 1\right]\right\} \qquad (10.52)$$

This was found to be a suitable equation for describing the progress of cumulative gas production obtained from the experiment, where $H(t)$ represents cumulative volume of hydrogen production (mL), P the gas production potential (mL), R_m (mL h^{-1}) the maximum production rate, λ (h) the lag time, t incubation time (h), and e the exp(1) = 2.718.
  Gompertz equation can be simplified as

$$Y = a \exp\{-\exp(b - cx)\} \qquad (10.53)$$

where

a = P
b = R_m e/P(\lambda + 1)
c = R_m e/P

The typical cumulative hydrogen production curve was nonlinearly modeled by the above equation. The values of P, R_m, and λ for each set of experiment were determined by best fitting the cumulative hydrogen production data for the above equation using a software program, Curve expert 1.3. The spent medium was obtained from the batch of dark fermentation by *E. cloacae* using MYG medium, originally containing 1% (w/v) of glucose. The values of P, R_m, and l were 615.65 mL, 16.15 mL h^{-1}, and 14.32 h, respectively, determined by best fitting the cumulative hydrogen production data for the above equation (Nath et al. 2008).

10.5.6 Luedeking–Piret Model

Product formation kinetics can be modeled on the basis of two-parameter kinetic expression, termed as Luedeking–Piret equation of the following form (Bailey and Ollis 1986):

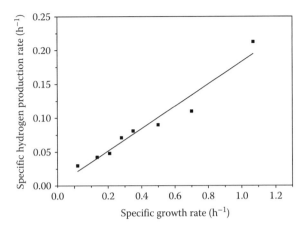

FIGURE 10.9
Plot of the specific growth rate of the cells versus the specific hydrogen production rate to justify that biohydrogen production is a growth-associated product.

$$r_{fp} = \alpha r_{fx} + \beta x \qquad (10.54)$$

Dividing both sides by X, we get,

$$\frac{1}{X} r_{fp} = \alpha \frac{1}{X} r_{fx} + \beta \qquad (10.55)$$

or

$$v = \alpha \mu + \beta \qquad (10.56)$$

v is the specific productivity = $(1/x)$ (dP/dt), where P = product formed after time interval t. α and β are constants. The volume of hydrogen produced was noted from time to time and the values of γ calculated. The values of γ and μ obtained at various time intervals were plotted on a graph. The nature of the best-fit curve tells us whether the product formation is growth-associated or not. In case of growth-associated product formation, the graph passes through the origin. Figure 10.9 indicates that the hydrogen is a purely growth-associated product. The value of growth associate coefficient (α) was 0.166. From Equation 10.33, values of dP/dt at different points of time can be found out from dx/dt. From dP/dt profile, a simulated P profile was obtained (Kumar et al. 2000).

10.6 Biological Reactors for the Improvement of Hydrogen Production

10.6.1 Dark Fermentation

Continuous H_2 production can be improved by increasing the cell concentration through whole cell immobilization. H_2 production using immobilized whole cells has been reported by several researchers (Seon et al. 1993; Yokoi et al. 1997). The major problems with immobilized whole cell systems are low substrate conversion efficiency and low rate of production due to mass transfer resistance between substrate and immobilized cells in a fixed bed reactor (Rachman et al. 1998). The solid matrices used for the immobilization of the whole cells are mostly synthetic polymers or inorganic materials (Yokoi et al. 1997). These materials impose disposal problems. So, environment friendly natural polymers such as lignocellulosic materials are selected as solid matrices in the present studies. Gas holdup is a major problem in gas generating system using packed bed reactor (Ghosh and Bandyopadyay 1980). The effect of different bioreactor configuration was taken into consideration to overcome this problem. Attempts were also made to increase the substrate conversion efficiency and to determine the cell growth kinetic parameters using immobilized *E. cloacae* IIT-BT 08. H_2 production in column bioreactor showed that the working volume was decreased with increase in dilution rate in a column bioreactor. This could be attributed to the fact that flow of substrate added convective transport contribution to the movement of substrate from the bulk solution to the external surface of the packed matrix where cells were immobilized. It was reported that the mass transfer coefficient increased with the increase of dilution rate in a packed bed bioreactor with immobilized cell (Seon et al. 1993). Therefore, mass transfer was more at higher dilution rate, which resulted in the production of large volumes of gas and hence higher gas holdup. The higher gas holdup resulted in decrease of working volume. To overcome the gas holdup problem, different bioreactor configurations were considered (Figure 10.10). SM-C was chosen as support matrix as it gave the best result in terms of cell density and H_2 production rate at dilution rate of $0.9\,h^{-1}$. The comparative studies on different bioreactor configurations showed that the rhomboid bioreactor with convergent divergent configuration had maximum H_2 production rate (Table 10.5) (Kumar and Das 2001).

10.6.2 Photobiological Process

Hydrogen production through biophotolysis or photofermentation is necessarily a two-stage process. Using CO_2 and light energy, the algal and cyanobacterial systems involve biomass production in the first stage followed by hydrogen production in anaerobic conditions through biophotolysis of water in second stage. For the stage of hydrogen production, while the algal biomass are harvested (by filtration, centrifugation, etc.) and resuspended in a sulfur-deprived media, the cyanobacterial cells are transferred to a separate unit in which the anaerobic conditions are maintained by sparging with an inert gas. In case of photofermentation, the organic acids produced in the first stage of dark fermentation are further metabolized to hydrogen in the second stage with the aid of light energy.

Based on the mode of operation, the bioreactors used for biomass production/hydrogen production can be broadly classified into batch, continuous, and fed batch. Open pond systems are examples of batch mode where closed tubular systems are operated in continuous mode. Large-scale production usually involves continuous mode during daylight, in which fresh media is fed at a constant rate and the same quantity of broth is removed simultaneously (Molina Grima et al. 1999). While feeding is stopped during night, agitation is kept high to prevent settling of biomass. For biomass production, the carbon dioxide is continuously fed during daylight hours.

The factors affecting the performance of a photobioreactor are

- Light penetration
- High surface-to-volume (S/V) ratio

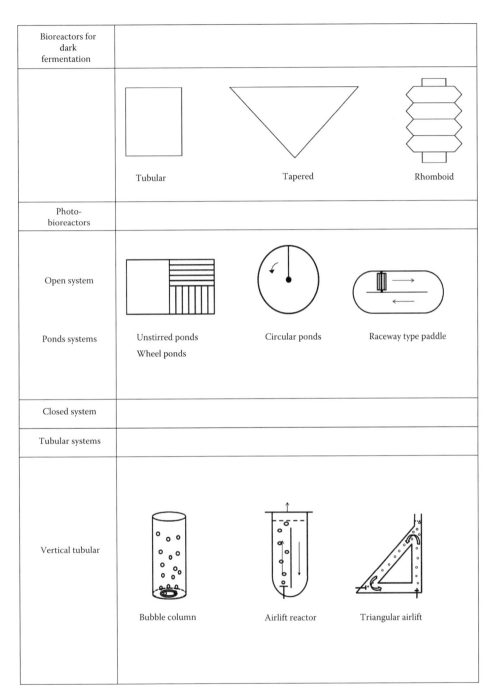

FIGURE 10.10
Configuration different bioreactors used for biomass and H_2 production.

(*continued*)

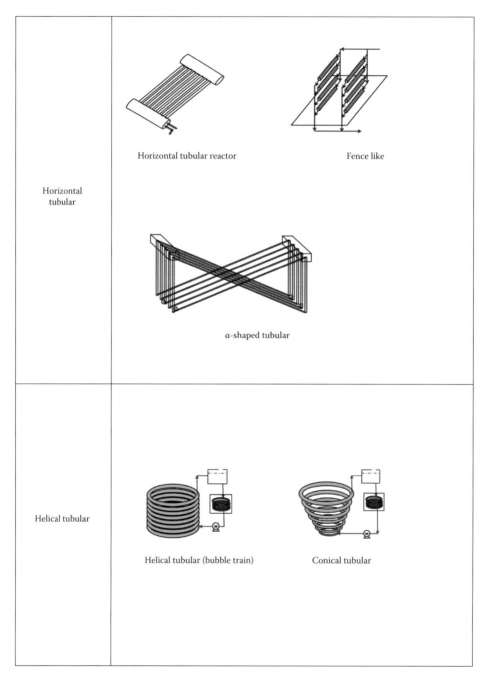

FIGURE 10.10 (continued)

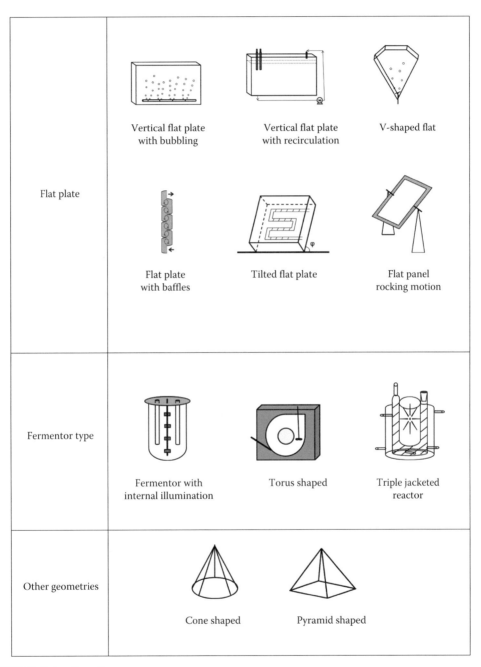

FIGURE 10.10 (continued)

(*continued*)

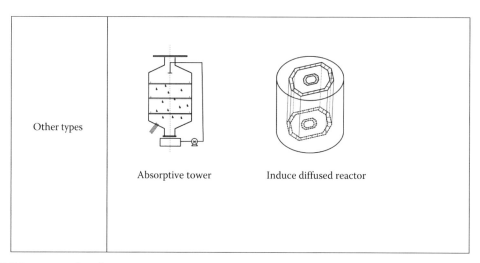

Other types

Absorptive tower   Induce diffused reactor

FIGURE 10.10 (continued)

TABLE 10.5

Effect of Different Bioreactor Configurations on H_2 Production

Reactor Type	Dilution Rate (h^{-1})	Rate of H_2 Production (mmol L^{-1} h^{-1})	Substrate Conversion Efficiency (%)	Relative Increase in H_2 Production (%)
Tubular	0.93	62.5	61	—
Tapered	0.93	65.2	65	6.5
Rhomboid	0.93	71.0	69	14.5

Note: *Packing material:* SM-C; media: MYG; temperature: 36°C; initial medium pH: 6.00; operational stability: 40 days; bioreactor volume: 380 mL; void fraction:0.65.

- Temperature control
- Transparency and durability of the material of construction
- Gas exchange
- Agitation system
- Harvesting efficiency

There are various types of photobioreactors designed and tested for biomass production and few have been successful in large-scale operation. Although these systems overcome the issues of S/V ratio, temperature control, etc., their suitability toward hydrogen production is still questionable due to factors like agitation and gas exchange. An additional problem faced with the algal and cyanobacterial system is that due to the basal activity of PSII, there is always some oxygen evolution that inhibits hydrogen production. Moreover, the hydrogen produced in the long tubes will be consumed back if the uptake hydrogenase activity is present. The photobioreactors can be broadly classified into two major types:

1. Open system—raceway ponds, lakes, etc.
2. Closed system—tubular, flat plate, conical, pyramidal, fermentor type, etc.

10.6.2.1 Open System

The open pond systems are of different types, viz., unstirred open ponds, circular ponds, raceway ponds with paddle wheel, and sloping cascade (Figure 10.10). These open pond systems are the simplest methods of algal cultivation system and offer advantage in lower construction and operation cost. The open pond systems have large surface area and are of necessary shallow depth to allow the maximum penetration of light. Contamination by other species is a major drawback. Hence, they are used to grow species like *Chlorella* and *Spirulina* that can survive extreme conditions like higher pH, nutrient concentrations, and salinity (Vonshak et al. 1983). Moreover, open systems are affected by rain that dilutes and causes contamination. Evaporation is another major concern.

10.6.2.2 Closed System

Open pond systems are suitable only for biomass production. They cannot provide the anaerobic conditions required in case of hydrogen production. Also the degree of control of parameters like temperature, pH, and nutrient concentrations is poor. Hence, it necessitates the need for closed systems (Figure 10.10). It was found that in closed systems the amount of biomass produced is about three times of that obtained with open systems (Chaumont 1993, Carvalho et al. 2006), hence reduced harvesting costs. The degree of control is very high and it is possible to control crucial parameters that influence the culture whether to produce biomass or hydrogen. The various types of closed system geometries that have been constructed and implemented have been compared in Table 10.6.

10.6.2.3 Material of Construction

The material used for construction of photobioreactors plays a major role in deciding the establishment and maintenance costs. The property of the material should be

TABLE 10.6

Performance of Different Photobioreactors

Type of Photobioreactor	Organism	H_2 Yield	Light Conversion Efficiency (%)	References
Helical tubular reactor (also called as bubble train)	*Anabaena azollae*	13 mL L^{-1} h^{-1}		Tsygankov (1998)
Flat plate tilted solar bioreactor	*R. sphaeroides* O.U.001	10 mL L^{-1} h^{-1}		Eroglu (2008)
Flat panel with gas recirculation	*Rhodopseudomonas* sp. HCC 2037	25 mL H_2 L^{-1} h^{-1}		Hoekema (2002)
Fermentor type	*R. rubrum*		8.67	Syahidah (2008)
Hollow fiber membrane bioreactor	*A. variabilis*	20 mL gdw h^{-1}		Markov (1995)
Induced diffuse PBR	*R. sphaeroides* RV	5.03 mM H_2 mM^{-1} lactate	9.23	El-Shishtawy (1998)
Floating-type PBR	*Rhodopseudomonas palustris* R-1	10–12 mL L^{-1} h^{-1}	0.308	Otsuki (1998)

such that it is stable and long lasting. The materials used should have the following characteristics:

- High transparency
- Flexible and durable
- Nontoxic
- Resistance to chemicals and metabolites produced by the microorganisms
- Resistance to weathering
- Low cost

The materials used for construction deteriorate over time on exposure to sunlight and various climatic conditions. For example, exposure of polyethylene to light and oxygen results in loss of strength and tear resistance. Zittelli and Tredici (Tredici 1999) carried out a preliminary experiment to determine the loss of transparency of six different materials—polycarbonate, PVC, polymethyl methyl acrylate, polyethylene, polypropylene, and Teflon in long-term exposure to natural climatic conditions. It was found that only PMMA and Teflon retained their transparency over months. By comparing the properties of the different types of materials (Table 10.6), we found that PMMA and PET are more advantageous for bioreactor construction.

10.6.2.4 Comparison of the Performance of the Photobioreactors

The performance of the reactors with respect to biomass productivity are usually compared based on one of the following parameters:

- Volumetric productivity (i.e., productivity per unit reactor volume per unit time)
- Areal productivity (i.e., productivity per unit of occupied land area per unit time)
- Productivity per illuminated surface area (i.e., productivity per unit of illuminated surface area per unit time)

Photosynthetic efficiency for biomass production (ψ) (Fernandez et al. 2003) is defined as the amount of energy stored in the generated biomass per unit of radiation energy absorbed by the culture. Mathematically, it is expressed as

$$\psi = \frac{P_b H_{biomass}}{F_{vol}} \tag{10.57}$$

where
P_b is the volumetric biomass productivity
$H_{biomass}$ is the combustion enthalpy of the biomass
F_{vol} is the photon flux absorbed in unit volume

Combustion enthalpy can be calculated from the biochemical composition or is measured directly. F_{vol} is calculated by the following equation:

$$F_{vol} = I_{av} K_a C_b \tag{10.58}$$

where
   I_{av} is the average irradiance inside the culture
   K_a is the absorption coefficient
   C_b is the biomass concentration

10.6.2.5 Light Conversion to Hydrogen

Of the broad spectrum of solar spectrum, green algae and cyanobacteria utilize only the region between 400 and 700 nm. Photosynthetic purple bacterium additionally requires light in the range of 700–1010 nm. It was calculated that for green algae 4 mol of photons in the range of 400–700 nm are required to generate 1 mol of hydrogen and for photosynthetic bacteria 15 mol of photons (400–600 and 800–1010 nm) are required to generate 1 mol of hydrogen. The light conversion efficiency is usually calculated on the broad spectrum, although the organism utilizes specific wavelength. Mathematically, the light conversion efficiency is defined as

$$\text{Light conversion efficiency (\%)} = \frac{H_2 \text{ production rate} \times H_2 \text{ energy content}}{\text{Absorbed light energy}} \tag{10.59}$$

The light conversion efficiency depends on the S/V ratio and the spatial distribution of light over the photobioreactor. It is also influenced by the thickness of the reactor as Miyake reports that only light in the wavelength range of 500–800 reaches the deeper regions of the reactor. The performance of the various photobioreactors are compared in Table 10.6.

10.7 Conclusion

Biohydrogen production has been established as a prospective alternative and integral component of green sustainable energy. These processes broadly divided in two types: Light dependent and light independent. Major advantages of these processes are low operation temperature and atmospheric pressure using renewable energy sources like different organic industrial wastes, municipal solid wastes, etc. Fundamental aspects of these processes for hydrogen production have been highlighted in this chapter. Major problems of these processes are low hydrogen yield as well as rate of hydrogen production. Overall hydrogen yield of the process may be improved by using two-stage process: Dark fermentation followed by photofermentation. Algae may also play major role for the carbon dioxide sequestration and hydrogen generation. After the fermentation, algae may be utilized for the extraction of several valuable products like pigment, biodegradable polymers, minerals, etc. The roles of two important enzymes hydrogenase and nitrogenase are discussed in detail. Possibilities of some genetical improvement of the microbial strains are also taken into consideration. Mathematical modeling of the process plays important roles for the scaling up of the process. Different mathematical models have also been discussed. A challenging problem in establishing biohydrogen as a source of energy from the renewable substrate is the improvement of reactor design. Several operational problems like gas holdup, light penetration, oxygen toxicity, etc. are to be overcome to make the process economically viable.

References

Aro, E., Virgin, I., and B. Andersson. 1993. Photoinhibition of photosystem II. Inactivation, protein damage and turnover. *Biochim. Biophys. Acta* 1143:113–134.

Asada, Y., Koike, Y., Schnackenberg, J., Miyake, M., Uemura, I., and J. Miyake. 2000. Heterologous expression of clostridial hydrogenase in the cyanobacterium *Synechococcus* PCC7942. *Biochim. Biophys. Acta* 1490:269–278.

Bailey, E. J. and Ollis, D. F. 1986. *Biochemical Engineering Fundamentals*, 2nd edn. New York: McGraw-Hill, pp. 380–405.

Berberoglu, H., Jay, J., and P. Laurent. 2008. Effect of nutrient media on photobiological hydrogen production by *Anabaena variabilis* ATCC 29413. *Int. J. Hydrogen Energy* 33:1172–1184.

Bernát, G., Waschewski, N., and M. Rögner. 2009. Towards efficient hydrogen production: The impact of antenna size and external factors on electron transport dynamics in *Synechocystis* PCC 6803. *Photosynth. Res.* 99:205–216.

Bishop, P. E. and R. Premakumar. 1992. Alternative nitrogen fixation systems. In *Biological Nitrogen Fixation*, eds. G. Stacey, R. H. Burris, and D. J. Evans, pp. 736–762. New York: Chapman & Hall.

Blankenship, R. E. 1985. Electron transport in green photosynthetic bacteria. *Photosynth. Res.* 6:317–333.

Blankenship, R. E. 1992. Origin and early evolution of photosynthesis. *Photosynth. Res.* 33:91–111.

Brosseau, J. D. and J. E. Zajik. 1982. Hydrogen gas production with *Citrobacter intermedius* and *Clostrium pasteurianum*. *J. Chem. Tech. Biotechnol.* 32:496–502.

Burrows, E. H., Chaplen, F. W. R., and R. L. Ely. 2008. Optimization of media nutrient composition for increased photofermentative hydrogen production by *Synechocystis* sp. PCC 6803. *Int. J. Hydrogen Energy* 33:6092–6099.

Cammack, R. 2001. The catalytic machinery. In *Hydrogen as a Fuel*, eds. R. Cammack, M. Frey, and R. Robson, pp. 159–180. Boca Raton, FL: CRC press.

Cammack, R., Fernandez, V. M., and E. C. Hatchikian. 1994. Nickel-iron hydrogenase. *Methods Enzymol.* 243:43–68.

Carlos, J., Camps, F., Frey, M. et al. 2001. Molecular architectures. In *Hydrogen as a Fuel*, eds. R. Cammack, M. Frey, and R. Robson, pp. 93–109. Boca Raton, FL: CRC Press.

Carvalho, A. P., Meireles, L. A., and F. X. Malcata. 2006. Microalgal reactors: A review of enclosed system designs and performances. *Biotechnol. Prog.* 22:1490–1506.

Chader, S., Haceneb, H., and S. N. Agathos. 2009. Study of hydrogen production by three strains of *Chlorella* isolated from the soil in the Algerian Sahara. *Int. J. Hydrogen Energy* 34:4941–4946.

Chaumont, D. 1993. Biotechnology of algal biomass production: A review of systems for outdoor mass culture. *J. Appl. Phycol.* 5:593–604.

Chen, C. C., Lin, C.-Y., and J. S. Chang. 2001. Kinetics of hydrogen production with continuous anaerobic cultures utilizing sucrose as the limiting substrate. *Appl. Microbiol. Biotechnol.* 57:56–64.

Chen, W. M., Tseng, Z. J., Lee, K. S., and J. S. Chan. 2005. Fermentative hydrogen production with *Clostridium butyricum* CGS5 isolated from anaerobic sewage sludge. *Int. J. Hydrogen Energy* 30:1063–1070.

Chen, H., Yokthongwattana, K., Newton, A., and A. Melis. 2003. SulP, a nuclear gene encoding a putative chloroplast-targeted sulfate permease in *Chlamydomonas reinhardtii*. *Planta* 218:98–106.

Claassen, P. A. M., de Vrije, T., and M. A. W. Budde. 2004. Biological hydrogen production from sweet sorghum by thermopilic bacteria. Paper presented at the *2nd World Conference on Biomass for Energy*, May 10–14, 2004, Rome, Italy.

Conrad, R. 1999. Contribution of hydrogen to methane production and control of hydrogen concentrations in methanogenic soils and sediments. *FEMS Microbiol. Ecol.* 28:193–202.

Cournac, L., Guedeney, G., Peltier, G., and P. M. Vignais. 2004. Sustained photoevolution of molecular hydrogen in a mutant of *Synechocystis sp.* Strain PCC 6803 deficient in the type I NADPH-dehydrogenase complex. *J. Bacteriol.* 186:1737–1746.

Das, D. 2009. Advances in biohydrogen production processes: An approach towards commercialization. *Int. J. Hydrogen Energy* 34:7349–7357.

Das, D. and T. N. Veziroğlu. 2001. Hydrogen production by biological processes: A survey of literature. *Int. J. Hydrogen Energy* 26:13–28.

Doebbe, A., Rupprecht, J., Beckmann, J. et al. 2007. Functional integration of the HUP1 hexose symporter gene into the genome of *C. reinhardtii*: Impacts on biological H_2 production. *J. Biotechnol.* 131:27–33.

Eroglu, I., Aslan, K., Gündüz, U., Yücel, M., and L. Türker. 1999. Substrate consumption rates for hydrogen production by *Rhodobacter sphaeroides* in a column photobioreactor. *J. Biotechnol.* 70:103–113.

Eroglu, I., Tabanoglu, A., Gunduz, U., Eroglu, E., and M. Yucel. 2008. Hydrogen production by *Rhodobacter sphaeroides* O.U.001 in a flat plate solar bioreactor. *Int. J. Hydrogen Energy* 33:531–541.

El-Shishtawy, R. M. A., Kitajima, Y., Otsuka, S., Kawasaki, S., and M. Morimoto. 1998. Study on the behavior of production and uptake of photobiohydrogen by photosyntheticbacterium *Rhodobacter sphaeroides* RV. In *Biohydrogen*, ed. O.R. Zaborsky, pp. 117–138. London, U.K.: Plenum Press.

Fan, Z., Yuan, L., and R. Chatterjee. 2009. Increased hydrogen production by genetic engineering of *Escherichia coli*. *PLoS ONE* 4(2):e4432. doi:10.1371/journal.pone.0004432.

Fascetti, E. and O. Todini. 1995. *Rhodobacter sphaeroides* RV cultivation and hydrogen production in a one- and two-stage chemostat. *Appl. Microbiol. Biotechnol.* 44:300–305.

Fernandez, F. G. A., Hall, D. O., Guerrero, E. C., Rao, K. K., and M. E. Grima. 2003. Outdoor production of *Phaeodactylum tricornutum* biomass in a helical reactor. *J. Biotechnol.* 103:137–152.

Flores, E. and A. Herrero. 1995. Assimilatory nitrogen metabolism and its regulation. In *The Molecular Biology of Cyanobacteria*, ed. D. A. Bryant, pp. 487–517. New York: Kluwer Academic Publishers.

Fodor, B., Rakhely, G., Kovacs, A. T., and K. L. Kovacs. 2001. Transposon mutagenesis in purple sulfur photosynthetic bacteria: identification of hypF, encoding a protein capable of processing [NiFe] hydrogenases in alpha, beta, and gamma subdivisions of the proteobacteria. *Appl. Environ. Microbiol.* 67:2476–2483.

Forestier, M., King, P., Zhang, L. et al. 2003. Expression of two [Fe]-hydrogenases in *Chlamydomonas reinhardtii* under anaerobic conditions. *Eur. J. Biochem.* 270:2750–2758.

Fox, G. E., Stackebrandt, E., Hespell, R. B. et al. 1980. The phylogeny of prokaryotes. *Science* 208:457–463.

Friedrich, C. G., Bardischewsky, F., Rother, D., Quentmeier, A., and J. Fischer. 2005. Prokaryotic sulfur oxidation. *Curr. Opin. Microbiol.* 8:253–259.

Gaffron, H. 1935. Über den Stoffwechsel der Purpurbakterien. *Biochem. Z.* 275:301–319.

Gaffron, H. and J. Rubin. 1942. Fermentative and photochemical production of hydrogen in algae. *J. Gen. Physiol.* 26:219–240.

Garcin, E., Vernede, X., Hatchikian, E., Volbeda, A., Frey, M., and J. Fontecilla-Camps. 1999. The crystal structure of a reduced [NiFeSe] hydrogenase provides an image of the activated catalytic center. *Structure* 7:557–566.

Gfeller, R. P. and M. Gibbs. 1984. Fermentative metabolism of *Chlamydomonas reinhardtii*: I. Analysis of fermentative products from starch in dark and light. *Plant Physiol.* 75:212–218.

Ghirardi, M. L., Zhang, L., Lee, J. W. et al. 2000. Microalgae: A green source of renewable H_2. *Trends Biotechnol.* 18:506–511.

Ghosh, T. K. and K. Bandyopadyay. 1980. Rapid ethanol fermentation in immobilized yeast cell reactor. *Biotechnol. Bioeng.* 12:1489–1496.

Gibson, J., Stackebrandt, E., Zablen, L. B., Gupta, R., and C. R. Woese. 1979. Phylogenetic analysis of purple photosynthetic bacteria. *Curr. Microbiol.* 3:59–64.

Gogotov, I. N., Zorin, N. A., and L. T. Serebriakova. 1991. Hydrogen production by model systems including hydrogenase from phototrophic bacteria. *Int. J. Hydrogen Energy* 16:393–396.

Gokhan, K., Ufuk, G., Gabor, R., Meral, Y., Inci, E., and L. K. Kornel. 2008. Improved hydrogen production by uptake hydrogenase deficient mutant strain of *Rhodobacter sphaeroides* O.U.001. *Int. J. Hydrogen Energy* 33:3056–3060.

Greenbaum, E. 1988. Energetic efficiency of hydrogen photoevolution by algal water splitting. *Biophys. J.* 54:365–368.

Guiral, M., Tron, P., Belle, V. et al. 2006. Hyperthermostable and oxygen resistant hydrogenases from a hyperthermophilic bacterium *Aquifex aeolicus*: Physicochemical properties. *Int. J. Hydrogen Energy* 31:1424–1431.

Gutekunst, K., Phunpruch, S., Schwarz, C., Schuchardt, S., Schulz-Friedrich, R., and J. Appel. 2005. LexA regulates the bidirectional hydrogenase in the cyanobacterium *Synechocystis sp.* PCC 6803 as a transcription activator. *Mol. Microbiol.* 58:810–823.

Hallenbeck, P. C. and J. R. Benemann. 2002. Biological hydrogen production; fundamentals and limiting processes. *Int. J. Hydrogen Energy* 27:1185–1193.

Happe, T., Hemschemeier, A., Winkler, M., and A. Kaminski. 2002. Hydrogenases in green algae: Do they save the algae's life and solve our energy problems? *Trends Plant Sci.* 7:246–250.

Happe, T. and A. Kaminski. 2002. Differential regulation of the Fe-hydrogenase during anaerobic adaptation in the green alga *Chlamydomonas reinhardtii. Eur. J. Biochem.* 269:1022–1032.

Happe, T. and J. D. Naber. 1993. Isolation, characterization and N-terminal amino acid sequence of hydrogenase from the green alga *Chlamydomonas reinhardtii. Eur. J. Biochem.* 214:475–481.

Hatch, M. D. and C. R. Slack. 1970. Photosynthetic CO_2-fixation pathways. *Annu. Rev. Plant Physiol.* 21:141–162.

Heda, G. D. and M. T. Madigan. 1986. Utilization of amino acids and lack of diazotrophy in the thermophilic anoxygenic phototroph *Chloroflexus aurantiacus. J. Gen. Microbiol.* 132:2469–2473.

Hemschemeier, A. and T. Happe. 2005. The exceptional photofermentative hydrogen metabolism of the green alga *Chlamydomonas reinhardtii. Biochem. Soc. Trans.* 33:39–41.

Heyer, H., Stal, L. J., and W. E. Krumbein. 1989. Simultaneous heterolatic and acetate fermentation in the marine cyanobacterium *Oscillatoria limosa* incubated anaerobically in the dark. *Arch. Microbiol.* 151:558–564.

Higuchi, Y., Yagi, T., and N. Yasuoka. 1997. Unusual ligand structure in Ni–Fe active center and an additional Mg site in hydrogenase revealed by high resolution x-ray structure analysis. *Structure* 5:1671–1680.

Himmelblau, D. M. 1996. *Principles and Calculation in Chemical Engineering*, 6th edn. New Delhi, India: Prentice Hall of India, pp. 700–703.

Hoekema, S., Bijmans, M., Janssen, M., Tramper, J., and R. H. Wijffels. 2003. A pneumatically agitated flat-panel photobioreactor with gas re-circulation: anaerobic photoheterotrophic cultivation of a purple non-sulfur bacterium. *Int. J. Hydrogen Energy* 27:1331–1338.

Holo, H. and R. Sirevåg. 1986. Autotrophic growth and CO_2 fixation of *Chloroflexus aurantiacus. Arch. Microbiol.* 145:173–180.

Horecker, B. L. 2002. The pentose phosphate pathway. *J. Biol. Chem.* 277:47965–47971.

Horwith, R., Orosz, T., Balint, B. et al. 2004. Application of gas separation to recover biohydrogen produced by *Thiocapsa roseopersicina. Desalination* 163:261–265.

Howarth, D. C. and G. A. Codd. 1985. The uptake and production of molecular hydrogen by unicellular cyanobacteria. *J. Gen. Microbiol.* 131:1561–1569.

Hügler, M., Huber, H., Stetter, K. O., and G. Fuchs. 2003. Autotrophic CO_2 fixation pathways in archaea (Crenarchaeota). *Arch. Microbiol.* 179:160–173.

Hügler, M., Wirsen, C. O., Fuchs, G., Taylor, C. D., and S. M. Sievert. 2005. Evidence for autotrophic CO_2 fixation via the reductive tricarboxylic acid cycle by members of the ε subdivision of proteobacteria. *J. Bacteriol.* 187:3020–3027.

Imhoff, J. 2004. Taxonomy and physiology of phototrophic purple bacteria and green sulfur bacteria. In *Anoxygenic Photosynthetic Bacteria*, eds. R. E. Blankenship, M. T. Madigan, and C. E. Bauer, pp. 1–15. New York: Kluwer Academic Publishers.

Kampf, C., and Pfennig, N. 1980. Capacity of chromatiaceae for chemotrophic growth. Specific respiration rates of *Thiocystis violacea* and *Chromatium vinosum*. *Arch. Microbiol.* 127:125–135.

Kashyap, A. K., Pandey, K. D., and S. Sarkar. 1996. Enhanced hydrogen photoproduction by nonheterocystous cyanobacterium *Plectonema boryanum*. *Int. J. Hydrogen Energy* 21:107–109.

Kennedy, R. A., Rumpho, M. E., and T. C. Fox. 1992. Anaerobic metabolism in plants. *Plant Physiol.* 100:1–6.

Kentemich, T., Danneberg, G., Hundeshagen, B., and H. Bothe. 1988. Evidence for the occurrence of the alternative, vanadium-containing nitrogenase in the cyanobacterium *Anabaena variabilis*. *FEMS Microbiol. Lett.* 51:19–24.

Kentemich, T., Haverkamp, G., and H. Bothe. 1991. The expression of a third nitrogenase in the cyanobacterium *Anabaena variabilis*. *Zeitschrift fur Naturforschung* 46:217–222.

Kim, E. J., Lee, M. K., Kim, M. S., and J. K. Lee. 2008a. Molecular hydrogen production by nitrogenase of *Rhodobacter sphaeroides* and by Fe-only hydrogenase of *Rhodospirillum rubrum*. *Int. J. Hydrogen Energy* 33:1516–1521.

Kim, S., Seol, E., Raj, S. M., Park, S., Oh, Y. K., and D. D. Y. Ryu. 2008b. Various hydrogenases and formate-dependent hydrogen production in *Citrobacter amalonaticus* Y19. *Int. J. Hydrogen Energy* 33:1509–1515.

Kimble, L. K. and Madigan, M. T. 1992. Evidence for an alternative nitrogenase in *Heliobacterium gestii*. *FEMS Microbiol. Lett.* 100:255–260.

King, P. W., Posewitz, M. C., Ghirardi, M. L., and M. Seibert. 2006. Functional studies of [FeFe] hydrogenase maturation in an *Escherichia coli* biosynthetic system. *J. Bacteriol.* 188:2163–2172.

Knappe, J. and G. Sawers. 1990. A radical-chemical route to acetyl-CoA: The anaerobically induced pyruvate formate-lyase system of *Escherichia coli*. *FEMS Microbiol. Lett.* 75:383–398.

Koku, H., Eroglu, I., Gündüz, U., Yücel, M., and L. Türker. 2002. Aspects of the metabolism of hydrogen production by *Rhodobacter sphaeroides*. *Int. J. Hydrogen Energy* 27:1315–1329.

Kondo, T., Arakawa, M., Hirai, T., Wakayama, T., Hara, M., and J. Miyake. 2002. Enhancement of hydrogen production by a photosynthetic bacterium mutant with reduced pigment. *J. Biosci. Bioeng.* 93:145–150.

Kondratieva, E. N. 1979. Interrelation between modes of carbon assimilation and energy production in phototrophic purple and green bacteria. In *Microbial Biochemistry*, ed. J. R. Quayle, pp. 117–175. Baltimore, MD: University Park Press.

Kruse, O., Rupprecht, J., Bader, K. et al. 2005a. Improved photobiological H_2 production in engineered green algal cells. *J. Biol. Chem.* 280:34170–34177.

Kruse, O., Rupprecht, J., Mussgnug, J. H., Dismukes, G. C., and B. Hankamer. 2005b. Photosynthesis: A blueprint for solar energy capture and biohydrogen production technologies. *Photochem. Photobiol. Sci.* 4:957–970.

Kumar, N. and Das, D. 2000. Enhancement of hydrogen production by *Enterobacter cloacae* IIT-BT 08. *Proc. Biochem.* 35:589–593.

Kumar, N. and Das, D. 2001. Continuous hydrogen production by immobilized *Enterobacter cloacae* IIT-BT 08 using lignocellulosic materials as solid matrices. *Enzyme Microb. Technol.* 29(4–5):280–287.

Kumar, N., Ghosh, A. K., and D. Das. 2001. Redirection of biochemical pathways for the enhancement of H_2 production by *Enterobacter cloacae*. *Biotechnol. Lett.* 23:537–541.

Kumar, N., Monga, P. S., Biswas, A. K., and D. Das. 2000. Modeling and simulation of clean fuel production by *Enterobacter cloacae* IIT-BT 08. *Int. J. Hydrogen Energy* 25:945–952.

Kumazawa, S. and Mitsui, A. 1982. Hydrogen metabolism of photosynthetic bacteria and algae. In *CRC Handbook of Biosolar Resources*, ed. O. R. Zaborsky, pp. 299–316. Boca Raton, FL: CRC press.

Kyle, D. J., Ohad, I., and C. J. Arntzen. 1984. Membrane protein damage and repair: Selective loss of a quinone–protein function in chloroplast membranes. *Proc. Natl. Acad. Sci. USA.* 81:4070–4074.

Lambert, G. R. and G. D. Smith. 1977. Hydrogen formation by marine blue-green algae. *FEBS Lett.* 83:159–162.

Laurinavichene, T. V., Fedorov, A. S., Ghirardi, M. L., Seibert, M., and A. A. Tsygankov. 2006. Demonstration of hydrogen photoproduction by immobilized, sulfur-deprived *Chlamydomonas reinhardtii* cells. *Int. J. Hydrogen Energy* 31:659–667.

Lay, J. J., Li, Y.-Y., and T. Noike. 1997. Influences of pH and moisture content on the methane production in high solids sludge digestion. *Water Res.* 3:1518–1524.

Lee, J. and Greenbaum, E. 2003. A new oxygen sensitivity and its potential application in photosynthetic H_2 production. *Appl. Biochem. Biotechnol.* 106:303–313.

Long, M., Liu, J., Chen, Z., Bleijlevens, B., Roseboom, W., and S. Albracht. 2007. Characterization of a HoxEFUYH type of [NiFe] hydrogenase from *Allochromatium vinosum* and some EPR and IR properties of the hydrogenase module. *J. Biol. Inorg. Chem.* 12:62–78.

Madigan, M. 2004. Microbiology of nitrogen fixation by anoxygenic photosynthetic bacteria. In *Anoxygenic Photosynthetic Bacteria*, eds. R. E. Blankenship, M. T. Madigan, and C. E. Bauer, pp. 915–928. New York: Kluwer Academic Publishers.

Madigan, M., Cox, S. S., and R. A. Stegeman. 1984. Nitrogen fixation and nitrogenase activities in members of the family *Rhodospirillaceae*. *J. Bacteriol.* 157(1):73–78.

Marin, M. R., 1999. Alcoholic fermentation modelling: Current state and perspectives. *Am. J. Enol. Vitic.* 50:166–178.

Markov, S. A., Thomas, A. D., Bazin, M. J., and D. O. Hall. 1997. Photoproduction of hydrogen by cyanobacteria under partial vacuum in batch culture or in a photobioreactor. *Int. J. Hydrogen Energy* 22:521–524.

Masepohl, B., Schoelisch, K., Goerlitz, K., Kutzki, C., and H. Boehme. 1997. The heterocyst-specific fdxH gene product of the cyanobacterium *Anabaena* sp. PCC 7120 is important but not essential for nitrogen fixation. *Mol. Gen. Genet.* 253:770–776.

McCord, J. M., Keele Jr., B. B., and I. Fridovich. 1971. An enzyme-based theory of obligate anaerobiosis: The physiological function of superoxide dismutase. *Proc. Natl. Acad. Sci. USA.* 68:1024–1027.

McEwan, A. G. 1994. Photosynthetic electron transport and anaerobic metabolism in purple non-sulfur phototrophic bacteria. *Antonie van Leeuwenhoek* 66:151–164.

Melis, A. 2002. Green alga hydrogen production: progress, challenges and prospects. *Int. J. Hydrogen Energy* 27:1217–1228.

Melis, A. 2007. Photosynthetic H_2 metabolism in *Chlamydomonas reinhardtii* (unicellular green algae). *Planta* 226:1075–1086.

Melis, A., Zhang, L., Forestier, M., Ghirardi, M. L., and M. Seibert. 2000. Sustained photobiological hydrogen gas production upon reversible inactivation of oxygen evolution in the green alga *Chlamydomonas reinhardtii*. *Plant Physiol.* 122:127–136.

Michel, H. and J. Deisenhofer. 1988. Relevance of the photosynthetic reaction center from purple bacteria to the structure of photosystem II. *Biochemistry* 27:1–7.

Minnan, L., Jinli, H., Xiaobin, W. et al. 2005. Isolation and characterization of a high H_2-producing strain *Klebsiella oxytoca* HP1 from a hot spring. *Res. Microbiol.* 156:76–81.

Mishra, J., Kumar, N., Ghosh, A. K., and D. Das. 2002. Isolation and molecular characterization of hydrogenase gene from a high rate of hydrogen-producing bacterial strain *Enterobacter cloacae* IIT-BT 08. *Int. J. Hydrogen Energy* 27:1475–1479.

Mitsui, A. 1975. The utilization of solar energy for hydrogen production by cell-free system of photosynthetic organisms. Paper presented at the *Miami Energy Conference*, March 18–20, 1975, Miami Beach, FL. New York: Plenum Press.

Miura, Y., Akano, T., Fukatsu, K. et al. 1997. Stably sustained hydrogen production by biophotolysis in natural day/night cycle. *Energy Convers. Manag.* 38:533–537.

Miyake, M. and Y. Asada. 1997. Direct electroporation of clostridial hydrogenase into cyanobacterial cells. *Biotechnol. Technol.* 11:787–790.

Moezelaar, R., Bijvank, S. M., and L. J. Stal. 1996. Fermentation and sulfur reduction in the mat-building cyanobacterium *Microcoleus chtonoplastes*. *Appl. Environ. Microbiol.* 62:1752–1758.

Molina Grima, E., Acién Fernández, F. G., García Camacho, F., Chisti, Y. 1999. Photobioreactors: Light regime, mass transfer, and scaleup. *J. Biotechnol.* 70:231–247.

Montet, Y., Amara, P., Volbeda, A. et al. 1997. Gas access to the active site of Ni–Fe hydrogenases probed by x-ray crystallography and molecular dynamics. *Nat. Struct. Biol.* 4:523–526.

Munekage, Y., Hashimoto, M., Miyake, C. et al. 2004. Cyclic electron flow around photosystem I is essential for photosynthesis. *Nature* 429:579–582.

Mus, F., Cournac, L., Cardettini, V., Caruana, A., and G. Peltier. 2005. Inhibitor studies on non-photochemical plastoquinone reduction and H$_2$ photoproduction in *Chlamydomonas reinhardtii. Biochim. Biophys. Acta* 1708:322–332.

Mussgnug, J. H., Thomas-Hall, S., Rupprecht, J. et al. 2007. Engineering photosynthetic light capture: Impacts on improved solar energy to biomass conversion. *Plant Biotechnol. J.* 5:802–814.

Nath, K. and D. Das. 2004. Improvement of fermentative hydrogen production: Various approaches. *Appl. Microbiol. Biotechnol.* 65:520–529.

Nath, K., Muthukumar, M., Kumar, A., and D. Das. 2008. Kinetics of two-stage fermentation process for the production of hydrogen. *Int. J. Hydrogen Energy* 33:1195–2203.

Nguyen, T. A. D., Kim, J. P., Kim, M. S., Oh, Y. K., and S. J. Sim. 2008. Optimization of hydrogen production by hyperthermophilic eubacteria, *Thermotoga maritime* and *Thermotoga neapolitana* in batch fermentation. *Int. J. Hydrogen Energy* 33:1483–1488.

Nicolet, Y., Lemon, B. J., Fontecilla-Camps, J. C., and J. W. Peters. 2000. A novel FeS cluster in Fe-only hydrogenases. *Trends Biochem. Sci.* 25:138–143.

Nicolet, Y., Piras, C., Legrand, P., Hatchikian, C. E., and J. C. Fontecilla-Camps. 1999. *Desulfovibrio desulfuricans* iron hydrogenase: The structure shows unusual coordination to an active site Fe binuclear center. *Structure* 7:13–23.

Nixon, P. J. 2000. Chlororespiration. *Philos. Trans. R. Soc. Lond. B Biol. Sci.* 355:1541–1547.

Nultsch, W. and G. Agel. 1986. Fluence rate and wavelength dependence of photobleaching in the cyanobacterium *Anabaena variabilis. Arch. Microbiol.* 144:268–271.

Oh, Y., Seol, E., Lee, E. Y., and S. Park. 2002. Fermentative hydrogen production by a new chemoheterotrophic bacterium *Rhodopseudomonas Palustris* P4. *Int. J. Hydrogen Energy* 27:1373–1379.

Orme-Johnson, W. H. 1992. Nitrogenase structure: Where to now? *Science* 257:1639–1640.

Otsuki, T., Uchiyama, S., Fujiki, K., and S. Fukunaga. 1998. Hydrogen production by a floating-type photobioreactor. In *Biohydrogen*, ed. O.R. Zaborsky, pp. 369–374. London, U.K.: Plenum Press.

O-Thong, S., Prasertsan, P., Karakashev, D., and I. Angelidaki. 2008. Thermophilic fermentative hydrogen production by the newly isolated *Thermoanaerobacterium thermosaccharolyticum* PSU-2. *Int. J. Hydrogen Energy* 33:1204–1214.

Penfold, D. W., Forster, C. F., and L. E. Macaskie. 2003. Increased hydrogen production by *Escherichia coli* strain HD701 in comparison with the wild-type parent strain MC4100. *Enzyme Microb. Technol.* 33:185–189.

Polle, J. E. W., Kanakagiri, S., and A. Melis. 2003. tla1, a DNA insertional transformant of the green alga *Chlamydomonas reinhardtii* with a truncated light-harvesting chlorophyll antenna size. *Planta* 217:49–59.

Prince, R. C. and H. S. Kheshgi. 2005. The Photobiological production of hydrogen: Potential efficiency and effectiveness as a renewable fuel. *Crit. Rev. Microbiol.* 31:19–31.

Rachman, M. A., Nkashimada, Y., Kakizono, T., and N. Nishio. 1998. Hydrogen production with high yield and high evolution rate by self-flocculated cells of *Enterobacter aerogenes* in a packed-bed reactor. *Appl. Microbiol. Biotechnol.* 49:450–454.

Rao, R. B. and S. K. Mahmood. 1997. Hydrogen production by phototrophic sulfur bacteria from waste waters. In *Recent Advanced in Ecobiological Research*, eds. M. P. Shinha and P. N. Mehrotra, pp. 227–232. New Delhi, India: A. P. H. Publishing Corporation.

Ruhle, T., Hemschemeier, A., Melis, A., and T. Happe. 2008. A novel screening protocol for the isolation of hydrogen producing *Chlamydomonas reinhardtii* strains. *BMC Plant Biol.* 8:107.

Ruiyan, Z., Di, W., Yaoping, Z., and Li. Jilun. 2006. Hydrogen production by *draTGB hupL* double mutant of *Rhodospirillum rubrum* under different light conditions. *Chin. Sci. Bull.* 51:2611–2618.

Sasikala, K., Ramana, C. V., Rao R. P., and K. L. Kovacs. 1993. Anoxygenic phototrophic bacteria: Physiology and advances in hydrogen production technology. *Adv. Appl. Microbiol.* 38:211–295.

Schmitz, O., Boison, G., Hilscher, R. et al. 1995. Molecular biological analysis of a bidirectional hydrogenase from cyanobacteria. *Eur. J. Biochem.* 233:266–276.

Schmitz, O., Boison, G., Salzmann, H. et al. 2002. HoxE: A subunit specific for the pentameric bidirectional hydrogenase complex (HoxEFUYH) of cyanobacteria. *Biochim. Biophys. Acta* 1554:66–74.

Schnackenberg, J., Ikemoto, H., and S. Miyachi. 1996. Photosynthesis and hydrogen evolution under stress conditions in a CO_2-tolerant marine green alga, *Chlorococcum littorale. J. Photochem. Photobiol. B: Biol.* 34:59–62.

Schütz, K., Happe, T., Troshina, O. et al. 2004. Cyanobacterial H_2 production: A comparative analysis. *Planta* 218:350–359.

Seabra, R., Santos, A., Pereira, S., Moradas-Ferreira, P., and P. Tamagnini. 2009. Immunolocalization of the uptake hydrogenase in themarine cyanobacterium *Lyngbyamajuscula* CCAP1446/4 and two *Nostoc* strains. *FEMS Microbiol. Lett.* 292:57–62.

Seol, E., Kim, S., Raj, S. M., and S. Park. 2008. Comparison of hydrogen-production capability of four different *Enterobacteriaceae* strains under growing and non-growing conditions. *Int. J. Hydrogen Energy* 33:5169–5175.

Seon, Y. H., Lee, C. G., Park, D. H., Hwang, K. Y., and Y. I. Joe. 1993. Hydrogen production by immobilized cells in the nozzle loop bioreactor. *Biotechnol. Lett.* 15:1275–1280.

Serebryakova, L., Sheremetieva, M., and A. Tsygankov. 1998. Reversible hydrogenase activity of *Gloeocapsa alpicola* in continuous culture. *FEMS Microbiol. Lett.* 166:89–94.

Sirevåg, R. and J. G. Ormerod. 1977. Synthesis, storage and degradation of polyglucose in *Chlorobium thiosulfatophilum. Arch. Microbiol.* 111:239–244.

Skjånes, K., Lindblad, P., and J. Muller. 2007. BioCO$_2$: A multidisciplinary, biological approach using solar energy to capture CO_2 while producing H_2 and high value products. *Biomol. Eng.* 24:405–413.

Stal, L. J. and W. E. Krumbein. 1985. Nitrogenase activity in the non-heterocystous cyanobacterium *Oscillatoria sp.* grown under alternating light-dark cycles. *Arch. Microbiol.* 143:67–71.

Stal, L. J. and W. E. Krumbein. 1987. Temporal separation of nitrogen fixation and photosynthesis in the filamentous, non-heterocystous cyanobacterium *Oscillatoria* sp. *Arch. Microbiol.* 149:76–80.

Stephenson, M. and L. H. Stickland. 1932. Hydrogenlyases. *Biochem. J.* 26:712–724.

Stewart, V. 1988. Nitrate respiration in relation to facultative metabolism in Enterobacteria. *Microbiol. Biol. Rev.* 52:190–232.

Suzuki, S. and I. Karube. 1981. Hydrogen production by immobilized whole cells of *Clostridium butyricum*. Paper presented at the *International Hydrogen Energy Progress*, June 23–26, 1980, Tokyo, Japan.

Syahidah, K., Ismail, K., Najafpour, G., Younesi, H., Mohamed, A. R., and A. H. Kamaruddin. 2008. Biological hydrogen production from CO: Bioreactor performance. *Biochem. Eng. J.* 39:468–477.

Tabita, F. R. 1999. Microbial ribulose 1,5-bisphosphate carboxylase/oxygenase: A different perspective. *Photosynth. Res.* 60:1–28.

Taguchi, F., Chang, J. D., Mizukami, N., Saito-Taki, T., Hasegawa, K., and M. Morimoto. 1993. Isolation of a hydrogen producing bacterium, *Clostridium beijerinckii* strain AM21B from termites. *Can. J. Microbio1.* 39:726–730.

Tamagnini, P., Axelsson, R., Lindberg, P., Oxelfelt, F., Wunschiers, R., and P. Lindblad. 2002. Hydrogenases and hydrogen metabolism of cyanobacteria. *Microbiol. Mol. Biol. Rev.* 66:1–20.

Tanisho, S., Kamiya, N., and N. Wakao. 1989. Hydrogen evolution of *Enterobacter aerogenes* depending on culture pH: Mechanism of hydrogen evolution from NADH by means of membrane-bound hydrogenase. *Biochim. Biophys. Acta* 973:1–6.

Thauer, R. K., Jungermann, K., and K. Decker. 1977. Energy conservation in chemotrophic anaerobic bacteria. *Bacteriol. Rev.* 41:100–180.

Thiel, T. 1993. Characterization of genes for an alternative nitrogenase in the cyanobacterium *Anabaena variabilis. J. Bacteriol.* 175:6276–6286.

Torzillo, G., Scoma, A., Faraloni, C., Ena, A., and U. Johanningmeier. 2009. Increased hydrogen photoproduction by means of a sulfur-deprived *Chlamydomonas reinhardtii* D1 protein mutant. *Int. J. Hydrogen Energy* 34:4529–4536.

Tredici, M. R. 1999. Bioreactors, photo. In *Encyclopedia of Bioprocess Technology: Fermentation, Biocatalysis and Bioseparation*, eds. M. C. Flickinger and S. W. Drew. New York: Wiley, Vol. 1, pp. 395–419.

Tsygankov, A. A., Serebryakova, L. T., Rao, K. K., and D. O. Hall. 1998. Acetylene reduction and hydrogen photoproduction by wild-type and mutant strains of *Anabaena* at different CO_2 and O_2 concentrations. *FEMS Microbiol. Lett.* 167:13–17.

Uyeda, K. and J. C. Rabinowitz. 1971. Pyruvate-ferredoxin oxidoreductase. *J. Biol. Chem.* 246:3111–3119.

van de Werken, H. J., Verhaart, M. R., VanFossen, A. L. et al. 2008. Hydrogenomics of the extremely thermophilic bacterium *Caldicellulosiruptor saccharolyticus*. *Appl. Environ. Microbiol.* 74(21):6720–6729.

Vasilyeva, L., Miyake, M., Khatipov, E., Wakayama, T., Sekine, M., Hara, M., Nakada, E., Asada, Y., and J. Miyake. 1999. Enhanced hydrogen production by a mutant of *Rhodobacter sphaeroides* having an altered light-harvesting system. *J. Biosci. Bioeng.* 87:619–624.

Vignais, P. M. and B. Billoud. 2007. Occurrence, classification, and biological function of hydrogenases: An overview. *Chem. Rev.* 107:4206–4272.

Vignais, P. M., Billoud, B., and J. Meyer. 2001. Classification and phylogeny of hydrogenases. *FEMS Microbiol. Rev.* 25:455–501.

Vignais, P. M. and A. Colbeau. 2004. Molecular biology of microbial hydrogenases. *Curr. Issues Mol. Biol.* 6:159–88.

Vignais, P. M., Magnin, J., and J. C. Willison. 2006. Increasing biohydrogen production by metabolic engineering. *Int. J. Hydrogen Energy* 31:1478–1483.

Vonshak, A., Boussiba, S., Abeliovich, A., and Richmond, A. 1983. On the production of Spirulina biomass: The maintenance of pure culture under outdoor conditions. *Biotechnol. Bioeng.* 25(2):341–351.

Warthmann, R., Cypionka, H., and N. Pfennig. 1992. Photoproduction of H_2 from acetate by syntrophic cocultures of green sulfur bacteria and sulfur-reducing bacteria. *Arch. Microbiol.* 157:343–348.

Wu, L. F. and M. A. Mandrand. 1993. Microbial hydrogenases: Primary structure, classification, signatures and phylogeny. *FEMS Microbiol. Rev.* 104:243–270.

Wykoff, D. D., Davies, J. P., Melis, A., and A. R. Grossman. 1998. The regulation of photosynthetic electron transport during nutrient deprivation in *Chlamydomonas reinhardtii*. *Plant Physiol.* 117:129–139.

Xu, Q., Yooseph, S., Smith, H. O., and C. J. Venter. 2005. Development of a novel recombinant cyanobacterial system for hydrogen production from water. Paper presented at *Genomics: GTL Program Projects*, December 7–9, 2005, Rockville, MD.

Yetis, M., Gunduz, U., Eroglu, I., Yucel, M., and L. Turker. 2000. Photoproduction of hydrogen from sugar refinery wastewater by *Rhodobacter sphaeroides* O.U.001. *Int. J. Hydrogen Energy* 25:1035–1041.

Yokoi, H., Tokushige, T., Hirose, J., Hayashi, S., and Y. Takasaki. 1997. Hydrogen production by immobilized cells of aciduric *E. aerogenes* strain HO-39. *J. Ferment. Bioeng.* 83:481–484.

Yoshida, A., Nishimura, T., Kawaguchi, H., Inui, M., and H. Yukawa. 2005. Enhanced hydrogen production from formic acid by formate hydrogen lyase-overexpressing *Escherichia coli* strains. *Appl. Environ. Microbiol.* 71:6762–6768.

Zeikus, J. G. 1977. The biology of methanogenic bacteria. *Bacteriol. Rev.* 41:514–541.

Zeikus, J. G., Hegge, P. W., and M. A. Anderson. 1979. *Thermoanaerobium brockii* gen. nov. and sp. nov., a new chemoorganotrophic, caldoactive, anaerobic bacterium. *Arch. Microbiol.* 122:41–48.

11

Photobiological and Photobiomimetic Production of Solar Fuels

Paul W. King, Katherine A. Brown, Kathleen Ratcliff, and Laura L. Beer

CONTENTS

11.1 Introduction

The solar energy that is adsorbed by the earth per year is estimated at 3.85 YJ, or 439 EJ on an hourly basis, which is enough to meet the global energy demand of 474 EJ for an entire year (see discussion in Refs. [1–3]). Solar energy capture into electricity with photovoltaics (PV) offers one alternative to the global dependence on nonrenewable fossil fuels (e.g., petroleum, coal, and natural gas) that is creating negative long-term environmental and socioeconomical effects [4,5]. One reality of a heavy reliance on fossil fuels is the rising levels of greenhouse gases (e.g., methane and CO_2), with myriad and complex consequences to climates that are causing widespread concerns [6–8]. From the need to address these realities, an increasing number of corporate and nation-sponsored research efforts are being organized to develop carbon-neutral energy sources, mitigate energy use with more energy-efficient technologies, and sequester CO_2 to ameliorate global warming [9–13].

Although PV-based electricity generation has numerous benefits for meeting global power needs, solar radiation is intermittent and insolence depends on geography [14], which together limit PV as a wide-scale practical alternative to conventional energy sources [1,15]. These challenges can be addressed with technologies that convert solar-generated electricity into gaseous or liquid fuels, providing storage (intermittency) and transport (geographical dependence) capabilities. Examples of target fuels include hydrogen gas that can be produced from water by photoelectrolysis, or carbon-based fuels produced from capture and conversion of CO_2. This chapter will summarize research efforts that support development of both photobiological and photobiomimetic technologies for capture and conversion of solar energy into both biohydrogen and carbon fuels (biofuels) (summarized in Figure 11.1). The section on photobiological research will encompass the progress toward developing photosynthetic microbes for direct production of biofuels. Section 11.3 on photobiomimetic approaches will summarize efforts to synthesize molecules and fabricate devices to capture solar energy for water splitting, hydrogen production and CO_2 conversion into biofuels. Although these two approaches may represent opposing philosophies toward addressing solutions to renewable energy, the physical and chemical challenges of capturing and converting light energy into the electrochemical potential required for chemical reactions are common. Thus, the examples that are highlighted below show how the structure and function of photoactive molecules and fuel-forming catalysts, whether of biological or chemical origin, share similar design themes.

11.2 Photobiological Biofuel and Biohydrogen Production

11.2.1 Background

The development of phototrophic microorganisms for the production of carbon-based biofuels [16–20] and biohydrogen (also referred to as H_2) [21–26] is being intensively explored on many basic and applied research fronts. When compared to enzymatic or cellular conversion of plant-derived biomass, photosynthetic microorganisms require less arable land, fewer resources, and have higher predicted sunlight-to-product conversion efficiencies [20,26–29]. For example, it is theoretically possible to gain a 100-fold improvement in lipid production per acre using algae compared to cultivation of soybeans for

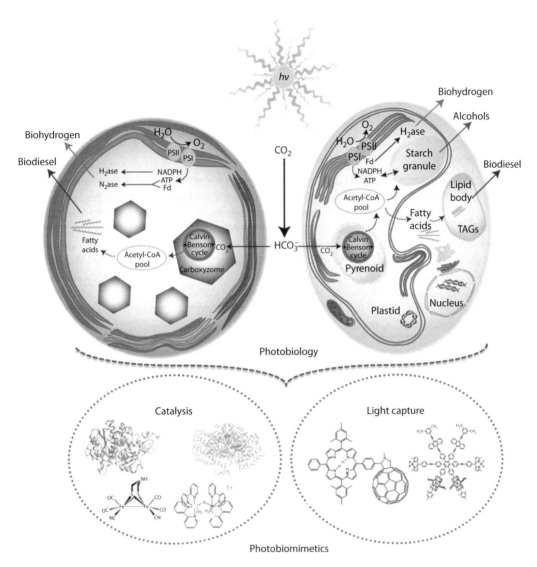

FIGURE 11.1
Schematics of photobiological reaction pathways and enzymes as inspirations for photobiomimetic designs. Models of cyanobacterial (left) and algal (right) cells showing the pathways that direct the capture of CO_2 and light into biofuels; fatty acids (FAs), triacylglycerols (TAGs), alcohols, and biohydrogen (H_2). Investigations of the molecular and atomic scale structures and functional mechanisms of chlorophyll and antenna molecules, hydrogenases, and the PSII Mn-center have led to the synthesis of molecular photobiomimetics for development of light capture, hydrogen production, water oxidation, and carbon capture. (Adapted from Gust, D. et al. *Acc. Chem. Res.*, 42(12), 1890, 2009; Capon, J.-F. et al. *Coord. Chem. Rev.*, 249(15–16), 1664, 2005; Brimblecombe, R. et al. *Dalton Trans.*, 43, 9374, 2009. With permission.)

biomass-derived lipid production [28]. However, the biofuel and biohydrogen production efficiencies of phototrophic microbes (e.g., diatoms, green algae, and cyanobacteria) are currently 10- to 20-fold lower than the corresponding calculated theoretical maximum values [16]. Ongoing research to engineer model species, to characterize the biofuel and biohydrogen production pathways, metabolisms, and physiologies, and explore natural microbial diversity to discover and isolate new strains is needed to support applied efforts toward attaining the efficiencies that are required for a cost-effective process [16,17,23,25,30–37]. Moreover, there are well-coordinated basic and applied research efforts in several countries that aim to understand factors that control solar-conversion efficiencies in phototrophs, regulate the reductant and carbon flux distributions, as well as how to improve enzymes through protein engineering (e.g., see Refs. [38,39]). These efforts collectively seek to optimize solar-conversion efficiencies toward achieving theoretical maxima.

11.2.2 Biofuels: Lipids for Biodiesel

The discussion of biofuels will address the use of phototrophic microorganisms for the photosynthetic production of lipids as biodiesel feedstocks. For discussions on the use of phototrophs for production of alcohols and other biofuels, please refer to the excellent reviews in Refs. [16,35,36,40,41]. Much of the research on lipid production in phototrophs evolved with the development of culture collections [38,39,42], screening individual species for lipid content using lipophilic dyes [43–45] or analytical techniques [46–48], and the development of model organisms to characterize genetic and physiological traits regulating lipid accumulation (for a summary of culture collections and associated research efforts, see Ref. [38]). However, only a few of the cultured organisms (e.g., *Botryococcus*, *Chlamydomonas*, *Chlorella*, and *Nanochloropsis*) naturally accumulate lipids, and not always in the ideal chemical form, nor at levels suitable for scalable, commercial production as a biodiesel feedstock [20,49]. Through isolation and curation efforts that have established culture collections [42,50], researchers have characterized a variety of new algal and cyanobacterial species and strains with the capacity for lipid production. Lipid contents and compositions are known to vary greatly among species [38,47,48], and there is no clear correlation of these properties with taxonomic groups. One noticeable trend, however, is that on a per-dry-cell-weight basis, species of both green algae and diatoms (summarized in Figure 11.2A and B, respectively) tend to produce more lipid than other algal groups (Figure 11.2C), whereas all three of these groups produce significantly more lipids than cyanobacteria (Figure 11.2D).

Oleaginous microorganisms may also accumulate hydrocarbons in the form of short-chain fatty acids (FAs) that are synthesized as precursors to longer-chain lipids. The FAs isolated from algae are predominantly saturated or mono-unsaturated, and 16–18 carbon (C) atoms in length [46,47,51]. However, lipid content and compositions do vary among species; for example, some species produce intermediate length, saturated lipids (10–14 C), whereas others produce poly-unsaturated and long-chain FAs (>20C) [46]. The compositions of the lipid fraction can also vary as cells are exposed to different growth conditions as discussed in the following.

11.2.2.1 Lipid Biosynthetic Pathways

Photosynthetic CO_2 fixation begins with the uptake of inorganic carbon (i.e., bicarbonate), which is converted to CO_2 by carbonic anhydrase (CA), and CO_2 fixation into organic carbon by the catalytic activity of ribulose-1,5-bisphosphate carboxylase oxygenase-5 (RuBisCo) via the Calvin-Benson-Bassham cycle (Figure 11.1) [52]. In green algae (e.g., *Chlamydomonas*), bicarbonate transport across the thylakoid membrane and CA activity in the plastid provide CO_2 to RuBisCo, located in the pyrenoid body [53,54]. In cyanobacteria (e.g., *Synechococcus* and *Synechocystis*), the bicarbonate transporter, CA and RuBisCo are part of the carboxysome, a multienzyme and protein complex [55,56]. Subsequent conversion of

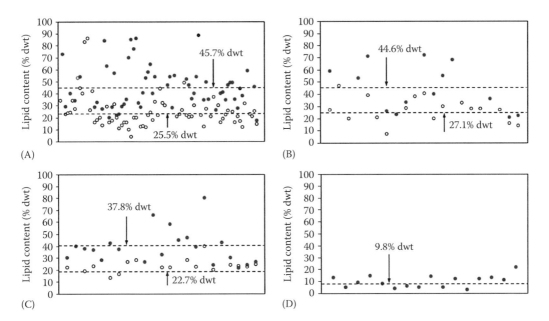

FIGURE 11.2
Cellular lipid content in various classes of microalgae and cyanobacteria under normal growth (open circles) and stress conditions (closed circles). The distribution along the *x*-axis in each graph refers to data points measured for various species and strains of (A) green algae; (B) diatoms; (C) oleaginous species and strains of other eukaryotic algae; and (D) cyanobacteria. (Hu, Q. et al.: Microalgal triacylglycerols as feedstocks for biofuel production: Perspectives and advances. *Plant J.* 2008. 54(4). 621–639. Copyright Wiley-VCH Verlag GmbH & Co. KGaA. Reprinted with permission.)

fixed carbon by central metabolism in algae results in the formation of glucose and starch, or glycogen, glucosylglycerol or polyhydroxybutyrate in cyanobacteria.

Lipids (or FAs) are biosynthesized in the plastid compartment in a process that initiates from glycolysis and production of acetyl coenzyme A (CoA). The exact origin and steady-state concentrations of the acetyl-CoA pool are not resolved, and based on flux analysis might be limiting for lipid biosynthesis [57–59]. FA synthesis in algae also occurs primarily in the chloroplast, where the committed step following glycolytic acetyl-CoA production is the conversion of acetyl-CoA to malonyl-CoA by acetyl-CoA carboxylase (ACC) [46,58] (Figure 11.3A). In higher plants, the levels and activities of ACC show a correlation with FA levels and it is postulated to be at least one step that controls the levels of lipid accumulation (see discussion in Refs. [60–62]). Similar to the process of FA synthesis in plants, the subsequent steps in algal FA synthesis take place in the plastid by type-II fatty acid synthase (FAS) enzymes [18,63,64] (Figure 11.3A). Unsaturated FAs are then generated through the introduction of double bonds by desaturases [65,66], and multiple desaturase genes have been identified in both cyanobacterial [67,68] and algal [69–71] genomes. Through these steps, biodiesel precursor FAs are produced directly from photosynthetic CO_2 fixation, or through autotrophic assimilation of carbon sources in the media. Free fatty acids (FFAs) are released from acyl carrier protein (ACP) during export by acyl-specific acyl-ACP thioesterases (see Ref. [72]).

Another major class of lipids for use as biofuels are triacylglycerols (TAGs) (reviewed in Ref. [73]) (Figure 11.3B). The biosynthesis of TAGs in plants can occur by the direct utilization of glycerol-3-phosphate (G3P) (the Kennedy pathway), which is a by-product of starch degradation [74]. During export of acyl-ACPs from the plastid pool, the acyl unit

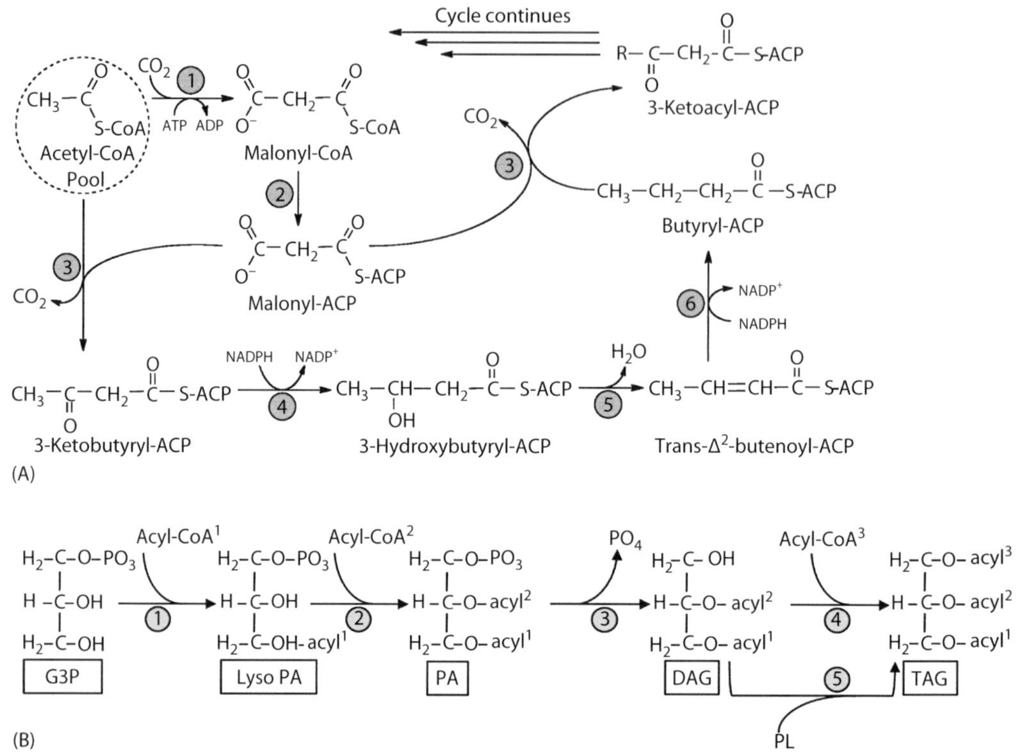

FIGURE 11.3
Fatty acid (FA) and triacylglycerol (TAG) biosynthetic pathways. (A) FA biosynthesis begins in the plastid compartment with the committed step of acetyl-CoA carboxylation to form malonyl-CoA by acetyl-CoA carboxylase (reaction 1). Subsequent steps occur via fatty acid synthase (FAS)-dependent reactions. Malonyl-CoA is converted to malonyl-ACP by malonyl-CoA:ACP transacylase (reaction 2), and eventually forms butyryl-ACP through reactions catalyzed by 3-ketoacyl-ACP synthase (reaction 3), 3-ketoacyl-ACP reductase (reaction 4), 3-hydroxyacyl-ACP dehydrase (reaction 5), and enoyl-ACP reductase (reaction 6). Condensation of butyryl-ACP with a second malonyl-ACP leads to chain elongation [63]. (B) TAG biosynthesis evolves from conversion of FFAs into acyl-CoA precursors [72], which are derived from the plastid acyl-ACP pool upon export. TAG pathway initiates with conversion of glycerol-3-phosphate (G3P) to lyso-phosphatidic acid (Lyso-PA) by glycerol-3-phosphate acyl transferase (reaction 1). Lyso-PA is converted to phosphatidic acid (PA) by lyso-phosphatidic acid acyltransferase (reaction 2). PA is dephosphorylated to diacylglycerol (DAG) by phosphatidic acid phosphatase (reaction 3), and a third acyl group added from acyl-CoA by acyl-CoA:diacylglycerol acyltransferase (reaction 4), or from phospholipids donors by phospholipid:diacylglycerol acyltransferase (reaction 5). (Hu, Q. et al.: Microalgal triacylglycerols as feedstocks for biofuel production: Perspectives and advances. *Plant J.* 2008. 54(4). 621–639. Copyright Wiley-VCH Verlag GmbH & Co. KGaA. Adapted with permission.)

is released from ACP at the inner membrane leaflet by acyl-specific thioesterases, which can be followed by the addition of CoA at the outer membrane leaflet by an acyl-CoA synthetase (see Refs. [61,72] and references therein). These comprise the acyl-CoA pool that is utilized for TAG biosynthesis. The reaction of G3P with two molecules of acyl-CoA to form phosphatidic acid (PA) is catalyzed by glycerol-3-phosphate acyltransferase (G3PAT) and acyl-CoA:lyso-phosphatidic acid acyltransferase (LPAT), respectively (see Figure 11.3B). In the green alga *Chlamydomonas reinhardtii*, G3PAT and LPAT activities are localized to the plastid [75]. The PA product is converted into diacylglycerol (DAG), which is subsequently converted into TAG through the addition of a third acyl-CoA by acyl-CoA:diacylglycerol acyltransferase (DGAT) (Figure 11.3B). Accumulation of TAGs in algae results in formation of lipid bodies (LBs) localized within the cytoplasmic compartment [76]. In plants, a second pathway for the utilization of phospholipids as acyl donors for the conversion of DAG directly to TAGs is catalyzed by the enzyme phospholipid:diacylglycerol acyltransferase (PDAT) [77]. The biochemical properties of PDAT have been studied for enzymes isolated from both *Saccharomyces cerevisiae* [77] and *Arabidopsis thaliana* [78], and have been shown to catalyze the transfer of an acyl group from the *sn*-2 position of a phospholipid into DAG to form TAG [77].

Orthologues of *PDAT* genes have been identified in the genome sequence of the green alga *Ostreococcus tauri*, but are absent in the genome of *C. reinhardtii* [18]. However, the *C. reinhardtii*, genome does encode orthologues of DGAT [18], and *C. reinhardtii* accumulates high levels of TAG-enriched LBs, when induced through prolonged nitrogen starvation in the presence of acetate [51]. Accumulation of LBs is also enhanced under photoautotrophic growth in *C. reinhardtii* mutants deficient in starch accumulation [51,79,80]. This starch metabolism effect contrasts to the attenuated levels of biohydrogen production observed in starchless mutants, and in the higher H_2 production rates observed in mutants that over-accumulate starch (see Refs. [81–84] for a more in-depth discussion). As will be discussed in more detail below, H_2 production and lipid accumulation compete for reductant and carbon pools, and the differential responses to changes in metabolism and physiological conditions must be considered in the selection and application of engineering strategies.

Once the process of LB accumulation is complete, the lipids must be harvested for conversion into biodiesel. The downstream process for the conversion of TAGs into biodiesel is relatively straightforward; TAGs are converted into FA methyl esters (FAME) and glycerol by methanolysis catalyzed in the presence of an acid or alkali (Figure 11.4). For this reason TAGs are an appealing biodiesel source [85].

One species of algae that is of particular interest is *Botryococcus braunii* [49,86,87]. This organism has numerous pathways to produce large quantities of lipids and accumulate long carbon chains of varying types. There are three distinct strains of *B. braunii*, A, B, and L. Strain A is defined by its ability to produce *n*-alkadiene and triene hydrocarbons, odd-carbon-numbered from 23 to 33 C in length. Strain B produces triterpenoid hydrocarbons, 30–37 C in length and 31–34 C long methylated squalenes. Finally, strain L only produces a single tetraterpenoid hydrocarbon, lycopadiene (Figure 11.5). In addition to these unique hydrocarbons, *B. braunii* makes FAs and TAGs chemically related to those discussed above, and the pathways for hydrocarbon production in each strain have been identified [86,87]. The lipid metabolism of *B. braunii* is just one example of the complexity that exists in biological organisms, and shows the benefit of further investigation of natural diversity in algae.

11.2.2.2 Growth Conditions and Lipid Accumulation

Nutrient levels (e.g., nitrogen, salts and silica) and environmental conditions (e.g., temperature, light intensity, salinity, pH, and water content) have been shown to significantly affect the amount and types of lipids that accumulate in algae, diatoms, and cyanobacteria. Nitrogen deprivation leads to the accumulation of lipids in both algal and cyanobacterial species, and conditions of drought (water stress) induce *Nostoc commune* to accumulate high lipid contents [88]. Nitrogen deprivation has been shown to have a strong affect on lipid accumulation in algae, with many species of algae specifically accumulating TAGs [89]. Deficiency in other nutrients, such as phosphates and sulfates, can also cause accumulation of lipids and TAGs [90], though the levels and types are highly dependent on the strain and species. Thus, the physiological and metabolic effects of nutrient depletion must

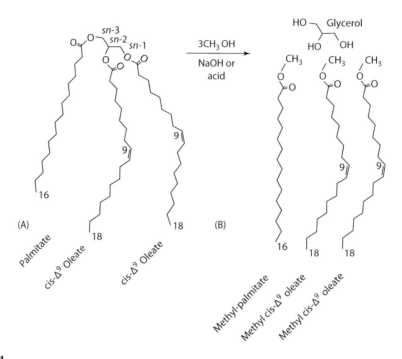

FIGURE 11.4

The production of biodiesel from triacylglycerol (TAG). TAGs (A) are converted to glycerol and fatty acid methyl esters (FAMEs) (B) by reacting with methanol in the presence of an acid or alkali (shown as NaOH in the figure) catalyst. (Durrett, T.P., Benning, C., and Ohlrogge, J.: Plant triacylglycerols as feedstocks for the production of biofuels. *Plant J.*, 2008. 54(4): 593–607. Reprinted with permission.)

FIGURE 11.5

Examples of the hydrocarbons produced and isolated from three strains of *B. braunii*. (A) C_{27} diene and C_{27} triene from strain A. (B) C30 botryococcene, squalene, and tetramethyl squalene from strain B. (C) *trs,trs*-Lycopadiene from strain L. (Reprinted with permission from Springer Science+Business Media: *Applied Microbiology and Biotechnology, Botryococcus braunii*: A rich source for hydrocarbons and related ether lipids, 66(5), 2005, 486–496, Metzger, P. and Largeau, C.)

be determined experimentally for each organism. The concentrations of salts and other osmolytes in culture media also affect carbon-fixation fluxes within phototrophs. Under salt stress or osmotic shock, cellular responses lead to elevated glycerol pathway gene expression and incorporation of fixed carbon into glycerol [91,92]. This effect is consistent with a mode of adaptation that has evolved in phototrophs that grow in high saline environments [93]. The strategy involves balancing osmotic strength by synthesizing organic molecules as solutes, which often consist of polyols such as glycerol.

In addition to nutrient depletion, environmental conditions, such as light intensity and temperature, can also alter the cellular accumulation levels of FAs and TAGs [94–96]. Most of the organisms discussed in this chapter are photosynthetic microorganisms that have evolved to survive under a range of temperatures. A noticeable bias toward accumulation of more unsaturated FAs has been observed in thermophilic algal species, or when mesophilic algae are cultured at higher temperatures [95,97]. This adaptation is most likely the result of an evolutionary bias toward structures that form a more packed lipid bilayer, which compensates for the increase in membrane fluidity that occurs as temperatures increase. At lower temperatures, more saturated FAs create "kinks" in the hydrocarbon chains, forming packing defects in the lipid bilayer, thereby increasing fluidity and counteracting the tendency of lipid membranes to become more rigid at low temperatures [96]. An increase in light intensities can cause algae to accumulate neutral storage lipids, such as TAGs [94]. This metabolic shift is accompanied by morphological, chemical, and physiological changes, which in turn can affect the yields from the harvesting process.

In summary, environmental factors strongly influence the cellular physiological state, and the characterization of this response from the molecular to the systems level will help optimize hydrocarbon formation in phototrophic organisms.

11.2.2.3 Strain Engineering for Lipid Production

Developments of high-throughput (HTP) sequencing [98,99] with advances in computational technologies have led to a proliferation of microbial genomic (whole-genome) and metagenomic data. From this sequence information and understanding of the biology of microbial diversity [100] as well as identification of new genes, enzymes, and pathways that support biofuels production is possible [33,40,101,102]. Complementary advances in strain engineering include genetic approaches to regulate pathway expression, alter enzyme composition or improve substrate fluxes, and enzyme engineering to improve pathway kinetics [49,101,103–106]. In practice successful strain engineering presents many challenges [107]. As shown in Figure 11.3, the generation of FAs and TAGs requires balancing substrate fluxes with other metabolic pathways that generate or utilize acetyl-CoA and the constant regeneration of reducing equivalents (e.g., nicotinamide adenine dinucleotide phosphate (NADPH)). A number of possible regulatory effects are also likely to be involved that compensate steady-state flux balances. These regulatory mechanisms involve both static and sensory responses, which act in combination to control metabolic fluxes [104]. As a result, altering gene expression (e.g., gene knockout) to remove a competitive pathway might trigger regulatory effects that result in little or no change in biofuel-production levels [107]. The fields of systems biology [26,108,109] and flux analysis [110] are mapping the metabolic fluxes within an organism to create models and predict outcomes of pathway engineering. It is possible now with the advent and refining of HTP techniques to thoroughly analyze the metabolic effects of either genetic mutations, changes in nutrient levels, or variations in culture conditions [111]. Validation of theoretical models by experimentation is required to generate accurate predictions. For example, it has been recently shown [112,113] in both bacterial and mammalian model organisms that acetylation of enzyme lysine groups, previously hypothesized to be a general mechanism of regulation [114], controls cellular carbon metabolism. Whether protein acetylation also regulates metabolism in phototrophs is a question that will likely be addressed by emerging HTP efforts. This and other regulatory mechanisms, in turn, are likely to have a functional role in cellular adaptations to stress (e.g., temperature, osmotic shock, and nutrient deprivation). Thus, a complete understanding of these metabolic complexities will improve the ability to successfully engineer photosynthetic microorganisms for biofuel-production efficiencies.

A complementary approach to altering endogenous metabolic pathways is to identify the key genetic components from one organism and express them in a more genetically tractable organism using synthetic biology techniques [101,115]. As discussed, elevating biofuel-production levels in a native host requires a significant amount of pathway optimization, and can lead to unpredictable effects. Expression of fuel-forming enzymes or whole pathways in heterologous hosts may alleviate unwanted regulatory steps and kinetic bottlenecks in a native organism [116]. Development of novel biofuel-production strains by this approach is an iterative process of pathway manipulation (e.g., expression, relative gene expression levels and enzyme compositions) and system-level assessments [102]. A successful example of this approach is the production of isobutanol in *Escherichia coli* [101]. Advances in genome sequencing and HTP techniques are making it possible to develop similar efforts in photosynthetic organisms such as the green alga *C. reinhardtii* [26,108,109,117] and cyanobacteria [118–120].

11.2.2.4 Harvesting and Lipid Extraction

Once the lipids are biosynthesized, the final steps in the production process are the separation of cells (dewatering) from the medium, followed by harvesting or extraction of lipids. Recovery of lipids is a significant challenge towards achieving a reliable and economical process, and several physical techniques are being developed based on centrifugation, filtration, flocculation, or milking [121,122]. A genetic approach to this problem has also been developed based on engineering of *Synechocystis* sp. PCC 6803 for inducible cell lysis [123]. The endolysin and holin genes from bacteriophages were introduced into the *Synechocystis* genome under the control of either constitutive or Ni-inducible promoters. By adding Ni^{2+} to the growth medium, cell-wall degradation by the expressed phage proteins resulted in cell lysis. As a model approach, controlled expression of cell-wall degrading enzymes, in conjunction with biofuel pathway engineering, can potentially improve both yield and ease of disruption in a single strain. In effect this approach also allows for precise control over the timing of cell disruption when production of biofuels is at its peak.

11.2.3 Background Biohydrogen

Hydrogen is a potential clean energy carrier that is produced naturally as a metabolic byproduct of photosynthetic or fermentative growth by a variety of microorganisms [22,124,125], but only under low O_2, or O_2-free, growth conditions [21,22,126,127]. A primary benefit of H_2 as a fuel is the low environmental impact when combined with O_2 (air) in a catalytic reaction to generate electrochemical potential (electricity) in polymer-electrolyte-membrane (PEM) fuel cells. The H_2O produced from this reaction can be recycled for the generation of H_2 by photosynthetic microbes to engineer a closed loop process for electricity generation [1,128–130]. A water cycle for H_2 generation avoids production of greenhouse gas emissions that arise from generating H_2 by natural gas reforming or electricity generation by conventional coal-fired power plants. Alternatively, H_2 can be produced by fermentation of carbon compounds where CO_2 is a coproduct. This is a carbon-neutral process provided the carbon substrates are derived from photosynthetic CO_2 fixation. Both of these light-dependent (photosynthetic) and light-independent (fermentative) pathways function in the photobiological production of H_2 [21,124,131–135], and details of the unique and fascinating enzymes [136] that catalyze H_2 production are discussed in more detail below. The capacity for H_2 production spans the diversity of phototrophic organisms, and has been studied in genera of filamentous (e.g., *Anabaena*, *Arthrospira*, and *Nostoc*) an non-filamentous (e.g., *Gloeocapsa*, *Synechococcus*, and *Synechocystis*) cyanobacteria and in species of green algae including *Scenedesmus*, *Chlamydomonas*, *Chlorella*, *Chlorococcum*, *Platymonas* and *Tetraselmis* among others [23,26,32,125,137,138].

11.2.3.1 Enzymes Catalyzing H₂ Production

The activation of H_2 in phototrophic microbes is a reaction mainly catalyzed by two classes of metalloenzymes, the hydrogenases (reviewed in Refs. [24,139–144]) and nitrogenases (reviewed in Refs. [145–148]). The first part of this section will review some of the fundamental properties of the hydrogenases from phototrophic organisms. These enzymes catalyze the reversible reaction shown in Equation 11.1:

$$2H^+ + 2e^- \leftrightarrow H_2 \quad \Delta G = 79.8 \, \text{kJ} \, \text{mol}^{-1} \quad (\text{pH}\,7) \tag{11.1}$$

$$\Delta E = -0.414 \, \text{V} \quad (\text{pH}\,7, 1\,\text{atm}\,H_2)$$

In some cases, the catalytic activity of hydrogenases can show an inherent bias for the H_2 oxidation reaction, whereas other enzymes show higher k_{cat} value for H_2 production and tend to have a neutral bias.

11.2.3.1.1 [FeFe]-Hydrogenases

The hydrogenases of phototrophic organisms belong to one of two phylogenetic groups that are differentiated by the metals that make up their catalytic sites as either [FeFe]-hydrogenases or [NiFe]-hydrogenases. Eukaryotic algae are known to only synthesize [FeFe]-hydrogenases (reviewed in Refs. [32,149]), which are monomeric, consist of only a catalytic domain and are localized to the chloroplast stroma [150–156] (Figure 11.6). The catalytic domain of all [FeFe]-hydrogenases contains three conserved cysteine-rich motifs that coordinate the catalytic H-cluster, as well as conserved proton and gas transport pathways [157–159]. The H-cluster is a unique organometallic [6Fe-6S]-cluster composed of a cubane [4Fe-4S]-subcluster linked by cysteine to a di-iron, di-thiolate, [2Fe]-subcluster that possesses Fe-CO and Fe-CN ligation and an organic bridging ligand [160,161] (Figure 11.6A). The chemical composition of the bridging ligand is a matter of current debate [159]. Recent studies that combine high-resolution crystallography with density functional theory (DFT) suggesting a dithiomethylether moiety [162], whereas earlier DFT [163] and more recent HYSCORE analysis of ^{14}N-labeled enzyme [164] support a dithiomethylamine assignment.

The physiological electron donor to algal [FeFe]-hydrogenases is [2Fe-2S] ferredoxin (Fd) [155,165,166], which is reduced photochemically by PSI or non-photochemically through oxidation of NAD(P)H [126,165,167,168]. The kinetics of the electron-transfer (ET) reaction between Fd and algal hydrogenase have been characterized [155,169] and the reaction modeled as a direct transfer from the Fd iron-sulfur cluster to the [FeFe]-hydrogenase H-cluster [150,170–172]. The algal [FeFe]-hydrogenases are known to have one of the highest sensitivities to inactivation by O_2 [124,158,173,174], but the biochemical

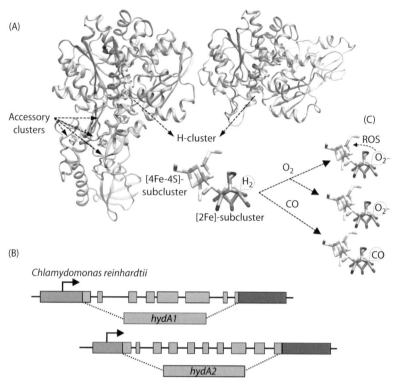

FIGURE 11.6
Gene organization, enzyme structures, and O_2 inactivation model of [FeFe]-hydrogenases. (A) The x-ray structure of [FeFe]-hydrogenase CpI from *Clostridium pasteurianum* (PDB ID 3C8Y [162]) on the left, and a homology model of the *Chlamydomonas reinhardtii* HydA2 on the right. Both CpI and HydA2 consist of a conserved catalytic domain that harbors the catalytic H-cluster (licorice format). In addition, CpI has an accessory domain that contains three [4Fe-4S]-clusters (distal, medial and proximal), and one [2Fe-2S]-cluster (distal). The structure of the catalytic H-cluster, shown in detail below, is comprised of a [4Fe-4S]-subcluster linked by cysteine to a [2Fe-2S]-subcluster with two CN (terminal), three CO (two terminal and one bridging) ligands and a bridging ligand between the two S atoms shown here as a dithiomethylamine. S = white; O = dark gray; N = black; Fe = light gray, at junctions; C = light gray (elsewhere). (B) Structural organization of the *hydA1* and *hydA2* nuclear-encoded genes from *C. reinhardtii* [153,484]. Medium gray boxes indicate the promoter regions, light gray—the exons (spaces indicate the introns), and dark gray—the terminator regions. (C) Models for the CO-inhibited [241] and O_2-inactivated H-cluster [238]. O_2 inactivation is hypothesized to initiate by reaction at the distal-Fe atom, where it is converted to a reactive O-species (ROS). Two consequences are proposed for degradation of the [4Fe-4S]-cluster: (i) generation and migration of ROS (top) or (ii) a through-bond oxidation (middle) [238]. CO inhibition occurs by binding of an exogenous CO to the distal Fe-atom, resulting in a terminally bound CO (bottom).

and structural origin for this high sensitivity is not characterized. Although a structure for algal [FeFe]-hydrogenase has not yet been solved, homology models based on the x-ray structure of [FeFe]-hydrogenase CpI from *Clostridium pasteurianum*, Figure 11.6A, have been used to identify the locations of the catalytic site, Fd binding site, and other key structural and functional features [150,157,158,170,175,176]. The lack of any evidence for a [FeFe]-hydrogenase in cyanobacteria suggests that acquisition of this class of hydrogenase among phototrophs is restricted to the eukaryotic algae [177].

11.2.3.1.2 [NiFe]-Hydrogenases

In contrast to algae, cyanobacteria are known to synthesize only [NiFe]-hydrogenases, enzymes that are phylogenetically distinct from the [FeFe]-hydrogenases [144]. Although structurally diverse, a distinguishing characteristic of [NiFe]-hydrogenases is the presence of a [NiFe]-cluster in the catalytic domain [142]. The [NiFe]-center is coordinated through conserved cysteines that bridge the Ni and Fe atoms [142]. Like [FeFe]-hydrogenases, the [NiFe]-hydrogenase catalytic domain also contains conserved gas transport [178] and proton transfer [179] regions, and both Fe-CO and Fe-CN ligation of the [NiFe]-center [180,181].

There are two forms of cyanobacterial [NiFe]-hydrogenases, a dimeric form that catalyzes H_2 oxidation (uptake) and a pentameric form that catalyzes reversible H_2 activation (bidirectional, NADP-dependent) coupled to NAD(P)H reduction-oxidation (Figure 11.7) [24]. The dimeric hydrogenase, HupSL, is found in N_2-fixing cyanobacteria and functions to recycle the H_2 generated by nitrogenase in heterocysts [182–186]. The HupSL enzyme

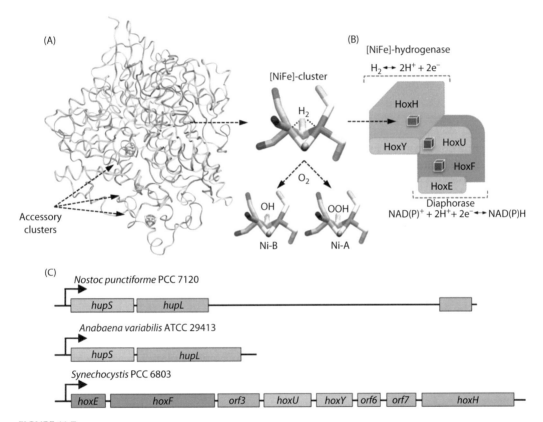

FIGURE 11.7
Gene organization and structural models of [NiFe]-hydrogenases. (A) The x-ray structure of the dimeric [NiFe]-hydrogenase isolated from *Desulfovibrio vulgaris* Miyazaki F, PDB 1WUL [485]. The large subunit (right half) contains the [NiFe]-cluster and the small subunit (left half, slightly lighter shade) contains the three accessory [Fe-S]-clusters, two [4Fe-4S] (distal and proximal) and one [3Fe-4S] (medial). The [NiFe]-cluster is shown in detail to the right with sidechains of four cysteine ligands and the product H_2 in a bridging position (Ni = medium gray on the right side of the structures, otherwise atom colors are as defined in Figure 11.6). Exposure of [NiFe]-hydrogenase to O_2 produces either the Ni-B (hydroxo) or Ni-A (peroxo) state (shown in detail below). (B) Model of the *Synechocystis* PCC 6803 pentameric Hox hydrogenase based on details from reference [194]. The [NiFe]-cluster is shown located in HoxH (upper box), with the assigned locations of accessory [Fe-S]-clusters in HoxU and HoxF as proposed in Ref. [194]. (C) Structural organization of the *hupSL* and *hox* genes from *Nostoc punctiforme* PCC 7120 (NCBI accession NC 003272) [486], *Anabaena variabilis* ATCC 29413 (NCBI accession NC 007413) [212], and *Synechocystis* sp. PCC 6803 [196]. The *hupL* gene in *N. punctiforme* (shown as two exons separated by an intron) undergoes a splicing rearrangement into a single exon during heterocyst development [24].

is a αβ-dimer composed of a large (HupL) and small (HupS) subunit. Coordination of the catalytic [NiFe]-center in HupL and the accessory [Fe-S]-clusters in HupS involves cysteine residues located in conserved motifs [187,188]. A phylogenetic analysis of [NiFe]-hydrogenase genes showed that the HupSL genes group with the cytoplasmic regulatory [NiFe]-hydrogenases that function as H_2 sensors in photosynthetic proteobacteria [143,144]. Interestingly, H_2-sensing [NiFe]-hydrogenases (e.g., *Rhodobacter capsulatus* HupUV) possess higher tolerances to O_2, which has been shown to be due to a more restricted gas channel that limits O_2 accessibility to the [NiFe]-center. A mutation of the HupUV channel that replaced a larger phenylalanine residue with a smaller valine resulted in an active enzyme that was more sensitive to O_2 [189–191]. Cyanobacterial HupSL share similar sequence features to HupUV, but display the faster inactivation kinetics of HupUV mutants when exposed to O_2 [192]. A theory for the mechanism of H_2 cleavage by HupSL has been suggested, and the residues critical to catalytic function and coordination of [Fe-S] groups have been identified [193].

Cyanobacterial bidirectional NAD(P)$^+$-dependent [NiFe]-hydrogenases consists of 5-subunits, HoxE, HoxF, HoxH, HoxU, and HoxY. The hydrogenase component is composed of a catalytic subunit HoxH that contains the [NiFe]-center and a small subunit HoxY. The diaphorase component is composed of the catalytic subunits HoxF, HoxE, and HoxU. It has been proposed that HoxU coordinates a [2Fe-2S]-cluster, and that either HoxU or HoxF coordinates one (or possibly two) [4Fe-4S]-cluster (Figure 11.7B) [194,195]. The HoxE and HoxY subunits co-purify with HoxH and HoxY, but their functions are not well understood [195]. The *hox* genes are encoded within a single operon [196], and co-transcribed as a single transcript from a single promoter [197,198].

11.2.3.1.3 Nitrogenases

The nitrogenases catalyze the irreversible, ATP-dependent reduction of N_2 to NH_3, producing H_2 as an obligatory by-product according to Equation 11.2 [147,199]:

$$N_2 + 8e^- + 8H^+ + 1MgATP \rightarrow H_2 + 2NH_3 + 16MgADP + 16P_i \qquad (11.2)$$

Although the reaction is energy intensive in terms of H_2 production (~1 mol H_2/16 mol ATP), the irreversibility means that in cyanobacteria lacking HupSL, there is no H_2 uptake. As a whole, the cyanobacteria synthesize both the Mo-type (Nif) and V-type (Vnf) nitrogenases [200,201], Although an Fe-type nitrogenase was reportedly isolated from *Anabaena variabilis* [202], there is no genetic evidence for the presence of *anf* in the genomes of sequenced cyanobacterial species [200,201]. Each enzyme class is phylogenetically distinct and can be distinguished based on peptide sequence conservation and catalytic sites that contain either Mo and Fe, V and Fe, or only Fe, respectively [203,204]. Some species of cyanobacteria (e.g., *Nostoc punctiforme* PCC 7120) synthesize only Mo-type nitrogenase, whereas others (e.g., *A. variabilis* ATCC 29413) synthesize both Mo and V-type enzymes [200]. The Mo-type nitrogenase is a $\alpha_2\beta_2$-tetramer composed of NifD (α-subunit) and NifK (β-subunit) (Figure 11.8), and contain the [8Fe-7S] P-cluster at the αβ-interface and the FeMo-co cluster of the catalytic site in the α-subunit [205]. The p-cluster mediates ET from the nitrogenase Fe-protein (NifH) to FeMo-co, following NifH reduction by reduced Fd. A series of proton-coupled ET (PCET) steps are catalyzed by FeMo-co and lead to the conversion of N_2 into NH_3 with coproduction of H_2. The exact mechanism of N_2-fixation remains under intense investigation [206–209]. Structural organization and transcription of the *nifH*, *nifD*, and *nifK* genes have been described for *Trichodesmium, N. punctiforme* and *A. variabilis* (Figure 11.8B) [125,210–212] and are typically, but not always, grouped into polycistronic operons.

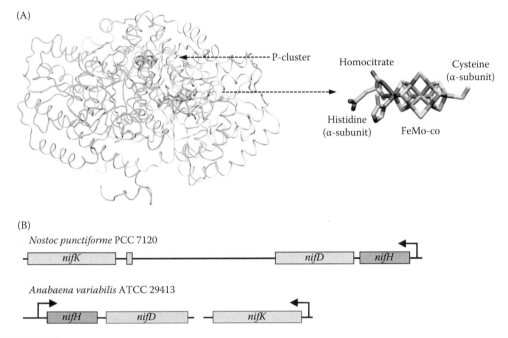

FIGURE 11.8
Gene organization and structural models of MoFe-nitrogenases. (A) The x-ray structure of the αβ-subunits of Mo-type nitrogenase from *Azotobacter vinelandii* (PDB 1M1N, [207]). The α-subunit (right-side lobe) and β-subunit (left-side lobe) are shown with the [8Fe-7S] P-cluster (at the αβ-interface) and FeMo-co catalytic cluster (α-subunit) in detail. Mo = dark gray, and lies on the left edge of FeMo-co as shown, coordinated by histidine and homocitrate, otherwise the coloring is the same as in Figure 11.6. The interstitial atom at the center is a light element, possibly N, O, or C [207]. (B) Structural organization of nitrogenase genes in *Nostoc punctiforme* PCC 7120 and *Anabaena variabilis* ATCC 29413. Nitrogenase iron-protein, *nifH* (dark gray); MoFe α-subunit, *nifD* (medium gray); and MoFe β-subunit, *nifK* (light gray). The nifD gene in *N. punctiforme* (shown as two exons separated by an intron) undergoes a splicing rearrangement into a single exon during heterocyst development [211].

11.2.3.2 Maturation of H_2-Producing Enzymes

Following the expression of hydrogenase and nitrogenase structural subunits, the accessory and catalytic clusters must be inserted to convert apo-proteins into active enzymes. This process of maturation involves enzyme-specific proteins, maturases, which are co-expressed with the structural proteins. For biotechnological applications involving the exchange of enzymes or structural modifications, maturation must be carefully considered in devising a successful strategy. All of the characterized maturation processes have been shown to require anaerobic, or reducing conditions, which stands in contrast to some examples noted below where enzymes are expressed and biosynthesized in otherwise oxygenic growth conditions.

11.2.3.2.1 Hydrogenases

Biosynthesis of the hydrogenase metallo-clusters involves both general (e.g., *iron-sulfur-cluster* [ISC], pathway; see Refs. [213–215]) and enzyme-specific biosynthetic pathways. Biosynthesis and insertion of accessory [4Fe-4S], [4Fe-3S], and [2Fe-2S]-clusters, and the H-cluster [4Fe-4S]-subcluster [216,217], are thought to occur by means of the general pathway. Biosynthesis of the H-cluster [2Fe]-subcluster, as well as the catalytic [NiFe]-clusters each occur by means of a specialized set of maturases. For each hydrogenase type and corresponding catalytic cluster (i.e., either [FeFe] or [NiFe]), a specific set of proteins has evolved to accomplish maturation [140]. Nevertheless, it has been observed that maturation of [FeFe]-hydrogenases by the HydE, HydF, and HydG proteins [218,219] is a less discriminating process than maturation of [NiFe]-hydrogenases [220]. There are several reported examples of successful [FeFe]-hydrogenase maturation using heterologous maturases and/or nonnative hosts [218,219,221–223], including cell-free maturation [224]. In contrast, the Hyp proteins that function in [NiFe]-cluster biosynthesis and enzyme maturation [225] have a greater degree of enzyme specificity [189,195,226–228]. This stringency is likely due to the greater diversity in catalytic subunit structure among [NiFe]-hydrogenases than for [FeFe]-hydrogenases [141,143], and differences in how maturation is supported by host-specific metabolic pathways [229–231]. Despite the presence of these constraints, there are several noted examples of successful heterologous expression of [NiFe]-hydrogenases in nonnative hosts [220]. These successes provide viable routes toward developing not only novel organisms for biohydrogen production, but biosynthetic systems to generate enzymes for biohybrid and other artificial applications described in Section 11.3.

11.2.3.2.2 Nitrogenases

Maturation of nitrogenase (restricted herein to a summary of the Mo-type nitrogenase) is an intricate process of in situ and ex situ assembly of the P-cluster and catalytic FeMo-co cluster, respectively. FeMo-co is assembled on NifEN through a series of steps involving NifS (cysteine desulfurase), NifU (scaffold for nascent [Fe-S] cluster), NifB (radical s-adenosylmethionine protein), NifX (nascent FeMo-co carrier protein), NifQ (Mo-binding), NifV (homocitrate synthase), NafY (delivers FeMo-co from NifEN to NifDK), and the iron-protein NifH [146,148,205,232,233]. Both NifU and NifS also assemble the P-cluster precursor on NifDK, and processing is completed by iron-protein NifH [205]. Much of the biochemical evidence for the functional role of the Nif proteins in maturation has been derived from in vitro reconstitution of FeMo-co biosynthesis using extracts and purified proteins [146,148,205]. In spite of this success, to date, the genetic engineering efforts with nitrogenase have primarily focused on the transfer of the *Nif* genes into plants to improve cultivation of crops in nitrogen poor soils, with less emphasis on bioengineering for biohydrogen or biofuel production [234].

11.2.3.3 Sensitivity of Enzymes to Oxygen

At this time, there is no natural or engineered organism that is capable of sustaining high rates of H_2 production under completely aerobic (photosynthetic) conditions. There are several basic issues that confront the development of a completely photolytic process and the inherent sensitivity of the H_2-forming enzymes to O_2 is one of them. The mechanism(s) by which O_2 inhibits or inactivates hydrogenases and nitrogenases has been the subject of decades of research. Nevertheless understanding of O_2 inactivation and the development of effective engineering strategies continues to progress [139,157,178,199,191,235–237].

11.2.3.3.1 [FeFe]-Hydrogenases

A barrier to the development and utilization of algal species for photolytic H_2 production is the high sensitivity of the endogenous [FeFe]-hydrogenases to inactivation by O_2 [158,174,236,238]. A model for the inactivation mechanism has been recently suggested based on x-ray adsorption spectroscopy of O_2-treated algal [FeFe]-hydrogenase that is based on a two-step process [238]. First, O_2 accesses the catalytic site through hydrophobic channels [157,158] and reacts with the distal Fe atom (Fe_D) of the H-cluster [2Fe]-subcluster to form reactive oxygen species (ROS) that either binds the [2Fe]-subcluster or oxidizes the [4Fe-4S]-subcluster into a [3Fe-4S]-cluster. This H-cluster oxidation renders the enzyme inactive. In a similar process, carbon monoxide (CO) reversibly inhibits [FeFe]-hydrogenase [239,240] by also binding to Fe_D [241]. In the CO-inhibited state, [FeFe]-hydrogenase is protected from inactivation by O_2 [174,236] due to competition for Fe_D, which favors CO presumably due to the faster binding kinetics of CO versus O_2 [236], and a binding mode that is resistant to enzyme oxidation. Early electron paramagnetic resonance (EPR) and biochemical studies of CpI showed oxidation by O_2 led to a loss of the enzyme EPR-active signal [139], consistent with more recent evidence that O_2 binds at Fe_D to result in oxidation of the H-cluster upon enzyme inactivation (Figure 11.6C) [238]. Thus, CO-protection of [FeFe]-hydrogenase from O_2 inactivation must limit any and all of the possible reaction schemes of O_2 with the enzyme catalytic iron-sulfur clusters [238].

11.2.3.3.2 *[NiFe]-Hydrogenases*

In contrast to [FeFe]-hydrogenases, exposure of [NiFe]-hydrogenases to O_2 results in the reversible O_2 inhibition of the catalytic [NiFe]-cluster [242], through the formation of a μ-oxo species [243]. Depending on the reduction state of the enzyme upon exposure to O_2, the species is either a hydroxo (Ni-B state) or peroxo (Ni-A state) that bridges the Ni and Fe atoms of the [NiFe]-center [244] (Figure 11.7A). Once returned to an anaerobic environment, reduction of the [NiFe]-center (e.g., sodium dithionite, NAD(P)H, H_2, or other physiological electron-donor molecules) reactivates the [NiFe]-hydrogenase via a fast, Ni-B, or slow, Ni-A, reaction scheme [245].

Nevertheless, O_2 competitively inhibits [NiFe]-hydrogenase, H_2-activation activity, with the effective inhibition rate determined by both O_2 diffusion into and its reaction with the [NiFe]-cluster [245]. The kinetics of these two processes vary from enzyme-to-enzyme, due to differences in the structures of the gas transport pathways and catalytic sites, and in enzyme redox properties [157,158,191,229,245–247]. As mentioned above, the HupSL enzymes of N_2-fixing cyanobacteria, group with the H_2-sensing [NiFe]-hydrogenases, HupUV, which possess a more restricted barrier for gas transport into the [NiFe]-center [189,191]. Two of the residues that form this barrier in HupUV, an isoleucine and phenylalanine, restrict O_2 access to the [NiFe]-center by virtue of the larger side-chain group and result in higher O_2 tolerance of these enzymes [189,191]. HupL of *Nostoc* and *Anabaena* share the isoleucine determinant, but have a smaller leucine in place of phenylalanine at the other position [212,248]. By comparison, HoxH also possesses the isoleucine–leucine pair, but HoxHY inactivates at an apparently faster rate than HupSL [249,250]. From this evidence, it appears that gas accessibility alone does not account for differences in the O_2 inhibition rate of cyanobacterial [NiFe]-hydrogenases. Spectroscopic features of purified HoxHY reacted with O_2 show formation of only a Ni-B-like state [194], which in the presence of H_2 and NAD(P)H undergoes a rapid (~min) reactivation [251]. Thus, HoxHY has evolved to be an ideal enzyme for disposing of excess reductant during the dark-to-light transition under microaerobic conditions [252] by switching between active and inactive states as the NADPH pool reduction levels change between anaerobic and oxygenic (photosynthetic) growth.

11.2.3.3.3 *Nitrogenases*

Like the hydrogenases, nitrogenases are also highly sensitive to O_2, and cyanobacteria have overcome this sensitivity by adaptations that protect nitrogenase and the N_2-fixing pathway from the deleterious effects of oxygenic photosynthesis [184]. Regimes that have evolved to segregate N_2-fixation and photosynthesis in cyanobacteria include (i) sequestration of nitrogenase into specialized cells, or heterocysts, that possess photosystem I (PSI) but lack photosystem II (PSII), have high rates of O_2 respiration and have cell wall structures that physically limit O_2 diffusion [253]; and (ii) temporal and spatial segregation of N_2-fixation and oxygenic photosynthesis [254]. That has been identified in N_2-fixing cyanobacterial strains from the open ocean is the complete [255]. For utilization of cyanobacteria for aerobic, solar-driven biohydrogen production, much of the research effort has focused on the heterocyst-forming cyanobacteria, and examples of recent progress with these organisms are summarized in more detail below.

Engineering to address enzyme O_2 sensitivity. The inactivation of H_2 production in phototrophs under oxygenic photosynthetic conditions involves both direct and indirect mechanisms. Direct effects of O_2 include the inhibition or inactivation of the H_2-producing enzyme, whereas indirect effects (discussed in Section 11.2.3.3) include changes in cellular metabolism that limit electron flux and the energetics of H_2 production. Examples of engineering strategies to address direct effects on hydrogenase include (i) whole-genome and site-directed mutagenesis to modify and control sensitivity and O_2 access into the catalytic site [32,124,256–258]; (ii) expression of more O_2-tolerant enzymes in oxygenic phototrophs [220,230,259]; and (iii) elucidating physiological and/or culture conditions that allow cells to maintain H_2 production under oxygenic photosynthesis [21,260,261]. Progress in engineering of hydrogenase for higher tolerance to O_2 and therefore continuous catalytic output under simultaneous O_2 evolution has been accomplished mainly with the [NiFe]-hydrogenases. By mutating the two barrier residues (see Section 11.2.3.3) several combinations were created in enzymes that showed slower oxidation kinetics due to restricted access of the [NiFe]-cluster [257], and possibly facilitating the reactivation of Ni-B and Ni-A states [256]. The expression of native hydrogenases with naturally high tolerances to O_2 in phototrophs has been limited by poor maturation efficiencies [220]. Recently, these efforts have benefited from advances in knowledge of the maturation process and host-specific factors required to produce active enzymes [24,218,220,230,231,259]. Each of these approaches has shown some promise, which may lead to a more robust hydrogenase for water splitting, biohydrogen production process.

11.2.3.4 Pathways to Biohydrogen Production

11.2.3.4.1 Cyanobacteria

There are two fundamental pathways for H_2 production in cyanobacteria that differ depending on whether nitrogenase or the NAD(P)-dependent, bidirectional [NiFe]-hydrogenase catalyzes the H_2 production reaction [262]. Nitrogenase-dependent H_2 production occurs in either nonfilamentous or filamentous cyanobacteria (e.g., *Anabaena, Gloeoethece, Nostoc,* and *Trichodesmium*). Enzyme sensitivity to O_2 is overcome by spatially and/or temporally segregating N_2-fixation from oxygenic photosynthesis. Spatial segregation can occur by cell differentiation into non-photosynthetic heterocysts [184,263] or localization of nitrogenase activity away from PSII [254], whereas temporal segregation occurs by conducting N_2-fixation during the dark phase of growth [264]. Nitrogenases' main function is to convert N_2 to ammonia (Equation 11.2), which co-produces one molecule of H_2. When cultivated in the absence of N_2, the enzyme catalyzes only H_2 production according to Equation 11.3 [31]:

$$2e^- + 2H^+ + 4ATP \rightarrow H_2 + 4ADP + 4P_i \tag{11.3}$$

As mentioned above, nitrogenases in heterocystic cyanobacteria are protected from O_2, thus are expressed under oxygenic conditions together with HupSL. Since HupSL catalyzes H_2 oxidation in vivo, its co-expression limits the overall H_2 rates and production efficiencies of these organisms [125,265,266]. Strategies to overcome this limitation include deletion of HupSL, which in either *Anabaena* [212] or *Nostoc* [267,268] led to higher rates of nitrogenase-catalyzed H_2 production. This increased rate was strongly coupled to photosynthesis and light intensity [269]. Nitrogenases from homocitrate-deficient strains are poor catalysts for N_2-fixation [233], but the same genetic defect combined with the lack of HupSL resulted in even higher yields of H_2 production from *Nostoc* [270]. These examples demonstrate the significant gains that can be made from strain engineering of nitrogenase-based H_2-producing organisms.

The direct photoevolution of H_2 from photosynthetic water splitting in cyanobacteria is catalyzed by the HoxEFUHY [NiFe]-hydrogenase [24]. Expression of HoxEFUHY is under both circadian control (higher expression in the dark phase), and O_2 regulation, and is induced under oxygenic conditions [198,271,272]. The enzyme functions to dispose of excess reducing equivalents from photosynthesis during the early stages of aerobic growth [252], and thus competes with other NADPH-dependent pathways including CO_2-fixation, NADPH-dependent respiration, and nitrate assimilation [251,273,274]. Competition from NADPH respiration was alleviated in a NADPH-dehydrogenase (NDH) mutant [251], and from aerobic respiration and nitrate assimilation in mutants defective in terminal cytochrome oxidase(s) and nitrate reductase [273]. In each of these cases, higher levels of sustained H_2 production were observed under oxygenic photosynthesis, however, accumulation of O_2 resulted in a gradual decline in rates [251]. Achieving sustained levels of photosynthetic H_2 production by the Hox-dependent pathway in cyanobacteria will require the application of additional engineering strategies that address the O_2 inhibition of Hox and elimination of other competing electron sinks [24].

11.2.3.4.2 Purple Non-Sulfur Bacteria

The purple non-sulfur bacteria (e.g., *Rhodopseudomonas palustris*), are not described in detail here, but are mentioned because they produce H_2 in a cyanobacterial-related light- and nitrogenase-dependent pathway. This pathway is functional during the non-oxygenic conversion of organic and inorganic substrates into CO_2 and NADPH under low N_2 growth conditions [262]. The reductant pool of NADPH is converted into a proton gradient for ATP formation and a reduced Fd pool through non-oxygenic, photosynthetic electron transport (PET). The resulting Fd and ATP provide the chemical reductant and energetic substrates for nitrogenase-catalyzed H_2 production (see Equation 11.3) [193].

11.2.3.4.3 Green Algae

The green algae produce H_2 under both photosynthetic and fermentative growth conditions involving the catalytic activity of [FeFe]-hydrogenases [126]. There are several biochemical, metabolic, and physiological factors that limit conversion efficiencies, and both basic and applied research efforts are addressing these limitations toward improving H_2 production efficiencies. These research efforts primarily support three approaches for achieving solar H_2 production in algae: (i) the direct photoevolution of both H_2 and O_2 from photosynthetic water splitting during CO_2 fixation; (ii) the temporal separation of photosynthetic water splitting and CO_2 fixation from anaerobic, light-driven H_2 production in a two-stage process; and (iii) the anaerobic H_2 production from dark fermentation of fixed carbon reserves (e.g., starch) [131,275–278].

The first approach of a direct photosynthetic water-splitting process is theoretically the most efficient pathway in algae, with an upper limit efficiency of ~10%–12% [29,279,280] for conversion of incident light into H_2. Under limiting light conditions, with a continuous flow of helium and a low CO_2 partial pressure, the coevolution of H_2 and O_2 occurs at a 2:1 ratio [281–283]. Light-dependent evolution of H_2 requires PSI [284], however, if light intensity is too high, the increased PSII rates lead to O_2 accumulation and inactivation of [FeFe]-hydrogenase [285]. Moreover, elevated CO_2 levels direct the partitioning of PET to Ferredoxin:NAD(P)$^+$ reductase (FNR) for NADPH production and CO_2 fixation by RuBisCo, the preferred electron sink [282]. Thus, the direct photoevolution of H_2 is currently an inefficient process due to the short H_2-production periods, and there are a number of research efforts around the world that are addressing this challenge.

The adverse effects of O_2 production and CO_2 fixation can be overcome in green algae (e.g., *C. reinhardtii*) by temporally separating oxygenic, CO_2 fixation from H_2 production in a two-stage process. This pathway cycles between physiological states by depriving cultures of sulfur [21,127,286], which decreases PSII-mediated O_2 production to the level of respiration, thereby creating an anaerobic environment in either sealed bioreactors [287,288] or in alginate films [260]. After an initial period of CO_2 fixation, cells transition to an anaerobic phase of growth. As a result, [FeFe]-hydrogenase expression is induced and catalyzes light-driven H_2 production [21,132,134]. Under sulfur deprivation, H_2 production can be sustained for periods lasting upto several weeks in continuous culture [289,290], or catalyzed in air (21% O_2) if sulfur-deprived cells are immobilized in an alginate film [260]. The accumulation of fixed carbon [288,291], for example as starch granules, has two roles in supporting H_2 production in cells under sulfur deprivation [21,84,135,292]. One is to provide the substrate for respiration to scavenge O_2 and maintain anaerobiosis [83,133,134,293]. A second is to provide electrons to the plastoquinone pool via NDH [294–296] that are transferred by light-induced PET to hydrogenase for catalytic H_2 production [81,84,133].

Finally, although it is not a direct solar-conversion process, H_2 production by dark fermentation in algae does require photosynthetic CO_2 fixation for biosynthesis of stored carbohydrates [110,261,292]. In this production mode, anaerobiosis is achived by sealing, or purging cultures with an inert carrier gas (e.g., argon). The H_2-evolution rates by dark production are typically low and depend on hydrogenase coupling to the electrochemical, rather than photoelectrochemical, generation of electron carrier pools (e.g., NADH, NADPH, and Fd). It has been proposed that pyruvate-ferredoxin oxidoreductase (PFOR) [131,167,293,297,298] or other redox-active enzymes and pathways catalyze the recycling of electron carriers into reduced Fd and H_2 production by hydrogenase.

11.2.4 Barriers to Efficient Solar Production of Biofuels and Biohydrogen

Optimization of photobiological production of biohydrogen and biofuels are currently limited by several factors: (i) oxygen inactivation of H_2-producing enzymes (addressed in Section 11.2.3); (ii) inefficient PET; (iii) competing sinks for reductant pools; and (iv) kinetically limited substrate fluxes, and inefficient light utilization by dense cultures. The conversion efficiency of PET is measured as the amount of photosynthetic productivity relative to a theoretical maximum of 8%–10% [17,29,278,279,299]. The efficiency varies among different organisms [20,28,29,300] and under different the growth conditions [27,132,293]. The main areas of investigation toward improving PET efficiencies include engineering chlorophyll antenna size [301,302] and the non-dissipation of the proton gradient during anaerobic photosynthesis [303], although neither has been demonstrated. A third area for addressing PET efficiencies of H_2-production is determining state-transition effects by mutagenesis to find strains that maintain linear electron flow (LEF or Z-scheme) under anaerobic conditions [83,304]. Under anaerobiosis cells normally transition from a LEF scheme of electron transport to a cyclic electron flow (CEF) scheme where electrons are cycled around PSI [305] to maintain optimal ATP/NADPH ratios for CO_2 fixation [305–308]. The compositional structure of the PSI-CEF complex is beginning to emerge from screens of *Arabidopsis* t-DNA insertional libraries [309], and isolation of PSI supercomplexes from CEF-induced *C. reinhardtii* cells [310]. In both cases, FNR is tightly associated with PSI-CEF, which strongly favors cycling of NADPH back to PSI as the major electron sink. In CEF the photoevolution of H_2 is low due to poor competition with FNR, but can be alleviated in state-transition mutants that are locked into a LEF mode [83]. In addition to PET-specific effects, competition among pathways is being characterized through HTP investigations and flux analysis, and in some cases these efforts are being combined with mutagenesis efforts targeting starch accumulation [79,81,84], glucose transport [311], and CO_2 fixation pathways [293,312] that seek to adjust carbon and electron flow toward accumulation of fuel end products [16,82,131,297,313].

11.3 Photobiomimetic Approaches to Solar H_2 Production and CO_2 Conversion

The light-harvesting and catalytic molecules of photosynthetic organisms detailed in Section 11.2 provide valuable structural and functional blueprints for developing synthetic counterparts for artificial production of fuels using light energy. From a design principle, artificial photosynthesis can potentially be more efficient than biological photosynthesis [314], which evolved to confer a competitive advantage to an organism rather than to optimize production of biofuels toward maximum efficiencies (see discussion in Section 11.2 and Refs. [29,128,315–317]). However, photosynthesis and biocatalysis utilize highly functional molecules composed of earth abundant elements, in contrast to currently available commercial technologies that mostly rely on Pt-based catalysts. A long-standing objective of engineering efforts in artificial photosynthesis, or photobiomimetics, is to design and synthesize molecular structures that integrate light capture, charge separation and transfer, and catalysis, to attain near ideal conversion efficiencies for sustained periods under full sunlight intensity [314,318]. For ease of discussion, the design principles guiding development of photobiomimetics are organized here into two general approaches: (i) modular, where functionally independent molecules are integrated into a multifunctional complex or device (e.g., a dye-solar cell) for light harvesting and catalysis; and (ii) integrated, where molecules incorporate all of the required functions as a single structure. For example, a modular approach may involve developing the coordination of an organometallic catalyst to a large organic light-adsorbing framework to generate a heterocomplex for light-driven catalysis. In this chapter, we will present an overview of current research efforts that are related to both of these types of approaches. The mechanistic aspects will address the themes that are fundamental to developing artificial photosynthesis including light capture and charge separation, as well as catalytic water oxidation, H_2 production, and carbon capture.

11.3.1 Light Capture and Charge Separation

The first steps of photosynthesis involve the absorption of light energy by chlorophylls (CHLs), carotenoids, and other light-adsorbing molecules. The photon energy is then transferred to either PSI or PSII by singlet-singlet energy transfer where a stable charge-separated electron-hole pair is formed. The supramolecular structure of the light absorbing pigments allows for transfer of the energy of absorbed photons from CHL to the photosynthetic reaction centers. Before relaxation of the excited state occurs, charge is transferred to bound quinone acceptors, which all occurs at picosecond timescales. Thus, photosystems have evolved to kinetically favor the forward reactions and charge-transfer steps over the back reactions and charge recombination (see Refs. [279,319–321]). This promotes energy conversion at a quantum yield (QY) of 1 for the formation of the initial excited state [279,322]. Downstream catalytic steps result in the conversion of the excited state into electrochemical potential through water oxidation and the reduction of electron carriers, as well as the formation of a transmembrane ΔpH. The kinetics of energy transfer, charge transfer, and the stability of the charge-separated state in photosystems are important parameters to the capacity to make efficient use of photon energy. Likewise, the challenges to synthesizing efficient artificial photoelectrochemical devices that reproduce photosynthetic-driven catalysis is to control the light-absorption properties, energy levels, and charge-transfer kinetics to achieve efficient energy conversion. Molecules that absorb light at the appropriate wavelengths are relatively easy to synthesize [323] and have been studied extensively [324–329]. However, synthesis of molecules in which recombination of the excited state is sufficiently slow to allow for ET, and the integration of such molecules into an antenna network with rapid energy transfer, are significantly more challenging.

11.3.1.1 Porphyrins as Light Absorbers

Porphyrins are one of the most common molecule families for use as light absorbers in photobiomimetic designs [323,329–331]. Synthetic porphyrins can be made to resemble the CHLa and CHLb of photosynthetic reaction centers [322], light-harvesting complexes [332,333], and macrocylic structures of other biological light absorbing porphyrinoid molecules. The basic structure of these molecules allow for p-to-p* transitions within the conjugated bond system upon photon absorption [330]. A general strategy in the design and synthesis of porphyrins has been to develop chemical variations of CHL molecules, due to the stability and synthetic routes for manipulating structures [328]. These synthetic porphyrins are effective as light collecting antenna, and are useful in artificial photosynthetic designs when coupled to electron acceptor moieties in dyad or triad complexes by helping to establish a more stable, longer-lived, charge-separated state [334].

11.3.1.1.1 Porphyrin–Fullerene Complexes

The molecular dyad shown in Figure 11.9 is an example of an initial effort toward a porphyrin–fullerene artificial antenna-reaction center (P-C_{60}) [335,336]. The covalent linkage between the two moieties allowed transfer of a photon of energy from the porphyrin to the fullerene on timescales that can compete with the photophysical relaxation pathways of the porphyrin excited states [337]. Photon absorption by the porphyrin resulted in a singlet excited state, which decays to a di-radical $P^{\bullet+}$-$C_{60}^{\bullet-}$ by photoinduced ET [335]. This decay occurs at a very high rate and successfully competes with other relaxation pathways in the molecule, resulting in a quantum yield for $P^{\bullet+}$-$C_{60}^{\bullet-}$ of 1 [335]. The oxidation potential of the di-radical has been estimated at +1.4 eV in polar solvents [335], making it a possible choice for an artificial antenna in a water-splitting scheme. However, the charge-separated state lifetimes are extremely short, ~50 ps, which makes radiative recombination compete effectively with charge transfer. This short exciton lifetime is an issue for many of these Dyad complexes [326].

In photosynthesis, the problem of short-lived charge-separated state lifetimes can be addressed by the physical (spatial) separation of the light absorbing molecule and the reaction center where charge separation occurs. This prevents energy dissipation that can occur via back transfer [279]. A large distance between absorber and acceptor requires multiple short-range ET steps in order to form a charge separated state, which makes fast, single step recombination prohibitively slow [338]. To incorporate this design, triad-based mimics were developed that incorporate the advantages of the conjugated P-C_{60} dyad, but

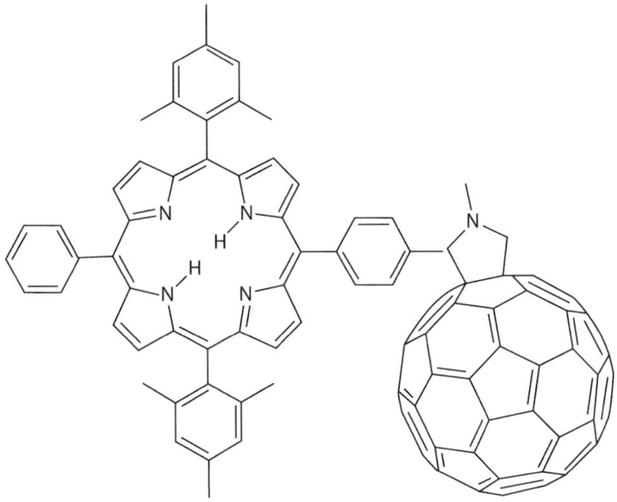

FIGURE 11.9
Structure of a porphyrin-fullerene, P-C_{60}, dyad complex. The porphyrin functions as a light absorber, which upon photoexcitation participates in ET to the conjugated fullerene creating a $P^{\bullet+}$-$C_{60}^{\bullet-}$ charge-separated state. (Reprinted with permission from Gust, D., Moore, T.A., and Moore, A.L., Solar fuels via artificial photosynthesis. *Acc. Chem. Res.*, 42(12), 1890–1898. Copyright 2009 American Chemical Society.)

include larger separation distances (Figure 11.9) [339–341]. The carotenoid (shown as "C" in the following schemes) moiety acts as a secondary electron donor, and the processes of photon absorption and ET to form C- $P^{\bullet+}$-$C_{60}^{\bullet-}$ is equivalent to those seen in the dyad. The additional presence of the carotenoid moiety creates a competing pathway to recombination. This arises from the carotenoid transferring an electron to the porphyrin to form $C^{\bullet+}$-P-$C_{60}^{\bullet-}$, in which the exciton pair are physically separated by a large enough distance to increase the lifetime of charge-separated state to ~50 ns [340]. Both dyad and triad complexes of porphyrin–fullerenes are effective mimics for light capture and charge separation, and provide excellent models for investigating the underlying mechanisms of these processes. The structures detailed here serve as examples of this class of antenna-reaction center mimics, of which numerous examples exist [339,340].

11.3.1.1.2 Porphyrin-Antenna Complexes

Although the major CHL molecules, CHLa and CHLb, are the primary light absorbers and excited-state electron donors to the photosynthetic reaction centers, phototrophic microorganisms also synthesize accessory pigments (e.g., carotenoids and phytochromes) [342]. The CHL absorbance has two maxima that lie at 430–450 and 640–660 nm, which represent only a portion of the incident solar spectrum [317]. However, the accessory pigments function to expand the light adsorption range into spectral regions where CHL is less active. Structural incorporation of accessory pigments into antenna structures has evolved to support energy transfer at ultrafast timescales, both between antenna molecules and to reaction centers [279,321,342]. Artificial configurations, in which a porphyrin or metalloporphyrin center is coupled to an antenna molecule, confer similar advantages of strong absorption across more of the visible spectrum [343–348]. The challenge of designing functional assemblies incorporating this molecular architecture lies in achieving timescales of energy transfer between the various moieties that allow for ET to compete with recombination.

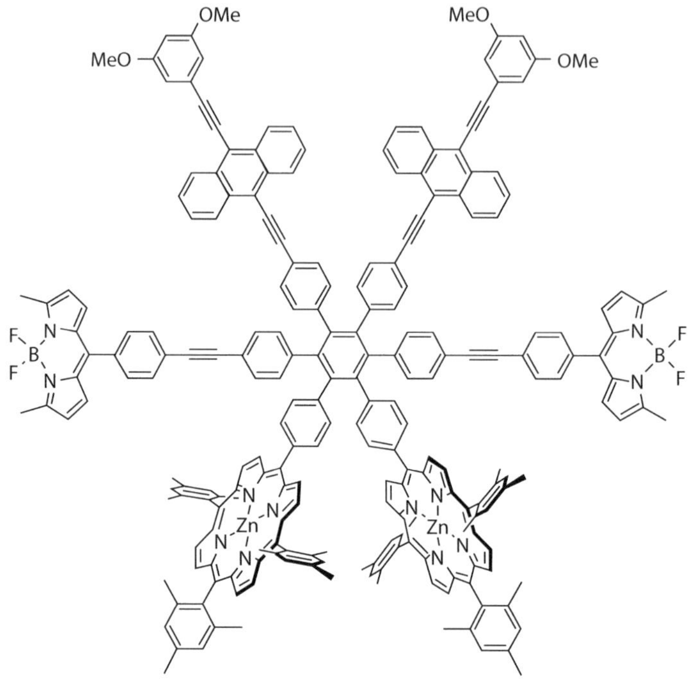

FIGURE 11.10
Structure of a molecular hexad antenna complex. The hexad incorporates three pairs of light absorbing antenna compounds that are conjugated to a hexaphenylbenzene core; bis(phenylethynyl)anthracene (top), borondipyrromethene (side), and zinc tetraarylporphyrin (bottom). (Reprinted with permission from Gust, D., Moore, T. A., and Moore, A. L., Solar fuels via artificial photosynthesis. *Acc. Chem. Res.*, 42(12), 1890–1898. Copyright 2009 American Chemical Society.)

Figure 11.10 shows an example of a porphyrin integrated into a more highly functional hexad antenna complex capable of a broad spectral response in the visible region (400–600 nm) and rapid energy transfer [349,350]. The structure consists of three light absorbing antenna, bis(phenylethynyl)anthracene (BPEA) that absorbs strongly in the blue region of the visible spectra (~450 nm), borondipyrromethene (BDPY) that absorbs in the green (~520 nm) and zinc tetraarylporphyrin that has two regions of absorption, the sorret (418 nm) and the orange-red region (575 nm). The emission and absorption spectra of the antennas overlap, allowing high efficiency energy transfer. In addition, the structure in Figure 11.10 self-assembles with fullerene to form an antenna-reaction center complex that forms a charge-separated state upon photoexcitation.

Porphyrin-antenna complexes have also been achieved by coupling imide and diimide dyes to metalloporphyrins [351–354]. These structures are not only capable of efficient energy transfer between moieties, but they also self-assemble into supermolecular structures with enhanced photon absorption, and absorption action spectra that extend beyond the visible into the ultraviolet and near infrared [330]. These antenna arrays generally take the form of a metalloporphyrin center with four imide or diimide arms that form an "X" configuration. Self-assembly occurs by π-stacking and hydrophobic–hydrophilic interactions [355]. The resulting supermolecular structures extend the energy transfer limits from nanometer to micrometer lengths in dyad and triad porphyrin complexes.

11.3.1.2 Inorganic Nanostructures

Nanostructured materials made from II to VI semiconductors have unique physical and optical properties that make them promising molecules for use in solar-driven water splitting, as well as for developing third generation PV materials [356,357]. In bulk, these materials have intrinsic valence (or, highest occupied molecular orbital [HOMO]) and conduction (or, lowest unoccupied molecular orbital [LUMO]) energy levels separated by a bandgap energy (E_g); photoexcitation with light energy above E_g results in a charge-separated state (electron-hole) that relaxes by recombination of electron-hole pairs or by charge transfer and chemical quenching [358]. When 3D length-scales are limited to dimensions smaller than the Bohr radius of the exciton (~nm), the resulting nanocrystals have quantized energy states [359,360], and recombination times are increased from the nanosecond to microsecond range. The nanocrystal E_g values scale inversely with size, and the HOMO-LUMO band edges show both size and material dependence [361,362]. For photobiomimetic applications, the band edge potentials of the photogenerated electron-hole pairs must be sufficient to couple the excited state to water-splitting and fuel-generating chemical reactions. Several II–VI nanocrystalline materials (e.g., CdTe, CdSe, and CdS) adsorb light with sufficient energy and at band-edges of appropriate redox potentials to couple water oxidation to proton reduction and other fuel-forming reactions [363–365]. Moreover, the transfer of photoexcited electrons and holes to surface-bound molecules has been demonstrated for CdS, CdSe, and CdTe [366–369]. Finally, there is great interest in quantum dots for their potential to produce multiple excitons from a single photon [370]. Multiple exciton generation (MEG) results from absorption of a photon energy greater than twice the E_g of the first excited-state transition [371–373]. In theory, MEG-based PV in a single or dual-bandgap scheme can exceed the maximal 31% or 42.4% limits [314] of traditional light-absorbing materials [370–373].

11.3.2 Water Oxidation

The reducing equivalents that drive both biological CO_2 fixation and H_2 production can be provided from the oxidation of water by photosynthesis, summarized in Equation 11.4:

$$2H_2O_2 \leftrightarrow +4e^- + 4H^+ \quad \Delta G = 314\,kJ\,mol^{-1}\,(pH\,7)$$

$$\Delta E = 0.816\,V\,(pH\,7)$$

$$(11.4)$$

This same reaction in an artificial or photobiomimetic context can be coupled to a H_2 production or CO_2 reduction catalyst to produce H_2 or biofuels using light energy. In PSII, water oxidation is catalyzed by a $CaMn_4O_X$-cluster composed of a cubic $CaMn_3O_4$ structure with a dangling Mn atom (Figure 11.11A) [374–378]. A conserved protein-encoded tyrosine residue functions as the photogenerated oxidant [322,374,378–380]. The design and synthesis of chemical catalysts that can drive this reaction is arguably the biggest current challenge to achieving efficient generation of solar fuels. There are several aspects of this reaction that make synthesis of an effective catalyst difficult. The half reaction has a high energetic requirement, and in vivo the potential necessary to split water is supplied by the excited state of the PSII reaction center. The redox poise of the 1.83 eV of energy created by PSII adsorption at 680 nm generates a tyrosine radical that lies ~0.34 V positive of the water oxidation half-reaction (Equation 11.4). Achieving the subsequent breaking of the water O–H bonds to form and oxygen O–O bond at high turnover rates presents a

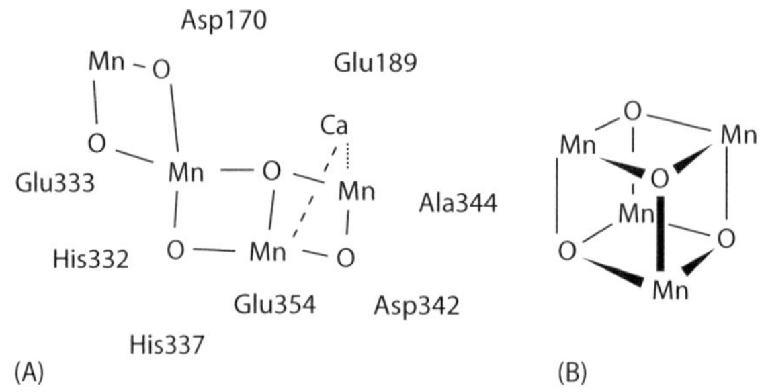

FIGURE 11.11

Biological and synthetic Mn-clusters for water oxidation. (A) Model of the Mn-cluster from PSII derived from x-ray spectroscopy [376] with putative coordination assignments designated as three-letter encoded amino acids with position numbers from the PSII x-ray structure [487]. (B) Structure of an oxo-Mn cubane core, which has been synthesized as a functional model for the Mn-cluster of PSII [385, 386]. (Reprinted with permission from Dismukes, G.C. et al., Development of bioinspired Mn_4O_4 cubane water oxidation catalysts: Lessons from photosynthesis. *Acc. Chem. Res.*, 42(12), 1935–1943. Copyright 2009 American Chemical Society.)

significant mechanistic challenge, and requires a catalyst that can undergo a series of coordinated structural modifications to accomplish multiple PCET reactions and redox leveling throughout the catalytic cycle [300]. Furthermore, the reaction coordinate for water oxidation requires formation of multiple (two per molecule of H_2O) oxidizing equivalents, a difficult prospect given that each successive equivalent can increase the reaction potential by 0.5–1.5 V [381]. Finally, synthetic water oxidation catalysts integrated into artificial, photoelectrochemical devices must have turnovers that can match average solar insolation (200–300 W m^{-2}) [1,3]. Based on Equations 11.1 and 11.4 and the addition of an overpotential for efficient charge transfer and catalysis, ~1.8 eV is needed for water splitting at a current density of between 10 and 40 mA cm^{-2} [381].

11.3.2.1 Manganese Catalysts

The $CaMn_4O_X$ core of PSII is ubiquitous across all O_2 evolving photosynthetic organisms [300]. The fact that no other alternative structure for water oxidation has been found in biology has lead many researchers to focus on organometallic Mn-clusters as catalysts for artificial photosynthesis. This section will review the development of catalyst designs that are based on the structure and composition of the PSII Mn-center. These are primarily Mn_4O_4 clusters with cubic topologies coordinated by organic ligands. Figure 11.11A shows one of several predicted structures for the PSII Mn-center (the exact structure of the water-oxidation complex is unresolved and remains a highly debated topic [382,383]) in comparison to a synthetic Mn_4O_4 cubane to highlight structural characteristics that are shared between natural and artificial complexes. The Mn-center of PSII is stabilized within a 3D-protein scaffold that also provides flexibility and hydration to accommodate the mechanistic and kinetic requirements of catalytic water oxidation [374,384]. This 3D structure is mimicked to an extent in artificial complexes by the addition of organic ligands to the Mn atoms. Like the Mn-center, some cubic Mn_4O_4 complexes can self-assemble with diphenylphosphate ligands [385,386], via a mechanism based on Mn-O bond formation in conjunction with π-π interactions between the phenyl rings.

There are two important mechanistic properties that have coevolved within the PSII Mn-center that together help to offset the high energetic requirement of water oxidation.

The first involves the accumulation of multiple oxidative equivalents at the Mn-center by the transfer of successive photon energies. This is accomplished by the control of Mn redox potential by the nature of the ligands and the chemistry of the secondary coordination sphere, which together allow the buildup of oxidative potential [387]. In synthetic versions of the Mn-center, the cubane cluster O-group ligands help to fulfill this requirement. Each Mn-atom is coordinated by three oxo and three phosphinate O-atoms, which allow oxidation of the cluster to a 7^+ state at redox potentials attainable by visible light absorption [388,389]. These potentials can be further tuned by varying the functionalization of phenyl groups of the ligands [388]. The second mechanism for energy offset in PSII Mn-center is synchronization of PCET reaction steps [338]. This achieves redox leveling and a lowering of the energetic cost of catalysis compared to uncoupled proton and electron transfer [390,391]. PCET by a PSII Mn-center biomimetic catalyst has been demonstrated for phosphinate-coordinated Mn_4O_4 cubane, which upon oxidation catalyzed a hydride-transfer $(2e^-, H^+)$ reaction [392].

In PSII, the protein structure not only acts to coordinate the metal atoms and tune redox potentials, but it also helps to stabilize the Mn-center and protect it from undesirable oxidation reactions [300]. This supporting environment can be mimicked in several ways, including the use of large polycyclic ligands [393], the embedding of catalysts in the pores of solid supports [394], and the incorporation of catalysts in polymer matrices. For example, cubane Mn_4O_4 complexes have been embedded in Nafion and studied as electrocatalysts for water oxidation by absorption to an electrode [395], or coupled to photoactive dye-sensitized titania film as photocatalysts [396]. In each of these cases however, the water-splitting reaction occurred at low turnovers and was less efficient compared to catalytic PSII water oxidation.

11.3.2.2 Ruthenium Catalytic Complexes

Ruthenium complexes were the early focus of research efforts into synthetic compounds for water oxidation. A di-Ru compound, the blue dimer, shown in Figure 11.12 was the first successful water oxidation catalyst synthesized [397], and numerous examples of structurally similar compounds now exist [398,399]. Many of these compounds achieve the high oxidative potentials required for water oxidation through successive PCET events to achieve a final transfer of four electrons and four protons [400]. In this manner, the mechanism resembles PSII water oxidation, with PCET offsetting the large energetic

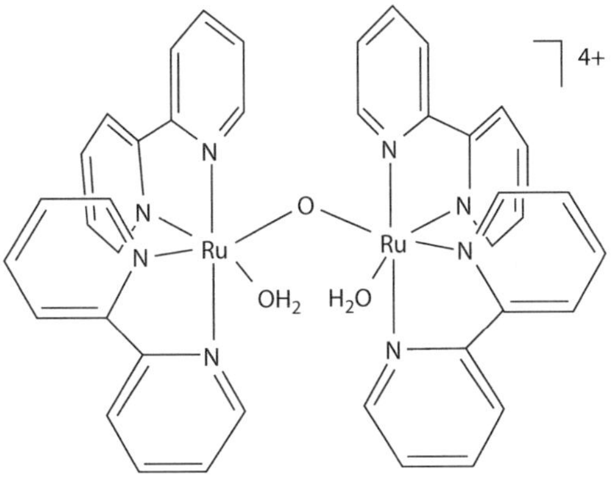

FIGURE 11.12
Structure of a di-ruthenium catalyst for water oxidation. The solution phase x-ray structure of $[(bipy)_2(H_2O)Ru^{III}ORu^{III}(H_2O)(bipy)_2]^{4+}$ (blue dimer) is shown. (Reprinted with permission from [Brimblecombe, R. et al., Molecular water-oxidation catalysts for photoelectrochemical cells. *Dalton Trans.*, 43, 9374–9384. Copyright 2009 Royal Society of Chemistry.)

cost of the loss of four electrons. Some limitations of Ru complexes are that they operate best under extreme pH conditions [300,381], require overpotentials, exhibit low turnovers in photoelectrochemical cells, and are composed of platinum-group metals of limited stock. These compounds also rely on the uses of cesium as a strong oxidant to drive the water oxidation reaction [381]. Mechanistic studies of this class of compounds have shown that O–O bond formation occurs at a single site [401,402]. This led to the development of catalysts containing a single Ru atom surrounded by organic frameworks [403–405]. Mono-Ru complexes are functional as water oxidation catalysts in acidic solutions, with turnover numbers in the hundreds. Like the blue dimer and its derivatives, the mono-Ru complexes are capable of catalyzing water oxidation using Ce(IV) as an oxidant in solution.

The mechanism of coupling ruthenium-complexes to the Ce(IV)–Ce(III) oxidant to drive the catalytic water oxidation highlights the importance of ET. The rate-limiting step for this reaction is the transfer of electrons from the catalyst to reduce Ce(IV)–Ce(III) [381]. The rate constant for this step can be substantially enhanced by modification of the Ru-coordinating organic ligands and substitution of the bipyridine moiety of the catalyst. Addition of ET mediators such as bipyrimidine or bipyrazine to the organic framework can increase the catalytic rate by up to 30-fold [406]. This rate enhancement effect is also observed for the electrocatalytic water oxidation by the complexes adsorbed onto indium-tin oxide (ITO) and fluorine-doped tin oxide (FTO) electrodes [407]. Electrodes modified by adsorption of the bipyridine Ru blue dimer showed currents for electrocatalytic water oxidation, whereas an unmodified electrode showed no detectable currents [381]. The electrodes functioned continuously for over 13 h at turnover frequencies of up to $0.6 s^{-1}$ [381]. In a photoelectrochemical cell, Ru complexes attached to nanocrystalline TiO_2 have catalyzed light-dependent dehydrogenation of isopropanol [408], and water oxidation with an IrO_2 co-catalyst [409].

11.3.2.3 Cobalt Catalysts

Cobalt can serve as an alternative to cesium as the oxidizer for Ru catalysts, however, this is an ineffective combination for sustained water oxidation catalysis due to eventual precipitation of Co [410,411]. With the loss of the oxidizer from solution, the catalytic yields of O_2 decreased. This precipitation can be exploited however, by using it to create catalysts in situ on an electrode. Aqueous solutions of Co^{2+} and phosphate form an oxidized amorphous layer on both ITO and FTO films under an applied potential of +1.3 V (vs. NHE) [412]. The electrodes sustained O_2 evolution for at least 8 h in aqueous solutions at neutral pH, with water oxidation occurring at the Co-phosphate layer. For water oxidation, the Co^{2+}-based catalysts represent an important advance over Ru-catalysts. Mechanistically, the pH dependence of the Co^{2+}/Co^{3+}-catalyzed reaction showed that in neutral solutions water oxidation involved a phosphate-dependant PCET step [413]. This observation suggested that in addition to mechanistic similarities, there might be structural similarities to the Mn-center of PSII. This assumption has been strengthened by x-ray absorption spectroscopy showing the Co film to be composed of a mix of Co_3O_4 and Co_4O_4 cubane units [414]. Potassium is also present in the film, and is thought to coordinate the Co_3O_4 subunits in a manner analogous to the role of Ca in the Mn-cluster of PSII. The capacity of the Co-based catalysts to self-assemble, self-repair, achieve redox leveling of intermediate reaction steps, and carry out sustained turnover for prolonged periods makes them a promising electrocatalyst for integrating with photoelectrochemical devices for coupling water oxidation to H_2 or biofuel production [129].

11.3.3 Artificial Hydrogen Production

Development of a H_2 fuel economy will require catalysts for production of H_2 to be used as the fuel, and for oxidization of H_2 coupled to O_2 reduction to generate electrical energy in a fuel cell. As discussed in Section 11.2.3, the natural structural and functional diversity that exists within the hydrogenases provide excellent examples of model structures for catalyzing both the H_2 oxidation and proton reduction reactions (Equation 11.1). The atomic structures of both [FeFe]-hydrogenase and [NiFe]-hydrogenase catalytic sites are well characterized, though the exact nature of their catalytic mechanisms remain unknown, and are inspiring the design of a number of synthetic catalysts. Examples of the progress in the synthesis of biomimetic catalysts are summarized in the following sections.

11.3.3.1 Hydrogenase-Inspired Biomimetic Catalysts

11.3.3.1.1 [FeFe]-Hydrogenase Mimics

The functional aspects and structural characteristics of the [FeFe]-hydrogenase H-cluster (see Section 11.2.3 for details) reveal a unique organometallic composition that has inspired synthetic designs of Fe-based catalysts. These efforts have principally focused on a variety of substituted di-iron di-thiolates that structurally mimic the composition of the H-cluster, [2Fe]-subcluster. Initial synthetic efforts conducted soon after the CpI and Ddh [FeFe]-hydrogenase structures [160,161] became available focused on synthesis of di-iron di-thiolates carbonyls as models of the [2Fe]-subcluster (Figure 11.13A) [415–417]. These structures are both water soluble and stable, but were unable to catalyze H_2 activation. Subsequent [2Fe]-subcluster analogs synthesized based on modifications of Figure 11.13A included substitutions of the bridgehead atom by amine (Figure 11.13B) [418] and ether (Figure 11.13C) [418,419] groups. In an effort to develop more biologically relevant chemical mimics, a [4Fe-4S]-cluster with a tripodal thiolate ligand was linked to a [2Fe]-subcluster with pentacarbonyl ligation and an ethyl bridging group [420]. This H-cluster analog was shown to function as an electrocatalyst for proton reduction, unfortunately at significant overpotentials. In general, the mechanism of H_2 production by [FeFe]-hydrogenases relies on a reaction between a proton and a hydride intermediate. Formation of the hydride is accomplished by the oxidative addition of a proton to a reduced di-iron center, followed by a second proton-transfer step and subsequent condensation of the proton and hydride to form H_2 [384,421]. However, the order of each protonation, oxidation, and reduction step depends on the redox chemistries of the catalyst, as well as the properties of the surrounding medium [106]. In order for H-cluster analogs to be most efficient as catalysts for solar

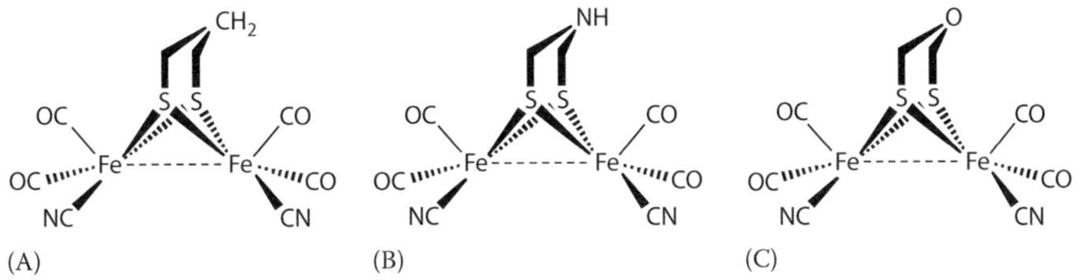

FIGURE 11.13
Structures of synthetic analogues of the H-cluster [2Fe]-subcluster. Each is based on a $[\mu\text{-}S_2C_2H_2X)Fe_2(CO)_4(CN)_2]^{2-}$ parent structure where in "X" is either; (A) methyl [415–417], (B) amine [418], or (C) an ether [418, 419] group. (Adapted from *Coordination Chemistry Reviews*, 249(15-16), Capon, J.-F. et al., Electrochemical H_2 evolution by di-iron sub-site models, 1664–1676, Copyright 2005, with permission from Elsevier.)

H_2 production, they must operate at or near to the reduction potential of the H^+/H_2 couple, and possess turnovers that can match timescales of charge recombination in the light capture material. Modifications to the bridging di-thiolate structure, and to the Fe atom ligands have produced a variety of synthetic compounds with a range of catalytic activities and reduction potentials, and the research progress in this field is nicely reviewed in Ref. [421].

The [2Fe]-subcluster analogs catalyze proton reduction but at potentials of around $-2.0\,V$ versus NHE [417,422,423], which equates to an overpotential of 1.6 V. This is well above the current densities that are achieved by visible light photosensitizers if the water oxidation reaction is included (Equations 11.1 and 11.4). Modification of the di-thiolate bridge with electron-withdrawing groups, specifically nitrogen, can shift the reduction potential to milder values (Figure 11.13B) [424], The shift in potential is the result of changes in the mechanism of proton addition and hydride formation. In di-iron structures without this modification, the proton addition reaction step occurs at one of the Fe atoms, where it is subsequently reduced. However, modification of the di-thiolate bridging group to an amine (Figure 11.13B) results in proton addition at the nitrogen atom, which is then transferred to the [2Fe]-center where it is reduced to form H_2. This modification is thought to improve the stability of the protonated form and facilitate ET of the added proton [424,425]. Also, it is thought that these modified structures bring the mechanism closer to that seen in the enzyme-active site, where the peptide structure surrounding the di-iron site contributes to the catalytic processes [426]. Improvements have also been achieved by modification of the terminal Fe ligands (CO and CN^- in Figure 11.13A) with additional electron-donating groups such as phosphine or CN^-. Like the di-thiolate modifications, these are thought to alter the kinetics of protonation at or near the Fe-Fe bond [423,427]. Also, it has been hypothesized that such modifications may destabilize the metal-hydride intermediate at the [2Fe]-center [426], leading to a faster protonation reaction toward H_2 formation. Combinations of both strategies have produced catalysts that operate at reduction potentials that although are above the H_2/H^+ couple, are within the potentials generated by the photoexcited state of existing photosensitizers [428]. These results are promising, but other weaknesses of these complexes must be addressed prior to development of a functional di-iron catalyst. These include sensitivity to O_2, and photostability of the complexes, both of which are also issues in the hydrogenase enzymes on which these structures are based. Also, despite successes in lowering the energetic requirement for H_2 production by di-iron mimetics, the required overpotentials are still not ideal. It should be noted that we have covered only the generalities of di-iron catalysts, and many complexes have been explored, which are well reviewed elsewhere [426,429–431].

11.3.3.1.2 [NiFe]-Hydrogenase Mimics

Together with synthetic efforts in Fe-based catalysts, there have also been efforts to develop synthetic complexes that mimic the structure and catalytic activity of the [NiFe]-hydrogenase [NiFe]-center. Though several [NiFe]-based catalysts have been synthesized [432–436], to date none of these has been shown to function as a catalyst for H_2 activation. Two types of Ni-based catalysts that have been synthesized that are functional include designs that can catalyze proton reduction [437] and those that can catalyze H_2 oxidation [438–440]. These catalysts take the general form of a Ni atom with four ligand atoms. In complexes that reduce protons, the ligands are generally composed of nitrogen and/or sulfur [441], while in complexes that oxidize H_2 the ligands are generally phosphates [442]. In either case, these ligands are held in place by an organic framework that arranges them in a tetrahedral structure. Complexes have been made in which the organic ligand framework is composed of a single large molecule while others are composed of two identical organic ligands form the tetrahedral framework. As with the H-cluster mimics, rates of proton and/or hydride binding and their respective binding constants greatly affect the reduction potentials and the rates of catalysis. These aspects of the complex can be influenced by the structure of the tetrahedral ligands, with a distorted tetrahedral structure yielding more active complexes [443]. This observation is explained by the increased accessibility of the Ni atom in a distorted structure. Despite some success at synthesizing catalytically active complexes, there are as yet no reports of successfully incorporating Ni-catalysts into functioning H_2 activating devices. Nevertheless, [NiFe]-hydrogenase mimics remain an active area of research. It is also important to note that the mono-atomic structures have also been synthesized using other metal atoms, including cobalt [444] and rhodium [445].

11.3.3.2 Hydrogenase Biohybrids

The integration of hydrogenases as H_2 production catalysts with photovoltaic materials and photochemical devices is a research area that seeks to guide and inspire the design of artificial, light-driven water-splitting devices for solar H_2 production [220,245,323,446–450]. Hydrogenases have high turnover numbers, operate at the equilibrium potential of the reaction shown in Equation 11.1, and can show high stabilities when immobilized [451–455]. Many of the current biomimetic catalysts suffer from low turnovers, overpotentials, and variable stability [106,421,426,443]. As previously discussed, the catalytic domain of hydrogenases is highly evolved to support many of the key functional requirements including tuning of redox potentials, electron transfer, proton transfer and H_2 diffusion, protection of the H-cluster, control of solubility, and interfacing with electron donors [426]. The challenge of designing functional synthetic catalysts with these functional attributes have benefitted from structural efforts that have elucidated how nature has designed chemical structures of enzyme catalytic sites to create highly functional catalysts. Another aspect to a biohybrid approach is to help establish the structural requirements for developing integration of biomimetic catalysts to high stability and efficient photoconversion. A further performance benefit that hydrogenases possess is a 3D-structure that is specific for activation of H_2, which can correspond to tolerances for O_2 and CO at concentrations that normally inhibit noble-metal electrocatalysts [246,455].

Successful integration of redox-active enzymes with artificial photosynthetic molecules and devices requires that each of the molecules (e.g., enzyme, photoactive molecule, and conductor) self-assembles into architectures that can support fast and efficient PCET [129,327,456]. For hydrogenases, oriented assembly with nanoparticles and electrodes can be mediated by electrostatics or van der Waals interactions [457,458]. Polar surfaces on hydrogenases and other redox-active enzymes assist in mediating the formation of ET complexes with donor molecules [169,170,459–461]. Incomplete bonding and surface defects on nanoparticles result in formation of functional groups, which can mediate intermolecular interactions with other molecules such as hydrogenase. Photodriven H_2 production by a biohybrid complex was first demonstrated in the 1980s with [FeFe]-hydrogenase and nanoparticulate TiO_2, with and without sensitizing dyes [462]. This work was expanded on through the use of [NiFe]-hydrogenases with both TiO_2 [463,464] and CdS [465]. More recently, nanoparticulate TiO_2 has been used as a conducting material to couple light-capture and charge-separation by dye molecules to catalytic H_2 production by [NiFe]-hydrogenases [454,466]. Assembly of these biohybrids relied on interactions between surface groups on semiconductor metals (i.e., hydroxyls or sulfur vacancies), and chemically compatible surfaces on the enzyme. As a result, orientations were not specifically

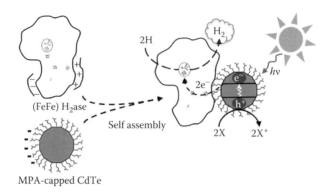

FIGURE 11.14
Schematic of a CdTe:[FeFe]-hydrogenase complex. Self-assembly is guided by the mutual electrostatic attraction between the negatively charged carboxylate groups capping the CdTe nanocrystal surface, and a positively charged region on the [FeFe]-hydrogenase surface composed of amine groups. Upon photoexcitation of the CdTe, the molecular orientation supports ET from CdTe to hydrogenase for H_2 production. (Reprinted with permission from Brown, K. et al., Controlled self assembly of hydrogenase-CdTe nanocrystal hybrids for solar hydrogen production. *J. Am. Chem. Soc.*, 132(28), 9672–9680. Copyright 2010 American Chemical Society.)

optimized for ET. Nanocrystal surfaces on the other hand, can be chemically tailored and made biocompatible through the attachment of amino acids as capping groups [467]. Controlled adsorption of [FeFe]-hydrogenase on nanocrystalline materials was achieved by mixing suspensions of mercapto-propionic acid (MPA)-capped CdTe nanocrystals with bacterial [FeFe]-hydrogenases (Figure 11.14) [458]. The MPA ligand chemi-adsorbs via the S-group to saturate CdTe surface defects improving quantum yields and charge-transfer properties. Through the solvent accessible carboxyl group, the MPA ligand can promote electrostatically controlled self-assembly with positively charged molecules. In this case, [FeFe]-hydrogenase has a high density of positively charged residues near an accessory [4Fe-4S]–cluster, which mediates ET with soluble electron donor Fd [176]. Thus, the MPA group induces CdTe to form ET complexes with [FeFe]-hydrogenase that functionally support light-driven H_2 production.

In addition to the process of assembly, the light-conversion efficiencies in biohybrids are also determined by the molecular compositions. The molecular distributions consist of multiple subcomplexes that differ in the relative number of the catalyst and light adsorber and are thought to contribute to a broad range of conversion efficiencies [454,458]. Isolation of individual biohybrid complexes of a defined stoichiometry has not yet been accomplished, nor have the interfacial structures been well characterized. These are central challenges to understanding the photoconversion process in any molecular complex, where the physical and chemical phenomena of the interface mediate ET and kinetics of photocatalysis.

In photochemical devices for solar H_2 production, hydrogenases have been utilized as electrocatalysts by immobilization on a conducting electrode that is electrically wired to a dye-sensitized photoanode [452,468,469]. Charge balance is maintained by regenerating the ground state of the photoanode with an electrolyte or reforming reaction [368,470,471]. In some cases the electron source can be a biofuel, such as ethanol, glucose or methanol, which can be produced using the systems detailed in Section 11.2 [453,469]. Hydrogenases are typically adsorbed onto conductive electrodes that are composed of metal (e.g., TiO_2 or gold) or carbon (e.g., glassy carbon, pyrolytic graphite edge, carbon felt, or carbon cloth) [245,451,452]. Electrodes have also been fabricated from nanostructured materials, for example single- or multi-wall carbon nanotubes (SWNT and MWNT, respectively) or

TiO_2 nanorods [457,472,473]. [NiFe]-hydrogenase was adsorbed onto MWNT electrodes under an applied electric field to induce oriented attachment, which resulted in high stability and H_2 oxidation current densities [457]. [FeFe]-hydrogenases and SWNTs self-assemble in solution as biohybrids, and steady-state ET within these complexes has been studied by photoluminescence and Raman spectroscopies [235,473]. The results showed efficient electron exchange with little or no thermodynamic driving force, indicating that SWNTs and other 2D materials can be highly efficient conductors for coupling redox catalysts to solar capture materials and devices. In order for enzymes to be viable as catalysts in artificial H_2 production schemes, developing scalable expression and purification processes are required, as well as the materials to stabilize and protect enzymes under applied conditions. None of these requirements have been completely addressed by current technologies.

11.3.4 Catalysts for CO_2 Reduction

The concentration and capture of CO_2 for production of biofuels is an important pathway for remediation of atmospheric CO_2 levels and the renewable production of energy carriers. Currently, the synthetic catalysts developed for CO_2 reduction typically are designed to catalyze either CO or formate production and fit into applied schemes as electrocatalysts in photoelectrochemical devices or as photocatalysts (similar to Ru-based water oxidation catalysts) by the addition of light absorbing metalloporphyrins [474]. Electrocatalysts that have specificity for CO_2 are often composed of tetraazo-macrocyclic compounds [475–478]. These structures, as the name implies, are composed of macrocylic molecules, which form a tetrahedral ligand cage of N atoms around a single metal atom, usually Co or Ni (Figure 11.15A). As a result, CO_2 binds to form a metal-formato complex at the central atom. The efficiency and turnover numbers of CO_2 reduction electrocatalysts vary widely, and the catalytic properties are tuned by alterations to the tetraazo-macrocycle structure to affect the CO_2 binding kinetics and the redox potential of the metal atom [474]. Covalently linked electrocatalyst and photosensitizer bifunctional complexes have been synthesized from several different catalytic macrocycles and light absorbers [479]. Though several examples of complexes that catalyze light-driven CO_2 reduction exist, improvements in overpotentials, efficiencies, and turnover are required. Multifunctional catalysts with both light absorbing and catalytic moieties for CO_2 reduction are generally in the form of metallomacrocycles (Figure 11.15B). The macrocyclic components are related to, and in some cases identical to, the porphyrin antennas discussed in Section 11.3.1. Structures include porphyrins, corrins, phtalocyanines, and corroles (Figure 11.13B) [480–482]. These tend to have planar geometries in which the metal atom, usually Fe or Co, is chelated by four nitrogen atoms [474]. Reduction of CO_2 under illumination has been measured with turnover numbers of up to 300. The catalytic performance of these compounds is a short lifetime of the photoexcited state that effectively competes with slow turnover timescales and requires characterizing the mechanism for CO_2 binding to improve efficiencies [443,474,483].

11.4 Perspectives

Current trends in biohydrogen and biofuel-production research should continue to emphasize application of HTP technologies, flux analysis, and systems biology toward developing more detailed and accurate models of cellular metabolism. Applying these models will no doubt assist in optimizing biofuel and biohydrogen production into scalable processes.

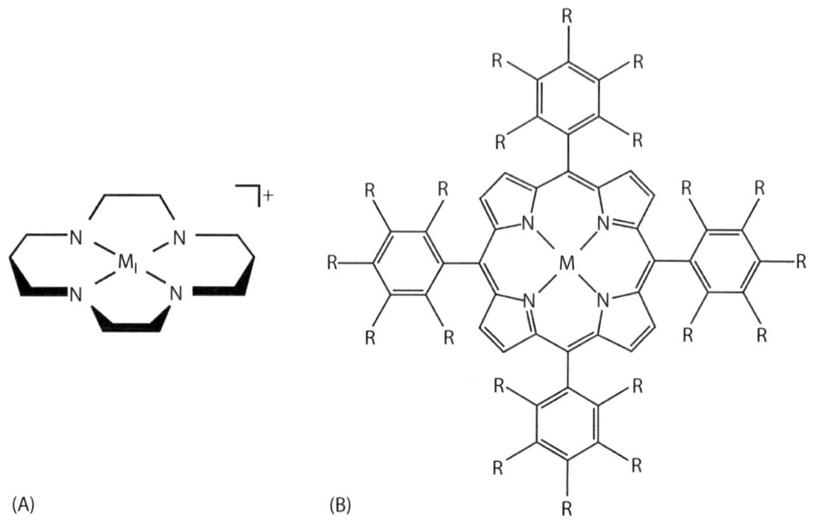

(A)   (B)

FIGURE 11.15

Chemical structures of catalytic metallomacrocycles for CO_2 reduction. (A) A tetraaza-metallomacrocycle, where M = Co or Ni (for review see Ref. [474]). The M is in the +1 (I) oxidation state and the compound has an overall charge of +1 (I^+). Putative intermediates in the PCET reduction of CO_2 to CO include stabilization of a C(OH)O (formato) group between macrocycle dimers [476]. Compounds like (A) can be indirectly reduced in an electrochemical or photochemical scheme. (B) A metalloprophyrin macrocycle, where M = Fe or Co [480, 481]. Under illumination, the phorphyrin moieties are photoexcited for the direct reduction of the metal center in the presence of a sacrificial donor, whereby CO_2 can be reduced to either CO or formate via multistep PCET. (Adapted with permission from Morris, A.J., Meyer, G.J., and Fujita, E., Molecular approaches to the photocatalytic reduction of carbon dioxide for solar fuels. *Acc. Chem. Res.*, 42(12), 1983–1994. Copyright 2009 American Chemical Society.)

For photobiomimetic schemes, the challenge lies in developing robust catalysts that are cheap and renewable to keep device designs and fuel production costs low. There have been a few significant breakthroughs in water oxidation catalysis toward this goal, and there will need to be similar breakthroughs in CO_2 reduction and H_2 production catalysis. Sunlight provides the energy necessary to sustain many lifeforms, which through evolution have developed highly functional molecular machines to accomplish solar conversion from abundant raw materials. Ongoing efforts to both understand and mimic these machines will continue to provide insights into advancing artificial photosynthesis and developing sustainable, cost-effective, and efficient renewable energy technologies.

Acknowledgments

The authors would like to thank Drs. Damien Carrieri and Michael Siebert of the NREL Photobiology Group for insightful comments during preparation of this chapter. Completion of this work was supported by the U.S. Department of Energy (DOE) under Contract No. DE-AC36-08-GO28308 with the National Renewable Energy Laboratory, the U.S. DOE Fuel Cell Technologies Program, and the U.S. DOE Office of Basic Energy Sciences, Photo- and Bio-Chemistry Program.

References

1. Lewis, N.S. and D.G. Nocera, Powering the planet: Chemical challenges in solar energy utilization. *Proceedings of the National Academy of Sciences*, 2006. **103**(43):15729–15735.
2. Smil, V., Energy at the crossroads, in *OECD Global Science Forum*, Paris, France. 2006. p. 27.
3. Program, U.N.D., *World Energy Assessment Report: Energy and the Challenge of Sustainability*, United Nations. 2003.
4. Chow, J., R.J. Kopp, and P.R. Portney, Energy resources and global development. *Science*, 2003. **302**(5650):1528–1531.
5. Hall, J. et al., Valuing the health benefits of clean air. *Science*, 1992. **255**(5046):812–817.
6. Hansen, J. et al., Global temperature change. *Proceedings of the National Academy of Sciences*, 2006. **103**(39):14288–14293.
7. Wigley, T.M.L., R. Richels, and J.A. Edmonds, Economic and environmental choices in the stabilization of atmospheric CO_2 concentrations. *Nature*, 1996. **379**(6562):240–243.
8. WMO, WMO World Data Center for Greenhouse Gases Data Summary, in *Greenhouse Gases and Other Atmospheric Gases*, Tokyo, Japan. 2010. p. 101.
9. Anastas, P. and N. Eghbali, Green chemistry: Principles and practice. *Chemical Society Reviews*, 2009. **39**(1):301–312.
10. Hoffert, M.I. et al., Advanced technology paths to global climate stability: Energy for a greenhouse planet. *Science*, 2002. **298**(5595):981–987.
11. LaVan, D.A. and J.N. Cha, Approaches for biological and biomimetic energy conversion. *Proceedings of the National Academy of Sciences*, 2006. **103**(14):5251–5255.
12. Mikkelsen, M., M. Jorgensen, and F.C. Krebs, The Teraton challenge. A review of fixation and transformation of carbon dioxide. *Energy and Environmental Science*, 2009. 3(1):43–81.
13. Song, C., Global challenges and strategies for control, conversion and utilization of CO_2 for sustainable development involving energy, catalysis, adsorption and chemical processing. *Catalysis Today*, 2006. **115**(1–4):2–32.
14. Batlles, F.J. et al., Determination of atmospheric parameters to estimate global radiation in areas of complex topography: Generation of global irradiation map. *Energy Conversion and Management*, 2008. **49**(2):336–345.
15. Gueymard, C.A., The Sun's total and spectral irradiance for solar energy applications and solar radiation models. *Solar Energy*, 2004. **76**(4):423–453.
16. Beer, L.L. et al., Engineering algae for biohydrogen and biofuel production. *Current Opinion Biotechnology*, 2009. **20**(3):264–271.
17. Dismukes, G.C. et al., Aquatic phototrophs: Efficient alternatives to land-based crops for biofuels. *Current Opinion Biotechnology*, 2008. **19**(3):235–240.
18. Lykidis, A. and N. Ivanova, Genomic prospecting for microbial biodiesel production, in *Bioenergy*, J.D. Wall, C.S. Harwood, and A. Demain, eds. 2008, ASM Press: Washington, DC.
19. Radakovits, R. et al., Genetic engineering of algae for enhanced biofuel production. *Eukaryotic Cell*, 2010. **9**(4):486–501.
20. Weyer, K. et al., Theoretical maximum algal oil production. *BioEnergy Research*, 2009. 3(2):204–213.
21. Ghirardi, M.L. et al., Microalgae: A green source of renewable H_2. *Trends Biotechnology*, 2000. **18**(12):506–511.
22. Weaver, P.F., S. Lien, and M. Seibert, Photobiological production of hydrogen. *Solar Energy*, 1980. **24**(1):3–45.
23. Boichenko, V.A., E. Greenbaum, and M. Seibert, Hydrogen production by photosynthetic microorganisms, in *Photoconversion of Solar Energy: Molecular to Global Photosynthesis*, M.D. Archer and J. Barber, eds. 2004, Imperial College Press: London, U.K.
24. Tamagnini, P. et al., Cyanobacterial hydrogenases: Diversity, regulation and applications. *FEMS Microbiology Reviews*, 2007. **31**(6):692–720.

25. Hemschemeier, A., A. Melis, and T. Happe, Analytical approaches to photobiological hydrogen production in unicellular green algae. *Photosynthesis Research*, 2009. **102**(2):523–540.

26. Rupprecht, J. et al., Perspectives and advances of biological H_2 production in microorganisms. *Applied Microbiology and Biotechnology*, 2006. **72**(3):442–449.

27. Hankamer, B. et al., Photosynthetic biomass and H_2 production by green algae: From bioengineering to bioreactor scale-up. *Physiologia Plantarum*, 2007. **131**(1):10–21.

28. Hu, Q. et al., Microalgal triacylglycerols as feedstocks for biofuel production: Perspectives and advances. *Plant Journal*, 2008. **54**(4):621–639.

29. Melis, A., Solar energy conversion efficiencies in photosynthesis: Minimizing the chlorophyll antennae to maximize efficiency. *Plant Science*, 2009. **177**(4):272–280.

30. Boyd, E.S., J.R. Spear, and J.W. Peters, [FeFe] hydrogenase genetic diversity provides insight into molecular adaptation in a saline microbial mat community. *Applied and Environmental Microbiology*, 2009. **75**(13):4620–4623.

31. Sakurai, H. and H. Masukawa, Promoting R & D in photobiological hydrogen production utilizing mariculture-raised cyanobacteria. *Marine Biotechnology*, 2007. **9**(2):128–145.

32. Seibert, M. et al., Photosynthetic water-splitting for hydrogen production, in *Bioenergy*, J.D. Wall, C.S. Harwood, and A. Demain, eds. 2008, ASM Press: Washington, DC.

33. Stephanopoulos, G., Challenges in engineering microbes for biofuels production. *Science*, 2007. **315**(5813):801–804.

34. Timmins, M. et al., Phylogenetic and molecular analysis of hydrogen-producing green algae. *Journal of Experimental Botany*, 2009. **60**(6):1691–1702.

35. Atsumi, S., T. Hanai, and J.C. Liao, Non-fermentative pathways for synthesis of branched-chain higher alcohols as biofuels. *Nature*, 2008. **451**(7174):86–89.

36. Atsumi, S., W. Higashide, and J.C. Liao, Direct photosynthetic recycling of carbon dioxide to isobutyraldehyde. *Nature Biotechnology*, 2009. **27**(12):1177–1180.

37. Hellingwerf, K.J. and M.J. Teixeira de Mattos, Alternative routes to biofuels: Light-driven biofuel formation from CO_2 and water based on the photanol approach. *Journal of Biotechnology*, 2009. **142**(1):87–90.

38. Sheehan, J. et al., *A Look Back at the U.S. Department if Energy's Aquatic Species Program-Biodeisel from Algae*, U.S.D.O. of Development, ed. 1998, National Renewable Energy Laboratory: Golden, CO.

39. Surek, B., Meeting report: Algal culture collections 2008: An international meeting at the Culture Collection of Algae and Protozoa (CCAP), Dunstaffnage Marine Laboratory, Dunbeg, Oban, United Kingdom; June 8–11, 2008. *Protist*, 2008. **159**(4):509–517.

40. Hellingwerf, K.J. and M.J. Teixeira de Mattos, Alternative routes to biofuels: Light-driven biofuel formation from CO_2 and water based on the photanol approach. *Journal of Biotechnology*, 2009. **142**(1):87–90.

41. Sheehan, J., Engineering direct conversion of CO_2 to biofuel. *Nature Biotechnology*, 2009. **27**(12):1128–1129.

42. Day, J.G., Cryopreservation technology transfer: Experiences from the COBRA project. *Cryoletters*, 2008. **29**(1):76–77.

43. Cooksey, K.E. et al., Fluormetric-determination of the neutral lipid-content of microalgal cells using Nile Red. *Journal of Microbiological Methods*, 1987. **6**(6):333–345.

44. Elsey, D. et al., Fluorescent measurement of microalgal neutral lipids. *Journal of Microbiological Methods*, 2007. **68**(3):639–642.

45. Guckert, J.B., K.E. Cooksey, and L.L. Jackson, Lipid solvent systems are not equivalent for analysis of lipid classes in the microeukaryotic green-alaga, *Chlorella*. *Journal of Microbiological Methods*, 1988. **8**(3):139–149.

46. Guschina, I.A. and J.L. Harwood, Lipids and lipid metabolism in eukaryotic algae. *Progress in Lipid Research*, 2006. **45**(2):160–186.

47. Kumari, P. et al., Tropical marine macroalgae as potential sources of nutritionally important PUFAs. *Food Chemistry*, 2010. **120**(3):749–757.

48. Paik, M.J. et al., Separation of triacylglycerols and free fatty acids in microalgal lipids by solid-phase extraction for separate fatty acid profiling analysis by gas chromatography. *Journal of Chromatography A*, 2009. **1216**(31):5917–5923.
49. Chisti, Y., Biodiesel from microalgae. *Biotechnology Advances*, 2007. **25**(3):294–306.
50. Wang, N. et al., The Hawaiian Algal Database: A laboratory LIMS and online resource for biodiversity data. *BMC Plant Biology*, 2009. **9**(117).
51. Wang, Z.T. et al., Algal lipid bodies: Stress induction, purification, and biochemical characterization in wild-type and starchless *Chlamydomonas reinhardtii. Eukaryotic Cell*, 2009. **8**(12):1856–1868.
52. Bassham, J.A., A.A. Benson, and M. Calvin, The path of carbon in photosynthesis. *Journal of Biological Chemistry*, 1950. **185**(2):781–787.
53. Spalding, M.H., Microalgal carbon-dioxide-concentrating mechanisms: *Chlamydomonas* inorganic carbon transporters. *Journal of Experimental Botany*, 2008. **59**(7):1463–1473.
54. Yamano, T. and H. Fukuzawa, Carbon-concentrating mechanism in a green alga, *Chlamydomonas reinhardtii*, revealed by transcriptome analyses. *Journal of Basic Microbiology*, 2009. **49**(1):42–51.
55. Savage, D.F. et al., Spatially ordered dynamics of the bacterial carbon fixation machinery. *Science*, 2010. **327**(5970):1258–1261.
56. Yeates, T.O. et al., Protein-based organelles in bacteria: Carboxysomes and related microcompartments. *Nature Reviews Microbiology*, 2008. **6**(9):681–691.
57. Bao, X.M. et al., Understanding in vivo carbon precursor supply for fatty acid synthesis in leaf tissue. *Plant Journal*, 2000. **22**(1):39–50.
58. Harwood, J.L., Recent advances in the biosynthesis of plant fatty acids. *Biochimica et Biophysica Acta (BBA): Lipids and Lipid Metabolism*, 1996. **1301**(1–2):7–56.
59. Ohlrogge, J. et al., Fatty acid synthesis: From CO_2 to functional genomics. *Biochemical Society Transactions*, 2000. **28**(6):567–573.
60. Page, R.A., S. Okada, and J.L. Harwood, Acetyl-CoA carboxylase exerts strong flux control over lipid synthesis in plants. *Biochimica et Biophysica Acta (BBA): Lipids and Lipid Metabolism*, 1994. **1210**(3): 369–372.
61. Ohlrogge, J.B. and J.G. Jaworski, Regulation of fatty acid synthesis. *Annual Review of Plant Physiology and Plant Molecular Biology*, 1997. **48**(1):109–136.
62. Rawsthorne, S., Carbon flux and fatty acid synthesis in plants. *Progress in Lipid Research*, 2002. **41**(2):182–196.
63. Ohlrogge, J. and J. Browse, Lipid biosynthesis. *Plant Cell*, 1995. **7**(7):957–970.
64. White, S.W. et al., The structural biology of type II fatty acid biosynthesis. *Annual Review of Biochemistry*, 2005. **74**(1):791–831.
65. Los, D.A. and N. Murata, Structure and expression of fatty acid desaturases. *Biochimica et Biophysica Acta (BBA): Lipids and Lipid Metabolism*, 1998. **1394**(1):3–15.
66. Shanklin, J. and E.B. Cahoon, Desaturation and related modifications of fatty acids. *Annual Review of Plant Physiology and Plant Molecular Biology*, 1998. **49**(1):611–641.
67. Chi, X.Y. et al., Comparative analysis of fatty acid desaturases in cyanobacterial genomes. *Comparative and Functional Genomics*, 2008. **2008**:25.
68. Wada, H., M.H. Avelange-Macherel, and N. Murata, The desA gene of the cyanobacterium *Synechocystis* sp. strain PCC6803 is the structural gene for delta 12 desaturase. *Journal of Bacteriology*, 1993. **175**(18):6056–6058.
69. Chi, X. et al., Fatty acid biosynthesis in eukaryotic photosynthetic microalgae: Identification of a microsomal delta 12 desaturase in *Chlamydomonas reinhardtii. The Journal of Microbiology*, 2008. **46**(2):189–201.
70. Giroud, C., A. Gerber, and W. Eichenberger, Lipids of *Chlamydomonas reinhardtii*. Analysis of molecular species and intracellular site(s) of biosynthesis. *Plant Cell Physiology*, 1988. **29**(4):587–595.
71. Tonon, T. et al., Fatty acid desaturases from the microalga *Thalassiosira pseudonana. FEBS Journal*, 2005. **272**(13):3401–3412.

72. Koo, A.J.K., J.B. Ohlrogge, and M. Pollard, On the export of fatty acids from the chloroplast. *Journal of Biological Chemistry*, 2004. **279**(16):16101–16110.

73. Yen, C.-L.E. et al., Thematic review series: Glycerolipids. DGAT enzymes and triacylglycerol biosynthesis. *Journal of Lipid Research*, 2008. **49**(11):2283–2301.

74. Klock, G. and K. Kreuzberg, Kinetic properties of a sn-glycerol-3-phosphate dehydrogenase purified from the unicellular alga *Chlamydomonas reinhardtii*. *Biochimica et Biophysica Acta (BBA): General Subjects*, 1989. **991**(2):347–352.

75. Jelsema, C. et al., Membrane lipid metabolism in *Chlamydomonas reinhardtii* 137+ and Y-1: I. Biochemical localization and characterization of acyltransferase activities. *Journal of Cell Science*, 1982. **58**(1):469–488.

76. Harris, E.H., *Chlamydomonas* as a model organism. *Annual Review of Plant Physiology and Plant Molecular Biology*, 2001. **52**(1):363–406.

77. Dahlqvist, A. et al., Phospholipid:diacylglycerol acyltransferase: An enzyme that catalyzes the acyl-CoA-independent formation of triacylglycerol in yeast and plants. *Proceedings of the National Academy of Sciences*, 2000. **97**(12):6487–6492.

78. Stahl, U. et al., Cloning and functional characterization of a phospholipid:diacylglycerol acyl-transferase from Arabidopsis. *Plant Physiology*, 2004. **135**(3):1324–1335.

79. Li, Y. et al., *Chlamydomonas* starchless mutant defective in ADP-glucose pyrophosphorylase hyper-accumulates triacylglycerol. *Metabolic Engineering*, 2010. **12**(4):387–391.

80. Work, V.H. et al., Increased lipid accumulation in the *Chlamydomonas reinhardtii* sta7–10 starch-less isoamylase mutant and increased carbohydrate synthesis in complemented strains. *Eukaryotic Cell*, 2010. **9**:1251–1261.

81. Chochois, V. et al., Hydrogen production in *Chlamydomonas*: Photosystem II-dependent and -independent pathways differ in their requirement for starch metabolism. *Plant Physiology*, 2009. **151**(2):631–640.

82. Grossman, A.R. et al., Novel metabolism in *Chlamydomonas* through the lens of genomics. *Current Opinion in Plant Biology*, 2007. **10**(2):190–198.

83. Kruse, O. et al., Improved photobiological H_2 production in engineered green algal cells. *Journal of Biological Chemistry*, 2005. **280**(40):34170–34177.

84. Posewitz, M.C. et al., Hydrogen photoproduction is attenuated by disruption of an isoamylase gene in *Chlamydomonas reinhardtii*. *Plant Cell*, 2004. **16**(8):2151–2163.

85. Durrett, T.P., C. Benning, and J. Ohlrogge, Plant triacylglycerols as feedstocks for the production of biofuels. *The Plant Journal*, 2008. **54**(4):593–607.

86. Banerjee, A. et al., *Botryococcus braunii*: A renewable source of hydrocarbons and other chemicals. *Critical Reviews Biotechnology*, 2002. **22**(3):245–279.

87. Metzger, P. and C. Largeau, *Botryococcus braunii*: A rich source for hydrocarbons and related ether lipids. *Applied Microbiology and Biotechnology*, 2005. **66**(5):486–496.

88. Taranto, P.A., T.W. Keenan, and M. Potts, Rehydration induces rapid onset of lipid biosynthesis in desiccated nostoc commune (cyanobacteria). *Biochimica et Biophysica Acta (BBA): Lipids and Lipid Metabolism*, 1993. **1168**(2):228–237.

89. Merzlyak, M.N. et al., Effect of nitrogen starvation on optical properties, pigments, and arachidonic acid content of the unicellular green alaga *Parietochloris Incisa* (Trebouxiophyceae, Chlorophyta). *Journal of Phycology*, 2007. **43**(4):833–843.

90. Khozin-Goldberg, I. and Z. Cohen, The effect of phosphate starvation on the lipid and fatty acid composition of the fresh water eustigmatophyte Monodus subterraneus. *Phytochemistry*, 2006. **67**(7):696–701.

91. Husic, H.D. and N.E. Tolbert, Effect of osmotic stress on carbon metabolism in *Chlamydomonas reinhardtii*: Accumulation of glycerol as an osmoregulatory solute. *Plant Physiology*, 1986. **82**(2):594–596.

92. Kanesaki, Y. et al., Salt stress and hyperosmotic stress regulate the expression of different sets of genes in *Synechocystis* sp. PCC 6803. *Biochemical and Biophysical Research Communications*, 2002. **290**(1):339–348.

93. Oren, A., Bioenergetic aspects of halophilism. *Microbiology Molecular Biology Reviews*, 1999. **63**(2):334–348.
94. Khotimchenko, S.V. and I.M. Yakovleva, Lipid composition of the red alga *Tichocarpus crinitus* exposed to different levels of photon irradiance. *Phytochemistry*, 2005. **66**(1):73–79.
95. Sato, N. and N. Murata, Temperature shift-induced responses in lipids in the blue-green alga, *Anabaena variabilis*: The central role of diacylmonogalactosylglycerol in thermo-adaptation. *Biochimica et Biophysica Acta (BBA): Lipids and Lipid Metabolism*, 1980. **619**(2):353–366.
96. Somerville, C., Direct tests of the role of membrane lipid-composition in low-temperature-induced photoinhibition and chilling sensitivity in plants and cyanobacteria. *Proceedings of the National Academy of Sciences*, 1995. **92**(14):6215–6218.
97. Sato, N. et al., Effect of growth temperature on lipid and fatty-acid compositions in the blue-green-algae, *Anabaena variabilis* and *Anacystis nidulans*. *Biochimica et Biophysica Acta*, 1979. **572**(1):19–28.
98. Margulies, M. et al., Genome sequencing in microfabricated high-density picolitre reactors. *Nature*, 2005. **437**(7057):376–380.
99. Wheeler, D.A. et al., The complete genome of an individual by massively parallel DNA sequencing. *Nature*, 2008. **452**(7189):872–876.
100. Wu, D. et al., A phylogeny-driven genomic encyclopaedia of Bacteria and Archaea. *Nature*, 2009. **462**(7276):1056–1060.
101. Lee, S.K. et al., Metabolic engineering of microorganisms for biofuels production: From bugs to synthetic biology to fuels. *Current Opinion in Biotechnology*, 2008. **19**(6):556–563.
102. Mukhopadhyay, A. et al., Importance of systems biology in engineering microbes for biofuel production. *Current Opinion in Biotechnology*, 2008. **19**(3):228–234.
103. Bar-Even, A. et al., Design and analysis of synthetic carbon fixation pathways. *Proceedings of the National Academy of Sciences*, 2010. **107**(19):8889–8894.
104. Holtz, W.J. and J.D. Keasling, Engineering static and dynamic control of synthetic pathways. *Cell*, 2010. **140**(1):19–23.
105. Khalil, A.S. and J.J. Collins, Synthetic biology: Applications come of age. *Nature Reviews Genetics*, 2010. **11**(5):367–379.
106. Magnuson, A. et al., Biomimetic and microbial approaches to solar fuel generation. *Accounts of Chemical Research*, 2009. **42**(12):1899–1909.
107. Ishii, N. et al., Multiple high-throughput analyses monitor the response of *E. coli* to perturbations. *Science*, 2007. **316**(5824):593–597.
108. Chang, C.H. et al., Photons, photosynthesis, and high-performance computing: Challenges, progress, and promise of modeling metabolism in green algae. *Journal of Physics: Conference Series*, 2008. **125**:012048.
109. Wienkoop, S. et al., Targeted proteomics for *Chlamydomonas reinhardtii* combined with rapid sub-cellular protein fractionation, metabolomics and metabolic flux analyses. *Molecular BioSystems*, 2010. **6**(6):1018–1031.
110. Boyle, N. and J. Morgan, Flux balance analysis of primary metabolism in *Chlamydomonas reinhardtii*. *BMC Systems Biology*, 2009. **3**(1):4.
111. Eberhard, S. et al., Generation of an oligonucleotide array for analysis of gene expression in *Chlamydomonas reinhardtii*. *Current Genetics*, 2006. **49**(2):106–124.
112. Wang, Q. et al., Acetylation of metabolic enzymes coordinates carbon source utilization and metabolic flux. *Science*, 2010. **327**(5968):1004–1007.
113. Zhao, S. et al., Regulation of cellular metabolism by protein lysine acetylation. *Science*, 2010. **327**(5968):1000–1004.
114. Kouzarides, T., Acetylation: A regulatory modification to rival phosphorylation? *EMBO Journal*, 2000. **19**(6):1176–1179.
115. Peralta-Yahya, P.P. and J.D. Keasling, Advanced biofuel production in microbes. *Biotechnology Journal*, 2010. **5**(2):147–162.
116. Keasling, J.D. and H. Chou, Metabolic engineering delivers next-generation biofuels. *Nature Biotechnology*, 2008. **26**(3):298–299.

117. May, P. et al., ChlamyCyc: An integrative systems biology database and web-portal for *Chlamydomonas reinhardtii. BMC Genomics*, 2009. **10**:11.
118. Burja, A.M., S. Dhamwichukorn, and P.C. Wright, Cyanobacterial postgenomic research and systems biology. *Trends Biotechnology*, 2003. **21**(11):504–511.
119. Demarsac, N.T. and J. Houmard, Adaptation of cyanobacteria to environmental stimuli: New steps towards molecular mechanisms. *FEMS Microbiology Reviews*, 1993. **104**(1–2):119–189.
120. Krall, L. et al., Assessment of sampling strategies for gas chromatography-mass spectrometry (GC-MS) based metabolomics of cyanobacteria. *Journal of Chromatography B: Analytical Technologies in the Biomedical and Life Sciences*, 2009. **877**(27):2952–2960.
121. Uduman, N. et al., Dewatering of microalgal cultures: A major bottleneck to algae-based fuels. *Journal of Renewable and Sustainable Energy*, 2010. **2**(1):15.
122. Sayre, R.T. and S. Pereira, Optimization of biofuel production from algae, U.S. Patent Office, 2008.
123. Liu, X.Y. and R. Curtiss, Nickel-inducible lysis system in *Synechocystis* sp PCC 6803. *Proceedings of the National Academy of Sciences*, 2009. **106**(51):21550–21554.
124. Ghirardi, M.L. et al., Hydrogenases and hydrogen photoproduction in oxygenic photosynthetic organisms. *Annual Review of Plant Biology*, 2007. **58**(1):71–91.
125. Tamagnini, P. et al., Hydrogenases and hydrogen metabolism of cyanobacteria. *Microbiology Molecular Biology Reviews*, 2002. **66**(1):1–20.
126. Gaffron, H. and J. Rubin, Fermentative and photochemical production of hydrogen in algae. *The Journal of General Physiology*, 1942. **26**(2):219–240.
127. Melis, A. et al., Sustained photobiological hydrogen gas production upon reversible inactivation of oxygen evolution in the green alga *Chlamydomonas reinhardtii. Plant Physiology*, 2000. **122**(1):127–136.
128. Hambourger, M. et al., Biology and technology for photochemical fuel production. *Chemical Society Reviews*, 2009. **38**(1):25–35.
129. Nocera, D.G., Chemistry of personalized solar energy. *Inorganic Chemistry*, 2009. **48**(21):10001–10017.
130. Ramage, M.P., *The Hydrogen Economy: Opportunities, Costs, Barriers, and R&D Needs*. 2004, National Academies Press: Washington, DC.
131. Mus, F. et al., Anaerobic acclimation in *Chlamydomonas reinhardtii. Journal of Biological Chemistry*, 2007. **282**(35):25475–25486.
132. Kosourov, S. et al., A comparison of hydrogen photoproduction by sulfur-deprived *Chlamydomonas reinhardtii* under different growth conditions. *Journal of Biotechnology*, 2007. **128**(4):776–787.
133. Kosourov, S., M. Seibert, and M.L. Ghirardi, Effects of extracellular pH on the metabolic pathways in sulfur-deprived, H_2-producing *Chlamydomonas reinhardtii* cultures. *Plant Cell Physiology*, 2003. **44**(2):146–155.
134. Makarova, V.V. et al., Photoproduction of hydrogen by sulfur-deprived *C. reinhardtii* mutants with impaired photosystem ii photochemical activity. *Photosynthesis Research*, 2007. **94**(1):79–89.
135. Zhang, L., T. Happe, and A. Melis, Biochemical and morphological characterization of sulfur-deprived and H_2-producing *Chlamydomonas reinhardtii* planta, 2002. **214**(4):552–561.
136. Rees, D.C., Great metalloclusters in enzymology. *Annual Review of Biochemistry*, 2002. **71**(1):221–246.
137. Brand, J.J., J.N. Wright, and S. Lien, Hydrogen production by eukaryotic algae. *Biotechnology and Bioengineering*, 1989. **33**(11):1482–1488.
138. Meuser, J.E. et al., Phenotypic diversity of hydrogen production in chlorophycean algae reflects distinct anaerobic metabolisms. *Journal of Biotechnology*, 2009. **142**(1):21–30.
139. Adams, M.W.W., L.E. Mortenson, and J.S. Chen, Hydrogenase. *Biochimica et Biophysica Acta*, 1980. **594**(2–3):105–176.
140. Bock, A. et al., Maturation of hydrogenases, in *Advances in Microbial Physiology*. 2006, Academic Press Ltd: London, U.K. pp. 1–71.
141. Meyer, J., [FeFe] hydrogenases and their evolution: A genomic perspective. *Cellular and Molecular Life Sciences*, 2007. **64**(9):1063–1084.

142. Przybyla, A.E. et al., Structure-function-relationships among the nickel-containing hydroge-nases. *FEMS Microbiology Reviews*, 1992. **88**(2):109–135.
143. Vignais, P.M. and B. Billoud, Occurrence, classification, and biological function of hydroge-nases: An overview. *Chemical Reviews*, 2007. **107**(10):4206–4272.
144. Vignais, P.M., B. Billoud, and J. Meyer, Classification and phylogeny of hydrogenases. *FEMS Microbiology Reviews*, 2001. **25**(4):455–501.
145. Burris, R.H., Nitrogenases. *Journal of Biological Chemistry*, 1991. **266**(15):9339–9342.
146. Dos Santos, P.C. et al., Formation and insertion of the nitrogenase iron¿½ molybdenum cofac-tor. *Chemical Reviews*, 2003. **104**(2):1159–1174.
147. Postgate, J., *Nitrogen Fixation*. 3rd edn. 1998, Cambridge University Press: Cambridge, U.K.
148. Rubio, L.M. and P.W. Ludden, Biosynthesis of the iron-molybdenum cofactor of nitrogenase. *Annual Review of Microbiology*, 2008. **62**(1):93–111.
149. Melis, A., M. Seibert, and T. Happe, Genomics of green algal hydrogen research. *Photosynthesis Research*, 2004. **82**(3):277–288.
150. Florin, L., A. Tsokoglou, and T. Happe, A novel type of iron hydrogenase in the green alga *Scenedesmus obliquus* is linked to the photosynthetic electron transport chain. *Journal of Biological Chemistry*, 2001. **276**(9):6125–6132.
151. Happe, T. and J.D. Naber, Isolation, characterization and N-terminal amino acid sequence of hydrogenase from the green alga *Chlamydomonas reinhardtii*. *European Journal of Biochemistry*, 1993. **214**(2):475–481.
152. Winkler, M. et al., Isolation and molecular characterization of the [Fe]-hydrogenase from the unicellular green alga *Chlorella fusca*. *Biochimica et Biophysica Acta (BBA): Gene Structure and Expression*, 2002. **1576**(3):330–334.
153. Forestier, M. et al., Expression of two [Fe]-hydrogenases in *Chlamydomonas reinhardtii* under anaerobic conditions. *European Journal of Biochemistry*, 2003. **270**(13):2750–2758.
154. Happe, T., B. Mosler, and J.D. Naber, Induction, localization and metal content of hydrogenase in the green alga *Chlamydomonas reinhardtii*. *European Journal of Biochemistry*, 1994. **222**(3):769–774.
155. Roessler, P.G. and S. Lien, Purification of hydrogenase from *Chlamydomonas reinhardtii*. *Plant Physiology*, 1984. **75**(3):705–709.
156. Wunschiers, R. et al., Molecular evidence for a Fe-hydrogenase in the green alga *Scenedesmus obliquus*. *Current Microbiology*, 2001. **42**(5):353–360.
157. Cohen, J. et al., Finding gas diffusion pathways in proteins: Application to O_2 and H_2 transport in CpI [FeFe]-hydrogenase and the role of packing defects. *Structure*, 2005. **13**(9):1321–1329.
158. Cohen, J. et al., Molecular dynamics and experimental investigation of H_2 and O_2 diffusion in [Fe]-hydrogenase. *Biochemical Society Transactions*, 2005. **33**(Pt 1):80–82.
159. Nicolet, Y. et al., A novel FeS cluster in Fe-only hydrogenases. *Trends in Biochemical Sciences*, 2000. **25**(3):138–143.
160. Nicolet, Y. et al., Desulfovibrio desulfuricans iron hydrogenase: The structure shows unusual coordination to an active site Fe binuclear center. *Structure with Folding and Design*, 1999. **7**(1):13–23.
161. Peters, J.W. et al., X-ray crystal structure of the Fe-Only hydrogenase (CpI) from *Clostridium pasteurianum* to 1.8 angstrom resolution. *Science*, 1998. **282**(5395):1853–1858.
162. Pandey, A.S. et al., Dithiomethylether as a ligand in the hydrogenase H-cluster. *Journal of the American Chemical Society*, 2008. **130**(13):4533–4540.
163. Fan, H.-J. and M.B. Hall, A capable bridging ligand for Fe-only hydrogenase: Density func-tional calculations of a low-energy route for heterolytic cleavage and formation of dihydrogen. *Journal of the American Chemical Society*, 2001. **123**(16):3828–3829.
164. Silakov, A. et al., ^{14}N HYSCORE investigation of the H-cluster of [FeFe] hydrogenase: Evidence for a nitrogen in the dithiol bridge. *Physical Chemistry Chemical Physics*, 2009. **11**(31):6592–6599.
165. Arnon, D.I. et al., Photoproduction of hydrogen, photofixation of nitrogen and a unified con-cept of photosynthesis. *Nature*, 1961. **190**(4776):601–606.
166. Tagawa, K. and D.I. Arnon, Ferredoxins as electron carriers in photosynthesis and in the bio-logical production and consumption of hydrogen gas. *Nature*, 1962. **195**(4841):537–543.

167. Healey, F.P., The mechanism of hydrogen evolution by *Chlamydomonas moewusii*. *Plant Physiology*, 1970. **45**(2):153–159.
168. Abeles, F.B., Cell-free hydrogenase from *Chlamydomonas*. *Plant Physiology*, 1964. **39**(2):169–176.
169. Winkler, M. et al., Characterization of the key step for light-driven hydrogen evolution in green algae. *Journal of Biological Chemistry*, 2009. **284**(52):36620–36627.
170. Chang, C.H. et al., Atomic resolution modeling of the ferredoxin:[FeFe] hydrogenase complex from *Chlamydomonas reinhardtii*. *Biophysics Journal*, 2007. **93**(9):3034–3045.
171. Long, H. et al., Hydrogenase/ferredoxin charge-transfer complexes: Effect of hydrogenase mutations on the complex association. *Journal of Physical Chemistry A*, 2009. **113**(16):4060–4067.
172. Long, H. et al., Brownian dynamics and molecular dynamics study of the association between hydrogenase and ferredoxin from *Chlamydomonas reinhardtii*. *Biophysics Journal*, 2008. **95**(8):3753–3766.
173. Flynn, T., M.L. Ghirardi, and M. Seibert, Accumulation of O_2-tolerant phenotypes in H_2-producing strains of *Chlamydomonas reinhardtii* by sequential applications of chemical mutagenesis and selection. *International Journal of Hydrogen Energy*, 2002. **27**(11–12):1421–1430.
174. Erbes, D.L., D. King, and M. Gibbs, Inactivation of hydrogenase in cell-free extracts and whole cells of *Chlamydomonas reinhardi* by oxygen. *Plant Physiology*, 1979. **63**(6):1138–1142.
175. Ghirardi, M.L. et al., Photobiological hydrogen-producing systems. *Chemical Society Reviews*, 2009. **38**(1):52–61.
176. Peters, J.W., Structure and mechanism of iron-only hydrogenases. *Current Opinion in Structural Biology*, 1999. **9**(6):670–676.
177. Ludwig, M., R. Schulz-Friedrich, and J. Appel, Occurrence of hydrogenases in cyanobacteria and anoxygenic photosynthetic bacteria: Implications for the phylogenetic origin of cyanobacterial and algal hydrogenases. *Journal of Molecular Evolution*, 2006. **63**(6):758–768.
178. Montet, Y. et al., Gas access to the active site of Ni-Fe hydrogenases probed by x-ray crystallography and molecular dynamics. *Nature Structural Biology*, 1997. **4**(7):523–526.
179. Dementin, S. et al., A glutamate is the essential proton transfer gate during the catalytic cycle of the [NiFe] hydrogenase. *Journal of Biological Chemistry*, 2004. **279**(11):10508–10513.
180. Bagley, K.A. et al., Infrared studies on the interaction of carbon-monoxide with divalent nickel in hydrogenase from *Chromatium vinosum*. *Biochemistry*, 1994. **33**(31):9229–9236.
181. Happe, R.P. et al., Biological activation of hydrogen. *Nature*, 1997. **385**(6612):126.
182. Bothe, H., E. Distler, and G. Eisbrenner, Hydrogen metabolism in blue-green algae. *Biochimie*, 1978. **60**(3):277–289.
183. Meyer, J., B.C. Kelley, and P.M. Vignais, Nitrogen fixation and hydrogen metabolism in photosynthetic bacteria. *Biochimie*, 1978. **60**(3):245–260.
184. Robson, R.L. and J.R. Postgate, Oxygen and hydrogen in biological nitrogen fixation. *Annual Review of Microbiology*, 1980. **34**(1):183–207.
185. Peterson, R.B. and C.P. Wolk, Localization of an uptake hydrogenase in *Anabaena*. *Plant Physiology*, 1978. **61**(4):688–691.
186. Tel-Or, E., L.W. Luijk, and L. Packer, Hydrogenase in N_2-fixing cyanobacteria. *Archives of Biochemistry and Biophysics*, 1978. **185**(1):185–194.
187. Oxelfelt, F., P. Tamagnini, and P. Lindblad, Hydrogen uptake in *Nostoc* sp. strain PCC 73102. Cloning and characterization of a hupSL homologue. *Archives of Microbiology*, 1998. **169**(4):267–274.
188. Schmitz, O. et al., Molecular biological analysis of a bidirectional hydrogenase from cyanobacteria. *European Journal of Biochemistry*, 1995. **233**(1):266–276.
189. Buhrke, T. et al., Oxygen tolerance of the H_2-sensing [NiFe] hydrogenase from *Ralstonia eutropha* H16 is based on limited access of oxygen to the active site. *Journal of Biological Chemistry*, 2005. **280**(25):23791–23796.
190. Duche, O. et al., Enlarging the gas access channel to the active site renders the regulatory hydrogenase HupUV of *Rhodobacter capsulatus* O_2 sensitive without affecting its transductory activity. *FEBS Journal*, 2005. **272**(15):3899–3908.

191. Volbeda, A. et al., High-resolution crystallographic analysis of *Desulfovibrio fructosovorans* [NiFe] hydrogenase. *International Journal of Hydrogen Energy*, 2002. **27**(11–12):1449–1461.

192. Houchins, J.P. and R.H. Burris, Comparative characterization of two distinct hydrogenases from *Anabaena* sp. strain 7120. *Journal of Bacteriology*, 1981. **146**(1):215–221.

193. Rey, F.E., E.K. Heiniger, and C.S. Harwood, Redirection of metabolism for biological hydrogen production. *Applied and Environmental Microbiology*, 2007. **73**(5):1665–1671.

194. Germer, F. et al., Overexpression, isolation, and spectroscopic characterization of the bidirectional [NiFe] hydrogenase from *Synechocystis* sp. PCC 6803. *Journal of Biological Chemistry*, 2009. **284**(52):36462–36472.

195. Massanz, C., S. Schmidt, and B. Friedrich, Subforms and in vitro reconstitution of the NAD-reducing hydrogenase of *Alcaligenes eutrophus*. *Journal of Bacteriology*, 1998. **180**(5):1023–1029.

196. Appel, J. and R. Schulz, Sequence Analysis of an operon of a NAD(P)-reducing nickel hydrogenase from the cyanobacterium *Synechocystis* sp PCC 6803 gives additional evidence for direct coupling of the enzyme to NAD(P)H-dehydrogenase (complex I). *Biochimica et Biophysica Acta: Protein Structure and Molecular Enzymology*, 1996. **1298**(2):141–147.

197. Gutekunst, K. et al., LexA regulates the bidirectional hydrogenase in the cyanobacterium *Synechocystis* sp. PCC 6803 as a transcription activator. *Molecular Microbiology*, 2005. **58**(3):810–823.

198. Oliveira, P. and P. Lindblad, LexA, a transcription regulator binding in the promoter region of the bidirectional hydrogenase in the cyanobacterium *Synechocystis* sp. PCC 6803. *FEMS Microbiology Letters*, 2005. **251**(1):59–66.

199. Burgess, B.K. and D.J. Lowe, Mechanism of molybdenum nitrogenase. *Chemical Reviews*, 1996. **96**(7):2983–3012.

200. Masukawa, H. et al., Survey of the distribution of different types of nitrogenases and hydrogenases in heterocyst-forming cyanobactera. *Marine Biotechnology*, 2009. **11**(3):397–409.

201. Tsygankov, A.A., Nitrogen-fixing cyanobacteria: A review. *Applied Biochemistry and Microbiology*, 2007. **43**(3):250–259.

202. Kentemich, T., G. Haverkamp, and H. Bothe, The expression of a 3RD nitrogenase in the cyanobacterium *Anabaena variabilis*. *Zeitschrift Fur Naturforschung C-A Journal of Biosciences*, 1991. **46**(3–4):217–222.

203. Glazer, A.N. and K.J. Kechris, Conserved amino acid sequence features in the subunits of MoFe, VFe, and FeFe nitrogenases. *PLoS ONE*, 2009. **4**(7):e6136.

204. Zhao, Y. et al., Diversity of nitrogenase systems in diazotrophs. *Journal of Integrative Plant Biology*, 2006. **48**(7):745–755.

205. Hu, Y. et al., Assembly of nitrogenase MoFe protein. *Biochemistry*, 2008. **47**(13):3973–3981.

206. Barney, B.M. et al., Trapping an intermediate of dinitrogen (N-2) reduction on nitrogenase. *Biochemistry*, 2009. **48**(38):9094–9102.

207. Einsle, O. et al., Nitrogenase MoFe-protein at 1.16 A resolution: A central ligand in the FeMo-cofactor. *Science*, 2002. **297**(5587):1696–1700.

208. Hoffman, B.M., D.R. Dean, and L.C. Seefeldt, Climbing nitrogenase: Toward a mechanism of enzymatic nitrogen fixation. *Accounts of Chemical Research*, 2009. **42**(5):609–619.

209. Pelmenschikov, V., D.A. Case, and L. Noodleman, Ligand-bound S=1/2 FeMo-cofactor of nitrogenase: Hyperfine interaction analysis and implication for the central ligand X identity. *Inorganic Chemistry*, 2008. **47**(14):6162–6172.

210. Dominic, B., Y.-B. Chen, and J.P. Zehr, Cloning and transcriptional analysis of the nifUHDK genes of *Trichodesmium* sp. IMS101 reveals stable nifD, nifDK and nifK transcripts. *Microbiology*, 1998. **144**(12):3359–3368.

211. Golden, J.W., S.J. Robinson, and R. Haselkorn, Rearrangement of nitrogen-fixation genes during heterocyst differentiation in the cyaobacterium *Anabaena*. *Nature*, 1985. **314**(6010):419–423.

212. Happe, T., K. Schutz, and H. Bohme, Transcriptional and mutational analysis of the uptake hydrogenase of the filamentous cyanobacterium *Anabaena variabilis* ATCC 29413. *Journal of Bacteriology*, 2000. **182**(6):1624–1631.

213. Bandyopadhyay, S., K. Chandramouli, and M.K. Johnson, Iron-sulfur cluster biosynthesis. *Biochemical Society Transactions*, 2008. **36**:1112–1119.
214. Fontecave, M. and S. Ollagnier-De-Choudens, Iron-sulfur cluster biosynthesis in bacteria: Mechanisms of cluster assembly and transfer. *Archives of Biochemistry and Biophysics*, 2008. **474**(2):226–237.
215. Lill, R. and U. Muhlenhoff, Iron-sulfur protein biogenesis in eukaryotes: Components and mechanisms. *Annual Review of Cell and Developmental Biology*, 2006. **22**:457–486.
216. Mulder, D.W. et al., Activation of HydAΔEFG requires a preformed [4Fe-4S] cluster. *Biochemistry*, 2009. **48**(26):6240–6248.
217. Mulder, D.W. et al., Stepwise [FeFe]-hydrogenase H-cluster assembly revealed in the structure of HydAEFG. *Nature*, 2010. **465**(7295):248–251.
218. King, P.W. et al., Functional studies of [FeFe] hydrogenase maturation in an *Escherichia coli* biosynthetic system. *Journal of Bacteriology*, 2006. **188**(6):2163–2172.
219. Posewitz, M.C. et al., Discovery of two novel radical S-adenosylmethionine proteins required for the assembly of an active [Fe] hydrogenase. *Journal of Biological Chemistry*, 2004. **279**(24):25711–25720.
220. English, C.M. et al., Recombinant and in vitro expression systems for hydrogenases: New frontiers in basic and applied studies for biological and synthetic H_2 production. *Dalton Transactions*, 2009. **45**:9970–9978.
221. Akhtar, M.K. and P.R. Jones, Engineering of a synthetic hydF-hydE-hydG-hydA operon for biohydrogen production. *Analytical Biochemistry*, 2008. **373**(1):170–172.
222. Girbal, L. et al., Homologous and heterologous overexpression in *Clostridium acetobutylicum* and characterization of purified clostridial and algal Fe-only hydrogenases with high specific activities. *Applied and Environmental Microbiology*, 2005. **71**:2777–2781.
223. Sybirna, K. et al., Shewanella oneidensis: A new and efficient system for expression and maturation of heterologous [Fe-Fe] hydrogenase from *Chlamydomonas reinhardtii*. *BMC Biotechnology*, 2008. **8**:73.
224. Boyer, M.E. et al., Cell-free synthesis and maturation of [FeFe] hydrogenases. *Biotechnology and Bioengineering*, 2008. **99**:59–67.
225. Forzi, L. and R. Sawers, Maturation of [NiFe]-hydrogenases in *Escherichia coli*. *BioMetals*, 2007. **20**(3):565–578.
226. Ludwig, M. et al., Oxygen-tolerant H_2 oxidation by membrane-bound [NiFe] hydrogenases of *Ralstonia* species: Coping with low level H_2 in air. *Journal of Biological Chemistry*, 2009. **284**(1):465–477.
227. Ludwig, M. et al., Concerted action of two novel auxiliary proteins in assembly of the active site in a membrane-bound [NiFe] hydrogenase. *Journal of Biological Chemistry*, 2009. **284**(4):2159–2168.
228. Winter, G. et al., A model system for [NiFe] hydrogenase maturation studies: Purification of an active site-containing hydrogenase large subunit without small subunit. *FEBS Letters*, 2005. **579**:4292–4296.
229. Burgdorf, T. et al., [NiFe]-Hydrogenases of *Ralstonia eutropha* H16: Modular enzymes for oxygen-tolerant biological hydrogen oxidation. *Journal of Molecular Microbiology and Biotechnology*, 2005. **10**(2–4):181–196.
230. Lenz, O. et al., Requirements for heterologous production of a complex metalloenzyme: The membrane-bound [NiFe] hydrogenase. *Journal of Bacteriology*, 2005. **187**:6590–6595.
231. Lenz, O. et al., Carbamoylphosphate serves as the source of CN^-, but not of the intrinsic CO in the active site of the regulatory [NiFe]-hydrogenase from *Ralstonia eutropha*. *FEBS Letters*, 2007. **581**(17):3322–3326.
232. Hoover, T.R. et al., Homocitrate is a component of the iron-molybdenum cofactor of nitrogenase. *Biochemistry*, 1989. **28**(7):2768–2771.
233. McLean, P.A. and R.A. Dixon, Requirement of nifV gene for production of wild-type nitrogenase enzyme in *Klebsiella pneumoniae*. *Nature*, 1981. **292**(5824):655–656.

234. Dixon, R. et al., Nif gene transfer and expression in chloroplasts: Prospects and problems. *Plant and Soil*, 1997. **194**(1–2):193–203.
235. Blackburn, J.L. et al., Raman spectroscopy of charge transfer interactions between single wall carbon nanotubes and [FeFe] hydrogenase. *Dalton Transactions*, 2008. **40**:5454–5461.
236. Goldet, G. et al., Electrochemical kinetic investigations of the reactions of [FeFe]-hydrogenases with carbon monoxide and oxygen: Comparing the importance of gas tunnels and active-site electronic/redox effects. *Journal of the American Chemical Society*, 2009. **131**(41):14979–14989.
237. Leroux, F et al., Experimental approaches to kinetics of gas diffusion in hydrogenase. *Proceedings of the National Academy of Sciences*, 2008. **105**(32):11188–11193.
238. Stripp, S.T. et al., How oxygen attacks [FeFe] hydrogenases from photosynthetic organisms. *Proceedings of the National Academy of Sciences*, 2009. **106**(41):17331–17336.
239. Adams, M.W.W., The structure and mechanism of iron-hydrogenases. *Biochimica et Biophysica Acta (BBA): Bioenergetics*, 1990. **1020**(2):115–145.
240. Thauer, R.K. et al., The reaction of the iron-sulfur protein hydrogenase with carbon monoxide. *European Journal of Biochemistry*, 1974. **42**(2):447–452.
241. Lemon, B.J. and J.W. Peters, Binding of exogenously added carbon monoxide at the active site of the iron-only hydrogenase (CpI) from *Clostridium pasteurianum. Biochemistry*, 1999. **38**(40):12969–12973.
242. Ogata, H., W. Lubitz, and Y. Higuchi, [NiFe] Hydrogenases: Structural and spectroscopic studies of the reaction mechanism. *Dalton Transactions*, 2009. **37**:7577–7587.
243. De Lacey, A.L. et al., Activation and inactivation of hydrogenase function and the catalytic cycle: Spectroelectrochemical studies. *Chemical Reviews*, 2007. **107**(10):4304–4330.
244. Volbeda, A. et al., Structural differences between the ready and unready oxidized states of [NiFe] hydrogenases. *Journal of Biological Inorganic Chemistry*, 2005. **10**(3):239–249.
245. Armstrong, F.A. et al., Dynamic electrochemical investigations of hydrogen oxidation and production by enzymes and implications for future technology. *Chemical Society Reviews*, 2009. **38**(1):36–51.
246. Vincent, K.A. et al., Electrochemical definitions of O_2 sensitivity and oxidative inactivation in hydrogenases. *Journal of the American Chemical Society*, 2005. **127**(51):18179–18189.
247. Volbeda, A. et al., Crystal-structure of the nickel-iron hydrogenase from desulfovibrio-gigas. *Nature*, 1995. **373**(6515):580–587.
248. Oliveira, P. et al., Characterization and transcriptional analysis of HupSLW in *Gloeothece* sp. ATCC 27152: An uptake hydrogenase from a unicellular cyanobacterium. *Microbiology*, 2004. **150**(11):3647–3655.
249. Houchins, J.P. and R.H. Burris, Physiological reactions of the reversible hydrogenase from *Anabaena* 7120. *Plant Physiology*, 1981. **68**(3):717–721.
250. Houchins, J.P. and R.H. Burris, Occurrence and localization of two distinct hydrogenases in the heterocystous cyanobacterium *Anabaena* sp. strain 7120. *Journal of Bacteriology*, 1981. **146**(1):209–214.
251. Cournac, L. et al., Sustained photoevolution of molecular hydrogen in a mutant of *Synechocystis* sp. strain PCC 6803 deficient in the type I NADPH-dehydrogenase complex. *Journal of Bacteriology*, 2004. **186**(6):1737–1746.
252. Appel, J. et al., The bidirectional hydrogenase of *Synechocystis* sp. PCC 6803 works as an electron valve during photosynthesis. *Archives of Microbiology*, 2000. **173**(5):333–338.
253. Wolk, C.P., Heterocyst formation. *Annual Review of Genetics*, 1996. **30**:59–78.
254. Berman-Frank, I. et al., Segregation of nitrogen fixation and oxygenic photosynthesis in the marine cyanobacterium *Trichodesmium. Science*, 2001. **294**(5546):1534–1537.
255. Zehr, J.P. et al., Globally distributed uncultivated oceanic N_2-fixing cyanobacteria lack oxygenic photosystem II. *Science*, 2008. **322**(5904):1110–1112.
256. Dementin, S. et al., Introduction of methionines in the gas channel makes [NiFe] hydrogenase aero-tolerant. *Journal of the American Chemical Society*, 2009. **131**(29):10156–10164.

257. Liebgott, P.P. et al., Relating diffusion along the substrate tunnel and oxygen sensitivity in hydrogenase. *Nature Chemical Biology*, 2010. **6**(1):63–70.

258. Nagy, L.E. et al., Application of gene-shuffling for the rapid generation of novel [FeFe]-hydrogenase libraries. *Biotechnology Letters*, 2007. **29**(3):421–430.

259. Maroti, G. et al., Discovery of [NiFe] hydrogenase genes in metagenomic DNA: Cloning and heterologous expression in *Thiocapsa roseopersicina*. *Applied and Environmental Microbiology*, 2009. **75**(18):5821–5830.

260. Kosourov, S.N. and M. Seibert, Hydrogen photoproduction by nutrient-deprived *Chlamydomonas reinhardtii* cells immobilized within thin alginate films under aerobic and anaerobic conditions. *Biotechnology and Bioengineering*, 2009. **102**(1):50–58.

261. Melis, A., M. Seibert, and M.L. Ghirardi, Hydrogen fuel production by transgenic microalgae, in *Transgenic Microalgae as Green Cell Factories*, R. Leon, A. Galvan and E. Fernandez, eds. 2007, Landes Bioscience: New York. pp. 110–121.

262. McKinlay, J.B. and C.S. Harwood, Photobiological production of hydrogen gas as a biofuel. *Current Opinion in Biotechnology*, 2010. **21**(3):244–251.

263. Fay, P., Oxygen relations of nitrogen fixation in cyanobacteria. *Microbiology and Molecular Biology Reviews*, 1992. **56**(2):340–373.

264. Stockel, J. et al., Global transcriptomic analysis of cyanothece 51142 reveals robust diurnal oscillation of central metabolic processes. *Proceedings of the National Academy of Sciences*, 2008. **105**(16):6156–6161.

265. Mikheeva, L.E. et al., Mutants of the cyanobacterium *Anabaena variabilis* is altered in hydrogenase activities. *Zeitschrift Fur Naturforschung C—A Journal of Biosciences*, 1995. **50**(7–8):505–510.

266. Markov, S.A., P.F. Weaver, and M. Seibert, Spiral tubular bioreactors for hydrogen production by photosynthetic microorganisms-design and operation. *Applied Biochemistry and Biotechnology*, 1997. **63–65**:577–584.

267. Lindberg, P. et al., A hydrogen-producing, hydrogenase-free mutant strain of *Nostoc punctiforme* ATCC 29133. *International Journal of Hydrogen Energy*, 2002. **27**(11–12):1291–1296.

268. Yoshino, F. et al., High photobiological hydrogen production activity of a *Nostoc* sp. PCC 7422 uptake hydrogenase-deficient mutant with high nitrogenase activity. *Marine Biotechnology*, 2007. **9**(1):101–112.

269. Lindberg, P., P. Lindblad, and L. Cournac, Gas exchange in the filamentous cyanobacterium *Nostoc punctiforme* strain ATCC 29133 and its hydrogenase-deficient mutant strain NHM5. *Applied and Environmental Microbiology*, 2004. **70**(4):2137–2145.

270. Masukawa, H., K. Inoue, and H. Sakurai, Effects of disruption of homocitrate synthase genes on *Nostoc* sp. strain PCC 7120 photobiological hydrogen production and nitrogenase. *Applied and Environmental Microbiology*, 2007. **73**(23):7562–7570.

271. Kucho, K.-i. et al., Global analysis of circadian expression in the cyanobacterium *Synechocystis* sp. strain PCC 6803. *Journal of Bacteriology*, 2005. **187**(6):2190–2199.

272. Schmitz, O., G. Boison, and H. Bothe, Quantitative analysis of expression of two circadian clock-controlled gene clusters coding for the bidirectional hydrogenase in the cyanobacterium *Synechococcus* sp. PCC7942. *Molecular Microbiology*, 2001. **41**(6):1409–1417.

273. Gutthann, F. et al., Inhibition of respiration and nitrate assimilation enhances photohydrogen evolution under low oxygen concentrations in *Synechocystis* sp. PCC 6803. *Biochimica et Biophysica Acta (BBA): Bioenergetics*, 2007. **1767**(2):161–169.

274. Ananyev, G., D. Carrieri, and G.C. Dismukes, Optimization of metabolic capacity and flux through environmental cues to maximize hydrogen production by the cyanobacterium Arthrospira (Spirulina) maxima. *Applied and Environmental Microbiology*, 2008. **74**(19):6102–6113.

275. Das, D. and T.N. Veziroğlu, Advances in biological hydrogen production processes. *International Journal of Hydrogen Energy*, 2008. **33**(21):6046–6057.

276. Hallenbeck, P.C. and D. Ghosh, Advances in fermentative biohydrogen production: The way forward? *Trends in Biotechnology*, 2009. **27**(5):287–297.

277. Lee, H.-S., W.F.J. Vermaas, and B.E. Rittmann, Biological hydrogen production: Prospects and challenges. *Trends in Biotechnology*, 2010. **28**(5):262–271.

278. Prince, R.C. and H.S. Kheshgi, The photobiological production of hydrogen: Potential efficiency and effectiveness as a renewable fuel. *Critical Reviews in Microbiology*, 2005. **31**(1):19–31.

279. Dau, H. and I. Zaharieva, Principles, efficiency, and blueprint character of solar-energy conversion in photosynthetic water oxidation. *Accounts of Chemical Research*, 2009. **42**(12):1861–1870.

280. Rao, K.K. and R. Cammack, Producing hydrogen as a fuel, in *Hydrogen as a Fuel*, R. Cammack, M. Frey, and R. Robson, eds. 2001, Taylor & Francis: New York. pp. 201–330.

281. Cinco, R.M., J.M. MacInnis, and E. Greenbaum, The role of carbon dioxide in light-activated hydrogen production by *Chlamydomonas reinhardtii*. *Photosynthesis Research*, 1993. **38**(1):27–33.

282. Graves, D.A., C.V. Tevault, and E. Greenbaum, Control of photosynthetic reductant: The role of light and temperature on sustained hydrogen photoevolution by *Chlamydomonas* sp. in an anoxic, carbon dioxide-containing atmosphere. *Photochemistry and Photobiology*, 1989. **50**(4):571–576.

283. Greenbaum, E., Photosynthetic hydrogen and oxygen production: Kinetic studies. *Science*, 1982. **215**(4530):291–293.

284. Redding, K. et al., Photosystem I is indispensable for photoautotrophic growth, CO_2 fixation, and H_2 photoproduction in *Chlamydomonas reinhardtii*. *Journal of Biological Chemistry*, 1999. **274**(15):10466–10473.

285. Ghirardi, M., R. Togasaki, and M. Seibert, Oxygen sensitivity of algal H_2-production. *Applied Biochemistry and Biotechnology*, 1997. **63–65**(1):141–151.

286. Zhang, L.P. and A. Melis, Probing green algal hydrogen production. *Philosophical Transactions of the Royal Society of London Series B-Biological Sciences*, 2002. **357**(1426):1499–1507.

287. Kosourov, S. et al., Sustained hydrogen photoproduction by *Chlamydomonas reinhardtii*: Effects of culture parameters. *Biotechnology and Bioengineering*, 2002. **78**(7):731–740.

288. Tsygankov, A. et al., Hydrogen photoproduction under continuous illumination by sulfur-deprived, synchronous *Chlamydomonas reinhardtii* cultures. *International Journal of Hydrogen Energy*, 2002. **27**(11–12):1239–1244.

289. Fedorov, A.S. et al., Continuous hydrogen photoproduction by *Chlamydomonas reinhardtii* using a novel two-stage, sulfate-limited chemostat system. *Applied Biochemistry and Biotechnology*, 2005. **121–124**:403–412.

290. Laurinavichene, T.V. et al., Prolongation of H_2 photoproduction by immobilized, sulfur-limited *Chlamydomonas reinhardtii* cultures. *Journal of Biotechnology*, 2008. **134**(3–4):275–277.

291. Winkler, M. et al., [Fe]-hydrogenases in green algae: Photo-fermentation and hydrogen evolution under sulfur deprivation. *International Journal of Hydrogen Energy*, 2002. **27**(11–12):1431–1439.

292. Gfeller, R.P. and M. Gibbs, Fermentative metabolism of *Chlamydomonas reinhardtii*: I. Analysis of fermentative products from starch in dark and light. *Plant Physiology*, 1984. **75**(1):212–218.

293. Hemschemeier, A. et al., Hydrogen production by *Chlamydomonas reinhardtii*: An elaborate interplay of electron sources and sinks. *Planta*, 2008. **227**(2):397–407.

294. Jans, F. et al., A type II NAD(P)H dehydrogenase mediates light-independent plastoquinone reduction in the chloroplast of *Chlamydomonas*. *Proceedings of the National Academy of Sciences*, 2008. **105**(51):20546–20551.

295. Antal, T.K. et al., The dependence of algal H_2 production on photosystem II and O_2 consumption activities in sulfur-deprived *Chlamydomonas reinhardtii* cells. *Biochimica et Biophysica Acta-Bioenergetics*, 2003. **1607**(2–3):153–160.

296. Mus, F. et al., Inhibitor studies on non-photochemical plastoquinone reduction and H_2 photoproduction in *Chlamydomonas reinhardtii*. *Biochimica et Biophysica Acta—Bioenergetics*, 2005. **1708**(3):322–332.

297. Dubini, A. et al., Flexibility in anaerobic metabolism as revealed in a mutant of *Chlamydomonas reinhardtii* lacking hydrogenase activity. *Journal of Biological Chemistry*, 2009. **284**(11):7201–7213.

298. Ohta, S., K. Miyamoto, and Y. Miura, Hydrogen evolution as a consumption mode of reducing equivalents in green algal fermentation. *Plant Physiology*, 1987. **83**(4):1022–1026.

299. Ghirardi, M.L., Hydrogen production by photosynthetic green algae. *Indian Journal of Biochemistry and Biophysics*, 2006. **43**(4):201–210.

300. Dismukes, G.C. et al., Development of bioinspired Mn$_4$O$_4$ cubane water oxidation catalysts: Lessons from photosynthesis. *Accounts of Chemical Research*, 2009. **42**(12):1935–1943.
301. Mitra, M. and A. Melis, Genetic and biochemical analysis of the TLA1 gene in *Chlamydomonas reinhardtii*. Planta, 2009. **231**(3):729–740.
302. Polle, J.E.W., S.-D. Kanakagiri, and A. Melis, tla1, a DNA insertional transformant of the green alga *Chlamydomonas reinhardtii* with a truncated light-harvesting chlorophyll antenna size. *Planta*, 2003. **217**(1):49–59.
303. Lee, J. and E. Greenbaum, A new oxygen sensitivity and its potential application in photosynthetic H$_2$ production. *Applied Biochemistry and Biotechnology*, 2003. **106**(1):303–313.
304. Eberhard, S., G. Finazzi, and F.-A. Wollman, The dynamics of photosynthesis. *Annual Review of Genetics*, 2008. **42**(1):463–515.
305. Forti, G. et al., In vivo changes of the oxidation-reduction state of NADP and of the ATP/ADP cellular ratio linked to the photosynthetic activity in *Chlamydomonas reinhardtii*. *Plant Physiology*, 2003. **132**(3):1464–1474.
306. Allen, J.F., Cyclic, pseudocyclic and noncyclic photophosphorylation: New links in the chain. *Trends in Plant Science*, 2003. **8**(1):15–19.
307. Shikanai, T., Cyclic electron transport around photosystem I: Genetic approaches. *Annual Review of Plant Biology*, 2007. **58**(1):199–217.
308. Cardol, P. et al., Impaired respiration discloses the physiological significance of state transitions in *Chlamydomonas*. *Proceedings of the National Academy of Sciences*, 2009. **106**(37):15979–15984.
309. DalCorso, G. et al., A complex containing PGRL1 and PGR5 is involved in the switch between linear and cyclic electron flow in arabidopsis. *Cell*, 2008. **132**(2):273–285.
310. Iwai, M. et al., Isolation of the elusive supercomplex that drives cyclic electron flow in photosynthesis. *Nature*, 2010. **464**:1210–1213.
311. Doebbe, A. et al., Functional integration of the HUP1 hexose symporter gene into the genome of *C. reinhardtii*: Impacts on biological H$_2$ production. *Journal of Biotechnology*, 2007. **131**(1):27–33.
312. White, A.L. and A. Melis, Biochemistry of hydrogen metabolism in *Chlamydomonas reinhardtii* wild type and a rubisco-less mutant. *International Journal of Hydrogen Energy*, 2006. **31**(4):455–464.
313. Matthew, T. et al., The metabolome of *Chlamydomonas reinhardtii* following induction of anaerobic H$_2$ production by sulfur depletion. *Journal of Biological Chemistry*, 2009. **284**(35):23415–23425.
314. Bolton, J.R., S.J. Strickler, and J.S. Connolly, Limiting and realizable efficiencies of solar photolysis of water. *Nature*, 1985. **316**(6028):495–500.
315. Gust, D. et al., Engineered and artificial photosynthesis: Human ingenuity enters the game. *MRS Bulletin*, 2008. **33**(4):383–387.
316. Long, S.P., S. Humphries, and P.G. Falkowski, Photoinhibition of photosynthesis in nature. *Annual Review Plant Physiology Plant Molecular Biology*, 1994. **45**(1):633–662.
317. Milo, R., What governs the reaction center excitation wavelength of photosystems I and II? *Photosynthesis Research*, 2009. **101**(1):59–67.
318. Bard, A.J. and M.A. Fox, Artificial photosynthesis-solar splitting of water to hydrogen and oxygen. *Accounts in Chemical Research*, 1995. **28**(3):141–145.
319. Brettel, K., Electron transfer and arrangement of the redox cofactors in photosystem I. *Biochimica et Biophysica Acta (BBA): Bioenergetics*, 1997. **1318**(3):322–373.
320. Golbeck, J.H., Structure and function of photosystem I. *Annual Review of Plant Physiology and Plant Molecular Biology*, 1992. **43**(1):293–324.
321. Schatz, G.H., H. Brock, and A.R. Holzwarth, Picosecond kinetics of fluorescence and absorbance changes in photosystem II particles excited at low photon density. *Proceedings of the National Academy of Sciences*, 1987. **84**(23):8414–8418.
322. Nelson, N. and C.F. Yocum, Structure and function of photosystems I and II. *Annual Review of Plant Biology*, 2006. **57**:521–565.
323. Gust, D., T.A. Moore, and A.L. Moore, Solar fuels via artificial photosynthesis. *Accounts of Chemical Research*, 2009. **42**(12):1890–1898.

324. Falkenstrom, M., O. Johansson, and L. Hammarstrom, Light-induced charge separation in ruthenium based triads-new variations on an old theme. *Inorganica Chimica Acta*, 2007. **360**(3):741–750.

325. Flamigni, L. et al., Photoinduced processes in dyads made of a porphyrin unit and a ruthenium complex. *The Journal of Physical Chemistry B*, 1997. **101**(31):5936–5943.

326. Gust, D., T.A. Moore, and A.L. Moore, Mimicking photosynthetic solar energy transduction. *Accounts of Chemical Research*, 2000. **34**(1):40–48.

327. Meyer, T.J., Chemical approaches to artificial photosynthesis. *Accounts of Chemical Research*, 1989. **22**(5):163–170.

328. Redmore, N.P., I.V. Rubtsov, and M.J. Therien, Synthesis, electronic structure, and electron transfer dynamics of (Aryl)ethynyl-bridged donor acceptor systems. *Journal of the American Chemical Society*, 2003. **125**(29):8769–8778.

329. Wasielewski, M.R., Photoinduced electron transfer in supramolecular systems for artificial photosynthesis. *Chemical Reviews*, 1992. **92**(3):435–461.

330. Wasielewski, M.R., Self-assembly strategies for integrating light harvesting and charge separation in artificial photosynthetic systems. *Accounts of Chemical Research*, 2009. **42**(12):1910–1921.

331. Yasuyuki, N., A. Naoki, and O. Atsuhiro, Cyclic porphyrin arrays as artificial photosynthetic antenna: Synthesis and excitation energy transfer. *Chemical Society Reviews*, 2007. **36**(6):831–845.

332. Koepke, J. et al., The crystal structure of the light-harvesting complex II (B800–850) from *Rhodospirillum molischianum*. *Structure*, 1996. **4**(5):581–597.

333. Roszak, A.W. et al., Crystal structure of the RC-LH1 core complex from *Rhodopseudomonas palustris*. *Science*, 2003. **302**(5652):1969–1972.

334. Wasielewski, M.R., Energy, charge, and spin transport in molecules and self-assembled nanostructures inspired by photosynthesis. *The Journal of Organic Chemistry*, 2006. **71**(14):5051–5066.

335. Kuciauskas, D. et al., Photoinduced electron transfer in carotenoporphyrin fullerene triads: Temperature and solvent effects. *The Journal of Physical Chemistry B*, 2000. **104**(18):4307–4321.

336. Liddell, P.A. et al., Preparation and photophysical studies of porphyrin-C_{60} dyads. *Photochemistry and Photobiology*, 1994. **60**(6):537–541.

337. Guldi, D., Fullerene–porphyrin architectures; photosynthetic antenna and reaction center models. *Chemical Society Reviews*, 2002. **31**:22–36.

338. Reece, S.Y. et al., Proton-coupled electron transfer: The mechanistic underpinning for radical transport and catalysis in biology. *Philosophical Transactions of the Royal Society B: Biological Sciences*, 2006. **361**(1472):1351–1364.

339. Gust, D. et al., Mimicking the photosynthetic triplet energy-transfer relay. *Journal of the American Chemical Society*, 1993. **115**(13):5684–5691.

340. Liddell, P.A. et al., Photoinduced charge separation and charge recombination to a triplet state in a carotene porphyrin fullerene triad. *Journal of the American Chemical Society*, 1997. **119**(6):1400–1405.

341. Moore, T.A. et al., Photodriven charge separation in a carotenoporphyrin quinone triad. *Nature*, 1984. **307**:630–632.

342. Melkozernov, A.N., J. Barber, and R.E. Blankenship, Light harvesting in photosystem I supercomplexes. *Biochemistry*, 2005. **45**(2):331–345.

343. Burrell, A.K. et al., Synthetic routes to multiporphyrin arrays. *Chemical Reviews*, 2001. **101**(9):2751–2796.

344. Davila, J., A. Harriman, and L.R. Milgrom, A light-harvesting array of synthetic porphyrins. *Chemical Physics Letters*, 1987. **136**:427–430.

345. Kodis, G. et al., Efficient energy transfer and electron transfer in an artificial photosynthetic antenna reaction center complex. *The Journal of Physical Chemistry A*, 2002. **106**(10):2036–2048.

346. Li, J. et al., Synthesis and properties of star-shaped multiporphyrin phthalocyanine light-harvesting arrays. *The Journal of Organic Chemistry*, 1999. **64**(25):9090–9100.

347. Morandeira, A. et al., Ultrafast excited state dynamics of tri- and hexaporphyrin arrays. *The Journal of Physical Chemistry A*, 2004. **108**(27):5741–5751.

348. Nakamura, Y. et al., Directly linked porphyrin rings: Synthesis, characterization, and efficient excitation energy hopping. *Journal of the American Chemical Society*, 2004. **127**(1):236–246.
349. Kodis, G. et al., Energy and photoinduced electron transfer in a wheel-shaped artificial photosynthetic antenna-reaction center complex. *Journal of the American Chemical Society*, 2006. **128**(6):1818–1827.
350. Terazono, Y. et al., Multiantenna artificial photosynthetic reaction center complex. *The Journal of Physical Chemistry B*, 2009. **113**(20):7147–7155.
351. Ahrens, M.J., M.J. Fuller, and M.R. Wasielewski, Cyanated perylene-3,4-dicarboximides and perylene-3,4:9,10-bis(dicarboximide): Facile chromophoric oxidants for organic photonics and electronics. *Chemistry of Materials*, 2003. **15**(14):2684–2686.
352. Ahrens, M.J. et al., Self-assembly of supramolecular light-harvesting arrays from covalent multi-chromophore perylene-3,4:9,10-bis(dicarboximide) building blocks. *Journal of the American Chemical Society*, 2004. **126**(26):8284–8294.
353. Schenning, A.P.H.J. et al., Photoinduced electron transfer in hydrogen-bonded oligo(p-phenylene vinylene) perylene bisimide chiral assemblies. *Journal of the American Chemical Society*, 2002. **124**(35):10252–10253.
354. Tomizaki, K.-Y. et al., Synthesis and photophysical properties of light-harvesting arrays comprised of a porphyrin bearing multiple perylene-monoimide accessory pigments. *The Journal of Organic Chemistry*, 2002. **67**(18):6519–6534.
355. Würthner, F., Perylene bisimide dyes as versatile building blocks for functional supramolecular architectures. *Chemical Communications*, 2004. **14**:1564–1579.
356. Green, M.A., Third generation photovoltaics: Ultra-high conversion efficiency at low cost. *Progress in Photovoltaics*, 2001. **9**(2):123–135.
357. Luque, A., A. Marti, and A.J. Nozik, Solar cells based on quantum dots: Multiple exciton generation and intermediate bands. *MRS Bulletin*, 2007. **32**(3):236–241.
358. Nozik, A.J. and R. Memming, Physical chemistry of semiconductor–liquid interfaces. *Journal of Physical Chemistry*, 1996. **100**(31):13061–13078.
359. Nirmal, M. and L. Brus, Luminescence photophysics in semiconductor nanocrystals. *Accounts of Chemical Research*, 1999. **32**(5):407–414.
360. Nozik, A.J., Spectroscopy and hot electron relaxation dynamics in semiconductor quantum wells and quantum dots. *Annual Review of Physical Chemistry*, 2001. **52**(1):193–231.
361. Bae, Y., N. Myung, and A.J. Bard, Electrochemistry and electrogenerated chemiluminescence of CdTe nanoparticles. *Nano Letters*, 2004. **4**(6):1153–1161.
362. Poznyak, S.K. et al., Size-dependent electrochemical behavior of thiol-capped CdTe nanocrystals in aqueous solution. *Journal of Physical Chemistry B*, 2005. **109**(3):1094–1100.
363. Rajh, T., O.I. Micic, and A.J. Nozik, Synthesis and characterization of surface-modified colloidal cadmium telluride quantum dots. *Journal of Physical Chemistry*, 1993. **97**(46):11999–12003.
364. Zhang, H. et al., The influence of carboxyl groups on the photoluminescence of mercaptocarboxylic acid-stabilized CdTe nanoparticles. *Journal of Physical Chemistry B*, 2003. **107**:8–13.
365. Zhang, Y.-H. et al., The influence of ligands on the preparation and optical properties of water-soluble CdTe quantum dots. *Applied Surface Science*, 2009. **255**(9):4747–4753.
366. Duonghong, D., J. Ramsden, and M. Graetzel, Dynamics of interfacial electron-transfer processes in colloidal semiconductor systems. *Journal of the American Chemical Society*, 1982. **104**(11):2977–2985.
367. Freeman, R. and I. Willner, NAD$^+$/NADH-sensitive quantum dots: Applications to probe NAD+-dependent enzymes and to sense the RDX explosive. *Nano Letters*, 2008. **9**(1):322–326.
368. Korgel, B.A. and H.G. Monbouquette, Quantum confinement effects enable photocatalyzed nitrate reduction at neutral pH using CdS nanocrystals. *The Journal of Physical Chemistry B*, 1997. **101**(25):5010–5017.
369. Warrier, M. et al., Photocatalytic reduction of aromatic azides to amines using CdS and CdSe nanoparticles. *Photochemical and Photobiological Sciences*, 2004. **3**:859–863.
370. Nozik, A., Multiple exciton generation in semiconductor quantum dots. *Chemical Physics Letters*, 2008. **457**(1–3):3–11.

371. Klimov, V.I., Mechanisms for photogeneration and recombination of multiexcitons in semiconductor nanocrystals: Implications for lasing and solar energy conversion. *The Journal of Physical Chemistry B*, 2006. **110**(34):16827–16845.

372. Schaller, R.D. and V.I. Klimov, High efficiency carrier multiplication in PbSe nanocrystals: Implications for solar energy conversion. *Physical Review Letters*, 2004. **92**(18):186601-1–186601-4.

373. Schaller, R.D., M.A. Petruska, and V.I. Klimov, Effect of electronic structure on carrier multiplication efficiency: Comparative study of PbSe and CdSe nanocrystals. *Applied Physics Letters*, 2005. **87**(25):253102-1–253102-3.

374. Barber, J., Crystal structure of the oxygen-evolving complex of photosystem II. *Inorganic Chemistry*, 2008. **47**(6):1700–1710.

375. Guskov, A. et al., Cyanobacterial photosystem II at 2.9-A resolution and the role of quinones, lipids, channels and chloride. *Nature Structural Molecular Biology*, 2009. **16**(3):334–342.

376. Yano, J. et al., High-resolution structure of the photosynthetic Mn_4Ca catalyst from x-ray spectroscopy. *Philosophical Transactions of the Royal Society B: Biological Sciences*, 2008. **363**(1494):1139–1147.

377. Yano, J. et al., Where water is oxidized to dioxygen: Structure of the photosynthetic Mn_4Ca cluster. *Science*, 2006. **314**(5800):821–825.

378. Zouni, A. et al., Crystal structure of photosystem II from *Synechococcus elongatus* at 3.8 A resolution. *Nature*, 2001. **409**(6821):739–743.

379. Iwata, S. and J. Barber, Structure of photosystem II and molecular architecture of the oxygen-evolving centre. *Current Opinion in Structural Biology*, 2004. **14**(4):447–453.

380. Siegbahn, P.E.M., Structures and energetics for O_2 formation in photosystem II. *Accounts of Chemical Research*, 2009. **42**(12):1871–1880.

381. Concepcion, J.J. et al., Making oxygen with ruthenium complexes. *Accounts of Chemical Research*, 2009. **42**(12):1954–1965.

382. Ferreira, K.N. et al., Architecture of the photosynthetic oxygen-evolving center. *Science*, 2004. **303**(5665):1831–1838.

383. Yano, J. et al., X-ray damage to the Mn4Ca complex in single crystals of photosystem II: A case study for metalloprotein crystallography. *Proceedings of the National Academy of Sciences*, 2005. **102**(34):12047–12052.

384. Lubitz, W., E.J. Reijerse, and J. Messinger, Solar water-splitting into H_2 and O_2: Design principles of photosystem II and hydrogenases. *Energy and Environmental Science*, 2008. **1**(1):15–31.

385. Carrell, T.G., S. Cohen, and G.C. Dismukes, Oxidative catalysis by $Mn_4O_4^{6+}$ cubane complexes. *Journal of Molecular Catalysis A: Chemical*, 2002. **187**(1):3–15.

386. Ruettinger, W.F., C. Campana, and G.C. Dismukes, Synthesis and characterization of $Mn_4O_4L_6$ complexes with cubane-like core structure: A new class of models of the active site of the photosynthetic water oxidase. *Journal of the American Chemical Society*, 1997. **119**(28):6670–6671.

387. Manchanda, R., G.W. Brudvig, and R.H. Crabtree, High-valent oxomanganese clusters: Structural and mechanistic work relevant to the oxygen-evolving center in photosystem II. *Coordination Chemistry Reviews*, 1995. **144**:1–38.

388. Brimblecombe, R. et al., Electrochemical investigation of Mn_4O_4-cubane water-oxidizing clusters. *Physical Chemistry Chemical Physics*, 2009. **11**:6441–6449.

389. Wu, J.-Z. et al., Tuning the photoinduced O_2-evolving reactivity of $Mn_4O_4^{7+}$, $Mn_4O_4^{6+}$, and $Mn_4O_3(OH)^{6+}$ manganese oxo cubane complexes. *Inorganic Chemistry*, 2005. **45**(1):189–195.

390. Haumann, M. et al., Photosynthetic O_2 formation tracked by time-resolved x-ray experiments. *Science*, 2005. **310**(5750):1019–1021.

391. Westphal, K. et al., Concerted hydrogen-atom abstraction in photosynthetic water oxidation. *Current Opinion in Plant Biology*, 2000. **3**(3):236–242.

392. Carrell, T.G. et al., Transition from hydrogen atom to hydride abstraction by $Mn_4O_4(O_2PPh_2)_6$ versus $[Mn_4O_4(O_2PPh_2)_6]^+$: OH bond dissociation energies and the formation of $Mn_4O_3(OH)(O_2PPh_2)_6$. *Inorganic Chemistry*, 2003. **42**(9):2849–2858.

393. Poulsen, A.K., A. Rompel, and C.J. McKenzie, Water oxidation catalyzed by a dinuclear Mn complex: A functional model for the oxygen-evolving center of photosystem II13. *Angewandte Chemie International Edition*, 2005. **44**(42):6916–6920.

394. Narita, K. et al., Characterization and activity analysis of catalytic water oxidation induced by hybridization of [(OH$_2$)(terpy)Mn(μ-O)$_2$Mn(terpy)(OH$_2$)]$^{3+}$ and clay compounds. *The Journal of Physical Chemistry B*, 2006. **110**(46):23107–23114.

395. Brimblecombe, R. et al., Sustained water oxidation photocatalysis by a bioinspired manganese cluster. *Angewandte Chemie International Edition*, 2008. **120**(38):7445–7448.

396. Service, R.A., New trick for splitting water with sunlight. *Science*, 2009. **325**:1200–1201.

397. Binstead, R.A. et al., Mechanism of water oxidation by the μ-oxo dimer [(bpy)$_2$(H$_2$O) RuIIIORuIII(OH$_2$)(bpy)$_2$]$^{4+}$. *Journal of the American Chemical Society*, 2000. **122**(35):8464–8473.

398. Kohl, S.W. et al., Consecutive thermal H$_2$ and light-induced O$_2$ evolution from water promoted by a metal complex. *Science*, 2009. **324**(5923):74–77.

399. Sala, X. et al., Molecular catalysts that oxidize water to dioxygen. *Angewandte Chemie International Edition*, 2009. **48**(16):2842–2852.

400. Liu, F. et al., Mechanisms of water oxidation from the blue dimer to photosystem II. *Inorganic Chemistry*, 2008. **47**(6):1727–1752.

401. Yang, X. and M.-H. Baik, Electronic structure of the water-oxidation catalyst [(bpy)2(OHx) RuORu(OHy)(bpy)2]z+: Weak coupling between the metal centers is preferred over strong coupling. *Journal of the American Chemical Society*, 2004. **126**(41):13222–13223.

402. Yang, X. and M.-H. Baik, cis,cis-[(bpy)2RuVO]2O4+ catalyzes water oxidation formally via in situ generation of radicaloid RuIV. *Journal of the American Chemical Society*, 2006. **128**(23):7476–7485.

403. Concepcion, J.J. et al., One site is enough. Catalytic water oxidation by [Ru(tpy)(bpm)(OH$_2$)]$^{2+}$ and [Ru(tpy)(bpz)(OH$_2$)]$^{2+}$. *Journal of the American Chemical Society*, 2008. **130**(49): 16462–16463.

404. Masaoka, S. and K. Sakai, Clear evidence showing the robustness of a highly active oxygen-evolving mononuclear ruthenium complex with an aqua ligand. *Chemistry Letters*, 2009. **38**(2):182–183.

405. Tseng, H.-W. et al., Mononuclear ruthenium(II) complexes that catalyze water oxidation. *Inorganic Chemistry*, 2008. **47**(24):11763–11773.

406. Concepcion, J.J. et al., Mediator-assisted water oxidation by the ruthenium blue dimer cis,cis-[(bpy)$_2$(H$_2$O)RuORu(OH$_2$)(bpy)$_2$]$^{4+}$. *Proceedings of the National Academy of Sciences*, 2008. **105**(46):17632–17635.

407. Jurss, J.W. et al., Surface catalysis of water oxidation by the blue ruthenium dimer. *Inorganic Chemistry*, 2010. **49**(9):3980–3982.

408. Treadway, J.A., J.A. Moss, and T.J. Meyer, Visible region photooxidation on TiO$_2$ with a chromophore catalyst molecular assembly. *Inorganic Chemistry*, 1999. **38**(20):4386–4387.

409. Youngblood, W.J. et al., Photoassisted overall water splitting in a visible light-absorbing dye-sensitized photoelectrochemical cell. *Journal of the American Chemical Society*, 2009. **131**(3):926–927.

410. Brunschwig, B.S. et al., Mechanisms of water oxidation to oxygen-cobalt(IV) as an intermediate in the aquocobalt(II)-catalyzed reaction. *Journal of the American Chemical Society*, 1983. **105**(14):4832–4833.

411. Shafirovich, V.Y., N.K. Khannanov, and V.V. Strelets, Chemical and light-induced catalytic water oxidation. *Nouveau Journal De Chimie—New Journal of Chemistry*, 1980. **4**(2):81–84.

412. Kanan, M.W. and D.G. Nocera, In situ formation of an oxygen-evolving catalyst in neutral water containing phosphate and Co^{2+}. *Science*, 2008. **321**(5892):1072–1075.

413. Kanan, M.W., Y. Surendranath, and D.G. Nocera, Cobalt-phosphate oxygen-evolving compound. *Chemical Society Reviews*, 2009. **38**(1):109–114.

414. Risch, M. et al., Cobalt oxo core of a water-oxidizing catalyst film. *Journal of the American Chemical Society*, 2009. **131**(20):6936–6937.

415. Le Cloirec, A. et al., A di-iron dithiolate possessing structural elements of the carbonyl/cyanide Sub-site of the H-centre of Fe-only hydrogenase. *Chemical Communications*, 1999. **22**:2285–2286.

416. Lyon, E. et al., Carbon monoxide and cyanide ligands in a classical organometallic complex model for Fe-only hydrogenase. *Angewandte Chemie International Edition*, 1999. **38**(21): 3178–3180.

417. Schmidt, M., S.M. Contakes, and T.B. Rauchfuss, First generation analogues of the binuclear site in the Fe-only hydrogenases: $Fe_2(\mu\text{-SR})_2(CO)_4(CN)_2^{2-}$. *Journal of the American Chemical Society*, 1999. **121**(41):9736–9737.

418. Li, H. and T.B. Rauchfuss, Iron carbonyl sulfides, formaldehyde, and amines condense to give the proposed azadithiolate cofactor of the Fe-only hydrogenases. *Journal of the American Chemical Society*, 2002. **124**(5):726–727.

419. Song, L.-C. et al., Novel single and double diiron oxadithiolates as models for the active site of [Fe]-only hydrogenases. *Organometallics*, 2004. **23**(13):3082–3084.

420. Tard, C. et al., Synthesis of the H-cluster framework of iron-only hydrogenase. *Nature*, 2005. **433**(7026):610–613.

421. Tard, C. and C.J. Pickett, Structural and functional analogues of the active sites of the [Fe]-, [NiFe]-, and [FeFe]-hydrogenases. *Chemical Reviews*, 2009. **109**(6):2245–2274.

422. Gloaguen, F., J.D. Lawrence, and T.B. Rauchfuss, Biomimetic hydrogen evolution catalyzed by an iron carbonyl thiolate. *Journal of the American Chemical Society*, 2001. **123**(38):9476–9477.

423. Gloaguen, F. et al., Bimetallic carbonyl thiolates as functional models for Fe-only hydrogenases. *Inorganic Chemistry*, 2002. **41**(25):6573–6582.

424. Ott, S. et al., A biomimetic pathway for hydrogen evolution from a model of the iron hydrogenase active site. *Angewandte Chemie International Edition*, 2004. **43**(8):1006–1009.

425. Capon, J.-F. et al., Electrochemical and theoretical investigations of the reduction of $[Fe_2(CO)$ $5L$ {m-SCH_2XCH_2S}] complexes related to [FeFe] hydrogenase. *New Journal of Chemistry*, 2007. **31**(12):2052–2064.

426. Capon, J.-F. et al., Catalysis of the electrochemical H_2 evolution by di-iron sub-site models. *Coordination Chemistry Reviews*, 2005. **249**(15–16):1664–1676.

427. Zhao, X. et al., H/D exchange reactions in dinuclear iron thiolates as activity assay models of Fe-H_2ase. *Journal of the American Chemical Society*, 2001. **123**(39):9710–9711.

428. Kluwer, A.M. et al., Self-assembled biomimetic [2Fe2S]-hydrogenase-based photocatalyst for molecular hydrogen evolution. *Proceedings of the National Academy of Sciences*, 2009. **106**(26):10460–10465.

429. Evans, D. and C. Pickett, Chemistry and the hydrogenases. *Chemical Society Reviews*, 2003. **32**:268–275.

430. Georgakaki, I.P. et al., Fundamental properties of small molecule models of Fe-only hydrogenase: Computations relative to the definition of an Entatic State in the active site. *Coordination Chemistry Reviews*, 2003. **238–239**:255–266.

431. Chong, I.P.G., R. Mejia-Rodriguez, J. Sanabria-Chinchilla, M.P. Soriaga, and M.Y. Darensbourg, Electrocatalysis of hydrogen production by active site analogues of the iron hydrogenase enzyme: Structure/function relationships. *Dalton Transactions*, 2003. **21**:4158–4163.

432. Colpas, G.J., R.O. Day, and M.J. Maroney, Synthesis and structure of a trinickeliron cluster featuring single and double thiolato bridges. *Inorganic Chemistry*, 1992. **31**(24):5053–5055.

433. Glaser, T. et al., Spin-dependent delocalization in three isostructural complexes $[LFeNiFeL]^{2+/3+/4+}$ (L=1,4,7-(4-tert-Butyl-2-mercaptobenzyl)-1,4,7-triazacyclononane). *Inorganic Chemistry*, 1999. **38**(4):722–732.

434. Mills, D.K. et al., Applications of the N_2S_2 ligand, N,N'-bis(mercaptoethyl)-1,5-diazacyclooctane (BME-DACO), toward the formation of bi- and heterometallics: [(BME-DACO)Fe]2 and [(BME-DACO)NiFeCl2]2. *Journal of the American Chemical Society*, 1991. **113**(4):1421–1423.

435. Steinfeld, G. and B. Kersting, Characterization of a triply thiolate-bridged Ni-Fe amine-thiolate complex: Insights into the electronic structure of the active site of [NiFe] hydrogenase. *Chemical Communications*, 2000. **3**:205–206.

436. Verhagen, J.A.W. et al., Synthesis and characterisation of new nickel-iron complexes with an S_4 coordination environment around the nickel centre. *European Journal of Inorganic Chemistry*, 2003. **21**:3968–3974.

437. Efros, L.L. et al., Towards a functional model of hydrogenase: Electrocatalytic reduction of protons to dihydrogen by a nickel macrocyclic complex. *Inorganic Chemistry*, 1992. **31**(9):1722–1724.

438. Barber, D.E. et al., Silane alcoholysis by a nickel(II) complex in a N, O, S ligand environment. *Inorganic Chemistry*, 1992. **31**(22):4709–4711.

439. Lu, Z. et al., Deprotonated thioamides as thiolate S-donor ligands with a high tendency to avoid M-S-M bridge formation: Crystal and molecular structure of bis(2-hydroxy-5-methylacetophenone N,N-dimethylthiosemicarbazonato)dinickel. *Inorganic Chemistry*, 1993. **32**(19):3991–3994.

440. Zimmer, M. et al., Functional-modeling of NiFe hydrogenases: A nickel-complex in an N,O,S environment. *Angewandte Chemie International Edition*, 1991. **30**:193–194.

441. Bouwman, E. and J. Reedijk, Structural and functional models related to the nickel hydrogenases. *Coordination Chemistry Reviews*, 2005. **249**(15–16):1555–1581.

442. Dubois, R. and D. Dubois, Development of molecular electrocatalysts for CO_2 reduction and H_2 production/oxidation. *Accounts of Chemical Research*, 2009. **42**(12):1974–1982.

443. Curtis, C.J. et al., [Ni(Et$_2$PCH$_2$NMeCH$_2$PEt$_2$)$_2$]$^{2+}$ as a functional model for hydrogenases. *Inorganic Chemistry*, 2002. **42**(1):216–227.

444. Dempsey, J.L. et al., Hydrogen evolution catalyzed by cobaloximes. *Accounts of Chemical Research*, 2009. **42**(12):1995–2004.

445. Mealli, C. and T.B. Rauchfuss, Models for the hydrogenases put the focus where it should be: Hydrogen. *Angewandte Chemie International Edition*, 2007. **46**(47):8942–8944.

446. Allakhverdiev, S.I. et al., Hydrogen photoproduction by use of photosynthetic organisms and biomimetic systems. *Photochemical and Photobiological Sciences*, 2009. **8**(2):148–156.

447. Cracknell, J.A., K.A. Vincent, and F.A. Armstrong, Enzymes as working or inspirational electrocatalysts for fuel cells and electrolysis. *Chemical Reviews*, 2008. **108**(7):2439–2461.

448. Hall, D.O. et al., Photolysis of water for H_2 production with the use of biological and artificial catalysts. *Philosophical Transactions of the Royal Society of London Series A: Mathematical Physical and Engineering Sciences*, 1980. **295**(1414): 473–476.

449. Klibanov, A.M., N.O. Kaplan, and M.D. Kamen, A rationale for stabilization of oxygen-labile enzymes: Application to a clostridial hydrogenase. *Proceedings of the National Academy of Sciences*, 1978. **75**(8):3640–3643.

450. Klibanov, M., Biotechnological potential of the enzyme hydrogenase. *Process Biochemistry*, 1983. **18**(4):13–16.

451. Bae, S. et al., Photoanodic and cathodic role of anodized tubular titania in light-sensitized enzymatic hydrogen production. *Journal of Power Sources*, 2008. **185**(1):439–444.

452. Hambourger, M. et al., [FeFe]-hydrogenase-catalyzed H_2 production in a photoelectrochemical biofuel cell. *Journal of the American Chemical Society*, 2008. **130**(6):2015–2022.

453. Hambourger, M. et al., Solar energy conversion in a photoelectrochemical biofuel cell. *Dalton Transactions*, 2009. **45**:9979–9989.

454. Reisner, E. et al., Visible light-driven H_2 production by hydrogenases attached to dye-sensitized TiO_2 nanoparticles. *Journal of the American Chemical Society*, 2009. **131**(51):18457–18466.

455. Vincent, K.A. et al., Electricity from low-level H_2 in still air: An ultimate test for an oxygen tolerant hydrogenase. *Chemical Communications*, 2006. **48**:5033–5035.

456. de Groot, H., Integration of catalysis with storage for the design of multi-electron photochemistry devices for solar fuel. *Applied Magnetic Resonance*, 2010. **37**(1):497–503.

457. Alonso-Lomillo, M.A. et al., Hydrogenase-coated carbon nanotubes for efficient H_2 oxidation. *Nano Letters*, 2007. **7**(6):1603–1608.

458. Brown, K. et al., Controlled self assembly of hydrogenase-CdTe nanocrystal hybrids for solar hydrogen production. *Journal of the American Chemical Society*, 2010. **132**(28):9672–9680.

459. Davidson, V.L., Protein control of true, gated, and coupled electron transfer reactions. *Accounts of Chemical Research*, 2008. **41**(6):730–738.

460. Nocek, J.M. et al., Theory and practice of electron transfer within protein-protein complexes: Application to the multidomain binding of cytochrome c by cytochrome c peroxidase. *Chemical Reviews*, 1996. **96**(7):2459–2489.

461. Schreiber, G., G. Haran, and H.X. Zhou, Fundamental aspects of protein-protein association kinetics. *Chemical Reviews*, 2009. **109**(3):839–860.

462. Cuendet, P. et al., Light induced H_2 evolution in a hydrogenase-TiO_2 particle system by direct electron transfer or via rhodium complexes. *Biochimie*, 1986. **68**(1):217–221.

463. Pedroni, P. et al., The hydrogenase from the hyperthermophilic archaeon *Pyrococcus furiosus*: From basic research to possible future applications. *International Journal of Hydrogen Energy*, 1996. **21**(10):853–858.

464. Selvaggi, A. et al., In vitro hydrogen photoproduction using *Pyrococcus furiosus* sulfhydrogenase and TiO_2. *Journal of Photochemistry and Photobiology*, 1999. **125**(1–3):107–112.

465. Shumilin, I.A. et al., Photogeneration of NADH under coupled action of CdS semiconductor and hydrogenase from *Alcaligenes eutrophus* without exogenous mediators. *FEBS Letters*, 1992. **306**(2–3):125–128.

466. Reisner, E., J.C. Fontecilla-Camps, and F.A. Armstrong, Catalytic electrochemistry of a [NiFeSe]-hydrogenase on TiO_2 and demonstration of its suitability for visible-light driven H_2 production. *Chemical Communications*, 2009. **5**:550–552.

467. Ai, X., et al. Photophysics of (CdSe) ZnS colloidal quantum dots in an aqueous environment stabilized with amino acids and genetically-modified proteins. *Photochemical and Photobiological Sciences*, 2007. **6**(9):1027–1033.

468. Hambourger, M. et al., Enzyme-assisted reforming of glucose to hydrogen in a photoelectrochemical cell. *Photochemistry and Photobiology*, 2009. **81**(4):1015–1020.

469. Hambourger, M. et al., Parameters affecting the chemical work output of a hybrid photoelectrochemical biofuel cell. *Photochemical and Photobiological Sciences*, 2007. **6**:431–437.

470. Clarke, T.M. and J.R. Durrant, Charge photogeneration in organic solar cells. *Chemical Reviews*, 2010. Epub Ahead of Print.

471. Peter, L., Sticky electrons, transport and interfacial transfer of electrons in the dye-sensitized solar cell. *Accounts of Chemical Research*, 2009. **42**(11):1839–1847.

472. Bae, B. et al., Performance evaluation of passive DMFC single cells. *Journal of Power Sources*, 2006. **158**(2):1256–1261.

473. McDonald, T.J. et al., Wiring-up hydrogenase with single-walled carbon nanotubes. *Nano Letters*, 2008. **7**(6):3528–3534.

474. Morris, A.J., G.J. Meyer, and E. Fujita, Molecular approaches to the photocatalytic reduction of carbon dioxide for solar fuels. *Accounts of Chemical Research*, 2009. **42**(12):1983–1994.

475. Creutz, C. et al., Thermodynamics and kinetics of carbon dioxide binding to two stereoisomers of a cobalt(I) macrocycle in aqueous solution. *Journal of the American Chemical Society*, 1991. **113**(9):3361–3371.

476. Fujita, E. et al., Carbon dioxide activation by cobalt(I) macrocycles: Factors affecting carbon dioxide and carbon monoxide binding. *Journal of the American Chemical Society*, 1991. **113**(1):343–353.

477. Fujita, E. et al., High electrocatalytic activity of RRSS-[NiIIHTIM]$(ClO_4)_2$ and [NiIIDMC]$(ClO_4)_2$ for carbon dioxide reduction (HTIM = 2,3,9,10-tetramethyl-1,4,8,11-tetraazacyclotetradecane, DMC = C-meso-5,12-dimethyl-1,4,8,11-tetraazacyclotetradecane). *Inorganic Chemistry*, 1994. **33**(21):4627–4628.

478. Kelly, C.A. et al., The thermodynamics and kinetics of CO_2 and H^+ binding to Ni(cyclam)$^+$ in aqueous solution. *Journal of the American Chemical Society*, 1995. **117**(17):4911–4919.

479. Gholamkhass, B. et al., Architecture of supramolecular metal complexes for photocatalytic CO_2 reduction: Ruthenium, rhenium bi- and tetranuclear complexes. *Inorganic Chemistry*, 2005. **44**(7):2326–2336.

480. Behar, D. et al., Cobalt porphyrin catalyzed reduction of CO_2. Radiation chemical, photochemical, and electrochemical studies. *The Journal of Physical Chemistry A*, 1998. **102**(17):2870–2877.

481. Grodkowski, J. et al., Iron porphyrin-catalyzed reduction of CO_2. Photochemical and radiation chemical studies. *The Journal of Physical Chemistry A*, 1997. **101**(3):248–254.
482. Grodkowski, J. et al., Reduction of cobalt and iron phthalocyanines and the role of the reduced species in catalyzed photoreduction of CO_2. *The Journal of Physical Chemistry A*, 2000. **104**(48):11332–11339.
483. Benson, E.E. et al., Electrocatalytic and homogeneous approaches to conversion of CO_2 to liquid fuels. *Chemical Society Reviews*, 2009. **38**(1):89–99.
484. Happe, T. and A. Kaminski, Differential regulation of the Fe-hydrogenase during anaerobic adaptation in the green alga *Chlamydomonas reinhardtii*. *European Journal of Biochemistry*, 2002. **269**(3):1022–1032.
485. Ogata, H. et al., Activation process of [NiFe] hydrogenase elucidated by high-resolution x-ray analyses: Conversion of the ready to the unready state. *Structure*, 2005. **13**(11):1635–1642.
486. Carrasco, C.D., J.A. Buettner, and J.W. Golden, Programmed DNA rearrangement of a cyanobacterial hupL gene in heterocysts. *Proceedings of the National Academy of Sciences*, 1995. **92**(3):791–795.
487. Loll, B. et al., Towards complete cofactor arrangement in the 3.0 A resolution structure of photosystem II. *Nature*, 2005. **438**(7070):1040–1044.
488. Brimblecombe, R. et al., Molecular water-oxidation catalysts for photoelectrochemical cells. *Dalton Transactions*, 2009. **43**:9374–9384.

12

Fermentative Biofuels: Prospects of Practical Application

David B. Levin, Carlo Carere, Umesh Ramachandran, Tom Rydzak, and Jessica Saunders

CONTENTS

12.1 Overview of Fermentative Biofuels

12.1.1 Rationale for Biofuels

In June, 2008, the Goldman Sachs Group predicted that crude oil prices could reach U.S. $200 per barrel by 2012. Within weeks of this assessment, the price of crude oil exceeded U.S. $140 per barrel, raising concerns that the unprecedented price of U.S. $200 per barrel may be reached well before 2012. Although the current economic downturn has resulted in lower crude oil prices, the global economy is expected to resurge in 2–3 years, and, with this, oil prices will again increase. Reserves of easy-to-reach, inexpensive oil (oil that was on land, near the surface, under pressure, light and "sweet"—meaning low sulfur content—and therefore easy to refine) are more or less depleted. The remaining oil reserves are offshore in deep waters, or in oil-sands, and of lower quality (i.e., high sulfur content). It therefore takes ever more money and energy to extract, refine, and transport. Under these conditions, the rate of production inevitably drops. Furthermore, all oil fields eventually reach a point where they become economically, and energetically, no longer viable (Kleykamp 2008; Rhodes 2008).

Increasingly higher costs of crude oil prices will have devastating effects on the world economy, which makes the development of alternative biofuels more critical than ever before. Despite large uncertainties about the quantity of oil that remains and its production path, the necessity for replacement fuels to represent a considerable portion of liquid fuel supply is likely to fall within a 22 year period, which is much earlier than previous estimates. The timeline of this transition suggests that a minimum of 10 million barrels per day of alternative fuels will be needed within a decade of the peak in production of conventional crude oil (Kaufman and Shiers 2008).

There is, therefore, an urgent need to develop viable, renewable, and sustainable energy systems that can displace global dependence on fossil fuel sources of energy. Biofuels such as ethanol can be utilized as transportation fuels with little change to current technologies and have the potential to improve sustainability and reduce greenhouse gas (GHG) emissions in the short term. Bioethanol, the most widely used liquid biofuel, is produced by converting sugars—directly from crops like sugarcane or sugar beets, indirectly through starch from corn, wheat, potatoes, or cassava, or through cellulose from biomass—into ethanol via fermentation followed by distillation. The worldwide production of bioethanol was 16.215 billion gallons (61.6 billion L) in 2008 (Market Research Analyst 2008). The largest producers in the world are the United States (38%), Brazil (31%), and China (6.6%) (Market Research Analyst 2008). Today, 18% of the U.S. corn crop is converted into 4.5 billion gallons (17.1 billion L) of ethanol, replacing only 1% of U.S. petroleum consumption (Patzek 2004). The U.S. government has set a production target of 36 billion gallons (136.8 billion L) of ethanol per year by 2022, with the goal of reducing America's gasoline consumption by 20% in 10 years. Over 40% of this ethanol (16 billion gallons = 60.8 billion L) is required to be made from cellulosic feedstocks (RFA, 2008a). Canada's current ethanol production is approximately 0.874 billion L/year and will double to more than 1.6 billion L/year by 2012 (Market Research Analyst 2008).

12.1.2 First-Generation Biofuels

Ethyl alcohol, also known as ethanol or grain alcohol, is a flammable, clear, and colorless liquid that melts at $-114.1°C$ and has a boiling point of $78.5°C$. Ethanol is widely used as a gasoline additive providing a cleaner-burning fuel with a higher octane number, and is expected to be one of the major renewable fuels for the transportation sector over the next 20 years. The largest ethanol-producing industries, representing more than 95% of world production, are located in Brazil and the United States. Ethanol production by fermentation has been utilized by human society for millennia. Traditional fermentation feedstock contains hexose sugars (sugarcane or sugar beet) or substrates from which sugars can be derived, such as starch or cellulose. There are numerous advantages of using ethanol as an energy carrier in today's transportation economy. The most significant is the potential for ethanol to serve as an essential renewable and domestic source of fuel acting to decrease the dependence on foreign oils. The addition of ethanol to gasoline also increases the fuel octane rating and results in cleaner and more complete combustion, offering mitigation to the growing global concerns regarding GHG emissions and climate change. Emissions from the combustion of 10% ethanol-blended gasoline have 30% less carbon monoxide (CO), 10% less carbon dioxide (CO_2), and 7% less NO_x/SO_x (Ethanol clean air facts, American coalition for ethanol 2004).

The present drivers for the development of the world's biofuel economy have little to do with the economic gain, as these industries do not, as of now, exist without heavy government subsidies. Subsidies per liter of ethanol are 60 times greater than those of gasoline (Pimentel and Marklein, 2009) and total more than $6 billion per year for U.S. corn ethanol (Koplow 2006). Both the European Union (EU) and the United States have made an economically inefficient choice, meaning that they are using significant government resources, with the introduction of biofuels in their economies (Slingerland and van Geuns 2005). Biofuels have been introduced not for economic reasons but for answering the pressing concerns raised by heavy reliance on fossil fuels. Global GHG emissions, fuel security, volatility in oil pricing, and the development of local, rural economies are used to justify an aggressive pursuit of biofuels development. Economic viability aside, first (1st)-generation technologies, which form the backbone of the current bioethanol and biodiesel industry, have several conspicuous limitations. Most notable of these drawbacks is their total reliance on cultivated biomass and associated issues relating to land-use patterns and the diversion of feedstock to biofuels production without disrupting food supply.

Many researchers have cited the benefits of biofuels in terms of their capacity to reduce GHG emissions. Plant biomass offers an attractive alternative (to petrochemical fuels), bypassing the need for fossil resources in chemical production and balancing the time constants of feedstock production and CO_2 fixation (van Maris et al. 2006). Biofuels are described as carbon neutral because of their ability to fix CO_2 within the same time frame as combustion liberates this gas. The rapid rise in global CO_2 levels has come about because of this imbalance between the fixation of CO_2 and the release; the former taking 350 million years to accumulate and the latter occurring during the last century. However, some argue that biofuels will cause dramatic changes in land-use patterns that could offset any CO_2 savings derived from the utilization of biomass. Searchinger et al. (2008) contend that land-use changes will cause a net increase in GHG emissions, doubling GHG emissions over 30 years and causing a net increase for 167 years. Carbon emissions occur as farmers worldwide respond to higher prices and convert forest and grassland to new cropland to replace the grain (or cropland) diverted to biofuels. There is a cost, in terms of carbon storage and sequestration, to diverting previously unused land or appropriating agricultural land to grow bioenergy crops. Searchinger et al. (2008) suggest that the only way to achieve carbon neutrality is to exclusively utilize agricultural wastes.

First-generation technologies pose environmental concerns both in regard to possible new GHG emissions related to a change in land-use patterns as well as suffering uncertainty in regard to possible GHG savings. Farrell (2006) summarized the findings of six well-to-wheel studies and surmised, "the impact of a switch from gasoline to ethanol has an ambiguous effect on GHG emissions, with the reported values ranging from a 20% increase to a decrease of 32%." Most researchers agree that a net decrease in GHG emissions (most suggest a modest decrease of 13%–18%) is likely to be observed with the incorporation of ethanol into the fuel supply (Farrell 2006; Kim and Dale 2004). Although ethanol, compared to petroleum, releases fewer GHGs upon combustion, the magnitude of the total GHG savings must be viewed in light of the primary sources of emission in the ethanol life cycle, which lie in agricultural practices (34%–44%) and petroleum inputs (45%–80%) (Farrell 2006).

Ethanol production, especially from corn, has additional environmental drawbacks centered around resource consumption and agricultural practice. It is estimated that a 50 million gallon per annum ethanol factory consumes 500 gallons of water a minute (*Economist* 2008). More than 1700 gall of water are required to produce 1 gallon of ethanol (Pimentel and Patzek 2007) and a total of 12 L of wastewater must be removed per liter of ethanol produced, and the sewage effluent has to be disposed of at an energy, economic, and environmental cost (Pimentel and Marklein 2009). Agriculturally, corn causes soil erosion, uses more nitrogen fertilizer than any other crop grown (NAS 2003), has significant phosphorus requirements (USDA 2007), and uses more insecticides (McLaughlin and Walsh 1998) and herbicides (Patzek 2004) than any other crop grown. Compounding these environmental issues is the reality of ethanol's lower energy density. Ethanol delivers only two-thirds the energy that petroleum does and therefore is required in greater quantity (Srinivasan 2009). From an environmental perspective, the veracity of incorporating biofuels into the existing fuel supply has to be questioned; the supposed gains tempered by the probable losses.

Diverting food crops for use as substrates in biofuel production has caused unceasing controversy since the inception of the biofuels movement. In 2006, 20% of the U.S. corn crop was diverted to fuel ethanol production. It is estimated that when the fuel ethanol plants under construction come online, increasing the present plant number from 118 to just more than 200, more than 50% of the U.S. corn crop will be consumed in the generation of bioenergy. Most authorities agree that biofuels have contributed to rising food prices. Much uncertainty exists in this regard and estimates of biofuel contribution range from 15% to 25% (Sims et al. 2008). The use of corn for ethanol production has increased the prices of U.S. beef, chicken, pork, eggs, breads, cereals, and milk by 10%–20% (Brown 2008). The effect of rising prices has, for instance, been seen in Mexico where thousands marched in protest of rising tortilla cost (Tan et al. 2008). The switch to fuel crops, from other nonenergy food crops, could cause additional food security issues. The projected corn ethanol production in 2016 would use 43% of the U.S. corn land harvested for grain in 2004 (Searchinger et al. 2008). The U.S. agricultural exports will decline sharply (compared to what they would otherwise be at the time; corn by 62%, soybeans by 28%, pork by 18%, and chicken by 12%), causing a myriad of problems for importing nations, who will be forced to become increasingly self-reliant, likely appropriating previously unused land for domestic and agricultural production. A large-scale switch to "energy plantation" is likely to "induce structural changes in agriculture and change the source, levels, and the variability of farm incomes" (Rajagopal and Sexton 2007). In addition, Jacques Diouf, Director General of the U.N. Food and Agriculture Organization, reports that the use of food grains to produce biofuels is already causing food shortages for the poor of the world (Diouf 2007).

The most salient of the counterarguments against 1st generation technologies, despite foment from environmental and food diversion concerns, comes with the reality of their limited supply. Even if you "dedicated all U.S. soybean and corn production to biofuels, you would only meet 12% of gasoline demand and 6% of diesel demand" (Srinivasan 2009). Globally, seven crops (wheat, rice, corn, sorghum, sugarcane, cassava, and sugar beet) account for 42% of the cropland; if all were dedicated to biofuels, this would satisfy just over half of the global gasoline consumption. The necessary cultivation of feedstock will always pose limitations to the amount of available biomass for processing. According to Charles et al. (2007), the heightened reliance on biofuels may potentially "inhibit the development and maturation of longer-term alternatives" that could mitigate fossil fuel dependence.

The first generation technologies are not a solution to the world's long-term energy needs. Adopting present processing technologies to utilize a feedstock, however, without the necessity of heavy cultivation and diversion of agricultural lands and foodstuffs, could represent a long-term solution to bioenergy generation and sustainable supply. Farrell (2006) claimed "large-scale use of ethanol for fuel will almost certainly require cellulosic (second (2nd)-generation) technology." First-generation technologies offer an imperfect solution whose utility must be viewed as a bridge between total reliance on petroleum and a future energy portfolio where biomass-derived alternatives feasibly contribute.

12.1.3 Second-Generation Biofuels

Second-generation biofuels are made from lignocellulosic biomass feedstock using advanced technological processes (Antizar-Ladislao and Turrion-Gomez 2008). Mature ethanol production relies on minimal preprocessing to liberate free glucose destined for uptake by yeast in conventional fermentation processes. Cellulosic ethanol technologies rely on a distinctly different substrate, that of cellulose, found in leaves, stems, and other structural elements of plants. The conversion of cellulose to ethanol involves the hydrolysis of lignocellulosic biomass to produce reducing sugars. The amount of cellulosic material available for potential use vastly outweighs the amount of available starch-based substrates. A conservative estimate is that presently, there is approximately 400 million tons of biomass available and this number is expected to grow to approximately 600 million by 2020 (Stephanopoulos 2008). Many agree this could represent a significant contribution to liquid transportation fuel demands when converted to liquid fuels (Rajagopal and Sexton 2007; Stephanopoulos 2008). Potentials of up to 130–410 EJ/year in 2050, equivalent to 33%–100% of the present energy production, might be available using only abandoned agricultural lands, low-productivity lands, and "rest lands" (Hoogwijk et al. 2005).

The cost of preprocessing cellulosic material to generate free glucose, however, is much higher than that for conventional feedstock, as both mechanical and thermochemical treatments are often required. Cellulosic ethanol technologies, therefore, are still in their infancy, as the cost of processing has been historically prohibitive. Figure 12.1 illustrates the process required to deconstruct lignocellulosic biomass and liberate fermentable sugars, which is often an expensive and time-consuming process (Moxley et al. 2008).

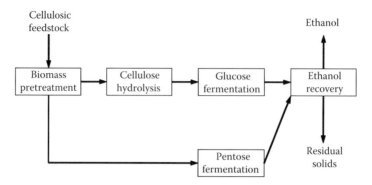

FIGURE 12.1
Cellulosic ethanol production process. (Adapted from Energy Information Administration. http://www.eia. doe.gov/oiaf/ieo/highlights.htlm, 2008.)

12.2 Cellulosic Bioethanol

12.2.1 Lignocellulose as a Feedstock for Bioethanol

Lignocellulosic biomass is a complex of biopolymers that makes up the structural components of plant material. The approximate composition of lignocellulose found in most biomass feedstock is roughly 45%–60% cellulose, 20%–40% hemicellulose, 25% lignin, and 1%–5% pectin (Demain et al. 2005; Desvaux 2005; Lynd et al. 2002). Cellulose consists of linear, insoluble polymers consisting of up to 25,000 repeating β-1,4-linked β-D-glucopyranose units. Cellulose is a highly ordered molecule consisting of 15–45 crystalline microfibril chains, which in turn associate to form cellulose fibers. In nature, cellulose is found primarily in plant cell walls and is associated with varying degrees of other bipolymers, including: (i) hemicellulose, a random, amorphous hetero-polysaccharide composed of typically β-1,3-linked xylans, arabinoxylan, glucomannan, and galactomannan; (ii) lignin, a complex hydrophobic network of phenylpropaniod units; (iii) pectins, composed of α-1-4-linked D-galacturonic acid; and (iv) proteins (Demain et al. 2005; Desvaux 2005; Lynd et al. 2002).

Cellulose is among the most abundant biopolymers on the planet Earth, with an estimated 7.5×10^{10} metric tons of cellulose produced yearly (Ljungdahl and Eriksson 1985). It is renewable, inexpensive, and constitutes a large fraction of waste biomass from municipal, agricultural, and forestry sectors, and thus offers excellent potential as a feedstock for renewable biofuels. Cellulose, however, is difficult to hydrolyze due to its crystalline structure. Current strategies that produce fuel ethanol from lignocellulosic biomass (or "second generation" biofuels) use simultaneous saccharification and fermentation (SSF) or simultaneous saccharification and cofermentation (SSCF) (Lynd et al. 2002, 2005). Both SSF and SSCF require extensive pretreatment of the cellulosic feedstock by steam explosion and/or acid treatment, followed by the addition of exogenously produced cocktails of cellulolytic enzymes to hydrolyze cellulose chains and release the glucose monomers required for fermentation. These pretreatments are costly, and some of the by-products generated, for example, furfurals, can inhibit downstream processes.

Biorefinery is a key concept used in the strategies and visions of many industrial countries for green processing and sustainable development. The objective of biorefineries is to convert agricultural residues and/or forestry-derived biomass feedstocks into a variety of value-added products including biofuels such as fuel ethanol, but also other value-added coproducts such as bio-plastics. The potential for Canadian biorefineries based on biomass derived from agricultural and/or agri-industrial by-products is enormous, but current biorefinery strategies are at an early phase of development and are focused on specific bioproduct generation based on sugars generated by hydrolysis of forestry-derived lignocellulosic biomass (Mabee et al. 2005).

Biofuels such as ethanol are high-volume, low-value commodities, and processes that produce only ethanol from lignocellulosics display poor economics. The general strategy to overcome this limitation is to construct large production facilities to gain economies of scale. An alternative strategy is to concurrently synthesize low-volume, but high-value coproducts. Biorefinery processes that maximize product generation from lignocellulosic substrates, and combine fuel ethanol production with the synthesis of value-added products, such as high-quality lignin for resins and adhesive production, offer greater economic viabilities (Kadam et al. 2008; Pan et al. 2005).

There are several processes by which lignocellulosic biomass may be converted to fuels and coproducts. These include the current paradigm of "second generation" cellulosic ethanol, which involves extensive pretreatment of the cellulosic feedstock by steam explosion and/or acid treatment, followed by the addition of exogenously produced cocktails of cellulolytic enzymes to hydrolyze cellulose chains and release the glucose monomers. The process also generates a mixture of five- and six-carbon sugars derived from hemicellulose that can be used to synthesize additional ethanol via pentose fermentation (Figure 12.1). The liberated sugars are used in fermentation reactions to synthesize fuel ethanol with yeast such as *Saccharomyces cerevisiae* or bacteria such as *Zymomonas mobilis*. This process is under intense investigation to improve efficiencies and reduce costs. Once the sugars are available, however, they may be used to synthesize fuels other than ethanol such as biobutanol or H_2, or to synthesize other value-added coproducts such as biopolymers for bioplastics.

Consolidated bioprocessing (CBP) is an alternative strategy in which cellulase production, substrate hydrolysis, and fermentation are accomplished in a single-step by microorganisms that express cellulolytic (and hemicellulolytic) enzymes (Demain et al. 2005; Lynd 1996; ; Lynd et al. 2002, 2005). CBP does not involve a dedicated process step for cellulase production, and in contrast to conventional (second generation biofuels) approaches of cellulose conversion that typically consist of four discrete steps (production of the saccharolytic cellulase and hemicellulase enzymes, hydrolysis of carbohydrates present in pretreated biomass to sugars, fermentation of glucose and other hexose sugars such as mannose and galactose, and fermentation of pentose sugars xylose and arabinose), CBP combines these four conversions into a single process. CBP offers the potential for lower biofuel production costs due to simpler feedstock processing, lower energy inputs (and therefore better energy balance), and higher conversion efficiencies than separate hydrolysis and fermentation-based processes. CBP is an economically attractive near-term goal for "third-generation" biofuel production (Demain et al. 2005; Lynd et al. 2002, 2005) (Figure 12.2).

12.2.2 Fermentation and Ethanol Synthesis

Alcohol fermentation is the formation of alcohol from sugar. The biochemical conversion of sugar to ethanol begins with the glycolysis pathway, the universal metabolic

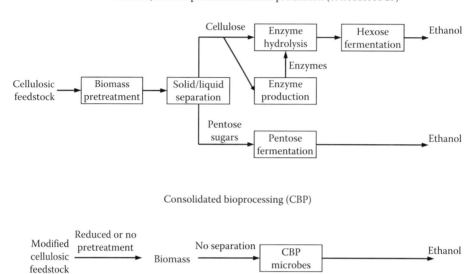

FIGURE 12.2

Comparison of multistep cellulosic biofuels production. Comparison of simultaneous hydrolysis and fermentation (SHF), simultaneous saccharification and fermentation (SSF), simultaneous saccharification and cofermentation (SSCF), and consolidated bioprocessing (CBP) strategies. (Adapted from GreenCar website http://www.greencar.com/articles/gm-nongrain-ethanol-1-per-gallon.php [accessed July 28, 2010].)

process whereby the six-carbon sugar, glucose, is split into two three-carbon molecules of pyruvate, adenosine triphosphate (ATP), and nicotinamide adenine dinucleotide (NADH) (Figure 12.3). Glycolysis can be carried out in the presence (aerobic conditions) or absence (anaerobic conditions) of oxygen (O_2) and is thus an important pathway for organisms that can ferment sugars. Pyruvate is a key metabolic branch point, and under anaerobic conditions, leads to the synthesis of alcohols like ethanol or butanol, and organic acids like lactate, acetate, or butyrate. These biochemical pathways, with their myriad reactions catalyzed by specific enzymes, all under genetic control, are extremely complex (Figure 12.3).

Aerobic microorganisms respire by converting pyruvate to CO_2, with O_2 as the terminal electron acceptor, and use the energy transformations for ATP production. When O_2 is present, the Kreb's cycle and electron transport chain produce a net of 38 ATP molecules per glucose molecule converted. In the absence of O_2 (anaerobic conditions), microorganisms must reoxidize NADH produced in glycolysis to NAD^+, which is needed for the glyceraldehyde-3-phosphate dehydrogenase reaction (Figure 12.3). Usually, NADH is reoxidized as pyruvate is converted to a more reduced compound. The complete pathway, including glycolysis and the reoxidation of NADH in the absence of O_2, is termed fermentation (Figure 12.4). During fermentation, a net of only two ATP molecules are generated per glucose molecule converted. In eukaryotes, these processes occur in the mitochondria, while in prokaryotes, they occur in the cytoplasm and cytoplasmic membrane. In most organisms, O_2 depletion controls the switch from respiration to fermentation.

12.2.3 Ethanol Synthesis Pathway in *Saccharomyces cerevisiae*

In facultative anaerobic microorganisms, like *S. cerevisiae*, ethanol production is controlled using both respiratory (aerobic) and fermentative (anaerobic) pathways. *S. cerevisiae* does

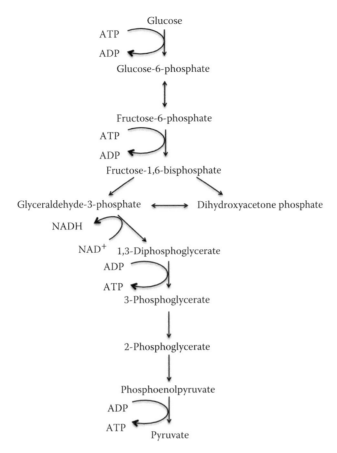

FIGURE 12.3
A schematic representation of glycolysis indicating the conversion of glucose to pyruvate.

not produce ethanol in the presence of low glucose concentrations and O_2. In *S. cerevisiae*, the switch from respiration to fermentation is controlled by external glucose levels (Otterstedt et al. 2004), and a mixed respiration and fermentation process occurs when the external glucose concentrations exceed 0.8 mM (Verduyn et al. 1984).

The three key enzymes that play essential roles in the controlled production of ethanol are pyruvate dehydrogenase (PDH), pyruvate decarboxylase (PDC), and alcohol dehydrogenase (ADH). PDH is the first component enzyme of the PDH complex. PDH acts as an essential enzyme for the decarboxylation of pyruvate followed by the reductive acetylation of lipoic acid. PDC is a homotetrameric enzyme that aids in the decarboxylation of pyruvate, in the presence of thiamine pyrophosphate, to acetaldehyde (nonacidic product) with the release of CO_2 (McMurry and Begley 2005). This reaction is very similar to the activity of the E1 subunit of PDH (Stoops et al. 1997). The difference between the two reactions is that PDC directly releases acetaldehyde, whereas PDH proceeds with the reaction by an oxidative attack of lipoamide resulting in acetyl-CoA synthesis (Figure 12.5) (McMurry and Begley 2005). Under anaerobic conditions, the PDC enzyme facilitates ethanol fermentation from acetaldehyde as a part of fermentative growth in *S. cerevisiae*. In the presence of high glucose concentrations, carbon flux through pyruvate exceeds the PDH reaction rate, shifting the PDC reaction toward ethanol synthesis. Aldehyde dehydrogenase (ALDH)

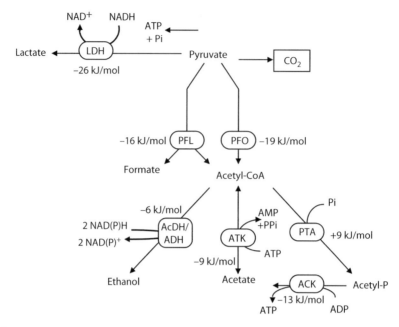

FIGURE 12.4

Metabolic pathways with corresponding standard free energies and end-products produced during anaerobic fermentation. The following enzymes are defined as lactate dehydrogenase (LDH), pyruvate ferredoxin oxidoreductase (POR), aldehyde/acetaldehyde dehydrogenase (ALDH/AcDH), alcohol dehydrogenase (ADH), phosphotransacetylase (PTA), acetate kinase (ACK), and pyruvate formate-lyase (PFL).

catalyzes the conversion of acetyl-CoA to acetaldehyde. Acetaldehyde is the two-carbon precursor of ethanol. The reduction of acetaldehyde to ethanol is catalyzed by ADH, which transfers two electrons between substrate and product, thereby consuming NADH and releasing NAD$^+$ so that glycolysis can continue (Figure 12.5).

PDC-constitutive mutants (Sharma and Tauro 1986) were used to determine the critical roles of PDC and ADH on the rate of ethanol production in *S. cerevisiae*. In this study, different *S. cerevisiae* strains with varied rates of ethanol production were examined. Strains 20 and 21 isolated from this laboratory were identified as slow and fast ethanol-producing strains, respectively. Both the strains showed similar invertase activity, whereas the levels of ADH and PDC were significantly different (Sharma and Tauro 1986). Strain 21 cells were collected by centrifugation during the batch fermentation, suspended in fresh medium, and the initial rate of ethanol production obtained. Strain 21 cells collected after 36 h showed a higher rate (55%) of ethanol production, with 48% more PDC activity, but only 12% more ADH activity than the cells collected after 24 h. The authors concluded that higher rates of ethanol synthesis are correlated with higher levels of PDC and ADH.

Sharma and Tauro (1987) further demonstrated the levels of other enzymes like ALDH in both fast (Strain 21) and slow (Strain 20) ethanol-producing strains of *S. cerevisiae*. Strain 21 showed higher levels of PDC and ADH and lower levels of ALDH, while Strain 20 had higher levels of ALDH and lower levels of PDC and ADH. High levels of ALDH and low levels of PDC resulted in the accumulation of intracellular pyruvate and increased synthesis of acetate during fermentation (Sharma and Tauro 1987). These data suggest that lower levels of PDC and ADH may be essential factors for the slower rates of ethanol production in *S. cerevisiae*, as carbon flux is diverted from acetaldehyde to acetate instead of to ethanol.

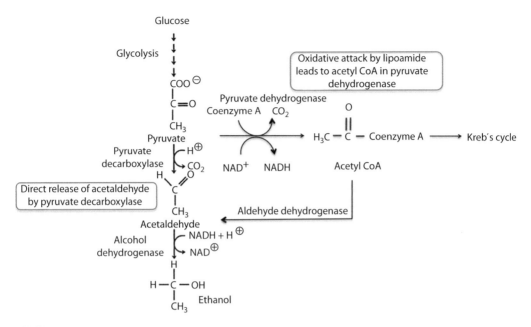

FIGURE 12.5
Mechanism of PDC, PDH, and ADH during ethanol fermentation in *S. cerevisiae*. In *S. cerevisiae*, the problem of recycling NADH is overcome through the enzyme pyruvate decarboxylase (PDC), which converts pyruvate to acetaldehyde by releasing CO_2, whereas pyruvate dehydrogenase continues the reaction forming acetyl-CoA due to the oxidation of lipoamide. Acetaldehyde is further reduced to ethanol in the presence of NADH by alcohol dehydrogenase (ADH).

12.2.4 Ethanol Synthesis Pathways and Enzymes in Other Microorganisms

Tables 12.1 and 12.2 show different mesophilic and thermophilic bacteria that produce ethanol as their main fermentation product. These bacteria have the capacity to convert sugars derived from hydrolysis of starch, cellulose, and/or hemicellulose into ethanol. The facultative anaerobic bacterium, *Zymomonas mobilis*, is a good candidate for industrial ethanol production (Stewart 1998). *Zymomonas mobilis* has been shown to have several advantages over *S. cerevisiae* for ethanol production (Table 12.1) and tolerance. *Z. mobilis* ferments glucose and fructose via Entner–Doudoroff (E-D) pathway that normally occurs in aerobic microorganisms (Montenecourt 1985). The cell-free extracts of *Zymomonas mobilis* showed rapid glucose fermentation and produced 15% more ethanol (w/v), due to higher ethanol tolerance of the E-D pathway enzymes (Algar and Scopes 1985). The cell membrane of *Zymomonas mobilis* consists of distinct fatty acids such as myristic acid, palmitic acid, and *cis*-vaccenic acid, and phospholipids like phosphotidyl ethanolamine. The presence of higher concentration of *cis*-vaccenic acid and specific hapanoids in the membrane confers higher ethanol tolerance in *Z. mobilis*, when compared to *Saccharomyces cerevisiae* (Buchholz et al. 1987). Some of the disadvantages of *Z. mobilis* include its inability to convert lignocellulosics to ethanol, formation of other end-products such as acetic acid, glycerol, acetoin, and sorbitol, and the formation of extracellular levan polymer (Gunasekaran and Chandra Raj 1999).

Increased yields of ethanol have been demonstrated using co-cultures of *S. cerevisiae* and *Z. mobilis*. The average ethanol yield in batch and continuous cultures of *S. cerevisiae* does not exceed 90%–95% of theoretical maximum. In a continuous fermentation using mixed

TABLE 12.1

Different Microorganisms That Produce Ethanol as Their Main Fermentation Product and the Ethanol Yield per Mole of Glucose Metabolized

Mesophilic Microorganisms	Ethanol Produced (mol) per Mole of Glucose Metabolized
Clostridium indolis	1.96[a]
Clostridium sphenoides	1.8[a]
Clostridium sordelli	1.7
Zymomonas mobilis (syn. *Anaerobica*)	1.9
Zymomonas mobilis ssp. *Pomaceas*	1.7
Spirochaeta aurantia	1.5
Erwinia amylovora	1.2
Leuconostoc mesenteroides	1.1
Streptococcus lactis	1.0
Sarcina ventriculi	1.0
Saccharomyces cereviseae	1.6

Source: Wiegel, J., *Experientia*, 36, 1434, 1980.

[a] In the presence of higher amounts of yeast extract.

TABLE 12.2

Different Thermophilic Bacteria That Produce Ethanol as Their Main Fermentation Product and the Ethanol Yield per Mole of Glucose Metabolized

Thermophilic Microorganisms	Ethanol Produced (mol) per Mole of Glucose Metabolized
Thermoanaerobacter ethanolicus	1.9
Clostridium thermohydrosulfuricum	1.6
Bacillus stearothermophilis	1.0 (above 55°C)
Thermoanaerobium brockii	0.95
Clostridium thermosaccharolyticum (Syn. *Tartarivorum*)	1.7
Clostridium thermocellum	1.0

Sources: Wiegel, J., *Experientia*, 36, 1434, 1980; Lamed, R. and Zeikus, J. G., *J. Bacteriol.*, 144, 569, 1980.

cultures of *Z. mobilis* and *S. cerevisiae*, higher ethanol productivity was obtained with a yield of 10.8 mM ethanol/mol glucose, with 98% theoretical yield and 99% substrate conversion (Gunasekaran and Chandra Raj, 1999).

ADH is the key ethanol synthesis enzyme in most species, including the solvent-producing Clostridia that can produce ethanol, butanol, and isopropanol (Chen 1995). The ADHs can be classified into three or four types based on their specificity in their primary structures. Ethanol- or butanol-producing Clostridia possess primary/secondary ADHs with Type 1 ADH zinc-binding domains, as well as Type 3 ADH domains (Reid and Fewson 1994). Coenzyme specificity, NAD(H) or NADP(H), is an essential factor for proper understanding of the roles of different ADHs in microbial metabolism. NAD(H)-dependent enzymes catalyze oxidative reactions for energy conservation, whereas NADP(H)-dependent enzymes catalyze reductive biosynthetic reactions. Pyruvate is converted to acetyl-CoA in the presence of pyruvate ferredoxin oxidoreductase (POR). Table 12.3 summarizes the different genes involved in ethanol synthesis within various microorganisms.

TABLE 12.3

Summary of Ethanol Synthesis Pathway Genes in Different Microorganisms

Organism	Designated Enzyme/Reaction/Gene/Cofactor					
	ADH-E	**ALDH**	**POR**	**PFL**	**PDH/PDC**	**BDH**
	Acetaldheyde to ethanol	Acetyl-CoA to ethanol			Pyruvate to ethanol	
S. cerevisiae	+ NADH		?	?	+	−
Clostridium acetobutylicum (ATCC 824)	+ aad/adhE NADH		+	?	?	− bdhA/bdhB
Clostridium beijerinckii (NRRL B592/B593)	+ adh1/adh2/adh3 NAD(P)H	Acetyl-CoA to butyryl-CoA NADH	?	?	+	−
Clostridium sp. (NCP 262)	+ adh-1 NADPH	+	?	?	+	−
Clostridium thermocellum (ATCC 27405)	+ Cthe 0423 NAD(P)H	Cthe 2238	+ Cthe 2794–2797 Cthe 2390–2393 Cthe 3120	+ Cthe 0505	−	
Clostridium cellulolyticum (H10)	+ Ccel 3198 NAD(P)H	−	+ Ccel 0016/1164	+ Ccel2582/2224	+	−
Zymomonas mobilis	+ adhA/adhB NADH	−	?	?	+ pdhA/pdhB NADH	−

+, presence of gene; −, absence of gene; ?, not determined.

ThDP, thiamine diphosphate; NADPH, nicotinamide adenine dinucleotide phosphate; NADH, nicotinamide adenine dinucleotide; ADH-E, alcohol dehydrogenase; ALDH, aldehyde dehydrogenase; POR, pyruvate-ferredoxin oxidoreductase; PFL, pyruvate formate lyase; PDH, pyruvate dehydrogenase complex; PDC, pyruvate decarboxylase; BDH, butanol dehydrogenase.

The detection of ADH and ALDH activity can be difficult due to the fact that their activities are often linked through the catalysis of consecutive reactions during ethanol synthesis (Yan and Chen 1990). Recently, however, six distinct ADHs were identified in different species and the NAD(H)- and NADP(H)-dependent activities under specific conditions were determined (Chen 1995; Johnson and Chen 1995). The six major primary ADHs include BDH I (BDH-B product of the *bdhB* gene), BDH II (BDH-A) from *Clostridium acetobutylicum* ATCC 824; NAD(P)H-dependent enzymes ADH-1, ADH-2, and ADH-3 from *Clostridium beijerinckii* NRRL B592; and NADPH-dependent ADH-1 (product of *adh-1* gene) from *Clostridium* sp. NCP 262 (Chen 1995).

Bioinformatic analyses of these enzymes revealed amino acid sequence similarities between the ADH domain of the alcohol/aldehyde dehydrogenase (ADH-E, product of *adhE/aad* gene) identified for butanol production (Nair et al. 1994) and all the major ADHs characterized from solvent-producing Clostridia, suggesting a similar function (Chen 1995). *C. acetobutylicum* ATCC 824/DSM 792 *aad* or *adhE* gene showed higher amino acid sequence similarity to the *Escherichia coli adhE* gene that encodes a multifunctional protein with pyruvate formate-lyase deactivase (PFL deactivase) and acetyl-CoA reductase (ethanol and acetaldehyde dehydrogenase) properties (Kessler et al. 1991).

C. beijerinckii NRRL B592 possesses three ADH isozymes that have high sequence similarity to *Clostridium* sp. NCP 262 (initially named as *C. acetobutylicum* P262) ADH-1 and to the alcohol/aldehyde dehydrogenase ADH domain of *C. acetobutylicum* ATCC 824/DSM 792 (Chen 1995). The type-3 ADH-1 of *Clostridium* sp. NCP 262 showed sequence similarity with type-3 iron-activated ADH-2 of *Zymomonas mobilis* and zinc-activated ADH-4 of *S. cerevisiae* (Conway et al. 1987; Reid and Fewson 1994). pH acts as an essential factor affecting the NADH-dependent activity of the ADHs from *C. beijerinckii* NRRL B592 and the butanol dehydrogenase (BDH) from *C. acetobutylicum* ATCC 824. Most of the solvent-forming enzymes, including primary ADH, ALDH, and 3-hydroxybutyryl-CoA dehydrogenase, show sequence similarity between various *Clostridium* sp. such as *C. beijerinckii*, *Clostrdium* sp. NCP2662, and NRRLS B643 (Chen 1995).

12.3 Methods of Increasing Ethanol Production

12.3.1 Genetic and Metabolic Engineering

Recent advancements in molecular biology and genetic engineering have resulted in the development of genetically engineered or modified microorganisms. One major approach of metabolic engineering focuses on expanding the pentose-utilizing capabilities of microorganisms, such as *S. cerevisiae* and *Z. mobilis*, which are predisposed to the efficient conversion of glucose to ethanol. Contrasting strategies seek to direct carbon flow away from competing products of fermentation and toward ethanol synthesis within bacteria such as *E. coli*, *Klebsiella oxytoca*, and *Erwinia* sp.; all of which can efficiently metabolize mixed sugars (Helle et al. 2004).

A xylose-fermenting *S. cerevisiae* strain was engineered by introducing genes encoding xylose-metabolizing enzymes, such as xylose isomerase from *Pichia stipitis* (yeast), *Thermus thermophilus* (bacteria), and *Piromyces* sp. (fungi) (Kotter and Ciriacy 1993; Walfridsson et al. 1996). A genetically engineered xylose-utilizing *Z. mobilis* strain was generated by introducing xylose-metabolizing pathway from *E. coli* (Zhang et al. 1995). *K. oxytoca* and *E. coli* naturally ferment arabinose and this property was used to integrate a functional arabinose-utilizing pathway into the diploid xylose-fermenting yeast *S. cerevisiae* strain TMB 3400 (Karhumaa et al. 2006).

Global transcription machinery engineering (gTME) is an approach that alters the essential proteins regulating the transcriptome. This approach (gTME) was used with *S. cerevisiae* to improve ethanol/glucose tolerance (Alper et al. 2006). In eukaryotes, RNA polymerase II is an important enzyme complex that consists of approximately 75 transcription factors or coactivators, and the loss of most of the components results in cell death. RNA polymerase II transcription factor D constitutes TATA-binding protein (SPT15) and 14 other associated factors that are considered to be important DNA-binding proteins regulating promoter specificity in yeast (Alper et al. 2006). Mutations in the *Spt15p* gene resulted in the substitution of three amino acids—S (117) → F; H (195) → Y; and R (218) → L, respectively—resulted in dominant mutations that conferred enhanced glucose conversion to ethanol, and increased ethanol tolerance of the yeast cell.

Respiration-deficient yeast strains have been extensively studied for the commercial synthesis of ethanol (Panoutsopoulou et al. 2001). Recently, Oner et al. (2005) developed a respiration-deficient amylolytic *S. cerevisiae* strain (NPB-G) through a mutation generated using polymerase chain reaction-mediated disruption of the *pet 191* gene with a kanamycin resistance (*kanMX4*) cassette that confers G418 sulfate (Genticin) resistance in yeast (Hutter and Oliver 1998). This recombinant yeast strain excreted a bifunctional fusion protein containing *Bacillus subtilis* α-amylase and *Aspergillus awamori* glucoamylase after transformation of the mutant FY23Δpet191 with the pPB-G plasmid-encoding genes for the two enzymes. Expression of the transgene was under transcriptional control of the PGK1 promoter (de Moraes et al. 1995). The NPB-G strain showed a 48% increase in both ethanol yield and productivity when compared to the parental respiration-sufficient WTPB-G strain. Growth measurement and end-products (ethanol and glucose) analysis of NPB-G mutant and WTPB-G parent strains grown in YEP-S medium showed lower biomass yields and concentrations from starch in the NPB-G strain as this nuclear petite strain derives its energy requirements using fermentation (Oner et al. 2005). Determination of maximum specific growth rates between these two strains indicated the insignificant effect of nuclear mutation on this parameter as it also depends on the pPB-G plasmid stability (Oner et al. 2005).

A specific recombinant *S. cerevisiae* strain 1400 (named pLNH32) was examined for fermentation of sugars like glucose, xylose, arabinose, and galactose that are primary monosaccharides present in corn fiber hydrolysates (Moniruzzaman et al. 1997). This strain was genetically engineered to ferment xylose by expressing genes coding for the enzymes xylitol dehydrogenase, xylose reductase, and xylulose kinase. The recombinant strain efficiently fermented xylose alone or in the presence of glucose. Also, it was shown that the highest production of ethanol was obtained with all sugars under aerobic conditions: 52 g/L (85% theoretical yield) of ethanol was produced from a mixture of 80 g/L glucose and 40 g/L xylose within 24 h.

Recently, advancements in the transport and regulation of glycerol synthesis resulted in increased ethanol production by reducing the glycerol yield. Osmotic stress conditions during cell growth can cause formation and accumulation of glycerol inside the cell where it acts as an osmolyte preventing the cell from lysis. The formation of glycerol and its concentration is controlled by its biosynthetic pathway and regulated by membrane transport system. Experiments in *S. cerevisiae* showed that glyceraldehyde-3-phosphate dehydrogenase acts as the rate-limiting factor for glycerol formation and not glycerol phosphatase (Remize et al. 2001). Also, studies showed that glyceraldehyde-3-phosphate dehydrogenase mutants showed higher rate of ethanol production when compared to the wild type

(Valadi et al. 1998). The *fps*1 gene acts as a glycerol facilitator and controls the transport of glycerol in both directions across the plasma membrane (Oliveira et al. 2003). Glycerol passes through the plasma membrane by passive diffusion through the MIP protein channel Fps1p, depending upon the osmotic stability. FPS1 deletion experiments were done to examine the effect of the gene on fermentation properties of *S. cerevisiae* and to determine whether the mutation caused in the gene could increase the yield of ethanol (Zhang et al. 2007). One-step gene replacement method was used to knock out the *fps*1 gene in *S. cerevisiae*. The *fps*1Δ mutant showed 10%±2% increase in the ethanol production, and an 18.8%±2% decrease in the glycerol yield, whereas acetic acid decreased by 5.4%±1% and pyruvate decreased by 58.6%±1%. Reducing glycerol yield by *fps*1 deletion in *S. cerevisiae* is a good approach to increase ethanol production without affecting the overall cost of carbon source as 5% of the carbon source is converted to glycerol during ethanol fermentation.

12.3.2 Ethanol Preadaption

Microorganisms are exposed to various unfavorable conditions during the industrial fermentations that can hinder growth and fermentation process. During alcohol fermentations using yeast, ethanol can inhibit cell growth. This factor is of economic importance and has been studied extensively (Jones 1989; Alexandre et al. 2001). Recently, more benefits have been achieved using preadaption techniques where microorganisms are exposed to a sublethal stress condition. A positive cumulative effect was obtained from preadaptation of the yeast cells with non-inhibitory concentrations of ethanol and then followed by the addition of 90 mg/L of acetaldehyde to the unadapted cultures. This method reduced the lag phase of ethanol-inhibited cultures of *S. cerevisiae* by 70% (Vriesekoop and Pamment 2005).

12.3.3 Addition of Boron

Recent studies involving addition of certain chemicals to the medium have been used to examine differences in growth and metabolic shifts due to pH changes in yeast cells. Bennett et al. (1999) showed that *S. cerevisiae* could be used as a valuable tool for investigating intracellular boron trafficking. Generally, boron is an essential element that is required for the growth and embryonic development in vascular plants and fishes, respectively. The molecular basis of boron and its role in metabolism, however, are still unknown. *S. cerevisiae* cultures were first grown in media containing low boron concentration (0.04 μmol B/L). After 24 h of incubation, the cells were subcultured in a new flask until early log phase (9 h), and then divided into two cultures. To one of the cultures, ultrapure boric acid was added to obtain a final concentration of 185 μmol B/L. To the other, the same amount of ultrapure water was added. Addition of 185 μmol B/L to the media resulted in a significant increase in the growth rate during the early log phase. Boron extends life during stationary phase enabling ethanol synthesis to continue for a longer period (Bennett et al. 1999).

12.3.4 Engineering of Hexose Transporters

Glucose uptake in *S. cerevisiae* is controlled by multiple hexose transporter proteins (Hxts) that display variable substrate affinity and specificity (Ozcan and Johnston 1999). Hxts may be expressed under different, but nonexclusive, conditions. Yeast strains that are devoid of Hxts are incapable of glucose uptake and this provides a way to study the role of glucose uptake in yeast glycolytic metabolism and glucose-induced signaling. Engineering yeast strains containing chimeric sugar transporters, namely, Hxt1 (low affinity) and Hxt7 (high affinity) could lead to new transporters with interesting properties. One of the chimeras showed respiratory metabolism at high-glucose concentration, when expressed as the glucose transporter. Also, this strain was able to ferment glucose even under aerobic conditions, thus indicating that modifying the glucose uptake process can cause changes in the mode of metabolic control.

12.3.5 Immobilization Techniques

Processes utilizing immobilized or flocculent yeast can improve the efficiency of ethanol production and reduce production costs (de Vasconcelos et al. 2004). Different immobilization methods include: (i) adhesion to a surface, (ii) matrix entrapment, and (iii) flocculation, including membrane techniques. All of these have been used for biocatalyst formation. Cell entrapment using a porous matrix is the most extensively used immobilization method for continuous ethanol production using yeast cells (Gódia et al. 1987). Calcium alginate is a common matrix used, but it is unstable in the presence of phosphates and is dissociated by the CO_2 liberated during the fermentation process.

Preformed cellulose beads have been used to immobilize yeast cells by adsorption (Szajáni et al. 1996). In this experiment, a vertical fluidized-bed reactor composed of glass and perforated plates to support the immobilized *S. cerevisiae* cells was used for the continuous production of ethanol from saccharose. The immobilized cells were independent of pH between 3.1 and 6.25. At 30°C, the final concentration of ethanol was 41.9 ± 0.1 g/L with a fermentation efficiency of $82.9\% \pm 2.1\%$ and a volumetric productivity of 3.94 ± 0.52 g/L/h.

de Vasconcelos et al. (2004) used sugarcane stalks as a support for immobilizing yeast cells for continuous alcoholic fermentation. The sugarcane stalks (2 cm long) used for the experiment were shown to be efficient for a period of 220 consecutive day cycles, much longer than the sugarcane harvest. Primarily, 38.5 L fermenters containing a bed of sugarcane stalks with 50% porosity were used and molasses was used in the medium formulation. Antibiotics such as penicillin (10 ppm) and tetracycline (10 ppm) were added to the medium and the ethanol yield and efficiency were calculated as 29.64 g/L/h, and 86.40%, respectively. 74.61% conversion of total reducing sugars was obtained at a dilution rate of $0.83\,h^{-1}$. The sugarcane stalks were stable even at variable dilution rates $0.05–3.0\,h^{-1}$. The immobilized yeast cell concentration was measured as 10^9 cells/g of the dry sugarcane stalk at the highest dilution rate $3.0\,h^{-1}$. Cells that are free in suspension have a larger surface area contact with the nutrient medium compared to immobilized cells. There are, however, several disadvantages to using free cell suspensions; a higher cost of microbial recycling and installation, increased chances of contamination, increased chances of cultural changes, and limitations to the dilution rate due to washout during the continuous fermentation process. On the other hand, cell immobilization technologies can limit the scale of industrial ethanol production (de Vasconcelos et al., 2004).

12.4 Biofuels beyond Ethanol: Biobutanol

12.4.1 Biobutanol

Butanol (IUPAC nomenclature 1-butanol) is a colorless, combustible liquid that has a wide range of industrial chemical uses such as acting as an intermediate for the production of

various butyl esters and butyl ethers and being used in the manufacture of pharmaceuticals, polymers, plastics, and resins. The worldwide demand for butanol as a chemical is
estimated to be 132 million liters per year, with a market size for of approximately U.S.
$130 million per year (based on a butanol wholesale chemical price of U.S. $0.99 per liter).

The refinery wholesale price of gasoline in the United States is U.S. $0.37 per liter (July
2009). Assuming that biobutanol fuel will need to sell for a comparable price to gasoline,
and taking into account that 1 L of biobutanol has 91% of the energy content of 1 L of gasoline, biobutanol fuel should conservatively sell for U.S. $0.33 per liter (ethanol currently
sells for U.S. $0.39 per liter). This gives a worldwide market size for butanol as a fuel of
about U.S. $61 billion (Festel 2008).

Like ethanol, butanol can be produced fermentatively. The French microbiologist,
Louis Pasteur, first described butanol as a product of fermentation in 1862 (Durre 2008).
The biological production of butanol is most commonly associated with Clostridial species (*C. acetobutylicum*, *C. beijerinckii*, *C. saccharoperbutylacetonicum*, and *C. saccharobuty-
licum*); however, production has also been reported among other Gram-positive species
including *Butyribacterium methylotrophicum*, and *Bacillus butylicus*, and within the archea,
Hyperthermus butylicus (Durre 2008). There are four isomeric structures of butanol with the
molecular formula C_4H_9OH (Figure 12.6).

Industrial production of butanol was initiated within the United Kingdom early in the
twentieth century as a means of biologically producing the precursors necessary for synthetic rubber production: amyl-alcohol, butanol, and acetone. During World War I, acetone
was required in high volumes as an essential chemical for the production of smokeless
ammunition (cordite). *Clostridium acetobutylicum*, originally isolated by Chaim Weizmann
in 1915, became the organism of choice for acetone synthesis, due to its high product yields,
and was used in all fermentation plants within the United Kingdom and United States
(Durre 2008). A by-product of fermentative acetone production, butanol produced during the World War I was stored until 1920 when prohibition banning drinkable alcoholic
beverages in the United States and the emerging automobile industry created a demand
for alternative industrial solvents and quick-drying lacquers. Butanol proved to be an
excellent substrate for the production of butyl acetate, a valuable solvent used in lacquer
production, and fermentative production via the acetone-butanol-ethanol (ABE) process
became widespread during the first half of the twentieth century within the United States,
South Africa, the former Soviet Union, and China. Increasing substrate costs (molasses)
and low crude oil prices in the early 1960s, however, led to the decline of the industrial
ABE process and the rise of chemical production. Until recently, butanol has been almost
exclusively produced petrochemically, via an oxidation reaction from propylene, yielding
the intermediate butyraldehyde, which is subsequently reduced to butanol via hydrogenation (Durre 2008).

The emergence of public discussion focusing on future petroleum reserves, peak-oil production, and climate change in conjunction with recent dramatic increases in crude oil
prices has led to a reintroduction of industrial butanol fermentation and stimulated scientific research as a means of developing alternative biofuels. In 2007, British Petroleum and

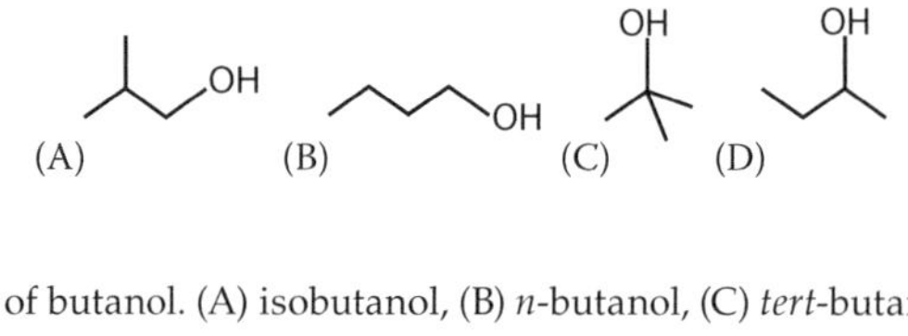

FIGURE 12.6
The isomeric configurations of butanol. (A) isobutanol, (B) *n*-butanol, (C) *tert*-butanol, and (D) *sec*-butanol.

DuPont announced an initiative to convert existing ethanol plants in the United Kingdom into facilities dedicated to the fermentative production of butanol biofuels using established Clostridial strains.

12.4.2 Butanol as Fuel

The utilization of butanol over ethanol as a biofuel offers a number of significant advantages: (i) Butanol can be used in any concentration with gasoline as a gasoline extender or may be used directly as a combustible fuel without requiring the modification of existing car engines. In contrast, ethanol can only be blended up to 85% with gasoline. (ii) Butanol is neither hygroscopic nor corrosive, and is therefore amenable to blending with gasoline on site at the refinery, whereas blending with ethanol must occur shortly before use. Not only does this illustrate that butanol is compatible with existing petroleum infrastructure (tanks, pipelines, pumps, filling stations, etc.), but it offers the added environmental benefit of minimizing contamination of groundwater in the case of accidental spills (Durre 2007). (iii) The energy density of butanol is greater than that of ethanol, 29.2 MJ/L compared to 21.1 MJ/L, and the dibutylether derivative produced during butanol production may be used as a substrate for the production of biodiesel (Durre 2007).

12.4.3 Fermentative Butanol Production: Metabolism of Solventogenic Bacteria

The metabolism of solventogenic bacteria, and hence the production of butanol, acetone, and ethanol, has been best studied in *Clostridium acetobutylicum* and is typically divided into acidogenic and solventogenic phases (Figure 12.7). During exponential growth, the

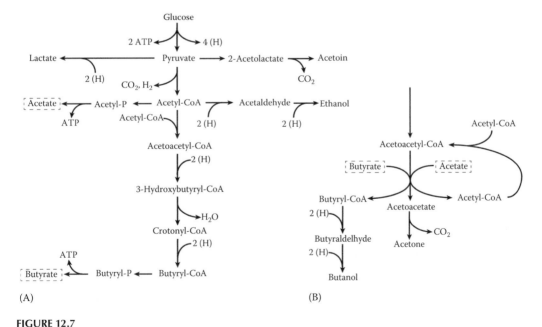

FIGURE 12.7
(A) Acidogenic and solventogenic (B) metabolic pathways utilized by *Clostridium acetobutylicum*. (Adapted from Durre, P., *Biotechnol. J.*, 2, 1525, 2007.)

bacterium typically undergoes acidogenesis where the major end-products are butyrate, acetate, CO_2, and H_2. Butyrate is produced from acetyl-CoA via the dimerization of two molecules into acetoacetyl-CoA that subsequently undergoes an enzyme-catalyzed reduction (four electrons) and dehydration (Fischer et al. 2008). Approximately, twice as much butyrate is produced relative to acetate during this time and low levels of ethanol and acetoin are typically observed. The production of lactate throughout growth may depend on specific growth conditions such as pH and carbon availability (Durre 2007).

As the cells enter into stationary phase, a major metabolic shift takes place where the excreted acids of acidogenesis (butyrate and acetate) are taken into the cell and converted into the neutral products butanol and acetone (in a typical ratio of 2:1). Here, butyrate undergoes a further four-electron reduction to *n*-butanol, whereas acetone is concomitantly produced via a decarboxylation of acetoacyl-CoA. In some organisms, hydrogen evolved via hydrogenase represents a competing outlet for reducing equivalents, and may affect yields of butanol and other reduced products. It is possible, however, for these reducing equivalents to be recovered via an uptake hydrogenase as demonstrated by decreased butanol yields following the downregulation of the *hupCBA* hydrogenase gene cluster in *Clostridium saccharoperbutylacetonicum* (Nakayama et al. 2008).

The shift from acidogenesis to solventogenesis occurs in response to decreasing extracellular pH. Anaerobic bacteria, such as *Clostridium acetobutylicum*, are unable to maintain cytoplasmic pH homeostasis (Durre 2008). At external pH values below 4.5, acids diffuse into the cytoplasm where they dissociate, causing a collapse of the proton gradient and cell death. Being able to convert acids into neutral solvents, thus increasing extracellular pH and preserving the proton gradient, provides solventogenic Clostridia with a distinct growth advantage over strictly acidogenic species. Solvents such as butanol and acetone, however, have a damaging effect on membranes and therefore the initiation of solventogenesis often occurs in conjunction with endospore synthesis (Durre 2007).

12.4.4 Organization and Regulation of Solventogenesis Genes

The genes related to acetone and butanol production within solventogenic bacteria have been studied in detail (Durre et al. 2002; Kosaka et al. 2007; Nakayama et al. 2008; Shi and Blaschek 2008). Elements encoding solventogenic enzymes within *C. acetobutylicum* are located both on the chromosome and episomally as clusters within a 192-kbp megaplasmid (pSOL1).

Briefly, the pSOL1 plasmid harbors three operons, two of which are organized as a cluster. The *sol* operon consists of four open reading frames (ORFs) transcribed from a single inducible promoter that encode: orfL (a peptide of unknown function), *adhE* (a bifunctional alcohol/aldehyde dehydrogenase), and *ctfA* and *ctfB* (encoding a two-subunit CoA transferase). The organization of the *sol* operon is unique within *C. acetobutylicum* as other solventogenic clostridia encode a butyraldehyde dehydrogenase in the place of *adhE* (Fischer et al. 1993). Immediately adjacent to the *sol* operon, but transcribed in the opposite orientation, is an ORF-encoding acetoacetate decarboxylase (*adc*). Both the *sol* operon and the *adc* exhibit transcriptional regulation, with induction coinciding with the onset of solventogenesis (Nakayama et al. 2008). In addition to transcriptional regulation, the *adc* protein also is co- or post-translationally modified in a still unknown way; however, given the enzyme remains active in the presence of O_2, the modification may contribute to stability of the protein (Durre et al. 2002).

The remaining solventogenesis genes reside on the chromosome and include *bdhA*, *bdhB*, and *adhE2*. *BdhA* encodes an ADH that is able to utilize both acetaldehyde and butyraldehyde resulting in minor ethanol and butanol synthesis during growth. *BdhB* encodes a butanol dehydrogenase that is responsible for the bulk of butanol production following the initiation of solvent production. *AdhE2* is another bifunctional acetaldehyde/ADH that is only expressed during growth on reduced substrates such as mixtures of glucose and glycerol. Growth under these conditions, and expression of *adhE2*, leads to "alcohologenic" fermentation in which both butanol and ethanol are produced, but no acetone (Durre 2007).

12.4.5 Strain Improvement for Biobutanol Synthesis

Strategies to enhance biobutanol synthesis are currently under development (Cascone 2008; Chemier et al. 2009; Durre 2007, 2008). Efforts to improve butanol yields during fermentation via metabolic engineering strategies have focused on redirecting carbon and electron flow and increasing solvent tolerance. Antisense RNA downregulation of the *C. acetobutylicum* CoA-transferase gene (*ctfB*) was applied in an effort to repress acetone production, and consequently increase butanol yields. Although acetone production was successfully reduced, butanol yields were similarly affected, suggesting that *adhE* expression was influenced by the antisense RNA construct (Tummala et al. 2003a,b). To overcome this problem, overexpression of *adhE* was combined with the downregulation of CoA-transferase. Although butanol levels increased, transformants still produced acetone, and increased quantities of ethanol were detected (Tummala et al. 2003c).

Alternative strategies for altering solvent ratios have focused on modifying electron flow by inhibiting the production of hydrogen. Butanol yields have been increased in cultures of *C. acetobutylicum* when grown under elevated hydrogen partial pressures (Doremus et al. 1985; Su et al. 1981), in the presence of CO (a hydrogenase inhibitor) or an artificial electron carrier (Datta and Zeikus 1985). Metabolic engineering strategies seeking to repress hydrogen production have been less successful. Downregulation of the *hupCBA* operon using antisense RNA in *Clostridium saccharoperbutylacetonicum* resulted in a 3.1-fold increase in hydrogen production and a 75.6% reduction in butanol yields relative to the control strain. Although the antisense RNA was effective at reducing expression of the encoded NAD-dependent hydrogenase, the physiological role of the *hupCBA* gene product was determined to be an uptake hydrogenase. Thus, downregulation of the *hupCBA* operon resulted in a smaller available pool of reduced electron carriers (NADH) poised for oxidation via butanol-producing pathways (Nakayama et al. 2008).

Attempts to engineer greater butanol tolerance in *C. acetobutylicum* have been partially successful. The cyclopropane fatty acid synthase gene (*cfa*) of *C. acetobutylicum* was overexpressed under the control of the clostridial *ptb* promoter. Constructs expressing *cfa* were introduced into *C. acetobutylicum* hosts and cultured in rich glucose broth in static flasks without pH control. Overexpression of the *cfa* gene in the wild type and in a butyrate kinase-deficient strain increased the cyclopropane fatty acid content of early-log-phase cells as well as initial acid and butanol resistance. However, solvent production in the *cfa*-overexpressing strain was considerably decreased, while acetate and butyrate levels remained high. The findings suggest that overexpression of *cfa* results in changes in membrane properties that affect full induction of solventogenesis (Zhao et al. 2003).

12.5 Biofuels beyond Ethanol: Long-Chain Alcohols, Alkanes, and Alkenes

12.5.1 Higher-Chain Alcohol Production from 2-Ketoacids

Synthetic biology has made significant progress in constructing recombinant organisms for the production of fuel chemicals beyond the scope of what microorganisms natively produce. These novel biofuels include longer-chain alcohols, fatty acid esters, and isoprenoids. Engineering "user-friendly" hosts, such as *E. coli* or *S. cerevisiae*, for biofuel production, ensures desirable characteristics such as fast growth rate and amenability to industrial scale-up, while avoiding the complex genetics characteristic of clostridial species.

Recently, *E. coli* was engineered to exploit existing amino acid biosynthesis pathways for the production of various higher alcohols. Specifically, six different natively produced 2-keto acids (2-ketobutyrate, 2-ketoisovalerate, 2-ketovalerate, 2-keto-3-methyl-valerate, 2-keto-4-methylpentanoate, and phenylpyruvate) were utilized as substrates for the non-fermentable production of long-chain alcohols (1-propanol, isobutanol, *n*-butanol, 2-methyl-1-butanol, 3-methyl-1-butanol, and 2-phenylethanol) following the heterologous introduction of a 2-keto-acid decarboxylase (*kdc*) and an ADH (Atsumi et al., 2008).

To test the capability of the endogenous 2-keto acids as substrates for KDC in *E. coli*, five different enzymes from *S. cerevisiae* (*Pdc6, Aro10, Thi3*), *Lactococcus lactis* (*Kivd*), and *C. acetobutylicum* (*Pdc*) were co-expressed with a *S. cerevisiae* ADH2. Cultures expressing either the *Kivd* or *Aro10* KDC produced all of the expected alcohols. The *Pdc*- and *Pdc10*-expressing cells were less versatile and *Thi3* did not have any observed activity. In all cultures, only trace amounts of aldehydes were detectable, indicating sufficient activity of the expressed ADH. Addition of various 2-keto acids to cultures expressing *Kivd* confirmed the specific production of corresponding alcohols by 2- to 23-fold, suggesting that increasing the flux to a specific 2-keto acid could improve both the productivity and the specificity of alcohol production (Atsumi et al. 2008). Subsequently, *E. coli* strains overexpressing the *ilvIHCD* genes, responsible for the synthesis of 2-keto isovalerate, and the heterologous alcohol-producing pathway (*Kivd, Adh2*) were able to achieve a fivefold increase (23 mM) in isobutanol yields relative to control strains.

By increasing flux toward specific 2-keto acid biosynthesis, it has been demonstrated that *E. coli* is capable of producing relatively high levels of some potentially valuable alcohols. Further pathways optimization, however, will be necessary to produce these chemicals at the low costs needed to compete with petroleum-based fuels (Keasling and Chou 2008).

12.5.2 Isoprenoids for the Production of Long-Chain Alcohols, Alkanes, and Alkenes

Isoprenoids (sometimes referred to as terpenes) are one of the most structurally diverse groups of natural compounds, with over 50,000 examples identified to date (Keasling and Chou 2008). They are all derived from the same basic units—isopentenyl diphosphate (IPP) and dimethylallyl diphosphate (DMADP), which are then combined to produce the precursors to monoterpenes, sesquiterpenes, triterpenes, and diterpenes. While isoprenoids have found use as neutraceuticals, flavors, fragrances, polymers, and antimalarial drugs, research has only recently focused on the exploitation of these compounds for the production of next-generation biofuels as they offer the potential to produce branched-chain and cyclic alkanes, alkenes, and alcohols with diverse structures and chemical properties

(Withers et al. 2007). Exploitation of isoprenoid biosynthetic pathways could produce fuels or precursors to gasoline, diesel, and jet fuel additives or substitutes (Keasling and Chou 2008).

Significant progress has already been made in engineering the native isoprenoid biosynthetic pathways of the microbial hosts *S. cerevisiae* and *E. coli*. Introduction of the mevalonate pathway into *E. coli* has proven to be one of the most effective routes for establishing a microbial host for isoprenoid synthesis. Initial work illustrated that expression of the mevalonate pathway in *E. coli* led to severe growth inhibition that was attributed to toxicity resulting from accumulation of high levels of the prenyl diphosphates (IPP, DMADP). Co-expression of a terpene synthase from the artemisinin pathway alleviated this toxicity and was used to evaluate IPP and DMADP production as the downstream amorphadiene product (Keasling 2008). Recently, the gene responsible for isoprene biosynthesis in *Bacillus subtilis* (*IspS*) was isolated and expressed in *E. coli*. Selection for clones resistant to the toxic affect of isoprene resulted in the discovery of two genes (*yhfR* and *nudF*) whose protein products were capable of catalyzing prenyl diphosphate precursors into isopentenol (Withers et al. 2007). Expression of *nudf* in *E. coli* engineered with the mevalonate-based isopentenyl pyrophosphate biosynthesis pathway resulted in the production of isopentenol. Further examination is needed to elucidate the genetic mechanisms responsible for the production of more complex isoprenoids and other candidate biofuels (Kirby and Keasling 2008; Kirby et al. 2008; Withers et al. 2007).

12.6 Biohydrogen

Hydrogen (H_2) is a versatile, clean burning, and renewable energy currency that can potentially displace the use of petroleum-based fuels in the transportation sector, which accounts for 74% of the total projected increase in liquid fuel consumption over the next 30 years (EIA, 2008). A number of alternative processes can be used for the production of hydrogen that do not result in a net increase in GHG emissions, and recent emphasis has been placed on the development of biological hydrogen production processes, which employ the use of hydrogen-producing microorganisms via light-dependant or fermentative processes, including direct biophotolysis (Green algae), indirect biophotolysis (Cyanobacteria), photofermentation (photosynthetic bacteria), and dark fermentation (fermentative bacteria) (Hallenbeck and Benemann 2002; Nath and Das 2004). Each process has its advantages and disadvantages, but there is a general consensus that light-dependent processes for biological hydrogen production have limited potential for economically viable hydrogen production (Levin et al. 2004). A thorough review of biological hydrogen production is provided in Chapter 10.

12.7 Concluding Remarks on Bioeconomy

The bioeconomy represents economic activity that uses renewable bioresources and bioprocesses to produce energy, industrial products, functional foods, and nutraceuticals. International market opportunities are the key to future biofuel development, but the economics of biofuel production will determine their long-term success and market penetration. Grain-based bioethanol is currently the only liquid transportation biofuel in the market, and biobutanol may be one of several "next-generation" biofuels currently under development. Although biohydrogen is attractive for a number of reasons, the prospects for commercial biohydrogen production are limited.

12.7.1 Economics of Bioethanol

The largest ethanol-producing industries, representing more than 90% of the global 65.7 billion L produced in 2008 (RFA 2008), are located in Brazil and the United States. Between 2000 and 2007, the average annual growth rate of U.S. ethanol production was about 23% and in 2006, the United States overtook Brazil to become the world's largest ethanol producer (Pryor 2009). Total production capacity in the United States is expected to reach about 90% of the 2015 goal of 56.2 billion L of corn ethanol production set in the 2007 Energy Independence and Security Act (Pryor 2009). In 2008, the United States produced 34 billion L of ethanol for use as fuel oxygenate in blend percentages of primarily 10% (E10) (Tao and Aden 2009). Brazil produced approximately 24.5 billion L in 2008 (Table 12.4) but mandates 20%–25% (E20–E25) blend ratios (Hahn 2008). According to Hahn (2008), the EU has also set targets of 5.75% for blended gas, as do India and Argentina (at least 5%), and Canada (5.75%) (CRFA 2009).

Biofuels, or any other petroleum alternative, are poised to contribute to energy supply when the price of oil is in the range of U.S. $20 a barrel, as it has been for most of the past three decades (Dale 2008). At U.S. $50 plus per barrel of oil, some biofuels, particularly cellulosic ethanol, begin to make economic sense (Dale 2008). Most commercial fuel ethanol now comes from corn (~97.5% in the United States) (Zhao et al. 2009) and the most commonly available blended gasoline contains 10% corn ethanol (Pryor 2009). Studies on the performance of ethanol/gasoline blends compared with unblended gasoline suggest that the mileage per gallon performance between 10% ethanol-blended gasoline (E10) and -unblended (100%) gasoline is not significantly different. For E20 and higher percentage ethanol-blended gasoline, however, significant declines in fuel economy were observed (Pryor 2009).

Corn or cereal grain feedstocks are the largest single cost contributor, ranging from 60% to 80% of total production cost. In the United States, during 2006/2007, net ethanol production costs of approximately $400/m^3 (~$0.396 per liter) were documented for corn

TABLE 12.4
Global Ethanol Production

Country	Billions of Liters
United States of America	34.1
Brazil	25.4
European Union	2.8
China	1.9
Canada	0.9

Sources: RFA. 2008b Renewable Fuels Association: Changing the Climate. Ethanol Industry Outlook. http://www.chemkeys.com/blog/wk-content/uploads/2008/12/changing-the-climate.pdf [accessed February 21, 2011].

purchased at $3.35 per bushel (Tao and Aden 2009). Average corn prices from 2002 to 2008 have ranged from $2 per bushel to $4.20 (USDA, 2008). Cost data suggests that a 10% price increase in corn alone would raise the production cost by 5.9% (Pryor 2009).

Production costs, however, are not the whole cost, and considering the larger picture for corn ethanol production, including capital depreciation, markup, shipping, and storing, and subsidies, the costs swell to $1.819 per gallon (~$0.48 per liter), considering the 2006/2007 baseline (Pryor 2009). Without subsidies, at corn prices of around $3.25 per bushel, ethanol as a high-octane fuel is competitive with oil at about $60 per barrel (Dale 2008). The retail price of gasoline in 2009 (January–October) has averaged $2.486 per gallon (~$0.656 per liter), with crude oil averaging $58.43/barrel (EIA 2009), and the average U.S. rack price of ethanol in November 2009 is listed as $2.2103/gal (~$0.584 per liter) (Fuel Ethanol 2009). Under the present pricing scenario, if E10 gasoline has equivalent fuel economy to gasoline, the incorporation of ethanol as E10 into the fuel supply has no negative effect on fuel price. In fact, the growth in ethanol production, according to an Iowa State University study, has caused retail gasoline prices to be $0.29–$0.40 per gallon lower than would otherwise have been the case (Du and Hayes 2008).

Wheat-based ethanol production, which predominates in the EU, has less favorable economics than corn-based production. In the EU, the 2006/2007 production cost of grain-based ethanol was reported to be $1.54/gallon (~$0.578 per liter) (Tao and Aden 2009). Wheat has numerous disadvantages as compared to corn for use as ethanol feedstock. Corn has historically been a cheaper commodity, costing roughly half that of wheat. Average corn yields are just below 8600 kg/ha of moist corn or 142.2 bushels/acre (Patzek 2004). The global average yield of wheat in 2000 was 2700 kg/ha (Koutinas et al. 2004). Wheat is 10%–15% lower in starch content than corn, which results in reduced concentrations of ethanol in the beer and substantially lower ethanol yields per unit weight (Sosulski and Sosulski 1994). It is estimated that bioethanol production from corn for a standard volume of ethanol produced costs roughly 60% of wheat-based production (Koutinas et al. 2004).

In contrast, sugarcane-based ethanol production has more favorable economics than corn-based production. Tao and Aden (2009) suggest that Brazilian sugarcane ethanol costs anywhere from $1.14/gal to $1.29/gal (~$0.30/gal to $0.34 per liter), making it less expensive than both corn- and wheat-based production. Sugarcane ethanol in Brazil is reported to offer higher energy return and GHG reductions per liter of ethanol than U.S.-made corn ethanol (Rajagopal and Sexton, 2007). Regional variability in agricultural conditions dictates the fuel crops that can feasibly be produced in the area. The economics of specific biofuels differ from one another depending on a variety of specific and local market and technological variations (Tao and Aden 2009).

Transportation biofuels such as cellulosic ethanol, if produced from low-input biomass grown on agriculturally marginal land or from waste biomass, could provide much greater supplies and environmental benefits than food-based biofuels (Hill et al. 2006). Potentials of 33%–100% of present energy production may be achievable using only abandoned agricultural lands, low-productivity/marginal lands, and/or "rest lands" (Hoogwijk et al. 2005). The U.S. government, under the Energy Independence and Security Act of 2007, has mandated 16 billion gallons (~60 billion L) of "cellulosic ethanol" be included into the renewable fuel supply (Tao and Aden 2009). Currently, the costs of preprocessing cellulosic for ethanol production are relatively large, about 70% of the total cellulosic ethanol production costs (Dale 2008). The ultimate goal is to produce cellulosic ethanol at $0.60–$0.70 per gallon ($0.158–$0.185 per liter), making cellulosic ethanol competitive with oil at U.S. $25–$30 per barrel (Dale 2008).

12.7.2 Economics of Biobutanol

The biological production of butanol declined from the 1940s through the 1950s, mainly because the price of petroleum-derived chemicals dropped below that of the carbohydrate-rich substrates used during fermentation (corn mash and molasses). The nature of labor-intensive batch fermentations and typically low solvent yields were contributing factors with production ceasing by the late 1950s. Current rising petroleum prices and developments in the field of biotechnology suggest that the biological production of butanol may once again become a cost-effective technology able to compete economically with petroleum-based and chemically manufactured fuels.

The cost of ABE production from corn, based on corn prices of U.S. $79.23 per ton, has been estimated to be U.S. $0.34 per kg (Qureshi and Blaschek 2001). Even under suboptimal conditions, with corn at U.S. $197.10 per ton, no gas capture, and a grass-rooted plant, the cost of butanol production would increase to U.S. $1.07 per kg. An estimation of the economics of butanol fermentation using a hyper-butanol-producing strain of *Clostridium beijerinckii* BA101 reported a production cost of U.S. $0.55–0.66 per kg, based on corn prices between U.S. $71 and 118 per ton (Qureshi and Blaschek 2000). The cost of petrochemically derived butanol in 2001 was U.S. $1.21 per kg. Thus, the fermentative production of butanol, even under suboptimal conditions, is already competitive with petrochemical processes, and may prove to be far less expensive if processes are properly optimized and substrate costs remain low (Qureshi and Blaschek 2001).

While technological advances have focused mainly on engineering strains for higher butanol productivity and yields, improved process design can help overcome the toxicity challenges typically observed. The National Renewable Energy Laboratory (Golden, CO) developed its own conceptual design for industrial ABE production from corn using an existing USDA corn dry mill ethanol model. According to the model, corn would be milled, liquefied, and the sugars would be fermented to acetone, butanol, and ethanol. The model assumed a total fermentor residence time of 72 h. The whole beer would be degassed (H_2 recovered) and then sent to distillation systems where the acetone, ethanol, and butanol would be recovered and dehydrated. The model also included further process options of extractive distillation, molecular sieve, or pervaporation membrane units (Tao and Aden 2009). Approximately 85% of the water could be removed during dehydration and recycled back to liquefaction. As with corn ethanol, Dried Distillers Grains with Solubels (DDGS) produced during this process could be sold as by-product for animal feed, although DDGS produced from bacteria would require animal-feeding trials and validation. According to this process description, a total of five products could be recovered: butanol, ethanol, acetone, hydrogen, and DDGS (Tao and Aden 2009).

In comparison to corn-based ethanol, using almost identical baseline economic assumptions, Tao and Aden (2009) calculated the yield of corn ethanol to be nearly double the yield of corn butanol. Following the addition of valuable coproducts formed from both processes, the production cost of butanol was calculated to be U.S. $1.96 per gallon butanol compared to U.S. $1.53/gal ethanol. The total project investment for a corn butanol plant was calculated to be nearly double the calculated cost for a corn ethanol plant, due primarily to the lower yields observed and more complicated separation technologies required (Tao and Aden 2009). While butanol produced from cellulosic substrates presents great promise to meet the goals of renewable fuels production, there is no available cost data for butanol derived from these substrates.

12.8 Fermentative Biofuels: Prospects for Practical Applications

A number of technologies that can produce liquid fuels for the transportation sector are now under intensive investigation. While it is clear that renewable fuels that can displace dependence on fossil fuels for transportation must be developed, it is not entirely clear which of the various alternatives will emerge as a winning technology. Considerable investments are currently focused on "second generation" cellulosic ethanol. Ethanol is the most easily attainable and immediately possible alternative transportation fuel, but deconstruction of lignocellulosic biomass to release fermentable sugars is a major obstacle that remains to be overcome. One of the first commercial cellulosic ethanol companies, Iogen Corporation, reports a yield of 340 L ethanol per ton biomass (http://www.iogen.ca/cellulose_ethanol/what_is_ethanol/process.html). By comparison, ethanol yields from grain (corn and/or wheat) are in the range of 370 L/ton (Husky Energy 2009). To become commercially viable, cellulosic ethanol processes must reduce both the costs and energy inputs of upstream pretreatment and increase the yields of ethanol per ton of feedstock.

Butanol has several advantages over ethanol such as higher energy content, lower water absorption, better blending ability, and use in conventional combustion engines without modification. Methods for enhanced synthesis of biobutanol are currently under development (Cascone 2008; Chemier et al. 2009; Durre 2007, 2008). Further advances, however, are necessary to make fermentation-based biobutanol production competitive to petrochemically produced butanol (Joseph et al. 2009).

Finally, the use of H_2 as a transportation fuel has great potential to displace fossil fuel consumption and drastically reduce GHG emissions generated by hydrocarbon combustion. There are, however, many technical and economic barriers to the implementation of a Hydrogen Economy (Edwards et al. 2008). Biohydrogen production via dark fermentation offers potential to generate renewable H_2 from inexpensive "waste" feedstocks. However, the rates and yields of H_2 production via fermentation are low and must be increased significantly if this technology is to become a viable method for generating usable H_2. A much more comprehensive understanding of the relationships between gene and gene product expression, end-product synthesis patterns, and the factors that regulate carbon and electron balance, within the context of the bioreactor conditions must be achieved if we are to improve molar yields of H_2 during cellulose fermentation. Strategies to increase yields of H_2 production include manipulation of carbon and electron flow via end-product inhibition (metabolic shift), metabolic engineering at the genetic level, synergistic co-cultures, and bioprocess engineering and bioreactor designs that maintain a neutral pH during fermentation and ensure rapid removal of H_2 and CO_2 from the aqueous phase (Levin et al. 2009).

References

Alexandre, H., V. Ansanay-Galeote, S. Dequin, and B. Blondin. 2001. Global gene expression during short term ethanol stress in *Saccharomyces cerevisiae*. *FEBS Letters* 498: 98–103.

Algar, E. M. and R. K. Scopes. 1985. Studies on cell-free metabolism: Ethanol production by extracts of *Zymomonas mobilis*. *Journal of Biotechnology* 2: 275–287.

Alper, H., J. Moxley, E. Nevoigt, G. R. Fink, and G. Stephanopoulos. 2006. Engineering yeast transcription machinery for improved ethanol tolerance and production. *Science* 314: 1565–1568.

Antizar-Ladislao, B. and J. L. Turrion-Gomez. 2008. Second-generation biofuels and local bioenergy systems. *Biofuels Bioproducts, & Biorefining* 2(5): 455–469.

Atsumi, S., T. Hanai, and J. C. Liao. 2008. Non-fermentative pathways for synthesis of branched-chain higher alcohols as biofuels. *Nature* 451: 86–89.

Bennett, A., R. I. Rowe, N. Soch, and C. D. Eckhert. 1999. Boron stimulates yeast (*Saccharomyces cerevisiae*) growth. *The Journal of Nutrition* 129: 2236–2238.

Brown, L. R. 2008. Why ethanol production will drive world food prices even higher in 2008. Earth Policy Institute, Washington, DC. January 24; http://www.earthpolicy.org/Updates/2008/Update69.htm

Buchholz, S. E., M. M. Dooley, and D. E. Eveleigh. 1987. *Zymomonas*—An alcoholic enigma. *Trends in Biotechnology* 5: 199–204.

Cascone, R. 2008. Biobutanol—A replacement for bioethanol? *Chemical Engineering Progress* 104(8): S4–S9.

Charles, M. B., R. Rachel, and R. Neal et al. 2007. Public policy and biofuels: The way forward? *Energy Policy* 35(11): 5737–5746.

Chemier, J. A., Z. L. Fowler, M. A. Koffas, and E. Leonard. 2009. Trends in microbial synthesis of natural products and biofuels. *Advances in Enzymology and Related Areas of Molecular Biology* 76: 151–217.

Chen, J. S. 1995. Alcohol-dehydrogenase—Multiplicity and relatedness in the solvent-producing clostridia. *FEMS Microbiology Reviews* 17: 263–273.

Conway, T., G. W. Sewell, Y. A. Osman, and L. O. Ingram. 1987. Cloning and sequencing of the alcohol dehydrogenase-Ii gene from *Zymomonas mobilis*. *Journal of Bacteriology* 169: 2591–2597.

CRFA. 2009. Canadian Renewable Fuels Association. Life cycle assessment of renewable fuel production from canadian biofuel plants for 2008–2009. http://www.greenfuels.org/en/resource-centre.aspx [accessed February 21, 2011].

Dale, B. 2008. Biofuels: Thinking clearly about the issues. *Journal of Agricultural and Food Chemistry* 56: 3885–3891.

Datta, R., and J. G. Zeikus. 1985. Modulation of acetone-butanol-ethanol fermentation by carbon monoxide and organic acids. *Applied and Environmental Microbiology* 49: 522–529.

Demain, A. L., M. Newcomb, and J. H. D. Wu. 2005. Cellulase, clostridia, and ethanol. *Microbiology and Molecular Biology Reviews* 69: 124–154.

de Moraes, L. M. P., S. Astolfi-Filho, and S. G. Oliver. 1995. Development of yeast strains for the efficient utilization of starch: Evaluation of constructs that express—Amylase and glucoamylase separately or as bifunctional fusion proteins. *Applied Microbiology Biotechnology* 43: 1067–1076.

Desvaux, M. 2005. *Clostridium cellulolyticum*: Model organism of mesophilic cellulolytic clostridia. *FEMS Microbiology Reviews* 29: 741–764.

de Vasconcelos, J. N., C. E. Lopes, and F. P. de França. 2004. Continuous ethanol production using yeast immobilized on sugarcane stalks. *Brazilian Journal of Chemical Engineering* 21: 357–365.

Diouf, J. 2007. Biofuels a disaster for world food. In *EU Coherence*. October 31; http://eucoherence.org/renderer.do/clearState/ false/menuld/227351/return

Doremus, M. G., J. C. Linden, and A. R. Moreira. 1985. Agitation and pressure effects on acetone-butanol fermentation. *Biotechnology and Bioengineering* 27: 852–860.

Du, X., D. J. Hayes. 2008. The impact of ethanol production on U.S. and regional gasoline prices and on the profitability of the U.S. oil refinery industry. In *Working Paper 08-WP 467*. Ames, IO: Center for Agricultural and Rural Development, Iowa State University.

Durre, P. 2007. Biobutanol: An attractive biofuel. *Biotechnology Journal* 2: 1525–1534.

Durre, P. 2008. Fermentative butanol production: Bulk chemical and biofuel. *Annals of the New York Academy of Sciences* 1125: 353–362.

Durre, P., M. Bohringer, S. Nakotte, S. Schaffer, K. Thormann, and B. Zickner. 2002. Transcriptional regulation of solventogenesis in *Clostridium acetobutylicum*. *Journal of Molecular Microbiology and Biotechnology* 4: 295–300.

Economist. 2008. Don't mix; The ethanol and water. March 1. http://www.economist.com/node/0766882 [accessed February 21, 2011].

Edwards, P. P., V. L. Kuznetsov, W. I. F. David, and N. P. Brandon. 2008. Hydrogen and fuel cells: Towards a sustainable energy future. *Energy Policy* 36(12): 4356–4362.

EIA. 2008. Energy Information Administration. Internal Energy Outlooks 2010-Highlights. http//www.eia.doe.gov/oiaf/ieo/highlights.html [accessed February 21, 2011].

EIA. 2009. Weekly U.S. All Grades All Formulations Retail Gasoline Prices. In *Energy Information Administration, Official Energy Statistics from the U.S. Government*. http://tonto.eia.doe.gov/dnav/pet/hist/LeafHandler.ashx?n=PET&s=MG_TT_US&f=W

Ethanol clean air facts, American coalition for ethanol—ACE. http://www.ethanol.org/ (Consulted 10 November, 2004).

Farrell, A. E. 2006. Ethanol can contribute to energy and environmental goals. *Science* 312(5781): 1748.

Festel, G. W. 2008. Biofuels—Economic aspects. *Chemical Engineering and Technology* 31(5): 715–720

Fischer, R. J., J. Helms, and P. Durre. 1993. Cloning, sequencing, and molecular analysis of the sol operon of *Clostridium acetobutylicum*, a chromosomal locus involved in solventogenesis. *Journal of Bacteriology* 175: 6959–6969.

Fischer, C. R., D. Klein-Marcuschamer, and G. Stephanopoulos. 2008. Selection and optimization of microbial hosts for biofuels production. *Metabolic Engineering* 10: 295–304.

Fuel Ethanol. 2009. Ethanol market weekly news and market report. In *Ethanol Market*. http://www.ethanolmarket.com/fuelethanol.html

Gódia, F., C. Casas, and C. Sola. 1987. A survey of continuous ethanol fermentation systems using immobilized cells. *Process Biochemistry* 22: 43–48.

Greencar. 2008. Nongrain-ethanol: Litres per gallon. http://www.greencar.com/articles/gm-nongrain-ethanol-1-per-gallon.php [accessed July 28, 2010].

Gunasekaran, P. and K. Chandra Raj. 1999. Ethanol Fermentation Technology—*Zymomonas mobilis*. *Current Science* 77: 56–68.

Hahn, R. W. 2008. Ethanol: Law, economics, and politics. In *Reg-Marktes Center Working Paper No. 08–02*. Available at SSRN: http://ssm.com/abstract=1082079

Hallenbeck, P. C. and J. R. Benemann. 2002. Biological hydrogen production: Fundamentals and limiting processes. *International Journal of Hydrogen Energy* 27: 1185–1193.

Helle, S. S., A. Murray, J. Lam, D. R. Cameron, and S. J. B. Duff. 2004. Xylose fermentation by genetically modified *Saccharomyces cerevisiae* 259ST in spent sulfite liquor. *Bioresource Technology* 92: 163–171.

Hill, J., E. Nelson, D. Tilman, S. Polasky, and D. Tiffany 2006 Environmental, economic, and energetic costs and benefits of biodiesel and ethanol biofuels. *Proceedings of the National Academy of Sciences of the United States of America* 103: 11206–11210.

Hoogwijk, M., A. Faaij, B. Eickhout, B. de Vries, and W. Turkenburg 2005 Potential of biomass energy out to 2100, for four IPCCSRES land-use scenarios. *Biomass & Bioenergy* 29: 225–257.

Husky Energy, 2009. Ethanol fact sheet. http://www.author-works.com/media/media-10163.pdf [accessed February 21, 2010].

Hutter, A. and S. G. Oliver. 1998. Ethanol production using nuclear petite yeast mutants. *Applied Microbiology and Biotechnology* 49: 511–516.

Jones, R. 1989. Biological principles for the effect of ethanol. *Enzyme Microbial Technology* 11: 130–153.

Johnson, J. L. and J. A. Chen. 1995. Taxonomic relationships among strains of *Clostridium acetobutylicum* and other phenotypically similar organisms. *FEMS Microbiology Reviews* 17: 233–240.

Joseph, A. C., L. F. Zachary, A. K. G. Mattheos, and L. Effendi. 2009. Trends in microbial synthesis of natural products and biofuels *Advances in Enzymology and Related Areas of Molecular Biology* 76: 151–217.

Kadam, K. L., C. Y. Chin, and L. W. Brown. 2008. Flexible biorefinery for producing fermentation sugars, lignin and pulp from corn stover. *Journal of Industrial Microbiology and Biotechnology* 35(5): 331–341.

Karhumaa, K., B. Wiedemann, E. Boles, B. Hahn-Hägerdal, and M. F. Gorwa-Grauslund. 2006. Co-utilization of L-arabinose and D-xylose by laboratory and industrial *Saccharomyces cerevisiae* strains. *Microbial Cell Factories* 5: 18.

Kaufman, R. K. and L. D. Shiers. 2008. Alternatives to conventional crude oil: When, how quickly, and market driven? *Ecological Economics* 67(3): 405–411.

Keasling, J. D. 2008. Synthetic biology for synthetic chemistry. *ACS Chemical Biology* 3: 64–76.

Keasling, J. D. and H. Chou. 2008. Metabolic engineering delivers next-generation biofuels. *Nature Biotechnology* 26: 298–299.

Kessler, D., I. Leibrecht, and J. Knappe. 1991. Pyruvate-formate-lyase-deactivase and acetyl-CoA reductase activities of *Escherichia coli* reside on a polymeric protein particle encoded by Adhe. *FEBS Letters* 281: 59–63.

Kim, S. and B. E. Dale. 2004. Global potential bioethanol production from wasted crops and crop residues. *Biomass & Bioenergy* 26: 361–375.

Kirby, J. and J. D. Keasling. 2008. Metabolic engineering of microorganisms for isoprenoid production. *Natural Products Report* 25: 656–661.

Kirby, J., D. W. Romanini, E. M. Paradise, and J. D. Keasling. 2008. Engineering triterpene production in *Saccharomyces cerevisiae*-beta-amyrin synthase from *Artemisia annua*. *FEBS Journal* 275: 1852–1859.

Kleykamp, D. 2008. Oil and the world economy. *Tamkang Journal of International Affairs* 12(2): 51–117.

Koplow, D. 2006. Biofuels—At what cost? Government support for ethanol and biodiesel in the United States. The Global Studies Initiative (GSI) of the International Institute for Sustainable development (IISD). February 16, 2007; http://www.globalsubsidies.org/IMG/pdf/biofuels_subsidies_us.pdf

Koutinas, A. A., R. Wang, and C. Webb. 2004. Evaluation of wheat as generic feedstock for chemical production. *Industrial Crops and Products* 20: 75–88.

Kosaka, T., S. Nakayama, K. Nakaya, S. Yoshino, and K. Furukawa. 2007. Characterization of the sol operon in butanol-hyperproducing *Clostridium saccharoperbutylacetonicum* strain N1-4 and its degeneration mechanism. *Biosciences Biotechnology and Biochemistry*. 71: 58–68.

Kotter, P. and M. Ciriacy. 1993. Xylose fermentation by *Saccharomyces cerevisiae*. *Applied Microbiology and Biotechnology* 38: 776–783.

Lamed, R. and J. G. Zeikus. 1980. Ethanol production by thermophilic bacteria: Relationship between fermentation product yields of and catabolic enzyme activities in *Clostridium thermocellum* and *Thermoanaerobium brockii*. *Journal of Bacteriology* 144: 569–578.

Levin, D. B., C. R. Carere, N. Cicek, and R. Sparling. 2009. Challenges for biohydrogen production via direct lignocellulose fermentation. *International Journal of Hydrogen Energy* 34: 7390–7403.

Levin, D. B., L. Pitt, and M. Love. 2004. Biohydrogen production: Prospects and limitations to practical application. *International Journal of Hydrogen Energy*. 29: 173–185.

Ljungdahl, L. G. and K. Eriksson. 1985. Ecology of microbial cellulose utilization. VIII. In *Advances in Microbial Ecology*, pp. 237–299. ed. K. C. Marshall. New York: Plenum.

Lynd, L. R. 1996. Overview and evaluation of fuel ethanol from cellulosic biomass: Technology, Economics, the Environment, and Policy. *Annual Review of Energy and the Environment* 21: 403–465.

Lynd, L. R., P. J. Weimer, W. H. van Zyl, and I. S. Pretorius. 2002. Microbial cellulose utilization: Fundamentals and biotechnology. *Microbiology and Molecular Biology Reviews* 66: 506–577.

Lynd, L. R., W. H. van Zyl, J. E. McBride, and M. Laser. 2005. Consolidated bioprocessing of cellulosic biomass: An update. *Current Opinion in Biotechnology*. 16: 577–583.

Mabee, W. E., D. J. Gregg, and J. N. Saddler. 2005. Assessing the emerging biorefinery sector in Canada. *Applied Biochemistry and Biotechnology* 121–124: 765–778.

Market Research Analyst. 2008. World ethanol production 2008–2012. http://www.marketresearch-analyst.com/2008/01/26/world-ethanol-production-forecast-2008–2012/ [accessed February 21, 2011].

McLaughlin, S. B. and M. E. Walsh, 1998. Evaluating environmental consequences of producing herbaceous crops for bioenergy. *Biomass & Bioenergy* 14: 4317–4324.

McMurry, J. and T. P. Begley. 2005. *The Organic Chemistry of Biological Pathways*. Greenwood Village, CO: Roberts and Co.

Moniruzzaman, M., B. S. Dien, C. D. Skory et al. 1997. Fermentation of corn fibre sugars by an engineered xylose utilizing *Saccharomyces* yeast strain. *World Journal of Microbiology and biotechnology* 13: 341–346.

Montenecourt, B. S. 1985. *Zymomonas*, a unique genus of bacteria. *Biology of Industrial Organisms*, pp. 261–289, eds. A. L. Demain and N. A. Soloman. CA: Benjamin–Cummings Publishing Co. Menlo Park, California.

Moxley, G. M., Z. Zhu, and Y. Zhang. 2008. Efficient sugar release by the cellulose solvent based lignocellulose fractionation technology and enzymatic cellulose hydrolysis *Journal OF Agricultural and Food Chemistry* 56 (17): 7885–7890.

Nair, R. V., G. N. Bennett, and E. T. Papoutsakis. 1994. Molecular characterization of an aldehyde/alcohol dehydrogenase gene from *Clostridium acetobutylicum* ATCC 824. *Journal of Bacteriology* 176: 871–885.

Nakayama, S., T. Kosaka, H. Hirakawa, K. Matsuura, S. Yoshino, and K. Furukawa. 2008. Metabolic engineering for solvent productivity by downregulation of the hydrogenase gene cluster hupCBA in *Clostridium saccharoperbutylacetonicum* strain N1-4. *Applied Microbiology and Biotechnology* 78: 483–493.

NAS. 2003. Frontiers in agricultural research: Food, health, environment, and communities. Washington, DC, National Academy of Sciences. November 5, 2004; http://dels.nas.edu/ rpt_briefs/frontiers_in_ag_final%20for%20print.pdf

Nath, K. and D. Das. 2004. Improvement of fermentative hydrogen production: Various approaches. *Applied Microbiology and Biotechnology* 65: 520–529.

Oliveira, R., F. Lages, M. Silva-Graca, and C. Lucas. 2003. Fps1p channel is the mediator of the major part of glycerol passive diffusion in *Saccharomyces cerevisiae*: Artefacts and re-definitions. *Biochemistry Biophysics Acta* 1613: 57–71.

Oner, E. T., S. G. Oliver, and B. Kırdar. 2005. Production of ethanol from starch by respiration-deficient recombinant *Saccharomyces cerevisiae*. *Applied and Environmental Microbiology* 71(10): 6443–6445.

Otterstedt, K., C. Larsson, and R. M. Bill. 2004. Switching the mode of metabolism in the yeast *Saccharomyces cerevisiae*. *EMBO Reports* 5: 532–537.

Ozcan, S. and M. Johnston. 1999. Function and regulation of yeast hexose transporters. *Microbiology and Molecular Biology Reviews* 63: 554–569.

Pan, X., C. Arato, N. Gilkes, D. Gregg, W. E. Mabee, K. Pye, et al., 2005. Biorefining of softwoods using ethanol organosolv pulping: Preliminary evaluation of process streams for manufacture of fuel-grade ethanol and co-products. *Biotechnology and Bioengineering* 90(4): 473–481.

Panoutsopoulou, B. K., A. Hutter, P. Jones, D. C. J. Gardner, and S. J. Oliver. 2001. Improvement of ethanol production by an industrial yeast strain via multiple gene deletions. *Journal of the Institute of Brewing* 107: 49–53.

Patzek, T. W. 2004. Thermodynamics of the corn–ethanol biofuel cycle. *Critical Review in Plant Sciences* 23: 6519–567.

Pimentel, D. and A. Marklein 2009. Food versus biofuels: Environmental and economic costs. *Human Ecology* 37: 1–12.

Pimentel, D. and T. W. Patzek, 2007. Ethanol production: Energy and economic issues related to U.S. and Brazilian sugarcane. *Natural Resources Research* 16: 3235–3242.

Pryor, F. L. 2009. The economics of gasohol. *Contemporary Economic Policy* 27: 523–537.

Qureshi, N. and H. P. Blaschek. 2000. Economics of butanol fermentation using hyper-butanol producing *Clostridium beijerinckii* BA101. *Food and Bioproducts Processing* 8: 139–144.

Qureshi, N. and H. P. Blaschek. 2001. ABE production from corn: A recent economic evaluation. *Journal of Industrial Microbiology and Biotechnology* 27: 292–297.

Rajagopal, D. and S. E. Sexton. 2007. Challenge of biofuel: Filling the tank without emptying the stomach? *Environmental Research Letters* 2(044004): 1–9.

Reid, M. F. and C. A. Fewson. 1994. Molecular characterization of microbial alcohol dehydrogenases. *Critical Reviews in Microbiology* 20:13–56.

Remize, F., L. Barnavon, and S. Dequin. 2001. Glycerol export and glycerol-3-phosphate dehydrogenase, but not glycerol phosphatase, are rate limiting for glycerol production in *Saccharomyces cerevisiae*. *Metabolic Engineering* 3: 301–312.

RFA. 2008a. Renewable Fuels Association. Ethanol production costs: A worldwide survey. A special study from FO Lichts and Agra CEAS Consulting. Tunbridge Wells, Kent, U.K.: Agra Informa. http://www.ceasc.com/Images/Content/Ethanol%20Production%20Costs.pdf [accessed February 21, 2011].

RFA. 2008b. Renewable Fuels Association: Changing the Climate. Ethanol Industry Outlook. http://www.chemkeys.com/blog/wp-content/uploads/2008/12/changing-the-climate.pdf [accessed February 21, 2011].

Rhodes, C.J. 2008. The oil question: Nature and prognosis. *Science Progress* 91(4): 317–375.

Searchinger, T., R. Heimlich and R. A. Houghton et al. 2008. Use of U.S. croplands for biofuels increases greenhouse gases through emissions from land-use change. *Science* 319: 1238–1240.

Sharma, S. and P. Tauro. 1986. Control of ethanol production by yeast: Role of pyruvate decarboxylase and alcohol dehydrogenase. *Biotechnology Letters* 8: 735–738.

Sharma, S. and P. Tauro. 1987. Control of ethanol production by yeast: Pyruvate accumulation in slow ethanol producing *Saccharomyces cerevisiae*. *Biotechnology Letters* 9:585–586.

Shi, Z. and H. P. Blaschek. 2008. Transcriptional analysis of *Clostridium beijerinckii* NCIMB 8052 and the hyper-butanol-producing mutant BA101 during the shift from acidogenesis to solventogenesis. *Applied and Environmental Microbiology* 74: 7709–7714.

Sims, R., M. Taylor, and J. Saddler. 2008. From 1st to 2nd Generation Biofuel Technologies: An overview of current industry and RD&D activities. *International Energy Agency*. http://www.iea.org/papers/2008/2nd_Biofuel_Gen.pdf [accessed February 21, 2011].

Slingerland, S. and L. van Geuns. 2005. Drivers for an international biofuel market. Discussion Paper *Clingendael International Energy Program Future Fuel Seminar*. http://www.clingendael.nl/publications/2005/20051209_ciep_misc_biofuelsmarket.pdf [accessed February 21, 2011].

Sosulski, K. and F. Sosulski. 1994. Wheat as a feedstock for fuel ethanol. *Applied Biochemistry and Biotechnology* 45: 169–180.

Srinivasan, S. 2009. The food v. fuel debate: A nuanced view of incentive structures. *Renewable Energy* 34(4): 950–54.

Stephanopoulos, G. 2008. Metabolic engineering: Enabling technology for biofuels production. *Metabolic Engineering* 10(6): 293–294.

Stewart, G. 1998. Twenty-five years of yeast research. *Journal of Industrial Microbiology* 3: 1–21.

Stoops, J. K., R. H. Cheng, and M. A. Yazdi. 1997. On the unique structural organization of the *Saccharomyces cerevisiae* pyruvate dehydrogenase complex. *The Journal of Biological Chemistry* 272: 5757–5764.

Su, T. M., R. Lamed, and J. H. Lobos. 1981. Effects of stirring and H_2 on ethanol production by thermophilic fermentation. *Proceedings of the 2nd World Congress on Chemical Engineering* 1: 353–356.

Szajáni, B., Z. Buza's, K. Dallmann, I. Gimesi, J. Krisch, and M. Toth. 1996. Continuous production of ethanol using yeast cells immobilized in preformed cellulose beads. *Applied Microbiology Biotechnology* 46: 122–125.

Tan, K. T., K. T. Lee, and A. R. Mohamed. 2008. Role of energy policy in renewable energy accomplishment: The case of second-generation bioethanol. *Energy Policy* 36(9): 3360–3365.

Tao, L. and A. Aden. 2009. The economics of current and future biofuels. *In Vitro Cellular and Developmental Biology—Plant* 45: 199–217.

Tummala, S. B., S. G. Junne, C. J. Paredes, and E. T. Papoutsakis. 2003a. Transcriptional analysis of product-concentration driven changes in cellular programs of recombinant *Clostridium acetobutylicum* strains. *Biotechnology and Bioengineering* 84: 842–854.

Tummala, S. B., N. E. Welker, and E. T. Papoutsakis. 2003b. Design of antisense RNA constructs for downregulation of the acetone formation pathway of *Clostridium acetobutylicum*. *Journal of Bacteriology* 185: 1923–1934.

Tummala, S. B., S. G. Junne, and E. T. Papoutsakis. 2003c. Antisense RNA downregulation of coenzyme A transferase combined with alcohol-aldehyde dehydrogenase overexpression leads to predominantly alcohologenic *Clostridium acetobutylicum* fermentations. *Journal of Bacteriology* 185: 3644–3653.

USDA. 2007. Major land uses. Economic Research Services. *United States Department of Agriculture.* http://www.ers.usda.gov/data/majorlanduses

USDA. 2008. National Agricultural Statistics Service—Corn, Field. In *United States Department of Agriculture.* http://www.nass.usda.gov/QuickStats/index2.jsp

Valadi, H., C. Larsson, and L. Gustafsson. 1998. Improved ethanol production by glycerol-3-phosphate dehydrogenase mutants of *Saccharomyces cerevisiae. Applied Microbiology and Biotechnology* 4: 434–439.

Van Maris, A., D. A. Abbott, and E. Bellissimi et al. 2006. Alcoholic fermentation of carbon sources in biomass hydrolysates by *Saccharomyces cerevisiae*: Current status. *Antonie Van Leeuwenhoek International Journal of General and Molecular Microbiology.* 90(4): 391–418.

Verduyn, C., T. P. L. Zomerdijk, J. P. Van Dijken, and W. A. Scheffers. 1984. Continuous measurement of ethanol production by aerobic yeast suspension with an enzyme electrode. *Applied Microbiology and Biotechnology* 19: 181–185.

Vriesekoop, F. and N. B. Pamment. 2005. Acetaldehyde addition and pre-adaptation to the stressor together virtually eliminate the ethanol-induced lag phase in *Saccharomyces cerevisiae. Letters in Applied Microbiology* 41: 424–427.

Walfridsson, M., X. Bao, M. Anderlund, G. Lilius, L. Bulow, and B. Hahn-Hägerdal. 1996. Ethanolic fermentation of xylose with *Saccharomyces cerevisiae* harboring the *Thermus thermophilus* xylA gene, which expresses an active xylose (glucose) isomerase. *Applied and Environmental Microbiology* 62: 4648–4651.

Wiegel, J. 1980. Formation of ethanol by bacteria. A pledge for the use of extreme thermophilic anaerobic bacteria in industrial ethanol fermentation processes. *Experientia* 36: 1434–1446.

Withers, S. T., S. S. Gottlieb, B. Lieu, J. D. Newman, and J. D. Keasling. 2007. Identification of isopentenol biosynthetic genes from *Bacillus subtilis* by a screening method based on isoprenoid precursor toxicity. *Applied and Environmental Microbiology* 73: 6277–6283.

Yan, R. T. and J. S. Chen. 1990. Coenzyme a-acylating aldehyde dehydrogenase from *Clostridium Beijerinckii* Nrrl-B592. *Applied and Environmental Microbiology* 56: 2591–2599.

Zhang, M., C. Eddy, K. Deanda, K. M. Finkelstein, and S. Picataggio. 1995. Metabolic engineering of a pentose metabolism pathway in ethanologenic *Zymomonas Mobilis. Science* 267: 240–243.

Zhang, A., Q. Kong, L. Cao, and X. Chen. 2007. Effect of FPS1 deletion on the fermentation properties of *Saccharomyces cerevisiae. Letters in Applied Microbiology* 44: 212–217.

Zhao, Y., L. A. Hindorff, and A. Chuang et al. 2003. Expression of a cloned cyclopropane fatty acid synthase gene reduces solvent formation in *Clostridium acetobutylicum* ATCC 824. *Applied and Environmental Microbiology* 69: 2831–2841.

Zhao, R., X. Wu, B. W. Seabourn, S. R. Bean, L. Guan, Y. C. Shi, J. D. Wilson, R. Madl, and D. Wang. 2009. Comparison of waxy vs. nonwaxy wheats in fuel ethanol fermentation. *Cereal Chemistry* 86: 145–156.

13

Biofuels from Oily Biomass

Maximino Manzanera

CONTENTS

13.1 Introduction

The combustion of fossil fuels is considered the major factor responsible for global warming due to large-scale carbon dioxide emissions. In addition, the recent steep rise in food-crop prices is caused mainly by the rising petroleum prices, threatening thousands of people with starvation. Oil prices have been climbing steadily since 2006 until May 2008, with a record high above of US$147 a barrel on July 11, 2008 due to the political instability in major oil-exporting regions, and due to the rapid demand growth in China,

India, and other developing countries (although oil prices have significantly declined since then due to the financial crisis). At oil prices above US$50 per barrel, it is profitable to produce some biofuels without subsidies. Although biofuels will add to supply growth, increasing from 1.35 million barrel per day in 2008 to 1.95 million by 2013, announced capacity additions may be difficult to achieve given available feedstock and growing concerns due to rising food prices. Current trends of crop-yield growth, greater fertilizer efficiency, and improvements in biofuel plant design and use of coproducts promise further increases in profit margins from biofuel production. In response, there is rapid expansion of biofuel production capacity from different crops in the Europe, United States, Brazil, and several Southeast Asian countries (Cassman and Liska, 2007). Using biofuels on a large scale will promote the planting of crops used to produce its feedstock. Applying appropriate farming policies would result in higher amounts of carbon-dioxide recycling by photosynthesis, thereby minimizing the impact on the greenhouse effect (Davis et al., 2009a). Biofuels from oily biomass are alternative energy sources and could be substitutes for petroleum-based diesel fuel. However, to be a viable alternative, a biofuel should provide a net energy gain, have environmental benefits, be economically competitive, and be producible in large quantities without reducing food supplies (Manzanera et al., 2008). This type of fuel offers a series of advantages over petroleum-based fuels and presents properties very close to the latter (Peterson et al., 1992). Therefore, biodiesel fuel can be used in diesel engines with little modifications. There are other advantages in using biodiesel since it has a higher cetane number (CN) than diesel fuel, no aromatics, no sulfur, and contains 10%–11% oxygen by weight. These characteristics of biodiesel are responsible for a reduction in the emissions of carbon monoxide (CO), hydrocarbon (HC), and particulate matter (PM) in the exhaust gas compared to diesel fuel (Chang et al., 1996). Alternative energy sources based on sustainable, renewable, and environmentally friendly processes are urgently needed to provide a net energy gain, be economically competitive, and be producible in large quantities without reducing food supplies. This chapter includes an overview of the oil-containing feedstocks and methods that can be used for biofuel production in order to fulfill these requirements.

13.2 Biodiesel as an Alternative Fuel

One of the most prominent alternative energy resources, attracting more and more interest in recent years, is the biofuel of oily vegetable origin, which is a possible substitute for petroleum-based diesel fuel, mainly in the form of biodiesel. Biodiesel fuels are defined as the alkyl monoesters of fatty acids from vegetable or microbial oils and animal fats for use in compression–ignition (diesel) engines. Many studies have shown that the properties of biodiesel are very close to diesel fuel (Alptekin and Canakci, 2009). Therefore, biodiesel fuel can be used in diesel engines with few modifications. Biodiesel has a higher CN than petroleum-based diesel, no aromatics, no sulfur, and contains 10%–11% oxygen by weight. Biodiesel has a relatively high flash point (150°C), which makes it less volatile and safer to transport or handle than petroleum diesel or petrodiesel (Tate et al., 2006). These characteristics of biodiesel are responsible for a reduction in the emissions of CO, HC, and PM in the exhaust gas compared to diesel fuel (Chang et al., 1996).

13.3 Direct Use of Vegetable Oils

Unprocessed oil can also be used in diesel engines, but some adjustments to the engine are required. Actually, the first diesel engine, developed by Rudolf Diesel in 1892, was run using groundnut oil as fuel. Despite the exclusive use of vegetable oil to run the early diesel engines, the feedstock shifted to petroleum distillates refined from crude oil during gasoline production in the 1920s. By then, diesel from fossil origin was more abundant and therefore cheaper than vegetable oil, and since petrodiesel is lighter and less viscous, carmakers modified their engine designs accordingly. Consequently, vegetable oil as a fuel source was practically ignored for decades. Then, as the Arab oil embargo in 1973 provoked a swift price increase, petrodiesel was suddenly four times more expensive than before and the interest in biofuels returned. Since the 1973 oil crisis, many different sources have been proposed in order to produce biodiesel preferably from vegetable oils.

Unlike fossil diesel fuel, vegetable oil consists mostly of triglycerides, these being glycerides in which the glycerols are esterified with three fatty acids. These fatty acids from vegetable oils consist of saturated HCs, and vary in their carbon-chain length and in the number of double bonds. Palmitic (16:0) and stearic (18:0) are the two most common saturated fatty acids found in at least small amounts in any vegetable oil (Murugesan et al., 2009).

An important fuel property is viscosity, which represents the level of thickness. This property affects the performance of the biofuel in a diesel engine. Due to the large molecular mass of the vegetable oils (in the range of 600–$900\,g\,mol^{-1}$), they have a very high kinematic viscosity (ratio of the viscous force to the inertial force) in the range of 30–$40\,mm^2\,s^{-1}$ at 38°C. This is about 20 times higher than that of petrodiesel fuel (Knothe and Steidley, 2007). On the other hand, the flash point (the lowest temperature at which this fuel can form an ignitable mixture in air) of vegetable oil is also very high (above 200°C). Due to the chemically bound oxygen, vegetable oils have lower heating values (39–$40\,MJ\,kg^{-1}$) than fossil diesel fuel (about $45\,MJ\,kg^{-1}$). The CNs for vegetable oils, a measurement of the combustion quality, range from 32 to 40 (Barnwal and Sharma, 2005). Generally, diesel engines run well with a CN from 40 to 55. In Europe, diesel CNs were set at a minimum of 38 in 1994 and 40 in 2000. The current standard for diesel sold in the European Union (EU) has to present a minimum cetane index of 46. In North America, diesel can be found in two CN ranges: 38–42 for regular diesel and 42–45 for premium. Premium diesel may have additives to improve CNs and lubricity and to reduce carbon deposits. Fuels with very low CNs can cause rough engine operation. They are more difficult to start, especially in cold weather or at high altitudes. They accelerate lube oil sludge formation. This high viscosity may lead to poor atomization of the fuel, incomplete combustion, choking of the injectors, ring carbonization, and accumulation of the fuel in the lubricating oils. Also, many low-cetane fuels increase engine wear. Furthermore, for the conventional diesel fuel, lower CNs have been correlated with increased nitrogen oxides (NO_x) exhaust emissions (Knothe, 2005). Hence, virgin vegetable oils are not the best fuels for diesel engines.

However, blending of vegetable oils with diesel fuel would solve some problems of diesel engine operation with pure vegetable oil.

13.4 Vegetable Oil Blends

Petrodiesel fuel dissolves quite well in vegetable oils. A way to avoid the problems caused by the higher viscosity of vegetable oils and to improve the performance is to reduce their viscosity by methods such as fuel blending. It has the advantages of improving the use of vegetable oil as fuel with minimum processing and engine modification. Using a Lister–Petter TS2 model, twin-cylinder air-cooled diesel engine, Murugesan et al. found that 25% vegetable oils mixed with 75% fossil diesel produced an exhaust temperature that was slightly higher than that for pure vegetable oils or pure petrodiesel and that the differences in brake-specific fuel consumption of the blend was very small at different engine loads (Murugesan et al., 2009). On the other hand, they showed that the CO emission for the vegetable-oil blends were lower than that of the conventional petrodiesel fuel at full load. This was also true for other gasses such as CO_2, NO_x, HCs, and polycyclic aromatic HCs, given that emissions were lower for vegetable blends than for petrodiesel fuels (Murugesan et al., 2009). However, there were drawbacks in using vegetable-oil blends. Schlick et al. (1988) evaluated the performance of a direct injection 2.59 L, three-cylinder 2600 series Ford diesel engine operating on a mixture of 25% soybean oil and sunflower oil blended with 75% number 2 diesel fuel. The power remained constant throughout 200 h of operation. However, the use of vegetable-oil blends for both direct and indirect diesel engines was not satisfactory and was considered impractical. The main reason remained within the composition of the vegetable oils consisting of polyunsaturated fatty acids that are highly susceptible to polymerization and to form gum caused by oxidation during storage or by complex oxidative and thermal polymerization at the higher temperatures and combustion pressures. Since gum does not combust completely, carbon deposits formed and lubricating oil thickened. Therefore, the high viscosity, acid composition, free fatty-acid (FFA) content, as well as gum formation during the combustion of these types of oils and its blends preclude the use of these fuel blends (Peterson et al., 1983; Harwood, 1984; Schlick et al., 1988). Three main processing techniques are used to convert vegetable oils to lower-viscosity fuel, namely, pyrolysis, microemulsification, and transesterification (Ma and Hanna, 1999).

13.5 Microemulsification

To solve the problem of the high viscosity of vegetable oils, microemulsions with alcohols have been proposed. These alcohols include mainly methanol, ethanol, and 1-butanol. Microemulsions consist of clear, stable, and uniform liquid mixtures of diesel fuel, vegetable oils, alcohols, surfactants, and cetane improvers in suitable proportions (Ma and Hanna, 1999). Alcohols such as methanol, ethanol, and propanol are used as viscosity-lowering additives, while higher alcohols are used as surfactants and alkyl nitrates are used as cetane improvers (Goering and Fry, 1984). Despite the viscosity reduction, increased

CN, and good spraying characteristics, the usage of microemulsions for extended periods of time still causes problems like injector-needle sticking, carbon deposits, and incomplete combustion (Pryde, 1984; Ziejewski et al., 1984).

13.6 Pyrolysis

For the reduction of oil viscosity, heat application to produce simpler compounds from triglycerides has been proposed. This mechanism, known as pyrolysis, refers to chemical changes caused by high temperatures or high temperatures in combination with the help of a catalyst, this being known also as thermal cracking. This conversion of oils involves heating in the absence of oxygen and cleavage of chemical bonds to yield small molecules (Sonntag, 1979). In 1947, a large scale of thermal cracking of tung-oil calcium soaps was reported (Chang and Wan, 1947). Tung oil was first saponified with lime and then thermally cracked to yield a crude oil, which was refined to produce diesel fuel and small amounts of gasoline and kerosene. In this process, 68 kg of the soap from the saponification of tung oil produced 50 L of crude oil. Therefore, vegetable oils can be cracked to reduce viscosity and improve CN. The products of cracking include alkanes, alkenes, and carboxylic acids. Different types of oils have successfully been cracked with appropriate catalysts to produce good-quality biofuels, including soybean oil, cottonseed oil, rapeseed oil, and other (Niehaus et al., 1986; Schwab et al., 1988; Ma and Hanna, 1999). This technique gave good flow characteristics by reducing viscosity (Ma and Hanna, 1999).

13.7 Transesterification: Biodiesel

The main alternative to reduce the viscosity of the vegetable oils is transesterification for biodiesel production. Biodiesel consists of long-chain alkyl (mainly methyl, propyl, or ethyl) esters also known as fatty acid methyl ester (FAME), fatty acid propyl ester (FAPE), or fatty acid ethyl ester (FAEE). Its production is performed mainly by the transesterification of a triglyceride with an alcohol in a chemical reaction also known as alcoholysis, as outlined in Figure 13.1. In the transesterification of different types of oils, triglycerides react with an alcohol, generally methanol or ethanol, to produce esters and glycerol.

Triglyceride   Glycerol   Biodiesel

Methanol
Catalyst

FIGURE 13.1
Chemical transesterification of triglyceride into glycerol and biodiesel. Large dark gray balls represent O atoms, medium dark gray balls represent C atoms corresponding to the glycerol backbone, medium light gray balls represent C atoms corresponding to the fatty acid backbone, and small dark gray balls correspond to H atoms.

The overall process is normally a sequence of three consecutive reversible reactions. First, triglycerides are transformed into diglycerides. Then these diglycerides are transformed into monoglycerides in a second reaction, and finally glycerol is produced from monoglycerides in the last reaction. In all these reactions, esters are produced to be used as biodiesel, and eventually as the resulting product have to be purified. The stoichiometric relation between alcohol and oil is 3:1. However, an excess of alcohol is usually added to improve the reaction toward the desired product. To make the reaction faster, a catalyst is also added. This transesterification reaction is affected by the alcohol type, molar ratio of alcohol to oil, temperature, purity of the reactants, and the type as well as amount of catalyst (Ma and Hanna, 1999). Different types of catalysts are used: basic ones, acidic ones, ion-exchange resins, lipases, and supercritical fluids (although this last method is also considered as non-catalytic method).

13.7.1 Alkali-Catalyzed Transesterification

For a basic catalyst, either sodium hydroxide (NaOH) or potassium hydroxide (KOH) is normally used with methanol or ethanol as well as any kind of oil, although sodium and potassium alkoxides, such as sodium methoxide, sodium ethoxide, sodium propoxide, and sodium butoxide, have been proposed as alkali catalysts as well. In 1945, Trent showed that in this process it is better to produce the alkoxide before the reaction to achieve a better overall efficiency (Trent, 1945). The alcohol-to-oil molar ratio that should be used varies from 1:1 to 6:1. However, 6:1 is the most widely used ratio, giving an important conversion for the alkali catalyst without using an excessive amount of alcohol. As stated above, the types of alcohol that are usually used are methanol and ethanol. The last one has fewer safety problems because it is less toxic, but methanol is most frequently used because of its lower cost and physical and chemical advantages (polar and shortest chain alcohol), although in some countries such as Brazil, there is a special interest in developing an appropriate technology for the use of ethanol instead of methanol, since the former is easily available due to the extensive fermentation industry in the country (Vieira et al., 2007). Additionally, since ethanol is derived from farm products (renewable sources) and is more environmentally friendly than methanol, ethanol should be the ideal candidate for the synthesis of a fully biogenerated fuel (Demirbas, 2003). Although ethanol produced from plant biomass is available at 95%, additional dehydration steps are required (e.g., by using a molecular sieve), before its use in biodiesel production with a consequent increase in the production costs (Van Gerpen, 2005; Solomon et al., 2007). On the other hand, there are also advantages for the use of alcohols with higher molecular weights, such as butanol. The usage of butanol contributes to a less pronounced initial mass-transfer-controlled regime since this alcohol has better miscibility with the lipid fraction than smaller alcohols. Moreover, the reaction can be accelerated by performing the process at higher temperatures and lower pressures due to the high boiling points of larger alcohols (Nye et al., 1983).

For this type of transesterification catalyzed by an alkali, reactants should meet certain specifications. Thus, starting materials (catalyst, oils, and alcohol) must be substantially free of water, the glycerides should have a pH above 1, and though the addition of more NaOH catalyst compensates for higher acidity, the resulting soaps increase viscosity and the formation of gels interfering the further separation of glycerol. As little as 0.3% water in the reaction mixture reduces biodiesel yields by consuming the catalyst (Wright et al., 1944). Therefore, catalysts such as sodium methoxide and NaOH should be maintained in an anhydrous state, since prolonged contact with air will diminish the effectiveness of these catalysts through interaction with moisture and carbon dioxide (Freedman et al.,

1984). Methoxide ions form by dissociation of methoxide salts in one case or when methanol reacts with hydroxyl ions from added alkaline hydroxides in the second situation (Equations 13.1 and 13.2):

$$CH_3O^-Na^+ \rightarrow CH_3O^- + Na^+ \qquad (13.1)$$

$$NaOH + CH_3OH \leftrightarrow CH_3O^- + H_2O + Na^+ \qquad (13.2)$$

Once formed, the methoxide ions are strong nucleophiles and attack the carbonyl moiety in glyceride molecules to produce the alkyl esters. Other investigators have also stressed the importance of a nearly dry, substantially fatty-acid-free oil (Feuge and Gros, 1949). With regard to the oils, an excessively high proportion of FFAs would inhibit the reaction by consuming the catalyst to form soaps, lowering the reaction yield and again making it more difficult to separate the products in a process known as saponification. FFAs are the saturated or unsaturated monocarboxylic acids that occur naturally in fats, oils, or greases and therefore are not attached to glycerol backbones. As a consequence, low-grade and degraded oils are not recommended for the alkali-catalyzed transesterification, since these types of oils contain FFAs decreasing the subsequent alkaline alcoholysis. Higher amounts of FFAs lead to higher acidity values. Therefore, oils should have a maximal FFA content for alkaline transesterification, beyond which either the reaction will not take place or the yield will be too low. The maximal level of FFAs present in the chosen oil for alkaline transesterification recommended varies depending on the author in a range from below 0.5% recommended by Zhang et al. (2003a) to below 3% recommended by Canakci and Van Gerpen (1999) as the maximal level of FFAs.

The reaction rates are also strongly influenced by the incubation temperatures. As a general rule, for any industrial reaction, the lowest possible temperature should be assayed for the reaction to be accomplished in the minimal time in order to save energy and production costs. The commonly employed temperature for the alkyl-catalyzed transesterification ranges from as low as room temperature to up to 65°C (Wright et al., 1944). Although the transesterification reactions are positively influenced by rising temperature, the maximal incubation temperature should be kept below 65°C. This is because the boiling point of methanol is 64.7°C and hence the transesterification reaction has to be carried out within this range, since a higher temperature than this enhances the explosion risk. In addition, higher temperatures also favor saponification and hence must be avoided (Figure 13.2) (Ramadhas et al., 2005).

13.7.2 Acid-Catalyzed Transesterification

The demanding feedstock specifications for base-catalyzed reactions have led researchers to seek catalytic and processing alternatives that could ease this difficulty and lower production costs. Methodologies based on acid-catalyzed reactions have the potential to achieve this since acid catalysts do not show measurable susceptibility to FFAs. The soap formation due to the presence of water in the alkali-catalyzed transesterification can be avoided by using an acid catalyst instead. Liquid acid catalysts such as sulfuric acid are less sensitive to FFAs and can simultaneously conduct esterification and transesterification (Goff et al., 2004).

This type of catalyst gives high yield in esters but needs higher reaction temperatures, which can be achieved at higher pressures. The reaction is very slow, requiring almost

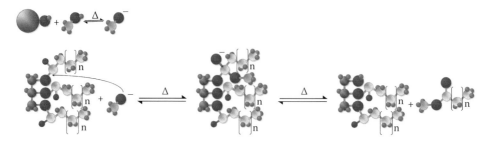

FIGURE 13.2

Base-catalyzed transesterification. The very large gray ball represents the metal atom of a base. Large dark gray balls represent O atoms, medium dark gray balls represent C atoms corresponding to the glycerol backbone, medium light gray balls represent C atoms corresponding to the fatty acid backbone, and small dark gray balls correspond to H atoms.

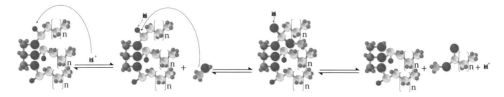

FIGURE 13.3

Acid-catalyzed transesterification. Large dark gray balls represent O atoms, medium dark gray balls represent C atoms corresponding to the glycerol backbone, medium light gray balls represent C atoms corresponding to the fatty acid backbone, and small dark gray balls correspond to H atoms.

always more than one day to finish, being 4000 times slower than the base-catalyzed reaction. Due to such requirements and their corrosive nature, these are not the most commonly used catalysts (Srivastava and Prasad, 2000; Lotero et al., 2005; Marchetti et al., 2007). Nonetheless, using an acid catalyst instead allows the use of cheaper feedstock such as yellow grease, instead of highly refined vegetable oils for which the price can account for 60%–75% of the final cost of biodiesel.

The acid-catalyzed transesterification is the second most commonly used method for producing biodiesel, using an acid as a catalyst that can be either a homogeneous acidic catalyst or a heterogeneous acidic catalyst (Figure 13.3).

13.7.2.1 Homogeneous Catalysts

Compounds such as sulfuric acid, hydrochloric acid, *para*-toluene sulfonic acid, and methane sulfonic acid are used as homogeneous acidic catalysts, although sulfuric acid is the preferred catalyst for this kind of reaction (Freedman et al., 1984, 1986; Harrington and Darcyevans, 1985).

In the mechanism for the transesterification of triglycerides in the homogeneous acid-catalyzed reaction, the key step is the protonation of the carbonyl oxygen. This in turn increases the electrophilicity of the adjoining carbon atom, making it more susceptible to nucleophilic attack. In contrast, base catalysis takes on a more direct route, creating first an alkoxide ion, which directly acts as a strong nucleophile, giving rise to a different chemical pathway for the reaction (Lotero et al., 2005).

As with the alkali reaction, if an excess of alcohol is used in the experiment, a better conversion of triglycerides is achieved, but glycerol recovery becomes more difficult and that is why the optimal relation between alcohol and the raw material should be determined experimentally while considering each process as a new problem. The usual operating condition is performed at a molar ratio of 30:1. For the optimum ratio of alcohol to oil to be established, two different factors must be considered. First, an increase in this molar ratio also translates as an increase in alcohol recovery and an increase in product-separation costs. Second, there is a greater and faster conversion of substrates at higher alcohol concentrations for the acid-catalyzed transesterification (Canakci and Van Gerpen, 1999; Crabbe et al., 2001).

The types of alcohol and oils that can be used are the same as those used in alkali catalyst reaction, although a broader range of oils can be employed as well, since oils with high FFAs content can be used too. Reaction rates in acid-catalyzed processes may also be boosted by the use of higher amounts of catalyst. The necessary concentration of catalyst to be added to the reaction mixture varies from 0.5% to 3.5%, and although the typical value is 1% sulfuric acid, some authors use 3.5% (Formo, 1954, Freedman et al., 1984, 1986, Aksoy et al., 1988, Zhang et al., 2003b).

Temperature is again another important factor in the acid-catalyzed synthesis of biodiesel. Higher temperatures could produce biodiesel with a lower specific gravity in a much shorter time. The lowest value of specific gravity (0.860) could be reached in the shortest time (1 h) at temperature of 200°C (Miao and Yao, 2009). Also, at higher temperatures, the extent of phase separation decreases and rate constants increase, as well as improved miscibility, leading to substantially shortened reaction times (Lotero et al., 2005).

Thus, the acid transesterification is a good way to make biodiesel if the sample has a relatively high FFA content. Although the acid-catalyzed reactions give very high yields in esters, the drawback of that method is the long time required for the reaction to take place, since the reaction usually requires longer than 24 h to finish.

Water presence is still a major issue for the acid-catalyzed transesterification. Canakci and Van Gerpen studied the effect of water on the sulfuric-acid-catalyzed formation of biodiesel and found that ester production was affected by as little as 0.1% water concentration and was almost totally inhibited when the water level reached 5%. These authors established the fact that the water content has to be kept under 0.5% to achieve a 90% ester yield under their reaction conditions (Canakci and Van Gerpen, 1999). Sulfuric acid has strong affinity for water and therefore it is likely that the acid will interact more strongly with water molecules than with alcohol molecules. Thus, if water is present in the feedstock or is produced during the reaction, the acid catalyst will be preferentially bound to water, leading to a reversible type of catalyst deactivation. The way that water affects the reaction is by binding to the protons in solution (H^+) more effectively than the alcohol, thereby producing a weaker acid (Rived et al., 2001). The additional molecules of water around the catalyst block the access of the nonpolar lipid molecules, consequently inhibiting the reaction.

13.7.2.2 Heterogeneous Catalysts

Despite of the high efficiency of the homogeneous acidic catalysts, the production cost is normally increased due to the extra cost of product purification. For the reduction of production costs, flow systems for continuous processing have been proposed. For this type of system, heterogeneous acidic catalysts are used (Lotero et al., 2005). In this way, using a packed-bed continuous flow reactor with solid heterogeneous acidic catalysts reduces the product-purification cost and minimizes the generation of wastes. However, due to the slow diffusion of triglycerides through the catalyst pores in heterogeneous catalysts such as supported metals, basic oxides, and zeolites, this type of transesterification reaction is even slower, so that a higher alcohol-to-glyceride molar ratio is required to achieve appreciable conversion of above 70% (Mbaraka et al., 2003). Mbaraka et al. also showed that using organosulfonic acid-functionalized mesoporous silicas can be translated into higher reactivity than commercially available solid acid esterification catalysts for the conversion of fatty acids into methyl esters. Consequently, by tailoring the textural properties of the catalyst structure and tuning the acidity of the active site, this group managed to enhance the performance of the mesoporous materials. In conclusion, they increased the pore diameter to diminish internal mass-transfer resistance by the appropriate choice of the surfactant template (Mbaraka et al., 2003). Additionally, in order to accelerate the reaction, the reactants can be exposed to microwaves or radio-frequency energy, followed by the transfer of these reactants over a heterogeneous catalyst at sufficiently high velocities to produce high shear conditions (Di Serio et al., 2007).

13.7.2.3 Two-Step Esterification

As indicated above, the main ways to produce biodiesel are the base and the acid chemically catalyzed reactions (by either homogeneous or heterogeneous processes) from refined oils. Despite the large amount of nonedible oils and fats available, the drawback with alkaline esterification of these oils is that they often contain large amounts of FFAs. As mentioned above, these FFA quickly react with the alkaline catalyst to produce soaps that inhibit the separation of the ester and glycerin. On the other hand, oils with a higher FFA content can be used for acid-catalyzed transesterification but at the cost of a much longer reaction time. To solve this problem, several groups have proposed the esterification of triglycerides in fats and oils with a high FFA content in a two-step process using an acidic catalyst in the first step (pre-esterification) followed by a second esterification using an alkaline catalyst (Liu, 1994, Ramadhas et al., 2005, Samios et al., 2009). In these works, the first step is the acid-catalyzed esterification, which reduces the FFA content of the oil and minimizes the soap formation in the second step (base-catalyzed reaction). In this way, by the application of the two-step procedure, less alcohol is required and the reaction time is much shorter than that in the acid-catalyzed transesterification. Also, a recent work has reported biodiesel production by a different two-step procedure based on the combination of consecutive base and acid catalyses known as transesterification double-step process (TDSP) (Samios et al., 2009). These authors claim that there is no initial acid attack and acid-esterification stage for the elimination of FFAs. Additionally, the second step is not a simple neutralization, but rather constitutes a catalytic stage (Samios et al., 2009). Therefore, this TDSP surpasses the need to use anhydrous materials, low-acidity vegetable oils, eliminates any soap- or emulsion-formation problem, requires far shorter times than the acid-catalysis transesterification, and results in high degrees of conversion. However, the TDSP compared to the single-step base catalysis requires two catalysts and approximately double the quantity of methanol, making the process more expensive (Figure 13.4).

13.7.3 Enzymatic Biodiesel

As stated above, either alkaline or acid chemical catalysts used for industrial biodiesel production can render a high transesterification rate of close to 99% under appropriate conditions. Nevertheless, to reach such high productivity yields, production has to be limited to conditions such as the necessity of using refined (free of FFAs) oils to avoid various

Step 1 (reactions 1 and 2)

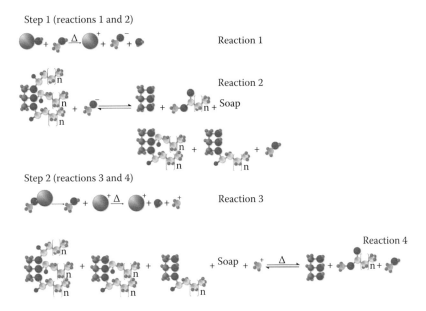

Step 2 (reactions 3 and 4)

FIGURE 13.4
Two-step transesterification. The very large gray balls represent the metal atom of a base. Large dark gray balls represent O atoms, medium dark gray balls represent C atoms corresponding to the glycerol backbone, medium light gray balls represent C atoms corresponding to the fatty acid backbone, and small dark gray balls correspond to H atoms.

problems, for example, those related to saponification, the recovery of pure glycerol, the presence of mono- and diacylglycerols, and the presence of pigments. If such residues are still present after the reaction product, the quality of final biodiesel will not satisfy with the standard requirements. Therefore, chemically catalyzed biodiesel requires several purification processes of the products comprising steps for the neutralization of the catalyst, deodorization, and the removal of pigments formed at an elevated temperature at which the process is conducted. Consequently, due to the requirements of the chemical reaction, such as a maximal concentration of FFAs of 0.5% in the reaction mixture, cheaper types of oil cannot be used as reactants in such processes. Hence, the necessity of using large amounts of alcohols and water-free reactants elevates production costs. As an alternative, enzymatic biodiesel production or biocatalysis has been proposed as better than chemically catalyzed reactions because less energy and little or no downstream processing are needed. In contrast to chemically catalysts, biocatalysts allow the synthesis of specific alkyl esters, easy recovery of glycerol, and transesterification of glycerides with a high FFA content (Nelson et al., 1996). Lipases are used to catalyze some reactions such as glycerol hydrolysis, alcoholysis, and acidolysis by living organisms, and it has also been discovered that they can be used as catalysts for transesterification and esterification reactions. Biocompatibility, biodegradability, and environmental acceptability of this biotechnological procedure are the desired properties in green technology such as biodiesel production. Lipases also known as triacylglycerol acylhydrolase (EC 3.1.1.3) are produced by bacteria (Jaeger and Eggert, 2002; Gupta et al., 2004; Sellappan and Akoh, 2005), fungi (Iwai and Tsujisaka, 1984), animals, and plants (Huang, 1993; Mukherjee and Hills, 1994). These enzymes catalyze the transformation of triacylglycerols into fatty acids and glycerol when present in aqueous conditions acting on the carboxyl ester bonds. Since their natural substrates are long-chain triacylglycerols, which have very low solubility in water, the reaction is catalyzed at the lipid–water interface. Under microaqueous conditions, lipases possess the unique ability to trigger the reverse reaction, leading to esterification, alcoholysis, and acidolysis (Jaeger et al., 1994, 1999; Gupta et al., 2004). Most commercial preparations are derived from microorganisms, and despite the large number of available lipases, only few of these have been found capable of efficient biodiesel synthesis (Jaeger et al., 1994; Palekar et al., 2000; Antczak et al., 2009). These commercial preparations can be found under the names of Lipozyme TL IM (for lipase from *Thermomyces lanuginose*), Lipozyme RM IM (for lipase from *Rhizomucor miehei*), Lipozyme IM 60 (*M. miehei*), Lipase PS-30 (from *Burkholderia cepacia*), and Novozyme SP435 (from *C. Antarctica*) to name a few.

Much work has been done on the lipase-catalyzed transesterification of triglyceride in the recent years (see Antczak et al. (2009) for a recent review). From the biotechnological standpoint, lipases are classified into extracellular and intracellular lipases. Both types of lipases are used for biodiesel synthesis, in the form of commercial preparations, although when extracellular ones are used they are normally used in their immobilized form. And both processes are highly efficient compared with using free enzymes (Hernandez-Martin and Otero, 2008; Ranganathan et al., 2008).

The optimum temperature for enzymatic transesterification results from the combination between the conformational stability of the lipase and the transesterification rate. Although the optimal temperature for the activity of various lipases ranges from 20°C to 65°C, this type of biodiesel production is performed at 30°C–55°C incubation temperature (Haas et al., 2002). In general, the transesterification reaction rate increases with temperature within the range, above which the conformation of the protein is compromised (Iso et al., 2001). In addition, when the molar ratio of alcohol to oil is higher, a rise in temperature has an adverse effect, promoting faster inactivation of the enzyme (Du et al., 2003; Xu et al., 2004). Therefore, this optimal temperature depends on the molar ratio of alcohol to oil, the type of organic solvent, and the thermostability of the enzymatic preparation.

Although for the chemically catalyzed biodiesel, water is a major issue, for the enzymatically catalyzed biodiesel, water is a plus. For these types of reactions, water is usually added to the system to enhance the catalytic efficiency of the enzyme and this water concentration needs to be optimized (Shimada et al., 2002; Noureddini et al., 2005; Shah and Gupta, 2007). This is of great interest since it gives the opportunity for the use of waste fats that are usually contaminated with water to a level that cannot be used for the production of chemically catalyzed biodiesel (Canakci, 2007). At the optimum water content, the hydrolysis of ester linkages is kept at the minimum level and this ensures the highest degree of transesterification and yield of biodiesel synthesis. For instance, if water is not added for jatropha oil transesterification by the lipase from *Burkholderia cepacia*, only 70% transformation is achieved after 5 h of reaction, while if 5% water is added then the transformation rate increases to 98% for the same time (Shah and Gupta, 2007). Consequently, the water concentration in the reaction mixture is one of the most important factors deciding the lipase-catalyzed transesterification reaction affecting the rate and yield of biodiesel synthesis. Thus, the determination of optimum water added to the reaction system for boosting the rate of ester synthesis and concomitantly maintaining the FFA concentration in the reaction mixture at the minimal level is of paramount importance (Antczak et al., 2009).

Another most important decision for this type of transesterification reactions is the type and concentration of alcohol, since these two factors affect the operational stability of enzymes. Several studies have reported the transesterification of oils with primary alcohols such as methanol, ethanol, and butanol using different lipases. Despite having high biodiesel yields with ethanol and butanol when methanol was used, only traces of methyl esters were found. This is attributed to the inhibitory effects caused by methanol on the lipases (Mittelbach, 1990; Nelson et al., 1996; Abigor et al., 2000; Iso et al., 2001). This was very discouraging since methanol is considered the main alcohol for industrial production because of its low cost and availability. Several attempts have been made to overcome this inhibitory effect of methanol, such as using 1,4-dioxane or *tert*-butanol as a solvent when methanol was used as the acyl acceptor (Iso et al., 2001; Li et al., 2006; Royon et al., 2007). Shimada et al. and Soumanou and Bornscheuer developed a stepwise methanolysis system with immobilized lipase from *Candida antarctica* in order to have an excess of alcohol and avoid the denaturing effect of methanol on the enzyme. This method consisted of a three-step flow reaction. In the first step, the substrates used consisted of waste oil and one-third molar equivalent of methanol. In the second step, the first-step eluate was mixed with one-third molar equivalent of methanol. Finally, in the third step, the second-step eluate and one-third molar equivalent of methanol were used. In this way, they converted waste oil into biodiesel fuel with a conversion rate higher than 90% in the two-reaction systems, and the lipase catalyst could be used for more than 100 days without diminished activity (Shimada et al., 2002; Soumanou and Bornscheuer, 2003).

Choice of the right solvent is also of great importance. Although lipases have high synthesis activity and good stability in hydrophobic solvents such as *n*-hexane, hydrophilic compounds such as glycerol also affect this activity, especially in continuous and repeated-batch processes. As the reaction proceeds, glycerol molecules are adsorbed onto the enzyme surface, forming a hydrophilic coating that impedes the access of the oils or any other hydrophobic substrates (Laane et al., 1987; Balcao et al., 1996; Dossat et al., 1999; Du et al., 2003). For this problem, different solutions have been proposed. One is the addition of another hydrophilic substance such as acetone or silica gel to the reaction system, since these compounds partially remove glycerol from the lipase surface, thereby restoring the lipase activity (Stevenson et al., 1994; Dossat et al., 1999). Another solution consists of intermittent washing of the immobilized lipase with a polar solvent such as isopropyl alcohol, butanol, or 2-methylbutan-2-ol or even the stepwise use of methanol to remove the attached glycerol (Chulalaksananukul et al., 1990; Du et al., 2003; Xu et al., 2004). Yet another solution proposed to recover the initial enzymatic rate involves using *n*-hexane mixed with a polar cosolvent in order to increase the medium polarity and thus improve glycerol solubility in the reaction medium (Colombie et al., 1998). Also, *in situ* glycerol removal by dialysis using isopropanol as a solvent has been suggested for removing this inhibitor (Belafi-Bako et al., 2002).

13.7.4 Supercritical Fluids

To avoid tedious purification and separation, steps need to be taken to remove saponified by-products and catalysts from biodiesel; and to speed up the reaction time, the supercritical method has been proposed. This supercritical method produces biodiesel by heating up the alcohol to its supercritical stage and letting it react with the oil (Saka and Kusdiana, 2001). In this way, a single phase of alcohol–oil mixture can be obtained instead of the two phases observed for the catalyst transesterification methods. This is because, at the supercritical stage, the dielectric constant of liquid alcohol tends to decrease and thereby increases the solubility of oil in alcohol, enabling the transesterification reaction to be completed in a very short reaction time. Since no catalyst is being used, this process is much simpler than the conventional catalytic transesterification process in terms of cost and purification of the product mixture. However, since this catalyst-free method requires high energy (pressures between 20 and 400 bar and temperatures between 150°C and 450°C) to reach the supercritical state, the production cost greatly rises in comparison to the chemically catalyzed methods, but still offers a potentially low-cost method with simpler technology for producing biodiesel (Kusdiana and Saka, 2001).

13.8 Biodiesel Feedstock

Biodiesel has traditionally been made from plants such as soybean and rapeseed. The traditional way uses several different types of mechanical extraction, although solvent extraction is more commonly used for a modern, faster, and more efficient way of extraction with a detrimental environmental impact since hexane is the common solvent of use. However, recently, enormous interest has emerged in using other types of feedstock including other edible and nonedible vegetable oils, or even used cooking oils as well as waste oils and fatty by-products from other industries such as the meat or diary industries. This source of energy can reduce our dependence on nonrenewable, potentially exhaustible, resources derived from fossil fuels, thereby reducing the environmental impact, the toxic problems associated with the use of petrodiesel, etc. Currently, great effort is being made to obtain cheap oils from microorganisms, particularly from microalgae in order to reduce production costs, reduce CO_2 inputs, lower the price of food products, and make biodiesel a real and attractive alternative.

13.8.1 First-Generation Biodiesel Feedstocks

Virgin oil from soybean and rapeseed oils are the most common feedstock used for biodiesel production, with soybean oil alone accounting for about 90% of all biofuel stocks in the United States. Over the last decade, there has been a swift surge in worldwide soybean production. While fuels produced from soybean oil in the United States amounted to barely a few thousand kilograms in 1994, 15 years later (2009), that volume is predicted to increase to 93.0 million tons, 16% rise from 2008. Soybean oil is extracted from the soybean plant (*Glycine max*), which has been used as a food crop for thousands of years, especially in China, where this plant originated. The oil content in the whole soybean represents some 20%, while 40% of the bean content is protein and 34% carbohydrates. Therefore, soybean is currently cultivated for oil and protein feed. In Europe, the most common biodiesel feedstock is rapeseed or canola oil, since this crop is widely adapted to grow in temperate climates and is, therefore, of greater strategic interest for the EU, where it is grown in increasingly large quantities. Updated information about rapeseed plantings for the marketing year 2008/2009 indicates that in the EU of 27 member states (EU-27), rapeseed-oil production grew to 19.7 million tons (Boshnakova et al., 2009). Nonetheless, vegetable oils are produced from other oil-seed crops affiliated with the temperate to warm climates in Southern Europe, such as cottonseed oil or sunflower oil, which can be very good candidates for biodiesel production. The EU-27 production estimate for the market year 2009/2010 is projected at 6.75 and 0.46 million tons for sunflower (*Helianthus annus*) and cottonseed (*Gossypium*) oil, respectively, where the sum of both types of oil represents nearly 10 times the European production of soybean oil.

Other important productions of virgin oils for biodiesel are palm and groundnuts oils. Palm oil is obtained from the fruit of the oil palm tree. It was the second most widely produced edible oil in the world, being the preferred oil in Southeast Asian cuisine.

In this sense, the Malaysian and Indonesian governments are refocusing on the use of palm oil for the production of biodiesel. The yearly production of crude palm oil is around 11 million tons. Palm oil is extracted from the palm (*Elaeis guineensis*) fruit, which produces two types of oil; the palm oil derived from the fruit mesocarp and the palm-kernel oil. The main disadvantage of palm-based biodiesel is that is unsuitable for cool-climate countries where average temperatures are below the oil pouring point, approximately 15°C for palm oil, due to its high content of saturated fatty compounds, and therefore its use is recommended only for tropical countries.

Groundnut oil was first used as a fuel during World War II due to the lack of fossil fuels and lubricants. Although its production per acre is nearly 2.5 times that of soybean, its use for biodiesel production is still economically impractical since groundnut oil on the world market is more valuable than soybean oil (Davis et al., 2009b). Although more than 350 crops have been identified as oil-producing plants, including almond (*Prunus dulcis*), andiroba (*Carapa guianensis*), babassu (*Orbignia phalerata*), camelina (*Camelina sativa*), coconut (*Cocos nucifera*), cumaru (*Dipteryx odorata*), and cardoon (*Cynara cardunculus*), the contribution to global biodiesel production from other crops is negligible (see Table 13.1) (Goering et al., 1982; Pinto et al., 2005; Demirbas, 2007).

The global oilseed production for 2009 is projected at 420.3 million tons since other important producers such as China, India, Argentina, and Brazil add up to the world production. Nevertheless, the oil used for biodiesel production is mainly high-quality food-grade vegetable oils, these increasing the cost biofuels and preventing a broad use of biodiesel, since the cost of these feedstocks account for 85% of the total cost of biodiesel fuel (Zhang et al., 2003a; Haas et al., 2006; Canakci and Sanli, 2008).

As we have indicated for palm oil, the selection of the oil feedstock depends not only on the production costs but also on its fatty-acid composition. Oils containing more saturated fatty acids than unsaturated ones tend to solidify and clog fuel lines. The fatty-acid composition of some biodiesel feedstock is shown in Table 13.2. Oils with high oleic-acid content are preferred because of the increased stability of their alkyl esters on storage and improved fuel properties (Pinto et al., 2005).

These plant-derived oils are renewable, inexhaustible, nontoxic, and biodegradable. The energy derived from them is comparable to that from petrodiesel. However, in order to compete economically with fossil diesel fuel, biodiesel must become a real alternative by

TABLE 13.1

Global Biodiesel Production from Different Feedstock

Type of Oil	2006/2007	2007/2008	2008/2009
	Million Tons		
Soybeans	235.9	220.6	238.0
Rapeseed	47.6	48.9	55.0
Cottonseed	44.6	44.1	42.4
Groundnuts	34.0	35.4	35.9
Sunflower seeds	30.2	28.5	31.8
Palm kernels	10.1	11.1	11.8
Copra	5.0	5.2	5.4
Total	407.4	393.8	420.3

Sources: From Pinto, A.C. et al., *J. Braz. Chem. Soc.*, 16, 1313, 2005; Goering, C.E. et al., *Trans. ASAE*, 25, 1472, 1982; Demirbas, A., *Energ. Policy*, 35, 4661, 2007.

TABLE 13.2

Fatty Acid Composition of Different Oils Used for Biodiesel Production

Oil	Fatty Acid Composition (%)									
	$C_{16:0}$	$C_{16:1}$	$C_{18:0}$	$C_{18:1}$	$C_{18:2}$	$C_{18:3}$	$C_{20:0}$	Others	SFA	UFA
Rapeseed	3.5		0.9	64.4	22.3	8.2			4.4	94.9
Soybean	11.4		4.4	20.8	53.8	9.3	0.3		16.1	83.9
Palm	42.6	0.3	4.4	40.5	10.1	0.2		1.1	47.0	51.1
Sunflower	7.1		4.7	25.5	62.4		0.3		12.1	87.9
Cottonseed	28.3		0.9	13.3	57.5				29.2	70.8
Groundnut	8.5		6.0	51.6	26.0				14.5	77.6
Jatropha	16.4	1.0	6.2	37.0	39.2		0.2		22.8	77.2

Sources: Akoh, C.C. et al., *J. Agric. Food Chem.*, 55, 8995, 2007; Aksoy, H.A., *J. Am. Oil Chem. Soc.*, 65, 936, 1988.
SFA, saturated fatty acids; UFA, unsaturated fatty acids.

reducing its production cost (Akoh et al., 2007). Today, the raw-material cost of biodiesel is already higher than the final cost of petrodiesel. Since biodiesel unit price is 1.5–3.0 times higher than that of petroleum-derived diesel fuel, depending on feedstock, cheaper feedstock needs to be used (Zhang et al., 2003a,b; Demirbas, 2007). More importantly, the use of edible oils for biofuel purposes may cause other significant problems such as starvation in developing countries and the deforestation of extensive areas, especially of rainforest in the Philippines, Brazil, and Indonesia (Gibbs et al., 2008). The environmental advantages of using biofuels for CO_2 balancing would be questioned if the clearance of such areas is driven for the cultivation of biodiesel crops. The release of CO_2 to the atmosphere when these areas are cleared, burned, and converted to biofuel crops is estimated to be up to 340 billion tons of carbon. This is the amount of carbon stored in tropical ecosystems, equivalent to more than 40 times the total emissions from fossil fuels used during the year (Canadell et al., 2007; Gibbs et al., 2007, 2008). Therefore, alternative feedstocks are needed for biodiesel production so that nonfood crops with lower or noncultivation requirements are used, avoiding higher food prices and greater deforestation. Such feedstocks are used for the second-generation biofuels.

13.8.2 Second-Generation Feedstocks

The main aim for the second-generation feedstock involves the sustainable use of biomass for biofuel production by using nonfood substrates. These second-generation feedstock consist of by-products derived from current crops such as stems, leaves, and husks that are normally considered as wastes once the food crop has been extracted, as well as other nonfood or feed crops, or in by-products from other industries such as chemicals or food (meat, diary, etc.).

Therefore, nonfood plants grown in unsuitable areas for food crops are considered as second-generation feedstock also. For example, the jatropha plant (*Jatropha curcas*) is a crop grown in certain parts of the world, promoted by governments and some corporations considering its ability to grow in hot and dry climates. Jatropha has attracted much attention to become one of the most promising sources of biodiesel. The plant can grow in wastelands, and it yields more than four times as much fuel per unit of area as soybean, and more than ten times that of corn. The fuel properties of the biodiesel derived from jatropha oil are similar to those of diesel and conform to European and American standards

(Tiwari et al., 2007). Due to its high content in unsaturated fatty acids, jatropha oil would be suitable for mixtures with palm oil in order to create a mixture with a lower oil-pouring point suitable for temperate countries. However, jatropha oil contains curcusones, diterpenoids of the tiglian (phorbol) type with levels between 0.03% and 3.4% of phorbol ester, which has potential irritant and toxic effects in humans (Joubert et al., 1984; Gressel, 2008). Similar considerations should be given to a group of oilseed-bearing shrubs from castor bean (*Ricinus communis*) to Indian Beech Tree (*Pongamia pinnata*) or Ballnut (*Calophyllum inophyllum*). The desert date plant (*Balanites aegyptiaca*) has recently been studied as a potential crop for hyperarid regions that are unsuitable for food crops. This plant is highly adapted to the drier parts of Africa and southern Asia and is distributed in most desert environments (Hall, 1992). Due to the high lipid content of their fruits, up to 46.7% (dry weight), and the quality of the fatty-acid composition (consisting mainly of palmitic, stearic, oleic, and linoleic acids), balanites has been proposed as a potential second-generation biodiesel feedstock (Chapagain et al., 2009).

Apart from plants, other sources of oil unsuitable for food include the residues or by-products generated from different industries such as meat, fish, or diary industries. These industries generate large amounts of wastes difficult to dispose off. Also, large amounts of oils and fats are generated during urban wastewater treatment, with which waste oils from cooking can work as second-generation feedstock.

In the case of meat and fish industries, rendering (the process whereby by-products of the food industry including animal fat, bone, offal, hides, feathers, and blood are recycled into usable products) provides products suitable for food, but, during this process, large amounts of oils and fats are generated that are unsuitable for food. In most cases, rendering operations are connected with slaughtering facilities, where large amounts of lipid-rich wastewaters are disposed off for appropriate treatment. Those waters in conjunction with inedible tallows and greases produced from rendered products can be used as biodiesel feedstock. In rendering facilities, these fat- and oil-rich residues are mixed with carcasses and other raw materials and ground into a uniform paste. This material is then heated to 121°C–135°C so that the water is removed and the fats are separated from the solids. Since the fat may still contain some water, it is common to reheat the fat with steam injection to ensure separation of the water. Depending on the water and FFA contents, these greases are classified into yellow (with FFA levels below 15%, and moisture, insolubles, and unsaponifiables below 2%) or brown grease (with FFAs levels above 15%). Renderers also filter out the spent cooking oil from restaurants. In Japan alone, more than 400,000 ton of lipid-containing wastewater are discharged each year (about 75% of this wastewater is from food industries and restaurants) (Matsumiya et al., 2007). The uses of theses fats and oils for transesterification purposes have been studied by chemical and enzymatic processes (Ma et al., 1998; Kulkarni and Dalai, 2006; Matsumiya et al., 2007).

Urban and diary wastewater is also a major source of lipids and fats. Wastewater-treatment facilities produce approximately 6.2 million metric tons (dry mass) of sludge in the United States annually. This amount is expected to increase in the future due to increasing urbanization and tighter legislation on wastewater policies in western countries (Mondala et al., 2009). The lipid fraction of the municipal wastewater sludge is between 17% and 30% dry weight depending on the authors. This lipid fraction is originated from the direct adsorption of lipids from domestic and industrial wastes to the sludge, and/or from the phospholipids in the cell membranes of microorganisms, their metabolites, and by-products of cell lysis (Boocock et al., 1992; Shen and Zhang, 2003; Jarde et al., 2005; Mondala et al., 2009). Despite representing a cheap and important source of oily biomass, due to the high content of FFAs among the lipid fraction, this feedstock is valid only for acid-catalyzed or enzymatic transesterification (Mondala et al., 2009).

13.8.3 Third-Generation Feedstocks

There are great advantages in using nonfood feedstock for biodiesel production, since more food and feed become available, thereby partially restricting rises in food prices. Nonetheless, there are more advantages in using alternative feedstock such as oils from algae, since aside from the fact that food is not being used as fuel, these organisms fix atmospheric CO_2, have rapid growth rates, and offer high per-acre yields (Roessler et al., 1994; Sawayama et al., 1995; Chisti, 2008). Since the current demand of transport fuel cannot be completely satisfied with first- and second-generation feedstock, additional sources or feedstock for production of biodiesel are needed. Microalgae used for oil could be included in the current biofuel scenario (Chisti, 2007). Microalgae are microscopic photosynthesizing organisms with dual heterotrophic and autotrophic metabolism, depending on the ambient conditions. Microalgae present enormous potential to convert water and carbon dioxide to biomass by using energy from the sunlight. Since many microalgae are highly rich in oil (sometimes with oil contents of more than 80% dry weight, although 20%–40% is more common), it seems to be the only source of biodiesel with the potential to contribute for an important partial replace of fossil diesel (Metting, 1996; Spolaore et al., 2006). Although not all microalgae are suitable for biodiesel production, many species produce appropriate kinds of oils (Banerjee et al., 2002; Metzger and Largeau, 2005; Guschina and Harwood, 2006).

It has been shown that oil production from algae reaches 100 and 200 times greater per acre than fuel production from oil palm or soybean, respectively (Metting, 1996; Chisti, 2008). This higher production rate derives from their ability to double their biomass within 24 h, growing extremely fast in comparison with other oil crops. For instance, microalgal doubling times during exponential growth can reach 3.5 h (Metting, 1996; Spolaore et al., 2006). Some authors estimate biomass production of 1.535 kg m^{-3} per day for systems located in tropical countries (Miron et al., 1999; Chisti, 2008). For large-scale production of microalgal biomass, continuous culture needs to be undertaken in raceway ponds or tubular photobioreactors (Figure 13.5). These raceway ponds use closed-loop open recirculation channels and tubular photobioreactors consisting of transparent tubes. The culture of a single microalga is possible when close tubular photobioreactors are used. In these cases, the advantage of using a single microalgal species allows the selection of the most appropriate microorganisms and provides much higher oil yield per hectare and reduces the cost of biomass recovery. Hence, this method produces nearly 30 times the typical biomass concentration (Terry and Raymond, 1985; Grima et al., 1999; Chisti, 2007). Raceway ponds, like any other open-culture system, are normally located in outdoor facilities, which quite often cause contamination problems with the outgrowth of undesired microalgae, the presence of predators such as viruses or rotifers, and poor control of nutrients (light, CO_2, N, or P), since the supply of these nutrients can be affected by weather changes (Fernandez et al., 1997; Pushparaj et al., 1997; Borowitzka, 1998; Carvalho et al., 2006; Xu et al., 2009). Tubular photobioreactors, on the other hand, are not as sensitive to such parameters, reduce contamination risk, and allow the cultivation of single microalgal species. This type of photobioreactor also minimizes CO_2 and water losses while boosting the light-use efficiency and improving the control of the process. Another advantage of tubular photobioreactors over raceway ponds is the lower area required, thus minimizing the cultivation space for the same biomass production (Grima et al., 1999; Carvalho et al., 2006). Therefore, microalgal

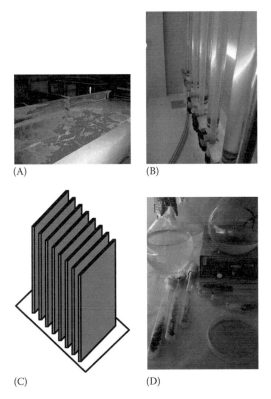

FIGURE 13.5

Isolation and growth of microalgae. Representation of different culturing systems for microalgae (A–C). (A) Raceway pond, (B) tubular photobioreactor, and (C) flat panel photobioreactor. (D) Materials used for the isolation of microalgal strains.

oil production using tubular photobioreactors is 13 times higher than for a raceway pond. Despite the clear advantages of the tubular photobioreactors, there are also disadvantages that need to be solved. One drawback is the selection of an appropriate balance between O_2 and CO_2. Due to their photosynthetic nature, microalgae fix CO_2 and produce O_2. By culturing the microorganisms in a close system, the CO_2 concentration has to be renewed in order to provide the carbon source for biomass production. On the other hand, O_2 accumulation can inhibit photosynthesis and therefore reduce oil production as this gas is toxic for most microalgae at concentrations above $35\,mg\,L^{-1}$ (Carvalho et al., 2006). The pumping of CO_2, O_2 removal using pumps or airlift technology, and degassing zones not only maintain the optimal gases concentration but also reduce the biomass sedimentation in the tubes and therefore restore the illumination area (Grima et al., 1999; Camacho et al., 2001; Fernandez et al., 2001; Miron et al., 2003; Carvalho et al., 2006). For these reasons, dimensions (length and thickness) of the tubes need to be maintained at certain levels. Tubes that are too thick (more than 0.1 m in diameter) decrease the surface-to-volume ratio so that microalgae at the center of the tube do not receive enough light (Jimenez et al., 2003). Excessively large tubes produce gradients in CO_2 and O_2 concentrations, leading to pH gradients above the aforementioned effect of imbalances in gas concentrations. In addition, mixing in the long tubes becomes complex and biomass sedimentation is difficult to avoid, leading to shaded areas that reduce the growth rate. Therefore, scaling up the tubular system is not as simple as extending the length of the tube or its diameter, as it involves the development of devices for appropriate mixing without damaging the microalgal cells as well as devices for gas balance or the use of modular reactor units (Grima et al., 1999; Molina et al., 2000, 2001; Eriksen, 2008). Alternative photobioreactors have been proposed, including flat panel photobioreactors, in which the microalgae are cultured in thin layers or media enclosed in a flat panel, and column photobioreactors, in which cells are cultured inside a tank where bubble columns aerate and mix the tank content from below (Hu et al., 1998; Miron et al., 1999; Doucha et al., 2005; Krichnavaruk et al., 2007).

More than 40,000 different species of microalgae have been identified, with marine microalgae in general having several advantages compared to other type of microalgae, given that they can be cultured on nonagricultural coastal areas without the need of freshwater. In addition, the need of high CO_2 concentration for microalgae to produce high oil yield makes the gas streams released from combustion at thermal-power plants an ideal place for both microalgal biomass production and CO_2 mitigation (Sawayama et al., 1995; Yun et al., 1997). From the 40,000 different species, more than 3,000 have shown special relevance for biodiesel production, species belonging to the *Bacillariophyceae, Chlorophyceae, Cyanophyceae, Prymnesiophyceae, Eustigmatophyceae,* and *Prasinophyceae* classes. A few of these algae have been cultivated on a large scale, such as in a photobioreactor or in raceway ponds. The isolation of new strains of microalgae has been proposed in order to increase their photosynthetic efficiency (PE), biomass yield, growth rate, temperature tolerance, and their oil content, as well as to reduce the photoinhibition, photooxidation damage, and the light saturation among the cells. These parameters are of great importance and are closely related. The amount of cells is normally expressed in form of biomass yield or productivity (either volumetric or areal) in $g·L^{-1}$ day^{-1}, or in $g·m^{-2}$ day^{-1}, although PE (the percentage of total solar energy that can be utilized for photosynthesis) is also a common parameter to define the best conditions to maximize production. PE is normally maximal at high light intensities if artificial light is used or at low light intensities when natural sunlight is used (morning and afternoon) (Carlozzi et al., 2006). Excessive light intensity can cause photoinhibition and photooxidation damage in photosystem II by promoting the formation of singlet oxygen (Melis, 1999). Camacho et al. showed that photoinhibition can be reduced by intermittently supplying the light (Rubio et al., 2003). Camp et al. have recently studied the use of different marine species as appropriate feedstock for biodiesel production (Fuentes-Grunewald et al., 2009). Apart from the isolation of new species, the improvement of current isolates through genetic modification and metabolic engineering would solve many of the current problems for microalgal biodiesel production (Gressel, 2008). Also, improved photobioreactor design and the identification of the other potential by-products generated from the microalgal under biorefinery-based cultivation could substantially lower production costs of microalgal biodiesel (Mata-Alvarez et al., 2000; Chisti, 2007; Raven and Gregersen, 2007).

13.8.4 Microdiesel

Heterotrophic microorganisms represent another potential source of diesel fuel. There are two ways to achieve heterotrophic microbial production of fuel. The first way is to employ microorganisms that accumulate high amounts of triglycerides and to use these triglycerides for any of the above-described types of transesterification. Despite that triglycerides can be found in most eukaryotic organisms as storage compounds for carbon and energy, most bacteria accumulate a different types of storage lipids, such as poly(3-hydroxybutyric acid), or polyhydroxyalkanoic acids (Anderson and Dawes, 1990; Steinbuchel and Valentin, 1995). Nevertheless, some bacteria such as *Mycobacterium, Streptomyces, Nocardia, Rhodococcus,* and *Gordonia* spp. accumulate triglycerides as reserve compounds. Some species of *Rhodococcus* and *Gordonia* are capable of accumulating up to 80% (dry weight) of triglycerides (Alvarez and Steinbuchel, 2002; Gouda et al., 2008). An alternative for using naturally oleaginous microbes is to adapt well-studied microorganisms such as *Escherichia coli* into oleaginous microorganisms by genetic engineering (Lu et al., 2008). The growth of these microorganisms represents a potentially good feedstock for biodiesel production, given an appropriate way for recovering of the triglycerides. The ideal scenario would be the production and secretion of a water-immiscible triglycerides (Rude and Schirmer, 2009). A second way of producing biodiesel from heterotrophic microorganisms is the direct production of biodiesel by the actual microorganism, also known as microdiesel. For microdiesel production uses, genetically modified microorganism are able to produce FAEEs. Kalscheuer et al. managed to obtain FAEEs from microorganisms by heterologous expression in *Escherichia coli* of *Zymomonas mobilis* pyruvate decarboxylase (pdc_{zm}) together with the alcohol dehydrogenase ($adhB_{zm}$) and the unspecific acyltransferase ($ws/dgata_{Ab}$) from *Acinetobacter baylyi* strain ADP1. With this approach, ethanol formation was combined with subsequent esterification of the ethanol with the acyl moieties of coenzyme A thioesters of fatty acids, with FAEE concentrations of $1.28\,g\,L^{-1}$ (Kalscheuer et al., 2006). This microorganism was improved by combining this second way with the first one for microbial production of biodiesel by deregulating and increasing de novo fatty-acid biosynthesis, which in combination with unspecific acyltransferase expression and alcohol feeding led to high titers of fatty esters without the need for adding fatty acids (Rude and Schirmer, 2009). Therefore, the broad metabolic potential of microorganisms can be used for producing biodiesel feedstocks or biodiesel itself as a low-cost and effective biofuel alternative.

13.9 Disadvantages of Biodiesel as Diesel Fuel

As cited above, despite the great advantages of using biodiesel as a substitute for petrodiesel, there are some disadvantages of biofuels of oily biomass such as higher viscosity, lower energy content, higher cloud point and pour point, lower engine speed and power, injector choking, engine incompatibility, high price, and higher engine wear. Each of these problems can be solved in different ways. However, there are additional disadvantages that include emissions produced from feedstock production, by-products made during production process, or emissions due to combustion of this fuel. Nevertheless, the main drawbacks for presenting biodiesel as a real fuel alternative are emissions and oxidation stability. With regard to the emissions, it has been shown that combustion of biodiesel and biodiesel blends increase mononitrogen oxides (NO_x) emissions by some 10% compared with petrodiesel fuel, particularly under high-temperature burning. These oxides are produced during combustion, and are believed to aggravate breathing conditions, to produce ozone at surface heights, which is also an irritant, and to react with the oxygen present in the air, eventually forming nitric acid when dissolved in water, and thereby producing acid rain (Hurley et al., 2007). These NO_x emissions can be reduced by different catalyst. Commonly used catalysts in diesel engine exhaust after treatment include precious metals, single metal oxides, mixed metal oxides, and zeolites (Peng et al., 2007).

Another emission of concern during biodiesel production is methyl bromide, a potent greenhouse gas. The production of this gas occurs mainly for natural sources, since algae, fungi, and higher plants produce these gasses when grown. However, some crops such as *Brassica* emit orders of magnitude more methyl bromide than all other crops, which is becoming a concern because rapeseed is currently one of the most widely used crop for biodiesel production, especially in Europe (Saini et al., 1995). In order to lower the levels of methyl bromide and its alkane halide analogues emitted by rapeseed, silencing of the methyl transferase that uses *S*-adenosyl-L-methionine to methyl halides and bisulfides to methanethiol has been proposed (Rhew et al., 2003; Gressel, 2008).

Also, a problem of oxidation stability arises from the deterioration of the fuel properties of biodiesel during storage. In general, biodiesel has relatively poorer oxidation stability compared with petrodiesel (Clark et al., 1984; Ali et al., 1995; Monyem and Van Gerpen, 2001; Xin et al., 2008). The different factors that affect biodiesel stability are closely related to its fatty-acid composition, storage conditions such as temperature or light, container composition, and water and metal content of the biodiesel itself (Falk and Meyer-Pittroff, 2004; Knothe, 2005). To enhance oxidation stability, antioxidants are employed such as aromatic derivatives including 3,4,5-trihydroxybenzoic acid *n*-propyl ester, 1,2,3-trihydroxybenzene, butylated hydroxyanisole, 2,6-di-*tert*-butyl-1-hydroxy-4-methylbenzene, α-tocopherol acetate, α-tocopherol, γ-tocopherol, δ-tocopherol, propyl gallate, dodecyl gallate, gallic acid, octyl gallate, ascorbyl palmitate, pyrogallol, α-naphthol, natural tocopherol, eugenol, or 6-ethoxy-1,2-dihydro-2,2,4-trimethylquinoline, although nonaromatic compounds are also used such as dilauryl thiodipropionate, isopropyl 2-hydroxy-4-methylthio butanoate, lecithin, stearyl citrate, palmityl citrate, chlorophyll, ascorbic acid, and citric acid. However, special interest is paid to the nonaromatic and the non-N- and S-containing compounds, such as antioxidants in an effort to avoid undermining the environmentally friendly nature of biodiesel (Mittelbach and Schober, 2003; Liang et al., 2006; Manzanera et al., 2008).

13.10 Concluding Remarks

The use of biofuels from oily biomass offers many advantages, and biodiesel is presented as the best alternative to fossil fuels. However, biodiesel has not yet been accepted because of its higher price than petrodiesel. Despite the recent increases in petroleum prices and uncertainties concerning petroleum availability, more research on biodiesel production fuels for diesel engines is imperative to lower production cost. This research on biodiesel production needs to promote the benefits and avoid the drawbacks mentioned above. With increasing interest in this subject, the scientific community is dedicating efforts to reach a good technological level for biodiesel production. The results indicate that when compared with petrodiesel, biodiesel offers overall increased benefits in almost all fields of study, although further research is needed for more appropriate methods and feedstocks to develop cheaper as well as more socially and environmentally acceptable technology that eventually can replace fossil fuels.

Acknowledgments

This work has been supported by Spanish Ministry of Science and Innovation (Ministerio de Ciencia e Innovación, MICINN, España) through the project reference CTM2009-09270. Maximino Manzanera was granted by Programa Ramón y Cajal by Spanish Ministry of Science and Innovation (Ministerio de Ciencia e Innovación, MICINN, Spain) and the European Regional Development Fund (ERDF, European Union).

References

Abigor RD, Uadia PO, Foglia TA et al. (2000). Lipase-catalysed production of biodiesel fuel from some Nigerian lauric oils. *Biochemical Society Transactions* **28**: 979–981.

Akoh CC, Chang SW, Lee GC, and Shaw JF (2007). Enzymatic approach to biodiesel production. *Journal of Agricultural and Food Chemistry* **55**: 8995–9005.

Aksoy HA, Kahraman I, Karaosmanoglu F, and Civelekoglu H (1988). Evaluation of Turkish sulfur olive oil as an alternative diesel fuel. *Journal of the American Oil Chemists Society* **65**: 936–938.

Ali Y, Hanna MA, and Cuppett SL (1995). Fuel properties of tallow and soybean oil esters. *Journal of the American Oil Chemists Society* **72**: 1557–1564.

Alptekin E and Canakci M (2009). Characterization of the key fuel properties of methyl ester-diesel fuel blends. *Fuel* **88**: 75–80.

Alvarez HM and Steinbuchel A (2002). Triacylglycerols in prokaryotic microorganisms. *Applied Microbiology and Biotechnology* **60**: 367–376.

Anderson AJ and Dawes EA (1990). Occurrence, metabolism, metabolic role, and industrial uses of bacterial polyhydroxyalkanoates. *Microbiological Reviews* **54**: 450–472.

Antczak MS, Kubiak A, Antczak T, and Bielecki S (2009). Enzymatic biodiesel synthesis—Key factors affecting efficiency of the process. *Renewable Energy* **34**: 1185–1194.

Balcao VM, Paiva AL, and Malcata FX (1996). Bioreactors with immobilized lipases: State of the art. *Enzyme and Microbial Technology* **18**: 392–416.

Banerjee A, Sharma R, Chisti Y, and Banerjee UC (2002). *Botryococcus braunii*: A renewable source of hydrocarbons and other chemicals. *Critical Reviews in Biotechnology* **22**: 245–279.

Barnwal BK and Sharma MP (2005). Prospects of biodiesel production from vegetables oils in India. *Renewable and Sustainable Energy Reviews* **9**: 363–378.

Belafi-Bako K, Kovacs F, Gubicza L, and Hancsok J (2002). Enzymatic biodiesel production from sunflower oil by *Candida antarctica* lipase in a solvent-free system. *Biocatalysis and Biotransformation* **20**: 437–439.

Boocock DGB, Konar SK, Leung A, and Ly LD (1992). Fuels and chemicals from sewage-sludge. 1. The solvent-extraction and composition of a lipid from a raw sewage-sludge. *Fuel* **71**: 1283–1289.

Borowitzka MA (1998). Commercial production of microalgae: Ponds, tanks, tubes and fermenters, pp. 313–321. Elsevier Science BV, Noordwijkerhout, the Netherlands.

Boshnakova M, Dobrescu M, Flach B, Henard MC, Krautgartner R, and Lieberz S (2009). *Oilseeds Crop Update*. Vol. E49059, Richey B, ed., USDA Foreign Agricultural Service, Washington, DC.

Camacho FG, Grima EM, Miron AS, Pascual VG, and Chisti Y (2001). Carboxymethyl cellulose protects algal cells against hydrodynamic stress. *Enzyme and Microbial Technology* **29**: 602–610.

Canadell JG, Le Quere C, Raupach MR et al. (2007). Contributions to accelerating atmospheric CO_2 growth from economic activity, carbon intensity, and efficiency of natural sinks. *Proceedings of the National Academy of Sciences of the United States of America* **104**: 18866–18870.

Canakci M (2007). The potential of restaurant waste lipids as biodiesel feedstocks. *Bioresource Technology* **98**: 183–190.

Canakci M and Sanli H (2008). Biodiesel production from various feedstocks and their effects on the fuel properties. *Journal of Industrial Microbiology and Biotechnology* **35**: 431–441.

Canakci M and Van Gerpen J (1999). Biodiesel production via acid catalysis. *Transactions of the ASAE* **42**: 1203–1210.

Carlozzi P, Pushparaj B, Degl'Innocenti A, and Capperucci A (2006). Growth characteristics of *Rhodopseudomonas palustris* cultured outdoors, in an underwater tubular photobioreactor, and investigation on photosynthetic efficiency. *Applied Microbiology and Biotechnology* **73**: 789–795.

Carvalho AP, Meireles LA, and Malcata FX (2006). Microalgal reactors: A review of enclosed system designs and performances. *Biotechnology Progress* **22**: 1490–1506.

Cassman KG and Liska AJ (2007). Food and fuel for all: Realistic or foolish? *Biofuels Bioproducts and Biorefining(Biofpr)* **1**: 18–23.

Chang DYZ, VanGerpen JH, Lee I, Johnson LA, Hammond EG, and Marley SJ (1996). Fuel properties and emissions of soybean oil esters as diesel fuel. *Journal of the American Oil Chemists Society* **73**: 1549–1555.

Chang CC and Wan SW (1947). China's motor fuels from tung oil. *Industrial and Engineering Chemistry* **39**: 1543–1548.

Chapagain BP, Yehoshua Y, and Wiesman Z (2009). Desert date (*Balanites aegyptiaca*) as an arid lands sustainable bioresource for biodiesel. *Bioresource Technology* **100**: 1221–1226.

Chisti Y (2007). Biodiesel from microalgae. *Biotechnology Advances* **25**: 294–306.

Chisti Y (2008). Biodiesel from microalgae beats bioethanol. *Trends in Biotechnology* **26**: 126–131.

Chulalaksananukul W, Condoret JS, Delorme P, and Willemot RM (1990). Kinetic-study of esterification by immobilized lipase in n-hexane. *FEBS Letters* **276**: 181–184.

Clark SJ, Wagner L, Schrock MD, and Piennaar PG (1984). Methyl and ethyl soybean esters as renewable fuels for diesel-engines. *Journal of the American Oil Chemists Society* **61**: 1632–1638.

Colombie S, Tweddell RJ, Condoret JS, and Marty A (1998). Water activity control: A way to improve the efficiency of continuous lipase esterification. *Biotechnology and Bioengineering* **60**: 362–368.

Crabbe E, Nolasco-Hipolito C, Kobayashi G, Sonomoto K, and Ishizaki A (2001). Biodiesel production from crude palm oil and evaluation of butanol extraction and fuel properties. *Process Biochemistry* **37**: 65–71.

Davis SC, Anderson-Teixeira KJ, and DeLucia EH (2009a). Life-cycle analysis and the ecology of biofuels. *Trends in Plant Science* **14**: 140–146.

Davis JP, Geller D, Faircloth WH, and Sanders TH (2009b). Comparisons of biodiesel produced from unrefined oils of different peanut cultivars. *Journal of the American Oil Chemists Society* **86**: 353–361.

Demirbas A (2003). Biodiesel fuels from vegetable oils via catalytic and non-catalytic supercritical alcohol transesterifications and other methods: A survey. *Energy Conversion and Management* **44**: 2093–2109.

Demirbas A (2007). Importance of biodiesel as transportation fuel. *Energy Policy* **35**: 4661–4670.

Di Serio M, Tesser R, Pengmei L, and Santacesaria E (2007). Heterogeneous catalysts for biodiesel production. *Energy and Fuels* **22**: 207–217.

Dossat V, Combes D, and Marty A (1999). Continuous enzymatic transesterification of high oleic sunflower oil in a packed bed reactor: Influence of the glycerol production. *Enzyme and Microbial Technology* **25**: 194–200.

Doucha J, Straka F, and Livansky K (2005). Utilization of flue gas for cultivation of microalgae (*Chlorella* sp.) in an outdoor open thin-layer photobioreactor. *Journal of Applied Phycology* **17**: 403–412.

Du W, Xu YY, and Liu DH (2003). Lipase-catalyzed transesterification of soya bean oil for biodiesel production during continuous batch operation. *Biotechnology and Applied Biochemistry* **38**: 103–106.

Eriksen NT (2008). The technology of microalgal culturing. *Biotechnology Letters* **30**: 1525–1536.

Falk O and Meyer-Pittroff R (2004). The effect of fatty acid composition on biodiesel oxidative stability. *European Journal of Lipid Science and Technology* **106**: 837–843.

Fernandez FGA, Camacho FG, Perez JAS, Sevilla JMF, and Grima EM (1997). A model for light distribution and average solar irradiance inside outdoor tubular photobioreactors for the microalgal mass culture. *Biotechnology and Bioengineering* **55**: 701–714.

Fernandez FGA, Sevilla JMF, Perez JAS, Grima EM, and Chisti Y (2001). Airlift-driven external-loop tubular photobioreactors for outdoor production of microalgae: Assessment of design and performance. *Chemical Engineering Science* **56**: 2721–2732.

Feuge RO and Gros AT (1949). Modification of vegetable oils. 7. Alkali catalyzed interesterification of peanut oil with ethanol. *Journal of the American Oil Chemists Society* **26**: 97–102.

Formo MW (1954). Ester reactions of fatty materials. *Journal of the American Oil Chemists Society* **31**: 548–559.

Freedman B, Butterfield RO, and Pryde EH (1986). Transesterification kinetics of soybean oil. *Journal of the American Oil Chemists Society* **63**: 1375–1380.

Freedman B, Pryde EH, and Mounts TL (1984). Variables affecting the yields of fatty esters from transesterified vegetable-oils. *Journal of the American Oil Chemists Society* **61**: 1638–1643.

Fuentes-Grunewald C, Garces E, Rossi S, and Camp J (2009). Use of the dinoflagellate Karlodinium veneficum as a sustainable source of biodiesel production. *Journal of Industrial Microbiology and Biotechnology* **36**: 1215–1224.

Gibbs HK, Brown S, Niles JO, and Foley JA (2007). Monitoring and estimating tropical forest carbon stocks: Making REDD a reality. *Environmental Research Letters* **2**: 13.

Gibbs HK, Johnston M, Foley JA, Holloway T, Monfreda C, Ramankutty N, and Zaks D (2008). Carbon payback times for crop-based biofuel expansion in the tropics: The effects of changing yield and technology. *Environmental Research Letters* **3**: 034001 (10 pp), doi:10.1088/1748-9326/3/3/034001.

Goering CE and Fry B (1984). Engine durability screening-test of a diesel oil soy oil alcohol micro-emulsion fuel. *Journal of the American Oil Chemists Society* **61**: 1627–1632.

Goering CE, Schwab AW, Daugherty MJ, Pryde EH, and Heakin AJ (1982). Fuel properties of 11 vegetable-oils. *Transactions of the ASAE* **25**: 1472–1477.

Goff MJ, Bauer NS, Lopes S, Sutterlin WR, and Suppes GJ (2004). Acid-catalyzed alcoholysis of soy-bean oil. *Journal of the American Oil Chemists Society* **81**: 415–420.

Gouda MK, Omar SH, and Aouad LM (2008). Single cell oil production by *Gordonia* sp DG using agro-industrial wastes. *World Journal of Microbiology and Biotechnology* **24**: 1703–1711.

Gressel J (2008). Transgenics are imperative for biofuel crops. *Plant Science* **174**: 246–263.

Grima EM, Fernandez FGA, Camacho FG, and Chisti Y (1999). Photobioreactors: Light regime, mass transfer, and scaleup. *Journal of Biotechnology* **70**: 231–247.

Gupta R, Gupta N, and Rathi P (2004). Bacterial lipases: An overview of production, purification and biochemical properties. *Applied Microbiology and Biotechnology* **64**: 763–781.

Guschina IA and Harwood JL (2006). Lipids and lipid metabolism in eukaryotic algae. *Progress in Lipid Research* **45**: 160–186.

Haas MJ, McAloon AJ, Yee WC, and Foglia TA (2006). A process model to estimate biodiesel produc-tion costs. *Bioresource Technology* **97**: 671–678.

Haas MJ, Piazza GJ, and Foglia TA (2002). Enzymatic approaches to the production of biodiesel fuels. *Lipid Biotechnology* 587–598.

Hall JB (1992). Ecology of a key african multipurpose tree species, *Balanites aegyptiaca* (*balanitaceae*)—The state-of-knowledge. *Forest Ecology and Management* **50**: 1–30.

Harrington KJ and Darcyevans C (1985). Trans-esterification in situ of sunflower seed oil. *Industrial and Engineering Chemistry Product Research and Development* **24**: 314–318.

Harwood HJ (1984). Oleochemicals as a fuel-mechanical and economic-feasibility. *Journal of the American Oil Chemists Society* **61**: 315–324.

Hernandez-Martin E and Otero C (2008). Different enzyme requirements for the synthesis of biodie-sel: Novozym (R) 435 and Lipozyme (R) TL IM. *Bioresource Technology* **99**: 277–286.

Hu Q, Kurano N, Kawachi M, Iwasaki I, and Miyachi S (1998). Ultrahigh-cell-density culture of a marine green alga *Chlorococcum littorale* in a flat-plate photobioreactor. *Applied Microbiology and Biotechnology* **49**: 655–662.

Huang AHC (1993). Lipases. In *Lipid Metabolism in Plants*, Moore TS et al., eds., pp. 473–502. CRC Press Inc., Boca Raton, FL.

Hurley MD, Ball JC, Wallington TJ, Toft A, Nielsen OJ, Bertman S, and Perkovic M (2007). Atmospheric chemistry of a model biodiesel fuel, CH3C(O)O(CH2)(2)OC(O)CH3: Kinetics, mechanisms, and products of Cl atom and OH radical initiated oxidation in the presence and absence of NOx. *Journal of Physical Chemistry A* **111**: 2547–2554.

Iso M, Chen BX, Eguchi M, Kudo T, and Shrestha S (2001). Production of biodiesel fuel from triglyc-erides and alcohol using immobilized lipase. *Journal of Molecular Catalysis B—Enzymatic* **16**: 53–58.

Iwai M and Tsujisaka Y (1984). Fungal lipase. In *Lipases*, Borgström B and Brockman HL, eds., pp. 443–469. Elsevier, Amsterdam, the Netherlands.

Jaeger KE, Dijkstra BW, and Reetz MT (1999). Bacterial biocatalysts: Molecular biology, three-dimensional structures, and biotechnological applications of lipases. *Annual Review of Microbiology* **53**: 315–351.

Jaeger KE and Eggert T (2002). Lipases for biotechnology. *Current Opinion in Biotechnology* **13**: 390–397.

Jaeger KE, Ransac S, Dijkstra BW, Colson C, Vanheuvel M, and Misset O (1994). Bacterial lipases. *FEMS Microbiology Reviews* **15**: 29–63.

Jarde E, Mansuy L, and Faure P (2005). Organic markers in the lipidic fraction of sewage sludges. *Water Research* **39**: 1215–1232.

Jimenez C, Cossio BR, and Niell FX (2003). Relationship between physicochemical variables and productivity in open ponds for the production of Spirulina: A predictive model of algal yield. *Aquaculture* **221**: 331–345.

Joubert PH, Brown JMM, Hay IT, and Sebata PDB (1984). Acute-poisoning with jatropha-curcas (purging nut tree) in children. *South African Medical Journal* **65**: 729–730.

Kalscheuer R, Stolting T, and Steinbuchel A (2006). Microdiesel: *Escherichia coli* engineered for fuel production. *Microbiology-Sgm* **152**: 2529–2536.

Knothe G (2005). Dependence of biodiesel fuel properties on the structure of fatty acid alkyl esters. *Fuel Processing Technology* **86**: 1059–1070.

Knothe G and Steidley KR (2007). Kinematic viscosity of biodiesel components (fatty acid alkyl esters) and related compounds at low temperatures. *Fuel* **86**: 2560–2567.

Krichnavaruk S, Powtongsook S, and Pavasant P (2007). Enhanced productivity of *Chaetoceros calcitrans* in airlift photobioreactors. *Bioresource Technology* **98**: 2123–2130.

Kulkarni MG and Dalai AK (2006). Waste cooking oil—An economical source for biodiesel: A review. *Industrial and Engineering Chemistry Research* **45**: 2901–2913.

Kusdiana D and Saka S (2001). Kinetics of transesterification in rapeseed oil to biodiesel fuel as treated in supercritical methanol. *Fuel* **80**: 693–698.

Laane C, Boeren S, Vos K, and Veeger C (1987). Rules for optimization of biocatalysis in organic-solvents. *Biotechnology and Bioengineering* **30**: 81–87.

Li LL, Du W, Liu DH, Wang L, and Li ZB (2006). Lipase-catalyzed transesterification of rapeseed oils for biodiesel production with a novel organic solvent as the reaction medium. *Journal of Molecular Catalysis B-Enzymatic* **43**: 58–62.

Liang YC, May CY, Foon CS, Ngan MA, Hock CC, and Basiron Y (2006). The effect of natural and synthetic antioxidants on the oxidative stability of palm diesel. *Fuel* **85**: 867–870.

Liu KS (1994). Preparation of fatty-acid methyl esters for gas-chromatographic analysis of lipids in biological-materials. *Journal of the American Oil Chemists Society* **71**: 1179–1187.

Lotero E, Liu YJ, Lopez DE, Suwannakarn K, Bruce DA, and Goodwin JG (2005). Synthesis of biodiesel via acid catalysis. *Industrial and Engineering Chemistry Research* **44**: 5353–5363.

Lu XF, Vora H, and Khosla C (2008). Overproduction of free fatty acids in *E. coli*: Implications for biodiesel production. *Metabolic Engineering* **10**: 333–339.

Ma FR, Clements LD, and Hanna MA (1998). Biodiesel fuel from animal fat. Ancillary studies on transesterification of beef tallow. *Industrial and Engineering Chemistry Research* **37**: 3768–3771.

Ma FR and Hanna MA (1999). Biodiesel production: A review. *Bioresource Technology* **70**: 1–15.

Manzanera M, Molina-Munoz ML, and Gonzalez-Lopez J (2008). Biodiesel: An alternative fuel. *Recent Patterns in Biotechnology* **2**: 25–34.

Marchetti JM, Miguel VU, and Errazu AF (2007). Possible methods for biodiesel production. *Renewable and Sustainable Energy Reviews* **11**: 1300–1311.

Mata-Alvarez J, Mace S, and Llabres P (2000). Anaerobic digestion of organic solid wastes. An overview of research achievements and perspectives. *Bioresource Technology* **74**: 3–16.

Matsumiya Y, Wakita D, Kimura A, Sanpa S, and Kubo M (2007). Isolation and characterization of a lipid-degrading bacterium and its application to lipid-containing wastewater treatment. *Journal of Bioscience and Bioengineering* **103**: 325–330.

Mbaraka IK, Radu DR, Lin VSY, and Shanks BH (2003). Organosulfonic acid-functionalized mesoporous silicas for the esterification of fatty acid. *Journal of Catalysis* **219**: 329–336.

Melis A (1999). Photosystem-II damage and repair cycle in chloroplasts: What modulates the rate of photodamage in vivo? *Trends in Plant Science* **4**: 130–135.

Metting FB (1996). Biodiversity and application of microalgae. *Journal of Industrial Microbiology and Biotechnology* **17**: 477–489.

Metzger P and Largeau C (2005). *Botryococcus braunii*: A rich source for hydrocarbons and related ether lipids. *Applied Microbiology and Biotechnology* **66**: 486–496.

Miao XLR and Yao H (2009). Effective acid-catalyzed transesterification for biodiesel production. *Energy Conversion and Management* 50: 2680–2684.

Miron AS, Garcia MCC, Gomez AC, Camacho FG, Grima EM, and Chisti Y (2003). Shear stress toler-ance and biochemical characterization of *Phaeodactylum tricornutum* in quasi steady-state con-tinuous culture in outdoor photobioreactors. *Biochemical Engineering Journal* **16**: 287–297.

Miron AS, Gomez AC, Camacho FG, Grima EM, and Chisti Y (1999). Comparative evaluation of compact photobioreactors for large-scale monoculture of microalgae. *Journal of Biotechnology* **70**: 249–270.

Mittelbach M (1990). Lipase catalyzed alcoholysis of sunflower oil. *Journal of the American Oil Chemists Society* **67**: 168–170.

Mittelbach M and Schober S (2003). The influence of antioxidants on the oxidation stability of biodie-sel. *Journal of the American Oil Chemists Society* **80**: 817–823.

Molina E, Fernandez J, Acien FG, and Chisti Y (2001). Tubular photobioreactor design for algal cul-tures. *Journal of Biotechnology* **92**: 113–131.

Molina EM, Fernandez FG, Camacho FG, Rubio FC, and Chisti Y (2000). Scale-up of tubular photo-bioreactors. *Journal of Applied Phycology* **12**: 355–368.

Mondala A, Liang KW, Toghiani H, Hernandez R, and French T (2009). Biodiesel production by in situ transesterification of municipal primary and secondary sludges. *Bioresource Technology* **100**: 1203–1210.

Monyem A and Van Gerpen JH (2001). The effect of biodiesel oxidation on engine performance and emissions. *Biomass and Bioenergy* **20**: 317–325.

Mukherjee KD and Hills MJ (1994). Lipases from plants. In *Lipases: Their Structure, Biochemistry and Application*, Woolley P and Petersen SB, eds., pp. 49–75. Cambridge University Press, Cambridge, U.K.

Murugesan A, Umarani C, Subramanian R, and Nedunchezhian N (2009). Bio-diesel as an alternative fuel for diesel engines—A review. *Renewable and Sustainable Energy Reviews* **13**: 653–662.

Nelson LA, Foglia TA, and Marmer WN (1996). Lipase-catalyzed production of biodiesel. *Journal of the American Oil Chemists Society* **73**: 1191–1195.

Niehaus RA, Goering CE, Savage LD and Sorenson SC (1986). Cracked soybean oil as a fuel for a diesel-engine. *Transactions of the ASAE* **29**: 683–689.

Noureddini H, Gao X, and Philkana RS (2005). Immobilized *Pseudomonas cepacia* lipase for biodiesel fuel production from soybean oil. *Bioresource Technology* **96**: 769–777.

Nye MJ, Williamson TW, Deshpande S, Schrader JH, Snively WH, Yurkewich TP, and French CL (1983). Conversion of used frying oil to diesel fuel by trans-esterification—Preliminary tests. *Journal of the American Oil Chemists Society* **60**: 1598–1601.

Palekar AA, Vasudevan PT, and Yan S (2000). Purification of lipase: A review. *Biocatalysis and Biotransformation* **18**: 177–200.

Peng XS, Lin H, Shangguan WF, and Huang Z (2007). A highly efficient and porous catalyst for simultaneous removal of NOx and diesel soot. *Catalysis Communications* **8**: 157–161.

Peterson CL, Auld DL, and Korus RA (1983). Winter rape oil fuel for diesel-engines—Recovery and utilization. *Journal of the American Oil Chemists Society* **60**: 1579–1587.

Peterson CL, Reece DL, Cruz R, and Thompson J (1992). A comparison of ethyl and methyl-esters of vegetable oil as diesel fuel substitutes. *Liquid Fuels from Renewable Resources* 99–110.

Pinto AC, Guarieiro LLN, Rezende MJC et al. (2005). Biodiesel: An overview. *Journal of the Brazilian Chemical Society* **16**: 1313–1330.

Pryde EH (1984). Vegetable-oils as fuel alternatives—Symposium overview. *Journal of the American Oil Chemists Society* **61**: 1609–1610.

Pushparaj B, Pelosi E, Tredici MR, Pinzani E, and Materassi R (1997). An integrated culture system for outdoor production of microalgae and cyanobacteria. *Journal of Applied Phycology* **9**: 113–119.

Ramadhas AS, Jayaraj S, and Muraleedharan C (2005). Biodiesel production from high FFA rubber seed oil. *Fuel* **84**: 335–340.

Ranganathan SV, Narasimhan SL, and Muthukumar K (2008). An overview of enzymatic production of biodiesel. *Bioresource Technology* **99**: 3975–3981.

Raven R and Gregersen KH (2007). Biogas plants in Denmark: Successes and setbacks. *Renewable and Sustainable Energy Reviews* **11**: 116–132.

Rhew RC, Ostergaard L, Saltzman ES, and Yanofsky MF (2003). Genetic control of methyl halide production in Arabidopsis. *Current Biology* **13**: 1809–1813.

Rived F, Canals I, Bosch E, and Roses M (2001). Acidity in methanol-water. *Analytica Chimica Acta* **439**: 315–333.

Roessler PG, Brown LM, Dunahay TG et al. (1994). Genetic-engineering approaches for enhanced production of biodiesel fuel from microalgae. *Enzymatic Conversion of Biomass for Fuels Production* **566**: 255–270.

Royon D, Daz M, Ellenrieder G, and Locatelli S (2007). Enzymatic production of biodiesel from cotton seed oil using *t*-butanol as a solvent. *Bioresource Technology* **98**: 648–653.

Rubio FC, Camacho FG, Sevilla JMF, Chisti Y, and Grima EM (2003). A mechanistic model of photosynthesis in microalgae. *Biotechnology and Bioengineering* **81**: 459–473.

Rude MA and Schirmer A (2009). New microbial fuels: A biotech perspective. *Current Opinion in Microbiology* **12**: 274–281.

Saini HS, Attieh JM, and Hanson AD (1995). Biosynthesis of halomethanes and methanethiol by higher-plants via a novel methyltransferase reaction. *Plant Cell and Environment* **18**: 1027–1033.

Saka S and Kusdiana D (2001). Biodiesel fuel from rapeseed oil as prepared in supercritical methanol. *Fuel* **80**: 225–231.

Samios D, Pedrotti F, Nicolau A, Reiznautt QB, Martini DD, and Dalcin FM (2009). A Transesterification Double Step Process—TDSP for biodiesel preparation from fatty acids triglycerides. *Fuel Processing Technology* **90**: 599–605.

Sawayama S, Inoue S, Dote Y, and Yokoyama SY (1995). CO_2 fixation and oil production through microalga. *Energy Conversion and Management* **36**: 729–731.

Schlick ML, Hanna MA, and Schinstock JL (1988). Soybean and sunflower oil performance in a diesel-engine. *Transactions of the ASAE* **31**: 1345–1349.

Schwab AW, Dykstra GJ, Selke E, Sorenson SC, and Pryde EH (1988). Diesel fuel from thermal-decomposition of soybean oil. *Journal of the American Oil Chemists Society* **65**: 1781–1786.

Sellappan S and Akoh CC (2005). Application of lipases in modification of food lipids. In *Handbook of Industrial Catalysis*, Hou CT et al., eds., pp. 9–39. Taylor & Francis, Boca Raton, FL.

Shah S and Gupta MN (2007). Lipase catalyzed preparation of biodiesel from jatropha oil in a solvent free system. *Process Biochemistry* **42**: 409–414.

Shen L and Zhang DK (2003). An experimental study of oil recovery from sewage sludge by low-temperature pyrolysis in a fluidized-bed. *Fuel* **82**: 465–472.

Shimada Y, Watanabe Y, Sugihara A, and Tominaga Y (2002). Enzymatic alcoholysis for biodiesel fuel production and application of the reaction to oil processing. *Journal of Molecular Catalysis B-Enzymatic* **17**: 133–142.

Solomon BD, Barnes JR, and Halvorsen KE (2007). Grain and cellulosic ethanol: History, economics, and energy policy. *Biomass and Bioenergy* **31**: 416–425.

Sonntag NOV (1979). Fat splitting. *Journal of the American Oil Chemists Society* **56**: A729–A732.

Soumanou MM and Bornscheuer UT (2003). Improvement in lipase-catalyzed synthesis of fatty acid methyl esters from sunflower oil. *Enzyme and Microbial Technology* **33**: 97–103.

Spolaore P, Joannis-Cassan C, Duran E, and Isambert A (2006). Commercial applications of microalgae. *Journal of Bioscience and Bioengineering* **101**: 87–96.

Srivastava A and Prasad R (2000). Triglycerides-based diesel fuels. *Renewable and Sustainable Energy Reviews* **4**: 111–133.

Steinbuchel A and Valentin HE (1995). Diversity of bacterial polyhydroxyalkanoic acids. *Fems Microbiology Letters* **128**: 219–228.

Stevenson DE, Stanley RA, and Furneaux RH (1994). Near-quantitative production of fatty-acid alkyl esters by lipase-catalyzed alcoholysis of fats and oils with adsorption of glycerol by silica-gel. *Enzyme and Microbial Technology* **16**: 478–484.

Tate RE, Watts KC, Allen CAW, and Wilkie KL (2006). The viscosities of three biodiesel fuels at temperatures up to 300 degrees C. *Fuel* **85**: 1010–1015.

Terry KL and Raymond LP (1985). System-design for the autotrophic production of microalgae. *Enzyme and Microbial Technology* **7**: 474–487.

Thomas ER and Sudborough JJ (1912). The direct esterification of saturated and unsaturated acids. *Journal of the Chemical Society* **101**: 317–328.

Tiwari AK, Kumar A, and Raheman H (2007). Biodiesel production from jatropha (*Jatropha curcas*) with high free fatty acids: An optimized process. *Biomass and Bioenergy* **31**: 569–575.

Trent W (1945). *Process for Treating Fatty Glycerides*. US Patent 2383633, issue date: August 28, 1945; assignee: Colgate Palmolive Peet Co.

Van Gerpen J (2005). Biodiesel processing and production. *Fuel Processing Technology* **86**: 1097–1107.

Vieira JAV, Sabba M, Dias BS et al. (2007). Preparing biodiesel from natural oils and/or fats comprises mixing and reacting oil/fat, alcohol and catalyst, separating the dense and light phases followed by purifying. Petrobras Petroleo Brasil Sa; Benson J E.

Wright H, Segur J, Clark H, Coburn S, Langdon E, and DuPuis R (1944). A report on ester interchange. *Journal of the American Oil Chemists' Society* **21**: 145–148.

Xin JY, Imahara H, and Saka S (2008). Oxidation stability of biodiesel fuel as prepared by supercritical methanol. *Fuel* **87**: 1807–1813.

Xu Y, Du W, Zeng J, and Liu D (2004). Conversion of soybean oil to biodiesel fuel using lipozyme TL IM in a solvent-free medium. *Biocatalysis and Biotransformation* **22**: 45–48.

Xu L, Weathers PJ, Xiong XR, and Liu CZ (2009). Microalgal bioreactors: Challenges and opportunities. *Engineering in Life Sciences* **9**: 178–189.

Yun YS, Lee SB, Park JM, Lee CI, and Yang JW (1997). Carbon dioxide fixation by algal cultivation using wastewater nutrients. *Journal of Chemical Technology and Biotechnology* **69**: 451–455.

Zhang Y, Dube MA, McLean DD, and Kates M (2003a). Biodiesel production from waste cooking oil: 1. Process design and technological assessment. *Bioresource Technology* **89**: 1–16.

Zhang Y, Dube MA, McLean DD, and Kates M (2003b). Biodiesel production from waste cooking oil: 2. Economic assessment and sensitivity analysis. *Bioresource Technology* **90**: 229–240.

Ziejewski M, Kaufman KR, Schwab AW, and Pryde EH (1984). Diesel-engine evaluation of a nonionic sunflower oil aqueous ethanol microemulsion. *Journal of the American Oil Chemists Society* **61**: 1620–1626.

14

Fossil Fuel Decarbonization: In the Quest
for Clean and Lasting Fossil Energy

Nazim Z. Muradov

CONTENTS

14.1 Introduction

Intensive worldwide efforts are under way to tackle anthropogenic greenhouse gas (GHG) emissions and their potentially devastating effect on our planet's biosphere and climate. There are numerous ways to deal with man-made CO_2 emissions and some measures are already underway. For example, several nations have already implemented carbon tax to discourage industries from generating and emitting massive quantities of CO_2 into the atmosphere. Widespread energy conservation, an increase in the efficiency of power plants and appliances, and an improvement in fuel economy of vehicles are also important first steps toward the carbon mitigation objectives. It is realized, however, that these measures, no matter how efficient, are unlikely to deliver the level of CO_2 emission reduction that would be necessary to stabilize the atmospheric CO_2 concentration against a backdrop of the ever growing global demand for energy. Thus, more radical approaches to the carbon mitigation problem are being sought, and one approach that many experts are pinning their hopes on is fossil fuel decarbonization, considered as a feasible and potentially cost-effective near-term solution for curbing man-made CO_2 emissions.

The objective of the fossil fuel decarbonization approach is to eliminate or drastically reduce the amount of CO_2 emitted from the use of primary fossil resources such as coal, petroleum, and natural gas (NG). The main rationale for the fossil decarbonization concept is that it potentially offers an extension of fossil fuel era by perhaps couple hundreds years (purportedly) without an adverse impact on our planet's ecosystem. Carbon (to be precise, CO_2) capture and storage (or sequestration) (CCS) is the most recognized and advanced fossil decarbonization strategy, particularly popular among the proponents of "clean" coal technology (note that in this chapter, the terms "CO_2 storage" and "CO_2 sequestration" are interchangeable). CCS is a way of preventing anthropogenic CO_2 emissions from reaching the atmosphere by capturing and securely storing CO_2 in such sinks as geological formations, the ocean floor, terrestrial ecosystems, etc. It is believed that as a result of fossil decarbonization, the concentration of CO_2 in the atmosphere could be stabilized at about 450 ppm level without abandoning the fossil fuel infrastructure. However, opponents of the CCS approach are concerned that it could provide only a temporary relief, and would make mankind even more dependent on fossil fuels, thus, making the necessary changes later even more difficult.

Most experts agree that the supply of primary energy will continue to be dominated by fossil fuels until at least the mid-century. Many energy scenarios project that known and emerging technological options could achieve a broad range of atmospheric CO_2

stabilization levels, but that the implementation of these measures would require a wide range of socioeconomic and institutional changes (IPCC CCS 2005). From this viewpoint, the availability of fossil decarbonization in the portfolio of carbon mitigation options could potentially facilitate achieving atmospheric CO_2 stabilization goals.

As of 2009, there were 275 active or planned CCS projects around the world, which is an indication of a growing commitment to this technological option. The importance of CCS as a carbon mitigation strategy was recently underscored by the Group of Eight (G8) at its 33rd convention in Hokkaido, Japan (July 2008). In particular, the G8 leaders expressed their support for initiating 20 new industrial-scale (larger than 1 Mt year^{-1}) CCS demonstration projects with the view of a widespread commercial deployment of the CCS technology by 2020 (Global CCS Institute 2009).

CCS is an extremely complex set of industrial-scale processes and operations comprising three major steps: CO_2 capture, transport/injection, and storage, and a diversity of technological options associated with each of these steps. Figure 14.1 provides a general outline of CCS system, which includes well-established and emerging technologies. The objective of this chapter is to assess the current state of knowledge and development with regard to scientific, technical, environmental, economic, and societal aspects of fossil decarbonization and its role as a carbon mitigation option.

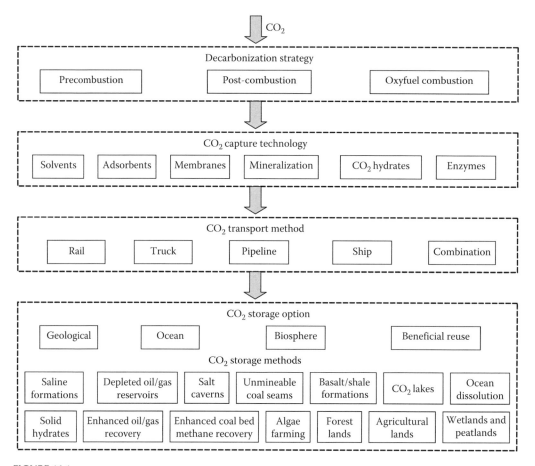

FIGURE 14.1
General outline of CO_2 capture and storage systems.

14.2 Fossil Fuel Decarbonization Strategies

Currently, several technological approaches to fossil fuel decarbonization are being developed, three of them, post-combustion carbon capture, precombustion carbon capture, and oxyfuel combustion, are well-developed technologies. Chemical looping combustion (CLC) and direct decarbonization technologies are still at pilot/demonstration and R&D stages of the development, respectively. The following is a brief discussion of the current status of the fossil decarbonization technologies.

14.2.1 Post-Combustion CO_2 Capture

In the post-combustion carbon capture option (Figure 14.2), CO_2 is captured after fossil fuel combustion. It is primarily applicable to conventional coal- and gas-fired power generation stations and other large industrial CO_2 point sources. In a typical coal- or gas-fired power generation system, the exhaust gas, or flue gas, consists mainly of nitrogen (N_2) and CO_2. CO_2 separation from the flue gas is challenging for the following reasons:

- CO_2 concentration in flue gases is relatively low: 13–15 vol.% and 3–4 vol.% for coal- and gas-fired power plants, respectively (U.S. DOE, NETL 2003).
- Pressure is low (1–1.7 ata), which necessitates the treatment of a large volume of gas.
- CO_2 has to be compressed from about atmospheric pressure to pipeline pressure of about 138 ata, which represents a significant auxiliary power load.
- The presence of impurities in the flue gas (e.g., SO_x, NO_x, and particulate matter) could potentially degrade some physical and chemical sorbents and adversely affect the performance of CO_2 capture processes.

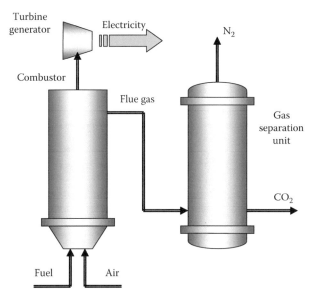

FIGURE 14.2
Schematic diagram of fossil fuel decarbonization via post-combustion CO_2 capture.

Despite the above challenges, the post-combustion capture systems offer certain advantages such as flexibility from the design and operation perspectives. Currently, this technology has niche market applications, and its use in power plants has been restricted to slipstream applications. However, it is projected that coal-fired power plants equipped with post-combustion capture will eventually dominate the field, in particular, a number of large-scale facilities are planned between 2012 and 2016 (IPCC CCS 2005; Global CCS Institute 2009). Preliminary analysis conducted at the U.S. DOE National Energy Technology Laboratory (NETL) indicates that CO_2 capture and compression to 152 ata could raise the cost of electricity (COE) by about 65% (U.S. DOE, NETL 2003).

14.2.2 Oxyfuel Combustion

In the oxyfuel combustion approach (Figure 14.3), fossil fuel is burned using oxygen instead of air along with recycled CO_2 and H_2O resulting in flue gas consisting mainly of water vapor and CO_2 (about 60 vol.% and more). After condensation of water (through the use of cooling, desiccant systems, and compression to a dew point of −40°C), almost pure CO_2 stream can be transported to a sequestration site. Oxy-combustion offers several advantages as follows:

- Applicability to existing and new coal-fired power plants (use of conventional steam cycle technology)
- Sixty to seventy percent reduction in NO_x emissions compared to air-fired combustion (U.S. DOE, NETL 2003)
- Possibility of co-capture of SO_2 (if co-disposal becomes technically and economically feasible)
- Increased mercury removal (due to oxidation of mercury followed by its downstream removal in the electrostatic precipitator)
- Potential cost savings (due to more compact units and the elimination of certain gas cleanup devices)

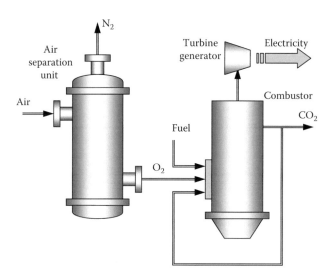

FIGURE 14.3
Block diagram of an oxyfuel combustion system.

The technology, however, needs to overcome several technical challenges to become competitive, in particular:

- The significant amount of oxygen required in the oxy-combustion process, increasing CO_2 capture costs. Oxygen is typically produced by energy-intensive cryogenic air separation or using adsorption techniques.
- Operation of an O_2-fired boiler (high-temperature-resistant materials, possible air leakage, etc.).
- Need for flue gas recycle (about 70%–80%) to approximate the combustion characteristics of air (in order to use a currently available combustion equipment).
- CO_2-NO_x-SO_x co-disposal problem (compressibility, corrosion issues during pipeline transport, environmental acceptability, economic viability, etc.).
- Environmental issues related to emissions of CO or unburned carbon, high concentration of acidic gases in the condensate, injection of toxic substances to sequestration sites, etc.

Although the technology could take advantage of the progress achieved in air separation and flue gas treatment processes, no full-scale commercial units are available today. Babcock & Wilcox Co. (Ohio) has successfully completed pulverized coal (PC) oxy-combustion testing at $1.5\,MW_{th}$ scale unit, and the technology is currently being evaluated at the $30\,MW_{th}$ pilot-scale unit. Preliminary results are encouraging: 80% reduction in the flue gas volume while achieving CO_2 concentration of 80 vol.% has been demonstrated (Ciferno et al. 2009). The pilot-scale testing has also demonstrated the possibility of a smooth transition between air- and oxygen-firing modes. Alstom Power also conducted pilot-scale ($3\,MW_{th}$) testing of the oxy-combustion process using a circulating fluidized bed combustor with coal and petroleum coke as fuels. Among other pilot-scale oxy-combustion systems, one should mention Vattenfall process (Schware Pumpe, Germany). Among planned full-scale commercial oxyfuel combustion units is a 200 MW coal-fired power plant in Meredosia (Illinois) (Johnson 2010). The oxyfuel combustion technology is also being actively pursued in Australia, where several existing coal power plants will be retrofitted with the oxyfuel and amine-based CO_2 capture systems. (Australia is particularly interested in the development of CCS technologies, because 80% of Australia's electricity comes from coal, and, besides, nuclear power has never been politically or socially acceptable in this country.) (Ritter 2010) Currently, the economic benefits of oxy-combustion compared to amine-based scrubbing systems is rather marginal (although certain improvements could be made to reduce its costs, for example, via the use of ion transport and chemical looping systems that are being developed).

14.2.3 Precombustion CO_2 Capture

A simplified block diagram of the precombustion capture technology is shown in Figure 14.4. In the gasification reactor, fuel (e.g., coal or NG) is gasified (or reformed) in the presence of oxidants (steam, oxygen, or oxygen-enriched air) into syngas (predominantly, H_2–CO mixture). The syngas is directed to a water gas shift (WGS) reactor where CO is converted to CO_2 and additional hydrogen in the presence of steam:

$$CO + H_2O \rightarrow H_2 + CO_2 \quad \Delta H^\circ = -41\,kJ\,mol^{-1} \tag{14.1}$$

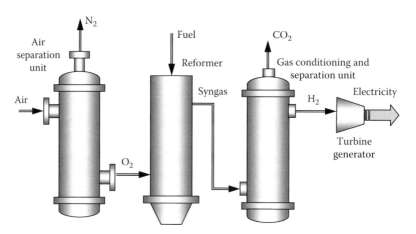

FIGURE 14.4
Simplified block diagram of precombustion CO_2 capture technology.

The CO_2 concentration in the resulting gaseous mixture varies in the range of 20–60 vol.% (on a dry basis, with balance being hydrogen, methane, CO, and minor impurities), depending on the feedstock composition and operational parameters. Separated CO_2 is then ready for transportation and sequestration. Essentially, carbon-free hydrogen fuel can be fired in a turbine-generating electricity (additional electricity can be produced by a steam turbine utilizing hot flue gas) or used in a variety of heaters and boilers. Note that in most applications, it is permissible for hydrogen fuel to contain low levels of methane, CO_x, and N_2. (More detailed discussion of coal gasification and hydrocarbon reforming technologies can be found in Section 14.6.3.)

Because CO_2 concentration in the shifted syngas is much higher than that in the postcombustion flue gas, CO_2 capture is less expensive for the precombustion than for the postcombustion capture technology. A state-of-the-art process for CO_2 capture from shifted syngas is based on physical absorption using glycol-based solvent called Selexol. A technoeconomic analysis conducted at NETL indicates that Selexol-based CO_2 capture raises the COE from a newly built coal-fired power plant by 30% (U.S. DOE, NETL 2003). U.S. DOE Gasification Research Program is expected to significantly improve coal gasification technology such that its cost will be comparable to that of PC combustion, potentially further reducing the cost of precombustion CO_2 capture in the future.

Precombustion strategy can also be utilized for the reduction in carbon content of fuels (the excess carbon is removed in the form of CO_2). For example, fuel with low H:C ratio (e.g., coal, where H:C ≈ 0.1) can be converted to fuels with significantly higher H:C ratio (e.g., hydrocarbons, where H:C ratio varies from 2 to 4). This can be accomplished by gasifying coal to syngas, followed by converting syngas to liquid hydrocarbons via Fischer–Tropsch synthesis, or methane via a methanation process.

As pointed out in several reports, for example, IPCC CCS (2005), the precombustion systems may be suited to implement CO_2 capture at a lower incremental cost compared to the same type of base technology without CO_2 capture (with a key driver being the absolute cost of the CO_2-emission-free product). Precombustion systems are flexible and strategically important due to their capacity to deliver a suitable mix of electricity, hydrogen, and low-carbon-containing fuels or chemical feedstocks with a relatively high efficiency.

14.2.4 Chemical Looping Combustion

The main principle of CLC is to oxidize carbonaceous fuels not directly by air, but rather by an oxygen carrier, which circulates between two reactors designed to combust fuel and regenerate the oxygen carrier. Oxides of iron (Fe_2O_3), copper (CuO), nickel (NiO), and manganese (Mn_2O_3) are among suitable oxygen carriers (Zafar et al. 2005). The advantage of CLC over conventional combustion approaches is that the use of oxygen carriers obviates the need for an air separation plant, and, as a result, CO_2 is formed in a highly concentrated ready for sequestration form (since it is not diluted with nitrogen).

Figure 14.5 depicts a simplified schematic diagram of a CLC system coupled with turbines. In the metal oxidation reactor, the reduced form of an oxygen carrier, for example, Ni, exothermically reacts with air yielding NiO:

$$2Ni + O_2(+N_2) \rightarrow 2NiO(+N_2) + heat \tag{14.2}$$

The exothermic reaction increases the temperature of air that enters a downstream expansion turbine producing electricity. The oxidized form of the oxygen carrier (e.g., NiO) is transported to a fuel oxidation (or metal oxide reduction) reactor, where it is reacted with carbonaceous fuel (e.g., NG), resulting in the reduction of metal oxide to its original (metallic) form with the release of heat:

$$4NiO + CH_4 \rightarrow 4Ni + CO_2 + 2H_2O + heat \tag{14.3}$$

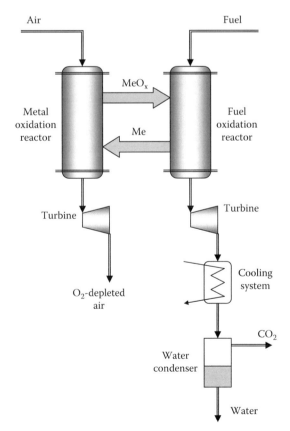

FIGURE 14.5
Schematic diagram of a CLC system.

Due to the strongly exothermic reaction (14.3), the exit stream can be expanded in a turbine for power generation. After water condensing, high-purity CO_2 can be compressed to necessary pressure for pipeline transmission. Thus, the overall reaction can be represented as CH_4 combustion with oxygen (not air):

$$CH_4 + 2O_2 \rightarrow CO_2 + 2H_2O + heat \tag{14.4}$$

The oxygen carrier particles (typically, 100–500 µm in size) circulate between two reactors in a fluidized form (which enables an efficient heat and mass transfer). Operational temperatures in the CLC process vary in the range of 800°C–1200°C. Thermodynamic analysis of the CLC system indicates that relatively high theoretical efficiencies could be achieved by the CLC-based power station (McGlashan 2008). It was estimated that the fuel-to-electricity (FTE) energy conversion efficiency of the CLC system running on NG fuel is in the range of 45%–50% (Brandvoll and Bolland 2004). Currently, the CLC process is still in pilot/demonstration stages of the development. Major challenges facing the CLC technology relate to the development of oxygen carriers (both in bulk and supported forms) that possess adequate long-term mechanical and chemical stabilities.

14.2.5 Direct Decarbonization

The direct decarbonization approach is based on the dissociation of hydrocarbons to hydrogen and elemental carbon in an air–water-free environment (the term "direct decarbonization" was suggested in the literature, because this approach involves direct extraction of carbon from fossil fuels, rather than its preliminary conversion to CO_2) (Muradov 2001a). Decarbonization of hydrocarbons is accomplished via a simple dissociation reaction:

$$C_nH_m \rightarrow nC + m/2H_2 \tag{14.5}$$

where $n \geq 1$ and $m \geq 2$.

Typically, this reaction is moderately endothermic and requires elevated temperatures (600°C–1000°C), depending on the type of hydrocarbon. Hydrogen produced in the process is used as a carbon-free energy carrier, for example, fuel for transportation and power generation, whereas carbon product can be utilized in a variety of traditional and novel application areas, or it can be securely stored in an environmentally safe manner (obviously, it is much safer to store solid carbon than gaseous or liquid CO_2). As a decarbonization strategy, decomposition of hydrocarbons is a relatively new approach in the portfolio of carbon mitigation options. More detailed discussion of this technological approach is given in Section 14.12 dedicated to alternative decarbonization methods.

14.3 CO_2 Capture Technologies

CO_2 capture is the first step in the multistage process of CO_2 separation from flue gases or shifted syngas and its storage (sequestration). In general, the CO_2 capture process is energy intensive and very costly (it accounts for the most of the CCS system cost). Therefore, CO_2 capture technology is a focus of intensive worldwide efforts aimed at improving the process efficiency and significantly reducing its cost. The overwhelming majority of these efforts are targeting CO_2 capture from coal-fired power plants because they are now the largest stationary sources of CO_2 emissions, and, most likely, will remain as such in the foreseeable future. However, the developed technology is also applicable to NG-fired power plants and the majority of large point sources of industrial CO_2 emissions, such as synthetic fuel plants, cement-manufacturing and metallurgical plants, hydrogen production plants, refineries, etc.

Technological options for CO_2 capture from CO_2-containing streams are summarized in Table 14.1. CO_2 capture processes are broadly divided into two classes: the technologies for CO_2 capturing from diluted streams (3–15 vol.%, e.g., from post-combustion flue gases) and from relatively concentrated streams (up to 40 vol.% and higher, e.g., from shifted syngas of precombustion capture processes or hydrogen-manufacturing plants) (although some

TABLE 14.1

CO_2 Capture Technologies for CCS Applications

Post-combustion capture (typically, from diluted CO_2 streams, 3–15 vol.% CO_2)	Chemical solvents	Amines: monoethanolamine (MEA); diglycolamine (DGA®)
		Advanced amines: stearically hindered amines KS-1, KS-2, and KS-3; Cansolv; HTC Purenergy; 2-amino-2-methyl-1-propanol (AMP)
		Ammonia: aqueous ammonia; chilled ammonia
	Physical solvents	Ionic liquids
	Chemical sorbents	Amine-enriched sorbents
		Metal-organic frameworks
	Physical sorbents	Metal-organic frameworks
	Membranes	N_2/CO_2 polymer membranes
		Membrane-amine hybrids
		Gas absorption membranes
	Enzymatic CO_2 capture processes	Carbozyme
Precombustion capture (typically, from relatively concentrated CO_2 streams up to 40 vol.% and higher)	Chemical solvents	Amines: diethanolamine (DEA); methyldiethanolamine (MDEA)
		Potassium carbonate (Benfield process)
	Physical solvents	Glycol: Selexol
		Methanol: Rectisol
		Propylene carbonates: Fluor process
		n-Methyl-2-pyrolidone (Purisol)
	Hybrid physical/chemical solvent absorption	Sulfinol (mixture of diisopropanolamine and tetrahydrothiophene oxide); Flexsorb® PS; Ucarsol® LE: Amisol
	Physical sorbents, PSA	Zeolites; activated carbon; metal-organic frameworks
	Membranes	Polymer membranes
		Ceramic membranes
		Hollow fiber membrane supports
	Cryogenic distillation	
	Hydrates	

Source: Data from U.S. DOE, NETL 2007; Global CCS Institute, Strategic analysis of the global status of carbon capture and storage, Final Report, Global CCS Institute, http://www/globalccsinstitute.com/downloads/reports/2009/worley/foundation-report-1-rev0.pdf (accessed August 3, 2010), 2009.

technologies are applicable to both cases). The technologies listed in this table fall into one of the following categories: chemical and physical absorption, chemical and physical adsorption, gas-separation membranes, cryogenic distillation, enzymatic systems, and hydrates. A brief discussion of some of the practically important CO_2 capture technologies follows.

14.3.1 Chemical and Physical Absorption

Chemical absorption systems are preferred for CO_2 capture from streams with low to moderate CO_2 concentrations. Since CO_2 is an acidic gas, most of the absorbing media for its capture are basic solvents, and the efficiency of CO_2 removal is controlled by acid–base neutralization reactions. Chemical absorption systems include aqueous amines, ammonia, and hot potassium carbonate. The most commonly used solvents for CO_2 capture in commercial systems are alkanolamines such as monoethanolamine (MEA), diethanolamine (DEA), and methyl-diethanolamine (MDEA). At low CO_2 partial pressures in the flue gases, alkanolamines are capable of achieving the high CO_2 recovery levels of 90% and more due to fast kinetics and high chemical reactivity (although the process capacity is equilibrium limited). Figure 14.6 shows a simplified schematic diagram of CO_2 recovery from diluted flue gases using MEA solvent.

   The use of the off-the-shelf amine-based solvents, however, is associated with a high energy penalty related to their regeneration via steam stripping (for MEA solvent, $\Delta H_R = 1.9\,MJ\,kg^{-1}\,CO_2$ captured). The minimum work required to separate CO_2 from coal-fired flue gas and compress it to 150 atm is 0.11 MWh t^{-1} CO_2 (Rochelle 2009). The presence in flue gases of such contaminants as SO_x, NO_x, hydrocarbons, and a particulate matter is undesirable, since these impurities may eventually reduce the absorption capacity of amine-based solvents and cause equipment corrosion problems. To prevent these problems, some commercial units practice various pretreatment options, which may add to the cost of CO_2 recovery. Other operating problems encountered in the amine solvent systems

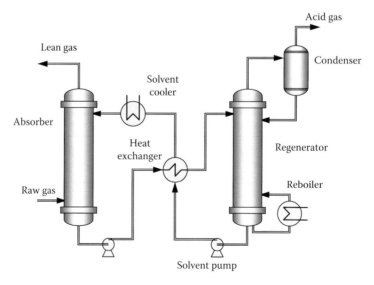

FIGURE 14.6
Schematic diagram of an amine-based CO_2 capture system.

are foaming, vapor entrainment of the solvent, and replenishment of the solvent; however, these factors have a small effect on the overall cost of the process.

Although MEA-based CO_2 scrubbing has been practiced for more than 60 years for NG and hydrogen purification and food-grade CO_2 production, it has not been demonstrated at a large-scale operation (e.g., 500 MW coal-fired plants where 10,000–15,000 ton day^{-1} of CO_2 would have to be removed) (Ciferno et al. 2009). In view of a considerable increase in the capacity of CCS systems in the near future, the amine-scrubbing technology developers such as Fluor Corporation and Mitsubishi Heavy Industries are in the process of optimizing the chemical scrubbing technology. In particular, the focus is on the improvement of solvent formulations, lowering stripping steam requirements, the thermal integration of the CO_2 capture system with a power plant, and so on.

Currently, DOE/NETL is investigating advanced solvents that have lower energy penalties for the regeneration step than MEA and are more resistant to flue gas impurities. Laboratory-scale studies of aqueous ammonia for capturing CO_2 (converting it to ammonium carbonate) show some promise. The system has the reduced heat requirement compared to amines, and it can potentially produce a fertilizer by-product by co-capturing SO_x and NO_x impurities in the flue gas. On the negative side, the system needs to be cooled down to 26.8°C for ammonium carbonate to be stable, and the reaction cycles involving ammonia reacting with CO_2 do not offer energy savings compared to amines. Other R&D efforts are focused on potassium carbonate promoted by piperazine, integrated vacuum carbonate absorption process, and novel oligomeric solvents (Ciferno et al. 2009).

Physical absorption methods are based on preferential absorption of gases from gaseous mixtures by inorganic or organic liquids. Physical absorption systems are governed by Henry's law, that is, low temperature and high pressure favor CO_2 capture. Thus, this method is preferred for CO_2 capture from the mixtures where CO_2 partial pressure is relatively high (greater than 500 kPa). Advantageously, the regeneration of physical solvents is less energy intensive than chemical solvent regeneration. Due to a high-pressure requirement, this technology is considered more practical for CO_2 capture from coal gasification gases in precombustion capture systems. Commercial processes for physical absorption of CO_2 include glycol-based compounds (e.g., dimethyl ether of polyethylene glycol), cold methanol, propylene carbonates, and others (Global CCS Institute 2009). In glycol-based systems, CO_2 recovery does not require heat for regeneration, and CO_2 and H_2S capture could be combined. However, they have relatively low carrying capacity, which requires circulating more than 20 kg of the glycol solution per kg of CO_2 captured. Other drawbacks of the glycol solvent use are that CO_2 pressure is lost during flash recovery, and some H_2 is lost with CO_2 stream. In methanol-based absorption systems (commercial process: Rectisol, by Lurgi), a CO_2-rich stream is cooled and contacted with liquid methanol, which readily dissolves CO_2. The process is capable of capturing the excess of 90% of CO_2 from a gaseous stream; however, the high cost of refrigeration hurts the process economics (e.g., compared with glycol-based and other solvents).

Currently, research efforts are under way to develop a new class of physical solvents based on reversible ionic liquids to capture CO_2 from low-pressure flue gases. Typically, ionic liquids contain an organic cation and either an inorganic or organic anion. Advantageously, the ionic liquids have very low vapor pressure and are thermally stable at temperatures up to several hundred degrees Centigrade (which helps to minimize the solvent loss during the operation). In the precombustion application, this would allow avoiding cooling down syngas and heating it back. Also, as a physical solvent, ionic liquid requires relatively low heat input for CO_2 recovery. However, on the negative side, most of the ionic liquids are very viscous liquids, which may make pumping of these solvents in a power plant application very difficult. Furthermore, since ionic liquids are not manufactured commercially, they are very expensive ($350–$2000 kg^{-1}) and their toxicity is unknown. Currently, Ion Engineering Co. (Colorado) is combining amines with ionic liquids to capture CO_2 from flue gases. A pilot project is under way to demonstrate the technology at 1 MW power plant scale (Thayer 2009).

14.3.2 Chemical and Physical Adsorption

Chemical adsorption systems based on regenerable solid sorbents are proposed for CO_2 removal at relatively high temperatures from post-combustion flue gases. Operating at high temperatures has the potential to reduce the efficiency penalties compared to wet-absorption systems. A simplified scheme of the regenerable sorbent-based chemical adsorption system is shown in Figure 14.7. In this system, flue gas is put in a contact with a sorbent to allow the solid–gas reaction of CO_2 with the sorbent (typically, carbonation of metal oxide). The solid sorbent then can be easily separated from the gas stream and sent for regeneration in a regenerator–reactor. Alternatively, the gas streams can be switched between the reactor and regenerator apparatuses. The key factors in the development of these systems are the adsorption capacity of sorbents and their cost.

As a regenerable CO_2 sorbent, CaO has been used for many decades. The carbonation step involving CaO sorbent at temperatures above 600°C is a very fast reaction (14.6), and its regeneration by calcining $CaCO_3$ to CaO and CO_2 is thermodynamically favorable at temperatures above 900°C and partial pressure of CO_2 of 0.1 MPa (reaction 14.7):

$$CaO + CO_2 \rightarrow CaCO_3 \tag{14.6}$$

$$CaCO_3 \rightarrow CaO + CO_2 \tag{14.7}$$

The use of the above CO_2 capture method was successfully tested at a pilot plant with the capacity of 40 t day^{-1} utilizing two interconnected fluidized bed reactors (acceptor coal

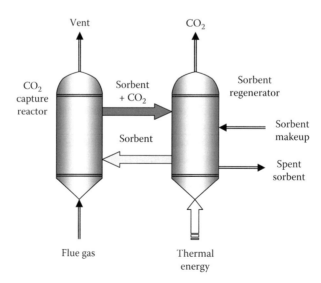

FIGURE 14.7
Schematic diagram of CO_2 capture system using a regenerable sorbent.

Gasification process), and in other projects (IPCC CCS 2005). The main drawback of these processes is that naturally occurring sorbents (limestone and dolomite) rapidly deactivate, and a large makeup flow of the sorbent is required to maintain the performance of the CO_2 capture–regeneration loop. Since the CaO sorbent is not expensive, and the spent sorbent could find an application (e.g., in cement industry), a wide range of R&D efforts are pursued worldwide to improve the efficiency of this method.

Several recently developed technologies are based on the chemical enhancement of physical sorption capacity. For example, researchers at DOE/NETL have developed amine-enriched adsorbents that are prepared by treating high surface area sorbents (e.g., zeolites) with various amine compounds (Ciferno et al. 2009). The amine immobilization on a solid substrate increases the surface contact area of the amine compound and facilitates CO_2 capture. Elimination of a water carrier in these systems has the potential to improve the energy efficiency and economics of the process relative to the MEA scrubbing technology. The amine-enriched sorbents have a high storage capacity (4 mol of CO_2 per kg of a solid sorbent), while the use of tertiary amines potentially allows for a lower energy requirement for CO_2 capture. The sorbents can be obtained by a simple spray-drying of the amine compound onto the clay sorbent. These sorbents have demonstrated at least 8 wt% CO_2 uptake and stand up to more than 250 operating cycles (Ritter 2010). The system drawbacks include the difficulty of lowering and raising temperature of the solid material (compared to liquid solvents), which may cause slow desorption rates. The use of small diameter particles can cause an undesirable significant pressure drop across the adsorber apparatus.

In another new development (dry carbonate process, by the Research Triangle Institute, North Carolina), researchers use supported sodium carbonate (Na_2CO_3) for scrubbing CO_2 from post-combustion flue gases. Sodium carbonate reacts with CO_2 and water forming sodium bicarbonate ($NaHCO_3$) via a reversible reaction that requires temperature swing from 60°C to 120°C for the sorbent to be regenerated. The economic advantages of this process over MEA scrubbing are the reduced capital costs, lower auxiliary power load, and lower material costs. However, the process faces the challenges related to the continuous circulation of large quantities of solids and requirements for contaminants. Currently, the technology is in a demonstration phase (1 t CO_2 per day), and a larger unit with the capacity of 102 t CO_2 per day is planned for 2012 (Global CCS Institute 2009).

CO_2 physical adsorption and separation processes are based on selective adsorption of CO_2 on high surface area solids such as zeolites, activated carbons, etc. Adsorption kinetics and capacities of the adsorbents are controlled by a number of factors including their surface area, pore size, volume, and the affinity of the adsorbed gas for the adsorbent. The most important technology in this category is a pressure swing adsorption (PSA) process based on the use of zeolite or activated carbon adsorbents. PSA is the method of choice for the separation of H_2–CO_2 mixtures in which high purity of hydrogen is required (e.g., 99.999% and higher). In the PSA cyclic operation using a plurality of adsorbent beds, gases are adsorbed at high pressure, isolated, and then desorbed at low pressure. Variants of the PSA process include vacuum swing adsorption (VSA), temperature swing adsorption (TSA), and electric swing adsorption (ESA) processes (Global CCS Institute 2009). PSA and TSA are commercially practiced technologies utilized in hydrogen-manufacturing plants (e.g., steam methane reforming) and in CO_2 removal from subquality NG, whereas VSA and ESA are still in pilot and demonstration stages of development (Global CCS Institute 2009). Although the adsorption-based processes are potentially applicable for capture and separation of CO_2 from large point sources, in general, they are not considered attractive for CO_2 removal from combustion gases due to their relatively low selectivity and capacity, and they are mostly limited to a low-temperature operation. The technical challenges

involve the development of adsorbents operating at higher temperatures in the presence of steam. Adsorbents with an increased adsorptive capacity and selectivity for CO_2 capture and improved kinetics and stability over thousands of cycles have to be developed. The pilot test results of coal-fired flue gas CO_2 recovery by the physical adsorption processes show that the energy consumption for capture has significantly improved from the original 708 kWh t^{-1} CO_2 to 560 kWh t^{-1} CO_2 (IPCC CCS 2005).

Novel physical adsorption concepts are also under development, for example, utilizing metal organic frameworks (MOF), carbon-based sorbents, etc. The main advantage of MOF is that they have very high porosity and adjustable chemical functionality that can be tailored to increase the CO_2 adsorption capacity. A wide variety of MOF have been synthesized and tested at laboratory-scale units. For example, MOF $Zn_3O_9(BTB)_2$ exhibited CO_2 sorption capacity of 1.4 g CO_2 per g of a sorbent material, which is an improvement over conventional zeolite-based sorbents (Millward and Yaghi 2005). The MOF technology is still at an early stage of development.

14.3.3 Cryogenic Distillation

In low-temperature (or cryogenic) distillation, a low boiling temperature liquid is separated from high-boiling temperature liquid via evaporation and condensation. Thus, the technology is most efficient and cost-effective when components in the gas feed have a significant difference in their boiling points. The process is widely used in the commercial processes involving liquefaction and purification of CO_2 from streams with relatively high CO_2 content (typically, higher than 90 vol.%). The advantages of the technology include good economy of scale and the possibility of direct production of liquid CO_2 that can be stored, transported, or sequestered at high pressure via liquid pumping. The major disadvantages of the process are that it is energy intensive, and it is necessary to remove the components having relatively high freezing point (e.g., water and NO_2), prior to cooling (to avoid a blockage of the equipment). These attributes make the cryogenic distillation method look less attractive than other routes to CO_2 capture and separation from lean CO_2 streams, especially, from post-combustion flue gases.

14.3.4 Gas Separation Membranes

Membrane separation of CO_2 utilizes permeable or semipermeable materials that allow selective transport of CO_2 from CO_2-containing gaseous mixtures, for example, flue gas (a partial pressure gradient of the permeable gas across the membrane is necessary to achieve a flow of that gas through the membrane). In principle, membranes have a promise of simplicity (no moving parts, a passive operation), compared to capital- and maintenance-intensive solvent-based separation systems, so they are expected to eventually become more reliable and more cost-effective. In general, gas separation is accomplished via physical or chemical interaction between a membrane and CO_2-containing gas. Polymer-based membranes, for example, transport gases by a solution-diffusion mechanism, which explains the relatively low gas transport flux. Polymeric membranes are quite effective (due to a large surface to volume ratio) and inexpensive; however, they are subject to a gradual degradation. Experimental studies have shown an apparent limit to the effectiveness of polymeric membranes. For example, a polymer composition can be changed to increase the membrane permeance, but this invariably decreases the separation factor (conversely, the increase in the separation factor reduces the membrane permeance).

Palladium membranes are widely used for the separation of H_2–CO_2 mixtures, but H_2 flux is typically low (besides, they are quite expensive and prone to degradation in the presence of sulfurous impurities). The use of porous inorganic membranes (ceramic or metallic) allows for a significant increase in the gas transport flux compared to nonporous (or solid) membranes. For example, porous ceramic membranes are two to four orders of magnitude more permeable than polymeric membranes; however, their cost is still high and the ratio of membrane area to the module volume is 100–1000 times smaller than that for the polymeric membranes. Inorganic membranes can be made of a wide range of materials with fine-tuned pore size in order to achieve the desired permeance and separation factors (e.g., the effective pore diameters of inorganic membranes can be made as small as 0.5 nm). Certain large molecules can be separated from smaller molecules by a so-called enhanced surface flow effect, which allows keeping the desired gas either on the high-pressure or on the low-pressure side of the membrane. Advantageously, inorganic membranes can operate at high temperatures and pressures and they are not sensitive to corrosive impurities in gas being separated. Although an individual inorganic membrane can be designed to separate CO_2 from any other gas, the separation of pure CO_2 from multiple gas mixtures may require the combination of several membranes with different characteristics. In general, the life cycle of inorganic membranes is expected to be much longer than that of polymeric membranes. One example of the inorganic membrane materials is zeolite-type membranes with pronounced molecular sieving characteristics. However, they are expensive and their permeance characteristics are substantially lower than the target values.

Considerable R&D efforts are required to implement the large-scale membrane separation of CO_2 from industrial gaseous streams. Although membranes are best suited for CO_2 separation from high-pressure gaseous mixtures (e.g., syngas from coal gasification), of particular practical interest are highly selective and permeable membranes for separating CO_2 from low partial pressure flue gas streams (e.g., post-combustion flue gas). New developments in this area include the gas absorption membranes, where CO_2 separation is achieved by a hybrid system combining a CO_2-permeable membrane with an absorption solvent (e.g., MEA) to selectively remove CO_2 from post-combustion flue gases (Ciferno et al. 2009). In this membrane-liquid sorbent hybrid system, flue gas is contacted with a membrane, and a sorbent solution on the permeate side absorbs CO_2 and creates a partial pressure differential to draw CO_2 across the membrane (see Figure 14.8). The advantages are threefold: (1) the membrane shields the amine compound from contaminants in flue gas, (2) attrition is reduced, and (3) higher loading differentials between lean and rich amine solutions can be achieved. The shortcomings of the hybrid system relate to high additional costs associated with the membrane and the need for CO_2 compression and inability of the membranes to keep out all unwanted contaminants. The hybrid membrane/liquid sorbent CO_2 capture system is still at laboratory-scale stage of the development and would require pilot-scale testing to draw the conclusion on the commercial potential of the technology.

MTR company is developing thin-film composite polymer membranes to increase the CO_2 flux across the membrane by using a novel countercurrent flow design. In this concept, a portion of incoming combustion air is utilized as a sweep gas to maximize the driving force for membrane permeation. According to preliminary estimates, 90% CO_2 capture at a 600 MW coal-fired power plant would require about 700,000 m^2 of membrane surface with a total footprint of about 2024 m^2 (Lin et al. 2007). Los Alamos National Laboratory (LANL) in partnership with Idaho National Energy and Engineering Laboratory, Pall Corp., and Shell Oil Co. are developing a new approach to CO_2 separation using thermally optimized

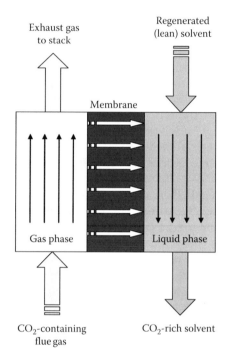

FIGURE 14.8
A membrane-liquid sorbent hybrid system for CO_2 separation.

membranes. The objective is to achieve an important combination of high selectivity, high permeability, and mechanical stability at temperatures significantly higher than that of conventional polymeric membranes (U.S. DOE, NETL 2003). Progress to date includes fabrication and testing of the polymeric–metallic membrane that is selective toward CO_2 at temperatures as high as 350°C. Further development of the high-temperature polymeric–metallic composite membranes will allow separating CO_2 at the range of temperatures of 100°C–450°C and pressures of 10–150 atm with associated economic and environmental benefits.

14.3.5 Enzymatic CO_2 Capture Systems

Enzyme-based CO_2 capture systems achieve CO_2 capture and release by mimicking a mammalian respiratory mechanism. The enzymatic sorbents enjoy fast kinetics at the lower system size and cost, and they can produce CO_2 at above atmospheric pressure. The system shortcomings include low-temperature resistance (38°C limit) and the requirement to cool the flue gas before sorption (since the CO_2 sorption process is exothermic). The enzymes could be sensitive to SO_x, NO_x, and other acid gases. In the enzymatic system developed by Carbozyme, carbonic anhydrase catalyzes the conversion of CO_2 to bicarbonate at the flue gas interface, and reverses the process at the CO_2 product side. The Carbozyme membrane system consists of two hollow-fiber microporous membranes separated by a thin liquid membrane. The laboratory-scale enzyme-facilitated membrane was validated recently on a $0.5\,m^3$ permeator demonstrating 85% removal of CO_2 from a $15.4\,vol.\%$ CO_2 containing feed stream (Trachtenberg et al. 2008). One practical consideration to take into account is the possibility of entrained solids in flue gas from coal boilers to block membrane channels.

14.3.6 CO$_2$ Hydrates

The use of CO$_2$ hydrates for CO$_2$ separation is based on the formation of solid supramolecular structures called CO$_2$ hydrates when liquid water is exposed to CO$_2$ at high pressure (10–70 atm) and low temperature (0°C–4°C), depending on the partial pressure of CO$_2$ in a gas stream. The solid hydrates are separated from the liquid stream and then heated to release CO$_2$. The method is particularly suitable for CO$_2$ recovery from high-pressure gaseous streams (e.g., precombustion streams) with minimal energy losses. The process drawbacks relate to high refrigeration energy requirements to counteract heat of hydrate formation ($\Delta H_F = 1.4$–3.3 MJ kg^{-1} CO$_2$ captured) (Carbon Sequestration Technology Roadmap 2006), and the possibility of the formation of cold spots in the hydrate formation reactor that may cause ice formation and associated operational problems. Research efforts in this area include the development of special additives to speed up hydrate formation while enabling 90% CO$_2$ capture, and the hydrate reactor design with the improved heat exchanger to ensure the uniform temperature distribution in the reactor.

DOE NETL in a partnership with Nexant and LANL have been developing a low-temperature/high-pressure SIMTECHE process for removing CO$_2$ from shifted synthesis gas containing H$_2$ (60%) and CO$_2$ (40%) via formation of CO$_2$ hydrates (U.S. DOE, NETL 2003). The technical feasibility of the continuous production of CO$_2$ hydrates has been demonstrated. Figure 14.9 provides a simplified block diagram of the SIMTECHE process for CO$_2$ separation from shifted syngas. If successfully implemented, the process will result in the production of a concentrated CO$_2$ stream for an industrial use or sequestration.

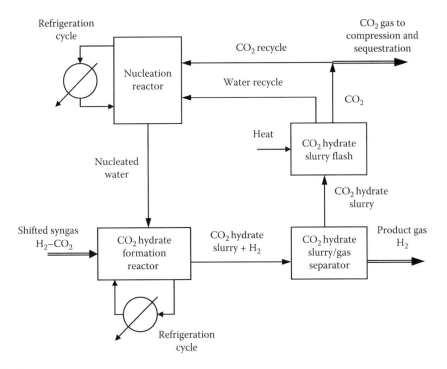

FIGURE 14.9
Simplified schematic block diagram of SIMTECHE CO$_2$ hydrate process for CO$_2$ separation from shifted syngas. (Modified from Carbon Sequestration Technology Roadmap, U.S. DOE, Office of Fossil Energy, Carbon Sequestration Technology Roadmap and Program Plan, NETL, 2006.)

14.3.7 Power Generation Efficiency Penalties due to CO_2 Capture

Practically, all CO_2 capture processes are energy intensive, which results in energy efficiency and fuel usage penalties on power generation cycles. These penalties originate from the following energy input requirements: (1) heat to regenerate solvents, (2) steam for stripping operations, and (3) electricity for fluids pumping and CO_2 compression. The reduction in an energy penalty is a focus of a variety of recent R&D and demonstration projects. The International Energy Agency (IEA) GHG Program has reported the evaluation of power plants with post- and precombustion CO_2 capture taking into consideration recent improvements in the technology (IEA GHG 2004). The comparative assessment of the overall net coal- and gas-based power plant efficiencies with and without CO_2 capture are depicted in Figure 14.10 (the percentage of CO_2 capture is 85%–90% for all plants except the NG oxyfuel plant that has 97% removal rate). Davidson (2005) estimated fuel usage due to CO_2 capture per $kW_{el}h$ electricity produced by coal- and gas-fired power plants and compared it to the same plants without capture (four categories were considered: CO_2 capture, fuel gas processing, CO_2 compression/purification, and O_2 production). It was found that the increase in the fuel usage for CO_2 separation is greater for the post-combustion capture than for precombustion capture option, because in the latter case, CO_2 is removed from more concentrated, higher-pressure stream, so a physical absorption rather than a chemical absorption method can be utilized. The fuel consumption for CO_2 compression also is lower in the precombustion capture than in the post-combustion capture option because a significant part of CO_2 leaves the separation unit at higher pressure.

Analyses of present-day and near-future post-combustion, precombustion, and oxyfuel CO_2 capture technologies indicate that a number of advanced systems that are currently under development have a potential to significantly reduce the fuel usage for CO_2 capture and increase the overall process energy efficiency, while still reducing CO_2 emissions by 90% or more.

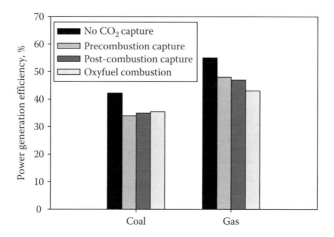

FIGURE 14.10
Thermal efficiencies of power plants with and without CO_2 capture. (Data from Davidson, J., CO_2 capture and storage and the IEA Greenhouse Gas Programme, *Workshop on CO₂ Issues*, Miffelfart, Denmark, May 2005, IEA Greenhouse Gas Programme, Cheltenham, U.K., 2005; IEA Report PH4/33, Potential for improvements in gasification combined cycle power generation with CO_2 capture, IEA Report PH4/33, IEA Greenhouse Gas R&D Programme, Cheltenham, U.K., 2004.)

14.3.8 Environmental Aspects of CO_2 Capture

According to IEA GHG report (2004), a power plant with CO_2 capture would produce a stream of concentrated CO_2 along with flue gas that, most likely, would be vented to the atmosphere (it is possible that certain amount of liquid and even solid wastes could also be produced) (IEA GHG 2004). Depending on the source of CO_2 and technology utilized for CO_2 capture, the CO_2 stream may contain certain impurities that would have practical implications on CO_2 transport and storage, as well as safety considerations and an environmental impact. The most common component of the captured CO_2 stream is moisture that has to be removed to avoid hydrate formation; this can be done by a number off-the-shelf technologies. CO_2 captured by solvent-scrubbing methods from post-combustion streams typically contains very low concentrations of impurities. For example, the total concentration of impurities in dried CO_2 from coal- and gas-fired plants with post-combustion capture is only 0.01 vol.% (the CO_2 purity is adequate for its use in food industry) (IEA GHG 2004). Precombustion CO_2 capture systems utilizing physical solvent-scrubbing processes produce less pure CO_2 streams. IEA reported typical concentrations of impurities in the precombustion CO_2 capture streams from coal-fired (so-called integrated gasification combined cycle, IGCC) plants as follows (vol.%): H_2 0.8–2.0, CO 0.03–0.4, CH_4 0.01 $N_2/O_2/Ar$ 0.03–0.6, H_2S 0.01–0.6, total of 2.1–2.7 (IEA GHG 2004; IEA GHG 2005). CO_2 streams from gas-fired power plants have somewhat a different slate of impurities (vol.%): H_2 1.0, CO 0.04, CH_4 2.0, $N_2/O_2/Ar$ 1.3, H_2S 0.01, total of 4.4. Production of combined CO_2/H_2S streams can potentially reduce the cost of the capture process (this option is only feasible if environmentally acceptable ways of transporting and storing the combined stream are available). CO_2 streams from oxyfuel combustion option contain a rather significant amount of $O_2/N_2/Ar$, along with smaller amounts of SO_2 and NO, totaling to 4.2 vol.% (IEA GHG 2005). These streams would be normally posttreated via cryogenic purification processes to reduce the level of impurities.

The concentration of harmful ingredients in flue (or vent) gases is quite similar to those from the plants with CO_2 capture, although the content of some substances could be lower due to their partial removal with the CO_2 stream. In the case of post-combustion capture by amine-based solvents, flue gases may contain traces of solvents and, possibly, ammonia (from the solvent decomposition). Liquid waste products of the CO_2 capture processes include degraded solvents, which are incinerated in specialized facilities (note that the post-combustion systems typically produce more liquid wastes than precombustion processes) (IPCC CCS 2005). The solid wastes contain deactivated reforming and CO shift catalysts from the precombustion capture systems (these wastes are either reprocessed or disposed in an environmentally safe manner).

14.3.9 Cost of CO_2 Capture

The cost of CO_2 capture depends on the choice of a CO_2-generating process and capture technology as well as underlying assumptions; thus, the reported cost estimates vary in a wide range. Four common measures of the CO_2 capture cost are (IPCC CCS 2005):

- Capital cost
- Incremental product cost (e.g., COE, hydrogen, fuels, etc.)
- Cost of CO_2 avoided (CCA)
- Cost of CO_2 captured (CCC) (or removed)

The entirety of these four measures represents a techno-economic perspective indicating the added cost of capturing CO_2 in a particular application. While the first two measures are commonly used in economic analyses and are self-explanatory, the third measure may require some explanation. The CCA reflects the average cost of reducing CO_2 emissions by one unit while providing the same amount of useful product as a reference plant without CO_2 capture (IPCC CCS 2005). For the case of a power plant, the CCA can be defined as follows:

$$CCA(\$t^{-1}CO_2) = \frac{[(COE)_{capture} - (COE)_{reference}]}{[(CO_2 \, kW \, h^{-1})_{reference} - (CO_2 \, kW \, h^{-1})_{capture}]} \quad (14.8)$$

where
  COE is the levelized cost of electricity ($ kW h^{-1})
  CO_2 kW h^{-1} is CO_2 emission rate (tons) per kWh generated

The subscripts "reference" and "capture" relate to the plant without and with CO_2 capture, respectively (the reference plant is assumed to be of the same type as the plant with CO_2 capture).

The CCC is defined as the mass of CO_2 captured (or removed) rather than emissions avoided. Again, for an electric power plant, the CCC measure can be defined as follows:

$$CCC(\$t^{-1}CO_2) = \frac{[(COE)_{capture} - (COE)_{reference}]}{(CO_2)_{captured} \, kW \, h^{-1}} \quad (14.9)$$

where
  $(CO_2)_{captured}$ kW h^{-1} is the total amount (in tons) of CO_2 captured per net kWh for a plant
    with CO_2 capture

CCC reflects the economic viability of the CO_2 capture system at a given market price for CO_2. Typically, the measure of CCC is lower than that of CCA because the energy input required to operate CO_2 capture systems results in an increase in the amount of CO_2 emitted per unit of product (IPCC CCS 2005). Figure 14.11 summarizes CO_2 capture costs for new coal- and gas-fired power plants and hydrogen plants based on current technologies. Figure 14.11 indicates that the current CO_2 capture systems reduce CO_2 emissions (per kWh) by about 85%–90% compared to similar plants without capture with an associated increase in the COE by 35%–70%, 40%–85%, and 20%–55% for NG combined cycle (NGCC), PC, and IGCC plants, respectively. The COE produced at new NGCC plants are typically lower compared to new PC and IGCC plants with or without capture. However, as NG prices are going up, it could be expected that the COE produced by NG-fired plants would become higher than that of coal-fired plants (especially, at lower capacity factors). Figure 14.11 shows that the lowest CO_2 capture costs relate to hydrogen production processes (from coal or NG) since they produce concentrated CO_2 streams as part of the technological chain of the process. These processes may represent earliest opportunities for the large-scale implementation of the CCS technology.

According to other sources, capturing and compressing CO_2 accounts for about two-thirds of the cost of CCS (Thayer 2009). The estimates for early commercial-scale coal-fired

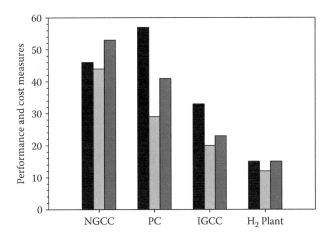

FIGURE 14.11
Summary of new plant performance and CO_2 capture costs based on current technology. NGCC is NG combined cycle; PC is pulverized coal. Black bars shows the increase in the product cost with CO_2 capture (%); light gray bars indicates CCC (U.S. $ per ton CO_2); dark gray bars shows CCA (U.S. $ per ton CO_2). All costs for CO_2 capture only and include compression, but do not include the cost of CO_2 transport and storage. PC and IGCC data are for bituminous coal at the cost of U.S. $1.0–$1.5 GJ^{-1}, NG price U.S. $2.8–$4.4 GJ^{-1} (LHV) (2002 basis). Power plant capacity: 400–800 MW without capture and 300–700 MW with capture. H_2 plant feedstock: NG cost at U.S. $4.7–$5.3 GJ^{-1}, coal cost at U.S. $0.9–$1.3 GJ^{-1}. Percent CO_2 reduction per kWh is 81%–91% for power plants and 72%–96% for H_2 plant. (Data from IPCC special report on carbon dioxide capture and storage, Prepared by Working Group III of the Intergovernmental Panel on Climate Change, B. Metz, O. Davidson, H. de Coninck, M. Loos, A. Meyer (eds.), Cambridge University Press, Cambridge, U.K., Table 3.15, p. 169, 2005.)

power plants expected at the year 2020 are projected at $50–$70 t^{-1} CO_2 abated. For nearer term demonstration projects, the cost will likely be in the range of $80–$130 t^{-1} CO_2. According to the same source, a power plant has to generate about 30% more power to cover the energy drain due to capturing 90% of CO_2 emissions.

The analyses of historical trends for the related energy and environmental technologies suggest that improvements to the current technologies can potentially reduce the cost of CO_2 capture by at least 20%–30% over the next decade (IPCC CCS 2005). It should be emphasized that despite numerous studies examining different technological options, there still exists a considerable uncertainty with regard to the magnitude of future cost reductions and the potential for cost increases above current estimates (especially for technologies in the early stages of development).

14.4 Transport of CO_2

Transport of CO_2 is an intermediate operation between CO_2 capture and its storage. CO_2 can be transported in three physical states: gaseous, liquid, and solid. In order to transport CO_2 in a gaseous phase, it has to be pressurized to a certain pressure to make the transport operation economical. CO_2 volume could be further reduced via its liquefaction, solidification, or hydration. CO_2 solidification is a more energy intensive and costly option compared to CO_2 liquefaction. Transport of CO_2 in the form of hydrates is at an early R&D

stage. In this section, commercial-scale systems for CO_2 transport are discussed: those include pipelines, tanks, and ships for transporting gaseous (G-CO_2) and liquid (L-CO_2) carbon dioxide.

14.4.1 CO_2 Compression and Dehydration

Captured CO_2 has to be compressed to pressure suitable for pipeline transport and subsequent storage. CO_2 is converted to a liquid or a dense-phase supercritical fluid by increasing the pressure above 7.4 MPa (the critical point pressure). CO_2 pipelines typically operate at pressures between 13.8 and 20.7 MPa, which allows for CO_2 to be pumped through the pipeline without further compression with the associated energy savings (Global CCS Institute 2009). The compression/pumping requires about 82 kWh of energy to raise the pressure of 1 t of CO_2 from atmospheric to 10.34 MPa. A compressor-type selection depends on the volumetric flow rates, starting and final pressures, and gas composition (e.g., for amine absorption process, starting pressure is 0.18 MPa). The three compressor types considered for CO_2 compression and pumping are

1. A reciprocating compressor
2. A multistage, integrally geared centrifugal compressor
3. A single-shaft, multistage centrifugal compressor

MAN Turbo AG is one of the leading suppliers of CO_2 compressors. During the staged compression of CO_2, its moisture content has to be reduced by cooling below its dew point and knocking out water. Finally, CO_2 stream is dehydrated by a treatment with glycol (typically, triethylene glycol, TEG) or solid adsorbents (e.g., molecular sieves). For example, CO_2 moisture content could be reduced to 20 ppmv via a TEG-based dehydration process (Global CCS Institute 2009). Molecular sieves can achieve even higher levels of CO_2 dehydration.

14.4.2 Pipeline Transport of CO_2

Pipeline transport of CO_2 is a well-developed technology: for example, in the United States, CO_2 pipelines extend over 2500 km carrying 50 million tons of CO_2 per year from industrial CO_2 sources to enhanced oil recovery (EOR) sites (in western Texas and other places). Before entering a pipeline, CO_2 has to be dried and cleaned of H_2S impurities in order to avoid possible pipeline corrosion problem. The following are the specifications for CO_2 pipeline transport for the Canyon Reef project (note that these specifications are for EOR applications, and some of them may not apply to CO_2 storage projects) (IPCC CCS 2005):

- Minimum CO_2 content of 95%
- No more than 1500 ppmw of H_2S and no more than 1450 ppmw of total sulfur
- Maximum temperature of 48.9°C
- No more than 4 mol% of nitrogen and 10 ppmw of oxygen
- No more than 5 mol% of hydrocarbons
- No more than 4×10^{-5} L m^{-3} of glycol

Typically, pipelines are made of carbon–manganese steel, and CO_2 does not corrode them as long as the relative humidity is below 60%. It was reported, for example, that at a temperature of 3°C–22°C, CO_2 pressure of 140 atm, and H_2S concentration of 800–1000 ppm, the corrosion rate for X-60 carbon steel was determined to be less than 0.5 μm year^{-1} (Seiersten 2001). The corrosion rates dramatically increase in the presence of free water (e.g., the corrosion rate of 0.7 mm year^{-1} was measured at 40°C in CO_2 equilibrated with water vapor under pressure of 95 atm). If corrosion-resistant pipelines will be developed in the future, the cost of CO_2 transportation may be reduced as CO_2 streams with water vapor, H_2S, and other impurities could be safely transported to a storage site (in principle, stainless steel pipelines are corrosion resistant, but their cost is prohibitively high to make this option economical).

During pipeline transport, the CO_2 fluid pressure slowly drops due to friction, so the provision has to be made to keep CO_2 in the dense phase. Intermediate pump stations are used to boost pressure to required levels (pressure reboosting requires significantly less energy than initial compression: 2.0 and 82 kWh t^{-1} CO_2, respectively) (Global CCS Institute 2009). Table 14.2 summarizes existing long-distance CO_2 pipelines. The composition of CO_2 stream carried by Weyburn pipeline at pressure of 15.2 MPa is as follows (vol.%): CO_2 – 96, H_2S – 0.9, CH_4 – 0.7, C_{2+} hydrocarbons – 2.3, CO – 0.1 (the content of N_2, O_2, and H_2O is less than 300, 50, and 20 ppm, respectively). McCoy and Rubin (2008) reported the cost of pipeline CO_2 transport as follows: U.S. \$1.16 t^{-1} of CO_2 transported via a 100 km pipeline constructed in the Midwest United States and handling 5 million tons CO_2 per year (the approximate output of an 800 MW coal-fired power plant with CCS) (for the same set of

TABLE 14.2

Existing and Planned Long-Distance CO_2 Pipelines

Pipeline	Location	Capacity, Mt CO_2 Year^{-1}	Length, km	Year
Canyon Reef Carriers	Texas	4.4–5.2	225	1970
Bravo Dome	New Mexico–Texas	7.3	350	1984
Cortez	Colorado–New Mexico–Texas	19.3–20.0	803	1982
Sheep Mountain	Colorado–New Mexico	9.2–9.5	772	
Bati Raman	Turkey	1.1	90	1983
Val Verde	Texas	2.5	130	1998
Weyburn	North Dakota–Canada	1.8–5	328	2000
PCOR Ft Nelson	British Columbia, Canada		78	2011
Kingsnorth	England		270	2014
Janschwalde	Germany		150	2015
Compostilla	Spain		80–90	2015
ZeroGen	Australia		100	2015
Rotterdam Afvang	The Netherlands		25	2015
FuturGas	Australia		80–200	2017

Sources: Data from Gale, J. and Davison, J., Transmission of CO_2—Safety and economic considerations, *Proceedings of the 6th International Conference on Greenhouse Gas Control Technology*, October 1–4, GHGT-6, Kyoto, Japan, 2002; IPCC special report on carbon dioxide capture and storage, Prepared by Working Group III of the Intergovernmental Panel on Climate Change, B. Metz, O. Davidson, H. de Coninck, M. Loos, A. Meyer (eds.), Cambridge University Press, Cambridge, U.K., 2005; Global CCS Institute, Strategic analysis of the global status of carbon capture and storage, Final Report, http://www/globalccsinstitute.com/downloads/reports/2009/worley/foundation-report-1-rev0.pdf (accessed August 3, 2010), 2009.

assumptions, the cost of CO_2 transport in the Central United States and the Northeast United States would be U.S. \$0.30 t^{-1} CO_2 and U.S. \$0.20 t^{-1} CO_2, respectively).

In many respects, CO_2 pipelines are no more technically challenging than those transporting NG, and the prior field experience indicates very few problems with the transport of high-pressure dry CO_2 in carbon steel pipelines. The pipelines are inspected internally (by piston-like devices) and externally by corrosion monitoring and leak detection systems. Although the incidence of CO_2 pipeline failure is relatively smalls 0.0002–0.001 km^{-1} year^{-1}, CO_2 leaking from pipelines may form a potential physiological hazard for humans and animals (Guijt 2004). If CO_2 pipelines run in a vicinity of densely populated areas, the number of people potentially exposed to the hazards of CO_2 leaking may be greater than the number of those exposed to potential risks from CO_2 storage. This may form a significant barrier to the large-scale use of CCS systems.

14.4.3 Land Transport of CO_2

In certain circumstances, for example, if the expected volumes of captured CO_2 cannot justify the construction of a new pipeline, or the point of CO_2 capture does not have an access to a pipeline facility, then mobile land transport of CO_2 by means of railroad, truck tankers, and trailers can be used. This choice would be particularly advantageous if an existing railway system is in a close proximity to a CO_2 point source. It is understood, however, that land transport of CO_2 is unlikely to be adopted as a transport option for long-term large-scale CCS projects (Global CCS Institute 2009).

14.4.4 Marine Transport of CO_2

Similar to liquid NG (L-NG) and liquefied petroleum gases (LPG) that are routinely transported by very large marine tankers, CO_2 is currently transported by ships, but on a much smaller scale (because of a current limited demand). The physical properties of liquefied CO_2 are not drastically different from those of liquefied light hydrocarbons, so the technology could be easily scaled up to extremely large L-CO_2 carriers for large-scale CCS projects. Liquid CO_2 can only exist at a condition of low temperature and elevated pressure. A semirefrigerated CO_2 tank (temperature from −50°C to −54°C, pressure 6–7 atm) is preferred by ship designers (these conditions are close to that of LPG carriers). Thus, CO_2 tankers could be constructed using the same technology as existing LPG carriers. Large L-NG carriers reach the capacity of 200,000 m^3, which could potentially transport 230,000 t of L-CO_2 (IPCC CCS 2005).

In contrast to continuous pipeline transmission of CO_2, transport by ships is discrete that requires temporary storage of CO_2 on land and a special dispensing facility. Furthermore, if the CO_2 delivery point is far from shore, CO_2 has to be unloaded from the ships to another temporary storage tanks. In case of the ocean storage option, L-CO_2 may be unloaded to a platform or a floating storage facility. Advantageously, all these issues are not specific to L-CO_2 transport and intermediate storage; similar operations are routinely handled during marine transportation of hydrocarbon products.

Currently, there are few relatively small ships in the world specifically designed to transport L-CO_2 (e.g., a relatively small 1250 m^3 Coral Carbonic tanker built in 1999 and operated by Anthony Veder) (Global CCS Institute 2009). These vessels are transporting food-grade L-CO_2 from ammonia plants in northern Europe to coastal distribution terminals, from where CO_2 is delivered to customers by tanker trucks or in pressurized cylinders (IPCC CCS 2005). There is a growing interest in using ship tankers to transport CO_2 for CCS applications. Companies in Norway and Japan are working on the design of large $L-CO_2$ carrier ships and associated infrastructure (i.e., liquefaction and intermediate storage facilities). For example, Statoil (Norway) plans to transport CO_2 extracted from flue gases in specialized tankers under pressure and at temperature of $-50°C$ to offshore oil and gas fields (it was estimated that tanker-based CO_2 transport will be more cost effective than pipeline one) (Global CCS Institute 2009). According to one planned project, a $22,000\,m^3$ tanker will carry LPG from an oil field to a shore terminal, where it would discharge the cargo and replace it with CO_2 for the return journey to an offshore field. Many major shipping companies expressed an interest in transporting CO_2. For example, Maersk Tankers (Denmark), one of the world's largest shipping companies, has announced its plans to enter the market to transport captured CO_2 (Global CCS Institute 2009). The company estimates that ships with the capacity of $20,000–35,000\,m^3$ that can hold about $25,000\,t$ of CO_2 would be best suited for the job.

It is noteworthy that marine transport of CO_2 may induce more CO_2 emissions than pipeline transport due to an additional energy-intensive step of CO_2 liquefaction and fuel usage in ships. IEA report estimated that 2.5% extra CO_2 emissions are produced during marine transport of CO_2 over the distance of $200\,km$ (18% extra emissions over $12,000\,km$), whereas 1%–2% extra CO_2 emissions are produced for each $1000\,km$ of pipeline transport (IEA GHG 2004). CO_2 could potentially leak during marine transport (about 3%–4% per $1000\,km$), which could be reduced to 1%–2% per $1000\,km$ by a carrier tank design optimization.

14.4.5 Cost of CO_2 Transport

Pipeline transport of CO_2 in a dense phase (pressure of $7.58\,MPa$ at ambient temperature) is by far the most cost-effective means of moving large quantities of CO_2 over long distances (Global CCS Institute 2009). The cost of pipeline transport includes (1) construction cost (material, equipment, booster station, installation, etc.), (2) operation and maintenance cost (monitoring, energy, etc.), and (3) other costs (management, insurance, regulatory fees, etc.). The pipeline material cost is a function of the pipeline length and diameter, the quantity and quality of transported CO_2 (in most of cost analyses, delivery pressure of CO_2 is assumed $10\,MPa$ and transport velocity $1–5\,m\,s^{-1}$) (IPCC CCS 2005). The cost of CO_2 pipeline transport also depends on a route and terrain: onshore pipelines running through heavily populated areas could be 50%–100% more expensive compared to less urbanized areas. Mountains, rivers, and other obstacles could also substantially add to the cost of CO_2 pipeline transport. Offshore pipelines, in general, are 40%–70% more expensive than onshore ones. Detailed analyses of economics of pipeline CO_2 transport can be found in IEA GHG (2002) and Hendriks et al. (2005).

Cost analysis of marine CO_2 transport is even more complicated than that of pipeline transport. The overall cost includes investments for ships, loading and unloading facilities, intermediate storage, liquefaction units, operation (labor, ship fuel cost, electricity, etc.), maintenance, harbor fees, extra facilities (for possible disruptions in the transport system), etc. Since no marine CO_2 transport systems have been implemented on a required scale of millions tons CO_2 per year, the cost of CO_2 transport by ships is still at the level of estimates. Several design studies have reported cost estimates varying in a wide range. For example, Statoil company estimated that a ship carrying $L-CO_2$ from a harbor to harbor may cost about 30%–50% more than a ship of about the same size carrying LPG, and the estimated cost of the ship of $20–30\,kt$ capacity is between U.S. \$50 million and U.S. \$70 million (Kaarstad 2004). Other estimates for a ship construction are as follows (in millions U.S. \$): 34, 60, and 85 for 10, 30, and $50\,kt$ capacities, respectively (IEA GHG PH 4/30 2004).

The estimates for the cost of a liquefaction facility are U.S. \$35–\$50 million for the capacity of 1 million tons per year (Kaarstad 2004) and U.S. \$80 million for the capacity of 6.2 million tons per year (IEA GHG PH 4/30 2004). The cost analysis of a marine transportation system carrying L-CO_2 for a distance of 7600 km reported the specific transport costs of U.S. \$35 t^{-1} for 30 kt ships and U.S. \$30 t^{-1} for 50 kt ships. Maersk (Denmark) estimated that shipping L-CO_2 by a tanker from Denmark to the North Sea would cost about U.S. \$12 t^{-1} CO_2 (Global CCS Institute 2009).

The economics of CO_2 ship transport is more favorable compared to pipeline transport for longer distances. The break-even distance for transporting 6 million tons of CO_2 per year by ships and offshore pipelines is about 1000 km; for an onshore pipeline, this distance is about 1600 km (IPCC CCS 2005). But distance is not the only factor determining the competitiveness of marine CO_2 transport: other factors include fuel cost, loading terminals, pipeline shore crossings, seabed stability, security, and others.

14.5 CO_2 Sequestration Options

CO_2 storage or sequestration is a final step in the multistep CCS process. Currently, the following technical options for CO_2 storage are under consideration:

- Geological storage (deep in the Earth's crust in rock, basalt, saline formations, salt caverns, etc.)
- Ocean storage (within deep ocean basins)
- Beneficial reuse (CO_2 storage is accompanied by the beneficial use of CO_2 in important industrial processes, e.g., EOR)
- Mineral carbonation (industrial fixation of CO_2 in the form of stable carbonates)
- Biological storage (CO_2 storage in terrestrial and aquatic systems)
- Industrial use of CO_2 (CO_2 conversion into useful products)

In the following sections, these options will be analyzed from the viewpoint of their advantages and limitations, technical maturity, and readiness for a large-scale commercial deployment, the status of current CO_2 storage projects, associated costs, as well as environmental and public perception issues. Since, in most cases, the beneficial reuse option is closely associated with geological storage, these two options are covered in Section 14.5.1.

14.5.1 Geological Storage

Geological storage of anthropogenic CO_2 as a carbon mitigation measure was first proposed in the 1970s by Marchetti and coworkers (Marchetti 1977), but it was not until early 1990s when the concept attracted a wide attention of scientists and environmentalists, and, eventually, gained credibility through the works of Kaarstad (1992), Koide et al. (1992), Holloway and Savage (1993), and others. The engineered injection of CO_2 and acid gases into subsurface geological formations was first undertaken in Texas, United States, and Alberta, Canada, as part of EOR and the disposal of acid gases (a by-product of oil production), respectively. In 1996, Statoil (Norway) and its partners initiated the world's first commercial-scale CO_2 storage project at the Sleipner Gas Field in the North Sea. This was followed by a number of research programs in United States, Europe, Canada, Japan, and Australia. Of particular practical interest were the geological CO_2 storage projects carried out by the oil and gas industries as applied to NG fields with a very high CO_2 content, for example, the Natuna field in Indonesia, In Salah in Algeria, and Gorgon in Australia (IPCC CCS 2005). Encouraged by the progress in these areas, more recently, coal mining companies and electric utility companies started showing an interest in geological storage as a carbon mitigation option. As the level of confidence in the technology increases, more and more industrial entities dealing with the large volumes of CO_2 are participating in commercial-scale projects.

Geological storage of CO_2 is carried out by the injection of CO_2 in deep porous rocks, basalt, or saline formations that are isolated from the atmosphere by the thick layers of an impermeable rock. The injected CO_2 is stored in the pore spaces between grains of rocks, for example, sandstone or limestone, or in the cavities and voids within rocks such as basalt or salt. Most importantly, an impermeable sealing layer (a caprock) should be located above the reservoir rocks such that it would not allow the stored CO_2 passing through it in appreciable quantities. The density of injected CO_2 increases with the depth, and it becomes practically constant at the depths below 1.5 km (Figure 14.12). This figure shows that proportionate with the increase in CO_2 density, its relative volume decreases. At a certain pressure corresponding to the depth of about 800 m, CO_2 turns in to a liquid-like form called the "supercritical" or "dense" phase that is much denser than gaseous CO_2 (consequently, most reservoirs that are considered for geological CO_2 storage have the depth of at least 800 m) (Global CCS Institute 2009). Advantageously, the CO_2 dense phase displays a high density like fluids and low viscosity like gases, which is beneficial from the viewpoint of an efficient utilization of the underground storage space, for example, within the pores of sedimentary rocks. Another peculiar feature of the dense-phase CO_2 is that, like oil, it is not mixed with water and forms a separate layer (although certain amount of CO_2 may be dissolved in water and form carbonic acid; as a result, carbonate ions may react with other elements within the formation and form rock minerals such as limestone).

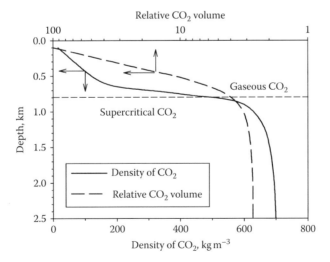

FIGURE 14.12
Density and relative volume of CO_2 as a function of depth. (Based on density data from Angus, S. et al., *International Thermodynamics Tables of the Fluid States, Vol. 3. Carbon Dioxide*, IUPAC, Pergamon Press, London, U.K., pp. 266–359, 1973.)

To put captured CO_2 into a deep geological reservoir, different types of injection wells are used, for example, horizontal or vertical wells. The well casing is cemented in place to prevent leakage. During injection of CO_2 into the reservoirs, it displaces naturally occurring fluids such as water, crude oil, or NG.

14.5.1.1 Options for Geological Storage

The most important questions to be addressed with regard to geological CO_2 storage are the following: (1) in what types of geological formations can CO_2 be safely stored and (2) is the overall storage capacity of these formations adequate to accommodate all anthropogenic CO_2 produced. Although there are many sedimentary regions spread all over the world, not all of them are suited for long-term CO_2 storage. The main requisites of a suitable geological formation are

- Adequate capacity and injectivity
- Satisfactory sealing caprock or a reliable confining unit
- Sufficiently stable geological environment to ensure the long-term integrity of the storage site (i.e., tectonic activity, sediment type, geothermal, and hydrodynamic regimes)
- Other factors such as basin characteristics and resources (oil, gas, coal, and salt), industry maturity and infrastructure, environmental concerns, and societal issues (level of development, economy, public acceptance, etc.) (Bachu 2003)

In order for geological CO_2 storage to be considered a viable carbon mitigation technology, it would be necessary (and it is expected) that once CO_2 injected into the deep geological formations at carefully selected sites, no less than 99% of injected CO_2 would be retained for at least 1000 years (IPCC CCS 2005). In general, CO_2 can remain confined underground as a result of

1. Trapping below an impermeable layer (a caprock)
2. Trapping in the pore spaces of the storage formation as an immobile phase
3. Dissolution in in situ formed fluids
4. Adsorption/absorption onto an organic matter in coal, shale, and other solid matter
5. Reaction with minerals in the storage formation and caprock to form carbonates

Advantageously, the injection of CO_2 in deep geological formations utilizes the technologies that have already been developed and practiced by oil and gas industry (e.g., well-drilling, injection, computer simulation, and others). Several types of geological formations have already been employed or under consideration for geological CO_2 storage:

- Deep saline formations
- Depleted oil and gas reservoirs
- Unmineable coal beds
- Other geological media (basalts, organic-rich shale, salt caverns, and abandoned mines)

Figure 14.13 depicts global CO_2 storage capacity estimates categorized by geological storage options.

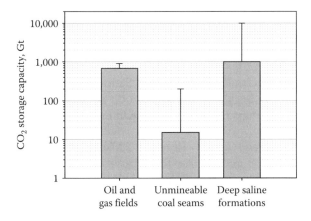

FIGURE 14.13
Lower and upper estimates of storage capacities of different geological CO_2 storage options. (Adapted from IPCC special report on carbon dioxide capture and storage, Prepared by Working Group III of the Intergovernmental Panel on Climate Change, B. Metz, O. Davidson, H. de Coninck, M. Loos, A. Meyer (eds.), Cambridge University Press, Cambridge, U.K., Table 5.2, p. 221, 2005.)

In the United States, the estimated CO_2 storage capacities of geological formations are as follows (in billions of tons) (U.S. DOE, NETL 2003):

Unmineable coal beds	15–20
Depleting oil reservoirs	40–50
Depleting gas reservoirs	80–100
Saline formations	130–500
Total	140–670

It can be seen that the storage option related to deep saline formations holds by far the largest capacity to store CO_2, and the basins are widely spread over the world both on shore and on continental shelves.

14.5.1.1.1 Saline Formations

Saline formations represent the layers of a porous rock saturated with highly mineralized brines (the level of salinity could be 10 times higher than in seawater). Water in the deep saline formations flows at an extremely slow rate of a few centimeters to meters per year in a response to pressure differences on a regional scale, and the formations can extend to hundreds to thousands kilometers (Global CCS Institute 2009). Thus, when CO_2 is stored in a deep saline formation, it is expected to be isolated from the near-surface layers for thousands of years. Due to the high concentration of minerals in the brine, CO_2 is likely to react with them forming solid carbonates.

Saline formations are widely occurring throughout the world. The geographical distribution of CO_2 storage capacity in the deep saline formations is as follows (in Gt) (IPCC CCS 2005):

• Europe	30–577
• United States	160–800
• Canada	4000
• Australia	740
• Japan	1.5–80

Advantageously, the potential storage sites are located not very far from major industrial CO_2 sources, and, most likely, their overall storage capacity will be adequate to accommodate significant part of the CO_2 emissions generated in the foreseeable future.

Examples of geological CO_2 storage in saline formations include the Statoil Sleipner Project (Norway), In Salah Gas Project (Algeria), Statoil Snøhvit Project in the Barents Sea (Norway), and several pilot-scale or demonstration projects such as the Ketzin Project (Germany), the Lacq CCS project (France), U.S. DOE Regional Carbon Storage Partnership Program, and others. Recently, U.S. and Chinese researchers reported on the estimates of China's geological CO_2 storage capacity (Johnson 2009b). The authors have found that most potential storage sites are in deep saline formations amounting to about 2300 Gt CO_2 of the total storage capacity. Beneficially, 90% of existing coal-fired power plants are located within 160 km of these formations. It is noteworthy that China has 1620 large stationary CO_2 sources generating 3.8 Gt year^{-1} CO_2, surpassing United States by almost 1 Gt year^{-1} (of those, 70% of emissions are originated from power plants). As of now, besides CO_2 sequestration, saline formations as a storage medium found very limited applications: just a few cases of chemical waste storage.

14.5.1.1.2 Depleted Oil and Gas Reservoirs

Depleted oil and gas reservoirs are considered excellent candidates for CO_2 storage due to following advantages:

1. The characteristics of these reservoirs are well known (based on history of their exploitation).
2. The infrastructure (e.g., wells and pipelines) is already in place and can be easily readjusted for CO_2 storage.
3. A proven natural trapping mechanism.
4. Large storage volumes due to previous extraction of fluids (oil or gas) from the reservoirs.

Presently, most commercial projects involving CO_2 storage in oil and gas fields relate to enhanced oil/gas recovery projects described in Section 14.5.1.2.

14.5.1.1.3 Unmineable Coal Seams

A significant advantage of coal-bed CO_2 sequestration over conventional gas reservoir storage is that coal seams can store six to seven times more CO_2 than a reservoir of an equivalent volume because solid coal contains natural fractures ("cleats"), pores, and micropores where CO_2 can diffuse and get tightly adsorbed. A coal seam becomes a suitable site for CO_2 storage if it is no longer economical to be mined for coal (which is determined by its geological conditions and world energy prices among many other factors). Typically, unmineable coal seams are likely to be several hundreds of meters or more in depth (Global CCS Institute 2009).

14.5.1.1.4 Shale and Basalt Formations

Shale and basalt are other types of the geological formations suitable for CO_2 storage. Shale is composed of thin layers of rock that in many cases contain organic matter (1%–2%) capable of absorbing CO_2. Basalt formations represent ancient volcanic rocks (lava) that have porosity and permeability in the fractures or cavities between blocks of a solid rock (Global CCS Institute 2009). A unique characteristic of basalt is that depending on its chemical composition it can react with CO_2 producing solid products—carbonates that would be easy to isolate from the atmosphere. Technical challenges facing storage in the basalt formations include difficulty of CO_2 injection into these heterogeneous formations, the presence of large porosity, and permeability of the formations that will make the sealing unreliable. Besides, little is known about CO_2 storage properties of basalt formations. Although basalts are widespread around the world, there are no current large-scale projects on CO_2 storage in the basalt formations.

14.5.1.1.5 Salt Caverns

Naturally occurring underground salt caverns can potentially store large quantities of CO_2 provided they have an adequate geological sealing layer and can support high pressures required for the storage of dense-phase CO_2. One feature of the salt caverns that distinguishes them from other geological formations is that they have an ability to deform and change volume until CO_2 pressure equalizes with surrounding pressure. The volume of salt caverns have to be large to accommodate industrial CO_2 sources, for example, a 500 MW coal-fired plant capturing 3 Mt year^{-1} CO_2 would require a cavern with the equivalent volume of a spherical cavern 150 m in diameter to store CO_2 produced by the plant just in one year (Global CCS Institute 2009). Although salt caverns have been used in the past for the temporary storage of NG and proved to be quite effective, currently, there are no CO_2 storage projects that make use of the salt caverns.

14.5.1.2 Beneficial CO_2 Reuse Applications

The term "beneficial reuse" refers to the CO_2 storage applications where captured CO_2 is used for generating a revenue from the sale of both CO_2 and raw materials that are obtained as a result of the CO_2 use (e.g., crude oil or NG).

14.5.1.2.1 Enhanced Oil Recovery

The most widely used beneficial CO_2 reuse application relates to EOR. EOR is a generic term for techniques for increasing the amount of extracted crude oil that would otherwise remain stranded (in order to distinguish CO_2-based EOR from other types of EOR, for example, chemical, microbial, and thermal EOR, hereinafter CO_2-EOR term will be used). CO_2-EOR offers a potential to increase oil production through CO_2 flooding of a well. It was estimated that only 5%–40% of original oil is recovered by a conventional primary production, and additional 10%–20% of oil is produced by a secondary recovery operation that utilizes water flooding (Holt et al. 1995). There are several estimates on the efficiency of oil recovery by CO_2 flooding: earlier estimates by Martin and Taber (1992), according to which the incremental tertiary oil recovery of 7%–23% of original oil could be achieved, and the latest one by Ferguson et al. (2009), who estimated that between 6.7% and 18.9% of original oil in place is recoverable by the CO_2-EOR method (the average for United States is 14.6%). For the United States with the estimated 595.7 billion barrels of oil in place, this translates into an additional 87.1 billion barrels of oil (Global CCS Institute 2009). About half of CO_2 used in CO_2-EOR is retained in the reservoir after oil production is ceased. (It is expected that the fraction of permanently retained CO_2 could be significantly increased by improving CO_2-EOR operation management and optimization.)

Figure 14.14 depicts a simplified sketch of CO_2-EOR. Captured CO_2 is compressed to dense-phase CO_2, which is injected into an oil reservoir through an injection well. Upon injection, CO_2 is mixed with crude oil in the reservoir making oil more mobile and forcing it to flow to the production wells (only one production well is shown in Figure 14.14,

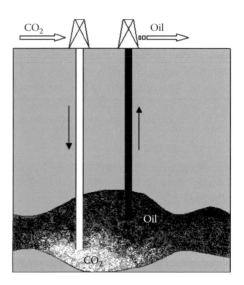

FIGURE 14.14
Schematic representation of EOR using CO_2.

although, typically, there are several production wells per each injection well). Typical examples of CO_2-EOR projects are the Rangely Project in Colorado and the Weyburn-Midale Project in Saskatchewan (Canada). Although the CO_2-EOR method is applicable to enhanced gas recovery (CO_2-EGR) projects, currently, there are no large-scale operations utilizing this approach.

14.5.1.2.2 Enhanced Coal Bed Methane Recovery

Most coal seams contain naturally occurring methane, and its content typically increases with the coal bed depth, coal rank, and pressure in the coal bed. Coal has much higher affinity to CO_2 than to methane (the ratio of CO_2:CH_4 adsorption capacities over coal surface varies in the range of 1–10, depending on the type of coal) (IPCC CCS 2005). Thus, when CO_2 is injected into coal seams, it displaces methane, which is the basis of CO_2-enhanced coal bed methane recovery (ECBMR) projects (Figure 14.15). An advantageous feature of

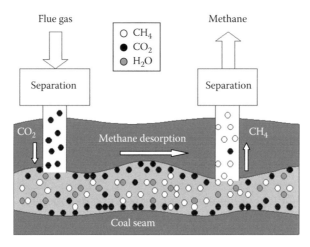

FIGURE 14.15
Schematic diagram of CO_2-ECBMR.

the ECBMR process is that it could utilize CO_2 in either dense or gaseous form depending on the coal seam depth. In order to be practical, an ECBMR project should meet a number of requirements including: (1) sufficient CO_2 permeability, (2) laterally extensive coal seams, (3) adequate depth up to 1.5 km, (4) adequate methane content in the coal deposit, and others (Global CCS Institute 2009). CO_2-ECBM operation can potentially increase the amount of recovered methane to about 90% of the total gas, compared to the conventional recovery of only 50% (by a pressure depletion method) (Stevens et al. 1996). Examples of the ECBMR projects are the Allison Unit (New Mexico) and the Recopol Project (Poland).

14.5.1.3 Geological CO_2 Storage Security

CO_2 storage security is one of the major issues that determine whether geological CO_2 storage is a feasible option for reducing CO_2 emissions from human activities. The security and effectiveness of geological CO_2 storage is determined by a combination of a variety of physical and geochemical trapping or confinement mechanisms and processes. The most preferred storage mechanisms would represent CO_2 conversion to solid stable minerals, or involve a thick, impermeable (or low-permeable) seal (a caprock) under which the immobile CO_2 phase is permanently trapped. Depending on the nature of the geological storage site, physical trapping of CO_2 can be accomplished via stratigraphic, structural, and hydrodynamic mechanisms. The first two mechanisms occur when CO_2 is trapped below low-permeability seals (e.g., shale and salt beds) or the traps formed by folded or fractured rocks. The hydrodynamic trapping mostly takes place in the deep saline formations where fluids migrate very slowly over long distances. After CO_2 is injected in a saline formation, it displaces saline water and then migrates to the top of the formation (since it is less dense than water) where it is trapped as a residual CO_2 saturation phase. Over longer period of time, CO_2 dissolves in saline formation water and migrates with groundwater. Since in most cases, the distance from a CO_2 injection site to the edge of the impermeable formation is in hundreds of kilometers, it would take millions of years for the CO_2 phase to reach surface (Bachu et al. 1994).

The geochemical trapping mechanism involves the chemical interaction of CO_2 with water and a rock formation; these interactions immensely increase the effectiveness of carbon trapping and enhance the storage security of this storage option. The trapping process consists of two stages:

1. CO_2 dissolution in saline formation water

$$CO_{2(gas)} + H_2O \leftrightarrow (HCO_3^-)_{aq} + H^+ \leftrightarrow CO_3^{2-} + 2H^+ \tag{14.10}$$

2. Conversion of ionic species to stable carbonate minerals

$$Ca^{2+} + CO_3^{2-} \rightarrow CaCO_3 \tag{14.11}$$

During the first stage, CO_2 moves from a gaseous to aqueous phase, and, as a result, it no longer exists in a separate phase, which prevents it from moving upward due to the buoyancy effect (therefore, the process is commonly called "solubility trapping"). A weak acid formed during CO_2 dissolution reacts with silicates of K, Na, Mg, Ca, and Fe (e.g., clays, micas, and feldspars present in the rock matrix) forming stable carbonate minerals (the process is called "mineral trapping").

Although the mineral trapping is the slowest process (thousands of years) among all geological storage options, it is considered the most preferred form of CO_2 storage due to the exceptional permanence it can provide coupled with a large CO_2 storage capacity. The adsorption-type CO_2 fixation occurs when CO_2 gets adsorbed onto coal or organic-rich shales. This type of CO_2 trapping has been observed in the field experiments at the Fenn Big Valley, Canada, and Alisson Unit CO_2-ECBMR project in United States (IPCC CCS 2005). It was estimated that the physical trapping mechanisms are effective in the relatively short time frame of tens to hundreds years, whereas the geochemical trapping mechanisms dominate from thousands to millions of years.

Modeling studies related to CO_2 storage at the Weyburn Oil Field site indicated that over 5000 years, all injected CO_2 would be dissolved and converted to carbonate minerals (Perkins et al. 2004). Another study provides a quantitative estimate of the probability of a CO_2 release from Weyburn Field as 1% in 5000 years (Casey 2008). It is projected that well-selected, designed, and managed geological storage sites are likely to release only 1% of injected CO_2 over 1000 years (IPCC CCS 2005). Data from natural systems such as NG fields and trapped CO_2 accumulations provide an indirect evidence of the storage permanence. For example, about 200 Mt CO_2 is trapped in the Pisgah Anticline (Mississippi) for more than 65 million years.

14.5.1.4 Cost of Geological CO_2 Sequestration

The cost of geological CO_2 storage is highly site specific, and it depends on the type of storage option (e.g., saline formation, depleted oil/gas reservoir, etc.) and its characteristics, storage depth, the site location, terrain, and other geographic factors. This leads to a high degree of variability in the CO_2 storage cost estimates. Typically, an offshore storage option is more costly compared to the onshore one since it involves expensive platforms and a subsea equipment coupled with higher operational costs. The cost of CO_2 storage may be substantially offset if it is combined with EOR or ECBM. The most significant components of the capital costs for CO_2 storage relate to wells, in-field pipelines, facilities, and infrastructure. The cost of a well could run from about U.S. $200,000 for an onshore site to U.S. $25 million for an offshore well (IPCC CCS 2005). The major operating costs include manpower, fuel, and maintenance. In most cases, the costs of geophysical and engineering feasibility studies, the reservoir evaluation and licensing are also included in the cost estimate.

Detailed cost estimates for geological CO_2 storage in United States, Europe, and Australia has been reported in a number of publications and reports (Hendriks et al. 2002; Allinson et al. 2003; Bock et al. 2003). Figure 14.16 summarizes the cost estimates (i.e., the ranges from low to high values) for geological CO_2 storage for different options, locations (e.g., onshore vs. offshore), and regions (Australia, Europe, and United States). The data are based on the statistics for multiple sites, representative ranges, and low/base/high cases. Figure 14.16 shows that the cost estimates for CO_2 storage in onshore saline formations, depleted oil field, and disused oil/gas fields are in a rather close range. It is noteworthy that the cost of offshore CO_2 storage in saline formations and disused oil/gas fields is about twice that of onshore storage. Figure 14.16 also includes the cost estimate for onshore CO_2 storage combined with CO_2-EOR project for European sites (Hendriks et al. 2002). The negative cost value indicates that the cost of CO_2 storage could be offset by additional revenues from oil production. For North American sites, the net cost of onshore CO_2 storage with EOR was estimated at a negative value of—U.S. $14.8 t^{-1} CO_2 stored (Bock et al. 2003). The

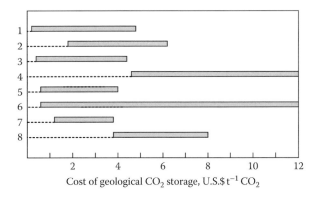

FIGURE 14.16
Cost estimates for geological CO_2 storage for different options. 1—saline formation, onshore (Australia), 2—saline formation, onshore (Europe), 3—saline formation, onshore (United States), 4—saline formation, offshore (North Sea), 5—depleted oil field, onshore (United States), 6—depleted gas field, onshore (United States), 7—depleted oil or gas field, onshore (Europe), 8—depleted oil or gas field, offshore (North Sea). (Data from IPCC special report on carbon dioxide capture and storage, Prepared by Working Group III of the Intergovernmental Panel on Climate Change, B. Metz, O. Davidson, H. de Coninck, M. Loos, A. Meyer (eds.), Cambridge University Press, Cambridge, U.K., Table 5.9, p. 260, 2005.)

values for offshore CO_2 storage cost with EOR are somewhat higher than that for onshore storage: from the negative value of -10.5 to $+21.0$ U.S. $ t^{-1} CO_2 stored (Hendriks et al. 2002). One should take into consideration that the potential benefit from coupling EOR with CO_2 storage would increase with the increase in oil prices; earlier cost estimates for CO_2 storage with EOR were based on relatively low oil price of U.S. $20 per barrel. According to IPCC 2005 report, the increase in the oil price from U.S. $20 to U.S. $50 per barrel would increase the potential benefit from EOR to CO_2 storage from U.S. $16 t^{-1} CO_2 to U.S. $30 t^{-1} CO_2 (high case value) (IPCC CCS 2005). The above data show that the economic benefits from oil production may provide incentives for the earlier implementation of CO_2-EOR for geological CO_2 storage compared to other options.

14.5.2 Ocean Storage of CO_2

Oceans cover about three quarters of the Earth's surface (precisely 71%) and as a natural sink for CO_2 they absorb the significant amount of anthropogenic and naturally occurring CO_2 (the ocean, the atmosphere, soil, and biosphere are part of the global carbon cycle, see Chapter 1 for details). In fact, because of the enormous volume of water and the relatively high CO_2 solubility in water coupled with the variety of chemical/ionic interactions, the ocean contains about 50 times more CO_2 than the atmosphere.

Since the beginning of industrial revolution, the oceans have absorbed about 500 Gt of anthropogenic CO_2 emissions of the total of 1300 Gt CO_2 (IPCC CCS 2005), and during this period, atmospheric CO_2 concentration increased from 280 to 392 ppm (CO2now 2010). Dissolved CO_2 is predominantly concentrated in the upper ocean layers resulting in a slight pH drop of about 0.1 pH units, compared with the preindustrial level (note that the average preindustrial pH of the ocean was about 8.2 pH units) (Key et al. 2004). Although no changes in the pH of deep ocean layers have been detected so far, the simulation studies and various models predict that over the next several centuries the upper ocean layers will be mixed with deep ocean waters, thus, decreasing its pH.

14.5.2.1 *CO₂ Storage in the Ocean as a Mitigation Strategy*

The ocean CO_2 sequestration approach is intended to deliberately inject CO_2 into the ocean at great depths (at least 1 km deep) where it is supposed to be retained for a prolonged period (centuries) isolated from the atmosphere. The idea was first proposed in mid-1970s by Marchetti, who hypothesized that if liquefied CO_2 is injected into the waters flowing from Mediterranean Sea to the mid-depth Atlantic Ocean, it would remain isolated from the atmosphere for centuries (Marchetti 1977). Due to the enormous volume of the Earth's oceans, it is assumed that practically unlimited amount of anthropogenic CO_2 can be stored in the ocean for at least a millennium (the timing will be dictated by the extent of the equilibrium between the ocean and the atmosphere). Most of analytical studies and models agree that CO_2 injected into the ocean will eventually become part of the global carbon cycle, and deeper injections would result in longer retention times. For example, it was estimated that 30%–85% of injected CO_2 will be retained after 500 years if stored at depths of 1000–3000 m (IPCC CCS 2005).

The modeling studies conducted by Kheshgi and Archer (2004) predict that in the timescale of about two millennia, the ocean storage approach would not significantly reduce the atmospheric CO_2 concentration. The authors compared atmospheric CO_2 concentrations for several scenarios: (1) emissions of 18,000 Gt CO_2 directly released to the atmosphere, (2) the same amount of CO_2 injected into the ocean at the depth of 3 km, and (3) release of all CO_2 to the atmosphere until the year 2050 followed by the release of 50% CO_2 to the atmosphere and remaining 50% CO_2 handled by other non-ocean-related mitigation technologies (e.g., permanent storage and the use of renewables) after 2050. The Kheshgi–Archer's model predicts that in the first case, the peak of atmospheric CO_2 concentration of about 1900 ppm will be reached by the year 2300 followed by a steep decline and leveling up at about 800–900 ppm beyond the year 3500. According to the second scenario, atmospheric CO_2 concentration will steadily increase until the year 3000–3200, and then it will level up at the value of about 700–800 ppm beyond the year 3500. In the third illustrative case, the atmospheric CO_2 concentration will reach a peak value of about 1000 ppm between the years 2200 and 2300 followed by a steep decline and leveling up at the value of 500 ppm beyond the year 3500. Thus, this and other studies conclude that comparing to the CO_2 release to the atmosphere, its injection into the ocean would substantially reduce the rate of atmospheric CO_2 build up over the time period of one to one and half millennia, but it will not make a big difference beyond the year 3500 because the atmosphere and the ocean would eventually equilibrate.

Figure 14.17 depicts the atmospheric CO_2 stabilization concentration (in ppm) as a function of the amount of anthropogenic CO_2 stored in the ocean at equilibrium conditions. The data show that the atmospheric CO_2 stabilization concentration increases exponentially with the amount of CO_2 stored in the deep ocean after atmosphere–ocean equilibration is reached. Simulation studies coupled with available experimental data and the ocean observations predict that CO_2 injected into the ocean will be isolated from the atmosphere for several hundreds of years with the fraction of CO_2 retained in the ocean increasing with the depth of injection (IPCC CCS 2005). Figure 14.18 shows the fraction of CO_2 retained in the ocean basin after 100 years of the continuous injection starting in the year 2000 as a function of the injection depth. If CO_2 is injected at a relatively shallow depth of 800 m, only roughly one-third of CO_2 will be retained in the ocean after 500 years. The CO_2 injection at a depth of 3000 m would result in the retention of more than 70% of injected CO_2.

In principle, the physical and chemical capacities of ocean CO_2 storage are immense with regard to conventional fossil fuel resources. The extent of utilizing the ocean storage

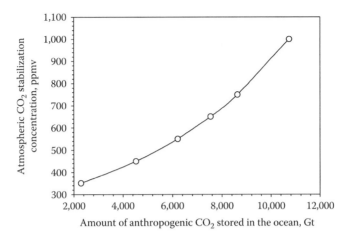

FIGURE 14.17
Atmospheric CO_2 stabilization concentration (in ppm by volume, ppmv) as a function of the amount of anthropogenic CO_2 stored in the ocean at equilibrium conditions. (Data from IPCC special report on carbon dioxide capture and storage, Prepared by Working Group III of the Intergovernmental Panel on Climate Change, B. Metz, O. Davidson, H. de Coninck, M. Loos, A. Meyer (eds.), Cambridge University Press, Cambridge, U.K., Table 5.9, p. 260, 2005.)

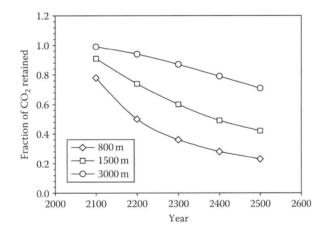

FIGURE 14.18
Fraction of CO_2 retained for the ocean storage after 100 years of continuous injections (starting in 2000) at three different depths of 800, 1500, and 3000 m. (Data from IPCC special report on carbon dioxide capture and storage, Prepared by Working Group III of the Intergovernmental Panel on Climate Change, B. Metz, O. Davidson, H. de Coninck, M. Loos, A. Meyer (eds.), Cambridge University Press, Cambridge, U.K., Table TS.7, p. 38, 2005.)

capacity will be determined by such factors as economics, its ecological impact, public acceptance, and others. The estimates indicate that roughly 2300–10,700 Gt CO_2 would be added to the ocean in equilibrium with the atmospheric CO_2 stabilization concentrations varying from 350 to 1000 ppm (depending on how CO_2 was initially released: to the ocean or the atmosphere) (IPCC CCS 2005). The overall capacity of the ocean to store CO_2 may be decreased by the increase in global temperature and increased with the increasing alkalinity of the ocean (e.g., via dissolution of minerals such as limestone).

14.5.2.2 CO₂ Storage in the Ocean: Technical Background

Before discussing technological approaches to storing CO_2 in the ocean, it would be worthwhile to touch upon physical and chemical properties of CO_2 that would govern its behavior at the atmosphere–ocean interface and during its long-term storage in the ocean. Atmospheric CO_2 is in chemical equilibrium with seawater according to the following equation:

$$CO_{2(gas)} + H_2O \leftrightarrow H_2CO_{3(aq)} \xleftrightarrow{pK_a=6.4} HCO_{3(aq)}^- + H^+ \xleftrightarrow{pK_a=10.3} CO_{3(aq)}^{2-} + 2H^+ \quad (14.12)$$

This equilibrium is governed by CO_2 partial pressure (i.e., its concentration in the atmosphere), seawater temperature, the rate of air/ocean exchange (mixing), the presence of other ionic species, chemistry of seawater, and other factors. Due to dissolution of minerals (e.g., $CaCO_3$) in seawater, the ocean pH is slightly above 7 (i.e., it is slightly alkaline), which favors dissolution of CO_2 in seawater. Since dissolved CO_2 is in dynamic equilibrium with atmospheric CO_2, the rise in CO_2 concentration in the atmosphere will cause the equilibrium shift resulting in the dissolution of additional CO_2 in the ocean and forming more bicarbonate ions:

$$CO_{2\,(gas)} + H_2O \leftrightarrow HCO_{3(aq)}^- + H^+ \quad (14.13)$$

$$CO_{2(gas)} + H_2O + CO_{3(aq)}^{2-} \leftrightarrow 2HCO_{3(aq)}^- \quad (14.14)$$

The combined effect of these reactions causes the ocean pH and carbonate ion concentration in seawater to drop. The physical and thermodynamic properties of pure CO_2 (in gaseous, liquid, and solid states) and the CO_2–H_2O system determine the fate of CO_2 upon its release into a relatively shallow and deep-sea environment. Of particular importance is to determine the conditions at which CO_2 would exist as gas, liquid, or hydrate and their relative density against seawater. Figure 14.19 shows how the densities of gaseous and liquid CO_2 and seawater change with the ocean depth.

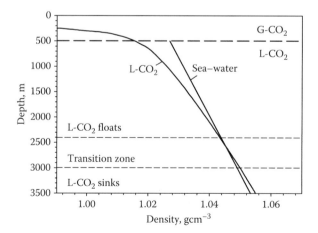

FIGURE 14.19
Changes in densities of gaseous and liquid CO_2 and seawater with the ocean depth.

14.5.2.2.1 *Gaseous CO₂*

At a typical temperature range in the ocean, CO_2 exists as gas at the depth of 0–500 m; the dissolution rate of gaseous CO_2 bubbles was estimated at 0.26–1.1 µmol cm^{-2} s^{-1} (Teng et al. 1996). By reducing the size of the CO_2 bubbles, it would be possible to dissolve them completely before they reach the surface.

14.5.2.2.2 *Liquid CO₂*

Below about 500 m, pressure is high enough to convert gaseous CO_2 to a liquid form. Interestingly, in the liquid phase, CO_2 is more compressible than seawater, the property that greatly affects the relative density of liquid CO_2 against seawater at different depths. At the depth shallower than about 2400–2500 m, liquid CO_2 is lighter than seawater and would float upward, whereas in the ocean layers deeper than 3000 m, liquid CO_2 is denser than seawater and it will sink. The rate of liquid CO_2 dissolution in seawater at the depth of roughly 2500 m has been reported to be ~3 µmol cm^{-2} s^{-1}; thus, if liquid CO_2 bubble with the size of 0.9 cm is released, about 90% of its mass would be dissolved in the first 200 m of its ascend to the surface (Brewer et al. 2002). Based on this and other relevant information, CO_2 diffusers should be designed such that the CO_2 bubbles are dissolved within the first 100–200 m of the depth of release, and they are not allowed to reach the depth of 500 m (where they will become gaseous CO_2 bubbles). The design of the diffusers for the release of liquid CO_2 at depths of 3000 m and more depends on the storage option: larger CO_2 bubbles would quickly drop to the sea floor, whereas small droplets would dissolve in seawater before reaching the seafloor. There is a transition zone between depths of about 2500 and 3000 m where buoyancy may vary depending on temperature, location, etc.

14.5.2.2.3 *CO₂ Hydrate*

At temperatures below 8°C–9°C and depths of about 400 m and more, CO_2 can react with water forming crystalline CO_2 hydrate ($CO_2 \cdot nH_2O$, where $6 < n < 8$), which is denser than seawater, and it sinks. It should be noted that CO_2 hydrate could form a skin on the liquid CO_2 bubble wall, thus affecting the rate of its dissolution. At depths of 500 m and greater, the fully formed crystalline CO_2 hydrate has the density of $r = 1.120 \pm 10$ g cm^{-3}, that is, about 10% greater than seawater (Fer and Haugan 2003). According to some estimates, CO_2 hydrate would dissolve in seawater at the rate of 0.2 cm h^{-1} (or 0.47–0.60 µmol cm^{-2} s^{-1}), which is significantly slower compared to the dissolution rate of liquid CO_2 (Rehder et al. 2004). Thus, the formation of CO_2 hydrate can potentially slow down the dissolution of CO_2 and, from this viewpoint, it is beneficial for ocean CO_2 storage. The formation of solid CO_2 hydrate may also create some technical difficulties during the CO_2 release into the ocean, in particular, by clogging pipelines and diffusers and impeding the CO_2 flow.

14.5.2.2.4 *Solid CO₂*

Similar to CO_2 hydrates, solid CO_2 is denser than seawater at any depth and, thus, would rapidly sink. Solid CO_2 is dissolved in seawater at roughly the same rate as CO_2 hydrate (0.2 cm h^{-1}) (Aya et al. 1997). Based on this rate of dissolution, relatively small chunks of solid CO_2 would be quickly dissolved before reaching the sea floor, whereas large masses of solid CO_2 could easily reach the bottom of the ocean.

14.5.2.3 *Ocean CO₂ Storage: Status of Development*

In contrast to geological CO_2 storage, the ocean storage technology has not yet reached a precommercial or even a large-scale demonstration stage of the development, and it is in

R&D phase that mostly involves modeling studies, laboratory, and small-scale field experiments. There have been two attempts to conduct large-scale in situ experiments involving the injection of liquid CO_2 into the ocean, but both of them did not even start due to a fierce opposition from a number of concerned organizations. The first project (Hawaii CO_2 Sequestration Field Experiment) involved an international consortium of experts from United States, Japan, Norway, and Canada who were planning to directly inject about 60 t of liquid CO_2 into the deep ocean from a ship near the Island of Hawaii (the total cost of the project was U.S. \$5 million (IPCC CCS 2005)). However, the project was canceled due to a stiff opposition from environmental groups and organizations. The second attempt to carry out a large-scale experiment would have involved the release of 5.4 t of liquid CO_2 at the depth of about 800 m off the coast of Norway. The research plans included monitoring CO_2 dispersion in the Norwegian Sea over a long period. This attempt, however, also proved to be unsuccessful due to the opposition from the Norwegian Environment Ministry (Giles 2002). A number of small-scale experiments (several dekaliters of liquid CO_2) were carried out in marine sanctuaries (Brewer et al. 2005). Thus, there is a very limited experience and practical knowledge with regard to ocean CO_2 storage as a carbon mitigation strategy.

14.5.2.4 Site Selection for Ocean CO_2 Storage

Among practical considerations related to ocean CO_2 storage is the selection of a suitable site. Many factors may influence the site selection process, for example, environmental considerations, the cost of storage, safety issues, laws and regulations. Since most of the large point sources of CO_2 are located far from oceans, the proximity to deep sea is an important factor determining the overall cost. Oceans with the depth of more than 1 and 3 km amount to about 88% and 75% of the total ocean surface area, respectively. Large CO_2 point sources located near these areas would be the most preferred and cost-effective settings for direct CO_2 injection into the ocean. In most cases, ocean storage would require liquid CO_2 transport by ships or deep-sea pipelines. The experience gained by oil and gas industries, especially with regard to offshore operations (e.g., drilling platforms, deep-sea pipelines), can be advantageously used during the realization of large-scale ocean CO_2 storage projects.

14.5.2.5 Technologies for Releasing Liquid CO_2 into the Ocean

In principle, ocean storage can be accomplished in four different ways:

- By dispersing dense-phase CO_2 into the ocean at relatively shallow depths and dissolving CO_2 in ocean layers (so-called *water column* option)
- By injecting liquid CO_2 into deep (about 3 km) ocean water, where it could form a plume of liquid CO_2 sinking to the bottom of the ocean (at this depth, liquid CO_2 is denser than seawater)
- By depositing the dense-phase CO_2 onto the sea floor at depths below 3 km where it would form a separate liquid phase ("CO_2 lake")
- By injecting dense-phase CO_2 at a depth where it would form CO_2 hydrate crystals sinking to the ocean floor

Some of the above technical options for ocean CO_2 storage are depicted in Figure 14.20. A brief discussion of various technical approaches with regard to the practical realization of CO_2 ocean storage follows.

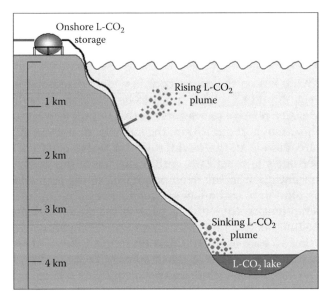

FIGURE 14.20
Various technical approaches to the practical realization of CO_2 ocean storage.

14.5.2.5.1 Water Column Release

Liquid CO_2 could be dispersed at depths of 1 km and more using available technologies. In particular, pipelines delivering liquid (or dense phase) CO_2 to these depths are commercially available and have been used for transporting CO_2 to EOR sites. It should be mentioned that oil and gas industries routinely lay pipelines on the sea floor at the depths of down to 1.6 km, and no technical obstacles are foreseen for placing the pipelines at depths of 3 km. It was estimated that a pipe with a diameter of 1 m would transport about 70,000 t CO_2 per day, which equals the amount of CO_2 produced by a coal-fired plant with the power output of $3\,GW_{el}$ (Ormerod et al. 2002).

Modeling studies indicate that it would be possible to diffuse CO_2 at these depths such that almost all CO_2 would be dissolved in seawater within about 100 m of the injection point (Liro et al. 1992). Once dispersed, CO_2-saturated seawater would be diluted and spread predominantly along the horizontal layers with the constant temperature and density. It is envisioned that the dilution would minimize the CO_2 acidic effect on ocean chemistry and biota. At the typical ocean pH of 7.5–8.5, the dominant form of dissolved CO_2 is bicarbonate ion. Eventually, CO_2 will reequilibrate with the atmosphere. For example, about 50% of CO_2 injected at a depth of 1 km would remain dissolved in the ocean after 200 years, and only 20% after 1000 years (Global CCS Institute 2009).

Injecting liquid CO_2 at a depth of about 1 km from a manifold would form a rising droplet plume (since at this depth liquid CO_2 is less dense than seawater, see Figure 14.19). Alternatively, liquid CO_2 could be transported by a ship or tanker to a stationary platform from which it would be injected into seawater by means of a vertical pipe, or it could also be released via a towed pipe. As mentioned, care should be exercised to avoid the formation of CO_2 hydrates (that could plug the pipes, nozzles, and diffusers) at depths below 500 m.

It should be noted that the United Nations Convention on the Law of the Sea (UNCLOS) restricts the projects dealing with ocean CO_2 storage, in particular, using the water column release method. Because of the restrictions, ocean CO_2 storage by the water column method is not currently practiced anywhere in the world.

14.5.2.5.2 CO$_2$ Lake on the Ocean Floor

The formation of CO_2 lakes on the ocean floor is widely discussed in the literature. In order to form a stable liquid CO_2 lake, it has to be released at depths of at least 3 km (due to its higher density relative to seawater at this depth) (see Figure 14.19). Since the average depth of the ocean is about 3.8 km, the suitable locations for CO_2 storage by the CO_2 lake method are abound in the world; thus, the ocean capacity for CO_2 storage in the form of seafloor lakes is practically unlimited. Modeling studies predict that the increase in CO_2 concentration in the deep ocean due to the formation of the CO_2 lake would promote dissolution of carbonate sediments (e.g., $CaCO_3$), which in turn would tend to prolong CO_2 retention time (this is especially true for Atlantic Ocean which is rich with $CaCO_3$ sediments inventory) (Archer et al. 1998). Numerical simulation calculations indicate that an increased stratification in the ocean and weaker overturning would tend to enhance the retention of ocean-stored CO_2, and, thus, favor the CO_2 lake option (Jain and Cao 2005). The techniques for forming CO_2 lakes include the usage of different types of pipes for discharging liquid CO_2 into the ocean floor and the formation of dense CO_2-containing slurries. The latter option was advocated by Aya et al. (2003) who proposed forming a slurry of liquid and solid CO_2 and injecting it into the ocean at a depth of 200–500 m, such that it would rapidly sink due to its relatively high density. This research group has conducted a field experiment off the coast of California using a slurry consisting of liquid CO_2 mixed with solid CO_2 pellets of about 8 cm size, and they found that the pellets melted in about 2 min while they traveled about 50 m deep into the ocean. The initial experiments demonstrated the importance of designing the slurry with the right dimension of solid CO_2 chunks making it possible to reach a depth of 3 km before they melt.

   Although liquid CO_2 forms a separate layer and is not directly mixed with seawater, CO_2 would gradually dissolve in water forming carbonate species (carbonic acid, carbonate and bicarbonate ions), thus, the depletion of the CO_2 pool is just a matter of time. It was estimated that 50 m thick CO_2 lake would dissolve in as low as 30 years (if ocean currents are nearby) and as long as 1000 years in stagnant areas where dissolution is controlled by diffusion (Global CCS Institute 2009). Once dissolved in seawater, CO_2 diffuses to the upper layers reaching the ocean–atmosphere interface and becomes part of the global carbon cycle. This process, however, is very slow: according to some estimates, 85% of CO_2 stored at the depth below 3 km would remain stored after 500 years (Global CCS Institute 2009).

14.5.2.5.3 CO$_2$ Hydrate

CO_2-releasing technologies have been proposed that take advantage of the formation of CO_2-hydrate, which is about 15% denser than seawater, and, therefore, sinks faster and dissolves slower over a broad range of seawater depths. In particular, Tsoiris et al. (2004) proposed a method for forming a CO_2–hydrate–seawater slurry and releasing it at the depths between 1 and 1.5 km, such that this sinking plume would dissolve as it moves downward, thus, distributing CO_2 over a large vertical distance (potentially, several kilometers). The idea has been tested in a field experiment at 1 km ocean depth where CO_2 was rapidly mixed with seawater using a capillary nozzle to form a composite paste that was neutrally buoyant at this depth. A rapidly sinking composite/slurry plume consisting of cold seawater and solid CO_2 with a hydrate skin has also been proposed with the aim of releasing it in shallow seawater, such that it moves by gravity to the deep ocean (this method allows avoiding the cost of a long pipeline) (Aya et al. 2004). It should be noted that the CO_2-hydrate formation process is reversible, and the hydrate crystals tend to redissolve in seawater when conditions are no longer favorable for their existence (e.g., if temperature increases, or pressure and CO_2 concentration decrease).

14.5.2.6 Carbonate Mineral Neutralization

The principle of carbonate neutralization strategy is based on enhancing the ocean CO_2 storage capacity via increasing seawater alkalinity by dissolving carbonate minerals in seawater. According to one approach, this can be accomplished by reacting CO_2 with calcium carbonate (limestone) in the seawater environment with the formation of bicarbonate anion and calcium cation as follows:

$$CaCO_3(s) + CO_2(gas) + H_2O \leftrightarrow Ca^{2+}(aq) + 2HCO_3^-(aq) \qquad (14.15)$$

This reaction neutralizes the effect of CO_2 on the pH of seawater (the pH around the release point rises by about 2 units, approaching the ambient pH value), which allows the ocean to absorb more CO_2 with less of a change in ocean acidity (it is assumed that the bicarbonate solution will have an indefinite sequestration time). Quantitatively, the dissolution of 1 mole of $CaCO_3$ would result in storage of additional 0.67–0.8 mole of CO_2 (or 2.8–3.5 t $CaCO_3$ per ton of CO_2) without an impact on CO_2 concentration in seawater (Kheshgi 1995; Caldeira and Rau 2000). The timescale for the increased ocean storage capacity due to carbonate neutralization would depend on the rate of natural $CaCO_3$ sedimentation, and it was estimated (via modeling studies) to last about 6000 years (Archer et al. 1998). The proponents of the carbonate neutralization strategy argue that this method provides a more permanent containment for disposed CO_2 compared with other ocean sequestration options, and allows avoiding its potentially negative environmental impact on marine biota (thus, this method was dubbed "environmentally benign sequestration," EBS, of CO_2).

Several technical approaches have been proposed to accomplish carbonate neutralization in seawater; however, all of them were conducted at the level of either modeling studies or laboratory-scale experiments, and no in situ field tests to prove the concept on a sufficiently large scale have been conducted so far. For example, Rau and Caldeira proposed a process combining the capture of waste CO_2 and carbonate neutralization methods (Rau and Caldeira 1999). According to the proposed approach, CO_2 is extracted from flue gas of a coal power plant by reacting it with a slurry of crushed limestone in seawater. The resulting seawater solution (containing mostly Ca^{2+} and HCO_3^- ions) would be released into the ocean, where it is further diluted by ocean water. The advantages of this approach are that it does not require a preliminary separation and transport of CO_2, and it does not involve the deep-sea injection via pipelines and nozzles (with an associated cost reduction). However, the extent of CO_2 removal from flue gases was not that great, since the reaction is limited by equilibrium. The authors of another study proposed to mix liquid CO_2 with pulverized limestone and water and release the resulting emulsion at the depth below 500 m (Golomb and Angelopoulos 2001). It was reported that carbonation of a widely occurring mineral olivine (Mg_2SiO_4) converts CO_2 into an environmentally benign mineral magnesite ($MgCO_3$) (Bearat et al. 2003). The techno-economic evaluation of the carbonate neutralization method, however, showed that its implementation would increase the cost of CO_2 disposal in the ocean by 150% (Golomb and Angelopoulos 2001) (the added cost could partially be offset by the use of shorter pipelines in order to reach the necessary depths for CO_2 disposal).

The main problem with the carbonate neutralization approach is the use of enormous quantities of carbonate minerals. Although the resources of these minerals far exceed that of carbon contained in all types of fossil fuels, the necessity of processing and handling such an immense amount of carbonates (up to 3.5 times the mass of CO_2 stored) may put certain limitations on the scope of the large-scale implementation of this approach. One should also take into consideration a possible negative environmental impact of this technology, which is associated with greatly expanded mining, transportation, and processing of the carbonate minerals, and potentially an adverse effect of impurities in carbonates on the ocean biota.

Among new EBS approaches, it would be worthwhile to mention CO_2 sequestration in the form of solid carbonic acid. It was proposed that under right conditions, carbonic acid could form solid oligomers $(H_2CO_3)_n$ (where $n = 2$–4) (Tossel 2009). The concept, however, is based exclusively on available spectral data and calculations, and is yet to be validated experimentally. Another new EBS concept deals with the use of micron-sized carbon particles to stabilize liquid CO_2 on the ocean floor by creating a protective carbon-CO_2-hydrate "skin" that would prevent liquid CO_2 from interacting with seawater and, subsequently, changing its pH (Muradov 2008). As mentioned, at high pressures (100 atm and higher) and ambient temperature, L-CO_2 and water are not miscible, and the two liquids would be separated into two phases similar to an oil–water system. The proposed concept is based on the assumption that if the L-CO_2–water system would behave like a two-phase oil–water system, it could be stabilized by micron-sized carbon particles; the stabilization of oil–water emulsions by a film of highly dispersed solid particles is well known in the production of crude oil (Golomb and Angelopoulos 2001). It was hypothesized that the presence of carbon particles could facilitate the formation of solid CO_2 hydrates that would form a layer between seawater and L-CO_2 phase (thus, L-CO_2 will be protected by a double "skin" of two solid phase layers: carbon and CO_2-hydrate). Computer simulations indicated that the CO_2-hydrate layer retards the CO_2 dissolution rate in water by a factor of 2.7 (Fer and Haugan 2003) (thus, the presence of an additional carbon layer, in all likelihood, would further slow down the dissolution process). It should be noted that neither of the above-discussed approaches to sequester CO_2 in an environmentally benign form has been tested on a sufficiently large-scale or long-duration experiments. EBS appears to be a promising approach; however, much fundamental and applied research and field testing will be needed to prove the cost-effectiveness and viability of this CO_2 disposal method.

14.5.2.7 Cost of Ocean CO_2 Disposal

There are fewer studies reported in the literature on the cost of ocean CO_2 storage compared to the geological storage option. Typically, the cost of storage includes the costs of handling of CO_2 and transport of CO_2 offshore (but not costs of onshore transport). The costs of capture and compression/liquefaction are considered to be the major cost components followed by transport costs (i.e., shipping and pipeline transport). Figure 14.21 depicts the costs of ocean storage for the following scenarios: (1) CO_2 transport and injection at the depth of 3 km from a floating platform, (2) CO_2 transport and injection at 2.0–2.5 km depth from a moving ship via a towing pipe, and (3) CO_2 injection via pipeline transport at 3.0 km depth (Akai et al. 2004). These scenarios assume that CO_2 is produced by three PC power

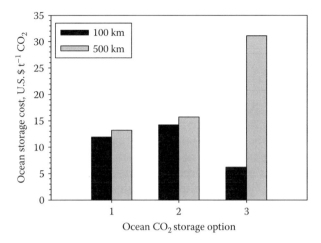

FIGURE 14.21
Ocean storage cost estimates for CO_2 transport and injection at different depths: 1—injection from a floating platform at 3 km depth, 2—injection from a moving ship at 2.0–2.5 km depth, 3—injection via pipeline transport at 3.0 km depth. (Data from IPCC special report on carbon dioxide capture and storage, Prepared by Working Group III of the Intergovernmental Panel on Climate Change, B. Metz, O. Davidson, H. de Coninck, M. Loos, A. Meyer (eds.), Cambridge University Press, Cambridge, U.K., Table 6.5, p. 310, 2005.)

plants with a net power generation capacity of $600\,MW_{el}$ and transported 100 or 500 km by a CO_2 tanker ship with the capacity of $80,000\,m^3$. In both scenarios, the cost of ocean storage includes the following costs: tank storage of CO_2 onshore, shipping of CO_2, and the injection of CO_2 (pipe, nozzle, etc.). The estimated costs for the first and second scenarios for the ship transport distance of 500 km are (in U.S. $ t^{-1} CO_2 net stored) 13.2 and 15.7, respectively.

The cost of ocean storage of CO_2 delivered by a pipeline to the ocean floor differs from that of the ship transport option. In particular, Akai et al. (2004) estimated that the cost of ocean storage of CO_2 produced at a $600\,MW_e$ coal-fired power plant transported 100 or 500 km by a CO_2 pipeline and injected at 3 km depth is (in U.S. $ t^{-1} CO_2 net stored) 6.2 and 31.1, respectively. No literature data is available on the cost of ocean CO_2 storage via CO_2 "lake" option.

14.5.3 Mineral Carbonation

Mineral carbonation (also called "mineral sequestration") is a CO_2 storage option related to the fixation of CO_2 in the form of insoluble carbonates. This relatively new approach is based on the reaction of CO_2 with complex metal oxides (preferably, silicates of Ca and Mg) to produce corresponding carbonates and silica (typically, these reactions are exothermic). The mineral carbonation approach is considered by many as a prospective EBS technology that could potentially provide an ecologically safe and geologically stable CO_2 disposal in the form of mineral carbonates.

The examples of such mineral carbonation reactions involve naturally occurring silicates of Ca and Mg such as olivine (Mg_2SiO_4), serpentine ($Mg_3Si_2O_5(OH)_4$), and wollastonite ($CaSiO_3$), as follows:

$$Mg_2SiO_4 + 2CO_2 \rightarrow 2MgCO_3 + SiO_2 \quad \Delta H° = -89\,kJ\,mol^{-1}\,CO_2 \tag{14.16}$$

$$\frac{1}{3}Mg_3Si_2O_5(OH)_4 + CO_2 \rightarrow MgCO_3 + \frac{2}{3}SiO_2 + \frac{2}{3}H_2O \quad \Delta H° = -64\,kJ\,mol^{-1}\,CO_2 \qquad (14.17)$$

$$CaSiO_3 + CO_2 \rightarrow CaCO_3 + SiO_2 \quad \Delta H° = -90\,kJ\,mol^{-1}\,CO_2 \qquad (14.18)$$

These reactions are thermodynamically favored at ambient conditions, and they spontaneously occur in nature on a timescale of thousands of years (the process is called "silicate weathering"). Mineral carbonates are the most thermodynamically stable forms of carbon, which ensures the permanent fixation of carbon (see Figure 14.22).

The material basis for the mineral carbonation process comprises abundant natural silicate rocks, as well as some types of industrial wastes, for example, a fly ash (CaO content of 35%), a slag from steel production (CaO and MgO content of 65%), waste cement, etc. (IPCC CCS 2005). Preliminary experimental studies of mineral carbonation conducted by NETL are encouraging. In particular, it was found that finely ground serpentine or olivine reacted with supercritical CO_2 to form magnesium carbonate ($MgCO_3$) (U.S. DOE, NETL 2003). At the temperature range of 150°C–250°C and pressure of 85–100 atm, 84% conversion of olivine was achieved in 6 h.

Depending on the specifics of the process, mineral sources, and targeted applications, mineral carbonation can be carried out in two modes: in situ and ex situ. In situ mineral carbonation involves the injection of a CO_2 stream into a silicate-rich formation, where the carbonation reaction will take place over certain period of time, resulting in permanent storage of CO_2 in the form of carbonates. This approach is very similar to geological CO_2 storage option (specifically, "mineral trapping"). In the ex situ approach, mineral carbonation is carried out in a special chemical processing plant, where the reaction between CO_2 and pretreated silicate minerals or inorganic industrial wastes takes place yielding carbonates, silica, and other solid products for subsequent disposal (see Figure 14.23).

It should be noted that mineral carbonation is an energy-intensive process since it requires a number of steps involving the preparation of the reactants (mining, grinding, and transport), the chemical processing, and the disposal of carbonates and other solid products with associated energy penalties. There have been significant difficulties

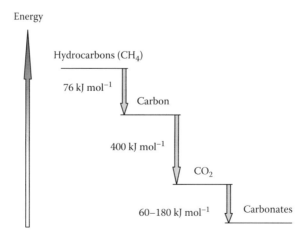

FIGURE 14.22
Thermodynamic stability of different carbon-containing products.

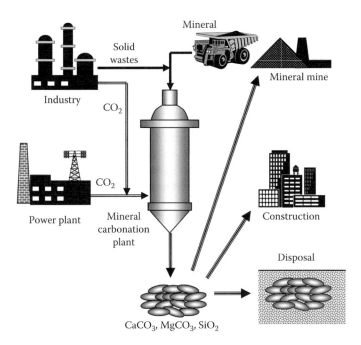

FIGURE 14.23
Schematic representation of ex situ mineral carbonation of silicate rocks and industrial residues.

encountered in an engineering arrangement of the process, which are attributed to very slow reaction kinetics and certain thermodynamic limitations lowering the products yield. In many cases, the mass of mineral matter required for CO_2 mineralization is many times larger than the mass of CO_2 to be stored. However, the selling points for this CO_2 storage option are twofold: the abundance and low cost of natural silicates suitable for this application, and the high degree of CO_2 storage permanence since CO_2 is permanently "locked" in the form of very stable insoluble carbonates. Despite these attractive features, mineral carbonation is still in a relatively early stage of development: most reported studies relate to laboratory-scale experiments that would not allow to thoroughly evaluate its commercial potential, or to make a reasonable techno-economic assessment of the technology and its potential environmental impact. U.S. DOE report provides some preliminary data on the energy consumption and economics of a wet mineral carbonation process using several types of minerals (olivine, lizardite, antigorite, and wollastonite with ore costs in the range of U.S. \$15–\$48 t^{-1} ore) (O'Connor et al. 2005). The authors estimated that the energy input requirements for conducting the mineral carbonation process varied in the range of 180–2300 kWh t^{-1} CO_2 stored, depending on the mineral used and pretreatment procedure (e.g., need for activation). Thus, between 30% and 50% of the total energy, output would be used for carrying out mineral carbonation with the associated reduction of power plant efficiency from original 35% to 25% and 18%, respectively. This would translate into an additional 60%–180% fuel usage in the power plant with CCS coupled with mineral carbonation compared to a plant without CCS (with the same energy output). The corresponding cost of CO_2 storage via mineral carbonation was estimated at the range of U.S. \$55–\$250 t^{-1} CO_2 stored.

The potential of mineral carbonation as a CO_2 mitigation strategy depends on a number of trade-offs between the costs of the mineral mining, processing, carbonate disposal, and the benefits of permanent CO_2 storage. If one takes into account that the fixation of 1 t of CO_2 in the form of carbonates would require processing of about 1.6–3.7 t of silicates and disposal of 2.6–4.7 t of solid products, it is clear that the large-scale implementation of the technology would require the expanding of a mining industry to the scale comparable with coal industry (IPCC CCS 2005). There may be the possibility of getting credits for by-products of the mineral carbonation processes. For example, during the mineral ores pretreatment stage, certain high-value minerals can be extracted from them, for example, magnetite from olivine, or minerals containing Cr, Ni, Mn, platinum group metals from periodite rocks, etc. It was reported recently on pilot-scale trials of a new mineralization process involving CO_2 reaction with bauxite residue (red mud) (Global CCS Institute 2009). The process addresses CO_2 storage needs while alleviating environmental concerns over the corrosive (caustic) waste product.

14.5.4 CO_2 Sequestration in Biosphere

This section addresses CO_2 sequestration in the Earth's biosphere and its potential as a carbon mitigation option. The objective of the enhanced CO_2 sequestration in biosphere is to achieve a relatively rapid withdrawal of CO_2 from the atmosphere over the next several decades to gain some time for the technological development and practical implementation of other CO_2 mitigation options. The advantage of CO_2 sequestration in biosphere over other carbon sequestration strategies is that it is applicable to all CO_2 sources including atmospheric CO_2, whereas geological and ocean sequestration is mainly limited to fossil-based power plants and large industrial point sources of CO_2. Three major components of the Earth's biosphere are terrestrial, aquatic, and ocean ecosystems, which are discussed below.

14.5.4.1 *CO_2 Sequestration in Terrestrial Biosphere*

Terrestrial ecosystems, which include vegetation and soil, are important biological CO_2 scrubbers and are considered as major natural sinks for removing CO_2 from the atmosphere. The terrestrial biosphere annually sequesters about 2 Gt of carbon (U.S. DOE, Carbon sequestration. State of the Science 1999). The objective of CO_2 sequestration in biosphere is to dramatically increase this rate while taking into the consideration all the possible ecological, economic, and social implications of this mitigation option. In principle, carbon sequestration in terrestrial biosphere can be achieved by either the enhanced net removal of CO_2 from the atmosphere or by reducing the net CO_2 emissions from the terrestrial ecosystems into the atmosphere, or by the combination of both approaches. The following is the list of possible measures to accomplish CO_2 storage in terrestrial biosphere:

- Increase photosynthetic carbon fixation
- Reduce decomposition of organic matter and other soil-derived CO_2 emissions
- Protect ecosystems that store carbon
- Reverse land use change trends that contribute to CO_2 emissions
- Create energy offsets by using biomass for production of fuels and other products
- Manipulate ecosystems to increase carbon sequestration beyond current conditions

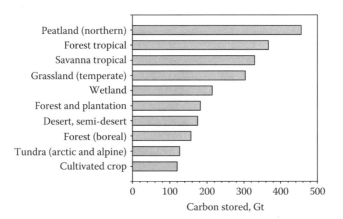

FIGURE 14.24
Global estimates of carbon stored in major terrestrial ecosystems and distribution of stored carbon between plant and soil ecosystems. (Data from U.S. DOE, Carbon sequestration, State of the Science, U.S. DOE, Office of Science, Office of Fossil Fuels, Working Paper on Carbon Sequestration Science and Technology, 1999.)

The Earth's terrestrial biosphere consists of a broad diversity of ecosystems that cumulatively store about 2542 Gt of carbon. Figure 14.24 depicts the global estimates of carbon stored in major terrestrial ecosystems (e.g., forest, tundra, grassland) and the distribution of stored carbon between the plant and soil ecosystems. As seen from this figure, peatland, tropical forest, and tropical savanna are the major carbon-storing terrestrial ecosystems of the world. The net primary productivity (NPP) of all ecosystems is about 59 Gt C year^{-1} (NPP is the rate at which all plants in the ecosystem produce net useful chemical energy, i.e., biomass). The values for the NPP vary in the range of 11 gC m^{-2} year^{-1} for extreme deserts to 925 gC m^{-2} year^{-1} for tropical forests and 1180 gC m^{-2} year^{-1} for wetlands (U.S. DOE, Carbon sequestration. State of the Science 1999). The carbon stock in soil by the factor of about 4 exceeds that of a plant matter. Thus, enhancing carbon storage in soil may potentially have a more pronounced impact on carbon sequestration in biosphere compared to that in a plant matter. However, many aspects of carbon sequestration in soil are much less known than those related to living plants.

Soil is a rather diverse ecosystem containing both organic and inorganic components. The organic components consist of decaying plant and animal matter, various microbial communities, different types of invertebrates and vertebrates, fungi, bacteria, etc. Soil microorganisms control organic carbon cycling and storage in soil by decomposing dead plant and animal matter and releasing CO_2 back to the atmosphere (USDA 2010). The inorganic part of soil mostly consists of minerals (e.g., carbonates) and absorbed CO_2. There are two main advantages to carbon sequestration in soil compared to that in the plant matter: (1) soil organic matter (SOM) has a longer residence time than most plant biomass (some stable compounds in SOM have the turnover times of hundreds to thousands of years), and (2) soils with high levels of SOM exhibit improved nutrient absorption, water retention, and texture, thus, increasing plant productivity and further enhancing atmospheric CO_2 absorption.

Figure 14.25 shows different soil processes that affect carbon fate and transport underground. It is evident that the dynamics of carbon transformations and transfer in soils are very complex and represent the balance of CO_2-absorbing and -releasing processes (thus, they may result in either net carbon sequestration or CO_2 emissions to the atmosphere). These transformations involve microbial conversion of organic matter to N-, S-,

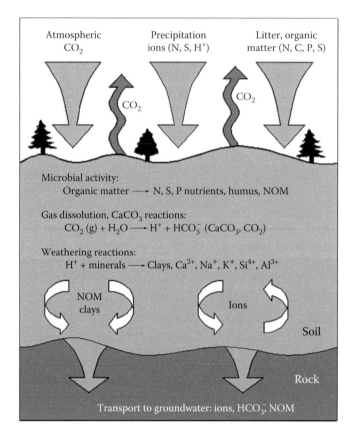

FIGURE 14.25
Carbon transformation and transport in soil. (Modified from U.S. DOE, Carbon sequestration, State of the Science, U.S. DOE, Office of Science, Office of Fossil Fuels, Working paper on Carbon Sequestration Science and Technology, 1999.)

and P-containing nutrients, oxidation-resistant humus, and natural organic matter (NOM). Atmospheric CO_2 dissolved in water produces bicarbonate ions (HCO_3^-) that could be sequestered if they enter the deep groundwater system that has residence time of hundreds to thousands of years. Carbon in the form of NOM could be sequestered by means of being transported to deep groundwater systems or deposited deeper in soil. Thus, one of the measures to enhance carbon sequestration in soil is to encourage the geohydrologic systems to promote deep transport of carbon into the groundwater systems (U.S. DOE, Carbon sequestration. State of the Science 1999).

Net removal of CO_2 from the atmosphere by terrestrial ecosystems ($\sim 2\,Gt$ C year^{-1}) is the balance of all processes involving plant photosynthesis, consumption, respiration, and decay. Plants, roots, and microbial biomass grow consuming CO_2 (carbon input) and return stored carbon to atmosphere through respiration and decay (carbon loss). Carbon sequestration in terrestrial ecosystems can be enhanced by increasing the amount of carbon stored in plant biomass, roots, and soil (both organic and inorganic carbon). Thus, biospheric carbon sequestration is and, most likely, will continue to be an immense natural scrubber for all the anthropogenic CO_2 emissions, and the methods relying on biological carbon transformation would potentially play a significant role in the management of carbon sequestration in the future (U.S. DOE, Carbon sequestration.

State of the Science 1999). However, the extent of the biosphere contribution to carbon mitigation cannot be adequately assessed due to a high uncertainty in the estimates of the balance between two immense fluxes of carbon related to photosynthesis and respiration. Although measuring annual changes in atmospheric CO_2 concentrations (and factoring in oceanic carbon dynamics) may provide some estimates of the net difference between global carbon uptake via photosynthesis and carbon release via respiration, this information cannot be reliably used for the prediction of the impact of biosphere on the future CO_2 concentration in the atmosphere. The major source of this uncertainty originates from the high sensitivity of the photosynthesis–respiration balance to a variety of environmental variables such as temperature, humidity, availability of nutrients, etc., and this balance varies widely among different ecosystems. Because the photosynthesis- and respiration- related carbon fluxes are so large, even small changes in their values (e.g., due to drastic increase in anthropogenic CO_2 emissions) could result in unpredictable consequences to global climate and may potentially cancel out carbon management measures.

Figure 14.26 depicts the estimates of potential carbon sequestration in major terrestrial ecosystems that might be sustained over a period of 25–50 years. The total capacity of carbon sequestration in terrestrial biosphere was estimated at 5.65–10.1 Gt C year^{-1}. The data presented in Figure 14.26, however, are rather optimistic since they do not adequately reflect many economic, energy, and environmental implications of biological carbon storage at such high rates. The complexities in estimating the potential for increasing biospheric carbon sequestration stem from many factors that are difficult to predict, for example, cost of petroleum fuels and fertilizers, availability of water, various socioeconomic issues, competition with other carbon management strategies, etc.

There are a number of approaches and strategies to achieve a considerable increase in the amount of carbon sequestered in the terrestrial ecosystems. The main objectives of these measures are to increase the amount of carbon in below-ground (i.e., soil and sediments) and above-ground biomass systems and to manage the land area. The following simple formula provides a means of quantifying the potential for carbon sequestration in terrestrial biosphere (U.S. DOE, Carbon sequestration. State of the Science 1999):

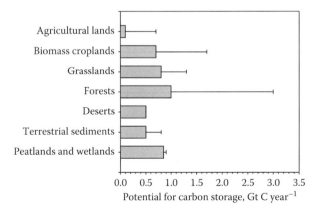

FIGURE 14.26

Global potential carbon sequestration rates in major terrestrial ecosystems that are likely to be sustained over the period of 25–50 years. (Data from U.S. DOE, Carbon sequestration, State of the Science, U.S. DOE, Office of Science, Office of Fossil Fuels, Working Paper on Carbon Sequestration Science and Technology, 1999.)

$$\text{PCS} = (a_i\text{AGC}_i + b_i\text{BGC}_i) \times c_i\text{LA}_i \tag{14.19}$$

where
  PCS is a potential for carbon sequestration
  a_i and b_i are the potential increase in above- and below-ground carbon, respectively, in the ith ecosystem
  AGC_i and BGC_i are the above-ground carbon (biomass) and below-ground carbon (root biomass + soil organic and inorganic carbon), respectively, in the ith ecosystem in the index year
  c_i is the potential change in the land area due to the carbon management in the ith ecosystem
  LA_i is the land area of each ecosystem in the index year

The options to increase carbon sequestration in the above-ground systems are:

- Increase in the rate of biomass accumulation
- Increase in the density of biomass per area
- Increase in the longevity of biomass carbon (by decreasing the decomposition rate)
- Increase in the beneficial use of biomass carbon in long-lived products (e.g., wood products, structural materials)

The ways to increase the amount of below-ground carbon are:

- Increase in the depth of soil carbon
- Increase in the density of both organic and inorganic carbon in soil
- Increase in mass and depth of roots
- Increase in the longevity of underground carbon (by decreasing the decomposition rate)
- Prevention of erosion

The potential for increased carbon sequestration via a land and crop management is substantial since it affects both above- and below-ground biomass systems. Although the land management strategies vary widely and are specific to each ecosystem, the following is a list of options to enhance carbon sequestration in terrestrial ecosystems (U.S. DOE, Carbon sequestration. State of the Science 1999):

- Forest management (reducing and reversing deforestation, fire management)
- Afforestation of marginal crops and pasture land
- Crop rotation, tillage, and residue management
- Range land management
- Improvement in cropping systems and precision farming
- Decrease in the pace of urbanization
- Decrease in the conversion of forest land to agricultural use
- Pest and invasive species control and disease management

According to some estimates, deforestation accounts for about 20% of the annual global emissions of CO_2 (U.S. DOE, NETL 2003). It was estimated that 12%–15% of fossil-based

CO_2 to be emitted in the years 1995 through 2050 could be offset by implementing the following measures: slowing tropical deforestation, allowing forests to regenerate, expanding plantations, and engaging in new plantings and other forms of agroforestry.

Other important strategies to support the objectives of carbon sequestration by terrestrial biosphere include

- *Soil improvement* for increasing (directly) below-ground and (indirectly) above-ground carbon stocks. This objective can be achieved through the following measures: irrigation and water retention, fertilization and nutrient acquisition, retention of soil carbon, erosion control, soil amendment, and creation of new soil.
- *Biotechnology, molecular genetics, and species selection.* This approach can directly affect both above- and below-ground carbon inventories, and enables the use of genetic engineering to produce plants with increased carbon sequestration capacity. The likely focus areas for this strategy are an improvement in photosynthetic efficiency, the extension of growing seasons of plants, increasing lignin content for longevity of woody biomass, and increasing pest and disease resistance.
- *Ecosystem dynamics* focuses on the whole terrestrial ecosystem behavior rather than on individual components of the system aiming at the optimization of carbon sequestration performance. The examples of specific actions related to this strategy include balancing biomass decomposition as a source of carbon loss to the atmosphere and as a source of nutrients essential to plant growth; designing strategies compatible with other human demands on land and natural resources; determining the potential feedback from carbon sequestration actions (e.g., the increased release of CH_4, CO, and N_2O to the atmosphere), etc.

In summary, carbon sequestration in the terrestrial ecosystems can potentially provide significant near- to mid-term (over the next 25–50 years) benefits as a carbon mitigation option. The estimated potential for carbon sequestration in terrestrial biosphere is in the order of 5–10 Gt C year^{-1}, although this value is rather speculative and more research is needed to determine this potential and assess all the economic, energy, and social implications of this option. It should be emphasized that carbon stored below ground (in both organic and inorganic forms) is more permanent than the one stored above ground (e.g., as plant biomass). From this point of view, the transformation of biomass into long-lived products is beneficial to enhance its potential as a carbon storage option. The following ecosystems offer significant opportunity for terrestrial carbon sequestration: forest lands, agricultural lands, biomass croplands, deserts and degraded lands, boreal wetlands, and peatlands.

14.5.4.2 Biological CO_2 Sequestration by Aquatic Systems

The potential of photoautotrophic organisms (algae) to efficiently capture CO_2 from both the atmosphere and CO_2-containing industrial streams (e.g., from power plants, cement-manufacturing plants, etc.) has recently attracted a worldwide attention. Numerous types of microalgae strains have biomass productivity superior to that of terrestrial plants (trees, grasses, etc.) by at least one order of magnitude. To achieve high CO_2 uptake rates and, consequently, biomass production rates, microalgae has to be grown under the optimal conditions of light, temperature, pH, nutrient, and CO_2 concentrations (Behrens 2005). The stoichiometry of photosynthesis-based cellular growth indicates that for every 106 mol of CO_2, 16 moles of nitrate and 1 mol of phosphate are consumed (Nakamura 2004):

$$106CO_2 + 16NO_3^- + H_2PO_4^- + 122H_2O + 17H^+ \rightarrow (C_{106}H_{263}O_{110}N_{16}P) + 138O_2 \qquad (14.20)$$

Technological implementation of this approach involves several factors:

- Selection of suitable microalgae species with high productivity and tolerance to environmental variations (temperature, pressure, pH, and CO_2 concentration) and impurities in CO_2 stream
- Development of an efficient photobioreactor and a light source (if necessary)
- Collection and handling of biomass products, etc.

In contrast to terrestrial plants that directly absorb CO_2 from air, microalgae uptakes its CO_2 feed from aqueous solutions, that is, in the form of ionic species HCO_3^- and CO_3^{2-}, as well as dissolved CO_2. These species exist in aqueous solutions in dynamic equilibrium (see Equation 14.12), with the formation of bicarbonate ion being a rate-limiting step. The equilibrium can be affected by the changes in the nutrient concentrations, pH of the aquatic system, and other factors. All these factors may influence the function of Rubisco (ribulose-1,5-biphosphate carboxylases/oxygenase), which is a key enzyme participating in CO_2 fixation (in microalgae, Rubisco is accompanied by carbon anhydrase).

It is important to emphasize that during the evolution, microalgae adapted to low concentration of atmospheric CO_2 (about 0.03 vol.%), therefore, the productivity of the system is a way below the values needed for its commercial use in industrial CO_2 sequestration systems. Thus, new microalgae species have to be developed to accommodate much higher concentrations of CO_2, for example, concentrations of 1–5 vol.% CO_2 that are typical of power plant emissions. In one study, 35 different microalgae strains were screened for their growth rate and the potential for the biofixation of CO_2 from thermal power plant emissions (Velea et al. 2008). The objective was to produce biomass with a high lipid content via controlled accelerated microalgal photosynthesis using CO_2-containing streams typical of flue gases generated by coal power plants. The authors demonstrated that the selected microalgae strains *Chlorella* sp., *Scenedesmus* sp., and *Chlorobotrus* sp. can be the basis for the development of a CO_2 biofixation system with the production of value-added products: horticultural oils, proteins, and carbohydrates. The reported productivity of the experimental system was 1 kg day^{-1} of *Chlorella* algal biomass obtained from 1.5 kg of CO_2 consumed by the algae strain.

Physical Sciences, Inc. (Massachusetts) in collaboration with Aquasearch and the University of Hawaii are jointly developing the technology for recovery and sequestration of CO_2 from stationary sources by photosynthesis of microalgae (Nakamura 2004). The main objective is to demonstrate the ability of selected microalgae species to efficiently fix carbon from typical power plant exhaust gases. The collaborators have selected several microalgal species (e.g., *Porphyridium* sp. and *Botryococcus braunii*) compatible with flue gases from coal and propane combustion gases and built a demonstration unit including a 2000 L outdoor photobioreactor. Figure 14.27 depicts the schematic diagram of the process for recovery and sequestration of CO_2 from stationary point sources by microalgae photosynthesis.

Numerous new companies have recently been formed that focus on transforming algae biomass into biodiesel and ethanol. For example, Green Fuel Technologies (Massachusetts)

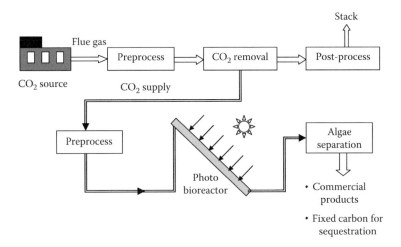

FIGURE 14.27
Recovery and sequestration of CO_2 from stationary CO_2 sources by microalgae photosynthesis. (Modified from Nakamura, T., Recovery and sequestration of CO_2 from stationary combustion systems by photosynthesis of microalgae, Technical Report to DOE, NETL, No. PSI-1356, December, 2004.)

has developed Emissions-to-Biofuels™ process that harnesses photosynthesis to grow algae by capturing CO_2 from retrofitted fossil-fired power plants and converting algae to transportation fuels (biodiesel and ethanol) (Biodiesel from algae oil 2008). The company is planning to build a \$92 million commercial-scale algae farm to capture and recycle CO_2 from power plants. It was estimated that it would take about 100 ac of algae ponds to consume 90% of CO_2 emitted by a 100 MW coal-burning plant (1 ac is equal to 4047 m^2). In June 2009, a partnership between Dow Chemical and Algenol Fuels on the development of a process for converting CO_2 and seawater to ethanol using a specific strain of algae was announced (Global CCS Institute 2009; Algenol Biofuels 2010). A pilot plant is under construction in Freeport (Texas) that will produce about 375,000 L year^{-1} of ethanol from 3100 tubular reactors. Other companies working on algae-based biofuels include (as of 2009) Solazyme, Inventure Chemical, Solena, Live Fuels, Solix Biofuels, Aurora Biofuels, Aquaflow Binomics, Petrosun, Bionavitas, Mighty Algae Biofuels, Bodega Algae, Seambiotic, Cellena, Petroalgae, Blue Marble Energy, AlgaeLink, Aquatic Energy, OriginOil, and many others.

The capacity of algae to capture CO_2 from industrial point sources and transform it into biomass that can be further processed into transportation fuels, foodstuff, or chemical products is of an immense practical interest and the idea is increasingly popular worldwide, receiving a sizable financial support from governments and entrepreneurs. While algae farming is an important part of the carbon mitigation portfolio, the widespread deployment of the algae growth technology might face challenges with regard to the availability of land and water. In particular, some studies point to a large land requirement of the algae farming process: about 40 ac of land per each MW_{el} produced by a coal-fired power plant (Global CCS Institute 2009). Currently, R&D efforts are focused on increasing the efficiency of algae production systems and chemical processes for its conversion to biofuels.

14.5.4.3 Biological CO_2 Sequestration in Ocean Ecosystem

Biological CO_2 sequestration in the ocean ecosystem is based on the enhancement of the natural process of CO_2 fixation by the ocean. Evidently, this approach drastically differs from the one discussed above that involves the direct ocean injection of CO_2 generated by large industrial point sources. Although the ocean's biomass amounts to only about 0.05% of the terrestrial biomass, it transforms about as much CO_2 to organic matter (about 50 Gt C year^{-1}) as the land ecosystems (U.S. DOE, Carbon sequestration. State of the Science 1999). The primary mechanism by which biological carbon sequestration in the ocean occurs is often referred to as a "biological pump" (Figure 14.28). The biological pump is a complex multistage process involving decomposition, gravitational settling, and the burial of biogenic debris formed in the upper layers of the ocean. CO_2 dissolved in the ocean is initially fixed by phytoplankton via photosynthesis. In the surface layers of the ocean, phytoplankton species are rapidly consumed by zooplankton, which, in turn, are grazed by higher trophic organisms, for example, fish. Organic carbon in the form of fecal pellets, decaying organisms, and other organic matter sinks to seafloor (portion of it is remineralized to CO_2 en route by bacteria). It was estimated that 70%–80% of the fixed carbon is recycled back into CO_2 in the surface layers of the ocean, and the remaining part settles as particulate organic carbon in the ocean depths, where it is slowly mineralized by bacteria (U.S. DOE, Carbon sequestration. State of the Science 1999).

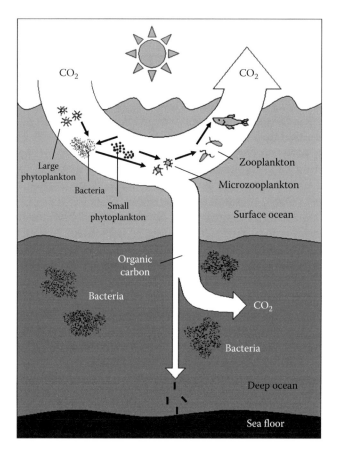

FIGURE 14.28
Schematic representation of a biological pump. (Adapted from U.S. DOE, NETL, Carbon sequestration, Technology Roadmap and Program Plan, U.S. DOE Office of Fossil Energy, National Energy Technology Laboratory, Morgantown, WV, 2003.)

14.5.4.4 *Iron Fertilization of the Ocean*

One controversial and highly debated strategy for increasing the rate and amount of CO_2 uptake by the ocean from the atmosphere relates to so-called ocean fertilization with iron and other nutrients such as nitrogen and phosphorus. It is assumed that this measure would drastically accelerate the action of the biological pump by shifting biological equilibrium and enhancing CO_2 absorption by the ocean from the atmosphere. The idea of ocean fertilization was originated from an observation that certain areas of the ocean (e.g., equatorial Pacific and southern oceans) have high levels of macronutrients (nitrogen and phosphorus) and low levels of micronutrients (iron), and in these particular areas, the phytoplankton population was very low (U.S. DOE, Carbon sequestration. State of the Science 1999). It was reasoned that the lack of iron limits the phytoplankton growth that explains why nitrogen and phosphorus are abundant in these regions. To test this hypothesis, several ocean fertilization experiments have been conducted in the Pacific Ocean (the earliest experiments: IRONEX I and II were conducted in 1993 and 1995, respectively). In the IRONEX II experiment, 500 kg of an iron compound was spread over the ocean surface of about 72 km^2. The effect of the iron addition was quick and dramatic: the photosynthesis yield rapidly increased, while nitrogen and phosphorus nutrients diminished, and phytoplankton biomass increased 30-fold within a week (U.S. DOE, Carbon sequestration. State of the Science 1999). As a result of the biological bloom, the partial pressure of CO_2 in the middle of the fertilized zone decreased and the production yield of dimethyl sulfide increased threefold. The fertilizing impact of this rather short (18 days) transient experiment dissipated shortly after the cease of iron injection. Since the IRONEX I and II trials, a number of international team projects have been conducted: SOIREE (1999), EisenEx (2000), SEEDS (2001), SOFeX (2002), SERIES (2002), SEEDS-II (2004), EIFEX (2004), CROZEX (2005), and LOHAFEX (2009). The most recent ocean iron fertilization experiment—LOHAFEX—was conducted during January–March 2009 in the Southern Atlantic Ocean despite a widespread international opposition (Eureka Alert 2009). In this experiment, iron sulfate was spread over a 30 × 30 km ocean patch, which triggered an intensive phytoplankton bloom. However, the experiment did not succeed in terms of enhanced carbon sequestration because the bloom was not accompanied by diatoms growth (which is essential for sustaining the complex bioecosystem responsible for the sinking of organic carbon) (apparently, the location of the experiment was not chosen correctly, since it had very low content of silicic acid).

The technical and economic aspects of the ocean iron fertilization concept sound quite reasonable. Calculations indicated and recent marine trials proved that 1 kg of iron can fix 83,000 kg of CO_2 and generate over 100,000 kg of plankton biomass (Sunda and Huntsman, 1995). The estimates of the required amount of iron to accomplish the ocean iron fertilization on a large scale (e.g., 3 Gt of CO_2 per year) vary in a wide range from 200 thousands to 4 millions tons per year, which could be shipped by about 16 supertanker loads of iron and would cost about $27 billion. Considering that the annual value of the global carbon credit is projected to exceed U.S. $1 trillion by 2012 (at the credit rate of $135 t^{-1} CO_2-equiv.), the ocean fertilization approach seems like a quite inexpensive strategy to offset about half of all anthropogenic CO_2 emissions (Sunda and Huntsman, 1995). For this reason, several private organizations are making plans to conduct larger-scale ocean fertilization projects to generate carbon offsets (Buesseler et al. 2008).

However, like many other geoengineering concepts (see Chapter 1), ocean iron fertilization remains a highly controversial issue and faces a stiff resistance from a majority of scientific and environmental communities. Many experts in the field emphasize that not only we poorly understand the unintended biogeochemical and ecological impacts of ocean fertilization but also the efficacy of this concept for atmospheric CO_2 sequestration is still in question (Buesseler et al. 2008). Possible environmental perturbations from ocean fertilization can spread over a large area by ocean circulations, which would make its long-term monitoring and verification practically impossible. In response to an increasingly strong opposition and concern about the possible long-term ecological consequences of large-scale iron fertilization, in May 2008, U.N. Convention on Biological Diversity issued a decree obligating member states to limit the ocean fertilization activities to small-scale scientific studies (Strong et al. 2009). Nevertheless, this issue remains a gray area and the recommendations are a subject to an interpretation (the LOHAFEX project was delayed by German government due to protests by environmental groups, but finally proceeded).

The main problem with the ocean fertilization concept is that its specific effects and the global ecological impact are difficult to predict, because the ocean's response strongly depends on the scale of the project. Unfortunately, small-scale experiments, although feasible from engineering and economical perspectives, are inherently inadequate to verify the model predictions, especially, those related to the long-term side effects of ocean fertilization. To resolve ecological uncertainties related to global ocean fertilization as a climate mitigation option, large-scale alteration of the ocean would be required with monitoring of its cumulative effect over extended period of time. It was estimated that it would take between decades and a century to unequivocally validate the effect of carbon sequestration by the ocean and to sort out various possible long-term ecological impacts of global fertilization, for example, ecosystem productivity, oxygen depletion, etc. To further complicate the issue, other factors associated with the dynamics of the ocean system may also interfere making the effect of ocean fertilization difficult to isolate from other ongoing effects.

Since testing the efficacy of the ocean fertilization approach on a global scale would require an unprecedented alteration of the ocean, and long-term ecological monitoring might not be feasible, robust dynamic models predicting the ocean ecosystem response are widely used. However, many modeling studies indicated that iron fertilization of the ocean as a mitigation option may not hold to its promise. For example, a model put forward by Zahariev et al. predicted that even if the entire Southern Ocean were fertilized with iron, less than 1 Gt carbon per year would be sequestered and only for a few years (Zahariev et al. 2008). In one modeling study reported in the *Science* magazine, the authors concluded that ocean iron fertilization in the areas of the high nutrient contents would unlikely sequester more than several hundred million tons of carbon per year (Buesseler et al. 2008). Based on the modeling and some field-test studies, many experts in oceanic and atmospheric sciences suggest to discontinue all activities on iron ocean fertilization. They argue that we already know enough about the ocean systems to conclude that iron fertilization on a global scale will be disruptive to the ocean ecosystems, and it is unlikely to be effective as a carbon mitigation strategy (Strong et al. 2009).

Despite the uncertainties in science of ocean fertilization and its potential as a mitigation option, the concept of ocean fertilization is being utilized for increasing fish harvest by some commercial ventures. For example, Ocean Farming, Inc., planned a large-scale fertilization project of the coastal waters of the Marshall Islands involving the use of iron-, silicon-, and phosphorus-based nutrients to increase the yield of tuna (U.S. DOE, Carbon sequestration. State of the Science 1999). A similar European project MARICULT pursues the commercial feasibility of fertilizing coastal waters to increase fish harvest. Although the primary goal of these projects is not carbon sequestration, the potential environmental impact of these processes would be the same (although at a smaller scale compared to using iron fertilization as a carbon mitigation option).

14.5.4.5 Advanced Biological Carbon Sequestration Systems

The objective of advanced biological processes as a carbon sequestration strategy is to enhance natural biological processes for CO_2 uptake from the atmosphere in terrestrial and marine ecosystems via the use of novel organisms, designed biological systems, and genetic improvements in microbial, plant, and animal species. Among the approaches to advance this strategy are (U.S. DOE, Carbon sequestration. State of the Science 1999)

- The development of faster-growing and more stress-resistant crops and plants
- New methods to enhance geological carbon sequestration via the use of microorganisms
- New processes to enhance carbon sequestration in the ocean ecosystem through transgenic and genetic manipulation of the members of the food chain
- Alternative microbial polymers or genetically improved plants as durable materials
- New microorganisms that do not rely on photosynthesis or carbon-based sources of energy

The progress in the area of genetic engineering could benefit the technology of carbon sequestration in all ecosystems including terrestrial, geological, and the ocean systems. In particular, by altering the structure of plants, carbon sequestration by soil could be significantly enhanced. New plant species could be designed to have a higher percentage of biomass below ground, which would make them more resistant to decay and promote the formation of carbonate minerals. The structure of above-ground plant biomass (e.g., its cell walls) could be engineered to facilitate the plant bioconversion processes and to make nonharvested biomass less degradable in the environment. Through genetic engineering, the metabolic networks of terrestrial and aquatic plants and algae could be designed to increase the share of products with desired characteristics. Bioengineering methods could be applied to slow down the decomposition rates of solid wastes in landfills, and, thus, reduce the CO_2 and methane emission rates and even trap, separate, and recycle waste products at landfill and sewage treatment facilities.

The application of advanced biological systems to carbon sequestration could lower the overall energy consumption, eliminate or reduce the need for elaborate chemical processing, and reduce the usage of CO_2-emitting fossil fuels. Due to the high cost of carbon capture and separation from point sources producing diluted CO_2 streams, the concept of integrated energy production and carbon capture by means of biological "energyplexes" has been proposed (National Laboratory Directors 1997). In particular, one of the energoplex schemes involves capturing CO_2 from flue gases of power plants by means of photosynthesis and storing the reduced carbon in the form of algal biomass. Other approaches involve the integration of power production with sewage and other waste treatment processes with the benefit of using nutrients and carbon in biological processes at the site. Thus, the energoplexes based on biological processes have the potential to produce energy, treat waste, sequester CO_2, and produce value-added products with a minimal environmental impact.

In geological systems, microbial processes could be engineered to greatly accelerate the conversion of CO_2 into carbonate rocks, for example, siderite ($FeCO_3$) using metal-reducing bacteria (or algae) and metal-containing fly ash or other low-value products from power plants and other industrial sources (U.S. DOE, Carbon sequestration. State of the Science 1999). The resulting materials can be used in roadbeds, as a filler or composite materials. The formation of the high-density carbonate rock enormously simplifies storage and disposal of CO_2 and enhances the storage permanence. Experimental studies on the use of metal-reducing thermophilic bacteria to produce siderite mineral proved the feasibility of this approach. Bioengineering methods could also be used to accelerate the formation of carbonates from natural silicate minerals, for example, serpentinite.

The use of bioengineering approaches could be used to produce refractory types of biomass from terrestrial plants that fix CO_2 into materials with recycling times much longer than wood. For example, the plant species that synthesize diterpenoid resins or natural rubber could be engineered to improve their efficiency for the conversion of CO_2 to plant biomass that is much more refractory that other species. Another approach involves the increase in lignin content of plants (lignin is relatively resistant to degradation) via manipulation of biochemical pathways to lignin biosynthesis (this has already been proved in case of aspen and poplar). Plants could be genetically engineered to increase the transfer of photosynthate to root systems that are less prone to decay compared to above-ground plant biomass.

Genetic modification of plants has the potential to significantly enhance their photosynthetic activity and ability to fix CO_2. The theoretical (maximum) efficiency of light energy conversion into chemical energy (plant biomass) is approximately 5%. In general, plants operate at lower efficiencies, which vary in a wide range depending on the ecosystem, species, and environmental conditions. For example, the photosynthetic efficiency of forests is in 0.05%–0.1% range, marsh grasses—in 2%–4% range, and corn and sugar cane—as high as 3.5%–4% (U.S. DOE, Carbon sequestration. State of the Science 1999). It is known that photosynthetic fixation of CO_2 is limited by the efficiency of two processes: conversion of captured light to chemical energy and primary carbon fixation catalyzed by the enzyme Rubisco. It is thought that both of these processes could be enhanced via advanced bioengineering approaches.

Production of durable and stable materials from biomass as a means of carbon fixation is an active area of development. Enzymes have been successfully used for the production of novel materials (e.g., biopolymers, bioplastics, and other materials) with highly specialized properties such as biodegradability, biocompatibility, chemical functionality, etc. In many areas, these materials are replacing petroleum-based polymers (e.g., polyethylene, polystyrene). One example of this product is polylactic acid (PLA)-based resins that are produced by conversion of starch to sugar followed by its fermentation to lactic acid (the process has been commercialized by Dow Chemical). The production of microbial cellulose is another example of the successful implementation of the biochemical approach. Since carbon originates from atmospheric CO_2 (via biomass), the net result of using these products is carbon sequestration.

14.5.5 Industrial Use of CO_2

The industrial use of CO_2 (e.g., as a chemical feedstock) provides an extremely attractive alternative to CO_2 storage options due to its potential of not only curbing CO_2 emissions but also producing value-added products, for example, fertilizers, polymers, chemicals, etc. This way, the industrial use of CO_2 provides a carbon sink, the extent of which would be dictated by the size of the market for the CO_2-derived products and lifetime of the products. The products produced from CO_2 as a feedstock have different life cycle characteristics that include their manufacture, use, and disposal, and, as such, they can function as CO_2 storage media for periods of time from months to centuries (i.e., until CO_2 is released back to the atmosphere via oxidation, decomposition, or decay processes). Evidently, only products with sufficiently long lifetime (e.g., centuries) can be considered as a means for the reduction of net CO_2 emissions to the atmosphere. From this viewpoint, it would be useful to examine existing and emerging industrial processes utilizing CO_2 and evaluate their potential for long-term CO_2 storage.

14.5.5.1 Existing and Emerging Industrial CO₂-Utilization Processes

Current industrial uses of CO_2 include a large number of important processes and applications such as production of urea, methanol, "dry" ice, the use in refrigeration systems (liquid CO_2), food and beverage industries, fire extinguishers, horticulture, EOR, water treatment, solvent (supercritical CO_2), and many others. Figure 14.29 summarizes most important industrial applications and yearly markets for CO_2-derived products, which include both existing and some emerging processes (only major products are shown in Figure 14.29). Currently, urea, a widely used nitrogen fertilizer and a chemical intermediate, is the major industrial product directly manufactured from CO_2 by reacting it with ammonia via intermediate formation of ammonium carbamate:

$$CO_2 + NH_3 \rightarrow H_2NCOOH \tag{14.21}$$

$$H_2NCOOH + NH_3 \rightarrow H_2NCOO^-NH_4^+ \tag{14.22}$$

$$H_2NCOO^-NH_4^+ \rightarrow CO(NH_2)_2 + H_2O \tag{14.23}$$

Most of CO_2 used in the process is recovered from waste streams of hydrogen and ammonia-manufacturing plants.

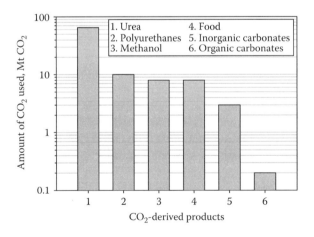

FIGURE 14.29
Amount of CO_2 needed to satisfy yearly markets for major CO_2-derived products. (Data from IPCC special report on carbon dioxide capture and storage, Prepared by Working Group III of the Intergovernmental Panel on Climate Change, B. Metz, O. Davidson, H. de Coninck, M. Loos, A. Meyer (eds.), Cambridge University Press, Cambridge, U.K., Table 7.2, p. 332, 2005; Methanol, www.methanol.org [accessed August 20, 2010], 2010; Ullmann's, *Ullmann's Encyclopedia of Industrial Chemistry*, Wiley-VCH Verlag, Weinheim, Germany, 2003.)

Methanol is another major product of chemical industry that could potentially utilize CO_2 as a feedstock during its manufacturing process. Currently, methanol is produced by passing H_2–CO–CO_2 mixture over alumina-supported Cu–Zn catalyst at 250°C–300°C and elevated pressure of 50–100 atm, resulting in the following exothermic reactions:

$$CO + 2H_2 \rightarrow CH_3OH \qquad (14.24)$$

$$CO_2 + 3H_2 \rightarrow CH_3OH + H_2O \quad \Delta H° = -90.7\,kJ\,mol^{-1} \qquad (14.25)$$

Typically, the H_2–CO–CO_2 mixture is predominantly produced by reforming of NG or gasification of coal, and, as such, the process cannot be considered as a CO_2 mitigation technology (because CO_2 is produced from the same source as H_2 and CO). In order to function as a sink for CO_2, the methanol production process has to rely on an external source of CO_2 (e.g., power plants or the atmosphere), which would increasingly replace CO in the reacting mixture. The technical feasibility of converting H_2–CO_2 mixtures to methanol with high selectivity has been demonstrated by a number of research groups worldwide. For example, high-performance Cu/ZnO-based multicomponent catalysts were developed for methanol synthesis from CO_2 and H_2 with the purity of 99.9% (Toyir et al. 2009). Ultimately, the hydrogen component of the H_2–CO_2 feedstock for methanol production has to be produced from nonfossil sources (e.g., nuclear-, solar-, or wind-powered water electrolysis) (as discussed in the Chapter 1).

Recently, new approaches to producing a number of important chemicals, pesticides, and polymers with the use of CO_2 as a chemical feedstock have been developed (or are at the different stages of the commercial development). In particular, in the polyurethane- and polycarbonates-manufacturing processes, CO_2 is a substitute for a traditionally used toxic compound—phosgene ($COCl_2$) (it should be noted that the primary motivation for the use of CO_2 was to solve the toxicity problem, not carbon mitigation). For example, polycarbonates can be produced by copolymerization of CO_2 and epoxy compounds in the presence of $Zn(CH_2CH_3)_2$ catalyst in benzene solution at room temperature and elevated pressure of 60 atm, as follows (Inoue et al. 1969):

$$H_2C(O)-CHR + CO_2 \rightarrow 1/n(-H_2C-CHR-O-CO-O-)_n \qquad (14.26)$$

Similarly, traditionally, the production of urea derivatives, carbamates (used in manufacturing of pesticides and pharmaceuticals), involved the use of phosgene and isocyanate (also a toxic compound) as starting materials. The use of CO_2 offers an alternative route to production of carbamates by reacting CO_2 with amines:

$$2RNH_2 + CO_2 \rightarrow RNH-C(O)-NHR + H_2O \qquad (14.27)$$

In addition to the above processes that are close to a practical realization, a large number of reactions involving CO_2 as a chemical feedstock are at R&D stage (Halmann and Steinberg 1999). One example is the reaction of CO_2 with methane to produce acetic acid—an important chemical product and an intermediate in the synthesis of a variety of pharmaceuticals, polymers, etc. In particular, it was discovered that the CH_4–CO_2 mixture charged into an autoclave containing $VO(acac)_2/K_2S_2O_8$ catalyst in CF_3COOH solvent at 20 atm and 80°C produced acetic acid with 97% yield (Taniguchi et al. 1997):

$$CH_4 + CO_2 \rightarrow CH_3COOH \tag{14.28}$$

Of particular practical interest is the production of liquid hydrocarbon fuels (gasoline, diesel, jet fuel, etc.) from CO_2 because of their high energy density and existing infrastructure:

$$nCO_2 + (m + 4n)H_2 \rightarrow C_nH_m + 2nH_2O \tag{14.29}$$

For example, hydrogenation of CO_2 to hydrocarbons over Fischer–Tropsch catalyst such as Co-Pt/Al$_2$O$_3$ has been recently reported (Dorner et al. 2009). The authors found that a reduction in the $H_2:CO_2$ ratio in the feed resulted in a shift toward higher-chain hydrocarbons. Kim et al. (2006) studied direct catalytic hydrogenation of CO_2 to liquid hydrocarbons (C_5+) over K-promoted iron catalysts (Fe-K/Al$_2$O$_3$) in fluidized bed and slurry reactors.

It should be noted that even if CO_2-based fuel production processes are based on renewable hydrogen, their products cannot be considered a carbon sink, because CO_2 is released during fuel combustion (these compounds and processes relate to "carbon-neutral" systems that are described in details in Chapter 1). More detailed information on chemical, photochemical, and electrochemical conversion of CO_2 to fuels and useful products could be found in the Chapters 4, 6, and 9. Additionally, we can recommend two recent review papers on CO_2 conversion to fuels and products by Ma et al. (2009) and Jiang et al. (2010).

Other current industrial uses of CO_2 include the following:

Food and beverages. CO_2 is used in the food and beverage industries in three main areas: beverage carbonation, foodstuff packaging, and in the form of dry ice in chilling and freezing operations. For example, the Prosint methanol plant in Brazil captures 32,850 t of CO_2 per year and uses it in soft drink manufacturing (Global CCS Institute 2009). Big Brown SkyMine in the United States captures 112,500 t of CO_2 per year and supplies it to food-processing facilities.

Metallurgical industry. CO_2 is used for fume suppression during charging of furnaces, tapping of electric arc furnaces, in Cu and Ni production processes, etc. (IPCC CCS 2005). The Kwinana carbonation plant in Australia uses captured CO_2 in the amount of 70,000 t of CO_2 per year in a carbonation of bauxite residue process (Global CCS Institute 2009). In this case, storage of CO_2 is accomplished via its mineralization (the caustic residue of the bauxite-to-alumina process is transformed to a stable carbonate product).

Pharmaceuticals. CO_2 is used in supercritical fluid extraction procedures, chemical synthesis, as inert gas, and as a cooling agent for storage and transportation of temperature-sensitive substances.

Pulp and paper. CO_2 is used for adjusting pH of recycled chemical pulps after an alkaline bleaching and for enhancing the performance of paper-manufacturing machines.

Electronics. CO_2 is used as a cooling medium during environmental testing of electronic devices, as a supercritical fluid for removing photoresists from wafers, and in the form of snow for abrasive cleaning of residues on wafers.

Waste water treatment. CO_2 injection enables to adjust and control the pH of aqueous effluents.

Fire extinguishers. Solid CO_2 in the form of snow is used in fire extinguishers.

Other applications. These include health care (an intra-abdominal insufflation during surgeries and other medical procedures), pH control, and regulation of waste waters in swimming pools (IPCC CCS 2005).

There is a number of emerging CO_2 utilization technologies specifically designed for carbon mitigation applications. For example, Calera Corp. (California) has developed a process by which they utilize CO_2 and calcium and magnesium in seawater to produce carbonates that can be used as a replacement for Portland cement (Biello 2008). The process essentially mimics "marine cement" production by corals when they make their shells and reefs. The company claims that for every ton of cement produced they could potentially sequester half a ton of CO_2 (note that in industrial manufacturing of cement, roughly 1 t of CO_2 is emitted per ton of cement produced). The technology is still in a pilot and demonstration stage of the development (a pilot plant with the capacity of 10 t day^{-1} uses flue gas from a nearby gas-fired power plant in Moss Landing, California). The California Department of Transportation is interested in testing the cement product. Other companies are pursuing similar concepts, for example, Carbon Sciences of Santa Barbara (California) uses flue gas and Ca, Mg-rich waste water from mining operations to produce similar cement. Halifax (Nova Scotia, Canada) based Carbon Sense Solutions company sequesters CO_2 by reacting it with a fresh batch of cement that readily absorbs CO_2.

14.5.5.2 *Mitigation Potential of Industrial CO_2 Utilization*

In order to determine the mitigation potential of industrial CO_2 utilization, it is very important to assess the overall volume of CO_2-derived products (i.e., its worldwide production and inventory) and the carbon-storing properties of the product (how long it would take for the product to degrade to CO_2 and release it to the atmosphere). If the worldwide production of polycarbonate and polyurethane products were switched to CO_2-based processes, the industrial CO_2 consumption would increase by about 3.3 Mt year^{-1} CO_2. The total industrial use of CO_2 is about 115 Mt year^{-1} CO_2, which is less than half a percent of the overall anthropogenic CO_2 emissions (IPCC CCS 2005). Thus, currently, the industrial use of CO_2 could potentially take up a very small portion of man-made CO_2 emissions, and no significant increase in the commercial production of CO_2-derived products is expected in the near future (most of the new developments aim mostly at replacing toxic feedstocks by CO_2).

Another important factor to consider relates to the long-term carbon-storing capacity of CO_2-based products. Figure 14.30 depicts typical life-time of CO_2-derived products before they degrade to CO_2. Evidently, the duration of CO_2 storage in these products varies in a wide range from a few hours-to-days for carbonated beverages and fuel-related applications to a few months-to-years for fertilizers and pesticides to several decades-to-centuries for construction materials (e.g., plastics and laminates). This fact would further diminish the capacity of CO_2-derived products to serve as a carbon sink since even in the best case scenarios (e.g., a very large market for the product), the duration of carbon storage is rather limited. Summarizing, it is unlikely that the industrial use of captured CO_2 would markedly contribute to the mitigation of anthropogenic CO_2 problem.

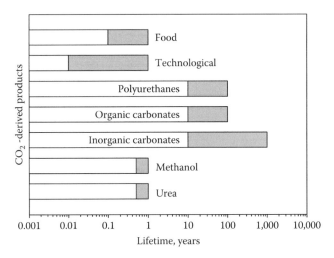

FIGURE 14.30
Typical lifetime of CO_2-derived products before they degrade to CO_2. (Data from IPCC special report on carbon dioxide capture and storage, Prepared by Working Group III of the Intergovernmental Panel on Climate Change, B. Metz, O. Davidson, H. de Coninck, M. Loos, A. Meyer (eds.), Cambridge University Press, Cambridge, U.K., Table 7.2, p. 332, 2005.)

14.6 Current Status of Carbon Capture and Storage Technologies

14.6.1 Current Status and Trends in the Development of CCS Technologies

The technical maturity of different CCS systems and components vary significantly (Figure 14.31). In this figure, the extent of the commercial maturity of the CCS systems is broken down to four phases that reflect the level of their technological advancement. This figure shows that while some CCS systems are still at R&D stage (e.g., ocean storage and mineral carbonation), other systems and components are mature technologies with a decades-long record of a successful industrial practice (e.g., EOR, amine-based CO_2 capture, CO_2 compression, H_2-fired turbines, and industrial use of CO_2). It should also be emphasized that while the precombustion CO_2 capture technology has been commercially deployed in an industrial sector, it has not been widely used in a power generation sector. The post-combustion and oxyfuel combustion technologies, with a few exceptions, are yet to be widely applied to both industrial and power generation sectors (at a large scale).

The analysis of the worldwide CCS-related R&D activities indicates that among main CO_2 capture technologies, post-combustion capture is the most active area of development followed by oxyfuel combustion and precombustion capture. The following data summarize the worldwide distribution of R&D efforts (by number of organizations) in the CCS area by technology (Global CCS Institute 2009):

Post-combustion capture	136
Oxyfuel combustion	94
Precombustion capture	72
CO_2 sequestration	132
CO_2 transport	65

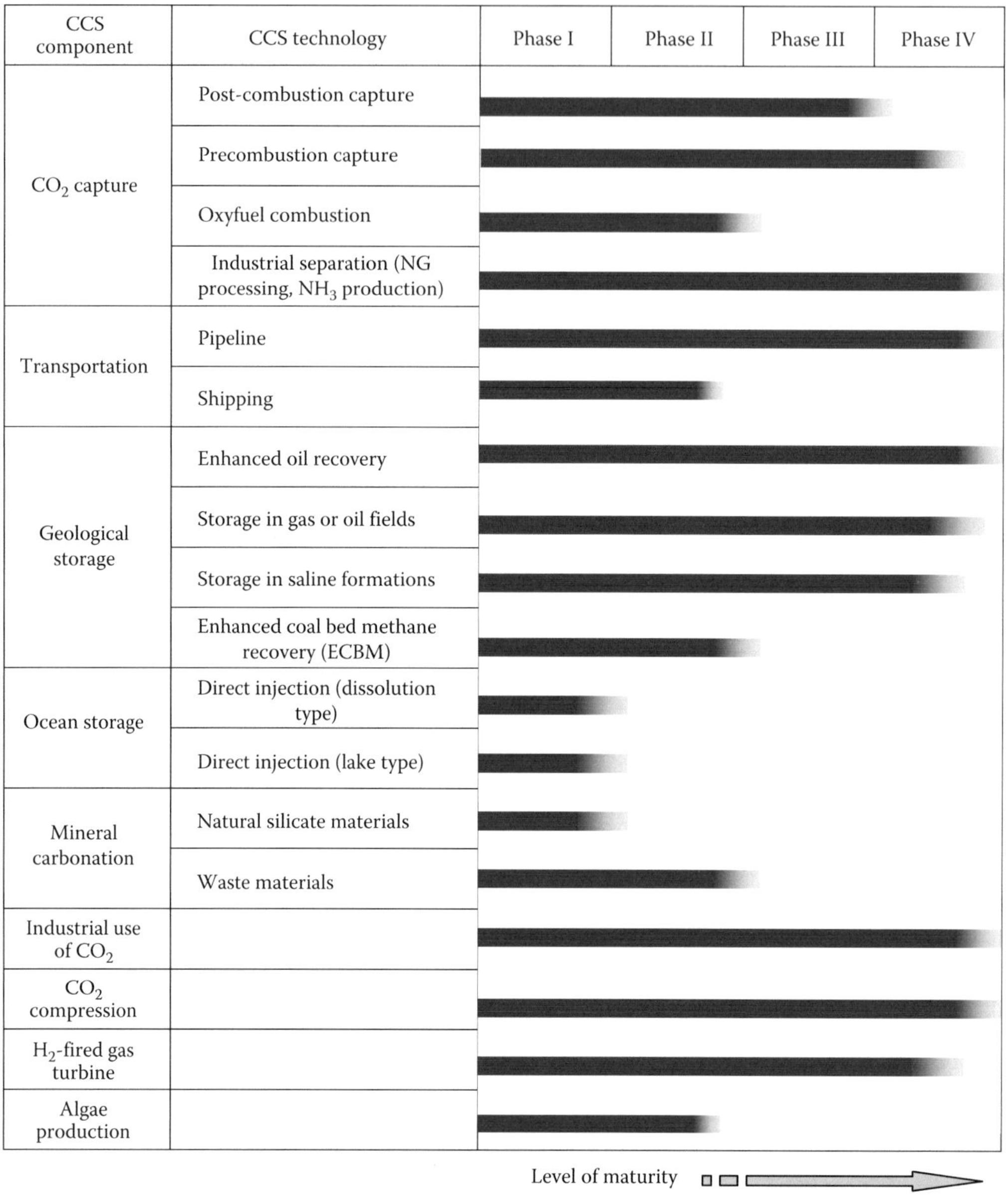

CCS component	CCS technology	Phase I	Phase II	Phase III	Phase IV
CO$_2$ capture	Post-combustion capture				
	Precombustion capture				
	Oxyfuel combustion				
	Industrial separation (NG processing, NH$_3$ production)				
Transportation	Pipeline				
	Shipping				
Geological storage	Enhanced oil recovery				
	Storage in gas or oil fields				
	Storage in saline formations				
	Enhanced coal bed methane recovery (ECBM)				
Ocean storage	Direct injection (dissolution type)				
	Direct injection (lake type)				
Mineral carbonation	Natural silicate materials				
	Waste materials				
Industrial use of CO$_2$					
CO$_2$ compression					
H$_2$-fired gas turbine					
Algae production					

Level of maturity ▫ ▢ ⟹

FIGURE 14.31
Current level of maturity of the CCS system components. (Data from Global CCS Institute, Strategic analysis of the global status of carbon capture and storage, Final Report, Global CCS Institute, http://www.globalccsinstitute.com/downloads/reports/2009/worley/foundation-report-1-rev0.pdf [accessed August 3, 2010], Table 1-2, p. 9, 2009; IPCC special report on carbon dioxide capture and storage, Prepared by Working Group III of the Intergovernmental Panel on Climate Change, B. Metz, O. Davidson, H. de Coninck, M. Loos, A. Meyer (eds.), Cambridge University Press, Cambridge, U.K., Table TS.1, p. 21, 2005.) Phase I: Basic science understood, research and development phase. Phase II: Demonstration (pilot plant). Phase III: Economically feasible under specific conditions (limited commercial applications, processing under 0.1 Mt CO$_2$ year^{-1} capacity). Phase IV: Mature market (the technology is in operation with multiple replications worldwide).

The current trends in the CCS systems development indicate that fossil-based power plants with post-combustion capture will dominate the field in the near-to-mid-term future, with a number of large-scale facilities already planned for the second decade of this century (IPCC CCS 2005). In view of the projected widespread commercial deployment of the CCS systems, several studies pointed to the advantages of building CO_2-capture ready (CCR) power plants (Lucquiaud and Gibbins 2009). This fact is especially important because a large number of coal-fired plants are planned to be built in developing countries and emerging economies such as China, India, and South Africa that are heavily reliant on coal for power generation. Since CCS is unlikely to be an immediate option for most of these plants, it would be feasible to make them CCR for the subsequent retrofit of the flue gas scrubbing systems for post-combustion capture.

The trends in CCS R&D activities indicate a shift away from deep ocean storage and mineralization options toward geological storage in saline formations, oil and gas reservoirs, unmineable coal seams, and basalt formations. The concerns over the potential adverse impact of the enormous quantities of stored CO_2 on marine biota is one of the main reasons why ocean storage is losing a favor. It is also realized that unrealistically large quantities of reactants would be required for the widespread deployment of the mineralization storage option (e.g., lime production would have to be increased by several orders of magnitude to meet the increased demand for this reagent). Biological CO_2 fixation and storage technologies, particularly, via algae growth are currently gaining momentum. The worldwide interest in this technology is fueled by the realization that algae could be further converted to fuels that are similar to conventional petroleum-based transportation fuels.

14.6.2 Overview of Active and Planned CCS Projects

As of 2009, there were 499 CCS-related activities around the world, which can be further refined to 275 CCS projects (Global CCS Institute 2009). Of the 275 CCS projects, 213 were active or planned projects, 34 (mostly, small scale) have been completed, 26 have been canceled or delayed, and 2 were of unknown status. Of 213 active/planned projects, 101 were commercial-scale projects, and among them, 62 were integrated CCS projects (due to their technological advantages, the integrated projects are of particular commercial interest). The following is a breakdown of the existing or planned CCS projects by geographical location:

United States	78
Western and central Europe	53
Australia and New Zealand	23
Canada	18
China	14
Eastern Europe	6
India	6
Japan	6
Middle East	3
East Asia (excluding Japan)	3
Africa	2
South America	1
Total	213

Among 62 commercial integrated CCS projects, Europe has the largest number, 23, thanks to a commitment by the EU (2007) to deploy 10–12 industrial-scale CCS demonstration units by 2015. United States has 15 active/planned integrated projects, followed by Australia (7), Canada (6), China (4), eastern Europe (4) (several remaining regions have only one project).

The analyses of active or planned CCS projects indicate that up until 2009, the post-combustion technology was a predominant decarbonization approach pursued by 48% of users, followed by precombustion (35%) and oxyfuel (9%) technologies (the remaining 8% of projects are of unspecified nature) (Global CCS Institute 2009). Within the 213 active/planned CCS projects, 129 involve some form of CO_2 storage; the categorization of these projects is as follows:

Geological storage	71
Beneficial reuse (EOR, EGR, and ECBM)	44
Terrestrial	6
Unspecified	8
Total	129

It can be seen that currently geological storage and beneficial reuse are the major means of storing CO_2 amounting to 89% of all projects. Among the geological storage related projects, the predominant fraction (53%) deals with the storage in saline formations, with 28% associated with depleted oil and gas fields (the balance is not specified). Enhanced oil and gas recovery as a carbon storage option provides an excellent economic incentive for the deployment of CCS projects. Among beneficial reuse projects, 55% relate to EOR, and 11% to EGR and ECBM, each. The remaining beneficial reuse projects involve the use of CO_2 in food and chemical industries.

CO_2 capture technologies currently being utilized or projected to be implemented cover a wide range of industrial applications are as follows (Global CCS Institute 2009):

Power generation	105
Gas processing	19
Oil and gas recovery	17
CO_2 sequestration plant	14
Fertilizer production	11
Chemical industry	8
Mining	7
Research facilities	4
Iron and steel production	3
Coal to liquids	3
Oil refining	2
Other areas	6
Not categorized	14
Total	213

CCS units associated with power generation plants represent the lion share (49%) of all existing or planned projects (most of them relate to coal-fired power plants). This is followed by gas and oil-processing facilities. Unfortunately, such large CO_2 emitters as cement-manufacturing plants or aluminum smelting plants have not shown much interest in the CCS technology so far, which is worrisome because these facilities contribute a significant fraction of man-made CO_2 emissions to the atmosphere.

The current annual rate of CO_2 injection (in thousand tons of CO_2 per year) into geological storage reservoirs categorized by their geographical location and carbon storage option is shown below:

Americas	Onshore saline formation (65), oil/gas reservoir (48,000), basalt (0.9), coal seam (70)
Europe	Onshore saline (125), offshore saline (1000), oil/gas reservoir (700)
Africa	Onshore saline (1300)
Asia	Onshore saline (10)
Australia	Oil/gas reservoir (65)

Most of the existing and proposed CCS projects in Europe relate to offshore storage, although they are more expensive compared with the onshore option. Some of the existing government- and privately funded geological storage projects are summarized in Table 14.3.

TABLE 14.3

Selected Existing and Planned Commercial and Demonstration Geological CO_2 Storage Projects (with Injection Rate More Than 5 t CO_2 per Day)

Project	Country	Daily Injection Rate, t day^{-1}	Storage Type	Start Date
Sleipner[a]	Norway	3,000	Aquifer	1996
Rangely[a]	United States	2,740	EOR	1986
Weyburn[a]	Canada	Up to 5,000	EOR	2000
In Salah[a]	Algeria	Up to 4,000	Depleted gas well	2004
Salt Creek[a]	United States	Up to 6,000	EOR	2004
Snohvit[a]	Norway	2,000	Saline formation	2007
Minami-Nagoaka	Japan	Up to 40	Aquifer	2002
Yubari	Japan	10	ECBM	2004
K12B	The Netherlands	Up to 1,000	EGR	2004
Ketzin	Germany	100	Saline formation	2006
Otway	Australia	160	Depleted gas field	2005
RECOPOL	EU	8,200	Coal seam	2004
CASTOR	EU	27	Geological storage	2004
Quinshi	China	6		2005
CO_2 SINK	EU	27	Saline formation	2006
Husky	Canada		EOR	2012
Masdar	UAE	11,780	EOR	2013
Aalborg	Denmark	5,205	Saline aquifer	2014
Gorgon	Australia	9,315	Saline aquifer	2015
RWE Goldenbergwek	Germany	7,671	Saline aquifer	2015
Futuregen	United States	2,740	Saline aquifer	2018

Sources: Data from IPCC special report on carbon dioxide capture and storage, Prepared by Working Group III of the Intergovernmental Panel on Climate Change, B. Metz, O. Davidson, H. de Coninck, M. Loos, A. Meyer (eds.), Cambridge University Press, Cambridge, U.K., 2005; Global CCS Institute, Strategic analysis of the global status of carbon capture and storage, Final Report, Global CCS Institute, http://www.globalccsinstitute.com/downloads/reports/2009/worley/foundation-report-1-rev0.pdf (accessed August 3, 2010), 2009; Carlson Easley, N., *Relay* Spring, 8, 2008.

[a] Commercial-scale projects.

Practically, all existing major commercial CCS projects relate to geological storage, mostly associated with NG processing. It should be noted, however, that originally, CO_2 capture was driven by the necessity of removing CO_2 from raw NG to meet the pipeline transport specifications (e.g., energy content). The following is a brief description of most important existing and planned geological CCS projects starting with dedicated CO_2 storage projects (i.e., the projects with the sole objective of capturing and storing CO_2 as a carbon mitigation measure), followed by beneficial reuse or enhanced fuel recovery type projects such as CO_2-EOR and ECBMR.

14.6.2.1 Dedicated CO₂ Storage Projects

The Sleipner Project successfully operated by the Statoil since 1996 in the North Sea (about 250 km off the coast of Norway) is the first commercial-scale project aiming at geological CO_2 storage in a saline aquifer formation (a schematic representation of the Sleipner project is shown in Figure 14.32). In this project, CO_2 is first separated from NG stream with CO_2 content of about 9 vol.%, and then compressed and injected into a deep sandstone saline formation located about 0.8–1.0 km below the seabed. On an average, about 1 million ton of CO_2 per year (or about 2700 t day^{-1}) is injected into the saline aquifer formation, and over the lifetime of the project, about 20 million tons of CO_2 is expected to be injected and stored underground. The saline formation represents a brine-saturated sandstone beneath the North Sea floor (called Utsira formation). The top of the formation and the overlaying primary seal consist of a fairly flat, extensive shale layer 75 m thick. It was estimated that the saline formation has a very large storage capacity of 1–10 Gt CO_2. Since the beginning of the Sleipner Project, the IEA GHG R&D Program has collaborated with Statoil to arrange the monitoring and research activities involving baseline data gathering and evaluation, simulation of reservoir geology,

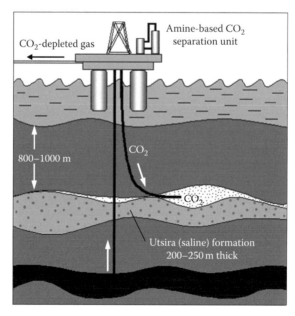

FIGURE 14.32
Sketch of the Sleipner CO_2 storage project.

geophysical modeling, data interpretation and model verification, and other activities. In particular, the fate and movement of CO_2 plume in saline formation has been successfully monitored by a seismic time-lapse survey, which showed that the CO_2 plume extends over the area of about $5\,km^2$, and its migration out of the storage site was effectively prevented by the caprock seal.

The In Salah Gas Project that is located in Saharan region of Algeria is a joint venture between Sonatrach, BP, and Statoil (operating since 2004) (Michael et al. 2009). In this project, CO_2 is stripped from NG that contains about 10 vol.% CO_2 and injected into a sandstone saline formation 1.8 km deep (Global CCS Institute 2009) (note that it is stated in some sources that in the In Salah project, CO_2 is injected into a depleted gas reservoir outside the boundaries of the gas field). CO_2 is injected into the sandstone formation via three horizontal wells, each 1.5 km long, at the rate of about 1.2 million tons CO_2 per year (over the lifetime of the project, 17 Mt CO_2 will be geologically stored), and the purified NG is delivered to European markets (IPCC CCS 2005). The top seal of the storage formation (caprock) is a layer of mudstones about 950 m thick. The storage site is under continuous surveillance and monitoring with regard to the risk assessment, CO_2 storage integrity, CO_2 migration from the injection wells, etc.

The Snøhvit project (Statoil, Norway) in the Barents Sea off the coast of northern Norway is a recent addition to the family of the CCS projects utilizing saline formations (Statoil 2009). Since 2008, CO_2 was extracted from NG and reinjected into a sandstone saline formation at a depth of 2.6 km beneath the seafloor (the formation is deeper than the gas reservoir; the caprock shale is 30 m thick). Since the beginning of its operation in 2008, the project stores about 0.7 Mt CO_2 per year.

The Gordon Project (off the north-western coast of Australia), a joint venture of Chevron, ExxonMobil, and Royal Dutch Shell, is projected to start operation in 2014–2015 (Global CCS Institute 2009). The project will extract CO_2 from NG containing 14 vol.% CO_2 and inject it into a saline formation at about 2 km depth (Flett et al. 2008). When the project will reach its full capacity of 3.5 Mt $year^{-1}$ of CO_2 injected, the Gordon Project will become the largest CO_2 storage project in the world.

Several demonstration and pilot-scale CO_2 storage projects are underway worldwide. For example, in the Ketzin Project (Germany), operated since 2008 by a consortium called CO2SINK, about 30,000 t $year^{-1}$ CO_2 is injected into a saline formation 650 m deep covered by a 210 m thick caprock made of mudstone and dolomite (Schilling et al. 2009). In the Lacq CCS project located near Pau, France, CO_2 captured from a gas-fired power plant is injected into a nearly depleted Rouse Formation carbonate reservoir (Leblond 2009). The process involves capturing CO_2 from an oxyfuel combustion unit, transporting it 27 km by pipeline to a site where it is injected into a reservoir at the depth of 4.5 km (Global CCS Institute 2009). During its 2 year long operation starting from 2009, the project will store over 120,000 t of CO_2. In the Netherlands' K-12-B project, CO_2 is injected into a nearly depleted offshore gas field. Since early 1990s, more than 40 relatively small-scale projects have been operating in western Canada with the objective of injecting CO_2–H_2S-rich acid gases into geological formations (Bachu and Haug 2005).

A new pilot-scale CO2CRC Otway Project with CO_2 storage in a depleted NG reservoir started in Victoria, Australia, in 2008 (Sharma et al. 2009). In this project, CO_2 is removed from NG containing up to 80% CO_2 and injected into a depleted sandstone gas reservoir at a depth of 2.05 km, located beneath the producing reservoir. Since the beginning of the operation, the project has injected 65,000 t of CO_2 (Global CCS Institute 2009).

14.6.2.2 Enhanced Fuel Recovery Projects

Oil and gas industries have been active in CO_2 geological storage through CO_2-EOR for about 35 years. It should be noted that the abundance of technical information and valuable experience gained from a large number of CO_2-EOR operations have been utilized in a variety of geological CO_2 storage projects.

The SACROC project in Texas was the first commercial-scale CO_2-EOR project in the world operating between 1972 and 1995 and using an industrial CO_2 source. The Rangely-Weber project (Colorado) is probably the longest operating CO_2-EOR project (since 1986). The project operated by Chevron injected CO_2 delivered by a 283 km long pipeline from a gas-processing plant in Wyoming with the average injection rate of 2.97 Mt year^{-1} CO_2 (IPCC CCS 2005; Global CCS Institute 2009). The Rangely-Weber CO_2-EOR project is projected to produce 129 million barrels of oil via EOR. As of 2006, 25 million tons of CO_2 has been pumped into the storage reservoir (Wackowski 2007).

Since 2000, a large CO_2-EOR operation has been established at the Weyburn Oil Field in Saskatchewan, Canada (the Weyburn-Midale CO_2-EOR project). With the injection rate of 1.75 Mt year^{-1} CO_2, this project remains the world's largest carbon storage project (Global CCS Institute, 2009). The CO_2 source for this CO_2-EOR project is the North Dakota's coal gasification plant producing substitute natural gas (SNG) with relatively pure stream of CO_2 as a by-product. CO_2 is delivered in a dense-phase form via a 320 km long pipeline. Currently, about 1600 m^3 day^{-1} of incremental oil is being recovered from the field. About 500 million barrels of oil had been recovered prior to the start of the CO_2-EOR project, and additional 215 million barrels (or 11.3% of the estimated total oil in place) is expected to be recovered over the life of the project (in the process, over 40 million tons of CO_2 will be stored in the reservoir) (PTRC 2007). It was estimated that the project would extend the oil field lifetime by 25 years (Moberg et al. 2003). The site is extensively monitored with high-resolution seismic surveys. Surface monitoring is also conducted to determine any potential leakage, which includes analysis of groundwater and soil (as of now, there was no CO_2 leakage to the surface). Besides the above commercial-scale CO_2-EOR projects, there is a number of demonstration projects worldwide, for example, in Trinidad, Turkey, Brazil, Germany, Australia, Japan, and Saudi Arabia.

ECBMR is another commercially significant method of geological CO_2 storage. The Allison Unit Project of Burlington Resources in New Mexico was the first ECBMR project. Injection of CO_2 for methane recovery at the Allison unit started in 1995 and lasted for 6 years during which a total of 322,000 t of natural CO_2 was pumped into the reservoir at a pressure of 10.4 MPa (of that amount, 277,000 t remained in the storage reservoir, with the rest produced with methane) (Global CCS Institute 2009). This resulted in an increase in methane recovery from about 77% to the estimated 95% of the original gas in place. The methane recovery ratio was one volume of methane recovered for every three volumes of CO_2 injected (Reeves et al. 2004).

Relatively small-scale projects testing CO_2 storage in coal beds have been carried in Europe and Japan. RECOPOL ECBMR project was launched in 2001 in Poland with the objective of evaluating the technical and economic feasibility of storing CO_2 in coal seams and producing methane. The U.S. DOE is sponsoring six CO_2-ECBMR pilot-scale projects. In one of them (SECARB project), about 1000 t of gaseous CO_2 was injected into two unmineable coal seams 490–670 m deep in Virginia in 2009 (Karmis 2009). CO_2-ECBMR projects are under consideration in China, Canada, Italy, and other countries. Other geological media for CO_2 storage such as basalts, oil and gas shale, salt caverns, and abandoned mines may provide local niche options for underground CO_2 storage (Gale 2003).

14.6.3 Enabling Technologies for Fossil Decarbonization Applications

In this section, the existing and emerging fossil-based hydrogen production technologies that could be coupled with CCS (e.g., for precombustion capture applications) are reported. In the case of coal as a feedstock, the hydrogen product is almost entirely originated from water (with the energy conversion efficiency of 50%–63%), whereas in the case of NG, about one half of hydrogen comes from water and another half from NG (with the energy conversion efficiency of 70%–76%).

14.6.3.1 Hydrocarbon Reforming

The concept of fossil hydrogen production coupled with CO_2 sequestration was advocated by a number of research groups worldwide with a special emphasis on steam-methane reforming (SMR) as the technology of choice for the production of hydrogen (Audus et al. 1996). SMR is by far the most important and widely used process for the industrial manufacture of hydrogen, amounting to about 80% of hydrogen produced in United States (40% in the world) (Roadmap for the hydrogen economy 2005). The overall SMR process could be presented by the following chemical equation:

$$CH_4 + 2H_2O_{liq.} \rightarrow 4H_2 + CO_2 \quad \Delta H^\circ = 253.1 \, kJ \, mol^{-1} \tag{14.30}$$

The SMR technology is well developed and commercially available in a wide range of hydrogen capacities, from less than $1 \, t \, h^{-1} \, H_2$ (small decentralized units) to about $100 \, t \, h^{-1}$ H_2 (large ammonia-manufacturing plants). The endothermic reforming reaction is accomplished over a Ni-based catalyst at high temperatures of 800°C–900°C. Heat is supplied to the reactor by combusting of part (about a third) of the NG feed. The produced syngas undergoes WGS (Reaction 14.1) and gas separation stages to produce pure hydrogen. Until about two decades ago, the CO_2 stream was separated from H_2 using chemical absorption methods (using amine or hot potassium carbonate solvents), resulting in pure CO_2 rejected to the atmosphere. Present-day SMR plants use physical adsorption technology, in particular, PSA units. The PSA process allows obtaining H_2 of very high purity (99.999%) at a pressure of 2.2 MPa by using a set of switching adsorbent beds. However, PSA does not selectively separate CO_2 from other waste gases (CH_4 and CO), thus, the off-gas from the PSA unit contains CO_2 (about 40%–50%), CH_4, CO, and small amounts of H_2, and it is used as fuel in the reforming reactor with CO_2 being vented to the atmosphere. Since the resulting off-gas is heavily diluted with N_2, the capture of CO_2 from modern SMR plant emissions would require one of the post-combustion CO_2 capture processes described in Section 14.3. In an alternative approach, the PSA process could be modified to recover both pure H_2 and CO_2, for example, by including an additional PSA section to remove CO_2 prior to the H_2 separation step (Air Products Gemini Process) (IPCC CCS 2005).

According to an SMR plant design study (IEA Report 1996), the overall efficiency of H_2 production (at the pressure of 6 MPa) at a modern SMR plant with the capacity of 720 t day^{-1} H_2 without CO_2 capture is estimated at 76% (on a lower heating value, LHV, basis) with overall CO_2 emissions of 9.1 kg CO_2 per kg H_2. If the process is modified such that a nearly pure CO_2 is produced as a coproduct (e.g., via combination of amine solvent scrubbing with PSA), the efficiency is reduced to 73%, while CO_2 removal rate also reduced to 8.0 kg CO_2 kg^{-1} H_2. Besides SMR, there are a number of other technological routes to hydrogen production from hydrocarbon feedstocks, for example, partial oxidation, autothermal reforming, combined reforming, gasification of residual oil, cracking, etc. A comprehensive overview of the state-of-the-art hydrocarbon-to-hydrogen technologies has been recently reported by Muradov (2009).

Some estimates have been reported in the literature on the economics of fossil hydrogen production coupled with CCS. Audus et al. (1996) analyzed the concept of hydrogen production by SMR process with CO_2 capture and disposal in a saline aquifer. It was estimated that CCS would add about 25%–30% to the cost of hydrogen production. The costs of avoiding CO_2 emissions would be about U.S. \$20 t^{-1} CO_2 avoided. The U.S. National Research Council (NRC) report provided the estimates of the cost of centralized hydrogen production by SMR without and with added CO_2 sequestration: \$1.03 kg^{-1} and \$1.22 kg^{-1} H_2, respectively (an increase in 18.5%) (NRC 2004). The additional cost of pipeline shipment and distribution of hydrogen is \$0.96 kg^{-1} H_2 (note that it is almost equal to the cost of hydrogen production) (NRC 2004).

The high cost of hydrogen delivery and distribution seems to favor the decentralized (or distributed) production of hydrogen (e.g., onsite SMR reformers at gas-filling stations). This approach is widely considered the lowest-cost option for hydrogen production during the transition to hydrogen economy. However, the cost of CO_2 capture and sequestration associated with distributed hydrogen production appears to be prohibitive (DiPietro 1997); thus, the SMR–CCS concept will most likely be limited to large centralized SMR plants preferably located near CO_2 disposal sites.

14.6.3.2 Coal Gasification

Gasification of coal aims at producing high-value commodities such as fuels, electricity, and chemicals from low-value "dirty" solid feedstocks such as coal or petroleum coke. The coal gasification process can be expressed by the following main reactions (for the sake of simplification, coal is presented as C):

$$C + H_2O \rightarrow H_2 + CO \quad \Delta H° = 132 \, kJ \, mol^{-1} \tag{14.31}$$

$$C + O_2 \rightarrow CO_2 \quad \Delta H° = -394 \, kJ \, mol^{-1} \tag{14.32}$$

The amount of O_2 is carefully controlled such that only a portion of fuel is burnt providing enough heat to run steam gasification of coal to syngas. The composition of syngas depends on the coal feedstock, the type of gasifier (e.g., entrained flow, fluidized bed, and moving bed) and operational parameters (pressure, temperature, H_2O/C, and O_2/C ratios), syngas cooling method (water quench vs. heat exchangers), gas cleanup system, etc. For example, Shell and Texaco (GE) type gasifiers produce syngas with the H_2–CO composition of 26.7–63.3 vol.% and 34–48 vol.%, respectively (Lin 2009). Besides main components (H_2, CO_2, CO, and H_2O), the gasifier output contains the appreciable amount of impurities, for example, COS, H_2S, NH_3, HCN, N_2, Hg, and volatile minerals that have to be captured and dealt with. Depending on the gasifier type, operational temperature could reach up to 1600°C and pressure up to 85 atm. The gasification stage is followed by WGS (Reaction 14.1) and gas separation stages. Typical thermal process efficiencies for coal gasification vary in the range of 51%–63% (Ball et al. 2009).

The coal gasification technology has been practiced for more than a century, and is common in chemical and fertilizer industries. Currently, there are 128 gasification plants operating worldwide with 366 gasifiers producing 42,700 MW$_{th}$ of syngas (NETL-DOE 2010).

Also, there is about 24,500 MW$_{th}$ of syngas capacity under construction or development, and 4000–5000 MW$_{th}$ of syngas capacity is added annually. Based on coal–gas interface and coal particles movement patterns, coal gasifiers belong to one of three categories (Wilson and Gerard 2007):

- Moving bed (e.g., Lurgi)
- Fluidized bed (e.g., KRW)
- Entrained flow bed (e.g., ChevronTexaco, Shell)

The operational conditions in the gasifiers depend on the gasifier type and vary in the wide range of temperatures (800°C–1350°C) and operating pressures (0.1–7 MPa). For the last 20 years, the predominant type of newly installed commercial gasifiers was of an entrained-flow type. A recent study (Stobs and Clark 2004), however, suggested that high-pressure fluidized bed gasifiers might be better suited for the use with low-rank coals (although there is very limited commercial-scale operation of such gasifiers).

Commercial coal gasification systems are mainly focused on the following application areas:

- Hydrogen production
- Ammonia production (many plants are operated in China)
- IGCC power plants
- SNG production (e.g., North Dakota plant)
- Fischer–Tropsch liquids production (e.g., Sasol technology in South Africa)
- Polygeneration (production of electricity, steam, and chemicals)

For example, in the North Dakota plant producing substitute (or synthetic) NG, 3.3 million tons CO_2 per year are captured using the cold methanol solvent scrubbing process (Rectisol process). Most of the captured CO_2 is vented, and about 1.5 million tons of CO_2 is pipelined to Weyburn, Canada, for EOR and CO_2 storage (IPCC CCS 2005).

To increase the efficiency of the CO_2 capture process, oxygen-blown gasification at high pressure is preferred because of high CO_2 partial pressure requirements. Thus, when considering a CO_2 separation stage, fossil decarbonization via the gasification route entails lesser energy penalties compared with the post-combustion systems, because CO_2 could be recovered at partial pressures up to three orders of magnitude higher. As a result, the CO_2 stripping energy requirements and solvent circulation rates are lower, with associated reduction in the equipment (e.g., CO_2 absorber) size. These advantages are offset by the additional energy penalties related to oxygen separation from air, CO shift, and other processes.

There are critical issues with regard to sulfur removal from syngas of the gasification process. During gasification, most of the sulfurous compounds in coal are converted to H_2S and, to a lesser extent, to COS. H_2S content in syngas must be reduced to about 10 ppm levels for power plants to be in a compliance with SO_2 emission regulations. For other applications, for example, ammonia or synfuels manufacturing, H_2S concentration in syngas has to be further reduced to 1 ppm level and below in order to prevent the permanent deactivation of industrial catalysts by the H_2S impurities. The specifics of H_2S/CO_2 capture and separation processes are dictated by CO_2 storage requirements. If combined storage of CO_2 and H_2S were permissible, it would be economically beneficial to recover both products in the same physical absorption unit (IEA Report PH4/33, 2004). Although H_2S/ CO_2 costorage is being practiced in western Canada for sour NG projects (about 0.48 million tons of CO_2 per year), it is not clear whether costorage would be feasible at much larger scales (a typical gasification plant would annually produce about 1–4 millions tons of CO_2 to be delivered to a storage site) (Bachu and Gunter 2004). However, if CO_2 has to be stored in a pure form, it is common to first recover H_2S (many physical solvents favor absorption of H_2S relative to CO_2), followed by conversion of H_2S to elemental sulfur (in a Claus plant) and tail gas clean up (CO_2 is recovered in a separate downstream unit).

The vast resources of coal in United States and worldwide could potentially last for about 250 years if consumed at the current consumption rate. However, if coal would be the basis for future large-scale production of hydrogen or synthetic fuels, its economically accessible resources would be depleted in about a century or even sooner.

14.6.3.3 Integrated Gasification Combined Cycle

Advanced coal gasification technologies, for example, IGCC, for the production of syngas and hydrogen have been under an intensive development for over several decades; for example, U.S. DOE's "Clean Coal Technology" concept envisions the development of a coal-processing technology, whereby CO_2 is captured and sequestered. In view of a relative abundance of coal reserves, the coal-to-hydrogen concept is an active area of research, especially, in countries rich in coal reserves (United States, China, and others).

Recently, IGCC has evolved as an ultralow emission power generation technology combining advanced coal gasification processes with highly efficient combined cycle power generation systems. Several IGCC plants are planned or under construction according to U.S. DOE Future Gen program aimed at a full-scale demonstration of this technology. Figure 14.33 provides a simplified block diagram of an IGCC plant. In the IGCC gasifier, coal feedstock is gasified with an oxygen–steam mixture to hot syngas, which is cooled and cleaned of particles. Cleaned syngas is further subjected to CO shift in sulfur-tolerant shift reactors to produce H_2–CO_2 mixture (containing gaseous impurities such as CH_4, CO,

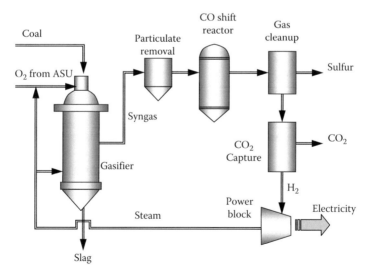

FIGURE 14.33
Simplified block diagram of an IGCC system.

H_2S, etc.). Typically, a double-stage Selexol unit is utilized for the removal of H_2S and CO_2 from syngas: in the first absorber, H_2S is preferentially scrubbed using a physical solvent, followed by the regeneration of the rich solution in a stripper by heat input (Wilson and Gerard 2007). The resulting gas (consisting of H_2S and CO_2) is treated in a Claus unit, where H_2S is oxidized to sulfur. H_2S-free ("sweet") syngas then enters a second absorber, where it is contacted with lean solvent resulting in the removal of remaining CO_2 from syngas. Fuel gas after the Selexol unit consisting mainly of H_2 is sent to a gas turbine. The CO_2 product is released at a relatively high pressure of 3.5 atm (it may require further pressurization before pipeline transport).

The advantages of the IGCC system are as follows:

- Syngas is produced and converted to electricity at the same site to avoid the high cost of pipeline transport (heating value of syngas is a third of that of NG).
- Extra electrical power can be produced onsite from steam generated during syngas cooling.
- Relatively low energy penalties due to high pressure and substantial concentration of CO_2 in the process gas.
- Significant reduction in the emission of criteria air pollutants such as SO_x and NO_x
- Production of value-added product sulfur.

The shortcomings of the IGCC system include

- Complexity of the system (the IGCC plants consist of a larger number of process units and a rather sophisticated equipment compared to conventional power plants).
- High capital/investment costs.
- The effect of coal quality on the process efficiency is not well known (most of the process development has been conducted using high-rank coals; it is not clear how the gasifier performance could be affected if widely available low-rank coals will be used).

The IGCC technology was initially demonstrated in the 1980s, and since then, several IGCC power plants fueled by oil, petroleum coke, or coal have been constructed. It should be noted, however, that the IGCC technology is yet to receive a wide deployment, mainly due to strong competition from NGCC, especially when NG is available at low prices. Moreover, coal-based IGCC plants are still less competitive than PC-fired steam-electric plants, and there are some reliability concerns with regard to this relatively new technology. Because of its environmental advantages over conventional coal-fired power plants, IGCC will be economically more attractive in a carbon-constrained world.

14.6.3.4 Production of Hydrogen and Liquid Fuels from Coal with CO_2 Capture

Although H_2 production from coal with CO_2 capture is technologically similar to IGCC, there are very few studies on this subject, for example, Kreutz et al. (2005) and Gray and Tomlinson (2003). For example, in a design study by Kreutz et al. (2005), a H_2-manufacturing plant using high-sulfur bituminous coal (3.4% sulfur) with the capacity of 1070 MW_{th} H_2 is analyzed. An entrained flow quench gasifier operating at the pressure of 7 MPa is assumed as a base case design. After syngas is cooled, purified of particulates, and CO shifted, H_2S and CO_2 are removed from the syngas using a physical solvent-based process (Selexol) in the first and second stages of the process, respectively. H_2S is further converted to sulfur in a Claus process, whereas CO_2 is dried and compressed to 150 atm for pipeline transport and underground storage. High-purity H_2 (99.999%) is extracted from CO shifted syngas via a PSA process. PSA purge gas is combusted in a gas turbine to generate 39 MW_{el} of electricity (if CO_2 is not captured, the amount of generated electricity would increase to 78 MW_{el}). For this base scenario with CO_2 capture, the efficiency of H_2 manufacturing and associated CO_2 emissions were estimated at 61% and 1.4 kg CO_2 kg^{-1} H_2, respectively (without CO_2 capture, these numbers would increase to 64% and 16.9 kg CO_2 kg^{-1} H_2). The authors of the study concluded that there are no economic advantages from increasing the electricity/H_2 output ratio, and this ratio will be determined by market demands for electricity and H_2.

Syngas produced by coal gasification, can be converted, besides H_2, to a number of other fuels and chemical products, for example, substitute NG, synthetic liquid hydrocarbons (FT gasoline, jet, and diesel fuels), methanol, and dimethyl ether (see Figure 14.34). Relatively pure stream of CO_2 is a by-product of these processes that can be conveniently captured and sequestered. Some of these processes have been commercially practiced for decades (e.g., Fischer–Tropsch-based Sasol technology for synfuel production in South Africa). Among other coal-based commercial technologies are coal-to-methanol plants in China and United States, an SNG-manufacturing plant in North Dakota.

Manufacturing of synthetic fuels from coal typically includes the stages of O_2-blown coal gasification to syngas, gas cleanup, CO shift, and acid gas (CO_2, H_2S) removal. Clean syngas enters a FT synthesis reactor, where it is converted to a wide range of liquid hydrocarbons. Unconverted syngas is either recycled back to the synthesis reactor, or used for electricity generation (for onsite use or for sale to electric grid). The latter single-pass option is considered more economical alternative to the syngas recycle approach (Larson and Ren 2003). For the single-pass systems, a WGS reactor and CO_2 scrubber are often placed upstream of the synthesis reactor to adjust the H_2/CO ratio and remove an excess of CO_2 from raw syngas.

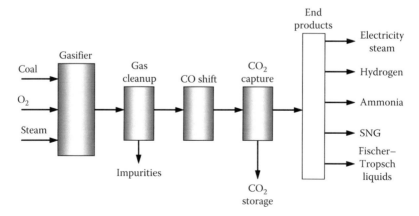

FIGURE 14.34
Block diagram of a coal gasification process with CO_2 capture and options for production of electricity, fuels, and chemicals.

14.6.4 Emerging Technologies

An increase in FTE energy conversion efficiency is beneficial from the viewpoints of both the efficient use of dwindling fossil resources and reduction in CO_2 emissions per kW of electrical power produced. Recent new developments in power generation technologies involving a combination of fuel processing (e.g., NG reforming or coal gasification) technologies with fuel cells (FCs) offer the advantage of a significant increase in the FTE energy conversion efficiency. In contrast to traditional power generation systems (e.g., turbines, diesel generators, internal combustion engines), FC efficiency is not controlled by the Carnot cycle heat-to-work limitations, since they directly convert the chemical energy of fuel to electricity via electrochemical reactions. Thus, the practical energy conversion efficiencies as high as 60% and 85% (without and with electricity-heat cogeneration, respectively) are possible (compared to 30%–40% efficiencies for the conventional power generators). Another advantage of FC is that fuel and air are not mixed (as in a turbine), but rather are introduced into separate compartments of the FC: fuel to an anode and air to a cathode compartment: this way, oxidation products are not diluted by nitrogen. The resulting anode outlet stream is rich with CO_2, which significantly simplifies its capture and storage.

The types of FC applicable to CCS applications include (FC temperature range is shown in parentheses)

- Polymer electrolyte membrane FC (PEM FC) (60°C–100°C)
- Alkaline (AFC) (90°C–100°C)
- Phosphoric (PAFC) (150°C–200°C)
- Molten carbonate FC (MCFC) (600°C–750°C)
- Solid oxide FC (SOFC) (650°C–1000°C)
- Direct carbon FC (DCFC) (650°C–950°C)

Depending on the type of FC used, three options for the FC-based power generation systems can be envisioned that utilize hydrogen, syngas, or fuel (directly). In the first option, fuel is first reformed to syngas, which is further conditioned via WGS to H_2–CO_2 mixture, and, in the final stage, H_2 is separated from this mix and fed to the anode compartment of FC (Figure 14.35). Practically pure CO_2 stream is ready for transportation and sequestration. This approach is applicable to both low- or high-temperature FC and to any type of primary fuel (NG, liquid hydrocarbons, coal, petroleum coke, etc.), and, conceptually, it is similar to the precombustion decarbonization option discussed in the Section 14.2.3. In the second approach, a high-temperature FC (e.g., MCFC or SOFC) utilizes syngas after the reforming and gas purification stages without the need for H_2 recovery from the reformate gas (Figure 14.36). The resulting anode outlet gaseous stream contains H_2O–CO_2 mixture, which can be easily separated by condensing water (knockout), thus leaving pure CO_2 as the only product of fuel oxidation (in practical systems, there may be unconverted H_2 and CO and some CH_4 impurity). This approach can also be applied to any type of gaseous, liquid, or solid fuel. The systems with the direct use of fuel (a third option) utilize internal reforming capabilities of high-temperature FC (e.g., SOFC). Due to the very high temperature in SOFC, internal reforming of gaseous fuels (e.g., NG) takes place in the anode compartment resulting in the H_2O–CO_2 stream as an oxidation product. This approach is mostly limited to gaseous fuels since the use of liquid fuels may result in excessive coking inside the device.

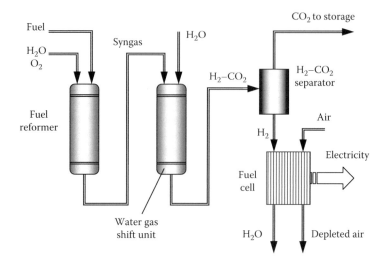

FIGURE 14.35
FC-based power generation system with upstream CO_2 capture.

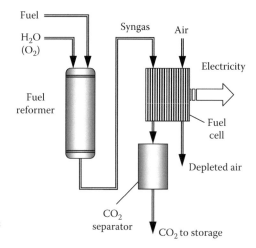

FIGURE 14.36
FC-based power generation system with downstream CO_2
capture.

Although most of FC-related activities are focused on hydrogen- and syngas-fueled FC, DCFC are increasingly attracting attention of researchers due to their potential to achieve much higher FTE efficiency compared to H_2-fed FC (H_2-FC). Besides, the DCFC-based power generators offer the advantages of direct utilization of coal as fuel (without prior gasification) and production of sequestration-ready CO_2. It is important to note that as fuel, carbon has the highest volumetric energy density (19 kWh L^{-1}) among all electrochemically active fuels, battery anodes, and transportation fuels (e.g., compared to Zn – 9.3, Li – 6.9, and diesel – 10 kWh L^{-1}). In DCFC, solid carbon is introduced into an anode compartment where it is electro-oxidized to CO_2, generating electricity:

$$C + O_2 \rightarrow CO_2 \quad E^\circ = 1.02\,V \tag{14.33}$$

The actual efficiency of DCFC (η) could be defined by the following equation:

$$\eta = \frac{\Delta G(T)}{\Delta H} \mu \frac{V}{V_o} = \frac{\Delta G(T)}{\Delta G + T\Delta S} \mu \frac{V}{V_o} \tag{14.34}$$

where
  η is the actual efficiency of FC
  ΔG, ΔH, and ΔS are free energy, enthalpy, and entropy, respectively, of reaction (14.33)
  m is a fuel utilization coefficient
  V and V_o are operating and open circuit voltages of FC, respectively

Since $\Delta S \approx 0$ for reaction (14.33), and assuming that for DCFC, $\mu \approx 0.9$–1.0 and $V/V_o \approx 0.8/0.9$, the FTE energy conversion efficiency of DCFC in practical systems could be as high as 80%–90%. Recently, different types of DCFC utilizing a variety of electrolytes at different temperature ranges have been under development; in particular, SOFC with yttria-stabilized zirconia as an electrolyte operating at temperatures of 900°C–1000°C (Saito et al. 2008), molten salt electrolyte FC operating at 600°C–800°C (Cherepy et al. 2005), and combined SOFC and MCFC operating at 525°C–700°C (Jain et al. 2007). Conversion efficiencies of 80% and higher have already been demonstrated on a laboratory-scale DCFC using different types of carbon (Steinberg et al. 2002). Thus, DCFC offers the following advantages over other types of FC: (1) the highest theoretical (and, potentially, practical) FTE energy conversion efficiency, (2) the highest fuel utilization efficiency (since fuel—carbon and the product—CO_2 exist in separate phases, allowing full conversion of carbon in a single pass), and (3) CO_2 is produced in the form of concentrated stream ready for transportation and sequestration.

Despite the advantages of DCFC over H_2-FC in terms of higher FTE energy conversion efficiency, the practical implementation of DCFC is hindered by several factors mostly related to the system sustainability, and a need for the supply of clean (i.e., sulfur- and ash-free) carbon fuel for DCFC. From this viewpoint, the use of hydrocarbon fuels as a source of pure carbon for DCFC would be advantageous. Recently, Muradov et al. (2010) reported on the development of a highly efficient integrated energy conversion system based on the decomposition of hydrocarbon fuels to hydrogen and carbon that are separately utilized by hydrogen- and carbon-powered FC. The schematic diagram of the integrated power generation system is shown in Figure 14.37. A hydrocarbon feedstock (e.g., methane) is catalytically decomposed to hydrogen-rich gas and carbon in a hydrocarbon decomposition reactor (HDR) in the temperature range of 800°C–900°C. Hydrogen-rich gas is used as fuel in H_2-FC, whereas the clean carbon product is withdrawn from the reactor and used in DCFC. Although any type of FC can be utilized in the scheme, the use of high-temperature FC such as SOFC is preferable since it allows for an efficient thermal integration of HDR and FC. It was estimated that overall CTE energy conversion efficiency of the integrated system could vary in the range of about 50%–80%, depending on the type of fuel and FC used, and CO_2 emission per $kW_{el}h$ produced is less than half of that from conventional power generation sources.

Although some types of FC are close to commercialization, in general, CCS-related applications of FC are still in R&D, demonstration, and validation stages. The largest existing FC-based demonstration units are at $1\,MW_{el}$ scale, and it will take, probably, another

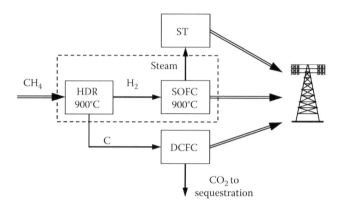

FIGURE 14.37
Schematic diagram of integrated power generation unit consisting of hydrogen and direct carbon FCs. HDR—hydrocarbon decomposition reactor and ST—steam turbine.

5–10 years before large-scale (10–100 MW$_{el}$) units will become commercially available. Due to high CO_2 concentration in off-gas, integrating FC-based power generation units with CO_2 capture systems could be relatively easy and, potentially, less costly compared to conventional power plants, and this may offer the prospect of reducing energy efficiency penalties due to CO_2 capture and improving the process economics. In a study by Jansen and Dijkstra (2003), the effect of incorporating CO_2 capture to a high-temperature FC-based power generation system is analyzed. The authors estimated that FTE efficiencies of 67% and higher can be achieved in the SOFC-based system without CO_2 capture; however, this value drops by about 7% if CO_2 capture is integrated into the system. It should be noted that since the cost of CO_2 transport is a function of CO_2 transmission capacity, combining several smaller FC systems into a group with the total capacity of 100 MW$_{el}$ and higher could reduce the cost of CO_2 to a more acceptable level (IEA GHG 2002).

14.7 Economics of CO_2 Capture and Storage

Although there is plenty of technical and economic analyses data in the literature related to individual components of CCS systems (mostly from relevant oil and gas exploration technologies), very little techno-economic data is available with regard to a full CCS system. In this section, the economic aspects of deploying an entire CCS system are discussed.

It was pointed out that although the costs of individual components of the CCS system provide a basis for determining the overall CCS cost, it is not a simple operation of summing up the costs of CO_2 capture, transport, and storage (IPCC CCS 2005). This is due to the difference between the categories of CO_2 captured and CO_2 avoided during generation of electricity or production of hydrogen. Therefore, the cost of CO_2 mitigation should be expressed either in terms of CO_2 captured or CO_2 avoided. Typically, the cost of CO_2 mitigation in terms of CO_2 avoided (in U.S. \$ t^{-1} CO_2 avoided) is greater than that of CO_2 captured (in U.S. \$ t^{-1} CO_2 captured). It is recognized that the carbon mitigation cost is best represented by the CCA.

U.S. DOE/NETL conducted extensive studies on the economic analysis of CCS systems with regard to their potential to meet the DOE CCS program's goal of 90% CO_2 capture

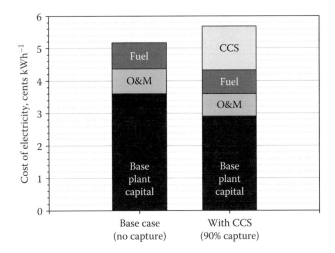

FIGURE 14.38
Cost analysis of CCS technology as applied to a coal-fired power plant with the following metrics: parasitic load (0.35); CO_2 captured at 30 atm; CO_2 capture and compression capital (0.80); capture plant $200 per kg CO_2 per hour; compressor $300 per kg CO_2 per hour; CO_2 compressor efficiency 80%; CO_2 capture O&M (0.08); chemical cost: $1 t^{-1} CO_2 captured; capital cost pipeline and injection wells (0.12); pipeline length 50 mi; pipeline pressure 148.6 atm; pipeline cost $375,000/km; $1.5 million/well; injection well capacity 1300 t CO_2 per day per well. (Data from Carbon Sequestration Technology Roadmap, U.S. DOE, Office of Fossil Energy, Carbon Sequestration Technology Roadmap and Program Plan, NETL, 2006.)

with 99% storage permanence at less than a 10% increase in the cost of energy services by 2012 (NETL 2006). Figure 14.38 shows the cost analysis of CCS technology as applied to a coal-fired power plant. The left bar relates to a reference case: a power plant without CO_2 capture with technology that is projected to be online in 2020. The estimated plant efficiency and capital cost are 47.4% and $1350 kW^{-1}, respectively (which represents an improvement over current technology). The right bar relates to a power plant with 90% CO_2 capture. According to the DOE program's targets, the lowering cost and improving the efficiency of the base power plant ($1100 kW^{-1} and 52%, respectively) would result in lower COE "making room" for the CCS technology, such that COE from the overall system is 10% higher compared to the reference scenario. In the DOE/NETL 2009 report, the updated R&D performance and cost goals for CO_2 capture applicable to new and existing coal-fired plant technologies were established as follows: minimum CO_2 captured—90%, maximum increase in the COE—35% (NETL 2009).

Figures 14.39 and 14.40 summarize the mitigation costs of CCS (with geological storage) (in U.S. $ t^{-1} CO_2 avoided) and the COE (U.S. $ MWh^{-1}) related to PC, NGCC, and IGCC power plants. The reference power plant is of the same type as the above plants, but without CCS. Two storage options are considered: geological storage and EOR. As expected, the inclusion of EOR significantly reduces the CO_2 mitigation cost and the COE (the lower range of carbon mitigation cost with EOR could actually have negative values). Figure 14.41 depicts the range of total costs of the CCS system for new hydrogen production plants (with geological storage and EOR options).

The results of the economic assessment of CCS technologies conducted by the Global CCS Institute (2009) for power generation and some industrial applications are summarized in Figure 14.42, which depicts the product cost increase over the base case without CO_2 capture. A caution must be exercised in drawing conclusions from this figure

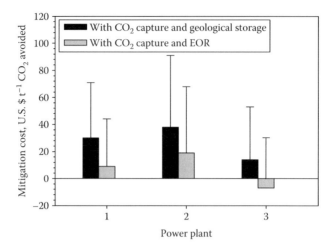

FIGURE 14.39
Range of CO_2 mitigation costs with CO_2 capture and geological storage and EOR for new power plants. 1—PC power plant, 2—NGCC power plant, and 3—IGCC power plant. (Data from IPCC special report on carbon dioxide capture and storage, Prepared by Working Group III of the Intergovernmental Panel on Climate Change, B. Metz, O. Davidson, H. de Coninck, M. Loos, A. Meyer (eds.), Cambridge University Press, Cambridge, U.K., Table 8.3a, p. 347, 2005.)

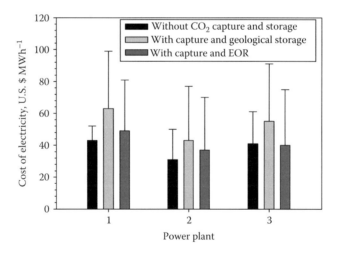

FIGURE 14.40
Range of total costs of electricity with and without CO_2 capture and geological storage based on current technology. 1—PC power plant, 2—NGCC power plant, and 3—IGCC power plant. (Data from IPCC special report on carbon dioxide capture and storage, Prepared by Working Group III of the Intergovernmental Panel on Climate Change, B. Metz, O. Davidson, H. de Coninck, M. Loos, A. Meyer (eds.), Cambridge University Press, Cambridge, U.K., Table 8.3a, p. 347, 2005.)

with regard to the advantages of one technology over another because of the margin of error and the significant impact of the project location and the CCS technology choice. The modeling studies indicated that the cost of CCS for power generation (based on the use of commercially available technology) was in the range of $62–$112 t^{-1} CO_2 avoided or $44–$90 $t^{-1}CO_2$ captured (the lowest values were for oxyfuel combustion and the highest

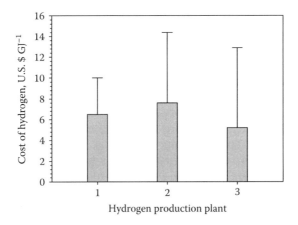

FIGURE 14.41
Range of total costs of CCS system for new hydrogen production plants (with geological storage and EOR options). 1—without CO_2 capture, 2—with CO_2 capture and geological storage, 3—with CO_2 capture and EOR. (Data from IPCC special report on carbon dioxide capture and storage, Prepared by Working Group III of the Intergovernmental Panel on Climate Change, B. Metz, O. Davidson, H. de Coninck, M. Loos, A. Meyer (eds.), Cambridge University Press, Cambridge, U.K., Table 8.3b, p. 348, 2005.)

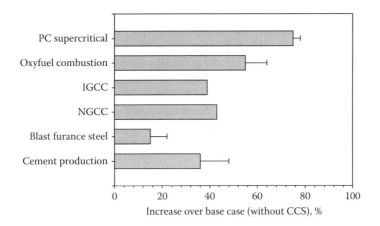

FIGURE 14.42
Product cost increase over the base case without CO_2 capture for different industrial applications. (Data from Global CCS Institute, Strategic analysis of the global status of carbon capture and storage, Final Report, Global CCS Institute, http://www.globalccsinstitute.com/downloads/reports/2009/worley/foundation-report-1-rev0.pdf [accessed August 3, 2010], 2009.)

for NGCC with post-combustion capture technology) (Global CCS Institute 2009). Among industrial applications, steel production and cement production show the highest percentage cost increase with the inclusion of CCS because the capture of CO_2 is not inherent in the design of these facilities. The analysis also indicated that across the integrated CCS system, the most significant contributor to the COE for power plants with CCS is CO_2 capture and compression followed by its transport and storage costs. The cost of CO_2 capture and compression (including fuel) is greater than 80% of the integrated CCS costs (Global CCS Institute 2009).

Considering a high degree of uncertainty, other cost predictions for CO_2 sequestration vary in a wide range. Lackner reported that the cost of CO_2 sequestration of $30 t^{-1} CO_2 (which is an equivalent of additional $13 bl^{-1} of oil or $0.25 gal^{-1} of gasoline) appears to be practically achievable in the long run (Lackner 2003). The author argues that initially niche markets (e.g., EOR) would help keep CO_2 disposal costs low; however, as cheap disposal sites are filled up and the requirements on CO_2 storage permanence and safety are tighten up, the cost of carbon sequestration may rise (even as the cost of CO_2 capture may drop due to technology improvements). Haszeldine (2009) estimates that in the United Kingdom, the deployment of CCS may cost each household an extra 10% per year for electricity cost. There is a possibility that in some application areas (e.g., aviation, road transportation), higher cost of CO_2 capture from air could be accommodated (since this would eliminate onsite capture and transport of CO_2) (Lackner 2003).

The general trend in the CCS technologies development is to lower its cost and increase a CO_2 storage capacity. Figure 14.43 summarizes the cost–capacity relation for the various CO_2 sequestration options. U.S. DOE considers terrestrial ecosystems the lowest cost option for CCS followed by storing CO_2 in unmineable coal seams and depleting oil and gas reservoirs (U.S. DOE, NETL 2003). Ocean sequestration is considered the highest storage cost option, although it has a huge potential as a carbon storage sink.

When discussing the economic aspects of CCS deployment, one should take into a consideration the indirect cost of existing fossil fuel infrastructure. In its recent report, U.S. NRC estimated that fossil-based energy production, mostly from coal and oil, causes U.S. $120 billion worth of health and other nonclimate-related damages each year in the United States only that are not included in the price of energy (Johnson 2009c). The cost numbers are primarily based on health impacts and premature deaths of nearly 20,000 people annually that relate to air pollution (primarily, due to SO_x, NO_x, and particulates) from coal-fired power plants (about $63 billion) and ground transportation (about $56 billion), over their full life cycles, with two-thirds of it from extraction and production of fuel (the remaining $1 billion is due to heating). The authors of the study emphasize that the estimate of the total cost of the damages is rather conservative because it does not account for the impact of climate change, a negative impact to ecosystems, or the adverse effect of toxic

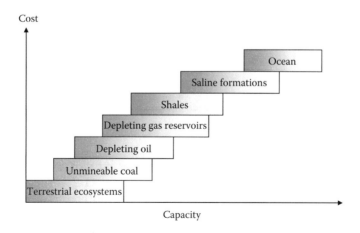

FIGURE 14.43
Cost–capacity relation for the various CO_2 sequestration options. (Data from U.S. DOE, NETL, Carbon sequestration, Technology Roadmap and Program Plan, U.S. DOE Office of Fossil Energy, National Energy Technology Laboratory, Morgantown, WV, 2003.)

air pollutants, such as mercury and lead. The report also shows that most of these damages come from coal-fired plants (United States has 407 of them); NG-fired power plants are responsible for only $1 billion in health and nonclimate-related damages (although they produce 20% of U.S. electricity). The NRC report estimates that the damage cost due to CO_2 emissions vary in the wide range of $1–$100 t^{-1} CO_2.

Summarizing, the economic impact of the CCS technology deployment is a subject to major uncertainties. There are estimates in the literature on the short- and long-term cost of CCS deployment. In particular, the World Energy Investment Outlook report provides an estimate of an upper limit for investment in the CCS system for the OECD (Organization for Economic Cooperation and Development) countries over the next three decades at about U.S. $350–$440 billion (WEIO 2003). To put this number in a perspective, it corresponds to the CCS share of about 3.5% of the total installed fossil-based power plant capacity in the OECD countries by 2020 (IPCC CCS 2005). The estimates of the long-term investment requirements for the CCS deployment suffer from even more uncertainties. Edmonds et al. reported that depending on the targeted CO_2 concentration stabilization levels of 750 or 450 ppmv, long-term savings associated with the CCS deployment would vary in the range of tens billions to trillions of U.S. dollars, respectively (Edmonds et al. 2000). Herzog (2002) estimated that a carbon tax of $100–$300 t^{-1} CO_2 would be necessary to make CO_2 sequestration viable with a current technology. In addition to an uncertain short- and long-term economic impact of CCS technologies, there are factors that are not well known, but can significantly affect the cost of the CCS deployment, such as costs related to a possible environmental damage, liability, monitoring cost, issues related to public acceptance, etc.

Due to the enormous potential cost of the CCS deployment, different policy options for providing sufficient funding for massive-scale CO_2 capture and storage have been discussed in the literature. It was proposed in one study that the World Trade Organization (WTO) should become an active player in the CCS market (Kithil 2007). The authors of the study assume that WTO could jumpstart the CO_2 sequestration market by issuing long-term contracts to purchase bona fide sequestration-derived CO_2 credits. This measure could bring forth the sequestration investment needed to achieve storage of up to 10 billion tons of CO_2 per year by 2025 (including 7 billion tons from biological ocean storage and 3 billion from geological and terrestrial storage). According to one set of assumptions, the net WTO subsidy would peak at U.S. $86 billion by 2022 (which is about 1.01% of the WTO's tariff on imports and exports). Thus, under this proposal, WTO would subsidize the CCS deployment in the near-to-medium term, and then recoups its investment and gains large profits over the long term.

14.8 Role of CCS in Portfolio of CO_2 Mitigation Options

The analyses of various CO_2 mitigation scenarios involve a large number of rather complex models (e.g., input–output, macroeconomic, energy-sector-based engineering, technology-specific and other models). Many modeling studies imply that CCS systems will be competitive with other CO_2 mitigation options, for example, an energy efficiency increase, the use of nuclear and renewable energy sources. IEA provides the following breakdown (in percentage of total) of the potential contributions of different carbon mitigating technologies targeting the reduction in CO_2 emissions by 50% by 2050 (IEA Energy Technology Perspectives 2008):

CCS industry and transformation	9
CCS power generation	10
Nuclear power	6
Renewables resources	21
Power generation efficiency and fuel switching	7
End-use fuel switching	11
End-use electricity efficiency	12
End-use fuel efficiency	24
Total	100%

It can be seen that the deployment of CCS technologies in power generation and industrial sectors can potentially contribute up to 19% reductions in CO_2 emissions by 2050. Conservation and energy efficiency increase measures along with renewable energy sources will also meet a significant fraction of the global energy demand. The share of nuclear-based technologies would be rather modest over the entire period. Overall, according to the IEA-2008 report, the application of all listed measures could reduce CO_2 emissions from 62 Gt year^{-1} (in a business-as-usual scenario) to 14 Gt year^{-1} by 2050.

The evaluation of different scenarios of the future deployment of CCS technologies indicate major uncertainties with regard to timing, the estimates of CO_2 emission reductions, and the implementation costs of these options. For example, Morita et al. (2000) reported the estimates of uncertainties associated with the deployment of CCS technologies and their role in total CO_2 emission reductions in order to stabilize atmospheric CO_2 concentrations between 450 and 750 ppmv. The IPCC report (2005) estimated that the average contribution of CCS to total CO_2 emission reductions would range from 15% to 54% for CO_2 concentration stabilization scenarios based on 750 and 450 ppm levels, respectively (with the extreme values ranging from 0% to 90%, respectively). The corresponding average cumulative CO_2 storage during the years 2000–2100 according to different mitigation scenarios could range from 377 to 2160 Gt CO_2 for 750 and 450 ppmv CO_2 stabilization scenarios, respectively (with the extreme values from zero to about 6000 Gt CO_2, for the stabilization target of 750–450 ppmv, respectively) (Figure 14.44).

The studies on the future deployment of CCS in North America and western Europe indicate that there is a large CO_2 storage potential in these regions (Dooley et al. 2004; Wildenborg et al. 2004). In particular, it was estimated that well over 80% of CO_2 emissions from present industrial CO_2 sources could be stored in suitable geological formations for about U.S. \$12–\$15 t^{-1} CO_2 and U.S. \$25 t^{-1} CO_2 in North America and western Europe, respectively. These studies show that the potential for the early CCS deployment would depend on such factors as the nature of CO_2 source (e.g., its purity), possibility of CCS coupling with EOR or ECBM, the proximity of a CO_2 point source, and geological and geographical conditions.

The majority of analytical studies agree that although the penetration of CCS technologies into the marketplace may occur within two to three decades, most of the global CCS deployment, in all likelihood, would take place in the second half of the century (Riahi et al. 2003). According to Morita et al. (2000) analysis, the value of cumulative CO_2 storage would average at about 185 Gt CO_2 in the first part of twenty-first century and at 1470 Gt CO_2 in the second part. It is also understood that timing of the CCS deployment would depend on a particular region and other special conditions. For example, there are countries such as United States, Canada, and Australia where the CCS deployment

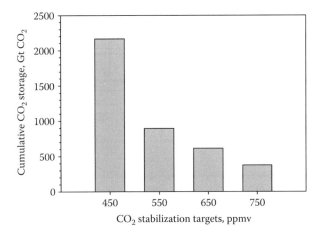

FIGURE 14.44
Average cumulative CO_2 storage during the years 2000–2100 according to the IPCC TAR mitigation scenario. (Data from IPCC special report on carbon dioxide capture and storage, Prepared by Working Group III of the Intergovernmental Panel on Climate Change, B. Metz, O. Davidson, H. de Coninck, M. Loos, A. Meyer (eds.), Cambridge University Press, Cambridge, U.K., Table 8.5, p. 356, 2005.)

would not face economic or physical constraints due to a relatively large geological storage capacity. These countries would be in a better position to start the early deployment of CCS technologies. However, some countries have rather limited reservoir capacity for geological CO_2 storage compared to a potential demand (e.g., Japan and South Korea), which would require using other mitigation options (e.g., ocean storage) (IPCC CCS 2005).

14.9 Risk Factors Associated with CCS

14.9.1 CO_2 Emissions and Leakage Sources during CCS

It is important to emphasize that CCS does not completely eliminate all CO_2 emissions, and CO_2 storage is not necessarily 100% permanent. Certain fraction of stored CO_2 is likely to return to the atmosphere each year via a variety of pathways such as molecular diffusion through the caprock, a gradual leakage through faults, fractures, or wells, an abrupt leakage, an injection well failure, etc. (Stone et al. 2009). In the case of ocean storage, the retention time is a strong function of the depth of injection with the trade-off between the cost of storage and the leakage rate: shallower injection is cheaper but is subject to faster leakage rates. Since CCS is a sequence of energy-intensive processes, capture, transportation, and storage, CO_2 emissions are likely at each stage of this sequence. Figure 14.45 summarizes the sources of CO_2 emissions during different stages of carbon capture with geological storage.

14.9.1.1 Emissions from CO_2 Capture

CO_2 capture processes produce two types of emissions: the emissions due to imperfect CO_2 capture and those resulting from an additional energy use due to CO_2 capture. The

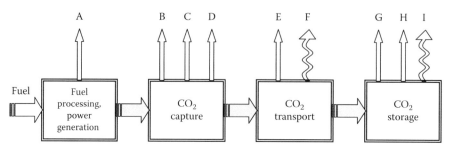

FIGURE 14.45

Flow diagram of potential CO_2 emission sources and leaks during fuel processing and CO_2 capture and storage. A—CO_2 from fuel processing (with precombustion capture, H_2 or low-carbon fuel production) or power generation (with post-combustion capture), for example, flue gas, vent gas, off-gas. B—CO_2 emissions from additional energy/fuel usage due to CO_2 capture. C—CO_2 emissions from imperfect capture. D—CO_2 emissions in the form of flue gases formed during CO_2 capture. E—CO_2 emissions from additional energy/fuel usage for CO_2 transport. F—CO_2 emissions during transport (fugitive). G—CO_2 emissions from additional energy/fuel usage for CO_2 injection. H—CO_2 leakage from storage. I—CO_2 emissions during injection (fugitive).

capture process can produce GHGs other than CO_2 (although in smaller quantities than CO_2) such as methane, NO_x, CO, and SO_2 (some of these gases could be produced as a result of the degradation of the solvents used in the CO_2 capture processes, for example, amine, or potassium carbonate scrubbing).

14.9.1.2 Emissions from CO₂ Transportation

CO_2 emissions associated with its transport result from additional energy requirements for this operation and fugitive CO_2 emissions released during CO_2 transport. Most of CO_2 is transported by pipelines and ships, and to a lesser extent, via railway and trucks. The pipeline transport consumes electrical energy used in compressors installed at a CO_2 capture site (longer distances would require more compressors resulting in more CO_2 emissions). The emissions related to CO_2 transportation by ships, rails, and trucks originate from mobile combustion sources, for example, vehicles. Other GHGs (e.g., NO_x, CO, SO_x, and VOCs) are also emitted during CO_2 transportation. Fugitive losses during CO_2 transportation could be insignificant.

14.9.1.3 Emissions from CO₂ Storage

CO_2 emissions generated during the CO_2 storage phase include emissions from an additional energy use for CO_2 injection (underground or the deep ocean), fugitive emissions during the injection operation, and CO_2 leakage from a storage reservoir. The assessment of CO_2 emissions due to additional energy requirements for its injection to some extent is similar to that related to CO_2 emissions due to its capture and transport. The quantification of fugitive CO_2 emissions during injection presents more difficulties and is less certain compared to other stages. Potential escape routes for CO_2 injected into saline formations include the passages of CO_2 through siltstone, the gap in a caprock, a poorly plugged abandoned well, and dissolution at CO_2–water interface.

The assessment of the extent of a physical leakage from geological and ocean storage sites is a focus of numerous studies. The amount of CO_2 stored over a given time period can be defined as follows (IPCC CCS 2005):

$$CO_2 \text{ stored} = \int_0^T (CO_2 \text{ injected}(t) - CO_2 \text{ emitted}(t)) dt \qquad (14.35)$$

where
   t is time
   T is the duration of assessment time period (note that this equation is valid for any CO_2 storage option: geological, the ocean, mineral carbonation, etc.)

Since the amount of injected CO_2 could be directly measured, this equation requires accurate measurements or rigorous source-specific estimates of the amount of CO_2 emitted from a storage site. A physical leakage from storage reservoirs could occur via either gradual long-term release or sudden rapid release due to disruption or perturbation of the storage reservoir. CO_2 leakage from a storage site can be defined by the following equation:

$$CO_2 \text{ emitted} = \int_0^T m(t) dt \qquad (14.36)$$

where
   $m(t)$ is the amount of CO_2 emitted to the atmosphere per unit of time
   T is the duration of the assessment time period

This equation determines the CO_2 leakage rate that might occur during a specific time frame in the future after the injection.

Although geological storage sites are meant to safely contain all injected CO_2 for thousands of years, the probability of CO_2 release even in small quantities is high and it has to be accurately assessed. Unfortunately, no studies have been reported on the underground CO_2 leakage monitoring over a sufficiently long period. The IPCC Report (2005) provided the quantitative estimates of achievable fractions of CO_2 retained in storage sites based on existing relevant engineered systems such as NG storage, CO_2-EOR, disposal of acid gases (CO_2 and H_2S), as well as modeling studies. About 470 NG storage facilities exist in United States containing about 160 Mt NG, and there have been documented nine incidents of a significant leakage mostly related to wellbore integrity and leaks in caprocks (in one incident at Kansas facility, about 3000 t of NG was released). The technical merits of this comparison, though, should not be overestimated due to different physical and chemical nature of CO_2 and methane (e.g., the difference in solubility in water, corrosiveness with regard to metal components, etc.).

Numerical simulations of CO_2 confinement in storage sites indicated that the probability of CO_2 release is rather insignificant. For example, a simulation study using a probabilistic model predicted that over 5000 years a statistical mean release of 0.001% and a maximum release of 0.14% of the total stored amount of CO_2 would occur (Zhou et al. 2005). Although estimates of the fraction of retained CO_2 in geological storage sites are highly site-specific, it is assumed (based on the available data) that for well-selected and monitored storage sites, the fraction of retained CO_2 is likely to be more than 99% over the first 1000 years (IPCC CCS 2005).

In a recent study by Stone et al. (2009), the authors used a set of relatively simple models involving carbon cycle to quantify the extent of a tolerable leakage from CO_2 storage reservoirs for a range of CCS engineering and implementation parameters. It was concluded that up to the year 2100, the CCS application is beneficial (although in some cases the benefit might be rather small). For a much longer time horizon (out to the year 2500), however, the models, quite unexpectedly, predicted that in many cases (depending on the leakage rate and the climate sensitivity assumptions), the deployment of CCS could lead to a greater negative impact on climate than had it not been applied at all (and the largest single controlling factor is the storage reservoir retention time).

14.9.1.4 CO_2 Emissions from Mineral Carbonation and CO_2 Industrial Uses

No data are available in the literature on the amount and rate of CO_2 leakage from mineral carbonation. It can be assumed that due to chemical nature of the carbonation process (i.e., formation of stable carbonates), the physical leakage from storage sites would be unlikely or minimal. However, fugitive CO_2 as well as a certain amount of CO_2 from an additional energy usage could be emitted during the carbonation process. A life cycle analysis of mineral carbonation process that involves mining, rock grinding, particle size reduction, waste disposal, and site restoration stages indicated additional CO_2 emissions of 0.05 t CO_2 per ton of CO_2 stored (Newall et al. 2000).

Processes involving industrial uses of CO_2 typically emit substantial amounts of CO_2 (e.g., methanol synthesis and urea production). The emissions could result from the use of fossil fuels for power/heat inputs to the technological processes, and may also include some fugitive CO_2 from cryogenic applications, for example, dry ice. In contrast to other CO_2 mitigation options (e.g., geological, ocean storage, and mineral carbonation), in this application, CO_2 is emitted after a relatively short period (typically, days to months). There is no discussion in the literature on the assessment of CO_2 emissions due to its industrial uses.

14.9.2 Health and Safety Issues Associated with CO_2 Exposure

Since CO_2 is present in the atmosphere (although in a very low concentration of about 390 ppm) and is a product of human metabolism, it is not considered a harmful (or toxic) substance. Furthermore, most healthy people can tolerate relatively high concentrations of CO_2 (up to 1.5 vol.%) for several hours without health complications. The exposure to higher CO_2 concentrations or for much longer durations could be harmful to human health. In particular, the exposure to CO_2 concentrations of 3 vol.% or higher in air negatively affects health of general population, potentially causing a hearing loss, a visual impairment, a headache, breathing difficulties, dizziness, etc. At the levels of 7–10 vol.%, CO_2 acts as an asphyxiant. Of particular danger is an increase in CO_2 concentration to the levels that would cause oxygen concentration in air to drop below 16 vol.%, which is considered a threshold level for sustaining human life (IPCC CCS 2005). CO_2 entering bloodstream tends to displace oxygen reducing the amount of oxygen taken during breathing, thus, resulting in a negative physiological effect. At CO_2 concentrations above 20 vol.%, death could occur in 20–30 min.

The potentially harmful effect of high concentrations of CO_2 on human health is reflected in the current U.S. occupational exposure standard (NIOSH 1996) limiting the maximum permissible CO_2 concentration in air to 0.5 vol.% for 8 h of a continuous exposure and 3 vol.% for a short period (15 min) of exposure (the CO_2 level of 5 vol.% is considered immediately dangerous to life and health).

Dangerous concentrations of CO_2 in air may arise from a number of sources both manmade and natural. Most typical man-made sources relate to different types of leaks (e.g., from pipework, storage tanks, industrial units, open pits, buildings, etc.). Since CO_2 is about 1.5 times denser than air, it tends to accumulate in low-lying confined or enclosed spaces (especially under stagnant conditions) leading to potentially hazardous situations that require an immediate attention. To make the matter worse, CO_2 is a colorless and odorless gas, which is almost impossible to detect by human senses until CO_2 intoxication already occurred. To avoid an overexposure to CO_2, monitors (or sensors) should be installed where accumulation of CO_2 in high concentrations is likely. Adequate ventilation must also be provided in the areas where CO_2 leak is possible or CO_2 could be vented into the air. Natural sources of CO_2 overexposure include sudden releases of CO_2: they are rarer, but, potentially, they are more deadly compared to man-made sources. The first known large-scale asphyxiation caused by naturally occurring CO_2 took place near the lake Nyos (Cameroon) in 1986 (see further for details).

14.10 Assessment of the Potential Environmental Impact of CCS

Long-term ecological consequences of CO_2 storage are largely unknown. The assessment of environmental risks is highly specific of the storage method (e.g., geological vs. ocean storage) and even of the techniques used to inject CO_2 into a storage site. Here, we outline some general considerations of possible hazards and a potential negative environmental impact of CO_2 storage with regard to the geological and ocean storage options.

14.10.1 Environmental Risks of Geological CO_2 Storage

The potential negative environmental impact of underground CO_2 storage mainly relates to possible hazards to groundwater, terrestrial and marine ecosystems, and to induced seismicity.

14.10.1.1 Hazards to Groundwater

Water suppliers understandably worry that underground CO_2 storage might adversely affect drinking water aquifers. They point to a possibility that in the presence of water CO_2 will form a weak acid, which may cause leaching of naturally occurring elements and minerals (containing metal ions such as Pb^{2+}, Cd^{2+}, and Hg^{2+}) contaminating groundwater. Moreover, CO_2 carries harmful impurities (e.g., H_2S) that may become mobile and further exacerbate the problem. At best, these compounds may alter the odor, color, and taste of water, but in worst cases, these may prevent the use of groundwater for drinking or even irrigation purposes. A computer simulation study conducted at the Lawrence Livermore National Laboratory (Livermore, California) indicated that CO_2 can cause leaching of naturally occurring trace elements such as As, Ba, Cd, Hg, Pb, Sb, Se, Zn, and even U from geological formations (however, only release of arsenic could really pose a problem due to its very high toxicity) (Johnson 2009a). Wang and Jaffe (2004) in their study used a chemical transport model to estimate the effect of dissolved CO_2 on the migration of trace metals in aquifers. It was determined that the release of CO_2 at the depth of 100 m into a shallow water formation containing high concentration of lead-based minerals caused the migration of weakly acidic lead-contaminated water over the distance of hundred meters from the CO_2 source (which could potentially pose a great health hazard for population drinking this water over a large area).

Recently, U.S. EPA has been struggling with the development of the regulations that would protect groundwater and allow some CO_2 sequestration projects. One of the proposed measures is to inject CO_2 very deep in the ground well below drinking water aquifers. However, some aquifers in many locations are already very deep, making CO_2 injection too costly. It is suggested that if CO_2 sequestration projects are to proceed, they have to include strong groundwater monitoring systems.

14.10.1.2 Hazards to Ecosystems

CO_2 stored in underground formations may have a negative impact on terrestrial and aquatic ecosystems (i.e., flora and fauna). In particular, microbial populations in deep subsurface and plants and animals in shallower subsurface and surface may be affected due to low pH or high concentration of CO_2. If CO_2 leaks upward to the surface, it would have a detrimental effect on soil causing a vegetation "die-off." CO_2 concentrations in soil above 5% are dangerous for plants, and the CO_2 concentrations of 20% and higher may be fatal for vegetation (normal CO_2 concentration in soil is 0.2%–4%) (IPCC CCS 2005). Although there are no evidences of the negative impact of a CO_2 leakage on terrestrial ecosystems from existing CO_2 storage projects (which could be due to relatively short period), some natural phenomena involving the underground seepage of CO_2 give some indications of a possible impact. For example, CO_2 seepage in a volcanic area near Mammoth Mountain (California, 1989–2001) caused an extended plant die-off (total CO_2 flux from underground in the affected areas was about 530 t day^{-1} CO_2 and CO_2 concentration in soil was 15%–90%). In Colli Albani (Italy) in 1999–2001, a natural CO_2 seepage from magmatic activity caused death of 29 cows and 8 sheep from asphyxiation (the average gas flux was about 60 t day^{-1}, and gas contained 98% CO_2 and 2% H_2S) (Carapeza et al. 2003).

Potential hazards to ecosystems can be exacerbated if other much more problematic gases such as H_2S, SO_2, and NO_2, are stored along with CO_2, thus, substantially increasing risk levels. These gases are routinely formed during combustion or gasification of coal, oil residues, and, depending on the level of CO_2 purification, can be present in CO_2 storage reservoirs in trace amounts or high concentrations (e.g., in the Weyburn project, injected CO_2 gas contains 2% H_2S). H_2S is a much more toxic gas than CO_2, so the level of tolerance to a leakage of H_2S-containing gas would be much lower compared to pure CO_2 due to a much higher risk of poisoning. Furthermore, SO_2 upon dissolution in groundwater would produce a much stronger acid compared to CO_2, resulting in a much higher concentration of toxic metal ions and minerals in groundwater, thus, potentially increasing the risk of exposure to hazardous compounds. There is insufficient information in the literature on the detailed quantitative assessment of risk levels related to the presence of these gaseous impurities in CO_2 storage sites.

Detrimental releases of CO_2 from deep subsurface caused by human activities are also well documented. For example, in Dixie Valley (Nevada), the flux of 570 g m^{-2} day^{-1} CO_2 originating from a 3 km deep geothermal source was measured (this is two orders of magnitude higher than the background level of CO_2 seepage) (Bergfeld et al. 2001). This caused a massive die-off of vegetation above the geothermal field. It is rather difficult to correlate the risks related to natural seepage phenomena with that from CO_2 leakage from storage reservoirs mostly due to differences in the scale of CO_2 fluxes (which is higher in the case of a natural seepage).

14.10.1.3 Induced Seismic Activity

The injection of CO_2 into deep subsurface formations (e.g., porous rock) at significant pressures (exceeding formation pressures) can potentially induce fracturing and fault activation (i.e., movement along faults). This may result in two detrimental consequences: (1) enhanced fracture permeability (enhancing undesirable CO_2 migration from a storage site to the surface) and (2) induced earthquakes (potentially, large enough to cause a damage). Although no direct evidences exist to point to the possibility of induced seismic activity related to CO_2 storage projects, it has been suggested that the deep-well injection of waste fluids were, in all likelihood, responsible for local earthquakes of moderate magnitudes (M_L) in a number of locations in United States, for example, in Denver (1967) with M_L of 5.3 and Ohio (1986, 1987) with M_L of 4.9 (IPCC CCS 2005).

The probability of an induced seismic activity is dependent on the extent of pore-fluid pressure increase at the epicenter of the seismic event and the magnitude of pore-fluid perturbations (thus, the quantity and rate of injection are more prevalent parameters than the type of fluid injected). Currently, besides EOR projects, there are a number of other projects involving an underground fluid injection, for example, the injection of brines for oil and gas recovery (>2 Gt year^{-1}), wastewater (>0.5 Gt year^{-1}), hazardous wastes (>30 Mt year^{-1}), but no significant seismic effects have been attributed to these projects (IPCC CCS 2005). However, most of these projects do not exactly represent the operational conditions (most importantly, pressure) under which large-scale CO_2 storage will be carried out. In particular, formation pressures in CO_2 storage reservoirs may exceed those found in existing CO_2-EOR projects; thus, more of large-scale and long-term experiments need to be conducted and more practical experience to be gained in order to assess the risks of induced seismicity.

14.10.2 Potential Environmental Impact of Ocean CO₂ Storage

There is limited knowledge of the possible environmental impact of CO_2 storage on the ocean ecosystem. Regardless of the techniques used to inject CO_2 in the ocean, a large-scale deployment of CO_2 storage projects would result in the production of immense volumes of seawater with an increased CO_2 concentration and, consequently, altered chemical balance. Conducting various simulation studies (using mathematical models) and making an extrapolation from small-scale field experiments, although very useful, are unlikely to predict all the ecological consequences of the worldwide implementation of ocean CO_2 storage. Although it is expected that CO_2 impact on the ocean ecosystem would increase with an increase in CO_2 levels in seawater, no threshold CO_2 concentrations have been determined. It is also unclear how ocean biota would adapt to sustained elevated CO_2 levels. Due to evolutionary selection (that eliminated species enduring ecological perturbation), deep-sea organisms are most likely more sensitive to environmental disturbance compared to shallow water relatives (IPCC CCS 2005).

As discussed above, the introduction of CO_2 into the ocean (regardless of the method used) would result in an increase in CO_2 concentration in seawater, drop in its pH, and decrease in CO_3^{2-} concentration. This would cause carbonates (e.g., $CaCO_3$) to dissolve producing bicarbonate (HCO_3^-) from carbonate (CO_3^{2-}) ions. The extent of pH decrease is a function of seawater volume to dilute the given quantity of CO_2 or the amount of CO_2 released into the ocean. Over the course of evolution, the ocean marine life was adopted to relatively low CO_2 concentrations, and its sensitivity to higher CO_2 concentration (either on a temporary or on a permanent basis) is unknown. CO_2 impact on marine life has primarily been studied in fish and invertebrates over the wide depth ranges; marine invertebrates appear to be more sensitive to the increased CO_2 levels than fish (Langenbuch and Portner 2004). However, no sufficiently long-duration studies (e.g., for intervals exceeding the duration of a reproduction cycle) have been conducted, so the long-term effects on the ecosystem could be easily overlooked.

To make the matters even more complex, the changes in seawater acidity (pH), CO_2, carbonate-ion, and bicarbonate-ion concentrations may have different and specific effects on marine life, for example, the impact of CO_2 accumulation could be more severe than the effects of pH reduction or an increase in carbonate-ion concentration. Furthermore, the changes in pH caused by CO_2 dissolution may affect marine organisms through mechanisms that do not directly relate to CO_2 presence. For example, decreasing pH value may affect many metabolic functions since enzymes and ion transporters efficiently function only over a narrow pH range. It was found that pH changes in the marine ecosystem affect nitrification and speciation of nutrients such as phosphate, silicate, and ammonia as well as speciation and uptake of essential trace elements (Huesemann et al. 2002).

Numerous studies indicated that the permanent exposure of marine animals with calcareous exoskeletons to elevated levels of CO_2 could lead to decalcification (i.e., dissolution of their skeleton) (Feely et al. 2004). Laboratory and filed studies indicated that the calcification rates in corals were reduced by 15%–85% with doubling of CO_2 concentration against the preindustrial level (to 560 ppm). Similar increase in CO_2 levels to 560 ppm caused reduction in growth rate and even survival of shelled marine animals (IPCC CCS 2005). These observations indicate that due to the increase in atmospheric CO_2 levels, the calcification rates in the world oceans may decrease by 50% over the next century, and this could cause a considerable shift in global biochemical cycles (Zondervan et al. 2001).

Long-term effects of CO_2 on marine life have been identified in cells and tissues as well as individual animal species and whole ecosystem levels. Different species may have different sensitivities toward elevated levels of CO_2, which would make defining a common threshold beyond which CO_2 cannot be tolerated very difficult. Some species may be able to withstand temporary or transient CO_2 level increases, but would not survive in an environment with permanent high CO_2 concentrations.

One often-overlooked issue related to the environmental impact of the ocean CO_2 storage and associated risks is concerned with harmful contaminants in CO_2. Although there are already some limitations in place for the permissible levels of impurities in CO_2 streams, this problem may arise if ocean disposal of CO_2 will be realized on a commercial scale. A typical impurity in CO_2 streams is H_2S. Although there are many examples of marine life adapted to the presence of high concentrations of naturally occurring H_2S and other sulfurous compounds (e.g., in Black Sea, the Cariaco Trench), the presence of H_2S in CO_2 would have a negative impact on marine organisms by locally reducing oxygen levels and, thus, affecting their respiration system.

Risk management issues related to intentional ocean carbon storage have not been extensively discussed in the literature. It is assumed that catastrophic degassing of CO_2 from the ocean floor similar to one occurred at Lake Nyos (Cameroon) is unlikely for the reasons discussed below. In 1986, about 2 million tons of CO_2 produced by volcanic activity and dissolved in Lake Nyos were suddenly released (apparently triggered by a landslide), causing death of more than 1700 people and 3500 livestock in surrounding villages (Kling et al. 1994). Such a rapid degassing was attributed to the spontaneous conversion of dissolved CO_2 into CO_2 bubbles that rose rapidly and dragged with them surrounding water producing even more bubbles. It is argued that most CO_2 injection techniques release it at the depths deeper than 500 m where bubble formation is not possible. However, if for some (currently, unknown) reason, large volumes of liquid or dissolved CO_2 were suddenly transported from the ocean depth above the gas–liquid phase boundary, there exists a possibility of the self-accelerating CO_2 degassing at the surface (IPCC CCS 2005).

14.10.3 Environmental Impact of Mineral Carbonation

The environmental impact of the mineral carbonation technology is directly associated with the issues related to large-scale mining, mineral processing, and the disposal of carbonates and solid waste products. These operations might lead to potential problems related to soil, water, and air pollution, land clearing, displacement of millions tons of soil and rock, leaching of metals, and others (IPCC CCS 2005). In particular, mining activities such as blasting, drilling, and earth moving would create dust and aerosol matter that may affect respiration and pollute local vegetation. Due to weak acidity of by-products, there is a possibility of leaching of metals and the contamination of groundwater. To minimize the negative impact of water and soil contamination, mine reclamation programs have to be implemented. On a positive side, mineral carbonation provides an almost 100% "leak-proof" method for long-term CO_2 storage, since there would be practically no CO_2 leakage from the disposal sites after CO_2 is chemically bound into mineral carbonates. This would practically eliminate the need for long-term storage site monitoring, which is an important issue with the geological and ocean options of CO_2 disposal.

14.10.4 Ecological Impact of Carbon Sequestration in Biosphere

One of the undesirable ecological consequences of carbon sequestration in biosphere relates to the increase in organic matter in wetland soil that may result in higher emissions of methane, which is about 20 times more potent GHG than CO_2. The drastic increase in grasslands (at the expense of croplands) may increase the emissions of nitrous oxide (N_2O)—another potent GHG (Marland et al. 1998). Furthermore, the land use change and sequestration actions could potentially alter the flow of nutrients (e.g., the changes in the fluxes of phosphorus- and nitrogen-based nutrients in aquatic systems in response to erosion control measures). An enhancement in carbon sequestration in deserts by growing drought-tolerant plants could result in the decrease in fluxes of iron-based wind-blown nutrients with a possible adverse effect on the ability of ocean to sequester carbon through iron-fertilized phytoplankton (U.S. DOE, Carbon sequestration. State of the Science. 1999).

14.10.4.1 Ocean Fertilization

Potential ecological consequences of large-scale ocean fertilization on the biosphere and biogeochemical cycling are not known. Such consequences could range from the changes in species diversity to the induction of anoxia and significant adverse effects on biocommunity structure function (U.S. DOE, Carbon sequestration. State of the Science 1999). It is not clear how sustained fertilization would affect the balance and structure of the ocean ecosystem, fluxes of GHGs, and transfer of carbon to the deep ocean. The fundamental question that should be addressed is whether any changes in the ocean ecosystem are justified relative to the benefits to society.

The knowledge gaps with regard to ocean fertilization as a strategy for carbon sequestration can be summarized as follows:

- The impact of long-term ocean fertilization on the structure and function of marine ecosystems is unknown (e.g., the changes in phytoplankton structure could have an impact through the food chain on fisheries). There is a possibility that fertilization with iron and phosphorus could lead to growth of toxin-producing cyanobacteria over other types of phytoplankton (such impact was observed in lake ecosystems).

- The impact of sustained fertilization on the natural biogeochemical cycles in the ocean is completely unknown (the biogeochemical cycles of carbon, nitrogen, phosphorus, silicon, and sulfur are intertwined, and the perturbation of one elemental cycle can adversely affect other elemental cycles to the extent that cannot be predicted).

- The ocean fertilization could potentially lead to eutrophication causing oxygen depletion, which could kill many species requiring oxygen (this may result in enhanced production of methane that is more potent GHG than CO_2).

- It is not well understood what kind of plankton will bloom after fertilization, so there is a possibility that a harmful algal bloom may occur causing poisoning of lagoons and other coastal ecosystems.

- Iron infusions may unpredictably alter surface ecosystems disturbing food chains and adversely impacting fisheries and whale populations.

- Currently, there is no good understanding of the potential effectiveness of ocean fertilization at a global scale. For example, it is not clear whether enhanced fixation of carbon in the surface ocean layers would result in an increase in the amount of carbon sequestered in the deep ocean. The existing models are based on simplified assumptions and have not been validated against real-world data (U.S. DOE, Carbon sequestration. State of the Science 1999).

The knowledge of the possible consequences of ocean fertilization is critical to the responsible use of oceans as a carbon sequestration option. The ocean plays an essential role in sustaining the Earth's biosphere, thus any alterations in the ocean ecosystem must be taken with an extreme caution.

14.10.5 Public Acceptance of Risks

Although the issues related to climate change and its implications are a hot topic of discussion nowadays, in general, there is an insufficient public knowledge of various carbon mitigation options and their potential environmental impact and practical ways of their realization. There are several studies with regard to the public perceptions and acceptance of risks related to CCS. For example, a survey conducted in United States indicated that on the scale from 1 (very negative) to 7 (very positive), the respondents rated ocean and geological storage 3.2 and 3.5, respectively (Palmgren et al. 2004). Interestingly, after receiving additional more detailed information on the CCS options, the respondents changed their ratings shifting to more negative side of the scale, that is, 2.4 for ocean and 3.0 for geological storage. In a similar survey in Japan, on a scale from 1 (negative) to 5 (positive), the respondents rated the CCS options as follows: dilution and lake-type ocean storage at 2.24 and 2.47, respectively, and onshore and offshore geological storage at 2.57 and 2.75, respectively (Itoaka et al. 2004). After receiving additional information, the ratings slightly increased for all categories (by about 0.1–0.3 points). In general, the public perception of ocean storage is more negative compared to geological storage. In particular, a considerable and widespread public opposition has been developed against the proposed CO_2 release in the Pacific Ocean.

One should not also underestimate the powerful NIMBY (not-in-my-backyard) factor, which resurfaces every time when local interests collide with a broader public purpose. There are many examples of the NIMBY factor intervening and sometimes killing some important initiatives with huge potential benefits to society based on the sentiments like this: "we strongly support clean alternative energy sources, but, please, put those wind turbines somewhere else." In 1987, U.S. Congress established the Yucca Mountain nuclear waster repository in Nevada, and billions of dollars have been spent since then to build and equip the facility, but even a single barrel of nuclear waste has not been received by the repository yet (Effron 2010). The latest example of the NYMBY factor as applied to the CCS technology relates to a 200 MW coal-fired power plant retrofitted with oxyfuel combustion and CO_2 capture technologies. Originally, CO_2 sequestration site was planned in nearby Mattoon (Illinois); however, the officials in Mattoon have announced that they are not interested in having the sequestration site in that location (Johnson 2010).

14.10.6 Legal Issues

In principle, the legal issues related to geological and ocean CO_2 storage are governed by general principles of an international law, according to which states exercise their sovereignty in their territories and have the responsibility to not cause damage to the environment of other states. A number of global and regional environmental treaties such as those related to climate change, the law of the sea and marine environment can, in principle, handle the issues related to ocean and offshore geological storage. It should be emphasized, however, that those treaties were not designed for CO_2 storage, but rather to prohibit a marine dumping. But there are some "gray" areas with regard to whether CO_2 storage constitutes dumping, or whether CO_2 containing H_2S and other impurities are considered industrial or even hazardous waste. Of particular ambiguity are legal issues related to ocean storage. It was indicated that while the U.N. Framework Convention of Climate Change encourages ocean storage of CO_2, the UNCLOS is ambiguous with regard to the use of the ocean as a reservoir for CO_2 storage (IPCC CCS 2005). In particular, according to one of the UNCLOS' provisions on protecting and preserving marine environment, CO_2 may be considered "a pollutant" (because of potential CO_2 impact on marine biota). The OSPAR Convention (an acronym for the Oslo and Paris Convention), which relates to the North Atlantic region was amended in 2007 to specifically prohibit ocean storage of CO_2 using a water column method. It stipulates that in order for storage to be allowed by the water column method, amendments to UNCLOS would be required including granting of authority to the International Seabed Authority (ISA) to regulate and set standards for ocean storage-related CCS activities (Global CCS Institute 2009).

14.10.7 Knowledge Gaps Associated with Geological and Ocean CO$_2$ Storage

The following are uncertainties and gaps in knowledge regarding geological and ocean CO$_2$ storage that contribute to the risk factors that have to be taken into consideration during a broad-scale implementation of these CCS options (IPCC CCS 2005):

1. Although existing projects evidence that geological CO$_2$ storage have a potential to be a long-term storage option, there is a lack of quantitative data on potential CO$_2$ leakage rates from different types of storage sites. There is a need for reliable hydrogeological–geochemical–geomechanical simulation models that can accurately predict long-term storage performance.

2. It is realized that current capacity estimates are imperfect: there are many gaps in capacity estimates at the global, regional, and local levels. In particular, there is a lack of reliable information on CO$_2$ storage capacity in the areas that are likely to experience the greatest growth in energy consumption, such as China, India, Southeast Asia, Middle East, Brazil, eastern Europe (including Russia), South Africa, and others.

3. There is a need for more studies on the mechanism and kinetics of geochemical trapping of CO$_2$ and the long-term impact of CO$_2$ trapping on reservoir fluids, minerals, and rocks.

4. There are a number of CO$_2$ leakage scenarios that have to be carefully assessed, for example, the risks of a leakage from abandoned wells caused by material and cement degradation, the spatial distribution of leaks that might arise from inadequate storage sites, possibility of subaquatic CO$_2$ seepage, and others. This would require the development of novel methods for monitoring, detection, remote sensing, and quantification of different forms of CO$_2$ leakage (especially dispersed leaks) and fracture detection. There is no track record of remediation measures for leaked CO$_2$. The methods for quantitative assessment of risks to human health and local environment from CO$_2$ leakage have to be developed.

5. The impact of biological (i.e., microbial) processes on CO$_2$ storage reservoirs in the deep subsurface are virtually unknown and have to be elucidated.

6. The response of biological systems in deep sea and marine seafloor to added CO$_2$ is unknown.

7. Very little is known about the long-term effect of corrosive environment (seawater, low pH) on the equipment operating in deep sea, for example, pipes, nozzles, and diffusers.

8. Reliable techniques, sensors, and equipments to detect CO$_2$ plumes and other spots with high CO$_2$ concentrations and their biological and geochemical impacts are yet to be developed.

9. Methodologies to estimate physical leakage from storage reservoirs and emission factors (including fugitive emissions) to determine emissions from capture, transport, and injection processes are not available.

10. Both geological and ocean storage options suffer from (1) uncertainties related to the permanence of the stored CO$_2$; (2) the lack of protocols on transboundary CO$_2$ transport and storage; and (3) the lack of accounting rules for CCS, long-term monitoring, timely detection, and liability/responsibility.

11. There is a lack of methodologies for estimating and dealing with the CO_2 leakage resulting from system failures due to seismic activities, pipeline disruptions, accidents, and other unexpected circumstances.

The long-term operation of existing commercial CO_2 storage projects and further development and field demonstrations in this area would decrease the uncertainties and knowledge gaps allowing to make a more informed decision on the prospects of geological and ocean storage as carbon mitigation options.

14.11 Alternative Routes to Fossil Fuel Decarbonization

14.11.1 Direct Fossil Decarbonization with Production of Solid Carbon

Due to long-term environmental uncertainties of the CO_2 sequestration approach, there have been proposals to decarbonize fossil fuels by sequestering solid carbon instead of gaseous CO_2 (Muradov 1993, 2001a; Gaudernack and Lynum 1996; Steinberg 1999; Muradov and Veziroğlu 2005). This alternative decarbonization approach is based on the dissociation of hydrocarbons to hydrogen and elemental carbon and is referred to in the literature as "direct decarbonization" (see also Section 14.2.5). Since NG is the most preferred feedstock for the present-day and, in all likelihood, near-future production of hydrogen, our discussion of this decarbonization strategy will be limited to NG (or methane) (although the concept is applicable to any gaseous or even liquid hydrocarbon fuel).

Direct decarbonization of methane is accomplished via a simple methane dissociation reaction:

$$CH_4 \rightarrow C + 2H_2 \quad \Delta H° = 75.6 \, kJ \, mol^{-1} \tag{14.37}$$

This reaction is much less endothermic than SMR: the energy requirement per mole of hydrogen produced is $37.8 \, kJ \, mol^{-1} \, H_2$ compared with $63.3 \, kJ \, mol^{-1} \, H_2$ for SMR. Although the methane decomposition reaction seems to be in a disadvantage to SMR (Reaction 14.30), that produces twice as much hydrogen per mole of CH_4, if one takes into account the amount of NG to fuel the strongly endothermic SMR process (i.e., almost a third of the total amount of NG) and the energy losses due to CO_2 sequestration, the overall energy efficiencies for the SMR–CCS and methane decomposition processes, according to Steinberg (1999), become comparable. One should also consider that in the latter option, the chemical energy of carbon product is not lost, but is safely stored for the possible future use (if ecological situation would permit). Figure 14.46 provides a simplified representation of the direct decarbonization concept.

The major problem with the methane dissociation reaction is that due to very strong C–H bonds ($E_{dis.} = 436 \, kJ$) and a lack of polarity, methane is one of the most stable organic molecules. Thus, reaction (14.37) would require an energy input in the form of either very high temperature heat (typically above 1200°C) or electrical discharge (or plasma) energy. There are several technological options for methane dissociation to hydrogen and carbon, which are summarized in Figure 14.47.

It should be noted that thermal decomposition of NG has been practiced since 1930s until recently for production of carbon black with hydrogen, a by-product, being a

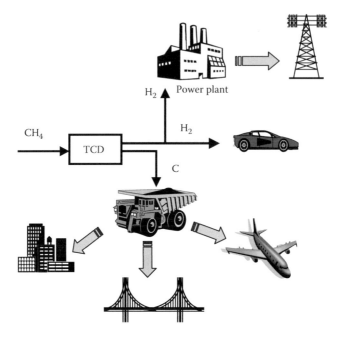

FIGURE 14.46
Schematic diagram of direct decarbonization concept. TCD, Thermocatalytic decomposition of hydrocarbons. (Modified from Muradov, N. and Veziroğlu, N., *Int. J. Hydrogen Energy*, 30, 225, 2005.)

supplementary fuel for the process (Thermal Black process). The process has been operated in a semicontinuous (cyclic) mode using two tandem decomposition–regeneration reactors at high operational temperatures (up to 1400°C). Kværner company of Norway has developed and operated on a limited commercial scale a thermal plasma process for decomposition of methane and other hydrocarbon feedstocks to hydrogen and carbon black (Lynum et al. 1998). Although technologically simple, the process is energy intensive: it was estimated that up to 1.9 kWh of electricity is consumed per normal cubic meter of hydrogen produced (Fulcheri and Schwob 1995).

The methane dissociation process lends itself to the use of catalysts in order to facilitate its kinetics and reduce maximum temperature of the reaction. Two types of catalysts have been developed: metal- and carbon-based catalysts for the use in thermocatalytic

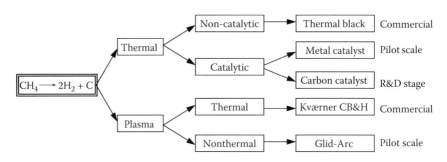

FIGURE 14.47
Major technological routes to decomposition of methane to hydrogen and carbon. (From Muradov, N. and Veziroğlu, N., *Int. J. Hydrogen Energy*, 33, 6804, 2008. With permission.)

decomposition (TCD) of methane. Metal catalysts such as Ni-, Fe-, and Co-based catalysts have been most commonly used for the methane TCD process (Callahan 1974; Poirier and Sapundzhiev 1998). However, there is a catalyst deactivation problem associated with carbon build up on the catalyst surface. In some processes, for example, Universal Oil Product's HYPRO process, catalyst was regenerated by combustion of carbon (which also provided a heat input to the process), resulting in considerable CO_2 emissions (CO_2 by-product, apparently, was not of a concern for the developers of the technology). Another serious problem arising from the oxidative regeneration of metal catalysts relates to an unavoidable contamination of hydrogen with carbon oxides, which would require an additional purification step (e.g., via methanation). It should be noted that at optimized operational conditions, high-value filamentous carbon (or multiwall carbon nanotubes, MWCN) could also be produced as a product of the metal-catalyzed methane decomposition process; Ni-, Co-, and Fe-based catalysts are most commonly used for this application. The prospect of CO_2-free production of two valuable products H_2 and MWCN currently drives very intensive research efforts in this area (Li et al. 2004; Mordkovich et al. 2007).

The use of carbon-based catalysts offers certain advantages over metal catalysts due to their durability and low cost. In contrast to metal-based catalysts, carbon catalysts are sulfur- and temperature resistant and would not require catalyst–carbon separation stage. Muradov (2001b) and Muradov et al. (2005) screened a variety of carbon materials and demonstrated that efficient catalytic methane decomposition can be accomplished over high-surface area disordered carbons at temperatures typical of SMR process (800°C–900°C). XRD studies of carbon catalysts indicated that after their exposure to hydrocarbons, the ordering in the "columnar" or stacking direction has evolved. Thus, carbons produced by decomposition of methane or other light hydrocarbons feature more ordered structure compared to amorphous carbons, but they are less structurally ordered than graphite (which is characteristic of the so-called *turbostratic* carbon). Technological aspects of hydrogen and carbon production via carbon-catalyzed decomposition of NG are discussed elsewhere (Muradov and Veziroğlu 2005). Recently, a series of novel carbon-based catalysts with improved stability has been reported (Serrano et al. 2008; Suelves et al. 2008; Muradov et al. 2009).

Techno-economic evaluation of the TCD of NG process indicates that the hydrogen production cost is a function of a carbon selling price (Muradov and Veziroğlu 2008). In particular, TCD process becomes competitive with SMR (without CCS) at a carbon-selling price of about \$350 t^{-1} carbon. Carbon produced in the carbon-catalyzed methane decomposition process is a sulfur- and ash-free product that could be marketed at even greater selling price. For example, as a high-quality substitute for petroleum coke, it could potentially be sold for \$310–\$460 t^{-1} (for manufacturing of electrodes in aluminum and ferro-alloy industries) (Gaudernack and Lynum 1996). Other forms of carbon (e.g., pyrolytic carbon, carbon black, carbon nanofibers) currently are of much greater value. It should be noted that the TCD process might become competitive with the SMR even without carbon credit if the cost CO_2 sequestration (about 30% of the H_2 product cost) is added to the cost of hydrogen manufactured by SMR (alternatively, imposing carbon tax on CO_2 emissions from the SMR process may also improve the competitiveness of TCD against the SMR process).

It should be emphasized that carbon credit could reduce the cost of the TCD hydrogen only if sufficiently large markets for the TCD-generated carbon products would be found. To put the numbers in a perspective, if 100 million tons of hydrogen were generated via the TCD process, close to 300 million tons of by-product carbon would be produced. However, annual worldwide consumption of all carbon products amounts to only 15–20 million tons, and it is unlikely that there will be any dramatic increase in carbon use in the traditional carbon utilization areas in the near future. Thus, for the TCD concept to become practically viable, very large new markets for carbon products would be necessary (Muradov and Veziroğlu 2005).

Building and construction materials area can potentially absorb immense amount of carbon products. There are plenty of examples where traditional construction materials are being substituted for advanced carbon-based materials, for example, carbon–carbon composites, manufactured graphites, etc. These materials are common in construction, aerospace, and automotive industries. Carbon fiber composites are used for specific automotive parts and components where the weight and strength properties outweigh cost considerations. The next generation of commercial airliners under development by Boeing and Airbus is expected to extensively use carbon fiber-based composites. Advancements in fabrication methods and cost reductions for carbon fiber composites in these areas may have implications for much broader applications of these materials. The advantages of using carbon composites over traditional materials (e.g., steel) are that they do not corrode, are five times lighter than steel, and can be installed without the use of a heavy construction equipment. The perspectives of using solid carbon products generated by decomposition of hydrocarbon feedstocks as structural materials capable of replacing steel and concrete were discussed by Muradov (2001a), Muradov and Veziroğlu (2005; 2008), and Halloran (2008).

14.11.2 Use of Carbon-Free Energy Sources for Fossil Decarbonization

It has been recognized that the use of alternative carbon-free (i.e., nonfossil) energy sources as a means of energy input to the precombustion type CCS processes would result in a substantial conservation of fossil resources and dramatic reduction in the amount of CO_2 by-product that would be required to store. From this viewpoint, the possibility of using high-temperature nuclear and solar heat sources has long attracted the interest of researchers.

14.11.2.1 Use of Nuclear Heat Input

A number of early studies on the technological use of nuclear heat pointed out that the SMR process has the greatest potential for a near-term integration into a nuclear process heat system (Jiacoletti 1977). According to a study conducted by General Atomic Co. researchers, the efficiency of a reformer heated by a high-temperature nuclear reactor is considerably higher than that of a conventional reformer (85% vs. 74%) (Peterma et al. 1975). In high-temperature gas-cooled nuclear reactors (HTGR), recycled helium is heated to temperatures up to 950°C, which is suitable for carrying out SMR process. Hot helium is circulated in indirectly heated heat exchangers countercurrent to methane and steam flowing through reformer tubes, releasing its sensible heat and being cooled from 950°C to 600°C. The preferred reformer tube design has an inner helical tube through which reformed gas is discharged to heat a catalyst-filled tube. Thermal energy is supplied to a helium heat carrier in the core of a high-temperature nuclear reactor. Such reactors, termed EVA reactors, have been under testing and pilot-scale operation since 1971 and are considered suitable for commercial syngas production. Since about one-third of the overall amount of NG used in SMR process is utilized as process fuel (to cover the process endothermicity and generate steam), the coupling of a reformer with a nuclear source would result in a substantial (about 30%) reduction in CO_2 by-product yield and proportionate savings of NG feed.

Thermal (catalytic) decomposition of methane (or NG) is another high-temperature (600°C–1000°C) endothermic process suitable for coupling with a nuclear source. The thermal energy for the process is typically provided by combusting some portion (about

15%) of methane. Thus, the process would greatly benefit from coupling with high-temperature nuclear reactors. Serban et al. proposed the concept of utilizing "waste" heat from Generation IV high-temperature nuclear reactors to produce hydrogen and carbon from methane or NG by direct contact pyrolysis (Serban et al. 2002). Certain types of the Generation IV nuclear reactors use heavy liquid metal coolant (Pb–Sn). According to the concept, high-temperature heat carrier transfers heat from the nuclear reactor to a thermal cracking unit. NG was bubbled through a bed of low-melting-point metals (Pb–Sn) at temperatures ranging from 600°C to 900°C. The main advantage of the proposed system is the ease of separation of the generated carbon by-product from a heat transfer media (by buoyancy, due to difference in densities). The conversion, though, was rather low due to poor catalytic activity of the melt (at 750°C methane conversion was about 9%). Two types of carbon were produced: soot and pyrocarbon. These experiments lay the groundwork for developing a technical expertise in producing pure H_2 cost effectively by utilizing heat energy contained in a liquid metal coolant in the Generation IV nuclear reactors. The state-of-the-art systems for the technological use of high-temperature nuclear reactors in the production of carbon-neutral fuels are presented in Chapter 3.

14.11.2.2 Solar Heat Input

Similar to a nuclear source, the use of concentrated solar radiation can potentially provide all the necessary energy input to endothermic CCS-related processes. Advantageously, solar concentrators can achieve much higher temperatures than nuclear reactors. The state-of-the-art solar concentrators can provide solar flux concentrations in the following ranges, depending on the type of a solar concentrator (Steinfeld and Meier 2004):

Trough concentrators: 30–100 suns

Tower systems: 500–5000 suns

Dish systems: 1000–10,000 suns

For a solar concentration ratio of 5000, optimum temperature of a solar receiver is about 1270°C, giving the maximum theoretical efficiency of 75% (i.e., the portion of solar energy that can be converted to chemical energy of fuels). This temperature is adequate to conduct high-temperature endothermic SMR or methane–CO_2 reforming (or "dry"-reforming) processes. Solar chemical reactors for highly concentrated solar systems typically utilize a cavity receiver-type configuration, that is, a well-insulated enclosure with a small opening (an aperture) to let in concentrated solar radiation.

Solar reforming of methane has been extensively studied in solar furnaces as well as in solar simulators using different reactor configurations and catalysts, for example, Tamme et al. (2001) and Berman et al. (2006). In his review paper, Steinfeld discusses an indirectly irradiated solar reforming reactor consisting of a pentagonal cavity receiver insulated with ceramic fibers and containing a set of Inconel tubes (the solar reactor was developed at the Weizmann Institute of Science, Israel). The solar reformer tubes were filled with Rh (2%)/ Al_2O_3 catalyst (Steinfeld and Meier 2004). Berman et al. (2006) reported the experimental results on the development of high-temperature steam reforming catalyst for "DIAPR-Kippod" volumetric-type reformer. The absorber in this reformer consisted of an array of ceramic pins. The authors have developed and tested alumina-supported Ru catalysts promoted with La and Mn oxides. The activity of catalysts in SMR and CO_2-methane reforming reactions was measured in the temperature range of 500°C–1100°C. Catalysts showed stable operation at 1100°C for 100 h.

In one study, the reforming reaction was conducted under direct illumination of catalyst by concentrated light in a reactor-receiver with a transparent wall (Aristov et al. 1997). In the reformer, both the specific rate of H_2 production and the specific power loading of the solar energy conversion device appeared to increase considerably compared to a conventional externally heated stainless steel reactor, reaching $130\,N\,dm^3\,h^{-1}\,g^{-1}$ of catalyst and 50–$100\,W$ cm^{-3}, respectively. It was proposed that the increase in the reaction rate is caused by a significant intensification of energy input into the catalyst bed due to absorption of light directly by the catalyst granules. Yokota et al. (2000) reported steam reforming of methane over Ni/Al_2O_3 catalyst using a solar simulator (Xe lamp). The reaction was conducted at $H_2O/CH_4 = 1/1$ ratio at the range of temperatures $650°C$–$950°C$. At $850°C$ and molar ratio of $H_2O/CH_4 = 1/1$, methane conversion was in excess of 85% under atmospheric pressure.

The use of solar source is especially advantageous for methane decomposition reaction because it could provide very high temperatures (up to $2000°C$) that could make the use of catalysts unnecessary (thus, avoiding catalyst deactivation and other related problems). Hirsch et al. discussed the perspectives of solar thermal decarbonization of NG (Hirsch et al. 2001). The authors studied methane decomposition into carbon and H_2 using concentrated solar energy as a model reaction for conducting the Second Law analysis of the solar decarbonization process, in which carbon is removed from fossil fuels prior to their use for power generation. It was determined that the theoretical maximum exergy efficiency of the system could be 35% for a blackbody solar reactor operating at $1223°C$ under a mean solar flux concentration ratio of 1000. The analysis indicated that power generation energy efficiency could exceed 65%, and the energy penalty for avoiding CO_2 emissions amounts to 30% of the electrical output.

Steinfeld et al. (1997) demonstrated the technical feasibility of solar decomposition of methane using a reactor with fluidized bed of catalyst particulates. The experiment was conducted at the Paul Scherrer Institute (PSI, Switzerland) solar furnace delivering up to $15\,kW$ with a peak concentration ratio of 3500 suns. A quartz reactor (diameter $2\,cm$) with a fluidized bed of Ni (90%)/Al_2O_3 catalyst and alumina grains was positioned in the focus of the solar furnace. The direct irradiation of the catalyst provided effective heat transfer to the reaction zone. The temperature was maintained below $577°C$ to prevent rapid deactivation of catalyst. The outlet gas composition corresponded to 40% conversion of methane to H_2 in a single pass. In the study reported by Meier et al. (1999), concentrated solar radiation was used as a source of high-temperature process heat for the production of hydrogen and filamentous carbon via decomposition of gaseous hydrocarbons CH_4 and C_4H_{10} in the presence of catalyst particles. Depending on the nature of the catalyst, two types of filamentous carbon were produced in the solar experiments: nanotubes were formed using Co/MgO catalyst and nanofibers were formed over Ni/Al_2O_3. The conversion of CH_4 to H_2 was approximately 30% as determined from the outlet gas composition during the solar experiments.

Kogan et al. (2005) investigated the production of hydrogen and carbon by solar decomposition of methane using a volumetric solar receiver. The authors demonstrated that the volumetric absorption of radiation in a solar reactor could be achieved by seeding gas in the reaction chamber with cloud of fine carbon particulates (carbon black). The experiments showed that loading of $0.1\,g$ of carbon black particles per cubic meter of irradiated gas would be sufficient for the attainment of adequate radiation absorption. Solar thermal dissociation of methane in a fluid-wall aerosol flow reactor was also reported by Dahl et al. (2004). A fluid-wall aerosol flow reactor was heated to temperatures in the excess of $1700°C$ by concentrated sunlight, which resulted in thermal decomposition of methane to carbon black particles and hydrogen-rich gas. It was shown that at the reactor wall temperature of $1860°C$ and an average residence time of $0.01\,s$, about 90% conversion of methane to products could be achieved. The carbon black product consisted of particles with the sizes in the range of 20–40 nm. A nozzle-type solar reactor for thermal methane splitting has been reported by Abanades and Flamant (2005). In the tested reactor, a graphite nozzle absorbed concentrated solar radiation provided by a solar furnace, heat was then transferred to reagents; carbon black product was recovered in a carbon trap. The researchers achieved methane-to-H_2 conversion in the range of 30%–90% using a 2 m diameter solar concentrator.

14.12 Conclusions

With energy demand climbing and the global ecological situation deteriorating, a variety of measures are being proposed to minimize potentially devastating climate-changing consequences of carbon-based economy. Two main pathways to curbing carbon emissions are currently being actively pursued: (1) *CO_2 reduction* through a variety of energy conservation measures and improvements in vehicles' mileage and power generation efficiency, and (2) *CO_2 rejection* through the use of renewable energy sources, carbon-neutral fuels, and the technologies that capture and safely store CO_2. Many experts believe that the last option can provide a potentially cost-effective near-term solution to the fossil fuel-related environmental problems. The CCS approach offers a promise of a reasonable compromise: the continuing use of fossil fuels, albeit at higher processing cost, but with the significantly reduced rate of CO_2 emissions (in average by 90%). It is acknowledged that as long as fossil fuels are used, CCS will be required to decouple economic growth from unacceptable levels of atmospheric CO_2 emissions. However, the global implementation of CO_2 capture and sequestration technologies faces three major challenges:

1. A significant reduction in its cost
2. Understanding the reservoir options, for example, its permanence, storage retention time, tolerable leakage rates, etc.
3. The assessment of key risk factors associated with the long-term ecological consequences of this approach (a potential environmental impact, possible health hazard, public acceptance, legal issues, etc.)

Deep-ocean CO_2 sequestration is actively debated in the literature, and it has many opponents because of its possible adverse impact on aquatic environment (i.e., ocean acidification, an effect on marine life and ecobalance, etc.), and in the long-term outlook, it still would not provide a permanent storage solution. The geological disposal of CO_2 seems to be a less expensive option than ocean sequestration, and currently, it is looked upon more favorably compared to ocean storage, but it has its own share of problems, for example, possible leakage, leaching of harmful trace elements in freshwater aquifers, etc. Although CO_2 injection into geological formations has been practiced for several decades by petroleum industry for EOR, it is not yet possible to predict with confidence the storage volumes, formation integrity, and storage permanence over long time periods (e.g., centuries or millennia). Another often overlooked fact is that CO_2 capture is technically feasible only for centralized sources such as power plants, large industrial point sources, etc. But those sources account for only about half of the total CO_2 emissions to the atmosphere, and another half comes from a myriad of small dispersed sources of CO_2 (e.g., vehicles, domestic heat sources) that would make CO_2 capture and disposal technically and economically prohibitive.

Despite the above challenges, CCS is an active area of research, development, and commercialization (especially, in conjunction with the beneficial reuse of CO_2). As of 2009, there were 275 CCS projects around the world, of which 213 are active or planned projects (among them, 101 projects are of a commercial scale). Some large-scale commercial projects, for example, Statoil's Sleipner project off the cost of Norway in the North Sea, have a good track record of successful operation over more than two decades (note that almost all commercial CCS projects are dealing with CO_2 captured from NG, and not from flue gases of coal-fired power plants). The industrialized countries lead the way, but even in developed nations, a large number of new power plants equipped with CCS technologies are expected to be built in coming decades. At the same time, new technological approaches such as a direct decarbonization concept are emerging and, most likely, they will play an increasing role in the future carbon mitigation scenarios. In contrast to conventional CCS technologies that handle *gaseous* or *liquefied* CO_2, the direct decarbonization approach is based on the extraction of *solid* carbon from hydrocarbons that can be used in a variety of applications (e.g., construction materials, commodity products, etc.), or stored for the future energy use in a less carbon-constrained environment.

It is realized that in order to stabilize atmospheric CO_2 concentration at the levels that would minimize its negative impact on ecosystems and climate, tens of billions of tons of CO_2 per year must be captured and safely stored. Advantageously, the global capacity for CO_2 storage is estimated at hundreds of billions of tons of CO_2, enough for several hundred years of carbon storage. Different geographic regions offer markedly different options for carbon sequestration, which include saline formations, depleted oil and gas formations, unmineable coal seams, basalts, and other reservoirs.

The majority of climate change scenarios and model predictions show that serious CO_2 reduction measures should start no later than 2020, because further delays may have unpredictable environmental and economical consequences. But we have yet to see a strong worldwide support from governments and policymakers and effective actions that are commensurate with the size of the problem. Despite the large number of ongoing CCS projects, there is a lack of cohesive policies and funds to support large-scale demonstration plants at the scale that will allow the widespread deployment of the technology by 2020. It is noteworthy that despite many years of commercial deployment in oil and gas industry, the concept of CCS is relatively new to general public, which, as many surveys showed, is practically unaware of its role as a carbon mitigation strategy (in fact, public opposition has halted several CCS projects in Europe). There is a clear opportunity for the worldwide deployment of the fossil fuel decarbonization technologies under GHG stabilization scenarios. At the same time, the technology has to be thoroughly evaluated for potential environmental impacts, and the provisions have to be made to ensure that humans' and our planet's health and well being are completely preserved. If successful, these measures can potentially defer the urgent environmental and climate change problems and buy time for the introduction of carbon-free fuels and energy sources and carbon-neutral technologies.

Acknowledgments

The author acknowledges the support of Florida Solar Energy Center for the preparation of the materials presented in this chapter. Special thanks to Dr. Ali T-Raissi for fruitful discussions.

Abbreviations

AFC	alkaline fuel cell
CCS	carbon capture and storage (sequestration)
CLC	chemical looping combustion
COE	cost of electricity
DCFC	direct carbon fuel cell
DOE	U.S. Department of Energy
EBS	environmentally benign sequestration
ECBMR	enhanced coal bed methane recovery
EGR	enhanced gas recovery
EOR	enhanced oil recovery
FC	fuel cell
FT	Fischer–Tropsch process
FTE	fuel to electricity
GHG	greenhouse gas
GT	gigaton
GW	gigawatt
HDR	hydrocarbon decomposition reactor
IEA	International Energy Agency
IGCC	integrated gasification combined cycle
IPCC	Intergovernmental Panel on Climate Change
LHV	lower heating value
MCFC	molten carbonate fuel cell
MEA	monoethanolamine
MW	megawatt
NETL	National Energy Technology Laboratory
NG	natural gas
NGCC	natural gas combined cycle
NOM	natural organic matter
NPP	net primary productivity
NRC	National Research Council
NREL	National Renewable Energy Laboratory
ppm	parts per million
PAFC	phosphoric acid fuel cell
PC	pulverized coal
PEM	proton exchange membrane
PLA	polylactic acid
PSA	pressure swing adsorption
SLH	synthetic liquid hydrocarbons
SMR	steam methane reforming
SNG	synthetic natural gas
SOFC	solid oxide fuel cell
SOM	soil organic matter
TCD	thermocatalytic decomposition
TSA	temperature swing adsorption
VSA	vacuum swing adsorption

References

Abanades, S., G. Flamant. 2005. Production of hydrogen by thermal methane splitting in a nozzle-type laboratory-scale solar reactor. *Int. J. Hydrogen Energy* 30, 843–853.

Akai, M., N. Nishio, M. Iijima, M. Ozaki, J. Minamiura, T. Tanaka. 2004. Performance and economic evaluation of CO_2 capture and sequestration technologies. *Proceedings of the 7th International Conference on Greenhouse Gas Control Technologies*, September 5–9, 2004, Vancouver, Canada, Elsevier Science, Oxford, U.K.

Algenol Biofuels. 2010. www.algenolbiofuels.com (accessed August 14, 2010).

Allinson, W., D. Nguen, J. Bradshaw. 2003. The economics of geological CO_2 storage in Australia. *APPEA J.* 43, 623–636.

Angus, S., B. Armstrong, K. De Reuck. 1973. *International Thermodynamics Tables of the Fluid States. Vol. 3. Carbon Dioxide*, IUPAC, Pergamon Press, London, U.K. pp. 266–359.

Archer, D., H. Kheshgi, E. Maier-Reimer. 1998. Dynamics of fossil fuel neutralization by marine $CaCO_3$. *Global Biochem. Cycles* 12, 259–276.

Aristov, Y., V. Fedoseev, V. Parmon. 1997. High-density conversion of light energy via direct illumination of catalyst. *Int. J. Hydrogen Energy* 22, 869–874.

Audus, H., O. Kaarstad, M. Kowal. 1996. Decarbonization of fossil fuels: Hydrogen as an energy carrier. *Proceedings of the 11th World Hydrogen Energy Conference*, June 23–28, 1996, Stuttgart, Germany.

Aya, I., R. Kojima, K. Yamane, P. Brewer, E. Peltzer. 2003. In situ experiments of cold CO_2 release in mid depth. *Proceedings of the 6th International Conference on Greenhouse Gas Control Technologies*, Kyoto, Japan.

Aya, I., R. Kojima, K. Yamane, P. Brewer, E. Peltzer. 2004. In situ experiments of cold CO_2 release in mid depth. *Energy* 29, 1499–1509.

Aya, I., K. Yamane, H. Hariai. 1997. Solubility of CO_2 and density of CO_2 hydrate at 30 MPa. *Energy* 22, 263–271.

Bachu, S. 2003. Screening and ranking of sedimentary basins for sequestration of CO_2 in geological media. *Environ. Geol.* 44, 277–289.

Bachu, S., W. Gunter. 2004. Overview of acid gas injection in western Canada. In E. Rubin, D. Keith, C. Gilroy (eds.), *Proceedings of the 7th International Conference on Greenhouse Gas Technologies*, Vol. I, September 5–9, Elsevier Science, Oxford, U.K., pp. 443–448.

Bachu, S., W. Gunter, E. Perkins. 1994. Aquifer disposal of CO_2: Hydrodynamic and mineral trapping. *Energy Convers. Manage.* 35, 269–279.

Bachu, S., K. Haug. 2005. In-situ characteristics of acid-gas injection operations in the Alberta basin, western Canada. In *Geologic Storage of Carbon Dioxide with Monitoring and Verification*, S. Benson (ed.), Elsevier, London, U.K., pp. 867–876.

Ball, M., W. Weindorf, U. Bunger. 2009. Hydrogen production. In *Hydrogen Economy. Opportunities and Challenges*, M. Ball, M. Wietschel (eds.), Cambridge University Press, Cambridge, U.K.

Bearat, H., M. McKelvy, A. Chizmeshya, R. Nunez, R. Carpenter. 2003. Investigation of the mechanisms that govern carbon dioxide sequestration via aqueous olivine mineral carbonation. *Proceedings of the International Technical Conference on Coal Utilization and Fuel System*, Clearwater, Florida. Vol. 1, p. 307.

Behrens, P. 2005. Photobioreactors and Fermentors: the light and dark sides of growing algae. *Algal Culturing Techniques*, R. Andersen (ed.), Elsevier, Amsterdam, the Netherlands, pp. 189–203.

Bergfeld, D., F. Goff, C. Janik. 2001. Elevated carbon dioxide flux at the Dixie Valley geothermal field, Nevada; relations between surface phenomena and the geothermal reservoir. *Chem. Geol.* 177, 43–66.

Berman, A., R. Karn, M. Epstein. 2006. Steam reforming of methane on Ru-catalysts for solar hydrogen production. *Abstracts of 232 ACS National Meeting*, Paper: FUEL-140, San Francisco, CA.

Biello, D. 2008. Cement from CO_2: A concrete cure for global warming? *Scientific American on-line*, August 7, 2008. www.scientificamerican.com/article.cfm?id=cement-from-carbon-dioxide (accessed August 11, 2010).

Biodiesel from algae oil. 2008. Pure Energy Systems Wiki. http://peswiki.com/index.php/directory:biodiesel_from_algae_oil (accessed May 27, 2009).

Bock, B., R. Rhudy, H. Herzog, M. Klett, J. Vavison, D. de la Torre Ugarte, D.Simbeck. 2003. Economic evaluation of CO_2 storage and sink options. DOE Research Report, DE-FC26-00NT40937.

Brandvoll, O., O. Bolland. 2004. Inherent CO_2 capture using chemical looping combustion in a natural gas fired power cycle. *ASME J. Eng. Gas Turbines Power* 126, 316–321 (ASME Paper GT-2002-30129).

Brewer, P., E. Peltzer, G. Friederich, G. Rehder. 2002. Experimental determination of the fate of a CO_2 plume in seawater. *Environ. Sci. Technol.* 36, 5441–5446.

Brewer, P., E. Peltzer, P. Walz, I. Aya, K. Yamane, R. Kojima, Y. Nakajima, N. Nakayama, P. Haugan, T. Johannessen. 2005. Deep ocean experiments with fossil fuel carbon dioxide: Creation and sensing of a controlled plume at 4 km depth. *J. Mar. Res.* 63, 9–33.

Buesseler, K., S. Doney, D. Karl, P. Boyd, K. Caldeira, F. Chai, K. Coale et al. 2008. Ocean iron fertilization—Moving forward in a sea of uncertainty. *Science* 319, 162.

Caldeira, K., G. Rau. 2000. Accelerating carbonate dissolution to sequester carbon dioxide in the ocean: Geochemical implications. *Geophys. Res. Lett.* 27, 225–228.

Callahan, M. 1974. Hydrocarbon fuel conditioner for a 1.5 kW fuel cell power plant. *Proceedings of the 26th Power Sources Symposium*, April 29–May 2, Red Bank, NJ, p. 181.

Carapeza, M., B. Badalamenti, L. Cavarra, A. Scalzo. 2003. Gas hazard assessment in a densely inhabited area of Colli Albani Volcano. *J. Volcanol. Geotherm. Res.* 123, 81–94.

Carbon Sequestration Technology Roadmap. 2006. Carbon Sequestration Technology Roadmap and Program Plan. U.S. DOE, Office of Fossil Energy, NETL.

Carlson Easley, N. 2008. The truth about carbon sequestration. *Relay* Spring, 8–12.

Casey, A. 2008. Carbon cemetery. *Can. Geogr. Mag.* January/February, p. 61.

Cherepy, N., R. Krueger, K. Fiet, A. Jankowski, J. Cooper. 2005. Direct conversion of carbon fuel in a molten carbonate fuel cell. *J. Electrochem. Soc.* 152, A80–A87.

Ciferno, J., T. Fout, A. Jones, J. Murphy. 2009. Capturing carbon from existing coal-fired power plants. *Chem. Eng. Prog.* April 2009, 33–47.

CO2now. Current data for atmospheric CO_2. 2010. http://CO2now.org (accessed July 9, 2010).

Dahl, J. et al. 2004. Solar thermal dissociation of methane in a fluid-wall aerosol flow reactor. *Int. J. Hydrogen Energy* 29, 725–736.

Davidson, J. 2005. CO_2 capture and storage and the IEA Greenhouse Gas Programme. *Workshop on CO_2 issues*, May 2005, Miffelfart, Denmark, IEA Greenhouse Gas Programme, Cheltenham, U.K.

DiPietro, P. 1997. Incorporating carbon dioxide sequestration into hydrogen energy systems. *Proceedings of the 1997 DOE Hydrogen Program Review*, National Renewable Energy Laboratory, Golden, CO.

Dooley, J., R. Dahowski, C. Davidson, S. Bachu, N. Gupta, H. Gale. 2004. A CO_2 storage supply curve for North America and its implications for the deployment of carbon dioxide capture and storage systems. In E. Rubin et al. (eds.), *Proceedings of the 7th International Conference on Greenhouse Gas Control Technologies*, Vol. 1, September 5–9, Vancouver, Canada, *IEA GHG Programme*, Cheltenham, U.K.

Dorner, R., D. Hardy, F. Williams, B. Davis, H. Willauer. 2009. Influence of gas feed composition and pressure on the catalytic conversion of CO2 to hydrocarbons using a traditional Cobalt-based Fischer-Tropsch catalyst. *Energy Fuels* 23, 4190–4195.

Edmonds, J., P. Feund, J. Dooley. 2000. The role carbon management technologies in addressing atmospheric stabilization of greenhouse gases. *Proceedings of the 5th International Conference on Greenhouse Gas Control Technologies*, Cairns, Australia, Sponsored by the IEA Greenhouse gas R&D Programme.

Effron, E. 2010. Comment. *The Week* 10(450), 7.

Eureka Alert. 2009. Amer. Assoc. for Advancement of Science. http://www.eurekaalert.org/pub_releases/2009-03/haog-lpn032409.php (accessed December 12, 2009).

Feely, R., C. Sabine, K. Lee et al. 2004. Impact of anthropogenic CO_2 on $CaCO_3$ system in the oceans. *Science* 305, 362–366.

Fer, I., P. Haugan. 2003. Dissolution from a liquid CO_2 lake disposed in the deep ocean. *Limnol. Oceanogr.* 48, 872–883.

Ferguson, R., C. Nichols, T. van Leeuwen, V. Kuuskraa. 2009. Storing CO_2 with enhanced oil recovery. Greenhouse Gas Technologies-9, Elsevier Science Direct. *Energy Procedia* 1, 1989–1996.

Flett, M., G. Beacher, J. Brantjes et al. 2008. Gordon Project: Subsurface evaluation of carbon dioxide disposal under Barrow island. *SPE Asia Pacific Oil and Gas Conference and Exhibition*, October, 20–22, 2008, Perth, Australia, Society of Petroleum Engineers, doi: 10.2118/116372-MS.

Fulcheri, L., Y. Schwob. 1995. From methane to hydrogen, carbon black and water. *Int. J. Hydrogen Energy* 20, 197–202.

Gale, J. 2003. Geological storage of CO_2: What's known, where are the gaps and what more needs to be done. In J. Gale, Y. Kaya (eds.), *Proceedings of the 6th International Conference on Greenhouse Gas Control Technology*, Vol. 1, October 1–4, 2002, Kyoto, Japan, Pergamon Press, Oxford, U.K., pp. 207–212.

Gale, J., J. Davison. 2002. Transmission of CO_2—Safety and economic considerations. *Proceedings of the 6th International Conference on Greenhouse Gas Control Technology*, October 1–4, GHGT-6, Kyoto, Japan.

Gaudernack, B., S. Lynum. 1996. Hydrogen from natural gas without release of CO_2 to the atmosphere. *Proceedings of the 11th World Hydrogen Energy Conference*, June 23–28, Stuttgart, Germany.

Giles, J. 2002. Norway sinks ocean carbon study. *Nature* 419, 6.

Global CCS Institute. 2009. Strategic analysis of the global status of carbon capture and storage. Final Report, Global CCS Institute. http://www/globalccsinstitute.com/downloads/reports/2009/worley/foundation-report-1-rev0.pdf (accessed August 3, 2010).

Golomb, D., A. Angelopoulos. 2001. A benign form of CO_2 sequestration in the ocean. *DOE NETL Workshop on Carbon Sequestration Science*, Pittsburg, PA.

Gray, D., G. Tomlinson. 2003. Hydrogen from coal. Mitretek Technical paper MTR-2003-13, prepared for NETL, U.S. DOE.

Guijt, W. 2004. Analyses of incident data show US, European pipelines becoming safer. *Oil Gas J.* 102, 68–73.

Halloran, J. 2008. Extraction of hydrogen from fossil fuels with production of solid carbon materials. *Int. J. Hydrogen Energy* 33, 2218–2224.

Halmann, M., M. Steinberg. 1999. *Greenhouse Gas Carbon Dioxide Mitigation. Science and Technology*, Lewis Publishers, Boca Raton, FL.

Haszeldine, R. 2009. Carbon capture and storage: How green can black be. *Science* 325, 1647–1652.

Hendriks, C., W. Graus, F. van Bergen. 2002. Global carbon dioxide storage potential and costs. Report Ecofys & The Netherlands Institute of Applied Geoscience TNO, Ecofys Report EEP, 63pp.

Hendriks, C., T. Wildenborg, P. Feron, W. Graus. 2005. Capture and storage, EC, DG-ENV. Ecofys Energy and Environment, report M70066.

Herzog, H. 2002. CO_2 sequestration. In R. Slott (ed.), *Proceedings of Greenhouse Gas Reduction Programs and Technologies Symposium*, Dedham, MA, pp. 11–12.

Hirsch, D., M. Epstein, A. Steinfeld. 2001. The solar thermal decarbonization of natural gas. *Int. J. Hydrogen Energy* 26, 1023–1033.

Holloway, S., D. Savage. 1993. The potential for aquifer disposal of carbon dioxide in UK. *Energy Convers. Manage.* 34, 925–932.

Holt, T., J. Jensen, E. Lindeberg. 1995. Underground of storage of CO_2 in aquifers and oil reservoirs. *Energy Convers. Manage.* 36, 535–538.

Huesemann, M., A. Skillman, E. Grecelius. 2002. The inhibition of marine nitrification by ocean disposal of carbon dioxide. *Mar. Pollut. Bull.* 44, 142–148.

IEA Clean Coal Center. 2005. *The World Coal-Fired Power Plants Database*, Gemini House, London, U.K.

IEA Energy Technology Perspectives. 2008. The role of RE in global energy scenarios. http://iedredt.org/files/REDT081022_102_Frankl.pdf (accessed March, 2011).

IEA GHG. 2002. Transmission of CO_2 and energy. IEA Greenhouse Gas R&D Programme. Report PH4/6, IEA Greenhouse gas R&D Programme, Cheltenham, U.K.

IEA GHG. 2004. Improvements in power generation with post combustion capture of CO_2. IEA Report PH4/33, IEA Greenhouse gas R&D Programme, Cheltenham, U.K.

IEA GHG. 2005. Retrofit of CO_2 capture to natural gas combined cycle power plants. Report 2005/1, IEA Greenhouse gas R&D Programme, Cheltenham, U.K.

IEA GHG PH 4/30. 2004. Ship transport of CO_2. IEA Greenhouse Gas R&D Programme. Report PH4/30, IEA Greenhouse gas R&D Programme, Cheltenham, U.K., 2004.

IEA Report. 1996. Decarbonization of fossil fuels. IEA Report PH2/2, March 1996, IEA Greenhouse Gas R&D Programme, Cheltenham, U.K.

IEA Report PH4/33. 2004. Potential for improvements in gasification combined cycle power generation with CO_2 capture. IEA Report PH4/33, IEA Greenhouse gas R&D Programme, Cheltenham, U.K., 2004.

Inoue, S., H. Koinuma, T. Tsuruta. 1969. Copolymerization of carbon dioxide and epoxide. *J. Polym. Sci. B* 7, 287–292.

IPCC CCS. 2005. IPCC special report on carbon dioxide capture and storage. Prepared by Working Group III of the Intergovernmental Panel on Climate Change. B. Metz, O. Davidson, H. de Coninck, M. Loos, A. Meyer (eds.), Cambridge University Press, Cambridge, U.K.

Itoaka, K., A. Saito, M. Akai. 2004. Public acceptance of CO_2 capture and storage technology: A survey of public opinion to explore influential factors. *Proceedings of the 7th International Conference on Greenhouse Gas Control Technology (GHGT-7)*, September 5–9, 2004, Vancouver, Canada.

Jain, A., L. Cao. 2005. Assessing the effectiveness of direct injection for ocean carbon sequestration under the influence of climate change. *Geophys. Res. Lett.* 32, L09609. doi: 10.1029/2005GL022818.

Jain, S., J. Lakerman, K. Pointon, J. Irvine. 2007. Carbon content in a direct carbon fuel cell. *ECS Trans.* 7, 829–836.

Jansen, D., J. Dijkstra. 2003. CO_2 capture in SOFC-GT systems. *Second Annual Conference on Carbon Sequestration*, May 7, Alexandria, VA.

Jiacoletti, R. 1977. Hydrogen from nuclear energy. In *Hydrogen: Its Technology and Applications*, K. Cox, K. Williamson, (eds.), CRC Press, Boca Raton, FL, Chapter 4.

Jiang, Z., T. Xiao, V. Kuznetsov, P. Edwards. 2010. Turning carbon dioxide into fuel. *Phil. Trans. R. Soc. A* 368, 3343–3364.

Johnson, J. 2009a. Water and CO_2 shouldn't mix. *Chem. Eng. News*, 87, 32, September 21, 2009.

Johnson, J. 2009b. Huge CO_2 storage potential in China. *Chem. Eng. News*, 87, 27, October 19, 2009.

Johnson, J. 2009c. Fossil-fuel costs. *Chem. Eng. News*, 87, 6, October 26, 2009.

Johnson, J. August 16, 2010. Gasification plant funds shifted by DOE. *Chem. Eng. News*, 88, 34.

Kaarstad, O. 1992. Emission-free fossil energy from Norway. *Energy Convers. Manage.* 33, 781–786.

Kaarstad, O. 2004. Statoil (Norway). Written communication to the Intergovernmental Panel on Climate Change (IPCC).

Karmis, M. 2009. SECARB Initiatives in central and southern Appalachia—Progress and future opportunities. *ECC Annual Meeting*, May 12, 2009, Kingsport, TN.

Key, R., A. Kozyr, C. Sabine et al. 2004. A global ocean carbon climatology: Results from GLODAP. *Global Biochem. Cycles* 18, GB4031.

Kheshgi, H. 1995. Sequestering atmospheric carbon dioxide by increasing ocean alkalinity. *Energy* 20, 915–922.

Kheshgi, H. 2004. Ocean carbon sink duration under stabilization of atmospheric CO_2: A 1000-year time scale. *Geophys. Res. Lett.* 31, L20204.

Kheshgi, H., D. Archer. 2004. A nonlinear convolution model for the evasion of CO_2 injected into the deep ocean. *J. Geophys. Res.* 109, C02007.1–C02007.13.

Kim, J.-S., S. Lee, S.-B. Lee, M.-J. Choi, K.-W. Lee. 2006. Performance of catalytic reactors for the hydrogenation of CO_2 to hydrocarbons. *Catal. Today* 115, 228–234.

Kithil, P. 2007. A policy option to provide sufficient funding for massive-scale sequestration of CO_2. *The Smithsonian/NASA Astrophysics Data System*. http://adsabs.harvard.edu/abs/2007AGUFM.U43C1410K (accessed January 14, 2010).

Kling, G., W. Evans, M. Tuttle, G. Tanyileke. 1994. Degassing of Lake Nyos. *Nature* 368, 405–406.

Kogan, A., M. Kogan, S. Barak. 2005. Production of hydrogen and carbon by solar thermal methane splitting. III. Fluidization, entrainment and seeding powder particles into a volumetric solar receiver. *Int. J. Hydrogen Energy* 30, 35–43.

Koide, H., Y. Tazaki, Y. Noguchi et al. 1992. Subterranean containment and long-term storage of carbon dioxide in unused aquifers and depleted natural gas reservoirs. *Energy Convers. Manage.* 33, 619–626.

Kreutz, T., P. Williams, P. Chiesa, S. Consonni. 2005. Co-production of hydrogen, electricity and CO_2 from coal with commercially ready technology. *Int. J. Hydrogen Energy* 30, 769–784.

Lackner, K. 2003. A guide to CO_2 sequestration. *Science* 300, 1677–1678.

Langenbuch, M., H. Portner. 2004. High sensitivity to chronically elevated CO_2 levels in a eurybathic marine sipunculid. *Aquat. Toxicol.* 70, 55–61.

Larson, E., T. Ren. 2003. Synthetic fuel production by indirect coal gasification. *Energy Sustain. Dev.* VII(4), 79–102.

Leblond, D. 2009. A pilot installation at Lacq; Total Corporate Website http://www.tota.com/en/corporate -socia-responsibility/special-reports/capture/carbon-dioxide-total-commitment/carbon-dioxide-lacq-pilot_11357.htm (accessed January 30, 2010).

Li, J., G. Lu, K. Li, W. Wang. 2004. Active Nb_2O_5-supported nickel and nickel-copper catalysts for methane decomposition to hydrogen and filamentous carbon. *J. Molec. Catal.* A: Chemical 221, 105–112.

Lin, S.-Y. 2009. Hydrogen production from coal. In *Hydrogen Fuel. Production, Transport and Storage*, R. Gupta (ed.), CRC Press, Boca Raton, FL.

Lin, H., T. Merkel, R. Baker. 2007. The membrane solution to global warming. *Sixth Annual Conference on Carbon Capture and Sequestration*, May 7, Pittsburgh, PA.

Liro, C., E. Adams, H. Herzog. 1992. Modeling the release of CO_2 in deep ocean. *Energy Convers. Manage.* 33, 667–674.

Lucquiaud, M., J. Gibbins. 2009. Retrofitting CO_2 capture ready fossil plants with post-combustion capture. Part 1: Requirements for supercritical pulverized coal plants using solvent-based flue gas scrubbing. *Proc. ImechE, Part A: J. Power Energy* 223, 213–226.

Lynum S., R. Hildrum, K. Hox, J. Hugdabl. 1998. Kværner based technologies for environmentally friendly energy and hydrogen production. *Proceedings of the12th World Hydrogen Energy Conference*, June 20–26, Buenos Aires, Argentina, p. 697.

Ma, J., N. Sun, X. Zhang et al. 2009. A short review of catalysis for CO_2 conversion. *Catal. Today* 148, 221–231.

Marchetti, C. 1977. On geoengineering and the CO_2 problem. *Clim. Change* 1, 59–68.

Marland, G., B. McCarl, U. Schneider. 1998. Soil carbon: Policy and economics. *Carbon Sequestration in Soils: Science, Monitoring and Beyond: Proceedings of the St. Michaelis Workshop*, December 1998, St. Michaelis, MD.

Martin, F., J. Taber. 1992. Carbon dioxide flooding. *J. Petrol. Technol.* 44, 396–400.

McCoy, S., E. Rubin. 2008. An engineering economic model of pipeline transport of CO_2 with application to carbon capture and storage. *Int. J. Greenhouse Gas Control* 2, 219–229.

McGlashan, N. 2008. Chemical looping combustion—A thermodynamic study. *J. Mech. Eng. Sci.* 222, 1005–1019.

Meier, A. et al. 1999. Solar thermal decomposition of hydrocarbons and carbon monoxide for the production of catalytic filamentous carbon. *Chem. Eng. Sci.* 54, 3341–3348.

Methanol. 2010. Methanol Institute. www.methanol.org (accessed August 20, 2010).

Michael, K., G. Allison, A. Golab, S. Sharma, V. Shulakova. 2009. CO_2 storage in saline aquifers II—Experience from existing storage operations. Greenhouse Gas Technologies-9 Proceedings, Elsevier Science Direct. *Energy Procedia* 1, 1973–1980.

Millward A., O. Yaghi. 2005. Metal-organic frameworks with exceptionally high capacity for storage of carbon dioxide at room temperature. *J. Am. Chem. Soc.* 127, 17998–17999.

Moberg, R., D. Stewart, D. Stachniak. 2003. The IEA Weyburn CO_2 monitoring and storage project. In J. Gale, Y. Kaya (eds.), *Proceedings of the 6th International Conference on Greenhouse Gas Control Technologies*, October 1–4, 2002, Kyoto, Japan, pp. 219–224.

Mordkovich, V., E. Dolgova, A. Karaeva et al. 2007. Synthesis of carbon nanotubes by catalytic conversion of methane: Competition between active components of catalyst. Carbon 45, 62–69.

Morita, T., N. Nakicenovic, J. Robinson. 2000. Overview of mitigation scenarios for global climate stabilization based on new IPCC emissions scenarios. *Environ. Econ. Policy* 3, 65–88.

Muradov, N. 1993. How to produce hydrogen from fossil fuels without CO_2 emission. *Int. J. Hydrogen Energy* 18, 211–215.

Muradov, N. 2001a. Hydrogen via methane decomposition: An application to decarbonization of fossil fuels. *Int. J. Hydrogen Energy* 26, 1165–1175.

Muradov, N. 2001b. Catalysis of methane decomposition over elemental carbon. *Catal. Commun.* 2, 89–94.

Muradov, N. 2008. Role of carbon in environmentally "benign" sequestration of CO_2. *Proceedings of the 17th World Hydrogen Energy Conference*, June 15–19, Brisbane, Australia.

Muradov, N. 2009. Production of hydrogen from hydrocarbons. In *Hydrogen Fuel. Production, Transport and Storage*, R. Gupta (ed.), CRC Press, Boca Raton, FL.

Muradov, N., P. Choi, F. Smith, G. Bokerman. 2010. Integration of direct carbon and hydrogen fuel cells for highly efficient power generation from hydrocarbon fuels. *J. Power Sources* 195, 1112–1121.

Muradov, N., F. Smith, G. Bokerman. 2009. Methane activation by non-thermal plasma generated carbon aerosols. *J. Phys. Chem. C* 113, 9737–9747.

Muradov, N., F. Smith, A. T-Raissi. 2005. Catalytic activity of carbons for methane decomposition reaction. *Catal. Today* 102–103, 225–233.

Muradov, N., N. Veziroğlu. 2005. From hydrocarbon to hydrogen-carbon to hydrogen economy. *Int. J. Hydrogen Energy* 30, 225–237.

Muradov, N., N. Veziroğlu. 2008. "Green" path from fossil to hydrogen economy. An overview of carbon-neutral technologies. *Int. J. Hydrogen Energy* 33, 6804–6839.

Nakamura, T. 2004. Recovery and sequestration of CO_2 from stationary combustion systems by photosynthesis of microalgae. Technical Report to DOE, NETL, No. PSI-1356, December.

National Laboratory Directors. 1997. Technology opportunities to reduce US greenhouse emissions, Oak Ridge National Laboratory, TN.

NETL. 2009. Existing plants, emissions and capture—Setting CO_2 program goals. U.S. DOE, Office of Fossil Energy, National Energy Technology Laboratory, April 20.

NETL-DOE. 2010. Worldwide gasification database online, Pittsburgh, PA. www.netl.doe.gov/coal-power/gasification/models/dtbs.pdf. [accessed December, 2010].

Newall, P., S. Clarke, H. Haywood et al. 2000. CO_2 storage as carbonate minerals. Report PH3/17 for IEA Greenhouse Gas Technologies R&D Programme, CSMA Consultants Ltd., Cornwall, U.K.

NIOSH. 1996. National Institute for Occupational Safety and Health. Criteria for a recommended standard, occupational exposure to carbon dioxide. 1976, 1965.

NRC. 2004. *The Hydrogen Economy. Opportunities, Costs, Barriers and R&D Needs*. National Research Council and National Academy of Engineering, National Academies Press, Washington, DC.

O'Connor, W.D. Dahlin, G. Rush, S. Gedermann, L. Penner, D. Nielsen. 2005. Aqueous mineral carbonation. Final Report. DOE/ARC-TR-04-002, March 15.

Ormerod, W., P. Freund, A. Smith, J. Davison. 2002. Ocean storage of CO_2. IEA, Greenhouse Gas Technologies R&D Programme. ISBN 1 898373302.

Palmgren, C., M. Granger Morgan, W. Bruine de Bruin, D. Keith. 2004. Initial public perception of deep geological and oceanic disposal of CO_2. *Environ. Sci. Technol.* 38, 6441–6450.

Perkins, E., I. Czernichowski-Lauriol, M. Azaroual, P. Durst. 2004. Long term predictions of CO_2 storage by mineral and solubility trapping in the Weybourn Midale Reservoir. *Proceedings of the 7th International Conference on Greenhouse Gas Control Technologies*, Vol. II, September 5–9, 2004, Vancouver, Canada, pp. 2093–2096.

Peterma, D. et al. 1975. Studies of the use of high temperature nuclear heat from an HTGR for hydrogen production. General Atomic Company Report GA-A11391, San Diego, CA.

Petroleum Technology Research Center (PTRC). 2007. IEA GHG Weyburn-Midale CO_2 monitoring & storage project. *Proceedings of the 6th Annual Conference on Carbon Capture and Sequestration*, May 7, 2007, Pittsburg, PA.

Poirier, M., C. Sapundzhiev. 1998. Catalytic decomposition of natural gas to hydrogen for fuel cell applications. *Int. J. Hydrogen Energy* 22, 429–433.

Rau, G., K. Caldeira. 1999. Enhanced carbonate dissolution: A means of sequestering waste CO_2 as ocean bicarbonate. *Energy Convers. Manage.* 40, 1803–1813.

Reeves, S., D. Davis, A. Oudinot. 2004. A technical and economic sensitivity study of enhanced coal-bed methane recovery and carbon sequestration in coal. DOE Topical Report, March.

Rehder, G., S. Kirby, W. Durham et al. 2004. Dissolution rates of pure methane hydrate and carbon dioxide hydrate in under-saturated seawater at 1000 m depth. *Geochim. Cosmochim. Acta* 68, 285–292.

Riahi, K., E. Rubin, L. Schrattenholzer. 2003. Prospects for carbon capture and sequestration technologies assuming their technological learning. In J. Gale, Y. Kaya (eds.), *Proceedings of the 6th International Conference on Greenhouse Gas Control Technologies*, October 1–4, 2002, Kyoto, Japan, Elsevier, Oxford, U.K., pp. 1095–1100.

Ritter, S. July 26, 2010. Carbon dioxide's unsettled future. *Chem. Eng. News*, 88, 36–37.

Roadmap for the hydrogen economy. 2005. Roadmap on manufacturing R&D for the hydrogen economy. *Workshop on Manufacturing R&D for the Hydrogen Economy*, Washington, DC.

Rochelle, G. 2009. Amine scrubbing for CO_2 capture. *Science* 325, 1652–1654.

Saito, H., S. Hasegawa, M. Ihara. 2008. Effective anode thickness in rechargeable direct carbon fuel cells using fuel charged by methane. *J. Electrochem. Soc.* 155, B443–B447.

Schilling, F., G. Borm, H. Würdemann, F. Möller, M. Kühn. 2009. CO2SINK Group, Status report on the first European on shore CO_2 storage site at Ketzin (Germany). *Proceedings of the 9th International Conference on Greenhouse Gas Control Technologies (GHGT-9)*, November 16–20, 2008, pp. 2029–2035.

Seiersten, M. 2001. Material selection for separation, transportation and disposal of CO_2. *Proceedings Corrosion 2001*, Houston, TX, National Association of Corrosion Engineers, paper 1042.

Serban, M. et al. 2002. Hydrogen production by direct contact pyrolysis of natural gas. *Prepr. Symp.—Am. Chem. Soc., Div. Fuel Chem.* 47, 746.

Serrano, D., J. Botas, P. Pizarro, R. Guil-Lopez, G. Gomez. 2008. Ordered mesoporous carbons as highly active catalysts for hydrogen production by CH_4 decomposition. *Chem. Commun.* 6585–6587.

Sharma, S., P. Cook, T. Berly, M. Lees. 2009. The CO2CRC Otway Project: Overcoming challenges from planning to execution of Australia's first CCS project. *Proceedings of the 9th International Conference on Greenhouse Gas Control Technologies*, Elsevier Science Direct. *Energy Procedia* 1, 1965–1972.

Statoil. 2009. Development solution. http://www.statoil.com/statoilcom/snohvit/svg02699.nsf?opendatabase&lang=en (accessed September 25, 2009).

Steinberg, M. 1999. Fossil fuel decarbonization technology for mitigating global warming. *Int. J. Hydrogen Energy* 24, 771–777.

Steinberg, M., J. Cooper, N. Cherepy. 2002. High efficiency direct carbon and hydrogen fuel cells for fossil fuel power generation. *Proceedings of the AIChE 2002 Spring Meeting*, New Orleans, LA, pp. 2112–2127.

Steinfeld, A., V. Kirillov, G. Kuvshinov et al. 1997. Production of filamentous carbon and hydrogen by solar thermal catalytic cracking of methane. *Chem. Eng. Sci.* 52, 3599–3603.

Steinfeld, A., A. Meier. 2004. Solar fuels and materials. In C.J. Cleveland (ed.), *Encyclopedia of Energy*, Vol. 5, Elsevier, Amsterdam, the Netherlands, pp. 623–637.

Stevens, S., J. Kuuskraa, R. Schraufnagel. 1996. Technology spurs growth of US coal bed methane. *Oil Gas J.* 94, 56–63.

Stobs, R., P. Clark. 2004. Canadian clean power coalition: The evaluation of options for CO_2 capture from existing and new coal-fired power plants. In M. Wilson, T. Morris, J. Gale, K. Thambimuthu (eds.), *Proceedings of the 7th International Conference on Greenhouse Gas Control Technology*, Vol. II, September 5–9, 2004, Vancouver, Canada, Elsevier Science, Oxford, U.K., pp. 1187–1192.

Stone, E., J. Lowe, K. Shine. 2009. The impact of carbon capture and storage on climate. *Energy Environ. Sci.* 2, 81–91.

Strong, A., S. Chisholm, C. Miller, J. Cullen. 2009. Ocean fertilization: Time to move on. *Nature* 461, 347–348.

Suelves, I., J. Pinilla, M. Lazaro, R. Moliner. 2008. Carbonaceous materials as catalysts for decomposition of methane. *Chem. Eng. J.* 140, 432–438.

Sunda, W. and S. Huntsman. 1995. Iron uptake and growth limitation in oceanic and coastal phytoplankton. *Mar. Chem.* 50, 189–206. doi: 10.1016/0304-4203(95)00035-P.

Tamme, R., R. Buck, M. Epstein et al. 2001. Solar upgrading of fuels for generation of electricity. *J. Solar Energy Eng.* 123, 160–163.

Taniguchi, Y., T. Kitamura, Y. Fujiwara. 1997. Vanadium-catalyzed acetic acid synthesis from methane and carbon dioxide. *Proceedings of the 4th International Conference on Carbon Dioxide Utilization,* Kyoto, Japan, P-030.

Teng, H., A. Yamasaki, Y. Shindo. 1996. The fate of liquid CO_2 disposed in the ocean. *Int. Energy* 21, 765–774.

Thayer, A. July 13, 2009. Chemicals to help coal come clean. *Chem. Eng. News,* 87, 18–20.

Tossel, J. March 30, 2009. Carbonic acid could be a fix for carbon. *Chem. Eng. News,* 87, 29.

Toyir, J., R. Milloua, N. Elkadri et al. 2009. Sustainable process for the production of methanol from CO_2 and H_2 using Cu/ZnO-based multicomponent catalyst. *Phys. Procedia* 2, 1075–1079.

Trachtenberg, M., R. Cowan, D. Smith et al. 2008. Membrane-based enzyme-facilitated efficient CO_2 capture. *Proceedings of the 9th International Conference on Greenhouse Gas Control Technologies,* November 16–20, 2008, Washington, DC, pp. 353–360.

Tsoiris, C., P. Brewer, E. Peltzer et al. 2004. Hydrate composite particles for ocean carbon sequestration: Field verification. *Environ. Sci. Technol.* 38, 2470–2475.

Ullmann's 2003. *Ullmann's Encyclopedia of Industrial Chemistry,* Wiley-VCH Verlag, Weinheim, Germany.

U.S. DOE, Carbon sequestration. State of the Science. 1999. U.S. DOE, Office of Science, Office of Fossil Fuels. Working Paper on Carbon Sequestration Science and Technology.

U.S. DOE, NETL. 2003. Carbon sequestration. Technology Roadmap and Program Plan. U.S. DOE, Office of Fossil Energy, National Energy Technology Laboratory, Morgantown, WV.

U.S. DOE, NETL, 2007. Carbon Sequestration Technology Roadmap and Program Plan. U.S. DOE Office of Fossil Energy, National Energy Technology Laboratory, Morgantown, WV.

USDA. 2010. Agricultural research service national program. http://www.ars.usda.gov/research/programs/programs.htm?np_code (accessed January 14, 2010).

Velea, S., N. Dragos, S. Serban et al. 2008. Biological sequestration of carbon dioxide from thermal power plant emissions by absorption in microalgal culture media. *Romanian Biotechnol. Lett.* 14, 4485–4500.

Wackowski, R. 2007. Rangely Weber Sand unit CO_2 flooding case study, a long history of CO_2 injection. *Proceedings of CO_2 Capture and Storage Conference,* February 7, 2007, The Canadian Institute, Calgary, Alberta, Canada.

Wang, S., P. Jaffe. 2004. Dissolution of trace metals in potable aquifers due to CO_2 releases from deep formations. *Energy Convers. Manage.* 45, 2833–2848.

WEIO. 2003. *World Energy Investment Outlook 2003,* OECD/IEA, Paris, France. ISBN: 92-64-01906-5.

Wildenborg, T., T. Gale, C. Hendrics et al. 2004. Cost curves for CO_2 storage: European sector. In E. Rubin et al. (eds.), *Proceedings of the 7th International Conference on Greenhouse Gas Control Technologies,* Vol. 1, September 5–9, 2004, Vancouver, Canada, IEA GHG Programme, Cheltenham, U.K.

Wilson, E., D. Gerard. 2007. *Carbon Capture and Sequestration. Integrating Technology, Monitoring and Regulation,* Blackwell Publishing, Ames, IA.

Yokota, O., Y. Oku, M. Arakawa et al. 2000. Steam reforming of methane by using a solar simulator controlled by $H_2O/CH_4 = 1/1$. *Appl. Organomet. Chem.* 14, 867–870.

Zafar, Q., T. Mattisson, B. Gevert. 2005. Integrated hydrogen and power production with CO_2 capture using chemical-looping reforming-redox reactivity of particles of CuO, Mn_2O_3, NiO and Fe_2O_3 using SiO_2 as a support. *Ind. Eng. Chem. Res.* 44, 3485–3496.

Zahariev, K., J. Christian, K. Denman. 2008. Preindustrial, historical and fertilization simulations using a global ocean carbon model with new parameterization of iron limitation, calcification and N_2 fixation. *Prog. Oceanogr.* 77, 56–82.

Zhou, W., M. Stenhouse, R. Arthur et al. 2005. The IEA Weybourn CO_2 monitoring and storage project—Modeling of long-term migration of CO_2 from Weybourn. *Proceedings of the 7th International Conference on Greenhouse Gas Control Technologies (GHGT-7)*, Vol. I, September 5–9, 2004, Vancouver, Canada, 2004, pp. 721–730, Elsevier, U.K.

Zondervan, I., R. Zeebe, B. Rost, U. Riebessel. 2001. Decreasing marine biogenic calcification: A negative feedback on rising atmospheric CO_2. *Global Biogeochem. Cycles* 15, 507–516.

15

Clean Car Options for the Twenty-First Century

C.E. (Sandy) Thomas

CONTENTS

15.1 Introduction

The global consumption of petroleum to make gasoline and diesel fuel for motor vehicles is not sustainable. If the developing world were to even *approach* our per capita oil consumption habits, the stress on the environment and the world energy security threats would be overwhelming. Even without these environmental and energy security concerns, the world would not be able to continue our voracious consumption of nonrenewable fossil fuels.

We must find alternatives to gasoline-powered cars with internal combustion engines (ICEs). Fortunately, there are many possible alternative vehicles and fuels to help reduce our dependence on fossil fuel including

- Hybrid electric vehicles (HEVs) such as the Toyota Prius that combine an electric motor with an ICE to improve vehicle efficiency
- Plug-in hybrid electric vehicles (PHEVs) like the Chevy Volt scheduled for production by GM in late 2010 or 2011 with a larger battery pack to allow some travel on electricity alone before the on-board ICE/generator begins running to recharge the batteries
- Biomass-fueled PHEVs
- Hydrogen-powered fuel cell electric vehicles (FCEVs)
- Battery-powered electric vehicles (BEVs)

We* developed a 100 year computer simulation program to compare the societal impacts of adopting some combination of these alternative vehicles and fuels in the United States spanning the twenty-first century.

With the discoveries of large quantities of natural gas in United States shale formations, some people are promoting natural gas to power our vehicles, so we have also analyzed the impact of using natural gas as a vehicle fuel. Strictly speaking, however, natural gas is not a sustainable option and could only serve as a transition fuel until one or more of the main alternatives listed above could provide a sustainable transportation system. In addition, as discussed in Section 15.3.9, natural gas, even if used in a PHEV (the most efficient option for burning natural gas in an ICE) would not allow us to achieve the 80% reduction in greenhouse gas (GHG) emissions recommended by the climate change community. Furthermore, converting natural gas to hydrogen for use in a FCEV would require less natural gas for a given number of vehicle miles traveled (VMT) than burning that natural gas directly in a PHEV. So a given quantity of natural gas could improve the environment and stretch our limited fossil fuel supplies if that natural gas were converted to hydrogen for FCEVs.

15.2 Simulation Assumptions

Many analysts have compared the environmental and energy security impacts of individual alternative fuel/vehicle combinations. These static comparisons of single vehicles

---

* This project was begun by some members of the National Hydrogen Association and resulted in the "Energy Evolution" report available on their web site at http://www.hydrogenassociation.org/general/evolution.asp; updates on this simulation program and more details can also be found at http://www.cleancaroptions.com

are useful but do not reveal the full societal impact of various options. In addition, we will most likely never again rely on a single fuel/vehicle combination like the gasoline- or diesel-powered ICE vehicle that has until now dominated most transportation around the world. Indeed, the world has already begun adding hybrids (HEVs) to the conventional vehicle fleet. We have hypothesized six main scenarios in this computer simulation program that, with the exception of the business-as-usual case, each analyzes a portfolio of conventional and alternative vehicles:

- A business-as-usual scenario assuming that all light duty vehicles (LDVs) and light duty trucks (pickup trucks, vans, and sport utility vehicle (SUVs)) continue to run on gasoline or diesel (nonhybrid) ICEs.*
- A base-case scenario that assumes a continuation of the current mixture of ICE vehicles and HEVs, with gasoline HEVs dominating new car sales by the end of the century.
- A PHEV scenario that assumes that large numbers of PHEVs are added to the existing mix of HEVs and internal combustion vehicles (ICVs).
- A biomass-fueled PHEV scenario that assumes that a liquid fuel such as cellulosic ethanol derived from biomass is used to power ICEs on the PHEVs.
- A hydrogen-powered FCEV scenario where electric vehicles deriving their electricity from hydrogen-powered fuel cells dominate vehicle sales by the end of the century.
- A BEV scenario.

15.2.1 Business-as-Usual with All ICE Vehicles

In this scenario (and all other scenarios), we assume that VMT in the United States increases over the century due to three factors:

- Population growth
- Slightly increased number of vehicles per person
- Slightly increased number of miles traveled per person

All these three factors had been increasing rapidly prior to the Great Recession of 2008/2009, but their rates of growth will presumably slow down in the future. Accordingly, we have assumed more modest growth in these factors, but U.S. VMT still increases from 2.7 trillion million miles traveled in 2007 up to 8.3 trillion million miles by 2100 (for details on these assumptions, see Refs. [1,2]).

15.2.2 Base Case Hybrid Electric Vehicle Scenario

The base case in this model assumes a continuation of current trends in new car sales with most car makers introducing gasoline-powered HEVs. We assume that the growth in HEV sales continues, reaching 98% of all new car sales by 2100, as shown in Figure 15.1.

---

* This business-as-usual scenario would require us to turn back the clock since HEVs have already entered the marketplace in significant numbers. Thus, this scenario is merely a theoretical benchmark to illustrate the impact of HEVs and other alternative vehicles.

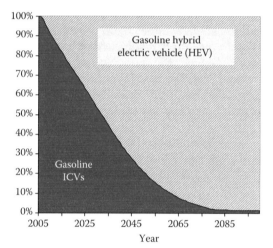

FIGURE 15.1
Fraction of light duty vehicle sales over the twenty-first century for the base case gasoline hybrid electric vehicle (HEV) scenario; the gasoline HEV is the only alternative vehicle in this scenario. (Reprinted from Thomas, C.E., *Int. J. Hydrogen Energy*, 34, 9279, 2009. With permission.)

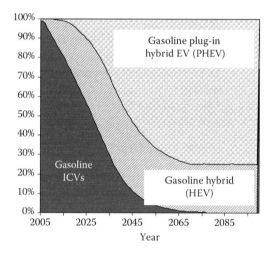

FIGURE 15.2
Fraction of light duty vehicle sales over the twenty-first century for the gasoline plug-in hybrid electric vehicle (PHEV) scenario; PHEVs replace HEVs and ICVs each year in proportion to their prevalence in the HEV scenario; PHEVs are limited to 75% of sales to simulate 25% of vehicles that do not have access to a charging outlet while parked at night. (Reprinted from Thomas, C.E., *Int. J. Hydrogen Energy*, 34, 9279, 2009. With permission.)

15.2.3 Plug-in Hybrid Electric Vehicle Scenario

In the PHEV scenario, we assume that gasoline-powered PHEVs are added to the HEV scenario beginning in 2010 as shown in Figure 15.2. This scenario assumes that most non-PHEVs are still HEVs. We assumed that PHEV sales are limited to 75% of all new cars sold, on the premise that not all car owners will have access to charging facilities. This 75% estimate may be high since Deloitte has since conducted a survey of potential PHEV owners and found that only 39% of the population have an access to a charging outlet in a garage, carport, or driveway [3]. Hence this model effectively assumes that 75% – 39% = 36% of potential PHEV owners have assigned off-street parking that could accommodate a controlled, secure electrical outlet to recharge their vehicles at night. It seems unlikely that apartment dwellers equal to 36% of all car owners would have assigned off-street parking places.

15.2.4 Biomass PHEV Scenario

In the biomass PHEV scenario, we assume the same new car sales as in the gasoline PHEV scenario, but with all PHEVs powered by liquid biomass-derived fuels such as cellulosic

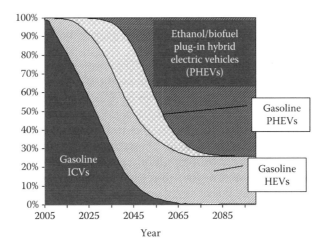

FIGURE 15.3
Fraction of light duty vehicle sales for the biofuels/cellulosic ethanol PHEV scenario. The total number of PHEV sold is equal to the number sold in the gasoline PHEV scenario, with biofuel-powered PHEVs taking over from gasoline-powered PHEVs as new biofuel plants begin operating and the biofuel distribution systems are put in place.

ethanol, butanol, or bio-diesel when the biofuel plants are built. As shown in Figure 15.3, gasoline-powered PHEVs enter the market at the same rate as the gasoline PHEV scenario described above. As new biofuel capacity is added, biofuel PHEVs take the place of gasoline PHEVs. This will be the most optimistic use of biofuels since the PHEV has higher fuel economy than either conventional ICVs or HEVs. In addition to the 75% charging outlet limitation for PHEV discussed above for gasoline PHEVs, the biofuel PHEVs would be limited by the availability of enough land without adversely affecting food crops to grow biomass to convert to biofuels such as cellulosic ethanol or butanol.

The estimates of the ultimate U.S. ethanol production potential vary widely, from 45 billion gallons per year up to 140 billion gallons per year. On the low end, the U.S. National Research Council (NRC) estimated in 2008 that the most probable cellulosic ethanol production limit would be 45 billion gallons per year by 2050, up from current production levels of 8 billion gallons per year of corn ethanol [4]. The 45 billion gallons estimate was based on a total biomass yield of 500 million dry tons per year, combined with an assumption that the cellulosic ethanol yield would be increased from 60 gallons per ton of biomass to 90 gallons/ton by 2050. The NRC also warned that water resources would likely be a "major issue" if ethanol production increased to these high levels and that the cost of cellulosic ethanol plants are likely to be two to three times that of current corn ethanol plants, which might limit actual ethanol production. The NRC also postulated an "upper-bound case" where they removed some of the water and cost limitations and estimated a potential annual biomass production rate of 700 million dry tons per year, yielding 63 billion gallons per year of cellulosic ethanol.

On the more optimistic side, a joint Sandia National Laboratory/General Motors assessment estimated that 90 billion gallons of ethanol per year might be feasible by 2030 [5]. A joint U.S. Department of Energy and Department of Agriculture assessment of the ultimate biomass production potential concluded that up to 1.37 billion dry tons of biomass (nearly 1 billion tons from agriculture and 370 million tons from forests) could be produced in the United States in a sustainable manner without adversely affecting soil fertility or food production [6]. If ethanol yields of 90 gal/ton of biomass could be achieved, then this would correspond to nearly 120 billion gallons per year or twice the NRC upper-bound estimate.

On the high side, a detailed 2004 assessment by Greene at the Natural Resources Defense Council showed that cellulosic biofuel production might displace up to 7.9 million barrels/day of crude oil by 2050 without unduly burdening U.S. food production or soil fertility [7]. They assumed a total biomass potential of 1.4 billion dry tons per year, based primarily on growing switchgrass, a fast growing native prairie grass. They also assumed that biofuel production yields would increase to 105 gal/dry ton, which translates into a cellulosic ethanol potential of more than 140 billion gallons per year, which we used as the upper limit in this study (it is noteworthy that this upper limit of 140 billion gallons per year of cellulosic ethanol coincidently supplies approximately 75% of all cars as PHEVs, so the 75% limit for PHEVs would be reduced if biofuel production fell below this highest estimate).

15.2.5 Hydrogen-Powered Fuel Cell Electric Vehicle Scenario

The hydrogen-powered FCEV scenario assumes that hydrogen infrastructure and FCEVs are introduced later than biofuels and PHEVs as shown in Figure 15.4. Thus, FCEV sales reach 50% of all new car sales by approximately 2048 while all other alternative vehicles achieve 50% market share earlier: HEVs reach 50% of all new car sales by approximately 2032; and biomass/ethanol PHEVs achieve the 50% sales goal by 2045. The FCEV sales are superimposed on the biomass PHEV sales scenario so that previously sold HEVs and PHEVs remain in the car fleet in the FCEV scenario.

15.2.6 Battery-Powered Electric Vehicle Scenario

The BEV scenario is basically identical to the FCEV scenario with BEVs replacing FCEVs. Thus, we are effectively assuming that the time to install charging outlets for BEVs and

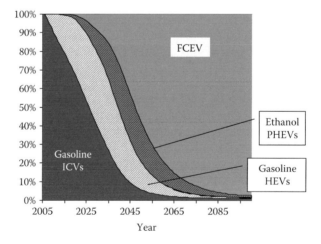

FIGURE 15.4
Fraction of light duty vehicle sales for the FCEV scenario; the long-range BEV scenario and the H_2 ICE HEV scenario use this same sales profile over time with the BEV or H_2 ICE HEV replacing the FCEV.

to develop BEV mass production capacity will be approximately the same as the time to install hydrogen fueling equipment (see Section 15.3.6 for a comparison of the electrical charging infrastructure cost and timing with hydrogen infrastructure cost and timing) and the time to build FCEV mass manufacturing capacity. This BEV scenario also assumes that *all* LDVs including pickup trucks, SUVs, and vans can meet customer expectations for range, acceleration, cost, and refueling time. Furthermore, we did not apply the 75% charging outlet availability limit that was applied to the PHEV cases.

15.2.7 Electricity Source Assumptions

One key input to the model is the mix of electricity generators, which will determine the GHGs emitted by charging car batteries in the PHEV and BEV scenarios. We have modeled the national electrical grid by scaling up the Western Electricity Coordinating Council (WECC) mix of western U.S. states as shown in Figure 15.5. These states have a lower fraction of coal-generated electric power (approximately 32%) compared to slightly over 50% for the average U.S. utility mix. Hence this model is more optimistic with respect to PHEV and BEV GHG emissions than would be the case for the entire country. The model assumes that low-carbon sources of electricity (renewables and nuclear power) will grow substantially over time. In particular, renewable electricity is assumed to provide approximately 40% of all U.S. electricity by 2100, which would require substantial bulk electrical storage capacity to smooth out the intermittency of solar and wind energy. Note that bulk electricity storage would require large banks of batteries or possibly bulk hydrogen storage to achieve 40% renewable capacity. Currently, intermittent renewables require fossil fuel generators (primarily natural gas turbines) to supply power when the sun is not shining or the wind is not blowing. As shown in Figure 15.5, there is very little natural gas generation capacity assumed by 2100 (to keep GHGs down). These battery banks would be extraordinarily expensive, so bulk hydrogen storage may be the

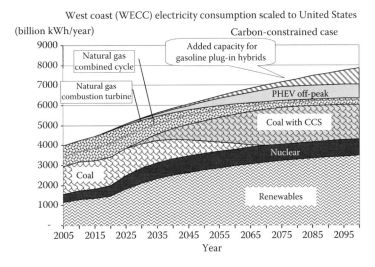

FIGURE 15.5
Sources of electricity based initially on the WECC grid mix expanded proportionately to the entire United States with increasing share of renewables, nuclear and coal carbon capture and storage over the century. (Reprinted from Thomas, C.E., *Int. J. Hydrogen Energy*, 34, 9279, 2009. With permission.)

least costly option to achieving large renewable energy generation. In other words, solar and wind energy would be used to generate hydrogen, which would be stored for later use in either FCEVs or to generate electricity with fuel cells, ICEs, turbines coupled to generators.

15.2.8 Hydrogen Source Assumptions

This model assumes that all hydrogen is made initially by reforming natural gas, which is the most common method of producing merchant hydrogen today. Over time, hydrogen production becomes "greener" by shifting to hydrogen made from biofuels or biomass and eventually be electrolyzing water using renewable or nuclear energy or electricity from coal with carbon capture and storage (CCS), as summarized in Figure 15.6.

15.2.9 Alternative Vehicle Fuel Economy Assumptions

The amount of pollution and oil consumption from each vehicle/fuel combination will depend in part on the assumed fuel economy for these alternative vehicles. In this model, we assume that the fuel economy of conventional (nonhybrid) ICE vehicles increases linearly over time, starting at the U.S. fleet average* of 20 miles per gallon of gasoline (mpg) (or 11.8 L/100 km), increasing to 34 mpg (or 6.9 L/100 km) by 2100.

Several studies have estimated the fuel economy of various alternative vehicle/fuel combinations compared to a gasoline ICV. We used the average relative fuel economy estimates from four sources: the Argonne National Laboratory GREET model [8], the Auto/Oil report

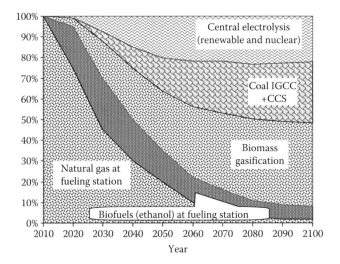

FIGURE 15.6
Sources of hydrogen over the century, beginning with distributed hydrogen generation by reforming natural gas at the fueling station, followed by reforming biofuels such as cellulosic ethanol at the fueling station, and central production by biomass gasification, coal integrated gasification combined cycle (IGCC) with CCS, and eventually electrolysis from zero-carbon electricity such as nuclear and renewables. (Reprinted from Thomas, C.E., *Int. J. Hydrogen Energy*, 34, 9279, 2009. With permission.)

---

* These fuel economy values apply to the *average* over all light duty vehicles including passenger cars and light-duty trucks (pickup trucks, vans, and SUVs), including both new and used vehicles that typically have lower fuel economy than new cars or trucks.

TABLE 15.1

Relative Fuel Economies Compared to Gasoline ICVs (Average Values in Last Column Used in Model)

Vehicle	Fuel	GREET	NRC	GM ANL	MIT	Ave.
SI ICV	Gasoline	1.00	1.00	1.00	1.00	1.00
SI ICV	Ethanol	1.00		1.00		1.00
SI ICV	H_2	1.20		1.20		1.20
CICI ICV	Diesel	1.20		1.21	1.16	1.19
SI IC HEV	Gasoline	1.48	1.45	1.24	1.79	1.49
SI IC HEV	Ethanol	1.48		1.24	1.79[a]	1.51
SI IC HEV	H_2	1.60		1.48	1.94[a]	1.67
CIDI IC HEV	Diesel	1.60		1.45	1.94[a]	1.66
FCEV	H_2	2.30	2.40	2.63	2.27	2.40

SI, spark ignition; ICV, internal combustion vehicle; CIDI, compression ignition direct injection; HEV, hybrid electric vehicle; FCEV, fuel cell electric vehicle.

[a] MIT numbers estimated by extrapolating MIT HEV data to keep relative averages realistic.

led by GM and Argonne [9], the MIT study on electric drive trains [10,11], and the NRC report on hydrogen [4], as summarized in Table 15.1. These are actual on-the-road fuel consumption numbers, not the uncorrected U.S. Environmental Protection Agency (EPA) ratings,* averaged over all LDVs including light duty trucks, vans, and SUVs, old and new.

Based on these literature averages, the hydrogen-powered FCEV would have an average fuel economy that is 2.4 times that of a conventional car. One recent road test of two Toyota Highlander SUV FCEVs provided a direct comparison of a FCEV with a virtually identical conventional ICV. The conventional (nonhybrid) Highlander has an EPA combined fuel economy rating of approximately 11.2 L/100 km (or 22 mpg). Engineers from the National Renewable Energy Laboratory (NREL) and the Savannah River National Laboratory instrumented the Highlander FCEVs and certified an actual on-the-road fuel economy of 3.4 L/100 km (on an energy equivalent basis) or 3.3 times the fuel economy of the ICV Highlander [12]. So the 2.4 times higher fuel economy used in this model is conservative based on this first direct comparison. The National Laboratories also certified an average range of 693 km (431 miles) between hydrogen refills for this SUV FCEV.

15.3 Simulation Results

15.3.1 Greenhouse Gas Emissions

The computer simulation model estimates the annual GHG emissions for each alternative vehicle scenario. We first show the 100 year long-term results to demonstrate which scenarios could achieve the goal of reducing GHGs by 80% below 1990 levels. We then show the near-term results for the decade between 2020 and 2030 to illustrate the benefits of moving to hydrogen-powered FCEVs now instead of later.

---

* The EPA drive cycles are very anemic and do not represent actual American driving habits that are much more aggressive.

15.3.2 Long-Term GHG Emissions Results

The long-term GHG pollution from each of the main scenarios is shown in Figure 15.7. This figure clearly demonstrates that each alternative vehicle option helps to reduce GHGs. Adding HEVs helps to offset the increasing GHGs due to increased VMT assumed in the model, nearly stabilizing GHG pollution from the LDV sector. Adding plug-in hybrids would reduce GHGs below the 1990 levels, and running PHEVs on biofuels such as cellulosic ethanol would cut GHGs even further. But to achieve our goal of reducing GHGs by 80% below 1990 levels, we must move toward either hydrogen-powered cars* or battery-powered vehicles. The basic message of this simulation is that we need to eliminate the ICE from most vehicles (unless they run on hydrogen) to achieve the 80% reduction: we must move to all-electric vehicles. We have two choices to power these all-electric vehicles: batteries or fuel cells (see Section 15.3.6 for a direct comparison of these two options).

15.3.3 Near-Term GHG Emissions Results

Some proponents of introducing PHEVs and BEVs now while postponing the introduction of FCEVs claim that PHEVs and BEVs would achieve GHG and oil reductions in the near

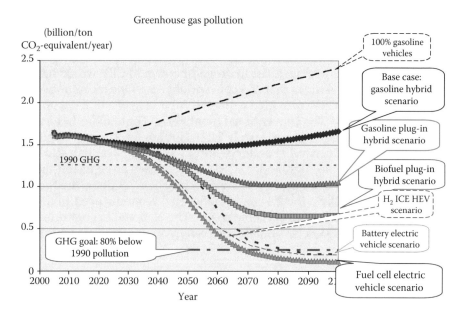

FIGURE 15.7

Primary model output showing the greenhouse gas pollution over the century for a reference case with no alternative vehicles, the four main vehicle scenarios and the two secondary scenarios; the upper dotted horizontal line corresponds to the 1990 light duty vehicle GHG pollution, and the lower dashed line represents an 80% reduction below the 1990 level. (Reprinted from Thomas, C.E., *Int. J. Hydrogen Energy*, 34, 9279, 2009. With permission.)

---

* Figure 15.6 includes both the BEV scenario and a scenario based on hydrogen-powered ICE hybrid electric vehicles that we have not discussed previously. While fuel cell vehicles cut GHGs more than hydrogen ICE HEVs due to higher fuel economy of FCEVs compared to H_2 ICE HEVs, hydrogen hybrids could approach the 80% reduction target by the end of the century.

term, while FCEVs are but long-term options. Their argument is based on the premise that PHEVs and BEVs can reap immediate results while FCEVs are but a long-term option. In fact, our research shows just the opposite. As shown in Figure 15.8, our model does assume that PHEVs enter the marketplace approximately 5 years before FCEVs: there are 10 million PHEVs on the road by 2024, while FCEVs do not achieve that level until 2029 in the model. Despite this 5 year head start, however, the FCEV would reduce GHGs more than the PHEV as shown in Figure 15.9 for the decade between 2020 and 2030. PHEVs would reduce GHGs by 2.2% relative to the HEV base case in 2030, while FCEVs would cut GHGs by 4.4% in 2030. Hence, FCEVs would reduce GHGs two times more than PHEVs, even though FCEVs enter the market 5 years later. This surprising result is due to the fact that hydrogen-powered FCEVs cut GHGs by 45%–50% immediately even when the hydrogen is made from natural gas, while PHEVs will not reduce GHGs much until the electrical grid

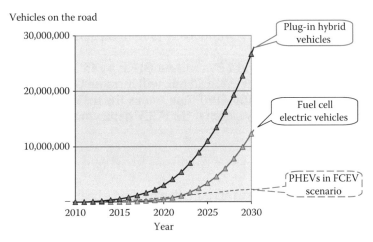

FIGURE 15.8
Number of vehicles on the road for the PHEV and FCEV scenarios in the near term.

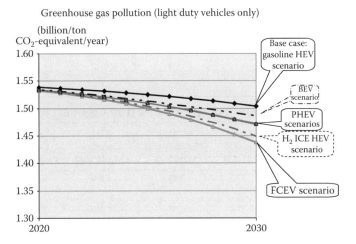

FIGURE 15.9
Greenhouse gas emissions in the decade between 2020 and 2030.

becomes cleaner toward the end of the century. Since most of the electricity sold in the United States comes from burning coal, running a PHEV on today's electricity will not cut GHGs, and in fact will *increase* GHGs in many regions of the United States.

For example, the Argonne National Laboratory has completed a very exhaustive evaluation of the GHG emissions from PHEVs in the United States [13]. They show that with the average U.S. grid mix today, plugging in an HEV will *increase* GHGs; thus, a gasoline HEV generates approximately 260 g/mile of GHGs according to the Argonne National Laboratory calculations, while a PHEV-40 like the proposed Chevy Volt will generate approximately 305 g/mile based on the average U.S. grid mix, a 17.3% *increase* (note that the Chevy Volt is designed to run up to 40 miles on electricity stored in its batteries before the gasoline-powered engine/generator needs to be turned on). From an environmental viewpoint, GHGs could be reduced by 17.3% by operating a PHEV-40 exclusively on gasoline as an HEV, and never recharging the batteries! The Argonne Laboratory report shows that in Illinois, with predominantly coal-based electricity, the GHG emissions from operating a PHEV-40 would reach 375 g/mile, which is a 47% increase in GHGs compared to a gasoline HEV without any plug-in capability at 255 g/mile running only on gasoline.

In summary, hydrogen-powered FCEVs will cut GHGs by 45%–50% immediately, while plugging in HEVs with the current U.S. grid mix will increase GHGs in many parts of the nation. To reduce GHGs in the 2020–2030 time frame, the nation will be best served by deploying FCEVs instead of PHEVs, even if the FCEV deployment lags the PHEV deployment by 5 years.

15.3.4 Local Air Pollution

We have monetized the cost of urban air pollution using the health costs or estimated costs of avoiding pollution using the average costs from a literature review of estimates for the primary criteria pollutants (VOCs, CO, and NO_x) as summarized in Table 15.2. As shown in Figure 15.10, the rankings of the various alternative vehicle scenarios for urban air pollution costs are very similar to the rankings for GHG emissions.

15.3.5 Oil Consumption

In the next two sections, we look at the oil consumed for each scenario in both long and short terms.

TABLE 15.2

Urban Air Pollution Costs ($/Metric Ton); Average Cost (Last Column) Used in Model

	Delucchi Average [16]	Litman [22]	EU AEA (Average of 4) [23]	EU [24] (Holland and Watkins)	ANL Damage Cost [25,26]	ANL Control Cost [25,26]	Average Air Pollution Costs
VOC	1,086	17,706	2,722	3,412	3,940	16,195	7,510
CO	76	534				4,420	1,677
NO_x	17,129	18,934	11,714	6,825	7,860	17,319	13,297
PM-10	138,257	6,565		22,750	10,599	6,005	36,835
PM-2.5	165,019		72,085				118,552
SO_2	69,094		15,506	8,450	4,733	11,581	21,873

Source: Reference numbers in this table are from Thomas, C.E., *Int. J. Hydrogen Energy*, 34, 9279, 2009. With permission.

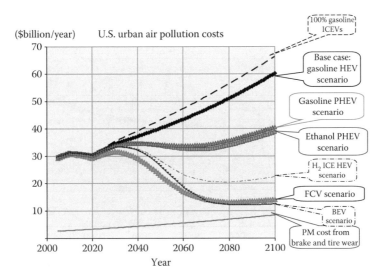

FIGURE 15.10
The costs of urban air pollution for the major alternative vehicle scenarios over the century; the bottom line shows the particulate matter (PM) costs from brake and tire wear that are common to all vehicles. (Reprinted from Thomas, C.E., *Int. J. Hydrogen Energy*, 34, 9279, 2009. With permission.)

15.3.5.1 Long-Term Oil Consumption

In the long term, the scenarios are ranked similar to the GHG chart, except that BEVs and FCEVs (and hydrogen ICE HEVs) have approximately the same result as FCEVs as shown in Figure 15.11.

15.3.5.2 Near-Term Oil Consumption

Finally, Figure 15.12 shows the oil consumption in the decade between 2020 and 2030. As with GHGs, oil consumption can be reduced more in the near term by introducing FCEVs than PHEVs, even though the FCEV deployments lag the PHEVs by 5 years. Thus, PHEVs would reduce oil consumption by 4.1% in 2030 compared to the base case, while FCEVs would cut oil use by 7.8%. This occurs because the FCEV eliminates all gasoline consumption,* while PHEVs still rely on gasoline for some of their travel. We conclude that introducing FCEVs even 5 years after PHEVs will still reduce oil consumption by a factor of 1.92 times the reduction from deploying PHEVs in 2030.

15.3.6 Net Societal Costs

We have monetized the costs of urban air pollution (Section 15.3.3), GHG emissions, and oil imports to estimate a total societal cost of running vehicles in the various scenarios. For GHG emissions, we used a linear fit from $25/metric ton of CO_2-equivalent in 2010 to $50/ton in 2100. In Ref. [1], we provide an explanation of how we arrived at a societal cost of $60/barrel for oil imports. (Note that this $60/barrel estimate is the societal *cost* of oil and is not necessarily related to the *price* of oil, which has been both higher and

---

* BEVs and hydrogen ICE HEVs would also achieve these early oil reductions if they could be introduced as fast as FCEVs (as assumed in this model).

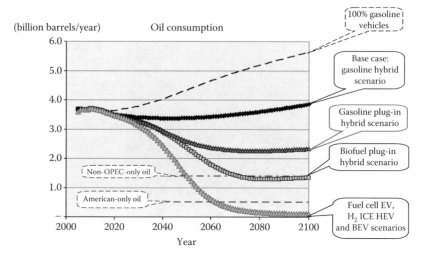

FIGURE 15.11
Oil consumption for the major alternative vehicle scenarios over the century; the upper horizontal dashed line indicates the oil consumption level that might be accommodated by non-OPEC nations, and the lower dashed line indicates the oil consumption level that might be provided in a crisis from the American continent—one representation of "energy independence." (Reprinted from Thomas, C.E. *Int. J. Hydrogen Energy*, 34, 6005, 2009. With permission.)

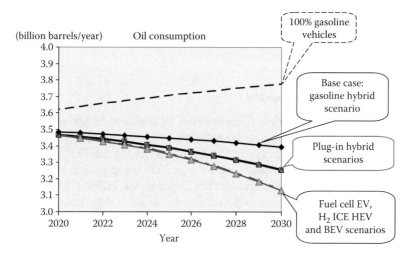

FIGURE 15.12
Expanded scale showing details of oil consumption in the near-term (2020–2030).

lower than $60/barrel over the last few years. This cost estimate includes an estimate of the fraction of military cost related to protecting our Middle Eastern oil pipeline, as well as the economic costs associated with imported oil; this economic cost includes three components as described by Greene and Leiby at Oak Ridge National Laboratory: the transfer of wealth, loss of potential to produce, and disruptive losses due to oil price fluctuations [14])

The resulting societal costs over the century for each vehicle scenario are summarized in Figure 15.13. For the base case scenario (HEVs only), the annual societal costs reach a

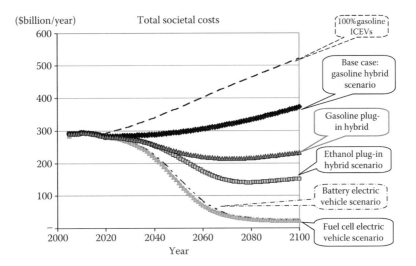

FIGURE 15.13
Estimate of the total societal costs of greenhouse gas pollution, urban air pollution, and the economic and military costs of imported oil for the major alternative vehicle scenarios (the hydrogen-powered ICE HEV scenario has approximately the same societal costs as the FCEV scenario). (Reprinted from Thomas, C.E., *Int. J. Hydrogen Energy*, 34, 9279, 2009. With permission.)

staggering $360 billion/year by 2100. The FCEV scenario reduces these annual costs to only $22 billion/year or a net societal *savings of $340 billion/year.* The next best option, the biofuels PHEV scenario would produce $157 billion in societal costs per year, or a savings of $205 billion compared to the base case.

15.3.7 Batteries and Fuel Cells Compared

The primary conclusion from this 100 year simulation program is that society must move from a nearly 100% dependence on the ICE for transportation to all-electric vehicles.* The main issue is how to provide the electricity for electric vehicles: batteries or fuel cells? Both batteries and fuel cell systems provide electricity to power the electric motor(s) connected to the wheels. The next few sections explore and compare batteries and fuel cells with respect to weight, volume, cost, and refueling time.

15.3.7.1 Battery versus Fuel Cell Weight

Weight or mass is very important on a motor vehicle. Every extra kilogram of mass requires more energy to propel the vehicle. One measure of weight or mass is the specific energy of an energy storage device as measured by energy per unit mass such as kWh/kg. As shown in Figure 15.14, the useful specific energy of a fuel cell system (including the hydrogen tanks, the fuel cell itself and a peak power battery[†]) is greater than the specific energy goal for advanced lithium ion batteries, the current preferred battery technology

---

* If ICEs are not eliminated, then they must be run on hydrogen to achieve our societal goals.
† All FCEVs will most likely have a battery bank in addition to the fuel cell system. This auxiliary battery provides peak power for vehicle acceleration and also permits the storage of regenerative braking when the FCEV is slowed down.

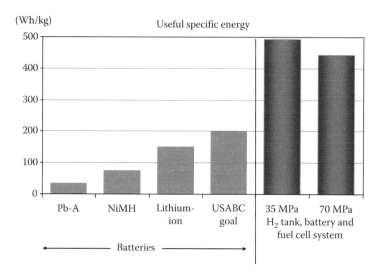

FIGURE 15.14
The useful specific energies of various battery systems compared to hydrogen-powered fuel cell systems.

for PHEVs and BEVs. Note that the specific energies shown in Figure 15.14 are the *useful* specific energies. That is, some batteries achieve high specific energy in the laboratory, but the "useful" specific energy must include only the energy that can be withdrawn from the battery, and the mass in the denominator of the specific energy calculation must include all the components of the full battery system including controls systems, the charger, cooling systems, etc. For example, the lithium-ion battery system for the proposed Chevy Volt PHEV-40 has a nameplate energy storage of 16 kWh. However, not all this energy can be withdrawn from the battery system due to limited state of charge (SOC) ranges. The useable energy from the VOLT battery is only 8 kWh or half the stored energy in the battery. The same applies to the fuel cell system, which must include only the energy that can be used, and the mass must include all auxiliary systems such as controls, cooling, air handling, humidification systems, etc., the useful energy for the fuel cell system must include only the hydrogen that can be withdrawn from the hydrogen tank, not the total stored hydrogen.

Thus, fuel cell systems can store more energy per unit mass than even advanced lithium ion battery systems. Figure 15.14 clearly shows the improvement in battery technology over the last century, from the lead-acid batteries used as starter batteries in most cars to the nickel metal hydride batteries used by most auto companies in their gasoline-powered HEVs, to the lithium-ion batteries used in laptops and cell phones and advocated for future PHEVs and BEVs. Note that the Chevy Volt lithium-ion battery has a specific energy of only 44.1 kWh/kg, compared to the 150 kWh/kg goal assumed in Figure 15.14 for lithium-ion batteries. However, PHEVs do not require as much energy storage as BEVs since the PHEV can turn on the ICE/generator to provide extra range.

In any case, existing fuel cell systems have more than twice the useful specific energy of even advanced lithium-ion batteries, which translates into less mass for the FCEV compared to the BEV as shown in Figure 15.15, which plots the vehicle test mass versus the range of a five-passenger vehicle based on the aluminum-intensive-vehicle (AIV) Mercury Sable developed by the Ford Motor Company under the Clinton administration's Partnership for a New Generation of Vehicles (PNGV) program beginning in

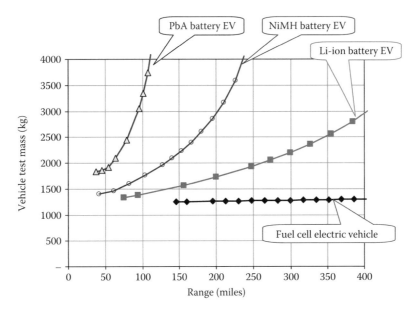

FIGURE 15.15
Vehicle test mass in kilogram as a function of vehicle range for BEVs and FCEVs assuming a five-passenger vehicle based on a light-weight Mercury Sable.

1993.* Note that the FCEV modeled in this program is based on the Ford/Mercury AIV Sable with the following characteristics: 1269 kg mass, 2.127 m² cross-sectional area, 0.33 drag coefficient, and 0.00092 rolling resistance. All vehicles were designed to achieve the following acceleration times: 0–60 miles/h (0–96.6 km/h) in 10 s, 5–20 miles/h (8.1–32.2 km/h) in 1.9 s, 40–60 miles/h (64.4–96.6 km/h) in 7 s, and 55–65 miles/h (88.6–104.7 km/h) in 6 s. As the alternative vehicles became heavier, they required larger motors to achieve these acceleration times and more stored energy to achieve the required range between refueling events. The lines in Figure 15.15 are curved due to the phenomenon of "mass-compounding" or "weight-compounding," whereby adding an extra kilogram of battery to increase BEV range, for example, requires extra vehicle structural support to hold the battery mass,† a slightly larger motor to accelerate that extra battery mass, a slightly larger motor inverter, and a slightly larger braking system to stop the vehicle, etc. Then this added mass in turn requires still more battery banks to achieve the specified vehicle range and acceleration in a nonlinear feedback loop. This is why the battery lines in Figure 15.15 are curved concave upward (without mass compounding, they would be straight lines‡). We developed a mass-compounding equation under contract to the Ford Motor Company in their FCEV development program [15].

---

* The PNGV program introduced by President Clinton on September 29, 1993 challenged the automobile companies to develop advanced vehicle technology to increase fuel economy by a factor of three. Several of the auto companies (including Ford) first developed FCEVs under this program to achieve the three-times fuel economy goal.
† Some analysts simply add a fixed structural mass, such as 15% of the battery mass to account for structure, and ignore the effects of mass compounding on other power train components on the car.
‡ The FCEV line is also curved in Figure 15.14, but this curvature is not pronounced due to the much higher specific energy of the fuel cell system compared to batteries.

15.3.7.2 Battery versus Fuel Cell Volume

Most people realize that batteries are very heavy. Many do not appreciate that batteries also tank up lots of space. Some analysts have been concerned about the large size of hydrogen tanks on FCEVs, sometimes stating that we need a "breakthrough" in hydrogen storage technology before FCEVs become feasible.

In fact, however, the space occupied by hydrogen tanks, the fuel cell system,* *and* the associated peak power battery is *less than* the space required to store an equivalent quantity of energy in batteries as shown in Figure 15.16, which compares the energy density in kWh/L of battery systems and fuel cell systems. Using today's hydrogen storage technology of tanks pressurized to either 35 MPa (or 350 bar) or 70 MPa (700 bar), Figure 15.16 demonstrates that hydrogen and fuel cell storage systems already take up less space than batteries for a given amount of stored energy; no "breakthrough" in hydrogen storage technology is required. (As another example, Toyota has developed a fuel cell version of their Highlander SUV that has been certified by the DOE's national laboratories [12] to achieve a range of 431 miles [or 693.9 km] on one tank of hydrogen, exceeding the often-stated benchmark of 300 miles [or 483 km] range for U.S. vehicles, again demonstrating that no "breakthrough" in hydrogen storage technology is required to meet even American standards of long-range passenger vehicles.)

The specific energies of Figure 15.16 translate into the vehicle energy storage volumes of Figure 15.17 as a function of the required vehicle range. Note again that the fuel cell system volume includes the fuel cell system itself and all auxiliary subsystems, the hydrogen

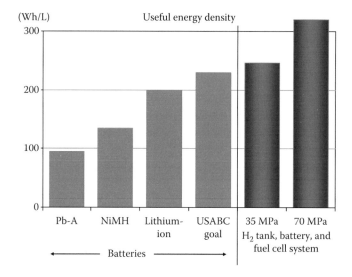

FIGURE 15.16
Comparison of useful specific energy for battery and fuel cell energy storage systems. (Reprinted from Thomas, C.E. *Int. J. Hydrogen Energy*, 34, 6005, 2009. With permission.)

---

* The fuel cell system-specific power in this model was assumed to be 0.94 W/kg based on GM's 2003 technology in their "Hy-Wire" FCEV. GM has since significantly reduced the weight and volume of their fuel cell stacks, but we were not able to find a quantitative estimate of the characteristics of their latest technology GM fuel cells. Hence these estimates of fuel cell-specific power should be considered conservative. We assumed a fuel cell power density of 1.91 kW/L based on the Nissan 130 kW fuel cell stack that occupies 69 L.

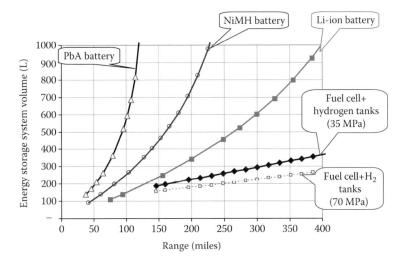

FIGURE 15.17
Volume of energy storage systems as a function of vehicle range required without refueling.

storage tank(s) and the peak power battery system. For a range of 300 miles (or 483 km), even an advanced lithium-ion battery system would take up two to three times more space on the vehicle than hydrogen tanks, fuel cell systems, and the peak power FCEV battery system combined.

15.3.7.3 Battery versus Fuel Cell Cost

The popular press makes much of the high cost of early FCEVs. These initial prototype FCEVs were often hand-built by researchers. The appropriate metric is the cost of alternative vehicles once they are mass produced. Kromer and Heywood at MIT have analyzed the expected mass-production costs of alternative vehicles, and they have concluded that FCEVs with 350 miles (or 579.6 km) will cost much less than BEVs, and even less the PHEVs with all-electric ranges greater than 20 miles (32.2 km), as summarized in Figure 15.18. In their analysis, MIT assumed that the fuel cell system would cost $50/kW in mass production (see Table 53 of Ref. [10]). Brian James and his colleagues at Directed Technologies, Inc. (DTI) reported at the DOE's 2010 Annual Merit Review of hydrogen and fuel cell technologies that fuel cell systems would cost only $39.45/kW if mass produced (500,000 units) using projected 2015 fuel cell technology [16]. If we assume this lower cost estimate for the fuel cell system (keeping all other costs the same as the MIT analysis), then the FCEV with 350 miles (or 563.5 km) range in mass production would cost less than a PHEV with only 10 miles (or 16.1 km) all-electric range as shown in Figure 15.19. We have also included the estimated cost of a BEV with 300 miles (or 483 km) range in Figure 15.19 (using the MIT estimates for BEV costs).

15.3.7.4 Battery versus Fuel Cell Refueling Time

Battery recharging time is one serious limitation of BEVs. To illustrate the difficulty of rapid recharging of batteries, consider the power flow for gasoline or hydrogen fueling events. Pumping 14 gal (or 53.2 L) of gasoline in 3 min corresponds to a power flow of

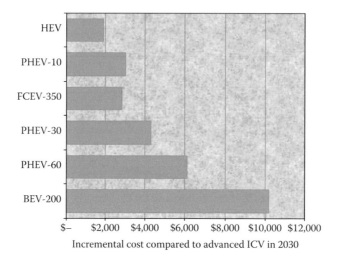

FIGURE 15.18
Incremental cost of alternative vehicles estimated by MIT compared to advanced gasoline ICVs in 2030. (Reprinted from Thomas, C.E. *Int. J. Hydrogen Energy*, 34, 6005, 2009. With permission; Kromer, M.A. and Heywood, J.B., Electric powertrains: Opportunities and challenges in the U.S. light-duty vehicle fleet, Sloan Automotive Laboratory Report # LFEE 2007-03 RP, Massachusetts Institute of Technology, May 2007. http://web. mit.edu/sloan-auto-lab/research/beforeh2/files/kromer_electric_powertrains.pdf [accessed August 2010].)

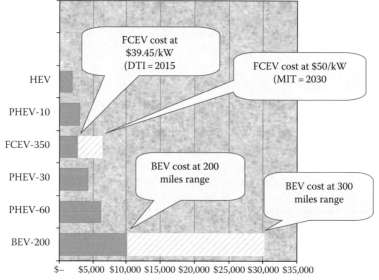

FIGURE 15.19
Estimated mass production incremental costs comparing MIT and DTI estimates for FCEVs.

10 million watts or 10 megawatts (MW). The NREL has monitored the flow of hydrogen in 14,000 separate hydrogen fueling events for FCEVs in actual on-the-road field testing and found an average hydrogen power flow rate of 1.61 MW [17].

For comparison, the maximum power that can be safely transmitted through a typical home 120 V/20 A circuit is 1.9 kilowatts (kW). As shown in Table 15.3, this is more than

TABLE 15.3

Comparison of Fuel Power Flow Rates for Gasoline, Hydrogen, and Electricity

		Fuel Power Flow (kW)	Ratio Gasoline to Alternatives	Ratio hydrogen to Alternatives
Gasoline	10 MW	10,000		
Hydrogen	1.61 MW	1,610	6	1
120 V/20 A circuit	1.9 kW	1.9	5,263	847
240 V/40 A circuit	7.7 kW	7.7	1,299	209

5200 times slower than pumping gasoline and 850 times slower than pumping hydrogen. Even if one installs a 240 V/40 A circuit, the maximum power flow of 7.7 kW is still 1300 times slower than pumping gasoline and 210 times slower than pumping hydrogen. This illustrates that it is much easier to move molecules of gasoline or molecules of hydrogen through a hose than to move electrons through a resistive circuit.

15.3.8 Natural Gas Vehicle Scenario

The discoveries of large amounts of natural gas stored in shale formations in the United States have led some to suggest that we should be converting vehicles to run on natural gas instead of gasoline or diesel fuel. While this may seem reasonable at first glance and will cut our dependence on foreign oil, we show in the following that burning natural gas in ICVs is not the most efficient way to utilize new natural gas resources and that both GHG emissions and oil consumption could be reduced more by converting that natural gas to hydrogen for use in FCEVs. We have analyzed three types of natural gas-powered vehicle:

- Conventional (nonhybrid) natural gas vehicles (NGVs)
- Hybrid electric vehicles running on natural gas (NG-HEV scenario)
- Plug-in hybrid electric vehicles running on natural gas (NG-PHEV scenario)

15.3.9 NGV Greenhouse Gas Emissions

As shown in Figure 15.20, replacing gasoline with natural gas does reduce GHGs in each case; thus, natural gas-powered HEVs generate less GHG than gasoline-powered HEVs; natural gas-powered PHEVs produce less GHGs than gasoline-powered PHEVs, etc. But even in the best case (lowest GHGs) of the NG PHEV scenario, the net GHGs could at best be reduced to only 12%–15% below 1990 levels, far above the goal of an 80% reduction below 1990 levels. Therefore, using natural gas in PHEVs will not achieve our GHG reduction goal. If, however, that natural gas were converted to hydrogen that was used in a FCEV, then we could achieve the 80% reduction goal by approximately 2076 with the assumptions in this model.

15.3.10 Natural Gas Utilization

Another related question is the efficiency of using natural gas in an ICE. In this model, we assume that FCEVs are 2.4 times more efficient than ICVs based on averages of literature

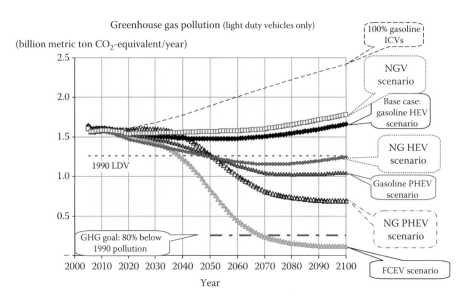

FIGURE 15.20
Greenhouse gas emissions including vehicles powered by natural gas.

estimates (although the first real-world comparison of the Toyota Highlander SUV showed more than three times higher fuel economy). Making hydrogen from natural gas in a steam reformer is approximately 75% efficient, meaning that the energy content of the hydrogen is only 75% of the energy content of the natural gas entering the reformer. So a FCEV running on hydrogen made from natural gas is $2.4 \times 0.75 = 1.8$ times more efficient than an ICV, assuming that the ICV running on natural gas (an NGV) has the same fuel economy as the gasoline version of that ICV. However, an ICE optimized to run on natural gas* can be up to 20% more efficient. In that case, the FCEV would be $1.8/1.2 = 1.5$ times more efficient in utilizing natural gas. In other words, the FCEV running on hydrogen made from natural gas would go 50% farther than an NGV running on natural gas even if the NGV was 20% more efficient than the gasoline version of the same car. A given quantity of natural gas would therefore support 50% more VMT if it was converted to hydrogen for a FCEV.

Another option would be to convert natural gas to electricity for use in a BEV. As shown in Figure 15.21, between 1.24 and 1.86 GJ of natural gas would be required to make enough electricity to propel a BEV for 300 miles (or 483 km), while only 0.85 GJ of natural gas would be sufficient to produce enough hydrogen to power a FCEV for 300 miles (or 483 km). In this case, the FCEV could travel between 1.44 and 2.19 times farther running on hydrogen made from natural gas compared to electricity made from natural gas powering a BEV.

15.3.11 Oil Consumption

Since natural gas could support from 44% to 119% more VMT if that natural gas were converted to hydrogen for a FCEV, then oil consumption would be reduced by these same factors; thus, more gasoline-powered cars could be taken off the road if the natural gas were converted to hydrogen.

---

* An internal combustion engine with higher compression ratio and leaner burn can be 20% more efficient than that same ICE running on gasoline.

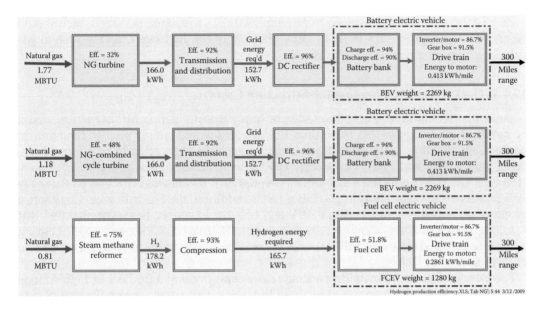

FIGURE 15.21
Comparison of natural gas required to make enough electricity to power a BEV for 300 miles (or 483 km) compared to the amount of natural gas required to make hydrogen for a FCEV to go 300 miles (or 483 km). (Reprinted from Thomas, C.E. *Int. J. Hydrogen Energy*, 34, 6005, 2009. With permission.)

15.4 Fuel Infrastructure Costs

Both BEVs and FCEVs will need new fueling infrastructure to reach a large fraction of the LDV fleet.

15.4.1 Hydrogen Infrastructure Costs

Hydrogen fueling facilities must be provided for FCEVs. The U.S. Department of Energy has estimated that the lowest cost hydrogen will come from steam methane reformers placed at the local fueling station; these reformers will convert natural gas and water to form hydrogen. Making hydrogen at the fueling station avoids the high cost of transporting hydrogen in gaseous or liquid form from a central production facility. This approach essentially relies on the existing natural gas and water pipeline systems to become the backbone of the hydrogen fueling infrastructure system. This distributed hydrogen fueling system also avoids high investment costs for hydrogen infrastructure before there are enough FCEVs on the road. Hydrogen fueling systems can be added when and where they are needed to match the introduction of FCEVs in each region of the country.

These on-site steam methane reformer systems will be expensive; initial stations may cost $3 million to $4 million each, including the necessary hydrogen gas compressors and hydrogen storage tanks and hydrogen dispensing systems. When there are very few FCEVs on the road, the cost per car will be excessive. Costs will fall as more stations are built; the DOE estimates a cost of $3.2 million if 500 systems are built (based on the DOE's H2A cost model) [18].

Each of these stations is designed to provide 1500 kg of hydrogen per day, which should be enough to support 2.013 FCEVs.* Thus, the cost of the hydrogen fueling system per vehicle will eventually reach $1391/FCEV.

15.4.2 Electric Vehicle Charging Infrastructure Costs

Some supporters of PHEVs and BEVs state or imply that no fueling infrastructure would be required: according to this reasoning, BEV owners would simply plug-in their cars at home at night, using lower cost off-peak electricity. This will certainly be a key advantage of early BEVs and PHEVs over early FCEVs, which must have a source of hydrogen. However, Deloitte conducted a survey of potential BEV owners and discovered that only 39% had access to home charging facilities [19]. In addition, the Electrification Coalition, a group of organizations supporting PHEV and BEV deployments, has recommended that to entice a large number of car owners to purchase PHEVs or BEVs, two public charging outlets be installed for every BEV initially decreasing to one public station for every two BEVs in the future [20].

The estimated costs for installing a single slow-charging outlet for BEV or PHEV range from $500 to $8043 as summarized in Table 15.4. (Note that these are just the cost of providing a safe electrical *outlet* and do not include the AC to DC rectifier or "charger" that is assumed to be installed on the vehicle and therefore part of the vehicle cost). The higher number, $8043 is the actual cost in 2010 of installing a Type 2 (240 V) public outlet provided by the installer Coulomb Technologies [21], based on their estimate of providing 4600 "free" Type 2 public outlets for $37 million (of which $15 million was supplied by the U.S. Government Recovery Act). The Idaho National Laboratory estimates given in Table 15.4 are based on their survey of actual contractor costs to install charging outlets [22]. If society decided to follow the Electrification Coalition recommendation of installing two public outlets for every BEV sold, then the costs of electrical charging outlets would vary between $1000 and $16,000 per BEV sold. At a minimum, one outlet could only accommodate one BEV since each BEV will require at least 4 h and more likely 8 h of recharging. GM estimates that their Chevy Volt will require 8 h of charging just to achieve a 40 mile range. Even if 4 h were sufficient, it is unlikely that multiple BEVs could be easily scheduled to share one charging outlet.

TABLE 15.4

Estimated Costs for Installing Electrical Outlets to Recharge Car Batteries for PHEVs and BEVs

	Electrification Coalition [20]	Idaho National Laboratory [22]	Coulomb Technologies [21]
Type 1 residential 120 V EVSE		$833–$878	
Type 2 residential 240 V EVSE	$500–$2,500	$1,520–$2,146	
Type 2 public 240 V EVSE	$2,000–$3,000	$1,853	$8,043
Type 3 public fast charger	$15,000–$50,000		

EVSE, electric vehicle supply equipment.

---

* Assuming that the FCEV travels an average of 13,000 miles (or 20,900 km) per year with a fuel economy of 68.3 kg/mile (42.4 kg/km) (the fuel economy measured for the Toyota Highlander FCEV by government engineers [11]) and an average fueling station capacity factor of 70%.

We conclude that the cost of installing electrical outlets will run between $500 and $16,000/BEV, or up to 11.5 times more than the cost per vehicle of installing a hydrogen infrastructure in a mature marketplace. These costs per vehicle apply to a mature market where there are thousands of vehicles per fueling station in the case of the FCEV.

The initial costs per vehicle will be much higher initially when there are very few vehicles on the road. We have also estimated the total government incentives required to jump-start the fueling infrastructure for a BEV and a FCEV scenario along with the associated industry investments required to install new fueling infrastructure. We assume in this model that the government pays 30% of the infrastructure cost* up to a maximum of $300,000/station, with industry picking up the rest of the cost by taking out a 10 year loan at 8% interest. We assume that industry charges enough for the electricity and for the hydrogen to make at least a 25% internal rate of return (IRR) on their infrastructure investments initially, dropping to 15% IRR once the alternative vehicle business (FCEVs or BEVs) becomes established.

The results of this analysis are summarized in Table 15.5. The total costs are large: over $80 billion for hydrogen infrastructure and over $ 276 billion for electrical charging outlets. But these costs are spread over 50 years in the model, from 2010 to 2060. The annual costs are not large compared to current government annual incentives or energy industry annual investments. For example, Figure 15.22 compares the estimated annual costs for installing a robust hydrogen infrastructure to support the FCEVs in this model with the projected government subsidies for ethanol. The ethanol subsidies could reach $16

TABLE 15.5

Estimated Government and Industry Incentives/Investments Necessary over 50 Years to Install Electrical and Hydrogen Fueling Infrastructure for BEVs and FCEVs

	Hydrogen Investments	Electrical Charging Outlet Investments
Government incentives required	$29.3 billion	$276 billion
Industry incentives required	$51.6 billion	$234 billion
Total incentives required	$80.9 billion	$276 billion

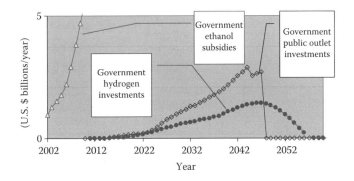

FIGURE 15.22
Estimated government annual hydrogen infrastructure investments compared to projected government subsidies for ethanol.

---

* The Federal government currently pays 30% of alternative fueling station costs under the current law.

billion per year by 2022 if the U.S. Congressional mandate of 36 billion gallons of ethanol is achieved, assuming a subsidy of 45 cents/gallon, down from the current subsidy of 51 cents/gallon. For comparison, the maximum annual hydrogen subsidy would be only $1.4 billion in 2045 [23].

Similarly, the government incentives for installing electrical outlets for BEVs would be small compared to ethanol subsidies as shown in Figure 15.23 (expanded scale) for both hydrogen and electrical outlets [24]. The maximum annual charging outlet subsidy would be $1.8 million in 2045.

The annual industry investments are also small compared to past investments by the oil industry to maintain our current gasoline and diesel infrastructure in the United States. We estimate that the U.S. energy industry has invested over $100 billion annually for fueling infrastructure in the United States only over the last 3 years, despite the Great Recession of 2008/2009 [25]. As shown in Figure 15.24, these past costs are huge compared to the projected industry annual investments required for either the electrical charging system or the hydrogen infrastructure. Thus, the maximum annual industry investments are $2.4 million for hydrogen in 2046 and $10.9 million for electrical outlets

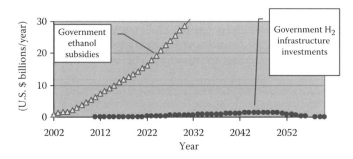

FIGURE 15.23
Estimated government investments required for hydrogen and electrical outlet investments compared to projected ethanol subsidies.

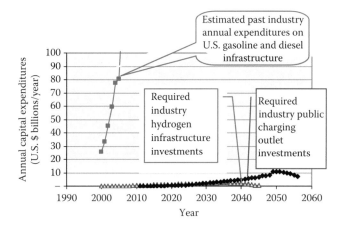

FIGURE 15.24
Annual industry capital investments to develop an electrical outlet infrastructure and a hydrogen infrastructure compared to past annual capital expenditures for U.S. gasoline and diesel fueling infrastructure (off the chart!).

in 2049, or just a small fraction of existing fueling infrastructure costs. These lower costs make sense: in the case of hydrogen production from natural gas, the costs of processing natural gas are less compared to the cost of oil refineries to make gasoline or diesel fuel, and the costs of transporting natural gas by pipeline are lesser than the costs of transporting crude oil and gasoline. In the case of charging car batteries, most of the electricity generation* and transmission capacity has already been paid for by electricity consumers. Some analyses have shown that many residential neighborhood distribution transformers are nearly at peak capacity and that adding even one PHEV could cause local distribution transformers to fail [26,27]. Hence utilities may have to upgrade local distribution transformers to accommodate many PHEV charging outlets at a cost of several thousand dollars per neighborhood; these costs were not included in the estimates reported above.

15.5 Conclusions

Greenhouse gas pollution

- In the long term, society must move toward all-electric vehicles (or hydrogen-powered ICE HEVs) to achieve our goal of cutting GHGs by 80% below 1990 levels.
- In the short term (2020–2030), deploying FCEVs would cut GHGs two times more than PHEVS, even if the FCEVs entered the market 5 years after PHEVs.

Oil consumption

- In the long term, society must move toward alternative fuels such as electricity or hydrogen to reduce our dependence on foreign oil to the degree that, in an emergency, all our nontransportation and remaining transportation needs could be supplied by oil from the American continent.[†]
- In the near-term (2020–2030), oil consumption would be reduced 1.9 times more by switching to hydrogen-powered FCEVs than by deploying PHEVs, even if the FCEVs entered the market 5 years after the PHEVs.

Total societal costs

- Introducing FCEVs would reduce total societal costs by more than $340 billion per year by 2100.
- The second-best option, biofuel-powered PHEVs, would reduce total societal costs by $205 billion per year.

---

* The United States has enormous unused electrical generation capacity at night, although the Electrification Coalition and EPRI both recommend that some charging take place during the day while BEVs and PHEVs are parked at work. To the degree that this daytime charging overlaps with utility peak loads, more production capacity may be required.

† Excluding Venezuela.

Batteries versus fuel cells

Fuel cells are superior to batteries in terms of:

- Weight. (A 300 mile, or 483 km, range FCEV would have a mass of approximately 1200 kg, while a 300 mile range BEV would have a mass of 2200 kg with advanced lithium-ion batteries).

- Space for energy storage on the vehicle (the hydrogen tanks, fuel cell system, and peak power battery would occupy approximately 200 L on a 300 mile range FCEV, while advanced lithium-ion battery systems would take up 540 L for a 300 mile BEV).

- Mass production cost (MIT estimates that a 350 mile (or 563.5 km) range FCEV would cost $3,600 more than a conventional car in mass production, while a 200 mile, or 322 km, range BEV would cost $10,200 more).

- Refueling time (charging car batteries will take 850–5200 times longer than pumping hydrogen or gasoline for a 120 V home circuit, and 210 times to 1300 times longer for a 240-V circuit).

- Initial emissions of GHGs.

- Long-term fueling infrastructure cost. (In a mature market, electrical outlets will cost $500 to $8000 per BEV, while hydrogen fueling stations will cost $1391/FCEV.).

Natural gas vehicles

- Running ICVs on natural gas would reduce GHGs and oil consumption more than running those vehicles on gasoline.

- But burning natural gas in ICEs, even natural gas-powered PHEVs, would not be the best way to utilize our finite natural gas resources.

- Converting natural gas into hydrogen for use in FCEVs would allow us to achieve our goal of cutting GHGs by 80% below 1990 levels, while at the same time multiplying the VMT on that natural gas resource by factors of between 1.29 and 1.44 compared to burning natural gas directly.

- Most importantly, relying on NGVs would defer the time when society moves to a truly sustainable transportation system built on zero-carbon fuels such as hydrogen or electricity derived from renewable or nuclear power.

References

1. Thomas, C.E. Transportation options in a carbon-constrained world: Hybrids, plug-in hybrids, biofuels, fuel cell electric vehicles, and battery electric vehicles. *International Journal of Hydrogen Energy*, 34 (2009), 9279–9296.
2. Thomas, C.E. Fuel cell and battery electric vehicles compared. *International Journal of Hydrogen Energy*, 34 (2009), 6005–6020.
3. Deloitte. Gaining traction: A customer view of electric vehicle mass adoption in the U.S. Automotive Market, June 17, 2010. http://www.deloitte.com/assets/Dcom-UnitedStates/Local%20Assets/Documents/us-automotive_gaining%20 Traction%20FINAL_061710.pdf.htm (accessed February 2011).

4. National Research Council of the National Academies, Ramage M.P., Chair. Committee on Assessment of Resource Needs for Fuel Cell and Hydrogen Technologies. *Transitions to Alternative Transportation Technologies: A Focus on Hydrogen*, 2008. http://books.nap.edu/catalog.php?record_id=12222 (accessed August 2010).

5. Sandia National Laboratory and General Motors. 90-Billion gallon deployment study, February 2009. http://hitectransportation.org/news/2009/Exec_Summary02-2009.pdf (accessed August 2010).

6. Perlack, R.D., Wright, L.L., Turhollow, A.F., Graham, R.L., Stokes, B.J., Erbach, D.C. Biomass as a feedstock for a bioenergy and bioproducts industry: The technical feasibility of a billion-ton annual supply, Oak Ridge National Laboratory Report TM-2005/66, Oak Ridge, TN, April 2005.

7. Greene, N. *Growing Energy: How Biofuels Can Help End America's Oil Dependence*, Natural Resources Defense Council, Washington, DC, December 2004.

8. Wang, M.Q. Greenhouse gases, Regulated Emissions, and Energy use in Transportation (GREET), The Argonne National Laboratory. http://www.transportation.anl.gov/modeling_simulation/GREET/index.html (accessed August 2010).

9. Ruselowski, G., Wallace, J.P., Choudhury, R., Wang, M., Weber, T., Finizza, A. *Well-to-Wheels Energy Use and Greenhouse Gas Emissions of Advanced Fuel/Vehicle Systems—North American Analysis*, Vol. 1, General Motors, The Argonne National Laboratory, BP, ExxonMobil, and Shell, June 2001.

10. Kromer, M.A., Heywood, J.B. Electric powertrains: Opportunities and challenges in the U.S. light-duty vehicle fleet, Sloan Automotive Laboratory Report # LFEE 2007-03 RP, Massachusetts Institute of Technology, May 2007. http://web.mit.edu/sloan-auto-lab/research/beforeh2/files/kromer_electric_powertrains.pdf (accessed August 2010).

11. Kromer, M.A., Heywood, J.B. A comparative assessment of electric propulsion systems in the 2030 US light-duty vehicle fleet. Society of Automotive Engineers: Warrendale, PA, Technical Paper Series 2008-01-0459, April 2008.

12. Wipke, K., Anton, D., Spirk, S. Evaluation of range estimates for Toyota FCHV-adv under open road driving conditions, National Renewable Energy Laboratory and Savannah River National Laboratory, 3 SRNS-STI-2009-00446, August 2009. http://www.nrel.gov/hydrogen/proj_learning_demo.html (accessed August 2010).

13. Elgowainy, A., Han, J., Poch, L., Wang, M., Vyas, A., Mahalik, M., Rousseau, A. *Well-to-Wheel Analysis of Energy Use and Greenhouse Gas Emissions of Plug-in Hybrid Electric Vehicles*. http://www.transportation.anl.gov/pdfs/TA/629.PDF (accessed August 2010).

14. Greene, D.L., Leiby, P.N. Oil independence: Realistic goal or empty slogan? Oak Ridge National Laboratory, March 2007. http://pzl1.ed.ornl.gov/GreeneAndLeiby2007%20Oil%20Independence-Realistic%20Goal%20or%20Empty%20Slogan.html (accessed September 2009).

15. Appendix G on Vehicle Weight Compounding in the final report of the Ford/DOE project entitled *Integrated Analysis of Hydrogen Passenger Vehicle Transportation Pathways*, Prepared by Directed Technologies, Inc. for the Ford Motor company and the National Renewable Energy Laboratory under subcontract No. AXE=6=166685-01, March 1998.

16. James, B., Kalinoski, J., Baum K. Mass-production cost estimation for automotive fuel cell systems, *U.S. Department of Energy Annual Hydrogen and Fuel Cell Program Merit Review*, June 9, 2010. http://www.hydrogen.energy.gov/pdfs/review10/fc018_james_2010_o_web.pdf (accessed February 2011).

17. Wipke, K., Sprik, S., Kurta, J., Ramsden, T. Controlled hydrogen fleet and infrastructure demonstration and Validation project, NREL/TP-560-4345451, March 2009. http://www.nrel.gov/hydrogen/pdfs/45451.pdf (accessed August 2010).

18. The U.S. Department of Energy's H2A Forecourt (distributed) natural gas production system. http://www.hydrogen.energy.gov/h2a_prod_studies.html (accessed August 2010).

19. Gardner, M., Hill, R., Hasegawa, M. Gaining traction: A customer view of electric vehicle mass adoption in the U.S. automotive market, Deloitte Consulting LLP, 2010. http://www.deloitte.com/view/en_US/us/Services/consulting/c3b1a4c65c948210VgnVCM100000ba42f00aRCRD.htm (accessed August 2010).

20. The Electrification Coalition Roadmap: Revolutionizing transportation and achieving energy security, November 2009. Available under Production case studies at http://www.electrification-coalition.org

21. Coulomb Technologies to Provide 4,600 Free EV Charging Stations, ConSensus newsletter, June 7, 2010. http://www.greencarcongress.com/2010/06/coulomb-20100602.html#more (accessed August 2010).

22. Morrow, K., Karner, D., Francfort, J. Plug-in hybrid electric vehicle charging infrastructure review. Final Report INL/EXT-08-15058, Idaho National Laboratory, November 2008. http://avt.inel.gov/pdf/phev/phevInfrastructureReport08.pdf (accessed August 2010).

23. Hydrogen infrastructure incentives required, from the Clean Car Options web site. http://www.cleancaroptions.com/html/hydrogen_infrastructure_incent.html (accessed August 2010).

24. BEV charging outlet incentives required from the Clean Car Options web site. http://www.cleancaroptions.com/html/charging_outlet_incentives.html (accessed August 2010).

25. Radler, M. Oil and gas capital spending to rise in US, fall in Canada, *Oil & Gas Journal*, 105(13), (April 02, 2007); Special Report: Capital budgets grow in US, drop in Canada, *Oil & Gas Journal*, (April 28, 2008), 20; and Economic slump to chill capital spending, *Oil & Gas Journal* (April 27, 2009), 26.

26. The Electrification Coalition Roadmap: Revolutionizing transportation and achieving energy security, November 2009, p. 102. Available under production case studies at http://www.electrificationcoalition.org

27. Taylor, J., Maitra, A., Alexander, D., Brooks, D., Duvall, M. *Evaluation of the Impact of Plug-in Electric Vehicles on Distribution System Operations* The Electric Power Research Institute, Palo Alto, CA, 2009. http://www.abve.org.br/destaques/2009/Evaluation_of_the_Impact_of_Plug-in_Electric_Vehicle_Loading_on_Distribution_System_Operations.pdf

Index

For Product Safety Concerns and Information please contact our EU representative GPSR@taylorandfrancis.com Taylor & Francis Verlag GmbH, Kaufingerstraße 24, 80331 München, Germany